STUDENT'S SOLUTIONS MANUAL

CINDY TRIMBLE & ASSOCIATES

BEGINNING & INTERMEDIATE ALGEBRA

FIFTH EDITION

Elayn Martin-Gay

University of New Orleans

PEARSON

Boston   Columbus   Indianapolis   New York   San Francisco   Upper Saddle River
Amsterdam   Cape Town   Dubai   London   Madrid   Milan   Munich   Paris   Montreal   Toronto
Delhi   Mexico City   Sao Paulo   Sydney   Hong Kong   Seoul   Singapore   Taipei   Tokyo

The author and publisher of this book have used their best efforts in preparing this book. These efforts include the development, research, and testing of the theories and programs to determine their effectiveness. The author and publisher make no warranty of any kind, expressed or implied, with regard to these programs or the documentation contained in this book. The author and publisher shall not be liable in any event for incidental or consequential damages in connection with, or arising out of, the furnishing, performance, or use of these programs.

Reproduced by Pearson from electronic files supplied by the author.

ISBN-13: 978-0-321-78574-9
ISBN-10: 0-321-78574-6

1 2 3 4 5 6 EBM 16 15 14 13 12

www.pearsonhighered.com

PEARSON

Contents

Chapter 1

Section 1.2 Practice

1. **a.** $5 < 8$ since 5 is to the left of 8 on the number line.

   b. $6 > 4$ since 6 is to the right of 4 on the number line.

   c. $16 < 82$ since 16 is to the left of 82 on the number line.

2. **a.** $9 \geq 3$ is true, since $9 > 3$ is true.

   b. $3 \geq 8$ is false, since neither $3 > 8$ nor $3 = 8$ is true.

   c. $25 \leq 25$ is true, since $25 = 25$ is true.

   d. $4 \leq 14$ is true, since $4 < 14$ is true.

3. **a.** $3 < 8$

   b. $15 \geq 9$

   c. $6 \neq 7$

4. The integer -10 represents 10 meters below sea level.

5. **a.** The natural number is 25.

   b. The whole number is 25.

   c. The integers are $25, -15, -99$.

   d. The rational numbers are $25, \dfrac{7}{3}, -15, \dfrac{-3}{4}, -3.7, 8.8, -99$.

   e. The irrational number is $\sqrt{5}$.

   f. The real numbers are $25, \dfrac{7}{3}, -15, \dfrac{-3}{4}, \sqrt{5}, -3.7, 8.8, -99$.

6. **a.** $0 < 3$ since 0 is to the left of 3 on a number line.

   b. $15 > -5$ since 15 is to the right of -5 on a number line.

   c. $3 = \dfrac{12}{4}$ since $\dfrac{12}{4}$ simplifies to 3.

7. **a.** $|-8| = 8$ since -8 is 8 units from 0 on a number line.

   b. $|9| = 9$ since 9 is 9 units from 0 on a number line.

   c. $|-2.5| = 2.5$ since -2.5 is 2.5 units from 0 on a number line.

   d. $\left|\dfrac{5}{11}\right| = \dfrac{5}{11}$ since $\dfrac{5}{11}$ is $\dfrac{5}{11}$ unit from 0 on a number line.

   e. $\left|\sqrt{3}\right| = \sqrt{3}$ since $\sqrt{3}$ is $\sqrt{3}$ units from 0 on a number line.

8. **a.** $|8| = |-8|$ since $8 = 8$.

   b. $|-3| > 0$ since $3 > 0$.

   c. $|-7| < |-11|$ since $7 < 11$.

   d. $|3| > |2|$ since $3 > 2$.

   e. $|0| < |-4|$ since $0 < 4$.

Vocabulary, Readiness & Video Check 1.2

1. The <u>whole</u> numbers are $\{0, 1, 2, 3, 4, ...\}$.

2. The <u>natural</u> numbers are $\{1, 2, 3, 4, 5, ...\}$.

3. The symbols $\neq$, $\leq$, and $>$ are called <u>inequality</u> symbols.

4. The <u>integers</u> are $\{..., -3, -2, -1, 0, 1, 2, 3, ...\}$.

5. The <u>real</u> numbers are {all numbers that correspond to points on the number line}.

6. The <u>rational</u> numbers are $\left\{ \dfrac{a}{b} \middle| a \text{ and } b \text{ are integers}, b \neq 0 \right\}$.

7. The <u>irrational</u> numbers are {nonrational numbers that correspond to points on the number line}.

8. The distance between a number b and 0 on a number line is <u>$|b|$</u>.

9. To form a true statement: $0 < 7$.

10. Five is greater than or equal to four; $5 \geq 4$

11. 0 belongs to the whole numbers, the integers, the rational numbers, and the real numbers; since 0 is a rational number, it cannot also be an irrational number.

12. absolute value

Exercise Set 1.2

1. $7 > 3$ since 7 is to the right of 3 on a number line.

3. $6.26 = 6.26$

5. $0 < 7$ since 0 is to the left of 7 on a number line.

7. $-2 < 2$ since -2 is to the left of 2 on a number line.

9. $32 < 212$ since 32 is to the left of 212 on a number line.

11. $30 \leq 45$ since 30 is to the left of 45 on a number line.

13. $11 \leq 11$ is true, since $11 = 11$.

15. $10 > 11$ is false, since 10 is to the left of 11 on a number line.

17. $3 + 8 \geq 3(8)$ is false, since 11 is to the left of 24 on a number line.

19. $9 > 0$ is true, since 9 is to the right of 0 on a number line.

21. $-6 > -2$ is false, since -6 is to the left of -2 on a number line.

23. Eight is less than twelve is written as $8 < 12$.

25. Five is greater than or equal to four is written as $5 \geq 4$.

27. Fifteen is not equal to negative two is written as $15 \neq -2$.

29. The integer 14,494 represents 14,494 feet above sea level. The integer -282 represents 282 feet below sea level.

31. The integer $-28,000$ represents 28,000 fewer students.

33. 350 represents a deposit of \$350. -126 represents a withdrawal of \$126.

35. The number 0 belongs to the sets of: whole numbers, integers, rational numbers, and real numbers.

37. The number -2 belongs to the sets of: integers, rational numbers, and real numbers.

39. The number 6 belongs to the sets of: natural numbers, whole numbers, integers, rational numbers, and real numbers.

41. The number $\dfrac{2}{3}$ belongs to the sets of: rational numbers and real numbers.

43. The number $-\sqrt{5}$ belongs to the sets of: irrational numbers and real numbers.

45. False; rational numbers may be non-integers.

47. True

49. True

51. True

53. False; an irrational number may not be written as a fraction.

55. $-10 > -100$ since -10 is to the right of -100 on the number line.

57. $32 > 5.2$ since 32 is to the right of 5.2 on the number line.

59. $\dfrac{18}{3} < \dfrac{24}{3}$ since $6 < 8$.

61. $-51 < -50$ since -51 is to the left of -50 on the number line.

63. $|-5| > -4$ since $5 > -4$.

65. $|-1| = |1|$ since $1 = 1$.

67. $|-2| < |-3|$ since $2 < 3$.

69. $|0| < |-8|$ since $0 < 8$.

71. The tallest bar represents the greatest number of visitors; 2009

73. Look for the bars that have heights greater than 280; 2009, 2010

75. In 2001, there were 280 million visitors. In 2010, there were 281 million visitors.
280 million < 281 million

77. The 2009 cranberry production in Oregon was 49 million pounds, while the 2009 cranberry production in Washington was 16 million pounds.
49 million > 16 million

79. The 2009 cranberry production in Washington was 16 million pounds, while the 2009 cranberry production in New Jersey was 54 million pounds.
54 − 16 = 38
The production in Washington was 38 million pounds less or −38 million.

81. −0.04 > −26.7 since −0.04 is to the right of −26.7 on the number line.

83. The sun is brighter since −26.7 < −0.04.

85. The sun is the brightest since −26.7 is to the left of all other numbers listed.

87. 20 ≤ 25 has the same meaning as 25 ≥ 20.

89. 6 > 0 has the same meaning as 0 < 6.

91. −12 < −10 has the same meaning as −10 > −12.

93. answers may vary

Section 1.3 Practice

1. a. $36 = 4 \cdot 9 = 2 \cdot 2 \cdot 3 \cdot 3$

   b. $200 = 2 \cdot 100 = 2 \cdot 4 \cdot 25 = 2 \cdot 2 \cdot 2 \cdot 5 \cdot 5$

2. a. $\dfrac{63}{72} = \dfrac{3 \cdot 3 \cdot 7}{2 \cdot 2 \cdot 2 \cdot 3 \cdot 3} = \dfrac{7}{2 \cdot 2 \cdot 2} = \dfrac{7}{8}$

   b. $\dfrac{64}{12} = \dfrac{2 \cdot 2 \cdot 2 \cdot 2 \cdot 2 \cdot 2}{2 \cdot 2 \cdot 3} = \dfrac{2 \cdot 2 \cdot 2 \cdot 2}{3} = \dfrac{16}{3}$

   c. $\dfrac{7}{25} = \dfrac{7}{5 \cdot 5}$
There are no common factors other than 1, so $\dfrac{7}{25}$ is already in lowest terms.

3. $\dfrac{3}{8} \cdot \dfrac{7}{9} = \dfrac{3 \cdot 7}{8 \cdot 9} = \dfrac{3 \cdot 7}{2 \cdot 2 \cdot 2 \cdot 3 \cdot 3} = \dfrac{7}{2 \cdot 2 \cdot 2 \cdot 3} = \dfrac{7}{24}$

4. a. $\dfrac{3}{4} \div \dfrac{4}{9} = \dfrac{3}{4} \cdot \dfrac{9}{4} = \dfrac{3 \cdot 9}{4 \cdot 4} = \dfrac{27}{16}$

   b. $\dfrac{5}{12} \div 15 = \dfrac{5}{12} \cdot \dfrac{1}{15} = \dfrac{5 \cdot 1}{12 \cdot 15} = \dfrac{5}{12 \cdot 3 \cdot 5} = \dfrac{1}{36}$

   c. $\dfrac{7}{6} \div \dfrac{7}{15} = \dfrac{7}{6} \cdot \dfrac{15}{7} = \dfrac{7 \cdot 15}{6 \cdot 7} = \dfrac{15}{6} = \dfrac{3 \cdot 5}{2 \cdot 3} = \dfrac{5}{2}$

5. a. $\dfrac{8}{5} - \dfrac{3}{5} = \dfrac{8-3}{5} = \dfrac{5}{5} = 1$

   b. $\dfrac{8}{5} - \dfrac{2}{5} = \dfrac{8-2}{5} = \dfrac{6}{5}$

   c. $\dfrac{3}{5} + \dfrac{1}{5} = \dfrac{3+1}{5} = \dfrac{4}{5}$

   d. $\dfrac{5}{12} + \dfrac{1}{12} = \dfrac{5+1}{12} = \dfrac{6}{12} = \dfrac{1}{2}$

6. $\dfrac{2}{3} = \dfrac{2}{3} \cdot \dfrac{7}{7} = \dfrac{2 \cdot 7}{3 \cdot 7} = \dfrac{14}{21}$

7. a. $\dfrac{5}{11} + \dfrac{1}{7} = \dfrac{5 \cdot 7}{11 \cdot 7} + \dfrac{1 \cdot 11}{7 \cdot 11}$
$= \dfrac{35}{77} + \dfrac{11}{77}$
$= \dfrac{35 + 11}{77}$
$= \dfrac{46}{77}$

   b.
$$\begin{array}{rl} \dfrac{5}{21} &= \dfrac{10}{42} \\[2mm] -\dfrac{1}{6} &= -\dfrac{7}{42} \\[1mm] \hline & \dfrac{3}{42} = \dfrac{1}{14} \end{array}$$

c. $\dfrac{1}{3}+\dfrac{29}{30}-\dfrac{4}{5}=\dfrac{10}{30}+\dfrac{29}{30}-\dfrac{4\cdot6}{5\cdot6}$

$=\dfrac{10+29}{30}-\dfrac{24}{30}$

$=\dfrac{39-24}{30}$

$=\dfrac{15}{30}$

$=\dfrac{1}{2}$

8. $5\dfrac{1}{6}\cdot4\dfrac{2}{5}=\dfrac{31}{6}\cdot\dfrac{22}{5}$

$=\dfrac{31\cdot22}{6\cdot5}$

$=\dfrac{31\cdot2\cdot11}{2\cdot3\cdot5}$

$=\dfrac{341}{15}$

$=22\dfrac{11}{15}$

9. $76\dfrac{1}{12}=76\dfrac{1}{12}=75\dfrac{13}{12}$

$-35\dfrac{1}{4}=-35\dfrac{3}{12}=-35\dfrac{3}{12}$

$40\dfrac{10}{12}=40\dfrac{5}{6}$

Vocabulary, Readiness & Video Check 1.3

1. A quotient of two numbers, such as $\dfrac{5}{8}$, is called a <u>fraction</u>.

2. In the fraction $\dfrac{3}{11}$, the number 3 is called the <u>numerator</u> and the number 11 is called the <u>denominator</u>.

3. To factor a number means to write it as a <u>product</u>.

4. A fraction is said to be <u>simplified</u> when the numerator and the denominator have no common factors other than 1.

5. In $7\cdot3=21$, the numbers 7 and 3 are called <u>factors</u> and the number 21 is called the <u>product</u>.

6. The fractions $\dfrac{2}{9}$ and $\dfrac{9}{2}$ are called <u>reciprocals</u>.

7. Fractions that represent the same quantity are called <u>equivalent</u> fractions.

8. 5, Fundamental Principle of Fractions

9. The division operation changes to multiplication and the second fraction $\dfrac{1}{20}$ changes to its reciprocal $\dfrac{20}{1}$.

10. Find the LCD; two fractions must have the same or common denominator before you can subtract (or add).

11. The number $4\dfrac{7}{6}$ is not in proper mixed number form as the fraction part, $\dfrac{7}{6}$, should not be an improper fraction.

Exercise Set 1.3

1. 3 of the 8 equal parts are shaded; $\dfrac{3}{8}$

3. 5 of the 7 equal parts are shaded; $\dfrac{5}{7}$

5. $33=3\cdot11$

7. $98=2\cdot49=2\cdot7\cdot7$

9. $20=4\cdot5=2\cdot2\cdot5$

11. $75=3\cdot25=3\cdot5\cdot5$

13. $45=9\cdot5=3\cdot3\cdot5$

15. $\dfrac{2}{4}=\dfrac{2}{2\cdot2}=\dfrac{1}{2}$

17. $\dfrac{10}{15}=\dfrac{2\cdot5}{3\cdot5}=\dfrac{2}{3}$

19. $\dfrac{3}{7}=\dfrac{3}{7}$

21. $\dfrac{18}{30} = \dfrac{2 \cdot 3 \cdot 3}{2 \cdot 3 \cdot 5} = \dfrac{3}{5}$

23. $\dfrac{120}{244} = \dfrac{2 \cdot 2 \cdot 2 \cdot 3 \cdot 5}{2 \cdot 2 \cdot 61} = \dfrac{2 \cdot 3 \cdot 5}{61} = \dfrac{30}{61}$

25. $\dfrac{1}{2} \cdot \dfrac{3}{4} = \dfrac{1 \cdot 3}{2 \cdot 4} = \dfrac{1 \cdot 3}{2 \cdot 2 \cdot 2} = \dfrac{3}{8}$

27. $\dfrac{2}{3} \cdot \dfrac{3}{4} = \dfrac{2 \cdot 3}{3 \cdot 4} = \dfrac{2 \cdot 3}{3 \cdot 2 \cdot 2} = \dfrac{1}{2}$

29. $\dfrac{1}{2} \div \dfrac{7}{12} = \dfrac{1}{2} \cdot \dfrac{12}{7} = \dfrac{1 \cdot 12}{2 \cdot 7} = \dfrac{1 \cdot 2 \cdot 2 \cdot 3}{2 \cdot 7} = \dfrac{2 \cdot 3}{7} = \dfrac{6}{7}$

31. $\dfrac{3}{4} \div \dfrac{1}{20} = \dfrac{3}{4} \cdot \dfrac{20}{1} = \dfrac{3 \cdot 20}{4 \cdot 1} = \dfrac{3 \cdot 2 \cdot 2 \cdot 5}{2 \cdot 2} = \dfrac{3 \cdot 5}{1} = 15$

33. $\dfrac{7}{10} \cdot \dfrac{5}{21} = \dfrac{7 \cdot 5}{10 \cdot 21} = \dfrac{7 \cdot 5}{2 \cdot 5 \cdot 3 \cdot 7} = \dfrac{1}{2 \cdot 3} = \dfrac{1}{6}$

35. $\dfrac{25}{9} \cdot \dfrac{1}{3} = \dfrac{25 \cdot 1}{9 \cdot 3} = \dfrac{5 \cdot 5 \cdot 1}{3 \cdot 3 \cdot 3} = \dfrac{25}{27}$

37. $\begin{aligned} \text{Area} = l \cdot w &= \dfrac{11}{12} \cdot \dfrac{3}{5} \\ &= \dfrac{11 \cdot 3}{12 \cdot 5} \\ &= \dfrac{11 \cdot 3}{2 \cdot 2 \cdot 3 \cdot 5} \\ &= \dfrac{11}{2 \cdot 2 \cdot 5} \\ &= \dfrac{11}{20} \text{ sq mi} \end{aligned}$

39. $\text{Area} = \dfrac{1}{2} bh = \dfrac{1}{2} \cdot \dfrac{7}{8} \cdot \dfrac{4}{9} = \dfrac{7 \cdot 4}{2 \cdot 2 \cdot 4 \cdot 9} = \dfrac{7}{36}$ sq ft

41. $\dfrac{4}{5} - \dfrac{1}{5} = \dfrac{4-1}{5} = \dfrac{3}{5}$

43. $\dfrac{4}{5} + \dfrac{1}{5} = \dfrac{4+1}{5} = \dfrac{5}{5} = 1$

45. $\dfrac{17}{21} - \dfrac{10}{21} = \dfrac{17-10}{21} = \dfrac{7}{21} = \dfrac{7}{3 \cdot 7} = \dfrac{1}{3}$

47. $\dfrac{23}{105} + \dfrac{4}{105} = \dfrac{23+4}{105} = \dfrac{27}{105} = \dfrac{3 \cdot 3 \cdot 3}{3 \cdot 5 \cdot 7} = \dfrac{3 \cdot 3}{5 \cdot 7} = \dfrac{9}{35}$

49. $\dfrac{7}{10} = \dfrac{7 \cdot 3}{10 \cdot 3} = \dfrac{21}{30}$

51. $\dfrac{2}{9} = \dfrac{2 \cdot 2}{9 \cdot 2} = \dfrac{4}{18}$

53. $\dfrac{4}{5} = \dfrac{4 \cdot 4}{5 \cdot 4} = \dfrac{16}{20}$

55. $\dfrac{2}{3} + \dfrac{3}{7} = \dfrac{2 \cdot 7}{3 \cdot 7} + \dfrac{3 \cdot 3}{7 \cdot 3} = \dfrac{14}{21} + \dfrac{9}{21} = \dfrac{23}{21}$

57. $\dfrac{4}{15} - \dfrac{1}{12} = \dfrac{16}{60} - \dfrac{5}{60} = \dfrac{16-5}{60} = \dfrac{11}{60}$

59. $\begin{aligned} \dfrac{5}{22} - \dfrac{5}{33} &= \dfrac{5 \cdot 3}{22 \cdot 3} - \dfrac{5 \cdot 2}{33 \cdot 2} \\ &= \dfrac{15}{66} - \dfrac{10}{66} \\ &= \dfrac{15-10}{66} \\ &= \dfrac{5}{66} \end{aligned}$

61. $\dfrac{12}{5} - 1 = \dfrac{12}{5} - \dfrac{5}{5} = \dfrac{12-5}{5} = \dfrac{7}{5}$

63. $\begin{aligned} 1 - \dfrac{3}{10} - \dfrac{5}{10} &= \dfrac{10}{10} - \dfrac{3}{10} - \dfrac{5}{10} \\ &= \dfrac{10-3-5}{10} \\ &= \dfrac{2}{10} \\ &= \dfrac{2}{2 \cdot 5} \\ &= \dfrac{1}{5} \end{aligned}$

The unknown part is $\dfrac{1}{5}$.

65. $1 - \dfrac{1}{4} - \dfrac{3}{8} = \dfrac{8}{8} - \dfrac{1 \cdot 2}{4 \cdot 2} - \dfrac{3}{8} = \dfrac{8-2-3}{8} = \dfrac{3}{8}$

The unknown part is $\dfrac{3}{8}$.

67. $1 - \dfrac{1}{2} - \dfrac{1}{6} - \dfrac{2}{9} = \dfrac{18}{18} - \dfrac{1 \cdot 9}{2 \cdot 9} - \dfrac{1 \cdot 3}{6 \cdot 3} - \dfrac{2 \cdot 2}{9 \cdot 2}$

$= \dfrac{18 - 9 - 3 - 4}{18}$

$= \dfrac{2}{18}$

$= \dfrac{1}{9}$

The unknown part is $\dfrac{1}{9}$.

69. $5\dfrac{1}{9} \cdot 3\dfrac{2}{3} = \dfrac{46}{9} \cdot \dfrac{11}{3} = \dfrac{506}{27} = 18\dfrac{20}{27}$

71. $8\dfrac{3}{5} \div 2\dfrac{9}{10} = \dfrac{43}{5} \div \dfrac{29}{10}$

$= \dfrac{43}{5} \cdot \dfrac{10}{29}$

$= \dfrac{43 \cdot 5 \cdot 2}{5 \cdot 29}$

$= \dfrac{86}{29}$

$= 2\dfrac{28}{29}$

73. $17\dfrac{2}{5} + 30\dfrac{2}{3} = \dfrac{87}{5} + \dfrac{92}{3}$

$= \dfrac{87 \cdot 3}{5 \cdot 3} + \dfrac{92 \cdot 5}{3 \cdot 5}$

$= \dfrac{261}{15} + \dfrac{460}{15}$

$= \dfrac{721}{15}$

$= 48\dfrac{1}{15}$

75. $\begin{array}{r} 8\frac{11}{12} \\ -1\frac{5}{6} \\ \hline \end{array}$ $\begin{array}{r} 8\frac{11}{12} \\ -1\frac{10}{12} \\ \hline 7\frac{1}{12} \end{array}$

77. $\dfrac{10}{21} + \dfrac{5}{21} = \dfrac{10+5}{21} = \dfrac{15}{21} = \dfrac{3 \cdot 5}{3 \cdot 7} = \dfrac{5}{7}$

79. $\dfrac{10}{3} - \dfrac{5}{21} = \dfrac{10 \cdot 7}{3 \cdot 7} - \dfrac{5}{21} = \dfrac{70}{21} - \dfrac{5}{21} = \dfrac{65}{21}$

81. $\dfrac{2}{3} \cdot \dfrac{3}{5} = \dfrac{2 \cdot 3}{3 \cdot 5} = \dfrac{2}{5}$

83. $\dfrac{2}{3} \div \dfrac{3}{5} = \dfrac{2}{3} \cdot \dfrac{5}{3} = \dfrac{2 \cdot 5}{3 \cdot 3} = \dfrac{10}{9}$

85. $5 + \dfrac{2}{3} = \dfrac{15}{3} + \dfrac{2}{3} = \dfrac{15+2}{3} = \dfrac{17}{3}$

87. $7\dfrac{2}{5} \div \dfrac{1}{5} = \dfrac{37}{5} \div \dfrac{1}{5} = \dfrac{37}{5} \cdot \dfrac{5}{1} = \dfrac{37 \cdot 5}{5 \cdot 1} = \dfrac{37}{1} = 37$

89. $\dfrac{1}{2} - \dfrac{14}{33} = \dfrac{1 \cdot 33}{2 \cdot 33} - \dfrac{14 \cdot 2}{33 \cdot 2}$

$= \dfrac{33}{66} - \dfrac{28}{66}$

$= \dfrac{33-28}{66}$

$= \dfrac{5}{66}$

91. $\dfrac{23}{105} - \dfrac{2}{105} = \dfrac{23-2}{105} = \dfrac{21}{105} = \dfrac{21}{21 \cdot 5} = \dfrac{1}{5}$

93. $1\dfrac{1}{2} + 3\dfrac{2}{3} = \dfrac{3}{2} + \dfrac{11}{3}$

$= \dfrac{3 \cdot 3}{2 \cdot 3} + \dfrac{11 \cdot 2}{3 \cdot 2}$

$= \dfrac{9}{6} + \dfrac{22}{6}$

$= \dfrac{9+22}{6}$

$= \dfrac{31}{6}$

$= 5\dfrac{1}{6}$

95. $\dfrac{2}{3} - \dfrac{5}{9} + \dfrac{5}{6} = \dfrac{2 \cdot 2 \cdot 3}{3 \cdot 2 \cdot 3} - \dfrac{5 \cdot 2}{9 \cdot 2} + \dfrac{5 \cdot 3}{6 \cdot 3}$

$= \dfrac{12}{18} - \dfrac{10}{18} + \dfrac{15}{18}$

$= \dfrac{12-10+15}{18}$

$= \dfrac{17}{18}$

97. $5 + 4\frac{1}{8} + 4\frac{1}{8} + 15\frac{3}{4} + 15\frac{3}{4} + 10\frac{1}{2}$

$= \frac{40}{8} + \frac{33}{8} + \frac{33}{8} + \frac{126}{8} + \frac{126}{8} + \frac{84}{8}$

$= \frac{40 + 33 + 33 + 126 + 126 + 84}{8}$

$= \frac{442}{8}$

$= 55\frac{1}{4}$ feet

99. answers may vary

101. $5\frac{1}{2} - 2\frac{1}{8} = \frac{11}{2} - \frac{17}{8}$

$= \frac{11 \cdot 4}{2 \cdot 4} - \frac{17}{8}$

$= \frac{44}{8} - \frac{17}{8}$

$= \frac{44 - 17}{8}$

$= \frac{27}{8}$

$= 3\frac{3}{8}$ miles

103. The graph shows $\frac{21}{100}$ are miniplexes.

105. The largest sector of the graph corresponds to multiplexes, so multiplexes have the greatest number of screens.

107. The work is incorrect.

$\frac{12}{24} = \frac{2 \cdot 2 \cdot 3}{2 \cdot 2 \cdot 2 \cdot 3} = \frac{1}{2}$

109. The work is incorrect.

$\frac{2}{7} + \frac{9}{7} = \frac{2 + 9}{7} = \frac{11}{7}$

Section 1.4 Practice

1. a. $1^3 = 1 \cdot 1 \cdot 1 = 1$

   b. $5^2 = 5 \cdot 5 = 25$

   c. $\left(\frac{1}{10}\right)^2 = \left(\frac{1}{10}\right)\left(\frac{1}{10}\right) = \frac{1}{100}$

   d. $9^1 = 9$

   e. $\left(\frac{2}{5}\right)^3 = \left(\frac{2}{5}\right)\left(\frac{2}{5}\right)\left(\frac{2}{5}\right) = \frac{8}{125}$

2. a. $6 + 3 \cdot 9 = 6 + 27 = 33$

   b. $4^3 \div 8 + 3 = 64 \div 8 + 3 = 8 + 3 = 11$

   c. $\left(\frac{2}{3}\right)^2 \cdot |-8| = \frac{4}{9} \cdot 8 = \frac{32}{9}$ or $3\frac{5}{9}$

   d. $\frac{9(14 - 6)}{|-2|} = \frac{9(8)}{2} = \frac{72}{2} = 36$

   e. $\frac{7}{4} \cdot \frac{1}{4} - \frac{1}{4} = \frac{7}{16} - \frac{4}{16} = \frac{3}{16}$

3. $\frac{6^2 - 5}{3 + |6 - 5| \cdot 8} = \frac{36 - 5}{3 + |1| \cdot 8} = \frac{31}{3 + 8} = \frac{31}{11}$

4. $4[25 - 3(5 + 3)] = 4[25 - 3(8)]$

$= 4[25 - 24]$

$= 4[1]$

$= 4$

5. $\frac{36 \div 9 + 5}{5^2 - 3} = \frac{4 + 5}{25 - 3} = \frac{9}{22}$

6. a. $2x + y = 2(2) + 5 = 4 + 5 = 9$

   b. $\frac{4x}{3y} = \frac{4(2)}{3(5)} = \frac{8}{15}$

   c. $\frac{3}{x} + \frac{x}{y} = \frac{3}{2} + \frac{2}{5} = \frac{15}{10} + \frac{4}{10} = \frac{19}{10}$

   d. $x^3 + y^2 = 2^3 + 5^2 = 8 + 25 = 33$

7. $9x - 6 = 7x$

$9(4) - 6 \overset{?}{=} 7(4)$

$36 - 6 \overset{?}{=} 28$

$30 = 28$ False

4 is not a solution of $9x - 6 = 7x$.

8. a. Six times a number is $6x$, since $6x$ denotes the product of 6 and x.

   b. A number decreased by 8 is $x - 8$ because "decreased by" means subtract.

   c. The product of a number and 9 is $x \cdot 9$ or $9x$.

   d. Two times a number is $2x$, plus 3 is $2x + 3$.

   e. The sum of 7 and a number x is $7 + x$.

9. a. A number x increased by 7 is $x + 7$, so $x + 7 = 13$.

   b. Two less than a number x is $x - 2$, so $x - 2 = 11$.

   c. Double a number x is $2x$, added to 9 is $2x + 9$, so $2x + 9 \neq 25$.

   d. Five times 11 is $5(11)$, so $5(11) \geq x$, where x is an unknown number.

Graphing Calculator Explorations

1. $5^4 = 625$

2. $7^4 = 2401$

3. $9^5 = 59,049$

4. $8^6 = 262,144$

5. $2(20 - 5) = 30$

6. $3(14 - 7) + 21 = 3(7) + 21 = 21 + 21 = 42$

7. $24(862 - 455) + 89 = 9857$

8. $99 + (401 + 962) = 1462$

9. $\dfrac{4623 + 129}{36 - 34} = 2376$

10. $\dfrac{956 - 452}{89 - 86} = 168$

Vocabulary, Readiness & Video Check 1.4

1. In the expression 5^2, the 5 is called the <u>base</u> and the 2 is called the <u>exponent</u>.

2. The symbols $(\ )$, $[\ ]$, and $\{\ \}$ are examples of <u>grouping</u> symbols.

3. A symbol that is used to represent a number is called a <u>variable</u>.

4. A collection of numbers, variables, operation symbols, and grouping symbols is called an <u>expression</u>.

5. A mathematical statement that two expressions are equal is called an <u>equation</u>.

6. A value for the variable that makes an equation a true statement is called a <u>solution</u>.

7. Deciding what values of a variable make an equation a true statement is called <u>solving</u> the equation.

8. The order in which we perform operations does matter! We came up with an order of operations to avoid getting more than one answer when evaluating an expression.

9. The replacement value for z is not used because it's not needed—there is no variable z in the given algebraic expression.

10. No; the variable was replaced with 0 in the equation to see if a true statement occurred, and it did not.

11. We translate phrases to mathematical expressions and sentences to mathematical equations.

Exercise Set 1.4

1. $3^5 = 3 \cdot 3 \cdot 3 \cdot 3 \cdot 3 = 243$

3. $3^3 = 3 \cdot 3 \cdot 3 = 27$

5. $1^5 = 1 \cdot 1 \cdot 1 \cdot 1 \cdot 1 = 1$

7. $5^1 = 5$

9. $7^2 = 7 \cdot 7 = 49$

11. $\left(\dfrac{2}{3}\right)^4 = \left(\dfrac{2}{3}\right)\left(\dfrac{2}{3}\right)\left(\dfrac{2}{3}\right)\left(\dfrac{2}{3}\right) = \dfrac{2 \cdot 2 \cdot 2 \cdot 2}{3 \cdot 3 \cdot 3 \cdot 3} = \dfrac{16}{81}$

13. $\left(\dfrac{1}{5}\right)^3 = \left(\dfrac{1}{5}\right)\left(\dfrac{1}{5}\right)\left(\dfrac{1}{5}\right) = \dfrac{1\cdot1\cdot1}{5\cdot5\cdot5} = \dfrac{1}{125}$

15. $(1.2)^2 = (1.2)(1.2) = 1.44$

17. $(0.04)^3 = (0.04)(0.04)(0.04) = 0.000064$

19. $5 + 6 \cdot 2 = 5 + 12 = 17$

21. $4 \cdot 8 - 6 \cdot 2 = 32 - 12 = 20$

23. $2(8 - 3) = 2(5) = 10$

25. $2 + (5 - 2) + 4^2 = 2 + 3 + 4^2 = 2 + 3 + 16 = 21$

27. $5 \cdot 3^2 = 5 \cdot 9 = 45$

29. $\dfrac{1}{4}\cdot\dfrac{2}{3} - \dfrac{1}{6} = \dfrac{2}{12} - \dfrac{1}{6} = \dfrac{1}{6} - \dfrac{1}{6} = 0$

31. $2[5 + 2(8 - 3)] = 2[5 + 2(5)]$
$= 2[5 + 10]$
$= 2[15]$
$= 30$

33. $\dfrac{19 - 3\cdot5}{6 - 4} = \dfrac{19 - 15}{6 - 4} = \dfrac{4}{2} = 2$

35. $\dfrac{|6 - 2| + 3}{8 + 2\cdot5} = \dfrac{|4| + 3}{8 + 2\cdot5} = \dfrac{4 + 3}{8 + 2\cdot5} = \dfrac{4 + 3}{8 + 10} = \dfrac{7}{18}$

37. $\dfrac{3 + 3(5 + 3)}{3^2 + 1} = \dfrac{3 + 3(8)}{3^2 + 1} = \dfrac{3 + 3(8)}{9 + 1} = \dfrac{3 + 24}{9 + 1} = \dfrac{27}{10}$

39. $\dfrac{6 + |8 - 2| + 3^2}{18 - 3} = \dfrac{6 + |6| + 3^2}{18 - 3}$
$= \dfrac{6 + 6 + 3^2}{18 - 3}$
$= \dfrac{6 + 6 + 9}{18 - 3}$
$= \dfrac{21}{15}$
$= \dfrac{3\cdot7}{3\cdot5}$
$= \dfrac{7}{5}$

41. $2 + 3[10(4\cdot5 - 16) - 30]$
$= 2 + 3[10(20 - 16) - 30]$
$= 2 + 3[10(4) - 30]$
$= 2 + 3[40 - 30]$
$= 2 + 3[10]$
$= 2 + 30$
$= 32$

43. $\left(\dfrac{2}{3}\right)^3 + \dfrac{1}{9} + \dfrac{1}{3}\cdot\dfrac{4}{3} = \dfrac{8}{27} + \dfrac{1}{9} + \dfrac{1}{3}\cdot\dfrac{4}{3}$
$= \dfrac{8}{27} + \dfrac{1}{9} + \dfrac{4}{9}$
$= \dfrac{8}{27} + \dfrac{3}{27} + \dfrac{12}{27}$
$= \dfrac{23}{27}$

45. a. $(6 + 2)\cdot(5 + 3) = 8\cdot8 = 64$

   b. $(6 + 2)\cdot5 + 3 = 8\cdot5 + 3 = 40 + 3 = 43$

   c. $6 + 2\cdot5 + 3 = 6 + 10 + 3 = 19$

   d. $6 + 2\cdot(5 + 3) = 6 + 2\cdot8 = 6 + 16 = 22$

47. Let $y = 3$.
$3y = 3(3) = 9$

49. Let $x = 1$ and $z = 5$.
$\dfrac{z}{5x} = \dfrac{5}{5(1)} = \dfrac{5}{5} = 1$

51. Let $x = 1$.
$3x - 2 = 3(1) - 2 = 3 - 2 = 1$

53. Let $x = 1$ and $y = 3$.
$|2x + 3y| = |2(1) + 3(3)| = |2 + 9| = |11| = 11$

55. Let $x = 1$, $y = 3$, and $z = 5$.
$xy + z = 1\cdot3 + 5 = 3 + 5 = 8$

57. Let $y = 3$.
$5y^2 = 5(3)^2 = 5(9) = 45$

59. Let $x = 12$, $y = 8$, and $z = 4$.
$\dfrac{x}{z} + 3y = \dfrac{12}{4} + 3(8) = 3 + 24 = 27$

61. Let $x = 12$ and $y = 8$.
$x^2 - 3y + x = (12)^2 - 3(8) + 12$
$= 144 - 24 + 12$
$= 132$

63. Let $x = 12$, $y = 8$, and $z = 4$.

$$\frac{x^2 + z}{y^2 + 2z} = \frac{(12)^2 + 4}{(8)^2 + 2(4)} = \frac{144 + 4}{64 + 8} = \frac{148}{72} = \frac{37}{18}$$

65. Evaluate $16t^2$ for each value of t.

$t = 1$: $16(1)^2 = 16(1) = 16$

$t = 2$: $16(2)^2 = 16(4) = 64$

$t = 3$: $16(3)^2 = 16(9) = 144$

$t = 4$: $16(4)^2 = 16(16) = 256$

Time t (in seconds)	Distance $16t^2$ (in feet)
1	16
2	64
3	144
4	256

67. Let $x = 5$.

$3x + 30 = 9x$

$3(5) + 30 \stackrel{?}{=} 9(5)$

$15 + 30 \stackrel{?}{=} 45$

$45 = 45$, true

5 is a solution of the equation.

69. Let $x = 0$.

$2x + 6 = 5x - 1$

$2(0) + 6 \stackrel{?}{=} 5(0) - 1$

$0 + 6 \stackrel{?}{=} 0 - 1$

$6 = -1$, false

0 is not a solution of the equation.

71. Let $x = 8$.

$2x - 5 = 5$

$2(8) - 5 \stackrel{?}{=} 5$

$16 - 5 \stackrel{?}{=} 5$

$9 = 5$, false

8 is not a solution of the equation.

73. Let $x = 2$.

$x + 6 = x + 6$

$2 + 6 \stackrel{?}{=} 2 + 6$

$8 = 8$, true

2 is a solution of the equation.

75. Let $x = 0$.

$x = 5x + 15$

$(0) \stackrel{?}{=} 5(0) + 15$

$0 \stackrel{?}{=} 0 + 15$

$0 = 15$, false

0 is not a solution of the equation.

77. Fifteen more than a number is written as $x + 15$.

79. Five subtracted from a number is written as $x - 5$.

81. The ratio of a number and 4 is written as $\frac{x}{4}$.

83. Three times a number, increased by 22 is written as $3x + 22$.

85. One increased by two equals the quotient of nine and three is written as $1 + 2 = 9 \div 3$.

87. Three is not equal to four divided by two is written as $3 \neq 4 \div 2$.

89. The sum of 5 and a number is 20 is written as $5 + x = 20$.

91. The product 7.6 and a number is 17 is written as $7.6x = 17$.

93. Thirteen minus three times a number is 13 is written as $13 - 3x = 13$.

95. To simplify the expression $1 + 3 \cdot 6$, first multiply.

97. To simplify the expression $(20 - 4) \cdot 2$, first subtract.

99. No; answers may vary.

	Length, l	Width, w	Perimeter of Rectangle: $2l + 2w$	Area of Rectangle: lw
101.	4 in.	3 in.	$2l + 2w$ $= 2(4 \text{ in.}) + 2(3 \text{ in.})$ $= 8 \text{ in.} + 6 \text{ in.}$ $= 14 \text{ in.}$	lw $= (4 \text{ in.})(3 \text{ in.})$ $= 12 \text{ sq in.}$
103.	5.3 in.	1.7 in.	$2l + 2w$ $= 2(5.3 \text{ in.}) + 2(1.7 \text{ in.})$ $= 10.6 \text{ in.} + 3.4 \text{ in.}$ $= 14 \text{ in.}$	lw $= (5.3 \text{ in.})(1.7 \text{ in.})$ $= 9.01 \text{ sq in.}$

105. Rectangles with the same perimeter can have different areas.

107. $(20 - 4) \cdot 4 \div 2 = 16 \cdot 4 \div 2 = 64 \div 2 = 32$

109. **a.** $5x + 6$ is an expression since it does not contain the equal symbol, "=."

      b. $2a = 7$ is an equation since it contains the equal symbol.

      c. $3a + 2 = 9$ is an equation since it contains the equal symbol.

      d. $4x + 3y - 8z$ is an expression since it does not contain the equal symbol.

      e. $5^2 - 2(6 - 2)$ is an expression since it does not contain the equal symbol.

111. answers may vary

113. answers may vary; for example, $-2(5) - 1$: $-2(5) - 1 = -10 - 1 = -11$

115. Let $l = 120$ and $w = 100$.
 $lw = (120)(100) = 12,000$ sq ft

117. Let $d = 432$ and $t = 8.5$.
 $\dfrac{d}{t} = \dfrac{432}{8.5} = 51$
 The rate is 51 mph.

Section 1.5 Practice

1.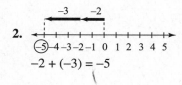
$2 + 4 = 6$

2.
$-2 + (-3) = -5$

3. **a.** $-5 + (-8)$
    Add the absolute values.
    $5 + 8 = 13$
    The common sign is negative, so
    $-5 + (-8) = -13.$

    b. $-31 + (-1)$
    Add the absolute values.
    $31 + 1 = 32$
    The common sign is negative, so
    $-31 + (-1) = -32.$

4.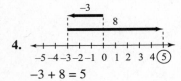
    $-3 + 8 = 5$

5. **a.** $15 + (-18)$
    Subtract the absolute values.
    $18 - 15 = 3$
    Use the sign of the number with the largest
    absolute value.
    $15 + (-18) = -3$

    b. $-19 + 20 = 20 - 19 = 1$

    c. $-0.6 + 0.4 = -(0.6 - 0.4) = -(0.2) = -0.2$

6. **a.** $-\dfrac{3}{5} + \left(-\dfrac{2}{5}\right) = -\dfrac{5}{5} = -1$

    b. $3 + (-9) = -6$

    c. $2.2 + (-1.7) = 0.5$

    d. $-\dfrac{2}{7} + \dfrac{3}{10} = -\dfrac{20}{70} + \dfrac{21}{70} = \dfrac{1}{70}$

7. **a.** $8 + (-5) + (-9) = 3 + (-9) = -6$

    b. $[-8+5] + \left[-5+|-2|\right] = [-3] + [-5+2]$
    $\qquad\qquad\qquad\qquad\qquad = -3 + [-3]$
    $\qquad\qquad\qquad\qquad\qquad = -6$

8. $-7 + 4 + 7 = -3 + 7 = 4$
    The temperature at 8 A.M. was 4°F.

9. **a.** The opposite of $-\dfrac{5}{9}$ is $\dfrac{5}{9}$.

    b. The opposite of 8 is -8.

c. The opposite of 6.2 is -6.2.

d. The opposite of -3 is 3.

10. **a.** Since $|-15| = 15,\ -|-15| = -15.$

    b. $-\left(-\dfrac{3}{5}\right) = \dfrac{3}{5}$

    c. $-(-5y) = 5y$

    d. $-(-8) = 8$

Vocabulary, Readiness & Video Check 1.5

1. Two numbers that are the same distance from 0
    but lie on opposite sides of 0 are called
    <u>opposites</u>.

2. If n is a number, then $n + (-n) = \underline{0}$.

3. If n is a number, then $-(-n) = \underline{n}$.

4. The sum of two negative numbers is always <u>a
    negative number</u>.

5. absolute values

6. Negative; when you add two numbers with
    different signs, the sign of the sum is the same as
    the sign of the number with the larger absolute
    value and -8.4 has a larger absolute value than
    6.3.

7. Negative temperatures; the high temperature for
    the day was -6°F.

8. Example 13 is an example of the opposite of the
    absolute value of $-a$, not the opposite of $-a$. The
    absolute value of $-a$ is positive, so its opposite is
    negative, therefore the answers to Examples 12
    and 13 have different signs.

Exercise Set 1.5

1. $6 + 3 = 9$

3. $-6 + (-8) = -14$

5. $8 + (-7) = 1$

7. $-14 + 2 = -12$

9. $-2 + (-3) = -5$

11. $-9 + (-3) = -12$

13. $-7 + 3 = -4$

15. $10 + (-3) = 7$

17. $5 + (-7) = -2$

19. $-16 + 16 = 0$

21. $27 + (-46) = -19$

23. $-18 + 49 = 31$

25. $-33 + (-14) = -47$

27. $6.3 + (-8.4) = -2.1$

29. $|-8| + (-16) = 8 + (-16) = -8$

31. $117 + (-79) = 38$

33. $-9.6 + (-3.5) = -13.1$

35. $-\dfrac{3}{8} + \dfrac{5}{8} = \dfrac{2}{8} = \dfrac{1}{4}$

37. $-\dfrac{7}{16} + \dfrac{1}{4} = -\dfrac{7}{16} + \dfrac{1 \cdot 4}{4 \cdot 4} = -\dfrac{7}{16} + \dfrac{4}{16} = -\dfrac{3}{16}$

39. $\begin{aligned}-\dfrac{7}{10} + \left(-\dfrac{3}{5}\right) &= -\dfrac{7}{10} + \left(-\dfrac{3 \cdot 2}{5 \cdot 2}\right) \\ &= -\dfrac{7}{10} + \left(-\dfrac{6}{10}\right) \\ &= -\dfrac{13}{10}\end{aligned}$

41. $-15 + 9 + (-2) = -6 + (-2) = -8$

43. $-21 + (-16) + (-22) = -37 + (-22) = -59$

45. $-23 + 16 + (-2) = -7 + (-2) = -9$

47. $|5 + (-10)| = |-5| = 5$

49. $6 + (-4) + 9 = 2 + 9 = 11$

51. $[-17 + (-4)] + [-12 + 15] = [-21] + [3] = -18$

53. $|9 + (-12)| + |-16| = |-3| + 16 = 3 + 16 = 19$

55. $\begin{aligned}-1.3 + [0.5 + (-0.3) + 0.4] &= -1.3 + [0.2 + 0.4] \\ &= -1.3 + [0.6] \\ &= -0.7\end{aligned}$

57. $-15 + 9 = -6$
The high temperature in Anoka was $-6°$.

59. $-512 + 658 = 146$
Your elevation is 146 feet.

61. $\begin{aligned}2.5 + 9 + (-14.2) + (-4.2) \\ = 11.5 + (-14.2) + (-4.2) \\ = -2.7 + (-4.2) \\ = -6.9\end{aligned}$
The total net income for fiscal year 2009 was $-\$6.9$ million.

63. $-6 + (-5) + (-3) + (-2) = -16$
Her score was -16 or 16 under par.

65. The opposite of 6 is -6.

67. The opposite of -2 is 2.

69. The opposite of 0 is 0.

71. Since $|-6|$ is 6, the opposite of $|-6|$ is -6.

73. $-|-2| = -2$

75. $-|0| = -0 = 0$

77. $-\left|-\dfrac{2}{3}\right| = -\dfrac{2}{3}$

79. Let $x = -4$.
$\begin{aligned}x + 9 &= 5 \\ (-4) + 9 &\overset{?}{=} 5 \\ 5 &= 5, \text{ true}\end{aligned}$
-4 is a solution of the equation.

81. Let $y = -1$.
$\begin{aligned}y + (-3) &= -7 \\ (-1) + (-3) &\overset{?}{=} -7 \\ -4 &= -7, \text{ false}\end{aligned}$
-1 is not a solution of the equation.

83. Look for the tallest bar. The temperature is the highest in July.

85. Look for the bar whose length has a positive value closest to 0; October

87. $\begin{aligned}[(-9.1) + 14.4 + 8.8] \div 3 &= [5.3 + 8.8] \div 3 \\ &= [14.1] \div 3 \\ &= 4.7\end{aligned}$
The average was $4.7°$F.

89. $7 + (-10) = -3$

91. $-10 + (-12) = -22$

93. Since a is a positive number, $-a$ is a <u>negative</u> number.

95. Since a is a positive number, $a + a$ is a <u>positive</u> number.

97. True

99. False; for example, $4 + (-2) = 2 > 0$.

101. answers may vary

103. answers may vary

Section 1.6 Practice

1. a. $-7 - 6 = -7 + (-6) = -13$

b. $-8 - (-1) = -8 + 1 = -7$

c. $9 - (-3) = 9 + 3 = 12$

d. $5 - 7 = 5 + (-7) = -2$

2. a. $8.4 - (-2.5) = 8.4 + 2.5 = 10.9$

b. $-\dfrac{5}{8} - \left(-\dfrac{1}{8}\right) = -\dfrac{5}{8} + \dfrac{1}{8} = -\dfrac{4}{8} = -\dfrac{1}{2}$

c. $-\dfrac{3}{4} - \dfrac{1}{5} = -\dfrac{3}{4} + \left(-\dfrac{1}{5}\right)$

$= -\dfrac{15}{20} + \left(-\dfrac{4}{20}\right)$

$= -\dfrac{19}{20}$

3. $-2 - 5 = -2 + (-5) = -7$

4. a. $-15 - 2 - (-4) + 7 = -15 + (-2) + 4 + 7 = -6$

b. $3.5 + (-4.1) - (-6.7) = 3.5 + (-4.1) + 6.7$

$= 6.1$

5. a. $-4 + [(-8 - 3) - 5] = -4 + [(-8 + (-3)) - 5]$

$= -4 + [(-11) - 5]$

$= -4 + [-11 + (-5)]$

$= -4 + [-16]$

$= -20$

b. $\left|-13\right| - 3^2 + [2 - (-7)] = 13 - 9 + [2 + 7]$

$= 13 - 9 + 9$

$= 13$

6. a. $\dfrac{7 - x}{2y + x} = \dfrac{7 - (-3)}{2(4) + (-3)} = \dfrac{7 + 3}{8 + (-3)} = \dfrac{10}{5} = 2$

b. $y^2 + x = (4)^2 + (-3) = 16 + (-3) = 13$

7. $282 - (-75) = 282 + 75 = 357$
The overall change was \$357.

8. a. $x = 90° - 62° = 28°$

b. $y = 180° - 43° = 137°$

Vocabulary, Readiness & Video Check 1.6

1. 7 minus a number <u>$7 - x$</u>

2. 7 subtracted from a number <u>$x - 7$</u>.

3. A number decreased by 7 <u>$x - 7$</u>

4. 7 less a number <u>$7 - x$</u>

5. A number less than 7 <u>$7 - x$</u>

6. A number subtracted from 7 <u>$7 - x$</u>

7. To evaluate $x - y$ for $x = -10$ and $y = -14$, we replace x with -10 and y with -14 and evaluate <u>$-10 - (-14)$</u>.   d

8. The expression $-5 - 10$ equals <u>$-5 + (-10)$</u>.   c

9. To subtract two real numbers, change the operation to <u>addition</u> and take the <u>opposite</u> of the second number.

10. $-10 + (8) + (-4) + (-20)$; it's rewritten to change the subtraction operations to addition and turn the problem into an addition of real numbers problem.

11. There's a minus sign in the numerator and the replacement value is negative (notice parentheses are used around the replacement value), and it's always good to be careful when working with negative signs.

12. This means that the overall vertical altitude change of the jet is actually a decrease in altitude from when the Example started.

13. In Example 9, you have two supplementary angles and know the measure of one of them. From the definition, you know that the two supplementary angles must sum to 180°. Therefore you can subtract the known angle measure from 180° to get the measure of the other angle.

Exercise Set 1.6

1. $-6 - 4 = -6 + (-4) = -10$

3. $4 - 9 = 4 + (-9) = -5$

5. $16 - (-3) = 16 + 3 = 19$

7. $\dfrac{1}{2} - \dfrac{1}{3} = \dfrac{1}{2} + \left(-\dfrac{1}{3}\right)$
$$= \dfrac{1 \cdot 3}{2 \cdot 3} + \left(-\dfrac{1 \cdot 2}{3 \cdot 2}\right)$$
$$= \dfrac{3}{6} + \left(-\dfrac{2}{6}\right)$$
$$= \dfrac{1}{6}$$

9. $-16 - (-18) = -16 + 18 = 2$

11. $-6 - 5 = -6 + (-5) = -11$

13. $7 - (-4) = 7 + 4 = 11$

15. $-6 - (-11) = -6 + 11 = 5$

17. $16 - (-21) = 16 + 21 = 37$

19. $9.7 - 16.1 = 9.7 + (-16.1) = -6.4$

21. $-44 - 27 = -44 + (-27) = -71$

23. $-21 - (-21) = -21 + 21 = 0$

25. $-2.6 - (-6.7) = -2.6 + 6.7 = 4.1$

27. $-\dfrac{3}{11} - \left(-\dfrac{5}{11}\right) = -\dfrac{3}{11} + \dfrac{5}{11} = \dfrac{2}{11}$

29. $-\dfrac{1}{6} - \dfrac{3}{4} = -\dfrac{1}{6} + \left(-\dfrac{3}{4}\right)$
$$= -\dfrac{1 \cdot 2}{6 \cdot 2} + \left(-\dfrac{3 \cdot 3}{4 \cdot 3}\right)$$
$$= -\dfrac{2}{12} + \left(-\dfrac{9}{12}\right)$$
$$= -\dfrac{11}{12}$$

31. $8.3 - (-0.62) = 8.3 + 0.62 = 8.92$

33. $8 - (-5) = 8 + 5 = 13$

35. $-6 - (-1) = -6 + 1 = -5$

37. $7 - 8 = 7 + (-8) = -1$

39. $-8 - 15 = -8 + (-15) = -23$

41. $-10 - (-8) + (-4) - 20 = -10 + 8 + (-4) + (-20)$
$$= -2 + (-4) + (-20)$$
$$= -6 + (-20)$$
$$= -26$$

43. $5 - 9 + (-4) - 8 - 8 = 5 + (-9) + (-4) + (-8) + (-8)$
$$= -4 + (-4) + (-8) + (-8)$$
$$= -8 + (-8) + (-8)$$
$$= -16 + (-8)$$
$$= -24$$

45. $-6 - (2 - 11) = -6 - (-9) = -6 + 9 = 3$

47. $3^3 - 8 \cdot 9 = 27 - 8 \cdot 9 = 27 - 72 = 27 + (-72) = -45$

49. $2 - 3(8 - 6) = 2 - 3(2) = 2 - 6 = 2 + (-6) = -4$

51. $(3 - 6) + 4^2 = [3 + (-6)] + 4^2$
$$= [-3] + 4^2$$
$$= [-3] + 16$$
$$= 13$$

53. $-2 + [(8 - 11) - (-2 - 9)]$
$$= -2 + [(8 + (-11)) - (-2 + (-9))]$$
$$= -2 + [(-3) - (-11)]$$
$$= -2 + [(-3) + 11]$$
$$= -2 + [8]$$
$$= 6$$

55. $|-3| + 2^2 + [-4 - (-6)] = 3 + 2^2 + [-4 + 6]$
$$= 3 + 2^2 + [2]$$
$$= 3 + 4 + [2]$$
$$= 7 + [2]$$
$$= 9$$

57. Let $x = -5$ and $y = 4$.
$x - y = -5 - 4 = -5 + (-4) = -9$

59. Let $x = -5$, $y = 4$, and $t = 10$.
$|x| + 2t - 8y = |-5| + 2(10) - 8(4)$
$$= 5 + 2(10) - 8(4)$$
$$= 5 + 20 - 32$$
$$= 25 - 32$$
$$= 25 + (-32)$$
$$= -7$$

61. Let $x = -5$ and $y = 4$.
$$\frac{9-x}{y+6} = \frac{9-(-5)}{4+6} = \frac{9+5}{4+6} = \frac{14}{10} = \frac{2 \cdot 7}{2 \cdot 5} = \frac{7}{5}$$

63. Let $x = -5$ and $y = 4$.
$y^2 - x = 4^2 - (-5) = 16 + 5 = 21$

65. Let $x = -5$ and $t = 10$.
$$\frac{|x-(-10)|}{2t} = \frac{|-5-(-10)|}{2(10)}$$
$$= \frac{|-5+10|}{2(10)}$$
$$= \frac{|5|}{2(10)}$$
$$= \frac{5}{20}$$
$$= \frac{5}{4 \cdot 5}$$
$$= \frac{1}{4}$$

67. The change in temperature is the difference between the last temperature and the first temperature.
$-56 - 44 = -56 + (-44) = -100$
The temperature dropped 100°F.

69. $136 - (-129) = 136 + 129 = 265$
136°F is 265°F warmer than −129°F.

71. $13{,}796 - (-21{,}857) = 13{,}796 + 21{,}857 = 35{,}653$
The difference in elevation is 35,653 feet.

73. $-250 + 120 - 178 = -250 + 120 + (-178)$
$$= -130 + (-178)$$
$$= -308$$
The overall vertical change is −308 feet.

75. $19{,}340 - (-512) = 19{,}340 + 512 = 19{,}852$
Mt. Kilimanjaro is 19,852 feet higher.

77. $y = 180 - 50 = 180 + (-50) = 130$
The supplementary angle is 130°.

79. $x = 90 - 60 = 90 + (-60) = 30$
The complementary angle is 30°.

81. Let $x = -4$.
$$x - 9 = 5$$
$$-4 - 9 \stackrel{?}{=} 5$$
$$-4 + (-9) \stackrel{?}{=} 5$$
$$-13 = 5, \text{ false}$$
−4 is not a solution of the equation.

83. Let $x = -2$.
$$-x + 6 = -x - 1$$
$$-(-2) + 6 \stackrel{?}{=} -(-2) - 1$$
$$2 + 6 \stackrel{?}{=} 2 + (-1)$$
$$8 = 1, \text{ false}$$
−2 is not a solution of the equation.

85. Let $x = 2$.
$$-x - 13 = -15$$
$$-2 - 13 \stackrel{?}{=} -15$$
$$-2 + (-13) \stackrel{?}{=} -15$$
$$-15 = -15, \text{ true}$$
2 is a solution of the equation.

87. The sum of −5 and a number is $-5 + x$.

89. Subtract a number from −20 is $-20 - x$.

91.

Month	Monthly Increase or Decrease
February	$-23.7 - (-19.3) = -23.7 + 19.3 = -4.4°$
March	$-21.1 - (-23.7) = -21.1 + 23.7 = 2.6°$
April	$-9.1 - (-21.1) = -9.1 + 21.1 = 12°$
May	$14.4 - (-9.1) = 14.4 + 9.1 = 23.5°$
June	$29.7 - 14.4 = 29.7 + (-14.4) = 15.3°$

93. The largest positive number corresponds to May.

95. answers may vary

97. $9 - (-7) = 9 + 7 = 16$

99. $10 - 30 = 10 + (-30) = -20$

101. true; answers may vary

103. false; answers may vary

105. Since 56,875 is less than 87,262, the answer is negative.
$56,875 - 87,262 = -30,387$

Integrated Review

1. The opposite of a positive number is a <u>negative</u> number.

2. The sum of two negative numbers is a <u>negative</u> number.

3. The absolute value of a negative number is a <u>positive</u> number.

4. The absolute value of zero is <u>0</u>.

5. The reciprocal of a positive number is a <u>positive</u> number.

6. The sum of a number and its opposite is <u>0</u>.

7. The absolute value of a positive number is a <u>positive</u> number.

8. The opposite of a negative number is a <u>positive</u> number.

	Number	Opposite	Absolute Value
9.	$\frac{1}{7}$	$-\frac{1}{7}$	$\frac{1}{7}$
10.	$-\frac{12}{5}$	$\frac{12}{5}$	$\frac{12}{5}$
11.	3	-3	3
12.	$-\frac{9}{11}$	$\frac{9}{11}$	$\frac{9}{11}$

13. $-19 + (-23) = -42$

14. $7 - (-3) = 7 + 3 = 10$

15. $-15 + 17 = 2$

16. $-8 - 10 = -8 + (-10) = -18$

17. $18 + (-25) = -7$

18. $-2 + (-37) = -39$

19. $-14 - (-12) = -14 + 12 = -2$

20. $5 - 14 = 5 + (-14) = -9$

21. $4.5 - 7.9 = 4.5 + (-7.9) = -3.4$

22. $-8.6 - 1.2 = -8.6 + (-1.2) = -9.8$

23. $-\dfrac{3}{4} - \dfrac{1}{7} = -\dfrac{21}{28} - \dfrac{4}{28} = -\dfrac{21}{28} + \left(-\dfrac{4}{28}\right) = -\dfrac{25}{28}$

24. $\dfrac{2}{3} - \dfrac{7}{8} = \dfrac{16}{24} - \dfrac{21}{24} = \dfrac{16}{24} + \left(-\dfrac{21}{24}\right) = -\dfrac{5}{24}$

25. $\begin{aligned} -9 - (-7) + 4 - 6 &= -9 + 7 + 4 - 6 \\ &= -9 + 7 + 4 + (-6) \\ &= -4 \end{aligned}$

26. $\begin{aligned} 11 - 20 + (-3) - 12 &= 11 + (-20) + (-3) + (-12) \\ &= -9 + (-3) + (-12) \\ &= -12 + (-12) \\ &= -24 \end{aligned}$

27. $\begin{aligned} 24 - 6(14 - 11) &= 24 - 6[14 + (-11)] \\ &= 24 - 6(3) \\ &= 24 - 18 \\ &= 24 + (-18) \\ &= 6 \end{aligned}$

28. $\begin{aligned} 30 - 5(10 - 8) &= 30 - 5[10 + (-8)] \\ &= 30 - 5(2) \\ &= 30 - 10 \\ &= 30 + (-10) \\ &= 20 \end{aligned}$

29. $(7 - 17) + 4^2 = [7 + (-17)] + 4^2 = (-10) + 16 = 6$

30. $\begin{aligned} 9^2 + (10 - 30) &= 9^2 + [10 + (-30)] \\ &= 81 + (-20) \\ &= 61 \end{aligned}$

31. $\begin{aligned} |-9| + 3^2 + (-4 - 20) &= 9 + 9 + [-4 + (-20)] \\ &= 9 + 9 + (-24) \\ &= 18 + (-24) \\ &= -6 \end{aligned}$

32. $\begin{aligned} |-4 - 5| + 5^2 + (-50) &= |-4 + (-5)| + 5^2 + (-50) \\ &= |-9| + 25 + (-50) \\ &= 9 + 25 + (-50) \\ &= 34 + (-50) \\ &= -16 \end{aligned}$

33. $\begin{aligned} -7 + [(1 - 2) + (-2 - 9)] &= -7 + [(-1) + (-11)] \\ &= -7 + [-12] \\ &= -19 \end{aligned}$

34. $\begin{aligned} -6 + [(-3 + 7) + (4 - 15)] &= -6 + [(4) + (-11)] \\ &= -6 + (-7) \\ &= -13 \end{aligned}$

35. $1 - 5 = 1 + (-5) = -4$

36. $-3 - (-2) = -3 + 2 = -1$

37. $\dfrac{1}{4} - \left(-\dfrac{2}{5}\right) = \dfrac{1}{4} + \dfrac{2}{5} = \dfrac{5}{20} + \dfrac{8}{20} = \dfrac{13}{20}$

38. $-\dfrac{5}{8} - \left(\dfrac{1}{10}\right) = -\dfrac{25}{40} - \dfrac{4}{40} = -\dfrac{25}{40} + \left(-\dfrac{4}{40}\right) = -\dfrac{29}{40}$

39. $\begin{aligned} 2(19 - 17)^3 &- 3(-7 + 9)^2 \\ &= 2[19 + (-17)]^3 - 3(-7 + 9)^2 \\ &= 2(2)^3 - 3(2)^2 \\ &= 2(8) - 3(4) \\ &= 16 - 12 \\ &= 16 + (-12) \\ &= 4 \end{aligned}$

40. $\begin{aligned} 3(10 - 9)^2 &+ 6(20 - 19)^3 \\ &= 3[10 + (-9)]^2 + 6[20 + (-19)]^3 \\ &= 3(1)^2 + 6(1)^3 \\ &= 3 + 6 \\ &= 9 \end{aligned}$

41. $x - y = -2 - (-1) = -2 + 1 = -1$

42. $x + y = -2 + (-1) = -3$

43. $y + z = -1 + 9 = 8$

44. $z - y = 9 - (-1) = 9 + 1 = 10$

45. $\dfrac{|5z-x|}{y-x} = \dfrac{|5(9)-(-2)|}{-1-(-2)} = \dfrac{|45+2|}{-1+2} = \dfrac{|47|}{1} = 47$

46. $\dfrac{|-x-y+z|}{2z} = \dfrac{|-(-2)-(-1)+9|}{2(9)}$

$\qquad = \dfrac{|2+1+9|}{18}$

$\qquad = \dfrac{|12|}{18}$

$\qquad = \dfrac{12}{18}$

$\qquad = \dfrac{2}{3}$

Section 1.7 Practice

1. a. $8(-5) = -40$

 b. $(-3)(-4) = 12$

 c. $(-6)(9) = -54$

2. a. $(-1)(-5)(-6) = 5(-6) = -30$

 b. $(-3)(-2)(4) = 6(4) = 24$

 c. $(-4)(0)(5) = 0(5) = 0$

 d. $(-2)(-3)-(-4)(5) = 6-(-20)$
$\qquad\qquad\qquad\qquad\;\; = 6+20$
$\qquad\qquad\qquad\qquad\;\; = 26$

3. a. $(0.23)(-0.2) = -[(0.23)(0.2)] = -0.046$

 b. $\left(-\dfrac{3}{5}\right)\cdot\left(\dfrac{4}{9}\right) = -\dfrac{3\cdot4}{5\cdot9} = -\dfrac{12}{45} = -\dfrac{4}{15}$

 c. $\left(-\dfrac{7}{12}\right)(-24) = \dfrac{7\cdot24}{12\cdot1} = 7\cdot2 = 14$

4. a. $(-6)^2 = (-6)(-6) = 36$

 b. $-6^2 = -(6\cdot6) = -(36) = -36$

 c. $(-4)^3 = (-4)(-4)(-4) = 16(-4) = -64$

 d. $-4^3 = -(4\cdot4\cdot4) = -[16(4)] = -64$

5. a. The reciprocal of $\dfrac{8}{3}$ is $\dfrac{3}{8}$ since $\dfrac{8}{3}\cdot\dfrac{3}{8} = 1$.

 b. The reciprocal of 15 is $\dfrac{1}{15}$ since $15\cdot\dfrac{1}{15} = 1$.

 c. The reciprocal of $-\dfrac{2}{7}$ is $-\dfrac{7}{2}$ since
$\left(-\dfrac{2}{7}\right)\left(-\dfrac{7}{2}\right) = 1$.

 d. The reciprocal of -5 is $-\dfrac{1}{5}$ since
$(-5)\left(-\dfrac{1}{5}\right) = 1$.

6. a. $\dfrac{16}{-2} = 16\left(-\dfrac{1}{2}\right) = -8$

 b. $24 \div (-6) = 24\left(-\dfrac{1}{6}\right) = -4$

 c. $\dfrac{-35}{-7} = \dfrac{35}{7} = \dfrac{5\cdot7}{7} = 5$

7. a. $\dfrac{-18}{-6} = \dfrac{18}{6} = \dfrac{3\cdot6}{6} = 3$

 b. $\dfrac{-48}{3} = -\dfrac{48}{3} = -\dfrac{3\cdot16}{3} = -16$

 c. $\dfrac{3}{5} \div \left(-\dfrac{1}{2}\right) = \dfrac{3}{5}\cdot(-2) = -\dfrac{6}{5}$

 d. $-\dfrac{4}{9} \div 8 = -\dfrac{4}{9}\cdot\dfrac{1}{8} = -\dfrac{4}{9\cdot4\cdot2} = -\dfrac{1}{9\cdot2} = -\dfrac{1}{18}$

8. a. $\dfrac{0}{-2} = 0$

 b. $\dfrac{-4}{0}$ is undefined.

 c. $\dfrac{-5}{6(0)} = \dfrac{-5}{0}$ is undefined.

9. a. $\dfrac{(-8)(-11)-4}{-9-(-4)} = \dfrac{88-4}{-9+4} = \dfrac{84}{-5} = -\dfrac{84}{5}$

b. $\dfrac{3(-2)^3-9}{-6+3} = \dfrac{3(-8)-9}{-3}$

$= \dfrac{-24-9}{-3}$

$= \dfrac{-33}{-3}$

$= 11$

10. a. $7y - x = 7(-2) - (-5) = -14 + 5 = -9$

b. $x^2 - y^3 = (-5)^2 - (-2)^3$

$= 25 - (-8)$

$= 25 + 8$

$= 33$

c. $\dfrac{2x}{3y} = \dfrac{2(-5)}{3(-2)} = \dfrac{-10}{-6} = \dfrac{5}{3}$

11. total score $= 4 \cdot (-13) = -52$
The card player's total score was −52.

Graphing Calculator Explorations

1. $-38(26 - 27) = 38$

2. $-59(-8) + 1726 = 2198$

3. $134 + 25(68 - 91) = -441$

4. $45(32) - 8(218) = -304$

5. $\dfrac{-50(294)}{175 - 265} = 163.\overline{3}$

6. $\dfrac{-444 - 444.8}{-181 - 324} = 1.76$

7. $9^5 - 4550 = 54,499$

8. $5^8 - 6259 = 384,366$

9. $(-125)^2 = 15,625$

10. $-125^2 = -15,625$

Vocabulary, Readiness & Video Check 1.7

1. If n is a real number, then $n \cdot 0 = \underline{0}$ and $0 \cdot n = \underline{0}$.

2. If n is a real number, but not 0, then $\dfrac{0}{n} = \underline{0}$ and we say $\dfrac{n}{0}$ is <u>undefined</u>.

3. The product of two negative numbers is a <u>positive</u> number.

4. The quotient of two negative numbers is a <u>positive</u> number.

5. The quotient of a positive number and a negative number is a <u>negative</u> number.

6. The product of a positive number and a negative number is a <u>negative</u> number.

7. The reciprocal of a positive number is a <u>positive</u> number.

8. The opposite of a positive number is a <u>negative</u> number.

9. The parentheses, or lack of them, determine the base of the expression. In Example 6, $(-2)^4$, the base is −2 and all of −2 is raised to the 4th power. In Example 7, -2^4, the base is 2 and only 2 is raised to the 4th power.

10. Remember, the product of a number and its reciprocal is 1, *not* −1. $\dfrac{2}{3} \cdot \dfrac{3}{2} = 1$, as needed.

11. Yes; because division of real numbers is defined in terms of multiplication.

12. The replacement values are negative and both will be squared. Therefore they must be placed in parentheses so the entire value, including the negative, is squared.

13. The football team lost 4 years on each play and a loss of yardage is represented by a negative number.

Exercise Set 1.7

1. $-6(4) = -24$

3. $2(-1) = -2$

5. $-5(-10) = 50$

7. $-3 \cdot 4 = -12$

9. $-7 \cdot 0 = 0$

11. $2(-9) = -18$

13. $-\dfrac{1}{2}\left(-\dfrac{3}{5}\right) = \dfrac{1 \cdot 3}{2 \cdot 5} = \dfrac{3}{10}$

15. $-\dfrac{3}{4}\left(-\dfrac{8}{9}\right) = \dfrac{3 \cdot 8}{4 \cdot 9} = \dfrac{24}{36} = \dfrac{2 \cdot 12}{3 \cdot 12} = \dfrac{2}{3}$

17. $5(-1.4) = -7$

19. $-0.2(-0.7) = 0.14$

21. $-10(80) = -800$

23. $4(-7) = -28$

25. $(-5)(-5) = 25$

27. $\dfrac{2}{3}\left(-\dfrac{4}{9}\right) = -\dfrac{2 \cdot 4}{3 \cdot 9} = -\dfrac{8}{27}$

29. $-11(11) = -121$

31. $-\dfrac{20}{25}\left(\dfrac{5}{16}\right) = -\dfrac{20 \cdot 5}{25 \cdot 16} = -\dfrac{100}{400} = -\dfrac{1}{4}$

33. $(-1)(2)(-3)(-5) = -2(-3)(-5) = 6(-5) = -30$

35. $(-2)(5) - (-11)(3) = -10 - (-33) = -10 + 33 = 23$

37. $(-6)(-1)(-2) - (-5) = -12 + 5 = -7$

39. True; example: $(-2)(-2)(-2) = -8$
False; example: $(-2)(-2)(-2)(-2) = 16$

41. False

43. $(-2)^4 = (-2)(-2)(-2)(-2)$
$\qquad = 4(-2)(-2)$
$\qquad = -8(-2)$
$\qquad = 16$

45. $-1^5 = -(1)(1)(1)(1)(1) = -1$

47. $(-5)^2 = (-5)(-5) = 25$

49. $-7^2 = -(7)(7) = -49$

51. Reciprocal of 9 is $\dfrac{1}{9}$ since $9 \cdot \dfrac{1}{9} = 1$.

53. Reciprocal of $\dfrac{2}{3}$ is $\dfrac{3}{2}$ since $\dfrac{2}{3} \cdot \dfrac{3}{2} = 1$.

55. Reciprocal of -14 is $-\dfrac{1}{14}$ since $-14 \cdot -\dfrac{1}{14} = 1$.

57. Reciprocal of $-\dfrac{3}{11}$ is $-\dfrac{11}{3}$ since $-\dfrac{3}{11} \cdot -\dfrac{11}{3} = 1$.

59. Reciprocal of 0.2 is $\dfrac{1}{0.2}$ since $0.2 \cdot \dfrac{1}{0.2} = 1$.

61. Reciprocal of $\dfrac{1}{-6.3}$ is -6.3 since
$\dfrac{1}{-6.3} \cdot -6.3 = 1$.

63. $\dfrac{18}{-2} = 18 \cdot -\dfrac{1}{2} = -9$

65. $\dfrac{-16}{-4} = -16 \cdot -\dfrac{1}{4} = 4$

67. $\dfrac{-48}{12} = -48 \cdot \dfrac{1}{12} = -4$

69. $\dfrac{0}{-4} = 0 \cdot -\dfrac{1}{4} = 0$

71. $-\dfrac{15}{3} = -15 \cdot \dfrac{1}{3} = -5$

73. $\dfrac{5}{0}$ is undefined.

75. $\dfrac{-12}{-4} = -12 \cdot -\dfrac{1}{4} = 3$

77. $\dfrac{30}{-2} = 30 \cdot -\dfrac{1}{2} = -15$

79. $\dfrac{6}{7} \div -\dfrac{1}{3} = \dfrac{6}{7} \cdot \left(-\dfrac{3}{1}\right) = -\dfrac{6 \cdot 3}{7 \cdot 1} = -\dfrac{18}{7}$

81. $-\dfrac{5}{9} \div \left(-\dfrac{3}{4}\right) = -\dfrac{5}{9} \cdot \left(-\dfrac{4}{3}\right) = \dfrac{5 \cdot 4}{9 \cdot 3} = \dfrac{20}{27}$

83. $-\dfrac{4}{9} \div \dfrac{4}{9} = -\dfrac{4}{9} \cdot \dfrac{9}{4} = -1$

85. $\dfrac{-9(-3)}{-6} = \dfrac{27}{-6} = -\dfrac{9}{2}$

87. $\dfrac{12}{9-12} = \dfrac{12}{-3} = -4$

89. $\dfrac{-6^2+4}{-2} = \dfrac{-36+4}{-2} = \dfrac{-32}{-2} = 16$

91. $\dfrac{8+(-4)^2}{4-12} = \dfrac{8+16}{4-12} = \dfrac{24}{-8} = -3$

93. $\dfrac{22+(3)(-2)}{-5-2} = \dfrac{22+(-6)}{-5-2} = \dfrac{16}{-7} = -\dfrac{16}{7}$

95. $\dfrac{-3-5^2}{2(-7)} = \dfrac{-3-25}{2(-7)} = \dfrac{-3+(-25)}{-14} = \dfrac{-28}{-14} = 2$

97. $\dfrac{6-2(-3)}{4-3(-2)} = \dfrac{6-(-6)}{4-(-6)} = \dfrac{6+6}{4+6} = \dfrac{12}{10} = \dfrac{6}{5}$

99. $\dfrac{-3-2(-9)}{-15-3(-4)} = \dfrac{-3-(-18)}{-15-(-12)} = \dfrac{-3+18}{-15+12} = \dfrac{15}{-3} = -5$

101. $\dfrac{|5-9|+|10-15|}{|2(-3)|} = \dfrac{|-4|+|-5|}{|-6|} = \dfrac{4+5}{6} = \dfrac{9}{6} = \dfrac{3}{2}$

103. Let $x = -5$ and $y = -3$.
$3x + 2y = 3(-5) + 2(-3) = -15 + (-6) = -21$

105. Let $x = -5$ and $y = -3$.
$$2x^2 - y^2 = 2(-5)^2 - (-3)^2$$
$$= 2(25) - 9$$
$$= 50 + (-9)$$
$$= 41$$

107. Let $x = -5$ and $y = -3$.
$x^3 + 3y = (-5)^3 + 3(-3) = -125 + (-9) = -134$

109. Let $x = -5$ and $y = -3$.
$\dfrac{2x-5}{y-2} = \dfrac{2(-5)-5}{-3-2} = \dfrac{-10-5}{-3-2} = \dfrac{-15}{-5} = 3$

111. Let $x = -5$ and $y = -3$.
$\dfrac{-3-y}{x-4} = \dfrac{-3-(-3)}{-5-4} = \dfrac{-3+3}{-5-4} = \dfrac{0}{-9} = 0$

113. The product of -71 and a number is $-71 \cdot x$ or $-71x$.

115. Subtract a number from -16 is $-16 - x$.

117. -29 increased by a number is $-29 + x$.

119. Divide a number by -33 is $\dfrac{x}{-33}$ or $x \div (-33)$.

121. A loss of 4 yards is represented by -4.
$3 \cdot (-4) = -12$
The team had a total loss of 12 yards.

123. Each move of 20 feet down is represented by -20.
$5 \cdot (-20) = -100$
The diver is at a depth of 100 feet.

125. Let $x = 7$.
$-5x = -35$
$-5(7) \overset{?}{=} -35$
$-35 = -35$, true
7 is a solution of the equation.

127. Let $x = -20$.
$\dfrac{x}{10} = 2$
$\dfrac{-20}{10} \overset{?}{=} 2$
$-2 = 2$, false
-20 is not a solution of the equation.

129. Let $x = 5$.
$-3x - 5 = -20$
$-3(5) - 5 \overset{?}{=} -20$
$-15 - 5 \overset{?}{=} -20$
$-20 = -20$, true
5 is a solution of the equation.

131. $2(-81) = -162$
The surface temperature of Jupiter is $-162°$F.

133. answers may vary

135. -1 and 1 are their own reciprocals.

137. Since q is negative, r is negative, and t is positive, then $\dfrac{q}{r \cdot t}$ is positive.

139. It is not possible to determine whether $q + t$ is positive or negative.

141. Since q is negative, r is negative, and t is positive, then $t(q + r)$ is negative.

143. $\begin{aligned} -2 + \dfrac{-15}{3} &= \dfrac{-2 \cdot 3}{1 \cdot 3} + \dfrac{-15}{3} \\ &= \dfrac{-6 + (-15)}{3} \\ &= \dfrac{-21}{3} \\ &= -7 \end{aligned}$

145. $2[-5 + (-3)] = 2(-8) = -16$

Section 1.8 Practice

1. a. $x \cdot 8 = \underline{8 \cdot x}$

    b. $x + 17 = \underline{17 + x}$

2. a. $(2 + 9) + 7 = \underline{2 + (9 + 7)}$

    b. $-4 \cdot (2 \cdot 7) = \underline{(-4 \cdot 2) \cdot 7}$

3. a. $(5 + x) + 9 = (x + 5) + 9 = x + (5 + 9) = x + 14$

    b. $5(-6x) = [5 \cdot (-6)]x = -30x$

4. a. $5(x - y) = 5(x) - 5(y) = 5x - 5y$

    b. $-6(4 + 2t) = -6(4) + (-6)(2t) = -24 - 12t$

    c. $\begin{aligned} 2(3x - 4y - z) &= 2(3x) + 2(-4y) + 2(-z) \\ &= 6x - 8y - 2z \end{aligned}$

    d. $(3 - y) \cdot (-1) = 3(-1) + (-y)(-1) = -3 + y$

    e. $\begin{aligned} -(x - 7 + 2s) &= (-1)(x - 7 + 2s) \\ &= (-1)x + (-1)(-7) + (-1)(2s) \\ &= -x + 7 - 2s \end{aligned}$

    f. $\begin{aligned} \dfrac{1}{2}(2x + 4) + 9 &= \dfrac{1}{2}(2x) + \dfrac{1}{2}(4) + 9 \\ &= x + 2 + 9 \\ &= x + 11 \end{aligned}$

5. a. $5 \cdot w + 5 \cdot 3 = 5(w + 3)$

    b. $9w + 9z = 9 \cdot w + 9 \cdot z = 9(w + z)$

6. a. $(7 \cdot 3x) \cdot 4 = (3x \cdot 7) \cdot 4$; commutative property of multiplication

    b. $6 + (3 + y) = (6 + 3) + y$; associative property of addition

    c. $8 + (t + 0) = 8 + t$; identity element for addition

    d. $-\dfrac{3}{4} \cdot \left(-\dfrac{4}{3}\right) = 1$; multiplicative inverse property

    e. $(2 + x) + 5 = 5 + (2 + x)$; commutative property of addition

    f. $3 + (-3) = 0$; additive inverse property

    g. $(-3b) \cdot 7 = (-3 \cdot 7) \cdot b$; commutative and associative properties of multiplication

Vocabulary, Readiness & Video Check 1.8

1. $x + 5 = 5 + x$ is a true statement by the <u>commutative property of addition</u>.

2. $x \cdot 5 = 5 \cdot x$ is a true statement by the <u>commutative property of multiplication</u>.

3. $3(y + 6) = 3 \cdot y + 3 \cdot 6$ is true by the <u>distributive property</u>.

4. $2 \cdot (x \cdot y) = (2 \cdot x) \cdot y$ is a true statement by the <u>associative property of multiplication</u>.

5. $x + (7 + y) = (x + 7) + y$ is a true statement by the <u>associative property of addition</u>.

6. The numbers $-\dfrac{2}{3}$ and $-\dfrac{3}{2}$ are called <u>reciprocals or multiplicative inverses</u>.

7. The numbers $-\dfrac{2}{3}$ and $\dfrac{2}{3}$ are called <u>opposites or additive inverses</u>.

8. order; grouping

9. 2 is outside the parentheses, so the point is made that you should only distribute the −9 to the terms within the parentheses and not also to the 2.

10. The identity element for addition is <u>0</u> because if we add <u>0</u> to any real number, the result is that real number.
The identity element for multiplication is <u>1</u> because any real number times <u>1</u> gives a result of that original real number.

Exercise Set 1.8

1. $x + 16 = 16 + x$

3. $-4 \cdot y = y \cdot (-4)$

5. $xy = yx$

7. $2x + 13 = 13 + 2x$

9. $(xy) \cdot z = x \cdot (yz)$

11. $2 + (a + b) = (2 + a) + b$

13. $4 \cdot (ab) = 4a \cdot (b)$

15. $(a + b) + c = a + (b + c)$

17. $8 + (9 + b) = (8 + 9) + b = 17 + b$

19. $4(6y) = (4 \cdot 6)y = 24y$

21. $\dfrac{1}{5}(5y) = \left(\dfrac{1}{5} \cdot 5\right)y = 1 \cdot y = y$

23. $(13 + a) + 13 = (a + 13) + 13$
$\qquad\qquad\qquad = a + (13 + 13)$
$\qquad\qquad\qquad = a + 26$

25. $-9(8x) = (-9 \cdot 8)x = -72x$

27. $\dfrac{3}{4}\left(\dfrac{4}{3}s\right) = \left(\dfrac{3}{4} \cdot \dfrac{4}{3}\right)s = 1s = s$

29. $\dfrac{2}{3} + \left(\dfrac{4}{3} + x\right) = \dfrac{2}{3} + \dfrac{4}{3} + x = \dfrac{6}{3} + x = 2 + x$

31. $4(x + y) = 4x + 4y$

33. $9(x - 6) = 9x - 9 \cdot 6 = 9x - 54$

35. $2(3x + 5) = 2(3x) + 2(5) = 6x + 10$

37. $7(4x - 3) = 7(4x) - 7(3) = 28x - 21$

39. $3(6 + x) = 3(6) + 3x = 18 + 3x$

41. $-2(y - z) = -2y - (-2)z = -2y + 2z$

43. $-7(3y + 5) = -7(3y) + (-7)(5) = -21y - 35$

45. $5(x + 4m + 2) = 5x + 5(4m) + 5(2)$
$\qquad\qquad\qquad\quad = 5x + 20m + 10$

47. $-4(1 - 2m + n) = -4(1) - (-4)(2m) + (-4)n$
$\qquad\qquad\qquad\quad = -4 + 8m - 4n$

49. $-(5x + 2) = -1(5x + 2)$
$\qquad\qquad\quad = -1(5x) + (-1)(2)$
$\qquad\qquad\quad = -5x - 2$

51. $-(r - 3 - 7p) = -1(r - 3 - 7p)$
$\qquad\qquad\qquad = -1r - (-1)(3) - (-1)(7p)$
$\qquad\qquad\qquad = -r + 3 + 7p$

53. $\dfrac{1}{2}(6x + 8) = \dfrac{1}{2}(6x) + \dfrac{1}{2}(8)$
$\qquad\qquad\quad = \left(\dfrac{1}{2} \cdot 6\right)x + \left(\dfrac{1}{2} \cdot 8\right)$
$\qquad\qquad\quad = 3x + 4$

55. $-\dfrac{1}{3}(3x - 9y) = -\dfrac{1}{3}(3x) - \left(-\dfrac{1}{3}\right)(9y)$
$\qquad\qquad\qquad = \left(-\dfrac{1}{3} \cdot 3\right)x - \left(-\dfrac{1}{3} \cdot 9\right)y$
$\qquad\qquad\qquad = -1 \cdot x + 3 \cdot y$
$\qquad\qquad\qquad = -x + 3y$

57. $3(2r + 5) - 7 = 3(2r) + 3(5) - 7$
$\qquad\qquad\qquad = 6r + 15 + (-7)$
$\qquad\qquad\qquad = 6r + 8$

59. $-9(4x + 8) + 2 = -9(4x) + (-9)(8) + 2$
$\qquad\qquad\qquad\quad = -36x - 72 + 2$
$\qquad\qquad\qquad\quad = -36x - 70$

61. $-4(4x + 5) - 5 = -4(4x) + (-4)(5) - 5$
$\qquad\qquad\qquad\quad = -16x + (-20) + (-5)$
$\qquad\qquad\qquad\quad = -16x - 25$

63. $4 \cdot 1 + 4 \cdot y = 4(1 + y)$

65. $11x + 11y = 11(x + y)$

67. $(-1) \cdot 5 + (-1) \cdot x = -1(5 + x) = -(5 + x)$

69. $30a + 30b = 30(a + b)$

71. $3 \cdot 5 = 5 \cdot 3$; commutative property of multiplication

73. $2 + (x + 5) = (2 + x) + 5$; associative property of addition

75. $9(3 + 7) = 9 \cdot 3 + 9 \cdot 7$; distributive property

77. $(4 \cdot y) \cdot 9 = 4 \cdot (y \cdot 9)$; associative property of multiplication

79. $0 + 6 = 6$; identity element of addition

81. $-4(y + 7) = -4 \cdot y + (-4) \cdot 7$; distributive property

83. $-4 \cdot (8 \cdot 3) = (8 \cdot -4) \cdot 3$; associative and commutative properties of multiplication

85.

Expression	Opposite	Reciprocal
8	-8	$\frac{1}{8}$

87.

Expression	Opposite	Reciprocal
x	$-x$	$\frac{1}{x}$

89.

Expression	Opposite	Reciprocal
$2x$	$-2x$	$\frac{1}{2x}$

91. False; the opposite of $-\dfrac{a}{2}$ is $\dfrac{a}{2}$. $-\dfrac{2}{a}$ is the reciprocal of $-\dfrac{a}{2}$.

93. "Taking a test" and "studying for the test" are not commutative, since the order in which they are performed affects the outcome.

95. "Putting on your left shoe" and "putting on your right shoe" are commutative, since the order in which they are performed does not affect the outcome.

97. "Mowing the lawn" and "trimming the hedges" are commutative, since the order in which they are performed does not affect the outcome.

99. "Dialing a number" and "turning on the cell phone" are not commutative, since the order in which they are performed affects the outcome.

101. a. The property illustrated is the commutative property of addition since the order in which they are added changed.

    b. The property illustrated is the commutative property of addition since the order in which they are added changed.

c. The property illustrated is the associative property of addition since the grouping of addition changed.

103. answers may vary

105. answers may vary

Chapter 1 Vocabulary Check

1. The symbols $\neq$, $<$, and $>$ are called <u>inequality symbols</u>.

2. A mathematical statement that two expressions are equal is called an <u>equation</u>.

3. The <u>absolute value</u> of a number is the distance between that number and 0 on the number line.

4. A symbol used to represent a number is called a <u>variable</u>.

5. Two numbers that are the same distance from 0 but lie on opposite sides of 0 are called <u>opposites</u>.

6. The number in a fraction above the fraction bar is called the <u>numerator</u>.

7. A <u>solution</u> of an equation is a value for the variable that makes the equation a true statement.

8. Two numbers whose product is 1 are called <u>reciprocals</u>.

9. In 2^3, the 2 is called the <u>base</u> and the 3 is called the <u>exponent</u>.

10. The number in a fraction below the fraction bar is called the <u>denominator</u>.

11. Parentheses and brackets are examples of <u>grouping symbols</u>.

12. A <u>set</u> is a collection of objects.

Chapter 1 Review

1. $8 < 10$ since 8 is to the left of 10 on the number line.

2. $7 > 2$ since 7 is to the right of 2 on the number line.

3. $-4 > -5$ since -4 is to the right of -5 on the number line.

4. $\dfrac{12}{2} > -8$ since $6 > -8$.

5. $|-7| < |-8|$ since $7 < 8$.

6. $|-9| > -9$ since $9 > -9$.

7. $-|-1| = -1$ since $-1 = -1$.

8. $|-14| = -(-14)$ since $14 = 14$.

9. $1.2 > 1.02$ since 1.2 is to the right of 1.02 on the number line.

10. $-\dfrac{3}{2} < -\dfrac{3}{4}$ since $-\dfrac{3}{2}$ is to the left of $-\dfrac{3}{4}$ on the number line.

11. Four is greater than or equal to negative three is written as $4 \geq -3$.

12. Six is not equal to five is written as $6 \neq 5$.

13. 0.03 is less than 0.3 is written as $0.03 < 0.3$.

14. $400 > 155$ or $155 < 400$

15. a. The natural numbers are 1 and 3.

    b. The whole numbers are 0, 1, and 3.

    c. The integers are -6, 0, 1, and 3.

    d. The rational numbers are -6, 0, 1, $1\dfrac{1}{2}$, 3, and 9.62.

    e. The irrational number is π.

    f. The real numbers are all numbers in the given set.

16. a. The natural numbers are 2 and 5.

    b. The whole numbers are 2 and 5.

    c. The integers are -3, 2, and 5.

    d. The rational numbers are -3, -1.6, 2, 5, $\dfrac{11}{2}$, and 15.1.

    e. The irrational numbers are $\sqrt{5}$ and 2π.

    f. The real numbers are all numbers in the given set.

17. Look for the negative number with the greatest absolute value. The greatest loss was on Friday.

18. Look for the largest positive number. The greatest gain was on Wednesday.

19. $36 = 4 \cdot 9 = 2 \cdot 2 \cdot 3 \cdot 3$

20. $120 = 8 \cdot 15 = 2 \cdot 2 \cdot 2 \cdot 3 \cdot 5$

21. $\dfrac{8}{15} \cdot \dfrac{27}{30} = \dfrac{8 \cdot 27}{15 \cdot 30} = \dfrac{2 \cdot 4 \cdot 3 \cdot 3 \cdot 3}{3 \cdot 5 \cdot 2 \cdot 3 \cdot 5} = \dfrac{12}{25}$

22. $\dfrac{7}{8} \div \dfrac{21}{32} = \dfrac{7}{8} \cdot \dfrac{32}{21} = \dfrac{7 \cdot 32}{8 \cdot 21} = \dfrac{7 \cdot 8 \cdot 4}{8 \cdot 3 \cdot 7} = \dfrac{4}{3}$

23.
$$\dfrac{7}{15} + \dfrac{5}{6} = \dfrac{7 \cdot 2}{15 \cdot 2} + \dfrac{5 \cdot 5}{6 \cdot 5}$$
$$= \dfrac{14}{30} + \dfrac{25}{30}$$
$$= \dfrac{14 + 25}{30}$$
$$= \dfrac{39}{30}$$
$$= \dfrac{3 \cdot 13}{3 \cdot 10}$$
$$= \dfrac{13}{10}$$

24.
$$\dfrac{3}{4} - \dfrac{3}{20} = \dfrac{3 \cdot 5}{4 \cdot 5} - \dfrac{3}{20}$$
$$= \dfrac{15}{20} - \dfrac{3}{20}$$
$$= \dfrac{15 - 3}{20}$$
$$= \dfrac{12}{20}$$
$$= \dfrac{3 \cdot 4}{5 \cdot 4}$$
$$= \dfrac{3}{5}$$

25. $2\dfrac{3}{4} + 6\dfrac{5}{8} = \dfrac{11}{4} + \dfrac{53}{8}$

$\qquad\qquad = \dfrac{11 \cdot 2}{4 \cdot 2} + \dfrac{53}{8}$

$\qquad\qquad = \dfrac{22}{8} + \dfrac{53}{8}$

$\qquad\qquad = \dfrac{22 + 53}{8}$

$\qquad\qquad = \dfrac{75}{8}$

$\qquad\qquad = 9\dfrac{3}{8}$

26. $7\dfrac{1}{6} - 2\dfrac{2}{3} = \dfrac{43}{6} - \dfrac{8}{3}$

$\qquad\qquad = \dfrac{43}{6} - \dfrac{8 \cdot 2}{3 \cdot 2}$

$\qquad\qquad = \dfrac{43}{6} - \dfrac{16}{6}$

$\qquad\qquad = \dfrac{43 - 16}{6}$

$\qquad\qquad = \dfrac{27}{6}$

$\qquad\qquad = \dfrac{9 \cdot 3}{2 \cdot 3}$

$\qquad\qquad = \dfrac{9}{2}$

$\qquad\qquad = 4\dfrac{1}{2}$

27. $5 \div \dfrac{1}{3} = 5 \cdot \dfrac{3}{1} = 15$

28. $2 \cdot 8\dfrac{3}{4} = 2 \cdot \dfrac{35}{4} = \dfrac{2 \cdot 35}{2 \cdot 2} = \dfrac{35}{2} = 17\dfrac{1}{2}$

29. $1 - \dfrac{1}{6} - \dfrac{1}{4} = \dfrac{12}{12} - \dfrac{1 \cdot 2}{6 \cdot 2} - \dfrac{1 \cdot 3}{4 \cdot 3}$

$\qquad\qquad = \dfrac{12}{12} - \dfrac{2}{12} - \dfrac{3}{12}$

$\qquad\qquad = \dfrac{12 - 2 - 3}{12}$

$\qquad\qquad = \dfrac{7}{12}$

The unknown part is $\dfrac{7}{12}$.

30. $1 - \dfrac{1}{2} - \dfrac{1}{5} = \dfrac{10}{10} - \dfrac{5}{10} - \dfrac{2}{10} = \dfrac{10 - 5 - 2}{10} = \dfrac{3}{10}$

The unknown part is $\dfrac{3}{10}$.

31. $P = 2l + 2w$

$P = 2\left(1\dfrac{1}{3}\right) + 2\left(\dfrac{7}{8}\right)$

$\quad = \dfrac{2}{1} \cdot \dfrac{4}{3} + \dfrac{2}{1} \cdot \dfrac{7}{8}$

$\quad = \dfrac{8}{3} + \dfrac{14}{8}$

$\quad = \dfrac{8 \cdot 8}{3 \cdot 8} + \dfrac{14 \cdot 3}{8 \cdot 3}$

$\quad = \dfrac{64}{24} + \dfrac{42}{24}$

$\quad = \dfrac{64 + 42}{24}$

$\quad = \dfrac{106}{24}$

$\quad = 4\dfrac{10}{24}$

$\quad = 4\dfrac{5}{12}$ meters

$A = lw$

$A = 1\dfrac{1}{3} \cdot \dfrac{7}{8}$

$\quad = \dfrac{4}{3} \cdot \dfrac{7}{8}$

$\quad = \dfrac{4 \cdot 7}{3 \cdot 2 \cdot 4}$

$\quad = \dfrac{7}{6}$

$\quad = 1\dfrac{1}{6}$ sq meters

32. $P = $ the sum of the lengths of the sides

$P = \dfrac{5}{11} + \dfrac{8}{11} + \dfrac{3}{11} + \dfrac{3}{11} + \dfrac{2}{11} + \dfrac{5}{11} = \dfrac{26}{11} = 2\dfrac{4}{11}$ in.

$A = $ the sum of the two areas, each given by lw

$A = \dfrac{5}{11} \cdot \dfrac{5}{11} + \dfrac{3}{11} \cdot \dfrac{3}{11} = \dfrac{25}{121} + \dfrac{9}{121} = \dfrac{34}{121}$ sq in.

33. $2\dfrac{1}{2}+3\dfrac{1}{16}+1\dfrac{3}{4}+2\dfrac{9}{16}+1\dfrac{13}{16}+2\dfrac{7}{16}$

$=\dfrac{5}{2}+\dfrac{49}{16}+\dfrac{7}{4}+\dfrac{41}{16}+\dfrac{29}{16}+\dfrac{39}{16}$

$=\dfrac{40}{16}+\dfrac{49}{16}+\dfrac{28}{16}+\dfrac{41}{16}+\dfrac{29}{16}+\dfrac{39}{16}$

$=\dfrac{226}{16}$

$=14\dfrac{2}{16}$

$=14\dfrac{1}{8}$ lb

34. $2\dfrac{1}{8}+2\dfrac{3}{16}=\dfrac{17}{8}+\dfrac{35}{16}=\dfrac{34}{16}+\dfrac{35}{16}=\dfrac{69}{16}=4\dfrac{5}{16}$ lb

35. Total weight = weight of boys + weight of girls

$=\dfrac{226}{16}+\dfrac{69}{16}$

$=\dfrac{295}{16}$

$=18\dfrac{7}{16}$ lb

36. Look for the largest number. Baby C weighed the most.

37. Look for the smallest number. Baby E weighed the least.

38. $3\dfrac{1}{16}-1\dfrac{3}{4}=\dfrac{49}{16}-\dfrac{7}{4}=\dfrac{49}{16}-\dfrac{28}{16}=\dfrac{21}{16}=1\dfrac{5}{16}$ lb

39. $6\cdot3^2+2\cdot8=6\cdot9+2\cdot8=54+16=70$
The answer is c.

40. $68-5\cdot2^3=68-5\cdot8=68-40=68+(-40)=28$
The answer is b.

41. $\left(\dfrac{2}{7}\right)^2=\dfrac{2}{7}\cdot\dfrac{2}{7}=\dfrac{4}{49}$

42. $\left(\dfrac{3}{4}\right)^3=\dfrac{3}{4}\cdot\dfrac{3}{4}\cdot\dfrac{3}{4}=\dfrac{27}{64}$

43. $3(1+2\cdot5)+4=3(1+10)+4$
$=3(11)+4$
$=33+4$
$=37$

44. $8+3(2\cdot6-1)=8+3(12-1)$
$=8+3(11)$
$=8+33$
$=41$

45. $\dfrac{4+|6-2|+8^2}{4+6\cdot4}=\dfrac{4+|4|+64}{4+24}$
$=\dfrac{4+4+64}{4+24}$
$=\dfrac{72}{28}$
$=\dfrac{4\cdot18}{4\cdot7}$
$=\dfrac{18}{7}$

46. $5[3(2+5)-5]=5[3(7)-5]$
$=5[21-5]$
$=5[16]$
$=80$

47. The difference of twenty and twelve is equal to the product of two and four is written as
$20-12=2\cdot4.$

48. The quotient of nine and two is greater than negative five is written as $\dfrac{9}{2}>-5.$

49. Let $x=6$ and $y=2$.
$2x+3y=2(6)+3(2)=12+6=18$

50. Let $x=6$, $y=2$, and $z=8$.
$x(y+2z)=6[2+2(8)]=6[2+16]=6[18]=108$

51. Let $x=6$, $y=2$, and $z=8$.
$\dfrac{x}{y}+\dfrac{z}{2y}=\dfrac{6}{2}+\dfrac{8}{2(2)}=\dfrac{6}{2}+\dfrac{8}{4}=3+2=5$

52. Let $x=6$ and $y=2$.
$x^2-3y^2=(6)^2-3(2)^2$
$=36-3(4)$
$=36-12$
$=36+(-12)$
$=24$

53. Replace a with 37 and b with 80.
$$\begin{aligned} 180 - a - b &= 180 - 37 - 80 \\ &= 180 + (-37) + (-80) \\ &= 143 + (-80) \\ &= 63 \end{aligned}$$
The measure of the unknown angle is 63°.

54. Replace a with 93, b with 80, and c with 82.
$$\begin{aligned} 360 - a - b - c &= 360 - 93 - 80 - 82 \\ &= 360 + (-93) + (-80) + (-82) \\ &= 267 + (-80) + (-82) \\ &= 187 + (-82) \\ &= 105 \end{aligned}$$
The measure of the unknown angle is 105°.

55. Let $x = 3$.
$$\begin{aligned} 7x - 3 &= 18 \\ 7(3) - 3 &\stackrel{?}{=} 18 \\ 21 - 3 &\stackrel{?}{=} 18 \\ 18 &= 18, \text{ true} \end{aligned}$$
3 is a solution to the equation.

56. Let $x = 1$.
$$\begin{aligned} 3x^2 + 4 &= x - 1 \\ 3(1)^2 + 4 &\stackrel{?}{=} 1 - 1 \\ 3 + 4 &\stackrel{?}{=} 0 \\ 7 &= 0, \text{ false} \end{aligned}$$
1 is not a solution to the equation.

57. The additive inverse of −9 is 9.

58. The additive inverse of $\dfrac{2}{3}$ is $-\dfrac{2}{3}$.

59. The additive inverse of $|-2|$ is −2 since $|-2| = 2$.

60. The additive inverse of $-|-7|$ is 7 since $-|-7| = -7$.

61. $-15 + 4 = -11$

62. $-6 + (-11) = -17$

63. $\dfrac{1}{16} + \left(-\dfrac{1}{4}\right) = \dfrac{1}{16} + \left(-\dfrac{1 \cdot 4}{4 \cdot 4}\right)$
$$\begin{aligned} &= \dfrac{1}{16} + \left(-\dfrac{4}{16}\right) \\ &= -\dfrac{3}{16} \end{aligned}$$

64. $-8 + |-3| = -8 + 3 = -5$

65. $-4.6 + (-9.3) = -13.9$

66. $-2.8 + 6.7 = 3.9$

67. $6 - 20 = 6 + (-20) = -14$

68. $-3.1 - 8.4 = -3.1 + (-8.4) = -11.5$

69. $-6 - (-11) = -6 + 11 = 5$

70. $4 - 15 = 4 + (-15) = -11$

71. $\begin{aligned}[t] -21 - 16 + 3(8 - 2) &= -21 + (-16) + 3[8 + (-2)] \\ &= -21 + (-16) + 3[6] \\ &= -21 + (-16) + 18 \\ &= -37 + 18 \\ &= -19 \end{aligned}$

72. $\dfrac{11 - (-9) + 6(8 - 2)}{2 + 3 \cdot 4} = \dfrac{11 + 9 + 6[8 + (-2)]}{2 + 3 \cdot 4}$
$$\begin{aligned} &= \dfrac{11 + 9 + 6[6]}{2 + 3 \cdot 4} \\ &= \dfrac{11 + 9 + 36}{2 + 12} \\ &= \dfrac{56}{14} \\ &= 4 \end{aligned}$$

73. Replace x with 3, y with −6, and z with −9.
$$\begin{aligned} 2x^2 - y + z &= 2(3)^2 - (-6) + (-9) \\ &= 2(9) + 6 + (-9) \\ &= 18 + 6 + (-9) \\ &= 24 + (-9) \\ &= 15 \end{aligned}$$
The answer is a.

74. Replace x with 3 and y with −6.
$$\begin{aligned} \dfrac{|y - 4x|}{2x} &= \dfrac{|-6 - 4(3)|}{2(3)} \\ &= \dfrac{|-6 - 12|}{6} \\ &= \dfrac{|-6 + (-12)|}{6} \\ &= \dfrac{|-18|}{6} \\ &= \dfrac{18}{6} \\ &= 3 \end{aligned}$$
The answer is a.

75. $50+1+(-2)+5+1+(-4)$
$= 51+(-2)+5+1+(-4)$
$= 49+5+1+(-4)$
$= 54+1+(-4)$
$= 55+(-4)$
$= 51$
The price at the end of the week is $51.

76. $50+1+(-2)+5 = 51+(-2)+5 = 49+5 = 54$
The price at the end of the day on Wednesday is $54.

77. The multiplicative inverse of -6 is $-\dfrac{1}{6}$ since

$-6 \cdot -\dfrac{1}{6} = 1.$

78. The multiplicative inverse of $\dfrac{3}{5}$ is $\dfrac{5}{3}$ since

$\dfrac{3}{5} \cdot \dfrac{5}{3} = 1.$

79. $6(-8) = -48$

80. $(-2)(-14) = 28$

81. $\dfrac{-18}{-6} = 3$

82. $\dfrac{42}{-3} = -14$

83. $\dfrac{4 \cdot (-3)+(-8)}{2+(-2)} = \dfrac{-12+(-8)}{2+(-2)} = \dfrac{-20}{0}$
The expression is undefined.

84. $\dfrac{3(-2)^2-5}{-14} = \dfrac{3(4)-5}{-14} = \dfrac{12-5}{-14} = \dfrac{7}{-14} = -\dfrac{1}{2}$

85. $\dfrac{-6}{0}$ is undefined.

86. $\dfrac{0}{-2} = 0$

87. $-4^2-(-3+5) \div (-1) \cdot 2 = -16-(2) \div (-1) \cdot 2$
$= -16+2 \cdot 2$
$= -16+4$
$= -12$

88. $-5^2-(2-20) \div (-3) \cdot 3 = -25-(-18) \div (-3) \cdot 3$
$= -25-6 \cdot 3$
$= -25-18$
$= -43$

89. Let $x = -5$ and $y = -2$.
$x^2-y^4 = (-5)^2-(-2)^4 = 25-16 = 9$

90. Let $x = -5$ and $y = -2$.
$x^2-y^3 = (-5)^2-(-2)^3 = 25-(-8) = 25+8 = 33$

91. $-7x$ or $-7 \cdot x$

92. $\dfrac{x}{13}$ or $x \div (-13)$

93. $-20-x$

94. $-1+x$

95. $-6+5 = 5+(-6)$; commutative property of addition

96. $6 \cdot 1 = 6$; multiplicative identity property

97. $3(8-5) = 3 \cdot 8 - 3 \cdot 5$; distributive property

98. $4+(-4) = 0$; additive inverse property

99. $2+(3+9) = (2+3)+9$; associative property of addition

100. $2 \cdot 8 = 8 \cdot 2$; commutative property of multiplication

101. $6(8+5) = 6 \cdot 8 + 6 \cdot 5$; distributive property

102. $(3 \cdot 8) \cdot 4 = 3 \cdot (8 \cdot 4)$; associative property of multiplication

103. $4 \cdot \dfrac{1}{4} = 1$; multiplicative inverse property

104. $8+0 = 8$; additive identity property

105. $5(y-2) = 5(y)+5(-2) = 5y-10$

106. $-3(z+y) = -3(z)+(-3)(y) = -3z-3y$

107. $-(7-x+4z) = (-1)(7)+(-1)(-x)+(-1)(4z)$
$= -7+x-4z$

108. $\frac{1}{2}(6z-10) = \frac{1}{2}(6z) + \frac{1}{2}(-10) = 3z - 5$

109. $-4(3x+5) - 7 = -4(3x) + (-4)(5) - 7$
$$= -12x - 20 - 7$$
$$= -12x - 27$$

110. $-8(2y+9) - 1 = -8(2y) + (-8)(9) - 1$
$$= -16y - 72 - 1$$
$$= -16y - 73$$

111. $-|-11| < |11.4|$ since $-|-11| = -11$ and $|11.4| = 11.4$.

112. $-1\frac{1}{2} > -2\frac{1}{2}$ since $-1\frac{1}{2}$ is to the right of $-2\frac{1}{2}$ on the number line.

113. $-7.2 + (-8.1) = -15.3$

114. $14 - 20 = 14 + (-20) = -6$

115. $4(-20) = -80$

116. $\frac{-20}{4} = -5$

117. $-\frac{4}{5}\left(\frac{5}{16}\right) = -\frac{4}{16} = -\frac{1}{4}$

118. $-0.5(-0.3) = 0.15$

119. $8 \div 2 \cdot 4 = 4 \cdot 4 = 16$

120. $(-2)^4 = (-2)(-2)(-2)(-2) = 16$

121. $\frac{-3 - 2(-9)}{-15 - 3(-4)} = \frac{-3 + 18}{-15 + 12} = \frac{15}{-3} = -5$

122. $5 + 2[(7-5)^2 + (1-3)] = 5 + 2[2^2 + (-2)]$
$$= 5 + 2[4 + (-2)]$$
$$= 5 + 2[2]$$
$$= 5 + 4$$
$$= 9$$

123. $-\frac{5}{8} \div \frac{3}{4} = -\frac{5}{8} \cdot \frac{4}{3} = -\frac{20}{24} = -\frac{5}{6}$

124. $\frac{-15 + (-4)^2 + |-9|}{10 - 2 \cdot 5} = \frac{-15 + 16 + 9}{10 - 10} = \frac{1 + 9}{0}$ is undefined.

125. $7\frac{1}{2} - 6\frac{1}{8} = \frac{15}{2} - \frac{49}{8}$
$$= \frac{15 \cdot 4}{2 \cdot 4} - \frac{49}{8}$$
$$= \frac{60}{8} - \frac{49}{8}$$
$$= \frac{60 - 49}{8}$$
$$= \frac{11}{8}$$
$$= 1\frac{3}{8} \text{ ft}$$

Chapter 1 Test

1. The absolute value of negative seven is greater than five is written as $|-7| > 5$.

2. The sum of nine and five is greater than or equal to four is written as $(9 + 5) \geq 4$.

3. $-13 + 8 = -5$

4. $-13 - (-2) = -13 + 2 = -11$

5. $12 \div 4 \cdot 3 - 6 \cdot 2 = 3 \cdot 3 - 6 \cdot 2 = 9 - 12 = -3$

6. $(13)(-3) = -39$

7. $(-6)(-2) = 12$

8. $\frac{|-16|}{-8} = \frac{16}{-8} = -2$

9. $\frac{-8}{0}$ is undefined.

10. $\frac{|-6| + 2}{5 - 6} = \frac{6 + 2}{5 + (-6)} = \frac{8}{-1} = -8$

11. $\frac{1}{2} - \frac{5}{6} = \frac{1 \cdot 3}{2 \cdot 3} - \frac{5}{6} = \frac{3}{6} - \frac{5}{6} = \frac{3 - 5}{6} = \frac{-2}{6} = -\frac{1}{3}$

12. $5\dfrac{3}{4} - 1\dfrac{1}{8} = \dfrac{23}{4} - \dfrac{9}{8}$

$= \dfrac{2 \cdot 23}{2 \cdot 4} - \dfrac{9}{8}$

$= \dfrac{46}{8} - \dfrac{9}{8}$

$= \dfrac{46 + (-9)}{8}$

$= \dfrac{37}{8}$

$= 4\dfrac{5}{8}$

13. $-0.6 + 1.875 = 1.275$

14. $3(-4)^2 - 80 = 3(16) - 80 = 48 + (-80) = -32$

15. $6[5 + 2(3 - 8) - 3] = 6\{5 + 2[3 + (-8)] + (-3)\}$
$= 6\{5 + 2[-5] + (-3)\}$
$= 6\{5 + (-10) + (-3)\}$
$= 6\{-5 + (-3)\}$
$= 6\{-8\}$
$= -48$

16. $\dfrac{-12 + 3 \cdot 8}{4} = \dfrac{-12 + 24}{4} = \dfrac{12}{4} = 3$

17. $\dfrac{(-2)(0)(-3)}{-6} = \dfrac{0(-3)}{-6} = \dfrac{0}{-6} = 0$

18. $-3 > -7$ since -3 is to the right of -7 on the number line.

19. $4 > -8$ since 4 is to the right of -8 on the number line.

20. $2 < |-3|$ since $2 < 3$.

21. $|-2| = -1 - (-3)$ since $|-2| = 2$ and $-1 - (-3) = -1 + 3 = 2$.

22. $2221 < 10{,}993$

23. a. The natural numbers are 1 and 7.

    b. The whole numbers are 0, 1 and 7.

    c. The integers are −5, −1, 0, 1, and 7.

    d. The rational numbers are $-5, -1, \dfrac{1}{4}, 0, 1, 7,$ and 11.6.

e. The irrational numbers are $\sqrt{7}$ and 3π.

f. The real numbers are all numbers in the given set.

24. Let $x = 6$ and $y = -2$.
$x^2 + y^2 = (6)^2 + (-2)^2 = 36 + 4 = 40$

25. Let $x = 6$, $y = -2$ and $z = -3$.
$x + yz = 6 + (-2)(-3) = 6 + 6 = 12$

26. Let $x = 6$ and $y = -2$.
$2 + 3x - y = 2 + 3(6) - (-2)$
$= 2 + 18 + 2$
$= 20 + 2$
$= 22$

27. Let $x = 6$, $y = -2$ and $z = -3$.
$\dfrac{y + z - 1}{x} = \dfrac{-2 + (-3) - 1}{6} = \dfrac{-5 + (-1)}{6} = \dfrac{-6}{6} = -1$

28. $8 + (9 + 3) = (8 + 9) + 3$; associative property of addition

29. $6 \cdot 8 = 8 \cdot 6$; commutative property of multiplication

30. $-6(2 + 4) = -6 \cdot 2 + (-6) \cdot 4$; distributive property

31. $\dfrac{1}{6}(6) = 1$; multiplicative inverse property

32. The opposite of −9 is 9.

33. The reciprocal of $-\dfrac{1}{3}$ is −3.

34. Look for the negative number that has the greatest absolute value. The second down had the greatest loss of yardage.

35. Gains: 5, 29
Losses: −10, −2
Total gain or loss $= 5 + (-10) + (-2) + 29$
$= (-5) + (-2) + 29$
$= -7 + 29$
$= 22$ yards gained
Yes, they scored a touchdown.

36. Since $-14 + 31 = 17$, the temperature at noon was 17°.

37. $356 + 460 + (-166) = 816 + (-166) = 650$
The net income was \$650 million.

38. Change in value per share $= -1.50$
Change in total value $= 280(-1.50) = -420$
She had a total loss of \$420.

Chapter 2

Section 2.1 Practice

1. a. The numerical coefficient of t is 1, since t is $1t$.

b. The numerical coefficient of $-7x$ is -7.

c. The numerical coefficient of $-\dfrac{w}{5}$ is $-\dfrac{1}{5}$, since $-\dfrac{w}{5}$ means $-\dfrac{1}{5} \cdot w$.

d. The numerical coefficient of $43x^4$ is 43.

e. The numerical coefficient of $-b$ is -1, since $-b$ is $-1b$.

2. a. $-4xy$ and $5yx$ are like terms, since $xy = yx$ by the commutative property.

b. $5q$ and $-3q^2$ are unlike terms, since the exponents on q are not the same.

c. $3ab^2$, $-2ab^2$, and $43ab^2$ are like terms, since each variable and its exponent match.

d. y^5 and $\dfrac{y^5}{2}$ are like terms, since the exponents on y are the same.

3. a. $-3y + 11y = (-3 + 11)y = 8y$

b. $4x^2 + x^2 = 4x^2 + 1x^2 = (4+1)x^2 = 5x^2$

c. $5x - 3x^2 + 8x^2 = 5x + (-3 + 8)x^2 = 5x + 5x^2$

d. $20y^2 + 2y^2 - y^2 = 20y^2 + 2y^2 - 1y^2$
$= (20 + 2 - 1)y^2$
$= 21y^2$

4. a. $3y + 8y - 7 + 2 = (3 + 8)y + (-7 + 2) = 11y - 5$

b. $6x - 3 - x - 3 = 6x - 1x + (-3 - 3)$
$= (6 - 1)x + (-3 - 3)$
$= 5x - 6$

c. $\dfrac{3}{4}t - t = \dfrac{3}{4}t - 1t = \left(\dfrac{3}{4} - 1\right)t = -\dfrac{1}{4}t$

d. $9y + 3.2y + 10 + 3 = (9 + 3.2)y + (10 + 3)$
$= 12.2y + 13$

e. $5z - 3z^4$
These two terms cannot be combined because they are unlike terms.

5. a. $3(2x - 7) = 3(2x) + 3(-7) = 6x - 21$

b. $-5(x - 0.5z - 5)$
$= -5(x) + (-5)(-0.5z) + (-5)(-5)$
$= -5x + 2.5z + 25$

c. $-(2x - y + z - 2)$
$= -1(2x - y + z - 2)$
$= -1(2x) - 1(-y) - 1(z) - 1(-2)$
$= -2x + y - z + 2$

6. a. $4(9x + 1) + 6 = 36x + 4 + 6 = 36x + 10$

b. $-7(2x - 1) - (6 - 3x) = -14x + 7 - 6 + 3x$
$= -11x + 1$

c. $8 - 5(6x + 5) = 8 - 30x - 25 = -30x - 17$

7. "Subtract $7x - 1$ from $2x + 3$" translates to
$(2x + 3) - (7x - 1) = 2x + 3 - 7x + 1 = -5x + 4$

8. a.

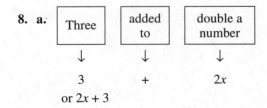

Three	added to	double a number
↓	↓	↓
3	+	2x

or $2x + 3$

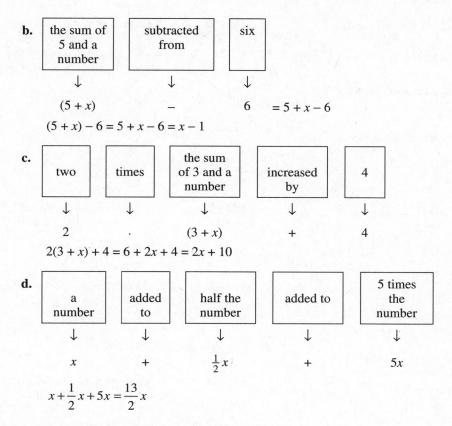

b.

the sum of 5 and a number	subtracted from	six

$\downarrow$ $\qquad$ $\downarrow$ $\qquad$ $\downarrow$

$(5 + x)$ $\qquad$ $-$ $\qquad$ 6 $\quad = 5 + x - 6$

$(5 + x) - 6 = 5 + x - 6 = x - 1$

c.

two	times	the sum of 3 and a number	increased by	4

$\downarrow$ $\qquad$ $\downarrow$ $\qquad$ $\downarrow$ $\qquad$ $\downarrow$ $\qquad$ $\downarrow$

2 $\qquad$ $\cdot$ $\qquad$ $(3 + x)$ $\qquad$ $+$ $\qquad$ 4

$2(3 + x) + 4 = 6 + 2x + 4 = 2x + 10$

d.

a number	added to	half the number	added to	5 times the number

$\downarrow$ $\qquad$ $\downarrow$ $\qquad$ $\downarrow$ $\qquad$ $\downarrow$ $\qquad$ $\downarrow$

x $\qquad$ $+$ $\qquad$ $\frac{1}{2}x$ $\qquad$ $+$ $\qquad$ $5x$

$x + \frac{1}{2}x + 5x = \frac{13}{2}x$

Vocabulary, Readiness & Video Check 2.1

1. $23y^2 + 10y - 6$ is called an <u>expression</u> while $23y^2$, $10y$, and -6 are each called a <u>term</u>.

2. To simplify $x + 4x$, we <u>combine like terms</u>.

3. The term y has an understood <u>numerical coefficient</u> of 1.

4. The terms $7z$ and $7y$ are <u>unlike</u> terms and the terms $7z$ and $-z$ are <u>like</u> terms.

5. For the term $-\frac{1}{2}xy^2$, the number $-\frac{1}{2}$ is the <u>numerical coefficient</u>.

6. $5(3x - y)$ equals $15x - 5y$ by the <u>distributive</u> property.

7. Although these terms have exactly the same variables, the exponents on each are not exactly the same—the exponents on x differ in each term.

8. distributive property

9. -1

10. The sum of 5 times a number and -2, plus 7 times the number; $5x + (-2) + 7x$; because there are like terms.

Exercise Set 2.1

1. The numerical coefficient of $-7y$ is -7.

3. The numerical coefficient of x is 1, since $x = 1x$.

5. The numerical coefficient of $17x^2y$ is 17.

7. $5y$ and $-y$ are like terms, since the variable and its exponent match.

9. $2z$ and $3z^2$ are unlike terms, since the exponents on z are not the same.

11. $8wz$ and $\frac{1}{7}zw$ are like terms, since $wz = zw$ by the commutative property.

13. $7y + 8y = (7+8)y = 15y$

15. $8w - w + 6w = (8 - 1 + 6)w = 13w$

17. $3b - 5 - 10b - 4 = 3b - 10b - 5 - 4$
$\qquad = (3 - 10)b - 9$
$\qquad = -7b - 9$

19. $m - 4m + 2m - 6 = (1 - 4 + 2)m - 6 = -m - 6$

21. $5g - 3 - 5 - 5g = (5g - 5g) + (-3 - 5)$
$\qquad = (5 - 5)g + (-8)$
$\qquad = 0g - 8$
$\qquad = -8$

23. $6.2x - 4 + x - 1.2 = 6.2x + x - 4 - 1.2$
$\qquad = (6.2 + 1)x - 5.2$
$\qquad = 7.2x - 5.2$

25. $6x - 5x + x - 3 + 2x = 6x - 5x + x + 2x - 3$
$\qquad = (6 - 5 + 1 + 2)x - 3$
$\qquad = 4x - 3$

27. $7x^2 + 8x^2 - 10x^2 = (7 + 8 - 10)x^2 = 5x^2$

29. $6x + 0.5 - 4.3x - 0.4x + 3$
$\qquad = 6x - 4.3x - 0.4x + 0.5 + 3$
$\qquad = (6 - 4.3 - 0.4)x + (0.5 + 3)$
$\qquad = 1.3x + 3.5$

31. $5(y - 4) = 5(y) - 5(4) = 5y - 20$

33. $-2(x + 2) = -2(x) + (-2)(2) = -2x - 4$

35. $7(d - 3) + 10 = 7d - 21 + 10 = 7d - 11$

37. $-5(2x - 3y + 6) = -5(2x) - (-5)(3y) + (-5)(6)$
$\qquad = -10x + 15y - 30$

39. $-(3x - 2y + 1) = -3x + 2y - 1$

41. $5(x + 2) - (3x - 4) = 5x + 10 - 3x + 4$
$\qquad = 2x + 14$

43. $\quad 6x + 7 \quad$ added to $\quad 4x - 10$
$\qquad \downarrow \qquad\qquad\quad \downarrow \qquad\qquad \downarrow$
$(6x + 7) \qquad + \qquad (4x - 10) = 6x + 4x + 7 - 10$
$\qquad\qquad\qquad\qquad\qquad\qquad = 10x - 3$

45. $\quad 3x - 8 \quad$ minus $\quad 7x + 1$
$\qquad \downarrow \qquad\quad \downarrow \qquad\quad \downarrow$
$(3x - 8) \quad - \quad (7x + 1) = 3x - 8 - 7x - 1$
$\qquad\qquad\qquad\qquad\qquad = 3x - 7x - 8 - 1$
$\qquad\qquad\qquad\qquad\qquad = -4x - 9$

47. $\quad m - 9 \quad$ minus $\quad 5m - 6$
$\qquad \downarrow \qquad\quad \downarrow \qquad\quad \downarrow$
$(m - 9) \quad - \quad (5m - 6) = m - 9 - 5m + 6$
$\qquad\qquad\qquad\qquad\qquad = m - 5m - 9 + 6$
$\qquad\qquad\qquad\qquad\qquad = -4m - 3$

49. $2k - k - 6 = (2 - 1)k - 6 = k - 6$

51. $-9x + 4x + 18 - 10x = -9x + 4x - 10x + 18$
$\qquad = (-9 + 4 - 10)x + 18$
$\qquad = -15x + 18$

53. $-4(3y - 4) + 12y = -4(3y) - (-4)(4) + 12y$
$\qquad = -12y + 16 + 12y$
$\qquad = -12y + 12y + 16$
$\qquad = 16$

55. $3(2x - 5) - 5(x - 4) = 6x - 15 - 5x + 20 = x + 5$

57. $-2(3x - 4) + 7x - 6 = -6x + 8 + 7x - 6 = x + 2$

59. $5k - (3k - 10) = 5k - 3k + 10 = 2k + 10$

61. $(3x + 4) - (6x - 1) = 3x + 4 - 6x + 1 = -3x + 5$

63. $3.4m - 4 - 3.4m - 7 = 3.4m - 3.4m - 4 - 7 = -11$

65. $\frac{1}{3}(7y-1)+\frac{1}{6}(4y+7)=\frac{7}{3}y-\frac{1}{3}+\frac{4}{6}y+\frac{7}{6}$

$=\frac{7}{3}y+\frac{2}{3}y-\frac{1}{3}+\frac{7}{6}$

$=\frac{9}{3}y-\frac{2}{6}+\frac{7}{6}$

$=3y+\frac{5}{6}$

67. $2+4(6x-6)=2+24x-24=-22+24x$

69. $0.5(m+2)+0.4m=0.5m+1+0.4m=0.9m+1$

71. $10-3(2x+3y)=10-6x-9y$

73. $6(3x-6)-2(x+1)-17x=18x-36-2x-2-17x$

$=18x-2x-17x-36-2$

$=-x-38$

75. $\frac{1}{2}(12x-4)-(x+5)=6x-2-x-5=5x-7$

77.
twice a number	decreased by	4
↓	↓	↓
$2x$	$-$	4

79.
seven	added to	double a number
↓	↓	↓
7	$+$	$2x$

81.
three-fourths of a number	increased by	12
↓	↓	↓
$\frac{3}{4}x$	$+$	12

83.
5 times a number	added to	-2	added to	7 times the number
↓	↓	↓	↓	↓
$5x$	$+$	-2	$+$	$7x$

$5x+(-2)+7x=12x-2$

85.
8	times	the sum of a number and 6
↓	↓	↓
8	$\cdot$	$(x+6)$

$8(x+6)=8x+48$

87.

double a number	minus	the sum of the number and 10
↓	↓	↓
$2x$	$-$	$(x+10)$

$2x - (x + 10) = 2x - x - 10 = x - 10$

89.

2	added to	3 times a number	added to	−9	added to	4 times the number
↓	↓	↓	↓	↓	↓	↓
2	+	$3x$	+	−9	+	$4x$

$2 + 3x + (-9) + 4x = 7x - 7$

91. $y - x^2 = 3 - (-1)^2 = 3 - 1 = 2$

93. $a - b^2 = 2 - (-5)^2 = 2 - 25 = -23$

95. $yz - y^2 = (-5)(0) - (-5)^2 = 0 - 25 = -25$

97. $5x + (4x - 1) + 5x + (4x - 1) = 5x + 4x - 1 + 5x + 4x - 1$
$$= 18x - 2$$
The perimeter is $(18x - 2)$ feet.

99. 1 cone + 1 cylinder $\stackrel{?}{=}$ 3 cubes
        1 cube + 2 cubes $\stackrel{?}{=}$ 3 cubes
            3 cubes = 3 cubes: Balanced

101. 2 cylinders + 1 cube $\stackrel{?}{=}$ 3 cones + 2 cubes
        $2 \cdot 2$ cubes + 1 cube $\stackrel{?}{=}$ 3 cubes + 2 cubes
            4 cubes + 1 cube $\stackrel{?}{=}$ 3 cubes + 2 cubes
                5 cubes = 5 cubes: Balanced

103. answers may vary

105. $12(x + 2) + (3x - 1) = 12x + 24 + 3x - 1 = 15x + 23$
The total length is $(15x + 23)$ inches.

107. answers may vary

109. $5b^2c^3 + 8b^3c^2 - 7b^3c^2 = 5b^2c^3 + b^3c^2$

111. $3x - (2x^2 - 6x) + 7x^2 = 3x - 2x^2 + 6x + 7x^2$
$$= 5x^2 + 9x$$

113. $-(2x^2y + 3z) + 3z - 5x^2y = -2x^2y - 3z + 3z - 5x^2y$
$$= -7x^2y$$

Section 2.2 Practice

1. $x + 3 = -5$
$x + 3 - 3 = -5 - 3$
$x = -8$
Check: $x + 3 = -5$
$-8 + 3 \stackrel{?}{=} -5$
$-5 = -5$
The solution is -8.

2. $y - 0.3 = -2.1$
$y - 0.3 + 0.3 = -2.1 + 0.3$
$y = -1.8$
Check: $y - 0.3 = -2.1$
$-1.8 - 0.3 \stackrel{?}{=} -2.1$
$-2.1 = -2.1$
The solution is -1.8.

3. $8x - 5x - 3 + 9 = x + x + 3 - 7$
$3x + 6 = 2x - 4$
$3x + 6 - 2x = 2x - 4 - 2x$
$x + 6 = -4$
$x + 6 - 6 = -4 - 6$
$x = -10$
Check:
$8x - 5x - 3 + 9 = x + x + 3 - 7$
$8(-10) - 5(-10) - 3 + 9 \stackrel{?}{=} -10 + (-10) + 3 - 7$
$-80 + 50 - 3 + 9 \stackrel{?}{=} -10 + (-10) + 3 - 7$
$-24 = -24$
The solution is -10.

4. $2 = 4(2a - 3) - (7a + 4)$
$2 = 4(2a) + 4(-3) - 7a - 4$
$2 = 8a - 12 - 7a - 4$
$2 = a - 16$
$2 + 16 = a - 16 + 16$
$18 = a$
Check by replacing a with 18 in the original equation.

5. $\dfrac{4}{5}x = 16$
$\dfrac{5}{4} \cdot \dfrac{4}{5}x = \dfrac{5}{4} \cdot 16$
$\left(\dfrac{5}{4} \cdot \dfrac{4}{5}\right)x = \dfrac{5}{4} \cdot 16$
$1x = 20$
$x = 20$

Check: $\dfrac{4}{5}x = 16$
$\dfrac{4}{5} \cdot 20 \stackrel{?}{=} 16$
$16 = 16$
The solution is 20.

6. $8x = -96$
$\dfrac{8x}{8} = \dfrac{-96}{8}$
$x = -12$
Check: $8x = -96$
$8(-12) \stackrel{?}{=} -96$
$-96 = -96$
The solution is -12.

7. $\dfrac{x}{5} = 13$
$5 \cdot \dfrac{x}{5} = 5 \cdot 13$
$x = 65$

Check: $\dfrac{x}{5} = 13$
$\dfrac{65}{5} \stackrel{?}{=} 13$
$13 = 13$
The solution is 65.

8. $6b - 11b = 18 + 2b - 6 + 9$
$-5b = 21 + 2b$
$-5b - 2b = 21 + 2b - 2b$
$-7b = 21$
$\dfrac{-7b}{-7} = \dfrac{21}{-7}$
$b = -3$
Check by replacing b with -3 in the original equation. The solution is -3.

9. **a.** The other number is $9 - 2 = 7$.

   b. The other number is $9 - x$.

   c. The other piece has length $(9 - x)$ feet.

10. Let x = first integer.
$x + 2$ = second even integer.
$x + 4$ = third even integer.
$x + (x + 2) + (x + 4) = 3x + 6$

Vocabulary, Readiness & Video Check 2.2

1. The difference between an equation and an expression is that an <u>equation</u> contains an equal sign, whereas an <u>expression</u> does not.

2. <u>Equivalent</u> equations are equations that have the same solution.

3. A value of the variable that makes the equation a true statement is called a <u>solution</u> of the equation.

4. The process of finding the solution of an equation is called <u>solving</u> the equation for the variable.

5. By the <u>addition</u> property of equality, $x = -2$ and $x + 10 = -2 + 10$ are equivalent equations.

6. By the <u>addition</u> property of equality, $x = -7$ and $x - 5 = -7 - 5$ are equivalent equations.

7. By the <u>multiplication</u> property of equality, $y = \dfrac{1}{2}$ and $5 \cdot y = 5 \cdot \dfrac{1}{2}$ are equivalent equations.

8. By the <u>multiplication</u> property of equality, $9x = -63$ and $\dfrac{9x}{9} = \dfrac{-63}{9}$ are equivalent equations.

9. The equations $x = \dfrac{1}{2}$ and $\dfrac{1}{2} = x$ are equivalent equations. The statement is true.

10. The equations $\dfrac{z}{4} = 10$ and $4 \cdot \dfrac{z}{4} = 10$ are not equivalent equations. The statement is false.

11. The addition property of equality means that if we have an equation, we can add the same real number to <u>both sides</u> of the equation and have an equivalent equation.

12. We can multiply both sides of an equation by <u>any</u> nonzero number and have an equivalent equation.

13. addition property; multiplication property; answers may vary

14. $(x + 1) + (x + 3) = 2x + 4$

Exercise Set 2.2

1. $x + 7 = 10$
    $x + 7 - 7 = 10 - 7$
    $x = 3$
    Check: $x + 7 = 10$
    $3 + 7 \overset{?}{=} 10$
    $10 = 10$
    The solution is 3.

3. $x - 2 = -4$
    $x - 2 + 2 = -4 + 2$
    $x = -2$
    Check: $x - 2 = -4$
    $-2 - 2 \overset{?}{=} -4$
    $-4 = -4$
    The solution is -2.

5. $3 + x = -11$
    $3 + x - 3 = -11 - 3$
    $x = -14$
    Check: $3 + x = -11$
    $3 + (-14) \overset{?}{=} -11$
    $-11 = -11$
    The solution is -14.

7. $r - 8.6 = -8.1$
    $r - 8.6 + 8.6 = -8.1 + 8.6$
    $r = 0.5$
    Check: $x - 8.6 = -8.1$
    $0.5 - 8.6 \overset{?}{=} -8.1$
    $-8.1 = -8.1$
    The solution is 0.5.

9. $8x = 7x - 3$
    $8x - 7x = 7x - 7x - 3$
    $x = -3$
    Check: $8x = 7x - 3$
    $8(-3) \overset{?}{=} 7(-3) - 3$
    $-24 \overset{?}{=} -21 - 3$
    $-24 = -24$
    The solution is -3.

11. $5b - 0.7 = 6b$
    $5b - 5b - 0.7 = 6b - 5b$
    $-0.7 = b$
    Check: $5b - 0.7 = 6b$
    $5(-0.7) - 0.7 \overset{?}{=} 6(-0.7)$
    $-3.5 - 0.7 \overset{?}{=} -4.2$
    $-4.2 = -4.2$
    The solution is -0.7.

13.
$$7x - 3 = 6x$$
$$7x - 6x - 3 = 6x - 6x$$
$$x - 3 = 0$$
$$x - 3 + 3 = 0 + 3$$
$$x = 3$$
Check: $7x - 3 = 6x$
$$7(3) - 3 \stackrel{?}{=} 6(3)$$
$$21 - 3 \stackrel{?}{=} 18$$
$$18 = 18$$
The solution is 3.

15.
$$2x + x - 6 = 2x + 5$$
$$3x - 6 = 2x + 5$$
$$3x - 2x - 6 = 2x + 5 - 2x$$
$$x - 6 = 5$$
$$x - 6 + 6 = 5 + 6$$
$$x = 11$$
The solution is 11.

17.
$$3t - t - 7 = t - 7$$
$$2t - 7 = t - 7$$
$$2t - t - 7 = t - t - 7$$
$$t - 7 = -7$$
$$t - 7 + 7 = -7 + 7$$
$$t = 0$$
The solution is 0.

19.
$$7x + 2x = 8x - 3$$
$$9x = 8x - 3$$
$$9x - 8x = 8x - 8x - 3$$
$$x = -3$$
The solution is -3.

21.
$$-2(x + 1) + 3x = 14$$
$$-2x - 2 + 3x = 14$$
$$-2 + x = 14$$
$$2 - 2 + x = 14 + 2$$
$$x = 16$$
The solution is 16.

23.
$$-5x = -20$$
$$\frac{-5x}{-5} = \frac{-20}{-5}$$
$$x = 4$$
The solution is 4.

25.
$$3x = 0$$
$$\frac{3x}{3} = \frac{0}{3}$$
$$x = 0$$
The solution is 0.

27.
$$-x = -12$$
$$\frac{-x}{-1} = \frac{-12}{-1}$$
$$x = 12$$
The solution is 12.

29.
$$3x + 2x = 50$$
$$5x = 50$$
$$\frac{5x}{5} = \frac{50}{5}$$
$$x = 10$$
The solution is 10.

31.
$$\frac{2}{3}x = -8$$
$$\frac{3}{2}\left(\frac{2}{3}x\right) = \frac{3}{2}(-8)$$
$$x = -12$$
The solution is -12.

33.
$$\frac{1}{6}d = \frac{1}{2}$$
$$6\left(\frac{1}{6}d\right) = 6\left(\frac{1}{2}\right)$$
$$d = 3$$
The solution is 3.

35.
$$\frac{a}{-2} = 1$$
$$-2\left(\frac{a}{-2}\right) = -2(1)$$
$$a = -2$$
The solution is -2.

37.
$$\frac{k}{7} = 0$$
$$7\left(\frac{k}{7}\right) = 7(0)$$
$$k = 0$$
The solution is 0.

39. Answers may vary

41.
$$2x - 4 = 16$$
$$2x - 4 + 4 = 16 + 4$$
$$2x = 20$$
$$\frac{2x}{2} = \frac{20}{2}$$
$$x = 10$$
Check: $2x - 4 = 16$
$$2(10) - 4 \stackrel{?}{=} 16$$
$$20 - 4 \stackrel{?}{=} 16$$
$$16 = 16$$
The solution is 10.

43.
$$-x + 2 = 22$$
$$-x + 2 - 2 = 22 - 2$$
$$-x = 20$$
$$x = -20$$
Check: $-x + 2 = 22$
$$-(-20) + 2 \stackrel{?}{=} 22$$
$$20 + 2 \stackrel{?}{=} 22$$
$$22 = 22$$
The solution is −20.

45.
$$6a + 3 = 3$$
$$6a + 3 - 3 = 3 - 3$$
$$6a = 0$$
$$\frac{6a}{6} = \frac{0}{6}$$
$$a = 0$$
Check: $6a + 3 = 3$
$$6(0) + 3 \stackrel{?}{=} 3$$
$$0 + 3 \stackrel{?}{=} 3$$
$$3 = 3$$
The solution is 0.

47.
$$6x + 10 = -20$$
$$6x + 10 - 10 = -20 - 10$$
$$6x = -30$$
$$\frac{6x}{6} = \frac{-30}{6}$$
$$x = -5$$

Check: $6x + 10 = -20$
$$6(-5) + 10 \stackrel{?}{=} -20$$
$$-30 + 10 \stackrel{?}{=} -20$$
$$-20 = -20$$
The solution is −5.

49.
$$5 - 0.3k = 5$$
$$5 - 5 - 0.3k = 5 - 5$$
$$-0.3k = 0$$
$$\frac{-0.3k}{-0.3} = \frac{0}{-0.3}$$
$$k = 0$$
Check: $5 - 0.3k = 5$
$$5 - 0.3(0) \stackrel{?}{=} 5$$
$$5 - 0 \stackrel{?}{=} 5$$
$$5 = 5$$
The solution is 0.

51.
$$-2x + \frac{1}{2} = \frac{7}{2}$$
$$-2x + \frac{1}{2} - \frac{1}{2} = \frac{7}{2} - \frac{1}{2}$$
$$-2x = \frac{6}{2}$$
$$-2x = 3$$
$$\frac{-2x}{-2} = \frac{3}{-2}$$
$$x = -\frac{3}{2}$$

Check: $-2x + \frac{1}{2} = \frac{7}{2}$
$$-2\left(-\frac{3}{2}\right) + \frac{1}{2} \stackrel{?}{=} \frac{7}{2}$$
$$\frac{6}{2} + \frac{1}{2} \stackrel{?}{=} \frac{7}{2}$$
$$\frac{7}{2} = \frac{7}{2}$$

The solution is $-\frac{3}{2}$.

53.
$$\frac{x}{3} + 2 = -5$$
$$\frac{x}{3} + 2 - 2 = -5 - 2$$
$$\frac{x}{3} = -7$$
$$3 \cdot \frac{x}{3} = 3 \cdot -7$$
$$x = -21$$

Check: $\dfrac{x}{3} + 2 = -5$

$\dfrac{-21}{3} + 2 \overset{?}{=} -5$

$-7 + 2 \overset{?}{=} -5$

$-5 = -5$

The solution is -21.

55. $10 = 2x - 1$

$10 + 1 = 2x - 1 + 1$

$11 = 2x$

$\dfrac{11}{2} = \dfrac{2x}{2}$

$\dfrac{11}{2} = x$

Check: $10 = 2x - 1$

$10 \overset{?}{=} 2\left(\dfrac{11}{2}\right) - 1$

$10 \overset{?}{=} 11 - 1$

$10 = 10$

The solution is $\dfrac{11}{2}$.

57. $6z - 8 - z + 3 = 0$

$5z - 5 = 0$

$5z - 5 + 5 = 0 + 5$

$5z = 5$

$\dfrac{5z}{5} = \dfrac{5}{5}$

$z = 1$

Check: $6z - 8 - z + 3 = 0$

$6 \cdot 1 - 8 - 1 + 3 \overset{?}{=} 0$

$6 - 8 - 1 + 3 \overset{?}{=} 0$

$0 = 0$

The solution is 1.

59. $10 - 3x - 6 - 9x = 7$

$4 - 12x = 7$

$4 - 4 - 12x = 7 - 4$

$-12x = 3$

$\dfrac{-12x}{-12} = \dfrac{3}{-12}$

$x = -\dfrac{1}{4}$

Check: $10 - 3x - 6 - 9x = 7$

$10 - 3\left(-\dfrac{1}{4}\right) - 6 - 9\left(-\dfrac{1}{4}\right) \overset{?}{=} 7$

$10 + \dfrac{3}{4} - 6 + \dfrac{9}{4} \overset{?}{=} 7$

$4 + \dfrac{12}{4} \overset{?}{=} 7$

$4 + 3 \overset{?}{=} 7$

$7 = 7$

The solution is $-\dfrac{1}{4}$.

61. $\dfrac{5}{6}x = 10$

$\dfrac{6}{5} \cdot \dfrac{5}{6}x = \dfrac{6}{5} \cdot 10$

$x = 12$

Check: $\dfrac{5}{6}x = 10$

$\dfrac{5}{6} \cdot 12 \overset{?}{=} 10$

$10 = 10$

The solution is 12.

63. $1 = 0.3x - 0.5x - 5$

$1 = -0.2x - 5$

$1 + 5 = -0.2x - 5 + 5$

$6 = -0.2x$

$\dfrac{6}{-0.2} = \dfrac{-0.2x}{-0.2}$

$-30 = x$

Check: $1 = 0.3x - 0.5x - 5$

$1 \overset{?}{=} 0.3(-30) - 0.5(-30) - 5$

$1 \overset{?}{=} -9 + 15 - 5$

$1 = 1$

The solution is -30.

65. $z - 5z = 7z - 9 - z$

$-4z = 6z - 9$

$-4z - 6z = 6z - 6z - 9$

$-10z = -9$

$\dfrac{-10z}{-10} = \dfrac{-9}{-10}$

$z = \dfrac{9}{10}$

Check: $z - 5z = 7z - 9 - z$

$$\frac{9}{10} - 5\left(\frac{9}{10}\right) \overset{?}{=} 7\left(\frac{9}{10}\right) - 9 - \frac{9}{10}$$

$$\frac{9}{10} - \frac{45}{10} \overset{?}{=} \frac{63}{10} - \frac{90}{10} - \frac{9}{10}$$

$$-\frac{36}{10} = -\frac{36}{10}$$

The solution is $\frac{9}{10}$.

67. $0.4x - 0.6x - 5 = 1$

$-0.2x - 5 = 1$

$-0.2x - 5 + 5 = 1 + 5$

$-0.2x = 6$

$$\frac{-0.2x}{-0.2} = \frac{6}{-0.2}$$

$x = -30$

Check: $0.4x - 0.6x - 5 = 1$

$0.4(-30) - 0.6(-30) - 5 \overset{?}{=} 1$

$-12 + 18 - 5 \overset{?}{=} 1$

$1 = 1$

The solution is -30.

69. $6 - 2x + 8 = 10$

$14 - 2x = 10$

$14 - 14 - 2x = 10 - 14$

$-2x = -4$

$$\frac{-2x}{-2} = \frac{-4}{-2}$$

$x = 2$

Check: $6 - 2x + 8 = 10$

$6 - 2 \cdot 2 + 8 \overset{?}{=} 10$

$6 - 4 + 8 \overset{?}{=} 10$

$10 = 10$

The solution is 2.

71. $-3a + 6 + 5a = 7a - 8a$

$6 + 2a = -a$

$6 + 2a - 2a = -a - 2a$

$6 = -3a$

$$\frac{6}{-3} = \frac{-3a}{-3}$$

$-2 = a$

Check: $-3a + 6 + 5a = 7a - 8a$

$-3(-2) + 6 + 5(-2) \overset{?}{=} 7(-2) - 8(-2)$

$6 + 6 - 10 \overset{?}{=} -14 + 16$

$2 = 2$

The solution is -2.

73. $20 = -3(2x + 1) + 7x$

$20 = -6x - 3 + 7x$

$20 = x - 3$

$20 + 3 = x - 3 + 3$

$23 = x$

Check: $20 = -3(2x + 1) + 7x$

$20 \overset{?}{=} -3(2 \cdot 23 + 1) + 7 \cdot 23$

$20 \overset{?}{=} -3(46 + 1) + 161$

$20 \overset{?}{=} -3(47) + 161$

$20 \overset{?}{=} -141 + 161$

$20 = 20$

The solution is 23.

75. The other number is $20 - p$.

77. The length of the other piece is $(10 - x)$ feet.

79. The supplement of the angle $x°$ is $(180 - x)°$.

81. April received $(n + 284)$ votes.

83. The length of the Golden Gate Bridge is $(m - 60)$ feet.

85. If the number of undergraduate students was n, and the number of graduate students was 28,000 fewer, then the number of graduate students is $n - 28,000$.

87. The area of the Sahara Desert is $7x$ square miles.

89. Sum = first integer + second integer
Sum = $x + (x + 2) = x + x + 2 = 2x + 2$

91. Sum = first integer + third integer
Sum = $x + (x + 2) = x + x + 2 = 2x + 2$

93. Let x = first room number.
$x + 2$ = second room number
$x + 4$ = third room number
$x + 6$ = fourth room number
$x + 8$ = fifth room number
$x + (x + 2) + (x + 4) + (x + 6) + (x + 8) = 5x + 20$

95. $5x + 2(x - 6) = 5x + 2x - 12 = 7x - 12$

97. $-(x - 1) + x = -x + 1 + x = 1$

99. $(-3)^2 = (-3)(-3) = 9$
$-3^2 = -3 \cdot 3 = -9$
$(-3)^2 > -3^2$

101. $(-2)^3 = (-2)(-2)(-2) = -8$
$-2^3 = -2 \cdot 2 \cdot 2 = -8$
$(-2)^3 = -2^3$

103. $180 - [x + (2x + 7)] = 180 - [x + 2x + 7]$
$\qquad\qquad\qquad\qquad = 180 - [3x + 7]$
$\qquad\qquad\qquad\qquad = 180 - 3x - 7$
$\qquad\qquad\qquad\qquad = 173 - 3x$
The third angle is $(173 - 3x)°$.

105. Answers may vary

107. $\qquad x - 4 = -9$
$x - 4 + (4) = -9 + (4)$
$\qquad\qquad x = -5$
The answer is 4.

109. Answers may vary

111. Answers may vary

113. $\quad 6x = \underline{\qquad}$
$\quad 6(-8) = \underline{\qquad}$
$\quad -48 = \underline{\qquad}$

115. $\quad 9x = 2100$
$\quad \dfrac{9x}{9} = \dfrac{2100}{9}$
$\qquad x = \dfrac{700}{3}$

Each dose should be $\dfrac{700}{3}$ milligrams.

117. Check $y = 1.2$: $8.13 + 5.85y = 20.05y - 8.91$
$\qquad\qquad 8.13 + 5.85(1.2) \overset{?}{=} 20.05(1.2) - 8.91$
$\qquad\qquad\quad 8.13 + 7.02 \overset{?}{=} 24.06 - 8.91$
$\qquad\qquad\qquad\quad 15.15 = 15.15$
Solution

119. $\quad -3.6x = 10.62$
$\quad \dfrac{-3.6x}{-3.6} = \dfrac{10.62}{-3.6}$
$\qquad\quad x = -2.95$

121. $\qquad 7x - 5.06 = -4.92$
$\quad 7x - 5.06 + 5.06 = -4.92 + 5.06$
$\qquad\qquad\quad 7x = 0.14$
$\qquad\qquad\quad \dfrac{7x}{7} = \dfrac{0.14}{7}$
$\qquad\qquad\qquad x = 0.02$

Section 2.3 Practice

1. $2(4a - 9) + 3 = 5a - 6$
$\quad 8a - 18 + 3 = 5a - 6$
$\qquad 8a - 15 = 5a - 6$
$\quad 8a - 15 - 5a = 5a - 6 - 5a$
$\qquad\quad 3a - 15 = -6$
$\quad 3a - 15 + 15 = -6 + 15$
$\qquad\qquad 3a = 9$
$\qquad\qquad \dfrac{3a}{3} = \dfrac{9}{3}$
$\qquad\qquad a = 3$
Check: $\quad 2(4a - 9) + 3 = 5a - 6$
$\qquad\quad 2[4(3) - 9] + 3 \overset{?}{=} 5(3) - 6$
$\qquad\qquad 2(12 - 9) + 3 \overset{?}{=} 15 - 6$
$\qquad\qquad\quad 2(3) + 3 \overset{?}{=} 9$
$\qquad\qquad\qquad 6 + 3 \overset{?}{=} 9$
$\qquad\qquad\qquad\qquad 9 = 1$
The solution is 3 or the solution set is $\{3\}$.

2. $\qquad\quad 7(x - 3) = -6x$
$\qquad\quad 7x - 21 = -6x$
$\quad 7x - 21 - 7x = -6x - 7x$
$\qquad\qquad -21 = -13x$
$\qquad\qquad \dfrac{-21}{-13} = \dfrac{-13x}{-13}$
$\qquad\qquad \dfrac{21}{13} = x$
Check: $\qquad\quad 7(x - 3) = -6x$
$\qquad 7\left(\dfrac{21}{13} - 3\right) \overset{?}{=} -6\left(\dfrac{21}{13}\right)$
$\qquad 7\left(\dfrac{21}{13} - \dfrac{39}{13}\right) \overset{?}{=} -\dfrac{126}{13}$
$\qquad\quad 7\left(-\dfrac{18}{13}\right) \overset{?}{=} -\dfrac{126}{13}$
$\qquad\qquad -\dfrac{126}{13} = -\dfrac{126}{13}$
The solution is $\dfrac{21}{13}$.

3.
$$\frac{3}{5}x - 2 = \frac{2}{3}x - 1$$
$$15\left(\frac{3}{5}x - 2\right) = 15\left(\frac{2}{3}x - 1\right)$$
$$15\left(\frac{3}{5}x\right) - 15(2) = 15\left(\frac{2}{3}x\right) - 15(1)$$
$$9x - 30 = 10x - 15$$
$$9x - 30 - 9x = 10x - 15 - 9x$$
$$-30 = x - 15$$
$$-30 + 15 = x - 15 + 15$$
$$-15 = x$$

Check: $\quad \dfrac{3}{5}x - 2 = \dfrac{2}{3}x - 1$
$$\frac{3}{5} \cdot -15 - 2 \stackrel{?}{=} \frac{2}{3} \cdot -15 - 1$$
$$-9 - 2 \stackrel{?}{=} -10 - 1$$
$$-11 = -11$$
The solution is -15.

4.
$$\frac{4(y+3)}{3} = 5y - 7$$
$$3 \cdot \frac{4(y+3)}{3} = 3 \cdot (5y - 7)$$
$$4(y+3) = 3(5y - 7)$$
$$4y + 12 = 15y - 21$$
$$4y + 12 - 4y = 15y - 21 - 4y$$
$$12 = 11y - 21$$
$$12 + 21 = 11y - 21 + 21$$
$$33 = 11y$$
$$\frac{33}{11} = \frac{11y}{11}$$
$$3 = y$$

To check, replace y with 3 in the original equation. The solution is 3.

5.
$$0.35x + 0.09(x+4) = 0.30(12)$$
$$100[0.35x + 0.09(x+4)] = 100[0.03(12)]$$
$$35x + 9(x+4) = 3(12)$$
$$35x + 9x + 36 = 36$$
$$44x + 36 = 36$$
$$44x + 36 - 36 = 36 - 36$$
$$44x = 0$$
$$\frac{44x}{44} = \frac{0}{44}$$
$$x = 0$$

To check, replace x with 0 in the original equation. The solution is 0.

6.
$$4(x+4) - x = 2(x+11) + x$$
$$4x + 16 - x = 2x + 22 + x$$
$$3x + 16 = 3x + 22$$
$$3x + 16 - 3x = 3x + 22 - 3x$$
$$16 = 22$$
There is no solution.

7.
$$12x - 18 = 9(x-2) + 3x$$
$$12x - 18 = 9x - 18 + 3x$$
$$12x - 18 = 12x - 18$$
$$12x - 18 + 18 = 12x - 18 + 18$$
$$12x = 12x$$
$$12x - 12x = 12x - 12x$$
$$0 = 0$$
The solution is all real numbers.

Graphing Calculator Explorations

1. Solution $(-24 = -24)$

2. Solution $(-4 = -4)$

3. Not a solution $(19.4 \neq 10.4)$

4. Not a solution $(-11.9 \neq -60.1)$

5. Solution $(17{,}061 = 17{,}061)$

6. Solution $(-316 = -316)$

Vocabulary, Readiness & Video Check 2.3

1. $x = -7$ is an equation.

2. $x - 7$ is an expression.

3. $4y - 6 + 9y + 1$ is an expression.

4. $4y - 6 = 9y + 1$ is an equation.

5. $\dfrac{1}{x} - \dfrac{x-1}{8}$ is an expression.

6. $\dfrac{1}{x} - \dfrac{x-1}{8} = 6$ is an equation.

7. $0.1x + 9 = 0.2x$ is an equation.

8. $0.1x^2 + 9y - 0.2x^2$ is an expression.

9. 3; distributive property, addition property of equality, multiplication property of equality

10. Because both sides have more than one term, you need to apply the distributive property to make sure you multiply every single term in the equation by the LCD.

11. The number of decimal places in each number helps you determine what power of 10 you can multiply through by so you are no longer dealing with decimals.

12. When solving a linear equation and all variable terms, subtract out:

    a. If you have a true statement, then the equation has <u>all real numbers</u> as a solution.

    b. If you have a false statement, then the equation has <u>no</u> solution.

Exercise Set 2.3

1.
$$-4y + 10 = -2(3y + 1)$$
$$-4y + 10 = -6y - 2$$
$$-4y + 10 - 10 = -6y - 2 - 10$$
$$-4y = -6y - 12$$
$$-4y + 6y = -6y - 12 + 6y$$
$$2y = -12$$
$$\frac{2y}{2} = \frac{-12}{2}$$
$$y = -6$$

3.
$$15x - 8 = 10 + 9x$$
$$15x - 8 + 8 = 10 + 9x + 8$$
$$15x = 18 + 9x$$
$$15x - 9x = 18 + 9x - 9x$$
$$6x = 18$$
$$\frac{6x}{6} = \frac{18}{6}$$
$$x = 3$$

5.
$$-2(3x - 4) = 2x$$
$$-6x + 8 = 2x$$
$$-6x + 6x + 8 = 2x + 6x$$
$$8 = 8x$$
$$\frac{8}{8} = \frac{8x}{8}$$
$$1 = x$$

7.
$$5(2x - 1) - 2(3x) = 1$$
$$10x - 5 - 6x = 1$$
$$4x - 5 = 1$$
$$4x - 5 + 5 = 1 + 5$$
$$4x = 6$$
$$\frac{4x}{4} = \frac{6}{4}$$
$$x = \frac{3}{2}$$

9.
$$-6(x - 3) - 26 = -8$$
$$-6x + 18 - 26 = -8$$
$$-6x - 8 = -8$$
$$-6x - 8 + 8 = -8 + 8$$
$$-6x = 0$$
$$\frac{-6x}{-6} = \frac{0}{-6}$$
$$x = 0$$

11.
$$8 - 2(a + 1) = 9 + a$$
$$8 - 2a - 2 = 9 + a$$
$$-2a + 6 = 9 + a$$
$$-2a + 6 - 6 = 9 + a - 6$$
$$-2a = 3 + a$$
$$-2a - a = 3 + a - a$$
$$-3a = 3$$
$$\frac{-3a}{-3} = \frac{3}{-3}$$
$$a = -1$$

13.
$$4x + 3 = -3 + 2x + 14$$
$$4x + 3 = 2x + 11$$
$$4x - 2x + 3 = 2x - 2x + 11$$
$$2x + 3 = 11$$
$$2x + 3 - 3 = 11 - 3$$
$$2x = 8$$
$$\frac{2x}{2} = \frac{8}{2}$$
$$x = 4$$

15.
$$-2y - 10 = 5y + 18$$
$$-2y - 5y - 10 = 5y - 5y + 18$$
$$-7y - 10 = 18$$
$$-7y - 10 + 10 = 18 + 10$$
$$-7y = 28$$
$$\frac{-7y}{-7} = \frac{28}{-7}$$
$$y = -4$$

17. $\dfrac{2}{3}x + \dfrac{4}{3} = -\dfrac{2}{3}$

$3\left(\dfrac{2}{3}x + \dfrac{4}{3}\right) = 3\left(-\dfrac{2}{3}\right)$

$2x + 4 = -2$

$2x + 4 - 4 = -2 - 4$

$2x = -6$

$\dfrac{2x}{2} = \dfrac{-6}{2}$

$x = -3$

19. $\dfrac{3}{4}x - \dfrac{1}{2} = 1$

$4\left(\dfrac{3}{4}x - \dfrac{1}{2}\right) = 4(1)$

$3x - 2 = 4$

$3x - 2 + 2 = 4 + 2$

$3x = 6$

$\dfrac{3x}{3} = \dfrac{6}{3}$

$x = 2$

21. $0.50x + 0.15(70) = 35.5$

$100[0.50x + 0.15(70)] = 100(35.5)$

$50x + 15(70) = 3550$

$50x + 1050 = 3550$

$50x + 1050 - 1050 = 3550 - 1050$

$50x = 2500$

$\dfrac{50x}{50} = \dfrac{2500}{50}$

$x = 50$

23. $\dfrac{2(x+1)}{4} = 3x - 2$

$4\left[\dfrac{2(x+1)}{4}\right] = 4(3x - 2)$

$2(x+1) = 12x - 8$

$2x + 2 = 12x - 8$

$2x - 12x + 2 = 12x - 12x - 8$

$-10x + 2 = -8$

$-10x + 2 - 2 = -8 - 2$

$-10x = -10$

$\dfrac{-10x}{-10} = \dfrac{-10}{-10}$

$x = 1$

25. $x + \dfrac{7}{6} = 2x - \dfrac{7}{6}$

$6\left(x + \dfrac{7}{6}\right) = 6\left(2x - \dfrac{7}{6}\right)$

$6x + 7 = 12x - 7$

$6x - 12x + 7 = 12x - 12x - 7$

$-6x + 7 = -7$

$-6x + 7 - 7 = -7 - 7$

$-6x = -14$

$\dfrac{-6x}{-6} = \dfrac{-14}{-6}$

$x = \dfrac{7}{3}$

27. $0.12(y - 6) + 0.06y = 0.08y - 0.70$

$100[0.12(y - 6) + 0.06y] = 100[0.08y - 0.70]$

$12(y - 6) + 6y = 8y - 70$

$12y - 72 + 6y = 8y - 70$

$18y - 72 = 8y - 70$

$18y - 8y - 72 = 8y - 8y - 70$

$10y - 72 = -70$

$10y - 72 + 72 = -70 + 72$

$10y = 2$

$\dfrac{10y}{10} = \dfrac{2}{10}$

$y = \dfrac{1}{5} = 0.2$

29. $4(3x + 2) = 12x + 8$

$12x + 8 = 12x + 8$

$12x + 8 - 12x = 12x + 8 - 12x$

$8 = 8$

All real numbers are solutions.

31. $\dfrac{x}{4} + 1 = \dfrac{x}{4}$

$4\left(\dfrac{x}{4} + 1\right) = 4\left(\dfrac{x}{4}\right)$

$x + 4 = x$

$x - x + 4 = x - x$

$4 = 0$

There is no solution.

33. $3x - 7 = 3(x + 1)$

$3x - 7 = 3x + 3$

$3x - 3x - 7 = 3x - 3x + 3$

$-7 = 3$

There is no solution.

35.
$$-2(6x-5)+4 = -12x+14$$
$$-12x+10+4 = -12x+14$$
$$-12x+14 = -12x+14$$
$$-12x+14+12x = -12x+14+12x$$
$$14 = 14$$
All real numbers are solutions.

37.
$$\frac{6(3-z)}{5} = -z$$
$$5\left[\frac{6(3-z)}{5}\right] = 5(-z)$$
$$6(3-z) = 5(-z)$$
$$18-6z = -5z$$
$$18-6z+6z = -5z+6z$$
$$18 = z$$

39.
$$-3(2t-5)+2t = 5t-4$$
$$-6t+15+2t = 5t-4$$
$$-4t+15 = 5t-4$$
$$-4t+15+4 = 5t-4+4$$
$$-4t+19 = 5t$$
$$-4t+19+4t = 5t+4t$$
$$19 = 9t$$
$$\frac{19}{9} = \frac{9t}{9}$$
$$\frac{19}{9} = t$$

41.
$$5y+2(y-6) = 4(y+1)-2$$
$$5y+2y-12 = 4y+4-2$$
$$7y-12 = 4y+2$$
$$7y-12+12 = 4y+2+12$$
$$7y = 4y+14$$
$$7y-4y = 4y+14-4y$$
$$3y = 14$$
$$\frac{3y}{3} = \frac{14}{3}$$
$$y = \frac{14}{3}$$

43.
$$\frac{3(x-5)}{2} = \frac{2(x+5)}{3}$$
$$6\left[\frac{3(x-5)}{2}\right] = 6\left[\frac{2(x+5)}{3}\right]$$
$$9(x-5) = 4(x+5)$$
$$9x-45 = 4x+20$$
$$9x-4x-45 = 4x-4x+20$$
$$5x-45 = 20$$
$$5x-45+45 = 20+45$$
$$5x = 65$$
$$\frac{5x}{5} = \frac{65}{5}$$
$$x = 13$$

45.
$$0.7x-2.3 = 0.5$$
$$10(0.7x-2.3) = 10(0.5)$$
$$7x-23 = 5$$
$$7x-23+23 = 5+23$$
$$7x = 28$$
$$\frac{7x}{7} = \frac{28}{7}$$
$$x = 4$$

47.
$$5x-5 = 2(x+1)+3x-7$$
$$5x-5 = 2x+2+3x-7$$
$$5x-5 = 5x-5$$
$$5x-5x-5 = 5x-5x-5$$
$$-5 = -5$$
All real numbers are solutions.

49.
$$4(2n+1) = 3(6n+3)+1$$
$$8n+4 = 18n+9+1$$
$$8n+4 = 18n+10$$
$$8n+4-4 = 18n+10-4$$
$$8n = 18n+6$$
$$8n-18n = 18n+6-18n$$
$$-10n = 6$$
$$\frac{-10n}{-10} = \frac{6}{-10}$$
$$n = -\frac{3}{5}$$

51.
$$x + \frac{5}{4} = \frac{3}{4}x$$
$$4\left(x + \frac{5}{4}\right) = 4\left(\frac{3}{4}x\right)$$
$$4x + 5 = 3x$$
$$4x + 5 - 4x = 3x - 4x$$
$$5 = -x$$
$$\frac{5}{-1} = \frac{-x}{-1}$$
$$-5 = x$$

53.
$$\frac{x}{2} - 1 = \frac{x}{5} + 2$$
$$10\left(\frac{x}{2} - 1\right) = 10\left(\frac{x}{5} + 2\right)$$
$$5x - 10 = 2x + 20$$
$$5x - 10 + 10 = 2x + 20 + 10$$
$$5x = 2x + 30$$
$$5x - 2x = 2x + 30 - 2x$$
$$3x = 30$$
$$\frac{3x}{3} = \frac{30}{3}$$
$$x = 10$$

55. $2(x+3) - 5 = 5x - 3(1+x)$
$$2x + 6 - 5 = 5x - 3 - 3x$$
$$2x + 1 = 2x - 3$$
$$2x - 2x + 1 = 2x - 2x - 3$$
$$1 = -3$$
There is no solution.

57.
$$0.06 - 0.01(x+1) = -0.02(2-x)$$
$$100[0.06 - 0.01(x+1)] = 100[-0.02(2-x)]$$
$$6 - (x+1) = -2(2-x)$$
$$6 - x - 1 = -4 + 2x$$
$$5 - x = -4 + 2x$$
$$5 - x - 2x = -4 + 2x - 2x$$
$$5 - 3x = -4$$
$$5 - 5 - 3x = -4 - 5$$
$$-3x = -9$$
$$\frac{-3x}{-3} = \frac{-9}{-3}$$
$$x = 3$$

59.
$$\frac{9}{2} + \frac{5}{2}y = 2y - 4$$
$$2\left(\frac{9}{2} + \frac{5}{2}y\right) = 2(2y-4)$$
$$9 + 5y = 4y - 8$$
$$9 + 5y - 4y = 4y - 8 - 4y$$
$$9 + y = -8$$
$$9 + y - 9 = -8 - 9$$
$$y = -17$$

61.
$$-2y - 10 = 5y + 18$$
$$-2y - 10 - 18 = 5y + 18 - 18$$
$$-2y - 28 = 5y$$
$$-2y - 28 + 2y = 5y + 2y$$
$$-28 = 7y$$
$$\frac{-28}{7} = \frac{7y}{7}$$
$$-4 = y$$

63.
$$0.6x - 0.1 = 0.5x + 0.2$$
$$10(0.6x - 0.1) = 10(0.5x + 0.2)$$
$$6x - 1 = 5x + 2$$
$$6x - 5x - 1 = 5x - 5x + 2$$
$$x - 1 = 2$$
$$x - 1 + 1 = 2 + 1$$
$$x = 3$$

65.
$$0.02(6t-3) = 0.12(t-2) + 0.18$$
$$100[0.02(6t-3)] = 100[0.12(t-2) + 0.18]$$
$$2(6t-3) = 12(t-2) + 18$$
$$12t - 6 = 12t - 24 + 18$$
$$12t - 6 = 12t - 6$$
$$12t - 12t - 6 = 12t - 12t - 6$$
$$-6 = -6$$
All real numbers are solutions.

67. -8 minus a number
$$\downarrow \quad \downarrow \quad \quad \downarrow$$
$$-8 \quad - \quad \quad x$$

69. -3 plus twice a number
$$\downarrow \quad \downarrow \quad \quad \downarrow$$
$$-3 \quad + \quad \quad 2x$$

71. 9 times a number plus 20
$$\downarrow \quad \downarrow \quad \quad \downarrow \quad \downarrow \quad \downarrow$$
$$9 \quad \cdot \quad (x \quad + \quad 20) = 9(x+20)$$

73. $x + (2x - 3) + (3x - 5) = x + 2x - 3 + 3x - 5$
$$= 6x - 8$$
The perimeter is $(6x - 8)$ meters.

75. a.
$$x + 3 = x + 3$$
$$x + 3 - x = x + 3 - x$$
$$3 = 3$$
$$3 - 3 = 3 - 3$$
$$0 = 0$$
All real numbers are solutions.

    b. answers may vary

    c. answers may vary

77.
$$5x + 1 = 5x + 1$$
$$5x + 1 - 5x = 5x + 1 - 5x$$
$$1 = 1$$
All real numbers are solutions. The answer is a.

79.
$$2x - 6x - 10 = -4x + 3 - 10$$
$$-4x - 10 = -4x - 7$$
$$-4x - 10 + 4x = -4x - 7 + 4x$$
$$-10 = -7$$
There is no solution. The answer is b.

81.
$$9x - 20 = 8x - 20$$
$$9x - 20 - 8x = 8x - 20 - 8x$$
$$x - 20 = -20$$
$$x - 20 + 20 = -20 + 20$$
$$x = 0$$
The answer is c.

83. answers may vary

85. a. Since the perimeter is the sum of the lengths of the sides, $x + x + x + 2x + 2x = 28$.

    b.
$$7x = 28$$
$$\frac{7x}{7} = \frac{28}{7}$$
$$x = 4$$

    c. $2x = 2(4) = 8$
The lengths are $x = 4$ centimeters and $2x = 8$ centimeters.

87. answers may vary

89.
$$1000(7x - 10) = 50(412 + 100x)$$
$$7000x - 10,000 = 20,600 + 5000x$$
$$7000x - 5000x - 10,000 = 20,600 + 5000x - 5000x$$
$$2000x - 10,000 = 20,600$$
$$2000x - 10,000 + 10,000 = 20,600 + 10,000$$
$$2000x = 30,600$$
$$\frac{2000x}{2000} = \frac{30,600}{2000}$$
$$x = 15.3$$

91.
$$0.035x + 5.112 = 0.010x + 5.107$$
$$1000(0.035x + 5.112) = 1000(0.010x + 5.107)$$
$$35x + 5112 = 10x + 5107$$
$$35x - 10x + 5112 = 10x - 10x + 5107$$
$$25x + 5112 = 5107$$
$$25x + 5112 - 5112 = 5107 - 5112$$
$$25x = -5$$
$$\frac{25x}{25} = \frac{-5}{25}$$
$$x = -\frac{1}{5} = -0.2$$

93.
$$x(x - 3) = x^2 + 5x + 7$$
$$x^2 - 3x = x^2 + 5x + 7$$
$$x^2 - x^2 - 3x = x^2 - x^2 + 5x + 7$$
$$-3x = 5x + 7$$
$$-3x - 5x = 5x - 5x + 7$$
$$-8x = 7$$
$$\frac{-8x}{-8} = \frac{7}{-8}$$
$$x = -\frac{7}{8}$$

95.
$$2z(z + 6) = 2z^2 + 12z - 8$$
$$2z^2 + 12z = 2z^2 + 12z - 8$$
$$2z^2 - 2z^2 + 12z = 2z^2 - 2z^2 + 12z - 8$$
$$12z = 12z - 8$$
$$12z - 12z = 12z - 12z - 8$$
$$0 = -8$$
There is no solution.

Integrated Review

1.
$$x - 10 = -4$$
$$x - 10 + 10 = -4 + 10$$
$$x = 6$$

2.
$$y + 14 = -3$$
$$y + 14 - 14 = -3 - 14$$
$$y = -17$$

3. $9y = 108$

$$\frac{9y}{9} = \frac{108}{9}$$

$$y = 12$$

4. $-3x = 78$

$$\frac{-3x}{-3} = \frac{78}{-3}$$

$$x = -26$$

5. $-6x + 7 = 25$

$$-6x + 7 - 7 = 25 - 7$$

$$-6x = 18$$

$$\frac{-6x}{-6} = \frac{18}{-6}$$

$$x = -3$$

6. $5y - 42 = -47$

$$5y - 42 + 42 = -47 + 42$$

$$5y = -5$$

$$\frac{5y}{5} = \frac{-5}{5}$$

$$y = -1$$

7. $\frac{2}{3}x = 9$

$$\frac{3}{2}\left(\frac{2}{3}x\right) = \frac{3}{2}(9)$$

$$x = \frac{27}{2}$$

8. $\frac{4}{5}z = 10$

$$\frac{5}{4}\left(\frac{4}{5}z\right) = \frac{5}{4}(10)$$

$$z = \frac{25}{2}$$

9. $\frac{r}{-4} = -2$

$$-4\left(\frac{r}{-4}\right) = -4(-2)$$

$$r = 8$$

10. $\frac{y}{-8} = 8$

$$-8\left(\frac{y}{-8}\right) = -8(8)$$

$$y = -64$$

11. $6 - 2x + 8 = 10$

$$-2x + 14 = 10$$

$$-2x + 14 - 14 = 10 - 14$$

$$-2x = -4$$

$$\frac{-2x}{-2} = \frac{-4}{-2}$$

$$x = 2$$

12. $-5 - 6y + 6 = 19$

$$-6y + 1 = 19$$

$$-6y + 1 - 1 = 19 - 1$$

$$-6y = 18$$

$$\frac{-6y}{-6} = \frac{18}{-6}$$

$$y = -3$$

13. $2x - 7 = 2x - 27$

$$2x - 2x - 7 = 2x - 2x - 27$$

$$-7 = -27$$

There is no solution.

14. $3 + 8y = 8y - 2$

$$3 + 8y - 8y = 8y - 8y - 2$$

$$3 = -2$$

There is no solution.

15. $-3a + 6 + 5a = 7a - 8a$

$$2a + 6 = -a$$

$$2a - 2a + 6 = -a - 2a$$

$$6 = -3a$$

$$\frac{6}{-3} = \frac{-3a}{-3}$$

$$-2 = a$$

16. $4b - 8 - b = 10b - 3b$

$$3b - 8 = 7b$$

$$3b - 3b - 8 = 7b - 3b$$

$$-8 = 4b$$

$$\frac{-8}{4} = \frac{4b}{4}$$

$$-2 = b$$

17. $-\frac{2}{3}x = \frac{5}{9}$

$$-\frac{3}{2}\left(-\frac{2}{3}x\right) = -\frac{3}{2}\left(\frac{5}{9}\right)$$

$$x = -\frac{5}{6}$$

18.
$$-\frac{3}{8}y = -\frac{1}{16}$$
$$-\frac{8}{3}\left(-\frac{3}{8}y\right) = -\frac{8}{3}\left(-\frac{1}{16}\right)$$
$$y = \frac{1}{6}$$

19.
$$10 = -6n + 16$$
$$10 - 16 = -6n + 16 - 16$$
$$-6 = -6n$$
$$\frac{-6}{-6} = \frac{-6n}{-6}$$
$$1 = n$$

20.
$$-5 = -2m + 7$$
$$-5 - 7 = -2m + 7 - 7$$
$$-12 = -2m$$
$$\frac{-12}{-2} = \frac{-2m}{-2}$$
$$6 = m$$

21.
$$3(5c - 1) - 2 = 13c + 3$$
$$15c - 3 - 2 = 13c + 3$$
$$15c - 5 = 13c + 3$$
$$15c - 13c - 5 = 13c - 13c + 3$$
$$2c - 5 = 3$$
$$2c - 5 + 5 = 3 + 5$$
$$2c = 8$$
$$\frac{2c}{2} = \frac{8}{2}$$
$$c = 4$$

22.
$$4(3t + 4) - 20 = 3 + 5t$$
$$12t + 16 - 20 = 3 + 5t$$
$$12t - 4 = 3 + 5t$$
$$12t - 5t - 4 = 3 + 5t - 5t$$
$$7t - 4 = 3$$
$$7t - 4 + 4 = 3 + 4$$
$$7t = 7$$
$$\frac{7t}{7} = \frac{7}{7}$$
$$t = 1$$

23.
$$\frac{2(z + 3)}{3} = 5 - z$$
$$3\left[\frac{2(z + 3)}{3}\right] = 3(5 - z)$$
$$2z + 6 = 15 - 3z$$
$$2z + 3z + 6 = 15 - 3z + 3z$$
$$5z + 6 = 15$$
$$5z + 6 - 6 = 15 - 6$$
$$5z = 9$$
$$\frac{5z}{5} = \frac{9}{5}$$
$$z = \frac{9}{5}$$

24.
$$\frac{3(w + 2)}{4} = 2w + 3$$
$$4\left[\frac{3(w + 2)}{4}\right] = 4(2w + 3)$$
$$3w + 6 = 8w + 12$$
$$3w - 8w + 6 = 8w - 8w + 12$$
$$-5w + 6 = 12$$
$$-5w + 6 - 6 = 12 - 6$$
$$-5w = 6$$
$$\frac{-5w}{-5} = \frac{6}{-5}$$
$$w = -\frac{6}{5}$$

25.
$$-2(2x - 5) = -3x + 7 - x + 3$$
$$-4x + 10 = -4x + 10$$
$$-4x + 4x + 10 = -4x + 4x + 10$$
$$10 = 10$$
All real numbers are solutions.

26.
$$-4(5x - 2) = -12x + 4 - 8x + 4$$
$$-20x + 8 = -20x + 8$$
$$-20x + 20x + 8 = -20x + 20x + 8$$
$$8 = 8$$
All real numbers are solutions.

27.
$$0.02(6t-3) = 0.04(t-2)+0.02$$
$$100[0.02(6t-3)] = 100[0.04(t-2)+0.02]$$
$$2(6t-3) = 4(t-2)+2$$
$$12t-6 = 4t-8+2$$
$$12t-6 = 4t-6$$
$$12t-4t-6 = 4t-4t-6$$
$$8t-6 = -6$$
$$8t-6+6 = -6+6$$
$$8t = 0$$
$$\frac{8t}{8} = \frac{0}{8}$$
$$t = 0$$

28.
$$0.03(m+7) = 0.02(5-m)+0.03$$
$$100[0.03(m+7)] = 100[0.02(5-m)+0.03]$$
$$3(m+7) = 2(5-m)+3$$
$$3m+21 = 10-2m+3$$
$$3m+21 = 13-2m$$
$$3m+2m+21 = 13-2m+2m$$
$$5m+21 = 13$$
$$5m+21-21 = 13-21$$
$$5m = -8$$
$$\frac{5m}{5} = \frac{-8}{5}$$
$$m = -\frac{8}{5} = -1.6$$

29.
$$-3y = \frac{4(y-1)}{5}$$
$$5(-3y) = 5\left[\frac{4(y-1)}{5}\right]$$
$$-15y = 4y-4$$
$$-15y-4y = 4y-4y-4$$
$$-19y = -4$$
$$\frac{-19y}{-19} = \frac{-4}{-19}$$
$$y = \frac{4}{19}$$

30.
$$-4x = \frac{5(1-x)}{6}$$
$$6(-4x) = 6\left[\frac{5(1-x)}{6}\right]$$
$$-24x = 5-5x$$
$$-24x+5x = 5-5x+5x$$
$$-19x = 5$$
$$\frac{-19x}{-19} = \frac{5}{-19}$$
$$x = -\frac{5}{19}$$

31.
$$\frac{5}{3}x - \frac{7}{3} = x$$
$$3\left(\frac{5}{3}x - \frac{7}{3}\right) = 3(x)$$
$$5x-7 = 3x$$
$$5x-5x-7 = 3x-5x$$
$$-7 = -2x$$
$$\frac{-7}{-2} = \frac{-2x}{-2}$$
$$\frac{7}{2} = x$$

32.
$$\frac{7}{5}n + \frac{3}{5} = -n$$
$$5\left(\frac{7}{5}n + \frac{3}{5}\right) = 5(-n)$$
$$7n+3 = -5n$$
$$7n-7n+3 = -5n-7n$$
$$3 = -12n$$
$$\frac{3}{-12} = \frac{-12n}{-12}$$
$$-\frac{1}{4} = n$$

33.
$$9(3x-1) = -4+49$$
$$27x-9 = 45$$
$$27x-9+9 = 45+9$$
$$27x = 54$$
$$\frac{27x}{27} = \frac{54}{27}$$
$$x = 2$$

34.
$$12(2x+1) = -6+66$$
$$24x+12 = 60$$
$$24x+12-12 = 60-12$$
$$24x = 48$$
$$\frac{24x}{24} = \frac{48}{24}$$
$$x = 2$$

35.
$$\frac{1}{10}(3x-7) = \frac{3}{10}x+5$$
$$10\left[\frac{1}{10}(3x-7)\right] = 10\left(\frac{3}{10}x+5\right)$$
$$3x-7 = 3x+50$$
$$3x-7-3x = 3x+50-3x$$
$$-7 = 50$$
There is no solution.

36. $\dfrac{1}{7}(2x-5)=\dfrac{2}{7}x+1$

$7\left[\dfrac{1}{7}(2x-5)\right]=7\left(\dfrac{2}{7}x+1\right)$

$2x-5=2x+7$

$2x-5-2x=2x+7-2x$

$-5=7$

There is no solution.

37. $5+2(3x-6)=-4(6x-7)$

$5+6x-12=-24x+28$

$6x-7=-24x+28$

$6x-7+24x=-24x+28+24x$

$30x-7=28$

$30x-7+7=28+7$

$30x=35$

$\dfrac{30x}{30}=\dfrac{35}{30}$

$x=\dfrac{7}{6}$

38. $3+5(2x-4)=-7(5x+2)$

$3+10x-20=-35x-14$

$10x-17=-35x-14$

$10x-17+35x=-35x-14+35x$

$45x-17=-14$

$45x-17+17=-14+17$

$45x=3$

$\dfrac{45x}{45}=\dfrac{3}{45}$

$x=\dfrac{1}{15}$

Section 2.4 Practice

1. Let x = the number.

$3x-6=2x+3$

$3x-6-2x=2x+3-2x$

$x-6=3$

$x-6+6=3+6$

$x=9$

The number is 9.

2. Let x = the number.

$3x-4=2(x-1)$

$3x-4=2x-2$

$3x-4-2x=2x-2-2x$

$x-4=-2$

$x-4+4=-2+4$

$x=2$

The number is 2.

3. Let x = the length of short piece, then $4x$ = the length of long piece.

$x+4x=45$

$5x=45$

$\dfrac{5x}{5}=\dfrac{45}{5}$

$x=9$

$4x=4(9)=36$

The short piece is 9 inches and the long piece is 36 inches.

4. Let x = number of Republican governors, then $x-9$ = number of Democratic governors.

$x+x-9=49$

$2x-9=49$

$2x-9+9=49+9$

$2x=58$

$\dfrac{2x}{2}=\dfrac{58}{2}$

$x=29$

$x-9=20$

There were 29 Republican and 20 Democratic governors.

5. x = degree measure of first angle
$3x$ = degree measure of second angle
$x+55$ = degree measure of third angle

$x+3x+(x+55)=180$

$5x+55=180$

$5x+55-55=180-55$

$5x=125$

$\dfrac{5x}{5}=\dfrac{125}{5}$

$x=25$

$3x=3(25)=75$

$x+55=25+55=80$

The measures of the angles are 25°, 75°, and 80°.

6. Let x = the first even integer, then
$x+2$ = the second even integer, and
$x+4$ = the third even integer.

$x+(x+2)+(x+4)=144$

$3x+6=144$

$3x+6-6=144-6$

$3x=138$

$\dfrac{3x}{3}=\dfrac{138}{3}$

$x=46$

$x+2=46+2=48$

$x+4=46+4=50$

The integers are 46, 48, and 50.

Vocabulary, Readiness & Video Check 2.4

1. $2x$; $2x - 31$

2. $3x$; $3x + 17$

3. $x + 5$; $2(x + 5)$

4. $x - 11$; $7(x - 11)$

5. $20 - y$; $\dfrac{20 - y}{3}$ or $(20 - y) \div 3$

6. $-10 + y$; $\dfrac{-10 + y}{9}$ or $(-10 + y) \div 9$

7. in the statement of the application

8. The original application asks for the measure of two supplementary angles. The solution of $x = 43$ only gives us the measure of one of the angles.

9. That the 3 angle measures are consecutive even integers and that they sum to $180°$.

Exercise Set 2.4

1. Let x = the number.
$$6x + 1 = 5x$$
$$6x + 1 - 5x = 5x - 5x$$
$$x + 1 = 0$$
$$x + 1 - 1 = 0 - 1$$
$$x = -1$$
The number is -1.

3. Let x = the number.
$$3x - 6 = 2x + 8$$
$$3x - 6 - 2x = 2x + 8 - 2x$$
$$x - 6 = 8$$
$$x - 6 + 6 = 8 + 6$$
$$x = 14$$
The number is 14.

5. Let x = the number.
$$2(x - 8) = 3(x + 3)$$
$$2x - 16 = 3x + 9$$
$$2x - 2x - 16 = 3x - 2x + 9$$
$$-16 = x + 9$$
$$-16 - 9 = x + 9 - 9$$
$$-25 = x$$
The number is -25.

7. Let x = the number.
$$2(-2 + x) = x - \frac{1}{2}$$
$$-4 + 2x = x - \frac{1}{2}$$
$$-4 + 2x - x = x - x - \frac{1}{2}$$
$$-4 + x = -\frac{1}{2}$$
$$-4 + 4 + x = -\frac{1}{2} + 4$$
$$x = -\frac{1}{2} + \frac{8}{2}$$
$$x = \frac{7}{2}$$
The number is $\dfrac{7}{2}$.

9. The sum of the three lengths is 25 inches.
$$x + 2x + 1 + 5x = 25$$
$$1 + 8x = 25$$
$$1 + 8x - 1 = 25 - 1$$
$$8x = 24$$
$$\frac{8x}{8} = \frac{24}{8}$$
$$x = 3$$
$2x = 2(3) = 6$
$1 + 5x = 1 + 5(3) = 1 + 15 = 16$
The lengths are 3 inches, 6 inches, and 16 inches.

11. Let x be the length of the first piece. Then the second piece is $2x$ and the third piece is $5x$. The sum of the lengths is 40 inches.
$$x + 2x + 5x = 40$$
$$8x = 40$$
$$\frac{8x}{8} = \frac{40}{8}$$
$$x = 5$$
$2x = 2(5) = 10$
$5x = 5(5) = 25$
The 1st piece is 5 inches, 2nd piece is 10 inches, and 3rd piece is 25 inches.

13.
$$x + x + 23{,}873 = 39{,}547$$
$$2x + 23{,}873 = 39{,}547$$
$$2x + 23{,}873 - 23{,}873 = 39{,}547 - 23{,}873$$
$$2x = 15{,}674$$
$$\frac{2x}{2} = \frac{15{,}674}{2}$$
$$x = 7837$$
In 2010, 7837 screens were 3D.

15. Let x be the measure of each of the two equal angles. Then $2x + 30$ is the measure of the third angle. Their sum is 180°.

$$x + x + 2x + 30 = 180$$
$$4x + 30 = 180$$
$$4x + 30 - 30 = 180 - 30$$
$$4x = 150$$
$$\frac{4x}{4} = \frac{150}{4}$$
$$x = 37.5$$

$2x + 30 = 2(37.5) + 30 = 75 + 30 = 105$

The 1st angle measures 37.5°, the 2nd angle measures 37.5°, and the 3rd angle measures 105°.

	First Integer	Next Integers			Indicated Sum
17. Three consecutive integers:	Integer: x	$x + 1$	$x + 2$		Sum of the three consecutive integers simplified: $(x + 1) + (x + 2) = 2x + 3$
19. Three consecutive even integers:	Even integer: x	$x + 2$	$x + 4$		Sum of the first and third even consecutive integers, simplified: $x + (x + 4) = 2x + 4$
21. Four consecutive integers:	Integer: x	$x + 1$	$x + 2$	$x + 3$	Sum of the four consecutive integers, simplified: $x + (x + 1) + (x + 2) + (x + 3)$ $= 4x + 6$
23. Three consecutive odd integers:	Odd integer: x	$x + 2$	$x + 4$		Sum of the second and third consecutive odd integers, simplified: $(x + 2) + (x + 4) = 2x + 6$

25. Let x = the number of the left page and $x + 1$ = the number of the right page.

$$x + x + 1 = 469$$
$$2x + 1 = 469$$
$$2x + 1 - 1 = 469 - 1$$
$$2x = 468$$
$$\frac{2x}{2} = \frac{468}{2}$$
$$x = 234$$

$x + 1 = 234 + 1 = 235$

The page numbers are 234 and 235.

27. Let x = the code for Belgium,
$x + 1$ = the code for France,
$x + 2$ = the code for Spain.

$$x + x + 1 + x + 2 = 99$$
$$3x + 3 = 99$$
$$3x + 3 - 3 = 99 - 3$$
$$3x = 96$$
$$\frac{3x}{3} = \frac{96}{3}$$
$$x = 32$$

$x + 1 = 32 + 1 = 33$
$x + 2 = 32 + 2 = 34$
The codes are Belgium: 32; France: 33; Spain: 34.

29. Let x represent the area of the Gobi Desert, in square miles. Then $7x$ represents the area of the Sahara Desert.
$$x + 7x = 4,000,000$$
$$8x = 4,000,000$$
$$\frac{8x}{8} = \frac{4,000,000}{8}$$
$$x = 500,000$$
$7x = 7(500,000) = 3,500,000$
The Gobi Desert's area is 500,000 square miles and the Sahara Desert's area is 3,500,000 square miles.

31. Let x be the length of the shorter piece. Then $2x + 2$ is the length of the longer piece. The measures sum to 17 feet.
$$x + 2x + 2 = 17$$
$$3x + 2 = 17$$
$$3x + 2 - 2 = 17 - 2$$
$$3x = 5$$
$$\frac{3x}{3} = \frac{15}{3}$$
$$x = 5$$
$2x + 2 = 2(5) + 2 = 10 + 2 = 12$
The pieces measure 5 feet and 12 feet.

33. Let x = the number.
$$10 - 5x = 3x$$
$$10 - 5x + 5x = 3x + 5x$$
$$10 = 8x$$
$$\frac{10}{8} = \frac{8x}{8}$$
$$\frac{5}{4} = x$$
The number is $\frac{5}{4}$.

35. Let x = carats in Angola, then $4x$ = carats in Botswana.
$$x + 4x = 40,000,000$$
$$5x = 40,000,000$$
$$\frac{5x}{5} = \frac{40,000,000}{5}$$
$$x = 8,000,000$$
$4x = 4(8,000,000) = 32,000,000$
Botswana produces 32,000,000 carats and Angola produces 8,000,000 carats.

37. Let x = the measure of the smallest angle, $x + 2$ = the measure of the second, and $x + 4$ = the measure of the third.
$$x + x + 2 + x + 4 = 180$$
$$3x + 6 = 180$$
$$3x + 6 - 6 = 180 - 6$$
$$3x = 174$$
$$\frac{3x}{3} = \frac{174}{3}$$
$$x = 58$$
$x + 2 = 58 + 2 = 60$
$x + 4 = 58 + 4 = 62$
The angles are 58°, 60°, and 62°.

39. Let x = first integer (South Korea), $x + 1$ = second integer (Russia), $x + 2$ = third integer (Austria).
$$x + (x + 1) + (x + 2) = 45$$
$$3x + 3 = 45$$
$$3 + 3 - 3 = 45 - 3$$
$$3x = 42$$
$$\frac{3x}{3} = \frac{42}{3}$$
$$x = 14$$
$x + 1 = 14 + 1 = 15$
$x + 2 = 14 + 2 = 16$
The number of medals for each country is South Korea: 14; Russia: 15; Austria: 16.

41. Let x = the number.
$$3(x + 5) = 2x - 1$$
$$3x + 15 = 2x - 1$$
$$3x + 15 - 2x = 2x - 1 - 2x$$
$$x + 15 = -1$$
$$x + 15 - 15 = -1 - 15$$
$$x = -16$$
The number is −16.

43. Let x = smaller angle, then $3x + 8$ = larger angle.
$$x + (3x + 8) = 180$$
$$4x + 8 = 180$$
$$4x + 8 - 8 = 180 - 8$$
$$4x = 172$$
$$\frac{4x}{4} = \frac{172}{4}$$
$$x = 43$$
$3x + 8 = 3(43) + 8 = 137$
The angles measure 43° and 137°.

45. Let x = the number.
$$\frac{x}{4}+\frac{1}{2}=\frac{3}{4}$$
$$4\left(\frac{x}{4}+\frac{1}{2}\right)=4\left(\frac{3}{4}\right)$$
$$x+2=3$$
$$x+2-2=3-2$$
The number is 1.

47.
$$\frac{2}{3}+4x=5x-\frac{5}{6}$$
$$6\cdot\left(\frac{2}{3}+4x\right)=6\cdot\left(5x-\frac{5}{6}\right)$$
$$4+24x=30x-5$$
$$4+24x-24x=30x-5-24x$$
$$4=6x-5$$
$$4+5=6x-5+5$$
$$9=6x$$
$$\frac{9}{6}=\frac{6x}{6}$$
$$\frac{3}{2}=x$$

The number is $\frac{3}{2}$.

49. Let x = speed of TGV, then
$x + 3.8$ = speed of Maglev.
$$x+x+3.8=718.2$$
$$2x+3.8=718.2$$
$$2x+3.8-3.8=718.2-3.8$$
$$2x=714.4$$
$$\frac{2x}{2}=\frac{714.4}{2}$$
$$x=357.2$$
$x + 3.8 = 357.2 + 3.8 = 361$
The speed of the TGV is 357.2 mph and the
speed of the Maglev is 361 mph.

51. Let x = the number.
$$\frac{1}{3}\cdot x=\frac{5}{6}$$
$$3\cdot\frac{1}{3}x=3\cdot\frac{5}{6}$$
$$x=\frac{5}{2}$$

The number is $\frac{5}{2}$.

53. Let x = number of counties in Montana and
$x + 2$ = number in California.
$$x+x+2=114$$
$$2x+2=114$$
$$2x+2-2=114-2$$
$$2x=112$$
$$\frac{2x}{2}=\frac{112}{2}$$
$$x=56$$
$x + 2 = 56 + 2 = 58$
There are 56 counties in Montana and
58 counties in California.

55. Let x = smaller angles, then
$x + 76.5$ = third angle.
$$x+x+(x+76.5)=180$$
$$3x+76.5=180$$
$$3x+76.5-76.5=180-76.5$$
$$3x=103.5$$
$$\frac{3x}{3}=\frac{103.5}{3}$$
$$x=34.5$$
$x + 76.5 = 34.5 + 76.5 = 111$
The third angle measures 111°.

57. Let x = length of first piece,
then $4x$ = length of second piece,
and $5x$ = length of third piece.
$$x+4x+5x=30$$
$$10x=30$$
$$\frac{10x}{10}=\frac{30}{10}$$
$$x=3$$
$4x = 4(3) = 12$
$5x = 5(3) = 15$
The first piece is 3 feet, the second piece is
12 feet, and the third piece is 15 feet.

59. The longest bar represents the album *Eagles:
Their Greatest Hits*, so *Eagles: Their Greatest
Hits* is the best selling album of all time.

61. Let x represent sales of *The Wall*. Then $x + 4$ is
the sales of *Thriller*.
$$x+x+4=50$$
$$2x+4=50$$
$$2x+4-4=50-4$$
$$2x=46$$
$$\frac{2x}{2}=\frac{46}{2}$$
$$x=23$$
$x + 4 = 23 + 4 = 27$
Thriller brought in $27 million and *The Wall*
brought in $23 million.

63. answers may vary

65. Replace W by 7 and L by 10.
$2W + 2L = 2(7) + 2(10) = 14 + 20 = 34$

67. Replace r by 15.
$\pi r^2 = \pi (15)^2 = \pi (225) = 225\pi$

69. Let x represent the width. Then $1.6x$ represents the length. The perimeter is
$2 \cdot$ length $+ 2 \cdot$ width.
$$2(1.6x) + 2x = 78$$
$$3.2x + 2x = 78$$
$$5.2x = 78$$
$$\frac{5.2x}{5.2} = \frac{78}{5.2}$$
$$x = 15$$
$1.6x = 1.6(15) = 24$
The dimensions of the garden are 15 feet by 24 feet.

71. 90 chirps every minute is $\dfrac{90 \text{ chirps}}{1 \text{ min}}$. There are 60 minutes in one hour.
$\dfrac{90 \text{ chirps}}{1 \text{ min}} \cdot 60 \text{ min} = 5400 \text{ chirps}$
At this rate, there are 5400 chirps each hour.
$24 \cdot 5400 = 129{,}600$
There are 129,600 chirps in one 24-hour day.
$365 \cdot 129{,}600 = 47{,}304{,}000$
There are 47,304,000 chirps in one year.

73. answers may vary

75. answers may vary

77. Measurements may vary. Rectangle (c) best approximates the shape of a golden rectangle.

Section 2.5 Practice

1. Let $d = 580$ and $r = 5$.
$$d = r \cdot t$$
$$580 = 5t$$
$$\frac{580}{5} = \frac{5t}{5}$$
$$116 = t$$
It takes 116 seconds or 1 minute 56 seconds.

2. Let $l = 40$ and $P = 98$.
$$P = 2l + 2w$$
$$98 = 2 \cdot 40 + 2w$$
$$98 = 80 + 2w$$
$$98 - 80 = 80 + 2w - 80$$
$$18 = 2w$$
$$\frac{18}{2} = \frac{2w}{2}$$
$$9 = w$$
The dog run is 9 feet wide.

3. Let $C = 8$.
$$F = \frac{9}{5}C + 32$$
$$F = \frac{9}{5} \cdot 8 + 32$$
$$F = \frac{72}{5} + \frac{160}{5}$$
$$F = \frac{232}{5} = 46.4$$
The equivalent temperature is $46.4°$F.

4. Let $w = $ width of sign, then
$5w + 3 = $ length of sign.
$$P = 2l + 2w$$
$$66 = 2(5w + 3) + 2w$$
$$66 = 10w + 6 + 2w$$
$$66 = 12w + 6$$
$$66 - 6 = 12w + 6 - 6$$
$$60 = 12w$$
$$\frac{60}{12} = \frac{12w}{12}$$
$$5 = w$$
$5w + 3 = 5(5) + 3 = 28$
The sign has length 28 inches and width 5 inches.

5. $$I = Prt$$
$$\frac{I}{Pt} = \frac{Prt}{Pt}$$
$$\frac{I}{Pt} = r \text{ or } r = \frac{I}{Pt}$$

6. $$H = 5as + 10a$$
$$H - 10a = 5as + 10a - 10a$$
$$H - 10a = 5as$$
$$\frac{H - 10a}{5a} = \frac{5as}{5a}$$
$$\frac{H - 10a}{5a} = s \text{ or } s = \frac{H - 10a}{5a}$$

7.
$$N = F + d(n-1)$$
$$N - F = F + d(n-1) - F$$
$$N - F = d(n-1)$$
$$\frac{N-F}{n-1} = \frac{d(n-1)}{n-1}$$
$$\frac{N-F}{n-1} = d \text{ or } d = \frac{N-F}{n-1}$$

8.
$$A = \frac{1}{2}a(b+B)$$
$$2 \cdot A = 2 \cdot \frac{1}{2}a(b+B)$$
$$2A = a(b+B)$$
$$2A = ab + aB$$
$$2A - ab = ab + aB - ab$$
$$2A - ab = aB$$
$$\frac{2A-ab}{a} = \frac{aB}{a}$$
$$\frac{2A-ab}{a} = B \text{ or } B = \frac{2A-ab}{a}$$

Vocabulary, Readiness & Video Check 2.5

1. A formula is an equation that describes known <u>relationships</u> among quantities.

2. This is a distance, rate, and time problem. The distance is given in miles and the time is given in hours, so the rate that we are finding must be in miles per hour (mph).

3. To show that the process of solving this equation for *x*—dividing both sides by 5, the coefficient of *x*—is the same process used to solve a formula for a specific variable. Treat whatever is multiplied by that specific variable as the coefficient—the coefficient is all the factors except that specific variable.

Exercise Set 2.5

1. Let $A = 45$ and $b = 15$.
$$A = bh$$
$$45 = 15h$$
$$\frac{45}{15} = \frac{15h}{15}$$
$$3 = h$$

3. Let $S = 102$, $l = 7$, and $w = 3$.
$$S = 4lw + 2wh$$
$$102 = 4(7)(3) + 2(3)h$$
$$102 = 84 + 6h$$
$$102 - 84 = 84 - 84 + 6h$$
$$18 = 6h$$
$$\frac{18}{6} = \frac{6h}{6}$$
$$3 = h$$

5. Let $A = 180$, $B = 11$, and $b = 7$.
$$A = \frac{1}{2}h(B+b)$$
$$180 = \frac{1}{2}h(11+7)$$
$$2(180) = 2\left[\frac{1}{2}h(18)\right]$$
$$360 = 18h$$
$$\frac{360}{18} = \frac{18h}{18}$$
$$20 = h$$

7. Let $P = 30$, $a = 8$, and $b = 10$.
$$P = a + b + c$$
$$30 = 8 + 10 + c$$
$$30 = 18 + c$$
$$30 - 18 = 18 - 18 + c$$
$$12 = c$$

9. Let $C = 15.7$, and $\pi \approx 3.14$.
$$C = 2\pi r$$
$$15.7 \approx 2(3.14)r$$
$$15.7 \approx 6.28r$$
$$\frac{15.7}{6.28} \approx \frac{6.28r}{6.28}$$
$$2.5 \approx r$$

11. Let $I = 3750$, $P = 25{,}000$, and $R = 0.05$.
$$I = PRT$$
$$3750 = 25{,}000(0.05)T$$
$$3750 = 1250T$$
$$\frac{3750}{1250} = \frac{1250T}{1250}$$
$$3 = T$$

13. Let $V = 565.2$, $r = 6$, and $\pi \approx 3.14$.

$$V = \frac{1}{3}\pi r^2 h$$

$$565.2 \approx \frac{1}{3}(3.14)(6)^2 h$$

$$565.2 \approx 37.68h$$

$$\frac{565.2}{37.68} \approx \frac{37.68h}{37.68}$$

$$15 \approx h$$

15. $f = 5gh$

$$\frac{f}{5g} = \frac{5gh}{5g}$$

$$\frac{f}{5g} = h$$

17. $V = lwh$

$$\frac{V}{lh} = \frac{lwh}{lh}$$

$$\frac{V}{lh} = w$$

19.
$$3x + y = 7$$
$$3x - 3x + y = 7 - 3x$$
$$y = 7 - 3x$$

21.
$$A = P + PRT$$
$$A - P = P - P + PRT$$
$$A - P = PRT$$
$$\frac{A-P}{PT} = \frac{PRT}{PT}$$
$$\frac{A-P}{PT} = R$$

23. $V = \frac{1}{3}Ah$

$$3V = 3\left(\frac{1}{3}Ah\right)$$

$$3V = Ah$$

$$\frac{3V}{h} = \frac{Ah}{h}$$

$$\frac{3V}{h} = A$$

25.
$$P = a + b + c$$
$$P - (b+c) = a + b + c - (b+c)$$
$$P - b - c = a + b + c - b - c$$
$$P - b - c = a$$

27.
$$S = 2\pi rh + 2\pi r^2$$
$$S - 2\pi r^2 = 2\pi rh + 2\pi r^2 - 2\pi r^2$$
$$S - 2\pi r^2 = 2\pi rh$$
$$\frac{S - 2\pi r^2}{2\pi r} = \frac{2\pi rh}{2\pi r}$$
$$\frac{S - 2\pi r^2}{2\pi r} = h$$

29. Use $A = lw$ when $A = 10,080$ and $w = 84$.
$$A = lw$$
$$10,080 = l(84)$$
$$\frac{10,080}{84} = \frac{84l}{84}$$
$$120 = l$$
The length (height) of the sign is 120 feet.

31. a.

$A = lw$	$P = 2l + 2w$
$A = 11.5(9)$	$P = 2(11.5) + 2(9)$
$A = 103.5$	$P = 23 + 18$
	$P = 41$

The area is 103.5 square feet and the perimeter is 41 feet.

b. Baseboards have to do with perimeter because they are installed around the edges. Carpet has to do with area because it is installed in the middle of the room.

33. a.

$A = \frac{1}{2}h(b_1 + b_2)$	$P = l_1 + l_2 + l_3 + l_4$
	$P = 24 + 20 + 56 + 20$
$A = \frac{1}{2}(12)(56 + 24)$	$P = 120$
$A = 6(80)$	
$A = 480$	

The area is 480 square inches and the perimeter is 120 inches.

b. The frame has to do with perimeter because it surrounds the edge of the picture. The glass has to do with area because it covers the entire picture.

35. Let $F = 14$.

$$14 = \frac{9}{5}C + 32$$

$$5(14) = 5\left(\frac{9}{5}\right)C + 5(32)$$

$$70 = 9C + 160$$

$$70 - 160 = 9C + 160 - 160$$

$$-90 = 9C$$

$$\frac{-90}{9} = \frac{9C}{9}$$

$$-10 = C$$

The equivalent temperature is $-10°C$.

37. Let $P = 260$ and $w = \frac{2}{3}l$.

$$P = 2l + 2w$$

$$260 = 2l + 2\left(\frac{2}{3}l\right)$$

$$260 = \frac{10}{3}l$$

$$3(260) = 3\left(\frac{10}{3}l\right)$$

$$780 = 10l$$

$$\frac{780}{10} = \frac{10l}{10}$$

$$78 = l$$

$$w = \frac{2}{3}l = \frac{2}{3}(78) = 52$$

The width is 52 feet and the length is 78 feet.

39. Let $P = 102$, a = the length of the shortest side, $b = 2a$, and $c = a + 30$.

$$P = a + b + c$$

$$102 = a + 2a + a + 30$$

$$102 = 4a + 30$$

$$102 - 30 = 4a + 30 - 30$$

$$72 = 4a$$

$$\frac{72}{4} = \frac{4a}{4}$$

$$18 = a$$

$$b = 2a = 2(18) = 36$$

$$c = a + 30 = 18 + 30 = 48$$

The lengths are 18 feet, 36 feet, and 48 feet.

41. Let $d = 138$ and $t = 2.5$.

$$d = rt$$

$$138 = r \cdot 2.5$$

$$\frac{138}{2.5} = \frac{r \cdot 2.5}{2.5}$$

$$55.2 = r$$

The speed is 55.2 mph.

43. Let $l = 8$, $w = 6$, and $h = 3$.

$$V = lwh$$

$$V = 8(6)(3) = 144$$

Let x = number of piranha and volume per fish = 1.5.

$$144 = 1.5x$$

$$\frac{144}{1.5} = \frac{1.5x}{1.5}$$

$$96 = x$$

96 piranhas can be placed in the tank.

45. Use $N = 86$.

$$T = 50 + \frac{N - 40}{4}$$

$$T = 50 + \frac{86 - 40}{4}$$

$$T = 50 + \frac{46}{4}$$

$$T = 50 + 11.5$$

$$T = 61.5$$

The temperature is $61.5°$ Fahrenheit.

47. Use $T = 55$.

$$T = 50 + \frac{N - 40}{4}$$

$$55 = 50 + \frac{N - 40}{4}$$

$$55 - 50 = 50 + \frac{N - 40}{4} - 50$$

$$5 = \frac{N - 40}{4}$$

$$4 \cdot 5 = 4 \cdot \frac{N - 40}{4}$$

$$20 = N - 40$$

$$20 + 40 = N - 40 + 40$$

$$60 = N$$

There are 60 chirps per minute.

49. As the number of cricket chirps per minute increases, the air temperature of their environment <u>increases</u>.

51. Let $h = 60$, $B = 130$, and $b = 70$.

$$A = \frac{1}{2}(B + b)h$$

$$A = \frac{1}{2}(130 + 70)60 = \frac{1}{2}(200)(60) = 6000$$

Let x = number of bags of fertilizer and the area per bag = 4000.

$$4000x = 6000$$

$$\frac{4000x}{4000} = \frac{6000}{4000}$$

$$x = 1.5$$

Two bags must be purchased.

53. Let $d = 16$, so $r = 8$.

$$A = \pi r^2 = \pi(8)^2 = 64\pi$$

Let $d = 10$, so $r = 5$.

$$A = 2\pi r^2 = 2\pi(5)^2 = 50\pi$$

One 16-inch pizza has more area and therefore gives more pizza for the price.

55. $x + x + x + 2.5x + 2.5x = 48$

$$8x = 48$$

$$\frac{8x}{8} = \frac{48}{8}$$

$$x = 6$$

$2.5x = 2.5(6) = 15$

Three sides measure 6 meters and two sides measure 15 meters.

57. $r = 0.5$ and $d = 11$.

$$d = rt$$

$$11 = 0.5t$$

$$\frac{11}{0.5} = \frac{0.5t}{0.5}$$

$$22 = t$$

It will take 22 hours.

59. Let $x =$ the length of a side of the square and $x + 5 =$ the length of a side of the triangle.

$$P(\text{triangle}) = P(\text{square}) + 7$$

$$3(x + 5) = 4x + 7$$

$$3x + 15 = 4x + 7$$

$$3x - 3x + 15 = 4x - 3x + 7$$

$$15 = x + 7$$

$$15 - 7 = x + 7 - 7$$

$$8 = x$$

$x + 5 = 8 + 5 = 13$

The side of the triangle is 13 inches.

61. Let $d = 135$ and $r = 60$.

$$d = rt$$

$$135 = 60t$$

$$\frac{135}{60} = \frac{60t}{60}$$

$$2.25 = t$$

It would take 2.25 hours.

63. Let $A = 1,813,500$ and $w = 150$.

$$A = lw$$

$$1,813,500 = l(150)$$

$$\frac{1,813,500}{150} = \frac{150l}{150}$$

$$12,090 = l$$

The length is 12,090 feet.

65. Let $F = 122$.

$$122 = \frac{9}{5}C + 32$$

$$5(122) = 5\left(\frac{9}{5}\right)C + 5(32)$$

$$610 = 9C + 160$$

$$610 - 160 = 9C + 160 - 160$$

$$450 = 9C$$

$$\frac{450}{9} = \frac{9C}{9}$$

$$50 = C$$

The equivalent temperature is 50°C.

67. Let $C = 167$.

$$F = \frac{9}{5}C + 32$$

$$= \frac{9}{5}(167) + 32$$

$$= 300.6 + 32$$

$$= 332.6$$

The equivalent temperature is 332.6°F.

69. Use $V = \frac{4}{3}\pi r^3$ when $r = \frac{9.5}{2} = 4.75$ and $\pi = 3.14$.

$$V = \frac{4}{3}\pi r^3 = \frac{4}{3}(3.14)(4.75)^3 \approx 449$$

The volume of the sphere is 449 cubic inches.

71. $32\% = 0.32$

73. $200\% = 2.00$ or 2

75. $0.17 = 0.17(100\%) = 17\%$

77. $7.2 = 7.2(100\%) = 720\%$

79. Use $V = lwh$. If the length is doubled, the new length is $2l$. If the width and height are doubled, the new width and height are $2w$ and $2h$, respectively.

$V = (2l)(2w)(2h) = 2 \cdot 2 \cdot 2lwh = 8lwh$

The volume of the box is multiplied by 8.

81. Replace T with N and solve for N.

$$T = 50 + \frac{N-40}{4}$$

$$N = 50 + \frac{N-40}{4}$$

$$N - 50 = 50 + \frac{N-40}{4} - 50$$

$$N - 50 = \frac{N-40}{4}$$

$$4(N-50) = 4 \cdot \frac{N-40}{4}$$

$$4N - 200 = N - 40$$

$$4N - 200 - N = N - 40 - N$$

$$3N - 200 = -40$$

$$3N - 200 + 200 = -40 + 200$$

$$3N = 160$$

$$\frac{3N}{3} = \frac{160}{3}$$

$$N = 53\frac{1}{3}$$

They are the same when the number of cricket chirps per minute is $53\frac{1}{3}$.

83.
$$N = R + \frac{V}{G}$$

$$N - R = R + \frac{V}{G} - R$$

$$N - R = \frac{V}{G}$$

$$G(N-R) = G \cdot \frac{V}{G}$$

$$G(N-R) = V$$

85. $\square - \bigcirc \cdot \square = \triangle$

$$-\bigcirc \cdot \square = \triangle - \square$$

$$\frac{-\bigcirc\square}{-\square} = \frac{\triangle - \square}{-\square}$$

$$\bigcirc = \frac{\square - \triangle}{\square}$$

87. Let $d = 93{,}000{,}000$ and $r = 186{,}000$.

$$d = rt$$

$$93{,}000{,}000 = 186{,}000t$$

$$\frac{93{,}000{,}000}{186{,}000} = \frac{186{,}000t}{186{,}000}$$

$$500 = t$$

It will take 500 seconds or $8\frac{1}{3}$ minutes.

89. Let $t = 365$ and $r = 20$.

$d = rt = 20(365) = 7300$ inches

$$\frac{7300 \text{ inches}}{1} \cdot \frac{1 \text{ foot}}{12 \text{ inch}} \approx 608.33 \text{ feet}$$

It moves about 608.33 feet.

91. Use $d = rt$, when $r = 581$ and $d = 42.8$.

$$d = rt$$

$$42.8 = 581t$$

$$\frac{42.8}{581} = \frac{581t}{581}$$

$$0.0737 \approx t$$

$$0.0737(60) \approx 4.42$$

It would last 4.42 minutes.

93. Use $d = rt$ when $d = 303$ and $t = 8\frac{1}{2}$.

$$d = rt$$

$$303 = r \cdot 8\frac{1}{2}$$

$$303 = \frac{17}{2}r$$

$$\frac{2}{17} \cdot 303 = \frac{2}{17} \cdot \frac{17}{2}r$$

$$\frac{606}{17} = r$$

$$35\frac{11}{17} = r$$

The average rate during the flight was $35\frac{11}{17}$ mph.

Section 2.6 Practice

1. Let $x =$ the unknown percent.

$$35 = x \cdot 56$$

$$\frac{35}{56} = \frac{56x}{56}$$

$$0.625 = x$$

The number 35 is 62.5% of 56.

2. Let $x =$ the unknown number.

$$198 = 55\% \cdot x$$

$$198 = 0.55x$$

$$\frac{198}{0.55} = \frac{0.55x}{0.55}$$

$$360 = x$$

The number 198 is 55% of 360.

3. a. From the circle graph, we see that 40% of pets owned are freshwater fish and 2% are saltwater fish; thus 40% + 2% = 42% of pets owned are freshwater fish or saltwater fresh.

   b. The circle graph percents have a sum of 100%; thus the percent of pets that are not equines is 100% − 2% = 98%.

   c. To find the number of dogs owned, we find
$$21\% \text{ of } 377.41 = (0.21)(377.41)$$
$$= 9.2561$$
$$\approx 79.3$$
Thus, about 79.3 million dogs are owned in the United States.

4. Let x = discount.
$x = 85\% \cdot 480$
$x = 0.85 \cdot 480$
$x = 408$
The discount is \$408.
New price = \$480 − \$408 = \$72

5. Increase = 2710 − 1900 = 810
Let x = percent increase.
$810 = x \cdot 1900$
$$\frac{810}{1900} = \frac{1900x}{1900}$$
$0.426 \approx x$
The percent increase is 42.6%.

6. Let x = number of digital 3D screens last year.
$x + 1.38x = 8459$
$2.38x = 8459$
$$\frac{2.38x}{2.38} = \frac{8459}{2.38}$$
$x \approx 3554$
There were 3554 screens last year.

7. Let x = number of liters of 2% solution.

Eyewash	No. of gallons	Acid Strength	Amt. of Acid
2%	x	2%	$0.02x$
5%	$6 - x$	5%	$0.05(6 - x)$
Mix: 3%	6	3%	$0.03(6)$

$0.02x + 0.05(6 - x) = 0.03(6)$
$0.02x + 0.3 - 0.05x = 0.18$
$-0.03x + 0.3 = 0.18$
$-0.03x + 0.3 - 0.3 = 0.18 - 0.3$
$-0.03x = -0.12$
$$\frac{-0.03x}{-0.03} = \frac{-0.12}{-0.03}$$
$x = 4$
$6 - x = 6 - 4 = 2$
She should mix 4 liters of 2% eyewash with 2 liters of 5% eyewash.

Vocabulary, Readiness & Video Check 2.6

1. No, $25\% + 25\% + 40\% = 90\% \neq 100\%$.

2. No, $30\% + 30\% + 30\% = 90\% \neq 100\%$.

3. Yes, $25\% + 25\% + 25\% + 25\% = 100\%$.

4. Yes, $40\% + 50\% + 10\% = 100\%$.

5. a. equals; =

   b. multiplication; ·

   c. Drop the percent symbol and move the decimal point two places to the left.

6. a. You also find a discount amount by multiplying the (discount) percent by the original price.

   b. For discount, the new price is the original price minus the discount amount, so you *subtract* from the original price rather than *add* as with mark-up.

7. You must first find the actual amount of increase in price by subtracting the original price from the new price.

8.

Alloy	Ounces	Copper Strength	Amount of Copper
10%	x	0.10	$0.10x$
30%	400	0.30	$0.30(400)$
20%	$x + 400$	0.20	$0.20(x + 400)$

Exercise Set 2.6

1. Let x be the unknown number.
$x = 16\% \cdot 70$
$x = 0.16 \cdot 70$
$x = 11.2$
11.2 is 16% of 70.

3. Let x be the unknown percent.
$28.6 = x \cdot 52$
$\dfrac{28.6}{52} = \dfrac{52x}{52}$
$0.55 = x$
$55\% = x$
The number 28.6 is 55% of 52.

5. Let x be the unknown number.
$45 = 25\% \cdot x$
$45 = 0.25 \cdot x$
$\dfrac{45}{0.25} = \dfrac{0.25x}{0.25}$
$180 = x$
45 is 25% of 180.

7. From the graph, 4% of adults spend more than 121 minutes on the phone each day.

9. 37% of adults talk 16–60 minutes on the phone each day.
37% of $27,000 = 37\% \cdot 27,000$
$= 0.37 \cdot 27,000$
$= 9990$
You would expect 9990 of the adults in Florence to talk 16–60 minutes each day.

11. Let $x =$ amount of discount.
$x = 8\% \cdot 18,500$
$x = 0.08 \cdot 18,500$
$x = 1480$
New price $= 18,500 - 1480 = 17,020$
The discount was \$1480 and the new price is \$17,020.

13. Let $x =$ tip.
$x = 15\% \cdot 40.50$
$x = 0.15 \cdot 40.5$
$x = 6.075 \approx 6.08$
Total $= 40.50 + 6.08 = 46.58$
The total cost is \$46.58.

15. Decrease $= 337 - 304 = 33$
Let $x =$ percent.
$33 = x \cdot 337$
$\dfrac{33}{337} = \dfrac{337x}{337}$
$0.098 \approx x$
The percent decrease is 9.8%.

17. Decrease $= 40 - 28 = 12$
Let $x =$ percent.
$12 = x \cdot 40$
$\dfrac{12}{40} = \dfrac{40x}{40}$
$0.3 = x$
The percent decrease is 30%.

19. Let x = the original price and
$0.25x$ = the discount.
$$x - 0.25x = 78$$
$$0.75x = 78$$
$$\frac{0.75x}{0.75} = \frac{78}{0.75}$$
$$x = 104$$
The original price was \$104.

21. Let x = last year's salary, and $0.04x$ = pay raise.
$$x + 0.04x = 44,200$$
$$1.04x = 44,200$$
$$\frac{1.04x}{1.04} = \frac{44,200}{1.04}$$
$$x = 42,500$$
Last year's salary was \$42,500.

23. Let x = the amount of pure acid.

	No. of gallons	· Strength	=	Amt. of Acid
100%	x	1.00		x
40%	2	0.4		2(0.4)
70%	$x + 2$	0.7		0.7(x + 2)

$$x + 2(0.4) = 0.7(x + 2)$$
$$x + 0.8 = 0.7x + 1.4$$
$$x - 0.7x + 0.8 = 0.7x - 0.7x + 1.4$$
$$0.3x + 0.8 = 1.4$$
$$0.3x + 0.8 - 0.8 = 1.4 - 0.8$$
$$0.3x = 0.6$$
$$\frac{0.3x}{0.3} = \frac{0.6}{0.3}$$
$$x = 2$$
Mix 2 gallons of pure acid.

25. Let x = the number of pounds at \$7/lb.

	No. of lb	· Cost/lb	=	Value
\$7/lb	x	7		$7x$
\$4/lb	14	4		4(14)
\$5/lb	$x + 14$	5		5(x + 14)

$$7x + 4(14) = 5(x + 14)$$
$$7x + 56 = 5x + 70$$
$$7x - 5x + 56 = 5x - 5x + 70$$
$$2x + 56 = 70$$
$$2x + 56 - 56 = 70 - 56$$
$$2x = 14$$
$$\frac{2x}{2} = \frac{14}{2}$$
$$x = 7$$
Add 7 pounds of \$7/pound coffee.

27. Let x = the number.
$$x = 23\% \cdot 20$$
$$x = 0.23 \cdot 20$$
$$x = 4.6$$
23% of 20 is 4.6.

29. Let x = the number.
$$40 = 80\% \cdot x$$
$$40 = 0.8x$$
$$\frac{40}{0.8} = \frac{0.8x}{0.8}$$
$$50 = x$$
40 is 80% of 50.

31. Let x = the percent.
$$144 = x \cdot 480$$
$$\frac{144}{480} = \frac{480x}{480}$$
$$0.3 = x$$
144 is 30% of 480.

33. From the graph, the height of the bar is about 23. Therefore, 23% of online purchases were in the category of electronic equipment.

35. 41% of $220,500 = 0.41 \cdot 220,500 \approx 90,405$
We predict 90,405 people will purchase books online.

37.

Top Cranberry-Producing States Forecast in 2010 (in millions of pounds)		
	Millions of Pounds	Percent of Total (rounded to nearest percent)
Wisconsin	435	$\frac{435}{736} \approx 0.5910 \approx 59\%$
Oregon	39	$\frac{39}{736} \approx 0.0529 \approx 5\%$
Massachusetts	195	$\frac{195}{736} \approx 0.2649 \approx 26\%$
Washington	14	$\frac{14}{736} \approx 0.0190 \approx 2\%$
New Jersey	53	$\frac{53}{736} \approx 7\%$
	736	99% due to rounding

39. Let x = the decrease in price.
$x = 0.25(256) = 64$
The decrease in price is \$64.
The sale price is $256 - 64 = \$192$.

41. $\text{percent decrease} = \dfrac{\text{amount of decrease}}{\text{original amount}}$

$= \dfrac{23.5 - 17.1}{23.5}$

$= \dfrac{6.4}{23.5}$

≈ 0.272

Head lettuce consumption decreased by 27.2%.

43. Let x = the number of vehicles in 2002.
$$x + 3\% \cdot x = 246$$
$$x + 0.03x = 246$$
$$1.03x = 246$$
$$\frac{1.03x}{1.03} = \frac{246}{1.03}$$
$$x \approx 239$$
There were about 239 registered vehicles in the United States in 2002.

45. $\text{percent increase} = \dfrac{\text{amount of increase}}{\text{original amount}}$

$= \dfrac{144 - 36}{36}$

$= \dfrac{108}{36}$

$= 3$

The area increased by 300%.

47. Let x be the ounces of alloy that is 20% copper.

	ounces	concentration	amount
20% copper	x	20%	$0.2x$
50% copper	200	50%	$0.5(200)$
30% copper	$200 + x$	30%	$0.3(200 + x)$

The amount of copper being combined must be the same as that in the mixture.

$$0.2x + 0.5(200) = 0.3(200 + x)$$
$$0.2x + 100 = 60 + 0.3x$$
$$0.2x + 100 - 0.2x = 60 + 0.3x - 0.2x$$
$$100 = 60 + 0.1x$$
$$100 - 60 = 60 + 0.1x - 60$$
$$40 = 0.1x$$
$$\frac{40}{0.1} = \frac{0.1x}{0.1}$$
$$400 = x$$

Thus 400 ounces should be used.

49. percent decrease $= \dfrac{\text{amount of decrease}}{\text{original amount}}$

$$= \frac{6.3 - 2.1}{6.3}$$
$$= \frac{4.2}{6.3}$$
$$\approx 0.667$$

The percent decrease in the number of farms is 66.7%.

51. Let x be the prior number of employees.

$$x - 0.35x = 78$$
$$0.65x = 78$$
$$\frac{0.65x}{0.65} = \frac{78}{0.65}$$
$$x = 120$$

There were 120 employees prior to the layoffs.

53. $42\% \cdot 860 = 0.42 \cdot 860 = 361.2$

You would expect 361 students to rank flexible hours as their top priority.

55. Let x be the ounces of self-tanning lotion.

	ounces	cost ($)	value
self-tanning	x	3	$3x$
everyday	800	0.30	$0.3(800)$
experimental	$800 + x$	1.20	$1.2(800 + x)$

The value of the lotions being combined must be the same as the value of the mixture.

$$3x + 0.3(800) = 1.2(800 + x)$$
$$3x + 240 = 960 + 1.2x$$
$$3x + 240 - 1.2x = 960 + 1.2x - 1.2x$$
$$1.8x + 240 = 960$$
$$1.8x + 240 - 240 = 960 - 240$$
$$1.8x = 720$$
$$\frac{1.8x}{1.8} = \frac{720}{1.8}$$
$$x = 400$$

Therefore, 400 ounces of the self-tanning lotion should be used.

57. increase $= 48\% \cdot 577 = 0.48 \cdot 577 = 276.96$
$577 + 276.96 = 853.96$
The Naga Jolokia pepper measures 854 thousand Scoville units.

59. $-5 > -7$

61. $|-5| = -(-5)$

63. $(-3)^2 = 9; \ -3^2 = -9$

$(-3)^2 > -3^2$

65. no; answers may vary

67. no; answers may vary

69. 230 mg is what percent of 2400 mg?
Let x represent the unknown percent.
$$x \cdot 2400 = 230$$
$$\frac{2400x}{2400} = \frac{230}{2400}$$
$$x = 0.0958\overline{3}$$
This food contains 9.6% of the daily value of sodium in one serving.

71. 35 is what percent of 130? Let x be the unknown percent.
$$35 = x \cdot 130$$
$$\frac{35}{130} = \frac{130x}{130}$$
$$0.269 \approx x$$
The percent calories from fat is 26.9%. Yes, this food satisfies the recommendation since $26.9\% \le 30\%$.

73. 12 g $\cdot$ 4 calories/gram = 48 calories
48 of the 280 calories come from protein.
$$\frac{48}{280} \approx 0.171$$
17.1% of the calories in this food come from protein.

Section 2.7 Practice

1. Let x = time down, then $x + 1$ = time up.

	Rate	· Time	= Distance
Up	1.5	$x + 1$	$1.5(x + 1)$
Down	4	x	$4x$

$$d = d$$
$$1.5(x + 1) = 4x$$
$$1.5x + 1.5 = 4x$$
$$1.5 = 2.5x$$
$$\frac{1.5}{2.5} = \frac{2.5x}{2.5}$$
$$0.6 = x$$

Total Time $= x + 1 + x = 0.6 + 1 + 0.6 = 2.2$
The entire hike took 2.2 hours.

2. Let x = speed of eastbound train, then $x - 10$ = speed of westbound train.

	r	· t	= d
East	x	1.5	$1.5x$
West	$x - 10$	1.5	$1.5(x - 10)$

$$1.5x + 1.5(x - 10) = 171$$
$$1.5x + 1.5x - 15 = 171$$
$$3x - 15 = 171$$
$$3x = 186$$
$$\frac{3x}{3} = \frac{186}{3}$$
$$x = 62$$
$x - 10 = 62 - 10 = 52$
The eastbound train is traveling at 62 mph and the westbound train is traveling at 52 mph.

3. Let x = the number of \$20 bills, then $x + 47$ = number of \$5 bills.

Denomination	Number	Value
\$5 bills	$x + 47$	$5(x + 47)$
\$20 bills	x	$20x$

$$5(x+47)+20x=1710$$
$$5x+235+20x=1710$$
$$235+25x=1710$$
$$25x=1475$$
$$x=59$$
$$x+47=59+47=106$$
There are 106 $5 bills and 59 $20 bills.

4. Let x = amount invested at 11.5%, then
$30,000 - x$ = amount invested at 6%.

	Principal $\cdot$	Rate $\cdot$	Time =	Interest
11.5%	x	0.115	1	$x(0.115)(1)$
6%	$30,000 - x$	0.06	1	$0.06(30,000 - x)(1)$
Total	30,000			2790

$$0.115x + 0.06(30,000 - x) = 2790$$
$$0.115x + 1800 - 0.06x = 2790$$
$$1800 + 0.055x = 2790$$
$$0.055x = 990$$
$$\frac{0.055x}{0.055} = \frac{990}{0.055}$$
$$x = 18,000$$
$$30,000 - x = 30,000 - 18,000 = 12,000$$
She invested $18,000 at 11.5% and $12,000 at 6%.

Vocabulary, Readiness & Video Check 2.7

1.

	r $\cdot$	t =	d
bus	55	x	$55x$
car	50	$x + 3$	$50(x + 3)$

$$55x = 50(x + 3)$$

2. The important thing is to remember the difference between the *number* of bills you have and the *value* of the bills.

3.

P $\cdot$	R $\cdot$	T =	I
x	0.06	1	$0.06x$
$36,000 - x$	0.04	1	$0.04(36,000 - x)$

$$0.06x = 0.04(36,000 - x)$$

Exercise Set 2.7

1. Let x = the time traveled by the jet plane.

	Rate	· Time	= Distance
Jet	500	x	$500x$
Prop	200	$x + 2$	$200(x + 2)$

$$d = d$$
$$500x = 200(x + 2)$$
$$500x = 200x + 400$$
$$300x = 400$$
$$\frac{300x}{300} = \frac{400}{300}$$
$$x = \frac{4}{3}$$

The jet traveled for $\frac{4}{3}$ hours.

$$d = rt$$
$$d = 500\left(\frac{4}{3}\right) = 666\frac{2}{3}$$

The planes are $666\frac{2}{3}$ miles from the starting point.

3. Let x = the average speed on the winding road and $x + 20$ on the level.

	Rate	· Time	= Distance
Winding	x	4	$4x$
Level	$x + 20$	3	$3(x + 20)$
Total			305

$$4x + 3(x + 20) = 305$$
$$4x + 3x + 60 = 305$$
$$7x + 60 = 305$$
$$7x = 245$$
$$\frac{7x}{7} = \frac{245}{7}$$
$$x = 35$$
$$x + 20 = 35 + 20 = 55$$

The average speed on level road was 55 mph.

5. The value of y dimes is $0.10y$.

7. The value of $x + 7$ nickels is $0.05(x + 7)$.

9. The value of $4y$ $20 bills is $20(4y)$ or $80y$.

11. The value of $35 - x$ $50 bills is $50(35 - x)$.

13. Let x = number of $10 bills, then $20 + x$ number of $5 bills.

	Number of Bills	Value of Bills
$5 bills	$20 + x$	$5(20 + x)$
$10 bills	x	$10x$
Total		280

$$5(20 + x) + 10x = 280$$
$$100 + 5x + 10x = 280$$
$$100 + 15x = 280$$
$$15x = 180$$
$$x = 12$$
$$20 + x = 32$$

There are 12 $10 bills and 32 $5 bills.

15. Let x = the amount invested at 9% for one year.

	Principal	· Rate	= Interest
9%	x	0.09	$0.09x$
8%	$25,000 - x$	0.08	$0.08(25,000 - x)$
Total	25,000		2135

$$0.09x + 0.08(25,000 - x) = 2135$$
$$0.09x + 2000 - 0.08x = 2135$$
$$0.01x + 2000 = 2135$$
$$0.01x = 135$$
$$\frac{0.01x}{0.01} = \frac{135}{0.01}$$
$$x = 13,500$$
$$25,000 - x = 25,000 - 13,500 = 11,500$$

She invested $11,500 at 8% and $13,500 at 9%.

17. Let x = the amount invested at 11% for one year.

	Principal	· Rate	= Interest
11%	x	0.11	$0.11x$
4%	$10,000 - x$	−0.04	$-0.04(10,000 - x)$
Total	10,000		650

$$0.11x - 0.04(10,000 - x) = 650$$
$$0.11x - 400 + 0.04x = 650$$
$$0.15x - 400 = 650$$
$$0.15x = 1050$$
$$\frac{0.15x}{0.15} = \frac{1050}{0.15}$$
$$x = 7000$$
$$10,000 - x = 10,000 - 7000 = 3000$$
He invested \$7000 at 11% and \$3000 at 4%.

19. Let x = the number of adult tickets, then $500 - x$ = the number of child tickets.

	Number ·	Rate =	Cost
Adult	x	43	$43x$
Child	$500 - x$	28	$28(500 - x)$
Total	500		16,805

$$43x + 28(500 - x) = 16,805$$
$$43x + 14,000 - 28x = 16,805$$
$$14,000 + 15x = 16,805$$
$$15x = 2805$$
$$x = 187$$
$$500 - x = 500 - 187 = 313$$
Sales included 187 adult tickets and 313 child tickets.

21. Let x = the time traveled.

	Rate ·	Time =	Distance
Car A	56	x	$56x$
Car B	47	x	$47x$

The total distance is 206 miles.
$$56x + 47x = 206$$
$$103x = 206$$
$$\frac{103x}{103} = \frac{206}{103}$$
$$x = 2$$
The two cars will be 206 miles apart in 2 hours.

23. Let x = the amount invested at 10% for one year.

	Principal ·	Rate =	Interest
10%	x	0.10	$0.10x$
8%	$54,000 - x$	0.08	$0.08(54,000 - x)$

$$0.10x = 0.08(54,000 - x)$$
$$0.10x = 4320 - 0.08x$$
$$0.18x = 4320$$
$$\frac{0.18x}{0.18} = \frac{4320}{0.18}$$
$$x = 24,000$$
$$54,000 - x = 54,000 - 24,000 = 30,000$$
Invest \$30,000 at 8% and \$24,000 at 10%.

25. Let x = the time they are able to talk.

	Rate ·	Time =	Distance
Alan	55	x	$55x$
Dave	65	$x - 1$	$65(x - 1)$
Total			250

$$55x + 65(x - 1) = 250$$
$$55x + 65x - 65 = 250$$
$$120x - 65 = 250$$
$$120x = 315$$
$$\frac{120x}{120} = \frac{315}{120}$$
$$x = 2\frac{5}{8}$$

They can talk for $2\frac{5}{8}$ hours or

2 hours $37\frac{1}{2}$ minutes.

27. Let x = the speed of the slower train.

	Rate ·	Time =	Distance
Train A	x	2.5	$2.5x$
Train B	$x + 10$	2.5	$2.5(x + 10)$

The total distance is 205 miles.
$$2.5x + 2.5(x + 10) = 205$$
$$2.5x + 2.5x + 25 = 205$$
$$5x + 25 = 205$$
$$5x = 180$$
$$\frac{5x}{5} = \frac{180}{5}$$
$$x = 36$$
$$x + 10 = 46$$
The speeds of the trains are 36 mph and 46 mph.

29. Let x = number of nickels, then
$3x$ = number of dimes.

	Number	Value
Nickels	x	$0.05x$
Dimes	$3x$	$0.10(3x)$
Total		56.35

$$0.05x + 0.10(3x) = 56.35$$
$$0.05x + 0.3x = 56.35$$
$$0.35x = 56.35$$
$$x = 161$$
$3x = 3(161) = 483$
They collected 161 nickels and 483 dimes.

31. Let x = time traveled.

	Rate	· Time	= Distance
Truck	52	x	$52x$
Van	63	x	$63x$

The total distance is 460 miles.
$$52x + 63x = 460$$
$$115x = 460$$
$$\frac{115x}{115} = \frac{460}{115}$$
$$x = 4$$
The truck and the van will be 460 miles apart in 4 hours.

33. Let x = time traveled.

	Rate	· Time	= Distance
Car A	70	x	$70x$
Car B	58	x	$58x$

They are traveling in the same direction, so find the difference of their distances.
$$70x - 58x = 30$$
$$12x = 30$$
$$\frac{12x}{12} = \frac{30}{12}$$
$$x = 2.5$$
They will be 30 miles apart in 2.5 hours.

35. Let x = the amount invested at 9% for one year.

	Principal ·	Rate	= Interest
9%	x	0.09	$0.09x$
6%	3000	0.06	$0.06(3000)$
Total			585

$$0.09x + 0.06(3000) = 585$$
$$0.09x + 180 = 585$$
$$0.09x = 405$$
$$\frac{0.09x}{0.09} = \frac{405}{0.09}$$
$$x = 4500$$
Should invest $4500 at 9%.

37. Let x = the rate of hiker 1.

	Rate	· Time	= Distance
Hiker 1	x	2	$2x$
Hiker 2	$x + 1.1$	2	$2(x + 1.1)$
Total			11

$$2x + 2(x + 1.1) = 11$$
$$2x + 2x + 2.2 = 11$$
$$4x + 2.2 = 11$$
$$4x = 8.8$$
$$\frac{4x}{4} = \frac{8.8}{4}$$
$$x = 2.2$$
$x + 1.1 = 2.2 + 1.1 = 3.3$
Hiker 1: 2.2 mph; Hiker 2: 3.3 mph

39. Let x = the time spent rowing upstream.

	Rate ·	Time	= Distance
Upstream	5	x	$5x$
Downstream	11	$4 - x$	$11(4 - x)$

$$5x = 11(4 - x)$$
$$5x = 44 - 11x$$
$$16x = 44$$
$$\frac{16x}{16} = \frac{44}{16}$$
$$x = 2.75$$
He rowed upstream for 2.75 hours.
$d = rt$
$d = 5(2.75) = 13.75$
He rowed 13.75 miles each way for a total of 27.5 miles.

41. $3 + (-7) = -4$

43. $\dfrac{3}{4} - \dfrac{3}{16} = \dfrac{4}{4} \cdot \dfrac{3}{4} - \dfrac{3}{16} = \dfrac{12}{16} - \dfrac{3}{16} = \dfrac{12-3}{16} = \dfrac{9}{16}$

45. $-5 - (-1) = -5 + 1 = -4$

47. Let x = number of \$100 bills, then
$x + 46$ = number of \$50 bills, and
$7x$ = number of \$20 bills.

	Number	Value
\$100 bills	x	$100x$
\$50 bills	$x + 46$	$50(x + 46)$
\$20 bills	$7x$	$20(7x)$
Total		9550

$$100x + 50(x+46) + 20(7x) = 9550$$
$$100x + 50x + 2300 + 140x = 9550$$
$$290x + 2300 = 9550$$
$$290x = 7250$$
$$x = 25$$

$x + 46 = 71$
$7x = 7(25) = 175$
There were 25 \$100 bills, 71 \$50 bills, and 175
\$20 bills.

49. $R = C$
$$24x = 100 + 20x$$
$$4x = 100$$
$$\frac{4x}{x} = \frac{100}{4}$$
$$x = 25$$
Should sell 25 skateboards to break even.

51. $R = C$
$$7.50x = 4.50x + 2400$$
$$3x = 2400$$
$$\frac{3x}{3} = \frac{2400}{3}$$
$$x = 800$$
Should sell 800 books to break even.

53. Answers may vary

Section 2.8 Practice

1. $x < 5$
Place a parenthesis at 5 since the inequality
symbol is $<$. Shade to the left of 5. The solution
set is $(-\infty, 5)$.

2. $x + 11 \geq 6$
$$x + 11 - 11 \geq 6 - 11$$
$$x \geq -5$$
The solution set is $[-5, \infty)$.

3. $-5x \geq -15$
$$\frac{-5x}{-5} \leq \frac{-15}{-5}$$
$$x \leq 3$$
The solution set is $(-\infty, 3]$.

4. $3x > -9$
$$\frac{3x}{3} > \frac{-9}{3}$$
$$x > -3$$
The solution set is $(-3, \infty)$.

5. $45 - 7x \leq -4$
$$45 - 7x - 45 \leq -4 - 45$$
$$-7x \leq -49$$
$$\frac{-7x}{-7} \geq \frac{-49}{-7}$$
$$x \geq 7$$
The solution set is $[7, \infty)$.

6. $3x + 20 \leq 2x + 13$
$$3x + 20 - 2x \leq 2x + 13 - 2x$$
$$x + 20 \leq 13$$
$$x + 20 - 20 \leq 13 - 20$$
$$x \leq -7$$
The solution set is $(-\infty, -7]$.

7.
$$6 - 5x > 3(x - 4)$$
$$6 - 5x > 3x - 12$$
$$6 - 5x - 3x > 3x - 12 - 3x$$
$$6 - 8x > -12$$
$$6 - 8x - 6 > -12 - 6$$
$$-8x > -18$$
$$\frac{-8x}{-8} < \frac{-18}{-8}$$
$$x < \frac{9}{4}$$

The solution set is $\left(-\infty, \frac{9}{4}\right)$.

8.
$$3(x - 4) - 5 \le 5(x - 1) - 12$$
$$3x - 12 - 5 \le 5x - 5 - 12$$
$$3x - 17 \le 5x - 17$$
$$3x - 17 - 5x \le 5x - 17 - 5x$$
$$-2x - 17 \le -17$$
$$-2x - 17 + 17 \le -17 + 17$$
$$-2x \le 0$$
$$\frac{-2x}{-2} \ge \frac{0}{-2}$$
$$x \ge 0$$

The solution set is $[0, \infty)$.

9. $-3 \le x < 1$

Graph all numbers greater than or equal to -3 and less than 1. Place a bracket at -3 and a parenthesis at 1.

The solution set is $[-3, 1)$.

10.
$$-4 < 3x + 2 \le 8$$
$$-4 - 2 < 3x + 2 - 2 \le 8 - 2$$
$$-6 < 3x \le 6$$
$$\frac{-6}{3} < \frac{3x}{3} \le \frac{6}{3}$$
$$-2 < x \le 2$$

The solution set is $(-2, 2]$.

11.
$$1 < \frac{3}{4}x + 5 < 6$$
$$4(1) < 4\left(\frac{3}{4}x + 5\right) < 4(6)$$
$$4 < 3x + 20 < 24$$
$$4 - 20 < 3x + 20 - 20 < 24 - 20$$
$$-16 < 3x < 4$$
$$\frac{-16}{3} < \frac{3x}{3} < \frac{4}{3}$$
$$-\frac{16}{3} < x < \frac{4}{3}$$

The solution set is $\left(-\frac{16}{3}, \frac{4}{3}\right)$.

12. Let x = the number.
$$35 - 2x > 15$$
$$35 - 2x - 35 > 15 - 35$$
$$-2x > -20$$
$$\frac{-2x}{-2} < \frac{-20}{-2}$$
$$x < 10$$

All numbers less than 10.

13. Let x = number of classes.
$$300 + 375x \le 1500$$
$$300 + 375x - 300 \le 1500 - 300$$
$$375x \le 1200$$
$$\frac{375x}{375} \le \frac{1200}{375}$$
$$x \le 3.2$$

Kasonga can afford at most 3 community college classes this semester.

Vocabulary, Readiness & Video Check 2.8

1. $6x - 7(x + 9)$ is an expression.

2. $6x = 7(x + 9)$ is an equation.

3. $6x < 7(x + 9)$ is an inequality.

4. $5y - 2 \ge -38$ is an inequality.

5. -5 is not a solution to $x \ge -3$.

6. $|-6| = 6$ is not a solution to $x < 6$.

7. The graph of Example 1 is shaded from $-\infty$ to and including -1, as indicated by a bracket. To write interval notation, you write down what is shaded for the inequality from left to right. A parenthesis is always used with $-\infty$, so from the graph, the interval notation is $(-\infty, -1]$.

8. Step 5 is where you apply the multiplication property of inequality. If a negative number is multiplied or divided when applying this property, you need to make sure you remember to reverse the direction of the inequality symbol.

9. You would divide the left, middle, and right by -3 instead of 3, which would reverse the directions of both inequality symbols.

10. no; greater than; $\leq$

Exercise Set 2.8

1. $[2, \infty), x \geq 2$

3. $(-\infty, -5), x < -5$

5. $x \leq -1, (-\infty, -1]$

7. $x < \dfrac{1}{2}, \left(-\infty, \dfrac{1}{2}\right)$

9. $y \geq 5, [5, \infty)$

11. $2x < -6$
$x < -3, (-\infty, -3)$

13. $x - 2 \geq -7$
$x \geq -5, [-5, \infty)$

15. $-8x \leq 16$
$\dfrac{-8x}{-8} \geq \dfrac{16}{-8}$
$x \geq -2, [-2, \infty)$

17. $3x - 5 > 2x - 8$
$x - 5 > -8$
$x > -3, (-3, \infty)$

19. $4x - 1 \leq 5x - 2x$
$4x - 1 \leq 3x$
$x - 1 \leq 0$
$x \leq 1, (-\infty, 1]$

21. $x - 7 < 3(x + 1)$
$x - 7 < 3x + 3$
$-2x - 7 < 3$
$-2x < 10$
$\dfrac{-2x}{-2} > \dfrac{10}{-2}$
$x > -5, (-5, \infty)$

23. $-6x + 2 \geq 2(5 - x)$
$-6x + 2 \geq 10 - 2x$
$-4x + 2 \geq 10$
$-4x \geq 8$
$\dfrac{-4x}{-4} \leq \dfrac{8}{-4}$
$x \leq -2, (\infty, -2]$

25. $4(3x - 1) \leq 5(2x - 4)$
$12x - 4 \leq 10x - 20$
$2x - 4 \leq -20$
$2x \leq -16$
$x \leq -8, (-\infty, -8]$

27. $3(x+2)-6 > -2(x-3)+14$
$3x+6-6 > -2x+6+14$
$3x > -2x+20$
$5x > 20$
$x > 4, (4, \infty)$

29. $-2x \le -40$
$\dfrac{-2x}{-2} \ge \dfrac{-40}{-2}$
$x \ge 20, [20, \infty)$

31. $-9+x > 7$
$x > 16, (16, \infty)$

33. $3x-7 < 6x+2$
$-3x-7 < 2$
$-3x < 9$
$\dfrac{-3x}{-3} > \dfrac{9}{-3}$
$x > -3, (-3, \infty)$

35. $5x-7x \ge x+2$
$-2x \ge x+2$
$-3x \ge 2$
$\dfrac{-3x}{-3} \le \dfrac{2}{-3}$
$x \le -\dfrac{2}{3}, \left(-\infty, -\dfrac{2}{3}\right]$

37. $\dfrac{3}{4}x > 2$
$x > \dfrac{8}{3}, \left(\dfrac{8}{3}, \infty\right)$

39. $3(x-5) < 2(2x-1)$
$3x-15 < 4x-2$
$-x-15 < -2$
$-x < 13$
$\dfrac{-x}{-1} > \dfrac{13}{-1}$
$x > -13, (-13, \infty)$

41. $4(2x+1) < 4$
$8x+4 < 4$
$8x < 0$
$x < 0, (-\infty, 0)$

43. $-5x+4 \ge -4(x-1)$
$-5x+4 \ge -4x+4$
$-x+4 \ge 4$
$-x \ge 0$
$\dfrac{-x}{-1} \le \dfrac{0}{-1}$
$x \le 0, (-\infty, 0]$

45. $-2(x-4)-3x < -(4x+1)+2x$
$-2x+8-3x < -4x-1+2x$
$-5x+8 < -2x-1$
$-3x+8 < -1$
$-3x < -9$
$\dfrac{-3x}{-3} > \dfrac{-9}{-3}$
$x > 3, (3, \infty)$

47. $\frac{1}{4}(x+4) < \frac{1}{5}(2x+3)$

$$20 \cdot \frac{1}{4}(x+4) < 20 \cdot \frac{1}{5}(2x+3)$$
$$5(x+4) < 4(2x+3)$$
$$5x + 20 < 8x + 12$$
$$5x + 20 - 5x < 8x + 12 - 5x$$
$$20 < 3x + 12$$
$$20 - 12 < 3x + 12 - 12$$
$$8 < 3x$$
$$\frac{8}{3} < \frac{3x}{3}$$
$$\frac{8}{3} < x, \left(\frac{8}{3}, \infty\right)$$

49. $-1 < x < 3$, $(-1, 3)$

51. $0 \le y < 2$, $[0, 2)$

53. $-3 < 3x < 6$

$-1 < x < 2$, $(-1, 2)$

55. $2 \le 3x - 10 \le 5$

$12 \le 3x \le 15$

$4 \le x \le 5$, $[4, 5]$

57. $-4 < 2(x-3) \le 4$

$-4 < 2x - 6 \le 4$

$2 < 2x \le 10$

$1 < x \le 5$, $(1, 5]$

59. $-2 < 3x - 5 < 7$

$3 < 3x < 12$

$1 < x < 4$, $(1, 4)$

61. $-6 < 3(x-2) \le 8$

$-6 < 3x - 6 \le 8$

$0 < 3x \le 14$

$0 < x \le \frac{14}{3}, \left(0, \frac{14}{3}\right]$

63. Let x be the number.

$$2x + 6 > -14$$
$$2x + 6 - 6 > -14 - 6$$
$$2x > -20$$
$$\frac{2x}{2} > \frac{-20}{2}$$
$$x > -10$$

All numbers greater than -10 make this statement true.

65. Use $P = 2l + 2w$ when $w = 15$ and $P \le 100$.

$$2l + 2(15) \le 100$$
$$2l + 30 \le 100$$
$$2l + 30 - 30 \le 100 - 30$$
$$2l \le 70$$
$$\frac{2l}{2} \le \frac{70}{2}$$
$$l \le 35$$

The maximum length of the rectangle is 35 cm.

67. Let x be the score in his third game.

$$\frac{146 + 201 + x}{3} \ge 180$$
$$\frac{347 + x}{3} \ge 180$$
$$3 \cdot \frac{347 + x}{3} \ge 3 \cdot 180$$
$$347 + x \ge 540$$
$$347 + x - 347 \ge 540 - 347$$
$$x \ge 193$$

He must bowl at least 193 in the third game.

69. Let x represent the number of people. Then the cost is $50 + 34x$.

$$50 + 34x \le 3000$$
$$50 + 34x - 50 \le 3000 - 50$$
$$34x \le 2950$$
$$\frac{34x}{34} \le \frac{2950}{34}$$
$$x \le \frac{2950}{34} \approx 86.76$$

They can invite at most 86 people.

71. Let x represent the number of minutes.
$$5.8x \geq 200$$
$$\frac{5.8x}{5.8} \geq \frac{200}{5.8}$$
$$x \geq \frac{200}{5.8} \approx 35$$
The person must walk at least 35 minutes.

73. Let x = the unknown number.
$$-5 < 2x + 1 < 7$$
$$-6 < 2x < 6$$
$$-3 < x < 3$$
All numbers between -3 and 3

75. $(2)^3 = (2)(2)(2) = 8$

77. $(1)^{12} = (1)(1)(1)(1)(1)(1)(1)(1)(1)(1)(1)(1) = 1$

79. $\left(\frac{4}{7}\right)^2 = \left(\frac{4}{7}\right)\left(\frac{4}{7}\right) = \frac{16}{49}$

81. Since $3 < 5$, $3(-4) > 5(-4)$.

83. If $m \leq n$, then $-2m \geq -2n$.

85. Reverse the direction of the inequality symbol when multiplying or dividing by a negative number.

87. Let x be the score on his final exam. Since the final counts as two tests, his final course average is $\frac{75 + 83 + 85 + 2x}{5}$.
$$\frac{75 + 83 + 85 + 2x}{5} \geq 80$$
$$\frac{243 + 2x}{5} \geq 80$$
$$5\left(\frac{243 + 2x}{5}\right) \geq 5(80)$$
$$243 + 2x \geq 400$$
$$243 + 2x - 243 \geq 400 - 243$$
$$2x \geq 157$$
$$\frac{2x}{2} \geq \frac{157}{2}$$
$$x \geq 78.5$$
His final exam score must be at least 78.5 for him to get a B.

89. answers may vary

91. answers may vary

93. $C = 3.14d$
$$2.9 \leq 3.14d \leq 3.1$$
$$0.924 \leq d \leq 0.987$$
The diameter must be between 0.924 cm and 0.987 cm.

95. $x(x + 4) > x^2 - 2x + 6$
$$x^2 + 4x > x^2 - 2x + 6$$
$$4x > -2x + 6$$
$$6x > 6$$
$$x > 1, (1, \infty)$$

97. $x^2 + 6x - 10 < x(x - 10)$
$$x^2 + 6x - 10 < x^2 - 10x$$
$$6x - 10 < -10x$$
$$16x - 10 < 0$$
$$16x < 10$$
$$x < \frac{10}{6}$$
$$x < \frac{5}{8}, \left(-\infty, \frac{5}{8}\right)$$

Chapter 2 Vocabulary Check

1. Terms with the same variables raised to exactly the same powers are called <u>like terms</u>.

2. If terms are not like terms, they are <u>unlike terms</u>.

3. A <u>linear equation in one variable</u> can be written in the form $ax + b = c$.

4. A <u>linear inequality in one variable</u> can be written in the form $ax + b < c$, (or $>$, $\leq$, $\geq$).

5. Inequalities containing two inequality symbols are called <u>compound inequalities</u>.

6. An equation that describes a known relationship among quantities is called a <u>formula</u>.

7. The <u>numerical coefficient</u> of a term is its numerical factor.

8. Equations that have the same solution are called <u>equivalent equations</u>.

9. The solutions to the equation $x + 5 = x + 5$ are <u>all real numbers</u>.

10. The solution to the equation $x + 5 = x + 4$ is <u>no solution</u>.

11. If both sides of an inequality are multiplied or divided by the same positive number, the direction of the inequality symbol is <u>the same</u>.

12. If both sides of an inequality are multiplied by the same negative number, the direction of the inequality symbol is <u>reversed</u>.

Chapter 2 Review

1. $5x - x + 2x = 6x$

2. $0.2z - 4.6x - 7.4z = -4.6x - 7.2z$

3. $\dfrac{1}{2}x + 3 + \dfrac{7}{2}x - 5 = \dfrac{8}{2}x - 2 = 4x - 2$

4. $\dfrac{4}{5}y + 1 + \dfrac{6}{5}y + 2 = \dfrac{10}{5}y + 3 = 2y + 3$

5. $2(n - 4) + n - 10 = 2n - 8 + n - 10 = 3n - 18$

6. $3(w + 2) - (12 - w) = 3w + 6 - 12 + w = 4w - 6$

7. $(x + 5) - (7x - 2) = x + 5 - 7x + 2 = -6x + 7$

8. $(y - 0.7) - (1.4y - 3) = y - 0.7 - 1.4y + 3$
$$= -0.4y + 2.3$$

9. Three times a number decreased by 7 is $3x - 7$.

10. Twice the sum of a number and 2.8 added to 3 times the number is $2(x + 2.8) + 3x$.

11. $\begin{aligned} 8x + 4 &= 9x \\ 8x + 4 - 8x &= 9x - 8x \\ 4 &= x \end{aligned}$

12. $\begin{aligned} 5y - 3 &= 6y \\ 5y - 3 - 5y &= 6y - 5y \\ -3 &= y \end{aligned}$

13. $\begin{aligned} \dfrac{2}{7}x + \dfrac{5}{7}x &= 6 \\ \dfrac{7}{7}x &= 6 \\ x &= 6 \end{aligned}$

14. $\begin{aligned} 3x - 5 &= 4x + 1 \\ -5 &= x + 1 \\ -6 &= x \end{aligned}$

15. $\begin{aligned} 2x - 6 &= x - 6 \\ x - 6 &= -6 \\ x &= 0 \end{aligned}$

16. $\begin{aligned} 4(x + 3) &= 3(1 + x) \\ 4x + 12 &= 3 + 3x \\ x + 12 &= 3 \\ x &= -9 \end{aligned}$

17. $\begin{aligned} 6(3 + n) &= 5(n - 1) \\ 18 + 6n &= 5n - 5 \\ 18 + n &= -5 \\ n &= -23 \end{aligned}$

18. $\begin{aligned} 5(2 + x) - 3(3x + 2) &= -5(x - 6) + 2 \\ 10 + 5x - 9x - 6 &= -5x + 30 + 2 \\ -4x + 4 &= -5x + 32 \\ x + 4 &= 32 \\ x &= 28 \end{aligned}$

19. $\begin{aligned} x - 5 &= 3 \\ x - 5 + \underline{5} &= 3 + \underline{5} \\ x &= 8 \end{aligned}$

20. $\begin{aligned} x + 9 &= -2 \\ x + 9 - \underline{9} &= -2 - \underline{9} \\ x &= -11 \end{aligned}$

21. $10 - x$; choice b.

22. $x - 5$; choice a.

23. Complementary angles sum to $90°$.
$(90 - x)°$; choice b.

24. Supplementary angles sum to $180°$.
$180 - (x + 5) = 180 - x - 5 = 175 - x$
$(175 - x)°$; choice c.

25. $\begin{aligned} \dfrac{3}{4}x &= -9 \\ \dfrac{4}{3}\left(\dfrac{3}{4}x\right) &= \dfrac{4}{3}(-9) \\ x &= -12 \end{aligned}$

26. $\begin{aligned} \dfrac{x}{6} &= \dfrac{2}{3} \\ 6 \cdot \dfrac{x}{6} &= 6 \cdot \dfrac{2}{3} \\ x &= 4 \end{aligned}$

27. $-5x = 0$

$\dfrac{-5x}{-5} = \dfrac{0}{-5}$

$x = 0$

28. $-y = 7$

$\dfrac{-y}{-1} = \dfrac{7}{-1}$

$y = -7$

29. $0.2x = 0.15$

$\dfrac{0.2x}{0.2} = \dfrac{0.15}{0.2}$

$x = 0.75$

30. $\dfrac{-x}{3} = 1$

$-3 \cdot \dfrac{-x}{3} = -3 \cdot 1$

$x = -3$

31. $-3x + 1 = 19$

$-3x = 18$

$\dfrac{-3x}{-3} = \dfrac{18}{-3}$

$x = -6$

32. $5x + 25 = 20$

$5x = -5$

$\dfrac{5x}{5} = \dfrac{-5}{5}$

$x = -1$

33. $7(x-1) + 9 = 5x$

$7x - 7 + 9 = 5x$

$7x + 2 = 5x$

$2 = -2x$

$\dfrac{2}{-2} = \dfrac{-2x}{-2}$

$-1 = x$

34. $7x - 6 = 5x - 3$

$2x - 6 = -3$

$2x = 3$

$\dfrac{2x}{2} = \dfrac{3}{2}$

$x = \dfrac{3}{2}$

35. $-5x + \dfrac{3}{7} = \dfrac{10}{7}$

$7\left(-5x + \dfrac{3}{7}\right) = 7 \cdot \dfrac{10}{7}$

$-35x + 3 = 10$

$-35x = 7$

$x = -\dfrac{7}{35}$

$x = -\dfrac{1}{5}$

36. $5x + x = 9 + 4x - 1 + 6$

$6x = 4x + 14$

$2x = 14$

$x = 7$

37. Let $x =$ the first integer, then
$x + 1 =$ the second integer, and
$x + 2 =$ the third integer.
sum $= x + (x + 1) + (x + 2) = 3x + 3$

38. Let $x =$ the first integer, then
$x + 2 =$ the second integer
$x + 4 =$ the third integer
$x + 6 =$ the fourth integer.
sum $= x + (x + 6) = 2x + 6$

39. $\dfrac{5}{3}x + 4 = \dfrac{2}{3}x$

$3\left(\dfrac{5}{3}x + 4\right) = 3\left(\dfrac{2}{3}x\right)$

$5x + 12 = 2x$

$12 = -3x$

$-4 = x$

40. $\dfrac{7}{8}x + 1 = \dfrac{5}{8}x$

$8\left(\dfrac{7}{8}x + 1\right) = 8\left(\dfrac{5}{8}x\right)$

$7x + 8 = 5x$

$8 = -2x$

$-4 = x$

41. $-(5x + 1) = -7x + 3$

$-5x - 1 = -7x + 3$

$2x - 1 = 3$

$2x = 4$

$x = 2$

42. $-4(2x+1) = -5x+5$
$-8x-4 = -5x+5$
$-3x-4 = 5$
$-3x = 9$
$x = -3$

43. $-6(2x-5) = -3(9+4x)$
$-12x+30 = -27-12x$
$30 = -27$
There is no solution.

44. $3(8y-1) = 6(5+4y)$
$24y-3 = 30+24y$
$-3 = 30$
There is no solution.

45. $\dfrac{3(2-z)}{5} = z$
$3(2-z) = 5z$
$6-3z = 5z$
$6 = 8z$
$\dfrac{6}{8} = z$
$\dfrac{3}{4} = z$

46. $\dfrac{4(n+2)}{5} = -n$
$4(n+2) = -5n$
$4n+8 = -5n$
$8 = -9n$
$-\dfrac{8}{9} = n$

47. $0.5(2n-3)-0.1 = 0.4(6+2n)$
$10[0.5(2n-3)-0.1] = 10[0.4(6+2n)]$
$5(2n-3)-1 = 4(6+2n)$
$10n-15-1 = 24+8n$
$10n-16 = 24+8n$
$2n-16 = 24$
$2n = 40$
$n = 20$

48. $-9-5a = 3(6a-1)$
$-9-5a = 18a-3$
$-9 = 23a-3$
$-6 = 23a$
$-\dfrac{6}{23} = a$

49. $\dfrac{5(c+1)}{6} = 2c-3$
$5(c+1) = 6(2c-3)$
$5c+5 = 12c-18$
$-7c+5 = -18$
$-7c = -23$
$c = \dfrac{23}{7}$

50. $\dfrac{2(8-a)}{3} = 4-4a$
$2(8-a) = 3(4-4a)$
$16-2a = 12-12a$
$10a+16 = 12$
$10a = -4$
$a = \dfrac{-4}{10}$
$a = -\dfrac{2}{5}$

51. $200(70x-3560) = -179(150x-19,300)$
$14,000x-712,000 = -26,850x+3,454,700$
$40,850x-712,000 = 3,454,700$
$40,850x = 4,166,700$
$x = 102$

52. $1.72y-0.04y = 0.42$
$1.68y = 0.42$
$y = 0.25$

53. Let x = length of a side of the square, then
$50.5 + 10x$ = the height.
$x+(50.5+10x) = 7327$
$11x+50.5 = 7327$
$11x = 7276.5$
$x = 661.5$
$50.5 + 10x = 50.5 + 10(661.5) = 6665.5$
The height is 6665.5 inches.

54. Let x = the length of the shorter piece and
$2x$ = the length of the other.
$x+2x = 12$
$3x = 12$
$x = 4$
$2x = 2(4) = 8$
The lengths are 4 feet and 8 feet.

55. Let x = number of Keebler plants, then
$2x - 1$ = number of Kellogg plants.
$$x + (2x - 1) = 53$$
$$3x - 1 = 53$$
$$3x = 54$$
$$x = 18$$
$$2x - 1 = 2(18) - 1 = 35$$
There were 18 Keebler plants and 35 Kellogg plants.

56. Let x = first integer, then
$x + 1$ = second integer, and
$x + 2$ = third integer.
$$x + (x + 1) + (x + 2) = -114$$
$$3x + 3 = -114$$
$$3x = -117$$
$$x = -39$$
$$x + 1 = -39 + 1 = -38$$
$$x + 2 = -39 + 2 = -37$$
The integers are $-39, -38, -37$.

57. Let x = the unknown number.
$$\frac{x}{3} = x - 2$$
$$3 \cdot \frac{x}{3} = 3(x - 2)$$
$$x = 3x - 6$$
$$-2x = -6$$
$$x = 3$$
The number is 3.

58. Let x = the unknown number.
$$2(x + 6) = -x$$
$$2x + 12 = -x$$
$$12 = -3x$$
$$-4 = x$$
The number is -4.

59. Let $P = 46$ and $l = 14$.
$$P = 2l + 2w$$
$$46 = 2(14) + 2w$$
$$46 = 28 + 2w$$
$$18 = 2w$$
$$9 = w$$

60. Let $V = 192$, $l = 8$, and $w = 6$.
$$V = lwh$$
$$192 = 8(6)h$$
$$192 = 48h$$
$$4 = h$$

61.
$$y = mx + b$$
$$y - b = mx$$
$$\frac{y - b}{x} = m$$

62.
$$r = vst - 5$$
$$r + 5 = vst$$
$$\frac{r + 5}{vt} = s$$

63. $2y - 5x = 7$
$$-5x = -2y + 7$$
$$x = \frac{-2y + 7}{-5}$$
$$x = \frac{2y - 7}{5}$$

64. $3x - 6y = -2$
$$-6y = -3x - 2$$
$$y = \frac{-3x - 2}{-6}$$
$$y = \frac{3x + 2}{6}$$

65. $C = \pi D$
$$\frac{C}{D} = \pi$$

66. $C = 2\pi r$
$$\frac{C}{2r} = \pi$$

67. Let $V = 900$, $l = 20$, and $h = 3$.
$$V = lwh$$
$$900 = 20w(3)$$
$$900 = 60w$$
$$15 = w$$
The width is 15 meters.

68. Let x = width, then $x + 6$ = length.
$$60 = 2x + 2(x + 6)$$
$$60 = 2x + 2x + 12$$
$$60 = 4x + 12$$
$$48 = 4x$$
$$12 = x$$
$$x + 6 = 12 + 6 = 18$$
The dimensions are 18 feet by 12 feet.

69. Let $d = 10{,}000$ and $r = 125$.

$$d = rt$$
$$10{,}000 = 125t$$
$$80 = t$$

It will take 80 minutes or 1 hour and 20 minutes.

70. Let $F = 104$.

$$C = \frac{5}{9}(F - 32)$$
$$= \frac{5}{9}(104 - 32)$$
$$= \frac{5}{9}(72)$$
$$= 40$$

The temperature was 40°C.

71. Let $x =$ the percent.

$$9 = x \cdot 45$$
$$\frac{9}{45} = \frac{45x}{45}$$
$$0.2 = x$$

9 is 20% of 45.

72. Let $x =$ the percent.

$$59.5 = x \cdot 85$$
$$\frac{59.5}{85} = \frac{85x}{85}$$
$$0.7 = x$$

59.5 is 70% of 85.

73. Let $x =$ the number.

$$137.5 = 125\% \cdot x$$
$$137.5 = 1.25x$$
$$\frac{137.5}{1.25} = \frac{1.25x}{1.25}$$
$$110 = x$$

137.5 is 125% of 110.

74. Let $x =$ the number.

$$768 = 60\% \cdot x$$
$$768 = 0.6x$$
$$\frac{768}{0.6} = \frac{0.6x}{0.6}$$
$$1280 = x$$

768 is 60% of 1280.

75. Let $x =$ mark-up.

$$x = 11\% \cdot 1900$$
$$x = 0.11 \cdot 1900$$
$$x = 209$$

New price $= 1900 + 209 = 2109$

The mark-up is $209 and the new price is $2109.

76. Find 79% of 76,000.
$0.79 \cdot 76{,}000 = 60{,}040$
We would expect 60,040 people in that city to use the Internet.

77. Let x = gallons of 40% solution.

Strength	gallons	Concentration	
40%	x	0.4	$0.4x$
10%	$30 - x$	0.1	$0.1(30 - x)$
20%	30	0.2	$0.2(30)$

$$0.4x + 0.1(30 - x) = 0.2(30)$$
$$0.4x + 3 - 0.1x = 6$$
$$0.3x + 3 = 6$$
$$0.3x = 3$$
$$x = 10$$
$30 - x = 30 - 10 = 20$
Mix 10 gallons of 40% acid solution with 20 gallons of 10% acid solution.

78. Increase $= 21.0 - 20.7 = 0.3$
Let x = percent.
$$0.3 = x \cdot 20.7$$
$$\frac{0.3}{20.7} = \frac{20.7x}{20.7}$$
$$0.0145 \approx x$$
The percent increase is 1.45%.

79. From the graph, the height of 'Almost hit a car' is 18%.

80. Choose the tallest bar. The most common effect is swerving into another lane.

81. Find 21% of 4600.
$0.21 \cdot 4600 = 966$
We would expect 966 customers to have cut someone off.

82. Find 41% of 4600.
$0.41 \cdot 4600 = 1886$
We would expect 1886 customers to have sped up.

83. $\text{percent decrease} = \dfrac{\text{amount of decrease}}{\text{original amount}}$
$$= \frac{250 - 170}{250}$$
$$= \frac{80}{250}$$
$$= 0.32$$
The percent decrease is 32%.

84. Let x = original price.

$$x - 0.20x = 19.20$$
$$0.80x = 19.20$$
$$\frac{0.80x}{0.80} = \frac{19.20}{0.80}$$
$$x = 24$$

The original price was \$24.

85. Let x = time up, then $3 - x$ = time down.

Rate $\cdot$ Time = Distance

	Rate	Time	Distance
Up	10	x	$10x$
Down	50	$3-x$	$50(3-x)$

$$d = d$$
$$10x = 50(3-x)$$
$$10x = 150 - 50x$$
$$60x = 150$$
$$x = 2.5$$

$$
\begin{aligned}
\text{Total distance} &= 10x + 50(3-x) \\
&= 10(2.5) + 50(3 - 2.5) \\
&= 25 + 50(0.5) \\
&= 25 + 25 \\
&= 50
\end{aligned}
$$

The distance traveled was 50 km.

86. Let x = the amount invested at 10.5% for one year.

Principal $\cdot$ Rate = Interest

	Principal	Rate	Interest
10.5%	x	0.105	0.105
8.5%	$50,000 - x$	0.085	$0.085(50,000 - x)$
Total	50,000		4550

$$0.105x + 0.085(50,000 - x) = 4550$$
$$0.105x + 4250 - 0.085x = 4550$$
$$0.02x + 4250 = 4550$$
$$0.02x = 300$$
$$x = 15,000$$

$50,000 - x = 50,000 - 15,000 = 35,000$
Invest \$35,000 at 8.5% and \$15,000 at 10.5%.

87. Let x = the number of dimes,
  $2x$ = the number of quarters, and
  $500 - x - 2x$ the number of nickels.

	No. of Coins ·	Value =	Amt. of Money
Dimes	x	0.1	$0.1x$
Quarters	$2x$	0.25	$0.25(2x)$
Nickels	$500 - 3x$	0.05	$0.05(500 - 3x)$
Total	500		88

$$0.1x + 0.25(2x) + 0.05(500 - 3x) = 88$$
$$0.1x + 0.5x + 25 - 0.15x = 88$$
$$0.45x + 25 = 88$$
$$0.45x = 63$$
$$x = 140$$
$$500 - 3x = 500 - 3(140) = 500 - 420 = 80$$
There are 80 nickels in the pay phone.

88. Let x = the time traveled by the Amtrak train.

	Rate ·	Time =	Distance
Amtrak	60	x	$60x$
Freight	45	$x + 1.5$	$45(x + 1.5)$

$$d = d$$
$$60x = 45(x + 1.5)$$
$$60x = 45x + 67.5$$
$$15x = 67.5$$
$$x = 4.5$$
It will take 4.5 hours.

89. $x > 0$, $(0, \infty)$

90. $x \le -2$, $(-\infty, -2]$

91. $0.5 \le y < 1.5$, $[0.5, 1.5)$

92. $-1 < x < 1$, $(-1, 1)$

93. $-3x > 12$

$$\frac{-3x}{-3} < \frac{12}{-3}$$

$$x < -4,\ (-\infty, -4)$$

94. $-2x \geq -20$

$$\frac{-2x}{-2} \leq \frac{-20}{-2}$$

$$x \leq 10,\ (-\infty, 10]$$

95. $x + 4 \geq 6x - 16$

$$-5x + 4 \geq -16$$

$$-5x \geq -20$$

$$\frac{-5x}{-5} \leq \frac{-20}{-5}$$

$$x \leq 4,\ (-\infty, 4]$$

96. $5x - 7 > 8x + 5$

$$-3x - 7 > 5$$

$$-3x > 12$$

$$\frac{-3x}{-3} < \frac{12}{-3}$$

$$x < -4,\ (-\infty, -4)$$

97. $-3 < 4x - 1 < 2$

$$-2 < 4x < 3$$

$$-\frac{1}{2} < x < \frac{3}{4},\ \left(-\frac{1}{2}, \frac{3}{4}\right)$$

98. $2 \leq 3x - 4 < 6$

$$6 \leq 3x < 10$$

$$2 \leq x < \frac{10}{3},\ \left[2, \frac{10}{3}\right)$$

99. $4(2x - 5) \leq 5x - 1$

$$8x - 20 \leq 5x - 1$$

$$3x - 20 \leq -1$$

$$3x \leq 19$$

$$x \leq \frac{19}{3},\ \left(-\infty, \frac{19}{3}\right]$$

100. $-2(x - 5) > 2(3x - 2)$

$$-2x + 10 > 6x - 4$$

$$-8x + 10 > -4$$

$$-8x > -14$$

$$\frac{-8x}{-8} < \frac{-14}{-8}$$

$$x < \frac{7}{4},\ \left(-\infty, \frac{7}{4}\right)$$

101. Let x = the amount of sales then
$0.05x$ = her commission.

$$175 + 0.05x \geq 300$$

$$0.05x \geq 125$$

$$x \geq 2500$$

Sales must be at least $2500.

102. Let x = her score on the fourth round.

$$\frac{76 + 82 + 79 + x}{4} < 80$$

$$237 + x < 320$$

$$x < 83$$

Her score must be less than 83.

103. $6x + 2x - 1 = 5x + 11$

$$8x - 1 = 5x + 11$$

$$3x - 1 = 11$$

$$3x = 12$$

$$x = 4$$

104. $2(3y - 4) = 6 + 7y$

$$6y - 8 = 6 + 7y$$

$$-8 = 6 + y$$

$$-14 = y$$

105. $4(3-a)-(6a+9)=-12a$
$$12-4a-6a-9=-12a$$
$$3-10a=-12a$$
$$3=-2a$$
$$-\frac{3}{2}=a$$

106. $\frac{x}{3}-2=5$
$$\frac{x}{3}=7$$
$$3\cdot\frac{x}{3}=3\cdot 7$$
$$x=21$$

107. $2(y+5)=2y+10$
$$2y+10=2y+10$$
$$10=10$$
All real numbers are solutions.

108. $7x-3x+2=2(2x-1)$
$$4x+2=4x-2$$
$$2=-2$$
There is no solution.

109. Let x = the number.
$$6+2x=x-7$$
$$6+x=-7$$
$$x=-13$$
The number is -13.

110. Let x = length of shorter piece, then $4x+3$ = length of longer piece.
$$x+(4x+3)=23$$
$$5x+3=23$$
$$5x=20$$
$$x=4$$
$4x+3=4(4)+3=19$
The shorter piece is 4 inches and the longer piece is 19 inches.

111. $V=\frac{1}{3}Ah$
$$3\cdot V=3\cdot\frac{1}{3}Ah$$
$$3V=Ah$$
$$\frac{3V}{A}=\frac{Ah}{A}$$
$$\frac{3V}{A}=h$$

112. Let x = the number.
$$x=26\%\cdot 85$$
$$x=0.26\cdot 85$$
$$x=22.1$$
22.1 is 26% of 85.

113. Let x = the number.
$$72=45\%\cdot x$$
$$72=0.45x$$
$$\frac{72}{0.45}=\frac{0.45x}{0.45}$$
$$160=x$$
72 is 45% of 160.

114. Increase $=282-235=47$
Let x = percent.
$$47=x\cdot 235$$
$$\frac{47}{235}=\frac{235x}{235}$$
$$0.2=x$$
The percent increase is 20%.

115. $4x-7>3x+2$
$$x-7>2$$
$$x>9,\ (9,\infty)$$

116. $-5x<20$
$$\frac{-5x}{-5}>\frac{20}{-5}$$
$$x>-4,\ (-4,\infty)$$

117. $-3(1+2x)+x\ge -(3-x)$
$$-3-6x+x\ge -3+x$$
$$-3-5x\ge -3+x$$
$$-5x\ge x$$
$$-6x\ge 0$$
$$\frac{-6x}{-6}\le\frac{0}{-6}$$
$$x\le 0,\ (-\infty,0]$$

Chapter 2 Test

1. $2y-6-y-4=y-10$

2. $2.7x+6.1+3.2x-4.9=5.9x+1.2$

3. $4(x-2)-3(2x-6)=4x-8-6x+18$
$$=-2x+10$$

4. $7 + 2(5y - 3) = 7 + 10y - 6 = 10y + 1$

5. $$-\frac{4}{5}x = 4$$
$$-\frac{5}{4}\cdot\left(-\frac{4}{5}x\right) = -\frac{5}{4}\cdot 4$$
$$x = -5$$

6. $4(n-5) = -(4-2n)$
$4n - 20 = -4 + 2n$
$2n - 20 = -4$
$2n = 16$
$n = 8$

7. $5y - 7 + y = -(y + 3y)$
$6y - 7 = -4y$
$-7 = -10y$
$\frac{7}{10} = y$

8. $4z + 1 - z = 1 + z$
$3z + 1 = 1 + z$
$2z + 1 = 1$
$2z = 0$
$z = 0$

9. $\frac{2(x+6)}{3} = x - 5$
$2(x+6) = 3(x-5)$
$2x + 12 = 3x - 15$
$12 = x - 15$
$27 = x$

10. $\frac{1}{2} - x + \frac{3}{2} = x - 4$
$2\left(\frac{1}{2} - x + \frac{3}{2}\right) = 2(x-4)$
$1 - 2x + 3 = 2x - 8$
$-2x + 4 = 2x - 8$
$-4x + 4 = -8$
$-4x = -12$
$x = 3$

11. $-0.3(x-4) + x = 0.5(3-x)$
$10[-0.3(x-4) + x] = 10[0.5(3-x)]$
$-3(x-4) + 10x = 5(3-x)$
$-3x + 12 + 10x = 15 - 5x$
$7x + 12 = 15 - 5x$
$12x + 12 = 15$
$12x = 3$
$x = \frac{3}{12} = \frac{1}{4} = 0.25$

12. $-4(a+1) - 3a = -7(2a-3)$
$-4a - 4 - 3a = -14a + 21$
$-7a - 4 = -14a + 21$
$7a - 4 = 21$
$7a = 25$
$a = \frac{25}{7}$

13. $-2(x-3) = x + 5 - 3x$
$-2x + 6 = -2x + 5$
$6 = 5$
There is no solution.

14. Let $y = -14$, $m = -2$, and $b = -2$.
$y = mx + b$
$-14 = -2x - 2$
$-12 = -2x$
$6 = x$

15. $V = \pi r^2 h$
$\frac{V}{\pi r^2} = \frac{\pi r^2 h}{\pi r^2}$
$\frac{V}{\pi r^2} = h$

16. $3x - 4y = 10$
$-4y = -3x + 10$
$y = \frac{-3x + 10}{-4}$
$y = \frac{3x - 10}{4}$

17. $3x - 5 \geq 7x + 3$
$-4x - 5 \geq 3$
$-4x \geq 8$
$\frac{-4x}{-4} \leq \frac{8}{-4}$
$x \leq -2$, $(-\infty, -2]$

18. $x + 6 > 4x - 6$
$-3x + 6 > -6$
$-3x > -12$
$\dfrac{-3x}{-3} < \dfrac{-12}{-3}$
$x < 4, \ (-\infty, 4)$

19. $-2 < 3x + 1 < 8$
$-3 < 3x < 7$
$-1 < x < \dfrac{7}{3}, \ \left(-1, \dfrac{7}{3}\right)$

20. $\dfrac{2(5x+1)}{3} > 2$
$2(5x + 1) > 6$
$10x + 2 > 6$
$10x > 4$
$x > \dfrac{4}{10} = \dfrac{2}{5}, \ \left(\dfrac{2}{5}, \infty\right)$

21. Let x = the number.
$x + \dfrac{2}{3}x = 35$
$3\left(x + \dfrac{2}{3}x\right) = 3(35)$
$3x + 2x = 105$
$5x = 105$
$x = 21$
The number is 21.

22. Let x = width, then $x + 2$ = length.
$P = 2w + 2l$
$252 = 2x + 2(x + 2)$
$252 = 2x + 2x + 4$
$252 = 4x + 4$
$252 - 4 = 4x + 4 - 4$
$248 = 4x$
$\dfrac{248}{4} = \dfrac{4x}{4}$
$62 = x$
$64 = x + 2$
The dimensions of the deck are 62 feet by 64 feet.

23. Let x = one area code, then
$2x$ = other area code.
$x + 2x = 1203$
$3x = 1203$
$\dfrac{3x}{3} = \dfrac{1203}{3}$
$x = 401$
$2x = 2(401) = 802$
The area codes are 401 and 802.

24. Let x = the amount invested at 10% for one year.

	Principal	· Rate	= Interest
10%	x	0.10	$0.1x$
12%	$2x$	0.12	$0.12(2x)$
Total			2890

$0.1x + 0.12(2x) = 2890$
$0.1x + 0.24x = 2890$
$0.34x = 2890$
$x = 8500$
$2x = 2(8500) = 17{,}000$
He invested \$8500 at 10% and \$17,000 at 12%.

25. Let x = the time they travel.

	Rate	· Time	= Distance
Train 1	50	x	$50x$
Train 2	64	x	$64x$
Total			285

$50x + 64x = 285$
$114x = 285$
$x = 2\dfrac{1}{2}$

They must travel for $2\dfrac{1}{2}$ hours.

26. From the graph, 69% are classified as weak.
Find 69% of 800.
$69\% \cdot 800 = 0.69 \cdot 800 = 552$
You would expect 552 of the 800 to be classified as weak.

27. Let x be the unknown percent.
$72 = x \cdot 180$
$\dfrac{72}{180} = \dfrac{180x}{180}$
$0.4 = x$
72 is 40% of 180.

28. $\text{percent decrease} = \dfrac{\text{amount of decrease}}{\text{original amount}}$

$$= \dfrac{225 - 189}{225}$$

$$= \dfrac{36}{225}$$

$$= 0.16$$

The percent decrease is 16%.

Chapter 2 Cumulative Review

1. a. the natural numbers are 11 and 112.

    b. The whole numbers are 0, 11, and 112.

    c. The integers are -3, -2, 0, 11, and 112.

    d. The rational numbers are -3, -2, -1.5, 0, $\dfrac{1}{4}$, 11, and 112.

    e. The irrational number is $\sqrt{2}$.

    f. All the numbers in the given set are real numbers.

2. a. The natural numbers are 2, 7, and 8.

    b. The whole numbers are 0, 2, 7, and 8.

    c. The integers are -185, 0, 2, 7, and 8.

    d. The rational numbers are -185, $-\dfrac{1}{5}$, 0, 2, 7, and 8.

    e. The irrational number is $\sqrt{3}$.

    f. All the numbers in the given set are real numbers.

3. a. $|4| = 4$

    b. $|-5| = 5$

    c. $|0| = 0$

    d. $\left|-\dfrac{1}{2}\right| = \dfrac{1}{2}$

    e. $|5.6| = 5.6$

4. a. $|5| = 5$

    b. $|-8| = 8$

    c. $\left|-\dfrac{2}{3}\right| = \dfrac{2}{3}$

5. a. $40 = 2 \cdot 2 \cdot 2 \cdot 5$

    b. $63 = 3 \cdot 3 \cdot 7$

6. a. $44 = 2 \cdot 2 \cdot 11$

    b. $90 = 2 \cdot 3 \cdot 3 \cdot 5$

7. $\dfrac{2}{5} = \dfrac{2}{5} \cdot \dfrac{4}{4} = \dfrac{8}{20}$

8. $\dfrac{2}{3} = \dfrac{2}{3} \cdot \dfrac{8}{8} = \dfrac{16}{24}$

9. $3[4 + 2(10 - 1)] = 3[4 + 2(9)]$
$$= 3[4 + 18]$$
$$= 3[22]$$
$$= 66$$

10. $5[16 - 4(2 + 1)] = 5[16 - 4(3)]$
$$= 5[16 - 12]$$
$$= 5[4]$$
$$= 20$$

11. Let $x = 2$.
$$3x + 10 = 8x$$
$$3(2) + 10 \overset{?}{=} 8(2)$$
$$6 + 10 \overset{?}{=} 16$$
$$16 = 16$$
2 is a solution of the equation.

12. Let $x = 3$.
$$5x - 2 = 4x$$
$$5(3) - 2 \overset{?}{=} 4(3)$$
$$15 - 2 \overset{?}{=} 12$$
$$13 \neq 12$$
3 is not a solution of the equation.

13. $-1 + (-2) = -3$

14. $(-2) + (-8) = -10$

15. $-4 + 6 = 2$

16. $-3 + 10 = 7$

17. a. $-(-10) = 10$

    b. $-\left(-\dfrac{1}{2}\right) = \dfrac{1}{2}$

c. $-(-2x) = 2x$

d. $-|-6| = -(6) = -6$

18. a. $-(-5) = 5$

b. $-\left(-\dfrac{2}{3}\right) = \dfrac{2}{3}$

c. $-(-a) = a$

d. $-|-3| = -(3) = -3$

19. a. $5.3 - (-4.6) = 5.3 + 4.6 = 9.9$

b.
$$-\dfrac{3}{10} - \dfrac{5}{10} = -\dfrac{3}{10} + \left(-\dfrac{5}{10}\right)$$
$$= \dfrac{-3-5}{10}$$
$$= -\dfrac{8}{10}$$
$$= -\dfrac{4}{5}$$

c.
$$-\dfrac{2}{3} - \left(-\dfrac{4}{5}\right) = -\dfrac{2}{3} \cdot \dfrac{5}{5} + \dfrac{4}{5} \cdot \dfrac{3}{3}$$
$$= -\dfrac{10}{15} + \dfrac{12}{15}$$
$$= \dfrac{2}{15}$$

20. a. $-2.7 - 8.4 = -2.7 + (-8.4) = -11.1$

b. $-\dfrac{4}{5} - \left(-\dfrac{3}{5}\right) = -\dfrac{4}{5} + \dfrac{3}{5} = \dfrac{-4+3}{5} = -\dfrac{1}{5}$

c. $\dfrac{1}{4} - \left(-\dfrac{1}{2}\right) = \dfrac{1}{4} + \dfrac{1}{2} \cdot \dfrac{2}{2} = \dfrac{1}{4} + \dfrac{2}{4} = \dfrac{3}{4}$

21. a. $x = 90 - 38 = 90 + (-38) = 52$
The complementary angle is 52°.

b. $y = 180 - 62 = 180 + (-62) = 118$
The supplementary angle is 118°.

22. a. $x = 90 - 72 = 90 + (-72) = 18$
The complementary angle is 18°.

b. $y = 180 - 47 = 180 + (-47) = 133$
The supplementary angle is 133°.

23. a. $(-1.2)(0.05) = -0.06$

b. $\dfrac{2}{3} \cdot \left(-\dfrac{7}{10}\right) = -\dfrac{2 \cdot 7}{3 \cdot 10} = -\dfrac{14}{30} = -\dfrac{7}{15}$

c. $\left(-\dfrac{4}{5}\right)(-20) = \dfrac{4 \cdot 20}{5} = \dfrac{80}{5} = 16$

24. a. $(4.5)(-0.08) = -0.36$

b. $-\dfrac{3}{4} \cdot \left(-\dfrac{8}{17}\right) = \dfrac{3 \cdot 8}{4 \cdot 17} = \dfrac{24}{68} = \dfrac{6}{17}$

25. a. $\dfrac{-24}{-4} = 6$

b. $\dfrac{-36}{3} = -12$

c. $\dfrac{2}{3} \div \left(-\dfrac{5}{4}\right) = \dfrac{2}{3}\left(-\dfrac{4}{5}\right) = -\dfrac{8}{15}$

d. $-\dfrac{3}{2} \div 9 = -\dfrac{3}{2} \div \dfrac{9}{1} = -\dfrac{3}{2} \cdot \dfrac{1}{9} = -\dfrac{3}{18} = -\dfrac{1}{6}$

26. a. $\dfrac{-32}{8} = -4$

b. $\dfrac{-108}{-12} = 9$

c. $-\dfrac{5}{7} \div \left(\dfrac{-9}{2}\right) = -\dfrac{5}{7}\left(-\dfrac{2}{9}\right) = \dfrac{10}{63}$

27. a. $x + 5 = 5 + x$

b. $3 \cdot x = x \cdot 3$

28. a. $y + 1 = 1 + y$

b. $y \cdot 4 = 4 \cdot y$

29. a. $8 \cdot 2 + 8 \cdot x = 8(2 + x)$

b. $7s + 7t = 7(s + t)$

30. a. $4 \cdot y + 4 \cdot \dfrac{1}{3} = 4\left(y + \dfrac{1}{3}\right)$

b. $0.10x + 0.10y = 0.10(x + y)$

31. $(2x - 3) - (4x - 2) = 2x - 3 - 4x + 2 = -2x - 1$

32. $(-5x+1)-(10x+3) = -5x+1-10x-3$
$$= -15x-2$$

33. $\quad y+0.6 = -1.0$
$$y+0.6-0.6 = -1.0-0.6$$
$$y = -1.6$$

34. $\qquad \dfrac{5}{6}+x = \dfrac{2}{3}$
$$6\left(\dfrac{5}{6}\right)+6(x) = 6\left(\dfrac{2}{3}\right)$$
$$5+6x = 4$$
$$6x = -1$$
$$x = -\dfrac{1}{6}$$

35. $\quad 7 = -5(2a-1)-(-11a+6)$
$$7 = -10a+5+11a-6$$
$$7 = a-1$$
$$7+1 = a-1+1$$
$$8 = a$$

36. $\quad -3x+1-(-4x-6) = 10$
$$-3x+1+4x+6 = 10$$
$$x+7 = 10$$
$$x = 3$$

37. $\dfrac{y}{7} = 20$
$$y = 140$$

38. $\dfrac{x}{4} = 18$
$$x = 72$$

39. $\quad 4(2x-3)+7 = 3x+5$
$$8x-12+7 = 3x+5$$
$$8x-5 = 3x+5$$
$$5x-5 = 5$$
$$5x = 10$$
$$x = 2$$

40. $\quad 6x+5 = 4(x+4)-1$
$$6x+5 = 4x+16-1$$
$$6x+5 = 4x+15$$
$$2x+5 = 15$$
$$2x = 10$$
$$x = 5$$

41. Let $x =$ a number.
$$2(x+4) = 4x-12$$
$$2x+8 = 4x-12$$
$$8 = 2x-12$$
$$20 = 2x$$
$$10 = x$$
The number is 10.

42. Let $x =$ a number.
$$x+4 = 3x-8$$
$$4 = 2x-8$$
$$12 = 2x$$
$$6 = x$$
The number is 6.

43. $\quad V = lwh$
$$\dfrac{V}{wh} = \dfrac{lwh}{wh}$$
$$\dfrac{V}{wh} = l$$

44. $\quad C = 2\pi r$
$$\dfrac{C}{2\pi} = \dfrac{2\pi r}{2\pi}$$
$$\dfrac{C}{2\pi} = r$$

45. $x+4 \le -6$
$$x \le -10, \ (-\infty, -10]$$

46. $x-3 > 2$
$$x > 5, \ (5, \infty)$$

Chapter 3

Section 3.1 Practice

1. a. We look for the shortest bar, which is the bar representing the Africa/Middle East region. We move from the right edge of this bar vertically downward to the Internet user axis. This region has approximately 145 million Internet users.

   b. The Asia/Oceania/Australia region has approximately 785 million Internet users. The Africa/Middle East region has approximately 145 million Internet users. We subtract $785 - 145 = 640$ or 640 million. The Asia/Oceania/Australia region has 640 million more Internet users than the Africa/Middle East region.

2. a. We locate the number 40 along the time axis and move vertically upward until the line is reached. From this point on the line, we move horizontally to the left until the pulse rate axis is reached. Reading the number of beats per minute, we find that the pulse rate is 70 beats per minute 40 minutes after a cigarette is lit.

   b. The number 0 on the time axis corresponds to the time when the cigarette is being lit. We move vertically upward to the point on the line and then horizontally to the left to the pulse rate axis. The pulse rate is 60 beats per minute when the cigarette is being lit.

   c. We find the highest point of the line graph, which represents the highest pulse rate. From this point, we move vertically downward to the time axis. We find the pulse rate is the highest at 5 minutes, which means 5 minutes after lighting a cigarette.

3. a. Point $(4, -3)$ lies in quadrant IV.

   b. Point $(-3, 5)$ lies in quadrant II.

   c. Point $(0, 4)$ lies on an axis, so it is not in any quadrant.

   d. Point $(-6, 1)$ lies in quadrant II.

   e. Point $(-2, 0)$ lies on an axis, so it is not in any quadrant.

   f. Point $(5, 5)$ lies in quadrant I.

   g. Point $\left(3\frac{1}{2}, 1\frac{1}{2}\right)$ lies in quadrant I.

   h. Point $(-4, -5)$ lies in quadrant III.

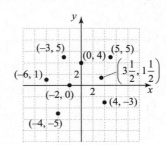

4. a. The ordered pairs are $(2004, 65)$, $(2005, 67)$, $(2006, 96)$, $(2007, 86)$, $(2008, 79)$, $(2009, 79)$, and $(2010, 72)$.

   b. We plot the ordered pairs. We label the horizontal axis "Year" and the vertical axis "Wildfires (in thousands)."

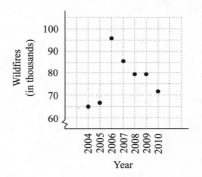

5. a. Let $x = 3$ and $y = 1$.
$$x + 3y = 6$$
$$3 + 3(1) = 6$$
$$3 + 3 = 6$$
$$6 = 6 \quad \text{true}$$
Yes, $(3, 1)$ is a solution.

   b. Let $x = 6$ and $y = 0$.
$$x + 3y = 6$$
$$6 + 3(0) = 6$$
$$6 + 0 = 6$$
$$6 = 6 \quad \text{true}$$
Yes, $(6, 0)$ is a solution.

c. Let $x = -2$ and $y = \dfrac{2}{3}$.

$$x + 3y = 6$$
$$-2 + 3\left(\dfrac{2}{3}\right) = 6$$
$$-2 + 2 = 6$$
$$0 = 6 \quad \text{false}$$

No, $\left(-2, \dfrac{2}{3}\right)$ is not a solution.

6. a. Let $x = 0$ and solve for y.

$$2x - y = 8$$
$$2(0) - y = 8$$
$$0 - y = 8$$
$$-y = 8$$
$$y = -8$$

The ordered pair is $(0, -8)$.

b. Let $y = 4$ and solve for x.

$$2x - y = 8$$
$$2x - 4 = 8$$
$$2x = 12$$
$$x = 6$$

The ordered pair is $(6, 4)$.

c. Let $x = -3$ and solve for y.

$$2x - y = 8$$
$$2(-3) - y = 8$$
$$-6 - y = 8$$
$$-y = 14$$
$$y = -14$$

The ordered pair is $(-3, -14)$.

7. a. Replace x with -2 in the equation and solve for y.

$$y = -4x$$
$$y = -4(-2)$$
$$y = 8$$

The ordered pair is $(-2, 8)$.

b. Replace y with -12 in the equation and solve for x.

$$y = -4x$$
$$-12 = -4x$$
$$3 = x$$

The ordered pair is $(3, -12)$.

c. Replace x with 0 in the equation and solve for y.

$$y = -4x$$
$$y = -4(0)$$
$$y = 0$$

The ordered pair is $(0, 0)$.

The completed table is shown below.

x	y
-2	8
3	-12
0	0

8. a. Let $x = -10$.

$$y = \dfrac{1}{5}x - 2$$
$$y = \dfrac{1}{5}(-10) - 2$$
$$y = -2 - 2$$
$$y = -4$$

Ordered pair: $(-10, -4)$

b. Let $x = 0$.

$$y = \dfrac{1}{5}x - 2$$
$$y = \dfrac{1}{5}(0) - 2$$
$$y = 0 - 2$$
$$y = -2$$

Ordered pair: $(0, -2)$

c. Let $y = 0$.

$$y = \dfrac{1}{5}x - 2$$
$$0 = \dfrac{1}{5}x - 2$$
$$2 = \dfrac{1}{5}x$$
$$10 = x$$

Ordered pair: $(10, 0)$

The completed table is shown below.

x	y
-10	-4
0	-2
10	0

9. When $x = 0$,
$y = -1800x + 12,000$
$y = -1800 \cdot 0 + 12,000$
$y = 0 + 12,000$
$y = 12,000$

When $x = 1$,
$y = -1800x + 12,000$
$y = -1800 \cdot 1 + 12,000$
$y = -1800 + 12,000$
$y = 10,200$

When $x = 2$,
$y = -1800x + 12,000$
$y = -1800 \cdot 2 + 12,000$
$y = -3600 + 12,000$
$y = 8400$

When $x = 3$,
$y = -1800x + 12,000$
$y = -1800 \cdot 3 + 12,000$
$y = -5400 + 12,000$
$y = 6600$

When $x = 4$,
$y = -1800x + 12,000$
$y = -1800 \cdot 4 + 12,000$
$y = -7200 + 12,000$
$y = 4800$

The completed table is shown below.

x	0	1	2	3	4
y	12,000	10,200	8400	6600	4800

Vocabulary, Readiness & Video Check 3.1

1. The horizontal axis is called the <u>*x*-axis</u> and the vertical axis is called the <u>*y*-axis</u>.

2. The intersection of the horizontal axis and the vertical axis is a point called the <u>origin</u>.

3. The axes divide the plane into regions called <u>quadrants</u>. There are <u>four</u> of these regions.

4. In the ordered pair of numbers (−2, 5), the number −2 is called the <u>*x*-coordinate</u> and the number 5 is called the <u>*y*-coordinate</u>.

5. Each ordered pair of numbers corresponds to <u>one</u> point in the plane.

6. An ordered pair is a <u>solution</u> of an equation in two variables if replacing the variables by the coordinates of the ordered pair results in a true statement.

7. horizontal: top tourist countries; vertical: number of arrivals (in millions) to these countries

8. Origin; left or right; up or down

9. Data occurring in pairs of numbers can be written as ordered pairs, called paired data, and then graphed on a coordinate system.

10. (7, 0) and (0, 7); since one of these points is a solution and one is not, it shows that it is very important to remember that the first number is the *x*-value and the second number is the *y*-value and not to mix them up.

11. a linear equation in one variable

Exercise Set 3.1

1. The tallest bar corresponds to France, so France is the most popular tourist destination.

3. Find the bars that have heights greater than 40. France, U.S., Spain, Italy, and China have more than 40 million tourists per year.

5. The height of the bar is near 43, so approximately 43 million tourists go to Italy each year.

7. The line is at about 71,000 for year 2009. Thus, the attendance was 71,000 in 2009.

9. The highest point corresponds to 2011, and is at a height of about 103,000.

11. From 2000 on the year axis, we move vertically up to the point on the line graph. When we move horizontally to the vertical axis. The number of students per teacher was approximately 15.9 in 2002.

13. The number of students per teacher shows the greatest decrease between 1998 and 2000. Notice that the line graph is steepest between 1998 and 2000.

15. The points on the line graph for 1998 through 2012 lie above the horizontal line at 15 on the vertical axis. The point for 2014 is the first that lies below this horizontal line. The first year shown that the number of students per teacher fell below 15 was 2014.

17. a. Point (1, 5) lies in quadrant I.

    b. Point (−5, −2) lies in quadrant III.

    c. Point (−3, 0) lies on the *x*-axis, so it is not in any quadrant.

    d. Point (0, −1) lies on the *y*-axis, so it is not in any quadrant.

    e. Point (2, −4) lies in quadrant IV.

    f. Point $\left(-1, 4\frac{1}{2}\right)$ lies in quadrant II.

    g. (3.7, 2.2) lies in quadrant I.

    h. Point $\left(\frac{1}{2}, -3\right)$ lies in quadrant IV.

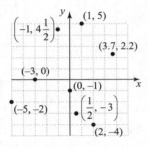

19. Point *A* lies at the origin. Its coordinates are given by the ordered pair (0, 0).

21. Point *C* lies three units to the right and two units above the origin. Its coordinates are given by the ordered pair (3, 2).

23. Point *E* lies two units to the left and two units below the origin. Its coordinates are given by the ordered pair (−2, −2).

25. Point *G* lies two units to the right and one unit below the origin. Its coordinates are given by the ordered pair (2, −1).

27. Point *B* lies on the *y*-axis three units below the origin. Its coordinates are given by the ordered pair (0, −3).

29. Point *D* lies one unit to the right and three units above the origin. Its coordinates are given by the ordered pair (1, 3).

31. Point *F* lies three units to the left and one unit below the origin. Its coordinates are given by the ordered pair (−3, −1).

33. a. The ordered pairs are (2006, 25.5), (2007, 26.3), (2008, 27.7), (2009, 29.4) and (2010, 31.8).

    b. The ordered pair (2010, 31.8) indicates that the worldwide box office in 2010 was $31.8 billion.

    c.

    d. The worldwide box office increased every year.

35. a. The ordered pairs are (0.50, 10), (0.75, 12), (1.00, 15), (1.25, 16), (1.50, 18), (1.50, 19), (1.75, 19), and (2.00, 20).

    b. The ordered pair (1.25, 16) indicates that when Minh studied 1.25 hours, her quiz score was 16.

    c.

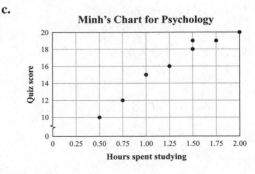

    d. answers may vary

37. a. The ordered pairs are (2313, 2), (2085, 1), (2711, 21), (2869, 39), (2920, 42), (4038, 99), (1783, 0), and (2493, 9).

b. We plot the ordered pairs. We label the horizontal axis "Distance from Equator (in miles)" and the vertical axis "Average Annual Snowfall (in inches)."

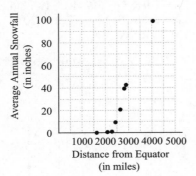

c. The farther from the equator, the more snowfall.

39. For (3, 1), let $x = 3$ and $y = 1$.
$$2x + y = 7$$
$$2(3) + 1 = 7$$
$$6 + 1 = 7$$
$$7 = 7 \quad \text{true}$$
Yes, (3, 1) is a solution.

For (7, 0), let $x = 7$ and $y = 0$.
$$2x + y = 7$$
$$2(7) + 0 = 7$$
$$14 + 0 = 7$$
$$14 = 7 \quad \text{false}$$
No, (7, 0) is not a solution.

For (0, 7), let $x = 0$ and $y = 7$.
$$2x + y = 7$$
$$2(0) + 7 = 7$$
$$0 + 7 = 7$$
$$7 = 7 \quad \text{true}$$
Yes, (0, 7) is a solution.

41. For (0, 0), let $x = 0$ and $y = 0$.
$$x = -\frac{1}{3}y$$
$$0 = -\frac{1}{3}(0)$$
$$0 = 0 \quad \text{true}$$
Yes, (0, 0) is a solution.

For (3, −9), let $x = 3$ and $y = -9$.

$$x = -\frac{1}{3}y$$
$$3 = -\frac{1}{3}(-9)$$
$$3 = 3 \quad \text{true}$$
Yes, (3, −9) is a solution.

43. For (4, 5), let $x = 4$ and $y = 5$.
$$x = 5$$
$$4 = 5 \quad \text{false}$$
No, (4, 5) is not a solution.

For (5, 4), let $x = 5$ and $y = 4$.
$$x = 5$$
$$5 = 5 \quad \text{true}$$
Yes, (5, 4) is a solution.

For (5, 0), let $x = 5$ and $y = 0$.
$$x = 5$$
$$5 = 5 \quad \text{true}$$
Yes, (5, 0) is a solution.

45. Replace y with −2 and solve for x.
$$x - 4y = 4$$
$$x - 4(-2) = 4$$
$$x + 8 = 4$$
$$x = -4$$
The ordered pair is (−4, −2).

Replace x with 4 and solve for y.
$$x - 4y = 4$$
$$4 - 4y = 4$$
$$-4y = 0$$
$$y = 0$$
The ordered pair is (4, 0).

47. Replace x with −8 and solve for y.
$$y = \frac{1}{4}x - 3$$
$$y = \frac{1}{4}(-8) - 3$$
$$y = -2 - 3$$
$$y = -5$$
The ordered pair is (−8, −5).

Replace y with 1 and solve for x.

$$y = \frac{1}{4}x - 3$$

$$1 = \frac{1}{4}x - 3$$

$$4 = \frac{1}{4}x$$

$$16 = x$$

The ordered pair is (16, 1).

49. Replace x with 0 and solve for y.

$$y = -7x$$
$$y = -7(0)$$
$$y = 0$$

The ordered pair is (0, 0).

Replace x with −1 and solve for y.

$$y = -7x$$
$$y = -7(-1)$$
$$y = 7$$

The ordered pair is (−1, 7).

Replace y with 2 and solve for x.

$$y = -7x$$
$$2 = -7x$$
$$-\frac{2}{7} = x$$

The ordered pair is $\left(-\frac{2}{7}, 2\right)$.

The completed table is shown below.

x	y
0	0
−1	7
$-\frac{2}{7}$	2

51. Replace x with 0 and solve for y.

$$y = -x + 2$$
$$y = -0 + 2$$
$$y = 2$$

The ordered pair is (0, 2).

Replace y with 0 and solve for x.

$$y = -x + 2$$
$$0 = -x + 2$$
$$x = 2$$

The ordered pair is (2, 0).

Replace x with −3 and solve for y.

$$y = -x + 2$$
$$y = -(-3) + 2$$
$$y = 3 + 2$$
$$y = 5$$

The ordered pair is (−3, 5).
The completed table is shown below.

x	y
0	2
2	0
−3	5

53. Replace x with 0 and solve for y.

$$y = \frac{1}{2}x$$

$$y = \frac{1}{2}(0)$$

$$y = 0$$

The ordered pair is (0, 0).

Replace x with −6 and solve for y.

$$y = \frac{1}{2}x$$

$$y = \frac{1}{2}(-6)$$

$$y = -3$$

The ordered pair is (−6, −3).

Replace y with 1 and solve for x.

$$y = \frac{1}{2}x$$

$$1 = \frac{1}{2}x$$

$$2 = x$$

The ordered pair is (2, 1).
The completed table is shown below.

x	y
0	0
−6	−3
2	1

55. Replace x with 0 and solve for y.

$$x + 3y = 6$$
$$0 + 3y = 6$$
$$3y = 6$$
$$y = 2$$

The ordered pair is (0, 2).

Replace y with 0 and solve for x.

$$x + 3y = 6$$
$$x + 3(0) = 6$$
$$x + 0 = 6$$
$$x = 6$$
The ordered pair is (6, 0).

Replace y with 1 and solve for x.
$$x + 3y = 6$$
$$x + 3(1) = 6$$
$$x + 3 = 6$$
$$x = 3$$
The ordered pair is (3, 1).
The completed table is shown below.

x	y
0	2
6	0
3	1

57. Replace x with 0 and solve for y.
$$y = 2x - 12$$
$$y = 2(0) - 12$$
$$y = 0 - 12$$
$$y = -12$$
The ordered pair is (0, −12).

Replace y with −2 and solve for x.
$$y = 2x - 12$$
$$-2 = 2x - 12$$
$$10 = 2x$$
$$5 = x$$
The ordered pair is (5, −2).

Replace x with 3 and solve for y.
$$y = 2x - 12$$
$$y = 2(3) - 12$$
$$y = 6 - 12$$
$$y = -6$$
The ordered pair is (3, −6).
The completed table is shown below.

x	y
0	−12
5	−2
3	−6

59. Replace x with 0 and solve for y.
$$2x + 7y = 5$$
$$2(0) + 7y = 5$$
$$7y = 5$$
$$y = \frac{5}{7}$$
The ordered pair is $\left(0, \frac{5}{7}\right)$.

Replace y with 0 and solve for x.
$$2x + 7y = 5$$
$$2x + 7(0) = 5$$
$$2x = 5$$
$$x = \frac{5}{2}$$
The ordered pair is $\left(\frac{5}{2}, 0\right)$.

Replace y with 1 and solve for x.
$$2x + 7y = 5$$
$$2x + 7(1) = 5$$
$$2x + 7 = 5$$
$$2x = -2$$
$$x = -1$$
The ordered pair is (−1, 1).
The completed table is shown below.

x	y
0	$\frac{5}{7}$
$\frac{5}{2}$	0
−1	1

61. Replace y with 0 and solve for x.
$$x = -5y$$
$$x = -5(0)$$
$$x = 0$$
The ordered pair is (0, 0).

Replace y with 1 and solve for x.
$$x = -5y$$
$$x = -5(1)$$
$$x = -5$$
The ordered pair is (−5, 1).

Replace x with 10 and solve for y.
$$x = -5y$$
$$10 = -5y$$
$$-2 = y$$

The ordered pair is (10, −2).
The completed table is shown below.

x	y
0	0
−5	1
10	−2

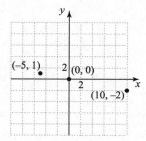

63. Replace x with 0 and solve for y.

$$y = \frac{1}{3}x + 2$$
$$y = \frac{1}{3}(0) + 2$$
$$y = 2$$

The ordered pair is (0, 2).

Replace x with −3 and solve for y.

$$y = \frac{1}{3}x + 2$$
$$y = \frac{1}{3}(-3) + 2$$
$$y = -1 + 2$$
$$y = 1$$

The ordered pair is (−3, 1).

Replace y with 0 and solve for x.

$$y = \frac{1}{3}x + 2$$
$$0 = \frac{1}{3}x + 2$$
$$-\frac{1}{3}x = 2$$
$$x = -6$$

The ordered pair is (−6, 0).

The completed table is shown below.

x	y
0	2
−3	1
−6	0

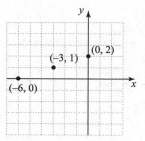

65. a. When $x = 100$,
$y = 80x + 5000$
$y = 80(100) + 5000$
$y = 8000 + 5000$
$y = 13,000$

When $x = 200$,
$y = 80x + 5000$
$y = 80(200) + 5000$
$y = 16,000 + 5000$
$y = 21,000$

When $x = 300$,
$y = 80x + 5000$
$y = 80(300) + 5000$
$y = 24,000 + 5000$
$y = 29,000$

The completed table is shown below.

x	100	200	300
y	13,000	21,000	29,000

b. Replace y with 8600 and solve for x.
$$y = 80x + 5000$$
$$8600 = 80x + 5000$$
$$3600 = 80x$$
$$45 = x$$

Thus, 45 computer desks can be produced for $8600.

67. a.

x	1	3	5
$y = 0.24x + 5.28$	$\begin{aligned} &0.24(1) + 5.28 \\ &= 0.24 + 5.28 \\ &= 5.52 \end{aligned}$	$\begin{aligned} &0.24(3) + 5.28 \\ &= 0.72 + 5.28 \\ &= 6.00 \end{aligned}$	$\begin{aligned} &0.24(5) + 5.28 \\ &= 1.2 + 5.28 \\ &= 6.48 \end{aligned}$

b. Find x when $y = 7.50$.
$$750 = 0.24x + 5.28$$
$$2.22 = 0.24x$$
$$9 \approx x$$
$$2000 + 9 = 2009$$
The average cinema admission price was \$7.50 in year 9 or 2009.

c. Find x when $y = 9.00$.
$$9.00 = 0.24x + 5.28$$
$$3.72 = 0.24x$$
$$15.5 = x$$
$$16 \approx x$$
$$2000 + 16 = 2016$$
The average cinema admission price is predicted to be \$9.00 in year 16 or 2016.

d. $(5, 6.48)$ means that in 2005, the average cinema price was \$6.48.

69. Ten years after 2000, or in 2010, there were 3755 Walmart stores.

71. In year 7, there appear to be approximately 3450 Walmart stores. In year 8 there appear to be approximately 3550 Walmart stores. The increase is approximately $3550 - 3450 = 100$ stores.

In year 8, there appear to be approximately 3550 Walmart stores. In year 9, there appear to be approximately 3655 Walmart stores. The increase is approximately $3655 - 3550 = 105$ stores.

In year 9, there appear to be approximately 3655 Walmart stores. In year 10, there appear to be approximately 3755 Walmart stores. The increase is approximately $3755 - 3655 = 100$ stores.

73. The coordinates of points whose graphs lie on the x-axis all have y-values of 0.

75. Subtract x from each side.
$$x + y = 5$$
$$y = 5 - x$$

77. Subtract $2x$ from each side. Then divide each side by 4.
$$2x + 4y = 5$$
$$4y = -2x + 5$$
$$y = -\frac{1}{2}x + \frac{5}{4}$$

79. Divide each side by -5.
$$10x = -5y$$
$$-2x = y$$
$$y = -2x$$

81. Subtract x from each side. Then divide each side by -3.
$$x - 3y = 6$$
$$-3y = -x + 6$$
$$y = \frac{1}{3}x - 2$$

83. False; the point $(-1, 5)$ lies in quadrant II.

85. True

87. In quadrant III, both coordinates are negative: (negative, negative).

89. In quadrant IV, the x-coordinate is positive and the y-coordinate is negative: (positive, negative).

91. At the origin, both coordinates are zero: $(0, 0)$.

93. A point of the form (0, number) is located on the y-axis.

95. No; answers may vary.

97. Answers may vary

99. The point four units to the right of the y-axis and seven units below the x-axis has ordered pair $(4, -7)$.

101. The length of the rectangle is $3 - (-1) = 4$ and the width of the rectangle is $5 - (-4) = 9$.
Perimeter $= 2(\text{length}) + 2(\text{width})$
$$= 2(4) + 2(9)$$
$$= 8 + 18$$
$$= 26$$
The perimeter is 26 units.

103.
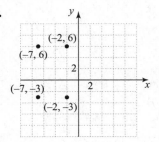

a. The fourth vertex is $(-2, 6)$. The rectangle is 9 units by 5 units.

b. The perimeter is $9 + 5 + 9 + 5 = 28$ units.

c. The area is $9 \times 5 = 45$ square units.

Section 3.2 Practice

1. a. $3x + 2.7y = -5.3$ is a linear equation in two variables because it is written in the form $Ax + By = C$ with $A = 3$, $B = 2.7$, and $C = -5.3$.

b. $x^2 + y = 8$ is not a linear equation in two variables because x is squared.

c. $y = 12$ is a linear equation in two variables because it can be written in the form $Ax + By = C$: $0x + y = 12$.

d. $5x = -3y$ is a linear equation in two variables because it can be written in the form $Ax + By = C$: $5x + 3y = 0$.

2. Find three ordered pair solutions.
Let $x = 0$.
$$x + 3y = 9$$
$$0 + 3y = 9$$
$$3y = 9$$
$$y = 3$$

Let $x = 3$.
$$x + 3y = 9$$
$$3 + 3y = 9$$
$$3y = 6$$
$$y = 2$$

Let $y = 1$.
$$x + 3y = 9$$
$$x + 3(1) = 9$$
$$x + 3 = 9$$
$$x = 6$$
The ordered pairs are $(0, 3)$, $(3, 2)$, and $(6, 1)$.

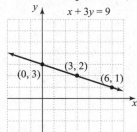

3. Find three ordered pair solutions.
Let $x = 0$.
$$3x - 4y = 12$$
$$3(0) - 4y = 12$$
$$-4y = 12$$
$$y = -3$$

Let $y = 0$.
$$3x - 4y = 12$$
$$3x - 4(0) = 12$$
$$3x = 12$$
$$x = 4$$

Let $x = 2$.
$$3x - 4y = 12$$
$$3(2) - 4y = 12$$
$$6 - 4y = 12$$
$$-4y = 6$$
$$y = -\frac{6}{4} = -\frac{3}{2}$$

The ordered pairs are $(0, -3)$, $(4, 0)$, and $\left(2, -\frac{3}{2}\right)$.

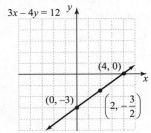

4. Find three ordered pair solutions.
If $x = 1$, $y = -2(1) = -2$.
If $x = 0$, $y = -2(0) = 0$.
If $x = -1$, $y = -2(-1) = 2$.

x	y
1	−2
0	0
−1	2

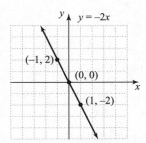

5. Find three ordered pair solutions.

If $x = 2$, $y = \frac{1}{2}(2) + 3 = 1 + 3 = 4$.

If $x = 0$, $y = \frac{1}{2}(0) + 3 = 0 + 3 = 3$.

If $x = -4$, $y = \frac{1}{2}(-4) + 3 = -2 + 3 = 1$.

x	y
2	4
0	3
−4	1

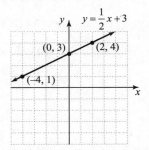

6. The equation $x = -2$ can be written in standard form as $x + 0y = -2$. No matter what value replaces y, x is always -2. It is a vertical line. Plot points $(-2, 2)$, $(-2, 0)$, and $(-2, -4)$, for example.

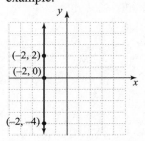

7. Find three ordered pair solutions.
If $x = 1$, $y = -2(1) + 3 = -2 + 3 = 1$.
If $x = 0$, $y = -2(0) + 3 = 0 + 3 = 3$.
If $x = 3$, $y = -2(3) + 3 = -6 + 3 = -3$.

x	y
1	1
0	0
3	−3

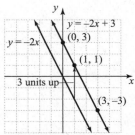

The graph of $y = -2x + 3$ is the same as the graph of $y = -2x$ except that the graph of $y = -2x + 3$ is moved three units upward.

8. **a.**   Find three ordered pair solutions.
             If $x = 0$, $y = 17.5(0) + 515 = 515$.
             If $x = 6$,
             $y = 17.5(6) + 515 = 620$.
             If $x = 10$,
             $y = 17.5(10) + 515 = 690$.

x	y
0	515
6	620
10	690

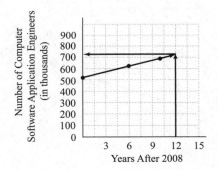

   b. $2020 - 2008 = 12$
            The graph shows that we predict approximately 725 thousand computer software application engineers in the year 2020.

Graphing Calculator Explorations

1. $y = -3x + 7$

2. $y = -x + 5$

3. $y = 2.5x - 7.9$

4. $y = -1.3x + 5.2$

5. $y = -\dfrac{3}{10}x + \dfrac{32}{5}$

6. $y = \dfrac{2}{9}x - \dfrac{22}{3}$

Vocabulary, Readiness & Video Check 3.2

1. In the definition, x and y both have an understood power of 1. Example 3 shows an equation where y has a power of 2, so it is not a linear equation in two variables.

2. Find 3 points in order to check your work. Make sure the points lie along one straight line—if not, an algebraic mistake was probably made.

3. An infinite number of points make up the line and each point corresponds to an ordered pair that is a solution of the linear equation in two variables.

Exercise Set 3.2

1. Yes; it can be written in the form $Ax + By = C$.

3. Yes; it can be written in the form $Ax + By = C$.

5. No; x is squared.

7. Yes; it can be written in the form $Ax + By = C$.

9. Let $y = 0$.
$$x - y = 6$$
$$x - 0 = 6$$
$$x = 6$$

Let $x = 4$.   Let $y = -1$
$$x - y = 6$$ $$x - y = 6$$
$$4 - y = 6$$ $$x - (-1) = 6$$
$$-y = 2$$ $$x + 1 = 6$$
$$y = -2$$ $$x = 5$$

x	y
6	0
4	-2
5	-1

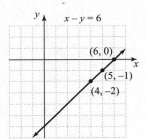

11. $y = -4x$
If $x = 1$, $y = -4(1) = -4$.
If $x = 0$, $y = -4(0) = 0$.
If $x = -1$, $y = -4(-1) = 4$.

x	y
1	-4
0	0
-1	4

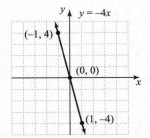

13. $y = \dfrac{1}{3}x$

If $x = 0$, $y = \dfrac{1}{3}(0) = 0$.

If $x = 6$, $y = \dfrac{1}{3}(6) = 2$.

If $x = -3$, $y = \dfrac{1}{3}(-3) = -1$.

x	y
0	0
6	2
-3	-1

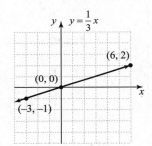

15. $y = -4x + 3$
If $x = 0$, $y = -4(0) + 3 = 0 + 3 = 3$.
If $x = 1$, $y = -4(1) + 3 = -4 + 3 = -1$.
If $x = 2$, $y = -4(2) + 3 = -8 + 3 = -5$.

x	y
0	3
1	-1
2	-5

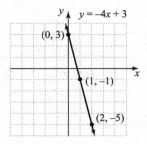

17. $x + y = 1$

x	y
0	1
1	0
2	−1

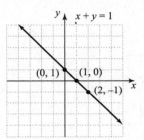

19. $x − y = −2$

x	y
−2	0
0	2
2	4

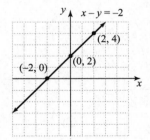

21. $x − 2y = 6$

x	y
−4	−5
0	−3
4	−1

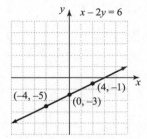

23. $y = 6x + 3$

x	y
−1	−3
0	3
1	9

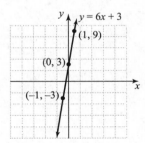

25. $x = −4$

x	y
−4	−1
−4	0
−4	2

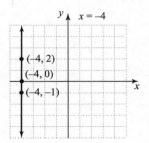

27. $y = 3$

x	y
-1	3
0	3
2	3

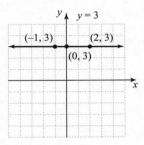

29. $y = x$

x	y
-1	-1
0	0
2	2

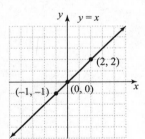

31. $x = -3y$

x	y
-6	2
0	0
6	-2

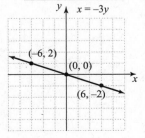

33. $x + 3y = 9$

x	y
-9	6
0	3
3	2

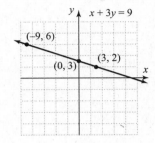

35. $y = \frac{1}{2}x + 2$

x	y
-4	0
0	2
4	4

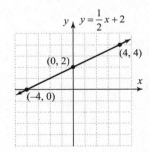

37. $3x - 2y = 12$

x	y
0	-6
2	-3
4	0

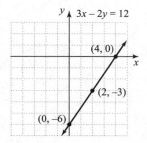

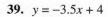

39. $y = -3.5x + 4$

x	y
0	4
1	0.5
2	-3

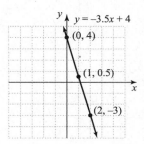

41.

$y = 5x$	
x	y
-1	-5
0	0
1	5

$y = 5x + 4$	
x	y
-1	-1
0	4
1	9

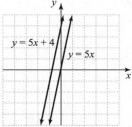

Answers may vary; possible answer: The graph of $y = 5x + 4$ is the same as the graph of $y = 5x$ except it is moved 4 units upward.

43.

$y = -2x$	
x	y
-2	4
0	0
2	-4

$y = -2x - 3$	
x	y
-2	1
0	-3
2	-7

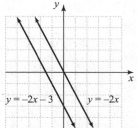

Answers may vary; possible answer: The graph of $y = -2x - 3$ is the same as the graph of $y = -2x$ except it is moved 3 units downward.

45.

$y = \frac{1}{2}x$	
x	y
-4	-2
0	0
4	2

$y = \frac{1}{2}x + 2$	
x	y
-4	0
0	2
4	4

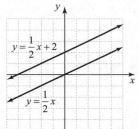

Answers may vary; possible answer: The graph of $y = \frac{1}{2}x + 2$ is the same as the graph of $y = \frac{1}{2}x$ except it is moved 2 units upward.

47. Comparing $y = 5x + 5$ to $y = mx + b$, we see that $b = 5$. We see that graph c crosses the y-axis at $(0, 5)$.

49. Comparing $y = 5x - 1$ to $y = mx + b$, we see that $b = -1$. We see that graph d crosses the y-axis at $(0, -1)$.

51. a. Using the equation, let $x = 8$.
$y = x + 23$
$y = 8 + 23 = 31$
The ordered pair is (8, 31).

b. Eight years after 2000, in 2008, there were 31 million joggers.

c. The year 2017 is 17 years after 2000, so let $x = 17$.
$y = x + 23$
$y = 17 + 23 = 40$
If the trend continues, there will be 40 million joggers in 2017.

53. a. $y = 2.2x + 190$
Let $x = 8$.
$y = 2.2(8) + 190 = 17.6 + 190 = 207.6$
The ordered pair is (8, 207.6).

b. In 2008, there were 207.6 million people with driver's licenses.

c. Let $x = 2016 - 2000 = 16$.
$y = 2.2(16) + 190 = 35.2 + 190 = 225.2$
In 2016, it is predicted that there will be 225.2 million people with driver's licenses.

55.

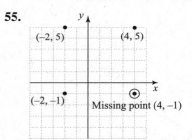

The fourth vertex is at (4, −1).

57. $x - y = -3$
$x = 0$: $0 - y = -3$
$\qquad\qquad y = 3$
$y = 0$: $x - 0 = -3$
$\qquad\qquad x = -3$

x	y
0	3
−3	0

59. $y = 2x$
$x = 0$: $y = 2(0) = 0$
$y = 0$: $0 = 2x$
$\qquad\quad 0 = x$

x	y
0	0
0	0

61. The equation is $y = x + 5$.

x	y
−2	3
0	5
2	7

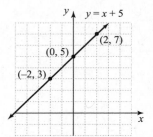

63. The equation is $2x + 3y = 6$.

x	y
3	0
0	2
−3	4

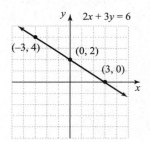

65. $x + y + 5 + 5 = 22$
$\quad x + y + 10 = 22$
$\qquad\quad x + y = 12$
Let $x = 3$.
$3 + y = 12$
$\quad\;\; y = 9$ centimeters

67. answers may vary

69. $y = x^2$

x	y
0	0
1	1
−1	1
2	4
−2	4

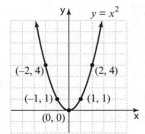

Section 3.3 Practice

1. The graph crosses the x-axis at the point (−4, 0). The x-intercept is (−4, 0). The graph crosses the y-axis at the point (0, −6). The y-intercept is (0, −6).

2. The graph crosses the x-axis at the point (−2, 0) and at the point (2, 0). The x-intercepts are (−2, 0) and (2, 0). The graph crosses the y-axis at the point (0, −3). The y-intercept is (0, −3).

3. The graph crosses both the x-axis and the y-axis at the point (0, 0). The x-intercept is (0, 0), and the y-intercept is (0, 0).

4. The graph does not cross the x-axis. There is no x-intercept. The graph crosses the y-axis at the point (0, 3). The y-intercept is (0, 3).

5. The graph crosses the x-axis at the point (−1, 0) and at the point (5, 0). The x-intercepts are (−1, 0) and (5, 0). The graph crosses the y-axis at the point (0, −2) and at the point (0, 2). The y-intercepts are (0, −2) and (0, 2).

6. Let $y = 0$.   Let $x = 0$.
   $x + 2y = -4$ $x + 2y = -4$
   $x + 2(0) = -4$ $0 + 2y = -4$
   $x + 0 = -4$ $2y = -4$
   $x = -4$ $y = -2$

   The x-intercept is (−4, 0), and the y-intercept is (0, −2).
   Let $x = 2$.
   $x + 2y = -4$
   $2 + 2y = -4$
   $2y = -6$
   $y = -3$

x	y
−4	0
0	−2
2	−3

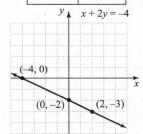

7. Let $y = 0$.   Let $x = 0$.
   $x = 3y$ $x = 3y$
   $x = 3(0)$ $0 = 3y$
   $x = 0$ $0 = y$

   Both the x-intercept and the y-intercept are (0, 0).
   Let $y = -1$ Let $y = 1$.
   $x = 3(-1)$ $x = 3(1)$
   $x = -3$ $x = 3$

x	y
0	0
3	1
−3	−1

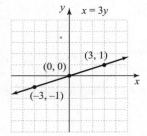

8. Let $y = 0$.   Let $x = 0$.
$\quad 3x = 2y + 4$ $\quad 3x = 2y + 4$
$\quad 3x = 2(0) + 4$ $\quad 3(0) = 2y + 4$
$\quad 3x = 4$ $\quad -4 = 2y$
$\quad x = \dfrac{4}{3}$ $\quad -2 = y$

Let $x = 2$.
$\quad 3x = 2y + 4$
$\quad 3(2) = 2y + 4$
$\quad 6 = 2y + 4$
$\quad 2 = 2y$
$\quad 1 = y$

x	y
0	−2
$\dfrac{4}{3}$	0
2	1

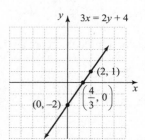

9. For any y-value chosen, notice that x is -2.

x	y
−2	−4
−2	0
−2	4

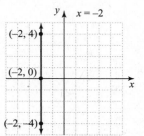

10. For any x-value chosen, notice that y is 2.

x	y
−5	2
0	2
5	2

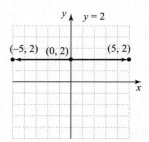

Graphing Calculator Explorations

1. $x = 3.78y$

$\quad y = \dfrac{x}{3.78}$

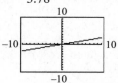

2. $-2.61y = x$

$\quad y = \dfrac{x}{-2.61}$

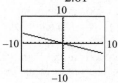

3. $3x + 7y = 21$

$7y = -3x + 21$

$y = -\dfrac{3}{7}x + 3$

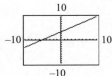

4. $-4x + 6y = 21$

$6y = 4x + 21$

$y = \dfrac{2}{3}x + \dfrac{7}{2}$

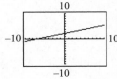

5. $-2.2x + 6.8y = 15.5$

$6.8y = 2.2x + 15.5$

$y = \dfrac{2.2}{6.8}x + \dfrac{15.5}{6.8}$

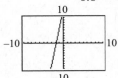

6. $5.9x - 0.8y = -10.4$

$-0.8y = -5.9x - 10.4$

$y = \dfrac{5.9}{0.8}x + \dfrac{10.4}{0.8}$

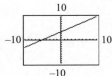

Vocabulary, Readiness & Video Check 3.3

1. An equation that can be written in the form $Ax + By = C$ is called a <u>linear</u> equation in two variables.

2. The form $Ax + By = C$ is called <u>standard</u> form.

3. The graph of the equation $y = -1$ is a <u>horizontal</u> line.

4. The graph of the equation $x = 5$ is a <u>vertical</u> line.

5. A point where a graph crosses the *y*-axis is called a <u>*y*-intercept</u>.

6. A point where a graph crosses the *x*-axis is called an <u>*x*-intercept</u>.

7. Given an equation of a line, to find the *x*-intercept (if there is one), let <u>*y*</u> = 0 and solve for <u>*x*</u>.

8. Given an equation of a line, to find the *y*-intercept (if there is one), let <u>*x*</u> = 0 and solve for <u>*y*</u>.

9. Because *x*-intercepts lie on the *x*-axis; because *y*-intercepts lie on the *y*-axis.

10. Using a third point as a check that your points lie along a straight line is always good practice.

11. For a horizontal line, the coefficient of *x* will be 0 and the coefficient of *y* will be 1; for a vertical line, the coefficient of *y* will be 0 and the coefficient of *x* will be 1.

Exercise Set 3.3

1. *x*-intercept: (−1, 0); *y*-intercept: (0, 1)

3. *x*-intercept: (−2, 0), (2, 0)

5. *x*-intercepts: (−2, 0), (1, 0), (3, 0)
y-intercept: (0, 3)

7. *x*-intercepts: (−1, 0), (1, 0)
y-intercepts: (0, 1), (0, −2)

9. Infinite; because the line could be vertical ($x = 0$) or horizontal ($y = 0$).

11. 0; because the circle could completely reside within one quadrant.

13. $x - y = 3$
$y = 0$: $x - 0 = 3$, $x = 3$
$x = 0$: $0 - y = 3$, $y = -3$
x-intercept: (3, 0); *y*-intercept: (0, −3)

x	y
3	0
0	−3

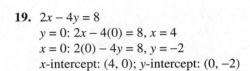

19. $2x - 4y = 8$
$y = 0$: $2x - 4(0) = 8$, $x = 4$
$x = 0$: $2(0) - 4y = 8$, $y = -2$
x-intercept: $(4, 0)$; y-intercept: $(0, -2)$

x	y
4	0
0	-2

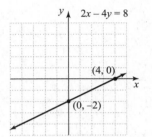

15. $x = 5y$
$y = 0$: $x = 5(0) = 0$
$x = 0$: $0 = 5y$, $y = 0$
x-intercept: $(0, 0)$; y-intercept: $(0, 0)$
$y = 1$: $x = 5(1) = 5$

x	y
0	0
5	1

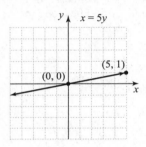

21. $y = 2x$
$y = 0$: $0 = 2x$, $0 = x$
$x = 0$: $y = 2(0)$, $y = 0$
x-intercept: $(0, 0)$; y-intercept: $(0, 0)$
$x = 1$: $y = 2(1)$, $y = 2$

x	y
0	0
1	2

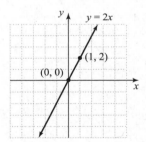

17. $-x + 2y = 6$
$y = 0$: $-x + 2(0) = 6$, $x = -6$
$x = 0$: $-0 + 2y = 6$, $y = 3$
x-intercept: $(-6, 0)$; y-intercept: $(0, 3)$

x	y
-6	0
0	3

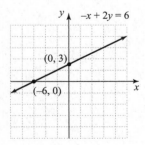

23. $y = 3x + 6$
$y = 0$: $0 = 3x + 6$, $-6 = 3x$, $-2 = x$
$x = 0$: $y = 3(0) + 6$, $y = 6$
x-intercept: $(-2, 0)$; y-intercept: $(0, 6)$

x	y
-2	0
0	6

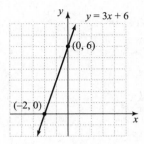

25. $x = -1$ for all values of y.

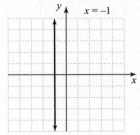

27. $y = 0$ for all values of x.

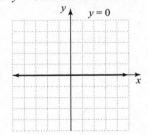

29. $y + 7 = 0$

$y = -7$ for all values of x.

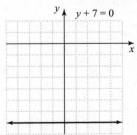

31. $x + 3 = 0$; $x = -3$ for all values of y.

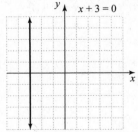

33. $x = y$

x-intercept: $(0, 0)$; y-intercept: $(0, 0)$

Second point: $(4, 4)$

x	y
4	4
0	0

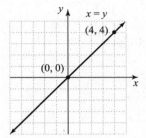

35. $x + 8y = 8$

x-intercept: $(8, 0)$; y-intercept: $(0, 1)$

x	y
8	0
0	1

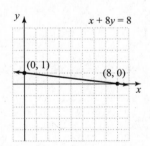

37. $5 = 6x - y$

x-intercept: $\left(\dfrac{5}{6}, 0\right)$; y-intercept: $(0, -5)$

x	y
$\dfrac{5}{6}$	0
0	-5

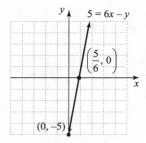

39. $-x + 10y = 11$

x-intercept: $(-11, 0)$; y-intercept: $\left(0, \dfrac{11}{10}\right)$

x	y
−11	0
0	$\dfrac{11}{10}$

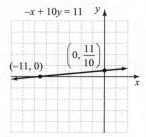

41. $x = -4\dfrac{1}{2}$ for all values of y.

x	y
$-4\dfrac{1}{2}$	0
$-4\dfrac{1}{2}$	3

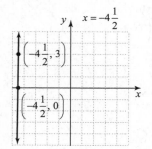

43. $y = 3\dfrac{1}{4}$ for all values of x.

x	y
0	$3\dfrac{1}{4}$
2	$3\dfrac{1}{4}$

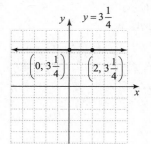

45. $y = -\dfrac{2}{3}x + 1$

x-intercept: $\left(\dfrac{3}{2}, 0\right)$; y-intercept: $(0, 1)$

x	y
$\dfrac{3}{2}$	0
0	1

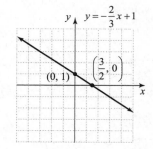

47. $4x - 6y + 2 = 0$

x-intercept: $\left(-\dfrac{1}{2}, 0\right)$; y-intercept: $\left(0, \dfrac{1}{3}\right)$

x	y
$-\dfrac{1}{2}$	0
0	$\dfrac{1}{3}$

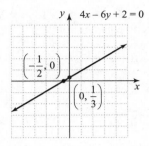

49. $y = 3$

The graph is a horizontal line with y-intercept $(0, 3)$.

C

51. $x = -1$

The graph is a vertical line with x-intercept $(-1, 0)$.

E

53. $y = 2x + 3$

The y-intercept is $(0, 3)$ and the x-intercept is $\left(-\dfrac{3}{2}, 0\right)$.

B

55. $\dfrac{-6-3}{2-8} = \dfrac{-9}{-6} = \dfrac{3}{2}$

57. $\dfrac{-8-(-2)}{-3-(-2)} = \dfrac{-6}{-1} = 6$

59. $\dfrac{0-6}{5-0} = \dfrac{-6}{5} = -\dfrac{6}{5}$

61. False; for example, the horizontal line $y = 2$ does not have an x-intercept.

63. True

65. $3x + 6y = 1200$

$x = 0: \ 3(0) + 6y = 1200$
$$6y = 1200$$
$$y = 200$$

The ordered pair $(0, 200)$ corresponds to manufacturing 0 chairs and 200 desks.

67. Manufacturing 50 desks corresponds to $y = 50$.

$3x + 6y = 1200$

$y = 50: \ 3x + 6(50) = 1200$
$$3x + 300 = 1200$$
$$3x = 900$$
$$x = 300$$

When 50 desks are manufactured, 300 chairs can be manufactured.

69. $y = -0.025x + 1.55$

a. $y = 0: \qquad 0 = -0.025x + 1.55$
$$0.025x = 1.55$$
$$x = 62$$

The x-intercept is $(62, 0)$.

b. 62 years after 2002 (2064); 0 people will attend movies at the theatre.

c. answers may vary

71. Parallel to $x = 5$ is vertical.
x-intercept is $(1, 0)$, so $x = 1$ for all values of y.
The equation is $x = 1$.

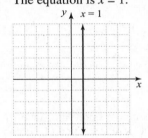

73. answers may vary

75. answers may vary

Section 3.4 Practice

1. If we let (x_1, y_1) be $(-4, 11)$, then $x_1 = -4$ and $y_1 = 11$. Also, let (x_2, y_2) be $(2, 5)$ so that $x_2 = 2$ and $y_2 = 5$.

$$m = \frac{y_2 - y_1}{x_2 - x_1} = \frac{5 - 11}{2 - (-4)} = \frac{-6}{6} = -1$$

The slope of the line is -1.

2. Let (x_1, y_1) be $(3, 1)$ and (x_2, y_2) be $(-3, -1)$.

$$m = \frac{y_2 - y_1}{x_2 - x_1} = \frac{-1 - 1}{-3 - 3} = \frac{-2}{-6} = \frac{1}{3}$$

3. $y = \dfrac{2}{3}x - 2$

   The equation is in slope-intercept form,

   $y = mx + b$. The coefficient of x, $\dfrac{2}{3}$, is the slope.

   The constant term, -2, is the y-value of the y-intercept, $(0, -2)$.

4. Write the equation in slope-intercept form by solving the equation for y.
   $$-y = -6x + 5$$
   $$\dfrac{-y}{-1} = \dfrac{-6x}{-1} + \dfrac{5}{-1}$$
   $$y = 6x - 5$$
   The coefficient of x, 6, is the slope. The constant term, -5, is the y-value of the y-intercept, $(0, -5)$.

5. Write the equation in slope-intercept form by solving the equation for y.
   $$5x + 2y = 8$$
   $$2y = -5x + 8$$
   $$\dfrac{2y}{2} = \dfrac{-5x}{2} + \dfrac{8}{2}$$
   $$y = -\dfrac{5}{2}x + 4$$

   The coefficient of x, $-\dfrac{5}{2}$, is the slope, and the y-intercept is $(0, 4)$.

6. Recall that $y = 3$ is a horizontal line. Two ordered pair solutions of $y = 3$ and $(1, 3)$ and $(3, 3)$.
   $$m = \dfrac{y_2 - y_1}{x_2 - x_1} = \dfrac{3 - 3}{3 - 1} = \dfrac{0}{2} = 0$$
   The slope of the line $y = 3$ is 0.

7. Recall that the graph of $x = -4$ is a vertical line. Two ordered pair solutions of $x = -4$ and $(-4, 1)$ and $(-4, 3)$.
   $$m = \dfrac{y_2 - y_1}{x_2 - x_1} = \dfrac{3 - 1}{-4 - (-4)} = \dfrac{2}{0}$$
   The slope of the vertical line $x = -4$ is undefined.

8. **a.** The slope of the line $y = -5x + 1$ is -5. We solve the second equation for y.
   $$x - 5y = 10$$
   $$-5y = -x + 10$$
   $$\dfrac{-5y}{-5} = \dfrac{-x}{-5} + \dfrac{10}{-5}$$
   $$y = \dfrac{1}{5}x - 2$$

   The slope of the second line is $\dfrac{1}{5}$. Since the

   product of the slopes is $\dfrac{1}{5}(-5) = -1$, the

   lines are perpendicular.

   b. Solve each equation for y.
   $$x + y = 11 \qquad\qquad 2x + y = 11$$
   $$y = -x + 11 \qquad\qquad y = -2x + 11$$
   The slopes are -1 and -2. The slopes are not the same, and their product is not -1. Thus, the lines are neither parallel nor perpendicular.

   c. Solve each equation for y.
   $$2x + 3y = 21 \qquad\qquad 6y = -4x - 2$$
   $$3y = -2x + 21 \qquad\qquad \dfrac{6y}{6} = \dfrac{-4x}{6} - \dfrac{2}{6}$$
   $$\dfrac{3y}{3} = \dfrac{-2x}{3} + \dfrac{21}{3} \qquad\qquad y = -\dfrac{2}{3}x - \dfrac{1}{3}$$
   $$y = -\dfrac{2}{3}x + 7$$

   The slopes are $-\dfrac{2}{3}$ and $-\dfrac{2}{3}$. Since the lines

   have the same slope and different y-intercepts, they are parallel.

9. $\text{grade} = \dfrac{\text{rise}}{\text{run}} = \dfrac{1794}{7176} = 0.25 = 25\%$

   The grade is 25%.

10. Use $(2, 2)$ and $(6, 5)$ to calculate slope.
    $$m = \dfrac{5 - 2}{6 - 2} = \dfrac{3}{4} = \dfrac{0.75 \text{ dollar}}{1 \text{ pound}}$$
    The Wash-n-Fold charges $0.75 per pound of laundry.

Graphing Calculator Explorations

1. $y_1 = 3.8x$
   $y_2 = 3.8x - 3$
   $y_3 = 3.8x + 9$

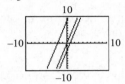

2. $y_1 = -4.9x$
   $y_2 = -4.9x + 1$
   $y_3 = -4.9x + 8$

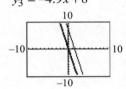

3. $y_1 = \dfrac{1}{4}x$
   $y_2 = \dfrac{1}{4}x + 5$
   $y_3 = \dfrac{1}{4}x - 8$

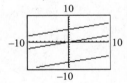

4. $y_1 = -\dfrac{3}{4}x$
   $y_2 = -\dfrac{3}{4}x - 5$
   $y_3 = -\dfrac{3}{4}x + 6$

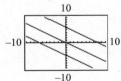

Vocabulary, Readiness & Video Check 3.4

1. The measure of the steepness or tilt of a line is called <u>slope</u>.

2. If an equation is written in the form $y = mx + b$, the value of the letter <u>m</u> is the value of the slope of the graph.

3. The slope of a horizontal line is <u>0</u>.

4. The slope of a vertical line is <u>undefined</u>.

5. If the graph of a line moves upward from left to right, the line has <u>positive</u> slope.

6. If the graph of a line moves downward from left to right, the line has <u>negative</u> slope.

7. Given two points of a line, slope $= \dfrac{\text{change in } \underline{y}}{\text{change in } \underline{x}}$.

8. Whatever y-value you decide to start with in the numerator, you *must* start with the corresponding x-value in the denominator.

9. Solve the equation for y; the slope is the coefficient of x.

10. Zero slope indicates $m = 0$ and a horizontal line; undefined slope indicates m is undefined and a vertical line; no slope refers to an undefined slope.

11. Slope-intercept form; this form makes the slope easy to see, and you need to compare slopes to determine if two lines are parallel or perpendicular.

12. Step 4: INTERPRET the results.

Exercise Set 3.4

1. $(x_1, y_1) = (-1, 5)$ and $(x_2, y_2) = (6, -2)$

   $$m = \frac{y_2 - y_1}{x_2 - x_1} = \frac{-2 - 5}{6 - (-1)} = \frac{-7}{7} = -1$$

3. $(x_1, y_1) = (-4, 3)$ and $(x_2, y_2) = (-4, 5)$

   $$m = \frac{y_2 - y_1}{x_2 - x_1} = \frac{5 - 3}{-4 - (-4)} = \frac{2}{0}$$

   The slope is undefined.

5. $(x_1, y_1) = (-2, 8)$ and $(x_2, y_2) = (1, 6)$

   $$m = \frac{y_2 - y_1}{x_2 - x_1} = \frac{6 - 8}{1 - (-2)} = \frac{-2}{3} = -\frac{2}{3}$$

7. $(x_1, y_1) = (5, 1)$ and $(x_2, y_2) = (-2, 1)$

$$m = \frac{y_2 - y_1}{x_2 - x_1} = \frac{1-1}{-2-5} = \frac{0}{-7} = 0$$

9. $(x_1, y_1) = (-1, 2)$ and $(x_2, y_2) = (2, -2)$

$$m = \frac{y_2 - y_1}{x_2 - x_1} = \frac{-2-2}{2-(-1)} = \frac{-4}{3} = -\frac{4}{3}$$

11. $(x_1, y_1) = (2, 3)$ and $(x_2, y_2) = (2, -1)$

$$m = \frac{y_2 - y_1}{x_2 - x_1} = \frac{-1-3}{2-2} = \frac{-4}{0}$$

The slope is undefined.

13. $(x_1, y_1) = (-3, -2)$ and $(x_2, y_2) = (-1, 3)$

$$m = \frac{y_2 - y_1}{x_2 - x_1} = \frac{3-(-2)}{-1-(-3)} = \frac{5}{2}$$

15. The line goes down. The slope is negative.

17. The line is vertical. The slope is undefined.

19. The slope is positive. The line is "upward."

21. The slope is 0. The line is horizontal.

23. The slope of line 1 is positive, and the slope of line 2 is negative. Thus, line 1 has the greater slope.

25. Both line 1 and line 2 have positive slopes, but line 2 is steeper than line 1. Thus, line 2 has the greater slope.

27. $(0, 0)$ and $(2, 2)$

$$m = \frac{y_2 - y_1}{x_2 - x_1} = \frac{2-0}{2-0} = \frac{2}{2} = 1$$

D

29. A vertical line has undefined slope.
B

31. $(2, 0)$ and $(4, -1)$

$$m = \frac{y_2 - y_1}{x_2 - x_1} = \frac{-1-0}{4-2} = -\frac{1}{2}$$

E

33. $x = 6$ is a vertical line, so it has an undefined slope.

35. $y = -4$ is a horizontal line, so it has a slope $m = 0$.

37. $x = -3$ is a vertical line, so it has an undefined slope.

39. $y = 0$ is a horizontal line, so it has a slope $m = 0$.

41. $y = 5x - 2$
The equation is in slope-intercept form. The coefficient of x, 5, is the slope.

43. $y = -0.3x + 2.5$
The equation is in slope-intercept form. The coefficient of x, -0.3, is the slope.

45. Solve for y.
$2x + y = 7$
$y = -2x + 7$
The coefficient of x, -2, is the slope.

47. Solve for y.
$2x - 3y = 10$
$-3y = -2x + 10$
$\frac{-3y}{-3} = \frac{-2x}{-3} + \frac{10}{-3}$
$y = \frac{2}{3}x - \frac{10}{3}$

The coefficient of x, $\frac{2}{3}$, is the slope.

49. The graph of $x = 1$ is a vertical line. The slope is undefined.

51. Solve for y.
$x = 2y$
$\frac{1}{2}x = y$ or $y = \frac{1}{2}x$

The coefficient of x, $\frac{1}{2}$, is the slope.

53. The graph of $y = -3$ is a horizontal line. The slope is 0.

55. Solve for y.
$-3x - 4y = 6$
$-4y = 3x + 6$
$\frac{-4y}{-4} = \frac{3x}{-4} + \frac{6}{-4}$
$y = -\frac{3}{4}x - \frac{3}{2}$

The coefficient of x, $-\frac{3}{4}$, is the slope.

57. Solve for y.

$$20x - 5y = 1.2$$
$$-5y = -20x + 1.2$$
$$\frac{-5y}{-5} = \frac{-20x}{-5} + \frac{1.2}{-5}$$
$$y = 4x - 0.24$$

The coefficient of x, 4, is the slope.

59. $(-3, -3)$ and $(0, 0)$

$$m = \frac{y_2 - y_1}{x_2 - x_1} = \frac{0 - (-3)}{0 - (-3)} = \frac{3}{3} = 1$$

a. $m = 1$

b. $m = -1$

61. $(-8, -4)$ and $(3, 5)$

$$m = \frac{y_2 - y_1}{x_2 - x_1} = \frac{5 - (-4)}{3 - (-8)} = \frac{9}{11}$$

a. $m = \frac{9}{11}$

b. $m = -\frac{11}{9}$

63. $y = \frac{2}{9}x + 3, \; y = -\frac{2}{9}x$

The slopes are $\frac{2}{9}$ and $-\frac{2}{9}$. The slopes are not the same, and their product is not -1. The lines are neither parallel nor perpendicular.

65. The slope of $y = 3x - 9$ is 3. Solve the other equation for y.

$$x - 3y = -6$$
$$-3y = -x - 6$$
$$\frac{-3y}{-3} = -\frac{x}{-3} - \frac{6}{-3}$$
$$y = \frac{1}{3}x + 2$$

The slope is $\frac{1}{3}$. The slopes are not the same, and their product is not -1. The lines are neither parallel nor perpendicular.

67. Solve the equations for y.

$$6x = 5y + 1 \qquad\qquad -12x + 10y = 1$$
$$6x - 1 = 5y \qquad\qquad 10y = 12x + 1$$
$$\frac{6x}{5} - \frac{1}{5} = \frac{5y}{5} \qquad\qquad \frac{10y}{10} = \frac{12x}{10} + \frac{1}{10}$$
$$y = \frac{6}{5}x - \frac{1}{5} \qquad\qquad y = \frac{6}{5}x + \frac{1}{10}$$

The lines have the same slope, $\frac{6}{5}$, but different y-intercepts. The lines are parallel.

69. Solve the equations for y.

$$6 + 4x = 3y \qquad\qquad 3x + 4y = 8$$
$$\frac{6}{3} + \frac{4x}{3} = \frac{3y}{3} \qquad\qquad 4y = -3x + 8$$
$$y = \frac{4}{3}x + 2 \qquad\qquad \frac{4y}{4} = -\frac{3x}{4} + \frac{8}{4}$$
$$\qquad\qquad\qquad\qquad y = -\frac{3}{4}x + 2$$

The slopes are $\frac{4}{3}$ and $-\frac{3}{4}$. Their product is -1, so the lines are perpendicular.

71. $\text{pitch} = \frac{6}{10} = \frac{3}{5}$

73. $\text{grade} = \dfrac{\text{rise}}{\text{run}} = \dfrac{2}{16} = 0.125 = 12.5\%$

75. $\text{grade} = \dfrac{\text{rise}}{\text{run}} = \dfrac{2580}{6450} = 0.40 = 40\%$

77. Canton Avenue:

$\text{grade} = \dfrac{\text{rise}}{\text{run}} = \dfrac{11 \text{ meters}}{30 \text{ meters}} \approx 0.37 = 37\%$

The grade of Canton Avenue is 37%.
Baldwin Street:

$\text{grade} = \dfrac{\text{rise}}{\text{run}} = \dfrac{1 \text{ meter}}{2.86 \text{ meters}} \approx 0.35 = 35\%$

The grade of Baldwin Street is 35%.

79. $m = \dfrac{y_2 - y_1}{x_2 - x_1} = \dfrac{115 - 110}{2010 - 2005} = \dfrac{5}{5} = \dfrac{1}{1}$ or 1

Every 1 year there are 1 million more U.S. households with televisions.

81. Use (5000, 2350) and (20,000, 9400) to calculate slope.

$$m = \frac{9400 - 2350}{20,000 - 5000} = \frac{7050}{15,000} = \frac{0.47 \text{ dollar}}{1 \text{ mile}}$$

It costs \$0.47 per 1 mile to own and operate a compact car.

83. $y - (-6) = 2(x - 4)$
$y + 6 = 2x - 8$
$y = 2x - 14$

85. $y - 1 = -6(x - (-2))$
$y - 1 = -6(x + 2)$
$y - 1 = -6x - 12$
$y = -6x - 11$

87. $(2, 1)$ and $(0, 0)$: $m = \dfrac{0 - 1}{0 - 2} = \dfrac{-1}{-2} = \dfrac{1}{2}$

$(2, 1)$ and $(-2, -1)$: $m = \dfrac{-1 - 1}{-2 - 2} = \dfrac{-2}{-4} = \dfrac{1}{2}$

$(2, 1)$ and $(-4, -2)$: $m = \dfrac{-2 - 1}{-4 - 2} = \dfrac{-3}{-6} = \dfrac{1}{2}$

$(0, 0)$ and $(-2, -1)$: $m = \dfrac{-1 - 0}{-2 - 0} = \dfrac{-1}{-2} = \dfrac{1}{2}$

$(0, 0)$ and $(-4, -2)$: $m = \dfrac{-2 - 0}{-4 - 0} = \dfrac{-2}{-4} = \dfrac{1}{2}$

$(-2, -1)$ and $(-4, -2)$: $m = \dfrac{-2 - (-1)}{-4 - (-2)} = \dfrac{-1}{-2} = \dfrac{1}{2}$

Since the slope of the line between each pair of points is the same, the points lie on the same line.

89. answers may vary

91. The line slopes down between 2005 and 2006, so there was a decrease in average fuel economy between 2005 and 2006.

93. The lowest point on the graph corresponds to 2000. The average fuel economy for that year was 28.5 miles per gallon.

95. Of the line segments listed, the segment from 2008 to 2009 is the steepest and therefore has the greatest slope.

97. pitch $= \dfrac{\text{rise}}{\text{run}}$
$\dfrac{1}{3} = \dfrac{x}{18}$
$3x = 18$
$x = 6$

99. a. $(2007, 2207)$ and $(2010, 2333)$

b. $m = \dfrac{y_2 - y_1}{x_2 - x_1}$
$= \dfrac{2333 - 2207}{2010 - 2007}$
$= \dfrac{126}{3}$
$= 42$
The slope is 42.

c. For the years 2007 through 2010, the number of heart transplants increased at a rate of 42 per year.

101. $(1, 1)$, $(-4, 4)$ and $(-3, 0)$
$m_1 = \dfrac{0 - 1}{-3 - 1} = \dfrac{1}{4}$, $m_2 = \dfrac{0 - 4}{-3 - (-4)} = -4$
$m_1 m_2 = -1$, so the sides are perpendicular.

103. $(2.1, 6.7)$ and $(-8.3, 9.3)$
$m = \dfrac{y_2 - y_1}{x_2 - x_1} = \dfrac{9.3 - 6.7}{-8.3 - 2.1} = \dfrac{2.6}{-10.4} = -0.25$

105. $(2.3, 0.2)$ and $(7.9, 5.1)$
$m = \dfrac{y_2 - y_1}{x_2 - x_1} = \dfrac{5.1 - 0.2}{7.9 - 2.3} = \dfrac{4.9}{5.6} = 0.875$

107. $y = -\dfrac{1}{3}x + 2$
$y = -2x + 2$
$y = -4x + 2$

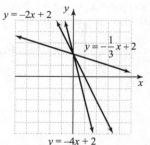

The line becomes steeper.

Integrated Review

1. $(0, 0)$ and $(2, 4)$
$m = \dfrac{y_2 - y_1}{x_2 - x_1} = \dfrac{4 - 0}{2 - 0} = \dfrac{4}{2} = 2$

2. Horizontal line, $m = 0$

3. $(0, 1)$ and $(3, -1)$

$$m = \frac{y_2 - y_1}{x_2 - x_1} = \frac{-1-1}{3-0} = -\frac{2}{3}$$

4. Vertical line, slope is undefined.

5. $y = -2x$
   $m = -2, b = 0$

x	y
0	0
1	-2
-1	2

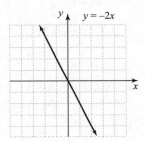

6. $x + y = 3$
   $y = -x + 3$
   $m = -1, b = 3$

x	y
0	3
3	0
1	2

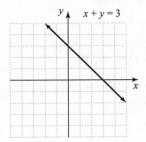

7. $x = -1$ for all values of y.
   Vertical line; slope is undefined.

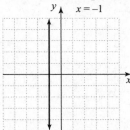

8. $y = 4$ for all values of x.
   Horizontal line; $m = 0$

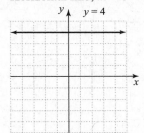

9. $x - 2y = 6$
   $-2y = -x + 6$
   $y = \frac{1}{2}x - 3$
   $m = \frac{1}{2}, b = -3$

x	y
0	-3
2	-2
4	-1

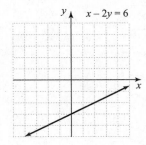

10. $y = 3x + 2$
  $m = 3, b = 2$

x	y
0	2
−1	−1
−2	−4

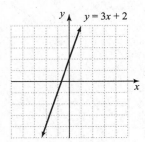

11. $5x + 3y = 15$

x	y
0	5
3	0

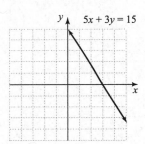

12. $2x - 4y = 8$

x	y
0	−2
4	0

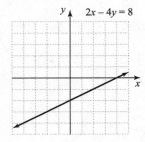

13. The slope of the first line is $-\dfrac{1}{5}$. Solve the second equation for y.
$$3x = -15y$$
$$\frac{3x}{-15} = \frac{-15y}{-15}$$
$$y = -\frac{1}{5}x$$

The slope of the second line is also $-\dfrac{1}{5}$. Since the lines have the same slope but different y-intercepts, the lines are parallel.

14. Solve the equations for y.

$$x - y = \frac{1}{2} \qquad\qquad 3x - y = \frac{1}{2}$$
$$-y = -x + \frac{1}{2} \qquad\quad -y = -3x + \frac{1}{2}$$
$$y = x - \frac{1}{2} \qquad\qquad y = 3x - \frac{1}{2}$$

The slopes are 1 and 3. Since the slopes are not equal and their product is not −1, the lines are neither parallel nor perpendicular.

15. a. Let $x = 0$.
  $y = 1.7(0) + 587 = 587$
  The y-intercept is (0, 587).

  b. In 2000, there were 587 thousand public bridges in the United States.

  c. The equation is in slope-intercept form. The coefficient of x, 1.7, is the slope.

  d. For the years 2000 through 2009, the number of public bridges increased at a rate of 1.7 thousand per year.

16. a. Let $x = 9$.
  $y = 3.3(9) - 3.1 = 29.7 - 3.1 = 26.6$
  The ordered pair is (9, 26.6).

  b. In 2009, the revenue for online advertising was $26.6 billion.

Section 3.5 Practice

1. $y = \dfrac{2}{3}x - 5$

The slope is $\dfrac{2}{3}$, and the y-intercept is $(0, -5)$.

We plot $(0, -5)$. From this point, we move up 2 units and then right 3 units. We stop at the point $(3, -3)$.

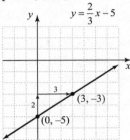

2. Solve the equation for y.

$3x - y = 2$

$-y = -3x + 2$

$y = 3x - 2$

The slope is 3, and the y-intercept is $(0, -2)$. We plot $(0, -2)$. From this point, we move up 3 units and then right 1 unit. We stop at the point $(1, 1)$.

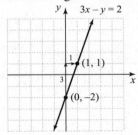

3. y-intercept: $(0, 7)$; slope: $\dfrac{1}{2}$

Let $m = \dfrac{1}{2}$ and $b = 7$.

$y = mx + b$

$y = \dfrac{1}{2}x + 7$

4. Line passing through $(2, 3)$ with slope 4

$y - y_1 = m(x - x_1)$

$y - 3 = 4(x - 2)$

$y - 3 = 4x - 8$

$-4x + y = -5$

$4x - y = 5$

5. Line through $(-1, 6)$ and $(3, 1)$

$m = \dfrac{1 - 6}{3 - (-1)} = \dfrac{-5}{4} = -\dfrac{5}{4}$

Use the slope $-\dfrac{5}{4}$ and the point $(3, 1)$.

$y - y_1 = m(x - x_1)$

$y - 1 = -\dfrac{5}{4}(x - 3)$

$4(y - 1) = 4\left(-\dfrac{5}{4}\right)(x - 3)$

$4y - 4 = -5(x - 3)$

$4y - 4 = -5x + 15$

$5x + 4y = 19$

6. The equation of a vertical line can be written in the form $x = c$, so an equation for a vertical line passing through $(3, -2)$ is $x = 3$.

7. Since the graph of $y = -2$ is a horizontal line, any line parallel to it is also vertical. The equation of a horizontal line can be written in the form $y = c$. An equation for the horizontal line passing through $(4, 3)$ is $y = 3$.

8. a. Write two ordered pairs, $(30, 150{,}000)$ and $(50, 120{,}000)$.

$m = \dfrac{120{,}000 - 150{,}000}{50 - 30}$

$\quad = \dfrac{-30{,}000}{20}$

$\quad = -1500$

Use the slope -1500 and the point $(30, 150{,}000)$.

$y - y_1 = m(x - x_1)$

$y - 150{,}000 = -1500(x - 30)$

$y - 150{,}000 = -1500x + 45{,}000$

$\qquad\qquad y = -1500x + 195{,}000$

b. Find y when $x = 60$.

$y = -1500x + 195{,}000$

$y = -1500(60) + 195{,}000$

$y = -90{,}000 + 195{,}000$

$y = 105{,}000$

To sell 60 condos per month, the price should be $105,000.

Graphing Calculator Explorations

1. $y_1 = x$, $y_2 = 6x$, $y_3 = -6x$

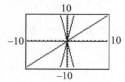

2. $y_1 = -x$, $y_2 = -5x$, $y_3 = -10x$

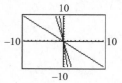

3. $y_1 = \frac{1}{2}x + 2$, $y_2 = \frac{3}{4}x + 2$, $y_3 = x + 2$

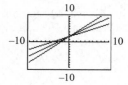

4. $y_1 = x + 1$, $y_2 = \frac{5}{4}x + 1$, $y_3 = \frac{5}{2}x + 1$

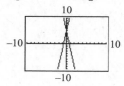

5. $y_1 = -7x + 5$, $y_2 = 7x + 5$

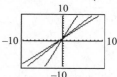

6. $y_1 = 3x - 1$, $y_2 = -3x - 1$

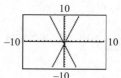

Vocabulary, Readiness & Video Check 3.5

1. The form $y = mx + b$ is called <u>slope-intercept</u> form. When a linear equation in two variables is written in this form, <u>m</u> is the slope of its graph and (0, <u>b</u>) is its y-intercept.

2. The form $y - y_1 = m(x - x_1)$ is called <u>point-slope</u> form. When a linear equation in two variables is written in this form, <u>m</u> is the slope of its graph and <u>(x_1, y_1)</u> is a point on the graph.

3. Start by graphing the <u>y-intercept</u>. Find another point by applying the slope to this point—rewrite the slope as a <u>fraction</u> if necessary.

4. $\left(0, -\frac{1}{6}\right)$

5. Write the equation with x- and y-terms on one side of the equal sign and a constant on the other side.

6. Yes, if one of the points given is the y-intercept. You will need to use the slope formula to find the slope, but then you'll have the slope and y-intercept for the slope-intercept form.

7. Example 6: $y = -3$; Example 7: $x = -2$

8. You need to know what your variables stand for in order to solve part b of the Example, and that depends on how you set up your ordered pairs in part a.

Exercise Set 3.5

1. $y = 2x + 1$

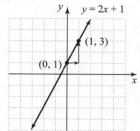

3. $y = \dfrac{2}{3}x + 5$

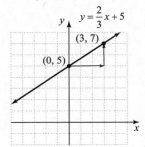

5. $y = -5x$

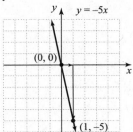

7. $4x + y = 6$
$y = -4x + 6$

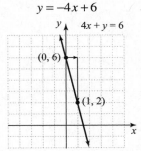

9. $4x - 7y = -14$
$-7y = -4x - 14$
$y = \dfrac{4}{7}x + 2$

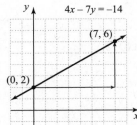

11. $x = \dfrac{5}{4}y$

$\dfrac{4}{5}x = y$

$y = \dfrac{4}{5}x$

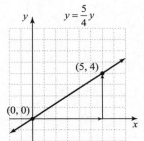

13. $m = 5, b = 3$
$y = mx + b$
$y = 5x + 3$

15. $m = -4, \; b = -\dfrac{1}{6}$

$y = mx + b$

$y = -4x + \left(-\dfrac{1}{6}\right)$

$y = -4x - \dfrac{1}{6}$

17. $m = \dfrac{2}{3}, \; b = 0$

$y = mx + b$

$y = \dfrac{2}{3}x + 0$

$y = \dfrac{2}{3}x$

19. $m = 0, b = -8$
$y = mx + b$
$y = 0x + (-8)$
$y = -8$

21. $m = -\dfrac{1}{5}, \; b = \dfrac{1}{9}$

$y = mx + b$

$y = -\dfrac{1}{5}x + \dfrac{1}{9}$

23. $m = 6$; $(2, 2)$
$$y - y_1 = m(x - x_1)$$
$$y - 2 = 6(x - 2)$$
$$y - 2 = 6x - 12$$
$$-6x + y = -10 \text{ or } 6x - y = 10$$

25. $m = -8$; $(-1, -5)$
$$y - y_1 = m(x - x_1)$$
$$y - (-5) = -8(x - (-1))$$
$$y + 5 = -8x - 8$$
$$8x + y = -13$$

27. $m = \dfrac{3}{2}$; $(5, -6)$
$$y - y_1 = m(x - x_1)$$
$$y - (-6) = \dfrac{3}{2}(x - 5)$$
$$2(y + 6) = 3(x - 5)$$
$$2y + 12 = 3x - 15$$
$$-3x + 2y = -27$$
$$3x - 2y = 27$$

29. $m = -\dfrac{1}{2}$; $(-3, 0)$
$$y - y_1 = m(x - x_1)$$
$$y - 0 = -\dfrac{1}{2}(x - (-3))$$
$$y = -\dfrac{1}{2}(x + 3)$$
$$-2y = x + 3$$
$$-x - 2y = 3$$
$$x + 2y = -3$$

31. $(3, 2)$ and $(5, 6)$
$$m = \dfrac{y_2 - y_1}{x_2 - x_1} = \dfrac{6 - 2}{5 - 3} = \dfrac{4}{2} = 2$$
$$m = 2; \ (3, 2)$$
$$y - y_1 = m(x - x_1)$$
$$y - 2 = 2(x - 3)$$
$$y - 2 = 2x - 6$$
$$-2x + y = -4$$
$$2x - y = 4$$

33. $(-1, 3)$ and $(-2, -5)$
$$m = \dfrac{y_2 - y_1}{x_2 - x_1} = \dfrac{-5 - 3}{-2 - (-1)} = \dfrac{-8}{-1} = 8$$
$$m = 8; \ (-1, 3)$$
$$y - y_1 = m(x - x_1)$$
$$y - 3 = 8(x - (-1))$$
$$y - 3 = 8x + 8$$
$$-8x + y = 11$$
$$8x - y = -11$$

35. $(2, 3)$ and $(-1, -1)$
$$m = \dfrac{y_2 - y_1}{x_2 - x_1} = \dfrac{-1 - 3}{-1 - 2} = \dfrac{-4}{-3} = \dfrac{4}{3}$$
$$m = \dfrac{4}{3}; \ (2, 3)$$
$$y - y_1 = m(x - x_1)$$
$$y - 3 = \dfrac{4}{3}(x - 2)$$
$$3(y - 3) = 4(x - 2)$$
$$3y - 9 = 4x - 8$$
$$-4x + 3y = 1$$
$$4x - 3y = -1$$

37. $(0, 0)$ and $\left(-\dfrac{1}{8}, \dfrac{1}{13}\right)$
$$m = \dfrac{\frac{1}{13} - 0}{-\frac{1}{8} - 0} = \dfrac{\frac{1}{13}}{-\frac{1}{8}} = \dfrac{1}{13}\left(-\dfrac{8}{1}\right) = -\dfrac{8}{13}$$
$$m = -\dfrac{8}{13}; \ (0, 0)$$
$$y - y_1 = m(x - x_1)$$
$$y - 0 = -\dfrac{8}{13}(x - 0)$$
$$y = -\dfrac{8}{13}x$$
$$13y = -8x$$
$$8x + 13y = 0$$

39. Vertical line, point $(0, 2)$
$$x = c$$
$$x = 0$$

41. Horizontal line, point $(-1, 3)$
$$y = c$$
$$y = 3$$

43. Vertical line, point $\left(-\dfrac{7}{3}, -\dfrac{2}{5}\right)$

$x = c$

$x = -\dfrac{7}{3}$

45. $y = 5$ is horizontal.

Parallel to $y = 5$ is horizontal; $y = c$.

Point $(1, 2)$

$y = 2$

47. $x = -3$ is vertical.

Perpendicular to $x = -3$ is horizontal; $y = c$.

Point $(-2, 5)$

$y = 5$

49. $x = 0$ is vertical.

Parallel to $x = 0$ is vertical; $x = c$.

Point $(6, -8)$

$x = 6$

51. $m = -\dfrac{1}{2}; \left(0, \dfrac{5}{3}\right)$

$y = mx + b$

$y = -\dfrac{1}{2}x + \dfrac{5}{3}$

53. $(10, 7)$ and $(7, 10)$

$m = \dfrac{y_2 - y_1}{x_2 - x_1} = \dfrac{10 - 7}{7 - 10} = \dfrac{3}{-3} = -1$

$m = -1; (10, 7)$

$y - y_1 = m(x - x_1)$

$y - 7 = -1(x - 10)$

$y - 7 = -x + 10$

$\quad\quad y = -x + 17$

55. Undefined slope, through $\left(-\dfrac{3}{4}, 1\right)$

A line with undefined slope is vertical. A vertical line has an equation of the form $x = c$.

$x = -\dfrac{3}{4}$

57. $m = 1; (-7, 9)$

$y - y_1 = m(x - x_1)$

$y - 9 = 1[x - (-7)]$

$y - 9 = x + 7$

$\quad\quad y = x + 16$

59. $m = -5, b = 7$

$y = mx + b$

$y = -5x + 7$

61. x-axis is horizontal.

Parallel to x-axis is horizontal; $y = c$.

Point $(6, 7)$

$y = 7$

63. $(2, 3)$ and $(0, 0)$

$m = \dfrac{y_2 - y_1}{x_2 - x_1} = \dfrac{3 - 0}{2 - 0} = \dfrac{3}{2}; \ b = 0$

$y = mx + b$

$y = \dfrac{3}{2}x + 0$

$y = \dfrac{3}{2}x$

65. y-axis is vertical.

Perpendicular to y-axis is horizontal; $y = c$.

Point $(-2, -3)$

$y = -3$

67. $m = -\dfrac{4}{7}; \ (-1, -2)$

$y - y_1 = m(x - x_1)$

$y - (-2) = -\dfrac{4}{7}[x - (-1)]$

$y + 2 = -\dfrac{4}{7}x - \dfrac{4}{7}$

$\quad\ y = -\dfrac{4}{7}x - \dfrac{4}{7} - 2$

$\quad\ y = -\dfrac{4}{7}x - \dfrac{18}{7}$

69. a. $(1, 32)$ and $(3, 96)$

$m = \dfrac{y_2 - y_1}{x_2 - x_1} = \dfrac{96 - 32}{3 - 1} = \dfrac{64}{2} = 32$

$m = 32; (1, 32)$

$s - s_1 = m(t - t_1)$

$s - 32 = 32(t - 1)$

$s - 32 = 32t - 32$

$\quad\ s = 32t$

b. If $t = 4$, then $s = 32(4) = 128$ ft/sec.

71. a. Use $(0, 356)$ and $(2, 290)$.

$m = \dfrac{290 - 356}{2 - 0} = \dfrac{-66}{2} = -33$

$b = 356$

$y = mx + b$

$y = -33x + 356$

b. Let $x = 2015 - 2007 = 8$

$$y = -33(8) + 356$$
$$= -264 + 356$$
$$= 92$$

We predict there will be 92 thousand or 92,000 hybrids sold in 2015.

73. a. Use $(0, 85)$ and $(5, 88)$.

$$m = \frac{88 - 85}{5 - 0} = \frac{3}{5} = 0.6$$
$$b = 85$$
$$y = mx + b$$
$$y = 0.6x + 85$$

b. Let $x = 2020 - 2006 = 14$.

$$y = 0.6(14) + 85 = 8.4 + 85 = 93.4$$

We predict there will be 93.4 persons per square mile in 2020.

75. a. Use $(0, 60)$ and $(8, 46)$.

$$m = \frac{46 - 60}{8 - 0} = \frac{-14}{8} = -1.75$$
$$b = 60$$
$$y = mx + b$$
$$y = -1.75x + 60$$

b. Let $x = 2017 - 2001 = 16$

$$y = -1.75(16) + 60 = -28 + 60 = 32$$

We predict that the newspaper circulation in 2017 will be 32 million.

77. a. The ordered pairs are $(3, 10{,}000)$ and $(5, 8000)$.

$$m = \frac{S_2 - S_1}{p_2 - p_1}$$
$$= \frac{8000 - 10{,}000}{5 - 3}$$
$$= \frac{-2000}{2}$$
$$= -1000$$

$$S - S_1 = m(p - p_1)$$
$$S - 10{,}000 = -1000(p - 3)$$
$$S - 10{,}000 = -1000p + 3000$$
$$S = -1000p + 13{,}000$$

b. $p = 3.50$: $S = -1000(3.50) + 13{,}000$
$$S = -3500 + 13{,}000$$
$$S = 9500$$

9500 Fun Noodles will be sold when the price is $3.50 each.

79. If $x = 2$, then

$$x^2 - 3x + 1 = (2)^2 - 3(2) + 1 = 4 - 6 + 1 = -1$$

81. If $x = -1$, then

$$x^2 - 3x + 1 = (-1)^2 - 3(-1) + 1 = 1 + 3 + 1 = 5$$

83. No

85. Yes

87. $y - 7 = 4(x + 3)$ is in <u>point-slope</u> form.

89. $y = \frac{3}{4}x - \frac{1}{3}$ is in <u>slope-intercept</u> form.

91. $y = \frac{1}{2}$ is a <u>horizontal</u> line.

93. answers may vary

95. $y = 3x - 1,\ m_1 = 3$

a. Parallel: $m_2 = m_1 = 3;\ (-1, 2)$

$$y - y_1 = m_2(x - x_1)$$
$$y - 2 = 3(x - (-1))$$
$$y - 2 = 3x + 3$$
$$-3x + y = 5$$
$$3x - y = -5$$

b. Perpendicular: $m_2 = -\frac{1}{m_1} = -\frac{1}{3};\ (-1, 2)$

$$y - y_1 = m_2(x - x_1)$$
$$y - 2 = -\frac{1}{3}(x - (-1))$$
$$3(y - 2) = -1(x + 1)$$
$$3y - 6 = -x - 1$$
$$x + 3y = 5$$

97. $3x + 2y = 7,\ y = -\frac{3}{2}x + \frac{7}{2},\ m_1 = -\frac{3}{2}$

a. Parallel: $m_2 = m_1 = -\frac{3}{2};\ (3, -5)$

$$y - y_1 = m_2(x - x_1)$$
$$y - (-5) = -\frac{3}{2}(x - 3)$$
$$2(y + 5) = -3(x - 3)$$
$$2y + 10 = -3x + 9$$
$$3x + 2y = -1$$

b. Perpendicular: $m_2 = -\dfrac{1}{m_1} = \dfrac{2}{3}$; $(3, -5)$

$$y - y_1 = m_2(x - x_1)$$
$$y - (-5) = \frac{2}{3}(x - 3)$$
$$3(y + 5) = 2(x - 3)$$
$$3y + 15 = 2x - 6$$
$$2x - 3y = 21$$

Section 3.6 Practice

1. The domain is the set of all x-values $\{0, 1, 5\}$.
   The range is the set of all y-values: $\{-2, 0, 3, 4\}$.

2. **a.** $\{(4, 1), (3, -2), (8, 5), (-5, 3)\}$
   Each x-value is assigned to only one
   y-value, so this set of ordered pairs is a
   function.

   b. $\{(1, 2), (-4, 3), (0, 8), (1, 4)\}$
   The x-value 1 is assigned to two y-values, 2
   and 4, so this set of ordered pairs is not a
   function.

3. **a.** This is the graph of the relation
   $\{(-2, 1), (3, -3), (3, 2)\}$. The x-coordinate 3
   is paired with two y-coordinates, -3 and 2,
   so this is not the graph of a function.

   b. This is the graph of the relation
   $\{(-2, 1), (0, 1), (1, -3), (3, 2)\}$. Each
   x-coordinate has exactly one y-coordinate,
   so this is the graph of a function.

4. **a.** This is the graph of a function since no
   vertical line will intersect this graph more
   than once.

   b. This is the graph of a function since no
   vertical line will intersect this graph more
   than once.

   c. This is the graph of a function since no
   vertical line will intersect this graph more
   than once.

   d. This is not the graph of a function. Vertical
   lines can be drawn that intersect the graph in
   two points. An example of one is shown.

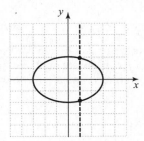

5. **a.** $y = 2x$ is a function because its graph is a
   nonvertical line.

   b. $y = -3x - 1$ is a function because its graph is
   a nonvertical line.

   c. $y = 8$ is a function because its graph is a
   nonvertical line.

   d. $x = 2$ is not a function because its graph is a
   vertical line.

6. **a.** Since June is the sixth month, we look for 6
   on the horizontal axis. From this point, we
   move vertically upward until the graph is
   reached. From the point on the graph, we
   move horizontally to the left to the vertical
   axis. The vertical axis there reads about
   69°F.

   b. We find 26°F on the temperature axis and
   move horizontally to the right. We
   eventually reach the point corresponding to
   2, or February.

   c. Yes, this is the graph of a function. It passes
   the vertical line test.

7. $h(x) = x^2 + 5$

   a. $h(2) = 2^2 + 5 = 4 + 5 = 9$
   $(2, 9)$

   b. $h(-5) = (-5)^2 + 5 = 25 + 5 = 30$
   $(-5, 30)$

   c. $h(0) = 0^2 + 5 = 0 + 5 = 5$
   $(0, 5)$

8. **a.** $h(x) = 6x + 3$
   In this function, x can be any real number.
   The domain of $h(x)$ is the set of all real
   numbers, or $(-\infty, \infty)$ in interval notation.

b. $f(x) = \dfrac{1}{x^2}$

Recall that we cannot divide by 0 so that the domain of $f(x)$ is the set of all real numbers except 0. In interval notation, we write $(-\infty, 0) \cup (0, \infty)$.

9. a.

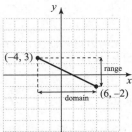

The domain is $[-4, 6]$.
The range is $[-2, 3]$.

b.

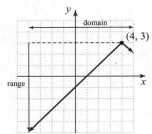

The domain is $(-\infty, \infty)$.
The range is $(-\infty, 3]$.

Vocabulary, Readiness & Video Check 3.6

1. A set of ordered pairs is called a <u>relation</u>.

2. A set of ordered pairs that assigns to each x-value exactly one y-value is called a <u>function</u>.

3. The set of all y-coordinates of a relation is called the <u>range</u>.

4. The set of all x-coordinates of a relation is called the <u>domain</u>.

5. All linear equations are functions except those whose graphs are <u>vertical</u> lines.

6. All linear equations are functions except those whose equations are of the form <u>$x = c$</u>.

7. A relation is a set of ordered pairs and an equation in two variables defines a set of ordered pairs. therefore, an equation in two variables can also define a relation.

8. Yes, this is a function. the definition restricts x-values to be assigned to exactly one y-value, but it makes no such restriction on the y-values.

9. A vertical line represents one x-value paired with many y-values. A function only allows an x-value paired with exactly one y-value, so if a vertical line intersects a graph more than once, there's an x-value paired with more than one y-value, and we don't have a function.

10. $f(-2) = 6$ corresponds to $(-2, 6)$, and $f(3) = 11$ corresponds to $(3, 11)$.

Exercise Set 3.6

1. $\{(2, 4), (0, 0), (-7, 10), (10, -7)\}$
   Domain: $\{-7, 0, 2, 10\}$
   Range: $\{-7, 0, 4, 10\}$

3. $\{(0, -2), (1, -2), (5, -2)\}$
   Domain: $\{0, 1, 5\}$
   Range: $\{-2\}$

5. Every point has a unique x-value: it is a function.

7. Two or more points have the same x-value: it is not a function.

9. No; two points have x-coordinate 1.

11. Yes; no two points have the same x-coordinate.

13. Yes; no vertical line can be drawn that intersects the graph more than once.

15. No; there are many vertical lines that intersect the graph twice, $x = 1$, for example.

17. Yes; $y = x + 1$ is a non-vertical line.

19. Yes; $y - x = 7$ is a non-vertical line.

21. Yes; $y = 6$ is a non-vertical line.

23. No; $x = -2$ is a vertical line.

25. No; does not pass the vertical line test.

27. The point on the graph above June corresponds to approximately 9:30 p.m. on the time axis.

29. The sunset is at approximately 3 p.m. twice, on January 1 and on December 1.

31. Yes; it passes the vertical line test.

33. $4.25 per hour; the segment representing dates before October 1996 corresponds to 4.25 on the vertical axis.

35. 2009; the first line segment above 7.00 on the vertical axis represents dates beginning July 24, 2009.

37. yes; answers may vary

39. According to the graph, the postage would be $1.50.

41. From the graph, it would cost $1 to mail a large envelope that weighs more than 1 ounce and less than or equal to 2 ounces.

43. yes; answers may vary

45. $f(x) = 2x - 5$
$f(-2) = 2(-2) - 5 = -4 - 5 = -9$
$f(0) = 2(0) - 5 = -5$
$f(3) = 2(3) - 5 = 6 - 5 = 1$

47. $f(x) = x^2 + 2$
$f(-2) = (-2)^2 + 2 = 4 + 2 = 6$
$f(0) = (0)^2 + 2 = 2$
$f(3) = (3)^2 + 2 = 9 + 2 = 11$

49. $f(x) = 3x$
$f(-2) = 3(-2) = -6$
$f(0) = 3(0) = 0$
$f(3) = 3(3) = 9$

51. $f(x) = |x|$
$f(-2) = |-2| = 2$
$f(0) = |0| = 0$
$f(3) = |3| = 3$

53. $h(x) = -5x$
$h(-1) = -5(-1) = 5$
$h(0) = -5(0) = 0$
$h(4) = -5(4) = -20$

55. $h(x) = 2x^2 + 3$
$h(-1) = 2(-1)^2 + 3 = 2 + 3 = 5$
$h(0) = 2(0)^2 + 3 = 3$
$h(4) = 2(4)^2 + 3 = 2 \cdot 16 + 3 = 32 + 3 = 35$

57. $f(3) = 6$ corresponds to the ordered pair (3, 6).

59. $g(0) = -\dfrac{1}{2}$ corresponds to the ordered pair $\left(0, -\dfrac{1}{2} \right)$.

61. $h(-2) = 9$ corresponds to the ordered pair (−2, 9).

63. $(-\infty, \infty)$

65. $x + 5 \neq 0 \Rightarrow x \neq -5$, therefore $(-\infty, -5) \cup (-5, \infty)$ or all real numbers except −5.

67. $(-\infty, \infty)$

69. D: $(-\infty, \infty)$, R: $x \geq -4$, $[-4, \infty)$

71. D: $(-\infty, \infty)$, R: $(-\infty, \infty)$

73. D: $(-\infty, \infty)$, R: $\{2\}$

75. When $x = 0$, $y = -1$, so the ordered-pair solution is (0, −1).

77. When $x = 0$, $y = -1$, so $f(0) = -1$.

79. When $y = 0$, $x = -1$ and $x = 5$.

81. (−2, 1)

83. (−3, −1)

85. $f(-5) = 12$

87. (3, −4)

89. $f(5) = 0$

91. $H(x) = 2.59x + 47.24$

　　a. $H(46) = 2.59(46) + 47.24 = 166.38$ cm

　　b. $H(39) = 2.59(39) + 47.24 = 148.25$ cm

93. Answers may vary

95. $y = x + 7$
$f(x) = x + 7$

97. $f(x) = 2x + 7$

　　a. $f(2) = 2(2) + 7 = 11$

　　b. $f(a) = 2(a) + 7 = 2a + 7$

99. $h(x) = x^2 + 7$

    a. $h(3) = (3)^2 + 7 = 16$

    b. $h(a) = (a)^2 + 7 = a^2 + 7$

Chapter 3 Vocabulary Check

1. An ordered pair is a <u>solution</u> of an equation in two variables if replacing the variables by the coordinates of the ordered pair results in a true statement.

2. The vertical number line in the rectangle coordinate system is called the <u>y-axis</u>.

3. A <u>linear</u> equation can be written in the form $Ax + By = C$.

4. An <u>x-intercept</u> is a point of the graph where the graph crosses the x-axis.

5. The form $Ax + By = C$ is called <u>standard</u> form.

6. A <u>y-intercept</u> is a point of the graph where the graph crosses the y-axis.

7. The equation $y = 7x - 5$ is written in <u>slope-intercept</u> form.

8. The equation $y + 1 = 7(x - 2)$ is written in <u>point-slope</u> form.

9. To find an x-intercept of a graph, let <u>y</u> = 0.

10. The horizontal number line in the rectangular coordinate system is called the <u>x-axis</u>.

11. To find a y-intercept of a graph, let <u>x</u> = 0.

12. The <u>slope</u> of a line measures the steepness or tilt of a line.

13. A set of ordered pairs that assigns to each x-value exactly one y-value is called a <u>function</u>.

14. The set of all x-coordinates of a relation is called the <u>domain</u> of the relation.

15. The set of all y-coordinates of a relation is called the <u>range</u> of the relation.

16. A set of ordered pairs is called a <u>relation</u>.

Chapter 3 Review

1–6.

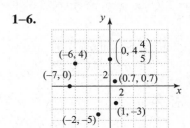

7. a. (5.00, 50), (8.50, 100), (20.00, 250), (27.00, 500)

    b.

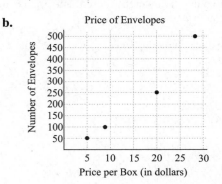

8. a. (2005, 13.8), (2006, 13.6), (2007, 14.1), (2008, 13.9), (2009, 14.6), (2010, 14.6)

    b.

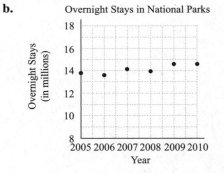

9. $7x - 8y = 56$
(0, 56)
$7(0) - 8(56) \overset{?}{=} 56$
$-448 \neq 0$ No

(8, 0)
$7(8) - 8(0) \overset{?}{=} 56$
$56 = 56$ Yes

10. $-2x + 5y = 10$
$(-5, 0)$
$-2(-5) + 5(0) \stackrel{?}{=} 10$
$\qquad\qquad 10 = 10$ Yes

$(1, 1)$
$-2(1) + 5(1) \stackrel{?}{=} 10$
$\qquad\qquad 3 \neq 10$ No

11. $x = 13$
$(13, 5)$
$(13) \stackrel{?}{=} 13$
$\quad 13 = 13$ Yes

$(13, 13)$
$(13) \stackrel{?}{=} 13$
$\quad 13 = 13$ Yes

12. $y = 2$
$(7, 2)$
$(2) \stackrel{?}{=} 2$
$\quad 2 = 2$ Yes

$(2, 7)$
$(7) \stackrel{?}{=} 2$
$\quad 7 \neq 2$ No

13. $-2 + y = 6x,\ x = 7$
$-2 + y = 6(7)$
$-2 + y = 42$
$\qquad y = 44$
$(7, 44)$

14. $y = 3x + 5,\ y = -8$
$-8 = 3x + 5$
$-13 = 3x$
$-\dfrac{13}{3} = x$
$\left(-\dfrac{13}{3}, -8\right)$

15. $9 = -3x + 4y$
$y = 0:\ 9 = -3x + 4(0),\ 9 = -3x,\ -3 = x$
$y = 3:\ 9 = -3x + 4(3),\ 9 = -3x + 12,\ -3 = -3x,$
$\qquad 1 = x$
$x = 9:\ 9 = -3(9) + 4y,\ 9 = -27 + 4y,\ 36 = 4y,$
$\qquad 9 = y$

x	y
-3	0
1	3
9	9

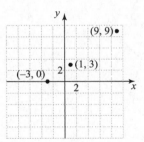

16. $x = 2y$
$y = 0:\ x = 2(0) = 0$
$y = 5:\ x = 2(5) = 10$
$y = -5:\ x = 2(-5) = -10$

x	y
0	0
10	5
-10	-5

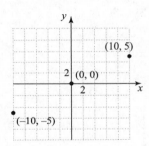

17. $y = 5x + 2000$
$x = 1:\ y = 5(1) + 2000 = 2005$
$x = 100:\ y = 5(100) + 2000 = 2500$
$x = 1000:\ y = 5(1000) + 2000 = 7000$

x	1	100	1000
y	2005	2500	7000

18. $y = 5x + 2000$
Let $y = 6430$.
$6430 = 5x + 2000$
$4430 = 5x$
$\ 886 = x$
886 compact disc holders can be produced.

19. $x - y = 1$

x	y
1	0
0	-1

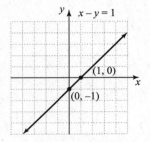

20. $x + y = 6$

x	y
6	0
0	6

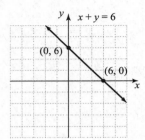

21. $x - 3y = 12$

x	y
12	0
0	-4

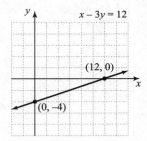

22. $5x - y = -8$

x	y
-2	-2
0	8

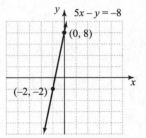

23. $x = 3y$

x	y
0	0
6	2

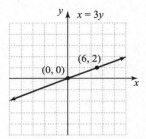

24. $y = -2x$

x	y
0	0
4	-8

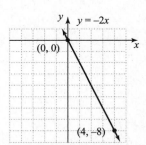

25. $2x - 3y = 6$

x	y
0	-2
3	0

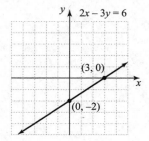

26. $4x - 3y = 12$

x	y
0	−4
3	0

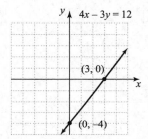

27. x-intercept: (4, 0)
y-intercept: (0, −2)

28. y-intercept: (0, −3)

29. x-intercepts: (−2, 0), (2, 0)
y-intercepts: (0, 2), (0, −2)

30. x-intercepts: (−1, 0), (2, 0), (3, 0)
y-intercept: (0, −2)

31. $x - 3y = 12$

x	y
0	−4
12	0

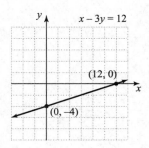

32. $-4x + y = 8$

x	y
0	8
−2	0

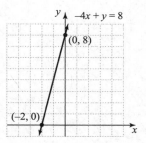

33. $y = -3$ for all x

x	y
0	−3

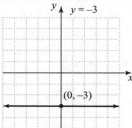

34. $x = 5$ for all y

x	y
5	0

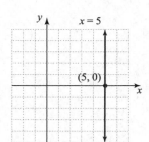

35. $y = -3x$
Find a second point.

x	y
0	0
3	-9

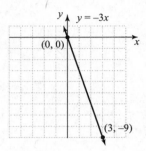

36. $x = 5y$
Find a second point.

x	y
0	0
5	1

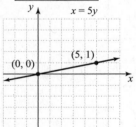

37. $x - 2 = 0$
$x = 2$ for all y

x	y
2	0

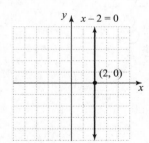

38. $y + 6 = 0$
$y = -6$ for all x

x	y
0	-6

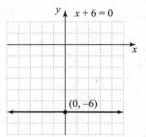

39. $(-1, 2)$, and $(3, -1)$
$$m = \frac{y_2 - y_1}{x_2 - x_1} = \frac{-1 - 2}{3 - (-1)} = -\frac{3}{4}$$

40. $(-2, -2)$ and $(3, -1)$
$$m = \frac{y_2 - y_1}{x_2 - x_1} = \frac{-1 - (-2)}{3 - (-2)} = \frac{1}{5}$$

41. $m = 0$
d

42. $m = -1$
b

43. Slope is undefined.
c

44. $m = 3$
a

45. $(2, 5)$ and $(6, 8)$
$$m = \frac{y_2 - y_1}{x_2 - x_1} = \frac{8 - 5}{6 - 2} = \frac{3}{4}$$

46. $(4, 7)$ and $(1, 2)$
$$m = \frac{y_2 - y_1}{x_2 - x_1} = \frac{2 - 7}{1 - 4} = \frac{-5}{-3} = \frac{5}{3}$$

47. $(1, 3)$ and $(-2, -9)$
$$m = \frac{y_2 - y_1}{x_2 - x_1} = \frac{-9 - 3}{-2 - 1} = \frac{-12}{-3} = 4$$

48. $(-4, 1)$, and $(3, -6)$
$$m = \frac{y_2 - y_1}{x_2 - x_1} = \frac{-6 - 1}{3 - (-4)} = \frac{-7}{7} = -1$$

49. $y = 3x + 7$
The equation is in slope-intercept form. The slope is the coefficient of x, or 3.

50. Solve for y.
$x - 2y = 4$
$\quad -2y = -x + 4$
$\qquad y = \dfrac{1}{2}x - 2$

The slope is $\dfrac{1}{2}$.

51. $y = -2$
This is the equation of a horizontal line. The slope is 0.

52. $x = 0$
This is the equation of a vertical line. The slope is undefined.

53. Solve the equations for y.
$\begin{array}{ll} x - y = 6 & x + y = 3 \\ \quad -y = -x + 6 & \quad y = -x + 3 \\ \qquad y = x - 6 & \end{array}$

The slopes are 1 and -1. Since their product is -1, the lines are perpendicular.

54. Solve the equations for y.
$\begin{array}{ll} 3x + y = 7 & -3x - y = 10 \\ \quad y = -3x + 7 & \quad -y = 3x + 10 \\ & \quad y = -3x - 10 \end{array}$

The slopes are both -3. Since the lines have the same slope but different y-intercepts, they are parallel.

55. The first line, $y = 4x + \dfrac{1}{2}$, has slope 4. Solve the second equation for y.
$4x + 2y = 1$
$\quad 2y = -4x + 1$
$\qquad y = -2x + \dfrac{1}{2}$

The second line has slope -2. Since the slopes are not the same and their product is not -1, the lines are neither parallel nor perpendicular.

56. $x = 4, y = -2$
The first equation's graph is a vertical line, and the second equation's graph is a horizontal line. These lines are perpendicular.

57. $m = \dfrac{y_2 - y_1}{x_2 - x_1} = \dfrac{785 - 565}{2009 - 2004} = \dfrac{220}{5} = 44$
Every 1 year, 44 thousand (44,000) more students graduate with an associate's degree.

58. $m = \dfrac{y_2 - y_1}{x_2 - x_1} = \dfrac{16,900 - 14,800}{2010 - 2004} = \dfrac{2100}{6} = 350$
Every 1 year, 350 more people get kidney transplants.

59. $3x + y = 7$
$\quad y = -3x + 7$
$y = mx + b$
$m = -3$, y-intercept $= (0, 7)$

60. $x - 6y = -1$
$\quad -6y = -x - 1$
$\qquad y = \dfrac{1}{6}x + \dfrac{1}{6}$
$y = mx + b$
$m = \dfrac{1}{6}$, y-intercept $= \left(0, \dfrac{1}{6}\right)$

61. $y = 2$
$y = mx + b$
$m = 0$, y-intercept $= (0, 2)$

62. $x = -5$
$y = mx + b$
m is undefined.
There is no y-intercept.

63. $y = 3x - 1$
$y = mx + b$
$m = 3$, $b = -1$

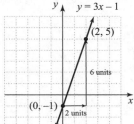

64. $y = -3x$
$y = mx + b$
$m = -3,\ b = 0$

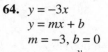

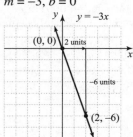

65. $5x - 3y = 15$
$\quad -3y = -5x + 15$
$\quad\quad\ y = \dfrac{5}{3}x - 5$
$y = mx + b$
$m = \dfrac{5}{3},\ b = -5$

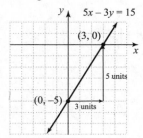

66. $-x + 2y = 8$
$\quad\ \ 2y = x + 8$
$\quad\quad y = \dfrac{1}{2}x + 4$
$y = mx + b$
$m = \dfrac{1}{2},\ b = 4$

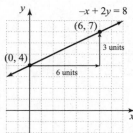

67. $m = -5,\ b = \dfrac{1}{2}$
$y = mx + b$
$y = -5x + \dfrac{1}{2}$

68. $m = \dfrac{2}{3},\ b = 6$
$y = mx + b$
$y = \dfrac{2}{3}x + 6$

69. $y = -4x$
$m = -4,\ b = 0$
c

70. $y = 2x$
$m = 2,\ b = 0$
a

71. $y = 2x - 1$
$m = 2,\ b = -1$
b

72. $y = -2x + 1$
$m = -2,\ b = 1$
d

73. $y = 56x + 1859$
$y = mx + b$
y-intercept $= (0,\ 1859)$

74. Since $x = 0$ represents 1995, then in 1995, the average cost for tuition and fees at a public two-year college was \$1859.

75. $m = -3;\ (0,\ -5)$
$\quad\quad y = mx + b$
$\quad\quad y = -3x - 5$
$3x + y = -5$

76. $m = \dfrac{1}{2};\ \left(0,\ -\dfrac{7}{2}\right)$
$\quad\quad y = mx + b$
$\quad\quad y = \dfrac{1}{2}x - \dfrac{7}{2}$
$\quad\ 2y = x - 7$
$x - 2y = 7$

77. Horizontal line, point $(-2,\ -3)$
$y = c$
$y = -3$

78. Horizontal line, point $(0,\ 0)$
$y = c$
$y = 0$

79. $m = -6; (2, -1)$

$$y - y_1 = m(x - x_1)$$
$$y - (-1) = -6(x - 2)$$
$$y + 1 = -6x + 12$$
$$6x + y = 11$$

80. $m = 12; \left(\dfrac{1}{2}, 5\right)$

$$y - y_1 = m(x - x_1)$$
$$y - 5 = 12\left(x - \dfrac{1}{2}\right)$$
$$y - 5 = 12x - 6$$
$$12x - y = 1$$

81. $(0, 6)$ and $(6, 0)$

$$m = \dfrac{y_2 - y_1}{x_2 - x_1} = \dfrac{0 - 6}{6 - 0} = \dfrac{-6}{6} = -1$$

$m = -1; (0, 6)$
$$y - y_1 = m(x - x_1)$$
$$y - 6 = -1(x - 0)$$
$$y - 6 = -x$$
$$x + y = 6$$

82. $(0, -4)$ and $(-8, 0)$

$$m = \dfrac{y_2 - y_1}{x_2 - x_1} = \dfrac{0 - (-4)}{-8 - 0} = \dfrac{4}{-8} = -\dfrac{1}{2}$$

$m = -\dfrac{1}{2}; (0, -4)$
$$y - y_1 = m(x - x_1)$$
$$y - (-4) = -\dfrac{1}{2}(x - 0)$$
$$y + 4 = -\dfrac{1}{2}x$$
$$2y + 8 = -x$$
$$x + 2y = -8$$

83. Vertical line, point $(5, 7)$
$$x = c$$
$$x = 5$$

84. Horizontal line, point $(-6, 8)$
$$y = c$$
$$y = 8$$

85. $y = 8$ is horizontal.
Perpendicular to $y = 8$ is vertical; $x = c$.
Point $(6, 0)$
$$x = 6$$

86. $x = -2$ is vertical.
Perpendicular to $x = -2$ is horizontal; $y = c$,
point $(10, 12)$
$$y = 12$$

87. Two points have the same *x*-value: it is not a function.

88. Every point has a unique *x*-value: it is a function.

89. Yes; $7x - 6y = 1$ is a non-vertical line.

90. Yes; $y = 7$ is a non-vertical line.

91. No; $x = 2$ is a vertical line.

92. Yes; for each value of *x* there is only one value of *y*.

93. No; the graph does not pass the vertical line test.

94. Yes; the graph passes the vertical line test.

95. $f(x) = -2x + 6$

    a. $f(0) = -2(0) + 6 = 6$

    b. $f(-2) = -2(-2) + 6 = 4 + 6 = 10$

    c. $f\left(\dfrac{1}{2}\right) = -2\left(\dfrac{1}{2}\right) + 6 = -1 + 6 = 5$

96. $h(x) = -5 - 3x$

    a. $h(2) = -5 - 3(2) = -11$

    b. $h(-3) = -5 - 3(-3) = 4$

    c. $h(0) = -5 - 3(0) = -5$

97. $g(x) = x^2 + 12x$

    a. $g(3) = (3)^2 + 12(3) = 45$

    b. $g(-5) = (-5)^2 + 12(-5) = -35$

    c. $g(0) = (0)^2 + 12(0) = 0$

98. $h(x) = 6 - |x|$

    a. $h(-1) = 6 - |-1| = 6 - 1 = 5$

    b. $h(1) = 6 - |1| = 6 - 1 = 5$

c. $h(-4) = 6 - |-4| = 6 - 4 = 2$

99. $(-\infty, \infty)$

100. $x - 2 \neq 0 \Rightarrow x \neq 2$, therefore $(-\infty, 2) \cup (2, \infty)$ or all real numbers except 2.

101. D: $[-3, 5]$, R: $[-4, 2]$

102. D: $(-\infty, \infty)$, R: $x \geq 0$, $[0, \infty)$

103. D: $\{3\}$, R: $(-\infty, \infty)$

104. D: $(-\infty, \infty)$, R: $x \leq 2$, $(-\infty, 2]$

105. $2x - 5y = 9$

Let $y = 1$.	Let $x = 2$.
$2x - 5(1) = 9$	$2(2) - 5y = 9$
$2x - 5 = 9$	$4 - 5y = 9$
$2x = 14$	$-5y = 5$
$x = 7$	$y = -1$

Let $y = -3$.
$2x - 5(-3) = 9$
$2x + 15 = 9$
$2x = -6$
$x = -3$

x	y
7	1
2	-1
-3	-3

106. $x = -3y$

Let $x = 0$.	Let $y = 1$.
$0 = -3y$	$x = -3(1)$
$0 = y$	$x = -3$

Let $x = 6$.
$6 = -3y$
$-2 = y$

x	y
0	0
-3	1
6	-2

107. $2x - 3y = 6$

Let $y = 0$.	Let $x = 0$.
$2x - 3(0) = 6$	$2(0) - 3y = 6$
$2x = 6$	$-3y = 6$
$x = 3$	$y = -2$

x-intercept: $(3, 0)$
y-intercept: $(0, -2)$

108. $-5x + y = 10$

Let $y = 0$.	Let $x = 0$.
$-5x + 0 = 10$	$-5(0) + y = 10$
$-5x = 10$	$y = 10$
$x = -2$	

x-intercept: $(-2, 0)$
y-intercept: $(0, 10)$

109. $x - 5y = 10$

x	y
10	0
0	-2

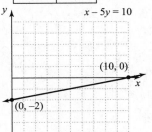

110. $x + y = 4$

x	y
4	0
0	4

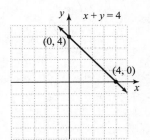

111. $y = -4x$

x	y
0	0
1	−4

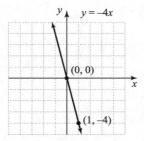

112. $2x + 3y = -6$

x	y
−3	0
0	−2

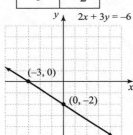

113. $x = 3$

This is the equation of a vertical line with x-intercept $(3, 0)$.

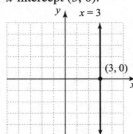

114. $y = -2$

This is the equation of a horizontal line with y-intercept $(0, -2)$.

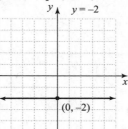

115. $(3, -5)$ and $(-4, 2)$

$$m = \frac{y_2 - y_1}{x_2 - x_1} = \frac{2 - (-5)}{-4 - 3} = \frac{7}{-7} = -1$$

116. $(1, 3)$ and $(-6, -8)$

$$m = \frac{y_2 - y_1}{x_2 - x_1} = \frac{-8 - 3}{-6 - 1} = \frac{-11}{-7} = \frac{11}{7}$$

117. $(0, -4)$ and $(2, 0)$

$$m = \frac{y_2 - y_1}{x_2 - x_1} = \frac{0 - (-4)}{2 - 0} = \frac{4}{2} = 2$$

118. $(0, 2)$ and $(6, 0)$

$$m = \frac{y_2 - y_1}{x_2 - x_1} = \frac{0 - 2}{6 - 0} = \frac{-2}{6} = -\frac{1}{3}$$

119. Solve for y.

$$-2x + 3y = -15$$
$$3y = 2x - 15$$
$$y = \frac{2}{3}x - 5$$

The slope is $\frac{2}{3}$. The y-intercept is $(0, -5)$.

120. Solve for y.

$$6x + y - 2 = 0$$
$$y = -6x + 2$$

The slope is -6. The y-intercept is $(0, 2)$.

121. $m = -5; (3, -7)$

$$y - y_1 = m(x - x_1)$$
$$y - (-7) = -5(x - 3)$$
$$y + 7 = -5x + 15$$
$$5x + y = 8$$

122. $m = 3$; $(0, 6)$
$y = mx + b$
$y = 3x + 6$
$3x - y = -6$

123. $(-3, 9)$ and $(-2, 5)$
$$m = \frac{y_2 - y_1}{x_2 - x_1} = \frac{5 - 9}{-2 - (-3)} = \frac{-4}{1} = -4$$
$m = -4$; $(-2, 5)$
$y - y_1 = m(x - x_1)$
$\quad y - 5 = -4(x - (-2))$
$\quad y - 5 = -4(x + 2)$
$\quad y - 5 = -4x - 8$
$\quad 4x + y = -3$

124. $(3, 1)$ and $(5, -9)$
$$m = \frac{y_2 - y_1}{x_2 - x_1} = \frac{-9 - 1}{5 - 3} = \frac{-10}{2} = -5$$
$m = -5$; $(3, 1)$
$y - y_1 = m(x - x_1)$
$\quad y - 1 = -5(x - 3)$
$\quad y - 1 = -5x + 15$
$\quad 5x + y = 16$

125. Use $(0, 2134)$ and $(7, 3800)$.
$$m = \frac{y_2 - y_1}{x_2 - x_1} = \frac{3800 - 2134}{7 - 0} = \frac{1666}{7} = 238$$
$b = 2134$
$y = mx + b$
$y = 238x + 2134$

126. Let $x = 2014 - 2002 = 12$
$y = 238(12) + 2134 = 2856 + 2134 = 4990$
In 2014, we predict the yogurt production to be 4990 million pounds.

Chapter 3 Test

1. $y = \frac{1}{2}x$

$m = \frac{1}{2}$; $b = 0$

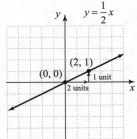

2. $2x + y = 8$

x	y
4	0
0	8

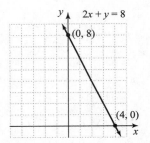

3. $5x - 7y = 10$

x	y
2	0
-5	-5

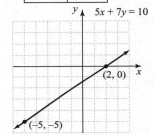

4. $y = -1$ for all values of x.

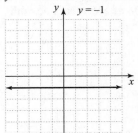

5. $x - 3 = 0$

     $x = 3$ for all values of y.

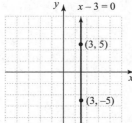

6. $(-1, -1)$ and $(4, 1)$

$$m = \frac{y_2 - y_1}{x_2 - x_1} = \frac{1 - (-1)}{4 - (-1)} = \frac{2}{5}$$

7. Horizontal line: $m = 0$

8. $(6, -5)$ and $(-1, 2)$

$$m = \frac{y_2 - y_1}{x_2 - x_1} = \frac{2 - (-5)}{-1 - 6} = \frac{7}{-7} = -1$$

9. $-3x + y = 5$

         $y = 3x + 5$

$y = mx + b$

$m = 3$

10. $x = 6$ is a vertical line. The slope is undefined.

11. $7x - 3y = 2$

     $-3y = -7x + 2$

         $y = \dfrac{7}{3}x - \dfrac{2}{3}$

$y = mx + b$

$m = \dfrac{7}{3}, \; b = -\dfrac{2}{3}, \left(0, -\dfrac{2}{3}\right)$

12. $y = 2x - 6, \; m_1 = 2$

     $-4x = 2y, \; -2x = y$

     $y = -2x, \; m_2 = -2$

     $m_1 \neq m_2$ and $m_1 m_2 \neq -1$, neither

13. $m = -\dfrac{1}{4}; \; (2, 2)$

       $y - y_1 = m(x - x_1)$

       $y - 2 = -\dfrac{1}{4}(x - 2)$

   $4(y - 2) = -(x - 2)$

     $4y - 8 = -x + 2$

     $x + 4y = 10$

14. $(0, 0)$ and $(6, -7)$

$$m = \frac{y_2 - y_1}{x_2 - x_1} = \frac{-7 - 0}{6 - 0} = -\frac{7}{6}$$

$$m = -\frac{7}{6}; \; (0, 0)$$

     $y - y_1 = m(x - x_1)$

       $y - 0 = -\dfrac{7}{6}(x - 0)$

        $6y = -7x$

  $7x + 6y = 0$

15. $(2, -5)$ and $(1, 3)$

$$m = \frac{y_2 - y_1}{x_2 - x_1} = \frac{3 - (-5)}{1 - 2} = \frac{8}{-1} = -8$$

$m = -8; \; (1, 3)$

     $y - y_1 = m(x - x_1)$

     $y - 3 = -8(x - 1)$

     $y - 3 = -8x + 8$

  $8x + y = 11$

16. $x = 7$ is vertical.

Parallel to $x = 7$ is vertical;

$x = c$, point $(-5, -1)$

$x = -5$

17. $m = \dfrac{1}{8}, \; b = 12$

       $y = mx + b$

       $y = \dfrac{1}{8}x + 12$

     $8y = x + 96$

  $x - 8y = -96$

18. Yes; it passes the vertical line test.

19. No; it does not pass the vertical line test.

20. $h(x) = x^3 - x$

   a. $h(-1) = (-1)^3 - (-1) = -1 + 1 = 0$

   b. $h(0) = (0)^3 - (0) = 0$

   c. $h(4) = (4)^3 - (4) = 64 - 4 = 60$

21. $x + 1 \neq 0 \Rightarrow x \neq -1$, therefore $(-\infty, -1) \cup (-1, \infty)$ or all real numbers except -1.

22. a. The graph crosses the *x*-axis at 0 and 4.
x-intercepts: (0, 0), (4, 0)
The graph crosses the *y*-axis at 0. The *y*-intercept is (0, 0).

  b. Domain: $(-\infty, \infty)$; range: $x \le 4$, $(-\infty, 4]$

23. a. The graph crosses the *x*-axis at 2, so the *x*-intercept is (2, 0). The graph crosses the *y*-axis at −2, so the
y-intercept is (0, −2).

  b. Domain: $(-\infty, \infty)$; range: $(-\infty, \infty)$

24. $f(7) = 20$ corresponds to the ordered pair (7, 20).

25. The bar for Denmark extends to about 210 on the horizontal axis. The average water use per person per day in
Denmark is approximately 210 liters.

26. The bar for Australia extends to about 490 on the horizontal axis. The average water use per person per day in
Australia is approximately 490 liters.

27. The highest point on the graph corresponds to 7 on the horizontal axis, denoting July. The average high
temperature is the greatest in July.

28. April corresponds to 4 on the horizontal axis. Moving horizontally to the left from the point on the graph above 4,
we reach approximately 63 on the vertical axis. The average high temperature for April is approximately 63°F.

29. The points for months 1, 2, 3, 11, and 12 lie below 60 on the vertical axis. Thus, the average high temperature is
below 60°F in January, February, March, November, and December.

30. a. The ordered pairs are (2003, 66.0), (2004, 65.4), (2005, 65.4), (2006, 65.6), (2007, 64.9), (2008, 63.7),
(2009, 62.1).

  b.

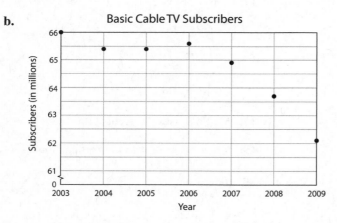

31. a. $m = \dfrac{\text{rise}}{\text{run}} = \dfrac{1340 - 1380}{2010 - 2005} = \dfrac{-40}{5} = \dfrac{-8}{1}$ or -8
For every 1 year, 8 million fewer tickets are sold.

  b. (0, 1380), (5, 1340)

c. $m = \dfrac{y_2 - y_1}{x_2 - x_1} = \dfrac{1340 - 1380}{5 - 0} = \dfrac{-40}{5} = -8$

$b = 1380$

$y = mx + b$

$y = -8x + 1380$

d. Let $x = 2015 - 2005 = 10$

$y = -8(10) + 1380 = -80 + 1380 = 1300$

In 2015, we predict that 1300 million movie tickets will be sold in the United States and Canada.

Chapter 3 Cumulative Review

1. a. $2 < 3$

   b. $7 > 4$

   c. $72 > 27$

2. $\dfrac{56}{64} = \dfrac{7 \cdot 8}{8 \cdot 8} = \dfrac{7}{8}$

3. $\dfrac{2}{15} \cdot \dfrac{5}{13} = \dfrac{2 \cdot 5}{3 \cdot 5 \cdot 13} = \dfrac{2}{39}$

4. $\dfrac{10}{3} + \dfrac{5}{21} = \dfrac{10 \cdot 7}{3 \cdot 7} + \dfrac{5}{21}$

$\qquad = \dfrac{70 + 5}{21}$

$\qquad = \dfrac{75}{21}$

$\qquad = \dfrac{3 \cdot 25}{3 \cdot 7}$

$\qquad = \dfrac{25}{7}$

$\qquad = 3\dfrac{4}{7}$

5. $\dfrac{3 + |4 - 3| + 2^2}{6 - 3} = \dfrac{3 + |1| + 2^2}{6 - 3} = \dfrac{3 + 1 + 4}{6 - 3} = \dfrac{8}{3}$

6. $16 - 3 \cdot 3 + 2^4 = 16 - 3 \cdot 3 + 16$

$\qquad\qquad = 16 - 9 + 16$

$\qquad\qquad = 23$

7. a. $-8 + (-11) = -19$

   b. $-5 + 35 = 30$

   c. $0.6 + (-1.1) = -0.5$

d. $-\dfrac{7}{10} + \left(-\dfrac{1}{10}\right) = -\dfrac{8}{10} = -\dfrac{4}{5}$

e. $11.4 + (-4.7) = 6.7$

f. $-\dfrac{3}{8} + \dfrac{2}{5} = -\dfrac{3 \cdot 5}{8 \cdot 5} + \dfrac{2 \cdot 8}{5 \cdot 8} = \dfrac{-15 + 16}{40} = \dfrac{1}{40}$

8. $|9 + (-20)| + |-10| = |-11| + |-10| = 11 + 10 = 21$

9. a. $-14 - 8 + 10 - (-6) = -14 + (-8) + 10 + 6$

$\qquad\qquad\qquad\qquad = -6$

   b. $1.6 - (-10.3) + (-5.6) = 1.6 + 10.3 + (-5.6)$

$\qquad\qquad\qquad\qquad\qquad = 6.3$

10. $-9 - (3 - 8) = -9 - (-5) = -9 + 5 = -4$

11. Let $x = -2$ and $y = -4$.

   a. $5x - y = 5(-2) - (-4) = -10 + 4 = -6$

   b. $x^4 - y^2 = (-2)^4 - (-4)^2 = 16 - 16 = 0$

   c. $\dfrac{3x}{2y} = \dfrac{3(-2)}{2(-4)} = \dfrac{-6}{-8} = \dfrac{3}{4}$

12. $\dfrac{x}{-10} = 2$

Let $x = -20$.

$\dfrac{-20}{-10} \overset{?}{=} 2$

$\quad 2 = 2$ True

-20 is a solution to the equation.

13. a. $10 + (x + 12) = 10 + x + 12 = x + 22$

    b. $-3(7x) = -21x$

14. $(12 + x) - (4x - 7) = 12 + x - 4x + 7 = 19 - 3x$

15. a. $-3y: -3$

    b. $22z^4: 22$

    c. $y = 1y: 1$

    d. $-x = -1x: -1$

    e. $\dfrac{x}{7} = \dfrac{1}{7}x: \dfrac{1}{7}$

16. $-5(x-7) = -5x - (-5)(7) = -5x + 35$

17. $\quad x - 7 = 10$
$\quad x - 7 + 7 = 10 + 7$
$\qquad\qquad x = 17$

18. $5(3+z) - (8z+9) = -4$
$\quad\; 15 + 5z - 8z - 9 = -4$
$\qquad\qquad -3z + 6 = -4$
$\qquad\qquad\quad\; -3z = -10$
$\qquad\qquad\qquad\; z = \dfrac{10}{3}$

19. $\qquad \dfrac{5}{2}x = 15$
$\quad \dfrac{2}{5}\left(\dfrac{5}{2}x\right) = \dfrac{2}{5}(15)$
$\qquad\qquad x = 6$

20. $\qquad \dfrac{x}{4} - 1 = -7$
$\quad 4\left(\dfrac{x}{4}\right) - 4(1) = 4(-7)$
$\qquad\quad x - 4 = -28$
$\qquad\qquad\; x = -24$

21. Sum
= first integer + second integer + third integer
Sum $= x + (x+1) + (x+2)$
$\qquad = x + x + 1 + x + 2$
$\qquad = 3x + 3$

22. $\qquad \dfrac{x}{3} - 2 = \dfrac{x}{3}$
$\quad 3\left(\dfrac{x}{3}\right) - 3(2) = 3\left(\dfrac{x}{3}\right)$
$\qquad\quad x - 6 = x$
$\qquad\qquad -6 = 0$
This is false. There is no solution.

23. $\quad \dfrac{2(a+3)}{3} = 6a + 2$
$\quad 2(a+3) = 3(6a+2)$
$\quad\; 2a + 6 = 18a + 6$
$\; -16a + 6 = 6$
$\qquad -16a = 0$
$\qquad\qquad a = 0$

24. $\qquad x + 2y = 6$
$\quad x - x + 2y = 6 - x$
$\qquad\quad\; 2y = 6 - x$
$\qquad\quad \dfrac{2y}{2} = \dfrac{6-x}{2}$
$\qquad\qquad y = \dfrac{6-x}{2}$

25. Let x = the number of Republican representatives, then $x - 49$ = the number of Democratic representatives.
$\quad x + x - 49 = 435$
$\qquad 2x - 49 = 435$
$\qquad\qquad 2x = 484$
$\qquad\qquad\; x = 242$
$x - 49 = 193$
There were 242 Republican representatives and 193 Democratic representatives.

26. $\quad 5(x+4) \geq 4(2x+3)$
$\quad 5x + 20 \geq 8x + 12$
$\; -3x + 20 \geq 12$
$\qquad -3x \geq -8$
$\qquad \dfrac{-3x}{-3} \leq \dfrac{-8}{-3}$
$\qquad\quad x \leq \dfrac{8}{3}, \left(-\infty, \dfrac{8}{3}\right]$

27. The perimeter of a rectangle is given by the formula $P = 2l + 2w$. Let l = the length of the garden.
$\quad P = 2l + 2w$
$\; 140 = 2l + 2w$
$\; 140 = 2l + 2(30)$
$\; 140 = 2l + 60$
$\quad 80 = 2l$
$\quad 40 = l$
The length of the garden is 40 feet.

28. $\quad -3 < 4x - 1 \leq 2$
$\quad\; -2 < 4x \leq 3$
$\; -\dfrac{1}{2} < x \leq \dfrac{3}{4}, \left(-\dfrac{1}{2}, \dfrac{3}{4}\right]$

29. $\qquad y = mx + b$
$\quad y - b = mx + b - b$
$\quad y - b = mx$
$\qquad \dfrac{y-b}{m} = \dfrac{mx}{m}$
$\qquad \dfrac{y-b}{m} = x$

30. $y = -5x$

x	y
0	0
−1	5
2	−10

31. Let x = the amount of 70% acid.
No. of liters · Strength = Amt of Acid

70%	x	0.7	$0.7x$
40%	$12 - x$	0.4	$0.4(12 - x)$
50%	12	0.5	$0.5(12)$

$$0.7x + 0.4(12 - x) = 0.5(12)$$
$$0.7x + 4.8 - 0.4x = 6$$
$$0.3x + 4.8 = 6$$
$$0.3x = 1.2$$
$$x = 4$$

$12 - x = 12 - 4 = 8$
Mix 4 liters of 70% acid with 8 liters of 40% acid.

32. $y = -3x + 5$

x	y
−1	8
0	5
1	2

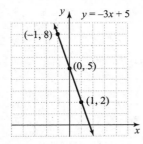

33. $x \geq -1$, $[-1, \infty)$

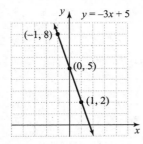

34. $2x + 4y = -8$
x-intercept, $y = 0$
$2x + 4(0) = -8 \Rightarrow x = -4$: $(-4, 0)$
y-intercept, $x = 0$
$2(0) + 4y = -8 \Rightarrow y = -2$: $(0, -2)$

35. $-1 \leq 2x - 3 < 5$
$2 \leq 2x < 8$
$1 \leq x < 4$, $[1, 4)$

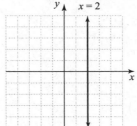

36. $x = 2$
$x = 2$ for all values of y.

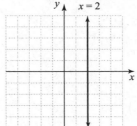

37. a. $x - 2y = 6$
$(6, 0)$
$(6) - 2(0) \overset{?}{=} 6$
$6 = 6$ Yes

   b. $x - 2y = 6$
$(0, 3)$
$(0) - 2(3) \overset{?}{=} 6$
$-6 \neq 6$ No

   c. $x - 2y = 6$
$\left(1, -\dfrac{5}{2}\right)$
$(1) - 2\left(-\dfrac{5}{2}\right) \overset{?}{=} 6$
$1 + 5 \overset{?}{=} 6$
$6 = 6$ Yes

38. $(0, 5)$ and $(-5, 4)$
$$m = \frac{y_2 - y_1}{x_2 - x_1} = \frac{4 - 5}{-5 - 0} = \frac{-1}{-5} = \frac{1}{5}$$

39. a. linear; because it can be written in the form $Ax + By = C$.

   b. linear; because it can be written in the form $Ax + By = C$.

   c. not linear; because y is squared.

   d. linear; because it can be written in the form $Ax + By = C$.

40. $x = -10$ is a vertical line. The slope is undefined.

41. $y = -1$ is horizontal, slope is 0.

42. $2x - 5y = 10$

$$-5y = -2x + 10$$

$$y = \frac{2}{5}x - 2$$

$y = mx + b$

$m = \frac{2}{5}, \ b = -2$

The slope is $\frac{2}{5}$.

The y-intercept is $(0, -2)$.

43. $m = \frac{1}{4}; \ b = -3$

$y = mx + b$

$y = \frac{1}{4}x + (-3)$

$y = \frac{1}{4}x - 3$

44. $(2, 3)$ and $(0, 0)$

$$m = \frac{y_2 - y_1}{x_2 - x_1} = \frac{0 - 3}{0 - 2} = \frac{-3}{-2} = \frac{3}{2}$$

Point: $(0, 0)$

$y - y_1 = m(x - x_1)$

$$y - 0 = \frac{3}{2}(x - 0)$$

$$2y = 3x$$

$3x - 2y = 0$

Chapter 4

Section 4.1 Practice

1. $\begin{cases} 4x - y = 2 \\ y = 3x \end{cases}$

   (4, 12)

   $4(4) - 12 \stackrel{?}{=} 2$

   $16 - 12 \stackrel{?}{=} 2$

   $4 = 2$ False

   (4, 12) is not a solution of the system.

2. $\begin{cases} x - 3y = -7 \\ 2x + 9y = 1 \end{cases}$

   (−4, 1)

   $-4 - 3(1) \stackrel{?}{=} -7$ $2(-4) + 9(1) \stackrel{?}{=} 1$

   $-4 - 3 \stackrel{?}{=} -7$ $-8 + 9 \stackrel{?}{=} 1$

   $-7 = -7$ True $1 = 1$ True

   (−4, 1) is a solution of the system.

3. $\begin{cases} x - y = 3 \\ x + 2y = 18 \end{cases}$

$x - y = 3$			$x + 2y = 18$	
x	y		x	y
−4	−7		−4	11
0	−3		0	9
4	1		4	7

   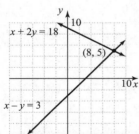

   The two lines appear to intersect at (8, 5).

   $x - y = 3$ $x + 2y = 18$

   $8 - 5 \stackrel{?}{=} 3$ $8 + 2(5) \stackrel{?}{=} 18$

   $3 = 3$ True $8 + 10 \stackrel{?}{=} 18$

                        $18 = 18$ True

   (8, 5) is the solution of the system.

4. $\begin{cases} -4x + 3y = -3 \\ y = -5 \end{cases}$

   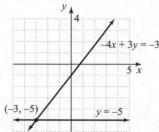

   The two lines appear to intersect at (−3, −5).

   Check.

       $-4x + 3y = -3$ $y = -5$

   $-4(-3) + 3(-5) \stackrel{?}{=} -3$ $-5 = -5$ True

        $12 - 15 \stackrel{?}{=} -3$

            $-3 = -3$ True

   (−3, −5) is the solution of the system.

5. $\begin{cases} 3y = 9x \\ 6x - 2y = 12 \end{cases}$

   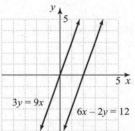

   The lines appear to be parallel. To confirm this, write both equations in slope-intercept form.

   $3y = 9x$ $6x - 2y = 12$

   $y = 3x$ $-2y = -6x + 12$

                       $y = 3x - 6$

   The slopes are the same, so the lines are parallel. Thus, there is no solution of the system and the system is inconsistent. The system has no solution or the solution set is { } or $\varnothing$.

6. $\begin{cases} x - y = 4 \\ -2x + 2y = -8 \end{cases}$

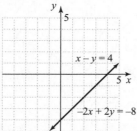

The graphs appear to be identical. To confirm this, write both equations in slope-intercept form.

$x - y = 4$ $-2x + 2y = -8$
$-y = -x + 4$ $-x + y = -4$
$y = x - 4$ $y = x - 4$

The equations are identical. Thus, there is an infinite number of solutions of the system; the system is consistent; the equations are dependent. The solution set is $\{(x, y) \mid x - y = 4\}$ or $\{(x, y) \mid -2x + 2y = -8\}$.

7. $\begin{cases} 5x + 4y = 6 \\ x - y = 3 \end{cases}$

Write each equation in slope-intercept form.
$5x + 4y = 6$ $x - y = 3$
$4y = -5x + 6$ $-y = -x + 3$
$y = -\dfrac{5}{4}x + \dfrac{3}{2}$ $y = x - 3$

The slopes are not equal, so the two lines are neither parallel nor identical and must intersect. Therefore, this system has one solution and is consistent.

8. $\begin{cases} -\dfrac{2}{3}x + y = 6 \\ 3y = 2x + 5 \end{cases}$

Write each equation in slope-intercept form.

$-\dfrac{2}{3}x + y = 6$ $3y = 2x + 5$

$y = \dfrac{2}{3}x + 6$ $y = \dfrac{2}{3}x + \dfrac{5}{3}$

The slope of each line is $\dfrac{2}{3}$, but they have different y-intercepts. Therefore, the lines are parallel. The system has no solution and is inconsistent.

Graphing Calculator Explorations

1. $\begin{cases} y = -2.68x + 1.21 \\ y = 5.22x - 1.68 \end{cases}$

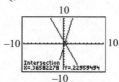

The approximate point of intersection is $(0.37, 0.23)$.

2. $\begin{cases} y = 4.25x + 3.89 \\ y = -1.88x + 3.21 \end{cases}$

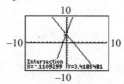

The approximate point of intersection is $(-0.11, 3.42)$.

3. $\begin{cases} 4.3x - 2.9y = 5.6 \\ 8.1x + 7.6y = -14.1 \end{cases}$

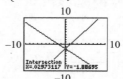

The approximate point of intersection is $(0.03, -1.89)$.

4. $\begin{cases} -3.6x - 8.6y = 10 \\ -4.5x + 9.6y = -7.7 \end{cases}$

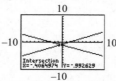

The approximate point of intersection is $(-0.41, -0.99)$.

Vocabulary, Readiness & Video Check 4.1

1. In a system of linear equations in two variables, if the graphs of the equations are the same, the equations are <u>dependent</u> equations.

2. Two or more linear equations are called a <u>system of linear equations</u>.

Copyright © 2013 Pearson Education, Inc.

3. A system of equations that has at least one solution is called a <u>consistent</u> system.

4. A <u>solution</u> of a system of two equations in two variables is an ordered pair of numbers that is a solution of both equations in the system.

5. A system of equations that has no solution is called an <u>inconsistent</u> system.

6. In a system of linear equations in two variables, if the graphs of the equations are different, the equations are <u>independent</u> equations.

7. The ordered pair must satisfy all equations of the system in order to be a solution of the system, so we must check that the ordered pair is a solution of both equations.

8. Graphing is not the most accurate method, especially if your graph is off just slightly, or the point of intersection does not have integer coordinates.

9. Writing the equations of a system in slope-intercept form lets you see their slope and y-intercept. Different slopes mean one solution; same slope with different y-intercepts means no solution; same slope with same y-intercept means infinite number of solutions.

Exercise Set 4.1

1. One solution, $(-1, 3)$

3. Infinite number of solutions

5. a. Let $x = 2$ and $y = 4$.

$$x + y = 8 \qquad\qquad 3x + 2y = 21$$
$$2 + 4 \stackrel{?}{=} 8 \qquad\quad 3(2) + 2(4) \stackrel{?}{=} 21$$
$$6 = 8 \;\; \text{False} \qquad\quad 6 + 8 \stackrel{?}{=} 21$$
$$14 = 21 \;\; \text{False}$$

(2, 4) is not a solution of the system.

b. Let $x = 5$ and $y = 3$.

$$x + y = 8 \qquad\qquad 3x + 2y = 21$$
$$5 + 3 \stackrel{?}{=} 8 \qquad\quad 3(5) + 2(3) \stackrel{?}{=} 21$$
$$8 = 8 \;\; \text{True} \qquad\quad 15 + 6 \stackrel{?}{=} 21$$
$$21 = 21 \;\; \text{True}$$

(5, 3) is a solution of the system.

7. a. Let $x = 3$ and $y = 4$.

$$3x - y = 5 \qquad\qquad x + 2y = 11$$
$$3(3) - 4 \stackrel{?}{=} 5 \qquad\quad 3 + 2(4) \stackrel{?}{=} 11$$
$$9 - 4 \stackrel{?}{=} 5 \qquad\qquad 3 + 8 \stackrel{?}{=} 11$$
$$5 = 5 \;\; \text{True} \qquad\qquad 11 = 11 \;\; \text{True}$$

(3, 4) is a solution of the system.

b. Let $x = 0$ and $y = -5$.

$$3x - y = 5 \qquad\qquad x + 2y = 11$$
$$3(0) - (-5) \stackrel{?}{=} 5 \qquad 0 + 2(-5) \stackrel{?}{=} 11$$
$$0 + 5 \stackrel{?}{=} 5 \qquad\qquad 0 - 10 \stackrel{?}{=} 11$$
$$5 = 5 \;\; \text{True} \qquad\quad -10 = 11 \;\; \text{False}$$

(0, −5) is not a solution of the system.

9. a. Let $x = -3$ and $y = -3$.

$$2y = 4x + 6 \qquad\qquad 2x - y = -3$$
$$2(-3) \stackrel{?}{=} 4(-3) + 6 \quad 2(-3) - (-3) \stackrel{?}{=} -3$$
$$-6 \stackrel{?}{=} -12 + 6 \qquad\quad -6 + 3 \stackrel{?}{=} -3$$
$$-6 = -6 \;\; \text{True} \qquad\quad -3 = -3 \;\; \text{True}$$

(−3, −3) is a solution of the system.

b. Let $x = 0$ and $y = 3$.

$$2y = 4x + 6 \qquad\qquad 2x - y = -3$$
$$2(3) \stackrel{?}{=} 4(0) + 6 \qquad 2(0) - 3 \stackrel{?}{=} -3$$
$$6 \stackrel{?}{=} 0 + 6 \qquad\qquad 0 - 3 \stackrel{?}{=} -3$$
$$6 = 6 \;\; \text{True} \qquad\quad -3 = -3 \;\; \text{True}$$

(0, 3) is a solution of the system.

11. a. Let $x = -2$ and $y = 0$.

$$-2 = x - 7y \qquad\qquad 6x - y = 13$$
$$-2 \stackrel{?}{=} -2 - 7(0) \quad 6(-2) - 0 \stackrel{?}{=} 13$$
$$-2 = -2 \;\; \text{True} \qquad\quad -12 = 13 \;\; \text{False}$$

(−2, 0) is not a solution of the system.

b. Let $x = \dfrac{1}{2}$ and $y = \dfrac{5}{14}$.

$$-2 = x - 7y \qquad\qquad 6x - y = 13$$
$$-2 \stackrel{?}{=} \frac{1}{2} - 7\left(\frac{5}{14}\right) \quad 6\left(\frac{1}{2}\right) - \left(\frac{5}{14}\right) \stackrel{?}{=} 13$$
$$2 \stackrel{?}{=} \frac{1}{2} - \frac{5}{2} = -\frac{4}{2} \quad 3 - \frac{5}{14} \stackrel{?}{=} 13$$
$$-2 = -2 \;\; \text{True} \qquad\qquad \frac{37}{14} = 13 \;\; \text{False}$$

$\left(\dfrac{1}{2}, \dfrac{5}{14}\right)$ is not a solution of the system.

13. $\begin{cases} x+y=4 \\ x-y=2 \end{cases}$

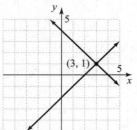

The solution of the system is (3, 1), consistent and independent.

15. $\begin{cases} x+y=6 \\ -x+y=-6 \end{cases}$

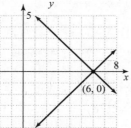

The solution of the system is (6, 0), consistent and independent.

17. $\begin{cases} y=2x \\ 3x-y=-2 \end{cases}$

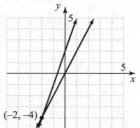

The solution of the system is (−2, −4), consistent and independent.

19. $\begin{cases} y=x+1 \\ y=2x-1 \end{cases}$

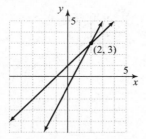

The solution of the system is (2, 3), consistent and independent.

21. $\begin{cases} 2x+y=0 \\ 3x+y=1 \end{cases}$

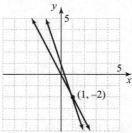

The solution of the system is (1, −2), consistent and independent.

23. $\begin{cases} y=-x-1 \\ y=2x+5 \end{cases}$

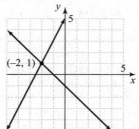

The solution of the system is (−2, 1), consistent and independent.

25. $\begin{cases} x+y=5 \\ x+y=6 \end{cases}$

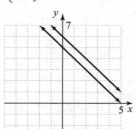

There is no solution, inconsistent and independent. The solution set is { } or ∅.

27. $\begin{cases} 2x - y = 6 \\ y = 2 \end{cases}$

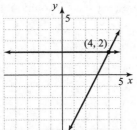

The solution of the system is (4, 2), consistent and independent.

29. $\begin{cases} x - 2y = 2 \\ 3x + 2y = -2 \end{cases}$

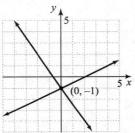

The solution of the system is (0, −1), consistent and independent.

31. $\begin{cases} 2x + y = 4 \\ 6x = -3y + 6 \end{cases}$

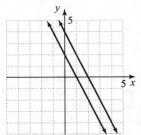

There is no solution, inconsistent and independent. The solution set is { } or ∅.

33. $\begin{cases} y - 3x = -2 \\ 6x - 2y = 4 \end{cases}$

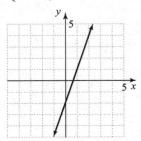

There is an infinite number of solutions, $\{(x, y) | y - 3x = -2\}$ or $\{(x, y) | 6x - 2y = 4\}$; consistent and dependent.

35. $\begin{cases} x = 3 \\ y = -1 \end{cases}$

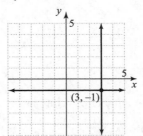

The solution of the system is (3, −1), consistent and independent.

37. $\begin{cases} y = x - 2 \\ y = 2x + 3 \end{cases}$

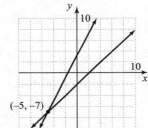

The solution of the system is (−5, −7), consistent and independent.

39. $\begin{cases} 2x - 3y = -2 \\ -3x + 5y = 5 \end{cases}$

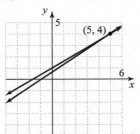

The solution of the system is (5, 4), consistent and independent.

41. $\begin{cases} 6x - y = 4 \\ \dfrac{1}{2}y = -2 + 3x \end{cases}$

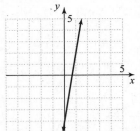

There is an infinite number of solutions,

$\{(x, y) | 6x - y = 4\}$ or $\left\{(x, y) \left| \dfrac{1}{2}y = -2 + 3x \right.\right\}$;

consistent and dependent.

43. $\begin{cases} 4x + y = 24 \\ x + 2y = 2 \end{cases} \rightarrow \begin{cases} y = -4x + 24 \\ y = -\dfrac{1}{2}x + 1 \end{cases}$

The lines are intersecting; there is one solution.

45. $\begin{cases} 2x + y = 0 \\ 2y = 6 - 4x \end{cases} \rightarrow \begin{cases} y = -2x \\ y = -2x + 3 \end{cases}$

The lines are parallel; there is no solution.

47. $\begin{cases} 6x - y = 4 \\ \dfrac{1}{2}y = -2 + 3x \end{cases} \rightarrow \begin{cases} y = 6x - 4 \\ y = 6x - 4 \end{cases}$

The lines are identical; there is an infinite number of solutions.

49. $\begin{cases} x = 5 \\ y = -2 \end{cases}$

The lines are intersecting; there is one solution.

51. $\begin{cases} 3y - 2x = 3 \\ x + 2y = 9 \end{cases} \rightarrow \begin{cases} y = \dfrac{2}{3}x + 1 \\ y = -\dfrac{1}{2}x + \dfrac{9}{2} \end{cases}$

The lines are intersecting; there is one solution.

53. $\begin{cases} 6y + 4x = 6 \\ 3y - 3 = -2x \end{cases} \rightarrow \begin{cases} y = -\dfrac{2}{3}x + 1 \\ y = -\dfrac{2}{3}x + 1 \end{cases}$

The lines are identical; there is an infinite number of solutions.

55. $\begin{cases} x + y = 4 \\ x + y = 3 \end{cases} \rightarrow \begin{cases} y = -x + 4 \\ y = -x + 3 \end{cases}$

The lines are parallel; there is no solution.

57. $5(x - 3) + 3x = 1$
$5x - 15 + 3x = 1$
$8x - 15 = 1$
$8x = 16$
$x = 2$
The solution is 2.

59. $4\left(\dfrac{y+1}{2}\right) + 3y = 0$
$2(y + 1) + 3y = 0$
$2y + 2 + 3y = 0$
$5y + 2 = 0$
$5y = -2$
$y = -\dfrac{2}{5}$
The solution is $-\dfrac{2}{5}$.

61. $8a - 2(3a - 1) = 6$
$8a - 6a + 2 = 6$
$2a + 2 = 6$
$2a = 4$
$a = 2$
The solution is 2.

63. answers may vary

65. answers may vary

67. answers may vary

69. The lines cross at a point between 1988 and 1989, and again between 2001 and 2002. The number of pounds of imported fishery products was equal to the domestic catch between 1988 and 1989, and also between 2001 and 2002.

71. The average attendance per game for the Texas Rangers was greater than the average attendance per game for the Minnesota Twins in 2003, 2004, 2005, 2006, and 2007.

73. answers may vary

75. a. (4, 9) appears in both tables, so it is a solution of the system.

b.

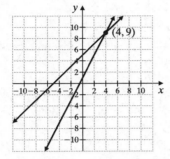

c. Yes; the two lines intersect at (4, 9).

77. answers may vary

Section 4.2 Practice

1. $\begin{cases} 2x - y = 9 \\ x = y + 1 \end{cases}$

Substitute $y + 1$ for x in the first equation.
$$2x - y = 9$$
$$2(y + 1) - y = 9$$
$$2y + 2 - y = 9$$
$$y + 2 = 9$$
$$y = 7$$

Let $y = 7$ in the second equation.
$$x = y + 1 = 7 + 1 = 8$$
The solution of the system is (8, 7).
Check.

$2x - y = 9$	$x = y + 1$
$2(8) - 7 \stackrel{?}{=} 9$	$8 \stackrel{?}{=} 7 + 1$
$16 - 7 \stackrel{?}{=} 9$	$8 = 8$ True
$9 = 9$ True	

The solution of the system is (8, 7).

2. $\begin{cases} 7x - y = -15 \\ y = 2x \end{cases}$

Substitute $2x$ for y in the first equation.
$$7x - y = -15$$
$$7x - 2x = -15$$
$$5x = -15$$
$$x = -3$$

Let $x = -3$ in the second equation.
$$y = 2x = 2(-3) = -6$$
The solution of the system is (−3, −6).

3. $\begin{cases} x + 3y = 6 \\ 2x + 3y = 10 \end{cases}$

Solve the first equation for x.
$$x + 3y = 6$$
$$x = -3y + 6$$

Substitute $-3y + 6$ for x in the second equation.

$$2x + 3y = 10$$
$$2(-3y + 6) + 3y = 10$$
$$-6y + 12 + 3y = 10$$
$$-3y + 12 = 10$$
$$-3y = -2$$
$$y = \frac{2}{3}$$

Let $y = \dfrac{2}{3}$ in the equation for x.

$$x = 3y + 6 = -3\left(\frac{2}{3}\right) + 6 = -2 + 6 = 4$$

The solution of the system is $\left(4, \dfrac{2}{3}\right)$.

4. $\begin{cases} 5x + 3y = -9 \\ -2x + y = 8 \end{cases}$

Solve the second equation for y.
$$-2x + y = 8$$
$$y = 2x + 8$$

Substitute $2x + 8$ for y in the first equation.
$$5x + 3y = -9$$
$$5x + 3(2x + 8) = -9$$
$$5x + 6x + 24 = -9$$
$$11x + 24 = -9$$
$$11x = -33$$
$$x = -3$$

Let $x = -3$ in the equation for y.
$$y = 2x + 8 = 2(-3) + 8 = -6 + 8 = 2$$
The solution of the system is (−3, 2).

5. $\begin{cases} \dfrac{1}{4}x - y = 2 \\ x = 4y + 8 \end{cases}$

Substitute $4y + 8$ for x in the first equation.

$$\frac{1}{4}x - y = 2$$

$$\frac{1}{4}(4y + 8) - y = 2$$

$$y + 2 - y = 2$$
$$2 = 2$$

The two linear equations are equivalent. Thus, the system has an infinite number of solutions,

$$\left\{ (x,\, y) \,\middle|\, \frac{1}{4}x - y = 2 \right\} \text{ or } \{(x,\, y) \mid x = 4y + 8\}.$$

6. $\begin{cases} 4x - 3y = 12 \\ -8x + 6y = -30 \end{cases}$

Solve the first equation for x.

$$4x - 3y = 12$$
$$4x = 3y + 12$$
$$x = \frac{3}{4}y + 3$$

Substitute $\frac{3}{4}y + 3$ for x in the second equation.

$$-8x + 6y = -30$$
$$-8\left(\frac{3}{4}y + 3\right) + 6y = -30$$
$$-6y - 24 + 6y = -30$$
$$-24 = -30$$

The false statement $-24 = -30$ indicates that the system has no solution and is inconsistent. The solution set is { } or $\varnothing$.

Vocabulary, Readiness & Video Check 4.2

1. Since $x = 1$, $y = 4x = 4(1) = 4$ and the solution is (1, 4).

2. There is no solution, since $0 = 34$ is a false statement.

3. There is an infinite number of solutions, since the statement $0 = 0$ is true for all values of the variables.

4. Since $y = 0$, $x = y + 5 = 0 + 5 = 5$ and the solution is (5, 0).

5. Since $x = 0$ and $x + y = 0$, $y = -x = -0 = 0$ and the solution is (0, 0).

6. There is an infinite number of solutions, since the statement $0 = 0$ is true for all values of the variables.

7. You solved one equation for a variable. Now be sure to substitute this expression for the variable into the *other* equation.

Exercise Set 4.2

1. $\begin{cases} x + y = 3 \\ x = 2y \end{cases}$

Substitute $2y$ for x in the first equation.
$$2y + y = 3$$
$$3y = 3$$
$$y = 1$$
Let $y = 1$ in the second equation.
$$x = 2(1) = 2$$
The solution is (2, 1).

3. $\begin{cases} x + y = 6 \\ y = -3x \end{cases}$

Substitute $-3x$ for y in the first equation.
$$x + (-3x) = 6$$
$$-2x = 6$$
$$x = -3$$
Let $x = -3$ in the second equation.
$$y = -3(-3) = 9$$
The solution is (−3, 9).

5. $\begin{cases} y = 3x + 1 \\ 4y - 8x = 12 \end{cases}$

Substitute $3x + 1$ for y in the second equation.
$$4(3x + 1) - 8x = 12$$
$$12x + 4 - 8x = 12$$
$$4x + 4 = 12$$
$$4x = 8$$
$$x = 2$$
Let $x = 2$ in the first equation.
$$y = 3(2) + 1 = 7$$
The solution is (2, 7).

7. $\begin{cases} y = 2x + 9 \\ y = 7x + 10 \end{cases}$

Substitute $2x + 9$ for y in the second equation.
$$2x + 9 = 7x + 10$$
$$-5x + 9 = 10$$
$$-5x = 1$$
$$x = -\frac{1}{5}$$

Let $x = -\frac{1}{5}$ in the first equation.

$$y = 2\left(-\frac{1}{5}\right) + 9 = -\frac{2}{5} + \frac{45}{5} = \frac{43}{5}$$

The solution is $\left(-\frac{1}{5}, \frac{43}{5}\right)$.

9. $\begin{cases} 3x - 4y = 10 \\ y = x - 3 \end{cases}$

Substitute $x - 3$ for y in the first equation.
$$3x - 4(x - 3) = 10$$
$$3x - 4x + 12 = 10$$
$$-x = -2$$
$$x = 2$$
Let $x = 2$ in the second equation
$$y = 2 - 3 = -1$$
The solution is (2, −1).

11. $\begin{cases} x+2y=6 \\ 2x+3y=8 \end{cases}$

Solve the first equation for x.

$x = 6 - 2y$

Substitute $6 - 2y$ for x in the second equation.

$2(6-2y)+3y=8$

$12-4y+3y=8$

$-y=-4$

$y=4$

Let $y = 4$ in $x = 6 - 2y$.

$x = 6 - 2(4) = -2$

The solution is $(-2, 4)$.

13. $\begin{cases} 3x+2y=16 \\ x=3y-2 \end{cases}$

Substitute $3y - 2$ for x in the first equation.

$3(3y-2)+2y=16$

$9y-6+2y=16$

$11y=22$

$y=2$

Let $y = 2$ in the second equation.

$x = 3(2) - 2 = 4$

The solution is $(4, 2)$.

15. $\begin{cases} 2x-5y=1 \\ 3x+y=-7 \end{cases}$

Solve the second equation for y.

$y = -7 - 3x$

Substitute $-7 - 3x$ for y in the first equation.

$2x-5(-7-3x)=1$

$2x+35+15x=1$

$17x=-34$

$x=-2$

Let $x = -2$ in $y = -7 - 3x$.

$y = -7 - 3(-2) = -1$

The solution is $(-2, -1)$.

17. $\begin{cases} 4x+2y=5 \\ -2x=y+4 \end{cases}$

Solve the second equation for y.

$y = -2x - 4$

Substitute $-2x - 4$ for y in the first equation.

$4x+2(-2x-4)=5$

$4x-4x-8=5$

$-8=5$ False

The system has no solution. The solution set is $\{\ \}$ or $\varnothing$.

19. $\begin{cases} 4x+y=11 \\ 2x+5y=1 \end{cases}$

Solve the first equation for y.

$y = 11 - 4x$

Substitute $11 - 4x$ for y in the second equation.

$2x+5(11-4x)=1$

$2x+55-20x=1$

$-18x=-54$

$x=3$

Let $x = 3$ in $y = 11 - 4x$.

$y = 11 - 4(3) = -1$

The solution is $(3, -1)$.

21. $\begin{cases} x+2y+5=-4+5y-x \\ \quad 2x+9=3y \\ \\ \quad 2x+x=y+4 \\ \quad -4+3x=y \end{cases}$

Substitute $3x - 4$ for y in $2x + 9 = 3y$.

$2x+9=3(3x-4)$

$2x+9=9x-12$

$21=7x$

$3=x$

Let $x = 3$ in $3x - 4 = y$.

$y = 3(3) - 4 = 5$

The solution of the system is $(3, 5)$.

23. $\begin{cases} 6x-3y=5 \\ x+2y=0 \end{cases}$

Solve the second equation for x.

$x = -2y$

Substitute $-2y$ for x in the first equation.

$6(-2y)-3y=5$

$-12y-3y=5$

$-15y=5$

$y=-\dfrac{1}{3}$

Let $y = -\dfrac{1}{3}$ in $x = -2y$.

$x = -2\left(-\dfrac{1}{3}\right) = \dfrac{2}{3}$

The solution is $\left(\dfrac{2}{3}, -\dfrac{1}{3}\right)$.

25. $\begin{cases} 3x-y=1 \\ 2x-3y=10 \end{cases}$

Solve the first equation for y.

$y = 3x - 1$

Substitute $3x - 1$ for y in the second equation.

$$2x - 3(3x - 1) = 10$$
$$2x - 9x + 3 = 10$$
$$-7x = 7$$
$$x = -1$$

Let $x = -1$ in $y = 3x - 1$.
$$y = 3(-1) - 1 = -4$$
The solution is $(-1, -4)$.

27. $\begin{cases} -x + 2y = 10 \\ -2x + 3y = 18 \end{cases}$

Solve the first equation for x.
$$x = 2y - 10$$
Substitute $2y - 10$ for x in the second equation.
$$-2(2y - 10) + 3y = 18$$
$$-4y + 20 + 3y = 18$$
$$-y = -2$$
$$y = 2$$

Let $y = 2$ in $x = 2y - 10$.
$$x = 2(2) - 10 = -6$$
The solution is $(-6, 2)$.

29. $\begin{cases} 5x + 10y = 20 \\ 2x + 6y = 10 \end{cases}$

Solve the first equation for x.
$$x + 2y = 4$$
$$x = 4 - 2y$$
Substitute $4 - 2y$ for x in the second equation.
$$2(4 - 2y) + 6y = 10$$
$$8 - 4y + 6y = 10$$
$$2y = 2$$
$$y = 1$$

Let $y = 1$ in $x = 4 - 2y$.
$$x = 4 - 2(1) = 2$$
The solution is $(2, 1)$.

31. $\begin{cases} 3x + 6y = 9 \\ 4x + 8y = 16 \end{cases}$

Solve the first equation for x.
$$x + 2y = 3$$
$$x = 3 - 2y$$
Substitute $3 - 2y$ for x in the second equation.
$$4(3 - 2y) + 8y = 16$$
$$12 - 8y + 8y = 16$$
$$12 = 16 \quad \text{False}$$
The system has no solution. The solution set is $\{\ \}$ or $\varnothing$.

33. $\begin{cases} \dfrac{1}{3}x - y = 2 \\ x - 3y = 6 \end{cases}$

Solve the second equation for x.
$$x = 6 + 3y$$
Substitute $6 + 3y$ for x in the first equation.
$$\frac{1}{3}(6 + 3y) - y = 2$$
$$2 + y - y = 2$$
$$2 = 2$$
The equations in the original system are equivalent and there is an infinite number of solutions, $\left\{(x,\ y)\ \middle|\ \dfrac{1}{3}x - y = 2\right\}$ or $\{(x, y) | x - 3y = 6\}$.

35. $\begin{cases} x = \dfrac{3}{4}y - 1 \\ 8x - 5y = -6 \end{cases}$

Substitute $\dfrac{3}{4}y - 1$ for x in the second equation.
$$8\left(\frac{3}{4}y - 1\right) - 5y = -6$$
$$6y - 8 - 5y = -6$$
$$y = 2$$
Let $y = 2$ in the first equation.
$$x = \frac{3}{4}(2) - 1 = \frac{1}{2}$$
The solution of the system is $\left(\dfrac{1}{2},\ 2\right)$.

37. $\begin{cases} -5y + 6y = 3x + 2(x - 5) - 3x + 5 \\ \quad\quad y = 3x + 2x - 10 - 3x + 5 \\ \quad\quad y = 2x - 5 \\ \ \\ 4(x + y) - x + y = -12 \\ \quad 4x + 4y - x + y = -12 \\ \quad\quad\quad 3x + 5y = -12 \end{cases}$

Substitute $2x - 5$ for y in the second equation.
$$3x + 5(2x - 5) = -12$$
$$3x + 10x - 25 = -12$$
$$13x = 13$$
$$x = 1$$
Let $x = 1$ in $y = 2x - 5$.
$$y = 2(1) - 5 = -3$$
The solution is $(1, -3)$.

39.
$$3x + 2y = 6$$
$$-2(3x + 2y) = -2(6)$$
$$-6x - 4y = -12$$

41.
$$-4x + y = 3$$
$$3(-4x + y) = 3(3)$$
$$-12x + 3y = 9$$

43.
$$\begin{array}{r} 3n + 6m \\ 2n - 6m \\ \hline 5n \end{array}$$

45.
$$\begin{array}{r} -5a - 7b \\ 5a - 8b \\ \hline -15b \end{array}$$

47. answers may vary

49. No; answers may vary.

51. **c**; answers may vary.

53. a.
$$\begin{cases} y = 3.9x + 443 \\ y = 14.2x + 314 \end{cases}$$

Substitute $14.2x + 314$ for y in the first equation.
$$14.2x + 314 = 3.9x + 443$$
$$10.3x = 129$$
$$x \approx 12.52$$
Let $x = 12.52$ in $y = 3.9x + 443$.
$$y \approx 3.9(12.52) + 443 \approx 491.828$$
The solution is (13, 492).

b. In 1970 + 13 = 1983, the number of men and the number of women receiving bachelor's degrees was the same.

c.

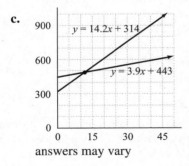

answers may vary

55.
$$\begin{cases} y = 5.1x + 14.56 \\ y = -2x - 3.9 \end{cases}$$

Substitute $-2x - 3.9$ for y in the first equation.

$$-2x - 3.9 = 5.1x + 14.56$$
$$-7.1x = 18.46$$
$$x = -2.6$$
Let $x = -2.6$ in $y = -2x - 3.9$.
$$y = -2(-2.6) - 3.9 = 1.3$$
The solution is (-2.6, 1.3).

57.
$$\begin{cases} 3x + 2y = 14.05 \\ 5x + y = 18.5 \end{cases}$$

Solve the second equation for y.
$$y = -5x + 18.5$$
Substitute $-5x + 18.5$ for y in the first equation.
$$3x + 2(-5x + 18.5) = 14.05$$
$$3x - 10x + 37 = 14.05$$
$$-7x = -22.95$$
$$x \approx 3.279$$
Let $x = 3.279$ in $y = -5x + 18.5$.
$$y \approx -5(3.279) + 18.5 \approx 2.105$$
The solution is approximately (3.28, 2.11).

Section 4.3 Practice

1.
$$\begin{cases} x - y = 2 \\ x + y = 8 \end{cases}$$

Add the left sides of the equations together and the right sides of the equations together.
$$\begin{array}{r} x - y = 2 \\ x + y = 8 \\ \hline 2x \quad\;\; = 10 \\ x = 5 \end{array}$$
Let $x = 5$ in the first equation.
$$x - y = 2$$
$$5 - y = 2$$
$$3 = y$$
The solution is (5, 3).
Check.

$x - y = 2$	$x + y = 8$
$5 - 3 \overset{?}{=} 2$	$5 + 3 \overset{?}{=} 8$
$2 = 2$ True	$8 = 8$ True

The solution of the system is (5, 3).

2.
$$\begin{cases} x - 2y = 11 \\ 3x - y = 13 \end{cases}$$

Multiply both sides of the first equation by -3 and add to the second equation.
$$\begin{array}{r} -3x + 6y = -33 \\ 3x - y = 13 \\ \hline 5y = -20 \\ y = -4 \end{array}$$
Let $y = -4$ in the first equation.

$$x - 2y = 11$$
$$x - 2(-4) = 11$$
$$x + 8 = 11$$
$$x = 3$$

The solution of the system is $(3, -4)$.

3. $\begin{cases} x - 3y = 5 \\ 2x - 6y = -3 \end{cases}$

Multiply both sides of the first equation by -2 and add to the second equation.

$$-2x + 6y = -10$$
$$\underline{2x - 6y = -3}$$
$$0 = -13 \quad \text{False}$$

The system has no solution. The solution set is $\{\ \}$ or $\varnothing$.

4. $\begin{cases} 4x - 3y = 5 \\ -8x + 6y = -10 \end{cases}$

Multiply the first equation by 2 and add to the second equation.

$$8x - 6y = 10$$
$$\underline{-8x + 6y = -10}$$
$$0 = 0 \quad \text{True}$$

The equations are equivalent, so the system has an infinite number of solutions,
$\{(x, y)|4x - 3y = 5\}$ or $\{(x, y)|-8x + 6y = -10\}$.

5. $\begin{cases} 4x + 3y = 14 \\ 3x - 2y = 2 \end{cases}$

Multiply the first equation by 2 and the second equation by 3 and add.

$$8x + 6y = 28$$
$$\underline{9x - 6y = 6}$$
$$17x \qquad = 34$$
$$x = 2$$

Let $x = 2$ in the second equation.

$$3x - 2y = 2$$
$$3(2) - 2y = 2$$
$$6 - 2y = 2$$
$$-2y = -4$$
$$y = 2$$

The solution of the system is $(2, 2)$.

6. $\begin{cases} -2x + \dfrac{3y}{2} = 5 \\ -\dfrac{x}{2} - \dfrac{y}{4} = \dfrac{1}{2} \end{cases}$

Clear fractions by multiplying the first equation by 2 and the second by 4.

$\begin{cases} -4x + 3y = 10 \\ -2x - y = 2 \end{cases}$

Multiply the second simplified equation by 3 and add.

$$-4x + 3y = 10$$
$$\underline{-6x - 3y = 6}$$
$$-10x \qquad = 16$$
$$x = -\frac{16}{10} = -\frac{8}{5}$$

Now multiply the second simplified equation by -2 and add.

$$-4x + 3y = 10$$
$$\underline{4x + 2y = -4}$$
$$5y = 6$$
$$y = \frac{6}{5}$$

The solution of the system is $\left(-\dfrac{8}{5}, \dfrac{6}{5}\right)$.

Vocabulary, Readiness & Video Check 4.3

1. $\begin{cases} 3x - 2y = -9 \\ x + 5y = 14 \end{cases}$

Multiply the second equation by -3, then add the resulting equations.

$$3x - 2y = -9$$
$$\underline{-3x - 15y = -42}$$
$$-17y = -51$$

The y's are not eliminated; the statement is false.

2. $\begin{cases} 3x - 2y = -9 \\ x + 5y = 14 \end{cases}$

Multiply the second equation by -3, then add the resulting equations.

$$3x - 2y = -9$$
$$\underline{-3x - 15y = -42}$$
$$-17y = -51$$

The statement is true.

3. $\begin{cases} 3x - 2y = -9 \\ x + 5y = 14 \end{cases}$

Multiply the first equation by 5 and the second equation by 2, then add the two new equations.

$$15x - 10y = -45$$
$$\underline{2x + 10y = 28}$$
$$17x \qquad = -17$$

The statement is true.

4. $\begin{cases} 3x - 2y = -9 \\ x + 5y = 14 \end{cases}$

Multiply the first equation by 5 and the second equation by −2, then add the two new equations.

$$15x - 10y = -45$$
$$\underline{-2x - 10y = -28}$$
$$13x - 20y = -73$$

The *y*'s are not eliminated; the statement is false.

5. The multiplication property of equality; be sure to multiply *both* sides of the equation by the number chosen.

Exercise Set 4.3

1. $\begin{cases} 3x + y = 5 \\ 6x - y = 4 \end{cases}$

$$3x + y = 5$$
$$\underline{6x - y = 4}$$
$$9x \quad\; = 9$$
$$x = 1$$

Let *x* = 1 in the first equation.
$$3(1) + y = 5$$
$$3 + y = 5$$
$$y = 2$$

The solution of the system is (1, 2).

3. $\begin{cases} x - 2y = 8 \\ -x + 5y = -17 \end{cases}$

$$x - 2y = 8$$
$$\underline{-x + 5y = -17}$$
$$3y = -9$$
$$y = -3$$

Let *y* = −3 in the first equation.
$$x - 2(-3) = 8$$
$$x + 6 = 8$$
$$x = 2$$

The solution of the system is (2, −3).

5. $\begin{cases} 3x + y = -11 \\ 6x - 2y = -2 \end{cases}$

Multiply the first equation by 2.
$$6x + 2y = -22$$
$$\underline{6x - 2y = \; -2}$$
$$12x \qquad = -24$$
$$x = -2$$

Let *x* = −2 in the first equation.

$$3(-2) + y = -11$$
$$-6 + y = -11$$
$$y = -5$$

The solution of the system is (−2, −5).

7. $\begin{cases} 3x + 2y = 11 \\ 5x - 2y = 29 \end{cases}$

$$3x + 2y = 11$$
$$\underline{5x - 2y = 29}$$
$$8x \qquad = 40$$
$$x = 5$$

Let *x* = 5 in the first equation.
$$3(5) + 2y = 11$$
$$15 + 2y = 11$$
$$2y = -4$$
$$y = -2$$

The solution of the system is (5, −2).

9. $\begin{cases} x + 5y = 18 \\ 3x + 2y = -11 \end{cases}$

Multiply the first equation by −3.
$$-3x - 15y = -54$$
$$\underline{3x + \; 2y = -11}$$
$$-13y = -65$$
$$y = 5$$

Let *y* = 5 in the first equation.
$$x + 5(5) = 18$$
$$x + 25 = 18$$
$$x = -7$$

The solution of the system is (−7, 5).

11. $\begin{cases} x + y = 6 \\ x - y = 6 \end{cases}$

$$x + y = 6$$
$$\underline{x - y = 6}$$
$$2x \quad\; = 12$$
$$x = 6$$

Let *x* = 6 in the first equation.
$$6 + y = 6$$
$$y = 0$$

The solution of the system is (6, 0).

13. $\begin{cases} 2x + 3y = 0 \\ 4x + 6y = 3 \end{cases}$

Multiply the first equation by −2.
$$-4x - 6y = 0$$
$$\underline{4x + 6y = 3}$$
$$0 = 3 \quad \text{False}$$

The system has no solution. The solution set is { } or ∅.

15. $\begin{cases} -x+5y=-1 \\ 3x-15y=3 \end{cases}$

Multiply the first equation by 3.

$$-3x+15y=-3$$
$$\underline{3x-15y=3}$$
$$0=0$$

There is an infinite number of solutions,
$\{(x, y) | -x + 5y = -1\}$ or $\{(x, y) | 3x - 15y = 3\}$.

17. $\begin{cases} 3x-2y=7 \\ 5x+4y=8 \end{cases}$

Multiply the first equation by 2.

$$6x-4y=14$$
$$\underline{5x+4y=8}$$
$$11x\qquad=22$$
$$x=2$$

Let $x = 2$ in the first equation.

$$3(2)-2y=7$$
$$6-2y=7$$
$$-2y=1$$
$$y=-\frac{1}{2}$$

The solution of the system is $\left(2, -\dfrac{1}{2}\right)$.

19. $\begin{cases} 8x=-11y-16 \\ 2x+3y=-4 \end{cases}$

Add $11y$ to both sides of the first equation and multiply the second equation by -4, then add.

$$8x+11y=-16$$
$$\underline{-8x-12y=16}$$
$$-y=0$$
$$y=0$$

Let $y = 0$ in the first equation.

$$8x=-11(0)-16$$
$$8x=-16$$
$$x=-2$$

The solution of the system is $(-2, 0)$.

21. $\begin{cases} 4x-3y=7 \\ 7x+5y=2 \end{cases}$

Multiply the first equation by 5 and the second equation by 3.

$$20x-15y=35$$
$$\underline{21x+15y=6}$$
$$41x\qquad=41$$
$$x=1$$

Let $x = 1$ in the first equation.

$$4x-3y=7$$
$$4(1)-3y=7$$
$$4-3y=7$$
$$-3y=3$$
$$y=-1$$

The solution of the system is $(1, -1)$.

23. $\begin{cases} 4x-6y=8 \\ 6x-9y=16 \end{cases}$

Multiply the first equation by 3 and the second equation by -2.

$$12x-18y=24$$
$$\underline{-12x+18y=-32}$$
$$0=-12 \text{ False}$$

The system has no solution. The solution set is $\{\ \}$ or $\varnothing$.

25. $\begin{cases} 2x-5y=4 \\ 3x-2y=4 \end{cases}$

Multiply the first equation by -3 and the second equation by 2.

$$-6x+15y=-12$$
$$\underline{6x-4y=8}$$
$$11y=-4$$
$$y=-\frac{4}{11}$$

Multiply the first equation by -2 and the second equation by 5.

$$-4x+10y=-8$$
$$\underline{15x-10y=20}$$
$$11x\qquad=12$$
$$x=\frac{12}{11}$$

The solution of the system is $\left(\dfrac{12}{11}, -\dfrac{4}{11}\right)$.

27. $\begin{cases} \dfrac{x}{3}+\dfrac{y}{6}=1 \\ \dfrac{x}{2}-\dfrac{y}{4}=0 \end{cases}$

Multiply the first equation by 6 and the second equation by 4.

$$2x+y=6$$
$$\underline{2x-y=0}$$
$$4x\qquad=6$$
$$x=\frac{3}{2}$$

Multiply the second equation of the simplified system by -1.

$$2x + y = 6$$
$$\underline{-2x + y = 0}$$
$$2y = 6$$
$$y = 3$$

The solution of the system is $\left(\dfrac{3}{2}, 3 \right)$.

29. $\begin{cases} \dfrac{10}{3}x + 4y = -4 \\ 5x + 6y = -6 \end{cases}$

Multiply the first equation by 3 and the second equation by -2.

$$10x + 12y = -12$$
$$\underline{-10x - 12y = 12}$$
$$0 = 0$$

The system has an infinite number of solutions,

$\left\{ (x, y) \Big| \dfrac{10}{3}x + 4y = -4 \right\}$ or $\{(x, y) | 5x + 6y = -6\}$.

31. $\begin{cases} x - \dfrac{y}{3} = -1 \\ -\dfrac{x}{2} + \dfrac{y}{8} = \dfrac{1}{4} \end{cases}$

Multiply the first equation by 3 and the second equation by 8.

$$3x - y = -3$$
$$\underline{-4x + y = 2}$$
$$-x = -1$$
$$x = 1$$

Multiply the first equation of the simplified system by 4 and the second equation by 3.

$$12x - 4y = -12$$
$$\underline{-12x + 3y = 6}$$
$$-y = -6$$
$$y = 6$$

The solution of the system is $(1, 6)$.

33. $\begin{cases} -4(x + 2) = 3y \\ 2x - 2y = 3 \end{cases} \rightarrow \begin{cases} -4x - 8 = 3y \\ 2x - 2y = 3 \end{cases}$

$\rightarrow \begin{cases} -4x - 3y = 8 \\ 2x - 2y = 3 \end{cases}$

Multiply the second equation by 2.

$$-4x - 3y = 8$$
$$\underline{4x - 4y = 6}$$
$$-7y = 14$$
$$y = -2$$

Let $y = -2$ in the second equation.

$$2x - 2(-2) = 3$$
$$2x + 4 = 3$$
$$2x = -1$$
$$x = -\dfrac{1}{2}$$

The solution of the system is $\left(-\dfrac{1}{2}, -2 \right)$.

35. $\begin{cases} \dfrac{x}{3} - y = 2 \\ -\dfrac{x}{2} + \dfrac{3y}{2} = -3 \end{cases}$

Multiply the first equation by 3 and the second equation by 2.

$$x - 3y = 6$$
$$\underline{-2x + 3y = -6}$$
$$0 = 0$$

The equations of the original system are equivalent and there is an infinite number of

solutions, $\left\{ (x, y) \Big| \dfrac{x}{3} - y = 2 \right\}$ or

$\left\{ (x, y) \Big| -\dfrac{x}{2} + \dfrac{3y}{2} = -3 \right\}$.

37. $\begin{cases} \dfrac{3}{5}x - y = -\dfrac{4}{5} \\ 3x + \dfrac{y}{2} = -\dfrac{9}{5} \end{cases}$

Multiply the first equation by 5 and the second equation by 10.

$$3x - 5y = -4$$
$$\underline{30x + 5y = -18}$$
$$33x = -22$$
$$x = -\dfrac{2}{3}$$

Let $x = -\dfrac{2}{3}$ in $30x + 5y = -18$.

$$30\left(-\dfrac{2}{3} \right) + 5y = -18$$
$$-20 + 5y = -18$$
$$5y = 2$$
$$y = \dfrac{2}{5}$$

The solution of the system is $\left(-\dfrac{2}{3}, \dfrac{2}{5} \right)$.

39. $\begin{cases} 3.5x + 2.5y = 17 \\ -1.5x - 7.5y = -33 \end{cases}$

Multiply the first equation by 6 and the second equation by 2.

$21x + 15y = 102$
$\underline{-3x - 15y = -66}$
$18x \quad\quad = 36$
$\quad\quad x = 2$

Let $x = 2$ in $-3x - 15y = -66$.
$-3(2) - 15y = -66$
$-6 - 15y = -66$
$-15y = -60$
$y = 4$

The solution of the system is $(2, 4)$.

41. $\begin{cases} 0.02x + 0.04y = 0.09 \\ -0.1x + 0.3y = 0.8 \end{cases}$

Multiply the first equation by 100 and the second equation by 20.

$2x + 4y = 9$
$\underline{-2x + 6y = 16}$
$10y = 25$
$y = \dfrac{5}{2} = 2.5$

Let $y = 2.5$ in $2x + 4y = 9$.
$2x + 4(2.5) = 9$
$2x + 10 = 9$
$2x = -1$
$x = -\dfrac{1}{2} = -0.5$

The solution of the system is $(-0.5, 2.5)$.

43. $\begin{cases} 2x - 3y = -11 \\ y = 4x - 3 \end{cases}$

Substitute $4x - 3$ for y in the first equation.
$2x - 3(4x - 3) = -11$
$2x - 12x + 9 = -11$
$-10x = -20$
$x = 2$

Let $x = 2$ in the second equation.
$y = 4(2) - 3 = 5$
The solution of the system is $(2, 5)$.

45. $\begin{cases} x + 2y = 1 \\ 3x + 4y = -1 \end{cases}$

Multiply the first equation by -2.
$-2x - 4y = -2$
$\underline{3x + 4y = -1}$
$x \quad\quad = -3$

Let $x = -3$ in the first equation.

$-3 + 2y = 1$
$2y = 4$
$y = 2$
The solution is $(-3, 2)$.

47. $\begin{cases} 2y = x + 6 \\ 3x - 2y = -6 \end{cases}$

Subtract x from both sides of the first equation.
$-x + 2y = 6$
$\underline{3x - 2y = -6}$
$2x \quad\quad = 0$
$\quad x = 0$

Let $x = 0$ in the first equation.
$2y = 0 + 6$
$2y = 6$
$y = 3$
The solution of the system is $(0, 3)$.

49. $\begin{cases} y = 2x - 3 \\ y = 5x - 18 \end{cases}$

Substitute $5x - 18$ for y in the first equation.
$5x - 18 = 2x - 3$
$3x = 15$
$x = 5$

Let $x = 5$ in the second equation.
$y = 5(5) - 18 = 7$
The solution of the system is $(5, 7)$.

51. $\begin{cases} x + \dfrac{1}{6}y = \dfrac{1}{2} \\ 3x + 2y = 3 \end{cases}$

Multiply the first equation by -12.
$-12x - 2y = -6$
$\underline{3x + 2y = 3}$
$-9x \quad\quad = -3$
$\quad\quad x = \dfrac{1}{3}$

Substitute $\dfrac{1}{3}$ for x in the second equation.

$3\left(\dfrac{1}{3}\right) + 2y = 3$
$1 + 2y = 3$
$2y = 2$
$y = 1$

The solution of the system is $\left(\dfrac{1}{3}, 1\right)$.

53. $\begin{cases} \dfrac{x+2}{2} = \dfrac{y+11}{3} \\[2mm] \dfrac{x}{2} = \dfrac{2y+16}{6} \end{cases}$

Multiply the first equation by 6 and the second equation by −6.
$$\begin{cases} 3(x+2) = 2(y+11) \\ 3x+6 = 2y+22 \\ 3x-2y = 16 \\[2mm] -3x = -2y-16 \\ -3x+2y = -16 \end{cases}$$

Add the two equations.
$$\begin{array}{r} 3x-2y=16 \\ -3x+2y=-16 \\ \hline 0=0 \end{array}$$

There is an infinite number of solutions,
$$\left\{(x,\,y)\,\middle|\,\dfrac{x+2}{2} = \dfrac{y+11}{3}\right\} \text{ or}$$
$$\left\{(x,\,y)\,\middle|\,\dfrac{x}{2} = \dfrac{2y+16}{6}\right\}.$$

55. $\begin{cases} 2x+3y = 14 \\ 3x-4y = -69.1 \end{cases}$

Multiply the first equation by 3 and the second equation by −2.
$$\begin{array}{r} 6x+9y=42 \\ -6x+8y=138.2 \\ \hline 17y=180.2 \\ y=10.6 \end{array}$$

Let $y = 10.6$ in the first equation.
$$\begin{aligned} 2x+3(10.6) &= 14 \\ 2x+31.8 &= 14 \\ 2x &= -17.8 \\ x &= -8.9 \end{aligned}$$
The solution of the system is (−8.9, 10.6).

57. Let x = a number.
$$2x + 6 = x - 3$$

59. Let x = a number.
$$20 - 3x = 2$$

61. Let n = a number.
$$4(n + 6) = 2n$$

63. $\begin{cases} 4x+2y = -7 \\ 3x-y = -12 \end{cases}$

To eliminate y, multiply the second equation by 2.
$$6x - 2y = -24$$

65. $\begin{cases} 3x+8y = -5 \\ 2x-4y = 3 \end{cases} = \begin{cases} 3x+8y = -5 \\ 4x-8y = 6 \end{cases}$

The correct answer is **b**; answers may vary

67. answers may vary

69. $\begin{cases} x+y = 5 \\ 3x+3y = b \end{cases}$

Multiply the first equation by −3.
$$\begin{array}{r} -3x-3y=-15 \\ 3x+3y=b \\ \hline 0=b-15 \end{array}$$

a. The system has an infinite number of solutions if this statement is true. $b = 15$

b. The system has no solution if this statement is false. b = any real number except 15.

71. $\begin{cases} 1.2x+3.4y = 27.6 \\ 7.2x-1.7y = -46.56 \end{cases}$

Multiply the second equation by 2.
$$\begin{array}{r} 1.2x+3.4y=27.6 \\ 14.4x-3.4y=-93.12 \\ \hline 15.6x =-65.52 \\ x=-4.2 \end{array}$$

Let $x = -4.2$ in the first equation.
$$\begin{aligned} 1.2(-4.2)+3.4y &= 27.6 \\ -5.04+3.4y &= 27.6 \\ 3.4y &= 32.64 \\ y &= 9.6 \end{aligned}$$

The solution of the system is (−4.2, 9.6).

73. a. $\begin{cases} 0.05x-y = -21.6 \\ 0.58x-y = -18.1 \end{cases}$

Multiply the first equation by −1.
$$\begin{array}{r} -0.05x+y=21.6 \\ 0.58x-y=-18.1 \\ \hline 0.53x =3.5 \\ x \approx 7 \end{array}$$

Let $x = 7$ in the first equation.

$$0.05(7) - y = -21.6$$
$$0.35 - y = -21.6$$
$$-y = -21.95$$
$$y \approx 22$$

The solution of the system is approximately (7, 22).

b. In 2015 (2008 + 7), the percent of workers age 25–34 and the percent of workers age 55 and older will be the same.

c. There will be approximately 22% of workers for each of these age groups.

Integrated Review

1. $\begin{cases} 2x - 3y = -11 \\ y = 4x - 3 \end{cases}$

Substitute $4x - 3$ for y in the first equation.
$$2x - 3(4x - 3) = -11$$
$$2x - 12x + 9 = -11$$
$$-10x = -20$$
$$x = 2$$
Let $x = 2$ in the second equation.
$$y = 4(2) - 3 = 5$$
The solution of the system is (2, 5).

2. $\begin{cases} 4x - 5y = 6 \\ y = 3x - 10 \end{cases}$

Substitute $3x - 10$ for y in the first equation.
$$4x - 5(3x - 10) = 6$$
$$4x - 15x + 50 = 6$$
$$-11x = -44$$
$$x = 4$$
Let $x = 4$ in the second equation.
$$y = 3(4) - 10 = 2$$
The solution of the system is (4, 2).

3. $\begin{cases} x + y = 3 \\ x - y = 7 \end{cases}$

$$\begin{array}{r} x + y = 3 \\ x - y = 7 \\ \hline 2x = 10 \\ x = 5 \end{array}$$
Let $x = 5$ in the first equation.
$$5 + y = 3$$
$$y = -2$$
The solution of the system is (5, –2).

4. $\begin{cases} x - y = 20 \\ x + y = -8 \end{cases}$

$$\begin{array}{r} x - y = 20 \\ x + y = -8 \\ \hline 2x = 12 \\ x = 6 \end{array}$$
Let $x = 6$ in the second equation.
$$6 + y = -8$$
$$y = -14$$
The solution of the system is (6, –14).

5. $\begin{cases} x + 2y = 1 \\ 3x + 4y = -1 \end{cases}$

Solve the first equation for x.
$$x = 1 - 2y$$
Substitute $1 - 2y$ for x in the second equation.
$$3(1 - 2y) + 4y = -1$$
$$3 - 6y + 4y = -1$$
$$-2y = -4$$
$$y = 2$$
Let $y = 2$ in $x = 1 - 2y$.
$$x = 1 - 2(2) = -3$$
The solution is (–3, 2).

6. $\begin{cases} x + 3y = 5 \\ 5x + 6y = -2 \end{cases}$

Solve the first equation for x.
$$x = 5 - 3y$$
Substitute $5 - 3y$ for x in the second equation.
$$5(5 - 3y) + 6y = -2$$
$$25 - 15y + 6y = -2$$
$$-9y = -27$$
$$y = 3$$
Let $y = 3$ in $x = 5 - 3y$.
$$x = 5 - 3(3) = -4$$
The solution is (–4, 3).

7. $\begin{cases} y = x + 3 \\ 3x - 2y = -6 \end{cases}$

Substitute $x + 3$ for y in the second equation.
$$3x - 2(x + 3) = -6$$
$$3x - 2x - 6 = -6$$
$$x = 0$$
Let $x = 0$ in the first equation.
$$y = 0 + 3 = 3$$
The solution is (0, 3).

8. $\begin{cases} y = -2x \\ 2x - 3y = -16 \end{cases}$

Substitute $-2x$ for y in the second equation.

$$2x - 3(-2x) = -16$$
$$2x + 6x = -16$$
$$8x = -16$$
$$x = -2$$

Let $x = -2$ in the first equation.
$$y = -2(-2) = 4$$
The solution is $(-2, 4)$.

9. $\begin{cases} y = 2x - 3 \\ y = 5x - 18 \end{cases}$

Substitute $5x - 18$ for y in the first equation.
$$5x - 18 = 2x - 3$$
$$3x = 15$$
$$x = 5$$

Let $x = 5$ in the second equation.
$$y = 5(5) - 18 = 7$$
The solution is $(5, 7)$.

10. $\begin{cases} y = 6x - 5 \\ y = 4x - 11 \end{cases}$

Substitute $6x - 5$ for y in the second equation.
$$6x - 5 = 4x - 11$$
$$2x = -6$$
$$x = -3$$

Let $x = -3$ in the first equation.
$$y = 6(-3) - 5 = -23$$
The solution is $(-3, -23)$.

11. $\begin{cases} x + \dfrac{1}{6}y = \dfrac{1}{2} \\ 3x + 2y = 3 \end{cases}$

Multiply the first equation by 6.
$\begin{cases} 6x + y = 3 \\ 3x + 2y = 3 \end{cases}$

Multiply the first equation of the simplified system by -2.
$$-12x - 2y = -6$$
$$\underline{3x + 2y = 3}$$
$$-9x = -3$$
$$x = \frac{1}{3}$$

Multiply the second equation of the simplified system by -2.
$\begin{cases} 6x + y = 3 \\ -6x - 4y = -6 \end{cases}$
$$\underline{}$$
$$-3y = -3$$
$$y = 1$$

The solution of the system is $\left(\dfrac{1}{3}, 1 \right)$.

12. $\begin{cases} x + \dfrac{1}{3}y = \dfrac{5}{12} \\ 8x + 3y = 4 \end{cases}$

Multiply the first equation by 12.
$\begin{cases} 12x + 4y = 5 \\ 8x + 3y = 4 \end{cases}$

Multiply the first equation of the simplified system by 2 and the second equation by -3.
$$24x + 8y = 10$$
$$\underline{-24x - 9y = -12}$$
$$-y = -2$$
$$y = 2$$

Multiply the first equation of the simplified system by 3 and the second equation by -4.
$$36x + 12y = 15$$
$$\underline{-32x - 12y = -16}$$
$$4x = -1$$
$$x = -\frac{1}{4}$$

The solution of the system is $\left(-\dfrac{1}{4}, 2 \right)$.

13. $\begin{cases} x - 5y = 1 \\ -2x + 10y = 3 \end{cases}$

Multiply the first equation by 2.
$$2x - 10y = 2$$
$$\underline{-2x + 10y = 3}$$
$$0 = 5 \quad \text{False}$$

The system has no solution. The solution set is $\{ \ \}$ or $\varnothing$.

14. $\begin{cases} -x + 2y = 3 \\ 3x - 6y = -9 \end{cases}$

Multiply the first equation by 3.
$$-3x + 6y = 9$$
$$\underline{3x - 6y = -9}$$
$$0 = 0$$

The equations in the original system are equivalent and there is an infinite number of solutions, $\{(x, y) | -x + 2y = 3\}$ or $\{(x, y) | 3x - 6y = -9\}$.

15. $\begin{cases} 0.2x - 0.3y = -0.95 \\ 0.4x + 0.1y = 0.55 \end{cases}$

Multiply both equations by 10.
$\begin{cases} 2x - 3y = -9.5 \\ 4x + y = 5.5 \end{cases}$

Multiply the first equation of the simplified system by -2.

$$-4x+6y=19$$
$$\underline{4x\ \ +y=5.5}$$
$$7y=24.5$$
$$y=3.5$$

Multiply the second equation of the simplified system by 3.
$$2x-3y=-9.5$$
$$\underline{12x+3y=16.5}$$
$$14x\ \ \ \ \ =7$$
$$x=0.5$$

The solution of the system is (0.5, 3.5).

16. $\begin{cases} 0.08x-0.04y=-0.11 \\ 0.02x-0.06y=-0.09 \end{cases}$

Multiply both equations by 100.
$$\begin{cases} 8x-4y=-11 \\ 2x-6y=-9 \end{cases}$$

Multiply the second equation of the simplified system by -4.
$$8x\ \ -4y=-11$$
$$\underline{-8x+24y=36}$$
$$20y=25$$
$$y=1.25$$

Multiply the first equation of the simplified system by -3 and the second equation by 2.
$$-24x+12y=33$$
$$\underline{4x-12y=-18}$$
$$-20x\ \ \ \ \ \ =15$$
$$x=-0.75$$

The solution of the system is $(-0.75, 1.25)$.

17. $\begin{cases} x=3y-7 \\ 2x-6y=-14 \end{cases}$

Substitute $3y-7$ for x in the second equation.
$$2(3y-7)-6y=-14$$
$$6y-14-6y=-14$$
$$-14=-14$$

The equations in the original system are equivalent and there is an infinite number of solutions, $\{x, y)|x=3y-7\}$ or $\{(x, y)|2x-6y=-14\}$.

18. $\begin{cases} y=\dfrac{x}{2}-3 \\ 2x-4y=0 \end{cases}$

Substitute $\dfrac{x}{2}-3$ for y in the second equation.

$$2x-4\left(\dfrac{x}{2}-3\right)=0$$
$$2x-2x+12=0$$
$$12=0\quad \text{False}$$

The system has no solution. The solution set is $\{\ \}$ or $\varnothing$.

19. $\begin{cases} 2x+5y=-1 \\ 3x-4y=33 \end{cases}$

Multiply the first equation by 4 and the second equation by 5.
$$8x+20y=-4$$
$$\underline{15x-20y=165}$$
$$23x\ \ \ \ \ \ =161$$
$$x=7$$

Let $x=7$ in the first equation.
$$2(7)+5y=-1$$
$$14+5y=-1$$
$$5y=-15$$
$$y=-3$$

The solution of the system is $(7, -3)$.

20. $\begin{cases} 7x-3y=2 \\ 6x+5y=-21 \end{cases}$

Multiply the first equation by 5 and the second equation by 3.
$$35x-15y=10$$
$$\underline{18x+15y=-63}$$
$$53x\ \ \ \ \ \ =-53$$
$$x=-1$$

Let $x=-1$ in the first equation.
$$7(-1)-3y=2$$
$$-7-3y=2$$
$$-3y=9$$
$$y=-3$$

The solution of the system is $(-1, -3)$.

21. answers may vary

22. answers may vary

Section 4.4 Practice

1. $\begin{cases} 3x+2y-z=0 & (1) \\ x-y+5z=2 & (2) \\ 2x+3y+3z=7 & (3) \end{cases}$

Multiply equation (2) by 2 and add to equation (1) to eliminate y.
$$\begin{cases} 3x+2y-z=0 \\ 2(x-y+5z)=2(2) \end{cases}$$

$$\begin{cases} 3x-2y-z=0 \\ 2x-2y+10z=4 \end{cases}$$
$$\overline{5x+9z=4}\quad(4)$$

Multiply equation (2) by 3 and add to equation (3) to eliminate y again.

$$\begin{cases} 3(x-y+5z)=3(2) \\ 2x+3y+3z=7 \end{cases}$$

$$\begin{cases} 3x-3y+15z=6 \\ 2x+3y+3z=7 \end{cases}$$
$$\overline{5x+18z=13}\quad(5)$$

Multiply equation (4) by -1 and add to equation (5) to eliminate x.

$$\begin{cases} -1(5x+9z)=-1(4) \\ 5x+18z=13 \end{cases}$$

$$\begin{cases} -5x-9z=-4 \\ 5x+18z=13 \end{cases}$$
$$\overline{9z=9}$$
$$z=1$$

Replace z with 1 in equation (4) or (5).
$$5x+9z=4$$
$$5x+9(1)=4$$
$$5x=-5$$
$$x=-1$$

Replace x with -1 and z with 1 in equation (1), (2), or (3).
$$x-y+5z=2$$
$$-1-y+5(1)=2$$
$$-y+4=2$$
$$-y=-2$$
$$y=2$$

The solution is $(-1, 2, 1)$. To check, let $x = -1$, $y = 2$, and $z = 1$ in all three original equations of the system.

2. $\begin{cases} 6x-3y+12z=4 & (1) \\ -6x+4y-2z=7 & (2) \\ -2x+y-4z=3 & (3) \end{cases}$

Multiply equation (3) by 3 and add to equation (1) to eliminate x.

$$\begin{cases} 6x-3y+12z=4 \\ 3(-2x+y-4z)=3(3) \end{cases}$$

$$\begin{cases} 6x-3y+12z=4 \\ -6x+3y-12z=9 \end{cases}$$
$$\overline{0=13}\quad\text{False}$$

Since the statement is false, this system is inconsistent and has no solution. The solution set is { } or $\varnothing$.

3. $\begin{cases} 3x+4y=0 & (1) \\ 9x-4z=6 & (2) \\ -2y+7z=1 & (3) \end{cases}$

Equation (2) has no term containing the variable y. Eliminate y using equations (1) and (3). Multiply equation (3) by 2 and add to equation (1).

$$\begin{cases} 3x+4y=0 \\ 2(-2y+7z)=2(1) \end{cases}$$

$$\begin{cases} 3x+4y=0 \\ -4y+14z=2 \end{cases}$$
$$\overline{3x+14z=2}\quad(4)$$

Multiply equation (4) by -3 and add to equation (2) to eliminate x.

$$\begin{cases} 9x-4z=6 \\ -3(3x+14z)=-3(2) \end{cases}$$

$$\begin{cases} 9x-4z=6 \\ -9x-52z=-6 \end{cases}$$
$$\overline{-56z=0}$$
$$z=0$$

Replace z with 0 in equation (2) and solve for x.
$$9x-4z=6$$
$$9x-4(0)=6$$
$$9x=6$$
$$x=\frac{6}{9}=\frac{2}{3}$$

Replace z with 0 in equation (3) and solve for y.
$$-2y+7z=1$$
$$-2y+7(0)=1$$
$$-2y=1$$
$$y=-\frac{1}{2}$$

The solution is $\left(\dfrac{2}{3}, -\dfrac{1}{2}, 0\right)$.

4. $\begin{cases} 2x+y-3z=6 & (1) \\ x+\dfrac{1}{2}y-\dfrac{3}{2}z=3 & (2) \\ -4x-2y+6z=-12 & (3) \end{cases}$

Multiply both sides of equation (2) by 2 to eliminate fractions, and multiply both sides of equation (3) by $-\dfrac{1}{2}$ since all coefficients in equation (3) are divisible by 2 and the coefficient of x is negative. The resulting system is

$$\begin{cases} 2x+y-3z=6 \\ 2x+y-3z=6 \\ 2x+y-3z=6 \end{cases}$$

Since the three equations are identical, there are infinitely many solutions of the system. The equations are dependent. The solution set can be written as $\{(x, y, z) | 2x + y - 3z = 6\}$.

5. $\begin{cases} x + 2y + 4z = 16 & (1) \\ x \quad\quad + 2z = -4 & (2) \\ \quad y - 3z = 30 & (3) \end{cases}$

Solve equation (2) for x and equation (3) for y.

$x + 2z = -4 \qquad\qquad y - 3z = 30$

$\quad x = -2z - 4 \qquad\qquad y = 3z + 30$

Substitute $-2z - 4$ for x and $3z + 30$ for y in equation (1) and solve for z.

$x + 2y + 4z = 16$

$(-2z - 4) + 2(3z + 30) + 4z = 16$

$-2z - 4 + 6z + 60 + 4z = 16$

$8z + 56 = 16$

$8z = -40$

$z = -5$

Use $x = -2z - 4$ to find x:

$x = -2(-5) - 4 = 10 - 4 = 6$.

Use $y = 3z + 30$ to find y:

$y = 3(-5) + 30 = -15 + 30 = 15$.

The solution is $(6, 15, -5)$.

Vocabulary, Readiness & Video Check 4.4

1. **a.** $x + y + z = 3$

$\quad -1 + 3 + 1 \overset{?}{=} 3$

$\quad\quad\quad\quad 3 = 3$ True

b. $-x + y + z = 5$

$\quad -(-1) + 3 + 1 \overset{?}{=} 5$

$\quad\quad 1 + 3 + 1 \overset{?}{=} 5$

$\quad\quad\quad\quad 5 = 5$ True

c. $-x + y + 2z = 0$

$\quad -(-1) + 3 + 2(1) \overset{?}{=} 0$

$\quad\quad 1 + 3 + 2 \overset{?}{=} 0$

$\quad\quad\quad\quad 6 = 0$ False

d. $x + 2y - 3z = 2$

$\quad -1 + 2(3) - 3(1) \overset{?}{=} 2$

$\quad\quad -1 + 6 - 3 \overset{?}{=} 2$

$\quad\quad\quad\quad 2 = 2$ True

$(-1, 3, 1)$ is a solution to the equations a, b, and d.

2. **a.** $x + y + z = -1$

$\quad 2 + 1 + (-4) \overset{?}{=} -1$

$\quad\quad\quad -1 = -1$ True

b. $x - y - z = -3$

$\quad 2 - 1 - (-4) \overset{?}{=} -3$

$\quad\quad 2 - 1 + 4 \overset{?}{=} -3$

$\quad\quad\quad\quad 5 = -3$ False

c. $2x - y + z = -1$

$\quad 2(2) - 1 + (-4) \overset{?}{=} -1$

$\quad\quad 4 - 1 - 4 \overset{?}{=} -1$

$\quad\quad\quad -1 = -1$ True

d. $-x - 3y - z = -1$

$\quad -2 - 3(1) - (-4) \overset{?}{=} -1$

$\quad\quad -2 - 3 + 4 \overset{?}{=} -1$

$\quad\quad\quad\quad -1 = -1$ True

$(2, 1, -4)$ is a solution to the equations a, c, and d.

3. yes; answers may vary

4. no; answers may vary

5. Once we have one equation in two variables, we need to get another equation in the *same* two variables, giving us a system of two equations in two variables. We solve this new system to find the values of two variables. We then substitute these values into an original equation to find the value of the third.

Exercise Set 4.4

1. $\begin{cases} x - y + z = -4 & (1) \\ 3x + 2y - z = 5 & (2) \\ -2x + 3y - z = 15 & (3) \end{cases}$

Add the equations (1) and (2) to eliminate z.

$\quad x - y + z = -4$

$\quad \underline{3x + 2y - z = 5}$

$\quad 4x + y \quad\quad = 1$ (4)

Add equations (1) and (3) to eliminate z.

$\quad x - y + z = -4$

$\quad \underline{-2x + 3y - z = 15}$

$\quad -x + 2y \quad = 11$ (5)

Multiply equation (5) by 4 and add it to equation (4).

$\quad 4x + y = 1$

$\quad \underline{-4x + 8y = 44}$

$\quad\quad\quad 9y = 45$

$\quad\quad\quad\quad y = 5$

Replace y with 5 in equation (4).

$\quad 4x + y = 1$

$\quad 4x + 5 = 1$

$\quad\quad 4x = -4$

$\quad\quad\quad x = -1$

Replace x with -1 and y with 5 in equation (1).

$x - y + z = -4$

$-1 - 5 + z = -4$

$-6 + z = -4$

$z = 2$

The solution of the system is $(-1, 5, 2)$.

3. $\begin{cases} x + y \quad\;\; = 3 & (1) \\ \quad\;\; 2y \quad = 10 & (2) \\ 3x + 2y - 3z = 1 & (3) \end{cases}$

Solve equation (2) for y.

$2y = 10$

$y = 5$

Replace y with 5 in equation (1).

$x + y = 3$

$x + 5 = 3$

$x = -2$

Replace x with -2 and y with 5 in equation (3).

$3x + 2y - 3z = 1$

$3(-2) + 2(5) - 3z = 1$

$-6 + 10 - 3z = 1$

$4 - 3z = 1$

$-3z = -3$

$z = 1$

The solution of the system is $(-2, 5, 1)$.

5. $\begin{cases} 2x + 2y + z = 1 & (1) \\ -x + y + 2z = 3 & (2) \\ x + 2y + 4z = 0 & (3) \end{cases}$

Add equations (2) and (3) to eliminate x.

$-x + y + 2z = 3$

$\underline{x + 2y + 4z = 0}$

$ 3y + 6z = 3$

$ y + 2z = 1 \quad (4)$

Multiply equation (2) by 2 and add it to equation (1) to eliminate x.

$2x + 2y + z = 1$

$\underline{-2x + 2y + 4z = 6}$

$ 4y + 5z = 7 \; (5)$

Multiply equation (4) by -4 and add it to equation (5).

$-4y - 8z = -4$

$\underline{4y + 5z = 7}$

$ -3z = 3$

$ z = -1$

Replace z with -1 in equation (4).

$y + 2z = 1$

$y + 2(-1) = 1$

$y - 2 = 1$

$y = 3$

Replace y with 3 and z with -1 in equation (3).

$x + 2y + 4z = 0$

$x + 2(3) + 4(-1) = 0$

$x + 6 - 4 = 0$

$x + 2 = 0$

$x = -2$

The solution of the system is $(-2, 3, -1)$.

7. $\begin{cases} x - 2y + z = -5 & (1) \\ -3x + 6y - 3z = 15 & (2) \\ 2x - 4y + 2z = -10 & (3) \end{cases}$

Multiply equation (1) by -3.

$-3x + 6y - 3z = 15$

This is equation (2).

Multiply equation (1) by 2.

$2x - 4y + 2z = -10$

This is equation (3).

The system is dependent and the solution set is $\{(x, y, z) | x - 2y + z = -5\}$.

9. $\begin{cases} 4x - y + 2z = 5 & (1) \\ 2y + z = 4 & (2) \\ 4x + y + 3z = 10 & (3) \end{cases}$

Multiply equation (1) by -1 and add it to equation (3) to eliminate x.

$-4x + y - 2z = -5$

$\underline{4x + y + 3z = 10}$

$ 2y + z = 5 \quad (4)$

Multiply equation (2) by -1 and add it to equation (4).

$-2y - z = -4$

$\underline{2y + z = 5}$

$ 0 = 1$

The statement $0 = 1$ is false, so the system has no solution.

11. $\begin{cases} x \quad\;\; + 5z = 0 & (1) \\ 5x + y \quad\;\; = 0 & (2) \\ \quad\;\; y - 3z = 0 & (3) \end{cases}$

Solve equation (1) for x.

$x + 5z = 0$

$x = -5z$

Solve equation (3) for y.

$y - 3z = 0$

$y = 3z$

Replace x with $-5z$ and y with $3z$ in equation (2).

$5x + y = 0$

$5(-5z) + 3z = 0$

$-25z + 3z = 0$

$-22z = 0$

$z = 0$

Replace z with 0 in equation (1).

$x + 5z = 0$
$x + 5(0) = 0$
$x + 0 = 0$
$x = 0$
Replace x with 0 in equation (2).
$5x + y = 0$
$5(0) + y = 0$
$0 + y = 0$
$y = 0$
The solution of the system is $(0, 0, 0)$.

13. $\begin{cases} 6x \quad -5z = 17 & (1) \\ 5x - y + 3z = -1 & (2) \\ 2x + y \quad = -41 & (3) \end{cases}$

Add equations (2) and (3) to eliminate y.
$5x - y + 3z = -1$
$\underline{2x + y \quad = -41}$
$7x \quad + 3z = -42$ (4)

Multiply equation (1) by 3 and equation (4) by 5, then add the resulting equations.
$18x - 15z = 51$
$\underline{35x + 15z = -210}$
$53x \quad = -159$
$x = -3$

Replace x with -3 in equation (1).
$6x - 5z = 17$
$6(-3) - 5z = 17$
$-18 - 5z = 17$
$-5z = 35$
$z = -7$

Replace x with -3 in equation (3).
$2x + y = -41$
$2(-3) + y = -41$
$-6 + y = -41$
$y = -35$

The solution of the system is $(-3, -35, -7)$.

15. $\begin{cases} x + y + z = 8 & (1) \\ 2x - y - z = 10 & (2) \\ x - 2y - 3z = 22 & (3) \end{cases}$

Add equations (1) and (2) to eliminate y and z.
$x + y + z = 8$
$\underline{2x - y - z = 10}$
$3x \quad = 18$
$x = 6$

Multiply equation (1) by 2 and add it to equation (3) to eliminate y.
$2x + 2y + 2z = 16$
$\underline{x - 2y - 3z = 22}$
$3x \quad - z = 38$ (4)

Replace x with 6 in equation (4).

$3x - z = 38$
$3(6) - z = 38$
$18 - z = 38$
$z = -20$
Replace x with 6 and z with -20 in equation (1).
$x + y + z = 8$
$6 + y + (-20) = 8$
$y - 14 = 8$
$y = 22$
The solution of the system is $(6, 22, -20)$.

17. $\begin{cases} x + 2y \quad - z = 5 & (1) \\ 6x + y \quad + z = 7 & (2) \\ 2x + 4y - 2z = 5 & (3) \end{cases}$

Add equations (1) and (2) to eliminate z.
$x + 2y - z = 5$
$\underline{6x + y + z = 7}$
$7x + 3y \quad = 12$ (4)

Multiply equation (2) by 2 and add to equation (3) to eliminate z.
$12x + 2y + 2z = 14$
$\underline{2x + 4y - 2z = 5}$
$14x + 6y \quad = 19$ (5)

Multiply equation (4) by -2 and add to equation (5).
$-14x - 6y = -24$
$\underline{14x + 6y = 19}$
$0 = -5$

The statement $0 = -5$ is false, so the system has no solution.

19. $\begin{cases} 2x - 3y + z = 2 & (1) \\ x - 5y + 5z = 3 & (2) \\ 3x + y - 3z = 5 & (3) \end{cases}$

Multiply equation (1) by -5 and add it to equation (2) to eliminate z.
$-10x + 15y - 5z = -10$
$\underline{x - 5y + 5z = 3}$
$-9x + 10y \quad = -7$ (4)

Multiply equation (1) by 3 and add it to equation (3) to eliminate z.
$6x - 9y + 3z = 6$
$\underline{3x + y - 3z = 5}$
$9x - 8y \quad = 11$ (5)

Add equations (4) and (5) to eliminate x.
$-9x + 10y = -7$
$\underline{9x - 8y = 11}$
$2y = 4$
$y = 2$

Replace y with 2 in equation (5).

$9x - 8y = 11$
$9x - 8(2) = 11$
$9x - 16 = 11$
$9x = 27$
$x = 3$

Replace x with 3 and y with 2 in equation (1).
$2x - 3y + z = 2$
$2(3) - 3(2) + z = 2$
$6 - 6 + z = 2$
$z = 2$

The solution of the system is $(3, 2, 2)$.

21. $\begin{cases} -2x - 4y + 6z = -8 & (1) \\ x + 2y - 3z = 4 & (2) \\ 4x + 8y - 12z = 16 & (3) \end{cases}$

Multiply equation (2) by -2.
$-2x - 4y + 6z = -8$
This is equation (1).
Multiply equation (2) by 4.
$4x + 8y - 12z = 16$
This is equation (3).
The system is dependent and the solution set is $\{(x, y, z) | x + 2y - 3z = 4\}$.

23. $\begin{cases} 2x + 2y - 3z = 1 & (1) \\ y + 2z = -14 & (2) \\ 3x - 2y = -1 & (3) \end{cases}$

Multiply equation (1) by -3 and equation (3) by 2, then add the results to eliminate x.
$-6x - 6y + 9z = -3$
$\underline{6x - 4y = -2}$
$-10y + 9z = -5$ (4)

Multiply equation (2) by 10 and add the result to equation (4) to eliminate y.
$10y + 20z = -140$
$\underline{-10y + 9z = -5}$
$29z = -145$
$z = -5$

Replace z with -5 in equation (2).
$y + 2z = -14$
$y + 2(-5) = -14$
$y - 10 = -14$
$y = -4$

Replace y with -4 in equation (3).
$3x - 2y = -1$
$3x - 2(-4) = -1$
$3x + 8 = -1$
$3x = -9$
$x = -3$

The solution of the system is $(-3, -4, -5)$.

25. $\begin{cases} x + 2y - z = 5 & (1) \\ -3x - 2y - 3z = 11 & (2) \\ 4x + 4y + 5z = -18 & (3) \end{cases}$

Add equations (1) and (2) to eliminate y.
$x + 2y - z = 5$
$\underline{-3x - 2y - 3z = 11}$
$-2x - 4z = 16$
$x + 2z = -8$ (4)

Multiply equation (2) by 2 and add to equation (3) to eliminate y.
$-6x - 4y - 6z = 22$
$\underline{4x + 4y + 5z = -18}$
$-2x - z = 4$ (5)

Multiply equation (5) by 2 and add to equation (4).
$x + 2z = -8$
$\underline{-4x - 2z = 8}$
$-3x = 0$
$x = 0$

Replace x with 0 in equation (4).
$x + 2z = -8$
$0 + 2z = -8$
$2z = -8$
$z = -4$

Replace x with 0 and z with -4 in equation (1).
$x + 2y - z = 5$
$0 + 2y - (-4) = 5$
$2y + 4 = 5$
$2y = 1$
$y = \dfrac{1}{2}$

The solution of the system is $\left(0, \dfrac{1}{2}, -4\right)$.

27. $\begin{cases} \dfrac{3}{4}x - \dfrac{1}{3}y + \dfrac{1}{2}z = 9 & (1) \\ \dfrac{1}{6}x + \dfrac{1}{3}y - \dfrac{1}{2}z = 2 & (2) \\ \dfrac{1}{2}x - y + \dfrac{1}{2}z = 2 & (3) \end{cases}$

Multiply equation (1) by 12, equation (2) by 6, and equation (3) by 2 to clear fractions.
$\begin{cases} 9x - 4y + 6z = 108 & (4) \\ x + 2y - 3z = 12 & (5) \\ x - 2y + z = 4 & (6) \end{cases}$

Multiply equation (5) by 2 and add it to equation (4) to eliminate y and z.
$9x - 4y + 6z = 108$
$\underline{2x + 4y - 6z = 24}$
$11x = 132$
$x = 12$

Add equations (5) and (6) to eliminate y.

$$x + 2y - 3z = 12$$
$$\underline{x - 2y + z = 4}$$
$$2x \qquad - 2z = 16 \quad (7)$$

Replace x with 12 in equation (7).

$$2x - 2z = 16$$
$$2(12) - 2z = 16$$
$$24 - 2z = 16$$
$$-2z = -8$$
$$z = 4$$

Replace x with 12 and z with 4 in equation (6).

$$x - 2y + z = 4$$
$$12 - 2y + 4 = 4$$
$$-2y + 16 = 4$$
$$-2y = -12$$
$$y = 6$$

The solution of the system is (12, 6, 4).

29. Let x = the first number, then
$2x$ = the second number.

$$x + 2x = 45$$
$$3x = 45$$
$$x = 15$$
$$2x = 2(15) = 30$$

The numbers are 15 and 30.

31.
$$2(x - 1) - 3x = x - 12$$
$$2x - 2 - 3x = x - 12$$
$$-x - 2 = x - 12$$
$$-2x = -10$$
$$x = 5$$

33.
$$-y - 5(y + 5) = 3y - 10$$
$$-y - 5y - 25 = 3y - 10$$
$$-6y - 25 = 3y - 10$$
$$-9y = 15$$
$$y = -\frac{15}{9} = -\frac{5}{3}$$

35. answers may vary

37. answers may vary

39. $\begin{cases} x + y + z = 1 & (1) \\ 2x - y + z = 0 & (2) \\ -x + 2y + 2z = -1 & (3) \end{cases}$

Add E1 and E3.

$$3y + 3z = 0 \text{ or } y + z = 0 \quad (4)$$

Add -2 times E1 to E2.

$$-2x - 2y - 2z = -2$$
$$\underline{2x - y + z = 0}$$
$$-3y - z = -2 \quad (5)$$

Add E4 and E5.

$$-2y = -2$$
$$y = 1$$

Replace y with 1 in E4.

$$1 + z = 0$$
$$z = -1$$

Replace y with 1 and z with -1 in E1.

$$x + 1 + (-1) = 1$$
$$x = 1$$

The solution is (1, 1, −1), and

$$\frac{x}{8} + \frac{y}{4} + \frac{z}{3} = \frac{1}{8} + \frac{1}{4} - \frac{1}{3}$$
$$= \frac{3}{24} + \frac{6}{24} - \frac{8}{24}$$
$$= \frac{1}{24}.$$

41. $\begin{cases} x + y \quad - w = 0 & (1) \\ y + 2z + w = 3 & (2) \\ x \quad - z \quad = 1 & (3) \\ 2x - y \quad - w = -1 & (4) \end{cases}$

Add E1 and E2.

$$x + 2y + 2z = 3 \quad (5)$$

Add E2 and E4.

$$2x + 2z = 2 \text{ or } x + z = 1 \quad (6)$$

Add E3 and E6.

$$x - z = 1$$
$$\underline{x + z = 1}$$
$$2x \quad = 2$$

Replace x with 1 in E3.

$$1 - z = 1$$
$$z = 0$$

Replace x with 1 and z with 0 in E5.

$$1 + 2y + 2(0) = 3$$
$$1 + 2y = 3$$
$$2y = 2$$
$$y = 1$$

Replace y with 1, and z with 0 in E2.

$$1 + 2(0) + w = 3$$
$$1 + w = 3$$
$$w = 2$$

The solution is (1, 1, 0, 2).

43. $\begin{cases} x+y+z+w=5 & (1) \\ 2x+y+z+w=6 & (2) \\ x+y+z=2 & (3) \\ x+y=0 & (4) \end{cases}$

Add -1 times E4 to E3.

$\begin{array}{r} -x-y=0 \\ \underline{x+y+z=2} \\ z=2 \end{array}$

Replace z with 2 in E1 and E2.

$\begin{cases} x+y+w=3 & (5) \\ 2x+y+w=4 & (6) \end{cases}$

Add -1 times E5 to E6.

$\begin{array}{r} -x-y-w=-3 \\ \underline{2x+y+w=4} \\ x=1 \end{array}$

Replace x with 1 in E4.

$1+y=0$
$y=-1$

Replace x with 1, y with -1, and z with 2 in E1.

$1+(-1)+2+w=5$
$2+w=5$
$w=3$

The solution is $(1, -1, 2, 3)$.

45. answers may vary

Section 4.5 Practice

1. a. $\begin{cases} y=-0.54x+110.6 \\ y=-0.36x+74.1 \end{cases}$

Use substitution.

$-0.54x+110.6=-0.36x+74.1$
$-0.18x=-36.5$
$x\approx 203$

The predicted year is 203 years from 2005 or in 2208.

b. yes; answers may vary

2. Let x be one number and y be the other.

The system is $\begin{cases} x+y=30 \\ x-y=6 \end{cases}$.

Add.

$\begin{array}{r} x+y=30 \\ \underline{x-y=6} \\ 2x=36 \\ x=18 \end{array}$

Let $x=18$ in the first equation.

$x+y=30$
$18+y=30$
$y=12$

The numbers are 18 and 12.

3. Let x = first number

y = second number

"A first number is five more than a second number" is translated as $x=y+5$. "Twice the first number is 2 less than 3 times the second number" is translated as $2x=3y-2$.

We solve the following system.

$\begin{cases} x=y+5 \\ 2x=3y-2 \end{cases}$

Since the first equation is solved for x, we use substitution. Substitute $y+5$ for x in the second equation.

$2(y+5)=3y-2$
$2y+10=3y-2$
$12=y$

Replace y with 12 in the equation $x=y+5$ and solve for x.

$x=12+5=17$

The numbers are 12 and 17.

4. Let x = price for adult admission

and y = price per child admission.

$\begin{cases} 3x+3y=75 \\ 2x+4y=62 \end{cases}$

Multiply the first equation by 2 and the second equation by -3.

$\begin{array}{r} 6x+6y=150 \\ \underline{-6x-12y=-186} \\ -6y=-36 \\ y=6 \end{array}$

Let $y=6$ in the second equation.

$2x+4y=62$
$2x+4(6)=62$
$2x+24=62$
$2x=38$
$x=19$

a. $x=19$, so the adult price is \$19.

b. $y=6$, so the child price is \$6.

c. $5(19)+15(6)=95+90=185<200$

No, the regular rates are less than the group rate.

5. Let x = speed of the V150
y = speed of the Atlantique
We summarize the information in a chart. Both trains have traveled two hours.

	Rate	•	Time	=	Distance
V150	x		2		$2x$
Atlantique	y		2		$2y$

The trains are 2150 kilometers apart, so the sum of the distances is 2150: $2x + 2y = 2150$.

The V150 is 75 kph faster than the Atlantique: $x = y + 75$.

We solve the following system.
$$\begin{cases} 2x + 2y = 2150 \\ x = y + 75 \end{cases}$$

Since the second equation is solved for x, we use substitution. Substitute $y + 75$ for x in the first equation.
$$2(y + 75) + 2y = 2150$$
$$2y + 150 + 2y = 2150$$
$$4y + 150 = 2150$$
$$4y = 2000$$
$$y = 500$$

To find x, we replace y with 500 in the second equation.
$$x = 500 + 75 = 575$$
The speed of the V150 is 575 kph, and the speed of the Atlantique is 500 kph.

6. Let x = amount of 99% acid
y = amount of water (0%)
Both x and y are measured in liters. We use a table to organize the given data.

	Amount	Acid Strength	Amount of Pure Acid
99% acid	x	99%	$0.99x$
Water	y	0%	$0y$

The amount of 99% acid and water combined must equal 1 liter, so $x + y = 1$.
The amount of pure acid in the mixture must equal the sum of the amounts of pure acid in the 99% acid and in the water, so
$0.99x + 0y = 0.05(1)$, which simplifies to
$0.99x = 0.05$.
We solve the following system.
$$\begin{cases} x + y = 1 \\ 0.99x = 0.05 \end{cases}$$

Since the second equation does not contain y, we solve it for x.
$$0.99x = 0.05$$
$$x = \frac{0.05}{0.99} \approx 0.05$$
To find y, we replace x with 0.05 in the first equation.
$$x + y = 1$$
$$0.05 + y = 1$$
$$y = 0.95$$
The teacher should use 0.05 liter of the 99% HCL solution and 0.95 liter of water.

7. Let x = the number of packages.
The firm charges the customer $4.50 for each package, so the revenue equation is $R(x) = 4.5x$.
Each package costs $2.50 to produce and the equipment costs $3000, so the cost equation is $C(x) = 2.5x + 3000$.
Since the break-even point is when $R(x) = C(x)$, we solve the equation $4.5x = 2.5x + 3000$.
$$4.5x = 2.5x + 3000$$
$$2x = 3000$$
$$x = 1500$$
The company must sell 1500 packages to break even.

8. Let x = measure of smallest angle
y = measure of largest angle
z = measure of third angle
The sum of the measures is 180°:
$x + y + z = 180$.

The measure of the largest angle is 40° more than the measure of the smallest angle:
$y = x + 40$.

The measure of the remaining angle is 20° more than the measure of the smallest angle:
$y = x + 20$.

We solve the following system.
$$\begin{cases} x + y + z = 1180 \\ y = x + 40 \\ z = x + 20 \end{cases}$$

We substitute $x + 40$ for y and $x + 20$ for z in the first equation.
$$x + (x + 40) + (x + 20) = 180$$
$$3x + 60 = 180$$
$$3x = 120$$
$$x = 40$$
Then $y = x + 40 = 40 + 40 = 80$ and $z = x + 20 = 40 + 20 = 60$.
The angle measures are 40°, 60°, and 80°.

Vocabulary, Readiness & Video Check 4.5

1. Up to now we've been choosing one variable/unknown and translating to one equation. To solve by a system of equations, we'll choose two variables to represent two unknowns and translate to two equations.

2. The break-even point occurs when revenue equals cost—money has not been lost or made; set the revenue function equal to the cost function and solve for the variable.

3. The ordered triple still needs to be interpreted in the context of the application. Each value actually represents the angle measure of a triangle, in degrees.

Exercise Set 4.5

1. **a.** $l - w = 8 - 5 = 3$
   $$P = 2l + 2w$$
   $$= 2(8) + 2(5)$$
   $$= 13 + 10$$
   $$= 23 \neq 30$$

   b. $l - w = 8 - 7 = 1 \neq 3$

   c. $l - w = 9 - 6 = 3$
   $$P = 2l + 2w = 2(9) + 2(6) = 18 + 12 = 30$$

   Choice **c** is correct.

3. **a.** $2d + 3n = 2(3) + 3(4) = 6 + 12 = 18 \neq 17$

   b. $2d + 3n = 2(4) + 3(3) = 8 + 9 = 17$
   $$5d + 4n = 5(4) + 4(3) = 20 + 12 = 32$$

   c. $2d + 3n = 2(2) + 3(5) = 4 + 15 = 19 \neq 17$

   Choice **b** is correct.

5. **a.** $80 + 20 = 100$
   $$80d + 20q = 80(0.10) + 20(0.25)$$
   $$= 8 + 5$$
   $$= 13$$

   b. $20 + 44 = 64 \neq 100$

   c. $60 + 40 = 100$
   $$60d + 40q = 60(0.10) + 40(0.25)$$
   $$= 6 + 10$$
   $$= 16 \neq 13$$

   Choice **a** is correct.

7. Let x = the larger number and y = the smaller number.
$$\begin{cases} x + y = 15 \\ x - y = 7 \end{cases}$$

9. Let x = the amount invested in the larger account and y = the amount invested in the smaller account.
$$\begin{cases} x + y = 6500 \\ x = y + 800 \end{cases}$$

11. Let x = the first number and y = the second number.
$$\begin{cases} x + y = 83 \\ x - y = 17 \end{cases}$$
$$\begin{array}{r} x + y = 83 \\ x - y = 17 \\ \hline 2x \quad\;\; = 100 \end{array}$$
$$x = 50$$
Let $x = 50$ in the first equation.
$$50 + y = 83$$
$$y = 33$$
The numbers are 50 and 33.

13. Let x = the first number, y = the second number.
$$\begin{cases} x = y + 2 \\ 2x = 3y - 4 \end{cases}$$
Substitute $x = y + 2$ in the second equation.
$$2(y + 2) = 3y - 4$$
$$2y + 4 = 3y - 4$$
$$y = 8$$
Replace y with 8 in the first equation.
$$x = 8 + 2 = 10$$
The numbers are 10 and 8.

15. Let x = the first number and y = the second number.
$$\begin{cases} x + 2y = 8 \\ 2x + y = 25 \end{cases}$$
Multiply the first equation by -2.
$$\begin{array}{r} -2x - 4y = -16 \\ 2x\; + y = 25 \\ \hline -3y = 9 \end{array}$$
$$y = -3$$
Let $y = -3$ in the first equation.
$$x + 2(-3) = 8$$
$$x - 6 = 8$$
$$x = 14$$
The numbers are 14 and -3.

17. Let x be the number of runs that Miguel Cabrera batted in and y be the number that Alex Rodriguez batted in.
$$\begin{cases} y = x - 1 \\ x + y = 251 \end{cases}$$
Substitute $x - 1$ for y in the second equation and solve for x.
$$x + y = 251$$
$$x + x - 1 = 251$$
$$2x = 252$$
$$x = 126$$
Now solve for y.
$$y = x - 1 = 126 - 1 = 125$$
Miguel Cabrera batted in 126 runs and Alex Rodriguez batted in 125 runs.

19. Let $x =$ the price of an adult's ticket and $y =$ the price of a child's ticket.
$$\begin{cases} 3x + 4y = 159 \\ 2x + 3y = 112 \end{cases}$$
Multiply the first equation by -2 and the second equation by 3.
$$-6x - 8y = -318$$
$$\underline{6x + 9y = 336}$$
$$y = 18$$
Let $y = 18$ in the first equation.
$$3x + 4(18) = 159$$
$$3x + 72 = 159$$
$$3x = 87$$
$$x = 29$$
An adult's ticket is \$29 and a child's ticket is \$18.

21. Let $x =$ the number of quarters and $y =$ the number of nickels.
$$\begin{cases} x + y = 80 \\ 0.25x + 0.05y = 14.6 \end{cases}$$
Solve the first equation for y.
$$y = 80 - x$$
Substitute $80 - x$ for y in the second equation.
$$0.25x + 0.05(80 - x) = 14.6$$
$$0.25x + 4 - 0.05x = 14.6$$
$$0.20x = 10.6$$
$$x = 53$$
Let $x = 53$ in $y = 80 - x$.
$$y = 80 - 53$$
$$y = 27$$
There are 53 quarters and 27 nickels.

23. Let x be the value of one McDonald's share and let y be the value of one Ohio Art Company share.
$$\begin{cases} 35x + 69y = 2814 \\ x = y + 70 \end{cases}$$
Substitute $y + 70$ for x in the first equation and solve for y.
$$35x + 69y = 2814$$
$$35(y + 70) + 69y = 2814$$
$$35y + 2450 + 69y = 2814$$
$$2450 + 104y = 2814$$
$$104y = 364$$
$$y = 3.5$$
Now solve for x.
$$x = y + 70 = 3.5 + 70 = 73.5$$
On that day, the closing price of the McDonald's stock was \$73.50 per share and the closing price of The Ohio Art Company stock was \$3.50 per share.

25. Let x be the daily fee and y be the mileage charge.
$$\begin{cases} 4x + 450y = 240.50 \\ 3x + 200y = 146.00 \end{cases}$$
Multiply the first equation by 3 and the second by -4.
$$\begin{cases} 3(4x + 450y) = 3(240.50) \\ -4(3x + 200y) = -4(146.00) \end{cases}$$
$$\rightarrow \begin{cases} 12x + 1350y = 721.5 \\ -12x - 800y = -584 \end{cases}$$
Add the equations to eliminate x and solve for y.
$$12x + 1350y = 721.5$$
$$\underline{-12x - 800y = -584}$$
$$550y = 137.5$$
$$y = 0.25$$
Now solve for x.
$$3x + 200y = 146$$
$$3x + 200(0.25) = 146$$
$$3x + 50 = 146$$
$$3x = 96$$
$$x = 32$$
There is a \$32 daily fee and a \$0.25 per mile mileage charge.

27.

	d	$=$	r	$\cdot$	t
Downstream	18		$x + y$		2
Upstream	18		$x - y$		$4\frac{1}{2}$

$$\begin{cases} 2(x+y) = 18 \\ \dfrac{9}{2}(x-y) = 18 \end{cases}$$

Multiply the first equation by $\dfrac{1}{2}$ and the second

equation by $\dfrac{2}{9}$.

$$\begin{array}{r} x + y = 9 \\ x - y = 4 \\ \hline 2x \quad\quad = 13 \\ x = 6.5 \end{array}$$

Let $x = 6.5$ in $x + y = 9$.
$6.5 + y = 9$
$\qquad y = 2.5$
Pratap can row 6.5 miles per hour in still water.
The rate of the current is 2.5 miles per hour.

29.

d	$=$	r	$\cdot$	t
With the wind	780	$x + y$		$1\frac{1}{2}$
Into the wind	780	$x - y$		2

$$\begin{cases} \dfrac{3}{2}(x+y) = 780 \\ 2(x-y) = 780 \end{cases}$$

Multiply the first equation by $\dfrac{2}{3}$ and the second

equation by $\dfrac{1}{2}$.

$$\begin{array}{r} x + y = 520 \\ x - y = 390 \\ \hline 2x \quad\quad = 910 \\ x = 455 \end{array}$$

Let $x = 455$ in $x + y = 520$.
$455 + y = 520$
$\qquad y = 65$
The plane can fly 455 miles per hour in still air.
The speed of the wind is 65 miles per hour.

31. Let x = the time spent walking and
y = the time spent on the bicycle.

	r	$\cdot$	t	$=$	d
Walking	4		x		$4x$
Biking	24		y		$24y$

$$\begin{cases} x + y = 6 \\ 4x + 24y = 114 \end{cases}$$

Multiply the first equation by -4.
$$\begin{array}{r} -4x - 4y = -24 \\ 4x + 24y = 114 \\ \hline 20y = 90 \\ y = 4.5 \end{array}$$

He spent $4\dfrac{1}{2}$ hours on the bicycle.

33. Let x = liters of 4% solution and
y = liters of 12% solution.

Concentration Rate	Liters of Solution	Liters of Pure Acid
0.04	x	$0.04x$
0.12	y	$0.12y$
0.09	12	$0.09(12)$

$$\begin{cases} x + y = 12 \\ 0.04x + 0.12y = 0.09(12) \end{cases}$$

Multiply the first equation by -4 and the second equation by 100.
$$\begin{array}{r} -4x - 4y = -48 \\ 4x + 12y = 108 \\ \hline 8y = 60 \\ y = 7.5 \end{array}$$

Let $y = 7.5$ in the first equation.
$x + 7.5 = 12$
$\qquad x = 4.5$

$4\dfrac{1}{2}$ liters of 4% solution and $7\dfrac{1}{2}$ liters of 12%

solution should be mixed.

35. Let x = pounds of $4.95 per pound beans and
y = pounds of $2.65 per pound beans.

	Cost Rate	Pounds of Beans	Dollars Cost
High Quality	4.95	x	$4.95x$
Low Quality	2.65	y	$2.65y$
Mixture	3.95	200	$3.95(200)$

$$\begin{cases} x + y = 200 \\ 4.95x + 2.65y = 3.95(200) \end{cases}$$

Solve the first equation for y.

$y = 200 - x$

Substitute $200 - x$ for y in the second equation.

$4.95x + 2.65(200 - x) = 3.95(200)$

$\quad 4.95x + 530 - 2.65x = 790$

$\qquad\qquad\qquad 2.30x = 260$

$\qquad\qquad\qquad\quad x \approx 113.04$

Let $x = 113.04$ in the first equation.

$113.04 + y = 200$

$\qquad\quad y \approx 86.96$

He needs 113 pounds of $4.95 per pound beans and 87 pounds of $2.65 per pound beans.

37. Let x = the first angle and y = the second angle.

$$\begin{cases} x + y = 90 \\ x = 2y \end{cases}$$

Substitute $2y$ for x in the first equation.

$2y + y = 90$

$\quad 3y = 90$

$\quad\ y = 30$

Let $y = 30$ in the second equation.

$x = 2(30) = 60$

The angles are 60° and 30°.

39. Let x = the first angle and y = the second angle.

$$\begin{cases} x + y = 90 \\ x = 3y + 10 \end{cases}$$

Substitute $3y + 10$ for x in the first equation.

$3y + 10 + y = 90$

$\qquad\quad 4y = 80$

$\qquad\quad\ y = 20$

Let $y = 20$ in the second equation.

$x = 3(20) + 10 = 70$

The angles are 70° and 20°.

41. Let x = the number sold at $9.50 and y = the number sold at $7.50.

$$\begin{cases} x + y = 90 \\ 9.5x + 7.5y = 721 \end{cases}$$

Solve the first equation for y.

$y = 90 - x$

Substitute $90 - x$ for y in the second equation.

$9.5x + 7.5(90 - x) = 721$

$\ 9.5x + 675 - 7.5x = 721$

$\qquad\qquad\qquad 2x = 46$

$\qquad\qquad\qquad\ x = 23$

Let $x = 23$ in $y = 90 - x$.

$y = 90 - 23 = 67$

They sold 23 at $9.50 and 67 at $7.50.

43. Let x = the rate of the faster group and y = the rate of the slower group.

	r	$\cdot$	t	$=$	d
Faster group	x		240		$240x$
Slower group	y		240		$240y$

$$\begin{cases} x = y + \dfrac{1}{2} \\ 240x + 240y = 1200 \end{cases}$$

Substitute $y + \dfrac{1}{2}$ for x in the second equation.

$240\left(y + \dfrac{1}{2}\right) + 240y = 1200$

$240y + 120 + 240y = 1200$

$\qquad\qquad\quad 480y = 1080$

$\qquad\qquad\qquad y = \dfrac{1080}{480} = 2\dfrac{1}{4}$

Let $y = 2\dfrac{1}{4}$ in the first equation.

$x = 2\dfrac{1}{4} + \dfrac{1}{2} = 2\dfrac{3}{4}$

The rate of the faster group is $2\dfrac{3}{4}$ miles per hour. The rate of the slower group is $2\dfrac{1}{4}$ miles per hour.

45. Let x = gallons of 30% solution and y = gallons of 60% solution.

Concentration Rate	Gallons of Solution	Gallons of Pure Fertilizer
0.30	x	$0.30x$
0.60	y	$0.60y$
0.50	150	$0.50(150)$

$$\begin{cases} x + y = 150 \\ 0.30x + 0.60y = 0.50(150) \end{cases}$$

Multiply the first equation by -3 and the second equation by 10.

Copyright © 2013 Pearson Education, Inc.

$$-3x - 3y = -450$$
$$\underline{3x + 6y = 750}$$
$$3y = 300$$
$$y = 100$$

Let $y = 100$ in the first equation.
$$x + 100 = 150$$
$$x = 50$$

50 gallons of 30% solution and 100 gallons of 60% solution.

47. Let $x =$ the width and $y =$ the length.
$$\begin{cases} 2x + 2y = 144 \\ y = x + 12 \end{cases}$$

Substitute $x + 12$ for y in the first equation.
$$2x + 2(x + 12) = 144$$
$$2x + 2x + 24 = 144$$
$$4x = 120$$
$$x = 30$$

Let $x = 30$ in the second equation.
$$y = 30 + 12 = 42$$

The width is 30 inches and the length is 42 inches.

49. a. $\begin{cases} y = -4.5x + 24 \\ y = 2x + 7 \end{cases}$

Replace y with $-4.5x + 24$ in the second equation.
$$y = 2x + 7$$
$$-4.5x + 24 = 2x + 7$$
$$-6.5x + 24 = 7$$
$$-6.5x = -17$$
$$x \approx 3$$
$$2007 + 3 = 2010$$

The predicted year was 2010 where the percent of adults under 30 and the percent of adults over 30 blogged at the same rate.

b. answers may vary

51. a. answers may vary

b. $\begin{cases} y = 0.06x + 9.7 \\ y = 0.21x + 9.3 \end{cases}$

Replace y with $0.21x + 9.3$ in the first equation.
$$y = 0.06x + 9.7$$
$$0.21x + 9.3 = 0.06x + 9.7$$
$$0.15x + 9.3 = 9.7$$
$$0.15x = 0.4$$
$$x \approx 2.7$$

The pounds of each cheese consumed were the same 3 years after 2000, or in 2003.

53. $\begin{cases} x = y - 30 \\ x + y = 180 \end{cases}$

Replace x with $y - 30$ in the second equation.
$$x + y = 180$$
$$y - 30 + y = 180$$
$$2y = 210$$
$$y = 105$$

Replace y with 105 in the first equation.
$$x = y - 30 = 105 - 30 = 75$$
The values are $x = 75$ and $y = 105$.

55. The break-even point is where $C(x) = R(x)$.
$$30x + 10,000 = 46x$$
$$10,000 = 16x$$
$$625 = x$$
625 units must be sold to break even.

57. The break-even point is where $C(x) = R(x)$.
$$1.2x + 1500 = 1.7x$$
$$1500 = 0.5x$$
$$3000 = x$$
3000 units must be sold to break even.

59. The break-even point is where $C(x) = R(x)$.
$$75x + 160,000 = 200x$$
$$160,000 = 125x$$
$$1280 = x$$
1280 units must be sold to break even.

61. a. Let x be the number of desks. The revenue from each desk is $450, so $R(x) = 450x$.

b. The cost is $6000 plus $200 for each desk, so $C(x) = 200x + 6000$.

c. $R(x) = C(x)$
$$450x = 200x + 6000$$
$$250x = 6000$$
$$x = 24$$
The break-even point is 24 desks.

63. Let $x =$ number of units of Mix A,
$y =$ number of units of Mix B,
$z =$ number of units of Mix C.
$$\begin{cases} 4x + 6y + 4z = 30 & (1) \\ 6x + y + z = 16 & (2) \\ 3x + 2y + 12z = 24 & (3) \end{cases}$$

Multiply equation (2) by -6 and add to equation (1).
$$-32x - 2z = -66 \text{ or}$$
$$16x + z = 33 \quad (4)$$
Multiply equation (2) by -2 and add to equation (3).

$-9x + 10z = -8$ (5)

Multiply equation (4) by -10 and add to equation (5).

$-169x = -338$

$x = 2$

Replace x with 2 in equation (4).

$16(2) + z = 33$

$32 + z = 33$

$z = 1$

Replace x with 2 and z with 1 in equation (2).

$6(2) + y + 1 = 16$

$12 + y + 1 = 16$

$y + 13 = 16$

$y = 3$

Combine 2 units of Mix A, 3 units of Mix B, and 1 unit of Mix C.

65. Let $x =$ length of shortest side,
$y =$ length of equal sides,
$z =$ length of longest side.

$$\begin{cases} x + 2y + z = 29 \\ z = 2x \\ y = x + 2 \end{cases}$$

Replace y with $x + 2$ and z with $2x$ in the first equation.

$x + 2(x + 2) + 2x = 29$

$x + 2x + 4 + 2x = 29$

$5x = 25$

$x = 5$

Replace x with 5 in the third equation.

$y = 5 + 2 = 7$

Replace x with 5 in the second equation.

$z = 2(5) = 10$

The four sides measure 5 inches, 7 inches, 7 inches, and 10 inches.

67. Let $x =$ the first number, $y =$ the second number, and $z =$ the third number.

$$\begin{cases} x + y + z = 40 \\ x = y + 5 \\ x = 2z \end{cases}$$

$$\begin{cases} x + y + z = 40 \\ y = x - 5 \\ z = \frac{1}{2}x \end{cases}$$

Substitute $y = x - 5$ and $z = \frac{1}{2}x$ in the first equation.

$x + x - 5 + \frac{1}{2}x = 40$

$\frac{5}{2}x - 5 = 40$

$\frac{5}{2}x = 45$

$x = \frac{2}{5}(45) = 18$

$y = x - 5 = 18 - 5 = 13$

$z = \frac{1}{2}x = \frac{1}{2}(18) = 9$

The three numbers are 18, 13, and 9.

69. Let $x =$ number of free throws,
$y =$ number of two-point field goals, and
$z =$ number of three-point field goals.

$$\begin{cases} x + 2y + 3z = 2161 \\ x = 4z + 14 \\ y = x - 28 \end{cases}$$

Substitute $4z + 14$ for x in the third equation.

$y = x - 28$

$y = (4z + 14) - 28$

$y = 4z - 14$

Substitute $4z + 14$ for x and $4z - 14$ for y in the first equation.

$x + 2y + 3z = 2161$

$(4z + 14) + 2(4z - 14) + 3z = 2161$

$4z + 14 + 8z - 28 + 3z = 2161$

$15z - 14 = 2161$

$15z = 2175$

$z = 145$

$x = 4z + 14 = 4(145) + 14 = 594$

$y = 4z - 14 = 4(145) - 14 = 566$

He scored 594 free throws, 566 2-point field goals, and 145 3-point field goals.

71. $\begin{cases} x + y + z = 180 \\ 2x + 5 + y = 180 \\ 2x - 5 + z = 180 \end{cases}$

$z = 185 - 2x$

$y = 175 - 2x$

Replace y with $175 - 2x$ and z with $185 - 2x$ in the first equation.

$x + (175 - 2x) + (185 - 2x) = 180$

$360 - 3x = 180$

$-3x = -180$

$x = 60$

$z = 185 - 2(60) = 185 - 120 = 65$

$y = 175 - 2(60) = 175 - 120 = 55$

The values are $x = 60$, $y = 55$, and $z = 65$.

73. $-3x < -9$

$$\frac{-3x}{-3} > \frac{-9}{-3}$$

$$x > 3, \ (3, \infty)$$

75. $4(2x - 1) \geq 0$

$$8x - 4 \geq 0$$

$$8x \geq 4$$

$$x \geq \frac{1}{2}, \ \left[\frac{1}{2}, \infty\right)$$

77. The minimum price is $0.49.
The maximum price is $0.65.
$0.72 > 0.65$ Impossible
$0.29 < 0.49$ Impossible
$0.49 < 0.58 < 0.65$ Possible
The answer is **a**.

79. Let x = the width and y = the length.

$$\begin{cases} 2x + y = 33 \\ y = 2x - 3 \end{cases}$$

Substitute $2x - 3$ for y in the first equation.

$$2x + 2x - 3 = 33$$

$$4x = 36$$

$$x = 9$$

Let $x = 9$ in the second equation.
$y = 2(9) - 3 = 15$
The width is 9 feet and the length is 15 feet.

81. $y = ax^2 + bx + c$

For $(1, 6)$, use $x = 1$ and $y = 6$.
$6 = a + b + c$ (1)
For $(-1, -2)$, use $x = -1$ and $y = -2$.
$-2 = a - b + c$ (2)
For $(0, -1)$, use $x = 0$ and $y = -1$.
$-1 = a \cdot 0 + b \cdot 0 + c$
$-1 = c$ (3)
The system is

$$\begin{aligned} 6 &= a + b + c & (1) \\ -2 &= a - b + c & (2) \\ -1 &= c & (3) \end{aligned}$$

From equation (3), we see that $c = -1$. Multiply equation (2) by -1 and add to equation (1).
$8 = 2b$
$4 = b$
Replace b with 4 and c with -1 in equation (1).
$6 = a + 4 - 1$
$6 = a + 3$
$3 = a$
The solution is $a = 3$, $b = 4$, and $c = -1$.

83. $y = ax^2 + bx + c$

For $(3, 927)$, $x = 3$ and $y = 927$.
$927 = a(3)^2 + b(3) + c$
$927 = 9a + 3b + c$ (1)
For $(11, 1179)$, $x = 11$ and $y = 1179$.
$1179 = a(11)^2 + b(11) + c$
$1179 = 121a + 11b + c$ (2)
For $(19, 1495)$, $x = 19$ and $y = 1495$.
$1495 = a(19)^2 + b(19) + c$
$1495 = 361a + 19b + c$ (3)
The system is

$$\begin{cases} 927 = 9a + 3b + c & (1) \\ 1179 = 121a + 11b + c & (2) \\ 1495 = 361a + 19b + c & (3) \end{cases}$$

Multiply equation (1) by -1 and add to equation (2).
$252 = 112a + 8b$ (4)
Multiply equation (2) by -1 and add to equation (3).
$316 = 240a + 8b$ (5)
Multiply equation (4) by -1 and add to equation (5).

$$\begin{aligned} -252 &= -112a - 8b \\ 316 &= 240a + 8b \\ \hline 64 &= 128a \\ \frac{1}{2} &= a \end{aligned}$$

Replace a with $\frac{1}{2} = 0.5$ in equation (4).
$252 = 112(0.5) + 8b$
$252 = 56 + 8b$
$196 = 8b$

$$b = \frac{49}{2} = 24.5$$

Replace a with 0.5 and b with 24.5 in equation (1).
$927 = 9(0.5) + 3(24.5) + c$
$927 = 4.5 + 73.5 + c$
$927 = 78 + c$
$849 = c$
The solution is $a = 0.5$, $b = 24.5$, and $c = 849$.
For 2015, $x = 25$.

$$y = 0.5(25)^2 + 24.5(25) + 849 = 1774$$

According to the model, 1774 thousand students will take the ACT in 2015.

85. $f(x) = -8.6x + 275$
$f(x) = 204.9x - 1217$
Use substitution.
$-8.6x + 275 = 204.9x - 1217$
$275 = 213.5x - 1217$
$1492 = 213.5x$
$7 \approx x$
$f(7) = -8.6(7) + 275$
$f(7) \approx 215$
The solution is (7, 215).

Chapter 4 Vocabulary Check

1. In a system of linear equations in two variables, if the graphs of the equations are the same, the equations are <u>dependent</u> equations.

2. Two or more linear equations are called a <u>system of linear equations</u>.

3. A system of equations that has at least one solution is called a <u>consistent</u> system.

4. A <u>solution</u> of a system of two equations in two variables is an ordered pair of numbers that is a solution of both equations in the system.

5. Two algebraic methods for solving systems of equations are <u>addition</u> and <u>substitution</u>.

6. A system of equations that has no solution is called an <u>inconsistent</u> system.

7. In a system of linear equations in two variables, if the graphs of the equations are different, the equations are <u>independent</u> equations.

Chapter 4 Review

1. a. Let $x = 12$ and $y = 4$.
$2x - 3y = 12$
$2(12) - 3(4) \stackrel{?}{=} 12$
$24 - 12 \stackrel{?}{=} 12$
$12 = 12$ True

$3x + 4y = 1$
$3(12) + 4(4) \stackrel{?}{=} 1$
$36 + 16 \stackrel{?}{=} 1$
$52 = 1$ False
(12, 4) is not a solution of the system.

b. Let $x = 3$ and $y = -2$.
$2x - 3y = 12$
$2(3) - 3(-2) \stackrel{?}{=} 12$
$6 + 6 \stackrel{?}{=} 12$
$2 = 12$ True

$3x + 4y = 1$
$3(3) + 4(-2) \stackrel{?}{=} 1$
$9 - 8 \stackrel{?}{=} 1$
$1 = 1$ True
(3, −2) is a solution of the system.

c. Let $x = -3$ and $y = 6$.
$2x - 3y = 12$
$2(-3) - 3(6) \stackrel{?}{=} 12$
$-6 - 18 \stackrel{?}{=} 12$
$-24 = 12$ False

$3x + 4y = 1$
$3(-3) + 4(6) \stackrel{?}{=} 1$
$-9 + 24 \stackrel{?}{=} 1$
$15 = 1$ False
(−3, 6) is not a solution of the system.

2. a. Let $x = \frac{3}{4}$ and $y = -3$.
$4x + y = 0$
$4\left(\frac{3}{4}\right) - 3 \stackrel{?}{=} 0$
$3 - 3 \stackrel{?}{=} 0$
$0 = 0$ True

$-8x - 5y = 9$
$-8\left(\frac{3}{4}\right) - 5(-3) \stackrel{?}{=} 9$
$-6 + 15 \stackrel{?}{=} 9$
$9 = 9$ True
$\left(\frac{3}{4}, -3\right)$ is a solution of the system.

b. Let $x = -2$ and $y = 8$.
$4x + y = 0$
$4(-2) + 8 \stackrel{?}{=} 0$
$-8 + 8 \stackrel{?}{=} 0$
$0 = 0$ True

$-8x - 5y = 9$
$-8(-2) - 5(8) \stackrel{?}{=} 9$
$16 - 40 \stackrel{?}{=} 9$
$-24 = 9$ False
(−2, 8) is not a solution of the system.

c. Let $x = \dfrac{1}{2}$ and $y = -2$.

$$4x + y = 0$$

$$4\left(\dfrac{1}{2}\right) - 2 \stackrel{?}{=} 0$$

$$2 - 2 \stackrel{?}{=} 0$$

$$0 = 0 \quad \text{True}$$

$$-8x - 5y = 9$$

$$-8\left(\dfrac{1}{2}\right) - 5(-2) \stackrel{?}{=} 9$$

$$-4 + 10 \stackrel{?}{=} 9$$

$$6 = 9 \quad \text{False}$$

$\left(\dfrac{1}{2}, -2\right)$ is not a solution of the system.

3. a. Let $x = -6$ and $y = -8$.

$$5x - 6y = 18$$

$$5(-6) - 6(-8) \stackrel{?}{=} 18$$

$$-30 + 48 \stackrel{?}{=} 18$$

$$18 = 18 \quad \text{True}$$

$$2y - x = -4$$

$$2(-8) - (-6) \stackrel{?}{=} -4$$

$$-16 + 6 \stackrel{?}{=} -4$$

$$-10 = -4 \quad \text{False}$$

$(-6, -8)$ is not a solution of the system.

b. Let $x = 3$ and $y = \dfrac{5}{2}$.

$$5x - 6y = 18$$

$$5(3) - 6\left(\dfrac{5}{2}\right) \stackrel{?}{=} 18$$

$$15 - 15 \stackrel{?}{=} 18$$

$$0 = 18 \quad \text{False}$$

$$2y - x = -4$$

$$2\left(\dfrac{5}{2}\right) - 3 \stackrel{?}{=} -4$$

$$5 - 3 \stackrel{?}{=} -4$$

$$2 = -4 \quad \text{False}$$

$\left(3, \dfrac{5}{2}\right)$ is not a solution of the system.

c. Let $x = 3$ and $y = -\dfrac{1}{2}$.

$$5x - 6y = 18$$

$$5(3) - 6\left(-\dfrac{1}{2}\right) \stackrel{?}{=} 18$$

$$15 + 3 \stackrel{?}{=} 18$$

$$18 = 18 \quad \text{True}$$

$$2y - x = -4$$

$$2\left(-\dfrac{1}{2}\right) - 3 \stackrel{?}{=} -4$$

$$-1 - 3 \stackrel{?}{=} -4$$

$$-4 = -4 \quad \text{True}$$

$\left(3, -\dfrac{1}{2}\right)$ is a solution of the system.

4. a. Let $x = 2$ and $y = 2$.

$$
\begin{array}{ll}
2x + 3y = 1 & 3y - x = 4 \\
2(2) + 3(2) \stackrel{?}{=} 1 & 3(2) - 2 \stackrel{?}{=} 4 \\
4 + 6 \stackrel{?}{=} 1 & 6 - 2 \stackrel{?}{=} 4 \\
10 = 1 \quad \text{False} & 4 = 4 \quad \text{True}
\end{array}
$$

$(2, 2)$ is not a solution of the system.

b. Let $x = -1$ and $y = 1$.

$$
\begin{array}{ll}
2x + 3y = 1 & 3y - x = 4 \\
2(-1) + 3(1) \stackrel{?}{=} 1 & 3(1) - (-1) \stackrel{?}{=} 4 \\
-2 + 3 \stackrel{?}{=} 1 & 3 + 1 \stackrel{?}{=} 4 \\
1 = 1 \quad \text{True} & 4 = 4 \quad \text{True}
\end{array}
$$

$(-1, 1)$ is a solution of the system.

c. Let $x = 2$ and $y = -1$.

$$2x + 3y = 1$$

$$2(2) + 3(-1) \stackrel{?}{=} 1$$

$$4 - 3 \stackrel{?}{=} 1$$

$$1 = 1 \quad \text{True}$$

$$3y - x = 4$$

$$3(-1) - 2 \stackrel{?}{=} 4$$

$$-3 - 2 \stackrel{?}{=} 4$$

$$-5 = 4 \quad \text{False}$$

$(2, -1)$ is not a solution of the system.

5. $\begin{cases} x + y = 5 \\ x - 1 = y \end{cases}$

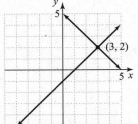

The solution of the system is (3, 2).

6. $\begin{cases} x + y = 3 \\ x - y = -1 \end{cases}$

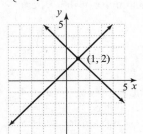

The solution of the system is (1, 2).

7. $\begin{cases} x = 5 \\ y = -1 \end{cases}$

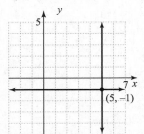

The solution of the system is (5, −1).

8. $\begin{cases} x = -3 \\ y = 2 \end{cases}$

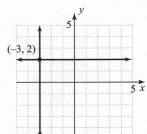

The solution of the system is (−3, 2).

9. $\begin{cases} 2x + y = 5 \\ x = -3y \end{cases}$

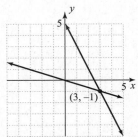

The solution of the system is (3, −1).

10. $\begin{cases} 3x + y = -2 \\ y = -5x \end{cases}$

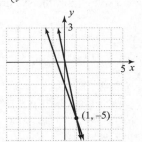

The solution of the system is (1, −5).

11. $\begin{cases} y = 3x \\ -6x + 2y = 6 \end{cases}$

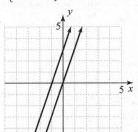

There is no solution. The solution set is { } or ∅.

12. $\begin{cases} x - 2y = 2 \\ -2x + 4y = -4 \end{cases}$

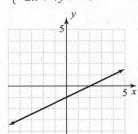

There is an infinite number of solutions, $\{(x, y)| x - 2y = 2\}$ or $\{(x, y)| -2x + 4y = -4\}$.

13. $\begin{cases} y = 2x + 6 \\ 3x - 2y = -11 \end{cases}$

Substitute $2x + 6$ for y in the second equation.
$3x - 2(2x + 6) = -11$
$3x - 4x - 12 = -11$
$-x = 1$
$x = -1$
Let $x = -1$ in the first equation.
$y = 2(-1) + 6 = 4$
The solution is $(-1, 4)$.

14. $\begin{cases} y = 3x - 7 \\ 2x - 3y = 7 \end{cases}$

Substitute $3x - 7$ for y in the second equation.
$2x - 3(3x - 7) = 7$
$2x - 9x + 21 = 7$
$-7x = -14$
$x = 2$
Let $x = 2$ in the first equation.
$y = 3(2) - 7 = -1$
The solution is $(2, -1)$.

15. $\begin{cases} x + 3y = -3 \\ 2x + y = 4 \end{cases}$

Solve the first equation for x.
$x = -3y - 3$
Substitute $-3y - 3$ for x in the second equation.
$2(-3y - 3) + y = 4$
$-6y - 6 + y = 4$
$-5y = 10$
$y = -2$
Let $y = -2$ in $x = -3y - 3$.
$x = -3(-2) - 3 = 3$
The solution is $(3, -2)$.

16. $\begin{cases} 3x + y = 11 \\ x + 2y = 12 \end{cases}$

Solve the first equation for y.
$y = 11 - 3x$
Substitute $11 - 3x$ for y in the second equation.
$x + 2(11 - 3x) = 12$
$x + 22 - 6x = 12$
$-5x = -10$
$x = 2$
Let $x = 2$ in $y = 11 - 3x$.
$y = 11 - 3(2) = 5$
The solution is $(2, 5)$.

17. $\begin{cases} 4y = 2x + 6 \\ x - 2y = -3 \end{cases}$

Solve the second equation for x.
$x = 2y - 3$
Substitute $2y - 3$ for x in the first equation,
$\{(x, y) | 4y = 2x + 6\}$ or $\{(x, y) | x - 2y = -3\}$.
$4y = 2(2y - 3) + 6$
$4y = 4y - 6 + 6$
$0 = 0$
The system has an infinite number of solutions,
$\{(x, y) | 4y = 2x + 6\}$ or $\{(x, y) | x - 2y = -3\}$.

18. $\begin{cases} 9x = 6y + 3 \\ 6x - 4y = 2 \end{cases}$

Solve the first equation for y.
$9x = 6y + 3$
$9x - 3 = 6y$
$\dfrac{3}{2}x - \dfrac{1}{2} = y$
Substitute $\dfrac{3}{2}x - \dfrac{1}{2}$ for y in the second equation.
$6x - 4\left(\dfrac{3}{2}x - \dfrac{1}{2}\right) = 2$
$6x - 6x + 2 = 2$
$2 = 2$
The system has an infinite number of solutions,
$\{(x, y) | 9x = 6y + 3\}$ or $\{(x, y) | 6x - 4y = 2\}$.

19. $\begin{cases} x + y = 6 \\ y = -x - 4 \end{cases}$

Substitute $-x - 4$ for y in the first equation.
$x + (-x - 4) = 6$
$x - x - 4 = 6$
$-4 = 6$ False
There is no solution. The solution set is $\{\ \}$ or $\varnothing$.

20. $\begin{cases} -3x + y = 6 \\ y = 3x + 2 \end{cases}$

Substitute $3x + 2$ for y in the first equation.
$-3x + (3x + 2) = 6$
$-3x + 3x + 2 = 6$
$2 = 6$ False
There is no solution. The solution set is $\{\ \}$ or $\varnothing$.

21. $\begin{cases} 2x+3y=-6 \\ x-3y=-12 \end{cases}$

$2x+3y=-6$

$\dfrac{x-3y=-12}{3x=-18}$

$x=-6$

Let $x=-6$ in the first equation.

$2(-6)+3y=-6$

$-12+3y=-6$

$3y=6$

$y=2$

The solution of the system is $(-6, 2)$.

22. $\begin{cases} 4x+y=15 \\ -4x+3y=-19 \end{cases}$

$4x+y=15$

$\dfrac{-4x+3y=-19}{4y=-4}$

$y=-1$

Let $y=-1$ in the first equation.

$4x+(-1)=15$

$4x-1=15$

$4x=16$

$x=4$

The solution of the system is $(4, -1)$.

23. $\begin{cases} 2x-3y=-15 \\ x+4y=31 \end{cases}$

Multiply the second equation by -2.

$2x-3y=-15$

$\dfrac{-2x-8y=-62}{-11y=-77}$

$y=7$

Let $y=7$ in the second equation.

$x+4(7)=31$

$x+28=31$

$x=3$

The solution of the system is $(3, 7)$.

24. $\begin{cases} x-5y=-22 \\ 4x+3y=4 \end{cases}$

Multiply the first equation by -4.

$-4x+20y=88$

$\dfrac{4x+3y=4}{23y=92}$

$y=4$

Let $y=4$ in the first equation.

$x-5(4)=-22$

$x-20=-22$

$x=-2$

The solution of the system is $(-2, 4)$.

25. $\begin{cases} 2x-6y=-1 \\ -x+3y=\dfrac{1}{2} \end{cases}$

Multiply the second equation by 2.

$2x-6y=-1$

$\dfrac{-2x+6y=1}{0=0}$

There is an infinite number of solutions,

$\{(x,\ y)|2x-6y=-1\}$ or $\left\{(x,\ y)\middle|-x+3y=\dfrac{1}{2}\right\}$.

26. $\begin{cases} 0.6x-0.3y=-1.5 \\ 0.04x-0.02y=-0.1 \end{cases}$

Multiply the first equation by 20 and the second equation by -300.

$12x-6y=-30$

$\dfrac{-12x+6y=30}{0=0}$

There is an infinite number of solutions,

$\{(x,\ y)|0.6x-0.3y=-1.5\}$ or

$\{(x,\ y)|0.04x-0.02y=-0.1\}$.

27. $\begin{cases} \dfrac{3}{4}x+\dfrac{2}{3}y=2 \\ x+\dfrac{y}{3}=6 \end{cases}$

Multiply the first equation by 12 and the second equation by 3.

$\begin{cases} 9x+8y=24 \\ 3x+y=18 \end{cases}$

Multiply the second equation in the simplified system by -3.

$9x+8y=24$

$\dfrac{-9x-3y=-54}{5y=-30}$

$y=-6$

Let $y=-6$ in $3x+y=18$.

$3x+(-6)=18$

$3x=24$

$x=8$

The solution of the system is $(8, -6)$.

28. $\begin{cases} 10x+2y=0 \\ 3x+5y=33 \end{cases}$

Multiply the first equation by -5 and the second equation by 2.

$$-50x-10y=0$$
$$\underline{6x+10y=66}$$
$$-44x\qquad\ =66$$
$$x=-\frac{3}{2}$$

Let $x=-\frac{3}{2}$ in the first equation.

$$10\left(-\frac{3}{2}\right)+2y=0$$
$$-15+2y=0$$
$$2y=15$$
$$y=\frac{15}{2}$$

The solution is $\left(-\frac{3}{2},\frac{15}{2}\right)$.

29. $\begin{cases} x+\quad z=4 \ (1) \\ 2x-y\quad =4 \ (2) \\ x+y-z=0 \ (3) \end{cases}$

Adding E2 and E3 gives $3x-z=4$ (4)
Adding E1 and E4 gives $4x=8$ or $x=2$
Replace x with 2 in E1.
$$2+z=4$$
$$z=2$$
Replace x with 2 and z with 2 in E3.
$$2+y-2=0$$
$$y=0$$
The solution is $(2, 0, 2)$.

30. $\begin{cases} 2x+5y\quad =4 \ (1) \\ x-5y+z=-1 \ (2) \\ 4x\quad -z=11 \ (3) \end{cases}$

Add E2 and E3.
$5x-5y=10$ (4)
Add E1 and E4.
$7x=14$
$x=2$
Replace x with 2 in E1.
$$2(2)+5y=4$$
$$4+5y=4$$
$$5y=0$$
$$y=0$$
Replace x with 2 in E3.

$$4(2)-z=11$$
$$8-z=11$$
$$z=-3$$
The solution is $(2, 0, -3)$.

31. $\begin{cases} \quad 4y+2z=5 \ (1) \\ 2x+8y\quad =5 \ (2) \\ 6x\quad +4z=1 \ (3) \end{cases}$

Multiply E1 by -2 and add to E2.
$$-8y-4z=-10$$
$$\underline{2x+8y\qquad =5}$$
$$2x\qquad -4z=-5 \ (4)$$
Add E3 and E4.
$$8x=-4$$
$$x=-\frac{1}{2}$$

Replace x with $-\frac{1}{2}$ in E2.

$$2\left(-\frac{1}{2}\right)+8y=5$$
$$-1+8y=5$$
$$8y=6$$
$$y=\frac{3}{4}$$

Replace x with $-\frac{1}{2}$ in E3.

$$6\left(-\frac{1}{2}\right)+4z=1$$
$$-3+4z=1$$
$$4z=4$$
$$z=1$$

The solution is $\left(-\frac{1}{2},\frac{3}{4},1\right)$.

32. $\begin{cases} 5x+7y\quad =9 \ (1) \\ \quad 14y-z=28 \ (2) \\ 4x\quad +2z=-4 \ (3) \end{cases}$

Dividing E3 by 2 gives $2x+z=-2$.
Add this equation to E2.
$$2x\qquad +z=-2$$
$$\underline{14y-z=28}$$
$$2x+14y\ =26 \text{ or } x+7y=13 \ (4)$$
Multiply E4 by -1 and add to E1.
$$-x-7y=-13$$
$$\underline{5x+7y=9}$$
$$4x\qquad =-4$$
$$x=-1$$
Replace x with -1 in E4.

$$-1 + 7y = 13$$
$$7y = 14$$
$$y = 2$$

Replace x with -1 in E3.
$$4(-1) + 2z = -4$$
$$-4 + 2z = -4$$
$$2z = 0$$
$$z = 0$$

The solution is $(-1, 2, 0)$.

33. $\begin{cases} 3x - 2y + 2z = 5 & (1) \\ -x + 6y + z = 4 & (2) \\ 3x + 14y + 7z = 20 & (3) \end{cases}$

Multiply E2 by 3 and add to E1.
$$3x - 2y + 2z = 5$$
$$\underline{-3x + 18y + 3z = 12}$$
$$16y + 5z = 17 \quad (4)$$

Multiply E3 by -1 and add to E1.
$$3x - 2y + 2z = 5$$
$$\underline{-3x - 14y - 7z = -20}$$
$$-16y - 5z = -15 \quad (5)$$

Add E4 and E5.
$$16y + 5z = 17$$
$$\underline{-16y - 5z = -15}$$
$$0 = 2 \quad \text{False}$$

The system is inconsistent. The solution is $\varnothing$.

34. $\begin{cases} x + 2y + 3z = 11 & (1) \\ y + 2z = 3 & (2) \\ 2x + 2z = 10 & (3) \end{cases}$

Multiply E2 by -2 and add to E1.
$$x + 2y + 3z = 11$$
$$\underline{-2y - 4z = -6}$$
$$x \quad - z = 5 \quad (4)$$

Multiply E4 by 2 and add to E3.
$$2x + 2z = 10$$
$$\underline{2x - 2z = 10}$$
$$4x = 20$$
$$x = 5$$

Replace x with 5 in E3.
$$2(5) + 2z = 10$$
$$10 + 2z = 10$$
$$2z = 0$$
$$z = 0$$

Replace z with 0 in E2.
$$y + 2(0) = 3$$
$$y + 0 = 3$$
$$y = 3$$

The solution is $(5, 3, 0)$.

35. $\begin{cases} 7x - 3y + 2z = 0 & (1) \\ 4x - 4y - z = 2 & (2) \\ 5x + 2y + 3z = 1 & (3) \end{cases}$

Multiply E2 by 2 and add to E1.
$$7x - 3y + 2z = 0$$
$$\underline{8x - 8y - 2z = 4}$$
$$15x - 11y = 4 \quad (4)$$

Multiply E2 by 3 and add to E3.
$$12x - 12y - 3z = 6$$
$$\underline{5x + 2y + 3z = 1}$$
$$17x - 10y = 7 \quad (5)$$

Solve the new system.
$$\begin{cases} 15x - 11y = 4 & (4) \\ 17x - 10y = 7 & (5) \end{cases}$$

Multiply E4 by -10, multiply E5 by 11, and add.
$$-150x + 110y = -40$$
$$\underline{187x - 110y = 77}$$
$$37x = 37$$
$$x = 1$$

Replace x with 1 in E4.
$$15(1) - 11y = 4$$
$$15 - 11y = 4$$
$$-11y = -11$$
$$y = 1$$

Replace x with 1 and y with 1 in E1.
$$7(1) - 3(1) + 2z = 0$$
$$4 + 2z = 0$$
$$2z = -4$$
$$z = -2$$

The solution is $(1, 1, -2)$.

36. $\begin{cases} x - 3y - 5z = -5 & (1) \\ 4x - 2y + 3z = 13 & (2) \\ 5x + 3y + 4z = 22 & (3) \end{cases}$

Multiply E1 by -4 and add to E2.
$$-4x + 12y + 20z = 20$$
$$\underline{4x - 2y + 3z = 13}$$
$$10y + 23z = 33 \quad (4)$$

Multiply E1 by -5 and add to E3.
$$-5x + 15y + 25z = 25$$
$$\underline{5x + 3y + 4z = 22}$$
$$18y + 29z = 47 \quad (5)$$

Solve the new system.
$$\begin{cases} 10y + 23z = 33 & (4) \\ 18y + 29z = 47 & (5) \end{cases}$$

Multiply E4 by 9, multiply E5 by -5 and add.

$$90y + 207z = 297$$
$$\underline{-90y - 145z = -235}$$
$$62z = 62$$
$$z = 1$$

Replace z with 1 in E4.
$$10y + 23(1) = 33$$
$$10y = 10$$
$$y = 1$$

Replace y with 1 and z with 1 in E1.
$$x - 3(1) - 5(1) = -5$$
$$x - 8 = -5$$
$$x = 3$$

The solution is $(3, 1, 1)$.

37. Let x = the larger number and
y = the smaller number.
$$\begin{cases} x + y = 16 \\ 3x - y = 72 \end{cases}$$

$$x + y = 16$$
$$\underline{3x - y = 72}$$
$$4x = 88$$
$$x = 22$$

Let $x = 22$ in the first equation.
$$22 + y = 16$$
$$y = -6$$

The numbers are -6 and 22.

38. Let x = the number of orchestra seats and
y = the number of balcony seats.
$$\begin{cases} x + y = 360 \\ 45x + 35y = 15{,}150 \end{cases}$$

Solve the first equation for x.
$$x = 360 - y$$

Substitute $360 - y$ for x in the second equation.
$$45(360 - y) + 35y = 15{,}150$$
$$16{,}200 - 45y + 35y = 15{,}150$$
$$-10y = -1050$$
$$y = 105$$

Let $y = 105$ in $x = 360 - y$.
$$x = 360 - 105 = 255$$

There are 255 orchestra seats and 105 balcony seats.

39. Let x = the riverboat's speed in still water and
y = the rate of the current.

	d	$=$	r	$\cdot$	t
Downriver	340		$x + y$		14
Upriver	340		$x - y$		19

$$\begin{cases} 14(x + y) = 340 \\ 19(x - y) = 340 \end{cases}$$

Multiply the first equation by $\dfrac{1}{14}$ and the second

equation by $\dfrac{1}{19}$.

$$x + y = \frac{340}{14} \approx 24.29$$
$$\underline{x - y = \frac{340}{19} \approx 17.89}$$
$$2x \approx 42.18$$
$$x \approx 21.09$$

Multiply the second equation of the simplified system by -1.
$$x + y \approx 24.29$$
$$\underline{-x + y \approx -17.89}$$
$$2y \approx 6.4$$
$$y \approx 3.2$$

The riverboat's speed in still water is 21.1 miles per hour. The rate of the current is 3.2 miles per hour.

40. Let x = amount of 6% solution and
y = amount of 14% solution.

Concentration Rate	Amount of Solution	Amount of Pure Acid
0.06	x	$0.06x$
0.14	y	$0.14y$
0.12	50	$0.12(50)$

$$\begin{cases} x + y = 50 \\ 0.06x + 0.14y = 0.12(50) \end{cases}$$

Multiply the first equation by -6 and the second equation by 100.
$$-6x - 6y = -300$$
$$\underline{6x + 14y = 600}$$
$$8y = 300$$
$$y = 37.5$$

Let $y = 37.5$ in the first equation.
$$x + 37.5 = 50$$
$$x = 12.5$$

$12\dfrac{1}{2}$ cc of 6% solution and $37\dfrac{1}{2}$ cc of 14% solution.

41. Let x = the cost of an egg and
y = the cost of a strip of bacon.
$$\begin{cases} 3x + 4y = 3.80 \\ 2x + 3y = 2.75 \end{cases}$$
Multiply the first equation by –2 and the second equation by 3.
$$\begin{aligned} -6x - 8y &= -7.60 \\ \underline{6x + 9y} &= \underline{8.25} \\ y &= 0.65 \end{aligned}$$
Let $y = 0.65$ in the first equation.
$$\begin{aligned} 3x + 4(0.65) &= 3.80 \\ 3x + 2.60 &= 3.80 \\ 3x &= 1.20 \\ x &= 0.40 \end{aligned}$$
An egg costs 40¢ and a strip of bacon costs 65¢.

42. Let x = the time spent walking and
y = the time spent jogging.

	r	$\cdot$ t	$=$ d
Walking	4	x	$4x$
Jogging	7.5	y	$7.5y$

$$\begin{cases} x + y = 3 \\ 4x + 7.5y = 15 \end{cases}$$
Multiply the first equation by –4.
$$\begin{aligned} -4x - 4y &= -12 \\ \underline{4x + 7.5y} &= \underline{15} \\ 3.5y &= 3 \\ y &\approx 0.857 \end{aligned}$$
Let $y = 0.857$ in the first equation.
$$\begin{aligned} x + 0.857 &= 3 \\ x &\approx 2.143 \end{aligned}$$
He spent 2.14 hours walking and 0.86 hours jogging.

43. Let x = the number of pennies,
y = the number of nickels, and
z = the number dimes.
$$\begin{cases} x + y + z = 53 \quad (1) \\ 0.01x + 0.05y + 0.10z = 2.77 \quad (2) \\ \qquad\qquad y = z + 4 \quad (3) \end{cases}$$
Clear the decimals from E2 by multiplying by 100.
$$x + 5y + 10z = 277 \quad (4)$$
Replace y with $z + 4$ in E1.
$$\begin{aligned} x + z + 4 + z &= 53 \\ x + 2z &= 49 \quad (5) \end{aligned}$$
Replace y with $z + 4$ in E4.

$$\begin{aligned} x + 5(z + 4) + 10z &= 277 \\ x + 15z &= 257 \quad (6) \end{aligned}$$
Solve the new system.
$$\begin{cases} x + 2z = 49 \quad (5) \\ x + 15z = 257 \quad (6) \end{cases}$$
Multiply E5 by –1 and add to E6.
$$\begin{aligned} -x - 2z &= -49 \\ \underline{x + 15z} &= \underline{257} \\ 13z &= 208 \\ z &= 16 \end{aligned}$$
Replace z with 16 in E3.
$$\begin{aligned} x + 2(16) &= 49 \\ x + 32 &= 49 \\ x &= 17 \end{aligned}$$
Replace z with 16 in E3.
$$y = 16 + 4 = 20$$
He has 17 pennies, 20 nickels, and 16 dimes in his jar.

44. Let c = pounds of chocolate used,
n = pounds of nuts used, and
r = pounds of raisins used.
$$\begin{cases} r = 2n \quad (1) \\ c + n + r = 45 \quad (2) \\ 3.00c + 2.70n + 2.25r = 2.80(45) \quad (3) \end{cases}$$
Replace r with $2n$ in E2.
$$\begin{aligned} c + n + 2n &= 45 \\ c + 3n &= 45 \\ c &= -3n + 45 \end{aligned}$$
Replace r with $2n$ and c with $-3n + 45$ in E3.
$$\begin{aligned} 3.00(-3n + 45) + 2.70n + 2.25(2n) &= 126 \\ -9n + 135 + 2.7n + 4.5n &= 126 \\ -1.8n + 135 &= 126 \\ -1.8n &= -9 \\ n &= 5 \end{aligned}$$
Replace n with 5 in E1.
$$r = 2(5) = 10$$
Replace n with 5 and r with 10 in E2.
$$\begin{aligned} c + 5 + 10 &= 45 \\ c + 15 &= 45 \\ c &= 30 \end{aligned}$$
She should use 30 pounds of creme-filled chocolates, 5 pounds of chocolate-covered nuts, and 10 pounds of chocolate-covered raisins.

45. Let x = length of the equal side and
y = length of the third side.
$$\begin{cases} 2x + y = 73 & (1) \\ y = x + 7 & (2) \end{cases}$$
Replace y with $x + 7$ in E1.
$2x + x + 7 = 73$
$3x = 66$
$x = 22$
Replace x with 22 in E2.
$y = 22 + 7 = 29$
Two sides of the triangle have length 22 cm and the third side has length 29 cm.

46. Let f = the first number, s = the second number, and t = the third number.
$$\begin{cases} f + s + t = 295 & (1) \\ f = s + 5 & (2) \\ f = 2t & (3) \end{cases}$$
Solve E2 for s and E3 for t.
$s = f - 5$
$t = \dfrac{f}{2}$

Replace s with $f - 5$ and t with $\dfrac{f}{2}$ in E1.

$f + f - 5 + \dfrac{f}{2} = 295$
$\dfrac{5}{2}f = 300$
$f = 120$
Replace f with 300 in the equation $s = f - 5$.
$s = 120 - 5 = 115$

Replace f with 120 the equation $\dfrac{f}{2}$.

$t = \dfrac{120}{2} = 60$
The first number is 120, the second number is 115, and the third number is 60.

47. $\begin{cases} x - 2y = 1 \\ 2x + 3y = -12 \end{cases}$

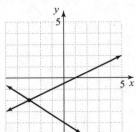

The solution is $(-3, -2)$.

48. $\begin{cases} 3x - y = -4 \\ 6x - 2y = -8 \end{cases}$

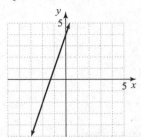

There is an infinite number of solutions.

49. $\begin{cases} x + 4y = 11 \\ 5x - 9y = -3 \end{cases}$
Solve the first equation for x.
$x = 11 - 4y$
Substitute $11 - 4y$ for x in the second equation.
$5(11 - 4y) - 9y = -3$
$55 - 20y - 9y = -3$
$-29y = -58$
$y = 2$
Let $y = 2$ in the first equation.
$x + 4(2) = 11$
$x + 8 = 11$
$x = 3$
The solution is $(3, 2)$.

50. $\begin{cases} x + 9y = 16 \\ 3x - 8y = 13 \end{cases}$
Solve the first equation for x.
$x = 16 - 9y$
Substitute $16 - 9y$ for x in the second equation.
$3(16 - 9y) - 8y = 13$
$48 - 27y - 8y = 13$
$-35y = -35$
$y = 1$
Let $y = 1$ in the first equation.
$x + 9(1) = 16$
$x + 9 = 16$
$x = 7$
The solution is $(7, 1)$.

51. $\begin{cases} y = -2x \\ 4x + 7y = -15 \end{cases}$
Substitute $-2x$ for y in the second equation.
$4x + 7(-2x) = -15$
$4x - 14x = -15$
$-10x = -15$
$x = \dfrac{3}{2} = 1\dfrac{1}{2}$

Let $x = \dfrac{3}{2}$ in the first equation.

$$y = -2\left(\dfrac{3}{2}\right) = -3$$

The solution is $\left(1\dfrac{1}{2}, -3\right)$.

52. $\begin{cases} 3y = 2x + 15 \\ -2x + 3y = 21 \end{cases}$

Solve the first equation for x.

$$3y = 2x + 15$$
$$3y - 15 = 2x$$
$$\dfrac{3}{2}y - \dfrac{15}{2} = x$$

Substitute $\dfrac{3}{2}y - \dfrac{15}{2}$ for x in the second equation.

$$-2\left(\dfrac{3}{2}y - \dfrac{15}{2}\right) + 3y = 21$$
$$-3y + 15 + 3y = 21$$
$$15 = 21 \quad \text{False}$$

The system has no solution. The solution set is $\{\ \}$ or $\varnothing$.

53. $\begin{cases} 3x - y = 4 \\ 4y = 12x - 16 \end{cases}$

Solve the first equation for y.

$$3x - 4 = y$$

Substitute $3x - 4$ for y in the second equation.

$$4(3x - 4) = 12x - 16$$
$$12x - 16 = 12x - 16$$
$$0 = 0$$

There is an infinite number of solutions, $\{(x, y) | 3x - y = 4\}$ or $\{(x, y) | 4y = 12x - 16\}$.

54. $\begin{cases} x + y = 19 \\ x - y = -3 \end{cases}$

$$\begin{array}{r} x + y = 19 \\ x - y = -3 \\ \hline 2x = 16 \\ x = 8 \end{array}$$

Let $x = 8$ in the first equation.

$$8 + y = 19$$
$$y = 11$$

The solution is $(8, 11)$.

55. $\begin{cases} x - 3y = -11 \\ 4x + 5y = -10 \end{cases}$

Solve the first equation for x.

$$x = 3y - 11$$

Substitute $3y - 11$ for x in the second equation.

$$4(3y - 11) + 5y = -10$$
$$12y - 44 + 5y = -10$$
$$17y = 34$$
$$y = 2$$

Let $y = 2$ in the first equation.

$$x - 3(2) = -11$$
$$x - 6 = -11$$
$$x = -5$$

The solution is $(-5, 2)$.

56. $\begin{cases} -x - 15y = 44 \\ 2x + 3y = 20 \end{cases}$

Solve the first equation for x.

$$-x - 15y = 44$$
$$-x = 15y + 44$$
$$x = -15y - 44$$

Substitute $-15y - 44$ for x in the second equation.

$$2(-15y - 44) + 3y = 20$$
$$-30y - 88 + 3y = 20$$
$$-27y = 108$$
$$y = -4$$

Let $y = -4$ in $x = -15y - 44$.

$$x = -15(-4) - 44 = 60 - 44 = 16$$

The solution is $(16, -4)$.

57. $\begin{cases} x - 3y + 2z = 0 & (1) \\ 9y - z = 22 & (2) \\ 5x + 3z = 10 & (3) \end{cases}$

Multiply E1 by 3 and add to E2.

$$\begin{array}{r} 3x - 9y + 6z = 0 \\ 9y - z = 22 \\ \hline 3x + 5z = 22 \quad (4) \end{array}$$

Multiply E3 by -5 and E4 by 3 and add the results.

$$\begin{array}{r} -25x - 15z = -50 \\ 9x + 15z = 66 \\ \hline -16x = 16 \\ x = -1 \end{array}$$

Replace x with -1 in E3.

$$5(-1) + 3z = 10$$
$$-5 + 3z = 10$$
$$3z = 15$$
$$z = 5$$

Replace z with 5 in E2.

$$9y - 5 = 22$$
$$9y = 27$$
$$y = 3$$

The solution is $(-1, 3, 5)$.

58. $\begin{cases} x - 4y = 4 \\ \dfrac{1}{8}x - \dfrac{1}{2}y = 3 \end{cases}$

Multiply the second by -8 and add to the first equation to eliminate x.

$\begin{cases} x - 4y = 4 \\ -x + 4y = -24 \end{cases}$

The equation $0 = -20$ is false. The system has no solution. The solution set is $\{\ \}$ or $\varnothing$.

59. Let x = the larger number and y = the smaller number.

$\begin{cases} x + y = 12 \\ x + 3y = 20 \end{cases}$

Multiply the first equation by -1.

$\begin{aligned} -x - y &= -12 \\ \underline{x + 3y} &= \underline{20} \\ 2y &= 8 \\ y &= 4 \end{aligned}$

Let $y = 4$ in the first equation.

$\begin{aligned} x + 4 &= 12 \\ x &= 8 \end{aligned}$

The numbers are 4 and 8.

60. Let x = the smaller number and y = the larger number.

$\begin{cases} x - y = -18 \\ 2x - y = -23 \end{cases}$

Multiply the first equation by -1.

$\begin{aligned} -x + y &= 18 \\ \underline{2x - y} &= \underline{-23} \\ x\ \ \ \ &= -5 \end{aligned}$

Let $x = -5$ in the first equation.

$\begin{aligned} -5 - y &= -18 \\ -y &= -13 \\ y &= 13 \end{aligned}$

The numbers are -5 and 13.

61. Let x = the number of nickels and y = the number of dimes.

$\begin{cases} x + y = 65 \\ 0.05x + 0.10y = 5.30 \end{cases}$

Multiply the first equation by -5 and the second equation by 100.

$\begin{aligned} -5x - 5y &= -325 \\ \underline{5x + 10y} &= \underline{530} \\ 5y &= 205 \\ y &= 41 \end{aligned}$

Let $y = 41$ in the first equation.

$\begin{aligned} x + 41 &= 65 \\ x &= 24 \end{aligned}$

There are 24 nickels and 41 dimes.

62. Let x = the number of 17¢ stamps and y = the number of 20¢ stamps.

$\begin{cases} x + y = 26 \\ 0.17x + 0.20y = 4.93 \end{cases}$

Multiply the first equation by -17 and the second equation by 100.

$\begin{aligned} -17x - 17y &= -442 \\ \underline{17x + 20y} &= \underline{493} \\ 3y &= 51 \\ y &= 17 \end{aligned}$

Let $y = 17$ in the first equation.

$\begin{aligned} x + 17 &= 26 \\ x &= 9 \end{aligned}$

They purchased 9 17¢ stamps and 17 20¢ stamps.

63. Let x = length of the shortest side
y = length of the second side
z = length of the third side
We solve the system

$\begin{cases} x + y + z = 126 \\ y = 2x \\ z = x + 14 \end{cases}$

We substitute $2x$ for y and $x + 14$ for z in the first equation.

$\begin{aligned} x + 2x + (x + 14) &= 126 \\ 4x + 14 &= 126 \\ 4x &= 112 \\ x &= 28 \end{aligned}$

Now we find y and z.
$y = 2x = 2(28) = 56$
$z = x + 14 = 28 + 14 = 42$
The lengths are 28 units, 42 units, and 56 units.

Chapter 4 Test

1. False; one solution, infinitely many solutions, or no solutions are the only possibilities.

2. False; a solution of a system of equations must be a solution of each equation in the system.

3. True

4. False; $x = 0$ is part of the solution.

5. Let $x = 1$ and $y = -1$.

$$2x - 3y = 5 \qquad\qquad 6x + y = 1$$
$$2(1) - 3(-1) \overset{?}{=} 5 \qquad 6(1) + (-1) \overset{?}{=} 1$$
$$2 + 3 \overset{?}{=} 5 \qquad\qquad 6 - 1 \overset{?}{=} 1$$
$$5 = 5 \quad \text{True} \qquad 5 = 1 \quad \text{False}$$

$(1, -1)$ is not a solution of the system.

6. Let $x = 3$ and $y = -4$.

$$4x - 3y = 24$$
$$4(3) - 3(-4) \overset{?}{=} 24$$
$$12 + 12 \overset{?}{=} 24$$
$$24 = 24 \quad \text{True}$$

$$4x + 5y = -8$$
$$4(3) + 5(-4) \overset{?}{=} -8$$
$$12 - 20 \overset{?}{=} -8$$
$$-8 = -8 \quad \text{True}$$

$(3, -4)$ is a solution of the system.

7. $\begin{cases} y - x = 6 \\ y + 2x = -6 \end{cases}$

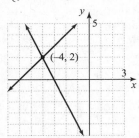

The solution is $(-4, 2)$.

8. $\begin{cases} 3x - 2y = -14 \\ x + 3y = -1 \end{cases}$

Solve the second equation for x.

$x = -3y - 1$

Substitute $-3y - 1$ for x in the first equation.

$$3(-3y - 1) - 2y = -14$$
$$-9y - 3 - 2y = -14$$
$$-11y = -11$$
$$y = 1$$

Let $y = 1$ in $x = -3y - 1$.

$x = -3(1) - 1 = -4$

The solution is $(-4, 1)$.

9. $\begin{cases} \dfrac{1}{2}x + 2y = -\dfrac{15}{4} \\ 4x = -y \end{cases}$

Solve the second equation for y.

$y = -4x$

Substitute $-4x$ for y in the first equation.

$$\frac{1}{2}x + 2(-4x) = -\frac{15}{4}$$
$$\frac{1}{2}x - 8x = -\frac{15}{4}$$
$$-\frac{15}{2}x = -\frac{15}{4}$$
$$x = \frac{1}{2}$$

Let $x = \dfrac{1}{2}$ in the equation $y = -4x$.

$$y = -4\left(\frac{1}{2}\right) = -2$$

The solution is $\left(\dfrac{1}{2}, -2\right)$.

10. $\begin{cases} 3x + 5y = 2 \\ 2x - 3y = 14 \end{cases}$

Multiply the first equation by 2 and the second equation by -3.

$$6x + 10y = 4$$
$$\underline{-6x + 9y = -42}$$
$$19y = -38$$
$$y = -2$$

Let $y = -2$ in the first equation.

$$3x + 5(-2) = 2$$
$$3x - 10 = 2$$
$$3x = 12$$
$$x = 4$$

The solution is $(4, -2)$.

11. $\begin{cases} 4x - 6y = 7 \\ -2x + 3y = 0 \end{cases}$

Multiply the second equation by 2.

$$4x - 6y = 7$$
$$\underline{-4x + 6y = 0}$$
$$0 = 7$$

The system is inconsistent. There is no solution. The solution set is { } or $\varnothing$.

12. $\begin{cases} 3x + y = 7 \\ 4x + 3y = 1 \end{cases}$

Solve the first equation for y.

$y = 7 - 3x$

Substitute $7 - 3x$ for y in the second equation.

$$4x + 3(7 - 3x) = 1$$
$$4x + 21 - 9x = 1$$
$$-5x = -20$$
$$x = 4$$

Let $x = 4$ in $y = 7 - 3x$.
$y = 7 - 3(4) = -5$
The solution is $(4, -5)$.

13. $\begin{cases} 3(2x + y) = 4x + 20 \\ 6x + 3y = 4x + 20 \\ 2x + 3y = 20 \\ \\ x - 2y = 3 \end{cases}$

Multiply the second equation by -2.

$\begin{aligned} 2x + 3y &= 20 \\ -2x + 4y &= -6 \\ \hline 7y &= 14 \\ y &= 2 \end{aligned}$

Let $y = 2$ in the second equation.
$\begin{aligned} x - 2(2) &= 3 \\ x - 4 &= 3 \\ x &= 7 \end{aligned}$
The solution of the system is $(7, 2)$.

14. $\begin{cases} \dfrac{x-3}{2} = \dfrac{2-y}{4} \\ \dfrac{7-2x}{3} = \dfrac{y}{2} \end{cases}$

Multiply the first equation by 4 and the second equation by 6.

$\begin{cases} 2(x-3) = 2 - y \\ 2x - 6 = 2 - y \\ 2x + y = 8 \\ \\ 2(7 - 2x) = 3y \\ 14 - 4x = 3y \\ 4x + 3y = 14 \end{cases}$

Multiply the first equation by -3.

$\begin{aligned} -6x - 3y &= -24 \\ 4x + 3y &= 14 \\ \hline -2x &= -10 \\ x &= 5 \end{aligned}$

Let $x = 5$ in the first equation.
$\begin{aligned} 2(5) + y &= 8 \\ 10 + y &= 8 \\ y &= -2 \end{aligned}$
The solution of the system is $(5, -2)$.

15. Let $x =$ the larger number and $y =$ the smaller number.
$\begin{cases} x + y = 124 \\ x - y = 32 \end{cases}$

$\begin{aligned} x + y &= 124 \\ x - y &= 32 \\ \hline 2x &= 156 \\ x &= 78 \end{aligned}$

Let $x = 78$ in the first equation.
$\begin{aligned} 78 + y &= 124 \\ y &= 46 \end{aligned}$
The numbers are 78 and 46.

16. Let $x =$ cc's of 12% solution and $y =$ cc's of 16% solution.

Concentration Rate	cc's of Solution	cc's of salt
12%	x	$0.12x$
22%	80	$0.22(80)$
16%	y	$0.16y$

$\begin{cases} x + 80 = y \\ 0.12x + 0.22(80) = 0.16y \end{cases}$

Multiply the first equation by -16 and the second equation by 100.

$\begin{aligned} -16x - 1280 &= -16y \\ 12x + 1760 &= 16y \\ \hline -4x + 480 &= 0 \\ -4x &= -480 \\ x &= 120 \end{aligned}$

Should add 120 cc's of 12% solution

17. Let $x =$ the number of thousands of farms in Texas and $y =$ the number of thousands of farms in Missouri.
$\begin{cases} x + y = 356 \\ x - y = 140 \end{cases}$

$\begin{aligned} x + y &= 356 \\ x - y &= 140 \\ \hline 2x &= 496 \\ x &= 248 \end{aligned}$

Let $x = 248$ in the first equation.
$\begin{aligned} 248 + y &= 356 \\ y &= 108 \end{aligned}$
There are 248 thousand farms in Texas and 108 thousand farms in Missouri.

18. Let x = the speed of the faster hiker and y = the speed of the slower hiker.

	r	$\cdot$ t	= d
Faster	x	4	$4x$
Slower	y	4	$4y$

$$\begin{cases} 4x+4y=36 \\ x=2y \end{cases}$$

Substitute $2y$ for x in the first equation.
$$4(2y)+4y=36$$
$$8y+4y=36$$
$$12y=36$$
$$y=3$$

Let $y = 3$ in the second equation.
$$x=2(3)=6$$
The speeds are 3 miles per hour and 6 miles per hour.

19. $$\begin{cases} 2x-3y \quad\;\;=4 \quad (1) \\ \quad\;\; 3y+2z=2 \quad (2) \\ \quad x \quad\;\; -z=-5 \quad (3) \end{cases}$$

Add E1 and E2.
$$2x+2z=6 \;\text{ or }\; x+z=3 \;\;(4)$$
Add E3 and E4.
$$\begin{array}{r} x+z=3 \\ x-z=-5 \\ \hline 2x\quad\;\;=-2 \\ x=-1 \end{array}$$
Replace x with -1 in E3.
$$-1-z=-5$$
$$-z=-4 \;\text{ so }\; z=4$$
Replace x with -1 in E1.
$$2(-1)-3y=4$$
$$-2-3y=4$$
$$-3y=6$$
$$y=-2$$
The solution is $(-1, -2, 4)$.

20. $$\begin{cases} 3x-2y-z=-1 \quad (1) \\ 2x-2y \quad\;\;=4 \quad\;\; (2) \\ 2x \quad\;\; -2z=-12 \;(3) \end{cases}$$

Multiply E2 by -1 and add to E1.
$$\begin{array}{r} 3x-2y-z=-1 \\ -2x+2y\quad\;\;=-4 \\ \hline x\quad\quad\;\; -z=-5 \;\;(4) \end{array}$$
Multiply E4 by -2 and add to E3.

$$\begin{array}{r} 2x-2z=-12 \\ -2x+2z=10 \\ \hline 0=-2 \;\text{ False} \end{array}$$
The system is inconsistent. The solution set is $\varnothing$.

21. Let x = measure of the smallest angle. Then the largest angle has a measure of $5x - 3$, and the remaining angle has a measure of $2x - 1$. The sum of the three angles must add to $180°$:
$$a+b+c=180$$
$$x+(5x-3)+(2x-1)=180$$
$$x+5x-3+2x-1=180$$
$$8x-4=180$$
$$8x=184$$
$$x=23$$
$$5x-3=5(23)-3=115-3=112$$
$$2x-1=2(23)-1=46-1=45$$
The angle measures are $23°$, $45°$, and $112°$.

Chapter 4 Cumulative Review

1. a. $-1<0$

  b. $7=\dfrac{14}{2}$

  c. $-5>-6$

2. a. $5^2=5\cdot5=25$

  b. $2^5=2\cdot2\cdot2\cdot2\cdot2=32$

3. a. commutative property of multiplication

  b. associative property of addition

  c. identity element for addition

  d. commutative property of multiplication

  e. multiplicative inverse property

  f. additive inverse property

  g. commutative and associative properties of multiplication

4. Let $x = 8$, $y = 5$.
$$y^2-3x=5^2-3(8)=25-24=1$$

5. $(2x-3)-(4x-2)=2x-3-4x+2=-2x-1$

6. $7 - 12 + (-5) - 2 + (-2)$
$= 7 + (-12) + (-5) + (-2) + (-2)$
$= 7 + (-21)$
$= -14$

7. $7 = -5(2a - 1) - (-11a + 6)$
$7 = -10a + 5 + 11a - 6$
$7 = a - 1$
$7 + 1 = a - 1 + 1$
$8 = a$

8. Let $x = -7$, $y = -3$.
$2y^2 - x^2 = 2(-3)^2 - (-7)^2$
$\quad\quad\quad\quad = 2(9) - 49$
$\quad\quad\quad\quad = 18 - 49$
$\quad\quad\quad\quad = -31$

9. $\dfrac{5}{2}x = 15$
$\dfrac{2}{5} \cdot \dfrac{5}{2}x = \dfrac{2}{5} \cdot 15$
$\quad\quad\quad x = 6$

10. $0.4y - 6.7 + y - 0.3 - 2.6y$
$= 0.4y + y + (-2.6y) + (-6.7) + (-0.3)$
$= -1.2y - 7$

11. $\dfrac{x}{2} - 1 = \dfrac{2}{3}x - 3$
$6\left(\dfrac{x}{2} - 1\right) = 6\left(\dfrac{2}{3}x - 3\right)$
$\quad 3x - 6 = 4x - 18$
$\quad -x - 6 = -18$
$\quad\quad -x = -12$
$\quad\quad\quad x = 12$

12. $7(x - 2) - 6(x + 1) = 20$
$7x - 14 - 6x - 6 = 20$
$\quad\quad\quad x - 20 = 20$
$\quad\quad\quad\quad\quad x = 40$

13. Let $x =$ the number.
$2(x + 4) = 4x - 12$
$2x + 8 = 4x - 12$
$-2x + 8 = -12$
$-2x = -20$
$x = 10$
The number is 10.

14. $5(y - 5) = 5y + 10$
$5y - 25 = 5y + 10$
$\quad -25 = 10$
False statement; there is no solution.

15. $\quad y = mx + b$
$y - b = mx + b - b$
$y - b = mx$
$\dfrac{y - b}{m} = \dfrac{mx}{m}$
$\dfrac{y - b}{m} = x$

16. Let $x =$ the number.
$5(x - 1) = 6x$
$5x - 5 = 6x$
$-x - 5 = 0$
$-x = 5$
$x = -5$
The number is -5.

17. $-2x \le -4$
$\dfrac{-2x}{-2} \ge \dfrac{-4}{-2}$
$x \ge 2, [2, \infty)$

18. $\quad P = a + b + c$
$P - a - c = a + b + c - a - c$
$P - a - c = b$

19. $x = -2y$

x	y
0	0
-4	2

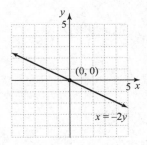

20. $3x + 7 \ge x - 9$
$2x + 7 \ge -9$
$2x \ge -16$
$x \ge -8, [-8, \infty)$

21. $(-1, 5)$ and $(2, -3)$

$$m = \frac{y_2 - y_1}{x_2 - x_1} = \frac{-3 - 5}{2 - (-1)} = \frac{-8}{3} = -\frac{8}{3}$$

22. $x - 3y = 3$

x	y
0	-1
3	0
9	2

23. $y = \frac{3}{4}x + 6$

$y = mx + b$

$m = \frac{3}{4}, b = 6$

Slope is $\frac{3}{4}$, y-intercept is $(0, 6)$.

24. $(-1, 3)$ and $(2, -8)$

$$m = \frac{y_2 - y_1}{x_2 - x_1} = \frac{-8 - 3}{2 - (-1)} = -\frac{11}{3}$$

A parallel line has the same slope.

Slope is $-\frac{11}{3}$.

25. $3x - 4y = 4$

$-4y = -3x + 4$

$y = \frac{-3x}{-4} + \frac{4}{-4}$

$y = \frac{3}{4}x - 1$

$y = mx + b$

$m = \frac{3}{4}, \ b = -1$

Slope is $\frac{3}{4}$, y-intercept is $(0, -1)$.

26. $y = 7x + 0$

$y = mx + b$

$m = 7, b = 0$

Slope is 7, y-intercept is $(0, 0)$.

27. $m = -2$, with point $(-1, 5)$

$y - y_1 = m(x - x_1)$

$y - 5 = -2[x - (-1)]$

$y - 5 = -2x - 2$

$y = -2x + 3$

$2x + y = 3$

28. Line: $y = 4x - 5 \Rightarrow m_1 = 4$

Line 2: $-4x + y = 7 \Rightarrow y = 4x + 7 \Rightarrow m_2 = 4$

$m_2 = m_1$

The lines are parallel.

29. A vertical line has an equation $x = c$.

Point, $(-1, 5)$

$x = -1$

30. $m = -5$, with point $(-2, 3)$

$y - y_1 = m(x - x_1)$

$y - 3 = -5[x - (-2)]$

$y - 3 = -5x - 10$

$y = -5x - 7$

31. Domain is $\{-1, 0, 3\}$

Range is $\{-2, 0, 2, 3\}$

32. $f(x) = 5x^2 - 6$

$f(0) = 5(0)^2 - 6 = -6$

$f(-2) = 5(-2)^2 - 6 = 5(4) - 6 = 14$

33. **a.** function

　　b. not a function

34. **a.** not a function

　　b. function

　　c. not a function

35. $\begin{cases} 3x - y = 4 \\ y = 3x - 4, \ m = 3 \\ \\ x + 2y = 8 \\ y = -\frac{1}{2}x + 4, \ m = -\frac{1}{2} \end{cases}$

Because they have different slopes, there is only one solution.

36. **a.** Let $x = 1$ and $y = -4$.

$2x - y = 6$

$2(1) - (-4) \stackrel{?}{=} 6$

$2 + 4 \stackrel{?}{=} 6$

$6 = 6$　True

$3x + 2y = -5$

$3(1) + 2(-4) \stackrel{?}{=} -5$

$3 - 8 \stackrel{?}{=} -5$

$-5 = -5$　True

$(1, -4)$ is a solution of the system.

b. Let $x = 0$ and $y = 6$.

$$2x - y = 6 \qquad\qquad 3x + 2y = -5$$
$$2(0) - (6) \overset{?}{=} 6 \qquad\qquad \text{Test not needed}$$
$$0 - 6 \overset{?}{=} 6$$
$$-6 = 6 \quad \text{False}$$

$(0, 6)$ is not a solution of the system.

c. Let $x = 3$ and $y = 0$.

$$2x - y = 6$$
$$2(3) - (0) \overset{?}{=} 6$$
$$6 - 0 \overset{?}{=} 6$$
$$6 = 6 \quad \text{True}$$

$$3x + 2y = -5$$
$$3(3) + 2(0) \overset{?}{=} -5$$
$$9 + 0 \overset{?}{=} -5$$
$$9 = -5 \quad \text{False}$$

$(3, 0)$ is not a solution of the system.

37. $\begin{cases} x + 2y = 7 \\ 2x + 2y = 13 \end{cases}$

Solve the first equation for x.
$$x = 7 - 2y$$
Substitute $7 - 2y$ for x in the second equation.
$$2(7 - 2y) + 2y = 13$$
$$14 - 4y + 2y = 13$$
$$-2y = -1$$
$$y = \frac{1}{2}$$
Let $y = \frac{1}{2}$ in $x = 7 - 2y$.
$$x = 7 - 2\left(\frac{1}{2}\right) = 6$$
The solution is $\left(6, \frac{1}{2}\right)$.

38. $\begin{cases} 3x - 4y = 10 \\ y = 2x \end{cases}$

Substitute $2x$ for y in the first equation.
$$3x - 4(2x) = 10$$
$$3x - 8x = 10$$
$$-5x = 10$$
$$x = -2$$
Let $x = -2$ in the second equation.
$$y = 2(-2) = -4$$
The solution is $(-2, -4)$.

39. $\begin{cases} x + y = 7 \\ x - y = 5 \end{cases}$

$$\begin{array}{r} x + y = 7 \\ x - y = 5 \\ \hline 2x = 12 \\ x = 6 \end{array}$$

Let $x = 6$ in the first equation.
$$6 + y = 7$$
$$y = 1$$
The solution to the system is $(6, 1)$.

40. $\begin{cases} x = 5y - 3 \\ x = 8y + 4 \end{cases}$

Substitute $8y + 4$ for x in the first equation.
$$8y + 4 = 5y - 3$$
$$3y + 4 = -3$$
$$3y = -7$$
$$y = -\frac{7}{3}$$

Let $y = -\frac{7}{3}$ in the second equation.
$$x = 8\left(-\frac{7}{3}\right) + 4$$
$$x = -\frac{56}{3} + \frac{12}{3}$$
$$x = -\frac{44}{3}$$

The solution is $\left(-\frac{44}{3}, -\frac{7}{3}\right)$.

41. $\begin{cases} 3x - y + z = -15 & (1) \\ x + 2y - z = 1 & (2) \\ 2x + 3y - 2z = 0 & (3) \end{cases}$

Add E1 and E2.
$$4x + y = -14 \quad (4)$$
Multiply E1 by 2 and add to E3.
$$\begin{array}{r} 6x - 2y + 2z = -30 \\ 2x + 3y - 2z = 0 \\ \hline 8x + y = -30 \quad (5) \end{array}$$
Solve the new system:
$$\begin{cases} 4x + y = -14 & (4) \\ 8x + y = -30 & (5) \end{cases}$$
Multiply E4 by -1 and add to E5.
$$\begin{array}{r} -4x - y = 14 \\ 8x + y = -30 \\ \hline 4x = -16 \\ x = -4 \end{array}$$
Replace x with -4 in E4.

$4(-4)+y=-14$
$-16+y=-14$
$y=2$
Replace x with -4 and y with 2 in E1.
$3(-4)-(2)+z=-15$
$-12-2+z=-15$
$-14+z=-15$
$z=-1$
The solution is $(-4, 2, -1)$.

42. $\begin{cases} x-2y+z=0 & (1) \\ 3x-y-2z=-15 & (2) \\ 2x-3y+3z=7 & (3) \end{cases}$

Multiply E1 by 2 and add to E2.
$2x-4y+2z=0$
$\underline{3x-y-2z=-15}$
$5x-5y=-15$ or $x-y=-3$ (4)
Multiply E1 by -3 and add to E3.
$-3x+6y-3z=0$
$\underline{2x-3y+3z=7}$
$-x+3y=7$ (5)
Add E4 and E5.
$2y=4$
$y=2$
Replace y with 2 in E4.
$x-2=-3$
$x=-1$
Replace x with -1 and y with 2 in E1.
$-1-2(2)+z=0$
$-5+z=0$
$z=5$
The solution is $(-1, 2, 5)$.

43. Let $x=$ the first number and
$y=$ the second number.
$\begin{cases} x=y-4 \\ 4x=2y+6 \end{cases}$

Substitute $y-4$ for x in the second equation.
$4(y-4)=2y+6$
$4y-16=2y+6$
$2y=22$
$y=11$
Let $y=11$ in $x=y-4$.
$x=11-4=7$
The numbers are 7 and 11.

44. Let $x=$ the first number and
$y=$ the second number.
$\begin{cases} x+y=37 \\ x-y=21 \end{cases}$

$x+y=37$
$\underline{x-y=21}$
$2x=58$
$x=29$
Let $x=29$ in the first equation.
$29+y=37$
$y=8$

The numbers are 29 and 8.

Chapter 5

1. a. $3^3 = 3 \cdot 3 \cdot 3 = 27$

   b. Use 4 as a factor once, $4^1 = 4$

   c. $(-8)^2 = (-8)(-8) = 64$

   d. $-8^2 = -(8 \cdot 8) = -64$

   e. $\left(\dfrac{3}{4}\right)^3 = \dfrac{3}{4} \cdot \dfrac{3}{4} \cdot \dfrac{3}{4} = \dfrac{27}{64}$

   f. $(0.3)^4 = (0.3)(0.3)(0.3)(0.3) = 0.0081$

   g. $3 \cdot 5^2 = 3 \cdot 25 = 75$

2. a. If x is 3, $\begin{aligned}3x^4 &= 3 \cdot (3)^4 \\ &= 3 \cdot (3 \cdot 3 \cdot 3 \cdot 3) \\ &= 3 \cdot 81 \\ &= 243\end{aligned}$

   b. If x is -4, $\dfrac{6}{x^2} = \dfrac{6}{(-4)^2} = \dfrac{6}{(-4)(-4)} = \dfrac{6}{16} = \dfrac{3}{8}$

3. a. $3^4 \cdot 3^6 = 3^{4+6} = 3^{10}$

   b. $y^3 \cdot y^2 = y^{3+2} = y^5$

   c. $z \cdot z^4 = z^1 \cdot z^4 = z^{1+4} = z^5$

   d. $x^3 \cdot x^2 \cdot x^6 = x^{3+2+6} = x^{11}$

   e. $(-2)^5 \cdot (-2)^3 = (-2)^{5+3} = (-2)^8$

   f. $b^3 \cdot t^5$, cannot be simplified because b and t are different bases.

4. $\begin{aligned}(-5y^3)(-3y^4) &= -5 \cdot y^3 \cdot -3 \cdot y^4 \\ &= -5 \cdot -3 \cdot y^3 \cdot y^4 \\ &= 15y^7\end{aligned}$

5. a. $\begin{aligned}(y^7 z^3)(y^5 z) &= (y^7 \cdot y^5) \cdot (z^3 \cdot z^1) \\ &= y^{12} \cdot z^4 \text{ or } y^{12}z^4\end{aligned}$

   b. $\begin{aligned}(-m^4 n^4)(7mn^{10}) &= (-1 \cdot 7) \cdot (m^4 \cdot m^1) \cdot (n^4 \cdot n^{10}) \\ &= (-7) \cdot (m^5) \cdot (n^{14}) \text{ or } -7m^5 n^{14}\end{aligned}$

6. a. $(z^3)^7 = z^{3 \cdot 7} = z^{21}$

   b. $(4^9)^2 = 4^{9 \cdot 2} = 4^{18}$

   c. $[(-2)^3]^5 = (-2)^{3 \cdot 5} = (-2)^{15}$

7. a. $(pr)^5 = p^5 \cdot r^5 = p^5 r^5$

   b. $(6b)^2 = 6^2 \cdot b^2 = 36b^2$

   c. $\begin{aligned}\left(\dfrac{1}{4} x^2 y\right)^3 &= \left(\dfrac{1}{4}\right)^3 \cdot (x^2)^3 \cdot y^3 \\ &= \dfrac{1}{64} \cdot x^6 \cdot y^3 \\ &= \dfrac{1}{64} x^6 y^3\end{aligned}$

   d. $\begin{aligned}(-3a^3 b^4 c)^4 &= (-3)^4 \cdot (a^3)^4 \cdot (b^4)^4 \cdot c^4 \\ &= 81a^{12} b^{16} c^4\end{aligned}$

8. a. $\left(\dfrac{x}{y^2}\right)^5 = \dfrac{x^5}{(y^2)^5} = \dfrac{x^5}{y^{10}},\ y \neq 0$

   b. $\left(\dfrac{2a^4}{b^3}\right)^5 = \dfrac{2^5 \cdot (a^4)^5}{(b^3)^5} = \dfrac{32a^{20}}{b^{15}},\ b \neq 0$

9. a. $\dfrac{z^8}{z^4} = z^{8-4} = z^4$

   b. $\dfrac{(-5)^5}{(-5)^3} = (-5)^{5-3} = (-5)^2 = 25$

   c. $\dfrac{8^8}{8^6} = 8^{8-6} = 8^2 = 64$

   d. $\dfrac{q^5}{t^2}$ cannot be simplified because q and t are different bases.

e. Begin by grouping common bases.

$$\frac{6x^3y^7}{xy^5} = 6 \cdot \frac{x^3}{x} \cdot \frac{y^7}{y^5} = 6 \cdot x^{3-1} \cdot y^{7-5} = 6x^2y^2$$

10. a. $-3^0 = -1 \cdot 3^0 = -1 \cdot 1 = -1$

b. $(-3)^0 = 1$

c. $8^0 = 1$

d. $(0.2)^0 = 1$

e. Assume that neither a nor y is zero.
$(7a^2y^4)^0 = 1$

f. $7y^0 = 7 \cdot y^0 = 7 \cdot 1 = 7$

11. a. $\left(\dfrac{z}{12}\right)^2 = \dfrac{z^2}{12^2} = \dfrac{z^2}{144}$

b. $(4x^6)^3 = 4^3 \cdot (x^6)^3 = 64x^{18}$

c. $y^{10} \cdot y^3 = y^{10+3} = y^{13}$

12. a. $8^2 - 8^0 = 64 - 1 = 63$

b. $(z^0)^6 + (4^0)^5 = 1^6 + 1^5 = 1 + 1 = 2$

c. $\left(\dfrac{5x^3}{15y^4}\right)^2 = \dfrac{5^2(x^3)^2}{15^2(y^4)^2} = \dfrac{25x^6}{225y^8} = \dfrac{x^6}{9y^8}$

d. $\dfrac{(2z^8x^5)^4}{-16z^2x^{20}} = \dfrac{2^4(z^8)^4(x^5)^4}{-16z^2x^{20}}$

$$= \dfrac{16z^{32}x^{20}}{-16z^2x^{20}}$$

$$= -1 \cdot (z^{32-2}) \cdot (x^{20-20})$$

$$= -1 \cdot z^{30} \cdot x^0$$

$$= -1 \cdot z^{30} \cdot 1$$

$$= -z^{30}$$

Vocabulary, Readiness & Video Check 5.1

1. Repeated multiplication of the same factor can be written using an <u>exponent</u>.

2. In 5^2, the 2 is called the <u>exponent</u> and the 5 is called the <u>base</u>.

3. To simplify $x^2 \cdot x^7$, keep the base and <u>add</u> the exponents.

4. To simplify $(x^3)^6$, keep the base and <u>multiply</u> the exponents.

5. The understood exponent on the term y is <u>1</u>.

6. If $x^{\square} = 1$, the exponent is <u>0</u>.

7. Example 4 can be written as $-4^2 = -1 \cdot 4^2$, which is similar to Example 7, $4 \cdot 3^2$, and shows why the negative sign should not be considered part of the base when there are no parentheses.

8. The properties allow us to reorder and regroup factors and put those with common bases together, making it easier to apply the product rule; yes, in Example 13.

9. Be careful not to confuse the power rule with the product rule. The power rule involves a power raised to a power (exponents are multiplied), and the product rule involves a product (exponents are added).

10. Remember to raise the -2 (or any number) to the power along with the variables.

11. the quotient rule

12. No, Example 30 is a fraction and does not use the quotient rule.

Exercise Set 5.1

1. In 3^2, the base is 3 and the exponent is 2.

3. In -4^2, the base is 4 and the exponent is 2.

5. In $5x^2$, the base 5 has exponent 1 and the base x has exponent 2.

7. $7^2 = 7 \cdot 7 = 49$

9. $(-5)^1 = -5$

11. $-2^4 = -2 \cdot 2 \cdot 2 \cdot 2 = -16$

13. $(-2)^4 = (-2)(-2)(-2)(-2) = 16$

15. $(0.1)^5 = (0.1)(0.1)(0.1)(0.1)(0.1)$
$= 0.00001$

17. $\left(\dfrac{1}{3}\right)^4 = \left(\dfrac{1}{3}\right)\left(\dfrac{1}{3}\right)\left(\dfrac{1}{3}\right)\left(\dfrac{1}{3}\right) = \dfrac{1}{81}$

19. $7 \cdot 2^5 = 7 \cdot 2 \cdot 2 \cdot 2 \cdot 2 \cdot 2 = 224$

21. $-2 \cdot 5^3 = -2 \cdot 5 \cdot 5 \cdot 5 = -250$

23. $x^2 = (-2)^2 = (-2)(-2) = 4$

25. $5x^3 = 5(3)^3 = 5 \cdot 3 \cdot 3 \cdot 3 = 135$

27. $2xy^2 = 2(3)(5)^2 = 2(3)(5)(5) = 150$

29. $\dfrac{2z^4}{5} = \dfrac{2(-2)^4}{5} = \dfrac{2(-2)(-2)(-2)(-2)}{5} = \dfrac{32}{5}$

31. $x^2 \cdot x^5 = x^{2+5} = x^7$

33. $(-3)^3 \cdot (-3)^9 = (-3)^{3+9} = (-3)^{12}$

35. $(5y^4)(3y) = 5(3)y^{4+1} = 15y^5$

37. $(x^9 y)(x^{10} y^5) = x^{9+10} y^{1+5} = x^{19} y^6$

39. $(-8mn^6)(9m^2 n^2) = --8(9)m^{1+2} n^{6+2}$
$= -72m^3 n^8$

41. $(4z^{10})(-6z^7)(z^3) = 4(-6)z^{10+7+3} = -24z^{20}$

43. $A = (4x^2) \cdot (5x^3)$
$= (4 \cdot 5) \cdot (x^2 \cdot x^3)$
$= 20x^{2+3}$
$= 20x^5$
The area is $20x^5$ square feet.

45. $(x^9)^4 = x^{9 \cdot 4} = x^{36}$

47. $(pq)^8 = p^8 q^8$

49. $(2a^5)^3 = 2^3 \cdot (a^5)^3 = 8 \cdot a^{5 \cdot 3} = 8a^{15}$

51. $(x^2 y^3)^5 = (x^2)^5 \cdot (y^3)^5 = x^{2 \cdot 5} \cdot y^{3 \cdot 5} = x^{10} y^{15}$

53. $(-7a^2 b^5 c)^2 = (-7)^2 \cdot (a^2)^2 \cdot (b^5)^2 \cdot c^2$
$= 49a^{2 \cdot 2} b^{5 \cdot 2} c^2$
$= 49a^4 b^{10} c^2$

55. $\left(\dfrac{r}{s}\right)^9 = \dfrac{r^9}{s^9}$

57. $\left(\dfrac{mp}{n}\right)^5 = \dfrac{(mp)^5}{n^5} = \dfrac{m^5 \cdot p^5}{n^5} = \dfrac{m^5 p^5}{n^5}$

59. $\left(\dfrac{-2xz}{y^5}\right)^2 = \dfrac{(-2)^2 x^2 z^2}{y^{5 \cdot 2}} = \dfrac{4x^2 z^2}{y^{10}}$

61. $A = (8z^5)^2 = 8^2 \cdot (z^5)^2 = 64 \cdot z^{5 \cdot 2} = 64z^{10}$
The area is $64z^{10}$ square decimeters.

63. $V = (3y^4)^3 = 3^3 y^{4 \cdot 3} = 27y^{12}$
The volume is $27y^{12}$ cubic feet.

65. $\dfrac{x^3}{x} = \dfrac{x^3}{x^1} = x^{3-1} = x^2$

67. $\dfrac{(-4)^6}{(-4)^3} = (-4)^{6-3} = (-4)^3 = -64$

69. $\dfrac{p^7 q^{20}}{pq^{15}} = p^{7-1} q^{20-15} = p^6 q^5$

71. $\dfrac{7x^2 y^6}{14x^2 y^3} = \dfrac{7}{14} x^{2-2} y^{6-3} = \dfrac{1}{2} x^0 y^3 = \dfrac{y^3}{2}$

73. $7^0 = 1$

75. $(2x)^0 = 1$

77. $-7x^0 = -7(1) = -7$

79. $5^0 + y^0 = 1 + 1 = 2$

81. $-9^2 = -9 \cdot 9 = -81$

83. $\left(\dfrac{1}{4}\right)^3 = \dfrac{1}{4} \cdot \dfrac{1}{4} \cdot \dfrac{1}{4} = \dfrac{1}{64}$

85. $b^4 b^2 = b^{4+2} = b^6$

87. $a^2 a^3 a^4 = a^{2+3+4} = a^9$

89. $(2x^3)(-8x^4) = (2 \cdot -8)(x^3 \cdot x^4)$
$\qquad = -16x^{3+4}$
$\qquad = -16x^7$

91. $(a^7 b^{12})(a^4 b^8) = a^7 a^4 \cdot b^{12} b^8$
$\qquad = a^{7+4} b^{12+8}$
$\qquad = a^{11} b^{20}$

93. $(-2mn^6)(-13m^8 n) = (-2)(-13)(m \cdot m^8)(n^6 \cdot n)$
$\qquad = 26m^{1+8} n^{6+1}$
$\qquad = 26m^9 n^7$

95. $(z^4)^{10} = z^{4 \cdot 10} = z^{40}$

97. $(4ab)^3 = (4)^3 a^3 b^3 = 64a^3 b^3$

99. $(-6xyz^3)^2 = (-6)^2 x^2 y^2 (z^3)^2$
$\qquad = 36x^2 y^2 z^{3 \cdot 2}$
$\qquad = 36x^2 y^2 z^6$

101. $\dfrac{3x^5}{x^4} = 3 \cdot \dfrac{x^5}{x^4} = 3x^{5-4} = 3x^1 = 3x$

103. $(9xy)^2 = 9^2 \cdot x^2 y^2 = 81x^2 y^2$

105. $2^3 + 2^0 = (2 \cdot 2 \cdot 2) + 1 = 8 + 1 = 9$

107. $\left(\dfrac{3y^5}{6x^4}\right)^3 = \left(\dfrac{y^5}{2x^4}\right)^3 = \dfrac{(y^5)^3}{2^3 (x^4)^3} = \dfrac{y^{5 \cdot 3}}{8x^{4 \cdot 3}} = \dfrac{y^{15}}{8x^{12}}$

109. $\dfrac{2x^3 y^2 z}{xyz} = 2 \cdot \dfrac{x^3}{x} \cdot \dfrac{y^2}{y} \cdot \dfrac{z}{z}$
$\qquad = 2x^{3-1} y^{2-1} z^{1-1}$
$\qquad = 2x^2 y^1 z^0$
$\qquad = 2x^2 y$

111. $(5^0)^3 + (y^0)^7 = 1^3 + 1^7 = 1 + 1 = 2$

113. $\left(\dfrac{5x^9}{10y^{11}}\right)^2 = \dfrac{5^2 (x^9)^2}{10^2 (y^{11})^2} = \dfrac{25x^{18}}{100y^{22}} = \dfrac{x^{18}}{4y^{22}}$

115. $\dfrac{(2a^5 b^3)^4}{-16a^{20} b^7} = \dfrac{2^4 (a^5)^4 (b^3)^4}{-16a^{20} b^7}$
$\qquad = \dfrac{16a^{20} b^{12}}{-16a^{20} b^7}$
$\qquad = -1 \cdot a^{20-20} \cdot b^{12-7}$
$\qquad = -1 \cdot a^0 \cdot b^5$
$\qquad = -1 \cdot 1 \cdot b^5$
$\qquad = -b^5$

117. $y - 10 + y = y + y - 10 = 2y - 10$

119. $7x + 2 - 8x - 6 = 7x - 8x + 2 - 6 = -x - 4$

121. $2(x - 5) + 3(5 - x) = 2x - 10 + 15 - 3x = -x + 5$

123. $(x^{14})^{23} = x^{14 \cdot 23} = x^{322}$
Multiply the exponents; choice c.

125. $x^{14} + x^{23}$ cannot be simplified further; choice e.

127. answers may vary

129. answers may vary

131. $V = x^3 = 7^3 = 7 \cdot 7 \cdot 7 = 343$
The volume is 343 cubic meters.

133. Volume; volume measures capacity.

135. answers may vary

137. answers may vary

139. $x^{5a} x^{4a} = x^{5a+4a} = x^{9a}$

141. $(a^b)^5 = a^{b \cdot 5} = a^{5b}$

143. $\dfrac{x^{9a}}{x^{4a}} = x^{9a-4a} = x^{5a}$

145. $A = P\left(1+\dfrac{r}{12}\right)^6$

$A = 1000\left(1+\dfrac{0.09}{12}\right)^6$

$= 1000(1.0075)^6$

≈ 1045.85

$1045.85 is needed to pay off the loan.

Section 5.2 Practice

1. a. The exponent on y is 3, so the degree of $5y^3$ is 3.

b. $10xy$ can be written as $10x^1y^1$. The degree of the term is the sum of the exponents, so the degree is $1 + 1 = 2$.

c. The degree of $z = z^1$ is 1.

d. $-3a^2b^5c$ can be written as $-3a^2b^5c^1$. The degree of the term is the sum of the exponents, so the degree is $2 + 5 + 1$ or 8.

e. The constant, 8, can be written as $8x^0$ (since $x^0 = 1$). The degree of 8 or $8x^0$ is 0.

2. a. The degree of the trinomial $5b^2 - 3b + 7$ is 2, the greatest degree of any of its terms.

b. Rewrite the binomial as $7t^1 + 3$, the degree is 1.

c. The degree of the polynomial $5x^2 + 3x - 6x^3 + 4$ is 3.

3.

Term	numerical coefficient	degree of term
$-3x^3y^2$	-3	5
$4xy^2$	4	3
$-y^2$	-1	2
$3x$	3	1
-2	-2	0

4. a. $P(x) = -2x^2 - x + 7$

$P(1) = -2(1)^2 - 1 + 7 = 4$

b. $P(x) = -2x^2 - x + 7$

$P(-4) = -2(-4)^2 - (-4) + 7 = -21$

5. To find each height, we evaluate $P(t)$ when $t = 1$ and when $t = 2$.

$P(t) = -16t^2 + 130$

$P(1) = -16(1)^2 + 130$

$= -16 + 130$

$= 114$

The height of the camera at 1 second is 114 feet.

$P(t) = -16t^2 + 130$

$P(2) = -16(2)^2 + 130$

$= -16(4) + 130$

$= -64 + 130$

$= 66$

The height of the camera at 2 seconds is 66 feet.

6. a. $-4y + 2y = (-4 + 2)y = -2y$

b. These terms cannot be combined because z and $5z^3$ are not like terms.

c. $15x^3 - x^3 = 15x^3 - 1x^3 = 14x^3$

d. $7a^2 - 5 - 3a^2 - 7 = 7a^2 - 3a^2 - 5 - 7$

$= 4a^2 - 12$

e. $\dfrac{3}{8}x^3 - x^2 + \dfrac{5}{6}x^4 + \dfrac{1}{12}x^3 - \dfrac{1}{2}x^4$

$= \left(\dfrac{5}{6} - \dfrac{1}{2}\right)x^4 + \left(\dfrac{3}{8} + \dfrac{1}{12}\right)x^3 - x^2$

$= \left(\dfrac{5}{6} - \dfrac{3}{6}\right)x^4 + \left(\dfrac{9}{24} + \dfrac{2}{24}\right)x^3 - x^2$

$= \dfrac{2}{6}x^4 + \dfrac{11}{24}x^3 - x^2$

$= \dfrac{1}{3}x^4 + \dfrac{11}{24}x^3 - x^2$

7. $9xy - 3x^2 - 4yx + 5y^2 = -3x^2 + (9-4)xy + 5y^2$

$= -3x^2 + 5xy + 5y^2$

8. $x \cdot x + 2 \cdot x + 2 \cdot 2 + 5 \cdot x + x \cdot 3x$

$= x^2 + 2x + 4 + 5x + 3x^2$

$= 4x^2 + 7x + 4$

9. a. $(4y^2 + x - 3y - 7) + (x + y^2 - 2)$
$= 4y^2 + x - 3y - 7 + x + y^2 - 2$
$= 4y^2 + y^2 - 3y + x + x - 7 - 2$
$= 5^2 - 3y + 2x - 9$

b. $(-8a^2b - ab^2 + 10) + (-2ab^2 - 10)$
$= -8a^2b - ab^2 + 10 - 2ab^2 - 10$
$= -8a^2b - ab^2 - 2ab^2 + 10 - 10$
$= -8a^2b - 3ab^2$

10. $(3x^2 - 9x + 11) + (-3x^2 + 7x^3 + 3x - 4)$
$= 7x^3 + 3x^2 - 3x^2 - 9x + 3x + 11 - 4$
$= 7x^3 - 6x + 7$

11.   First, change the sign of each term of the second polynomial and then add.
$(3x^3 - 5x^2 + 4x) - (x^3 - x^2 + 6)$
$= (3x^3 - 5x^2 + 4x) + (-x^3 + x^2 - 6)$
$= 3x^3 - x^3 - 5x^2 + x^2 + 4x - 6$
$= 2x^3 - 4x^2 + 4x - 6$

12. $[(8x - 11) + (2x + 5)] - (3x + 5)$
$= 8x - 11 + 2x + 5 - 3x - 5$
$= 8x + 2x - 3x - 11 + 5 - 5$
$= 7x - 11$

13. a. $(3a^2 - 4ab + 7b^2) + (-8a^2 + 3ab - b^2)$
$= 3a^2 - 4ab + 7b^2 - 8a^2 + 3ab - b^2$
$= -5a^2 - ab + 6b^2$

b. $(5x^2y^2 - 6xy - 4xy^2)$
$\qquad\qquad - (2x^2y^2 + 4xy - 5 + 6y^2)$
$= 5x^2y^2 - 6xy - 4xy^2 - 2x^2y^2$
$\qquad\qquad - 4xy + 5 - 6y^2$
$= 3x^2y^2 - 10xy - 4xy^2 - 6y^2 + 5$

Graphing Calculator Explorations

1. $(2x^2 + 7x + 6) + (x^3 - 6x^2 - 14)$
$= x^3 - 4x^2 + 7x - 8$

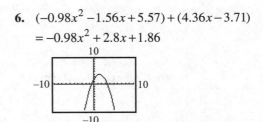

2. $(-14x^3 - x + 2) + (-x^3 + 3x^2 + 4x)$
$= -15x^3 + 3x^2 + 3x + 2$

3. $(1.8x^2 - 6.8x - 1.7) - (3.9x^2 - 3.6x)$
$= -2.1x^2 - 3.2x - 1.7$

4. $(-4.8x^2 + 12.5x - 7.8) - (3.1x^2 - 7.8x)$
$= -7.9x^2 + 20.3x - 7.8$

5. $(1.29x - 5.68) + (7.69x^2 - 2.55x + 10.98)$
$= 7.69x^2 - 1.26x + 5.3$

6. $(-0.98x^2 - 1.56x + 5.57) + (4.36x - 3.71)$
$= -0.98x^2 + 2.8x + 1.86$

Vocabulary, Readiness & Video Check 5.2

1.   A <u>binomial</u> is a polynomial with exactly 2 terms.

2.   A <u>monomial</u> is a polynomial with exactly one term.

3.   A <u>trinomial</u> is a polynomial with exactly three terms.

4. The numerical factor of a term is called the coefficient.

5. A number term is also called a constant.

6. The degree of a polynomial is the greatest degree of any term of the polynomial.

7. The degree of the polynomial is the greatest degree of any of its terms, so we need to find the degree of each term first.

8. Substitute the replacement value for x throughout the function, then simplify.

9. simplifying it

10. Addition; no, we subtract in Examples 9–11. To subtract, we first change the signs of the polynomial being subtracted and then add.

Exercise Set 5.2

1. $x + 2$ is a binomial because it has two terms. The degree is 1 since x is x^1.

3. $9m^3 - 5m^2 + 4m - 8$ is neither a monomial, a binomial, nor a trinomial because it has more than three terms. The degree is 3, the greatest degree of any of its terms.

5. $12x^4y - x^2y^2 - 12x^2y^4$ is a trinomial because it has three terms. The degree is 6, the greatest degree of any of its terms.

7. $3 - 5x^8$ is a binomial because it has two terms. The degree is 8, the greatest degree of any of its terms.

	Polynomial	*Degree*
9.	$3xy^2 - 4$	3
11.	$5a^2 - 2a + 1$	2

13. $P(x) = x^2 + x + 1$
$P(7) = 7^2 + 7 + 1 = 49 + 7 + 1 = 57$

15. $Q(x) = 5x^2 - 1$
$$Q(-10) = 5(-10)^2 - 1$$
$$= 5(100) - 1$$
$$= 500 - 1$$
$$= 499$$

17. $P(x) = x^2 + x + 1$
$P(0) = 0^2 + 0 + 1 = 0 + 1 = 1$

19. $Q(x) = 5x^2 - 1$
$$Q\left(\frac{1}{4}\right) = 5\left(\frac{1}{4}\right)^2 - 1 = 5\left(\frac{1}{16}\right) - 1 = \frac{5}{16} - \frac{16}{16} = -\frac{11}{16}$$

21. $P(t) = -16t^2 + 1150$
$P(1) = -16(1)^2 + 1150 = -16 + 1150 = 1134$
After 1 second, the height is 1134 feet.

23. $P(t) = -16t^2 + 1150$
$P(3) = -16(3)^2 + 1150 = -144 + 1150 = 1006$
After 3 seconds, the height is 1006 feet.

25. $P(x) = -7.5x^2 + 93x - 100$
$$P(8) = -7.5(8)^2 + 93(8) - 100$$
$$= -7.5(64) + 93(8) - 100$$
$$= -480 + 744 - 100$$
$$= 164$$
There were 164 thousand visitors in 2008.

27. $14x^2 + 9x^2 = (14 + 9)x^2 = 23x^2$

29. $15x^2 - 3x^2 - y = (15 - 3)x^2 - y = 12x^2 - y$

31. $8s - 5s + 4s = (8 - 5 + 4)s = 7s$

33. $0.1y^2 - 1.2y^2 + 6.7 - 1.9$
$$= (0.1 - 1.2)y^2 + (6.7 - 1.9)$$
$$= -1.1y^2 + 4.8$$

35. $\frac{2}{5}x^2 - \frac{1}{3}x^3 + x^2 - \frac{1}{4}x^3 + 6$
$$= \left(-\frac{1}{3} - \frac{1}{4}\right)x^3 + \left(\frac{2}{5} + 1\right)x^2 + 6$$
$$= \left(-\frac{4}{12} - \frac{3}{12}\right)x^3 + \left(\frac{2}{5} + \frac{5}{5}\right)x^2 + 6$$
$$= -\frac{7}{12}x^3 + \frac{7}{5}x^2 + 6$$

37. $6a^2 - 4ab + 7b^2 - a^2 - 5ab + 9b^2$
$= (6-1)a^2 + (-4-5)ab + (7+9)b^2$
$= 5a^2 - 9ab + 16b^2$

39. $(-7x+5) + (-3x^2 + 7x + 5)$
$= -7x + 5 - 3x^2 + 7x + 5$
$= -3x^2 + (-7x + 7x) + (5+5)$
$= -3x^2 + 10$

41. $(2x^2 + 5) - (3x^2 - 9) = 2x^2 + 5 - 3x^2 + 9$
$= (2x^2 - 3x^2) + (5+9)$
$= -x^2 + 14$

43. $3x - (5x - 9) = 3x - 5x + 9$
$= (3x - 5x) + 9$
$= -2x + 9$

45. $(2x^2 + 3x - 9) - (-4x + 7)$
$= 2x^2 + 3x - 9 + 4x - 7$
$= 2x^2 + (3x + 4x) + (-9 - 7)$
$= 2x^2 + 7x - 16$

47. $\begin{array}{r} 3t^2 + 4 \\ + \; 5t^2 - 8 \\ \hline 8t^2 - 4 \end{array}$

49. $\begin{array}{r} 4z^2 - 8z + 3 \\ - \; (6z^2 + 8z - 3) \\ \hline \end{array}$ $\begin{array}{r} 4z^2 - 8z + 3 \\ + \; (-6z^2 - 8z + 3) \\ \hline -2z^2 - 16z + 6 \end{array}$

51. $\begin{array}{r} 5x^3 - 4x^2 + 6x - 2 \\ - \; (3x^3 - 2x^2 \;\; - x - 4) \\ \hline \end{array}$ $\begin{array}{r} 5x^3 - 4x^2 + 6x - 2 \\ + \; (-3x^3 + 2x^2 \;\; + x + 4) \\ \hline 2x^3 - 2x^2 + 7x + 2 \end{array}$

53. $(-3y^2 - 4y) + (2y^2 + y - 1)$
$= -3y^2 - 4y + 2y^2 + y - 1$
$= -y^2 - 3y - 1$

55. $(5x+8) - (-2x^2 - 6x + 8)$
$= 5x + 8 + 2x^2 + 6x - 8$
$= 2x^2 + 11x$

57. $(-8x^4 + 7x) + (-8x^4 + x + 9)$
$= -8x^4 + 7x - 8x^4 + x + 9$
$= -16x^4 + 8x + 9$

59. $(3x^2 + 5x - 8) + (5x^2 + 9x + 12) - (x^2 - 14)$
$= 3x^2 + 5x - 8 + 5x^2 + 9x + 12 - x^2 + 14$
$= 7x^2 + 14x + 18$

61. $(7x - 3) - 4x = 7x - 3 - 4x = 3x - 3$

63. $(4x^2 - 6x + 1) + (3x^2 + 2x + 1)$
$= 4x^2 - 6x + 1 + 3x^2 + 2x + 1$
$= 7x^2 - 4x + 2$

65. $(81x^2 + 10) - (19x^2 + 5) = 81x^2 + 10 - 19x^2 - 5$
$= 62x^2 + 5$

67. $(7x^2 + 3x + 9) - (5x + 7)$
$= (7x^2 + 3x + 9) + (-5x - 7)$
$= 7x^2 + 3x + 9 - 5x - 7$
$= 7x^2 - 2x + 2$

69. $[(8x+1) + (6x+3)] - (2x+2)$
$= 8x + 1 + 6x + 3 - 2x - 2$
$= 8x + 6x - 2x + 1 + 3 - 2$
$= 12x + 2$

71. $[(8y^2 + 7) + (6y+9)] - (4y^2 - 6y - 3)$
$= (8y^2 + 7) + (6y+9) + (-4y^2 + 6y + 3)$
$= 8y^2 + 7 + 6y + 9 - 4y^2 + 6y + 3$
$= 4y^2 + 12y + 19$

73. $2x \cdot 2x + x \cdot 7 + x \cdot x + x \cdot 5 = 4x^2 + 7x + x^2 + 5x$
$= 5x^2 + 12x$

75. $(9a + 6b - 5) + (-11a - 7b + 6)$
$= 9a + 6b - 5 - 11a - 7b + 6$
$= -2a - b + 1$

77. $(4x^2 + y^2 + 3) - (x^2 + y^2 - 2)$
$= 4x^2 + y^2 + 3 - x^2 - y^2 + 2$
$= 3x^2 + 5$

79. $(x^2 + 2xy - y^2) + (5x^2 - 4xy + 20y^2)$
$= x^2 + 2xy - y^2 + 5x^2 - 4xy + 20y^2$
$= 6x^2 - 2xy + 19y^2$

81. $(11r^2s + 16rs - 3 - 2r^2s^2) - (3sr^2 + 5 - 9r^2s^2)$
$= 11r^2s + 16rs - 3 - 2r^2s^2 - 3sr^2 - 5 + 9r^2s^2$
$= 8r^2s + 16rs - 8 + 7r^2s^2$

83. $7.75x + 9.16x^2 - 1.27 - 14.58x^2 - 18.34$
$= (9.16 - 14.58)x^2 + 7.75x + (-1.27 - 18.34)$
$= -5.42x^2 + 7.75x - 19.61$

85. $[(7.9y^4 - 6.8y^3 + 3.3y) + (6.1y^3 - 5)]$
$\quad -(4.2y^4 + 1.1y - 1)$
$= 7.9y^4 - 6.8y^3 + 3.3y + 6.1y^3 - 5 - 4.2y^4$
$\quad -1.1y + 1$
$= 3.7y^4 - 0.7y^3 + 2.2y - 4$

87. $3x(2x) = 3 \cdot 2 \cdot x \cdot x = 6x^2$

89. $(12x^3)(-x^5) = (12x^3)(-1x^5)$
$\qquad\qquad = (12)(-1)(x^3)(x^5)$
$\qquad\qquad = -12x^8$

91. $10x^2(20xy^2) = 10 \cdot 20x^2 \cdot x \cdot y^2 = 200x^3y^2$

93. $9x + 10 + 3x + 12 + 4x + 15 + 2x + 7$
$= (9x + 3x + 4x + 2x) + (10 + 12 + 15 + 7)$
$= 18x + 44$

95. $(-x^2 + 3x) + (2x^2 + 5) + (4x - 1)$
$= -x^2 + 3x + 2x^2 + 5 + 4x - 1$
$= x^2 + 7x + 4$
The perimeter is $(x^2 + 7x + 4)$ feet.

97. $(4y^2 + 4y + 1) - (y^2 - 10)$
$= 4y^2 + 4y + 1 - y^2 + 10$
$= 3y^2 + 4y + 11$
The length of the remaining piece is
$(3y^2 + 4y + 11)$ meters.

99. $x = 2009 - 2004 = 5$
$P(x) = 0.16x^2 + 0.29x + 1.88$
$P(5) = 0.16(5)^2 + 0.29(5) + 1.88$
$\qquad = 4 + 1.45 + 1.88$
$\qquad = 7.33$
The number of cell phones recycled in 2009 is estimated at 7.33 million.

101. answers may vary

103. answers may vary

105. $10y - 6y^2 - y = (10 - 1)y - 6y^2 = 9y - 6y^2$
choice b

107. $(5x - 3) + (5x - 3) = (5x + 5x) + (-3 - 3)$
$\qquad\qquad\qquad\qquad = (5 + 5)x - 6$
$\qquad\qquad\qquad\qquad = 10x - 6$
choice e

109. a. $z + 3z = 1z + 3z = 4z$

   b. $z \cdot 3z = z^1 \cdot 3z^1 = 3z^{1+1} = 3z^2$

   c. $-z - 3z = -1z - 3z = -4z$

   d. $(-z)(-3z) = (-z^1)(-3z^1) = 3z^{1+1} = 3z^2$

   answers may vary

111. a. $m \cdot m \cdot m = m^1 \cdot m^1 \cdot m^1 = m^{1+1+1} = m^3$

   b. $m + m + m = 1m + 1m + 1m = (1 + 1 + 1)m = 3m$

   c. $(-m)(-m)(-m) = (-1 \cdot m^1)(-1 \cdot m^1)(-1 \cdot m^1)$
$\qquad\qquad\qquad = (-1)(-1)(-1)(m \cdot m \cdot m)$
$\qquad\qquad\qquad = -1m^3$
$\qquad\qquad\qquad = -m^3$

   answers may vary

113. $(4x^{2a} - 3x^a + 0.5) - (x^{2a} - 5x^a - 0.2)$
$= 4x^{2a} - 3x^a + 0.5 - x^{2a} + 5x^a + 0.2$
$= 4x^{2a} - x^{2a} - 3x^a + 5x^a + 0.5 + 0.2$
$= 3x^{2a} + 2x^a + 0.7$

115. $(8x^{2y} - 7x^y + 3) + (-4x^{2y} + 9x^y - 14)$
$= 8x^{2y} - 7x^y + 3 - 4x^{2y} + 9x^y - 14$
$= 8x^{2y} - 4x^{2y} - 7x^y + 9x^y + 3 - 14$
$= 4x^{2y} + 2x^y - 11$

117. $P(x) + Q(x) = (3x + 3) + (4x^2 - 6x + 3)$
$= 4x^2 + 3x - 6x + 3 + 3$
$= 4x^2 - 3x + 6$

119. $Q(x) - R(x) = (4x^2 - 6x + 3) - (5x^2 - 7)$
$= 4x^2 - 5x^2 - 6x + 3 + 7$
$= -x^2 - 6x + 10$

121. $2[Q(x)] - R(x) = 2(4x^2 - 6x + 3) - (5x^2 - 7)$
$= 2(4x^2) - 2(6x) + 2(3) - 5x^2 + 7$
$= 8x^2 - 12x + 6 - 5x^2 + 7$
$= 3x^2 - 12x + 13$

123. $P(x) = 2x - 3$

　a. $P(a) = 2a - 3$

　b. $P(-x) = 2(-x) - 3 = -2x - 3$

　c. $P(x + h) = 2(x + h) - 3 = 2x + 2h - 3$

125. $P(x) = 4x$

　a. $P(a) = 4a$

　b. $P(-x) = 4(-x) = -4x$

　c. $P(x + h) = 4(x + h) = 4x + 4h$

127. Since $3 + 4 = 7$, $3x^2 + 4x^2 = 7x^2$ is a true statement.

129. Since $2 + 4 = 6$ and $3 - 5 = -2$,
$2x^4 + 3x^3 - 5x^3 + 4x^4 = 6x^4 - 2x^3$ is a true statement.

131. $x^2 + x^2 + xy + xy + xy + xy = 2x^2 + 4xy$

Section 5.3 Practice

　1. $5y \cdot 2y = (5 \cdot 2)(y \cdot y) = 10y^2$

　2. $(5z^3) \cdot (-0.4z^5) = (5 \cdot -0.4)(z^3 \cdot z^5) = -2z^8$

　3. $\left(-\dfrac{1}{9}b^6\right)\left(-\dfrac{7}{8}b^3\right) = \left(-\dfrac{1}{9} \cdot -\dfrac{7}{8}\right)(b^6 \cdot b^3) = \dfrac{7}{72}b^9$

　4. a. $3x(9x^5 + 11) = 3x(9x^5) + 3x(11)$
　　　　　　$= 27x^6 + 33x$

　b. $-6x^3(2x^2 - 9x + 2)$
　　$= -6x^3(2x^2) + (-6x^3)(-9x) + (-6x^2)(2)$
　　$= -12x^5 + 54x^4 - 12x^3$

　5. Multiply each term of the first binomial by each term of the second.
　$(5x - 2)(2x + 3)$
　$= 5x(2x) + 5x(3) + (-2)(2x) + (-2)(3)$
　$= 10x^2 + 15x - 4x - 6$
　$= 10x^2 + 11x - 6$

　6. Recall that $a^2 = a \cdot a$, so
　$(5x - 3y)^2 = (5x - 3y)(5x - 3y)$. Multiply each term of the first binomial by each term of the second.
　$(5x - 3y)(5x - 3y)$
　$= 5x(5x) + 5x(-3y) + (-3y)(5x) + (-3y)(-3y)$
　$= 25x^2 - 15xy - 15xy + 9y^2$
　$= 25x^2 - 30xy + 9y^2$

　7. Multiply each term of the first polynomial by each term of the second.
　$(y + 4)(2y^2 - 3y + 5)$
　$= y(2y^2) + y(-3y) + y(5) + 4(2y^2)$
　　$+ 4(-3y) + 4(5)$
　$= 2y^3 - 3y^2 + 5y + 8y^2 - 12y + 20$
　$= 2y^3 + 5y^2 - 7y + 20$

　8. Write $(s + 2t)^3$ as $(s + 2t)(s + 2t)(s + 2t)$.
　$(s + 2t)(s + 2t)(s + 2t)$
　$= (s^2 + 2st + 2st + 4t^2)(s + 2t)$
　$= (s^2 + 4st + 4t^2)(s + 2t)$
　$= (s^2 + 4st + 4t^2)s + (s^2 + 4st + 4t^2)(2t)$
　$= s^3 + 4s^2t + 4st^2 + 2s^2t + 8st^2 + 8t^3$
　$= s^3 + 6s^2t + 12st^2 + 8t^3$

9.
$$
\begin{array}{r}
5x^2 - 3x + 5 \\
\times \quad\quad\quad x - 4 \\
\hline
-20x^2 + 12x - 20 \\
5x^3 - 3x^2 + 5x \quad\quad\quad \\
\hline
5x^3 - 23x^2 + 17x - 20
\end{array}
$$

10.
$$
\begin{array}{r}
x^3 - 2x^2 + 1 \\
\times \quad\quad\quad x^2 + 2 \\
\hline
2x^3 - 4x^2 + 2 \\
x^5 - 2x^4 \quad\quad + x^2 \quad\quad \\
\hline
x^5 - 2x^4 + 2x^3 - 3x^2 + 2
\end{array}
$$

11.
$$
\begin{array}{r}
5x^2 + 2x - 2 \\
x^2 - x + 3 \\
\hline
15x^2 + 6x - 6 \\
-5x^3 - 2x^2 + 2x \quad\quad \\
5x^4 + 2x^3 - 2x^2 \quad\quad\quad\quad \\
\hline
5x^4 - 3x^3 + 11x^2 + 8x - 6
\end{array}
$$

Vocabulary, Readiness & Video Check 5.3

1. The expression $5x(3x + 2)$ equals $5x \cdot 3x + 5x \cdot 2$ by the <u>distributive</u> property.

2. The expression $(x + 4)(7x - 1)$ equals $x(7x - 1) + 4(7x - 1)$ by the <u>distributive</u> property.

3. The expression $(5y - 1)^2$ equals <u>$(5y - 1)(5y - 1)$</u>.

4. The expression $9x \cdot 3x$ equals <u>$27x^2$</u>.

5. No; the monomials are unlike terms.

6. distributive property, product rule

7. Yes; the parentheses have been removed for the vertical format, but every term in the first polynomial is still distributed to every term in the second polynomial.

Exercise Set 5.3

1. $-4n^3 \cdot 7n^7 = (-4 \cdot 7)(n^3 \cdot n^7) = -28n^{10}$

3. $(-3.1x^3)(4x^9) = (-3.1 \cdot 4)(x^3 \cdot x^9) = -12.4x^{12}$

5. $\left(-\dfrac{1}{3}y^2\right)\left(\dfrac{2}{5}y\right) = \left(-\dfrac{1}{3} \cdot \dfrac{2}{5}\right)(y^2 \cdot y) = -\dfrac{2}{15}y^3$

7. $(2x)(-3x^2)(4x^5) = (2 \cdot -3 \cdot 4)(x \cdot x^2 \cdot x^5) = -24x^8$

9. $3x(2x + 5) = 3x(2x) + 3x(5) = 6x^2 + 15x$

11. $-2a(a + 4) = -2a(a) + (-2a)(4) = -2a^2 - 8a$

13. $3x(2x^2 - 3x + 4) = 3x(2x^2) + 3x(-3x) + 3x(4)$
$$= 6x^3 - 9x^2 + 12x$$

15. $-2a^2(3a^2 - 2a + 3)$
$$= -2a^2(3a^2) + (-2a^2)(-2a) + (-2a^2)(3)$$
$$= -6a^4 + 4a^3 - 6a^2$$

17. $-y(4x^3 - 7x^2y + xy^2 + 3y^3)$
$$= -y(4x^3) + (-y)(-7x^2y) + (-y)(xy^2)$$
$$\quad\quad\quad + (-y)(3y^3)$$
$$= -4x^3y + 7x^2y^2 - xy^3 - 3y^4$$

19. $\dfrac{1}{2}x^2(8x^2 - 6x + 1)$
$$= \dfrac{1}{2}x^2(8x^2) + \dfrac{1}{2}x^2(-6x) + \dfrac{1}{2}x^2(1)$$
$$= 4x^4 - 3x^3 + \dfrac{1}{2}x^2$$

21. $(x + 4)(x + 3) = x(x) + x(3) + 4(x) + 4(3)$
$$= x^2 + 3x + 4x + 12$$
$$= x^2 + 7x + 12$$

23. $(a + 7)(a - 2) = a(a) + a(-2) + 7(a) + 7(-2)$
$$= a^2 - 2a + 7a - 14$$
$$= a^2 + 5a - 14$$

25. $\left(x + \dfrac{2}{3}\right)\left(x - \dfrac{1}{3}\right)$
$$= x(x) + x\left(-\dfrac{1}{3}\right) + \dfrac{2}{3}(x) + \dfrac{2}{3}\left(-\dfrac{1}{3}\right)$$
$$= x^2 - \dfrac{1}{3}x + \dfrac{2}{3}x - \dfrac{2}{9}$$
$$= x^2 + \dfrac{1}{3}x - \dfrac{2}{9}$$

27. $(3x^2+1)(4x^2+7)$
$= 3x^2(4x^2)+3x^2(7)+1(4x^2)+1(7)$
$= 12x^4+21x^2+4x^2+7$
$= 12x^4+25x^2+7$

29. $(2y-4)^2$
$= (2y-4)(2y-4)$
$= 2y(2y)+2y(-4)+(-4)(2y)+(-4)(-4)$
$= 4y^2-8y-8y+16$
$= 4y^2-16y+16$

31. $(4x-3)(3x-5)$
$= 4x(3x)+4x(-5)+(-3)(3x)+(-3)(-5)$
$= 12x^2-20x-9x+15$
$= 12x^2-29x+15$

33. $(3x^2+1)^2 = (3x^2+1)(3x^2+1)$
$= 3x^2(3x^2)+3x^2(1)+1(3x^2)+1(1)$
$= 9x^4+3x^2+3x^2+1$
$= 9x^4+6x^2+1$

35. a. $4y^2(-y^2)=4(-1)y^{2+2}=-4y^4$

   b. $4y^2-y^2=4y^2-1y^2=(4-1)y^2=3y^2$

   c. answers may vary

37. $(x-2)(x^2-3x+7)$
$= x(x^2)+x(-3x)+x(7)+(-2)(x^2)$
$\qquad +(-2)(-3x)+(-2)(7)$
$= x^3-3x^2+7x-2x^2+6x-14$
$= x^3-5x^2+13x-14$

39. $(x+5)(x^3-3x+4)$
$= x(x^3)+x(-3x)+x(4)+5(x^3)+5(-3x)+5(4)$
$= x^4-3x^2+4x+5x^3-15x+20$
$= x^4+5x^3-3x^2-11x+20$

41. $(2a-3)(5a^2-6a+4)$
$= 2a(5a^2)+2a(-6a)+2a(4)+(-3)(5a^2)$
$\qquad +(-3)(-6a)+(-3)(4)$
$= 10a^3-12a^2+8a-15a^2+18a-12$
$= 10a^3-27a^2+26a-12$

43. $(x+2)^3 = (x+2)(x+2)(x+2)$
$= (x^2+2x+2x+4)(x+2)$
$= (x^2+4x+4)(x+2)$
$= (x^2+4x+4)x+(x^2+4x+4)2$
$= x^3+4x^2+4x+2x^2+8x+8$
$= x^3+6x^2+12x+8$

45. $(2y-3)^3$
$= (2y-3)(2y-3)(2y-3)$
$= (4y^2-6y-6y+9)(2y-3)$
$= (4y^2-12y+9)(2y-3)$
$= (4y^2-12y+9)2y+(4y^2-12y+9)(-3)$
$= 8y^3-24y^2+18y-12y^2+36y-27$
$= 8y^3-36y^2+54y-27$

47.
$$\begin{array}{r} 2x-11 \\ \times \quad 6x+1 \\ \hline 2x-11 \\ 12x^2-66x \quad\;\; \\ \hline 12x^2-64x-11 \end{array}$$

49.
$$\begin{array}{r} 2x^2+4x-1 \\ \times \quad 5x+1 \\ \hline 2x^2+4x-1 \\ 10x^3+20x^2-5x \quad\;\; \\ \hline 10x^3+22x^2\;-x-1 \end{array}$$

51.
$$\begin{array}{r} 2x^2-7x-9 \\ \times \quad x^2+5x-7 \\ \hline -14x^2+49x+63 \\ 10x^3-35x^2-45x \quad\;\; \\ 2x^4-7x^3\;-9x^2 \quad\;\;\;\;\; \\ \hline 2x^4+3x^3-58x^2\;+4x+63 \end{array}$$

53. $-1.2y(-7y^6)=-1.2(-7)(y\cdot y^6)=8.4y^7$

55. $-3x(x^2+2x-8)$
$= -3x(x^2)+(-3x)(2x)+(-3x)(-8)$
$= -3x^3-6x^2+24x$

57. $(x+19)(2x+1)=x(2x)+x(1)+19(2x)+19(1)$
$= 2x^2+x+38x+19$
$= 2x^2+39x+19$

59. $\left(x+\dfrac{1}{7}\right)\left(x-\dfrac{3}{7}\right)$

$= x(x)+x\left(-\dfrac{3}{7}\right)+\dfrac{1}{7}(x)+\dfrac{1}{7}\left(-\dfrac{3}{7}\right)$

$= x^2-\dfrac{3}{7}x+\dfrac{1}{7}(x)-\dfrac{3}{49}$

$= x^2-\dfrac{2}{7}x-\dfrac{3}{49}$

61. $(3y+5)^2 = (3y+5)(3y+5)$

$\qquad = 3y(3y)+3y(5)+5(3y)+5(5)$

$\qquad = 9y^2+15y+15y+25$

$\qquad = 9y^2+30y+25$

63. $(a+4)(a^2-6a+6)$

$= a(a^2)+a(-6a)+a(6)+4(a^2)+4(-6a)+4(6)$

$= a^3-6a^2+6a+4a^2-24a+24$

$= a^3-2a^2-18a+24$

65. $(2x-5)^3$

$= (2x-5)(2x-5)(2x-5)$

$= (4x^2-10x-10x+25)(2x-5)$

$= (4x^2-20x+25)(2x-5)$

$= (4x^2-20x+25)2x+(4x^2-20x+25)(-5)$

$= 8x^3-40x^2+50x-20x^2+100x-125$

$= 8x^3-60x^2+150x-125$

67. $(4x+5)(8x^2+2x-4)$

$= 4x(8x^2)+4x(2x)+4x(-4)+5(8x^2)$
$\qquad\qquad +5(2x)+5(-4)$

$= 32x^3+8x^2-16x+40x^2+10x-20$

$= 32x^3+48x^2-6x-20$

69.

$$
\begin{array}{r}
3x^2+2x-4 \\
\times \qquad 2x^2-4x+3 \\
\hline
9x^2\ +6x-12 \\
-12x^3-8x^2+16x \\
6x^4+4x^3-8x^2 \qquad\qquad \\
\hline
6x^4-8x^3-7x^2+22x-12
\end{array}
$$

71. $(2x-5)(2x+5)$

$= 2x(2x)+2x(5)+(-5)(2x)+(-5)(5)$

$= 4x^2+10x-10x-25$

$= 4x^2-25$

The area is $(4x^2-25)$ square yards.

73. $\dfrac{1}{2}(3x-2)(4x) = 2x(3x-2)$

$\qquad\qquad\qquad = 2x(3x)+2x(-2)$

$\qquad\qquad\qquad = 6x^2-4x$

The area is $(6x^2-4x)$ square inches.

75. Add: $5a+15a = (5+15)a = 20a$

Subtract: $5a-15a = (5-15)a = -10a$

Multiply: $5a\cdot 15a = 5\cdot 15\cdot a^1\cdot a^1 = 75a^{1+1} = 75a^2$

Divide: $\dfrac{5a}{15a} = \dfrac{1}{3}a^{1-1} = \dfrac{1}{3}a^0 = \dfrac{1}{3}\cdot 1 = \dfrac{1}{3}$

77. Add: $-3y^5+9y^4$ cannot be simplified.

Subtract: $-3y^5-9y^4$ cannot be simplified.

Multiply: $-3y^5\cdot 9y^4 = -3\cdot 9\cdot y^5\cdot y^4$

$\qquad\qquad\qquad = -27y^{5+4}$

$\qquad\qquad\qquad = -27y^9$

Divide: $\dfrac{-3y^5}{9y^4} = -\dfrac{3}{9}\cdot y^{5-4} = -\dfrac{1}{3}y^1 = -\dfrac{y}{3}$

79. a. $(3x+5)+(3x+7) = 3x+5+3x+7$
$\qquad\qquad\qquad\qquad = 6x+12$

b. $(3x+5)(3x+7)$

$= 3x(3x+7)+5(3x+7)$

$= 3x(3x)+3x(7)+5(3x)+5(7)$

$= 9x^2+21x+15x+35$

$= 9x^2+36x+35$

answers may vary

81. $(3x-1)+(10x-6) = 3x-1+10x-6 = 13x-7$

83. $(3x-1)(10x-6)$

$= 3x(10x-6)+(-1)(10x-6)$

$= 3x(10x)+3x(-6)+(-1)(10x)+(-1)(-6)$

$= 30x^2-18x-10x+6$

$= 30x^2-28x+6$

85. $(3x-1)-(10x-6) = (3x-1)+(-10x+6)$
$= 3x-1-10x+6$
$= -7x+5$

87. left rectangle: $x \cdot x = x^2$
right rectangle: $x \cdot 3 = 3x$
left rectangle + right rectangle: $x^2 + 3x$

89. top left rectangle: $x \cdot x = x^2$
top right rectangle: $x \cdot 3 = 3x$
bottom left rectangle: $2 \cdot x = 2x$
bottom right rectangle: $2 \cdot 3 = 6$
entire figure: $x^2 + 3x + 2x + 6 = x^2 + 5x + 6$

91. $5a + 6a = (5+6)a = 11a$

93. $(5x)^2 + (2y)^2 = (5x)(5x) + (2y)(2y)$
$= 25x^2 + 4y^2$

95. a. $(a+b)(a-b) = a(a) + a(-b) + b(a) + b(-b)$
$= a^2 - ab + ab - b^2$
$= a^2 - b^2$

b. $(2x+3y)(2x-3y)$
$= 2x(2x) + 2x(-3y) + 3y(2x) + 3y(-3y)$
$= 4x^2 - 6xy + 6xy - 9y^2$
$= 4x^2 - 9y^2$

c. $(4x+7)(4x-7)$
$= 4x(4x) + 4x(-7) + 7(4x) + 7(-7)$
$= 16x^2 - 28x + 28x - 49$
$= 16x^2 - 49$

d. answers may vary

97. larger square: $(x+3)^2 = (x+3)(x+3)$
$= x(x) + x(3) + 3(x) + 3(3)$
$= x^2 + 3x + 3x + 9$
$= x^2 + 6x + 9$
smaller square: $2^2 = 2 \cdot 2 = 4$
shaded region: $x^2 + 6x + 9 - 4 = x^2 + 6x + 5$
The area of the shaded region is
$(x^2 + 6x + 5)$ square units.

Section 5.4 Practice

1. $(x+2)(x-5)$
$= (x)(x) + (x)(-5) + (2)(x) + (2)(-5)$
$= x^2 - 5x + 2x - 10$
$= x^2 - 3x - 10$

2. $(4x-9)(x-1)$
$= 4x(x) + 4x(-1) + (-9)(x) + (-9)(-1)$
$= 4x^2 - 4x - 9x + 9$
$= 4x^2 - 13x + 9$

3. $3(x+5)(3x-1) = 3(3x^2 - x + 15x - 5)$
$= 3(3x^2 + 14x - 5)$
$= 9x^2 + 42x - 15$

4. $(4x-1)^2$
$= (4x-1)(4x-1)$
$= (4x)(4x) + (4x)(-1) + (-1)(4x) + (-1)(-1)$
$= 16x^2 - 4x - 4x + 1$
$= 16x^2 - 8x + 1$

5. a. $(b+3)^2 = b^2 + 2(b)(3) + 3^2 = b^2 + 6b + 9$

b. $(x-y)^2 = x^2 - 2(x)(y) + y^2 = x^2 - 2xy + y^2$

c. $(3y+2)^2 = (3y)^2 + 2(3y)(2) + 2^2$
$= 9y^2 + 12y + 4$

d. $(a^2 - 5b)^2 = (a^2)^2 - 2(a^2)(5b) + (5b)^2$
$= a^4 - 10a^2b + 25b^2$

6. a. $3(x+5)(x-5) = 3(x^2 - 5^2)$
$= 3x(x^2 - 25)$
$= 3x^2 - 75$

b. $(4b-3)(4b+3) = (4b)^2 - 3^2 = 16b^2 - 9$

c. $\left(x+\dfrac{2}{3}\right)\left(x-\dfrac{2}{3}\right) = x^2 - \left(\dfrac{2}{3}\right)^2 = x^2 - \dfrac{4}{9}$

d. $(5s+t)(5s-t) = (5s)^2 - t^2 = 25s^2 - t^2$

e. $(2y-3z^2)(2y+3z^2) = (2y)^2 - (3z^2)^2$
$= 4y^2 - 9z^4$

7. a. $(4x+3)(x-6) = 4x^2 - 24x + 3x - 18$
$$= 4x^2 - 21x - 18$$

b. $(7b-2)^2 = (7b)^2 - 2(7b)(2) + 2^2$
$$= 49b^2 - 28b + 4$$

c. $(x+0.4)(x-0.4) = x^2 - (0.4)^2 = x^2 - 0.16$

d. $\left(x^2 - \dfrac{3}{7}\right)\left(3x^4 + \dfrac{2}{7}\right)$
$$= 3x^6 + \frac{2}{7}x^2 - \frac{9}{7}x^4 - \frac{6}{49}$$
$$= 3x^6 - \frac{9}{7}x^4 + \frac{2}{7}x^2 - \frac{6}{49}$$

e. $(x+1)(x^2 + 5x - 2)$
$$= x(x^2 + 5x - 2) + 1(x^2 + 5x - 2)$$
$$= x^3 + 5x^2 - 2x + x^2 + 5x - 2$$
$$= x^3 + 6x^2 + 3x - 2$$

Vocabulary, Readiness & Video Check 5.4

1. $(x+4)^2 = x^2 + 2(x)(4) + 4^2$
$$= x^2 + 8x + 16 \neq x^2 + 16$$
The statement is false.

2. $(x+6)(2x-1) = 2x^2 - x + 12x - 6$
$$= 2x^2 + 11x - 6$$
The statement is true.

3. $(x+4)(x-4) = x^2 - 4^2 = x^2 - 16 \neq x^2 + 16$
The statement is false.

4. $(x-1)(x^3 + 3x - 1)$
$$= x(x^3 + 3x - 1) - 1(x^3 + 3x - 1)$$
$$= x^4 + 3x^2 - x - x^3 - 3x + 1$$
$$= x^4 - x^3 + 3x^2 - 4x + 1$$
This is a polynomial of degree 4; the statement is false.

5. a binomial times a binomial

6. FOIL order for multiplication, distributive property

7. Multiplying gives you four terms, and the two like terms will always subtract out.

8. No; the FOIL method is used for multiplying a binomial and a binomial.

Exercise Set 5.4

1. $(x+3)(x+4) = x^2 + 4x + 3x + 12 = x^2 + 7x + 12$

3. $(x-5)(x+10) = x^2 + 10x - 5x - 50$
$$= x^2 + 5x - 50$$

5. $(5x-6)(x+2) = 5x^2 + 10x - 6x - 12$
$$= 5x^2 + 4x - 12$$

7. $5(y-6)(4y-1) = 5(4y^2 - 1y - 24y + 6)$
$$= 5(4y^2 - 25y + 6)$$
$$= 20y^2 - 125y + 30$$

9. $(2x+5)(3x-1) = 6x^2 - 2x + 15x - 5$
$$= 6x^2 + 13x - 5$$

11. $\left(x - \dfrac{1}{3}\right)\left(x + \dfrac{2}{3}\right) = x^2 + \dfrac{2}{3}x - \dfrac{1}{3}x - \dfrac{2}{9}$
$$= x^2 + \frac{1}{3}x - \frac{2}{9}$$

13. $(x+2)^2 = x^2 + 2(x)(2) + 2^2 = x^2 + 4x + 4$

15. $(2x-1)^2 = (2x)^2 - 2(2x)(1) + (1)^2$
$$= 4x^2 - 4x + 1$$

17. $(3a-5)^2 = (3a)^2 - 2(3a)(5) + 5^2$
$$= 9a^2 - 30a + 25$$

19. $(5x+9)^2 = (5x)^2 + 2(5x)(9) + 9^2$
$$= 25x^2 + 90x + 81$$

21. $(a-7)(a+7) = a^2 - 7^2 = a^2 - 49$

23. $(3x-1)(3x+1) = (3x)^2 - 1^2 = 9x^2 - 1$

25. $\left(3x - \dfrac{1}{2}\right)\left(3x + \dfrac{1}{2}\right) = (3x)^2 - \left(\dfrac{1}{2}\right)^2 = 9x^2 - \dfrac{1}{4}$

27. $(9x+y)(9x-y) = (9x)^2 - y^2 = 81x^2 - y^2$

29. $(2x+0.1)(2x-0.1) = (2x)^2 - (0.1)^2 = 4x^2 - 0.01$

31. $(a+5)(a+4) = a^2 + 4a + 5a + 20 = a^2 + 9a + 20$

33. $(a+7)^2 = a^2 + 2(a)(7) + 7^2 = a^2 + 14a + 49$

35. $(4a+1)(3a-1) = 12a^2 - 4a + 3a - 1$
$$= 12a^2 - a - 1$$

37. $(x+2)(x-2) = x^2 - 2^2 = x^2 - 4$

39. $(3a+1)^2 = (3a)^2 + 2(3a)(1) + 1^2 = 9a^2 + 6a + 1$

41. $(x^2 + y)(4x - y^4) = 4x^3 - x^2 y^4 + 4xy - y^5$

43. $(x+3)(x^2 - 6x + 1)$
$$= x(x^2 - 6x + 1) + 3(x^2 - 6x + 1)$$
$$= x^3 - 6x^2 + x + 3x^2 - 18x + 3$$
$$= x^3 - 3x^2 - 17x + 3$$

45. $(2a-3)^2 = (2a)^2 - 2(2a)(3) + (3)^2$
$$= 4a^2 - 12a + 9$$

47. $(5x-6z)(5x+6z) = (5x)^2 - (6z)^2 = 25x^2 - 36z^2$

49. $(x^5 - 3)(x^5 - 5) = x^{10} - 5x^5 - 3x^5 + 15$
$$= x^{10} - 8x^5 + 15$$

51. $(x+0.8)(x-0.8) = x^2 - 0.8^2 = x^2 - 0.64$

53. $(a^3 + 11)(a^4 - 3) = a^7 - 3a^3 + 11a^4 - 33$

55. $3(x-2)^2 = 3[x^2 - 2(x)(2) + 2^2]$
$$= 3(x^2 - 4x + 4)$$
$$= 3x^2 - 12x + 12$$

57. $(3b+7)(2b-5) = 6b^2 - 15b + 14b - 35$
$$= 6b^2 - b - 35$$

59. $(7p-8)(7p+8) = (7p)^2 - (8)^2 = 49p^2 - 64$

61. $\left(\dfrac{1}{3}a^2 - 7\right)\left(\dfrac{1}{3}a^2 + 7\right) = \left(\dfrac{1}{3}a^2\right)^2 - (7)^2$
$$= \dfrac{1}{9}a^4 - 49$$

63. $5x^2(3x^2 - x + 2) = 5x^2(3x^2) + 5x^2(-x) + 5x^2(2)$
$$= 15x^4 - 5x^3 + 10x^2$$

65. $(2r-3s)(2r+3s) = (2r)^2 - (3s)^2 = 4r^2 - 9s^2$

67. $(3x-7y)^2 = (3x)^2 - 2(3x)(7y) + (7y)^2$
$$= 9x^2 - 42xy + 49y^2$$

69. $(4x+5)(4x-5) = (4x)^2 - 5^2 = 16x^2 - 25$

71. $(8x+4)^2 = (8x)^2 + 2(8x)(4) + (4)^2$
$$= 64x^2 + 64x + 16$$

73. $\left(a - \dfrac{1}{2}y\right)\left(a + \dfrac{1}{2}y\right) = a^2 - \left(\dfrac{1}{2}y\right)^2 = a^2 - \dfrac{1}{4}y^2$

75. $\left(\dfrac{1}{5}x - y\right)\left(\dfrac{1}{5}x + y\right) = \left(\dfrac{1}{5}x\right)^2 - y^2 = \dfrac{1}{25}x^2 - y^2$

77. $(a+1)(3a^2 - a + 1)$
$$= a(3a^2 - a + 1) + 1(3a^2 - a + 1)$$
$$= 3a^3 - a^2 + a + 3a^2 - a + 1$$
$$= 3a^3 + 2a^2 + 1$$

79. $(2x+1)^2 = (2x)^2 + 2(2x)(1) + 1^2 = 4x^2 + 4x + 1$

The area is $(4x^2 + 4x + 1)$ square feet.

81. $\dfrac{50b^{10}}{70b^5} = \dfrac{50}{70}b^{10-5} = \dfrac{5b^5}{7}$

83. $\dfrac{8a^{17}b^{15}}{-4a^7b^{10}} = \dfrac{8}{-4}a^{17-7}b^{15-10} = -2a^{10}b^5$

85. $\dfrac{2x^4 y^{12}}{3x^4 y^4} = \dfrac{2}{3}x^{4-4}y^{12-4} = \dfrac{2y^8}{3}$

87. $(-1, 1)$ and $(2, 2)$
$$m = \dfrac{y_2 - y_1}{x_2 - x_1} = \dfrac{2-1}{2-(-1)} = \dfrac{1}{3}$$

89. $(-1, -2)$ and $(1, 0)$
$$m = \dfrac{y_2 - y_1}{x_2 - x_1} = \dfrac{0-(-2)}{1-(-1)} = \dfrac{2}{2} = 1$$

91. $(a-b)^2 = a^2 - 2ab + b^2$
Choice c.

93. $(a+b)^2 = a^2 + 2ab + b^2$
Choice d.

95. From FOIL, the first term in the result is
$(x^{\square})^2 = x^{2\square}$. Thus, $2\square = 4$ so $\square = 2$.

97. $(x^2 - 1)^2 - x^2 = ((x^2)^2 - 2(x^2)(1) + 1^2) - x^2$
$\qquad = (x^4 - 2x^2 + 1) - x^2$
$\qquad = x^4 - 2x^2 + 1 - x^2$
$\qquad = x^4 - 3x^2 + 1$
The area is $(x^4 - 3x^2 + 1)$ square meters.

99. $(5x-3)^2 - (x+1)^2$
$= (25x^2 - 30x + 9) - (x^2 + 2x + 1)$
$= 25x^2 - 30x + 9 - x^2 - 2x - 1$
$= (24x^2 - 32x + 8)$
The shaded area is
$(24x^2 - 32x + 8)$ square meters.

101. $(x+5)(x+5) = (x+5)^2$
$\qquad = x^2 + 2(x)(5) + 5^2$
$\qquad = x^2 + 10x + 25$
The area is $(x^2 + 10x + 25)$ square units.

103. answers may vary

105. answers may vary

107. answers may vary

109. $[(x+y)-3][(x+y)+3] = (x+y)^2 - 3^2$
$\qquad\qquad\qquad\qquad = x^2 + 2xy + y^2 - 9$

111. $[(a-3)+b][(a-3)-b] = (a-3)^2 - b^2$
$\qquad\qquad\qquad\qquad = a^2 - 6a + 9 - b^2$

Integrated Review

1. $(5x^2)(7x^3) = (5 \cdot 7)(x^2 \cdot x^3) = 35x^5$

2. $(4y^2)(8y^7) = (4 \cdot 8)(y^2 \cdot y^7) = 32y^9$

3. $-4^2 = -(4 \cdot 4) = -16$

4. $(-4)^2 = (-4)(-4) = 16$

5. $(x-5)(2x+1) = 2x^2 + x - 10x - 5$
$\qquad\qquad\qquad = 2x^2 - 9x - 5$

6. $(3x-2)(x+5) = 3x^2 + 15x - 2x - 10$
$\qquad\qquad\qquad = 3x^2 + 13x - 10$

7. $(x-5)+(2x+1) = x - 5 + 2x + 1 = 3x - 4$

8. $(3x-2)+(x+5) = 3x - 2 + x + 5 = 4x + 3$

9. $\dfrac{7x^9 y^{12}}{x^3 y^{10}} = 7x^{9-3} y^{12-10} = 7x^6 y^2$

10. $\dfrac{20a^2 b^8}{14a^2 b^2} = \dfrac{20}{14} a^{2-2} b^{8-2} = \dfrac{10b^6}{7}$

11. $(12m^7 n^6)^2 = 12^2 m^{7 \cdot 2} n^{6 \cdot 2} = 144 m^{14} n^{12}$

12. $(4y^9 z^{10})^3 = 4^3 y^{9 \cdot 3} z^{10 \cdot 3} = 64 y^{27} z^{30}$

13. $3(4y-3)(4y+3) = 3[(4y)^2 - 3^2]$
$\qquad\qquad\qquad = 3(16y^2 - 9)$
$\qquad\qquad\qquad = 48y^2 - 27$

14. $2(7x-1)(7x+1) = 2[(7x)^2 - 1^2]$
$\qquad\qquad\qquad = 2(49x^2 - 1)$
$\qquad\qquad\qquad = 98x^2 - 2$

15. $(x^7 y^5)^9 = x^{7 \cdot 9} y^{5 \cdot 9} = x^{63} y^{45}$

16. $(3^1 x^9)^3 = 3^{1 \cdot 3} x^{9 \cdot 3} = 3^3 x^{27} = 27x^{27}$

17. $(7x^2 - 2x + 3) - (5x^2 + 9)$
$= 7x^2 - 2x + 3 - 5x^2 - 9$
$= 2x^2 - 2x - 6$

18. $(10x^2 + 7x - 9) - (4x^2 - 6x + 2)$
$= 10x^2 + 7x - 9 - 4x^2 + 6x - 2$
$= 6x^2 + 13x - 11$

19. $0.7y^2 - 1.2 + 1.8y^2 - 6y + 1 = 2.5y^2 - 6y - 0.2$

20. $7.8x^2 - 6.8x + 3.3 + 0.6x^2 - 9$
$= 8.4x^2 - 6.8x - 5.7$

21. $(x + 4y)^2 = x^2 + 2(x)(4y) + (4y)^2$
$\qquad = x^2 + 8xy + 16y^2$

22. $(y - 9z)^2 = y^2 - 2(y)(9z) + (9z)^2$
$\qquad = y^2 - 18yz + 81z^2$

23. $(x + 4y) + (x + 4y) = x + 4y + x + 4y = 2x + 8y$

24. $(y - 9z) + (y - 9z) = y - 9z + y - 9z = 2y - 18z$

25. $7x^2 - 6xy + 4(y^2 - xy) = 7x^2 - 6xy + 4y^2 - 4xy$
$\qquad = 7x^2 - 10xy + 4y^2$

26. $5a^2 - 3ab + 6(b^2 - a^2) = 5a^2 - 3ab + 6b^2 - 6a^2$
$\qquad = -a^2 - 3ab + 6b^2$

27. $(x - 3)(x^2 + 5x - 1)$
$= x(x^2 + 5x - 1) - 3(x^2 + 5x - 1)$
$= x^3 + 5x^2 - x - 3x^2 - 15x + 3$
$= x^3 + 2x^2 - 16x + 3$

28. $(x + 1)(x^2 - 3x - 2)$
$= x(x^2 - 3x - 2) + 1(x^2 - 3x - 2)$
$= x^3 - 3x^2 - 2x + x^2 - 3x - 2$
$= x^3 - 2x^2 - 5x - 2$

29. $(2x^3 - 7)(3x^2 + 10)$
$= 2x^3(3x^2) + 2x^3(10) - 7(3x^2) - 7(10)$
$= 6x^5 + 20x^3 - 21x^2 - 70$

30. $(5x^3 - 1)(4x^4 + 5)$
$= 5x^3(4x^4) + 5x^3(5) - 1(4x^4) - 1(5)$
$= 20x^7 + 25x^3 - 4x^4 - 5$

31. $(2x - 7)(x^2 - 6x + 1)$
$= 2x(x^2 - 6x + 1) - 7(x^2 - 6x + 1)$
$= 2x^3 - 12x^2 + 2x - 7x^2 + 42x - 7$
$= 2x^3 - 19x^2 + 44x - 7$

32. $(5x - 1)(x^2 + 2x - 3)$
$= 5x(x^2 + 2x - 3) - 1(x^2 + 2x - 3)$
$= 5x^3 + 10x^2 - 15x - x^2 - 2x + 3$
$= 5x^3 + 9^2 - 17x + 3$

33. $5x^3 + 5y^3$ cannot be simplified.

34. $(5x^3)(5y^3) = 5 \cdot 5x^3 y^3 = 25x^3 y^3$

35. $(5x^3)^3 = 5^3 x^{3 \cdot 3} = 125x^9$

36. $\dfrac{5x^3}{5y^3} = \dfrac{x^3}{y^3}$

37. $x + x = 2x$

38. $x \cdot x = x^2$

Section 5.5 Practice

1. a. $5^{-3} = \dfrac{1}{5^3} = \dfrac{1}{125}$

    b. $3y^{-4} = 3 \cdot \dfrac{1}{y^4} = \dfrac{3}{y^4}$

    c. $3^{-1} + 2^{-1} = \dfrac{1}{3} + \dfrac{1}{2} = \dfrac{2}{6} + \dfrac{3}{6} = \dfrac{5}{6}$

    d. $(-5)^{-2} = \dfrac{1}{(-5)^2} = \dfrac{1}{(-5)(-5)} = \dfrac{1}{25}$

    e. $x^{-5} = \dfrac{1}{x^5}$

2. a. $\dfrac{1}{s^{-5}} = \dfrac{s^5}{1} = s^5$

    b. $\dfrac{1}{2^{-3}} = \dfrac{2^3}{1} = 8$

    c. $\dfrac{x^{-7}}{y^{-5}} = \dfrac{y^5}{x^7}$

    d. $\dfrac{4^{-3}}{3^{-2}} = \dfrac{3^2}{4^3} = \dfrac{9}{64}$

3. a. $\dfrac{x^{-3}}{x^2} = x^{-3-2} = x^{-5} = \dfrac{1}{x^5}$

b. $\dfrac{5}{y^{-7}} = 5 \cdot \dfrac{1}{y^{-7}} = 5 \cdot y^7 = 5y^7$

c. $\dfrac{z}{z^{-4}} = \dfrac{z^1}{z^{-4}} = z^{1-(-4)} = z^5$

d. $\left(\dfrac{5}{9}\right)^{-2} = \dfrac{5^{-2}}{9^{-2}} = \dfrac{9^2}{5^2} = \dfrac{81}{25}$

4. a. $(a^4 b^{-3})^{-5} = a^{-20} b^{15} = \dfrac{b^{15}}{a^{20}}$

b. $\dfrac{x^2 (x^5)^3}{x^7} = \dfrac{x^2 \cdot x^{15}}{x^7}$
$= \dfrac{x^{2+15}}{x^7}$
$= \dfrac{x^{17}}{x^7}$
$= x^{17-7}$
$= x^{10}$

c. $\left(\dfrac{5p^8}{q}\right)^{-2} = \dfrac{5^{-2}(p^8)^{-2}}{q^{-2}}$
$= \dfrac{5^{-2} p^{-16}}{q^{-2}}$
$= \dfrac{q^2}{5^2 p^{16}}$
$= \dfrac{q^2}{25 p^{16}}$

d. $\dfrac{6^{-2} x^{-4} y^{-7}}{6^{-3} x^3 y^{-9}} = 6^{-2-(-3)} x^{-4-3} y^{-7-(-9)}$
$= 6^1 x^{-7} y^2$
$= \dfrac{6y^2}{x^7}$

e. $\left(\dfrac{-3x^4 y}{x^2 y^{-2}}\right)^3 = \dfrac{(-3)^3 x^{12} y^3}{x^6 y^{-6}}$
$= \dfrac{-27 x^{12} y^3}{x^6 y^{-6}}$
$= -27 x^{12-6} y^{3-(-6)}$
$= -27 x^6 y^9$

5. a. $0.000007 = 7 \times 10^{-6}$
The decimal point is moved 6 places, and the original number is less than 1, so the count is −6.

b. $20,700,000 = 2.07 \times 10^7$
The decimal point is moved 7 places, and the original number is 10 or greater, so the count is 7.

c. $0.0043 = 4.3 \times 10^{-3}$
The decimal point is moved 3 places, and the original number is less than 1, so the count is −3.

d. $812,000,000 = 8.12 \times 10^8$
The decimal point is moved 8 places, and the original number is 10 or greater, so the count is 8.

6. a. Move the decimal point 4 places to the left.
$3.67 \times 10^{-4} = 0.000367$

b. Move the decimal point 6 places to the right.
$8.954 \times 10^6 = 8,954,000$

c. Move the decimal point 5 places to the left.
$2.009 \times 10^{-5} = 0.00002009$

d. Move the decimal point 3 places to the right.
$4.054 \times 10^3 = 4054$

7. a. $(5 \times 10^{-4})(8 \times 10^6) = (5 \cdot 8) \times (10^{-4} \cdot 10^6)$
$= 40 \times 10^2$
$= 4000$

b. $\dfrac{64 \times 10^3}{32 \times 10^{-7}} = \dfrac{64}{32} \times 10^{3-(-7)}$
$= 2 \times 10^{10}$
$= 20,000,000,000$

Graphing Calculator Explorations

1. $5.31 \times 10^3 = 5.31 \text{ EE } 3$

2. $-4.8 \times 10^{14} = -4.8 \text{ EE } 14$

3. $6.6 \times 10^{-9} = 6.6 \text{ EE } -9$

4. $-9.9811 \times 10^{-2} = -9.9811 \text{ EE } -2$

5. $3,000,000 \times 5,000,000 = 1.5 \times 10^{13}$

6. $230,000 \times 1000 = 2.3 \times 10^8$

7. $(3.26 \times 10^6)(2.5 \times 10^{13}) = 8.15 \times 10^{19}$

8. $(8.76 \times 10^{-4})(1.237 \times 10^9) = 1.083612 \times 10^6$

Vocabulary, Readiness & Video Check 5.5

1. The expression x^{-3} equals $\dfrac{1}{\underline{x^3}}$.

2. The expression 5^{-4} equals $\dfrac{1}{\underline{625}}$.

3. The number 3.021×10^{-3} is written in <u>scientific notation</u>.

4. The number 0.0261 is written in <u>standard form</u>.

5. A negative exponent has nothing to do with the sign of the simplified result.

6. power of a product rule, power rule for exponents, negative exponent definition, quotient rule for exponents

7. When you move the decimal point to the left, the sign of the exponent will be positive; when you move the decimal point to the right, the sign of the exponent will be negative.

8. the exponent on 10

9. the quotient rule

Exercise Set 5.5

1. $4^{-3} = \dfrac{1}{4^3} = \dfrac{1}{64}$

3. $(-3)^{-4} = \dfrac{1}{(-3)^4} = \dfrac{1}{81}$

5. $7x^{-3} = 7 \cdot \dfrac{1}{x^3} = \dfrac{7}{x^3}$

7. $\left(\dfrac{1}{2}\right)^{-5} = \dfrac{1^{-5}}{2^{-5}} = \dfrac{2^5}{1^5} = 32$

9. $\left(-\dfrac{1}{4}\right)^{-3} = \dfrac{(-1)^{-3}}{(4)^{-3}} = \dfrac{4^3}{(-1)^3} = \dfrac{64}{-1} = -64$

11. $3^{-1} + 5^{-1} = \dfrac{1}{3} + \dfrac{1}{5} = \dfrac{5}{15} + \dfrac{3}{15} = \dfrac{8}{15}$

13. $\dfrac{1}{p^{-3}} = p^3$

15. $\dfrac{p^{-5}}{q^{-4}} = \dfrac{q^4}{p^5}$

17. $\dfrac{x^{-2}}{x} = x^{-2-1} = x^{-3} = \dfrac{1}{x^3}$

19. $\dfrac{z^{-4}}{z^{-7}} = z^{-4-(-7)} = z^3$

21. $3^{-2} + 3^{-1} = \dfrac{1}{3^2} + \dfrac{1}{3} = \dfrac{1}{9} + \dfrac{1}{3} = \dfrac{1}{9} + \dfrac{3}{9} = \dfrac{4}{9}$

23. $\dfrac{-1}{p^{-4}} = -1(p^4) = -p^4$

25. $-2^0 - 3^0 = -1(1) - 1 = -1 - 1 = -2$

27. $\dfrac{x^2 x^5}{x^3} = x^{2+5-3} = x^4$

29. $\dfrac{p^2 p}{p^{-1}} = p^{2+1-(-1)} = p^4$

31. $\dfrac{(m^5)^4 m}{m^{10}} = \dfrac{m^{20} m}{m^{10}} = m^{20+1-10} = m^{11}$

33. $\dfrac{r}{r^{-3}r^{-2}} = r^{1-(-3)-(-2)} = r^6$

35. $(x^5 y^3)^{-3} = x^{5(-3)} y^{3(-3)} = x^{-15} y^{-9} = \dfrac{1}{x^{15} y^9}$

37. $\dfrac{(x^2)^3}{x^{10}} = \dfrac{x^6}{x^{10}} = x^{6-10} = x^{-4} = \dfrac{1}{x^4}$

39. $\dfrac{(a^5)^2}{(a^3)^4} = \dfrac{a^{10}}{a^{12}} = a^{10-12} = a^{-2} = \dfrac{1}{a^2}$

41. $\dfrac{8k^4}{2k} = \dfrac{8}{2} \cdot k^{4-1} = 4k^3$

43. $\dfrac{-6m^4}{-2m^3} = \dfrac{-6}{-2} \cdot m^{4-3} = 3m$

45. $\dfrac{-24a^6 b}{6ab^2} = \dfrac{-24}{6} \cdot a^{6-1} b^{1-2} = -4a^5 b^{-1} = -\dfrac{4a^5}{b}$

47. $(-2x^3 y^{-4})(3x^{-1} y) = -2(3)x^{3+(-1)} y^{-4+1}$

$= -6x^2 y^{-3}$

$= -\dfrac{6x^2}{y^3}$

49. $(a^{-5} b^2)^{-6} = a^{-5(-6)} b^{2(-6)} = a^{30} b^{-12} = \dfrac{a^{30}}{b^{12}}$

51. $\left(\dfrac{x^{-2} y^4}{x^3 y^7}\right)^2 = \dfrac{x^{-2(2)} y^{4(2)}}{x^{3(2)} y^{7(2)}}$

$= \dfrac{x^{-4} y^8}{x^6 y^{14}}$

$= x^{-4-6} y^{8-14}$

$= x^{-10} y^{-6}$

$= \dfrac{1}{x^{10} y^6}$

53. $\dfrac{4^2 z^{-3}}{4^3 z^{-5}} = 4^{2-3} z^{-3-(-5)} = 4^{-1} z^2 = \dfrac{z^2}{4}$

55. $\dfrac{2^{-3} x^{-4}}{2^2 x} = 2^{-3-2} x^{-4-1}$

$= 2^{-5} x^{-5}$

$= \dfrac{1}{2^5 x^5}$

$= \dfrac{1}{32x^5}$

57. $\dfrac{7ab^{-4}}{7^{-1} a^{-3} b^2} = 7^{1-(-1)} a^{1-(-3)} b^{-4-2}$

$= 7^2 a^4 b^{-6}$

$= \dfrac{49a^4}{b^6}$

59. $\left(\dfrac{a^{-5} b}{ab^3}\right)^{-4} = \dfrac{a^{-5(-4)} b^{-4}}{a^{-4} b^{3(-4)}}$

$= \dfrac{a^{20} b^{-4}}{a^{-4} b^{-12}}$

$= a^{20-(-4)} b^{-4-(-12)}$

$= a^{24} b^8$

61. $\dfrac{(xy^3)^5}{(xy)^{-4}} = \dfrac{x^5 y^{3(5)}}{x^{-4} y^{-4}}$

$= \dfrac{x^5 y^{15}}{x^{-4} y^{-4}}$

$= x^{5-(-4)} y^{15-(-4)}$

$= x^9 y^{19}$

63. $\dfrac{(-2xy^{-3})^{-3}}{(xy^{-1})^{-1}} = \dfrac{(-2)^{-3} x^{-3} y^9}{x^{-1} y^1}$

$= (-2)^{-3} x^{-3-(-1)} y^{9-1}$

$= (-2)^{-3} x^{-2} y^8$

$= \dfrac{y^8}{(-2)^3 x^2}$

$= -\dfrac{y^8}{8x^2}$

65. $\dfrac{6x^2 y^3}{-7xy^5} = -\dfrac{6}{7} x^{2-1} y^{3-5} = -\dfrac{6}{7} x^1 y^{-2} = -\dfrac{6x}{7y^2}$

67. $\dfrac{(a^4 b^{-7})^{-5}}{(5a^2 b^{-1})^{-2}} = \dfrac{(a^4)^{-5}(b^{-7})^{-5}}{5^{-2}(a^2)^{-2}(b^{-1})^{-2}}$

$= \dfrac{a^{-20} b^{35}}{5^{-2} a^{-4} b^2}$

$= 5^2 a^{-20-(-4)} b^{35-2}$

$= 5^2 a^{-16} b^{33}$

$= \dfrac{25 b^{33}}{a^{16}}$

69. $78{,}000 = 7.8 \times 10^4$

71. $0.00000167 = 1.67 \times 10^{-6}$

73. $0.00635 = 6.35 \times 10^{-3}$

75. $1{,}160{,}000 = 1.16 \times 10^6$

77. $2{,}000{,}000{,}000 = 2 \times 10^9$

79. $2400 = 2.4 \times 10^3$

81. $8.673 \times 10^{-10} = 0.0000000008673$

83. $3.3 \times 10^{-2} = 0.033$

85. $2.032 \times 10^4 = 20{,}320$

87. $7.0 \times 10^8 = 700{,}000{,}000$

89. $9.460 \times 10^{12} = 9{,}460{,}000{,}000{,}000$

91. $184{,}000{,}000{,}000 = 1.84 \times 10^{11}$

93. $1.55 \times 10^{11} = 155{,}000{,}000{,}000$

95. $3.5 \times 10^4 = 35{,}000$

97. $(1.2 \times 10^{-3})(3 \times 10^{-2}) = (1.2 \cdot 3) \times (10^{-3} \cdot 10^{-2})$

$= 3.6 \times 10^{-5}$

$= 0.000036$

99. $(4 \times 10^{-10})(7 \times 10^{-9}) = (4 \cdot 7) \times (10^{-10} \cdot 10^{-9})$

$= 28 \times 10^{-19}$

$= 2.8 \times 10^{-18}$

$= 0.0000000000000000028$

101. $\dfrac{8 \times 10^{-1}}{16 \times 10^5} = \dfrac{8}{16} \times 10^{-1-5}$

$= 0.5 \times 10^{-6}$

$= 5 \times 10^{-7}$

$= 0.0000005$

103. $\dfrac{1.4 \times 10^{-2}}{7 \times 10^{-8}} = \dfrac{1.4}{7} \times 10^{-2-(-8)}$

$= 0.2 \times 10^6$

$= 2.0 \times 10^5$

$= 200{,}000$

105. $\dfrac{5x^7}{3x^4} = \dfrac{5}{3} \cdot x^{7-4} = \dfrac{5x^3}{3}$

107. $\dfrac{15z^4 y^3}{21zy} = \dfrac{15}{21} z^{4-1} y^{3-1} = \dfrac{5z^3 y^2}{7}$

109. $\dfrac{1}{y}(5y^2 - 6y + 5) = \dfrac{1}{y}(5y^2) + \dfrac{1}{y}(-6y) + \dfrac{1}{y}(5)$

$= 5y - 6 + \dfrac{5}{y}$

111. $\left(\dfrac{3x^{-2}}{z}\right)^3 = \dfrac{3^3 x^{-6}}{z^3} = \dfrac{27}{x^6 z^3}$

The volume is $\dfrac{27}{x^6 z^3}$ cubic inches.

113. $(2a^3)^3 a^4 + a^5 a^8 = 2^3 (a^3)^3 a^4 + a^{5+8}$

$= 8a^9 a^4 + a^{13}$

$= 8a^{13} + a^{13}$

$= 9a^{13}$

115. $x^{-5} = \dfrac{1}{x^5}$

117. answers may vary

119. a. $9.7 \times 10^{-2} = 0.097$

$1.3 \times 10^1 = 130$

1.3×10^1 is larger.

b. $8.6 \times 10^5 = 860,000$

$4.4 \times 10^7 = 44,000,000$

4.4×10^7 is larger.

c. $6.1 \times 10^{-2} = 0.061$

$5.6 \times 10^{-4} = 0.00056$

6.1×10^{-2} is larger.

121. answers may vary

123. $a^{-4m} \cdot a^{5m} = a^{-4m+5m} = a^m$

125. $(3y^{2z})^3 = 3^3 y^{2z \cdot 3} = 27 y^{6z}$

127. $(2.63 \times 10^{12})(-1.5 \times 10^{-10}) = -394.5$

129. $t = \dfrac{d}{r}$

$t = \dfrac{238,857}{1.86 \times 10^5} = \dfrac{238,857}{186,000} \approx 1.3$

It takes 1.3 seconds for the reflected light of the moon to reach Earth.

Section 5.6 Practice

1. $\dfrac{8t^3 + 4t^2}{4t^2} = \dfrac{8t^3}{4t^2} + \dfrac{4t^2}{4t^2} = 2t + 1$

Check: $4t^2(2t+1) = 4t^2(2t) + 4t^2(1)$

$\qquad\qquad = 8t^3 + 4t^2$

2. $\dfrac{16x^6 + 20x^3 - 12x}{4x^2} = \dfrac{16x^6}{4x^2} + \dfrac{20x^3}{4x^2} - \dfrac{12x}{4x^2}$

$\qquad\qquad\qquad\qquad = 4x^4 + 5x - \dfrac{3}{x}$

Check: $4x^2\left(4x^4 + 5x - \dfrac{3}{x}\right)$

$\qquad = 4x^2(4x^4) + 4x^2(5x) - 4x^2\left(\dfrac{3}{x}\right)$

$\qquad = 16x^6 + 20x^3 - 12x$

3. $\dfrac{15x^4y^4 - 10xy + y}{5xy} = \dfrac{15x^4y^4}{5xy} - \dfrac{10xy}{5xy} + \dfrac{y}{5xy}$

$\qquad\qquad\qquad\qquad = 3x^3y^3 - 2 + \dfrac{1}{5x}$

Check: $5xy\left(3x^3y^3 - 2 + \dfrac{1}{5x}\right)$

$\qquad = 5xy(3x^3y^3) - 5xy(2) + 5xy\left(\dfrac{1}{5x}\right)$

$\qquad = 15x^4y^4 - 10xy + y$

4.
$$\begin{array}{r} x+3 \\ x+2 \overline{\smash{\big)}\ x^2+5x+6} \\ \underline{x^2+2x} \\ 3x+6 \\ \underline{3x+6} \\ 0 \end{array}$$

Check: $(x+2) \cdot (x+3) + 0 = x^2 + 5x + 6$

The quotient checks.

5.
$$\begin{array}{r} 2x+3 \\ 2x+1 \overline{\smash{\big)}\ 4x^2+8x-7} \\ \underline{4x^2+2x} \\ 6x-7 \\ \underline{6x+3} \\ -10 \end{array}$$

$\dfrac{4x^2 + 8x - 7}{2x+1} = 2x + 3 + \dfrac{-10}{2x+1}$ or $2x+3 - \dfrac{10}{2x+1}$

Check:

$(2x+1)(2x+3) + (-10) = (4x^2 + 8x + 3) - 10$

$\qquad\qquad\qquad\qquad\qquad = 4x^2 + 8x - 7$

The quotient checks.

6. Rewrite $11x - 3 + 9x^3$ as $9x^3 + 0x^2 + 11x - 3$.

$$\begin{array}{r} 3x^2 - 2x + 5 \\ 3x+2 \overline{\smash{\big)}\ 9x^3 + 0x^2 + 11x - 3} \\ \underline{9x^3 + 6x^2} \\ -6x^2 + 11x \\ \underline{-6x^2 - 4x} \\ 15x - 3 \\ \underline{15x + 10} \\ -13 \end{array}$$

$\dfrac{11x - 3 + 9x^3}{3x+2} = 3x^2 - 2x + 5 + \dfrac{-13}{3x+2}$ or

$3x^2 - 2x + 5 - \dfrac{13}{3x+2}$

7. Rewrite $x^2 + 2$ as $x^2 + 0x + 2$.

$$
\begin{array}{r}
3x^2 - 2x - 9 \\
x^2+0x+2{\overline{\smash{\big)}\,3x^4-2x^3-3x^2\ \ +x+4}} \\
\underline{3x^4+0x^3+6x^2} \\
-2x^3-9x^2\ \ +x \\
\underline{-2x^3+0x^2-4x} \\
-9x^2+5x\ +4 \\
\underline{-9x^2+0x-18} \\
5x+22
\end{array}
$$

$$\frac{3x^4-2x^3-3x^2+x+4}{x^2+2}=3x^2-2x-9+\frac{5x+22}{x^2+2}$$

8. Rewrite $x^3 + 27$ as $x^3 + 0x^2 + 0x + 27$.

$$
\begin{array}{r}
x^2-3x+9 \\
x+3{\overline{\smash{\big)}\,x^3+0x^2+0x+27}} \\
\underline{x^3+3x^2} \\
-3x^2+0x \\
\underline{-3x^2-9x} \\
9x+27 \\
\underline{9x+27} \\
0
\end{array}
$$

$$\frac{x^3+27}{x+3}=x^2-3x+9$$

Vocabulary, Readiness & Video Check 5.6

1. In $6\overline{\smash{\big)}\,18}^{\,3}$, the 18 is the <u>dividend</u>, the 3 is the <u>quotient</u> and the 6 is the <u>divisor</u>.

2. In $x+1\overline{\smash{\big)}\,x^2+3x+2}^{\,x+2}$, the $x+1$ is the <u>divisor</u>, the x^2+3x+2 is the <u>dividend</u> and the $x+2$ is the <u>quotient</u>.

3. $\dfrac{a^6}{a^4}=a^{6-4}=2$

4. $\dfrac{p^8}{p^3}=p^{8-3}=p^5$

5. $\dfrac{y^2}{y}=\dfrac{y^2}{y^1}=y^{2-1}=y$

6. $\dfrac{a^3}{a}=\dfrac{a^3}{a^1}=a^{3-1}=a^2$

7. the common denominator

8. Filling in missing powers helps you keep like terms lined up and your work clear and neat.

Exercise Set 5.6

1. $\dfrac{12x^4+3x^2}{x}=\dfrac{12x^4}{x}+\dfrac{3x^2}{x}=12x^3+3x$

3. $\dfrac{20x^3-30x^2+5x+5}{5}=\dfrac{20x^3}{5}-\dfrac{30x^2}{5}+\dfrac{5x}{5}+\dfrac{5}{5}$
$=4x^3-6x^2+x+1$

5. $\dfrac{15p^3+18p^2}{3p}=\dfrac{15p^3}{3p}+\dfrac{18p^2}{3p}=5p^2+6p$

7. $\dfrac{-9x^4+18x^5}{6x^5}=\dfrac{-9x^4}{6x^5}+\dfrac{18x^5}{6x^5}=-\dfrac{3}{2x}+3$

9. $\dfrac{-9x^5+3x^4-12}{3x^3}=\dfrac{-9x^5}{3x^3}+\dfrac{3x^4}{3x^3}-\dfrac{12}{3x^3}$
$=-3x^2+x-\dfrac{4}{x^3}$

11. $\dfrac{4x^4-6x^3+7}{-4x^4}=\dfrac{4x^4}{-4x^4}-\dfrac{6x^3}{-4x^4}+\dfrac{7}{-4x^4}$
$=-1+\dfrac{3}{2x}-\dfrac{7}{4x^4}$

13. $x+3\overline{\smash{\big)}\,x^2+4x+3}^{\,x+1}$
$$
\begin{array}{r}
\underline{x^2+3x} \\
x+3 \\
\underline{x+3} \\
0
\end{array}
$$
$$\frac{x^2+4x+3}{x+3}=x+1$$

15.

$$x+5\overline{\smash{\big)}\,2x^2+13x+15} \quad\begin{array}{r}2x+3\end{array}$$
$$\underline{2x^2+10x}$$
$$3x+15$$
$$\underline{3x+15}$$
$$0$$

$$\frac{2x^2+13x+15}{x+5}=2x+3$$

17.

$$x-4\overline{\smash{\big)}\,2x^2-7x+3} \quad\begin{array}{r}2x+1\end{array}$$
$$\underline{2x^2-8x}$$
$$x+3$$
$$\underline{x-4}$$
$$7$$

$$\frac{2x^2-7x+3}{x-4}=2x+1+\frac{7}{x-4}$$

19.

$$3a+2\overline{\smash{\big)}\,9a^3-3a^2-3a+4} \quad\begin{array}{r}3a^2-3a+1\end{array}$$
$$\underline{9a^3+6a^2}$$
$$-9a^2-3a$$
$$\underline{-9a^2-6a}$$
$$3a+4$$
$$\underline{3a+2}$$
$$2$$

$$\frac{9a^3-3a^2-3a+4}{3a+2}=3a^2-3a+1+\frac{2}{3a+2}$$

21.

$$2x+1\overline{\smash{\big)}\,8x^2+10x+1} \quad\begin{array}{r}4x+3\end{array}$$
$$\underline{8x^2+4x}$$
$$6x+1$$
$$\underline{6x+3}$$
$$-2$$

$$\frac{8x^2+10x+1}{2x+1}=4x+3-\frac{2}{2x+1}$$

23.

$$x-2\overline{\smash{\big)}\,2x^3+2x^2\ -17x+8} \quad\begin{array}{r}2x^2+6x-5\end{array}$$
$$\underline{2x^3-4x^2}$$
$$6x^2-17x$$
$$\underline{6x^2-12x}$$
$$-5x\ +8$$
$$\underline{-5x+10}$$
$$-2$$

$$\frac{2x^3+2x^2-17x+8}{x-2}=2x^2+6x-5-\frac{2}{x-2}$$

25. Rewrite x^2-36 as $x^2+0x-36$.

$$x-6\overline{\smash{\big)}\,x^2+0x-36} \quad\begin{array}{r}x+6\end{array}$$
$$\underline{x^2-6x}$$
$$6x-36$$
$$\underline{6x-36}$$
$$0$$

$$\frac{x^2-36}{x-6}=x+6$$

27. Rewrite x^3-27 as $x^3+0x^2+0x-27$.

$$x-3\overline{\smash{\big)}\,x^3+0x^2+0x-27} \quad\begin{array}{r}x^2+3x+9\end{array}$$
$$\underline{x^3-3x^2}$$
$$3x^2+0x$$
$$\underline{3x^2-9x}$$
$$9x-27$$
$$\underline{9x-27}$$
$$0$$

$$\frac{x^3-27}{x-3}=x^2+3x+9$$

29. Rewrite $1-3x^2$ as $-3x^2+0x+1$.

$$x+2\overline{\smash{\big)}\,-3x^2+0x\ +1} \quad\begin{array}{r}-3x+6\end{array}$$
$$\underline{-3x^2-6x}$$
$$6x+\ 1$$
$$\underline{6x+12}$$
$$-11$$

$$\frac{1-3x^2}{x+2}=-3x+6-\frac{11}{x+2}$$

31. Rewrite $-4b + 4b^2 - 5$ as $4b^2 - 4b - 5$.

$$
\begin{array}{r}
2b - 1 \\
2b - 1 \overline{\smash{\big)}\ 4b^2 - 4b - 5} \\
\underline{4b^2 - 2b} \\
-2b - 5 \\
\underline{-2b + 1} \\
-6
\end{array}
$$

$$\frac{-4b + 4b^2 - 5}{2b - 1} = 2b - 1 - \frac{6}{2b - 1}$$

33. $\dfrac{a^2 b^2 - ab^3}{ab} = \dfrac{a^2 b^2}{ab} - \dfrac{ab^3}{ab} = ab - b^2$

35.
$$
\begin{array}{r}
4x + 9 \\
2x - 3 \overline{\smash{\big)}\ 8x^2 + 6x - 27} \\
\underline{8x^2 - 12x} \\
18x - 27 \\
\underline{18x - 27} \\
0
\end{array}
$$

$$\frac{8x^2 + 6x - 27}{2x - 3} = 4x + 9$$

37. $\dfrac{2x^2 y + 8x^2 y^2 - xy^2}{2xy} = \dfrac{2x^2 y}{2xy} + \dfrac{8x^2 y^2}{2xy} - \dfrac{xy^2}{2xy}$

$$= x + 4xy - \frac{y}{2}$$

39.
$$
\begin{array}{r}
2b^2 + b + 2 \\
b + 4 \overline{\smash{\big)}\ 2b^3 + 9b^2 + 6b - 4} \\
\underline{2b^3 + 8b^2} \\
b^2 + 6b \\
\underline{b^2 + 4b} \\
2b - 4 \\
\underline{2b + 8} \\
-12
\end{array}
$$

$$\frac{2b^3 + 9b^2 + 6b - 4}{b + 4} = 2b^2 + b + 2 - \frac{12}{b + 4}$$

41.
$$
\begin{array}{r}
5x - 2 \\
x + 6 \overline{\smash{\big)}\ 5x^2 + 28x - 10} \\
\underline{5x^2 + 30x} \\
-2x - 10 \\
\underline{-2x - 12} \\
2
\end{array}
$$

$$\frac{5x^2 + 28x - 10}{x + 6} = 5x - 2 + \frac{2}{x + 6}$$

43. $\dfrac{10x^3 - 24x^2 - 10x}{10x} = \dfrac{10x^3}{10x} - \dfrac{24x^2}{10x} - \dfrac{10x}{10x}$

$$= x^2 - \frac{12x}{5} - 1$$

45.
$$
\begin{array}{r}
6x - 1 \\
x + 3 \overline{\smash{\big)}\ 6x^2 + 17x - 4} \\
\underline{6x^2 + 18x} \\
-x - 4 \\
\underline{-x - 3} \\
-1
\end{array}
$$

$$\frac{6x^2 + 17x - 4}{x + 3} = 6x - 1 - \frac{1}{x + 3}$$

47.
$$
\begin{array}{r}
6x - 1 \\
5x - 2 \overline{\smash{\big)}\ 30x^2 - 17x + 2} \\
\underline{30x^2 - 12x} \\
-5x + 2 \\
\underline{-5x + 2} \\
0
\end{array}
$$

$$\frac{30x^2 - 17x + 2}{5x - 2} = 6x - 1$$

49. $\dfrac{3x^4 - 9x^3 + 12}{-3x} = \dfrac{3x^4}{-3x} - \dfrac{9x^3}{-3x} + \dfrac{12}{-3x}$

$$= -x^3 + 3x^2 - \frac{4}{x}$$

51.
$$\begin{array}{r} x^2+3x+9 \\ x+3\overline{)\,x^3+6x^2+18x+27} \\ \underline{x^3+3x^2} \\ 3x^2+18x \\ \underline{3x^2+9x} \\ 9x+27 \\ \underline{9x+27} \\ 0 \end{array}$$

$$\frac{x^3+6x^2+18x+27}{x+3}=x^2+3x+9$$

53. Rewrite y^3+3y^2+4 as y^3+3y^2+0y+4.
$$\begin{array}{r} y^2+5y+10 \\ y-2\overline{)\,y^3+3y^2\;+0y+4} \\ \underline{y^3-2y^2} \\ 5y^2+0y \\ \underline{5y^2-10y} \\ 10y+4 \\ \underline{10y-20} \\ 24 \end{array}$$

$$\frac{y^3+3y^2+4}{y-2}=y^2+5y+10+\frac{24}{y-2}$$

55. Rewrite $5-6x^2$ as $-6x^2+0x+5$.
$$\begin{array}{r} -6x-12 \\ x-2\overline{)\,-6x^2\;+0x\;+5} \\ \underline{-6x^2+12x} \\ -12x+5 \\ \underline{-12x+24} \\ -19 \end{array}$$

$$\frac{5-6x^2}{x-2}=-6x-12-\frac{19}{x-2}$$

57. Rewrite x^5+x^2 as $x^5+0x^4+0x^3+x^2$.
$$\begin{array}{r} x^3-x^2+x \\ x^2+x\overline{)\,x^5+0x^4+0x^3+x^2} \\ \underline{x^5+x^4} \\ -x^4+0x^3 \\ \underline{-x^4-x^3} \\ x^3+x^2 \\ \underline{x^3+x^2} \\ 0 \end{array}$$

$$\frac{x^5+x^2}{x^2+x}=x^3-x^2+x$$

59. $2a(a^2+1)=2a(a^2)+2a(1)=2a^3+2a$

61. $2x(x^2+7x-5)=2x(x^2)+2x(7x)+2x(-5)$
$$=2x^3+14x^2-10x$$

63. $-3xy(xy^2+7x^2y+8)$
$$=-3xy(xy^2)-3xy(7x^2y)-3xy(8)$$
$$=-3x^2y^3-21x^3y^2-24xy$$

65. $9ab(ab^2c+4bc-8)$
$$=9ab(ab^2c)+9ab(4bc)+9ab(-8)$$
$$=9a^2b^3c+36ab^2c-72ab$$

67. $P=4s,\ s=\dfrac{P}{4}$
$$\frac{12x^3+4x-16}{4}=\frac{12x^3}{4}+\frac{4x}{4}-\frac{16}{4}$$
$$=3x^3+x-4$$
Each side is $(3x^3+x-4)$ feet.

69. $\dfrac{a+7}{7}=\dfrac{a}{7}+\dfrac{7}{7}=\dfrac{a}{7}+1;$
choice c

71. answers may vary

73. $A = bh$, $h = \dfrac{A}{b}$

$$h = \frac{10x^2 + 31x + 15}{5x + 3}$$

$$\begin{array}{r}
2x + 5 \\
5x+3{\overline{\smash{\big)}\,10x^2 + 31x + 15}} \\
\underline{10x^2 + 6x} \\
25x + 15 \\
\underline{25x + 15} \\
0
\end{array}$$

The height is $(2x + 5)$ meters.

75. $\dfrac{18x^{10a} - 12x^{8a} + 14x^{5a} - 2x^{3a}}{2x^{3a}}$

$$= \frac{18x^{10a}}{2x^{3a}} - \frac{12x^{8a}}{2x^{3a}} + \frac{14x^{5a}}{2x^{3a}} - \frac{2x^{3a}}{2x^{3a}}$$

$$= 9x^{7a} - 6x^{5a} + 7x^{2a} - 1$$

Section 5.7 Practice

1. Since $x - c = x - 1$, c is 1.

$$\begin{array}{r|rrrr}
1 & 4 & -3 & 6 & 5 \\
  &   & 4 & 1 & 7 \\
\hline
  & 4 & 1 & 7 & 12
\end{array}$$

$$4x^2 + x + 7 + \frac{12}{x-1}$$

2. Since $x - c = x + 3 = x - (-3)$, c is -3.

$$\begin{array}{r|rrrrr}
-3 & 1 & 3 & -5 & 6 & 12 \\
   &   & -3 & 0 & 15 & -63 \\
\hline
   & 1 & 0 & -5 & 21 & -51
\end{array}$$

$$x^3 - 5x + 21 - \frac{51}{x+3}$$

3. a. $P(x) = x^3 - 5x - 2$

$P(2) = 2^3 - 5(2) - 2$

$= 8 - 10 - 2$

$= -4$

b. Since $x - c = x - 2$, c is 2.

$$\begin{array}{r|rrrr}
2 & 1 & 0 & -5 & -2 \\
  &   & 2 & 4 & -2 \\
\hline
  & 1 & 2 & -1 & -4
\end{array}$$

The remainder is -4.

4.

$$\begin{array}{r|rrrrrr}
3 & 2 & -18 & 0 & 90 & 59 & 0 \\
  &   & 6 & -36 & -108 & -54 & 15 \\
\hline
  & 2 & -12 & -36 & -18 & 5 & 15
\end{array}$$

$P(3) = 15$

Vocabulary, Readiness & Video Check 5.7

1. The last number n is the remainder and the other numbers are the coefficients of the variables in the quotient; the degree of the quotient is one less than the degree of the dividend.

2. When the calculations for substituting in c might be tedious, that is, if $P(x)$ is a many-termed, high-degree polynomial and/or if the coefficients are very large numbers, etc.

Exercise Set 5.7

1. $x - 5 = x - c$ where $c = 5$.

$$\begin{array}{r|rrr}
5 & 1 & 3 & -40 \\
  &   & 5 & 40 \\
\hline
  & 1 & 8 & 0
\end{array}$$

$$\frac{x^2 + 3x - 40}{x - 5} = x + 8$$

3. $x + 6 = x - c$ where $c = -6$.

$$\begin{array}{r|rrr}
-6 & 1 & 5 & -6 \\
   &   & -6 & 6 \\
\hline
   & 1 & -1 & 0
\end{array}$$

$$\frac{x^2 + 5x - 6}{x + 6} = x - 1$$

5. $x - 2 = x - c$ where $c = 2$.

$$\begin{array}{r|rrr}
2 & 1 & -7 & -13 & 5 \\
  &   & 2 & -10 & -46 \\
\hline
  & 1 & -5 & -23 & -41
\end{array}$$

$$\frac{x^3 - 7x^2 - 13x + 5}{x - 2} = x^2 - 5x - 23 - \frac{41}{x - 2}$$

7. $x - 2 = x - c$ where $c = 2$.

$$\begin{array}{r|rrr} 2 & 4 & 0 & -9 \\ & & 8 & 16 \\ \hline & 4 & 8 & 7 \end{array}$$

$$\frac{4x^2 - 9}{x - 2} = 4x + 8 + \frac{7}{x - 2}$$

9. $P(x) = 3x^2 - 4x - 1$

   a. $P(2) = 3(2)^2 - 4(2) - 1$
             $= 3(4) - 8 - 1$
             $= 12 - 8 - 1$
             $= 3$

   b.
$$\begin{array}{r|rrr} 2 & 3 & -4 & -1 \\ & & 6 & 4 \\ \hline & 3 & 2 & 3 \end{array}$$

        Thus, $P(2) = 3$.

11. $P(x) = 4x^4 + 7x^2 + 9x - 1$
            $= 4x^4 + 0x^3 + 7x^2 + 9x - 1$

   a. $P(-2) = 4(-2)^4 + 7(-2)^2 + 9(-2) - 1$
               $= 4(16) + 7(4) - 18 - 1$
               $= 64 + 28 - 18 - 1$
               $= 73$

   b.
$$\begin{array}{r|rrrrr} -2 & 4 & 0 & 7 & 9 & -1 \\ & & -8 & 16 & -46 & 74 \\ \hline & 4 & -8 & 23 & -37 & 73 \end{array}$$

        Thus, $P(-2) = 73$.

13. $P(x) = x^5 + 3x^4 + 3x - 7$
            $= x^5 + 3x^4 + 0x^3 + 0x^2 + 3x - 7$

   a. $P(-1) = (-1)^5 + 3(-1)^4 + 3(-1) - 7$
               $= -1 + 3 - 3 - 7$
               $= -8$

   b.
$$\begin{array}{r|rrrrrr} -1 & 1 & 3 & 0 & 0 & 3 & -7 \\ & & -1 & -2 & 2 & -2 & -1 \\ \hline & 1 & 2 & -2 & 2 & 1 & -8 \end{array}$$
        Thus, $P(-1) = -8$.

15. $x^3 - 3x^2 + 2 = x^3 - 3x^2 + 0x + 2$

$$\begin{array}{r|rrrr} 3 & 1 & -3 & 0 & 2 \\ & & 3 & 0 & 0 \\ \hline & 1 & 0 & 0 & 2 \end{array}$$

$$\frac{x^3 - 3x^2 + 2}{x - 3} = x^2 + \frac{2}{x - 3}$$

17. $x + 1 = x - (-1)$

$$\begin{array}{r|rrr} -1 & 6 & 13 & 8 \\ & & -6 & -7 \\ \hline & 6 & 7 & 1 \end{array}$$

$$\frac{6x^2 + 13x + 8}{x + 1} = 6x + 7 + \frac{1}{x + 1}$$

19.
$$\begin{array}{r|rrrrr} 5 & 2 & -13 & 16 & -9 & 20 \\ & & 10 & -15 & 5 & -20 \\ \hline & 2 & -3 & 1 & -4 & 0 \end{array}$$

$$\frac{2x^4 - 13x^3 + 16x^2 - 9x + 20}{x - 5} = 2x^3 - 3x^2 + x - 4$$

21. $3x^2 - 15 = 3x^2 + 0x - 15;\ x + 3 = x - (-3)$

$$\begin{array}{r|rrr} -3 & 3 & 0 & -15 \\ & & -9 & 27 \\ \hline & 3 & -9 & 12 \end{array}$$

$$\frac{3x^2 - 15}{x + 3} = 3x - 9 + \frac{12}{x + 3}$$

23.
$$\begin{array}{r|rrrr} \frac{1}{2} & 3 & -6 & 4 & 5 \\ & & \frac{3}{2} & -\frac{9}{4} & \frac{7}{8} \\ \hline & 3 & -\frac{9}{2} & \frac{7}{4} & \frac{47}{8} \end{array}$$

$$\frac{3x^3 - 6x^2 + 4x + 5}{x - \frac{1}{2}} = 3x^2 - \frac{9}{2}x + \frac{7}{4} + \frac{47}{8\left(x - \frac{1}{2}\right)}$$

25.

$$
\begin{array}{r|rrrr}
\frac{1}{3} & 3 & 2 & -4 & 1 \\
 & & 1 & 1 & -1 \\
\hline
 & 3 & 3 & -3 & 0
\end{array}
$$

$$\frac{3x^3 + 2x^2 - 4x + 1}{x - \frac{1}{3}} = 3x^2 + 3x - 3$$

27. $7x^2 - 4x + 12 + 3x^3 = 3x^3 + 7x^2 - 4x + 12$

$x + 1 = x - (-1)$

$$
\begin{array}{r|rrrr}
-1 & 3 & 7 & -4 & 12 \\
 & & -3 & -4 & 8 \\
\hline
 & 3 & 4 & -8 & 20
\end{array}
$$

$$\frac{7x^2 - 4x + 12 + 3x^3}{x + 1} = 3x^2 + 4x - 8 + \frac{20}{x + 1}$$

29. $x^3 - 1 = x^3 + 0x^2 + 0x - 1$

$$
\begin{array}{r|rrrr}
1 & 1 & 0 & 0 & -1 \\
 & & 1 & 1 & 1 \\
\hline
 & 1 & 1 & 1 & 0
\end{array}
$$

$$\frac{x^3 - 1}{x - 1} = x^2 + x + 1$$

31.

$$
\begin{array}{r|rrrr}
-6 & 2 & 12 & -3 & -20 \\
 & & -12 & 0 & 18 \\
\hline
 & 2 & 0 & -3 & -2
\end{array}
$$

$$\frac{2x^3 - 12x^2 - 3x - 20}{x + 6} = 2x^2 - 3 - \frac{2}{x + 6}$$

33.

$$
\begin{array}{r|rrrr}
1 & 1 & 3 & -7 & 4 \\
 & & 1 & 4 & -3 \\
\hline
 & 1 & 4 & -3 & 1
\end{array}
$$

Thus, $P(1) = 1$.

35.

$$
\begin{array}{r|rrrr}
-3 & 3 & -7 & -2 & 5 \\
 & & -9 & 48 & -138 \\
\hline
 & 3 & -16 & 46 & -133
\end{array}
$$

Thus, $P(-3) = -133$.

37.

$$
\begin{array}{r|rrrrr}
-1 & 4 & 0 & 1 & 0 & -2 \\
 & & -4 & 4 & -5 & 5 \\
\hline
 & 4 & -4 & 5 & -5 & 3
\end{array}
$$

Thus, $P(-1) = 3$.

39.

$$
\begin{array}{r|rrrrr}
\frac{1}{3} & 2 & 0 & -3 & 0 & -2 \\
 & & \frac{2}{3} & \frac{2}{9} & -\frac{25}{27} & -\frac{25}{81} \\
\hline
 & 2 & \frac{2}{3} & -\frac{25}{9} & -\frac{25}{27} & -\frac{187}{181}
\end{array}
$$

Thus, $P\left(\frac{1}{3}\right) = -\frac{187}{81}$.

41.

$$
\begin{array}{r|rrrrrr}
\frac{1}{2} & 1 & 1 & -1 & 0 & 0 & 3 \\
 & & \frac{1}{2} & \frac{3}{4} & -\frac{1}{8} & -\frac{1}{16} & -\frac{1}{32} \\
\hline
 & 1 & \frac{3}{2} & -\frac{1}{4} & -\frac{1}{8} & -\frac{1}{16} & \frac{95}{32}
\end{array}
$$

Thus, $P\left(\frac{1}{2}\right) = \frac{95}{32}$.

43. answers may vary

45. $7x + 2 = x - 3$

$7x - x = -3 - 2$

$6x = -5$

$x = -\frac{5}{6}$

The solution is $-\frac{5}{6}$.

47. $\quad \frac{x}{3} - 5 = 13$

$3\left(\frac{x}{3} - 5\right) = (13) \cdot 3$

$x - 15 = 39$

$x = 54$

The solution is 54.

49. $2^3 = 2 \cdot 2 \cdot 2 = 8$

51. $(-2)^5 = (-2)(-2)(-2)(-2)(-2) = -32$

53. $3 \cdot 4^2 = 3 \cdot 16 = 48$

55. Let $x = -5$.

$x^2 = (-5)^2 = (-5)(-5) = 25$

57. Let $x = -1$.

$2x^3 = 2(-1)^3 = 2(-1) = -2$

59. $(5x^2 - 3x + 2) \div (x + 2)$ is a candidate for synthetic division since $x + 2$ is in the form $x - c$, where $c = -2$.

61. $(x^7 - 2) \div (x^5 + 1)$ is not a candidate for synthetic division since $x^5 + 1$ does not have the form $x - c$.

63. $A = bh$ so $h = \dfrac{A}{b} = \dfrac{x^4 - 23x^2 + 9x - 5}{x + 5}$

$$\begin{array}{r|rrrrr} -5 & 1 & 0 & -23 & 9 & -5 \\ & & -5 & 25 & -10 & 5 \\ \hline & 1 & -5 & 2 & -1 & 0 \end{array}$$

The height is $(x^3 - 5x^2 + 2x - 1)$ cm.

65.

$$x - 1 \overline{\smash{\big)}\, x^4 + \tfrac{2}{3}x^3 + 0x^2 + x + 0}$$

quotient: $x^3 + \tfrac{5}{3}x^2 + \tfrac{5}{3}x + \tfrac{8}{3}$

$$\underline{x^4 - x^3}$$
$$\tfrac{5}{3}x^3 - 0x^2$$
$$\underline{\tfrac{5}{3}x^3 - \tfrac{5}{3}x^2}$$
$$\tfrac{5}{3}x^2 + x$$
$$\underline{\tfrac{5}{3}x^2 - \tfrac{5}{3}x}$$
$$\tfrac{8}{3}x + 0$$
$$\underline{\tfrac{8}{3}x - \tfrac{8}{3}}$$
$$\tfrac{8}{3}$$

Answer: $x^3 + \dfrac{5}{3}x^2 + \dfrac{5}{3}x + \dfrac{8}{3} + \dfrac{8}{3(x-1)}$

67.

$$\begin{array}{r|rrr} -3 & 1 & 3 & 4 & 12 \\ & & -3 & 0 & -12 \\ \hline & 1 & 0 & 4 & 0 \end{array}$$

Remainder = 0 and
$(x + 3)(x^2 + 4) = x^3 + 3x^2 + 4x + 12$

69. $P(c)$ is equal to the remainder when $P(x)$ is divided by $x - c$. Therefore, $P(c) = 0$.

71. Multiply $(x^2 - x + 10)$ by $(x + 3)$ and add the remainder, –2.
$(x^2 - x + 10)(x + 3) - 2$
$= (x^3 + 3x^2 - x^2 - 3x + 10x + 30) - 2$
$= x^3 + 2x^2 + 7x + 28$

Chapter 5 Vocabulary Check

1. A <u>term</u> is a number or the product of numbers and variables raised to powers.

2. The <u>FOIL</u> method may be used when multiplying two binomials.

3. A polynomial with exactly 3 terms is called a <u>trinomial</u>.

4. The <u>degree of polynomial</u> is the greatest degree of any term of the polynomial.

5. A polynomial with exactly 2 terms is called a <u>binomial</u>.

6. The <u>coefficient</u> of a term is its numerical factor.

7. The <u>degree of a term</u> is the sum of the exponents on the variables in the term.

8. A polynomial with exactly 1 term is called a <u>monomial</u>.

9. Monomials, binomials, and trinomials are all examples of <u>polynomials</u>.

10. The <u>distributive</u> property is used to multiply $2x(x - 4)$.

Chapter 5 Review

1. In 7^9, the base is 7 and the exponent is 9.

2. In $(-5)^4$, the base is -5 and the exponent is 4.

3. In -5^4, the base is 5 and the exponent is 4.

4. In x^6, the base is x and the exponent is 6.

5. $8^3 = 8 \cdot 8 \cdot 8 = 512$

6. $(-6)^2 = (-6)(-6) = 36$

7. $-6^2 = -6 \cdot 6 = -36$

8. $-4^3 - 4^0 = -64 - 1 = -65$

9. $(3b)^0 = 1$

10. $\dfrac{8b}{8b} = 1$

11. $y^2 \cdot y^7 = y^{2+7} = y^9$

12. $x^9 \cdot x^5 = x^{9+5} = x^{14}$

13. $(2x^5)(-3x^6) = (2 \cdot -3)(x^5 \cdot x^6) = -6x^{11}$

14. $(-5y^3)(4y^4) = (-5 \cdot 4)(y^3 \cdot y^4) = -20y^7$

15. $(x^4)^2 = x^{4 \cdot 2} = x^8$

16. $(y^3)^5 = y^{3 \cdot 5} = y^{15}$

17. $(3y^6)^4 = 3^4(y^6)^4 = 81y^{24}$

18. $2^3(x^3)^3 = 8x^9$

19. $\dfrac{x^9}{x^4} = x^{9-4} = x^5$

20. $\dfrac{z^{12}}{z^5} = z^{12-5} = z^7$

21. $\dfrac{a^5 b^4}{ab} = a^{5-1}b^{4-1} = a^4 b^3$

22. $\dfrac{x^4 y^6}{xy} = x^{4-1}y^{6-1} = x^3 y^5$

23. $\dfrac{3x^4 y^{10}}{12xy^6} = \dfrac{3}{12}x^{4-1}y^{10-6} = \dfrac{1}{4}x^3 y^4 = \dfrac{x^3 y^4}{4}$

24. $\dfrac{2x^7 y^8}{8xy^2} = \dfrac{2}{8}x^{7-1}y^{8-2} = \dfrac{x^6 y^6}{4}$

25. $5a^7(2a^4)^3 = 5a^7(2^3)(a^4)^3$
$\qquad = (5 \cdot 8)(a^7 \cdot a^{12})$
$\qquad = 40a^{19}$

26. $(2x)^2(9x) = (2^2 \cdot x^2)(9x)$
$\qquad = (4 \cdot 9)(x^2 \cdot x)$
$\qquad = 36x^3$

27. $(-5a)^0 + 7^0 + 8^0 = 1 + 1 + 1 = 3$

28. $8x^0 + 9^0 = 8(1) + 1 = 9$

29. $\left(\dfrac{3x^4}{4y}\right)^3 = \dfrac{3^3 x^{4 \cdot 3}}{4^3 y^3} = \dfrac{27x^{12}}{64y^3}$, choice b.

30. $\left(\dfrac{5a^6}{b^3}\right)^2 = \dfrac{5^2 a^{6 \cdot 2}}{b^{3 \cdot 2}} = \dfrac{25a^{12}}{b^6}$, choice c.

31. The degree of $-5x^4 y^3$ is $4 + 3 = 7$.

32. The degree of $10x^3 y^2 z$ is $3 + 2 + 1 = 6$.

33. The degree of $35a^5 bc^2$ is $5 + 1 + 2 = 8$.

34. The degree of $95xyz$ is $1 + 1 + 1 = 3$.

35. The degree is 5 because y^5 is the term with the highest degree.

36. The degree is 2 because $9y^2$ is the term with the highest degree.

37. The degree is 5 because $-28x^2 y^3$ is the term with the highest degree.

38. The degree is 6 because $6x^2 y^2 z^2$ is the term with the highest degree.

39. $P(t) = -16t^2 + 4000$
$P(0) = -16(0)^2 + 4000$
$\qquad = 0 + 4000$
$\qquad = 4000$
$P(1) = -16(1)^2 + 4000$
$\qquad = -16 + 4000$
$\qquad = 3984$

$$P(3) = -16(3)^2 + 4000$$
$$= -144 + 4000$$
$$= 3856$$
$$P(5) = -16(5)^2 + 4000$$
$$= -400 + 4000$$
$$= 3600$$

t	0 seconds	1 second	3 seconds	5 seconds
$P(t) = -16t^2 + 4000$	4000 feet	3984 feet	3856 feet	3600 feet

40. $P(x) = 2x^2 + 20x$

$$P(1) = 2(1)^2 + 20(1) = 22$$
$$P(3) = 2(3)^2 + 20(3) = 78$$
$$P(5.1) = 2(5.1)^2 + 20(5.1) = 154.02$$
$$P(10) = 2(10)^2 + 20(10) = 400$$

41. $6a^2 + 4a + 9a^2 = (6+9)a^2 + 4a$
$$= 15a^2 + 4a$$

42. $21x^2 + 3x + x^2 + 6 = (21+1)x^2 + 3x + 6$
$$= 22x^2 + 3x + 6$$

43. $4a^2b - 3b^2 - 8q^2 - 10a^2b + 7q^2 = (4a^2b - 10a^2b) - 3b^2 + (-8q^2 + 7q^2)$
$$= -6a^2b - 3b^2 - q^2$$

44. $2s^{14} + 3s^{13} + 12s^{12} - s^{10}$ cannot be combined.

45. $(3x^2 + 2x + 6) + (5x^2 + x) = 3x^2 + 2x + 6 + 5x^2 + x$
$$= 8x^2 + 3x + 6$$

46. $(2x^5 + 3x^4 + 4x^3 + 5x^2) + (4x^2 + 7x + 6) = 2x^5 + 3x^4 + 4x^3 + 5x^2 + 4x^2 + 7x + 6$
$$= 2x^5 + 3x^4 + 4x^3 + 9x^2 + 7x + 6$$

47. $(-5y^2 + 3) - (2y^2 + 4) = -5y^2 + 3 - 2y^2 - 4$
$$= -7y^2 - 1$$

48. $(3x^2 - 7xy + 7y^2) - (4x^2 - xy + 9y^2) = 3x^2 - 7xy + 7y^2 - 4x^2 + xy - 9y^2$
$$= -x^2 - 6xy - 2y^2$$

49. $(7x - 14y) - (3x - y) = 7x - 14y - 3x + y$
$$= 4x - 13y$$

50. $[(x^2 + 7x + 9) + (x^2 + 4)] - (4x^2 + 8x - 7)$
$= x^2 + 7x + 9 + x^2 + 4 - 4x^2 - 8x + 7$
$= -2x^2 - x + 20$

51. $P(x) = 9x^2 - 7x + 8$
$P(6) = 9(6)^2 - 7(6) + 8$
$\quad = 9(36) - 42 + 8$
$\quad = 324 - 42 + 8$
$\quad = 290$

52. $P(x) = 9x^2 - 7x + 8$
$P(-2) = 9(-2)^2 - 7(-2) + 8$
$\quad = 9(4) + 14 + 8$
$\quad = 36 + 14 + 8$
$\quad = 58$

53. $(x^2 y + 5) + (2x^2 y - 6x + 1) + (x^2 y + 5)$
$\quad\quad + (2x^2 y - 6x + 1)$
$= x^2 y + 2x^2 y + x^2 y + 2x^2 y - 6x - 6x$
$\quad\quad + 5 + 1 + 5 + 1$
$= 6x^2 y - 12x + 12$

The perimeter is $(6x^2 y - 12x + 12)$ cm.

54. Let $x = 8$.
$f(8) = 754(8)^2 - 228(8) + 80,134$
$\quad = 126,566$
Revenues from software sales in 2009 were to be $126,566 million.

55. $4(2a + 7) = 4(2a) + 4(7) = 8a + 28$

56. $9(6a - 3) = 9(6a) - 9(3) = 54a - 27$

57. $-7x(x^2 + 5) = -7(x^2) - 7x(5) = -7x^3 - 35x$

58. $-8y(4y^2 - 6) = -8y(4y^2) - 8y(-6)$
$\quad\quad = -32y^3 + 48y$

59. $(3a^3 - 4a + 1)(-2a)$
$= 3a^3(-2a) - 4a(-2a) + 1(-2a)$
$= -6a^4 + 8a^2 - 2a$

60. $(6b^3 - 4b + 2)(7b) = 6b^3(7b) - 4b(7b) + 2(7b)$
$\quad\quad = 42b^4 - 28b^2 + 14b$

61. $(2x + 2)(x - 7) = 2x^2 - 14x + 2x - 14$
$\quad\quad = 2x^2 - 12x - 14$

62. $(2x - 5)(3x + 2) = 6x^2 + 4x - 15x - 10$
$\quad\quad = 6x^2 - 11x - 10$

63. $(x - 9)^2 = (x - 9)(x - 9)$
$\quad\quad = x^2 - 9x - 9x + 81$
$\quad\quad = x^2 - 18x + 81$

64. $(x - 12)^2 = (x - 12)(x - 12)$
$\quad\quad = x^2 - 12x - 12x + 144$
$\quad\quad = x^2 - 24x + 144$

65. $(4a - 1)(a + 7) = 4a^2 + 28a - a - 7$
$\quad\quad = 4a^2 + 27a - 7$

66. $(6a - 1)(7a + 3) = 42a^2 + 18a - 7a - 3$
$\quad\quad = 42a^2 + 11a - 3$

67. $(5x + 2)^2 = (5x + 2)(5x + 2)$
$\quad\quad = 25x^2 + 10x + 10x + 4$
$\quad\quad = 25x^2 + 20x + 4$

68. $(3x + 5)^2 = (3x + 5)(3x + 5)$
$\quad\quad = 9x^2 + 15x + 15x + 25$
$\quad\quad = 9x^2 + 30x + 25$

69. $(x + 7)(x^3 + 4x - 5)$
$= x(x^3 + 4x - 5) + 7(x^3 + 4x - 5)$
$= x^4 + 4x^2 - 5x + 7x^3 + 28x - 35$
$= x^4 + 7x^3 + 4x^2 + 23x - 35$

70. $(x + 2)(x^5 + x + 1) = x(x^5 + x + 1) + 2(x^5 + x + 1)$
$\quad\quad = x^6 + x^2 + x + 2x^5 + 2x + 2$
$\quad\quad = x^6 + 2x^5 + x^2 + 3x + 2$

71. $(x^2 + 2x + 4)(x^2 + 2x - 4)$
$= x^2(x^2 + 2x - 4) + 2x(x^2 + 2x - 4)$
$\quad\quad + 4(x^2 + 2x - 4)$
$= x^4 + 2x^3 - 4x^2 + 2x^3 + 4x^2 - 8x$
$\quad\quad + 4x^2 + 8x - 16$
$= x^4 + 4x^3 + 4x^2 - 16$

72. $(x^3 + 4x + 4)(x^3 + 4x - 4)$
$= x^3(x^3 + 4x - 4) + 4x(x^3 + 4x - 4)$
$\qquad + 4(x^3 + 4x - 4)$
$= x^6 + 4x^4 - 4x^3 + 4x^4 + 16x^2 - 16x + 4x^3$
$\qquad + 16x - 16$
$= x^6 + 8x^4 + 16x^2 - 16$

73. $(x + 7)^3 = (x + 7)(x + 7)(x + 7)$
$= (x^2 + 7x + 7x + 49)(x + 7)$
$= (x^2 + 14x + 49)(x + 7)$
$= (x^2 + 14x + 49)x + (x^2 + 14x + 49)7$
$= x^3 + 14x^2 + 49x + 7x^2 + 98x + 343$
$= x^3 + 21x^2 + 147x + 343$

74. $(2x - 5)^3$
$= (2x - 5)(2x - 5)(2x - 5)$
$= (4x^2 - 10x - 10x + 25)(2x - 5)$
$= (4x^2 - 20x + 25)(2x - 5)$
$= (4x^2 - 20x + 25)(2x) + (4x^2 - 20x + 25)(-5)$
$= 8x^3 - 40x^2 + 50x - 20x^2 + 100x - 125$
$= 8x^3 - 60x^2 + 150x - 125$

75. $(x + 7)^2 = x^2 + 2(x)(7) + 7^2 = x^2 + 14x + 49$

76. $(x - 5)^2 = x^2 - 2(x)(5) + 5^2 = x^2 - 10x + 25$

77. $(3x - 7)^2 = (3x)^2 - 2(3x)(7) + 7^2$
$= 9x^2 - 42x + 49$

78. $(4x + 2)^2 = (4x)^2 + 2(4x)(2) + 2^2$
$= 16x^2 + 16x + 4$

79. $(5x - 9)^2 = (5x)^2 - 2(5x)(9) + 9^2$
$= 25x^2 - 90x + 81$

80. $(5x + 1)(5x - 1) = (5x)^2 - 1^2 = 25x^2 - 1$

81. $(7x + 4)(7x - 4) = (7x)^2 - 4^2 = 49x^2 - 16$

82. $(a + 2b)(a - 2b) = a^2 - (2b)^2 = a^2 - 4b^2$

83. $(2x - 6)(2x + 6) = (2x)^2 - 6^2 = 4x^2 - 36$

84. $(4a^2 - 2b)(4a^2 + 2b) = (4a^2)^2 - (2b)^2$
$= 16a^4 - 4b^2$

85. $(3x - 1)^2 = (3x)^2 - 2(3x)(1) + 1^2$
$= 9x^2 - 6x + 1$
The area is $(9x^2 - 6x + 1)$ square meters.

86. $(5x + 2)(x - 1) = 5x^2 - 5x + 2x - 2$
$= 5x^2 - 3x - 2$
The area is $(5x^2 - 3x - 2)$ square miles.

87. $7^{-2} = \dfrac{1}{7^2} = \dfrac{1}{49}$

88. $-7^{-2} = -\dfrac{1}{7^2} = -\dfrac{1}{49}$

89. $2x^{-4} = \dfrac{2}{x^4}$

90. $(2x)^{-4} = \dfrac{1}{(2x)^4} = \dfrac{1}{16x^4}$

91. $\left(\dfrac{1}{5}\right)^{-3} = \dfrac{1^{-3}}{5^{-3}} = \dfrac{5^3}{1^3} = 125$

92. $\left(\dfrac{-2}{3}\right)^{-2} = \dfrac{(-2)^{-2}}{3^{-2}} = \dfrac{3^2}{(-2)^2} = \dfrac{9}{4}$

93. $2^0 + 2^{-4} = 1 + \dfrac{1}{2^4} = \dfrac{16}{16} + \dfrac{1}{16} = \dfrac{17}{16}$

94. $6^{-1} - 7^{-1} = \dfrac{1}{6} - \dfrac{1}{7} = \dfrac{7}{42} - \dfrac{6}{42} = \dfrac{1}{42}$

95. $\dfrac{x^5}{x^{-3}} = x^{5-(-3)} = x^8$

96. $\dfrac{z^4}{z^{-4}} = z^{4-(-4)} = z^8$

97. $\dfrac{r^{-3}}{r^{-4}} = r^{-3-(-4)} = r$

98. $\dfrac{y^{-2}}{y^{-5}} = y^{-2-(-5)} = y^3$

99. $\left(\dfrac{bc^{-2}}{bc^{-3}}\right)^4 = \dfrac{b^4 c^{-8}}{b^4 c^{-12}} = b^{4-4} c^{-8-(-12)} = c^4$

100. $\left(\dfrac{x^{-3} y^{-4}}{x^{-2} y^{-5}}\right)^{-3} = \dfrac{x^9 y^{12}}{x^6 y^{15}}$

$= x^{9-6} y^{12-15}$

$= x^3 y^{-3}$

$= \dfrac{x^3}{y^3}$

101. $\dfrac{x^{-4} y^{-6}}{x^2 y^7} = x^{-4-2} y^{-6-7}$

$= x^{-6} y^{-13}$

$= \dfrac{1}{x^6 y^{13}}$

102. $\dfrac{a^5 b^{-5}}{a^{-5} b^5} = a^{5-(-5)} b^{-5-5} = a^{10} b^{-10} = \dfrac{a^{10}}{b^{10}}$

103. $a^{6m} a^{5m} = a^{6m+5m} = a^{11m}$

104. $\dfrac{(x^{5+h})^3}{x^5} = \dfrac{x^{3(5+h)}}{x^5}$

$= \dfrac{x^{15+3h}}{x^5}$

$= x^{15+3h-5}$

$= x^{10+3h}$

105. $(3xy^{2z})^3 = 3^3 x^3 y^{2z(3)} = 27 x^3 y^{6z}$

106. $a^{m+2} a^{m+3} = a^{(m+2)+(m+3)} = a^{2m+5}$

107. $0.00027 = 2.7 \times 10^{-4}$

108. $0.8868 = 8.868 \times 10^{-1}$

109. $80{,}800{,}000 = 8.08 \times 10^7$

110. $868{,}000 = 8.68 \times 10^5$

111. $91{,}000{,}000 = 9.1 \times 10^7$

112. $150{,}000 = 1.5 \times 10^5$

113. $8.67 \times 10^5 = 867{,}000$

114. $3.86 \times 10^{-3} = 0.00386$

115. $8.6 \times 10^{-4} = 0.00086$

116. $8.936 \times 10^5 = 893{,}600$

117. $1.43128 \times 10^{15} = 1{,}431{,}280{,}000{,}000{,}000$

118. $1 \times 10^{-10} = 0.0000000001$

119. $(8 \times 10^4)(2 \times 10^{-7}) = (8 \cdot 2) \times (10^4 \cdot 10^{-7})$

$= 16 \times 10^{-3}$

$= 0.016$

120. $\dfrac{8 \times 10^4}{2 \times 10^{-7}} = \dfrac{8}{2} \times 10^{4-(-7)}$

$= 4 \times 10^{11}$

$= 400{,}000{,}000{,}000$

121. $\dfrac{x^2 + 21x + 49}{7x^2} = \dfrac{x^2}{7x^2} + \dfrac{21x}{7x^2} + \dfrac{49}{7x^2}$

$= \dfrac{1}{7} + \dfrac{3}{x} + \dfrac{7}{x^2}$

122. $\dfrac{5a^3 b - 15ab^2 + 20ab}{-5ab} = \dfrac{5a^3 b}{-5ab} - \dfrac{15ab^2}{-5ab} + \dfrac{20ab}{-5ab}$

$= -a^2 + 3b - 4$

123.

$$
\begin{array}{r}
a+1 \\
a-2 \overline{\smash{\big)}\, a^2 - a + 4} \\
\underline{a^2 - 2a} \\
a + 4 \\
\underline{a - 2} \\
6
\end{array}
$$

$(a^2 - a + 4) \div (a - 2) = a + 1 + \dfrac{6}{a-2}$

124.

$$x+5\overline{\smash{\big)}4x^2+20x+7}$$
$$\underline{4x^2+20x}$$
$$7$$

$$(4x^2+20x+7)\div(x+5)=4x+\frac{7}{x+5}$$

125.

$$a-2\overline{\smash{\big)}a^3+a^2+2a+6}$$
$$\underline{a^3-2a^2}$$
$$3a^2+2a$$
$$\underline{3a^2-6a}$$
$$8a+6$$
$$\underline{8a-16}$$
$$22$$

$$\frac{a^3+a^2+2a+6}{a-2}=a^2+3a+8+\frac{22}{a-2}$$

126.

$$3b-2\overline{\smash{\big)}9b^3-18b^2+8b-1}$$
$$\underline{9b^3-6b^2}$$
$$-12b^2+8b$$
$$\underline{-12b^2+8b}$$
$$-1$$

$$\frac{9b^3-18b^2+8b-1}{3b-2}=3b^2-4b-\frac{1}{3b-2}$$

127.

$$2x-1\overline{\smash{\big)}4x^4-4x^3+x^2+4x-3}$$
$$\underline{4x^4-2x^3}$$
$$-2x^3+x^2$$
$$\underline{-2x^3+x^2}$$
$$4x-3$$
$$\underline{4x-2}$$
$$-1$$

$$\frac{4x^4-4x^3+x^2+4x-3}{2x-1}=2x^3-x^2+2-\frac{1}{2x-1}$$

128. Rewrite $-10x^2-x^3-21x+18$ as
$-x^3-10x^2-21x+18$.

$$x-6\overline{\smash{\big)}-x^3-10x^2-21x+18}$$
$$\underline{-x^3+6x^2}$$
$$-16x^2-21x$$
$$\underline{-16x^2+96x}$$
$$-117x+18$$
$$\underline{-117x+702}$$
$$-684$$

$$\frac{-10x^2-x^3-21x+18}{x-6}=-x^2-16x-117-\frac{684}{x-6}$$

129. $\dfrac{15x^3-3x^2+60}{3x^2}=\dfrac{15x^3}{3x^2}-\dfrac{3x^2}{3x^2}+\dfrac{60}{3x^2}$

$$=5x-1+\frac{20}{x^2}$$

The width is $\left(5x-1+\dfrac{20}{x^2}\right)$ feet.

130. $\dfrac{21a^3b^6+3a-3}{3}=\dfrac{21a^3b^6}{3}+\dfrac{3a}{3}-\dfrac{3}{3}$

$$=7a^3b^6+a-1$$

The length of a side is $(7a^3b^6+a-1)$ units.

131. $3x^3+12x-4=3x^3+0x^2+12x-4$

2	3	0	12	−4
		6	12	48
	3	6	24	44

$$\frac{3x^3+12x-4}{x-2}=3x^2+6x+24+\frac{44}{x-2}$$

132. $x+\dfrac{3}{2}=x-\left(-\dfrac{3}{2}\right)$

$-\frac{3}{2}$	3	2	−4	−1
		$-\frac{9}{2}$	$\frac{15}{4}$	$\frac{3}{8}$
	3	$-\frac{5}{2}$	$-\frac{1}{4}$	$-\frac{5}{8}$

$$\frac{3x^3+2x^2-4x-1}{x+\frac{3}{2}}=3x^2-\frac{5}{2}x-\frac{1}{4}-\frac{5}{8\left(x+\frac{3}{2}\right)}$$

133. $x^5 - 1 = x^5 + 0x^4 + 0x^3 + 0x^2 + 0x - 1;$
$x + 1 = x - (-1)$

$$
\begin{array}{r|rrrrrr}
-1 & 1 & 0 & 0 & 0 & 0 & -1 \\
   &   & -1 & 1 & -1 & 1 & -1 \\
\hline
   & 1 & -1 & 1 & -1 & 1 & -2
\end{array}
$$

$$\frac{x^5 - 1}{x + 1} = x^4 - x^3 + x^2 - x + 1 - \frac{2}{x+1}$$

134. $x^3 - 81 = x^3 + 0x^2 + 0x - 81$

$$
\begin{array}{r|rrrr}
3 & 1 & 0 & 0 & -81 \\
  &   & 3 & 9 & 27 \\
\hline
  & 1 & 3 & 9 & -54
\end{array}
$$

$$\frac{x^3 - 81}{x - 3} = x^2 + 3x + 9 - \frac{54}{x - 3}$$

135. $x^3 - x^2 + 3x^4 - 2 = 3x^4 + x^3 - x^2 + 0x - 2$

$$
\begin{array}{r|rrrrr}
4 & 3 & 1 & -1 & 0 & -2 \\
  &   & 12 & 52 & 204 & 816 \\
\hline
  & 3 & 13 & 51 & 204 & 814
\end{array}
$$

$$\frac{x^3 - x^2 + 3x^4 - 2}{x - 4}$$
$$= 3x^3 + 13x^2 + 51x + 204 + \frac{814}{x - 4}$$

136. $3x^4 - 2x^2 + 10 = 3x^4 + 0x^3 - 2x^2 + 0x + 10$
$x + 2 = x - (-2)$

$$
\begin{array}{r|rrrrr}
-2 & 3 & 0 & -2 & 0 & 10 \\
   &   & -6 & 12 & -20 & 40 \\
\hline
   & 3 & -6 & 10 & -20 & 50
\end{array}
$$

$$\frac{3x^4 - 2x^2 + 10}{x + 2} = 3x^3 - 6x^2 + 10x - 20 + \frac{50}{x + 2}$$

137. $P(x) = 3x^5 + 0x^4 + 0x^3 + 0x^2 - 9x + 7$

$$
\begin{array}{r|rrrrrr}
4 & 3 & 0 & 0 & 0 & -9 & 7 \\
  &   & 12 & 48 & 192 & 768 & 3036 \\
\hline
  & 3 & 12 & 48 & 192 & 759 & 3043
\end{array}
$$

Thus, $P(4) = 3043$.

138. $P(x) = 3x^5 + 0x^4 + 0x^3 + 0x^2 - 9x + 7$

$$
\begin{array}{r|rrrrrr}
-5 & 3 & 0 & 0 & 0 & -9 & 7 \\
   &   & -15 & 75 & -375 & 1875 & -9330 \\
\hline
   & 3 & -15 & 75 & -375 & 1866 & -9323
\end{array}
$$

Thus, $P(-5) = -9323$.

139. $\left(-\dfrac{1}{2}\right)^3 = \left(-\dfrac{1}{2}\right)\left(-\dfrac{1}{2}\right)\left(-\dfrac{1}{2}\right) = -\dfrac{1}{8}$

140. $(4xy^2)(x^3 y^5) = 4(x \cdot x^3)(y^2 \cdot y^5)$
$\qquad\qquad = 4x^{1+3} y^{2+5}$
$\qquad\qquad = 4x^4 y^7$

141. $\dfrac{18x^9}{27x^3} = \dfrac{18}{27} x^{9-3} = \dfrac{2x^6}{3}$

142. $\left(\dfrac{3a^4}{b^2}\right)^3 = \dfrac{3^3(a^4)^3}{(b^2)^3} = \dfrac{27a^{12}}{b^6}$

143. $(2x^{-4} y^3)^{-4} = 2^{-4}(x^{-4})^{-4}(y^3)^{-4}$
$\qquad\qquad = \dfrac{1}{2^4} x^{16} y^{-12}$
$\qquad\qquad = \dfrac{x^{16}}{16 y^{12}}$

144. $\dfrac{a^{-3} b^6}{9^{-1} a^{-5} b^{-2}} = 9a^{-3-(-5)} b^{6-(-2)} = 9a^2 b^8$

145. $(6x + 2) + (5x - 7) = 6x + 2 + 5x - 7 = 11x - 5$

146. $(-y^2 - 4) + (3y^2 - 6) = -y^2 - 4 + 3y^2 - 6$
$\qquad\qquad\qquad = 2y^2 - 10$

147. $(8y^2 - 3y + 1) - (3y^2 + 2) = 8y^2 - 3y^2 - 3y + 1 - 2$
$\qquad\qquad\qquad\qquad = 5y^2 - 3y - 1$

148. $(5x^2 + 2x - 6) - (-x - 4) = 5x^2 + 2x - 6 + x + 4$
$\qquad\qquad\qquad\qquad = 5x^2 + 3x - 2$

149. $4x(7x^2 + 3) = 4x(7x^2) + 4x(3)$
$\qquad\qquad = 28x^3 + 12x$

150. $(2x+5)(3x-2) = 6x^2 - 4x + 15x - 10$
$$= 6x^2 + 11x - 10$$

151. $(x-3)(x^2+4x-6)$
$$= x(x^2+4x-6) - 3(x^2+4x-6)$$
$$= x^3 + 4x^2 - 6x - 3x^2 - 12x + 18$$
$$= x^3 + x^2 - 18x + 18$$

152. $(7x-2)(4x-9) = 28x^2 - 63x - 8x + 18$
$$= 28x^2 - 71x + 18$$

153. $(5x+4)^2 = (5x)^2 + 2(5x)(4) + 4^2$
$$= 25x^2 + 40x + 16$$

154. $(6x+3)(6x-3) = (6x)^2 - (3)^2 = 36x^2 - 9$

155. $\dfrac{8a^4 - 2a^3 + 4a - 5}{2a^3} = \dfrac{8a^4}{2a^3} - \dfrac{2a^3}{2a^3} + \dfrac{4a}{2a^3} - \dfrac{5}{2a^3}$
$$= 4a - 1 + \dfrac{2}{a^2} - \dfrac{5}{2a^3}$$

156.
$$
\begin{array}{r}
x - 3 \\
x+5 \overline{)\,x^2 + 2x + 10} \\
\underline{x^2 + 5x} \\
-3x + 10 \\
\underline{-3x - 15} \\
25
\end{array}
$$

$$\dfrac{x^2 + 2x + 10}{x+5} = x - 3 + \dfrac{25}{x+5}$$

157.
$$
\begin{array}{r}
2x^2 + 7x + 5 \\
2x-3 \overline{)\,4x^3 + 8x^2 - 11x + 4} \\
\underline{4x^3 - 6x^2} \\
14x^2 - 11x \\
\underline{14x^2 - 21x} \\
10x + 4 \\
\underline{10x - 15} \\
19
\end{array}
$$

$$\dfrac{4x^3 + 8x^2 - 11x + 4}{2x-3} = 2x^2 + 7x + 5 + \dfrac{19}{2x-3}$$

Chapter 5 Test

1. $2^5 = 2 \cdot 2 \cdot 2 \cdot 2 \cdot 2 = 32$

2. $(-3)^4 = (-3)(-3)(-3)(-3) = 81$

3. $-3^4 = -3 \cdot 3 \cdot 3 \cdot 3 = -81$

4. $4^{-3} = \dfrac{1}{4^3} = \dfrac{1}{64}$

5. $(3x^2)(-5x^9) = (3)(-5)(x^2 \cdot x^9) = -15x^{11}$

6. $\dfrac{y^7}{y^2} = y^{7-2} = y^5$

7. $\dfrac{r^{-8}}{r^{-3}} = r^{-8-(-3)} = r^{-5} = \dfrac{1}{r^5}$

8. $\left(\dfrac{x^2 y^3}{x^3 y^{-4}}\right)^2 = \dfrac{x^4 y^6}{x^6 y^{-8}}$
$$= x^{4-6} y^{6-(-8)}$$
$$= x^{-2} y^{14}$$
$$= \dfrac{y^{14}}{x^2}$$

9. $\left(\dfrac{6^2 x^{-4} y^{-1}}{6^3 x^{-3} y^7}\right) = 6^{2-3} x^{-4-(-3)} y^{-1-7}$
$$= 6^{-1} x^{-1} y^{-8}$$
$$= \dfrac{1}{6xy^8}$$

10. $563,000 = 5.63 \times 10^5$

11. $0.0000863 = 8.63 \times 10^{-5}$

12. $1.5 \times 10^{-3} = 0.0015$

13. $6.23 \times 10^4 = 62,300$

14. $(1.2 \times 10^5)(3 \times 10^{-7}) = (1.2)(3) \times 10^{5-7}$
$$= 3.6 \times 10^{-2}$$
$$= 0.036$$

15. a.

Term	Numerical Coefficient	Degree of Term
$4xy^2$	4	3
$7xyz$	7	3
x^3y	1	4
-2	-2	0

    b.   The degree is 4.

16. $5x^2 + 4xy - 7x^2 + 11 + 8xy$
$= (5x^2 - 7x^2) + (4xy + 8xy) + 11$
$= -2x^2 + 12xy + 11$

17. $(8x^3 + 7x^2 + 4x - 7) + (8x^3 - 7x - 6)$
$= 8x^3 + 7x^2 + 4x - 7 + 8x^3 - 7x - 6$
$= 16x^3 + 7x^2 - 3x - 13$

18. $\quad 5x^3 \;\; + x^2 + 5x - 2$
$\quad \underline{-(8x^3 - 4x^2 \;\; + x - 7)}$

$\quad\quad 5x^3 \;\; + x^2 + 5x - 2$
$\quad \underline{- 8x^3 + 4x^2 \;\; - x + 7}$
$\quad -3x^3 + 5x^2 + 4x + 5$

19. $[(8x^2 + 7x + 5) + (x^3 - 8)] - (4x + 2)$
$= 8x^2 + 7x + 5 + x^3 - 8 - 4x - 2$
$= x^3 + 8x^2 + 3x - 5$

20. $(3x + 7)(x^2 + 5x + 2)$
$= 3x(x^2 + 5x + 2) + 7(x^2 + 5x + 2)$
$= 3x^3 + 15x^2 + 6x + 7x^2 + 35x + 14$
$= 3x^3 + 22x^2 + 41x + 14$

21. $3x^2(2x^2 - 3x + 7)$
$= 3x^2(2x^2) + 3x^2(-3x) + 3x^2(7)$
$= 6x^4 - 9x^3 + 21x^2$

22. $(x + 7)(3x - 5) = 3x^2 - 5x + 21x - 35$
$= 3x^2 + 16x - 35$

23. $\left(3x - \dfrac{1}{5}\right)\left(3x + \dfrac{1}{5}\right) = (3x)^2 - \left(\dfrac{1}{5}\right)^2 = 9x^2 - \dfrac{1}{25}$

24. $(4x - 2)^2 = (4x)^2 - 2(4x)(2) + 2^2$
$= 16x^2 - 16x + 4$

25. $(8x + 3)^2 = (8x)^2 + 2(8x)(3) + (3)^2$
$= 64x^2 + 48x + 9$

26. $(x^2 - 9b)(x^2 + 9b) = (x^2)^2 - (9b)^2 = x^4 - 81b^2$

27. $P(t) = -16t^2 + 1001$
$P(0) = -16(0)^2 + 1001 = 1001$ ft
$P(1) = -16(1)^2 + 1001 = 985$ ft
$P(3) = -16(3)^2 + 1001 = 857$ ft
$P(5) = -16(5)^2 + 1001 = 601$ ft

28. $(2x + 3)(2x - 3) = (2x)^2 - (3)^2 = 4x^2 - 9$
The area is $(4x^2 - 9)$ square inches.

29. $\dfrac{4x^2 + 24xy - 7x}{8xy} = \dfrac{4x^2}{8xy} + \dfrac{24xy}{8xy} - \dfrac{7x}{8xy}$
$= \dfrac{x}{2y} + 3 - \dfrac{7}{8y}$

30.

$$
\begin{array}{r}
x + 2 \\
x + 5 \overline{)\, x^2 + 7x + 10} \\
\underline{x^2 + 5x} \\
2x + 10 \\
\underline{2x + 10} \\
0
\end{array}
$$

$\dfrac{x^2 + 7x + 10}{x + 5} = x + 2$

31. Rewrite $27x^3 - 8$ as $27x^3 + 0x^2 + 0x - 8$.

$$
\begin{array}{r}
9x^2 - 6x + 4 \\
3x+2\overline{\smash{\big)}\,27x^3 + 0x^2 + 0x - 8} \\
\underline{27x^3 + 18x^2} \\
-18x^2 + 0x \\
\underline{-18x^2 - 12x} \\
12x - 8 \\
\underline{12x + 8} \\
-16
\end{array}
$$

$$\frac{27x^3 - 8}{3x + 2} = 9x^2 - 6x + 4 - \frac{16}{3x+2}$$

32. $h(t) = -16t^2 + 96t + 880$

   a. $h(1) = -16(1)^2 + 96(1) + 880$
$$= -16 + 96 + 880$$
$$= 960$$
The height of the pebble is 960 feet when $t = 1$.

   b. $h(5.1) = -16(5.1)^2 + 96(5.1) + 880$
$$= -16(26.01) + 489.6 + 880$$
$$= -416.16 + 489.6 + 880$$
$$= 953.44$$
The height of the pebble is 953.44 feet when $t = 5.1$.

33. $4x^4 - 3x^3 - x - 1 = 4x^4 - 3x^3 + 0x^2 - x - 1$
$x + 3 = x - (-3)$

$$
\begin{array}{r|rrrrr}
-3 & 4 & -3 & 0 & -1 & -1 \\
   &   & -12 & 45 & -135 & 408 \\
\hline
   & 4 & -15 & 45 & -136 & 407
\end{array}
$$

$$\frac{4x^4 - 3x^3 - x - 1}{x + 3}$$
$$= 4x^3 - 15x^2 + 45x - 136 + \frac{407}{x+3}$$

34. $P(x) = 4x^4 + 0x^3 + 7x^2 - 2x - 5$

$$
\begin{array}{r|rrrrr}
-2 & 4 & 0 & 7 & -2 & -5 \\
   &   & -8 & 16 & -46 & 96 \\
\hline
   & 4 & -8 & 23 & -48 & 91
\end{array}
$$

Thus, $P(-2) = 91$.

Chapter 5 Cumulative Review

1. a. $8 \geq 8$ is true since $8 = 8$.

   b. $8 \leq 8$ is true since $8 = 8$.

   c. $23 \leq 0$ is false.

   d. $23 \geq 0$ is true

2. a. $|-7.2| = 7.2$

   b. $|0| = 0$

   c. $\left|-\dfrac{1}{2}\right| = \dfrac{1}{2}$

3. a. $\dfrac{4}{5} \div \dfrac{5}{16} = \dfrac{4}{5} \cdot \dfrac{16}{5} = \dfrac{64}{25}$

   b. $\dfrac{7}{10} \div 14 = \dfrac{7}{10} \div \dfrac{14}{1} = \dfrac{7}{10} \cdot \dfrac{1}{14} = \dfrac{7}{10 \cdot 7 \cdot 2} = \dfrac{1}{20}$

   c. $\dfrac{3}{8} \div \dfrac{3}{10} = \dfrac{3}{8} \cdot \dfrac{10}{3} = \dfrac{3 \cdot 2 \cdot 5}{2 \cdot 4 \cdot 3} = \dfrac{5}{4}$

4. a. $\dfrac{3}{4} \cdot \dfrac{7}{21} = \dfrac{3 \cdot 7}{4 \cdot 3 \cdot 7} = \dfrac{1}{4}$

   b. $\dfrac{1}{2} \cdot 4\dfrac{5}{6} = \dfrac{1}{2} \cdot \dfrac{29}{6} = \dfrac{29}{12} = 2\dfrac{5}{12}$

5. a. $3^2 = 3 \cdot 3 = 9$

   b. $5^3 = 5 \cdot 5 \cdot 5 = 125$

   c. $2^4 = 2 \cdot 2 \cdot 2 \cdot 2 = 16$

   d. $7^1 = 7$

   e. $\left(\dfrac{3}{7}\right)^2 = \left(\dfrac{3}{7}\right)\left(\dfrac{3}{7}\right) = \dfrac{9}{49}$

6. Let $x = 5$ and $y = 1$.
$$\frac{2x - 7y}{x^2} = \frac{2(5) - 7(1)}{5^2}$$
$$= \frac{10 - 7}{25}$$
$$= \frac{3}{25}$$

7. a. $-3 + (-7) = -10$

b. $-1 + (-20) = -21$

c. $-2 + (-10) = -12$

8. $8 + 3(2 \cdot 6 - 1) = 8 + 3(12 - 1)$
$$= 8 + 3(11)$$
$$= 8 + 33$$
$$= 41$$

9. $-4 - 8 = -4 + (-8) = -12$

10. $x = 1$
$$5x^2 + 2 = x - 8$$
$$5(1)^2 + 2 \overset{?}{=} 1 - 8$$
$$5 + 2 \overset{?}{=} -7$$
$$7 \overset{?}{=} -7 \quad \text{False}$$
$x = 1$ is not a solution.

11. a. The reciprocal of 22 is $\dfrac{1}{22}$.

b. The reciprocal of $\dfrac{3}{16}$ is $\dfrac{16}{3}$.

c. The reciprocal of -10 is $-\dfrac{1}{10}$.

d. The reciprocal of $-\dfrac{9}{13}$ is $-\dfrac{13}{9}$.

12. a. $7 - 40 = 7 + (-40) = -33$

b. $-5 - (-10) = -5 + 10 = 5$

13. a. $5 + (4 + 6) = (5 + 4) + 6$

b. $(-1 \cdot 2) \cdot 5 = -1 \cdot (2 \cdot 5)$

14. $\dfrac{4(-3) + (-8)}{5 + (-5)} = \dfrac{-12 + (-8)}{0}$ is undefined.

15. a. $10 + (x + 12) = 10 + (12 + x)$
$$= (10 + 12) + x$$
$$= 22 + x$$

b. $-3(7x) = (-3 \cdot 7)x = -21x$

16. $-2(x + 3y - z) = -2(x) + (-2)(3y) - (-2)(z)$
$$= -2x - 6y + 2z$$

17. a. $5(3x + 2) = 5(3x) + 5(2) = 15x + 10$

b. $-2(y + 0.3z - 1)$
$$= -2(y) + (-2)(0.3z) - (-2)(1)$$
$$= -2y - 0.6z + 2y$$

c. $-(9x + y - 2z + 6)$
$$= -1(9x + y - 2z + 6)$$
$$= -1(9x) + (-1)(y) - (-1)(2z) + (-1)(6)$$
$$= -9x - y + 2z - 6$$

18. $2(6x - 1) - (x - 7) = 12x - 2 - x + 7$
$$= 11x + 5$$

19. $\quad x - 7 = 10$
$$x - 7 + 7 = 10 + 7$$
$$x = 17$$

20. Let $x =$ a number.
$$(x + 7) - 2x$$

21. $\quad \dfrac{5}{2}x = 15$
$$\dfrac{2}{5} \cdot \dfrac{5}{2}x = \dfrac{2}{5} \cdot 15$$
$$x = 6$$

22. $2x + \dfrac{1}{8} = x - \dfrac{3}{8}$
$$x + \dfrac{1}{8} = -\dfrac{3}{8}$$
$$x = -\dfrac{4}{8}$$
$$x = -\dfrac{1}{2}$$

23. Let $x =$ a number.
$$7 + 2x = x - 3$$
$$7 + x = -3$$
$$x = -10$$
The number is -10.

24. $10 = 5j - 2$
$$12 = 5j$$
$$\dfrac{12}{5} = j$$

25. Let x = a number.
$$2(x+4) = 4x - 12$$
$$2x + 8 = 4x - 12$$
$$-2x + 8 = -12$$
$$-2x = -20$$
$$x = 10$$
The number is 10.

26.
$$\frac{7x+5}{3} = x + 3$$
$$3\left(\frac{7x+5}{3}\right) = 3(x+3)$$
$$7x + 5 = 3x + 9$$
$$4x + 5 = 9$$
$$4x = 4$$
$$x = 1$$

27. Let x = the width and $3x - 2$ = the length.
$$2L + 2W = P$$
$$2(3x-2) + 2x = 28$$
$$6x - 4 + 2x = 28$$
$$8x - 4 = 28$$
$$8x = 32$$
$$x = 4$$
$3x - 2 = 3(4) - 2 = 10$
The width is 4 feet and the length is 10 feet.

28. $x < 5$, $(-\infty, 5)$

29.
$$F = \frac{9}{5}C + 32$$
$$F - 32 = \frac{9}{5}C$$
$$\frac{5}{9}(F - 32) = C$$
$$\frac{5F - 160}{9} = C$$

30. a. $x = -1$ is a vertical line and the slope is undefined.

    b. $y = 7$ is a horizontal line and the slope is zero.

31. $2 < x \le 4$

32. $m = \dfrac{y_2 - y_1}{x_2 - x_1} = \dfrac{2}{20} = \dfrac{1}{10} \cdot 100\% = 10\%$

33. $3x + y = 12$

    a. $(0, \ )$: $3(0) + y = 12$
$$y = 12, \ (0, 12)$$

    b. $(\ , 6)$: $3x + 6 = 12$
$$3x = 6$$
$$x = 2, \ (2, 6)$$

    c. $(-1, \ )$: $3(-1) + y = 12$
$$-3 + y = 12$$
$$y = 15, \ (-1, 15)$$

34. $\begin{cases} 3x + 2y = -8 \\ 2x - 6y = -9 \end{cases}$

Multiply the first equation by 3 and add.
$$9x + 6y = -24$$
$$\underline{2x - 6y = -9}$$
$$11x \quad\quad = -33$$
$$x = -3$$

Replace x with -3 in the first equation.
$$3(-3) + 2y = -8$$
$$-9 + 2y = -8$$
$$2y = 1$$
$$y = \frac{1}{2}$$

The solution to the system is $\left(-3, \dfrac{1}{2}\right)$.

35. $2x + y = 5$

x	y
0	5
$\frac{5}{2}$	0

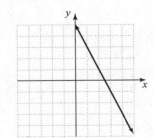

36. $\begin{cases} x = -3y + 3 \\ 2x + 9y = 5 \end{cases}$

Replace x with $-3y + 3$ in the second equation.
$2(-3y + 3) + 9y = 5$
$\qquad -6y + 6 + 9y = 5$
$\qquad\qquad 3y + 6 = 5$
$\qquad\qquad\quad 3y = -1$
$\qquad\qquad\quad\ y = -\dfrac{1}{3}$

Replace y with $-\dfrac{1}{3}$ in the first equation.

$x = -3\left(-\dfrac{1}{3}\right) + 3 = 1 + 3 = 4$

The solution to the system is $\left(4, -\dfrac{1}{3}\right)$.

37.

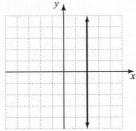

38. a. $(-5)^2 = (-5)(-5) = 25$

   b. $-5^2 = -(5)(5) = -25$

   c. $2 \cdot 5^2 = 2 \cdot 5 \cdot 5 = 50$

39. $x = 5$ is a vertical line and the slope is undefined.

40. $\dfrac{(z^2)^3 \cdot z^7}{z^9} = \dfrac{z^6 \cdot z^7}{z^9} = z^{6+7-9} = z^4$

41. $(2x^3 + 8x^2 - 6x) - (2x^3 - x^2 + 1)$
$= 2x^3 + 8x^2 - 6x - 2x^3 + x^2 - 1$
$= 2x^3 - 2x^3 + 8x^2 + x^2 - 6x - 1$
$= 9x^2 - 6x - 1$

42. $(5y^2 - 6) - (y^2 + 2) = 5y^2 - 6 - y^2 - 2 = 4y^2 - 8$

43. $(2x^2)(-3x^5) = (2 \cdot -3)(x^2 \cdot x^5) = -6x^{2+5} = -6x^7$

44. $-x^2$

   a. $-(2)^2 = -4$

   b. $-(-2)^2 = -4$

45. $(11x^3 - 12x^2 + x - 3) + (x^3 - 10x + 5)$
$= 11x^3 - 12x^2 + x - 3 + x^3 - 10x + 5$
$= 11x^3 + x^3 - 12x^2 + x - 10x - 3 + 5$
$= 12x^3 - 12x^2 - 9x + 2$

46. $(10x^2 - 3)(10x^2 + 3) = (10x^2)^2 - 3^2$
$\qquad\qquad\qquad\qquad\quad = 100x^4 - 9$

47. $(2x - y)^2 = (2x)^2 - 2(2x)(y) + (y)^2$
$\qquad\qquad\quad = 4x^2 - 4xy + y^2$

48. $(10x^2 + 3)^2 = (10x^2)^2 + 2(10x^2)(3) + 3^2$
$\qquad\qquad\qquad\ = 100x^4 + 60x^2 + 9$

49. $\dfrac{6m^2 + 2m}{2m} = \dfrac{6m^2}{2m} + \dfrac{2m}{2m} = 3m + 1$

50. a. $5^{-1} = \dfrac{1}{5}$

   b. $7^{-2} = \dfrac{1}{7^2} = \dfrac{1}{49}$

Chapter 6

Section 6.1 Practice

1. a. $36 = 2 \cdot 2 \cdot 3 \cdot 3 = 2^2 \cdot 3^2$
$42 = 2 \cdot 3 \cdot 7$
$\text{GCF} = 2 \cdot 3 = 6$

b. $35 = 5 \cdot 7$
$44 = 2 \cdot 2 \cdot 11$
$\text{GCF} = 1$

c. $12 = 2 \cdot 2 \cdot 3 = 2^2 \cdot 3$
$16 = 2 \cdot 2 \cdot 2 \cdot 2 = 2^4$
$40 = 2 \cdot 2 \cdot 2 \cdot 5 = 2^3 \cdot 5$
$\text{GCF} = 2^2 = 4$

2. a. The GCF is y^4 since 4 is the smallest exponent to which y is raised.

b. The GCF is x^1 or x, since 1 is the smallest exponent on x.

3. a. $5y^4 = 5 \cdot y^4$
$15y^2 = 3 \cdot 5 \cdot y^2$
$-20y^3 = -1 \cdot 2 \cdot 2 \cdot 5 \cdot y^3$
$\text{GCF} = 5 \cdot y^2 = 5y^2$

b. $4x^2 = 2 \cdot 2 \cdot x^2$
$x^3 = x^3$
$3x^8 = 3 \cdot x^8$
$\text{GCF} = x^2$

c. The GCF of a^4, a^3, and a^2 is a^2.
The GCF of b^2, b^5, and b^3 is b^2.
Thus, the GCF of $a^4 b^2$, $a^3 b^5$, and $a^2 b^3$ is $a^2 b^2$.

4. a. $4t + 12; \text{GCF} = 4$
$4t + 12 = 4 \cdot t + 4 \cdot 3 = 4(t + 3)$

b. $y^8 + y^4; \text{GCF} = y^4$
$y^8 + y^4 = y^4 \cdot y^4 + y^4 \cdot 1 = y^4(y^4 + 1)$

5. $-8b^6 + 16b^4 - 8b^2$
$= -8b^2(b^4) - 8b^2(-2b^2) - 8b^2(1)$
$= -8b^2(b^4 - 2b^2 + 1) \text{ or } 8b^2(-b^4 + 2b^2 - 1)$

6. $5x^4 - 20x = 5x(x^3 - 4)$

7. $\dfrac{5}{9}z^5 + \dfrac{1}{9}z^4 - \dfrac{2}{9}z^3 = \dfrac{1}{9}z^3(5z^2 + z - 2)$

8. $8a^2b^4 - 20a^3b^3 + 12ab^3 = 4ab^3(2ab - 5a^2 + 3)$

9. $8(y - 2) + x(y - 2) = (y - 2)(8 + x)$

10. $7xy^3(p + q) - (p + q) = 7xy^3(p + q) - 1(p + q)$
$= (p + q)(7xy^3 - 1)$

11. $xy + 3y + 4x + 12 = (xy + 3y) + (4x + 12)$
$= y(x + 3) + 4(x + 3)$
$= (x + 3)(y + 4)$
Check: $(x + 3)(y + 4) = xy + 3y + 4x + 12$

12. $40x^3 - 24x^2 + 15x - 9 = 8x^2(5x - 3) + 3(5x - 3)$
$= (5x - 3)(8x^2 + 3)$

13. $2xy + 3y^2 - 2x - 3y = (2xy + 3y^2) + (-2x - 3y)$
$= y(2x + 3y) - 1(2x + 3y)$
$= (2x + 3y)(y - 1)$

14. $7a^3 + 5a^2 + 7a + 5 = (7a^3 + 5a^2) + (7a + 5)$
$= a^2(7a + 5) + 1(7a + 5)$
$= (7a + 5)(a^2 + 1)$

15. $4xy + 15 - 12x - 5y = 4xy - 12x - 5y + 15$
$= (4xy - 12x) + (-5y + 15)$
$= 4x(y - 3) - 5(y - 3)$
$= (y - 3)(4x - 5)$

16. $9y - 18 + y^3 - 4y^2 = 9(y - 2) + y^2(y - 4)$
There is no common binomial factor, so it cannot be factored by grouping.

17. $3xy - 3ay - 6ax + 6a^2 = 3(xy - ay - 2ax + 2a^2)$
$= 3[y(x - a) - 2a(x - a)]$
$= 3(x - a)(y - 2a)$

Vocabulary, Readiness & Video Check 6.1

1. Since $5 \cdot 4 = 20$, the numbers 5 and 4 are called <u>factors</u> of 20.

2. The <u>greatest common factor</u> of a list of integers is the largest integer that is a factor of all the integers in the list.

3. The greatest common factor of a list of common variables raised to powers is the variable raised to the <u>least</u> exponent in the list.

4. The process of writing a polynomial as a product is called <u>factoring</u>.

5. $7(x + 3) + y(x + 3)$ is a sum, not a product. The statement is false.

6. $3x^3 + 6x + x^2 + 2 = 3x(x^2 + 2) + (x^2 + 2)$
   $$= (x^2 + 2)(3x + 1)$$
   The statement is false.

7. The GCF of a list of numbers is the largest number that is a factor of all numbers in the list.

8. The GCF of common variable factors is the variable raised to the smallest exponent.

9. When factoring out a GCF, the number of terms in the other factor should have the same number of terms as your original polynomial.

10. Look for a GCF other than 1 or −1; if you have a simplified four-term polynomial.

Exercise Set 6.1

1. $32 = 2 \cdot 2 \cdot 2 \cdot 2 \cdot 2 = 2^5$
   $36 = 2 \cdot 2 \cdot 3 \cdot 3 = 2^2 \cdot 3^2$
   GCF $= 2 \cdot 2 = 4$

3. $18 = 2 \cdot 3 \cdot 3 = 2 \cdot 3^2$
   $42 = 2 \cdot 3 \cdot 7$
   $84 = 2 \cdot 2 \cdot 3 \cdot 7 = 2^2 \cdot 3 \cdot 7$
   GCF $= 2 \cdot 3 = 6$

5. $24 = 2 \cdot 2 \cdot 2 \cdot 3 = 2^3 \cdot 3$
   $14 = 2 \cdot 7$
   $21 = 3 \cdot 7$
   GCF $= 1$

7. The GCF of y^2, y^4, and y^7 is y^2.

9. The GCF of z^7, z^9, and z^{11} is z^7.

11. The GCF of x^{10}, x, and x^3 is x.
    The GCF of y^2, y^2, and y^3 is y^2.
    Thus the GCF of $x^{10}y^2$, xy^2, and x^3y^3 is xy^2.

13. $14x = 2 \cdot 7 \cdot x$
    $21 = 3 \cdot 7$
    GCF $= 7$

15. $12y^4 = 2 \cdot 2 \cdot 3 \cdot y^4$
    $20y^3 = 2 \cdot 2 \cdot 5 \cdot y^3$
    GCF $= 2 \cdot 2 \cdot y^3 = 4y^3$

17. $-10x^2 = -1 \cdot 2 \cdot 5 \cdot x^2$
    $15x^3 = 3 \cdot 5 \cdot x^3$
    GCF $= 5 \cdot x^2 = 5x^2$

19. $12x^3 = 2 \cdot 2 \cdot 3 \cdot x^3$
    $-6x^4 = -1 \cdot 2 \cdot 3 \cdot x^4$
    $3x^5 = 3 \cdot x^5$
    GCF $= 3 \cdot x^3 = 3x^3$

21. $-18x^2 y = -1 \cdot 2 \cdot 3 \cdot 3 \cdot x^2 \cdot y$
    $9x^3 y^3 = 3 \cdot 3 \cdot x^3 \cdot y^3$
    $36x^3 y = 2 \cdot 2 \cdot 3 \cdot 3 \cdot x^3 \cdot y$
    GCF $= 3 \cdot 3 \cdot x^2 \cdot y = 9x^2 y$

23. $20a^6 b^2 c^8 = 2 \cdot 2 \cdot 5 \cdot a^6 \cdot b^2 \cdot c^8$
    $50a^7 b = 2 \cdot 5 \cdot 5 \cdot a^7 \cdot b$
    GCF $= 2 \cdot 5 \cdot a^6 \cdot b = 10a^6 b$

25. $3a + 6 = 3(a + 2)$

27. $30x - 15 = 15(2x - 1)$

29. $x^3 + 5x^2 = x^2(x + 5)$

31. $6y^4 + 2y^3 = 2y^3(3y + 1)$

33. $4x - 8y + 4 = 4(x - 2y + 1)$

35. $6x^3 - 9x^2 + 12x = 3x(2x^2 - 3x + 4)$

Copyright © 2013 Pearson Education, Inc.

37. $a^7b^6 - a^3b^2 + a^2b^5 - a^2b^2$
$= a^2b^2(a^5b^4 - a + b^3 - 1)$

39. $8x^5 + 16x^4 - 20x^3 + 12 = 4(2x^5 + 4x^4 - 5x^3 + 3)$

41. $\dfrac{1}{3}x^4 + \dfrac{2}{3}x^3 - \dfrac{4}{3}x^5 + \dfrac{1}{3}x$
$= \dfrac{1}{3}x(x^3 + 2x^2 - 4x^4 + 1)$

43. $y(x^2 + 2) + 3(x^2 + 2) = (x^2 + 2)(y + 3)$

45. $z(y + 4) - 3(y + 4) = (y + 4)(z - 3)$

47. $r(z^2 - 6) + (z^2 - 6) = r(z^2 - 6) + 1(z^2 - 6)$
$= (z^2 - 6)(r + 1)$

49. $-2x - 14 = -2(x + 7)$

51. $-2x^5 + x^7 = -x^5(2 - x^2)$

53. $-6a^4 + 9a^3 - 3a^2 = -3a^2(2a^2 - 3a + 1)$

55. $x^3 + 2x^2 + 5x + 10 = x^2(x + 2) + 5(x + 2)$
$= (x + 2)(x^2 + 5)$

57. $5x + 15 + xy + 3y = 5(x + 3) + y(x + 3)$
$= (x + 3)(5 + y)$

59. $6x^3 - 4x^2 + 15x - 10 = 2x^2(3x - 2) + 5(3x - 2)$
$= (3x - 2)(2x^2 + 5)$

61. $5m^3 + 6mn + 5m^2 + 6n$
$= m(5m^2 + 6n) + 1(5m^2 + 6n)$
$= (5m^2 + 6n)(m + 1)$

63. $2y - 8 + xy - 4x = 2(y - 4) + x(y - 4)$
$= (y - 4)(2 + x)$

65. $2x^3 - x^2 + 8x - 4 = x^2(2x - 1) + 4(2x - 1)$
$= (2x - 1)(x^2 + 4)$

67. $3x - 3 + x^3 - 4x^2 = 3(x - 1) + x^2(x - 4)$
The polynomial is not factorable by grouping.

69. $4x^2 - 8xy - 3x + 6y = 4x(x - 2y) - 3(x - 2y)$
$= (x - 2y)(4x - 3)$

71. $5q^2 - 4pq - 5q + 4p = q(5q - 4p) - 1(5q - 4p)$
$= (5q - 4p)(q - 1)$

73. $2x^4 + 5x^3 + 2x^2 + 5x = x(2x^3 + 5x^2 + 2x + 5)$
$= x[x^2(2x + 5) + 1(2x + 5)]$
$= x(2x + 5)(x^2 + 1)$

75. $12x^2y - 42x^2 - 4y + 14$
$= 2(6x^2y - 21x^2 - 2y + 7)$
$= 2[3x^2(2y - 7) - 1(2y - 7)]$
$= 2(2y - 7)(3x^2 - 1)$

77. $32xy^2 - 18x^2 = 2x(16y - 9x)$

79. $y(x + 2) - 3(x + 2) = (x + 2)(y - 3)$

81. $14x^3y + 7x^2y - 7xy = 7xy(2x^2 + x - 1)$

83. $28x^3 - 7x^2 + 12x - 3 = 7x^2(4x - 1) + 3(4x - 1)$
$= (4x - 1)(7x^2 + 3)$

85. $-40x^8y^6 - 16x^9y^5 = -8x^8y^5(5y + 2x)$

87. $6a^2 + 9ab^2 + 6ab + 9b^3$
$= 3(2a^2 + 3ab^2 + 2ab + 3b^3)$
$= 3[a(2a + 3b^2) + b(2a + 3b^2)]$
$= 3(2a + 3b^2)(a + b)$

89. $(x + 2)(x + 5) = x^2 + 5x + 2x + 10 = x^2 + 7x + 10$

91. $(b + 1)(b - 4) = b^2 - 4b + b - 4 = b^2 - 3b - 4$

	Two Numbers	Their Product	Their Sum
93.	2, 6	12	8
95.	−1, −8	8	−9
97.	−2, 5	−10	3

99. a. $8 \cdot a - 24 = 8a - 24$

    b. $8(a - 3) = 8a - 24$

    c. $4(2a - 12) = 8a - 48$

    d. $8 \cdot a - 2 \cdot 12 = 8a - 24$

The answer is b.

101. $(a + 6)(a + 2)$ is factored.

103. $5(2y + z) - b(2y + z)$ is not factored.

105. answers may vary

107. answers may vary

109. a. $-20x^2 + 300x + 120$
$= -20(6)^2 + 300(6) + 120$
$= -720 + 1800 + 120$
$= 1200$
There were 1200 million single digital downloads in 2010.

b. Let $x = 2014 - 2004 = 10$.
$-20x^2 + 300x + 120$
$= -20(10)^2 + 300(10) + 120$
$= -2000 + 3000 + 120$
$= 1120$
There were 1120 million single downloads in 2009.

c. $-20x^2 + 300x + 120$
$= -20 \cdot x^2 + (-20)(-15x) + (-20)(-6)$
$= -20(x^2 - 15x - 6)$

111. Subtract the area of the inner rectangle from the area of the outer rectangle.
Outer rectangle: $A = l \cdot w$
$$A = 12x \cdot x^2 = 12x^3$$
Inner rectangle: $A = l \cdot w$
$$A = 2 \cdot x = 2x$$
The area of the shaded region is given by the expression $12x^3 - 2x = 2x(6x^2 - 1)$.

113. Area $= 4n^4 - 24n = 4n(n^3 - 6)$
Since the width is $4n$ units, the length is $(n^3 - 6)$ units.

115. $x^{2n} + 2x^n + 3x^n + 6 = x^n(x^n + 2) + 3(x^n + 2)$
$= (x^n + 2)(x^n + 3)$

117. $3x^{2n} + 21x^n - 5x^n - 35$
$= 3x^n(x^n + 7) - 5(x^n + 7)$
$= (x^n + 7)(3x^n - 5)$

Section 6.2 Practice

1.

Positive Factors of 6	Sum of Factors
1, 6	7
2, 3	5

$x^2 + 5x + 6 = (x + 2)(x + 3)$

2.

Negative Factors of 70	Sum of Factors
$-1, -70$	-71
$-2, -35$	-37
$-5, -14$	-19
$-7, -10$	-17

$x^2 - 17x + 70 = (x - 7)(x - 10)$

3.

Factors of -14	Sum of Factors
$-1, 14$	13
$1, -14$	-13
$-2, 7$	5
$2, -7$	-5

$x^2 + 5x - 14 = (x - 2)(x + 7)$

4. The first term of each binomial is p. Then look for two numbers whose product is -63 and whose sum is -2.
$p^2 - 2p - 63 = (p - 9)(p + 7)$

5. The first term of each binomial is b. Then look for two numbers whose product is 1 and whose sum is 5. There are no such numbers.
$b^2 + 5b + 1$ is a prime polynomial.

6. The first term of each polynomial is x. Then look for two terms whose product is $12y^2$ and whose sum is $7y$.
$x^2 + 7xy + 12y^2 = (x + 3y)(x + 4y)$

7. The first term of each polynomial is x^2. Then look for two numbers whose product is 12 and whose sum is 13.
$x^4 + 13x^2 + 12 = (x^2 + 1)(x^2 + 12)$

8. $48 - 14x + x^2 = x^2 - 14x + 48$

The first term of each binomial is x. Then look for two factors whose product is 48 and whose sum is -14.

$x^2 - 14x + 48 = (x - 6)(x - 8)$

9. $4x^2 - 24x + 36 = 4(x^2 - 6x + 9)$

The first term of each binomial is x. Then look for two factors whose product is 9 and whose sum is -6.

$4(x^2 - 6x + 9) = 4(x - 3)(x - 3)$ or $4(x - 3)^2$

10. $3y^4 - 18y^3 - 21y^2 = 3y^2(y^2 - 6y - 7)$

The first term of each binomial is y. Then look for two factors whose product is -7 and whose sum is -6.

$3y^2(y^2 - 6y - 7) = 3y^2(y - 7)(y + 1)$

Vocabulary, Readiness & Video Check 6.2

1. The statement is true.

2. The statement is true.

3. Since $4x - 12 = 4(x - 3)$, the statement is false.

4. $(x + 2y)^2 = (x + 2y)(x + 2y) \neq (x + 2y)(x + y)$
The statement is false.

5. $x^2 + 9x + 20 = (x + 4)(x + \underline{5})$

6. $x^2 + 12x + 35 = (x + 5)(x + \underline{7})$

7. $x^2 - 7x + 12 = (x - 4)(x - \underline{3})$

8. $x^2 - 13x + 22 = (x - 2)(x - \underline{11})$

9. $x^2 + 4x + 4 = (x + 2)(x + \underline{2})$

10. $x^2 + 10x + 24 = (x + 6)(x + \underline{4})$

11. 15 is positive, so its factors would have to either be both positive or both negative. Since the factors need to sum to -8, both factors must be negative.

12. Since the sum of the factors is 3, the factors are -2 and 5, $(-2 + 5 = 3)$. If you accidentally choose factors whose sum is -3, simply "switch" the signs of the factors.

Exercise Set 6.2

1. $x^2 + 7x + 6 = (x + 6)(x + 1)$

3. $y^2 - 10y + 9 = (y - 9)(y - 1)$

5. $x^2 - 6x + 9 = (x - 3)(x - 3)$ or $(x - 3)^2$

7. $x^2 - 3x - 18 = (x - 6)(x + 3)$

9. $x^2 + 3x - 70 = (x + 10)(x - 7)$

11. $x^2 + 5x + 2$ is a prime polynomial.

13. $x^2 + 8xy + 15y^2 = (x + 5y)(x + 3y)$

15. $a^4 - 2a^2 - 15 = (a^2 - 5)(a^2 + 3)$

17. $13 + 14m + m^2 = m^2 + 14m + 13 = (m + 13)(m + 1)$

19. $10t - 24 + t^2 = t^2 + 10t - 24 = (t - 2)(t + 12)$

21. $a^2 - 10ab + 16b^2 = (a - 2b)(a - 8b)$

23. $2z^2 + 20z + 32 = 2(z^2 + 10z + 16)$
$= 2(z + 8)(z + 2)$

25. $2x^3 - 18x^2 + 40x = 2x(x^2 - 9x + 20)$
$= 2x(x - 5)(x - 4)$

27. $x^2 - 3xy - 4y^2 = (x - 4y)(x + y)$

29. $x^2 + 15x + 36 = (x + 12)(x + 3)$

31. $x^2 - x - 2 = (x - 2)(x + 1)$

33. $r^2 - 16r + 48 = (r - 12)(r - 4)$

35. $x^2 + xy - 2y^2 = (x + 2y)(x - y)$

37. $3x^2 + 9x - 30 = 3(x^2 + 3x - 10) = 3(x + 5)(x - 2)$

39. $3x^2 - 60x + 108 = 3(x^2 - 20x + 36)$
　　　　　　　　　　　$= 3(x - 18)(x - 2)$

41. $x^2 - 18x - 144 = (x - 24)(x + 6)$

43. $r^2 - 3r + 6$ is a prime polynomial.

45. $x^2 - 8x + 15 = (x - 5)(x - 3)$

47. $6x^3 + 54x^2 + 120x = 6x(x^2 + 9x + 20)$
　　　　　　　　　　　　$= 6x(x + 4)(x + 5)$

49. $4x^2 y + 4xy - 12y = 4y(x^2 + x - 3)$

51. $x^2 - 4x - 21 = (x - 7)(x + 3)$

53. $x^2 + 7xy + 10y^2 = (x + 5y)(x + 2y)$

55. $64 + 24t + 2t^2 = 2t^2 + 24t + 64$
　　　　　　　　　　　$= 2(t^2 + 12t + 32)$
　　　　　　　　　　　$= 2(t + 8)(t + 4)$

57. $x^3 - 2x^2 - 24x = x(x^2 - 2x - 24)$
　　　　　　　　　　　$= x(x - 6)(x + 4)$

59. $2t^5 - 14t^4 + 24t^3 = 2t^3(t^2 - 7t + 12)$
　　　　　　　　　　　　$= 2t^3(t - 4)(t - 3)$

61. $5x^3 y - 25x^2 y^2 - 120xy^3 = 5xy(x^2 - 5xy - 24y^2)$
　　　　　　　　　　　　　　$= 5xy(x - 8y)(x + 3y)$

63. $162 - 45m + 3m^2 = 3m^2 - 45m + 162$
　　　　　　　　　　　$= 3(m^2 - 15m + 54)$
　　　　　　　　　　　$= 3(m - 9)(m - 6)$

65. $-x^2 + 12x - 11 = -1(x^2 - 12x + 11)$
　　　　　　　　　　　$= -1(x - 11)(x - 1)$

67. $\dfrac{1}{2}y^2 - \dfrac{9}{2}y - 11 = \dfrac{1}{2}(y^2 - 9y - 22)$
　　　　　　　　　　　$= \dfrac{1}{2}(y - 11)(y + 2)$

69. $x^3 y^2 + x^2 y - 20x = x(x^2 y^2 + xy - 20)$
　　　　　　　　　　　$= x(xy - 4)(xy + 5)$

71. $(2x + 1)(x + 5) = 2x^2 + 10x + x + 5$
　　　　　　　　　$= 2x^2 + 11x + 5$

73. $(5y - 4)(3y - 1) = 15y^2 - 5y - 12y + 4$
　　　　　　　　　$= 15y^2 - 17y + 4$

75. $(a + 3b)(9a - 4b) = 9a^2 - 4ab + 27ab - 12b^2$
　　　　　　　　　$= 9a^2 + 23ab - 12b^2$

77. $(x - 3)(x + 8) = x^2 + 8x - 3x - 3(8) = x^2 + 5x - 24$

79. Answers may vary

81. $P = 2l + 2w$
　　$l = x^2 + 10x$ and $w = 4x + 33$, so
　　$P = 2(x^2 + 10x) + 2(4x + 33)$
　　　$= 2x^2 + 20x + 8x + 66$
　　　$= 2x^2 + 28x + 66$
　　　$= 2(x^2 + 14x + 33)$
　　　$= 2(x + 11)(x + 3)$
　　The perimeter of the rectangle is given by the polynomial $2x^2 + 28x + 66$ which factors as $2(x + 11)(x + 3)$.

83. $-16t^2 + 64t + 80 = -16(t^2 - 4t - 5)$
　　　　　　　　　　　$= -16(t - 5)(t + 1)$

85. $x^2 + \dfrac{1}{2}x + \dfrac{1}{16} = \left(x + \dfrac{1}{4}\right)\left(x + \dfrac{1}{4}\right)$ or $\left(x + \dfrac{1}{4}\right)^2$

87. $z^2(x + 1) - 3z(x + 1) - 70(x + 1)$
　　$= (x + 1)(z^2 - 3z - 70)$
　　$= (x + 1)(z - 10)(z + 7)$

89. $x^{2n} + 8x^n - 20 = (x^n + 10)(x^n - 2)$

91. c must be the product of positive numbers that sum to 6.
　　$6 = 1 + 5; 1 \cdot 5 = 5$
　　$6 = 2 + 4; 2 \cdot 4 = 8$
　　$6 = 3 + 3; 3 \cdot 3 = 9$
　　$x^2 + 6x + c$ if factorable when c is 5, 8, or 9.

93. c must be the product of negative numbers that sum to -4.
　　$-4 = -1 + (-3); -1 \cdot -3 = 3$
　　$-4 = -2 + (-2); -2 \cdot -2 = 4$
　　$y^2 - 4y + c$ if factorable when c is 3 or 4.

95. b must be the sum of positive numbers whose product is 15.
$15 = 1 \cdot 15; \; 1 + 15 = 16$
$15 = 3 \cdot 5; \; 3 + 5 = 8$
$x^2 + bx + 15$ is factorable when b is 8 or 16.

97. b must be the positive sum of a positive number and a negative number whose product is -27.
$-27 = 27 \cdot -1; \; 27 + (-1) = 26$
$-27 = 9 \cdot -3; \; 9 + (-3) = 6$
$m^2 + bm - 27$ is factorable when b is 6 or 26.

Section 6.3 Practice

1. Factors of $2x^2$: $2x^2 = 2x \cdot x$
Factors of 15: $15 = 1 \cdot 15, \; 15 = 3 \cdot 5$
Try possible combinations.
Factored form: $2x^2 + 11x + 15 = (2x + 5)(x + 3)$

2. Factors of $15x^2$: $15x^2 = 15x \cdot x, \; 15x^2 = 5x \cdot 3x$
Factors of 8: $8 = -1 \cdot -8, \; 8 = -2 \cdot -4$
Try possible combinations.
Factored form: $15x^2 - 22x + 8 = (5x - 4)(3x - 2)$

3. Factors of $4x^2$: $4x^2 = 4x \cdot x, \; 4x^2 = 2x \cdot 2x$
Factors of -3: $-3 = -1 \cdot 3, \; -3 = 1 \cdot -3$
Try possible combinations.
Factored form: $4x^2 + 11x - 3 = (4x - 1)(x + 3)$

4. Factors of $21x^2$: $21x^2 = 21x \cdot x, \; 21x^2 = 3x \cdot 7x$
Factors of
$-2y^2$: $-2y^2 = -2y \cdot y, \; -2y^2 = 2y \cdot -y$
Try possible combinations.
Factored form:
$21x^2 + 11xy - 2y^2 = (7x - y)(3x + 2y)$

5. Factors of $2x^4$: $2x^4 = 2x^2 \cdot x^2$
Factors of -7: $-7 = -7 \cdot 1, \; -7 = 7 \cdot -1$
Try possible combinations.
$2x^4 - 5x^2 - 7 = (2x^2 - 7)(x^2 + 1)$

6. $3x^3 + 17x^2 + 10x = x(3x^2 + 17x + 10)$
Factors of $3x^2$: $3x^2 = 3x \cdot x$
Factors of 10: $10 = 1 \cdot 10, \; 10 = 2 \cdot 5$
Try possible combinations:
$3x^3 + 17x^2 + 10x = x(3x^2 + 17x + 10)$
$\qquad\qquad\qquad\quad = x(3x + 2)(x + 5)$

7. $-8x^2 + 2x + 3 = -1(8x^2 - 2x - 3)$
$\qquad\qquad\qquad = -1(4x - 3)(2x + 1)$

8. $x^2 = (x)^2$ and $49 = 7^2$
Is $2 \cdot x \cdot 7 = 14x$ the middle term? Yes.
$x^2 + 14x + 49 = (x + 7)^2$

9. $4x^2 = (2x)^2$ and $9y^2 = (3y)^2$
Is $2 \cdot 2x \cdot 3y = 12xy$ the middle term? No.
Try other possibilities.
$4x^2 + 20xy + 9y^2 = (2x + 9y)(2x + y)$

10. $36n^4 = (6n^2)^2$ and $1 = 1^2$
Is $2 \cdot 6n^2 \cdot 1 = 12n^2$ the middle term? Yes, the opposite of the middle term.
$36n^4 - 12n^2 + 1 = (6n^2 - 1)^2$

11. $12x^3 - 84x^2 + 147x = 3x(4x^2 - 28x + 49)$
$\qquad\qquad\qquad\qquad = 3x[(2x)^2 - 2 \cdot 2x \cdot 7 + 7^2]$
$\qquad\qquad\qquad\qquad = 3x(2x - 7)^2$

Vocabulary, Readiness & Video Check 6.3

1. A <u>perfect square trinomial</u> is a trinomial that is the square of a binomial.

2. The term $25y^2$ written as a square is $\underline{(5y)^2}$.

3. The expression $x^2 + 10xy + 25y^2$ is called a <u>perfect square trinomial</u>.

4. The factorization $(x + 5y)(x + 5y)$ may also be written as $\underline{(x + 5y)^2}$.

5. $2x^2 + 5x + 3$ factors as $(2x + 3)(x + 1)$, which is choice d.

6. $7x^2 + 9x + 2$ factors as $(7x + 2)(x + 1)$, which is choice b.

7. Consider the factors of the first and last terms and the signs of the trinomial. Continue to check by multiplying until you get the middle term of the trinomial.

8. If the GCF has been factored out, then neither binomial can contain a common factor other than 1 or −1. This helps limit your choice of factors for one or both binomials since you cannot choose factors that would give the terms in either binomial a common factor.

9. The first and last terms are squares, a^2 and b^2, and the middle term is $2 \cdot a \cdot b$ or $-2 \cdot a \cdot b$.

Exercise Set 6.3

1. $5x^2 + 22x + 8 = (5x + 2)(x + 4)$

3. $50x^2 + 15x - 2 = (5x + 2)(10x - 1)$

5. $25x^2 - 20x + 4 = (5x - 2)(5x - 2)$

7. $2x^2 + 13x + 15 = (2x + 3)(x + 5)$

9. $8y^2 - 17y + 9 = (y - 1)(8y - 9)$

11. $2x^2 - 9x - 5 = (2x + 1)(x - 5)$

13. $20r^2 + 27r - 8 = (4r - 1)(5r + 8)$

15. $10x^2 + 31x + 3 = (10x + 1)(x + 3)$

17. $2m^2 + 17m + 10$ is prime.

19. $6x^2 - 13xy + 5y^2 = (3x - 5y)(2x - y)$

21. $15m^2 - 16m - 15 = (3m - 5)(5m + 3)$

23. $12x^3 + 11x^2 + 2x = x(12x^2 + 11x + 2)$
$= x(3x + 2)(4x + 1)$

25. $21b^2 - 48b - 45 = 3(7b^2 - 16b - 15)$
$= 3(7b + 5)(b - 3)$

27. $7z + 12z^2 - 12 = 12z^2 + 7z - 12 = (3z + 4)(4z - 3)$

29. $6x^2 y^2 - 2xy^2 - 60y^2 = 2y^2(3x^2 - x - 30)$
$= 2y^2(3x - 10)(x + 3)$

31. $4x^2 - 8x - 21 = (2x - 7)(2x + 3)$

33. $-x^2 + 2x + 24 = -1(x^2 - 2x - 24)$
$= -1(x - 6)(x + 4)$

35. $4x^3 - 9x^2 - 9x = x(4x^2 - 9x - 9)$
$= x(4x + 3)(x - 3)$

37. $24x^2 - 58x + 9 = (4x - 9)(6x - 1)$

39. $x^2 + 22x + 121 = x^2 + 2 \cdot x \cdot 11 + 11^2 = (x + 11)^2$

41. $x^2 - 16x + 64 = x^2 - 2 \cdot x \cdot 8 + 8^2 = (x - 8)^2$

43. $16a^2 - 24a + 9 = (4a)^2 - 2 \cdot 4a \cdot 3 + 3^2 = (4a - 3)^2$

45. $x^4 + 4x^2 + 4 = (x^2)^2 + 2 \cdot x^2 \cdot 2 + 2^2 = (x^2 + 2)^2$

47. $2n^2 - 28n + 98 = 2(n^2 - 14n + 49)$
$= 2(n^2 - 2 \cdot n \cdot 7 + 7^2)$
$= 2(n - 7)^2$

49. $16y^2 + 40y + 25 = (4y)^2 + 2 \cdot 4y \cdot 5 + 5^2$
$= (4y + 5)^2$

51. $2x^2 - 7x - 99 = (2x + 11)(x - 9)$

53. $24x^2 + 41x + 12 = (8x + 3)(3x + 4)$

55. $3a^2 + 10ab + 3b^2 = (3a + b)(a + 3b)$

57. $-9x + 20 + x^2 = x^2 - 9x + 20 = (x - 4)(x - 5)$

59. $p^2 + 12pq + 36q^2 = p^2 + 2 \cdot p \cdot 6q + (6q)^2$
$= (p + 6q)^2$

61. $x^2 y^2 - 10xy + 25 = (xy)^2 - 2 \cdot xy \cdot 5 + 5^2$
$= (xy - 5)^2$

63. $40a^2 b + 9ab - 9b = b(40a^2 + 9a - 9)$
$= b(8a - 3)(5a + 3)$

65. $30x^3 + 38x^2 + 12x = 2x(15x^2 + 19x + 6)$
$= 2x(3x + 2)(5x + 3)$

67. $6y^3 - 8y^2 - 30y = 2y(3y^2 - 4y - 15)$
$= 2y(3y + 5)(y - 3)$

69. $10x^4 + 25x^3y - 15x^2y^2 = 5x^2(2x^2 + 5xy - 3y^2)$
$$= 5x^2(2x - y)(x + 3y)$$

71. $-14x^2 + 39x - 10 = -1(14x^2 - 39x + 10)$
$$= -1(2x - 5)(7x - 2)$$

73. $16p^4 - 40p^3 + 25p^2 = p^2(16p^2 - 40p + 25)$
$$= p^2[(4p)^2 - 2 \cdot 4p \cdot 5 + 5^2]$$
$$= p^2(4p - 5)^2$$

75. $x + 3x^2 - 2 = 3x^2 + x - 2 = (3x - 2)(x + 1)$

77. $8x^2 + 6xy - 27y^2 = (4x + 9y)(2x - 3y)$

79. $1 + 6x^2 + x^4 = x^4 + 6x^2 + 1$ is prime.

81. $9x^2 - 24xy + 16y^2 = (3x)^2 - 2 \cdot 3x \cdot 4y + (4y)^2$
$$= (3x - 4y)^2$$

83. $18x^2 - 9x - 14 = (6x - 7)(3x + 2)$

85. $-27t + 7t^2 - 4 = 7t^2 - 27t - 4 = (7t + 1)(t - 4)$

87. $49p^2 - 7p - 2 = (7p + 1)(7p - 2)$

89. $m^3 + 18m^2 + 81m = m(m^2 + 18m + 81)$
$$= m(m^2 + 2 \cdot m \cdot 9 + 9^2)$$
$$= m(m + 9)^2$$

91. $5x^2y^2 + 20xy + 1$ is prime.

93. $6a^5 + 37a^3b^2 + 6ab^4 = a(6a^4 + 37a^2b^2 + 6b^4)$
$$= a(6a^2 + b^2)(a^2 + 6b^2)$$

95. $(x - 2)(x + 2) = x^2 + 2x - 2x - 4 = x^2 - 4$

97. $(a + 3)(a^2 - 3a + 9)$
$$= a^3 - 3a^2 + 9a + 3a^2 - 9a + 27$$
$$= a^3 + 27$$

99. Look for the tallest bar. The age range is 25–34.

101. answers may vary

103. no

105. answers may vary

107. $P = (3x^2 + 1) + (6x + 4) + (x^2 + 15x)$
$$= 3x^2 + 1 + 6x + 4 + x^2 + 15x$$
$$= 4x^2 + 21x + 5$$
$$= (4x + 1)(x + 5)$$

109. $4x^2 + 2x + \dfrac{1}{4} = (2x)^2 + 2 \cdot 2x \cdot \dfrac{1}{2} + \left(\dfrac{1}{2}\right)^2$
$$= \left(2x + \dfrac{1}{2}\right)^2$$

111. $4x^2(y-1)^2 + 10x(y-1)^2 + 25(y-1)^2$
$$= (y-1)^2(4x^2 + 10x + 25)$$

113. $16 = 4^2$; $2 \cdot x \cdot 4 = 8x$; 8

115. $(a + b)^2 = a^2 + 2ab + b^2$

117. $b = 2$: $3x^2 + 2x - 5 = (3x + 5)(x - 1)$
$b = 14$: $3x^2 + 14x - 5 = (3x - 1)(x + 5)$

119. $c = 2$: $5x^2 + 7x + 2 = (5x + 2)(x + 1)$

121. $-12x^3y^2 + 3x^2y^2 + 15xy^2$
$$= -3xy^2(4x^2 - x - 5)$$
$$= -3xy^2(4x - 5)(x + 1)$$

123. $4x^2(y-1)^2 + 20x(y-1)^2 + 25(y-1)^2$
$$= (y-1)^2(4x^2 + 20x + 25)$$
$$= (y-1)^2[(2x)^2 + 2 \cdot 2x \cdot 5 + 5^2]$$
$$= (y-1)^2(2x + 5)^2$$

125. $3x^{2n} + 17x^n + 10 = (3x^n + 2)(x^n + 5)$

127. answers may vary

Section 6.4 Practice

1.

Factors of $ac = 60$	Sum of Factors
1, 60	61
2, 30	32
3, 20	23
4, 15	19
5, 12	17
6, 10	16

$\leftarrow$ correct sum $b = 61$.

$$5x^2 + 61x + 12 = 5x^2 + 1x + 60x + 12$$
$$= x(5x+1) + 12(5x+1)$$
$$= (5x+1)(x+12)$$

2.

Factors of $ac = 60$	Sum of Factors
−1, −60	−61
−2, −30	−32
−3, −20	−23
−4, −15	−19
−5, −12	−17
−6, −10	−60

$\leftarrow$ Correct sum $b = -19$

$$12x^2 - 19x + 5 = 12x^2 - 15x - 4x + 5$$
$$= 3x(4x-5) - 1(4x-5)$$
$$= (4x-5)(3x-1)$$

3. $30x^2 - 14x - 4 = 2(15x^2 - 7x - 2)$

Find two numbers whose product is $ac = 15(-2) = -30$ and whose sum is b, −7. The numbers are −10 and 3.

$$2(15x^2 - 7x - 2) = 2(15x^2 - 10x + 3x - 2)$$
$$= 2[5x(3x-2) + 1(3x-2)]$$
$$= 2(3x-2)(5x+1)$$

4. $40m^4 + 5m^3 - 35m^2 = 5m^2(8m^2 + m - 7)$

Find two numbers whose product is $ac = 8(-7) = -56$ and whose sum is b, 1. The numbers are 8 and −7.

$$5m^2(8m^2 + m - 7) = 5m^2(8m^2 + 8m - 7m - 7)$$
$$= 5m^2[8m(m+1) - 7(m+1)]$$
$$= 5m^2(m+1)(8m-7)$$

5. Find two numbers whose product is $ac = 16 \cdot 9 = 144$ and whose sum is b, 24. The numbers are 12 and 12.

$$16x^2 + 24x + 9 = 16x^2 + 12x + 12x + 9$$
$$= 4x(4x+3) + 3(4x+3)$$
$$= (4x+3)(4x+3)$$
$$= (4x+3)^2$$

Vocabulary, Readiness & Video Check 6.4

1. $a = 1, b = 6, c = 8$
$a \cdot c = 1 \cdot 8 = 8$
$4 \cdot 2 = 8$ and $4 + 2 = 6$; choice a.

2. $a = 1, b = 11, c = 24$
$a \cdot c = 1 \cdot 24 = 24$
$8 \cdot 3 = 24$ and $8 + 3 = 11$; choice c.

3. $a = 2, b = 13, c = 6$
$a \cdot c = 2 \cdot 6 = 12$
$12 \cdot 1 = 12$ and $12 + 1 = 13$; choice b.

4. $a = 4, b = 8, c = 3$
$a \cdot c = 4 \cdot 3 = 12$
$2 \cdot 6 = 12$ and $2 + 6 = 8$; choice d.

5. This gives us a four-term polynomial which may be factored by grouping.

Exercise Set 6.4

1. $x^2 + 3x + 2x + 6 = x(x+3) + 2(x+3)$
$$= (x+3)(x+2)$$

3. $y^2 + 8y - 2y - 16 = y(y+8) - 2(y+8)$
$$= (y+8)(y-2)$$

5. $8x^2 - 5x - 24x + 15 = x(8x-5) - 3(8x-5)$
$$= (8x-5)(x-3)$$

7. $5x^4 - 3x^2 + 25x^2 - 15 = x^2(5x^2 - 3) + 5(5x^2 - 3)$
$$= (5x^2 - 3)(x^2 + 5)$$

9. a. $9 \cdot 2 = 18; 9 + 2 = 11; 9, 2$

b. $11x = 9x + 2x$

c. $6x^2 + 11x + 3 = 6x^2 + 9x + 2x + 3$
$$= 3x(2x + 3) + 1(2x + 3)$$
$$= (3x + 1)(2x + 3)$$

11. a. $-20 \cdot (-3) = 60; -20 + (-3) = -23; -20, -3$

b. $-23x = -20x - 3x$

c. $15x^2 - 23x + 4 = 15x^2 - 20x - 3x + 4$
$$= 5x(3x - 4) - 1(3x - 4)$$
$$= (3x - 4)(5x - 1)$$

13. $ac = 21 \cdot 2 = 42; b = 17;$ two numbers: 14, 3
$$21y^2 + 17y + 2 = 21y^2 + 14y + 3y + 2$$
$$= 7y(3y + 2) + 1(3y + 2)$$
$$= (3y + 2)(7y + 1)$$

15. $ac = 7 \cdot (-11) = -77; b = -4;$
two numbers: $-11, 7$
$$7x^2 - 4x - 11 = 7x^2 - 11x + 7x - 11$$
$$= x(7x - 11) + 1(7x - 11)$$
$$= (7x - 11)(x + 1)$$

17. $ac = 10 \cdot 2 = 20; b = -9;$ two numbers: $-4, -5$
$$10x^2 - 9x + 2 = 10x^2 - 4x - 5x + 2$$
$$= 2x(5x - 2) - 1(5x - 2)$$
$$= (5x - 2)(2x - 1)$$

19. $ac = 2 \cdot 5 = 10; b = -7;$ two numbers: $-5, -2$
$$2x^2 - 7x + 5 = 2x^2 - 5x - 2x + 5$$
$$= x(2x - 5) - 1(2x - 5)$$
$$= (2x - 5)(x - 1)$$

21. $12x + 4x^2 + 9 = 4x^2 + 12x + 9$
$ac = 4 \cdot 9 = 36; b = 12;$ two numbers: 6, 6
$$4x^2 + 12x + 9 = 4x^2 + 6x + 6x + 9$$
$$= 2x(2x + 3) + 3(2x + 3)$$
$$= (2x + 3)(2x + 3)$$
$$= (2x + 3)^2$$

23. $ac = 4 \cdot (-21) = -84; b = -8;$
two numbers: 6, -14
$$4x^2 - 8x - 21 = 4x^2 + 6x - 14x - 21$$
$$= 2x(2x + 3) - 7(2x + 3)$$
$$= (2x + 3)(2x - 7)$$

25. $ac = 10 \cdot 12 = 120; b = -23;$
two numbers: $-8, -15$
$$10x^2 - 23x + 12 = 10x^2 - 8x - 15x + 12$$
$$= 2x(5x - 4) - 3(5x - 4)$$
$$= (5x - 4)(2x - 3)$$

27. $2x^3 + 13x^2 + 15x = x(2x^2 + 13x + 15)$
$ac = 2 \cdot 15 = 30; b = 13;$ two numbers: 3, 10
$$x(2x^2 + 13x + 15) = x(2x^2 + 3x + 10x + 15)$$
$$= x[x(2x + 3) + 5(2x + 3)]$$
$$= x(2x + 3)(x + 5)$$

29. $16y^2 - 34y + 18 = 2(8y^2 - 17y + 9)$
$ac = 8(9) = 72; b = -17;$ two numbers: $-9, -8$
$$2(8y^2 - 17y + 9) = 2(8y^2 - 9y - 8y + 9)$$
$$= 2[y(8y - 9) - 1(8y - 9)]$$
$$= 2(8y - 9)(y - 1)$$

31. $-13x + 6 + 6x^2 = 6x^2 - 13x + 6$
$ac = 6 \cdot 6 = 36; b = -13;$ two numbers: $-9, -4$
$$6x^2 - 13x + 6 = 6x^2 - 9x - 4x + 6$$
$$= 3x(2x - 3) - 2(2x - 3)$$
$$= (2x - 3)(3x - 2)$$

33. $54a^2 - 9a - 30 = 3(18a^2 - 3a - 10)$
$ac = 18(-10) = -180; b = -3;$
two numbers: 12, -15
$$3(18a^2 - 3a - 10) = 3(18a^2 + 12a - 15a - 10)$$
$$= 3[6a(3a + 2) - 5(3a + 2)]$$
$$= 3(3a + 2)(6a - 5)$$

35. $20a^3 + 37a^2 + 8a = a(20a^2 + 37a + 8)$
$ac = 20(8) = 160; b = 37;$ two numbers: 5, 32
$$a(20a^2 + 37a + 8) = a(20a^2 + 5a + 32a + 8)$$
$$= a[5a(4a + 1) + 8(4a + 1)]$$
$$= a(4a + 1)(5a + 8)$$

37. $12x^3 - 27x^2 - 27x = 3x(4x^2 - 9x - 9)$
$ac = 4(-9) = -36; b = -9;$ two numbers: 3, -12
$$3x(4x^2 - 9x - 9) = 3x(4x^2 + 3x - 12x - 9)$$
$$= 3x[x(4x + 3) - 3(4x + 3)]$$
$$= 3x(4x + 3)(x - 3)$$

39. $3x^2y + 4xy^2 + y^3 = y(3x^2 + 4xy + y^2)$

$ac = 3 \cdot 1 = 3;\ b = 4;$ two numbers: 1, 3

$$y(3x^2 + 4xy + y^2) = y(3x^2 + xy + 3xy + y^2)$$
$$= y[x(3x + y) + y(3x + y)]$$
$$= y(3x + y)(x + y)$$

41. $ac = 20 \cdot 1 = 20;\ b = 7;$ there are no two numbers.

$20z^2 + 7z + 1$ is prime.

43. $5x^2 + 50xy + 125y^2 = 5(x^2 + 10xy + 25y^2)$

$ac = 1 \cdot 25 = 25;\ b = 10;$ two numbers: 5, 5

$$5(x^2 + 10xy + 25y^2) = 5(x^2 + 5xy + 5xy + 25y^2)$$
$$= 5[x(x + 5y) + 5y(x + 5y)]$$
$$= 5(x + 5y)(x + 5y)$$
$$= 5(x + 5y)^2$$

45. $24a^2 - 6ab - 30b^2 = 6(4a^2 - ab - 5b^2)$

$ac = 4 \cdot (-5) = -20;\ b = -1;$ two numbers: 4, −5

$$6(4a^2 - ab - 5b^2) = 6(4a^2 + 4ab - 5ab - 5b^2)$$
$$= 6[4a(a + b) - 5b(a + b)]$$
$$= 6(a + b)(4a - 5b)$$

47. $15p^4 + 31p^3q + 2p^2q^2 = p^2(15p^2 + 31pq + 2q^2)$

$ac = 15(2) = 30;\ b = 31;$ two numbers: 1, 30

$$p^2(15p^2 + 31pq + 2q^2)$$
$$= p^2(15p^2 + pq + 30pq + 2q^2)$$
$$= p^2[p(15p + q) + 2q(15p + q)]$$
$$= p^2(15p + q)(p + 2q)$$

49. $162a^4 - 72a^2 + 8 = 2(81a^4 - 36a^2 + 4)$

$ac = 81 \cdot 4 = 324;\ b = -36;$

two numbers: −18, −18

$$2(81a^4 - 36a^2 + 4)$$
$$= 2(81a^4 - 18a^2 - 18a^2 + 4)$$
$$= 2[9a^2(9a^2 - 2) - 2(9a^2 - 2)]$$
$$= 2(9a^2 - 2)(9a^2 - 2)$$
$$= 2(9a^2 - 2)^2$$

51. $35 + 12x + x^2 = x^2 + 12x + 35$

$ac = 1 \cdot 35 = 35;\ b = 12;$ two numbers: 5, 7

$$x^2 + 12x + 35 = x^2 + 5x + 7x + 35$$
$$= x(x + 5) + 7(x + 5)$$
$$= (x + 5)(x + 7)$$

53. $6 - 11x + 5x^2 = 5x^2 - 11x + 6$

$ac = 5 \cdot 6 = 30;\ b = -11;$ two numbers: −6, −5

$$5x^2 - 11x + 6 = 5x^2 - 6x - 5x + 6$$
$$= x(5x - 6) - 1(5x - 6)$$
$$= (5x - 6)(x - 1)$$

55. $(x - 2)(x + 2) = x^2 - 2^2 = x^2 - 4$

57. $(y + 4)(y + 4) = y^2 + 2 \cdot y \cdot 4 + 4^2 = y^2 + 8y + 16$

59. $(9z + 5)(9z - 5) = (9z)^2 - 5^2 = 81z^2 - 25$

61. $(x - 3)(x^2 + 3x + 9) = x^3 - 3^3 = x^3 - 27$

63. $5(2x^2 + 9x + 9) = 10x^2 + 45x + 45$

$ac = 2 \cdot 9 = 18;\ b = 9;$ two numbers: 3, 6

$$5(2x^2 + 9x + 9) = 5(2x^2 + 3x + 6x + 9)$$
$$= 5[x(2x + 3) + 3(2x + 3)]$$
$$= 15(2x + 3)(x + 3)$$

65. $x^{2n} + 2x^n + 3x^n + 6 = x^n(x^n + 2) + 3(x^n + 2)$
$$= (x^n + 2)(x^n + 3)$$

67. $ac = 3 \cdot (-35) = -105;\ b = 16;$

two numbers: −5, 21

$$3x^{2n} + 16x^n - 35 = 3x^{2n} - 5x^n + 21x^n - 35$$
$$= x^n(3x^n - 5) + 7(3x^n - 5)$$
$$= (3x^n - 5)(x^n + 7)$$

69. answers may vary

Section 6.5 Practice

1. $x^2 - 81 = x^2 - 9^2 = (x + 9)(x - 9)$

2. a. $9x^2 - 1 = (3x)^2 - 1^2 = (3x + 1)(3x - 1)$

b. $36a^2 - 49b^2 = (6a)^2 - (7b)^2$
$$= (6a + 7b)(6a - 7b)$$

c. $p^2 - \dfrac{25}{36} = p^2 - \left(\dfrac{5}{6}\right)^2 = \left(p + \dfrac{5}{6}\right)\left(p - \dfrac{5}{6}\right)$

3. $p^4 - q^{10} = (p^2)^2 - (q^5)^2 = (p^2 + q^5)(p^2 - q^5)$

4. a. $z^4 - 81 = (z^2)^2 - 9^2$

$\qquad\qquad = (z^2 + 9)(z^2 - 9)$

$\qquad\qquad = (z^2 + 9)(z + 3)(z - 3)$

b. $m^2 + 49$ is a prime polynomial.

5. $36y^3 - 25y = y(36y^2 - 25)$

$\qquad\qquad = y[(6y)^2 - 5^2]$

$\qquad\qquad = y(6y + 5)(6y - 5)$

6. $80y^4 - 5 = 5(16y^2 - 1)$

$\qquad\qquad = 5[(4y)^2 - 1^2]$

$\qquad\qquad = 5(4y + 1)(4y - 1)$

7. $-9x^2 + 100 = -1(9x^2 - 100)$

$\qquad\qquad = -1[(3x)^2 - 10^2]$

$\qquad\qquad = -1(3x + 10)(3x - 10)$

or $-9x^2 + 100 = 100 - 9x^2$

$\qquad\qquad = 10^2 - (3x)^2$

$\qquad\qquad = (10 + 3x)(10 - 3x)$

8. $x^3 + 64 = x^3 + 4^3$

$\qquad\qquad = (x + 4)(x^2 - x \cdot 4 + 4^2)$

$\qquad\qquad = (x + 4)(x^2 - 4x + 16)$

9. $x^3 - 125 = x^3 - 5^3$

$\qquad\qquad = (x - 5)(x^2 + x \cdot 5 + 5^2)$

$\qquad\qquad = (x - 5)(x^2 + 5x + 25)$

10. $27y^3 + 1 = (3y)^3 + 1^3$

$\qquad\qquad = (3y + 1)[(3y)^2 - 3y \cdot 1 + 1^2]$

$\qquad\qquad = (3y + 1)(9y^2 - 3y + 1)$

11. $32x^3 - 500y^3$

$\quad = 4(8x^3 - 125y^3)$

$\quad = 4[(2x)^3 - (5y)^3]$

$\quad = 4(2x - 5y)[(2x)^2 + 2x \cdot 5y + (5y)^2]$

$\quad = 4(2x - 5y)(4x^2 + 10xy + 25y^2)$

Graphing Calculator Explorations

x	$x^2 - 2x + 1$	$x^2 - 2x - 1$	$(x-1)^2$
5	16	14	16
−3	16	14	16
2.7	2.89	0.89	2.89
−12.1	171.61	169.61	171.61
0	1	−1	1

Vocabulary, Readiness & Video Check 6.5

1. The expression $x^3 - 27$ is called a <u>difference of two cubes</u>.

2. The expression $x^2 - 49$ is called a <u>difference of two squares</u>.

3. The expression $z^3 + 1$ is called a <u>sum of two cubes</u>.

4. The binomial $y^2 + 9$ is prime. The statement is false.

5. $49x^2 = (7x)^2$

6. $25y^4 = (5y^2)^2$

7. $8y^3 = (2y)^3$

8. $x^6 = (x^2)^3$

9. In order to recognize the binomial as a difference of squares and also to identify the terms to use in the special factoring formula

10. A prime polynomial is one that can't be factored further.

11. First rewrite the original binomial with terms written as cubes. Answers will then vary depending on your interpretation.

Exercise Set 6.5

1. $x^2 - 4 = x^2 - 2^2 = (x + 2)(x - 2)$

3. $81p^2 - 1 = (9p)^2 - 1^2 = (9p + 1)(9p - 1)$

5. $25y^2 - 9 = (5y)^2 - 3^2 = (5y + 3)(5y - 3)$

7. $121m^2 - 100n^2 = (11m)^2 - (10n)^2$
$$= (11m + 10n)(11m - 10n)$$

9. $x^2y^2 - 1 = (xy)^2 - 1^2 = (xy + 1)(xy - 1)$

11. $x^2 - \dfrac{1}{4} = x^2 - \left(\dfrac{1}{2}\right)^2 = \left(x + \dfrac{1}{2}\right)\left(x - \dfrac{1}{2}\right)$

13. $-4r^2 + 1 = -1(4r^2 - 1)$
$$= -1[(2r)^2 - 1^2]$$
$$= -1(2r + 1)(2r - 1)$$

15. $16r^2 + 1$ is the sum of two squares, $(4r)^2 + 1^2$, not the difference of two squares. $16r^2 + 1$ is a prime polynomial.

17. $-36 + x^2 = -1(36 - x^2)$
$$= -1(6^2 - x^2)$$
$$= -1(6 + x)(6 - x) \text{ or } (-6 + x)(6 + x)$$

19. $m^4 - 1 = (m^2)^2 - 1^2$
$$= (m^2 + 1)(m^2 - 1)$$
$$= (m^2 + 1)(m + 1)(m - 1)$$

21. $m^4 - n^{18} = (m^2)^2 - (n^9)^2$
$$= (m^2 + n^9)(m^2 - n^9)$$

23. $x^3 + 125 = x^3 + 5^3$
$$= (x + 5)(x^2 - x \cdot 5 + 5^2)$$
$$= (x + 5)(x^2 - 5x + 25)$$

25. $8a^3 - 1 = (2a)^3 - 1^3$
$$= (2a - 1)[(2a)^2 + 2a \cdot 1 + 1^2]$$
$$= (2a - 1)(4a^2 + 2a + 1)$$

27. $m^3 + 27n^3 = m^3 + (3n)^3$
$$= (m + 3n)[m^2 - m \cdot 3n + (3n)^2]$$
$$= (m + 3n)(m^2 - 3mn + 9n^2)$$

29. $5k^3 + 40 = 5(k^3 + 8)$
$$= 5(k^3 + 2^3)$$
$$= 5(k + 2)[k^2 - k \cdot 2 + 2^2]$$
$$= 5(k + 2)(k^2 - 2k + 4)$$

31. $x^3y^3 - 64 = (xy)^3 - 4^3$
$$= (xy - 4)[(xy)^2 + xy \cdot 4 + 4^2]$$
$$= (xy - 4)(x^2y^2 + 4xy + 16)$$

33. $250r^3 - 128t^3 = 2(125r^3 - 64t^3)$
$$= 2[(5r)^3 - (4t)^3]$$
$$= 2(5r - 4t)[(5r)^2 + 5r \cdot 4t + (4t)^2]$$
$$= 2(5r - 4t)(25r^2 + 20rt + 16t^2)$$

35. $r^2 - 64 = r^2 - 8^2 = (r + 8)(r - 8)$

37. $x^2 - 169y^2 = x^2 - (13y)^2 = (x + 13y)(x - 13y)$

39. $27 - t^3 = 3^3 - t^3$
$$= (3 - t)(3^2 + 3 \cdot t + t^2)$$
$$= (3 - t)(9 + 3t + t^2)$$

41. $18r^2 - 8 = 2(9r^2 - 4)$
$$= 2[(3r)^2 - 2^2]$$
$$= 2(3r + 2)(3r - 2)$$

43. $9xy^2 - 4x = x(9y^2 - 4)$
$$= x[(3y)^2 - 2^2]$$
$$= x(3y + 2)(3y - 2)$$

45. $8m^3 + 64 = 8(m^3 + 8)$
$$= 8(m^3 + 2^3)$$
$$= 8(m + 2)(m^2 - m \cdot 2 + 2^2)$$
$$= 8(m + 2)(m^2 - 2m + 4)$$

47. $xy^3 - 9xyz^2 = xy(y^2 - 9z^2)$
$$= xy[y^2 - (3z)^2]$$
$$= xy(y + 3z)(y - 3z)$$

49. $36x^2 - 64y^2 = 4(9x^2 - 16y^2)$
$$= 4[(3x)^2 - (4y)^2]$$
$$= 4(3x + 4y)(3x - 4y)$$

51. $144 - 81x^2 = 9(16 - 9x^2)$
$$= 9[4^2 - (3x)^2]$$
$$= 9(4 + 3x)(4 - 3x)$$

53. $x^3 y^3 - z^6 = (xy)^3 - (z^2)^3$
$$= (xy - z^2)[(xy)^2 + xy \cdot z^2 + (z^2)^2]$$
$$= (xy - z^2)(x^2 y^2 + xyz^2 + z^4)$$

55. $49 - \dfrac{9}{25}m^2 = 7^2 - \left(\dfrac{3}{5}m\right)^2 = \left(7 + \dfrac{3}{5}m\right)\left(7 - \dfrac{3}{5}m\right)$

57. $t^3 + 343 = t^3 + 7^3$
$$= (t + 7)(t^2 - t \cdot 7 + 7^2)$$
$$= (t + 7)(t^2 - 7t + 49)$$

59. $n^3 - 49n = n(n^2 + 49)$

61. $x^6 - 81x^2 = x^2(x^4 - 81)$
$$= x^2[(x^2)^2 - 9^2]$$
$$= x^2(x^2 + 9)(x^2 - 9)$$
$$= x^2(x^2 + 9)(x + 3)(x - 3)$$

63. $64p^3 q - 81pq^3 = pq(64p^2 - 81q^2)$
$$= pq[(8p)^2 - (9q)^2]$$
$$= pq(8p + 9q)(8p - 9q)$$

65. $27x^2 y^3 + xy^2 = xy^2(27xy + 1)$

67. $125a^4 - 64ab^3$
$$= a(125a^3 - 64b^3)$$
$$= a[(5a)^3 - (4b)^3]$$
$$= a(5a - 4b)[(5a)^2 + 5a \cdot 4b + (4b)^2]$$
$$= a(5a - 4b)(25a^2 + 20ab + 16b^2)$$

69. $16x^4 - 64x^2 = 16x^2(x^2 - 4)$
$$= 16x^2(x^2 - 2^2)$$
$$= 16x^2(x + 2)(x - 2)$$

71. $x - 6 = 0$
$$x - 6 + 6 = 0 + 6$$
$$x = 6$$

73. $2m + 4 = 0$
$$2m + 4 - 4 = 0 - 4$$
$$2m = -4$$
$$\dfrac{2m}{2} = \dfrac{-4}{2}$$
$$m = -2$$

75. $5z - 1 = 0$
$$5z - 1 + 1 = 0 + 1$$
$$5z = 1$$
$$\dfrac{5z}{5} = \dfrac{1}{5}$$
$$z = \dfrac{1}{5}$$

77. $(x + 2)^2 - y^2 = (x + 2 + y)(x + 2 - y)$

79. $a^2(b - 4) - 16(b - 4) = (b - 4)(a^2 - 16)$
$$= (b - 4)(a^2 - 4^2)$$
$$= (b - 4)(a + 4)(a - 4)$$

81. $(x^2 + 6x + 9) - 4y^2 = (x + 3)^2 - 4y^2$
$$= (x + 3)^2 - (2y)^2$$
$$= [(x + 3) + 2y][(x + 3) - 2y]$$
$$= (x + 3 + 2y)(x + 3 - 2y)$$

83. $x^{2n} - 100 = (x^n)^2 - 10^2 = (x^n + 10)(x^n - 10)$

85. $x + 6$ since
$$(x + 6)(x - 6) = x^2 - 6x + 6x - 36$$
$$= x^2 - 36$$
$$= x^2 - 6^2$$

87. answers may vary

89. a. $2704 - 16t^2 = 2704 - 16(3)^2$
$$= 2704 - 16 \cdot 9$$
$$= 2704 - 144$$
$$= 2560$$
After 3 seconds, the filter is 2560 feet above the river.

b. $2704 - 16t^2 = 2704 - 16(7)^2$
$$= 2704 - 16 \cdot 49$$
$$= 2704 - 784$$
$$= 1920$$
After 7 seconds, the filter is 1920 feet above the river.

c. The filter lands in the river when its height is 0 feet.
$$2704 - 16t^2 = 0$$
$$(52 + 4t)(52 - 4t) = 0$$
$$52 + 4t = 0 \quad \text{or} \quad 52 - 4t = 0$$
$$4t = -52 \qquad\qquad -4t = -52$$
$$t = -13 \qquad\qquad t = 13$$
Discard $t = -13$ since time cannot be negative. The filter lands in the river after 13 seconds.

d. $2704 - 16t^2 = 16(169 - t^2)$
$$= 16[(13)^2 - t^2]$$
$$= 16(13 - t)(13 + t)$$

91. a. Let $t = 3$.
$$1600 - 16t^2 = 1600 - 16(3)^2 = 1456$$
After 3 seconds the height is 1456 feet.

b. Let $t = 7$.
$$1600 - 16t^2 = 1600 - 16(7)^2 = 816$$
After 7 seconds the height is 816 feet.

c. When it hits the ground, the height is 0.
Let $0 = 1600 - 16t^2$.
$$16t^2 = 1600$$
$$t^2 = 100$$
$$t = \sqrt{100}$$
$$t = 10$$
Thus, it will hit the ground after 10 seconds.

d. $1600 - 16t^2 = 16(100 - t^2)$
$$= 16(10^2 - t^2)$$
$$= 16(10 + t)(10 - t)$$

Integrated Review Practice

1. $6x^2 - 11x + 3$
$ac = 6 \cdot 3 = 18$; $b = -11$; two numbers: $-2, -9$
$6x^2 - 11x + 3 = 6x^2 - 2x - 9x + 3$
$$= 2x(3x - 1) - 3(3x - 1)$$
$$= (3x - 1)(2x - 3)$$

2. $3x^3 + x^2 - 12x - 4 = (3x^3 + x^2) + (-12x - 4)$
$$= x^2(3x + 1) - 4(3x + 1)$$
$$= (3x + 1)(x^2 - 4)$$
$$= (3x + 1)(x + 2)(x - 2)$$

3. $27x^2 - 3y^2 = 3(9x^2 - y^2)$
$$= 3[(3x)^2 - y^2]$$
$$= 3(3x + y)(3x - y)$$

4. $8a^3 + b^3 = (2a)^3 + b^3$
$$= (2a + b)[(2a)^2 - 2a \cdot b + b^2]$$
$$= (2a + b)(4a^2 - 2ab + b^2)$$

5. $60x^3y^2 - 66x^2y^2 - 36xy^2$
$$= 6xy^2(10x^2 - 11x - 6)$$
$$= 6xy^2(5x + 2)(2x - 3)$$

Integrated Review

1. $x^2 + 2xy + y^2 = (x + y)(x + y) = (x + y)^2$

2. $x^2 - 2xy + y^2 = (x - y)(x - y) = (x - y)^2$

3. $a^2 + 11a - 12 = (a + 12)(a - 1)$

4. $a^2 - 11a + 10 = (a - 10)(a - 1)$

5. $a^2 - a - 6 = (a - 3)(a + 2)$

6. $a^2 - 2a + 1 = (a - 1)(a - 1) = (a - 1)^2$

7. $x^2 + 2x + 1 = (x + 1)(x + 1) = (x + 1)^2$

8. $x^2 + x - 2 = (x + 2)(x - 1)$

9. $x^2 + 4x + 3 = (x + 3)(x + 1)$

10. $x^2 + x - 6 = (x + 3)(x - 2)$

11. $x^2 + 7x + 12 = (x + 4)(x + 3)$

12. $x^2 + x - 12 = (x + 4)(x - 3)$

13. $x^2 + 3x - 4 = (x + 4)(x - 1)$

14. $x^2 - 7x + 10 = (x - 5)(x - 2)$

15. $x^2 + 2x - 15 = (x + 5)(x - 3)$

16. $x^2 + 11x + 30 = (x + 6)(x + 5)$

17. $x^2 - x - 30 = (x-6)(x+5)$

18. $x^2 + 11x + 24 = (x+8)(x+3)$

19. $2x^2 - 98 = 2(x^2 - 49)$
$\qquad\qquad = 2(x^2 - 7^2)$
$\qquad\qquad = 2(x+7)(x-7)$

20. $3x^2 - 75 = 3(x^2 - 25)$
$\qquad\qquad = 3(x^2 - 5^2)$
$\qquad\qquad = 3(x+5)(x-5)$

21. $x^2 + 3x + xy + 3y = x(x+3) + y(x+3)$
$\qquad\qquad\qquad\quad = (x+3)(x+y)$

22. $3y - 21 + xy - 7x = 3(y-7) + x(y-7)$
$\qquad\qquad\qquad\quad = (y-7)(3+x)$

23. $x^2 + 6x - 16 = (x+8)(x-2)$

24. $x^2 - 3x - 28 = (x-7)(x+4)$

25. $4x^3 + 20x^2 - 56x = 4x(x^2 + 5x - 14)$
$\qquad\qquad\qquad\quad = 4x(x+7)(x-2)$

26. $6x^3 - 6x^2 - 120x = 6x(x^2 - x - 20)$
$\qquad\qquad\qquad\quad = 6x(x-5)(x+4)$

27. $12x^2 + 34x + 24 = 2(6x^2 + 17x + 12)$
$\qquad\qquad\qquad = 2(6x^2 + 9x + 8x + 12)$
$\qquad\qquad\qquad = 2[3x(2x+3) + 4(2x+3)]$
$\qquad\qquad\qquad = 2(2x+3)(3x+4)$

28. $8a^2 + 6ab - 5b^2 = 8a^2 + 10ab - 4ab - 5b^2$
$\qquad\qquad\qquad = 2a(4a+5b) - b(4a+5b)$
$\qquad\qquad\qquad = (4a+5b)(2a-b)$

29. $4a^2 - b^2 = (2a)^2 - b^2 = (2a+b)(2a-b)$

30. $28 - 13x - 6x^2 = 28 - 21x + 8x - 6x^2$
$\qquad\qquad\qquad = 7(4-3x) + 2x(4-3x)$
$\qquad\qquad\qquad = (4-3x)(7+2x)$

31. $20 - 3x - 2x^2 = 20 - 8x + 5x - 2x^2$
$\qquad\qquad\qquad = 4(5-2x) + x(5-2x)$
$\qquad\qquad\qquad = (5-2x)(4+x)$

32. $x^2 - 2x + 4$ is a prime polynomial.

33. $a^2 + a - 3$ is a prime polynomial.

34. $6y^2 + y - 15 = 6y^2 + 10y - 9y - 15$
$\qquad\qquad\qquad = 2y(3y+5) - 3(3y+5)$
$\qquad\qquad\qquad = (3y+5)(2y-3)$

35. $4x^2 - x - 5 = 4x^2 - 5x + 4x - 5$
$\qquad\qquad\qquad = x(4x-5) + 1(4x-5)$
$\qquad\qquad\qquad = (4x-5)(x+1)$

36. $x^2 y - y^3 = y(x^2 - y^2) = y(x-y)(x+y)$

37. $4t^2 + 36 = 4(t^2 + 9)$

38. $x^2 + x + xy + y = x(x+1) + y(x+1)$
$\qquad\qquad\qquad = (x+1)(x+y)$

39. $ax + 2x + a + 2 = x(a+2) + 1(a+2)$
$\qquad\qquad\qquad = (a+2)(x+1)$

40. $18x^3 - 63x^2 + 9x = 9x(2x^2 - 7x + 1)$

41. $12a^3 - 24a^2 + 4a = 4a(3a^2 - 6a + 1)$

42. $x^2 + 14x - 32 = (x+16)(x-2)$

43. $x^2 - 14x - 48$ is prime.

44. $16a^2 - 56ab + 49b^2 = (4a)^2 - 2(4a)(7b) + (7b)^2$
$\qquad\qquad\qquad\qquad = (4a - 7b)^2$

45. $25p^2 - 70pq + 49q^2 = (5p)^2 - 2(5p)(7q) + (7q)^2$
$\qquad\qquad\qquad\qquad = (5p - 7q)^2$

46. $7x^2 + 24xy + 9y^2 = 7x^2 + 3xy + 21xy + 9y^2$
$\qquad\qquad\qquad\qquad = x(7x+3y) + 3y(7x+3y)$
$\qquad\qquad\qquad\qquad = (7x+3y)(x+3y)$

47. $125 - 8y^3 = 5^3 - (2y)^3$
$\qquad\qquad = (5-2y)[5^2 + 5 \cdot 2y + (2y)^2]$
$\qquad\qquad = (5-2y)(25 + 10y + 4y^2)$

48. $64x^3 + 27 = (4x)^3 + 3^3$
$= (4x + 3)[(4x)^2 - 4x \cdot 3 + 3^2]$
$= (4x + 3)(16x^2 - 12x + 9)$

49. $-x^2 - x + 30 = -1(x^2 + x - 30) = -(x + 6)(x - 5)$

50. $-x^2 + 6x - 8 = -1(x^2 - 6x + 8) = -(x - 2)(x - 4)$

51. $14 + 5x - x^2 = (7 - x)(2 + x)$

52. $3 - 2x - x^2 = (3 + x)(1 - x)$

53. $3x^4 y + 6x^3 y - 72x^2 y = 3x^2 y(x^2 + 2x - 24)$
$= 3x^2 y(x + 6)(x - 4)$

54. $2x^3 y + 8x^2 y^2 - 10xy^3 = 2xy(x^2 + 4xy - 5y^2)$
$= 2xy(x + 5y)(x - y)$

55. $5x^3 y^2 - 40x^2 y^3 + 35xy^4 = 5xy^2 - 8xy + 7y^2)$
$= 5xy^2(x - 7y)(x - y)$

56. $4x^4 y - 8x^3 y - 60x^2 y = 4x^2 y(x^2 - 2x - 15)$
$= 4x^2 y(x - 5)(x + 3)$

57. $12x^3 y + 243xy = 3xy(4x^2 + 81)$

58. $6x^3 y^2 + 8xy^2 = 2xy^2(3x^2 + 4)$

59. $4 - x^2 = 2^2 - x^2 = (2 + x)(2 - x)$

60. $9 - y^2 = 3^2 - y^2 = (3 + y)(3 - y)$

61. $3rs - s + 12r - 4 = s(3r - 1) + 4(3r - 1)$
$= (3r - 1)(s + 4)$

62. $x^3 - 2x^2 + 3x - 6 = x^2(x - 2) + 3(x - 2)$
$= (x - 2)(x^2 + 3)$

63. $4x^2 - 8xy - 3x + 6y = 4x(x - 2y) - 3(x - 2y)$
$= (x - 2y)(4x - 3)$

64. $4x^2 - 2xy - 7yz + 14xz$
$= 2x(2x - y) + 7z(-y + 2x)$
$= (2x - y)(2x + 7z)$

65. $6x^2 + 18xy + 12y^2 = 6(x^2 + 3xy + 2y^2)$
$= 6(x + 2)(x + y)$

66. $12x^2 + 46xy - 8y^2 = 2(6x^2 + 23xy - 4y^2)$
$= 2(6x^2 + 24xy - xy - 4y^2)$
$= 2[6x(x + 4y) - y(x + 4y)]$
$= 2(x + 4y)(6x - y)$

67. $xy^2 - 4x + 3y^2 - 12 = x(y^2 - 4) + 3(y^2 - 4)$
$= (y^2 - 4)(x + 3)$
$= (y^2 - 2^2)(x + 3)$
$= (y + 2)(y - 2)(x + 3)$

68. $x^2 y^2 - 9x^2 + 3y^2 - 27 = x^2(y^2 - 9) + 3(y^2 - 9)$
$= (y^2 - 9)(x^2 + 3)$
$= (y^2 - 3^2)(x^2 + 3)$
$= (y - 3)(y + 3)(x^2 + 3)$

69. $5(x + y) + x(x + y) = (x + y)(5 + x)$

70. $7(x - y) + y(x - y) = (x - y)(7 + y)$

71. $14t^2 - 9t + 1 = 14t^2 - 7t - 2t + 1$
$= 7t(2t - 1) - 1(2t - 1)$
$= (2t - 1)(7t - 1)$

72. $3t^2 - 5t + 1$ is a prime polynomial.

73. $3x^2 + 2x - 5 = 3x^2 + 5x - 3x - 5$
$= x(3x + 5) - 1(3x + 5)$
$= (3x + 5)(x - 1)$

74. $7x^2 + 19x - 6 = 7x^2 + 21x - 2x - 6$
$= 7x(x + 3) - 2(x + 3)$
$= (x + 3)(7x - 2)$

75. $x^2 + 9xy - 36y^2 = (x + 12y)(x - 3y)$

76. $3x^2 + 10xy - 8y^2 = 3x^2 - 2xy + 12xy - 8y^2$
$= x(3x - 2y) + 4y(3x - 2y)$
$= (3x - 2y)(x + 4y)$

77. $1 - 8ab - 20a^2 b^2 = 1 - 10ab + 2ab - 20a^2 b^2$
$= 1(1 - 10ab) + 2ab(1 - 10ab)$
$= (1 - 10ab)(1 + 2ab)$

78. $1 - 7ab - 60a^2b^2 = 1 - 12ab + 5ab - 60a^2b^2$
$$= 1(1 - 12ab) + 5ab(1 - 12ab)$$
$$= (1 - 12ab)(1 + 5ab)$$

79. $9 - 10x^2 + x^4 = (9 - x^2)(1 - x^2)$
$$= (3^2 - x^2)(1^2 - x^2)$$
$$= (3 + x)(3 - x)(1 + x)(1 - x)$$

80. $36 - 13x^2 + x^4 = (9 - x^2)(4 - x^2)$
$$= (3^2 - x^2)(2^2 - x^2)$$
$$= (3 + x)(3 - x)(2 + x)(2 - x)$$

81. $x^4 - 14x^2 - 32 = (x^2 + 2)(x^2 - 16)$
$$= (x^2 + 2)(x^2 - 4^2)$$
$$= (x^2 + 2)(x + 4)(x - 4)$$

82. $x^4 - 22x^2 - 75 = (x^2 + 3)(x^2 - 25)$
$$= (x^2 + 3)(x^2 - 5^2)$$
$$= (x^2 + 3)(x + 5)(x - 5)$$

83. $x^2 - 23x + 120 = (x - 15)(x - 8)$

84. $y^2 + 22y + 96 = (y + 16)(y + 6)$

85. $6x^3 - 28x^2 + 16x = 2x(3x^2 - 14x + 8)$
$$= 2x(3x - 2)(x - 4)$$

86. $6y^3 - 8y^2 - 30y = 2y(3y^2 - 4y - 15)$
$$= 2y(3y + 5)(y - 3)$$

87. $27x^3 - 125y^3 = (3x)^3 - (5y)^3$
$$= (3x - 5y)[(3x)^2 + 3x \cdot 5y + (5y)^2]$$
$$= (3x - 5y)(9x^2 + 15xy + 25y^2)$$

88. $216y^3 - z^3 = (6y)^3 - z^3$
$$= (6y - z)[(6y)^2 + 6y \cdot z + z^2]$$
$$= (6y - z)(36y^2 + 6yz + z^2)$$

89. $x^3y^3 + 8z^3 = (xy)^3 + (2z)^3$
$$= (xy + 2z)[(xy)^2 - xy \cdot 2z + (2z)^2]$$
$$= (xy + 2z)(x^2y^2 - 2xyz + 4z^2)$$

90. $27a^3b^3 + 8 = (3ab)^3 + 2^3$
$$= (3ab + 2)[(3ab)^2 - 3ab \cdot 2 + 2^2]$$
$$= (3ab + 2)(9a^2b^2 - 6ab + 4)$$

91. $2xy - 72x^3y = 2xy(1 - 36x^2)$
$$= 2xy[1^2 - (6x)^2]$$
$$= 2xy(1 + 6x)(1 - 6x)$$

92. $2x^3 - 18x = 2x(x^2 - 9)$
$$= 2x(x^2 - 3^2)$$
$$= 2x(x + 3)(x - 3)$$

93. $x^3 + 6x^2 - 4x - 24 = x^2(x + 6) - 4(x + 6)$
$$= (x + 6)(x^2 - 4)$$
$$= (x + 6)(x^2 - 2^2)$$
$$= (x + 6)(x + 2)(x - 2)$$

94. $x^3 - 2x^2 - 36x + 72 = x^2(x - 2) - 36(x - 2)$
$$= (x - 2)(x^2 - 36)$$
$$= (x - 2)(x^2 - 6^2)$$
$$= (x - 2)(x + 6)(x - 6)$$

95. $6a^3 + 10a^2 = 2a^2(3a + 5)$

96. $4n^2 - 6n = 2n(2n - 3)$

97. $a^2(a + 2) + 2(a + 2) = (a + 2)(a^2 + 2)$

98. $a - b + x(a - b) = (a - b)(1 + x)$

99. $x^3 - 28 + 7x^2 - 4x = x^3 + 7x^2 - 28 - 4x$
$$= x^2(x + 7) - 4(7 + x)$$
$$= (x + 7)(x^2 - 4)$$
$$= (x + 7)(x^2 - 2^2)$$
$$= (x + 7)(x + 2)(x - 2)$$

100. $a^3 - 45 - 9a + 5a^2 = a^3 + 5a^2 - 9a - 45$
$$= a^2(a + 5) - 9(a + 5)$$
$$= (a + 5)(a^2 - 9)$$
$$= (a + 5)(a^2 - 3^2)$$
$$= (a + 5)(a + 3)(a - 3)$$

101. $(x - y)^2 - z^2 = (x - y + z)(x - y - z)$

102. $(x+2y)^2 - 9 = (x+2y)^2 - 3^2$
$$= (x+2y+3)(x+2y-3)$$

103. $81 - (5x+1)^2 = 9^2 - (5x+1)^2$
$$= [9 + (5x+1)][9 - (5x+1)]$$
$$= (9 + 5x + 1)(9 - 5x - 1)$$

104. $b^2 - (4a+c)^2$
$$= [b + (4a+c)][b - (4a+c)]$$
$$= (b + 4a + c)(b - 4a - c)$$

105. answers may vary

106. Yes; $9x^2 + 81y^2 = 9(x^2 + 9y^2)$

107. $(x+10)(x-7) = (x-7)(x+10)$
$$= -1(x+10)(7-x);$$
a, c

108. $(x-2)(x-5) = (x-5)(x-2) = (5-x)(2-x)$; b, c

Section 6.6 Practice

1. $(x+4)(x-5) = 0$
$x + 4 = 0$ or $x - 5 = 0$
     $x = -4$ $x = 5$
Check:
Let $x = -4$.
    $(x+4)(x-5) = 0$
$(-4+4)(-4-5) \stackrel{?}{=} 0$
             $0(-9) = 0$ True
Let $x = 5$.
$(x+4)(x-5) = 0$
$(5+4)(5-5) \stackrel{?}{=} 0$
           $9(0) = 0$ True
The solutions are -4 and 5.

2. $(x - 12)(4x + 3) = 0$
$x - 12 = 0$ or $4x + 3 = 0$
    $x = 12$ $4x = -3$
                  $x = -\dfrac{3}{4}$
Check:
Let $x = 12$.
    $(x-12)(4x+3) = 0$
$(12-12)(4(12)+3) \stackrel{?}{=} 0$
            $0(51) \stackrel{?}{=} 0$
               $0 = 0$ True
Let $x = -\dfrac{3}{4}$.

$(x-12)(4x+3) = 0$
$\left(-\dfrac{3}{4} - 12\right)\left[4\left(-\dfrac{3}{4}\right) + 3\right] \stackrel{?}{=} 0$
$\left(-\dfrac{3}{4} - 12\right)(0) \stackrel{?}{=} 0$
            $0 = 0$ True
The solutions are 12 and $-\dfrac{3}{4}$.

3. $x(7x - 6) = 0$
$x = 0$ or $7x - 6 = 0$
                 $7x = 6$
                   $x = \dfrac{6}{7}$
Check:
Let $x = 0$.
   $x(7x - 6) = 0$
$0(7 \cdot 0 - 6) \stackrel{?}{=} 0$
    $0(-6) = 0$ True
Let $x = \dfrac{6}{7}$.
    $x(7x - 6) = 0$
$\dfrac{6}{7}\left(7 \cdot \dfrac{6}{7} - 6\right) \stackrel{?}{=} 0$
    $\dfrac{6}{7}(6 - 6) \stackrel{?}{=} 0$
      $\dfrac{6}{7}(0) = 0$ True

The solutions are 0 and $\dfrac{6}{7}$.

4. $x^2 - 8x - 48 = 0$
$(x+4)(x-12) = 0$
$x + 4 = 0$ or $x - 12 = 0$
    $x = -4$ $x = 12$
Check:
Let $x = -4$.
      $x^2 - 8x - 48 = 0$
$(-4)^2 - 8(-4) - 48 \stackrel{?}{=} 0$
     $16 + 32 - 48 \stackrel{?}{=} 0$
          $48 - 48 \stackrel{?}{=} 0$
               $0 = 0$ True
Let $x = 12$.

$$x^2 - 8x - 48 = 0$$
$$12^2 - 8\cdot 12 - 48 \stackrel{?}{=} 0$$
$$144 - 96 - 48 \stackrel{?}{=} 0$$
$$48 - 48 \stackrel{?}{=} 0$$
$$0 = 0 \quad \text{True}$$

The solutions are -4 and 12.

5.
$$9x^2 - 24x = -16$$
$$9x^2 - 24x + 16 = 0$$
$$(3x - 4)(3x - 4) = 0$$
$$3x - 4 = 0$$
$$3x = 4$$
$$x = \frac{4}{3}$$

The solution is $\frac{4}{3}$.

6.
$$x(3x + 7) = 6$$
$$3x^2 + 7x = 6$$
$$3x^2 + 7x - 6 = 0$$
$$(3x - 2)(x + 3) = 0$$
$$3x - 2 = 0 \quad \text{or} \quad x + 3 = 0$$
$$3x = 2 \qquad\qquad x = -3$$
$$x = \frac{2}{3}$$

The solutions are $\frac{2}{3}$ and -3.

7.
$$-3x^2 - 6x + 72 = 0$$
$$-3(x^2 + 2x - 24) = 0$$
$$-3(x + 6)(x - 4) = 0$$
$$x + 6 = 0 \quad \text{or} \quad x - 4 = 0$$
$$x = -6 \qquad\qquad x = 4$$

The solutions are -6 and 4.

8.
$$7x^3 - 63x = 0$$
$$7x(x^2 - 9) = 0$$
$$7x(x + 3)(x - 3) = 0$$
$$7x = 0 \quad \text{or} \quad x + 3 = 0 \quad \text{or} \quad x - 3 = 0$$
$$x = 0 \qquad\qquad x = -3 \qquad\qquad x = 3$$

The solutions are 0, -3, and 3.

9.
$$(3x - 2)(2x^2 - 13x + 15) = 0$$
$$(3x - 2)(2x - 3)(x - 5) = 0$$
$$3x - 2 = 0 \quad \text{or} \quad 2x - 3 = 0 \quad \text{or} \quad x - 5 = 0$$
$$3x = 2 \qquad\qquad 2x = 3 \qquad\qquad x = 5$$
$$x = \frac{2}{3} \qquad\qquad x = \frac{3}{2}$$

The solutions are $\frac{2}{3}$, $\frac{3}{2}$, and 5.

10.
$$5x^3 + 5x^2 - 30x = 0$$
$$5x(x^2 + x - 6) = 0$$
$$5x(x + 3)(x - 2) = 0$$
$$5x = 0 \quad \text{or} \quad x + 3 = 0 \quad \text{or} \quad x - 2 = 0$$
$$x = 0 \qquad\qquad x = -3 \qquad\qquad x = 2$$

The solutions are 0, -3, and 2.

11.
$$y = x^2 - 6x + 8$$
$$0 = x^2 - 6x + 8$$
$$0 = (x - 4)(x - 2)$$
$$x - 4 = 0 \quad \text{or} \quad x - 2 = 0$$
$$x = 4 \qquad\qquad x = 2$$

The x-intercepts of the graph of $y = x^2 - 6x + 8$ are $(2, 0)$ and $(4, 0)$.

Graphing Calculator Explorations

1. -0.9, 2.2

2. -2.5, 3.5

3. no real solution

4. no real solution

5. -1.8, 2.8

6. -0.9, 0.3

Vocabulary, Readiness & Video Check 6.6

1. An equation that can be written in the form $ax^2 + bx + c = 0$, (with $a \ne 0$), is called a <u>quadratic</u> equation.

2. If the product of two numbers is 0, then at least one of the numbers must be <u>0</u>.

3. The solutions to $(x - 3)(x + 5) = 0$ are <u>3, −5</u>.

4. If $a \cdot b = 0$, then <u>$a = 0$ or $b = 0$</u>.

5. One side of the equation must be a factored polynomial and the other side must be zero.

6. Because no matter how many factors you have in a multiplication problem, it's still true that for a zero product, at least one of the factors must be zero.

7. To find the x-intercepts of any graph in two variables we let $y = 0$. Doing this with our quadratic equation gives us an equation $= 0$ which we can try to solve by factoring.

Exercise Set 6.6

1. $(x - 6)(x - 7) = 0$
$x - 6 = 0$ or $x - 7 = 0$
$\quad x = 6 \qquad\qquad x = 7$
The solutions are 6 and 7.

3. $(x - 2)(x + 1) = 0$
$x - 2 = 0$ or $x + 1 = 0$
$\quad x = 2 \qquad\qquad x = -1$
The solutions are 2 and -1.

5. $(x + 9)(x + 17) = 0$
$x + 9 = 0$ or $x + 17 = 0$
$\quad x = -9 \qquad\qquad x = -17$
The solutions are -9 and -17.

7. $x(x + 6) = 0$
$x = 0$ or $x + 6 = 0$
$\qquad\qquad\qquad x = -6$
The solutions are 0 and -6.

9. $3x(x - 8) = 0$
$3x = 0$ or $x - 8 = 0$
$\quad x = 0 \qquad\qquad x = 8$
The solutions are 0 and 8.

11. $(2x + 3)(4x - 5) = 0$
$2x + 3 = 0$ or $4x - 5 = 0$
$\quad 2x = -3 \qquad\qquad 4x = 5$
$\quad x = -\dfrac{3}{2} \qquad\qquad x = \dfrac{5}{4}$
The solutions are $-\dfrac{3}{2}$ and $\dfrac{5}{4}$.

13. $(2x - 7)(7x + 2) = 0$
$2x - 7 = 0$ or $7x + 2 = 0$
$\quad 2x = 7 \qquad\qquad 7x = -2$
$\quad x = \dfrac{7}{2} \qquad\qquad x = -\dfrac{2}{7}$
The solutions are $\dfrac{7}{2}$ and $-\dfrac{2}{7}$.

15. $\left(x - \dfrac{1}{2}\right)\left(x + \dfrac{1}{3}\right) = 0$
$x - \dfrac{1}{2} = 0$ or $x + \dfrac{1}{3} = 0$
$\quad x = \dfrac{1}{2} \qquad\qquad x = -\dfrac{1}{3}$
The solutions are $\dfrac{1}{2}$ and $-\dfrac{1}{3}$.

17. $(x + 0.2)(x + 1.5) = 0$
$x + 0.2 = 0$ or $x + 1.5 = 0$
$\quad x = -0.2 \qquad\qquad x = -1.5$
The solutions are -0.2 and -1.5

19. $x^2 - 13x + 36 = 0$
$(x - 9)(x - 4) = 0$
$x - 9 = 0$ or $x - 4 = 0$
$\quad x = 9 \qquad\qquad x = 4$
The solutions are 9 and 4.

21. $x^2 + 2x - 8 = 0$
$(x + 4)(x - 2) = 0$
$x + 4 = 0$ or $x - 2 = 0$
$\quad x = -4 \qquad\qquad x = 2$
The solutions are -4 and 2.

23. $x^2 - 7x = 0$
$x(x - 7) = 0$
$x = 0$ or $x - 7 = 0$
$\qquad\qquad\qquad x = 7$
The solutions are 0 and 7.

25. $x^2 - 4x = 32$
$x^2 - 4x - 32 = 0$
$(x - 8)(x + 4) = 0$
$x - 8 = 0$ or $x + 4 = 0$
$\quad x = 8 \qquad\qquad x = -4$
The solutions are 8 and -4.

27.
$$x^2 = 16$$
$$x^2 - 16 = 0$$
$$(x+4)(x-4) = 0$$
$$x + 4 = 0 \quad \text{or} \quad x - 4 = 0$$
$$x = -4 \qquad\qquad x = 4$$
The solutions are -4 and 4.

29. $(x+4)(x-9) = 4x$
$$x^2 - 5x - 36 = 4x$$
$$x^2 - 9x - 36 = 0$$
$$(x-12)(x+3) = 0$$
$$x - 12 = 0 \quad \text{or} \quad x + 3 = 0$$
$$x = 12 \qquad\qquad x = -3$$
The solutions are 12 and -3.

31.
$$x(3x-1) = 14$$
$$3x^2 - x = 14$$
$$3x^2 - x - 14 = 0$$
$$(3x-7)(x+2) = 0$$
$$3x - 7 = 0 \quad \text{or} \quad x + 2 = 0$$
$$3x = 7 \qquad\qquad x = -2$$
$$x = \frac{7}{3}$$
The solutions are $\frac{7}{3}$ and -2.

33.
$$-3x^2 + 75 = 0$$
$$-3(x^2 - 25) = 0$$
$$-3(x+5)(x-5) = 0$$
$$x + 5 = 0 \quad \text{or} \quad x - 5 = 0$$
$$x = -5 \qquad\qquad x = 5$$
The solutions are -5 and 5.

35.
$$24x^2 + 44x = 8$$
$$24x^2 + 44x - 8 = 0$$
$$4(6x^2 + 11x - 2) = 0$$
$$4(6x-1)(x+2) = 0$$
$$6x - 1 = 0 \quad \text{or} \quad x + 2 = 0$$
$$6x = 1 \qquad\qquad x = -2$$
$$x = \frac{1}{6}$$
The solutions are $\frac{1}{6}$ and -2.

37. $x^3 - 12x^2 + 32x = 0$
$$x(x^2 - 12x + 32) = 0$$
$$x(x-8)(x-4) = 0$$
$$x = 0 \quad \text{or} \quad x - 8 = 0 \quad \text{or} \quad x - 4 = 0$$
$$x = 8 \qquad\qquad x = 4$$
The solutions are 0, 8, and 4.

39. $(4x-3)(16x^2 - 24x + 9) = 0$
$$(4x-3)(4x-3)^2 = 0$$
$$(4x-3)^3 = 0$$
$$4x - 3 = 0$$
$$4x = 3$$
$$x = \frac{3}{4}$$
The solution is $\frac{3}{4}$.

41.
$$4x^3 - x = 0$$
$$x(4x^2 - 1) = 0$$
$$x(2x+1)(2x-1) = 0$$
$$x = 0 \quad \text{or} \quad 2x + 1 = 0 \quad \text{or} \quad 2x - 1 = 0$$
$$2x = -1 \qquad\qquad 2x = 1$$
$$x = -\frac{1}{2} \qquad\qquad x = \frac{1}{2}$$
The solutions are 0, $-\frac{1}{2}$, and $\frac{1}{2}$.

43. $32x^3 - 4x^2 - 6x = 0$
$$2x(16x^2 - 2x - 3) = 0$$
$$2x(2x-1)(8x+3) = 0$$
$$2x = 0 \quad \text{or} \quad 2x - 1 = 0 \quad \text{or} \quad 8x + 3 = 0$$
$$x = 0 \qquad\qquad 2x = 1 \qquad\qquad 8x = -3$$
$$x = \frac{1}{2} \qquad\qquad x = -\frac{3}{8}$$
The solutions are 0, $\frac{1}{2}$, and $-\frac{3}{8}$.

45. $(x+3)(x-2) = 0$
$$x + 3 = 0 \quad \text{or} \quad x - 2 = 0$$
$$x = -3 \qquad\qquad x = 2$$
The solutions are -3 and 2.

47. $x^2 + 20x = 0$
$$x(x+20) = 0$$
$$x = 0 \quad \text{or} \quad x + 20 = 0$$
$$x = -20$$
The solutions are 0 and -20.

49. $4(x-7)=6$

$4x-28=6$

$4x=34$

$x=\dfrac{34}{4}$

$x=\dfrac{17}{2}$

The solution is $\dfrac{17}{2}$.

51. $\quad 4y^2-1=0$

$(2y+1)(2y-1)=0$

$2y+1=0 \quad$ or $\quad 2y-1=0$

$2y=-1 \qquad\qquad 2y=1$

$y=-\dfrac{1}{2} \qquad\qquad y=\dfrac{1}{2}$

The solutions are $-\dfrac{1}{2}$ and $\dfrac{1}{2}$.

53. $(2x+3)(2x^2-5x-3)=0$

$(2x+3)(2x+1)(x-3)=0$

$2x+3=0 \quad$ or $\quad 2x+1=0 \quad$ or $\quad x-3=0$

$2x=-3 \qquad\qquad 2x=-1 \qquad\qquad x=3$

$x=-\dfrac{3}{2} \qquad\qquad x=-\dfrac{1}{2}$

The solutions are $-\dfrac{3}{2}$, $-\dfrac{1}{2}$, and 3.

55. $\qquad x^2-15=-2x$

$x^2+2x-15=0$

$(x+5)(x-3)=0$

$x+5=0 \quad$ or $\quad x-3=0$

$x=-5 \qquad\qquad x=3$

The solutions are -5 and 3.

57. $30x^2-11x-30=0$

$(6x+5)(5x-6)=0$

$6x+5=0 \quad$ or $\quad 5x-6=0$

$6x=-5 \qquad\qquad 5x=6$

$x=-\dfrac{5}{6} \qquad\qquad x=\dfrac{6}{5}$

The solutions are $-\dfrac{5}{6}$ and $\dfrac{6}{5}$.

59. $\quad 5x^2-6x-8=0$

$(5x+4)(x-2)=0$

$5x+4=0 \quad$ or $\quad x-2=0$

$5x=-4 \qquad\qquad x=2$

$x=-\dfrac{4}{5}$

The solutions are $-\dfrac{4}{5}$ and 2.

61. $\quad 6y^2-22y-40=0$

$2(3y^2-11y-20)=0$

$2(3y+4)(y-5)=0$

$3y+4=0 \quad$ or $\quad y-5=0$

$3y=-4 \qquad\qquad y=5$

$y=-\dfrac{4}{3}$

The solutions are $-\dfrac{4}{3}$ and 5.

63. $(y-2)(y+3)=6$

$y^2+y-6=6$

$y^2+y-12=0$

$(y+4)(y-3)=0$

$y+4=0 \quad$ or $\quad y-3=0$

$y=-4 \qquad\qquad y=3$

The solutions are -4 and 3.

65. $3x^3+19x^2-72x=0$

$x(3x^2+19x-72)=0$

$x(3x-8)(x+9)=0$

$x=0 \quad$ or $\quad 3x-8=0 \quad$ or $\quad x+9=0$

$3x=8 \qquad\qquad x=-9$

$x=\dfrac{8}{3}$

The solutions are 0, $\dfrac{8}{3}$, and -9.

67. $x^2+14x+49=0$

$(x+7)^2=0$

$x+7=0$

$x=-7$

The solution is -7.

69. $12y = 8y^2$

$0 = 8y^2 - 12y$

$0 = 4y(2y - 3)$

$4y = 0$ or $2y - 3 = 0$

$y = 0$ $2y = 3$

                  $y = \dfrac{3}{2}$

The solutions are 0 and $\dfrac{3}{2}$.

71. $7x^3 - 7x = 0$

       $7x(x^2 - 1) = 0$

$7x(x + 1)(x - 1) = 0$

$7x = 0$ or $x + 1 = 0$ or $x - 1 = 0$

$x = 0$ $x = -1$ $x = 1$

The solutions are 0, −1, and 1.

73. $3x^2 + 8x - 11 = 13 - 6x$

    $3x^2 + 14x - 24 = 0$

    $(3x - 4)(x + 6) = 0$

    $3x - 4 = 0$ or $x + 6 = 0$

       $3x = 4$ $x = -6$

        $x = \dfrac{4}{3}$

The solutions are $\dfrac{4}{3}$ and −6.

75. $3x^2 - 20x = -4x^2 - 7x - 6$

   $7x^2 - 13x + 6 = 0$

   $(7x - 6)(x - 1) = 0$

   $7x - 6 = 0$ or $x - 1 = 0$

       $7x = 6$ $x = 1$

        $x = \dfrac{6}{7}$

The solutions are $\dfrac{6}{7}$ and 1.

77. Let $y = 0$ and solve for x.

$y = (3x + 4)(x - 1)$

$0 = (3x + 4)(x - 1)$

$3x + 4 = 0$ or $x - 1 = 0$

   $3x = -4$ $x = 1$

    $x = -\dfrac{4}{3}$

The intercepts are $\left(-\dfrac{4}{3}, 0 \right)$ and (1, 0).

79. Let $y = 0$ and solve for x.

$y = x^2 - 3x - 10$

$0 = x^2 - 3x - 10$

$0 = (x - 5)(x + 2)$

$x - 5 = 0$ or $x + 2 = 0$

   $x = 5$ $x = -2$

The x-intercepts are (5, 0) and (−2, 0).

81. Let $y = 0$ and solve for x.

$y = 2x^2 + 11x - 6$

$0 = 2x^2 + 11x - 6$

$0 = (2x - 1)(x + 6)$

$2x - 1 = 0$ or $x + 6 = 0$

   $2x = 1$ $x = -6$

    $x = \dfrac{1}{2}$

The x-intercepts are $\left(\dfrac{1}{2}, 0 \right)$ and (−6, 0).

83. e; x-intercepts are (−2, 0), (1, 0)

85. b; x-intercepts are (0, 0), (−3, 0)

87. c; $y = 2x^2 - 8 = 2(x - 2)(x + 2)$

x-intercepts are (2, 0), (−2, 0).

89. $\dfrac{3}{5} + \dfrac{4}{9} = \dfrac{3 \cdot 9}{5 \cdot 9} + \dfrac{4 \cdot 5}{9 \cdot 5}$

        $= \dfrac{27}{45} + \dfrac{20}{45}$

        $= \dfrac{27 + 20}{45}$

        $= \dfrac{47}{45}$

91. $\dfrac{7}{10} - \dfrac{5}{12} = \dfrac{7 \cdot 6}{10 \cdot 6} - \dfrac{5 \cdot 5}{12 \cdot 5}$

         $= \dfrac{42}{60} - \dfrac{25}{60}$

         $= \dfrac{42 - 25}{60}$

         $= \dfrac{17}{60}$

93. $\dfrac{7}{8} \div \dfrac{7}{15} = \dfrac{7}{8} \cdot \dfrac{15}{7} = \dfrac{15}{8}$

95. $\dfrac{4}{5} \cdot \dfrac{7}{8} = \dfrac{4 \cdot 7}{5 \cdot 8} = \dfrac{4 \cdot 7}{5 \cdot 2 \cdot 4} = \dfrac{7}{10}$

97. Didn't write the equation in standard form; standard form should be:

$$x(x-2) = 8$$
$$x^2 - 2x = 8$$
$$x^2 - 2x - 8 = 0$$
$$(x-4)(x+2) = 0$$
$$x - 4 = 0 \quad \text{or} \quad x + 2 = 0$$
$$x = 4 \qquad\qquad x = -2$$

99. Answers may vary. Possible answer: If the solutions are $x = 6$ and $x = -1$, then, by the zero factor property,

$$x = 6 \quad \text{or} \quad x = -1$$
$$x - 6 = 0 \qquad x + 1 = 0$$
$$(x-6)(x+1) = 0$$

101. Answers may vary. Possible answer: If the solutions are $x = 5$ and $x = 7$, then, by the zero factor property,

$$x = 5 \quad \text{or} \quad x = 7$$
$$x - 5 = 0 \qquad x - 7 = 0$$
$$(x-5)(x-7) = 0$$
$$x^2 - 7x - 5x + 35 = 0$$
$$x^2 - 12x + 35 = 0$$

103. $y = -16x^2 + 20x + 300$

   a.

time x	0	1	2	3	4	5	6
height y	300	304	276	216	124	0	−156

   b. The compass strikes the ground after 5 seconds, when the height, y, is zero feet.

   c. The maximum height was approximately 304 feet.

   d.

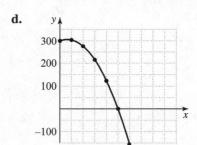

105. $(x-3)(3x+4) = (x+2)(x-6)$

$$3x^2 - 5x - 12 = x^2 - 4x - 12$$
$$2x^2 - x = 0$$
$$x(2x-1) = 0$$
$$2x - 1 = 0 \quad \text{or} \quad x = 0$$
$$x = \frac{1}{2}$$

The solutions are $\frac{1}{2}$ and 0.

107. $(2x-3)(x+8) = (x-6)(x+4)$

$2x^2 + 13x - 24 = x^2 - 2x - 24$

$x^2 + 15x = 0$

$x(x+15) = 0$

$x + 15 = 0$　　or　$x = 0$

$x = -15$

The solutions are -15 and 0.

Section 6.7 Practice

1. Find t when $h = 0$.

$h = -16t^2 + 64$

$0 = -16t^2 + 64$

$0 = -16(t^2 - 4)$

$0 = -16(t-2)(t+2)$

$t - 2 = 0$　or　$t + 2 = 0$

$t = 2$　　　$t = -2$

Since time cannot be negative, the diver will reach the pool in 2 seconds.

2. Let x = the number.

$x^2 - 8x = 48$

$x^2 - 8x - 48 = 0$

$(x-12)(x+4) = 0$

$x - 12 = 0$　　or　$x + 4 = 0$

$x = 12$　　　　$x = -4$

There are two numbers. They are -4 and 12.

3. Let x = height, then $3x - 1$ = base.

$A = \dfrac{1}{2}bh$

$210 = \dfrac{1}{2}(3x-1)(x)$

$420 = (3x-1)(x)$

$420 = 3x^2 - x$

$0 = 3x^2 - x - 420$

$0 = (3x+35)(x-12)$

$3x + 35 = 0$　　　or　$x - 12 = 0$

$x = -\dfrac{35}{3}$　　　　$x = 12$

Since height cannot be negative, the height is 12 feet and the base is $3(12) - 1 = 35$ feet.

4. Let x = first integer, then
$x + 1$ = next consecutive integer.

$x(x+1) = x + (x+1) + 41$

$x^2 + x = 2x + 42$

$x^2 - x - 42 = 0$

$(x-7)(x+6) = 0$

$x - 7 = 0$　or　$x + 6 = 0$

$x = 7$　　　　$x = -6$

The numbers are 7 and 8 or -6 and -5.

5. Let x = first leg, then $2x - 1$ = second leg, and $2x + 1$ = hypotenuse.

$x^2 + (2x-1)^2 = (2x+1)^2$

$x^2 + 4x^2 - 4x + 1 = 4x^2 + 4x + 1$

$x^2 - 8x = 0$

$x(x-8) = 0$

$x = 0$　or　$x - 8 = 0$

$x = 8$

Since the length cannot be 0, the legs have lengths 8 units and $2(8) - 1 = 15$ units and the hypotenuse has length $2(8) + 1 = 17$ units.

Vocabulary, Readiness & Video Check 6.7

1. In applications, the context of the problem needs to be considered. Each exercise resulted in both a positive and a negative solution, and a negative solution is not appropriate for any of the problems.

Exercise Set 6.7

1. Let x = the width, then $x + 4$ = the length.

3. Let x = the first odd integer, then
$x + 2$ = the next consecutive odd integer.

5. Let x = the base, then $4x + 1$ = the height.

7. Let x = the length of one side.

$A = x^2$

$121 = x^2$

$0 = x^2 - 121$

$0 = x^2 - 11^2$

$0 = (x+11)(x-11)$

$x + 11 = 0$　　or　$x - 11 = 0$

$x = -11$　　　　$x = 11$

Since the length cannot be negative, the sides are 11 units long.

9. The perimeter is the sum of the lengths of the sides.
$$120 = (x+5)+(x^2-3x)+(3x-8)+(x+3)$$
$$120 = x+5+x^2-3x+3x-8+x+3$$
$$120 = x^2+2x$$
$$0 = x^2+2x-120$$
$$x^2+2x-120=0$$
$$(x+12)(x-10)=0$$
$$x+12=0 \quad \text{or} \quad x-10=0$$
$$x=-12 \qquad\qquad x=10$$
Since the dimensions cannot be negative, the lengths of the sides are:
$10+5=15$ cm, $10^2-3(10)=70$ cm,
$3(10)-8=22$ cm, and $10+3=13$ cm.

11. $x+5$ = the base and $x-5$ = the height.
$$A = bh$$
$$96 = (x+5)(x-5)$$
$$96 = x^2-25$$
$$0 = x^2-121$$
$$x^2-121=0$$
$$(x+11)(x-11)=0$$
$$x+11=0 \quad \text{or} \quad x-11=0$$
$$x=-11 \qquad\qquad x=11$$
Since the dimensions cannot be negative, $x=11$. The base is $11+5=16$ miles, and the height is $11-5=6$ miles.

13. Find t when $h=0$.
$$h = -16t^2+64t+80$$
$$0 = -16t^2+64t+80$$
$$0 = -16(t^2-4t-5)$$
$$0 = -16(t-5)(t+1)$$
$$t-5=0 \quad \text{or} \quad t+1=0$$
$$t=5 \qquad\qquad t=-1$$
Since the time t cannot be negative, the object hits the ground after 5 seconds.

15. Let x = the length then $2x-7$ = the width.
$$A = lw$$
$$30 = (x)(2x-7)$$
$$30 = 2x^2-7x$$
$$0 = 2x^2-7x-30$$
$$0 = (2x+5)(x-6)$$
$$2x+5=0 \quad \text{or} \quad x-6=0$$
$$x=-\frac{5}{2} \qquad\qquad x=6$$
Since the dimensions cannot be negative, the length is 6 cm and the width is $2(6)-7=5$ cm.

17. Let $n=12$.
$$D = \frac{1}{2}n(n-3)$$
$$D = \frac{1}{2}\cdot 12(12-3) = 6(9) = 54$$
A polygon with 12 sides has 54 diagonals.

19. Let $D=35$ and solve for n.
$$D = \frac{1}{2}n(n-3)$$
$$35 = \frac{1}{2}n(n-3)$$
$$70 = n^2-3n$$
$$0 = n^2-3n-70$$
$$0 = (n-10)(n+7)$$
$$n-10=0 \quad \text{or} \quad n+7=0$$
$$n=10 \qquad\qquad n=-7$$
The polygon has 10 sides.

21. Let x = the unknown number.
$$x+x^2 = 132$$
$$x^2+x-132=0$$
$$(x+12)(x-11)=0$$
$$x+12=0 \quad \text{or} \quad x-11=0$$
$$x=-12 \qquad\qquad x=11$$
The two numbers are -12 and 11.

23. Let x = the first room number, then $x+1$ = next room number.
$$x(x+1) = 210$$
$$x^2+x = 210$$
$$x^2+x-210=0$$
$$(x-14)(x+15)=0$$
$$x-14=0 \quad \text{or} \quad x+15=0$$
$$x=14 \qquad\qquad x=-15$$
Since the room number is not negative, the room numbers are 14 and 15.

25. Let x = hypotenuse, then $x-1$ = height.
$$a^2+b^2 = c^2$$
$$5^2+(x-1)^2 = x^2$$
$$25+x^2-2x+1 = x^2$$
$$26-2x = 0$$
$$26 = 2x$$
$$13 = x$$
The length of the ladder is 13 feet.

27. Let x = the length of a side of the original square. Then $x + 3$ = the length of a side of the larger square.

$$64 = (x+3)^2$$
$$64 = x^2 + 6x + 9$$
$$0 = x^2 + 6x - 55$$
$$0 = (x+11)(x-5)$$
$$x + 11 = 0 \quad \text{or} \quad x - 5 = 0$$
$$x = -11 \qquad\qquad x = 5$$

Since the length cannot be negative, the sides of the original square are 5 inches long.

29. Let x = the length of the shorter leg. Then $x + 4$ = the length of the longer leg and $x + 8$ = the length of the hypotenuse. By the Pythagorean theorem,

$$x^2 + (x+4)^2 = (x+8)^2$$
$$x^2 + x^2 + 8x + 16 = x^2 + 16x + 64$$
$$x^2 - 8x - 48 = 0$$
$$(x-12)(x+4) = 0$$
$$x - 12 = 0 \quad \text{or} \quad x + 4 = 0$$
$$x = 12 \qquad\qquad x = -4$$

Since the length cannot be negative, the sides of the triangle are 12 mm, 12 + 4 = 16 mm, and 12 + 8 = 20 mm.

31. Let x = the height of the triangle, then $2x$ = the base.

$$A = \frac{1}{2}bh$$
$$100 = \frac{1}{2}(2x)(x)$$
$$100 = x^2$$
$$0 = x^2 - 100$$
$$0 = (x+10)(x-10)$$
$$x + 10 = 0 \quad \text{or} \quad x - 10 = 0$$
$$x = -10 \qquad\qquad x = 10$$

Since the height cannot be negative, the height of the triangle is 10 km.

33. Let x = the length of the shorter leg, then $x + 12$ = the length of the longer leg and $2x - 12$ = the length of the hypotenuse. By the Pythagorean theorem,

$$x^2 + (x+12)^2 = (2x-12)^2$$
$$x^2 + x^2 + 24x + 144 = 4x^2 - 48x + 144$$
$$0 = 2x^2 - 72x$$
$$0 = 2x(x-36)$$

$$2x = 0 \quad \text{or} \quad x - 36 = 0$$
$$x = 0 \qquad\qquad x = 36$$

Since the length cannot be zero feet, the shorter leg is 36 feet long.

35. Find t when $h = 0$.

$$h = -16t^2 + 1444$$
$$0 = -16t^2 + 1444$$
$$0 = -4(4t^2 - 361)$$
$$0 = -4(2t-19)(2t+19)$$
$$2t - 19 = 0 \quad \text{or} \quad 2t + 19 = 0$$
$$t = \frac{19}{2} \qquad\qquad t = -\frac{19}{2}$$

Since time cannot be negative, the object reaches the ground in $\frac{19}{2} = 9.5$ seconds.

37. Let $P = 100$ and $A = 144$.

$$A = P(1+r)^2$$
$$144 = 100(1+r)^2$$
$$144 = 100 + 200r + 100r^2$$
$$0 = 100r^2 + 200r - 44$$
$$0 = 4(25r^2 + 50r - 11)$$
$$0 = 4(5r-1)(5r+11)$$
$$5r - 1 = 0 \quad \text{or} \quad 5r + 11 = 0$$
$$5r = 1 \qquad\qquad 5r = -11$$
$$r = \frac{1}{5} \qquad\qquad r = -\frac{11}{5}$$
$$r = 0.2 \qquad\qquad r = -2.2$$

Since the interest rate cannot be negative $r = 0.2$ and the rate is 20%.

39. Let x = the length and $x - 7$ = the width.

$$A = lw$$
$$120 = (x-7)(x)$$
$$120 = x^2 - 7x$$
$$0 = x^2 - 7x - 120$$
$$0 = (x+8)(x-15)$$
$$x + 8 = 0 \quad \text{or} \quad x - 15 = 0$$
$$x = -8 \qquad\qquad x = 15$$

Since the length cannot be negative, the length is 15 miles. The width is 15 − 7 = 8 miles.

41. Let $C = 9500$.

$$C = x^2 - 15x + 50$$
$$9500 = x^2 - 15x + 50$$
$$0 = x^2 - 15x - 9450$$
$$0 = (x+90)(x-105)$$

$x + 90 = 0$ or $x - 105 = 0$
$x = -90$ $x = 105$
Since the number of units cannot be negative the solution is 105 units.

43. From the graph, there were approximately 2 million visitors to Glacier National Park in 2009.

45. From the graph, there were approximately 1.9 million visitors to Glacier National Park in 2005.

47. From the graph, the lines intersect at approximately 2003.

49. answers may vary

51. $\dfrac{20}{35} = \dfrac{2 \cdot 2 \cdot 5}{5 \cdot 7} = \dfrac{4}{7}$

53. $\dfrac{27}{18} = \dfrac{3 \cdot 3 \cdot 3}{2 \cdot 3 \cdot 3} = \dfrac{3}{2}$

55. $\dfrac{14}{42} = \dfrac{2 \cdot 7}{2 \cdot 3 \cdot 7} = \dfrac{1}{3}$

57. Let x = the rate (in mph) of the slower boat, then $x + 7$ = the rate (in mph) of the faster boat. After one hour, the slower boat has traveled x miles and the faster boat has traveled $x + 7$ miles. By the Pythagorean theorem,
$$x^2 + (x+7)^2 = 17^2$$
$$x^2 + x^2 + 14x + 49 = 289$$
$$2x^2 + 14x + 49 = 289$$
$$2x^2 + 14x - 240 = 0$$
$$2(x^2 + 7x - 120) = 0$$
$$2(x + 15)(x - 8) = 0$$
$x + 15 = 0$ or $x - 8 = 0$
$x = -15$ $x = 8$
Since the rate cannot be negative, the slower boat travels at 8 mph. The faster boat travels at 8 + 7 = 15 mph.

59. Let x = the first number, then $20 - x$ = the other number.
$$x^2 + (20 - x)^2 = 218$$
$$x^2 + 400 - 40x + x^2 = 218$$
$$2x^2 - 40x + 400 = 218$$
$$2x^2 - 40x + 182 = 0$$
$$2(x^2 - 20x + 91) = 0$$
$$2(x - 13)(x - 7) = 0$$
$x - 13 = 0$ or $x - 7 = 0$
$x = 13$ $x = 7$
The numbers are 13 and 7.

61. Pool: width = x and length = $x + 6$
Total Area: width = $x + 8$ and length = $x + 14$
Total area = 576 + Pool area
$$(x + 14)(x + 8) = 576 + (x + 6)(x)$$
$$x^2 + 22x + 112 = 576 + x^2 + 6x$$
$$16x + 112 = 576$$
$$16x = 464$$
$$x = 29$$
$x + 6 = 29 + 6 = 35$
The pool has length 35 meters and width 29 meters.

63. answers may vary

Chapter 6 Vocabulary Check

1. An equation that can be written in the form $ax^2 + bx + c = 0$ (with a not 0) is called a quadratic equation.

2. Factoring is the process of writing an expression as a product.

3. The greatest common factor of a list of terms is the product of all common factors.

4. A trinomial that is the square of some binomial is called a perfect square trinomial.

5. The expression $a^2 - b^2$ is called a difference of two squares.

6. The expression $a^3 - b^3$ is called a difference of two cubes.

7. The expression $a^3 + b^3$ is called a sum of two cubes.

8. By the zero factor property, if the product of two numbers is 0, then at least one of the numbers must be <u>0</u>.

9. In a right triangle, the side opposite the right angle is called the <u>hypotenuse</u>.

10. In a right triangle, each side adjacent to the right angle is called a <u>leg</u>.

11. The Pythagorean theorem states that
$(\text{leg})^2 + (\text{leg})^2 = (\underline{\text{hypotenuse}})^2$.

Chapter 6 Review

1. $6x^2 - 15x = 3x(2x - 5)$

2. $2x^3y + 6x^2y^2 + 8xy^3 = 2xy(x^2 + 3xy + 4y^2)$

3. $20x^2 + 12x = 4x(5x + 3)$

4. $6x^2y^2 - 3xy^3 = 3xy^2(2x - y)$

5. $3x(2x + 3) - 5(2x + 3) = (2x + 3)(3x - 5)$

6. $5x(x + 1) - (x + 1) = (x + 1)(5x - 1)$

7. $3x^2 - 3x + 2x - 2 = 3x(x - 1) + 2(x - 1)$
$= (x - 1)(3x + 2)$

8. $3a^2 + 9ab + 3b^2 + ab = 3a(a + 3b) + b(3b + a)$
$= (a + 3b)(3a + b)$

9. $10a^2 + 5ab + 7b^2 + 14ab$
$= 5a(2a + b) + 7b(b + 2a)$
$= (2a + b)(5a + 7b)$

10. $6x^2 + 10x - 3x - 5 = 2x(3x + 5) - 1(3x + 5)$
$= (3x + 5)(2x - 1)$

11. $x^2 + 6x + 8 = (x + 4)(x + 2)$

12. $x^2 - 11x + 24 = (x - 8)(x - 3)$

13. $x^2 + x + 2$ is prime.

14. $x^2 - x + 2$ is prime.

15. $x^2 + 4xy - 12y^2 = (x + 6y)(x - 2y)$

16. $x^2 + 8xy + 15y^2 = (x + 5y)(x + 3y)$

17. $72 - 18x - 2x^2 = 2(36 - 9x - x^2)$
$= 2(3 - x)(12 + x)$
or
$72 - 18x - 2x^2 = -2x^2 - 18x + 72$
$= -2(x^2 + 9x - 36)$
$= -2(x - 3)(x + 12)$

18. $32 + 12x - 4x^2 = 4(8 + 3x - x^2)$
or
$32 + 12x - 4x^2 = -4x^2 + 12x + 32$
$= -4(x^2 - 3x - 8)$

19. $10a^3 - 110a^2 + 100a = 10a(a^2 - 11a + 10a)$
$= 10a(a - 1)(a - 10)$

20. $5y^3 - 50y^2 + 120y = 5y(y^2 - 10y + 24)$
$= 5y(y - 6)(y - 4)$

21. To factor $x^2 + 2x - 48$, think of two numbers whose product is <u>−48</u> and whose sum is <u>2</u>.

22. The first step in factoring $3x^2 + 15x + 30$ is to factor out the GCF, 3.

23. Factors of $2x^2$: $2x \cdot x$
Factors of 6: $6 = 1 \cdot 6, 6 = 2 \cdot 3$
$2x^2 + 13x + 6 = (2x + 1)(x + 6)$

24. Factors of $4x^2$: $4x^2 = 4x \cdot x, 4x^2 = 2x \cdot 2x$
Factors of -3: $-3 = -1 \cdot 3, -3 = 1 \cdot -3$
$4x^2 + 4x - 3 = (2x + 3)(2x - 1)$

25. Factors of $6x^2$: $6x^2 = 6x \cdot x, 6x^2 = 3x \cdot 2x$
Factors of $-4y^2$: $-4y^2 = -4y \cdot y$,
$-4y^2 = 4y \cdot -y, -4y^2 = -2y \cdot 2y$
$6x^2 + 5xy - 4y^2 = (3x + 4y)(2x - y)$

26. $18 \cdot -20y^2 = -360y^2$

$15y \cdot -24y = -360y^2$

$15y + (-24y) = -9y$

$18x^2 - 9xy - 20y^2 = 18x^2 + 15xy - 24xy - 20y^2$
$= 3x(6x + 5y) - 4y(6x + 5y)$
$= (6x + 5y)(3x - 4y)$

27. $10y^3 + 25y^2 - 60y = 5y(2y^2 + 5y - 12)$

$2 \cdot -12 = -24$

$-3 \cdot 8 = -24$

$-3 + 8 = 5$

$10y^3 + 25y^2 - 60y = 5y(2y^2 + 5y - 12)$
$= 5y(2y^2 - 3y + 8y - 12)$
$= 5y[y(2y - 3) + 4(2y - 3)]$
$= 5y(2y - 3)(y + 4)$

28. $60y^3 - 39y^2 + 6y = 3y(20y^2 - 13y + 2)$

$20 \cdot 2 = 40$

$-5 \cdot -8 = 40$

$-5 + (-8) = -13$

$60y^3 - 39y^2 + 6y = 3y(20y^2 - 13y + 2)$
$= 3y(20y^2 - 5y - 8y + 2)$
$= 3y[5y(4y - 1) - 2(4y - 1)]$
$= 3y(4y - 1)(5y - 2)$

29. $18x^2 - 60x + 50 = 2(9x^2 - 30x + 25)$
$= 2[(3x)^2 - 2 \cdot 3x \cdot 5 + 5^2]$
$= 2(3x - 5)^2$

30. $4x^2 - 28xy + 49y^2 = [(2x)^2 - 2 \cdot 2x \cdot 7y + (7y)^2]$
$= (2x - 7y)^2$

31. $4x^2 - 9 = (2x)^2 - 3^2 = (2x + 3)(2x - 3)$

32. $9t^2 - 25s^2 = (3t)^2 - (5s)^2 = (3t + 5s)(3t - 5s)$

33. $16x^2 + y^2$ is a prime polynomial.

34. $x^3 - 8y^3 = x^3 - (2y)^3$
$= (x - 2y)[x^2 + x \cdot 2y + (2y)^2]$
$= (x - 2y)(x^2 + 2xy + 4y^2)$

35. $8x^3 + 27 = (2x)^3 + 3^3$
$= (2x + 3)[(2x)^2 - 2x \cdot 3 + 3^2]$
$= (2x + 3)(4x^2 - 6x + 9)$

36. $2x^3 + 8x = 2x(x^2 + 4)$

37. $54 - 2x^3y^3 = 2(27 - x^3y^3)$
$= 2[3^3 - (xy)^3]$
$= 2(3 - xy)[3^2 + 3 \cdot xy + (xy)^2]$
$= 2(3 - xy)(9 + 3xy + x^2y^2)$

38. $9x^2 - 4y^2 = (3x)^2 - (2y)^2 = (3x - 2y)(3x + 2y)$

39. $16x^4 - 1 = (4x^2)^2 - 1^2$
$= (4x^2 + 1)(4x^2 - 1)$
$= (4x^2 + 1)[(2x)^2 - 1^2]$
$= (4x^2 + 1)(2x + 1)(2x - 1)$

40. $x^4 + 16$ is a prime polynomial.

41. $(x + 6)(x - 2) = 0$

$x + 6 = 0$ or $x - 2 = 0$

$x = -6$ $x = 2$

The solutions are -6 and 2.

42. $3x(x + 1)(7x - 2) = 0$

$3x = 0$ or $x + 1 = 0$ or $7x - 2 = 0$

$x = 0$ $x = -1$ $7x = 2$

$x = \dfrac{2}{7}$

The solutions are 0, -1, and $\dfrac{2}{7}$.

43. $4(5x + 1)(x + 3) = 0$

$5x + 1 = 0$ or $x + 3 = 0$

$5x = -1$ $x = -3$

$x = -\dfrac{1}{5}$

The solutions are $-\dfrac{1}{5}$ and -3.

44. $x^2 + 8x + 7 = 0$

$(x + 7)(x + 1) = 0$

$x + 7 = 0$ or $x + 1 = 0$

$x = -7$ $x = -1$

The solutions are -7 and -1.

45. $x^2 - 2x - 24 = 0$

$(x-6)(x+4) = 0$

$x - 6 = 0$ or $x + 4 = 0$

$x = 6$ $\qquad x = -4$

The solutions are 6 and –4.

46. $x^2 + 10x = -25$

$x^2 + 10x + 25 = 0$

$(x+5)(x+5) = 0$

$x + 5 = 0$ or $x + 5 = 0$

$x = -5$ $\qquad x = -5$

The solution is –5.

47. $x(x - 10) = -16$

$x^2 - 10x = -16$

$x^2 - 10x + 16 = 0$

$(x-8)(x-2) = 0$

$x - 8 = 0$ or $x - 2 = 0$

$x = 8$ $\qquad x = 2$

The solutions are 8 and 2.

48. $(3x-1)(9x^2 - 6x + 1) = 0$

$(3x-1)(3x-1)(3x-1) = 0$

$3x - 1 = 0$ or $3x - 1 = 0$ or $3x - 1 = 0$

$3x = 1$ $\qquad 3x = 1$ $\qquad 3x = 1$

$x = \dfrac{1}{3}$ $\qquad x = \dfrac{1}{3}$ $\qquad x = \dfrac{1}{3}$

The solution is $\dfrac{1}{3}$.

49. $56x^2 - 5x - 6 = 0$

$56x^2 + 16x - 21x - 6 = 0$

$8x(7x+2) - 3(7x+2) = 0$

$(7x+2)(8x-3) = 0$

$7x + 2 = 0$ or $8x - 3 = 0$

$7x = -2$ $\qquad 8x = 3$

$x = -\dfrac{2}{7}$ $\qquad x = \dfrac{3}{8}$

The solutions are $-\dfrac{2}{7}$ and $\dfrac{3}{8}$.

50. $20x^2 - 7x - 6 = 0$

$(4x-3)(5x+2) = 0$

$4x - 3 = 0$ or $5x + 2 = 0$

$4x = 3$ $\qquad 5x = -2$

$x = \dfrac{3}{4}$ $\qquad x = -\dfrac{2}{5}$

The solutions are $\dfrac{3}{4}$ and $-\dfrac{2}{5}$.

51. $5(3x + 2) = 4$

$15x + 10 = 4$

$15x = -6$

$x = -\dfrac{6}{15} = -\dfrac{2}{5}$

The solution is $-\dfrac{2}{5}$.

52. $6x^2 - 3x + 8 = 0$

The equation has no real solution.

53. $12 - 5t = -3$

$-5t = -15$

$t = 3$

The solution is 3.

54. $5x^3 + 20x^2 + 20x = 0$

$5x(x^2 + 4x + 4) = 0$

$5x(x+2)(x+2) = 0$

$x + 2 = 0$ or $5x = 0$

$x = -2$ $\qquad x = 0$

The solutions are –2 and 0.

55. $4t^3 - 5t^2 - 21t = 0$

$t(4t^2 - 5t - 21) = 0$

$t(4t+7)(t-3) = 0$

$t = 0$ or $4t + 7 = 0$ or $t - 3 = 0$

$\qquad\qquad 4t = -7$ $\qquad t = 3$

$\qquad\qquad t = -\dfrac{7}{4}$

The solutions are 0, $-\dfrac{7}{4}$, and 3.

56. Answers may vary. Possible answer:

$(x-4)(x-5) = 0$

$x^2 - 9x + 20 = 0$

57. a. $7 \neq 2 \cdot 5$

b. $10 = 2 \cdot 5$
$$\begin{aligned}
P &= 2l + 2w \\
&= 2(10) + 2(5) \\
&= 20 + 10 \\
&= 30 \neq 24
\end{aligned}$$

c. $8 = 2 \cdot 4$
$$P = 2l + 2w = 2(8) + 2(4) = 16 + 8 = 24$$

d. $10 \neq 2 \cdot 2$

Choice **c** gives the correct dimensions.

58. a. $3 \cdot 8 + 1 = 25 \neq 10$

b. $3 \cdot 4 + 1 = 13$
$$A = lw = 13(4) = 52 \neq 80$$

c. $3 \cdot 4 + 1 = 13 \neq 20$

d. $3 \cdot 5 + 1 = 16$
$$A = lw = 5(16) = 80$$

Choice **d** gives the correct dimensions.

59.
$$\begin{aligned}
x^2 &= 81 \\
x^2 - 81 &= 0 \\
(x - 9)(x + 9) &= 0
\end{aligned}$$
$$x - 9 = 0 \quad \text{or} \quad x + 9 = 0$$
$$x = 9 \qquad\qquad x = -9$$
Since length is not negative, the length of the side is 9 units.

60. $(2x + 3) + (3x + 1) + (x^2 - 3x) + (x + 3) = 47$
$$\begin{aligned}
x^2 + 3x + 7 &= 47 \\
x^2 + 3x - 40 &= 0 \\
(x - 5)(x + 8) &= 0
\end{aligned}$$
$$x - 5 = 0 \quad \text{or} \quad x + 8 = 0$$
$$x = 5 \qquad\qquad x = -8$$
Length is not negative, so $x = 5$. The lengths are:
$x + 3 = 5 + 3 = 8$ units
$2x + 3 = 2(5) + 3 = 13$ units
$3x + 1 = 3(5) + 1 = 16$ units
$x^2 - 3x = 5^2 - 3(5) = 10$ units

61.   Let x = the width of the flag. Then
$2x - 15$ = the length of the flag.
$$\begin{aligned}
A &= lw \\
500 &= (2x - 15)(x) \\
500 &= 2x^2 - 15x \\
0 &= 2x^2 - 15x - 500 \\
0 &= (2x + 25)(x - 20)
\end{aligned}$$
$$2x + 25 = 0$$
$$2x = -25 \quad \text{or} \quad x - 20 = 0$$
$$x = -\frac{25}{2} \qquad\qquad x = 20$$
Since the dimensions cannot be negative, the width is 20 inches and the length is $2(20) - 15 = 25$ inches.

62.   Let x = the height of the sail, then $4x$ = the base of the sail.
$$A = \frac{1}{2}bh$$
$$162 = \frac{1}{2}(4x)(x)$$
$$162 = 2x^2$$
$$0 = 2x^2 - 162$$
$$0 = 2(x^2 - 81)$$
$$0 = 2(x + 9)(x - 9)$$
$$x + 9 = 0 \quad \text{or} \quad x - 9 = 0$$
$$x = -9 \qquad\qquad x = 9$$
Since the dimensions cannot be negative, the height is 9 yards and the base is $4 \cdot 9 = 36$ yards.

63.   Let x = the first integer. Then $x + 1$ = the next consecutive integer.
$$x(x + 1) = 380$$
$$x^2 + x = 380$$
$$x^2 + x - 380 = 0$$
$$(x + 20)(x - 19) = 0$$
$$x + 20 = 0 \quad \text{or} \quad x - 19 = 0$$
$$x = -20 \qquad\qquad x = 19$$
The integers are 19 and 20.

64.   Let x be the first positive even integer. Then $x + 2$ is the next consecutive even integer.
$$x(x + 2) = 440$$
$$x^2 + 2x = 440$$
$$x^2 + 2x - 440 = 0$$
$$(x + 22)(x - 20) = 0$$
$$x + 22 = 0 \quad \text{or} \quad x - 20 = 0$$
$$x = -22 \qquad\qquad x = 20$$
Discard $x = -22$ since it is not positive. The integers are 20 and $20 + 2 = 22$.

65. a. Let $h = 2800$ and solve for t.

$$h = -16t^2 + 440t$$
$$2800 = -16t^2 + 440t$$
$$0 = -16t^2 + 440t - 2800$$
$$0 = -8(2t^2 - 55t + 350)$$
$$0 = -8(2t - 35)(t - 10)$$
$$2t - 35 = 0 \quad \text{or} \quad t - 10 = 0$$
$$2t = 35 \qquad\qquad t = 10$$
$$t = \frac{35}{2}$$
$$t = 17.5$$

The solutions are 17.5 sec and 10 sec. There are two answers because the rocket reaches a height of 2800 feet on its way up and on its way back down.

b. Let $h = 0$ and solve for t.

$$h = -16t^2 + 440t$$
$$0 = -16t^2 + 440t$$
$$0 = -8t(2t - 55)$$
$$-8t = 0 \quad \text{or} \quad 2t - 55 = 0$$
$$t = 0 \qquad\qquad 2t = 55$$
$$t = \frac{55}{2}$$
$$t = 27.5$$

$t = 0$ is when the rocket is launched, so it reaches the ground again after 27.5 seconds.

66. Let x = the length of the longer leg. Then $x + 8$ = the length of the hypotenuse and $x - 8$ = the length of the shorter leg.

$$(x+8)^2 = (x-8)^2 + x^2$$
$$x^2 + 16x + 64 = x^2 - 16x + 64 + x^2$$
$$0 = x^2 - 32x$$
$$0 = x(x - 32)$$
$$x = 0 \quad \text{or} \quad x - 32 = 0$$
$$x = 32$$

The longer leg is 32 centimeters.

67. $7x - 63 = 7(x - 9)$

68. $11x(4x - 3) - 6(4x - 3) = (4x - 3)(11x - 6)$

69. $m^2 - \dfrac{4}{25} = m^2 - \left(\dfrac{2}{5}\right)^2 = \left(m + \dfrac{2}{5}\right)\left(m - \dfrac{2}{5}\right)$

70. $3x^3 - 4x^2 + 6x - 8 = x^2(3x - 4) + 2(3x - 4)$
$$= (3x - 4)(x^2 + 2)$$

71. $xy + 2x - y - 2 = x(y + 2) - 1(y + 2)$
$$= (y + 2)(x - 1)$$

72. $2x^2 + 2x - 24 = 2(x^2 + x - 12) = 2(x + 4)(x - 3)$

73. $3x^3 - 30x^2 + 27x = 3x(x^2 - 10x + 9)$
$$= 3x(x - 9)(x - 1)$$

74. $4x^2 - 81 = (2x)^2 - 9^2 = (2x + 9)(2x - 9)$

75. $2x^2 - 18 = 2(x^2 - 9)$
$$= 2(x^2 - 3^2)$$
$$= 2(x + 3)(x - 3)$$

76. $16x^2 - 24x + 9 = (4x)^2 - 2 \cdot 4x \cdot 3 + 3^2$
$$= (4x - 3)^2$$

77. $5x^2 + 20x + 20 = 5(x^2 + 4x + 4)$
$$= 5(x^2 + 2 \cdot x \cdot 2 + 2^2)$$
$$= 5(x + 2)^2$$

78. $2x^2 + 5x - 12 = (2x - 3)(x + 4)$

79. $4x^2y - 6xy^2 = 2xy(2x - 3y)$

80. $125x^3 + 27 = (5x)^3 + 3^3$
$$= (5x + 3)[(5x)^2 - 5x \cdot 3 + 3^2]$$
$$= (5x + 3)(25x^2 - 15x + 9)$$

81. $24x^2 - 3x - 18 = 3(8x^2 - x - 6)$

82. $(x + 7)^2 - y^2 = [(x + 7) + y][(x + 7) - y]$
$$= (x + 7 + y)(x + 7 - y)$$

83. $x^2(x + 3) - 4(x + 3) = (x + 3)(x^2 - 4)$
$$= (x + 3)(x^2 - 2^2)$$
$$= (x + 3)(x - 2)(x + 2)$$

84. $54a^3b - 2b = 2b(27a^3 - 1)$
$$= 2b[(3a)^3 - 1^3]$$
$$= 2b(3a - 1)[(3a)^2 + 3a \cdot 1 + 1^2]$$
$$= 2b(3a - 1)(9a^2 + 3a + 1)$$

85. $(x^2 - 2) + (x^2 - 4x) + (3x^2 - 5x)$
$= x^2 + x^2 + 3x^2 - 4x - 5x - 2$
$= 5x^2 - 9x - 2$
$= (5x + 1)(x - 2)$

86. $2(2x^2 + 3) + 2(6x^2 - 14x)$
$= 4x^2 + 6 + 12x^2 - 28x$
$= 16x^2 - 28x + 6$
$= 2(8x^2 - 14x + 3)$
$= 2(4x - 1)(2x - 3)$

87. $2x^2 - x - 28 = 0$
$(2x + 7)(x - 4) = 0$
$2x + 7 = 0$ or $x - 4 = 0$
$\quad x = -\dfrac{7}{2} \qquad\qquad x = 4$

The solutions are $-\dfrac{7}{2}$ and 4.

88. $x^2 - 2x = 15$
$x^2 - 2x - 15 = 0$
$(x + 3)(x - 5) = 0$
$x + 3 = 0$ or $x - 5 = 0$
$\quad x = -3 \qquad\qquad x = 5$
The solutions are -3 and 5.

89. $2x(x + 7)(x + 4) = 0$
$2x = 0$ or $x + 7 = 0$ or $x + 4 = 0$
$\;x = 0 \qquad\quad x = -7 \qquad\qquad x = -4$
The solutions are 0, -7, and -4.

90. $x(x - 5) = -6$
$\quad x^2 - 5x = -6$
$\;x^2 - 5x + 6 = 0$
$(x - 3)(x - 2) = 0$
$x - 3 = 0$ or $x - 2 = 0$
$\quad x = 3 \qquad\qquad x = 2$
The solutions are 3 and 2.

91. $x^2 = 16x$
$x^2 - 16x = 0$
$x(x - 16) = 0$
$x = 0$ or $x - 16 = 0$
$\qquad\qquad\qquad x = 16$
The solutions are 0 and 16.

92. $(x^2 + 3) + (4x + 5) + 2x = 48$
$\qquad\quad x^2 + 6x + 8 = 48$
$\qquad\quad x^2 + 6x - 40 = 0$
$\qquad\quad (x - 4)(x + 10) = 0$
$x - 4 = 0$ or $x + 10 = 0$
$\quad x = 4 \qquad\qquad x = -10$
Since the length cannot be negative, $x = 4$.
The lengths are:
$x^2 + 3 = 4^2 + 3 = 19$ inches
$4x + 5 = 4 \cdot 4 + 5 = 21$ inches
$2x = 2 \cdot 4 = 8$ inches

93. Let $x = $ length, then $x - 4 = $ width.
$\quad A = lw$
$\;12 = x(x - 4)$
$\;12 = x^2 - 4x$
$\quad 0 = x^2 - 4x - 12$
$\quad 0 = (x - 6)(x + 2)$
$x - 6 = 0$ or $x + 2 = 0$
$\quad x = 6 \qquad\qquad x = -2$
Since length cannot be negative, the length is 6 inches and the width is $6 - 4 = 2$ inches.

94. Find t when $h = 0$.
$\quad h = -16t^2 + 729$
$\quad 0 = -16t^2 + 729$
$\quad 0 = -(16t^2 - 729)$
$\quad 0 = -[(4t)^2 - (27)^2]$
$\quad 0 = -(4t + 27)(4t - 27)$
$4t + 27 = 0$ or $4t - 27 = 0$
$\quad t = -\dfrac{27}{4} \qquad\qquad t = \dfrac{27}{4}$
Since time cannot be negative, the object reaches
the ground in $\dfrac{27}{4} = 6.75$ seconds.

95. Area of large figure – Area of circle
$= [(6x)(5x) - 2x^2] - \pi x^2$
$= 30x^2 - 2x^2 - \pi x^2$
$= 28x^2 - \pi x^2$
$= x^2(28 - \pi)$

Chapter 6 Test

1. $x^2 + 11x + 28 = (x + 7)(x + 4)$

2. $49 - m^2 = (7^2 - m^2) = (7 - m)(7 + m)$

3. $y^2 + 22y + 121 = y^2 + 2 \cdot y \cdot 11 + 11^2 = (y+11)^2$

4. $4(a+3) - y(a+3) = (a+3)(4-y)$

5. $x^2 + 4$ is the sum of two perfect squares (not the difference). The polynomial is prime.

6. $y^2 - 8y - 48 = (y-12)(y+4)$

7. $x^2 + x - 10$ is a prime polynomial.

8. $9x^3 + 39x^2 + 12x = 3x(3x^2 + 13x + 4)$
$\qquad\qquad\qquad\quad = 3x(3x+1)(x+4)$

9. $3a^2 + 3ab - 7a - 7b = 3a(a+b) - 7(a+b)$
$\qquad\qquad\qquad\qquad\quad = (a+b)(3a-7)$

10. $3x^2 - 5x + 2 = (3x-2)(x-1)$

11. $x^2 + 14xy + 24y^2 = (x+12y)(x+2y)$

12. $180 - 5x^2 = 5(36 - x^2)$
$\qquad\qquad\quad = 5(6^2 - x^2)$
$\qquad\qquad\quad = 5(6+x)(6-x)$

13. $6t^2 - t - 5 = (6t+5)(t-1)$

14. $xy^2 - 7y^2 - 4x + 28 = y^2(x-7) - 4(x-7)$
$\qquad\qquad\qquad\qquad\quad = (x-7)(y^2 - 4)$
$\qquad\qquad\qquad\qquad\quad = (x-7)(y^2 - 2^2)$
$\qquad\qquad\qquad\qquad\quad = (x-7)(y+2)(y-2)$

15. $x - x^5 = x(1 - x^4)$
$\qquad\qquad = x[1 - (x^2)^2]$
$\qquad\qquad = x(1 + x^2)(1 - x^2)$
$\qquad\qquad = x(1 + x^2)(1^2 - x^2)$
$\qquad\qquad = x(1 + x^2)(1 + x)(1 - x)$

16. $-xy^3 - x^3 y = xy(y^2 + x^2)$

17. $64x^3 - 1 = (4x)^3 - 1^3$
$\qquad\qquad\quad = (4x-1)[(4x)^2 + 4x \cdot 1 + 1^2]$
$\qquad\qquad\quad = (4x-1)(16x^2 + 4x + 1)$

18. $8y^3 - 64 = 8(y^3 - 8)$
$\qquad\qquad\quad = 8(y^3 - 2^3)$
$\qquad\qquad\quad = 8(y-2)(y^2 + y \cdot 2 + 2^2)$
$\qquad\qquad\quad = 8(y-2)(y^2 + 2y + 4)$

19. $(x-3)(x+9) = 0$
$\quad x - 3 = 0 \quad$ or $\quad x + 9 = 0$
$\qquad x = 3 \qquad\qquad\quad x = -9$
The solutions are 3 and -9.

20. $\qquad\quad x^2 + 5x = 14$
$\qquad\quad x^2 + 5x - 14 = 0$
$\qquad\quad (x+7)(x-2) = 0$
$\quad x + 7 = 0 \quad$ or $\quad x - 2 = 0$
$\qquad x = -7 \qquad\qquad x = 2$
The solutions are -7 and 2.

21. $\qquad\quad x(x+6) = 7$
$\qquad\qquad x^2 + 6x = 7$
$\qquad\quad x^2 + 6x - 7 = 0$
$\qquad\quad (x+7)(x-1) = 0$
$\quad x + 7 = 0 \quad$ or $\quad x - 1 = 0$
$\qquad x = -7 \qquad\qquad x = 1$
The solutions are -7 and 1.

22. $3x(2x-3)(3x+4) = 0$
$\quad 3x = 0 \quad$ or $\quad 2x - 3 = 0 \quad$ or $\quad 3x + 4 = 0$
$\quad x = 0 \qquad\qquad 2x = 3 \qquad\qquad\quad 3x = -4$
$\qquad\qquad\qquad\qquad x = \dfrac{3}{2} \qquad\qquad\quad x = -\dfrac{4}{3}$
The solutions are 0, $\dfrac{3}{2}$, and $-\dfrac{4}{3}$.

23. $\qquad\quad 5t^3 - 45t = 0$
$\qquad\qquad 5t(t^2 - 9) = 0$
$\qquad 5t(t+3)(t-3) = 0$
$\quad 5t = 0 \quad$ or $\quad t + 3 = 0 \quad$ or $\quad t - 3 = 0$
$\quad t = 0 \qquad\qquad t = -3 \qquad\qquad t = 3$
The solutions are 0, -3, and 3.

24. $t^2 - 2t - 15 = 0$
$\quad (t-5)(t+3) = 0$
$\quad t - 5 = 0 \quad$ or $\quad t + 3 = 0$
$\qquad t = 5 \qquad\qquad t = -3$
The solutions are 5 and -3.

25.
$$6x^2 = 15x$$
$$6x^2 - 15x = 0$$
$$3x(2x - 5) = 0$$
$$3x = 0 \quad \text{or} \quad 2x - 5 = 0$$
$$x = 0 \qquad\qquad 2x = 5$$
$$x = \frac{5}{2}$$

The solutions are 0 and $\frac{5}{2}$.

26. Let x = the altitude of the triangle, then $x + 9$ = the base.
$$A = \frac{1}{2}bh$$
$$68 = \frac{1}{2}(x + 9)(x)$$
$$136 = x^2 + 9x$$
$$0 = x^2 + 9x - 136$$
$$0 = (x + 17)(x - 8)$$
$$x + 17 = 0 \quad \text{or} \quad x - 8 = 0$$
$$x = -17 \qquad\qquad x = 8$$
Since the length of the base cannot be negative, the base is $8 + 9 = 17$ feet.

27. Let x = the first number, then $17 - x$ = the other number.
$$x^2 + (17 - x)^2 = 145$$
$$x^2 + 289 - 34x + x^2 = 145$$
$$2x^2 - 34x + 144 = 0$$
$$2(x^2 - 17x + 72) = 0$$
$$2(x - 9)(x - 8) = 0$$
$$x - 9 = 0 \quad \text{or} \quad x - 8 = 0$$
$$x = 9 \qquad\qquad x = 8$$
The numbers are 8 and 9.

28. Find t when $h = 0$.
$$h = -16t^2 + 784$$
$$0 = -16t^2 + 784$$
$$0 = -16(t^2 - 49)$$
$$0 = -16(t + 7)(t - 7)$$
$$t + 7 = 0 \quad \text{or} \quad t - 7 = 0$$
$$t = -7 \qquad\qquad t = 7$$
Since the time cannot be negative, the object reaches the ground after 7 seconds.

29. Let x = length of the shorter leg, then $x + 10$ = length of hypotenuse, and $x + 5$ = length of longer leg.
$$x^2 + (x + 5)^2 = (x + 10)^2$$
$$x^2 + x^2 + 10x + 25 = x^2 + 20x + 100$$
$$x^2 - 10x - 75 = 0$$
$$(x + 5)(x - 15) = 0$$
$$x + 5 = 0 \quad \text{or} \quad x - 15 = 0$$
$$x = -5 \qquad\qquad x = 15$$
Since length cannot be negative, the lengths of the triangle sides are:
shorter leg = 15 cm, longer leg = 20 cm, hypotenuse = 25 cm.

Chapter 6 Cumulative Review

1. a. $9 \le 11$

 b. $8 > 1$

 c. $3 \ne 4$

2. a. $|-5| > |-3|$

 b. $|0| < |-2|$

3. a. $\frac{42}{49} = \frac{6 \cdot 7}{7 \cdot 7} = \frac{6}{7}$

 b. $\frac{11}{27} = \frac{11}{3 \cdot 3 \cdot 3} = \frac{11}{27}$

 c. $\frac{88}{20} = \frac{4 \cdot 22}{4 \cdot 5} = \frac{22}{5}$

4. Let $x = 20$ and $y = 10$.
$$\frac{x}{y} + 5x = \frac{20}{10} + 5(20) = 2 + 100 = 102$$

5. $\frac{8 + 2 \cdot 3}{2^2 - 1} = \frac{8 + 6}{4 - 1} = \frac{14}{3}$

6. Let $x = -20$ and $y = 10$.
$$\frac{x}{y} + 5x = \frac{-20}{10} + 5(-20) = -2 - 100 = -102$$

7. a. $3 + (-7) + (-8) = 3 + (-15) = -12$

 b. $[7 + (-10)] + [-2 + |-4|] = -3 + (-2 + 4)$
$$= -3 + 2$$
$$= -1$$

8. Let $x = -20$ and $y = -10$.

$$\frac{x}{y} + 5x = \frac{-20}{-10} + 5(-20) = 2 - 100 = -98$$

9. a. $(-8)(4) = -32$

b. $14(-1) = -14$

c. $(-9)(-10) = 90$

10. $5 - 2(3x - 7) = 5 - 6x + 14 = -6x + 19$

11. a. $7x - 3x = (7 - 3)x = 4x$

b. $10y^2 + y^2 = (10 + 1)y^2 = 11y^2$

c. $8x^2 + 2x - 3x = 8x^2 + (2 - 3)x = 8x^2 - x$

d. $9n^2 - 5n^2 + n^2 = (9 - 5 + 1)n^2 = 5n^2$

12. $0.8y + 0.2(y - 1) = 1.8$
$0.8y + 0.2y - 0.2 = 1.8$
$1.0y - 0.2 = 1.8$
$y = 2.0$

13. $\dfrac{y}{7} = 20$

$y\left(\dfrac{y}{7}\right) = 7(20)$

$y = 140$

14. $\dfrac{x}{-7} = -4$

$-7\left(\dfrac{x}{-7}\right) = -7(-4)$

$x = 28$

15. $-3x = 33$
$\dfrac{-3x}{-3} = \dfrac{33}{-3}$
$x = -11$

16. $-\dfrac{2}{3}x = -22$

$\left(-\dfrac{3}{2}\right)\left(-\dfrac{2}{3}\right)x = \left(-\dfrac{3}{2}\right)(-22)$

$x = 33$

17. $8(2 - t) = -5t$
$16 - 8t = -5t$
$16 - 8t + 5t = -5t + 5t$
$16 - 3t = 0$
$16 - 16 - 3t = -16$
$-3t = -16$
$\dfrac{-3t}{-3} = \dfrac{-16}{-3}$
$t = \dfrac{16}{3}$

18. $-z = \dfrac{7z + 3}{5}$

$5(-z) = 5\left(\dfrac{7z + 3}{5}\right)$

$-5z = 7z + 3$
$-5z - 7z = 7z - 7z + 3$
$-12z = 3$
$\dfrac{-12z}{-12} = \dfrac{3}{-12}$
$z = -\dfrac{1}{4}$

19. Let $x =$ the length of the shorter piece and
$3x =$ the length of the longer piece.
$x + 3x = 48$
$4x = 48$
$x = 12$
$3x = 3(12) = 36$
The pieces are 12 inches and 36 inches in length.

20. $3x + 9 \le 5(x - 1)$
$3x + 9 \le 5x - 5$
$-2x + 9 \le -5$
$2x \le -14$
$\dfrac{-2x}{-2} \ge \dfrac{-14}{-2}$
$x \ge 7, [7, \infty)$

21. $y = -\dfrac{1}{3}x + 2$

x	y
0	2
-3	3
3	1

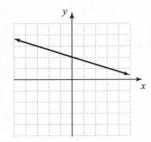

22. $-7x - 8y = -9$

$(-1, 2): -7(-1) - 8(2) \overset{?}{=} -9$

$\qquad\qquad 7 - 16 \overset{?}{=} -9$

$\qquad\qquad\qquad -9 = -9 \quad$ True

$(-1, 2)$ is a solution of the equation.

23. $3x - 4y = 4$

$\qquad -4y = -3x + 4$

$\qquad\quad y = \dfrac{3}{4}x - 1$

$\qquad\quad y = mx + b$

slope $= \dfrac{3}{4}$; y-intercept $= (0, -1)$

24. $(5, -6)$ and $(5, 2)$

$m = \dfrac{y_2 - y_1}{x_2 - x_1} = \dfrac{2 - (-6)}{5 - 5} = \dfrac{8}{0}$

The slope is undefined.

25. a. If $x = 5$, $2x^3 = 2(5)^3 = 2(125) = 250$.

b. If $x = -3$, $\dfrac{9}{x^2} = \dfrac{9}{(-3)^2} = \dfrac{9}{9} = 1$.

26. $7x - 3y = 2$

$\qquad -3y = -7x + 2$

$\qquad\quad y = \dfrac{-7x}{-3} + \dfrac{2}{-3}$

$\qquad\quad y = \dfrac{7}{3}x - \dfrac{2}{3}$

$\qquad\quad y = mx + b$

slope $= \dfrac{7}{3}$; y-intercept $= \left(0, -\dfrac{2}{3}\right)$

27. a. $3x^2$ has degree 2.

b. $-2^3 x^5 = -8x^5$ has degree 5.

c. $y = y^1$ has degree 1.

d. $12x^2 yz^3$ has degree $2 + 1 + 3 = 6$.

e. 5 has degree 0.

28. Vertical line has equation $x = c$.
Point $(0, 7)$
$x = 0$

29. $(2x^3 + 8x^2 - 6x) - (2x^3 - x^2 + 1)$

$= 2x^3 + 8x^2 - 6x - 2x^3 + x^2 - 1$

$= 9x^2 - 6x - 1$

30. $m = 4$, $b = \dfrac{1}{2}$

$\qquad y = mx + b$

$\qquad y = 4x + \dfrac{1}{2}$

$\qquad 2y = 8x + 1$

$\qquad 8x - 2y = -1$

31. $(3x + 2)(2x - 5)$

$= 3x(2x) + 3x(-5) + 2(2x) + 2(-5)$

$= 6x^2 - 15x + 4x - 10$

$= 6x^2 - 11x - 10$

32. $(-4, 0)$ and $(6, -1)$

$m = \dfrac{y_2 - y_1}{x_2 - x_1} = \dfrac{-1 - 0}{6 - (-4)} = -\dfrac{1}{10}$

$m = -\dfrac{1}{10}$, point $(-4, 0)$

$\quad y - y_1 = m(x - x_1)$

$\quad y - 0 = -\dfrac{1}{10}[x - (-4)]$

$\qquad\quad y = -\dfrac{1}{10}x - \dfrac{4}{10}$

$\qquad 10y = -x - 4$

$\qquad x + 10y = -4$

33. $(3y + 1)^2 = (3y)^2 + 2(3y)(1) + 1^2 = 9y^2 + 6y + 1$

34. $\begin{cases} -x + 3y = 18 \\ -3x + 2y = 19 \end{cases}$

Multiply the first equation by -3.

$\quad 3x - 9y = -54$

$\underline{-3x + 2y = 19}$

$\qquad\quad -7y = -35$

$\qquad\qquad y = 5$

Substitute 5 for y in the first equation.

$$-x + 3(5) = 18$$
$$-x + 15 = 18$$
$$-x = 3$$
$$x = -3$$

The solution to the system is $(-3, 5)$.

35. a. $3^{-2} = \dfrac{1}{3^2} = \dfrac{1}{9}$

b. $2x^{-3} = \dfrac{2}{x^3}$

c. $2^{-1} + 4^{-1} = \dfrac{1}{2} + \dfrac{1}{4}$
$$= \dfrac{1 \cdot 2}{2 \cdot 2} + \dfrac{1}{4}$$
$$= \dfrac{2}{4} + \dfrac{1}{4}$$
$$= \dfrac{2+1}{4}$$
$$= \dfrac{3}{4}$$

d. $(-2)^{-4} = \dfrac{1}{(-2)^4} = \dfrac{1}{16}$

e. $y^{-4} = \dfrac{1}{y^4}$

36. $\dfrac{(5a^7)^2}{a^5} = \dfrac{5^2 a^{14}}{a^5} = 25a^{14-5} = 25a^9$

37. a. $367,000,000 = 3.67 \times 10^8$

b. $0.000003 = 3.0 \times 10^{-6}$

c. $20,520,000,000 = 2.052 \times 10^{10}$

d. $0.00085 = 8.5 \times 10^{-4}$

38. $(3x - 7y)^2 = (3x)^2 - 2(3x)(7y) + (7y)^2$
$$= 9x^2 - 42xy + 49y^2$$

39.
$$\require{enclose}\begin{array}{r} x+4 \\ x+3 \enclose{longdiv}{x^2 + 7x + 12} \\ \underline{x^2 + 3x} \\ 4x + 12 \\ \underline{4x + 12} \\ 0 \end{array}$$

$$\dfrac{x^2 + 7x + 12}{x + 3} = x + 4$$

40. $\dfrac{(xy)^{-3}}{(x^5 y^6)^3} = \dfrac{x^{-3} y^{-3}}{x^{15} y^{18}}$
$$= x^{-3-15} y^{-3-18}$$
$$= x^{-18} y^{-21}$$
$$= \dfrac{1}{x^{18} y^{21}}$$

41. a. x^3, x^7, x^5: GCF $= x^3$

b. y, y^4, y^7: GCF $= y$

42. $z^3 + 7z + z^2 + 7 = z(z^2 + 7) + 1(z^2 + 7)$
$$= (z^2 + 7)(z + 1)$$

43. $x^2 + 7x + 12 = (x + 4)(x + 3)$

44. $2x^3 + 2x^2 - 84x = 2x(x^2 + x - 42)$
$$= 2x(x + 7)(x - 6)$$

45. $8x^2 - 22x + 5 = 8x^2 - 20x - 2x + 5$
$$= 4x(2x - 5) - 1(2x - 5)$$
$$= (2x - 5)(4x - 1)$$

46. $-4x^2 - 23x + 6 = -1(4x^2 + 23x - 6)$
$$= -(4x^2 - x + 24x - 6)$$
$$= -[x(4x - 1) + 6(4x - 1)]$$
$$= -(4x - 1)(x + 6)$$

47. $25a^2 - 9b^2 = (5a)^2 - (3b)^2 = (5a + 3b)(5a - 3b)$

48. $9xy^2 - 16x = x(9y^2 - 16)$
$$= x[(3y)^2 - 4^2]$$
$$= x(3y + 4)(3y - 4)$$

49. $(x - 3)(x + 1) = 0$

$x - 3 = 0$ or $x + 1 = 0$

$x = 3$ $x = -1$

The solutions are 3 and -1.

50. $x^2 - 13x = -36$

$x^2 - 13x + 36 = 0$

$(x - 9)(x - 4) = 0$

$x - 9 = 0$ or $x - 4 = 0$

$x = 9$ $x = 4$

The solutions are 9 and 4.

Chapter 7

Section 7.1 Practice

1. a. The denominator of $f(x)$ is never 0.
Domain: $\{x | x \text{ is a real number}\}$

   b. Undefined values when
$x + 3 = 0$, or $x = -3$
Domain: $\{x | x \text{ is a real number and } x \neq -3\}$

   c. Undefined values when
$$x^2 - 5x + 6 = 0$$
$$(x-3)(x-2) = 0$$
$$x - 3 = 0 \quad \text{or} \quad x - 2 = 0$$
$$x = 3 \quad \text{or} \quad x = 2$$
Domain:
$\{x | x \text{ is a real number and } x \neq 2, x \neq 3\}$

2. a.
$$\frac{5z^4}{10z^5 - 5z^4} = \frac{5z^4 \cdot 1}{5z^4(2z-1)}$$
$$= 1 \cdot \frac{1}{2z-1} = \frac{1}{2z-1}$$

   b.
$$\frac{5x^2 + 13x + 6}{6x^2 + 7x - 10} = \frac{(5x+3)(x+2)}{(6x-5)(x+2)}$$
$$= \frac{5x+3}{6x-5} \cdot 1$$
$$= \frac{5x+3}{6x-5}$$

3. a.
$$\frac{x+3}{3+x} = \frac{x+3}{x+3} = 1$$

   b.
$$\frac{3-x}{x-3} = \frac{-1(-3+x)}{x-3} = \frac{-1(x-3)}{x-3} = \frac{-1}{1} = -1$$

4.
$$\frac{20 - 5x^2}{x^2 + x - 6} = \frac{5(4 - x^2)}{(x+3)(x-2)}$$
$$= \frac{5(2+x)(2-x)}{(x+3)(x-2)}$$
$$= \frac{5(2+x) \cdot (-1)(x-2)}{(x+3)(x-2)}$$
$$= -\frac{5(2+x)}{x+3}$$

5. a.
$$\frac{x^3 + 64}{4 + x} = \frac{(x+4)(x^2 - 4x + 16)}{x+4}$$
$$= x^2 - 4x + 16$$

   b.
$$\frac{5z^2 + 10}{z^3 - 3z^2 + 2z - 6} = \frac{5(z^2 + 2)}{(z^3 - 3z^2) + (2z - 6)}$$
$$= \frac{5(z^2 + 2)}{z^2(z-3) + 2(z-3)}$$
$$= \frac{5(z^2 + 2)}{(z-3)(z^2 + 2)}$$
$$= \frac{5}{z-3}$$

6.
$$-\frac{x+3}{6x-11} = \frac{-(x+3)}{6x-11} = \frac{-x-3}{6x-11}$$
Also,
$$-\frac{x+3}{6x-11} = \frac{x+3}{-(6x-11)} = \frac{x+3}{-6x+11} \text{ or } \frac{x+3}{11-6x}$$
Thus, some equivalent forms of $-\dfrac{x+3}{6x-11}$ are
$$\frac{-(x+3)}{6x-11}, \frac{-x-3}{6x-11}, \frac{x+3}{-(6x-11)}, \frac{x+3}{-6x+11}, \text{ and }$$
$$\frac{x+3}{11-6x}.$$

7. a.
$$C(100) = \frac{3.2(100) + 400}{100} = \frac{720}{100} = 7.2$$
$7.20 per tee shirt

   b.
$$C(1000) = \frac{3.2(1000) + 400}{1000} = \frac{3600}{1000} = 3.6$$
$3.60 per tee shirt

Graphing Calculator Explorations

1.
$$x^2 - 4 = 0$$
$$(x+2)(x-2) = 0$$
$$x + 2 = 0 \quad \text{or} \quad x - 2 = 0$$
$$x = -2 \quad \text{or} \quad x = 2$$
Domain: $\{x | x \text{ is a real number and } x \neq -2, x \neq 2\}$

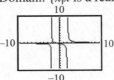

2.
$$x^2 - 9 = 0$$
$$(x+3)(x-3) = 0$$
$$x + 3 = 0 \quad \text{or} \quad x - 3 = 0$$
$$x = -3 \quad \text{or} \quad x = 3$$
Domain: $\{x | x$ is a real number and $x \neq -3, x \neq 3\}$

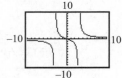

3.
$$2x^2 + 7x - 4 = 0$$
$$(2x-1)(x+4) = 0$$
$$2x - 1 = 0 \quad \text{or} \quad x + 4 = 0$$
$$2x = 1 \quad \text{or} \quad x = -4$$
$$x = \frac{1}{2}$$
Domain:
$$\left\{ x \middle| x \text{ is a real number and } x \neq -4, \ x \neq \frac{1}{2} \right\}$$

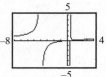

4.
$$4x^2 - 19x - 5 = 0$$
$$(4x+1)(x-5) = 0$$
$$4x + 1 = 0 \quad \text{or} \quad x - 5 = 0$$
$$4x = -1 \quad \text{or} \quad x = 5$$
$$x = -\frac{1}{4}$$
Domain:
$$\left\{ x \middle| x \text{ is a real number and } x \neq -\frac{1}{4}, \ x \neq 5 \right\}$$

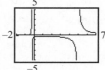

Vocabulary, Readiness & Video Check 7.1

1. A <u>rational</u> expression is an expression that can be written as the quotient $\dfrac{P}{Q}$ of two polynomials P and Q as long as $Q \neq 0$.

2. A rational expression is undefined if the denominator is <u>0</u>.

3. The <u>domain</u> of the rational function $f(x) = \dfrac{2}{x}$ is $\{x | x$ is a real number and $x \neq 0\}$.

4. A rational expression is <u>simplified</u> if the numerator and denominator have no common factors other than 1 or −1.

5. The expression $\dfrac{x^2 + 2}{2 + x^2}$ simplifies to <u>1</u>.

6. The expression $\dfrac{y - z}{z - y}$ simplifies to <u>−1</u>.

7. For a rational expression, $-\dfrac{a}{b} = \dfrac{-a}{\underline{b}} = \dfrac{a}{\underline{-b}}$.

8. The statement $\dfrac{a-6}{a+2} = \dfrac{-(a-6)}{-(a+2)} = \dfrac{-a+6}{-a-2}$ is true.

9. Rational expressions are fractions and are therefore undefined if the denominator is zero; the domain of a rational function is all real numbers except those that make the denominator of the related rational expression equal to 0. If a denominator contains variables, set it equal to zero and solve.

10. Although x is a factor in the numerator, it is not a factor in the denominator—factor means write as a product and the denominator is a difference, not a product.

11. We insert parentheses around the numerator or denominator if it has more than one term because the negative sign needs to apply to the entire numerator or denominator.

12. For $R(x)$, x is defined as the number of years since publication. Part b. gives us the revenue from publication to the end of the second year, thus two years of revenue. Since part c. asks for the second year revenue only, we subtract part a. from part b.

Exercise Set 7.1

1. 4 is never 0, so the domain of $f(x) = \dfrac{5x - 7}{4}$ is $\{x | x$ is a real number$\}$.

3. $2t = 0$

$\quad t = 0$

The domain of $s(t) = \dfrac{t^2 + 1}{2t}$ is

$\{t | t$ is a real number and $t \neq 0\}$.

5. $7 - x = 0$

$\quad 7 = x$

The domain of $f(x) = \dfrac{3x}{7-x}$ is

$\{x | x$ is a real number and $x \neq 7\}$.

7. $3x - 1 = 0$

$\quad 3x = 1$

$\quad x = \dfrac{1}{3}$

The domain of $f(x) = \dfrac{x}{3x-1}$ is

$\left\{ x \middle| x \text{ is a real number and } x \neq \dfrac{1}{3} \right\}$.

9. $x^3 + x^2 - 2x = 0$

$\quad x(x^2 + x - 2) = 0$

$\quad x(x+2)(x-1) = 0$

$\quad x = 0 \quad$ or $\quad x + 2 = 0 \quad$ or $\quad x - 1 = 0$

$\quad x = 0 \quad$ or $\quad\quad\ x = -2 \quad$ or $\quad\quad\ x = 1$

The domain of $R(x) = \dfrac{3 + 2x}{x^3 + x^2 - 2x}$ is

$\{x | x$ is a real number and $x \neq -2, x \neq 0,$

$\quad\quad x \neq 1\}$.

11. $\quad\quad\ x^2 - 4 = 0$

$\quad (x+2)(x-2) = 0$

$\quad x + 2 = 0 \quad$ or $\quad x - 2 = 0$

$\quad\quad x = -2 \quad$ or $\quad\quad x = 2$

The domain of $C(x) = \dfrac{x+3}{x^2 - 4}$ is

$\{x | x$ is a real number and $x \neq 2, x \neq -2\}$.

13. $-\dfrac{x-10}{x+8} = \dfrac{-(x-10)}{x+8} = \dfrac{-x+10}{x+8}$ or $\dfrac{10-x}{x+8}$

$\quad -\dfrac{x-10}{x+8} = \dfrac{x-10}{-(x+8)} = \dfrac{x-10}{-x-8}$

15. $-\dfrac{5y-3}{y-12} = \dfrac{-(5y-3)}{y-12} = \dfrac{-5y+3}{y-12}$ or $\dfrac{3-5y}{y-12}$

$\quad -\dfrac{5y-3}{y-12} = \dfrac{5y-3}{-(y-12)} = \dfrac{5y-3}{-y+12}$ or $\dfrac{5y-3}{12-y}$

17. $\dfrac{x+7}{7+x} = \dfrac{x+7}{x+7} = 1$

19. $\dfrac{x-7}{7-x} = \dfrac{x-7}{-1(x-7)} = \dfrac{1}{-1} = -1$

21. $\dfrac{2}{8x+16} = \dfrac{2}{8(x+2)} = \dfrac{2(1)}{2(4)(x+2)} = \dfrac{1}{4(x+2)}$

23. $\dfrac{-5a-5b}{a+b} = \dfrac{-5(a+b)}{a+b} = -5$

25. $\dfrac{7x+35}{x^2+5x} = \dfrac{7(x+5)}{x(x+5)} = \dfrac{7}{x}$

27. $\dfrac{x+5}{x^2-4x-45} = \dfrac{x+5}{(x+5)(x-9)} = \dfrac{1}{x-9}$

29. $\dfrac{5x^2+11x+2}{x+2} = \dfrac{(5x+1)(x+2)}{x+2} = 5x+1$

31. $\dfrac{x^3+7x^2}{x^2+5x-14} = \dfrac{x^2(x+7)}{(x+7)(x-2)} = \dfrac{x^2}{x-2}$

33. $\dfrac{2x^2-8}{4x-8} = \dfrac{2(x^2-4)}{4(x-2)} = \dfrac{2(x+2)(x-2)}{2\cdot 2(x-2)} = \dfrac{x+2}{2}$

35. $\dfrac{4-x^2}{x-2} = \dfrac{(-1)(x^2-4)}{x-2}$

$\quad\quad = -\dfrac{(x+2)(x-2)}{x-2}$

$\quad\quad = -(x+2)$ or $-x-2$

37. $\dfrac{11x^2-22x^3}{6x-12x^2} = \dfrac{11x^2(1-2x)}{6x(1-2x)} = \dfrac{11x}{6}$

39. $\dfrac{x^2+xy+2x+2y}{x+2} = \dfrac{x(x+y)+2(x+y)}{x+2}$

$\quad\quad = \dfrac{(x+y)(x+2)}{x+2}$

$\quad\quad = x+y$

41. $\dfrac{x^3+8}{x+2} = \dfrac{(x+2)(x^2-2x+4)}{x+2} = x^2-2x+4$

43. $\dfrac{x^3-1}{1-x} = \dfrac{(x-1)(x^2+x+1)}{-1(x-1)}$

$\qquad = -1(x^2+x+1)$

$\qquad = -x^2-x-1$

45. $\dfrac{2xy+5x-2y-5}{3xy+4x-3y-4} = \dfrac{x(2y+5)-1(2y+5)}{x(3y+4)-1(3y+4)}$

$\qquad = \dfrac{(2y+5)(x-1)}{(3y+4)(x-1)}$

$\qquad = \dfrac{2y+5}{3y+4}$

47. $\dfrac{3x^2-5x-2}{6x^3+2x^2+3x+1} = \dfrac{(3x+1)(x-2)}{2x^2(3x+1)+1(3x+1)}$

$\qquad = \dfrac{(3x+1)(x-2)}{(3x+1)(2x^2+1)}$

$\qquad = \dfrac{x-2}{2x^2+1}$

49. $\dfrac{9x^2-15x+25}{27x^3+125} = \dfrac{9x^2-15x+25}{(3x+5)(9x^2-15x+25)}$

$\qquad = \dfrac{1}{3x+5}$

51. $\dfrac{9-x^2}{x-3} = \dfrac{-1(x^2-9)}{x-3}$

$\qquad = -\dfrac{(x+3)(x-3)}{x-3}$

$\qquad = -(x+3)$ or $-x-3$

The given answer is correct.

53. $\dfrac{7-34x-5x^2}{25x^2-1} = \dfrac{-1(5x^2+34x-7)}{(5x+1)(5x-1)}$

$\qquad = -\dfrac{(x+7)(5x-1)}{(5x+1)(5x-1)}$

$\qquad = -\dfrac{x+7}{5x+1}$

$\qquad = \dfrac{x+7}{-(5x+1)}$ or $\dfrac{x+7}{-5x-1}$

The given answer is correct.

55. $f(x) = \dfrac{x+8}{2x-1}$

$f(2) = \dfrac{2+8}{2(2)-1} = \dfrac{10}{4-1} = \dfrac{10}{3}$

$f(0) = \dfrac{0+8}{2(0)-1} = \dfrac{8}{0-1} = \dfrac{8}{-1} = -8$

$f(-1) = \dfrac{-1+8}{2(-1)-1} = \dfrac{7}{-2-1} = \dfrac{7}{-3} = -\dfrac{7}{3}$

57. $g(x) = \dfrac{x^2+8}{x^3-25x}$

$g(3) = \dfrac{3^2+8}{3^3-25(3)} = \dfrac{9+8}{27-75} = \dfrac{17}{-48} = -\dfrac{17}{48}$

$g(-2) = \dfrac{(-2)^2+8}{(-2)^3-25(-2)} = \dfrac{4+8}{-8+50} = \dfrac{12}{42} = \dfrac{2}{7}$

$g(1) = \dfrac{1^2+8}{1^3-25(1)} = \dfrac{1+8}{1-25} = \dfrac{9}{-24} = -\dfrac{3}{8}$

59. $R(x) = \dfrac{1000x^2}{x^2+4}$

a. $R(1) = \dfrac{1000\cdot1^2}{1^2+4} = \dfrac{1000}{5} = 200$

The revenue at the end of the first year is $200 million.

b. $R(2) = \dfrac{1000\cdot2^2}{2^2+4}$

$\qquad = \dfrac{1000\cdot4}{4+4}$

$\qquad = \dfrac{4000}{8}$

$\qquad = 500$

The revenue at the end of the second year is $500 million.

c. The revenue during the second year is equal to the revenue at the end of the second year minus the revenue at the end of the first year.

$500 - 200 = 300$

The revenue during the second year is $300 million.

d. $x^2 + 4 = 0$
$$x^2 = -4$$
This equation has no solutions. Thus there are no values of x that would make the denominator equal to zero and the function undefined. The domain of $R(x)$ is $\{x \mid x \text{ is a real number}\}$.

61. Let $D = 1000$ and $A = 8$.
$$C = \frac{DA}{A+12} = \frac{1000(8)}{8+12} = \frac{8000}{20} = 400$$
The child should receive 400 mg.

63. $C = \dfrac{100W}{L}$; $W = 5$, $L = 6.4$
$$C = \frac{100(5)}{6.4} = \frac{500}{6.4} = 78.125$$
The skull is medium.

65. Use $S = \dfrac{h + d + 2t + 3r}{b}$ with $h = 183$, $d = 39$, $t = 1$, $r = 42$, and $b = 587$.
$$S = \frac{183 + 39 + 2(1) + 3(42)}{587} = \frac{350}{587} \approx 0.5963$$
Pujols' slugging percentage was about 59.6%.

67. $\dfrac{1}{3} \cdot \dfrac{9}{11} = \dfrac{1 \cdot 9}{3 \cdot 11} = \dfrac{3 \cdot 3}{3 \cdot 11} = \dfrac{3}{11}$

69. $\dfrac{1}{3} \div \dfrac{1}{4} = \dfrac{1}{3} \cdot \dfrac{4}{1} = \dfrac{4}{3}$

71. $\dfrac{13}{20} \div \dfrac{2}{9} = \dfrac{13}{20} \cdot \dfrac{9}{2} = \dfrac{13 \cdot 9}{20 \cdot 2} = \dfrac{117}{40}$

73. $\dfrac{5a - 15}{5} = \dfrac{5(a-3)}{5} = a - 3$
The statement is correct.

75. $\dfrac{1+2}{1+3} = \dfrac{3}{4}$
The statement is incorrect.

77. $\dfrac{x}{x+7}$ cannot be simplified.

79. Since $3 + x = x + 3$, $\dfrac{3+x}{x+3}$ can be simplified to 1.

81. Since $5 - x = -1(x - 5)$, $\dfrac{5-x}{x-5}$ can be simplified.

83. No; answers may vary

85. answers may vary

87. $f(x) = \dfrac{20x}{100 - x}$

x	0	10	30	50	70	90	95	99
y	0	$\dfrac{20}{9}$	$\dfrac{60}{7}$	20	$\dfrac{140}{3}$	180	380	1980

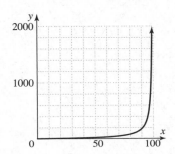

89. $y = \dfrac{x^2 - 16}{x - 4} = \dfrac{(x+4)(x-4)}{x-4} = x + 4,\ x \neq 4$

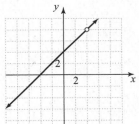

91. $y = \dfrac{x^2 - 6x + 8}{x - 2} = \dfrac{(x-2)(x-4)}{x-2} = x - 4,\ x \neq 2$

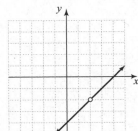

Section 7.2 Practice

1. a. $\dfrac{4a}{5}\cdot\dfrac{3}{b^2}=\dfrac{4a\cdot 3}{5\cdot b^2}=\dfrac{12a}{5b^2}$

  b. $\dfrac{-3p^4}{q^2}\cdot\dfrac{2q^3}{9p^4}=\dfrac{-3p^4\cdot 2q^3}{q^2\cdot 9p^4}$

  $=\dfrac{-1\cdot 3\cdot p^4\cdot 2\cdot q\cdot q^2}{q^2\cdot 3\cdot 3\cdot p^4}$

  $=-\dfrac{2q}{3}$

2. $\dfrac{x^2-x}{5x}\cdot\dfrac{15}{x^2-1}=\dfrac{x(x-1)}{5x}\cdot\dfrac{3\cdot 5}{(x+1)(x-1)}$

  $=\dfrac{x(x-1)\cdot 3\cdot 5}{5x\cdot(x+1)(x-1)}$

  $=\dfrac{3}{x+1}$

3. $\dfrac{6-3x}{6x+6x^2}\cdot\dfrac{3x^2-2x-5}{x^2-4}$

  $=\dfrac{3(2-x)}{2\cdot 3\cdot x(1+x)}\cdot\dfrac{(x+1)(3x-5)}{(x+2)(x-2)}$

  $=\dfrac{3(2-x)(x+1)(3x-5)}{2\cdot 3x(1+x)(x+2)(x-2)}$

  $=\dfrac{-1(x-2)(x+1)(3x-5)}{2x(x+1)(x+2)(x-2)}$

  $=-\dfrac{3x-5}{2x(x+2)}$

4. $\dfrac{5a^3b^2}{24}\div\dfrac{10a^5}{6}=\dfrac{5a^3b^2}{24}\cdot\dfrac{6}{10a^5}$

  $=\dfrac{5a^3b^2\cdot 6}{4\cdot 6\cdot 2\cdot 5\cdot a^2\cdot a^3}$

  $=\dfrac{b^2}{8a^2}$

5. $\dfrac{(x-5)^2}{3}\div\dfrac{4x-20}{9}=\dfrac{(x-5)(x-5)}{3}\cdot\dfrac{9}{4x-20}$

  $=\dfrac{(x-5)(x-5)\cdot 3\cdot 3}{3\cdot 4(x-5)}$

  $=\dfrac{3(x-5)}{4}$

6. $\dfrac{10x-2}{x^2-9}\div\dfrac{5x^2-x}{x+3}=\dfrac{10x-2}{x^2-9}\cdot\dfrac{x+3}{5x^2-x}$

  $=\dfrac{2(5x-1)(x+3)}{(x+3)(x-3)\cdot x(5x-1)}$

  $=\dfrac{2}{x(x-3)}$

7. $\dfrac{3x^2-11x-4}{2x-8}\div\dfrac{9x+3}{6}=\dfrac{3x^2-11x-4}{2x-8}\cdot\dfrac{6}{9x+3}$

  $=\dfrac{(3x+1)(x-4)\cdot 2\cdot 3}{2(x-4)\cdot 3(3x+1)}$

  $=\dfrac{1}{1}$ or 1

8. a. $\dfrac{y+9}{8x}\cdot\dfrac{y+9}{2x}=\dfrac{(y+9)\cdot(y+9)}{8x\cdot 2x}=\dfrac{(y+9)^2}{16x^2}$

  b. $\dfrac{y+9}{8x}\div\dfrac{y+9}{2}=\dfrac{y+9}{8x}\cdot\dfrac{2}{y+9}$

  $=\dfrac{(y+9)\cdot 2}{2\cdot 4\cdot x\cdot(y+9)}$

  $=\dfrac{1}{4x}$

  c. $\dfrac{35x-7x^2}{x^2-25}\cdot\dfrac{x^2+3x-10}{x^2+4x}$

  $=\dfrac{7x(5-x)}{(x+5)(x-5)}\cdot\dfrac{(x-2)(x+5)}{x(x+4)}$

  $=\dfrac{7x\cdot(-1)(x-5)\cdot(x-2)(x+5)}{(x+5)(x-5)\cdot x(x+4)}$

  $=-\dfrac{7(x-2)}{x+4}$

9. $288\text{ sq in.}=\dfrac{288\text{ sq in.}}{1}\cdot\dfrac{1\text{ sq ft}}{144\text{ sq in.}}=2\text{ sq ft}$

10. $3.5\text{ sq ft}=\dfrac{3.5\text{ sq ft}}{1}\cdot\dfrac{144\text{ sq in.}}{1\text{ sq ft}}=504\text{ sq in.}$

11. $61{,}000\text{ sq yd}=61{,}000\text{ sq yd}\cdot\dfrac{9\text{ sq ft}}{1\text{ sq yd}}$

  $=549{,}000\text{ sq ft}$

12. 102.7 feet/second

$$= \frac{102.7 \text{ feet}}{1 \text{ second}} \cdot \frac{3600 \text{ seconds}}{1 \text{ hour}} \cdot \frac{1 \text{ mile}}{5280 \text{ feet}}$$

$$= \frac{102.7 \cdot 3600}{5280} \text{ miles/hour}$$

$$\approx 70.0 \text{ miles/hour}$$

Vocabulary, Readiness & Video Check 7.2

1. The expressions $\dfrac{x}{2y}$ and $\dfrac{2y}{x}$ are called
   <u>reciprocals</u>.

2. $\dfrac{a}{b} \cdot \dfrac{c}{d} = \dfrac{a \cdot c}{\underline{b \cdot d}}$ or $\dfrac{ac}{\underline{bd}}$

3. $\dfrac{a}{b} \div \dfrac{c}{d} = \dfrac{a \cdot d}{\underline{b \cdot c}}$ or $\dfrac{ad}{\underline{bc}}$

4. $\dfrac{x}{7} \cdot \dfrac{x}{6} = \dfrac{x^2}{\underline{42}}$

5. $\dfrac{x}{7} \div \dfrac{x}{6} = \dfrac{6}{\underline{7}}$

6. Yes, multiplying and simplifying rational
   expressions often requires polynomial factoring.
   Example 2 alone involves factoring out a GCF,
   factoring a trinomial with $a \neq 1$, and factoring a
   difference of squares.

7. Dividing rational expressions is exactly like
   dividing <u>fractions</u>. Therefore, to divide by a
   rational expression, multiply by its <u>reciprocal</u>.

8. Multiplication and division of rational
   expressions are performed similarly—both
   involve multiplication—but there are important
   differences. Note the operation first to see
   whether you multiply by the reciprocal or not.

9. The units in the unit fraction consist of
   $\dfrac{\text{units converting to}}{\text{original units}}$.

Exercise Set 7.2

1. $\dfrac{3x}{y^2} \cdot \dfrac{7y}{4x} = \dfrac{3 \cdot x \cdot 7 \cdot y}{y \cdot y \cdot 4 \cdot x} = \dfrac{21}{4y}$

3. $\dfrac{8x}{2} \cdot \dfrac{x^5}{4x^2} = \dfrac{8x \cdot x^5}{2 \cdot 4x^2} = \dfrac{2 \cdot 4 \cdot x \cdot x \cdot x^4}{2 \cdot 4 \cdot x \cdot x} = x^4$

5. $-\dfrac{5a^2 b}{30a^2 b^2} \cdot b^3 = \dfrac{5a^2 b \cdot b^3}{30a^2 b^2}$

$$= -\dfrac{5 \cdot a^2 \cdot b \cdot b \cdot b^2}{5 \cdot 6 \cdot a^2 \cdot b^2}$$

$$= -\dfrac{b \cdot b}{6}$$

$$= -\dfrac{b^2}{6}$$

7. $\dfrac{x}{2x-14} \cdot \dfrac{x^2 - 7x}{5} = \dfrac{x \cdot (x^2 - 7x)}{(2x-14) \cdot 5}$

$$= \dfrac{x \cdot x(x-7)}{2(x-7) \cdot 5}$$

$$= \dfrac{x \cdot x}{2 \cdot 5}$$

$$= \dfrac{x^2}{10}$$

9. $\dfrac{6x+6}{5} \cdot \dfrac{10}{36x+36} = \dfrac{(6x+6) \cdot 10}{5 \cdot (36x+36)}$

$$= \dfrac{6(x+1) \cdot 2 \cdot 5}{5 \cdot 36(x+1)}$$

$$= \dfrac{6 \cdot 5 \cdot 2 \cdot (x+1)}{6 \cdot 5 \cdot 2 \cdot 3 \cdot (x+1)}$$

$$= \dfrac{1}{3}$$

11. $\dfrac{(m+n)^2}{m-n} \cdot \dfrac{m}{m^2 + mn} = \dfrac{(m+n)(m+n) \cdot m}{(m-n) \cdot m(m+n)} = \dfrac{m+n}{m-n}$

13. $\dfrac{x^2 - 25}{x^2 - 3x - 10} \cdot \dfrac{x+2}{x} = \dfrac{(x^2 - 25) \cdot (x+2)}{(x^2 - 3x - 10) \cdot x}$

$$= \dfrac{(x-5)(x+5) \cdot (x+2)}{(x-5)(x+2) \cdot x}$$

$$= \dfrac{x+5}{x}$$

15. $\dfrac{x^2+6x+8}{x^2+x-20}\cdot\dfrac{x^2+2x-15}{x^2+8x+16}$

$=\dfrac{(x+2)(x+4)}{(x+5)(x-4)}\cdot\dfrac{(x+5)(x-3)}{(x+4)(x+4)}$

$=\dfrac{(x+2)(x+4)\cdot(x+5)(x-3)}{(x+5)(x-4)\cdot(x+4)(x+4)}$

$=\dfrac{(x+2)(x-3)}{(x-4)(x+4)}$

17. $\dfrac{5x^7}{2x^5}\div\dfrac{15x}{4x^3}=\dfrac{5x^7}{2x^5}\cdot\dfrac{4x^3}{15x}$

$=\dfrac{5\cdot x^2\cdot x^5\cdot 2\cdot 2\cdot x\cdot x^2}{2\cdot x^5\cdot 3\cdot 5\cdot x}$

$=\dfrac{2x^4}{3}$

19. $\dfrac{8x^2}{y^3}\div\dfrac{4x^2y^3}{6}=\dfrac{8x^2}{y^3}\cdot\dfrac{6}{4x^2y^3}=\dfrac{2\cdot 4\cdot x^2\cdot 6}{y^3\cdot 4x^2y^3}=\dfrac{12}{y^6}$

21. $\dfrac{(x-6)(x+4)}{4x}\div\dfrac{2x-12}{8x^2}$

$=\dfrac{(x-6)(x+4)}{4x}\cdot\dfrac{8x^2}{2x-12}$

$=\dfrac{(x-6)(x+4)\cdot 2\cdot 4\cdot x\cdot x}{4x\cdot 2(x-6)}$

$=x(x+4)$

23. $\dfrac{3x^2}{x^2-1}\div\dfrac{x^5}{(x+1)^2}=\dfrac{3x^2}{x^2-1}\cdot\dfrac{(x+1)^2}{x^5}$

$=\dfrac{3x^2\cdot(x+1)(x+1)}{(x-1)(x+1)\cdot x^2\cdot x^3}$

$=\dfrac{3(x+1)}{x^3(x-1)}$

25. $\dfrac{m^2-n^2}{m+n}\div\dfrac{m}{m^2+nm}=\dfrac{m^2-n^2}{m+n}\cdot\dfrac{m^2+nm}{m}$

$=\dfrac{(m-n)(m+n)\cdot m(m+n)}{(m+n)\cdot m}$

$=(m-n)(m+n)$

$=m^2-n^2$

27. $\dfrac{x+2}{7-x}\div\dfrac{x^2-5x+6}{x^2-9x+14}=\dfrac{x+2}{7-x}\cdot\dfrac{x^2-9x+14}{x^2-5x+6}$

$=\dfrac{(x+2)\cdot(x-7)(x-2)}{-1(x-7)\cdot(x-3)(x-2)}$

$=-\dfrac{x+2}{x-3}$

29. $\dfrac{x^2+7x+10}{x-1}\div\dfrac{x^2+2x-15}{x-1}$

$=\dfrac{x^2+7x+10}{x-1}\cdot\dfrac{x-1}{x^2+2x-15}$

$=\dfrac{(x+5)(x+2)\cdot(x-1)}{(x-1)\cdot(x+5)(x-3)}$

$=\dfrac{x+2}{x-3}$

31. $\dfrac{5x-10}{12}\div\dfrac{4x-8}{8}=\dfrac{5x-10}{12}\cdot\dfrac{8}{4x-8}$

$=\dfrac{5(x-2)\cdot 2\cdot 4}{6\cdot 2\cdot 4(x-2)}$

$=\dfrac{5}{6}$

33. $\dfrac{x^2+5x}{8}\cdot\dfrac{9}{3x+15}=\dfrac{x(x+5)\cdot 3\cdot 3}{8\cdot 3(x+5)}=\dfrac{3x}{8}$

35. $\dfrac{7}{6p^2+q}\div\dfrac{14}{18p^2+3q}=\dfrac{7}{6p^2+q}\cdot\dfrac{18p^2+3q}{14}$

$=\dfrac{7\cdot 3(6p^2+q)}{(6p^2+q)\cdot 7\cdot 2}$

$=\dfrac{3}{2}$

37. $\dfrac{3x+4y}{x^2+4xy+4y^2}\cdot\dfrac{x+2y}{2}=\dfrac{(3x+4y)\cdot(x+2y)}{(x+2y)(x+2y)\cdot 2}$

$=\dfrac{3x+4y}{2(x+2y)}$

39. $\dfrac{(x+2)^2}{x-2}\div\dfrac{x^2-4}{2x-4}=\dfrac{(x+2)^2}{x-2}\cdot\dfrac{2x-4}{x^2-4}$

$=\dfrac{(x+2)(x+2)\cdot 2(x-2)}{(x-2)\cdot(x+2)(x-2)}$

$=\dfrac{2(x+2)}{x-2}$

41. $\dfrac{x^2-4}{24x} \div \dfrac{2-x}{6xy} = \dfrac{x^2-4}{24x} \cdot \dfrac{6xy}{2-x}$

$= \dfrac{(x+2)(x-2)\cdot 6x\cdot y}{4\cdot 6x\cdot(-1)(x-2)}$

$= -\dfrac{y(x+2)}{4}$

43. $\dfrac{a^2+7a+12}{a^2+5a+6} \cdot \dfrac{a^2+8a+15}{a^2+5a+4}$

$= \dfrac{(a+3)(a+4)\cdot(a+5)(a+3)}{(a+3)(a+2)\cdot(a+4)(a+1)}$

$= \dfrac{(a+5)(a+3)}{(a+2)(a+1)}$

45. $\dfrac{5x-20}{3x^2+x} \cdot \dfrac{3x^2+13x+4}{x^2-16}$

$= \dfrac{5(x-4)}{x(3x+1)} \cdot \dfrac{(3x+1)(x+4)}{(x+4)(x-4)}$

$= \dfrac{5(x-4)\cdot(3x+1)(x+4)}{x(3x+1)\cdot(x+4)(x-4)}$

$= \dfrac{5}{x}$

47. $\dfrac{8n^2-18}{2n^2-5n+3} \div \dfrac{6n^2+7n-3}{n^2-9n+8}$

$= \dfrac{8n^2-18}{2n^2-5n+3} \cdot \dfrac{n^2-9n+8}{6n^2+7n-3}$

$= \dfrac{2(2n+3)(2n-3)\cdot(n-8)(n-1)}{(n-1)(2n-3)\cdot(2n+3)(3n-1)}$

$= \dfrac{2(n-8)}{3n-1}$

49. $\dfrac{x^2-9}{2x} \div \dfrac{x+3}{8x^4} = \dfrac{x^2-9}{2x} \cdot \dfrac{8x^4}{x+3}$

$= \dfrac{(x+3)(x-3)}{2x} \cdot \dfrac{8x^4}{x+3}$

$= \dfrac{2x\cdot 4x^3\cdot(x+3)(x-3)}{2x\cdot(x+3)}$

$= 4x^3(x-3)$

51. $\dfrac{a^2+ac+ba+bc}{a-b} \div \dfrac{a+c}{a+b}$

$= \dfrac{a(a+c)+b(a+c)}{a-b} \cdot \dfrac{a+b}{a+c}$

$= \dfrac{(a+c)(a+b)}{a-b} \cdot \dfrac{a+b}{a+c}$

$= \dfrac{(a+c)\cdot(a+b)\cdot(a+b)}{(a-b)\cdot(a+c)}$

$= \dfrac{(a+b)^2}{a-b}$

53. $\dfrac{3x^2+8x+5}{x^2+8x+7} \cdot \dfrac{x+7}{x^2+4} = \dfrac{(3x+5)(x+1)}{(x+7)(x+1)} \cdot \dfrac{x+7}{x^2+4}$

$= \dfrac{(3x+5)\cdot(x+1)\cdot(x+7)}{(x+7)\cdot(x+1)\cdot(x^2+4)}$

$= \dfrac{3x+5}{x^2+4}$

55. $\dfrac{x^3+8}{x^2-2x+4} \cdot \dfrac{4}{x^2-4}$

$= \dfrac{(x+2)(x^2-2x+4)}{x^2-2x+4} \cdot \dfrac{4}{(x+2)(x-2)}$

$= \dfrac{4\cdot(x+2)\cdot(x^2-2x+4)}{(x+2)(x-2)(x^2-2x+4)}$

$= \dfrac{4}{x-2}$

57. $\dfrac{a^2-ab}{6a^2+6ab} \div \dfrac{a^3-b^3}{a^2-b^2}$

$= \dfrac{a^2-ab}{6a^2+6ab} \cdot \dfrac{a^2-b^2}{a^3-b^3}$

$= \dfrac{a(a-b)}{6a(a+b)} \cdot \dfrac{(a-b)(a+b)}{(a-b)(a^2+ab+b^2)}$

$= \dfrac{a\cdot(a-b)\cdot(a-b)\cdot(a+b)}{6\cdot a\cdot(a+b)\cdot(a-b)\cdot(a^2+ab+b^2)}$

$= \dfrac{a-b}{6(a^2+ab+b^2)}$

59. 10 square feet

$= \dfrac{10\text{ square feet}}{1} \cdot \dfrac{144\text{ square inches}}{1\text{ square foot}}$

$= 1440$ square inches

61. $45 \text{ square feet} = \dfrac{45 \text{ square feet}}{1} \cdot \dfrac{1 \text{ square yard}}{9 \text{ square feet}}$

$= 5 \text{ square yards}$

63. $3 \text{ cubic yards} = \dfrac{3 \text{ cubic yards}}{1} \cdot \dfrac{27 \text{ cubic feet}}{1 \text{ cubic yard}}$

$= 81 \text{ cubic feet}$

65. $\dfrac{50 \text{ miles}}{1 \text{ hour}} = \dfrac{50 \text{ miles}}{1 \text{ hour}} \cdot \dfrac{5280 \text{ feet}}{1 \text{ mile}} \cdot \dfrac{1 \text{ hour}}{3600 \text{ seconds}}$

$\approx 73 \text{ feet per second}$

67. 6.3 square yards

$= \dfrac{6.3 \text{ square yards}}{1} \cdot \dfrac{9 \text{ square feet}}{1 \text{ square yard}}$

$= 56.7 \text{ square feet}$

69. $133,500 \text{ square yards}$

$= \dfrac{133,500 \text{ square yards}}{1} \cdot \dfrac{9 \text{ square feet}}{1 \text{ square yard}}$

$= 1,201,500 \text{ square feet}$

71. $359.2 \text{ feet per second}$

$= \dfrac{359.2 \text{ feet}}{1 \text{ second}} \cdot \dfrac{3600 \text{ seconds}}{1 \text{ hour}} \cdot \dfrac{1 \text{ mile}}{5280 \text{ feet}}$

$\approx 244.9 \text{ miles per hour}$

73. $\dfrac{1}{5} + \dfrac{4}{5} = \dfrac{5}{5} = 1$

75. $\dfrac{9}{9} - \dfrac{19}{9} = -\dfrac{10}{9}$

77. $\dfrac{6}{5} + \left(\dfrac{1}{5} - \dfrac{8}{5}\right) = \dfrac{6}{5} + \left(-\dfrac{7}{5}\right) = -\dfrac{1}{5}$

79. $x - 2y = 6$

x	y
0	−3
6	0

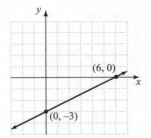

81. $\dfrac{4}{a} \cdot \dfrac{1}{b} = \dfrac{4 \cdot 1}{a \cdot b} = \dfrac{4}{ab}$

The statement is true.

83. $\dfrac{x}{5} \cdot \dfrac{x+3}{4} = \dfrac{x \cdot (x+3)}{5 \cdot 4} = \dfrac{x^2 + 3x}{20}$

The statement is false.

85. $\text{Area} = \text{length} \cdot \text{width}$

$\dfrac{x+5}{9x} \cdot \dfrac{2x}{x^2 - 25} = \dfrac{(x+5) \cdot 2 \cdot x}{9 \cdot x \cdot (x+5)(x-5)} = \dfrac{2}{9(x-5)}$

The area of the rectangle is $\dfrac{2}{9(x-5)}$ square feet.

87. $\left(\dfrac{x^2 - y^2}{x^2 + y^2} \div \dfrac{x^2 - y^2}{3x}\right) \cdot \dfrac{x^2 + y^2}{6}$

$= \dfrac{x^2 - y^2}{x^2 + y^2} \cdot \dfrac{3x}{x^2 - y^2} \cdot \dfrac{x^2 + y^2}{6}$

$= \dfrac{(x^2 - y^2) \cdot 3x \cdot (x^2 + y^2)}{(x^2 + y^2) \cdot (x^2 - y^2) \cdot 2 \cdot 3}$

$= \dfrac{x}{2}$

89. $\left(\dfrac{2a+b}{b^2} \cdot \dfrac{3a^2 - 2ab}{ab + 2b^2}\right) \div \dfrac{a^2 - 3ab + 2b^2}{5ab - 10b^2}$

$= \dfrac{2a+b}{b^2} \cdot \dfrac{3a^2 - 2ab}{ab + 2b^2} \cdot \dfrac{5ab - 10b^2}{a^2 - 3ab + 2b^2}$

$= \dfrac{(2a+b) \cdot (3a^2 - 2ab) \cdot (5ab - 10b^2)}{b^2 \cdot (ab + 2b^2) \cdot (a^2 - 3ab + 2b^2)}$

$= \dfrac{(2a+b) \cdot a(3a - 2b) \cdot 5b(a - 2b)}{b^2 \cdot b(a + 2b) \cdot (a - 2b)(a - b)}$

$= \dfrac{5a(2a + b)(3a - 2b)}{b^2(a + 2b)(a - b)}$

91. answers may vary

Section 7.3 Practice

1. $\dfrac{7a}{4b}+\dfrac{a}{4b}=\dfrac{7a+a}{4b}=\dfrac{8a}{4b}=\dfrac{2a}{b}$

2. $\dfrac{3x}{3x-2}-\dfrac{2}{3x-2}=\dfrac{3x-2}{3x-2}=\dfrac{1}{1}$ or 1

3. $\dfrac{4x^2+15x}{x+3}-\dfrac{8x+15}{x+3}=\dfrac{(4x^2+15x)-(8x+15)}{x+3}$

   $=\dfrac{4x^2+15x-8x-15}{x+3}$

   $=\dfrac{4x^2+7x-15}{x+3}$

   $=\dfrac{(x+3)(4x-5)}{x+3}$

   $=4x-5$

4. **a.** Find the prime factorization of each denominator.
   $14=2\cdot7$
   $21=3\cdot7$
   The greatest number of times that the factor 2 appears is 1. The greatest number of times that the factor 3 appears is 1. The greatest number of times that the factor 7 appears is 1.
   $LCD=2^1\cdot3^1\cdot7^1=42$

   b. Factor each denominator.
   $9y=3\cdot3\cdot y=3^2\cdot y$
   $15y^3=3\cdot5\cdot y^3$
   The greatest number of times that the factor 3 appears is 2. The greatest number of times that the factor 5 appears is 1. The greatest number of times that the factor y appears is 3.
   $LCD=3^2\cdot5^1\cdot y^3=9\cdot5\cdot y^3=45y^3$

5. **a.** The denominators $y-5$ and $y-4$ are completely factored already. The factor $y-5$ appears once and the factor $y-4$ appears once.
   $LCD=(y-5)(y-4)$

   b. The denominators a and $a+2$ cannot be factored further. The factor a appears once and the factor $a+2$ appears once.
   $LCD=a(a+2)$

6. Factor each denominator.
   $(2x-1)^2=(2x-1)^2$
   $6x-3=3(2x-1)$
   The greatest number that the factor $2x-1$ appears in any one denominator is 2.
   The greatest number of times that the factor 3 appears is 1.
   $LCD=3(2x-1)^2$

7. Factor each denominator.
   $x^2+5x+4=(x+1)(x+4)$
   $x^2-16=(x-4)(x+4)$
   $LCD=(x+1)(x+4)(x-4)$

8. The denominators $3-x$ and $x-3$ are opposites. That is, $3-x=-1(x-3)$. Use $x-3$ or $3-x$ as the LCD.
   $LCD=x-3$ or $LCD=3-x$

9. **a.** Since $5y(7xy)=35xy^2$, multiply by 1 in the form of $\dfrac{7xy}{7xy}$.

   $\dfrac{3x}{5y}=\dfrac{3x}{5y}\cdot1=\dfrac{3x}{5y}\cdot\dfrac{7xy}{7xy}=\dfrac{3x(7xy)}{5y(7xy)}=\dfrac{21x^2y}{35xy^2}$

   b. First, factor the denominator on the right.
   $\dfrac{9x}{4x+7}=\dfrac{}{2(4x+7)}$
   To obtain the denominator on the right from the denominator on the left, multiply by 1 in the form of $\dfrac{2}{2}$.
   $\dfrac{9x}{4x+7}=\dfrac{9x}{4x+7}\cdot\dfrac{2}{2}$

   $=\dfrac{9x\cdot2}{(4x+7)\cdot2}$

   $=\dfrac{18x}{2(4x+7)}$ or $\dfrac{18x}{8x+14}$

10. First, factor the denominator $x^2 - 2x - 15$ as $(x + 3)(x - 5)$. If we multiply the original denominator $(x + 3)(x - 5)$ by $x - 2$, the result is the new denominator $(x - 2)(x + 3)(x - 5)$. Thus, we multiply by 1 in the form $\dfrac{x-2}{x-2}$.

$$\frac{3}{x^2-2x-15} = \frac{3}{(x+3)(x-5)}$$
$$= \frac{3}{(x+3)(x-5)} \cdot \frac{x-2}{x-2}$$
$$= \frac{3(x-2)}{(x+3)(x-5)(x-2)}$$
$$= \frac{3x-6}{(x-2)(x+30(x-5)}$$

Vocabulary, Readiness & Video Check 7.3

1. $\dfrac{7}{11} + \dfrac{2}{11} = \dfrac{9}{11}$

2. $\dfrac{7}{11} - \dfrac{2}{11} = \dfrac{5}{11}$

3. $\dfrac{a}{b} + \dfrac{c}{b} = \dfrac{a+c}{b}$

4. $\dfrac{a}{b} - \dfrac{c}{b} = \dfrac{a-c}{b}$

5. $\dfrac{5}{x} - \dfrac{6+x}{x} = \dfrac{5-(6+x)}{x}$

6. In order to carry out the subtraction properly—parentheses make sure each term in the numerator is affected by the subtraction and not just the first term.

7. We factor denominators into the smallest factors—including coefficients—so we can determine the most number of times each unique factor occurs in any one denominator for the LCD.

8. To write an equivalent rational expression, you multiply the <u>numerator</u> of a rational expression by the same expression as the denominator. This means you're multiplying the original rational expression by a factor of <u>one</u> and therefore not changing the <u>value</u> of the original expression.

Exercise Set 7.3

1. $\dfrac{a+1}{13} + \dfrac{8}{13} = \dfrac{a+1+8}{13} = \dfrac{a+9}{13}$

3. $\dfrac{4m}{3n} + \dfrac{5m}{3n} = \dfrac{4m+5m}{3n} = \dfrac{9m}{3n} = \dfrac{3m}{n}$

5. $\dfrac{4m}{m-6} - \dfrac{24}{m-6} = \dfrac{4m-24}{m-6} = \dfrac{4(m-6)}{m-6} = 4$

7. $\dfrac{9}{3+y} + \dfrac{y+1}{3+y} = \dfrac{9+y+1}{3+y} = \dfrac{10+y}{3+y}$

9. $\dfrac{5x^2+4x}{x-1} - \dfrac{6x+3}{x-1} = \dfrac{5x^2+4x-(6x+3)}{x-1}$
$$= \frac{5x^2+4x-6x-3}{x-1}$$
$$= \frac{5x^2-2x-3}{x-1}$$
$$= \frac{(5x+3)(x-1)}{x-1}$$
$$= 5x+3$$

11. $\dfrac{4a}{a^2+2a-15} - \dfrac{12}{a^2+2a-15} = \dfrac{4a-12}{a^2+2a-15}$
$$= \frac{4(a-3)}{(a+5)(a-3)}$$
$$= \frac{4}{a+5}$$

13. $\dfrac{2x+3}{x^2-x-30} - \dfrac{x-2}{x^2-x-30} = \dfrac{2x+3-(x-2)}{x^2-x-30}$
$$= \frac{2x+3-x+2}{x^2-x-30}$$
$$= \frac{x+5}{x^2-x-30}$$
$$= \frac{x+5}{(x-6)(x+5)}$$
$$= \frac{1}{x-6}$$

15. $\dfrac{2x+1}{x-3} + \dfrac{3x+6}{x-3} = \dfrac{2x+1+3x+6}{x-3} = \dfrac{5x+7}{x-3}$

17. $\dfrac{2x^2}{x-5} - \dfrac{25+x^2}{x-5} = \dfrac{2x^2 - (25+x^2)}{x-5}$

$\qquad\qquad = \dfrac{2x^2 - 25 - x^2}{x-5}$

$\qquad\qquad = \dfrac{x^2 - 25}{x-5}$

$\qquad\qquad = \dfrac{(x+5)(x-5)}{x-5}$

$\qquad\qquad = x+5$

19. $\dfrac{5x+4}{x-1} - \dfrac{2x+7}{x-1} = \dfrac{5x+4-(2x+7)}{x-1}$

$\qquad\qquad = \dfrac{5x+4-2x-7}{x-1}$

$\qquad\qquad = \dfrac{3x-3}{x-1}$

$\qquad\qquad = \dfrac{3(x-1)}{x-1}$

$\qquad\qquad = 3$

21. $2x = 2 \cdot x$

$4x^3 = 2^2 \cdot x^3$

$\text{LCD} = 2^2 \cdot x^3 = 4x^3$

23. $8x = 2^3 \cdot x$

$2x + 4 = 2(x+2)$

$\text{LCD} = 2^3 \cdot x \cdot (x+2) = 8x(x+2)$

25. $x + 3 = x + 3$

$x - 2 = x - 2$

$\text{LCD} = (x+3)(x-2)$

27. $x + 6 = x + 6$

$3x + 18 = 3(x+6)$

$\text{LCD} = 3(x+6)$

29. $(x-6)^2 = (x-6)^2$

$5x - 30 = 5(x-6)$

$\text{LCD} = 5(x-6)^2$

31. $3x + 3 = 3 \cdot (x+1)$

$2x^2 + 4x + 2 = 2(x^2 + 2x + 1) = 2 \cdot (x+1)^2$

$\text{LCD} = 2 \cdot 3(x+1)^2 = 6(x+1)^2$

33. $x - 8 = x - 8$

$8 - x = -(x-8)$

$\text{LCD} = x - 8 \text{ or } 8 - x$

35. $x^2 + 3x - 4 = (x-1)(x+4)$

$x^2 + 2x - 3 = (x-1)(x+3)$

$\text{LCD} = (x-1)(x+4)(x+3)$

37. $3x^2 + 4x + 1 = (3x+1)(x+1)$

$2x^2 - x - 1 = (x-1)(2x+1)$

$\text{LCD} = (3x+1)(x+1)(x-1)(2x+1)$

39. $x^2 - 16 = (x+4)(x-4)$

$2x^3 - 8x^2 = 2x^2(x-4)$

$\text{LCD} = 2x^2(x+4)(x-4)$

41. $\dfrac{3}{2x} = \dfrac{3(2x)}{2x(2x)} = \dfrac{6x}{4x^2}$

43. $\dfrac{6}{3a} = \dfrac{6(4b^2)}{3a(4b^2)} = \dfrac{24b^2}{12ab^2}$

45. $\dfrac{9}{2x+6} = \dfrac{9}{2(x+3)} = \dfrac{9(y)}{2(x+3)(y)} = \dfrac{9y}{2y(x+3)}$

47. $\dfrac{9a+2}{5a+10} = \dfrac{9a+2}{5(a+2)} = \dfrac{(9a+2)(b)}{5(a+2)(b)} = \dfrac{9ab+2b}{5b(a+2)}$

49. $\dfrac{x}{x^3 + 6x^2 + 8x} = \dfrac{x}{x(x+4)(x+2)}$

$\qquad\qquad = \dfrac{x(x+1)}{x(x+4)(x+2)(x+1)}$

$\qquad\qquad = \dfrac{x^2 + x}{x(x+4)(x+2)(x+1)}$

51. $\dfrac{9y-1}{15x^2 - 30} = \dfrac{(9y-1)(2)}{(15x^2-30)2} = \dfrac{18y-2}{30x^2 - 60}$

53. $\dfrac{5x}{7} + \dfrac{9x}{7} = \dfrac{5x+9x}{7} = \dfrac{14x}{7} = \dfrac{2x}{1} = 2x$

55. $\dfrac{x+3}{4} \div \dfrac{2x-1}{4} = \dfrac{x+3}{4} \cdot \dfrac{4}{2x-1}$

$\qquad\qquad = \dfrac{(x+3) \cdot 4}{4 \cdot (2x-1)}$

$\qquad\qquad = \dfrac{x+3}{2x-1}$

57.
$$\frac{x^2}{x-6} - \frac{5x+6}{x-6} = \frac{x^2-(5x+6)}{x-6}$$
$$= \frac{x^2-5x-6}{x-6}$$
$$= \frac{(x+1)(x-6)}{x-6}$$
$$= x+1$$

59.
$$\frac{-2x}{x^3-8x} + \frac{3x}{x^3-8x} = \frac{-2x+3x}{x^3-8x}$$
$$= \frac{x}{x(x^2-8)}$$
$$= \frac{1}{x^2-8}$$

61.
$$\frac{12x-6}{x^2+3x} \cdot \frac{4x^2+13x+3}{4x^2-1}$$
$$= \frac{6(2x-1)\cdot(x+3)(4x+1)}{x(x+3)\cdot(2x+1)(2x-1)}$$
$$= \frac{6(4x+1)}{x(2x+1)}$$

63. LCD = 21
$$\frac{2}{3} + \frac{5}{7} = \frac{2(7)}{3(7)} + \frac{5(3)}{7(3)} = \frac{14}{21} + \frac{15}{21} = \frac{29}{21}$$

65. $6 = 2 \cdot 3$
$4 = 2^2$
LCD $= 2^2 \cdot 3 = 12$
$$\frac{1}{6} - \frac{3}{4} = \frac{1(2)}{6(2)} - \frac{3(3)}{4(3)} = \frac{2}{12} - \frac{9}{12} = \frac{2-9}{12} = -\frac{7}{12}$$

67. $12 = 2 \cdot 2 \cdot 3 = 2^2 \cdot 3$
$20 = 2 \cdot 2 \cdot 5 = 2^2 \cdot 5$
LCD $= 2^2 \cdot 3 \cdot 5 = 60$
$$\frac{1}{12} + \frac{3}{20} = \frac{1(5)}{12(5)} + \frac{3(3)}{20(3)} = \frac{5}{60} + \frac{9}{60} = \frac{14}{60} = \frac{7}{30}$$

69. $4a - 20 = 4(a-5)$
$(a-5)^2 = (a-5)^2$
LCD $= 4(a-5)^2$
The correct choice is d.

71. answers may vary

73.
$$\frac{3}{x} + \frac{y}{x} = \frac{3+y}{x}$$
The correct choice is c.

75.
$$\frac{3}{x} \cdot \frac{y}{x} = \frac{3 \cdot y}{x \cdot x} = \frac{3y}{x^2}$$
The correct choice is b.

77.
$$\frac{5}{2-x} = \frac{5(-1)}{(2-x)(-1)} = -\frac{5}{x-2}$$

79.
$$-\frac{7+x}{2-x} = \frac{7+x}{(-1)(2-x)} = \frac{7+x}{x-2}$$

81.
$$P = \frac{5}{x-2} + \frac{5}{x-2} + \frac{5}{x-2} + \frac{5}{x-2}$$
$$= \frac{5+5+5+5}{x-2}$$
$$= \frac{20}{x-2}$$

The perimeter is $\dfrac{20}{x-2}$ meters.

83. answers may vary

85. $88 = 2^3 \cdot 11$
$4332 = 2^3 \cdot 3 \cdot 19^2$
LCM $= 2^3 \cdot 3 \cdot 11 \cdot 19^2 = 95,304$
They will align again in 95,304 Earth days.

87. answers may vary

89. answers may vary

Section 7.4 Practice

1. a. Since $5 = 5$ and $15 = 3 \cdot 5$, the
LCD $= 3 \cdot 5 = 15$.
$$\frac{2x}{5} - \frac{6x}{15} = \frac{2x(3)}{5(3)} - \frac{6x}{15}$$
$$= \frac{6x}{15} - \frac{6x}{15}$$
$$= \frac{6x-6x}{15}$$
$$= \frac{0}{15}$$
$$= 0$$

b. Since $8a = 2^3 \cdot a$ and $12a^2 = 2^2 \cdot 3 \cdot a^2$, the LCD $= 2^3 \cdot 3 \cdot a^2 = 24a^2$.

$$\frac{7}{8a} + \frac{5}{12a^2} = \frac{7(3a)}{8a(3a)} + \frac{5(2)}{12a^2(2)}$$
$$= \frac{21a}{24a^2} + \frac{10}{24a^2}$$
$$= \frac{21a+10}{24a^2}$$

2. Since $x^2 - 25 = (x+5)(x-5)$, the LCD $= (x+5)(x-5)$.

$$\frac{12x}{x^2-25} - \frac{6}{x+5} = \frac{12x}{(x+5)(x-5)} - \frac{6(x-5)}{(x+5)(x-5)}$$
$$= \frac{12x - 6(x-5)}{(x+5)(x-5)}$$
$$= \frac{12x - 6x + 30}{(x+5)(x-5)}$$
$$= \frac{6x+30}{(x+5)(x-5)}$$
$$= \frac{6(x+5)}{(x+5)(x-5)}$$
$$= \frac{6}{x-5}$$

3. The LCD is $5y(y+1)$.

$$\frac{3}{5y} + \frac{2}{y+1} = \frac{3(y+1)}{5y(y+1)} + \frac{2(5y)}{(y+1)(5y)}$$
$$= \frac{3(y+1) + 2(5y)}{5y(y+1)}$$
$$= \frac{3y+3+10y}{5y(y+1)}$$
$$= \frac{13y+3}{5y(y+1)}$$

4. $x - 5$ and $5 - x$ are opposites. Write the denominator $5 - x$ as $-(x - 5)$ and simplify.

$$\frac{6}{x-5} - \frac{7}{5-x} = \frac{6}{x-5} - \frac{7}{-(x-5)}$$
$$= \frac{6}{x-5} - \frac{-7}{x-5}$$
$$= \frac{6-(-7)}{x-5}$$
$$= \frac{13}{x-5}$$

5. Note that 2 is the same as $\frac{2}{1}$. The LCD of $\frac{2}{1}$ and $\frac{b}{b+3}$ is $b + 3$.

$$2 + \frac{b}{b+3} = \frac{2}{1} + \frac{b}{b+3}$$
$$= \frac{2(b+3)}{1(b+3)} + \frac{b}{b+3}$$
$$= \frac{2(b+3)+b}{b+3}$$
$$= \frac{2b+6+b}{b+3}$$
$$= \frac{3b+6}{b+3} \text{ or } \frac{3(b+2)}{b+3}$$

6. First, factor the denominators.

$$\frac{5}{2x^2+3x} - \frac{3x}{4x+6} = \frac{5}{x(2x+3)} - \frac{3x}{2(2x+3)}$$

The LCD is $2x(2x+3)$.

$$\frac{5}{2x^2+3x} - \frac{3x}{4x+6} = \frac{5(2)}{x(2x+3)(2)} - \frac{3x(x)}{2(2x+3)(x)}$$
$$= \frac{10-3x^2}{2x(2x+3)}$$

7. First, factor the denominators.

$$x^2 + 7x + 12 = (x+4)(x+3)$$
$$x^2 - 9 = (x+3)(x-3)$$
$$\text{LCD} = (x+4)(x+3)(x-3)$$

$$\frac{2x}{x^2+7x+12} + \frac{3x}{x^2-9}$$
$$= \frac{2x}{(x+4)(x+3)} + \frac{3x}{(x+3)(x-3)}$$
$$= \frac{2x(x-3)}{(x+4)(x+3)(x-3)} + \frac{3x(x+4)}{(x+3)(x-3)(x+4)}$$
$$= \frac{2x(x-3) + 3x(x+4)}{(x+4)(x+3)(x-3)}$$
$$= \frac{2x^2 - 6x + 3x^2 + 12x}{(x+4)(x+3)(x-3)}$$
$$= \frac{5x^2+6x}{(x+4)(x+3)(x-3)} \text{ or } \frac{x(5x+6)}{(x+4)(x+3)(x-3)}$$

Vocabulary, Readiness & Video Check 7.4

1. The first step to perform on $\frac{3}{4} - \frac{y}{4}$ is to subtract the numerators and place the difference over the common denominator; choice d.

2. The first step to perform on $\dfrac{2}{a} \cdot \dfrac{3}{a+6}$ is to multiply the numerators and multiply the denominators; choice c.

3. The first step to perform on $\dfrac{x+1}{x} \div \dfrac{x-1}{x}$ is to multiply the first rational expression by the reciprocal of the second rational expression; choice a.

4. The first step to perform on $\dfrac{9}{x-2} - \dfrac{x}{x+2}$ is to find the LCD and write each expression as an equivalent expression with the LCD as denominator; choice b.

5. The problem adds two rational expressions with denominators that are opposites of each other. Recognizing this special case can save you time and effort. If you recognize that one denominator is −1 times the other denominator, you may save time.

Exercise Set 7.4

1. LCD $= 2 \cdot 3 \cdot x = 6x$

$$\frac{4}{2x} + \frac{9}{3x} = \frac{4(3)}{2x(3)} + \frac{9(2)}{3x(2)}$$
$$= \frac{12}{6x} + \frac{18}{6x}$$
$$= \frac{30}{6x}$$
$$= \frac{5(6)}{6x}$$
$$= \frac{5}{x}$$

3. LCD $= 5b$

$$\frac{15a}{b} - \frac{6b}{5} = \frac{15a(5)}{b(5)} - \frac{6b(b)}{5(b)}$$
$$= \frac{75a}{5b} - \frac{6b^2}{5b}$$
$$= \frac{75a - 6b^2}{5b}$$

5. LCD $= 2x^2$

$$\frac{3}{x} + \frac{5}{2x^2} = \frac{3(2x)}{x(2x)} + \frac{5}{2x^2} = \frac{6x}{2x^2} + \frac{5}{2x^2} = \frac{6x+5}{2x^2}$$

7. $2x + 2 = 2(x + 1)$
  LCD $= 2(x + 1)$

$$\frac{6}{x+1} + \frac{10}{2x+2} = \frac{6}{x+1} + \frac{10}{2(x+1)}$$
$$= \frac{6(2)}{(x+1)2} + \frac{10}{2(x+1)}$$
$$= \frac{12}{2(x+1)} + \frac{10}{2(x+1)}$$
$$= \frac{12+10}{2(x+1)}$$
$$= \frac{22}{2(x+1)}$$
$$= \frac{2(11)}{2(x+1)}$$
$$= \frac{11}{x+1}$$

9. $x^2 - 4 = (x+2)(x-2)$
  LCD $= (x + 2)(x - 2)$

$$\frac{3}{x+2} - \frac{2x}{x^2-4} = \frac{3(x-2)}{(x+2)(x-2)} - \frac{2x}{(x+2)(x-2)}$$
$$= \frac{3(x-2)-2x}{(x+2)(x-2)}$$
$$= \frac{3x-6-2x}{(x+2)(x-2)}$$
$$= \frac{x-6}{(x+2)(x-2)}$$

11. LCD $= 4x(x - 2)$

$$\frac{3}{4x} + \frac{8}{x-2} = \frac{3(x-2)}{4x(x-2)} + \frac{8(4x)}{(x-2)(4x)}$$
$$= \frac{3x-6}{4x(x-2)} + \frac{32x}{4x(x-2)}$$
$$= \frac{3x-6+32x}{4x(x-2)}$$
$$= \frac{35x-6}{4x(x-2)}$$

13. $3 - x = -(x - 3)$

$$\frac{6}{x-3} + \frac{8}{3-x} = \frac{6}{x-3} + \frac{8}{-(x-3)}$$
$$= \frac{6}{x-3} + \frac{-8}{x-3}$$
$$= \frac{6+(-8)}{x-3}$$
$$= -\frac{2}{x-3}$$

15. $3 - x = -(x-3)$

$$\frac{9}{x-3} + \frac{9}{3-x} = \frac{9}{x-3} + \frac{9}{-(x-3)}$$

$$= \frac{9}{x-3} + \frac{-9}{x-3}$$

$$= \frac{9+(-9)}{x-3}$$

$$= \frac{0}{x-3}$$

$$= 0$$

17. $1 - x^2 = -(x^2 - 1)$

$$\frac{-8}{x^2-1} - \frac{7}{1-x^2} = \frac{8}{-(x^2-1)} - \frac{7}{1-x^2}$$

$$= \frac{8}{1-x^2} - \frac{7}{1-x^2}$$

$$= \frac{8-7}{1-x^2}$$

$$= \frac{1}{1-x^2} \text{ or } -\frac{1}{x^2-1}$$

19. LCD = x

$$\frac{5}{x} + 2 = \frac{5}{x} + \frac{2}{1} = \frac{5}{x} + \frac{2(x)}{1(x)} = \frac{5+2x}{x}$$

21. LCD = $x - 2$

$$\frac{5}{x-2} + 6 = \frac{5}{x-2} + \frac{6}{1}$$

$$= \frac{5}{x-2} + \frac{6(x-2)}{1(x-2)}$$

$$= \frac{5}{x-2} + \frac{6x-12}{x-2}$$

$$= \frac{5+6x-12}{x-2}$$

$$= \frac{6x-7}{x-2}$$

23. LCD = $y + 3$

$$\frac{y+2}{y+3} - 2 = \frac{y+2}{y+3} - \frac{2}{1}$$

$$= \frac{y+2}{y+3} - \frac{2(y+3)}{y+3}$$

$$= \frac{y+2}{y+3} - \frac{2y+6}{y+3}$$

$$= \frac{y+2-(2y+6)}{y+3}$$

$$= \frac{y+2-2y-6}{y+3}$$

$$= \frac{-y-4}{y+3}$$

$$= \frac{-(y+4)}{y+3}$$

$$= -\frac{y+4}{y+3}$$

25. LCD = $4x$

$$\frac{-x+2}{x} - \frac{x-6}{4x} = \frac{(-x+2)(4)}{x(4)} - \frac{x-6}{4x}$$

$$= \frac{4(-x+2)-(x-6)}{4x}$$

$$= \frac{-4x+8-x+6}{4x}$$

$$= \frac{-5x+14}{4x} \text{ or } -\frac{5x-14}{4x}$$

27.

$$\frac{5x}{x+2} - \frac{3x-4}{x+2} = \frac{5x-(3x-4)}{x+2}$$

$$= \frac{5x-3x+4}{x+2}$$

$$= \frac{2x+4}{x+2}$$

$$= \frac{2(x+2)}{x+2}$$

$$= 2$$

29. LCD = 21

$$\frac{3x^4}{7} - \frac{4x^2}{21} = \frac{3x^4(3)}{7(3)} - \frac{4x^2}{21}$$

$$= \frac{3(3x^4)-4x^2}{21}$$

$$= \frac{9x^4-4x^2}{21}$$

31. $LCD = (x+3)^2$

$$\frac{1}{x+3} - \frac{1}{(x+3)^2} = \frac{1(x+3)}{(x+3)(x+3)} - \frac{1}{(x+3)^2}$$
$$= \frac{x+3}{(x+3)^2} - \frac{1}{(x+3)^2}$$
$$= \frac{x+3-1}{(x+3)^2}$$
$$= \frac{x+2}{(x+3)^2}$$

33. $LCD = 5b(b-1)$

$$\frac{4}{5b} + \frac{1}{b-1} = \frac{4(b-1)}{5b(b-1)} + \frac{1(5b)}{(b-1)(5b)}$$
$$= \frac{4b-4}{5b(b-1)} + \frac{5b}{5b(b-1)}$$
$$= \frac{4b-4+5b}{5b(b-1)}$$
$$= \frac{9b-4}{5b(b-1)}$$

35. $LCD = m$

$$\frac{2}{m} + 1 = \frac{2}{m} + \frac{1}{1} = \frac{2}{m} + \frac{1(m)}{1(m)} = \frac{2+m}{m}$$

37. $LCD = (x-7)(x-2)$

$$\frac{2x}{x-7} - \frac{x}{x-2} = \frac{2x(x-2)}{(x-7)(x-2)} - \frac{x(x-7)}{(x-2)(x-7)}$$
$$= \frac{2x(x-2) - x(x-7)}{(x-7)(x-2)}$$
$$= \frac{2x^2 - 4x - x^2 + 7x}{(x-7)(x-2)}$$
$$= \frac{x^2 + 3x}{(x-7)(x-2)} \text{ or } \frac{x(x+3)}{(x-7)(x-2)}$$

39. $2x - 1 = -(1 - 2x)$

$$\frac{6}{1-2x} - \frac{4}{2x-1} = \frac{6}{1-2x} - \frac{4}{-(1-2x)}$$
$$= \frac{6}{1-2x} - \frac{-4}{1-2x}$$
$$= \frac{6-(-4)}{1-2x}$$
$$= \frac{10}{1-2x}$$

41. $LCD = (x-1)(x+1)^2$

$$\frac{7}{(x+1)(x-1)} + \frac{8}{(x+1)^2}$$
$$= \frac{7(x+1)}{(x+1)(x-1)(x+1)} + \frac{8(x-1)}{(x+1)^2(x-1)}$$
$$= \frac{7x+7}{(x+1)^2(x-1)} + \frac{8x-8}{(x+1)^2(x-1)}$$
$$= \frac{7x+7+8x-8}{(x+1)^2(x-1)}$$
$$= \frac{15x-1}{(x+1)^2(x-1)}$$

43. $x^2 - 1 = (x+1)(x-1)$

$x^2 - 2x + 1 = (x-1)^2$

$LCD = (x+1)(x-1)^2$

$$\frac{x}{x^2-1} - \frac{2}{x^2-2x+1}$$
$$= \frac{x(x-1)}{(x-1)(x+1)(x-1)} - \frac{2(x+1)}{(x-1)^2(x+1)}$$
$$= \frac{x^2-x}{(x-1)^2(x+1)} - \frac{2x+2}{(x-1)^2(x+1)}$$
$$= \frac{x^2-x-(2x+2)}{(x-1)^2(x+1)}$$
$$= \frac{x^2-x-2x-2}{(x-1)^2(x+1)}$$
$$= \frac{x^2-3x-2}{(x-1)^2(x+1)}$$

45. $2a + 6 = 2(a+3)$

$LCD = 2(a+3)$

$$\frac{3a}{2a+6} - \frac{a-1}{a+3} = \frac{3a}{2(a+3)} - \frac{(a-1)(2)}{(a+3)(2)}$$
$$= \frac{3a}{2(a+3)} - \frac{2a-2}{2(a+3)}$$
$$= \frac{3a-(2a-2)}{2(a+3)}$$
$$= \frac{3a-2a+2}{2(a+3)}$$
$$= \frac{a+2}{2(a+3)}$$

47. $\text{LCD} = (2y+3)^2$

$$\frac{y-1}{2y+3} + \frac{3}{(2y+3)^2} = \frac{(y-1)(2y+3)}{(2y+3)(2y+3)} + \frac{3}{(2y+3)^2}$$

$$= \frac{(y-1)(2y+3)+3}{(2y+3)^2}$$

$$= \frac{2y^2 + y - 3 + 3}{(2y+3)^2}$$

$$= \frac{2y^2 + y}{(2y+3)^2} \text{ or } \frac{y(2y+1)}{(2y+3)^2}$$

49. $2 - x = -(x-2)$

$2x - 4 = 2(x-2)$

$\text{LCD} = 2(x-2)$

$$\frac{5}{2-x} + \frac{x}{2x-4} = \frac{5}{-(x-2)} + \frac{x}{2(x-2)}$$

$$= \frac{-5}{x-2} + \frac{x}{2(x-2)}$$

$$= \frac{-5(2)}{(x-2)(2)} + \frac{x}{2(x-2)}$$

$$= \frac{-10}{2(x-2)} + \frac{x}{2(x-2)}$$

$$= \frac{x-10}{2(x-2)}$$

51. $x^2 + 6x + 9 = (x+3)^2$

$\text{LCD} = (x+3)^2$

$$\frac{15}{x^2+6x+9} + \frac{2}{x+3} = \frac{15}{(x+3)^2} + \frac{2(x+3)}{(x+3)(x+3)}$$

$$= \frac{15 + 2(x+3)}{(x+3)^2}$$

$$= \frac{15 + 2x + 6}{(x+3)^2}$$

$$= \frac{2x+21}{(x+3)^2}$$

53. $x^2 - 5x - 6 = (x-3)(x-2)$

$\text{LCD} = (x-3)(x-2)$

$$\frac{13}{x^2-5x+6} - \frac{5}{x-3}$$

$$= \frac{13}{(x-3)(x-2)} - \frac{5(x-2)}{(x-3)(x-2)}$$

$$= \frac{13-(5x-10)}{(x-3)(x-2)}$$

$$= \frac{13-5x+10}{(x-3)(x-2)}$$

$$= \frac{-5x+23}{(x-3)(x-2)}$$

55. $m^2 - 100 = (m+10)(m-10)$

$\text{LCD} = 2(m+10)(m-10)$

$$\frac{70}{m^2-100} + \frac{7}{2(m+10)}$$

$$= \frac{70(2)}{(m+10)(m-10)(2)} + \frac{7(m-10)}{2(m+10)(m-10)}$$

$$= \frac{70(2)+7(m-10)}{2(m+10)(m-10)}$$

$$= \frac{140+7m-70}{2(m+10)(m-10)}$$

$$= \frac{7m+70}{2(m+10)(m-10)}$$

$$= \frac{7(m+10)}{2(m+10)(m-10)}$$

$$= \frac{7}{2(m-10)}$$

57. $x^2 - 5x - 6 = (x-6)(x+1)$

$x^2 - 4x - 5 = (x-5)(x+1)$

$\text{LCD} = (x-6)(x+1)(x-5)$

$$\frac{x+8}{x^2-5x-6} + \frac{x+1}{x^2-4x-5}$$

$$= \frac{(x+8)(x-5)}{(x-6)(x+1)(x-5)} + \frac{(x+1)(x-6)}{(x-5)(x+1)(x-6)}$$

$$= \frac{x^2+3x-40+x^2-5x-6}{(x-6)(x+1)(x-5)}$$

$$= \frac{2x^2-2x-46}{(x-6)(x+1)(x-5)}$$

$$\text{or } \frac{2(x^2-x-23)}{(x-6)(x+1)(x-5)}$$

59. $4n^2 - 12n + 8 = 4(n-1)(n-2)$

$3n^2 - 6n = 3n(n-2)$

$\text{LCD} = 4 \cdot 3n(n-1)(n-2) = 12n(n-1)(n-2)$

$$\frac{5}{4n^2 - 12n + 8} - \frac{3}{3n^2 - 6n}$$

$$= \frac{5(3n)}{4(n-1)(n-2)(3n)} - \frac{3(4)(n-1)}{3n(n-2)(4)(n-1)}$$

$$= \frac{5(3n) - 3(4)(n-1)}{12n(n-1)(n-2)}$$

$$= \frac{15n - 12n + 12}{12n(n-1)(n-2)}$$

$$= \frac{3n + 12}{12n(n-1)(n-2)}$$

$$= \frac{3(n+4)}{12n(n-1)(n-2)}$$

$$= \frac{n+4}{4n(n-1)(n-2)}$$

61. $\dfrac{15x}{x+8} \cdot \dfrac{2x+16}{3x} = \dfrac{15x}{x+8} \cdot \dfrac{2(x+8)}{3x}$

$$= \frac{2 \cdot 5 \cdot 3x \cdot (x+8)}{3x \cdot (x+8)}$$

$$= 10$$

63. $\dfrac{8x+7}{3x+5} - \dfrac{2x-3}{3x+5} = \dfrac{8x+7-(2x-3)}{3x+5}$

$$= \frac{8x+7-2x+3}{3x+5}$$

$$= \frac{6x+10}{3x+5}$$

$$= \frac{2(3x+5)}{3x+5}$$

$$= 2$$

65. $\dfrac{5a+10}{18} \div \dfrac{a^2-4}{10a} = \dfrac{5a+10}{18} \cdot \dfrac{10a}{a^2-4}$

$$= \frac{5(a+2) \cdot 2 \cdot 5a}{2 \cdot 9 \cdot (a-2)(a+2)}$$

$$= \frac{25a}{9(a-2)}$$

67. $x^2 - 3x + 2 = (x-2)(x-1)$

$\text{LCD} = (x-2)(x-1)$

$$\frac{5}{x^2-3x+2} + \frac{1}{x-2}$$

$$= \frac{5}{(x-2)(x-1)} + \frac{1}{(x-2)} \cdot \frac{(x-1)}{(x-1)}$$

$$= \frac{5}{(x-2)(x-1)} + \frac{x-1}{(x-2)(x-1)}$$

$$= \frac{5+x-1}{(x-2)(x-1)}$$

$$= \frac{x+4}{(x-2)(x-1)}$$

69. $3x+5 = 7$

$3x+5-5 = 7-5$

$3x = 2$

$\dfrac{3x}{3} = \dfrac{2}{3}$

$x = \dfrac{2}{3}$

71. $2x^2 - x - 1 = 0$

$(2x+1)(x-1) = 0$

$2x+1 = 0 \quad \text{or} \quad x-1 = 0$

$\quad 2x = -1 \qquad\qquad x = 1$

$\quad x = -\dfrac{1}{2}$

The solutions are $x = -\dfrac{1}{2}$ and $x = 1$.

73. $4(x+6) + 3 = -3$

$4x + 24 + 3 = -3$

$4x + 27 = -3$

$4x = -30$

$x = \dfrac{-30}{4} = -\dfrac{15}{2}$

75. $x^2 - 1 = (x+1)(x-1)$

LCD $= x(x+1)(x-1)$

$$\frac{3}{x} - \frac{2x}{x^2-1} + \frac{5}{x+1} = \frac{3(x+1)(x-1)}{x(x+1)(x-1)} - \frac{2x(x)}{(x+1)(x-1)(x)} + \frac{5(x)(x-1)}{(x+1)(x)(x-1)}$$

$$= \frac{3(x+1)(x-1) - 2x(x) + 5x(x-1)}{x(x+1)(x-1)}$$

$$= \frac{3x^2 - 3 - 2x^2 + 5x^2 - 5x}{x(x+1)(x-1)}$$

$$= \frac{6x^2 - 5x - 3}{x(x+1)(x-1)}$$

77. $x^2 - 4 = (x+2)(x-2)$

$x^2 - 4x + 4 = (x-2)^2$

$x^2 - x - 6 = (x-3)(x+2)$

LCD $= (x+2)(x-2)^2(x-3)$

$$\frac{5}{x^2-4} + \frac{2}{x^2-4x+4} - \frac{3}{x^2-x-6} = \frac{5(x-2)(x-3)}{(x-2)(x+2)(x-2)(x-3)} + \frac{2(x+2)(x-3)}{(x-2)^2(x+2)(x-3)} - \frac{3(x-2)^2}{(x-3)(x+2)(x-2)^2}$$

$$= \frac{5(x^2-5x+6)}{(x-2)^2(x+2)(x-3)} + \frac{2(x^2-x-6)}{(x-2)^2(x+2)(x-3)} - \frac{3(x^2-4x+4)}{(x-2)^2(x+2)(x-3)}$$

$$= \frac{5x^2-25x+30}{(x-2)^2(x+2)(x-3)} + \frac{2x^2-2x-12}{(x-2)^2(x+2)(x-3)} - \frac{3x^2-12x+12}{(x-2)^2(x+2)(x-3)}$$

$$= \frac{5x^2-25x+30+2x^2-2x-12-3x^2+12x-12}{(x-2)^2(x+2)(x-3)}$$

$$= \frac{4x^2-15x+6}{(x-2)^2(x+2)(x-3)}$$

79. $x^2 + 9x + 14 = (x+2)(x+7)$

$x^2 + 10x + 21 = (x+3)(x+7)$

$x^2 + 5x + 6 = (x+2)(x+3)$

LCD $= (x+2)(x+7)(x+3)$

$$\frac{9}{x^2+9x+14} - \frac{3x}{x^2+10x+21} + \frac{x+4}{x^2+5x+6} = \frac{9(x+3)}{(x+2)(x+7)(x+3)} - \frac{3x(x+2)}{(x+3)(x+7)(x+2)} + \frac{(x+4)(x+7)}{(x+2)(x+3)(x+7)}$$

$$= \frac{9(x+3) - 3x(x+2) + (x+4)(x+7)}{(x+2)(x+7)(x+3)}$$

$$= \frac{9x+27-3x^2-6x+x^2+11x+28}{(x+2)(x+7)(x+3)}$$

$$= \frac{-2x^2+14x+55}{(x+2)(x+7)(x+3)}$$

81. The length of the other board is

$\left(\dfrac{3}{x+4} - \dfrac{1}{x-4}\right)$ inches.

LCD $= (x+4)(x-4)$

$\dfrac{3}{x+4} - \dfrac{1}{x-4} = \dfrac{3(x-4)}{(x+4)(x-4)} - \dfrac{1(x+4)}{(x-4)(x+4)}$

$= \dfrac{3(x-4) - (x+4)}{(x+4)(x-4)}$

$= \dfrac{3x-12-x-4}{(x+4)(x-4)}$

$= \dfrac{2x-16}{(x+4)(x-4)}$

The length of the other board is

$\dfrac{2x-16}{(x+4)(x-4)}$ inches.

83. $1 - \dfrac{G}{P} = \dfrac{1}{1} - \dfrac{G}{P} = \dfrac{1(P)}{1(P)} - \dfrac{G}{P} = \dfrac{P-G}{P}$

85. answers may vary

87. $90° - \left(\dfrac{40}{x}\right)° = \left(90 - \dfrac{40}{x}\right)°$

LCD $= x$

$\left(90 \cdot \dfrac{x}{x} - \dfrac{40}{x}\right)° = \left(\dfrac{90x}{x} - \dfrac{40}{x}\right)° = \left(\dfrac{90x-40}{x}\right)°$

89. answers may vary

Section 7.5 Practice

1. The LCD of 3, 5, and 15 is 15.

$\dfrac{x}{3} + \dfrac{4}{5} = \dfrac{12}{5}$

$15\left(\dfrac{x}{3} + \dfrac{4}{5}\right) = 15\left(\dfrac{2}{15}\right)$

$15\left(\dfrac{x}{3}\right) + 15\left(\dfrac{4}{5}\right) = 15\left(\dfrac{2}{15}\right)$

$5 \cdot x + 12 = 2$

$5x = -10$

$x = -2$

Check: $\dfrac{x}{3} + \dfrac{4}{5} = \dfrac{2}{15}$

$\dfrac{-2}{3} + \dfrac{4}{5} \overset{?}{=} \dfrac{2}{15}$

$\dfrac{2}{15} = \dfrac{2}{15}$ True

This number checks, so the solution is −2.

2. The LCD of 4, 3, and 12 is 12.

$\dfrac{x+4}{4} - \dfrac{x-3}{3} = \dfrac{11}{12}$

$12\left(\dfrac{x+4}{4} - \dfrac{x-3}{3}\right) = 12\left(\dfrac{11}{12}\right)$

$12\left(\dfrac{x+4}{4}\right) - 12\left(\dfrac{x-3}{3}\right) = 12\left(\dfrac{11}{12}\right)$

$3(x+4) - 4(x-3) = 11$

$3x+12-4x+12 = 11$

$-x+24 = 11$

$-x = -13$

$x = 13$

Check: $\dfrac{x+4}{4} - \dfrac{x-3}{3} = \dfrac{11}{12}$

$\dfrac{13+4}{4} - \dfrac{13-3}{3} \overset{?}{=} \dfrac{11}{12}$

$\dfrac{17}{4} - \dfrac{10}{3} \overset{?}{=} \dfrac{11}{12}$

$\dfrac{11}{12} = \dfrac{11}{12}$ True

The solution is 13.

3. In this equation, 0 cannot be a solution. The LCD is x.

$8 + \dfrac{7}{x} = x + 2$

$x\left(8 + \dfrac{7}{x}\right) = x(x+2)$

$x(8) + x\left(\dfrac{7}{x}\right) = x \cdot x + x \cdot 2$

$8x + 7 = x^2 + 2x$

$0 = x^2 - 6x - 7$

$0 = (x+1)(x-7)$

$x + 1 = 0$ or $x - 7 = 0$

$x = -1$ $x = 7$

Neither −1 nor 7 makes the denominator in the original equation equal to 0.

Check:

$x = -1$

$8 + \dfrac{7}{x} = x + 2$

$8 + \dfrac{7}{-1} \overset{?}{=} -1 + 2$

$8 + (-7) \overset{?}{=} 1$

$1 = 1$ True

$x = 7$

$8 + \dfrac{7}{7} \stackrel{?}{=} 7 + 2$

$8 + 1 \stackrel{?}{=} 9$

$9 = 9$ True

Both -1 and 7 are solutions.

4. $x^2 - 5x - 14 = (x+2)(x-7)$

The LCD is $(x+2)(x-7)$.

$$\frac{6x}{x^2 - 5x - 14} - \frac{3}{x+2} = \frac{1}{x-7}$$

$$(x+2)(x-7)\left(\frac{6x}{x^2 - 5x - 14} - \frac{3}{x+2}\right) = (x+2)(x-7)\left(\frac{1}{x-7}\right)$$

$$(x+2)(x-7)\cdot\frac{6x}{x^2 - 5x - 14} - (x+2)(x-7)\cdot\frac{3}{x+2} = (x+2)(x-7)\cdot\frac{1}{x-7}$$

$$6x - 3(x-7) = x+2$$

$$6x - 3x + 21 = x+2$$

$$3x + 21 = x+2$$

$$2x = -19$$

$$x = -\frac{19}{2}$$

Check by replacing x with $-\dfrac{19}{2}$ in the original equation. The solution is $-\dfrac{19}{2}$.

5. The LCD is $x - 2$.

$$\frac{7}{x-2} = \frac{3}{x-2} + 4$$

$$(x-2)\left(\frac{7}{x-2}\right) = (x-2)\left(\frac{3}{x-2} + 4\right)$$

$$(x-2)\cdot\frac{7}{x-2} = (x-2)\cdot\frac{3}{x-2} + (x-2)\cdot 4$$

$$7 = 3 + 4x - 8$$

$$7 = 4x - 5$$

$$12 = 4x$$

$$3 = x$$

Check by replacing x with 3 in the original equation. The solution is 3.

6. From the denominators in the equation, 5 can't be a solution. The LCD is $x - 5$.

$$x + \frac{x}{x-5} = \frac{5}{x-5} - 7$$

$$(x-5)\left(x + \frac{x}{x-5}\right) = (x-5)\left(\frac{5}{x-5} - 7\right)$$

$$(x-5)(x) + (x-5)\left(\frac{x}{x-5}\right) = (x-5)\left(\frac{5}{x-5}\right) - (x-5)(7)$$

$$x^2 - 5x + x = 5 - 7x + 35$$

$$x^2 - 4x = 40 - 7x$$

$$x^2 + 3x - 40 = 0$$

$$(x+8)(x-5) = 0$$

$x+8=0$ or $x-5=0$
$x=-8$ $x=5$

Since 5 can't be a solution, check by replacing x with -8 in the original equation. The only solution is -8.

7. The LCD is abx.

$$\frac{1}{a}+\frac{1}{b}=\frac{1}{x}$$

$$abx\left(\frac{1}{a}+\frac{1}{b}\right)=abx\left(\frac{1}{x}\right)$$

$$abx\left(\frac{1}{a}\right)+abx\left(\frac{1}{b}\right)=abx\cdot\frac{1}{x}$$

$$bx+ax=ab$$

$$ax=ab-bx$$

$$ax=b(a-x)$$

$$\frac{ax}{a-x}=b$$

Graphing Calculator Explorations

1. $y_1=\dfrac{x-4}{2}-\dfrac{x-3}{9},\ y_2=\dfrac{5}{18}$

Use INTERSECT

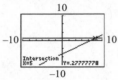

The solution of the equation is 5.

2. $y_1=3-\dfrac{6}{x},\ y_2=x+8$

Use INTERSECT

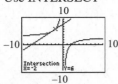

One solution is -3.

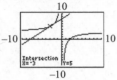

The other solution is -2.

3. $y_1=\dfrac{2x}{x-4},\ y_2=\dfrac{8}{x-4}+1$

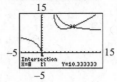

Using TRACE and ZOOM, it is clear that the curves never intersect. The equation has no solution.

4. $y_1=x+\dfrac{14}{x-2},\ y_2=\dfrac{7x}{x-2}+1$

Use INTERSECT

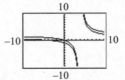

The solution is 8.

Vocabulary, Readiness & Video Check 7.5

1.
$$4\left(\frac{3x}{2}+5\right)=4\left(\frac{1}{4}\right)$$
$$4\left(\frac{3x}{2}\right)+4\cdot5=4\left(\frac{1}{4}\right)$$
$$6x+20=1$$
The correct choice is c.

2.
$$5x\left(\frac{1}{x}-\frac{3}{5x}\right)=5x(2)$$
$$5x\left(\frac{1}{x}\right)-5x\left(\frac{3}{5x}\right)=5x(2)$$
$$5-3=10x$$
The correct choice is b.

3. The LCD of $\dfrac{9}{x},\dfrac{3}{4}$, and $\dfrac{1}{12}=\dfrac{1}{3\cdot4}$ is $12x$; b.

4. The LCD of $\dfrac{8}{3x},\dfrac{1}{x}$, and $\dfrac{7}{9}=\dfrac{7}{3\cdot3}$ is $9x$; d.

5. The LCD of $\dfrac{9}{x-1}$ and $\dfrac{7}{(x-1)^2}$ is $(x-1)^2$; a.

6. The LCD of $\dfrac{1}{x-2},\dfrac{3}{x^2-4}=\dfrac{3}{(x+2)(x-2)}$, and $8=\dfrac{8}{1}$ is x^2-4; c.

7. These equations are solved in very different ways, so you need to determine the next correct move to make. For a linear equation, you first "move" variables terms on one side and numbers on the other; for a quadratic equation, you first set the equation equal to 0.

8. If there are variables in any denominator, you should first check to see if the proposed solutions make these denominators zero in the original equation, giving you an undefined rational expression. If so, that solution is an extraneous solution and is not a solution to the equation.

9. the steps for solving an equation containing rational expressions; as if it's the only variable in the equation.

Exercise Set 7.5

1. The LCD is 5.

$$\frac{x}{5} + 3 = 9$$

$$5\left(\frac{x}{5} + 3\right) = 5(9)$$

$$5\left(\frac{x}{5}\right) + 5(3) = 5(9)$$

$$x + 15 = 45$$

$$x = 30$$

Check: $\dfrac{x}{5} + 3 = 9$

$$\frac{30}{5} + 3 \stackrel{?}{=} 9$$

$$6 + 3 \stackrel{?}{=} 9$$

$$9 = 9 \quad \text{True}$$

The solution is 30.

3. The LCD is 12.

$$\frac{x}{2} + \frac{5x}{4} = \frac{x}{12}$$

$$12\left(\frac{x}{2} + \frac{5x}{4}\right) = 12\left(\frac{x}{12}\right)$$

$$12\left(\frac{x}{2}\right) + 12\left(\frac{5x}{4}\right) = 12\left(\frac{x}{12}\right)$$

$$6x + 15x = x$$

$$21x = x$$

$$20x = 0$$

$$x = 0$$

Check: $\dfrac{x}{2} + \dfrac{5x}{4} = \dfrac{x}{12}$

$$\frac{0}{2} + \frac{5 \cdot 0}{4} \stackrel{?}{=} \frac{0}{12}$$

$$0 + \frac{0}{4} \stackrel{?}{=} 0$$

$$0 = 0 \quad \text{True}$$

The solution is 0.

5. The LCD is x.

$$2 - \frac{8}{x} = 6$$

$$x\left(2 - \frac{8}{x}\right) = x(6)$$

$$x \cdot 2 - x \cdot \frac{8}{x} = x \cdot 6$$

$$2x - 8 = 6x$$

$$-8 = 4x$$

$$-2 = x$$

Check: $2 - \dfrac{8}{x} = 6$

$$2 - \frac{8}{-2} \stackrel{?}{=} 6$$

$$2 - (-4) \stackrel{?}{=} 6$$

$$6 = 6 \quad \text{True}$$

The solution is -2.

7. The LCD is x.

$$2 + \frac{10}{x} = x + 5$$

$$x\left(2 + \frac{10}{x}\right) = x(x + 5)$$

$$x(2) + x\left(\frac{10}{x}\right) = x(x + 5)$$

$$2x + 10 = x^2 + 5x$$

$$0 = x^2 + 3x - 10$$

$$0 = (x + 5)(x - 2)$$

$$x + 5 = 0 \quad \text{or} \quad x - 2 = 0$$

$$x = -5 \qquad\qquad x = 2$$

Check:

$$x = -5: \quad 2 + \frac{10}{x} = x + 5$$

$$2 + \frac{10}{-5} \stackrel{?}{=} -5 + 5$$

$$2 + (-2) \stackrel{?}{=} -5 + 5$$

$$0 = 0 \quad \text{True}$$

$x = 2:$ $2 + \dfrac{10}{x} = x + 5$

$2 + \dfrac{10}{2} \stackrel{?}{=} 2 + 5$

$2 + 5 \stackrel{?}{=} 2 + 5$

$7 = 7$ True

Both -5 and 2 are solutions.

9. The LCD is 10.

$\dfrac{a}{5} = \dfrac{a-3}{2}$

$10\left(\dfrac{a}{5}\right) = 10\left(\dfrac{a-3}{2}\right)$

$2a = 5(a-3)$

$2a = 5a - 15$

$-3a = -15$

$a = 5$

Check: $\dfrac{a}{5} = \dfrac{a-3}{2}$

$\dfrac{5}{5} \stackrel{?}{=} \dfrac{5-3}{2}$

$\dfrac{5}{5} \stackrel{?}{=} \dfrac{2}{2}$

$1 = 1$ True

The solution is 5.

11. The LCD is 10.

$\dfrac{x-3}{5} + \dfrac{x-2}{2} = \dfrac{1}{2}$

$10\left(\dfrac{x-3}{5} + \dfrac{x-2}{2}\right) = 10\left(\dfrac{1}{2}\right)$

$10\left(\dfrac{x-3}{5}\right) + 10\left(\dfrac{x-2}{2}\right) = 10\left(\dfrac{1}{2}\right)$

$2(x-3) + 5(x-2) = 5$

$2x - 6 + 5x - 10 = 5$

$7x - 16 = 5$

$7x = 21$

$x = 3$

Check:

$\dfrac{x-3}{5} + \dfrac{x-2}{2} = \dfrac{1}{2}$

$\dfrac{3-3}{5} + \dfrac{3-2}{2} \stackrel{?}{=} \dfrac{1}{2}$

$\dfrac{0}{5} + \dfrac{1}{2} \stackrel{?}{=} \dfrac{1}{2}$

$0 + \dfrac{1}{2} \stackrel{?}{=} \dfrac{1}{2}$

$\dfrac{1}{2} = \dfrac{1}{2}$ True

The solution is 3.

13. The LCD is $2a - 5$.

$\dfrac{3}{2a-5} = -1$

$(2a-5)\left(\dfrac{3}{2a-5}\right) = (2a-5)(-1)$

$3 = -2a + 5$

$-2 = -2a$

$1 = a$

Check: $\dfrac{3}{2a-5} = -1$

$\dfrac{3}{2(1)-5} \stackrel{?}{=} -1$

$\dfrac{3}{-3} \stackrel{?}{=} -1$

$-1 = -1$ True

The solution is 1.

15. The LCD is $y - 4$.

$\dfrac{4y}{y-4} + 5 = \dfrac{5y}{y-4}$

$(y-4)\left(\dfrac{4y}{y-4} + 5\right) = (y-4)\left(\dfrac{5y}{y-4}\right)$

$(y-4)\left(\dfrac{4y}{y-4}\right) + (y-4)(5) = (y-4)\left(\dfrac{5y}{y-4}\right)$

$4y + 5y - 20 = 5y$

$9y - 20 = 5y$

$4y - 20 = 0$

$4y = 20$

$y = 5$

Check: $\dfrac{4y}{y-4} + 5 = \dfrac{5y}{y-4}$

$\dfrac{4(5)}{5-4} + 5 \stackrel{?}{=} \dfrac{5(5)}{5-4}$

$\dfrac{20}{1} + 5 \stackrel{?}{=} \dfrac{25}{1}$

$25 = 25$ True

The solution is 5.

17. The LCD is $a - 3$.

$$2 + \frac{3}{a-3} = \frac{a}{a-3}$$

$$(a-3)\left(2 + \frac{3}{a-3}\right) = (a-3)\left(\frac{a}{a-3}\right)$$

$$(a-3)(2) + (a-3)\left(\frac{3}{a-3}\right) = a$$

$$2a - 6 + 3 = a$$

$$2a - 3 = a$$

$$-3 = a - 2a$$

$$-3 = -a$$

$$\frac{-3}{-1} = a$$

$$3 = a$$

When a is 3, a denominator equals zero. The equation has no solution.

19. $x^2 - 9 = (x+3)(x-3)$

The LCD is $(x + 3)(x - 3)$.

$$\frac{1}{x+3} + \frac{6}{x^2-9} = 1$$

$$(x+3)(x-3)\left(\frac{1}{x+3} + \frac{6}{(x+3)(x-3)}\right) = (x+3)(x-3)(1)$$

$$(x+3)(x-3) \cdot \frac{1}{x+3} + (x+3)(x-3) \cdot \frac{6}{(x+3)(x-3)} = (x+3)(x-3) \cdot 1$$

$$x - 3 + 6 = x^2 - 9$$

$$x + 3 = x^2 - 9$$

$$0 = x^2 - x - 12$$

$$0 = (x+3)(x-4)$$

$x + 3 = 0 \quad$ or $\quad x - 4 = 0$

$x = -3 \qquad\qquad x = 4$

When x is -3, a denominator equals zero. Check $x = 4$.

Check: $\dfrac{1}{x+3} + \dfrac{6}{x^2-9} = 1$

$$\frac{1}{4+3} + \frac{6}{4^2-9} \overset{?}{=} 1$$

$$\frac{1}{7} + \frac{6}{7} \overset{?}{=} 1$$

$$1 = 1 \quad \text{True}$$

The solution is 4.

21. The LCD is $y + 4$.

$$\frac{2y}{y+4} + \frac{4}{y+4} = 3$$

$$(y+4)\left(\frac{2y}{y+4} + \frac{4}{y+4}\right) = (y+4)(3)$$

$$(y+4)\cdot\frac{2y}{y+4} + (y+4)\cdot\frac{4}{y+4} = (y+4)\cdot 3$$

$$2y + 4 = 3y + 12$$

$$4 = y + 12$$

$$-8 = y$$

Check: $\dfrac{2y}{y+4} + \dfrac{4}{y+4} = 3$

$$\frac{2(-8)}{-8+4} + \frac{4}{-8+4} \stackrel{?}{=} 3$$

$$\frac{-16}{-4} + \frac{4}{-4} \stackrel{?}{=} 3$$

$$4 - 1 \stackrel{?}{=} 3$$

$$3 = 3 \quad \text{True}$$

The solution is -8.

23. The LCD is $(x + 2)(x - 2)$.

$$\frac{2x}{x+2} - 2 = \frac{x-8}{x-2}$$

$$(x+2)(x-2)\left(\frac{2x}{x+2} - 2\right) = (x+2)(x-2)\left(\frac{x-8}{x-2}\right)$$

$$(x+2)(x-2)\cdot\frac{2x}{x+2} - (x+2)(x-2)(2) = (x+2)(x-2)\cdot\frac{x-8}{x-2}$$

$$2x(x-2) - 2(x^2 - 4) = (x+2)(x-8)$$

$$2x^2 - 4x - 2x^2 + 8 = x^2 - 6x - 16$$

$$-4x + 8 = x^2 - 6x - 16$$

$$0 = x^2 - 2x - 24$$

$$0 = (x+4)(x-6)$$

$$x + 4 = 0 \quad \text{or} \quad x - 6 = 0$$
$$x = -4 \qquad\qquad x = 6$$

Check $x = -4$: $\dfrac{2x}{x+2} - 2 = \dfrac{x-8}{x-2}$

$$\frac{2(-4)}{-4+2} - 2 \stackrel{?}{=} \frac{-4-8}{-4-2}$$

$$\frac{-8}{-2} - 2 \stackrel{?}{=} \frac{-12}{-6}$$

$$4 - 2 \stackrel{?}{=} 2$$

$$2 = 2 \quad \text{True}$$

Check $x = 6$: $\dfrac{2x}{x+2} - 2 = \dfrac{x-8}{x-2}$

$$\dfrac{2(6)}{6+2} - 2 \overset{?}{=} \dfrac{6-8}{6-2}$$

$$\dfrac{12}{8} - 2 \overset{?}{=} \dfrac{-2}{4}$$

$$\dfrac{3}{2} - 2 \overset{?}{=} -\dfrac{1}{2}$$

$$-\dfrac{1}{2} = -\dfrac{1}{2} \quad \text{True}$$

The solutions are -4 an 6.

25. The LCD is $2y$.

$$\dfrac{2}{y} + \dfrac{1}{2} = \dfrac{5}{2y}$$

$$2y\left(\dfrac{2}{y} + \dfrac{1}{2}\right) = 2y\left(\dfrac{5}{2y}\right)$$

$$2y\left(\dfrac{2}{y}\right) + 2y\left(\dfrac{1}{2}\right) = 2y\left(\dfrac{5}{2y}\right)$$

$$4 + y = 5$$

$$y = 1$$

The solution is 1.

27. The LCD is $(a-6)(a-1)$.

$$\dfrac{a}{a-6} = \dfrac{-2}{a-1}$$

$$(a-6)(a-1)\left(\dfrac{a}{a-6}\right) = (a-6)(a-1)\left(\dfrac{-2}{a-1}\right)$$

$$a(a-1) = -2(a-6)$$

$$a^2 - a = -2a + 12$$

$$a^2 + a - 12 = 0$$

$$(a+4)(a-3) = 0$$

$$a+4 = 0 \quad \text{or} \quad a-3 = 0$$

$$a = -4 \qquad\qquad a = 3$$

The solutions are -4 and 3.

29. The LCD is $6x$.

$$\dfrac{11}{2x} + \dfrac{2}{3} = \dfrac{7}{2x}$$

$$6x\left(\dfrac{11}{2x} + \dfrac{2}{3}\right) = 6x\left(\dfrac{7}{2x}\right)$$

$$6x \cdot \dfrac{11}{2x} + 6x \cdot \dfrac{2}{3} = 6x \cdot \dfrac{7}{2x}$$

$$33 + 4x = 21$$

$$4x = -12$$

$$x = -3$$

The solution is -3.

31. The LCD is $(x + 2)(x - 2)$.

$$\frac{2}{x-2} + 1 = \frac{x}{x+2}$$

$$(x+2)(x-2)\left(\frac{2}{x-2} + 1\right) = (x+2)(x-2)\left(\frac{x}{x+2}\right)$$

$$(x+2)(x-2) \cdot \frac{2}{x-2} + (x+2)(x-2) \cdot 1 = (x+2)(x-2) \cdot \frac{x}{x+2}$$

$$2(x+2) + (x+2)(x-2) = x(x-2)$$

$$2x + 4 + x^2 - 4 = x^2 - 2x$$

$$x^2 + 2x = x^2 - 2x$$

$$2x = -2x$$

$$4x = 0$$

$$x = 0$$

The solution is 0.

33. The LCD is 6.

$$\frac{x+1}{3} - \frac{x-1}{6} = \frac{1}{6}$$

$$6\left(\frac{x+1}{3} - \frac{x-1}{6}\right) = 6\left(\frac{1}{6}\right)$$

$$6\left(\frac{x+1}{3}\right) - 6\left(\frac{x-1}{6}\right) = 6\left(\frac{1}{6}\right)$$

$$2(x+1) - (x-1) = 1$$

$$2x + 2 - x + 1 = 1$$

$$x + 3 = 1$$

$$x = -2$$

The solution is -2.

35. The LCD is $6(t - 4)$.

$$\frac{t}{t-4} = \frac{t+4}{6}$$

$$6(t-4)\left(\frac{t}{t-4}\right) = 6(t-4)\left(\frac{t+4}{6}\right)$$

$$6t = (t-4)(t+4)$$

$$6t = t^2 - 16$$

$$0 = t^2 - 6t - 16$$

$$0 = (t-8)(t+2)$$

$$t + 2 = 0 \quad \text{or} \quad t - 8 = 0$$

$$t = -2 \qquad\qquad t = 8$$

The solutions are -2 and 8.

37. $2y + 2 = 2(y + 1)$
$4y + 4 = 2 \cdot 2(y + 1)$
The LCD is $4(y + 1)$.

$$\frac{y}{2y+2}+\frac{2y-16}{4y+4}=\frac{2y-3}{y+1}$$

$$4(y+1)\left(\frac{y}{2(y+1)}+\frac{2y-16}{4(y+1)}\right)=4(y+1)\left(\frac{2y-3}{y+1}\right)$$

$$4(y+1)\left(\frac{y}{2(y+1)}\right)+4(y+1)\left(\frac{2y-16}{4(y+1)}\right)=4(y+1)\left(\frac{2y-3}{y+1}\right)$$

$$2y+2y-16=4(2y-3)$$

$$4y-16=8y-12$$

$$-4y=4$$

$$y=-1$$

In the original equation, -1 makes a denominator 0. This equation has no solution.

39. $r^2+5r-14=(r+7)(r-2)$

The LCD is $(r+7)(r-2)$.

$$\frac{4r-4}{r^2+5r-14}+\frac{2}{r+7}=\frac{1}{r-2}$$

$$(r+7)(r-2)\left(\frac{4r-4}{(r+7)(r-2)}+\frac{2}{r+7}\right)=(r+7)(r-2)\left(\frac{1}{r-2}\right)$$

$$(r+7)(r-2)\left(\frac{4r-4}{(r+7)(r-2)}\right)+(r+7)(r-2)\left(\frac{2}{r+7}\right)=(r+7)(r-2)\left(\frac{1}{r-2}\right)$$

$$4r-4+2(r-2)=(r+7)(1)$$

$$4r-4+2r-4=r+7$$

$$6r-8=r+7$$

$$5r=15$$

$$r=3$$

The solution is 3.

41. $x^2+x-6=(x+3)(x-2)$

The LCD is $(x+3)(x-2)$.

$$\frac{x+1}{x+3}=\frac{x^2-11x}{x^2+x-6}-\frac{x-3}{x-2}$$

$$(x+3)(x-2)\left(\frac{x+1}{x+3}\right)=(x+3)(x-2)\left(\frac{x^2-11x}{(x+3)(x-2)}-\frac{x-3}{x-2}\right)$$

$$(x+3)(x-2)\cdot\frac{x+1}{x+3}=(x+3)(x-2)\cdot\frac{x^2-11x}{(x+3)(x-2)}-(x+3)(x-2)\cdot\frac{x-3}{x-2}$$

$$(x-2)(x+1)=x^2-11x-(x+3)(x-3)$$

$$x^2-x-2=x^2-11x-(x^2-9)$$

$$x^2-x-2=x^2-11x-x^2+9$$

$$x^2-x-2=-11x+9$$

$$x^2+10x-11=0$$

$$(x+11)(x-1)=0$$

$$x+11=0 \quad \text{or} \quad x-1=0$$

$$x=-11 \qquad\qquad x=1$$

The solutions are -11 and 1.

43.
$$R = \frac{E}{I}$$
$$I(R) = I\left(\frac{E}{I}\right)$$
$$IR = E$$
$$I = \frac{E}{R}$$

45.
$$T = \frac{2U}{B+E}$$
$$(B+E)(T) = (B+E)\left(\frac{2U}{B+E}\right)$$
$$BT + ET = 2U$$
$$BT = 2U - ET$$
$$B = \frac{2U - ET}{T}$$

47.
$$B = \frac{705w}{h^2}$$
$$h^2(B) = h^2\left(\frac{705w}{h^2}\right)$$
$$Bh^2 = 705w$$
$$\frac{Bh^2}{705} = w$$

49.
$$N = R + \frac{V}{G}$$
$$G(N) = G\left(R + \frac{V}{G}\right)$$
$$GN = GR + V$$
$$GN - GR = V$$
$$G(N - R) = V$$
$$G = \frac{V}{N - R}$$

51.
$$\frac{C}{\pi r} = 2$$
$$\pi r\left(\frac{C}{\pi r}\right) = \pi r(2)$$
$$C = 2\pi r$$
$$\frac{C}{2\pi} = \frac{2\pi r}{2\pi}$$
$$\frac{C}{2\pi} = r$$

53.
$$\frac{1}{y} + \frac{1}{3} = \frac{1}{x}$$
$$3xy\left(\frac{1}{y} + \frac{1}{3}\right) = 3xy\left(\frac{1}{x}\right)$$
$$3xy \cdot \frac{1}{y} + 3xy \cdot \frac{1}{3} = 3xy \cdot \frac{1}{x}$$
$$3x + xy = 3y$$
$$x(3 + y) = 3y$$
$$x = \frac{3y}{3 + y}$$

55. The reciprocal of x is $\dfrac{1}{x}$.

57. The reciprocal of x, added to the reciprocal of 2 is $\dfrac{1}{x} + \dfrac{1}{2}$.

59. If a tank is filled in 3 hours, then $\dfrac{1}{3}$ of the tank is filled in one hour.

61. The graph crosses the x-axis at $x = 2$. It crosses the y-axis at $y = -2$. The x-intercept is $(2, 0)$ and the y-intercept is $(0, -2)$.

63. The graph crosses the x-axis at $x = -4$, $x = -2$ and $x = 3$. It crosses the y-axis at $y = 4$. The x-intercepts are $(-4, 0)$, $(-2, 0)$ and $(3, 0)$, and the y-intercept is $(0, 4)$.

65. answers may vary

67. expression
$$\frac{1}{x} + \frac{5}{9} = \frac{1(9)}{x(9)} + \frac{5x}{9x} = \frac{5x + 9}{9x}$$

69. equation
$$\frac{5}{x-1} - \frac{2}{x} = \frac{5}{x(x-1)}$$
$$x(x-1)\left(\frac{5}{x-1}\right) - x(x-1)\left(\frac{2}{x}\right) = x(x-1)\left(\frac{5}{x(x-1)}\right)$$
$$5x - 2(x-1) = 5$$
$$5x - 2x + 2 = 5$$
$$3x = 3$$
$$x = 1$$
1 makes a denominator zero. There is no solution.

71.
$$\frac{20x}{3} + \frac{32x}{6} = 180$$
$$6\left(\frac{20x}{3} + \frac{32x}{6}\right) = 6(180)$$
$$6\left(\frac{20x}{3}\right) + 6\left(\frac{32x}{6}\right) = 6(180)$$
$$40x + 32x = 1080$$
$$72x = 1080$$
$$\frac{72x}{72} = \frac{1080}{72}$$
$$x = 15$$

$$\frac{20x}{3} = \frac{20(15)}{3} = 100$$

$$\frac{32x}{6} = \frac{32(15)}{6} = 80$$

The angles are 100° and 80°.

73.
$$\frac{150}{x} + \frac{450}{x} = 90$$
$$x\left(\frac{150}{x} + \frac{450}{x}\right) = x(90)$$
$$x\left(\frac{150}{x}\right) + x\left(\frac{450}{x}\right) = x(90)$$
$$150 + 450 = 90x$$
$$600 = 90x$$
$$\frac{600}{90} = \frac{90x}{90}$$
$$\frac{20}{3} = x$$

$$\frac{150}{x} = \frac{150}{\frac{20}{3}} = 150\left(\frac{3}{20}\right) = \frac{45}{2} = 22.5$$

$$\frac{450}{x} = \frac{450}{\frac{20}{3}} = 450\left(\frac{3}{20}\right) = \frac{135}{2} = 67.5$$

The angles are 22.5° and 67.5°.

75.

$$\frac{5}{a^2+4a+3}+\frac{2}{a^2+a-6}-\frac{3}{a^2-a-2}=0$$

$$\frac{5}{(a+3)(a+1)}+\frac{2}{(a+3)(a-2)}-\frac{3}{(a-2)(a+1)}=0$$

$$(a+3)(a+1)(a-2)\left(\frac{5}{(a+3)(a+1)}+\frac{2}{(a+3)(a-2)}-\frac{3}{(a-2)(a+1)}\right)=(a+3)(a+1)(a-2)(0)$$

$$(a+3)(a+1)(a-2)\left(\frac{5}{(a+3)(a+1)}\right)+(a+3)(a+1)(a-2)\left(\frac{2}{(a+3)(a-2)}\right)$$

$$-(a+3)(a+1)(a-2)\left(\frac{3}{(a-2)(a+1)}\right)=0$$

$$5(a-2)+2(a+1)-3(a+3)=0$$

$$5a-10+2a+2-3a-9=0$$

$$4a-17=0$$

$$4a=17$$

$$a=\frac{17}{4}$$

The solution is $\frac{17}{4}$.

Integrated Review

1. expression

$$\frac{1}{x}+\frac{2}{3}=\frac{1(3)}{x(3)}+\frac{2(x)}{3(x)}=\frac{3}{3x}+\frac{2x}{3x}=\frac{3+2x}{3x}$$

2. expression

$$\frac{3}{a}+\frac{5}{6}=\frac{3(6)}{a(6)}+\frac{5(a)}{6(a)}=\frac{18}{6a}+\frac{5a}{6a}=\frac{18+5a}{6a}$$

3. equation

$$\frac{1}{x}+\frac{2}{3}=\frac{3}{x}$$

$$3x\left(\frac{1}{x}+\frac{2}{3}\right)=3x\left(\frac{3}{x}\right)$$

$$3x\left(\frac{1}{x}\right)+3x\left(\frac{2}{3}\right)=3x\left(\frac{3}{x}\right)$$

$$3+2x=9$$

$$2x=6$$

$$x=3$$

The solution is 3.

4. equation

$$\frac{3}{a}+\frac{5}{6}=1$$

$$6a\left(\frac{3}{a}+\frac{5}{6}\right)=6a(1)$$

$$6a\left(\frac{3}{a}\right)+6a\left(\frac{5}{6}\right)=6a$$

$$18+5a=6a$$

$$18=a$$

The solution is 18.

5. expression

$$\frac{2}{x-1}-\frac{1}{x}=\frac{2(x)}{(x-1)(x)}-\frac{1(x-1)}{x(x-1)}$$

$$=\frac{2x-(x-1)}{x(x-1)}$$

$$=\frac{x+1}{x(x-1)}$$

6. expression

$$\frac{4}{x-3}-\frac{1}{x}=\frac{4(x)}{(x-3)(x)}-\frac{1(x-3)}{x(x-3)}$$

$$=\frac{4x-(x-3)}{x(x-3)}$$

$$=\frac{4x-x+3}{x(x-3)}$$

$$=\frac{3x+3}{x(x-3)}$$

$$=\frac{3(x+1)}{x(x-3)}$$

7. equation

$$\frac{2}{x+1}-\frac{1}{x}=1$$

$$x(x+1)\left(\frac{2}{x+1}-\frac{1}{x}\right)=x(x+1)(1)$$

$$x(x+1)\left(\frac{2}{x+1}\right)-x(x+1)\left(\frac{1}{x}\right)=x(x+1)$$

$$2x-(x+1)=x(x+1)$$

$$2x-x-1=x^2+x$$

$$x-1=x^2+x$$

$$-1=x^2$$

There is no real number solution.

8. equation

$$\frac{4}{x-3}-\frac{1}{x}=\frac{6}{x(x-3)}$$

$$x(x-3)\left(\frac{4}{x-3}-\frac{1}{x}\right)=x(x-3)\left(\frac{6}{x(x-3)}\right)$$

$$x(x-3)\left(\frac{4}{x-3}\right)-x(x-3)\left(\frac{1}{x}\right)=6$$

$$4x-(x-3)=6$$

$$4x-x+3=6$$

$$3x+3=6$$

$$3x=3$$

$$x=1$$

The solution is 1.

9. expression

$$\frac{15x}{x+8}\cdot\frac{2x+16}{3x}=\frac{15x\cdot(2x+16)}{(x+8)\cdot 3x}$$

$$=\frac{3\cdot 5\cdot x\cdot 2\cdot(x+8)}{(x+8)\cdot 3\cdot x}$$

$$=5\cdot 2$$

$$=10$$

10. expression

$$\frac{9z+5}{15}\cdot\frac{5z}{81z^2-25}=\frac{(9z+5)\cdot 5z}{15\cdot(81z^2-25)}$$

$$=\frac{(9z+5)\cdot 5\cdot z}{5\cdot 3\cdot(9z+5)(9z-5)}$$

$$=\frac{z}{3(9z-5)}$$

11. expression

$$\frac{2x+1}{x-3}+\frac{3x+6}{x-3}=\frac{2x+1+3x+6}{x-3}=\frac{5x+7}{x-3}$$

12. expression

$$\frac{4p-3}{2p+7}+\frac{3p+8}{2p+7}=\frac{4p-3+3p+8}{2p+7}=\frac{7p+5}{2p+7}$$

13. equation

$$\frac{x+5}{7}=\frac{8}{2}$$

$$14\left(\frac{x+5}{7}\right)=14\left(\frac{8}{2}\right)$$

$$2(x+5)=56$$

$$2x+10=56$$

$$2x=46$$

$$x=23$$

The solution is 23.

14. equation

$$\frac{1}{2} = \frac{x-1}{8}$$

$$8\left(\frac{1}{2}\right) = 8\left(\frac{x-1}{8}\right)$$

$$4 = x - 1$$

$$5 = x$$

The solution is 5.

15. expression

$$\frac{5a+10}{18} \div \frac{a^2-4}{10a} = \frac{5a+10}{18} \cdot \frac{10a}{a^2-4}$$

$$= \frac{5(a+2) \cdot 2 \cdot 5 \cdot a}{2 \cdot 9(a+2)(a-2)}$$

$$= \frac{5 \cdot 5 \cdot a}{9(a-2)}$$

$$= \frac{25a}{9(a-2)}$$

16. expression

$$\frac{9}{x^2-1} + \frac{12}{3x+3} = \frac{9(3)}{(x+1)(x-1)(3)} + \frac{12(x-1)}{3(x+1)(x-1)}$$

$$= \frac{27+12x-12}{3(x-1)(x+1)}$$

$$= \frac{15+12x}{3(x+1)(x-1)}$$

$$= \frac{3(5+4x)}{3(x+1)(x-1)}$$

$$= \frac{4x+5}{(x+1)(x-1)}$$

17. expression

$$\frac{x+2}{3x-1} + \frac{5}{(3x-1)^2} = \frac{(x+2)(3x-1)}{(3x-1)(3x-1)} + \frac{5}{(3x-1)^2}$$

$$= \frac{3x^2+5x-2+5}{(3x-1)^2}$$

$$= \frac{3x^2+5x+3}{(3x-1)^2}$$

18. expression

$$\frac{4}{(2x-5)^2} + \frac{x+1}{2x-5} = \frac{4}{(2x-5)^2} + \frac{(x+1)(2x-5)}{(2x-5)(2x-5)}$$

$$= \frac{4+2x^2-3x-5}{(2x-5)^2}$$

$$= \frac{2x^2-3x-1}{(2x-5)^2}$$

19. expression

$$\frac{x-7}{x} - \frac{x+2}{5x} = \frac{(x-7)(5)}{x(5)} - \frac{x+2}{5x}$$

$$= \frac{5x-35-x-2}{5x}$$

$$= \frac{4x-37}{5x}$$

20. expression

$$\frac{10x-9}{x} - \frac{x-4}{3x} = \frac{(10x-9)(3)}{x(3)} - \frac{x-4}{3x}$$

$$= \frac{30x-27-x+4}{3x}$$

$$= \frac{29x-23}{3x}$$

21. equation

$$\frac{3}{x+3} = \frac{5}{x^2-9} - \frac{2}{x-3}$$

$$(x^2-9)\left(\frac{3}{x+3}\right) = (x^2-9)\left(\frac{5}{x^2-9}\right) - (x^2-9)\left(\frac{2}{x-3}\right)$$

$$(x-3)(3) = 5 - (x+3)(2)$$

$$3x-9 = 5 - 2x - 6$$

$$3x-9 = -2x - 1$$

$$5x-9 = -1$$

$$5x = 8$$

$$x = \frac{8}{5}$$

The solution is $\frac{8}{5}$.

22. equation

$$\frac{9}{x^2-4} + \frac{2}{x+2} = \frac{-1}{x-2}$$

$$(x^2-4)\left(\frac{9}{x^2-4}\right) + (x^2-4)\left(\frac{2}{x+2}\right) = (x^2-4)\left(\frac{-1}{x-2}\right)$$

$$9 + (x-2)(2) = (x+2)(-1)$$

$$9 + 2x - 4 = -x - 2$$

$$2x + 5 = -x - 2$$

$$3x + 5 = -2$$

$$3x = -7$$

$$x = -\frac{7}{3}$$

The solution is $-\frac{7}{3}$.

23. answers may vary

24. answers may vary

Section 7.6 Practice

1. Solve the equation as a rational equation.

$$\frac{36}{x} = \frac{4}{11}$$

$$11x \cdot \frac{36}{x} = 11x \cdot \frac{4}{11}$$

$$11 \cdot 36 = x \cdot 4$$

$$396 = 4x$$

$$\frac{396}{4} = \frac{4x}{4}$$

$$99 = x$$

Solve the proportion using cross products.

$$\frac{36}{x} = \frac{4}{11}$$

$$36 \cdot 11 = x \cdot 4$$

$$396 = 4x$$

$$\frac{396}{4} = \frac{4x}{4}$$

$$99 = x$$

Check: Both methods give a solution of 99. To check, substitute 99 for x in the original proportion. The solution is 99.

2. $$\frac{3x+2}{9} = \frac{x-1}{2}$$

$$2(3x+2) = 9(x-1)$$

$$6x+4 = 9x-9$$

$$6x = 9x-13$$

$$-3x = -13$$

$$\frac{-3x}{-3} = \frac{-13}{-3}$$

$$x = \frac{13}{3}$$

Check: Verify that $\frac{13}{3}$ is the solution.

3. Let x = price of seven 2-liter bottles of Diet Pepsi.

$$\frac{4 \text{ bottles}}{7 \text{ bottles}} = \frac{\text{price of 4 bottles}}{\text{price of 7 bottles}}$$

$$\frac{4}{7} = \frac{5.16}{x}$$

$$4x = 7(5.16)$$

$$4x = 36.12$$

$$x = 9.03$$

Check: Verify that 4 bottles is to 7 bottles as $5.16 is to $9.03.
Seven 2-liter bottles of Diet Pepsi cost $9.03.

4. Since the triangles are similar, their corresponding sides are in proportion.

$$\frac{20}{8} = \frac{15}{x}$$
$$20x = 8 \cdot 15$$
$$20x = 120$$
$$x = 6$$

Check: To check, replace x with 6 in the original proportion and see that a true statement results. The missing length is 6 meters.

5. Let x = the unknown number.

In words	the quotient of x and 5	minus	$\frac{3}{2}$	is	the quotient of x and 10
	↓	↓	↓	↓	↓
Translate:	$\frac{x}{5}$	−	$\frac{3}{2}$	=	$\frac{x}{10}$

The LCD is 10.

$$10\left(\frac{x}{5} - \frac{3}{2}\right) = 10\left(\frac{x}{10}\right)$$
$$10\left(\frac{x}{5}\right) - 10\left(\frac{3}{2}\right) = 10\left(\frac{x}{10}\right)$$
$$2x - 15 = x$$
$$x - 15 = 0$$
$$x = 15$$

Check: To check, verify that "the quotient of 15 and 5 minus $\frac{3}{2}$ is the quotient of 15 and 10," or $\frac{15}{5} - \frac{3}{2} = \frac{15}{10}$.

6. Let x = the time in hours it takes Cindy and Mary to complete the job together. Then $\frac{1}{x}$ = the part of the job they complete in 1 hour.

	Hours to Complete Total Job	Part of Job Completed in 1 Hour
Cindy	3	$\frac{1}{3}$
Mary	4	$\frac{1}{4}$
Together	x	$\frac{1}{x}$

The part of the job Cindy completes in 1 hour, added to the part of the job Mary completes in 1 hour is equal to the part of the job they complete together in 1 hour.

$$\frac{1}{3}+\frac{1}{4}=\frac{1}{x}$$

$$12x\left(\frac{1}{3}\right)+12x\left(\frac{1}{4}\right)=12x\left(\frac{1}{x}\right)$$

$$4x+3x=12$$

$$7x=12$$

$$x=\frac{12}{7} \text{ or } 1\frac{5}{7}$$

Check: The proposed solution is reasonable since $1\frac{5}{7}$ hours is more than half of Cindy's time and less than half of Mary's time. Check $1\frac{5}{7}$ hours in the originally stated problem.

Cindy and Mary can complete the garden planting in $1\frac{5}{7}$ hours.

7. Let x = the speed of the bus. Then since the car's speed is 15 mph faster than that of the bus, the speed of the car is $x + 15$.
Since distance = rate · time, or $d = r \cdot t$, then

$$t=\frac{d}{r}.$$

The bus travels 180 miles in the same time that the car travels 240 miles.

	Distance =	Rate ·	Time
Bus	180	x	$\frac{180}{x}$
Car	240	$x + 15$	$\frac{240}{x+15}$

Since the car and the bus traveled the same amount of time, $\frac{180}{x}=\frac{240}{x+15}$.

$$\frac{180}{x}=\frac{240}{x+15}$$

$$180(x+15)=240x$$

$$180x+2700=240x$$

$$2700=60x$$

$$45=x$$

The speed of the bus is 45 miles per hour. The speed of the car must then be $x + 15$ or 60 miles per hour.
Check: Find the time it takes the car to travel 240 miles and the time it takes the bus to travel 180 miles.

Car: $t=\frac{d}{r}=\frac{240}{60}=4$ hours

Bus: $t=\frac{d}{r}=\frac{180}{45}=4$ hours

Since the times are the same, the proposed solution is correct. The speed of the bus is 45 miles per hour and the speed of the car is 60 miles per hour.

Vocabulary, Readiness & Video Check 7.6

1. If both people work together, they can complete the job in less time than either person working alone. That is, in less than 5 hours; choice c.

2. If both inlet pipes are on, they can fill the pond in less time than either pipe alone. That is, in less than 25 hours; choice a.

3. A number: x
The reciprocal of the number: $\frac{1}{x}$
The reciprocal of the number, decreased by 3: $\frac{1}{x}-3$

4. A number: y
The reciprocal of the number: $\frac{1}{y}$
The reciprocal of the number, increased by 2: $\frac{1}{y}+2$

5. A number: z
The sum of the number and 5: $z + 5$
The reciprocal of the sum of the number and 5: $\frac{1}{z+5}$

6. A number: x
The difference of the number and 1: $x - 1$
The reciprocal of the difference of the number and 1: $\frac{1}{x-1}$

7. A number: y
Twice the number: $2y$
Eleven divided by twice the number: $\frac{11}{2y}$

8. A number: z

Triple the number: $3z$

Negative 10 divided by triple the number: $\dfrac{-10}{3y}$

9. No; proportions are actually equations containing rational expressions, so they can also be solved by using the steps to solve those equations.

10. There are also many ways to set up an incorrect proportion, so just checking your solution in your proportion isn't enough. You need to determine if your solution is reasonable from the relationships given in the problem.

11. divided by, quotient

12. Two machines (or people) take different amounts of time to complete the task, one faster and one slower than the other. When working together, they will complete the task in less time than the faster machine, so your answer must be less than the time of the faster machine.

13.

d	$=$	r	$\cdot$	t
car	325	$x + 7$		$\dfrac{325}{x+7}$
motorcycle	290	x		$\dfrac{290}{x}$

$$\frac{325}{x+7} = \frac{290}{x}$$

Exercise Set 7.6

1. $\dfrac{2}{3} = \dfrac{x}{6}$

$12 = 3x$

$4 = x$

3. $\dfrac{x}{10} = \dfrac{5}{9}$

$9x = 50$

$x = \dfrac{50}{9}$

5. $\dfrac{x+1}{2x+3} = \dfrac{2}{3}$

$3(x+1) = 2(2x+3)$

$3x + 3 = 4x + 6$

$3 = x + 6$

$-3 = x$

7. $\dfrac{9}{5} = \dfrac{12}{3x+2}$

$9(3x+2) = 5(12)$

$27x + 18 = 60$

$27x = 42$

$x = \dfrac{42}{27} = \dfrac{14}{9}$

9. Let $x =$ the elephant's weight on Pluto.

$\dfrac{100}{3} = \dfrac{4100}{x}$

$100x = 3(4100)$

$100x = 12,300$

$x = 123$

The elephant's weight is 123 pounds.

11. Let $x =$ the number of calories in 43.2 grams.

$\dfrac{110}{28.8} = \dfrac{x}{43.2}$

$110(43.2) = 28.8x$

$4752 = 28.8x$

$165 = x$

There are 165 calories in 43.2 grams.

13. $\dfrac{16}{10} = \dfrac{34}{y}$

$16y = 340$

$y = 21.25$

15. $\dfrac{28}{20} = \dfrac{8}{y}$

$28y = 160$

$y = \dfrac{160}{28} = \dfrac{40}{7}$

$y = 5\dfrac{5}{7}$ feet

17.
$$3 \cdot \frac{1}{x} = 9 \cdot \frac{1}{6}$$
$$\frac{3}{x} = \frac{9}{6}$$
$$6x\left(\frac{3}{x}\right) = 6x\left(\frac{9}{6}\right)$$
$$18 = 9x$$
$$x = 2$$
The unknown number is 2.

19.
$$\frac{3+2x}{x+1} = \frac{3}{2}$$
$$2(x+1)\left(\frac{3+2x}{x+1}\right) = 2(x+1)\left(\frac{3}{2}\right)$$
$$2(3+2x) = 3(x+1)$$
$$6+4x = 3x+3$$
$$x = -3$$
The unknown number is −3.

21. Let x be the number of hours for the two surveyors to survey the roadbed together.

	Hours to Complete Total Job	Part of Job Completed in 1 Hour
Experienced	4	$\frac{1}{4}$
Apprentice	5	$\frac{1}{5}$
Together	x	$\frac{1}{x}$

$$\frac{1}{4} + \frac{1}{5} = \frac{1}{x}$$
$$20x\left(\frac{1}{4}\right) + 20x\left(\frac{1}{5}\right) = 20x\left(\frac{1}{x}\right)$$
$$5x + 4x = 20$$
$$9x = 20$$
$$x = \frac{20}{9} \text{ or } 2\frac{2}{9}$$

The experienced surveyor and apprentice surveyor, working together, can survey the road in $2\frac{2}{9}$ hours.

23. Let x be the number of minutes it takes the belts working together.

	Minutes to Complete Total Job	Part of Job Completed in 1 Minute
Larger belt	2	$\frac{1}{2}$
Smaller belt	6	$\frac{1}{6}$
Both belts	x	$\frac{1}{x}$

$$\frac{1}{2} + \frac{1}{6} = \frac{1}{x}$$
$$6x\left(\frac{1}{2}\right) + 6x\left(\frac{1}{6}\right) = 6x\left(\frac{1}{x}\right)$$
$$3x + x = 6$$
$$4x = 6$$
$$x = \frac{6}{4} = \frac{3}{2} = 1\frac{1}{2}$$

Both belts together can move the cans to the storage area in $1\frac{1}{2}$ minutes.

25. Let r be the jogger's rate. Then, since distance = rate · time, or $d = r \cdot t$, then $t = \frac{d}{r}$.

	Distance =	Rate ·	Time
Trip to Park	12	r	$\frac{12}{r}$
Return Trip	18	r	$\frac{18}{r}$

Since her time on the return trip is 1 hour longer than on the trip to the park, $\frac{18}{r} = \frac{12}{r} + 1$.
$$r\left(\frac{18}{r}\right) = r\left(\frac{12}{r}\right) + r(1)$$
$$18 = 12 + r$$
$$6 = r$$
She jogs at 6 miles per hour.

27. Let r be his speed on the first portion. Then his speed on the cooldown portion is $r - 2$.

	Distance =	Rate	·	Time
1st portion	20	r		$\frac{20}{r}$
Cooldown portion	16	$r-2$		$\frac{16}{r-2}$

$$\frac{20}{r} = \frac{16}{r-2}$$
$$20(r-2) = 16r$$
$$20r - 40 = 16r$$
$$-40 = -4r$$
$$r = 10$$

and $r - 2 = 10 - 2 = 8$

His speed was 10 miles per hour during the first portion and 8 miles per hour during the cooldown portion.

29. Let $x =$ the minimum floor space needed by 40 students.
$$\frac{1}{9} = \frac{40}{x}$$
$$1x = 9(40)$$
$$x = 360$$
40 students need 360 square feet.

31.
$$\frac{1}{4} = \frac{x}{8}$$
$$8\left(\frac{1}{4}\right) = 8\left(\frac{x}{8}\right)$$
$$2 = x$$
The unknown number is 2.

33. Let x be the amount of time it takes Marcus and Tony working together.

	Hours to Complete Total Job	Part of Job Completed in 1 Hour
Marcus	6	$\frac{1}{6}$
Tony	4	$\frac{1}{4}$
Together	x	$\frac{1}{x}$

$$\frac{1}{6} + \frac{1}{4} = \frac{1}{x}$$
$$12x\left(\frac{1}{6}\right) + 12x\left(\frac{1}{4}\right) = 12x\left(\frac{1}{x}\right)$$
$$2x + 3x = 12$$
$$5x = 12$$
$$x = \frac{12}{5} = 2\frac{2}{5}$$
$$45\left(\frac{12}{5}\right) = 108$$

Together Marcus and Tony work for $2\frac{2}{5}$ hours at \$45 per hour. The labor estimate should be \$108.

35. Let w be the speed of the wind.

	Distance =	Rate	·	Time
With wind	400	$230 + w$		$\frac{400}{230+w}$
Against wind	336	$230 - w$		$\frac{336}{230-w}$

Since the time with the wind is the same as the time against the wind, $\frac{336}{230-w} = \frac{400}{230+w}$.

$$\frac{336}{230-w} = \frac{400}{230+w}$$
$$336(230 + w) = 400(230 - w)$$
$$77,280 + 336w = 92,000 - 400w$$
$$736w = 14,720$$
$$w = 20$$

The speed of the wind is 20 miles per hour.

37.
$$\frac{y}{25} = \frac{3}{2}$$
$$y \cdot 2 = 25 \cdot 3$$
$$y \cdot 2 = 75$$
$$y = \frac{75}{2}$$
$$y = 37\frac{1}{2}$$

The unknown length is $37\frac{1}{2}$ feet.

39. Let x be the speed of the slower train. In 3.5 hours, the slower train travels $3.5x$ miles and the faster train travels $3.5(x + 10)$ miles.

$$3.5x + 3.5(x + 10) = 322$$
$$3.5x + 3.5x + 35 = 322$$
$$7x + 35 = 322$$
$$7x = 287$$
$$x = 41$$

The slower train travels 41 mph and the faster train travels $41 + 10 = 51$ mph.

41.
$$\frac{2}{x-3} - \frac{4}{x+3} = 8 \cdot \frac{1}{x^2 - 9}$$
$$(x-3)(x+3)\left(\frac{2}{x-3} - \frac{4}{x+3}\right) = (x-3)(x+3)\left(\frac{8}{x^2-9}\right)$$
$$(x-3)(x+3)\left(\frac{2}{x-3}\right) - (x-3)(x+3)\left(\frac{4}{x+3}\right) = 8$$
$$2(x+3) - 4(x-3) = 8$$
$$2x + 6 - 4x + 12 = 8$$
$$-2x = -10$$
$$x = 5$$

The unknown number is 5.

43. Let r be the rate of the plane in still air.

	Distance =	Rate	$\cdot$	Time
With wind	630	$r + 35$		$\frac{630}{r+35}$
Against wind	455	$r - 35$	$\cdot$	$\frac{455}{r-35}$

$$\frac{630}{r+35} = \frac{455}{r-35}$$
$$630(r-35) = 455(r+35)$$
$$630r - 22{,}050 = 455r + 15{,}925$$
$$175r = 37{,}975$$
$$r = 217$$

The speed in still air is 217 mph.

45. Let x = the number of gallons of water needed.
$$\frac{8}{2} = \frac{36}{x}$$
$$8x = 2(36)$$
$$8x = 72$$
$$x = 9$$
Nine gallons of water are needed for the entire box.

47.

	r $\times$	t $=$	d
With wind	$16 + x$	$\dfrac{48}{16+x}$	48
Into Wind	$16 - x$	$\dfrac{16}{16-x}$	16

Since the times are the same, $\dfrac{48}{16+x} = \dfrac{16}{16-x}$.

$$\frac{48}{16+x} = \frac{16}{16-x}$$
$$48(16-x) = 16(16+x)$$
$$768 - 48x = 256 + 16x$$
$$512 = 64x$$
$$8 = x$$

The rate of the wind is 8 miles per hour.

49. Let x be the rate of the slower hiker. Then the rate of the faster hiker is $x + 1.1$. In 2 hours, the slower hiker walks $2x$ miles, while the faster hiker walks $2(x + 1.1)$ miles.

$$2x + 2(x+1.1) = 11$$
$$2x + 2x + 2.2 = 11$$
$$4x + 2.2 = 11$$
$$4x = 8.8$$
$$x = 2.2$$
$$x + 1.1 = 2.2 + 1.1 = 3.3$$

The hikers walk 2.2 miles per hour and 3.3 miles per hour.

51. Let x be the amount of time it takes the second worker to do the job alone.

	Hours to Complete Total Job	Part of Job Completed in 1 Hour
Custodian	3	$\dfrac{1}{3}$
2nd Worker	x	$\dfrac{1}{x}$
Together	$1\dfrac{1}{2}$ or $\dfrac{3}{2}$	$\dfrac{2}{3}$

$$\frac{1}{3} + \frac{1}{x} = \frac{2}{3}$$
$$3x\left(\frac{1}{3}\right) + 3x\left(\frac{1}{x}\right) = 3x\left(\frac{2}{3}\right)$$
$$x + 3 = 2x$$
$$3 = x$$

It takes the second worker 3 hours to do the job alone.

53. Let x be the missing dimension.

$$\frac{x}{8} = \frac{20}{6}$$
$$6x = 8 \cdot 20$$
$$x = \frac{160}{6}$$
$$x = \frac{80}{3} = 26\frac{2}{3}$$

The side is $26\dfrac{2}{3}$ feet long.

55.

$$\frac{3}{2} = \frac{324}{x}$$
$$3 \cdot x = 2 \cdot 324$$
$$3x = 648$$
$$x = \frac{648}{3} = 216$$

There should be 216 other nuts in the can.

57. Let t be the time in hours that the jet plane travels.

	distance	rate	time
jet plane	$500t$	500	t
propeller plane	$200(t + 2)$	200	$t + 2$

$$500t = 200(t + 2)$$
$$500t = 200t + 400$$
$$300t = 400$$
$$t = \frac{400}{300}$$
$$t = \frac{4}{3}$$

$$\text{distance} = 500t = 500\left(\frac{4}{3}\right) = 666\frac{2}{3}$$

The planes are $666\dfrac{2}{3}$ miles from the starting point.

59. Let x be the time that it takes the third pipe to fill the pool alone.

$$\frac{1}{20} + \frac{1}{15} + \frac{1}{x} = \frac{1}{6}$$
$$60x\left(\frac{1}{20} + \frac{1}{15} + \frac{1}{x}\right) = 60x\left(\frac{1}{6}\right)$$
$$3x + 4x + 60 = 10x$$
$$7x + 60 = 10x$$
$$60 = 3x$$
$$20 = x$$

It will take the third pump 20 hours to do the job alone.

61. Let r be the motorcycle's speed.

	distance	rate	time
car	280	$r + 10$	$\frac{280}{r+10}$
motorcycle	240	r	$\frac{240}{r}$

$$\frac{280}{r+10} = \frac{240}{r}$$
$$280r = 240(r+10)$$
$$280r = 240r + 2400$$
$$40r = 2400$$
$$r = 60$$
$$r + 10 = 60 + 10 = 70$$

The motorcycle's speed was 60 miles per hour and the car's speed was 70 miles per hour.

63. Let x be the time for the third cook to prepare the same number of pies.

$$\frac{1}{6} + \frac{1}{7} + \frac{1}{x} = \frac{1}{2}$$
$$42x\left(\frac{1}{6} + \frac{1}{7} + \frac{1}{x}\right) = 42x\left(\frac{1}{2}\right)$$
$$7x + 6x + 42 = 21x$$
$$13x + 42 = 21x$$
$$42 = 8x$$
$$\frac{42}{8} = x$$

$$\frac{42}{8} = \frac{21}{4} = 5\frac{1}{4}$$

It will take the third cook $5\frac{1}{4}$ hours to prepare the pies working alone.

65. Let x be the number.

$$\frac{x}{3} - 1 = \frac{5}{3}$$
$$3\left(\frac{x}{3} - 1\right) = 3\left(\frac{5}{3}\right)$$
$$x - 3 = 5$$
$$x = 8$$

The number is 8.

67. Let x be the speed of the second car.

	distance	rate	time
first car	224	$x + 14$	$\frac{224}{x+14}$
second car	175	x	$\frac{175}{x}$

$$\frac{224}{x+14} = \frac{175}{x}$$
$$224x = 175(x+14)$$
$$224x = 175x + 2450$$
$$49x = 2450$$
$$x = 50$$
$$x + 14 = 50 + 14 = 64$$

The speed of the first car is 64 miles per hour and the speed of the second car is 50 miles per hour.

69. Let x be the speed of the plane in still air.

	distance	rate	time
with wind	2160	$x + 30$	$\frac{2160}{x+30}$
against wind	1920	$x - 30$	$\frac{1920}{x-30}$

$$\frac{2160}{x+30} = \frac{1920}{x-30}$$
$$2160(x-30) = 1920(x+30)$$
$$2160x - 64,800 = 1920x + 57,600$$
$$240x = 122,400$$
$$x = 510$$

The speed of the plane in still air is 510 miles per hour.

71. $$\frac{9}{12} = \frac{3.75}{x}$$
$$9x = 45$$
$$x = 5$$

The missing length is 5.

73. $$\frac{16}{24} = \frac{9}{x}$$
$$16x = 216$$
$$x = 13.5$$

The missing length is 13.5.

75. $(-2, 5), (4, -3)$

$$m = \frac{-3-5}{4-(-2)} = \frac{-8}{6} = -\frac{4}{3}$$

Since the slope is negative, the line moves downward.

77. $(-3, -6), (1, 5)$

$$m = \frac{5-(-6)}{1-(-3)} = \frac{11}{4}$$

Since the slope is positive, the line moves upward.

79. $(3, 7), (3, -2)$

$$m = \frac{-2-7}{3-3} = \frac{-9}{0}$$

The slope is undefined. Since the slope is undefined, the line is vertical.

81. The capacity in 2010 was approximately 40,200 megawatts.

83. The capacity in 2010 was approximately 40,200 megawatts, or 40.2(1000 megawatts).
$40.2(560,000) = 22,512,000$
In 2010, the number of megawatts generated from wind served the electricity needs of 22,512,000 people.

85. Yes, since each side of the equation is one quotient.

87. Let x be the number of minutes it takes the first pump to fill the tank. Then it takes $3x$ minutes for the second pump to fill the tank.

	Minutes to Complete Total Job	Part of Job Completed in 1 Minute
1st Pump	x	$\frac{1}{x}$
2nd Pump	$3x$	$\frac{1}{3x}$
Together	21	$\frac{1}{21}$

$$\frac{1}{x} + \frac{1}{3x} = \frac{1}{21}$$

$$21x\left(\frac{1}{x}\right) + 21x\left(\frac{1}{3x}\right) = 21x\left(\frac{1}{21}\right)$$

$$21 + 7 = x$$
$$28 = x$$

$3x + 3(28) = 84$
The 1st pump takes 28 minutes and the 2nd takes 84 minutes.

89. none; answers may vary

91. answers may vary

93. $D = RT$

$$\frac{D}{T} = \frac{RT}{T}$$

$$\frac{D}{T} = R \text{ or } R = \frac{D}{T}$$

Section 7.7 Practice

1. a.
$$\frac{\frac{5k}{36m}}{\frac{15k}{9}} = \frac{5k}{36m} \div \frac{15k}{9}$$
$$= \frac{5k}{36m} \cdot \frac{9}{15k}$$
$$= \frac{5k \cdot 9}{36m \cdot 15k}$$
$$= \frac{1}{12m}$$

b.
$$\frac{\frac{8x}{x-4}}{\frac{3}{x+4}} = \frac{8x}{x-4} \div \frac{3}{x+4}$$
$$= \frac{8x}{x-4} \cdot \frac{x+4}{3}$$
$$= \frac{8x(x+4)}{3(x-4)}$$

c.
$$\frac{\frac{5}{a} + \frac{b}{a^2}}{\frac{5a}{b^2} + \frac{1}{b}} = \frac{\frac{5 \cdot a}{a \cdot a} + \frac{b}{a^2}}{\frac{5a}{b^2} + \frac{1 \cdot b}{b \cdot b}}$$
$$= \frac{\frac{5a+b}{a^2}}{\frac{5a+b}{b^2}}$$
$$= \frac{5a+b}{a^2} \cdot \frac{b^2}{5a+b}$$
$$= \frac{b^2(5a+b)}{a^2(5a+b)}$$
$$= \frac{b^2}{a^2}$$

2. a. The LCD is $(x-4)(x+4)$.
$$\frac{\frac{8x}{x-4}}{\frac{3}{x+4}} = \frac{\left(\frac{8x}{x-4}\right) \cdot (x-4)(x+4)}{\left(\frac{3}{x+4}\right) \cdot (x-4)(x+4)}$$
$$= \frac{8x(x+4)}{3(x-4)}$$

b. The LCD is a^2b^2.

$$\frac{\frac{b}{a^2}+\frac{1}{a}}{\frac{a}{b^2}+\frac{1}{b}}=\frac{\left(\frac{b}{a^2}+\frac{1}{a}\right)\cdot a^2b^2}{\left(\frac{a}{b^2}+\frac{1}{b}\right)\cdot a^2b^2}$$

$$=\frac{\frac{b}{a^2}\cdot a^2b^2+\frac{1}{a}\cdot a^2b^2}{\frac{a}{b^2}\cdot a^2b^2+\frac{1}{b}\cdot a^2b^2}$$

$$=\frac{b^3+ab^2}{a^3+a^2b}$$

$$=\frac{b^2(b+a)}{a^2(a+b)}$$

$$=\frac{b^2}{a^2}$$

3. $\dfrac{3x^{-1}+x^{-2}y^{-1}}{y^{-2}+xy^{-1}}=\dfrac{\frac{3}{x}+\frac{1}{x^2y}}{\frac{1}{y^2}+\frac{x}{y}}$

The LCD is x^2y^2.

$$=\frac{\left(\frac{3}{x}+\frac{1}{x^2y}\right)\cdot x^2y^2}{\left(\frac{1}{y^2}+\frac{x}{y}\right)\cdot x^2y^2}$$

$$=\frac{\frac{3}{x}\cdot x^2y^2+\frac{1}{x^2y}\cdot x^2y^2}{\frac{1}{y^2}\cdot x^2y^2+\frac{x}{y}\cdot x^2y^2}$$

$$=\frac{3xy^2+y}{x^2+x^3y}\text{ or }\frac{y(3xy+1)}{x^2(1+xy)}$$

4. $\dfrac{(3x)^{-1}-2}{5x^{-1}+2}=\dfrac{\frac{1}{3x}-2}{\frac{5}{x}+2}$

$$=\frac{\left(\frac{1}{3x}-2\right)\cdot 3x}{\left(\frac{5}{x}+2\right)\cdot 3x}$$

$$=\frac{\frac{1}{3x}\cdot 3x-2\cdot 3x}{\frac{5}{x}\cdot 3x+2\cdot 3x}$$

$$=\frac{1-6x}{15+6x}\text{ or }\frac{1-6x}{3(5+2x)}$$

Vocabulary, Readiness & Video Check 7.7

1. $\dfrac{\frac{7}{x}}{\frac{1}{x}+\frac{z}{x}}=\dfrac{x\left(\frac{7}{x}\right)}{x\left(\frac{1}{x}\right)+x\left(\frac{z}{x}\right)}=\dfrac{7}{1+z}$

2. $\dfrac{\frac{x}{4}}{\frac{x^2}{2}+\frac{1}{4}}=\dfrac{4\left(\frac{x}{4}\right)}{4\left(\frac{x^2}{2}\right)+4\left(\frac{1}{4}\right)}=\dfrac{x}{2x^2+1}$

3. $x^{-2}=\dfrac{1}{x^2}$

4. $y^{-3}=\dfrac{1}{y^3}$

5. $2x^{-1}=\dfrac{2}{x}$

6. $(2x)^{-1}=\dfrac{1}{2x}$

7. $(9y)^{-1}=\dfrac{1}{9y}$

8. $9y^{-2}=\dfrac{9}{y^2}$

9. a single fraction in the numerator and in the denominator

10. In method 2, you find the LCD of all fractions in both the numerator and the denominator in order to clear the complex fraction of fractions in the numerator and denominator; in method 1, you find the LCD of the fractions only in the numerator and/or only in the denominator in order to get single fractions in the numerator and denominator.

11. Since a negative exponent moves its base from a numerator to a denominator of the expression only, a rational expression containing negative exponents can become a complex fraction when rewritten with positive exponents.

Exercise Set 7.7

1. $\dfrac{\frac{10}{3x}}{\frac{5}{6x}}=\dfrac{10}{3x}\cdot\dfrac{6x}{5}=\dfrac{60x}{15x}=4$

3. $\dfrac{1+\frac{2}{5}}{2+\frac{3}{5}}=\dfrac{5\left(1+\frac{2}{5}\right)}{5\left(2+\frac{3}{5}\right)}=\dfrac{5+2}{10+3}=\dfrac{7}{13}$

5. $\dfrac{\frac{4}{x-1}}{\frac{x}{x-1}} = \dfrac{4}{x-1}\cdot\dfrac{x-1}{x} = \dfrac{4}{x}$

7. $\dfrac{1-\frac{2}{x}}{x+\frac{4}{9x}} = \dfrac{9x\left(1-\frac{2}{x}\right)}{9x\left(x+\frac{4}{9x}\right)} = \dfrac{9x-18}{9x^2+4} = \dfrac{9(x-2)}{9x^2+4}$

9. $\dfrac{\frac{4x^2-y^2}{xy}}{\frac{2}{y}-\frac{1}{x}} = \dfrac{\left(\frac{4x^2-y^2}{xy}\right)\cdot xy}{\left(\frac{2}{y}-\frac{1}{x}\right)\cdot xy}$

$= \dfrac{4x^2-y^2}{2x-y}$

$= \dfrac{(2x-y)(2x+y)}{2x-y}$

$= 2x+y$

11. $\dfrac{\frac{x+1}{3}}{\frac{2x-1}{6}} = \dfrac{x+1}{3}\cdot\dfrac{6}{2x-1} = \dfrac{2(x+1)}{2x-1}$

13. $\dfrac{\frac{2}{x}+\frac{3}{x^2}}{\frac{4}{x^2}-\frac{9}{x}} = \dfrac{\left(\frac{2}{x}+\frac{3}{x^2}\right)x^2}{\left(\frac{4}{x^2}-\frac{9}{x}\right)x^2} = \dfrac{2x+3}{4-9x}$

15. $\dfrac{\frac{1}{x}+\frac{2}{x^2}}{x+\frac{8}{x^2}} = \dfrac{x^2\left(\frac{1}{x}+\frac{2}{x^2}\right)}{x^2\left(x+\frac{8}{x^2}\right)}$

$= \dfrac{x+2}{x^3+8}$

$= \dfrac{x+2}{(x+2)(x^2-2x+4)}$

$= \dfrac{1}{x^2-2x+4}$

17. $\dfrac{\frac{4}{5-x}+\frac{5}{x-5}}{\frac{2}{x}+\frac{3}{x-5}} = \dfrac{-\frac{4}{x-5}+\frac{5}{x-5}}{\frac{2(x-5)+3x}{x(x-5)}}$

$= \dfrac{\frac{1}{x-5}}{\frac{2x-10+3x}{x(x-5)}}$

$= \dfrac{1}{x-5}\cdot\dfrac{x(x-5)}{5x-10}$

$= \dfrac{x}{5x-10} \text{ or } \dfrac{x}{5(x-2)}$

19. $\dfrac{\frac{x+2}{x}-\frac{2}{x-1}}{\frac{x+1}{x}+\frac{x+1}{x-1}} = \dfrac{\frac{(x+2)(x-1)-2x}{x(x-1)}}{\frac{(x+1)(x-1)+(x+1)(x)}{x(x-1)}}$

$= \dfrac{\frac{x^2+x-2-2x}{x(x-1)}}{\frac{x^2-1+x^2+x}{x(x-1)}}$

$= \dfrac{x^2-x-2}{x(x-1)}\cdot\dfrac{x(x-1)}{2x^2+x-1}$

$= \dfrac{(x-2)(x+1)}{x(x-1)}\cdot\dfrac{x(x-1)}{(2x-1)(x+1)}$

$= \dfrac{x-2}{2x-1}$

21. $\dfrac{\frac{2}{x}+3}{\frac{4}{x^2}-9} = \dfrac{\left(\frac{2}{x}+3\right)\cdot x^2}{\left(\frac{4}{x^2}-9\right)\cdot x^2}$

$= \dfrac{2x+3x^2}{4-9x^2}$

$= \dfrac{x(2+3x)}{(2+3x)(2-3x)}$

$= \dfrac{x}{2-3x}$

23. $\dfrac{1-\frac{x}{y}}{\frac{x^2}{y^2}-1} = \dfrac{\left(1-\frac{x}{y}\right)\cdot y^2}{\left(\frac{x^2}{y^2}-1\right)\cdot y^2}$

$= \dfrac{y^2-xy}{x^2-y^2}$

$= \dfrac{y(y-x)}{(x+y)(x-y)}$

$= \dfrac{-y(x-y)}{(x+y)(x-y)}$

$= -\dfrac{y}{x+y}$

25. $\dfrac{\frac{-2x}{x-y}}{\frac{y}{x^2}} = \dfrac{-2x}{x-y}\cdot\dfrac{x^2}{y} = -\dfrac{2x^3}{y(x-y)}$

27. $\dfrac{\frac{2}{x}+\frac{1}{x^2}}{\frac{y}{x^2}} = \dfrac{\left(\frac{2}{x}+\frac{1}{x^2}\right)x^2}{\left(\frac{y}{x^2}\right)x^2} = \dfrac{2x+1}{y}$

29.
$$\frac{\frac{x}{9}-\frac{1}{x}}{1+\frac{3}{x}}=\frac{\left(\frac{x}{9}-\frac{1}{x}\right)\cdot 9x}{\left(1+\frac{3}{x}\right)\cdot 9x}$$
$$=\frac{x^2-9}{9x+27}$$
$$=\frac{(x+3)(x-3)}{9(x+3)}$$
$$=\frac{x-3}{9}$$

31.
$$\frac{\frac{x-1}{x^2-4}}{1+\frac{1}{x-2}}=\frac{\frac{x-1}{x^2-4}}{\frac{x-2+1}{x-2}}=\frac{\frac{x-1}{x^2-4}}{\frac{x-1}{x-2}}$$
$$=\frac{x-1}{x^2-4}\cdot\frac{x-2}{x-1}$$
$$=\frac{x-1}{(x+2)(x-2)}\cdot\frac{x-2}{x-1}$$
$$=\frac{1}{x+2}$$

33.
$$\frac{\frac{2}{x+5}+\frac{4}{x+3}}{\frac{3x+13}{x^2+8x+15}}=\frac{\frac{2}{x+5}+\frac{4}{x+3}}{\frac{3x+13}{(x+5)(x+3)}}$$
$$=\frac{\left(\frac{2}{x+5}+\frac{4}{x+3}\right)(x+5)(x+3)}{\frac{3x+13}{(x+5)(x+3)}(x+5)(x+3)}$$
$$=\frac{2(x+3)+4(x+5)}{3x+13}$$
$$=\frac{2x+6+4x+20}{3x+13}$$
$$=\frac{6x+26}{3x+13}$$
$$=\frac{2(3x+13)}{3x+13}$$
$$=2$$

35.
$$\frac{x^{-1}}{x^{-2}+y^{-2}}=\frac{\frac{1}{x}}{\frac{1}{x^2}+\frac{1}{y^2}}$$
$$=\frac{x^2y^2\left(\frac{1}{x}\right)}{x^2y^2\left(\frac{1}{x^2}+\frac{1}{y^2}\right)}$$
$$=\frac{xy^2}{y^2+x^2}$$
$$=\frac{xy^2}{x^2+y^2}$$

37.
$$\frac{2a^{-1}+3b^{-2}}{a^{-1}-b^{-1}}=\frac{\frac{2}{a}+\frac{3}{b^2}}{\frac{1}{a}-\frac{1}{b}}$$
$$=\frac{ab^2\left(\frac{2}{a}+\frac{3}{b^2}\right)}{ab^2\left(\frac{1}{a}-\frac{1}{b}\right)}$$
$$=\frac{2b^2+3a}{b^2-ab}$$
$$=\frac{2b^2+3a}{b(b-a)}$$

39.
$$\frac{1}{x-x^{-1}}=\frac{1}{x-\frac{1}{x}}$$
$$=\frac{x(1)}{x\left(x-\frac{1}{x}\right)}$$
$$=\frac{x}{x^2-1}$$
$$=\frac{x}{(x+1)(x-1)}$$

41.
$$\frac{a^{-1}+1}{a^{-1}-1}=\frac{\frac{1}{a}+1}{\frac{1}{a}-1}=\frac{a\left(\frac{1}{a}+1\right)}{a\left(\frac{1}{a}-1\right)}=\frac{1+a}{1-a}$$

43.
$$\frac{3x^{-1}+(2y)^{-1}}{x^{-2}}=\frac{\frac{3}{x}+\frac{1}{2y}}{\frac{1}{x^2}}$$
$$=\frac{2x^2y\left(\frac{3}{x}+\frac{1}{2y}\right)}{2x^2y\left(\frac{1}{x^2}\right)}$$
$$=\frac{6xy+x^2}{2y}$$
$$=\frac{x(x+6y)}{2y}$$

45.
$$\frac{2a^{-1}+(2a)^{-1}}{a^{-1}+2a^{-2}}=\frac{\frac{2}{a}+\frac{1}{2a}}{\frac{1}{a}+\frac{2}{a^2}}$$
$$=\frac{2a^2\left(\frac{2}{a}+\frac{1}{2a}\right)}{2a^2\left(\frac{1}{a}+\frac{2}{a^2}\right)}$$
$$=\frac{4a+a}{2a+4}$$
$$=\frac{5a}{2(a+2)}$$

47. $\dfrac{5x^{-1}+2y^{-1}}{x^{-2}y^{-2}} = \dfrac{\frac{5}{x}+\frac{2}{y}}{\frac{1}{x^2y^2}}$

$= \dfrac{x^2y^2\left(\frac{5}{x}+\frac{2}{y}\right)}{x^2y^2\left(\frac{1}{x^2y^2}\right)}$

$= 5xy^2 + 2x^2y$

$= xy(5y+2x)$

49. $\dfrac{5x^{-1}-2y^{-1}}{25x^{-2}-4y^{-2}} = \dfrac{\frac{5}{x}-\frac{2}{y}}{\frac{25}{x^2}-\frac{4}{y^2}}$

$= \dfrac{x^2y^2\left(\frac{5}{x}-\frac{2}{y}\right)}{x^2y^2\left(\frac{25}{x^2}-\frac{4}{y^2}\right)}$

$= \dfrac{5xy^2 - 2x^2y}{25y^2 - 4x^2}$

$= \dfrac{xy(5y-2x)}{(5y+2x)(5y-2x)}$

$= \dfrac{xy}{5y+2x}$ or $\dfrac{xy}{2x+5y}$

51. $\dfrac{3x^3y^2}{12x} = \dfrac{3x\cdot x^2y^2}{3x\cdot 4} = \dfrac{x^2y^2}{4}$

53. $\dfrac{144x^5y^5}{-16x^2y} = \dfrac{16x^2y\cdot 9x^3y^4}{16x^2y\cdot(-1)} = -9x^3y^4$

55. $P(x) = -x^2$

$P(-3) = -(-3)^2 = -9$

57. $\dfrac{\frac{x+1}{9}}{\frac{y-2}{5}} = \dfrac{x+1}{9} \div \dfrac{y-2}{5} = \dfrac{x+1}{9}\cdot\dfrac{5}{y-2}$

Both a and c are equivalent to the original expression.

59. $\dfrac{a}{1-\frac{s}{770}} = \dfrac{770(a)}{770\left(1-\frac{s}{770}\right)} = \dfrac{770a}{770-s}$

61. $\dfrac{\frac{1}{x}}{\frac{3}{y}} = \dfrac{1}{x} \div \dfrac{3}{y} = \dfrac{1}{x}\cdot\dfrac{y}{3}$

Both a and b are equivalent to the original expression.

63. answers may vary

65. $\dfrac{1}{1+(1+x)^{-1}} = \dfrac{1}{1+\frac{1}{1+x}}$

$= \dfrac{(1+x)\cdot 1}{(1+x)\left(1+\frac{1}{1+x}\right)}$

$= \dfrac{1+x}{1+x+1}$

$= \dfrac{1+x}{2+x}$

67. $\dfrac{x}{1-\frac{1}{1+\frac{1}{x}}} = \dfrac{x}{1-\frac{1}{\frac{x+1}{x}}}$

$= \dfrac{x}{1-\frac{x}{x+1}}$

$= \dfrac{(x+1)(x)}{(x+1)\left(1-\frac{x}{x+1}\right)}$

$= \dfrac{x(x+1)}{x+1-x}$

$= \dfrac{x(x+1)}{1}$

$= x(x+1)$

69. $\dfrac{\frac{2}{y^2}-\frac{5}{xy}-\frac{3}{x^2}}{\frac{2}{y^2}+\frac{7}{xy}+\frac{3}{x^2}} = \dfrac{x^2y^2\left(\frac{2}{y^2}-\frac{5}{xy}-\frac{3}{x^2}\right)}{x^2y^2\left(\frac{2}{y^2}+\frac{7}{xy}+\frac{3}{x^2}\right)}$

$= \dfrac{2x^2-5xy-3y^2}{2x^2+7xy+3y^2}$

$= \dfrac{(2x+y)(x-3y)}{(2x+y)(x+3y)}$

$= \dfrac{x-3y}{x+3y}$

71. $\dfrac{3(a+1)^{-1}+4a^{-2}}{(a^3+a^2)^{-1}} = \dfrac{\frac{3}{a+1}+\frac{4}{a^2}}{\frac{1}{a^3+a^2}}$

$= \dfrac{\frac{3a^2+4(a+1)}{a^2(a+1)}}{\frac{1}{a^2(a+1)}}$

$= \dfrac{3a^2+4a+4}{a^2(a+1)}\cdot\dfrac{a^2(a+1)}{1}$

$= 3a^2+4a+4$

73. $f(x) = \dfrac{1}{x}$

a. $f(a+h) = \dfrac{1}{a+h}$

b. $f(a) = \dfrac{1}{a}$

c. $\dfrac{f(a+h)-f(a)}{h} = \dfrac{\frac{1}{a+h}-\frac{1}{a}}{h}$

d. $\dfrac{\frac{1}{a+h}-\frac{1}{a}}{h} = \dfrac{a(a+h)\left(\frac{1}{a+h}-\frac{1}{a}\right)}{a(a+h)\cdot h}$

$= \dfrac{a-(a+h)}{ah(a+h)}$

$= \dfrac{-h}{ah(a+h)}$

$= -\dfrac{1}{a(a+h)}$

75. $f(x) = \dfrac{3}{x+1}$

a. $f(a+h) = \dfrac{3}{a+h+1}$

b. $f(a) = \dfrac{3}{a+1}$

c. $\dfrac{f(a+h)-f(a)}{h} = \dfrac{\frac{3}{a+h+1}-\frac{3}{a+1}}{h}$

d. $\dfrac{\frac{3}{a+h+1}-\frac{3}{a+1}}{h}$

$= \dfrac{\left(\frac{3}{a+h+1}-\frac{3}{a+1}\right)\cdot(a+h+1)(a+1)}{h\cdot(a+h+1)(a+1)}$

$= \dfrac{3(a+1)-3(a+h+1)}{h(a+h+1)(a+1)}$

$= \dfrac{3a+3-3a-3h-3}{h(a+h+1)(a+1)}$

$= \dfrac{-3h}{h(a+h+1)(a+1)}$

$= -\dfrac{3}{(a+h+1)(a+1)}$

Chapter 7 Vocabulary Check

1. A <u>rational expression</u> is an expression that can be written in the form $\dfrac{P}{Q}$, where P and Q are polynomials and Q is not 0.

2. In a <u>complex fraction</u>, the numerator or denominator or both may contain fractions.

3. For a rational expression, $-\dfrac{a}{b} = \dfrac{-a}{\underline{b}} = \dfrac{a}{\underline{-b}}$.

4. A rational expression is undefined when the <u>denominator</u> is 0.

5. The process of writing a rational expression in lowest terms is called <u>simplifying</u>.

6. The expressions $\dfrac{2x}{7}$ and $\dfrac{7}{2x}$ are called <u>reciprocals</u>.

7. The <u>least common denominator</u> of a list of rational expressions is a polynomial of least degree whose factors include all factors of the denominators in the list.

8. A <u>ratio</u> is the quotient of two numbers.

9. $\dfrac{x}{2} = \dfrac{7}{16}$ is an example of a <u>proportion</u>.

10. If $\dfrac{a}{b} = \dfrac{c}{d}$, then ad and bc are called <u>cross products</u>.

11. The <u>domain</u> of the rational function $f(x) = \dfrac{1}{x-3}$ is $\{x|x \text{ is a real number } x \neq 3\}$.

Chapter 7 Review

1. 7 is never 0 so the domain of $f(x) = \dfrac{3-5x}{7}$ is $\{x|x \text{ is a real number}\}$.

2. 11 is never 0 so the domain of $g(x) = \dfrac{2x+4}{11}$ is $\{x|x \text{ is a real number}\}$.

3. $x - 5 = 0$
$ x = 5$

The domain of $F(x) = \dfrac{-3x^2}{x-5}$ is

$\{x | x \text{ is a real number and } x \neq 5\}$.

4. $3x - 12 = 0$
$ 3x = 12$
$ x = 4$

The domain of $h(x) = \dfrac{4x}{3x-12}$ is

$\{x | x \text{ is a real number and } x \neq 4\}$.

5. $x^2 + 8x = 0$
$x(x+8) = 0$
$x = 0 \quad \text{or} \quad x + 8 = 0$
$x = 0 \quad \text{or} \quad x = -8$

The domain of $f(x) = \dfrac{x^3 + 2}{x^2 + 8x}$ is

$\{x | x \text{ is a real number and } x \neq 0, x \neq -8\}$.

6. $3x^2 - 48 = 0$
$ 3(x^2 - 16) = 0$
$3(x+4)(x-4) = 0$
$x + 4 = 0 \quad \text{or} \quad x - 4 = 0$
$ x = -4 \quad \text{or} \quad x = 4$

The domain of $G(x) = \dfrac{20}{3x^2 - 48}$ is

$\{x | x \text{ is a real number and } x \neq -4, x \neq 4\}$.

7. $\dfrac{x-12}{12-x} = \dfrac{x-12}{-(x-12)} = -1$

8. $\dfrac{2x}{2x^2 - 2x} = \dfrac{2x}{2x(x-1)} = \dfrac{1}{x-1}$

9. $\dfrac{x+7}{x^2 - 49} = \dfrac{x+7}{(x-7)(x+7)} = \dfrac{1}{x-7}$

10. $\dfrac{2x^2 + 4x - 30}{x^2 + x - 20} = \dfrac{2(x^2 + 2x - 15)}{(x+5)(x-4)}$
$= \dfrac{2(x+5)(x-3)}{(x+5)(x-4)}$
$= \dfrac{2(x-3)}{x-4}$

11. $\dfrac{x^2 + xa + xb + ab}{x^2 - xc + bx - bc} = \dfrac{x(x+a) + b(x+a)}{x(x-c) + b(x-c)}$
$= \dfrac{(x+a)(x+b)}{(x-c)(x+b)}$
$= \dfrac{x+a}{x-c}$

12. $\dfrac{x^2 + 5x - 2x - 10}{x^2 - 3x - 2x + 6} = \dfrac{x(x+5) - 2(x+5)}{x(x-3) - 2(x-3)}$
$= \dfrac{(x+5)(x-2)}{(x-3)(x-2)}$
$= \dfrac{x+5}{x-3}$

13. $\dfrac{4-x}{x^3 - 64} = -\dfrac{x-4}{x^3 - 64}$
$= -\dfrac{x-4}{(x-4)(x^2 + 4x + 16)}$
$= -\dfrac{1}{x^2 + 4x + 16}$

14. $\dfrac{x^2 - 4}{x^3 + 8} = \dfrac{(x+2)(x-2)}{(x+2)(x^2 - 2x + 4)} = \dfrac{x-2}{x^2 - 2x + 4}$

15. $C(x) = \dfrac{35x + 4200}{x}$

$C(50) = \dfrac{35(50) + 4200}{50}$
$= \dfrac{1750 + 4200}{50}$
$= \dfrac{5950}{50}$
$= 119$

The average cost is \$119.

16. $C(x) = \dfrac{35x + 4200}{x}$

$C(100) = \dfrac{35(100) + 4200}{100}$
$= \dfrac{3500 + 4200}{100}$
$= \dfrac{7700}{100}$
$= 77$

The average cost is \$77.

17. $\dfrac{15x^3y^2}{z} \cdot \dfrac{z}{5xy^3} = \dfrac{15x^3y^2 \cdot z}{z \cdot 5xy^3}$

$= \dfrac{3 \cdot 5 \cdot x^2 \cdot x \cdot y^2 \cdot z}{z \cdot 5 \cdot x \cdot y^2 \cdot y}$

$= \dfrac{3x^2}{y}$

18. $\dfrac{-y^3}{8} \cdot \dfrac{9x^2}{y^3} = -\dfrac{y^3 \cdot 9x^2}{8 \cdot y^3} = -\dfrac{9x^2}{8}$

19. $\dfrac{x^2-9}{x^2-4} \cdot \dfrac{x-2}{x+3} = \dfrac{(x^2-9) \cdot (x-2)}{(x^2-4) \cdot (x+3)}$

$= \dfrac{(x-3)(x+3)(x-2)}{(x+2)(x-2)(x+3)}$

$= \dfrac{x-3}{x+2}$

20. $\dfrac{2x+5}{x-6} \cdot \dfrac{2x}{-x+6} = \dfrac{2x+5}{x-6} \cdot \dfrac{2x}{-(x-6)}$

$= \dfrac{2x+5}{x-6} \cdot \dfrac{-2x}{x-6}$

$= \dfrac{(2x+5) \cdot (-2x)}{(x-6) \cdot (x-6)}$

$= \dfrac{-2x(2x+5)}{(x-6)^2}$

21. $\dfrac{x^2-5x-24}{x^2-x-12} \div \dfrac{x^2-10x+16}{x^2+x-6}$

$= \dfrac{x^2-5x-24}{x^2-x-12} \cdot \dfrac{x^2+x-6}{x^2-10x+16}$

$= \dfrac{(x-8)(x+3) \cdot (x+3)(x-2)}{(x-4)(x+3) \cdot (x-8)(x-2)}$

$= \dfrac{x+3}{x-4}$

22. $\dfrac{4x+4y}{xy^2} \div \dfrac{3x+3y}{x^2y} = \dfrac{4x+4y}{xy^2} \cdot \dfrac{x^2y}{3x+3y}$

$= \dfrac{4(x+y) \cdot x \cdot x \cdot y}{x \cdot y \cdot y \cdot 3(x+y)}$

$= \dfrac{4x}{3y}$

23. $\dfrac{x^2+x-42}{x-3} \cdot \dfrac{(x-3)^2}{x+7}$

$= \dfrac{(x+7)(x-6) \cdot (x-3)(x-3)}{(x-3) \cdot (x+7)}$

$= (x-6)(x-3)$

24. $\dfrac{2a+2b}{3} \cdot \dfrac{a-b}{a^2-b^2} = \dfrac{2(a+b) \cdot (a-b)}{3 \cdot (a+b)(a-b)} = \dfrac{2}{3}$

25. $\dfrac{2x^2-9x+9}{8x-12} \div \dfrac{x^2-3x}{2x} = \dfrac{2x^2-9x+9}{8x-12} \cdot \dfrac{2x}{x^2-3x}$

$= \dfrac{(2x-3)(x-3) \cdot 2x}{4(2x-3) \cdot x(x-3)}$

$= \dfrac{2}{4}$

$= \dfrac{1}{2}$

26. $\dfrac{x^2-y^2}{x^2+xy} \div \dfrac{3x^2-2xy-y^2}{3x^2+6x}$

$= \dfrac{x^2-y^2}{x^2+xy} \cdot \dfrac{3x^2+6x}{3x^2-2xy-y^2}$

$= \dfrac{(x-y)(x+y) \cdot 3x(x+2)}{x(x+y) \cdot (3x+y)(x-y)}$

$= \dfrac{3(x+2)}{3x+y}$

27. $\dfrac{x-y}{4} \div \dfrac{y^2-2y-xy+2x}{16x+24}$

$= \dfrac{x-y}{4} \cdot \dfrac{16x+24}{y^2-2y-xy+2x}$

$= \dfrac{x-y}{4} \cdot \dfrac{8(2x+3)}{y(y-2)-x(y-2)}$

$= \dfrac{x-y}{4} \cdot \dfrac{8(2x+3)}{(y-2)(y-x)}$

$= -\dfrac{y-x}{4} \cdot \dfrac{8(2x+3)}{(y-2)(y-x)}$

$= -\dfrac{2 \cdot 4(y-x)(2x+3)}{4(y-2)(y-x)}$

$= -\dfrac{2(2x+3)}{y-2}$

28. $\dfrac{5+x}{7} \div \dfrac{xy+5y-3x-15}{7y-35}$

$= \dfrac{5+x}{7} \cdot \dfrac{7y-35}{xy+5y-3x-15}$

$= \dfrac{(5+x) \cdot 7(y-5)}{7 \cdot (x+5)(y-3)}$

$= \dfrac{y-5}{y-3}$

29. $\dfrac{x}{x^2+9x+14} + \dfrac{7}{x^2+9x+14} = \dfrac{x+7}{x^2+9x+14}$

$= \dfrac{x+7}{(x+7)(x+2)}$

$= \dfrac{1}{x+2}$

30. $\dfrac{x}{x^2+2x-15} + \dfrac{5}{x^2+2x-15} = \dfrac{x+5}{x^2+2x-15}$

$= \dfrac{x+5}{(x+5)(x-3)}$

$= \dfrac{1}{x-3}$

31. $\dfrac{4x-5}{3x^2} - \dfrac{2x+5}{3x^2} = \dfrac{4x-5-(2x+5)}{3x^2}$

$= \dfrac{4x-5-2x-5}{3x^2}$

$= \dfrac{2x-10}{3x^2}$

32. $\dfrac{9x+7}{6x^2} - \dfrac{3x+4}{6x^2} = \dfrac{9x+7-(3x+4)}{6x^2}$

$= \dfrac{9x+7-3x-4}{6x^2}$

$= \dfrac{6x+3}{6x^2}$

$= \dfrac{3(2x+1)}{3 \cdot 2x^2}$

$= \dfrac{2x+1}{2x^2}$

33. $2x = 2 \cdot x$

$7x = 7 \cdot x$

$\text{LCD} = 2 \cdot 7 \cdot x = 14x$

34. $x^2 - 5x - 24 = (x-8)(x+3)$

$x^2 + 11x + 24 = (x+8)(x+3)$

$\text{LCD} = (x-8)(x+3)(x+8)$

35. $\dfrac{5}{7x} = \dfrac{5}{7x} \cdot \dfrac{2x^2y}{2x^2y} = \dfrac{5 \cdot 2x^2y}{7x \cdot 2x^2y} = \dfrac{10x^2y}{14x^3y}$

36. $\dfrac{9}{4y} = \dfrac{9}{4y} \cdot \dfrac{4y^2x}{4y^2x} = \dfrac{9 \cdot 4y^2x}{4y \cdot 4y^2x} = \dfrac{36y^2x}{16y^3x}$

37. $\dfrac{x+2}{x^2+11x+18} = \dfrac{x+2}{(x+9)(x+2)}$

$= \dfrac{(x+2)(x-5)}{(x+9)(x+2)(x-5)}$

$= \dfrac{x^2-3x-10}{(x+2)(x-5)(x+9)}$

38. $\dfrac{3x-5}{x^2+4x+4} = \dfrac{3x-5}{(x+2)^2}$

$= \dfrac{(3x-5)(x+3)}{(x+2)^2(x+3)}$

$= \dfrac{3x^2+4x-15}{(x+2)^2(x+3)}$

39. $\dfrac{4}{5x^2} - \dfrac{6}{y} = \dfrac{4(y)}{5x^2(y)} - \dfrac{6(5x^2)}{y(5x^2)} = \dfrac{4y-30x^2}{5x^2y}$

40. $\dfrac{2}{x-3} - \dfrac{4}{x-1} = \dfrac{2(x-1)}{(x-3)(x-1)} - \dfrac{4(x-3)}{(x-1)(x-3)}$

$= \dfrac{2(x-1)-4(x-3)}{(x-3)(x-1)}$

$= \dfrac{2x-2-4x+12}{(x-3)(x-1)}$

$= \dfrac{-2x+10}{(x-3)(x-1)}$

41. $\dfrac{4}{x+3} - 2 = \dfrac{4}{x+3} - \dfrac{2(x+3)}{x+3}$

$= \dfrac{4-2(x+3)}{x+3}$

$= \dfrac{4-2x-6}{x+3}$

$= \dfrac{-2x-2}{x+3}$

42. $\dfrac{3}{x^2+2x-8}+\dfrac{2}{x^2-3x+2}$

$=\dfrac{3}{(x+4)(x-2)}+\dfrac{2}{(x-1)(x-2)}$

$=\dfrac{3(x-1)}{(x+4)(x-2)(x-1)}+\dfrac{2(x+4)}{(x-1)(x-2)(x+4)}$

$=\dfrac{3(x-1)+2(x+4)}{(x+4)(x-2)(x-1)}$

$=\dfrac{3x-3+2x+8}{(x+4)(x-2)(x-1)}$

$=\dfrac{5x+5}{(x+4)(x-2)(x-1)}$

43. $\dfrac{2x-5}{6x+9}-\dfrac{4}{2x^2+3x}=\dfrac{2x-5}{3(2x+3)}-\dfrac{4}{x(2x+3)}$

$=\dfrac{(2x-5)(x)}{3(2x+3)(x)}-\dfrac{4(3)}{x(2x+3)(3)}$

$=\dfrac{2x^2-5x-12}{3x(2x+3)}$

$=\dfrac{(2x+3)(x-4)}{3x(2x+3)}$

$=\dfrac{x-4}{3x}$

44. $\dfrac{x-1}{x^2-2x+1}-\dfrac{x+1}{x-1}=\dfrac{x-1}{(x-1)^2}-\dfrac{x+1}{x-1}$

$=\dfrac{1}{x-1}-\dfrac{x+1}{x-1}$

$=\dfrac{1-(x+1)}{x-1}$

$=\dfrac{1-x-1}{x-1}$

$=\dfrac{-x}{x-1}$

$=-\dfrac{x}{x-1}$

45. $P=2l+2w$

$P=2\left(\dfrac{x}{8}\right)+2\left(\dfrac{x+2}{4x}\right)$

$=\dfrac{x}{4}+\dfrac{2(x+2)}{4x}$

$=\dfrac{x\cdot x}{4\cdot x}+\dfrac{2x+4}{4x}$

$=\dfrac{x^2+2x+4}{4x}$

$A=l\cdot w$

$A=\dfrac{x}{8}\cdot\dfrac{x+2}{4x}=\dfrac{x\cdot(x+2)}{8\cdot 4x}=\dfrac{x+2}{32}$

The perimeter is $\dfrac{x^2+2x+4}{4x}$ units and the area

is $\dfrac{x+2}{32}$ square units.

46. $P=\dfrac{3x}{4x-4}+\dfrac{2x}{3x-3}+\dfrac{x}{x-1}$

$=\dfrac{3x}{4(x-1)}+\dfrac{2x}{3(x-1)}+\dfrac{x}{x-1}$

$=\dfrac{3x(3)}{4(x-1)(3)}+\dfrac{2x(4)}{3(x-1)(4)}+\dfrac{x(12)}{(x-1)(12)}$

$=\dfrac{9x+8x+12x}{12(x-1)}$

$=\dfrac{29x}{12(x-1)}$

$A=\dfrac{1}{2}\cdot b\cdot h$

$A=\dfrac{1}{2}\cdot\dfrac{x}{x-1}\cdot\dfrac{6y}{5}=\dfrac{1\cdot x\cdot 2\cdot 3y}{2\cdot(x-1)\cdot 5}=\dfrac{3xy}{5(x-1)}$

The perimeter is $\dfrac{29x}{12(x-1)}$ units and the area is

$\dfrac{3xy}{5(x-1)}$ square units.

47. $\dfrac{n}{10}=9-\dfrac{n}{5}$

$10\left(\dfrac{n}{10}\right)=10\left(9-\dfrac{n}{5}\right)$

$10\left(\dfrac{n}{10}\right)=10(9)-10\left(\dfrac{n}{5}\right)$

$n=90-2n$

$3n=90$

$n=30$

48.

$$\frac{2}{x+1}-\frac{1}{x-2}=-\frac{1}{2}$$

$$2(x+1)(x-2)\left(\frac{2}{x+1}-\frac{1}{x-2}\right)=2(x+1)(x-2)\left(-\frac{1}{2}\right)$$

$$2(x+1)(x-2)\left(\frac{2}{x+1}\right)-2(x+1)(x-2)\left(\frac{1}{x-2}\right)=2(x+1)(x-2)\left(-\frac{1}{2}\right)$$

$$4(x-2)-2(x+1)=-(x+1)(x-2)$$

$$4x-8-2x-2=-(x^2-x-2)$$

$$2x-10=-x^2+x+2$$

$$x^2+x-12=0$$

$$(x+4)(x-3)=0$$

$$x+4=0 \quad \text{or} \quad x-3=0$$
$$x=-4 \qquad\qquad x=3$$

49.

$$\frac{y}{2y+2}+\frac{2y-16}{4y+4}=\frac{y-3}{y+1}$$

$$\frac{y}{2(y+1)}+\frac{2y-16}{4(y+1)}=\frac{y-3}{y+1}$$

$$4(y+1)\left(\frac{y}{2(y+1)}+\frac{2y-16}{4(y+1)}\right)=4(y+1)\left(\frac{y-3}{y+1}\right)$$

$$4(y+1)\left(\frac{y}{2(y+1)}\right)+4(y+1)\left(\frac{2y-16}{4(y+1)}\right)=4(y+1)\left(\frac{y-3}{y+1}\right)$$

$$2y+2y-16=4(y-3)$$

$$4y-16=4y-12$$

$$-16=-12 \quad \text{False}$$

This equation has no solution.

50.

$$\frac{2}{x-3}-\frac{4}{x+3}=\frac{8}{x^2-9}$$

$$(x-3)(x+3)\left(\frac{2}{x-3}-\frac{4}{x+3}\right)=(x-3)(x+3)\left(\frac{8}{(x-3)(x+3)}\right)$$

$$(x-3)(x+3)\left(\frac{2}{x-3}\right)-(x-3)(x+3)\left(\frac{4}{x+3}\right)=8$$

$$2(x+3)-4(x-3)=8$$

$$2x+6-4x+12=8$$

$$-2x+18=8$$

$$-2x=-10$$

$$x=5$$

51.

$$\frac{x-3}{x+1}-\frac{x-6}{x+5}=0$$

$$(x+1)(x+5)\left(\frac{x-3}{x+1}-\frac{x-6}{x+5}\right)=(x+1)(x+5)(0)$$

$$(x+1)(x+5)\left(\frac{x-3}{x+1}\right)-(x+1)(x+5)\left(\frac{x-6}{x+5}\right)=0$$

$$(x+5)(x-3)-(x+1)(x-6)=0$$

$$x^2+2x-15-(x^2-5x-6)=0$$

$$x^2+2x-15-x^2+5x+6=0$$

$$7x-9=0$$

$$7x=9$$

$$x=\frac{9}{7}$$

52.

$$x+5=\frac{6}{x}$$

$$x(x+5)=x\left(\frac{6}{x}\right)$$

$$x^2+5x=6$$

$$x^2+5x-6=0$$

$$(x+6)(x-1)=0$$

$$x+6=0 \quad \text{or} \quad x-1=0$$

$$x=-6 \qquad\qquad x=1$$

53.

$$\frac{4A}{5b}=x^2$$

$$4A=5bx^2$$

$$\frac{4A}{5x^2}=\frac{5bx^2}{5x^2}$$

$$\frac{4A}{5x^2}=b$$

54.

$$\frac{x}{7}+\frac{y}{8}=10$$

$$56\left(\frac{x}{7}\right)+56\left(\frac{y}{8}\right)=56(10)$$

$$8x+7y=560$$

$$7y=560-8x$$

$$y=\frac{560-8x}{7}$$

55.

$$\frac{x}{2}=\frac{12}{4}$$

$$4x=24$$

$$x=6$$

56. $\dfrac{20}{1} = \dfrac{x}{25}$

$500 = x$

57. $\dfrac{2}{x-1} = \dfrac{3}{x+3}$

$2(x+3) = 3(x-1)$

$2x+6 = 3x-3$

$6 = x-3$

$9 = x$

58. $\dfrac{4}{y-3} = \dfrac{2}{y-3}$

$4(y-3) = 2(y-3)$

$4y-12 = 2y-6$

$2y-12 = -6$

$2y = 6$

$y = 3$

$y = 3$ doesn't check, so this equation has no solution.

59. Let x = the number of parts processed in 45 minutes.

$\dfrac{300}{20} = \dfrac{x}{45}$

$13,500 = 20x$

$675 = x$

675 parts can be processed in 45 minutes.

60. Let x = the charge for 3 hours.

$\dfrac{90.00}{8} = \dfrac{x}{3}$

$270.00 = 8x$

$33.75 = x$

He charges \$33.75 for 3 hours.

61. $5 \cdot \dfrac{1}{x} = \dfrac{3}{2} \cdot \dfrac{1}{x} + \dfrac{7}{6}$

$\dfrac{5}{x} = \dfrac{3}{2x} + \dfrac{7}{6}$

$6x\left(\dfrac{5}{x}\right) = 6x\left(\dfrac{3}{2x}\right) + 6x\left(\dfrac{7}{6}\right)$

$30 = 9 + 7x$

$21 = 7x$

$x = 3$

The unknown number is 3.

62. $\dfrac{1}{x} = \dfrac{1}{4-x}$

$4-x = x$

$4 = 2x$

$2 = x$

The unknown number is 2.

63. Let r be the rate of the faster car. Then the rate of the slower car is $r - 10$.

	Distance	= Rate	· Time
Fast car	90	r	$\dfrac{90}{r}$
Slow car	60	$r-10$	$\dfrac{60}{r-10}$

$\dfrac{90}{r} = \dfrac{60}{r-10}$

$90(r-10) = 60r$

$90r - 900 = 60r$

$-900 = -30r$

$30 = r$

$r - 10 = 30 - 10 = 20$

The rate of the fast car is 30 miles per hour and the rate of the slower car is 20 miles per hour.

64. Let r be the speed of the boat in still water.

	Distance	= Rate	· Time
Upstream	48	$r-4$	$\dfrac{48}{r-4}$
Downstream	72	$r+4$	$\dfrac{72}{r+4}$

$\dfrac{48}{r-4} = \dfrac{72}{r+4}$

$48(r+4) = 72(r-4)$

$48r + 192 = 72r - 288$

$480 = 24r$

$r = 20$

The speed of the boat in still water is 20 miles per hour.

65. Let x be the time it takes Maria working alone.

	Hours to Complete Total Job	Part of Job Completed in 1 Hour
Mark	7	$\frac{1}{7}$
Maria	x	$\frac{1}{x}$
Together	5	$\frac{1}{5}$

$$\frac{1}{7} + \frac{1}{x} = \frac{1}{5}$$

$$35x\left(\frac{1}{7}\right) + 35x\left(\frac{1}{x}\right) = 35x\left(\frac{1}{5}\right)$$

$$5x + 35 = 7x$$

$$35 = 2x$$

$$x = \frac{35}{2} \text{ or } 17\frac{1}{2}$$

It takes Maria $17\frac{1}{2}$ hours to complete the job alone.

66. Let x be the number of days it takes the pipes to fill the pond together.

	Days to Complete Total Job	Part of Job Completed in 1 Day
Pipe A	20	$\frac{1}{20}$
Pipe B	15	$\frac{1}{15}$
Together	x	$\frac{1}{x}$

$$\frac{1}{20} + \frac{1}{25} = \frac{1}{x}$$

$$60x\left(\frac{1}{20}\right) + 60x\left(\frac{1}{15}\right) = 60x\left(\frac{1}{x}\right)$$

$$3x + 4x = 60$$

$$7x = 60$$

$$x = \frac{60}{7} = 8\frac{4}{7}$$

Both pipes fill the pond in $8\frac{4}{7}$ days.

67.
$$\frac{2}{3} = \frac{10}{x}$$
$$2x = 30$$
$$x = 15$$
The missing length is 15.

68.
$$\frac{12}{4} = \frac{18}{x}$$
$$12x = 72$$
$$x = 6$$
The missing length is 6.

69. $\dfrac{\frac{5x}{27}}{-\frac{10xy}{21}} = \dfrac{5x}{27} \cdot -\dfrac{21}{10xy} = -\dfrac{5x \cdot 3 \cdot 7}{3 \cdot 9 \cdot 5 \cdot 2 \cdot x \cdot y} = -\dfrac{7}{18y}$

70. $\dfrac{\frac{3}{5} + \frac{2}{7}}{\frac{1}{5} + \frac{5}{6}} = \dfrac{\frac{21}{35} + \frac{10}{35}}{\frac{6}{30} + \frac{25}{30}} = \dfrac{\frac{31}{35}}{\frac{31}{30}} = \dfrac{31}{35} \cdot \dfrac{30}{31} = \dfrac{31 \cdot 5 \cdot 6}{5 \cdot 7 \cdot 31} = \dfrac{6}{7}$

71. $\dfrac{3 - \frac{1}{y}}{2 - \frac{1}{y}} = \dfrac{y\left(3 - \frac{1}{y}\right)}{y\left(2 - \frac{1}{y}\right)} = \dfrac{y(3) - y\left(\frac{1}{y}\right)}{y(2) - y\left(\frac{1}{y}\right)} = \dfrac{3y - 1}{2y - 1}$

72. $\dfrac{\frac{6}{x+2} + 4}{\frac{8}{x+2} - 4} = \dfrac{(x+2)\left(\frac{6}{x+2} + 4\right)}{(x+2)\left(\frac{8}{x+2} - 4\right)}$

$$= \dfrac{(x+2)\left(\frac{6}{x+2}\right) + (x+2)(4)}{(x+2)\left(\frac{8}{x+2}\right) - (x+2)(4)}$$

$$= \dfrac{6 + 4x + 8}{8 - 4x - 8}$$

$$= \dfrac{4x + 14}{-4x}$$

$$= -\dfrac{2(2x + 7)}{2 \cdot 2x}$$

$$= -\dfrac{2x + 7}{2x}$$

73. $\dfrac{\frac{x-3}{x+3}+\frac{x+3}{x-3}}{\frac{x-3}{x+3}-\frac{x+3}{x-3}}=\dfrac{(x+3)(x-3)\left(\frac{x-3}{x+3}+\frac{x+3}{x-3}\right)}{(x+3)(x-3)\left(\frac{x-3}{x+3}-\frac{x+3}{x-3}\right)}$

$=\dfrac{(x-3)^2+(x+3)^2}{(x-3)^2-(x+3)^2}$

$=\dfrac{x^2-6x+9+x^2+6x+9}{x^2-6x+9-(x^2+6x+9)}$

$=\dfrac{2x^2+18}{x^2-6x+9-x^2-6x-9}$

$=\dfrac{2(x^2+9)}{-12x}$

$=-\dfrac{x^2+9}{6x}$

74. $\dfrac{\frac{3}{x-1}-\frac{2}{1-x}}{\frac{2}{x-1}-\frac{2}{x}}=\dfrac{\frac{3}{x-1}+\frac{2}{x-1}}{\frac{2}{x-1}-\frac{2}{x}}$

$=\dfrac{\frac{5}{x-1}}{\frac{2}{x-1}-\frac{2}{x}}$

$=\dfrac{x(x-1)\frac{5}{x-1}}{x(x-1)\left(\frac{2}{x-1}-\frac{2}{x}\right)}$

$=\dfrac{5x}{2x-2(x-1)}$

$=\dfrac{5x}{2x-2x+2}$

$=\dfrac{5x}{2}$

75. $\dfrac{x+y^{-1}}{\frac{x}{y}}=\dfrac{x+\frac{1}{y}}{\frac{x}{y}}=\dfrac{y\left(x+\frac{1}{y}\right)}{x\left(\frac{x}{y}\right)}=\dfrac{xy+1}{x}$

76. $\dfrac{x-xy^{-1}}{\frac{1+x}{y}}=\dfrac{x-\frac{x}{y}}{\frac{1+x}{y}}=\dfrac{y\left(x-\frac{x}{y}\right)}{y\left(\frac{1+x}{y}\right)}=\dfrac{xy-x}{1+x}$

77. $\dfrac{4x+12}{8x^2+24x}=\dfrac{4(x+3)}{2\cdot4\cdot x(x+3)}=\dfrac{1}{2x}$

78. $\dfrac{x^3-6x^2+9x}{x^2+4x-21}=\dfrac{x(x-3)^2}{(x+7)(x-3)}=\dfrac{x(x-3)}{x+7}$

79. $\dfrac{x^2+9x+20}{x^2-25}\cdot\dfrac{x^2-9x+20}{x^2+8x+16}$

$=\dfrac{(x+4)(x+5)\cdot(x-4)(x-5)}{(x+5)(x-5)\cdot(x+4)(x+4)}$

$=\dfrac{x-4}{x+4}$

80. $\dfrac{x^2-x-72}{x^2-x-30}\div\dfrac{x^2+6x-27}{x^2-9x+18}$

$=\dfrac{x^2-x-72}{x^2-x-30}\cdot\dfrac{x^2-9x+18}{x^2+6x-27}$

$=\dfrac{(x-9)(x+8)\cdot(x-3)(x-6)}{(x+5)(x-6)\cdot(x+9)(x-3)}$

$=\dfrac{(x-9)(x+8)}{(x+5)(x+9)}$

81. $\dfrac{x}{x^2-36}+\dfrac{6}{x^2-36}=\dfrac{x+6}{x^2-36}$

$=\dfrac{x+6}{(x+6)(x-6)}$

$=\dfrac{1}{x-6}$

82. $\dfrac{5x-1}{4x}-\dfrac{3x-2}{4x}=\dfrac{5x-1-(3x-2)}{4x}$

$=\dfrac{5x-1-3x+2}{4x}$

$=\dfrac{2x+1}{4x}$

83. $\dfrac{4}{3x^2+8x-3}+\dfrac{2}{3x^2-7x+2}$

$=\dfrac{4}{(x+3)(3x-1)}+\dfrac{2}{(x-2)(3x-1)}$

$=\dfrac{4(x-2)}{(x+3)(3x-1)(x-2)}+\dfrac{2(x+3)}{(x-2)(3x-1)(x+3)}$

$=\dfrac{4(x-2)+2(x+3)}{(x+3)(3x-1)(x-2)}$

$=\dfrac{4x-8+2x+6}{(x+3)(3x-1)(x-2)}$

$=\dfrac{6x-2}{(x+3)(3x-1)(x-2)}$

$=\dfrac{2(3x-1)}{(x+3)(3x-1)(x-2)}$

$=\dfrac{2}{(x+3)(x-2)}$

84. $\dfrac{3x}{x^2+9x+14} - \dfrac{6x}{x^2+4x-21}$

$= \dfrac{3x}{(x+7)(x+2)} - \dfrac{6x}{(x+7)(x-3)}$

$= \dfrac{3x(x-3)}{(x+7)(x+2)(x-3)} - \dfrac{6x(x+2)}{(x+7)(x-3)(x+2)}$

$= \dfrac{3x(x-3) - 6x(x+2)}{(x+7)(x+2)(x-3)}$

$= \dfrac{3x^2 - 9x - 6x^2 - 12x}{(x+7)(x+2)(x-3)}$

$= \dfrac{-3x^2 - 21x}{(x+7)(x+2)(x-3)}$

$= \dfrac{-3x(x+7)}{(x+7)(x+2)(x-3)}$

$= -\dfrac{3x}{(x+2)(x-3)}$

85. $\dfrac{4}{a-1} + 2 = \dfrac{3}{a-1}$

$(a-1)\left(\dfrac{4}{a-1}\right) + (a-1)(2) = (a-1)\left(\dfrac{3}{a-1}\right)$

$4 + 2(a-1) = 3$

$4 + 2a - 2 = 3$

$2 + 2a = 3$

$2a = 1$

$a = \dfrac{1}{2}$

86. $\dfrac{x}{x+3} + 4 = \dfrac{x}{x+3}$

$(x+3)\left(\dfrac{x}{x+3}\right) + (x+3)(4) = (x+3)\left(\dfrac{x}{x+3}\right)$

$x + 4(x+3) = x$

$x + 4x + 12 = x$

$5x + 12 = x$

$12 = -4x$

$-3 = x$

Since $x = -3$ makes a denominator 0, the solution does not check. This equation has no solution.

87. $\dfrac{2x}{3} - \dfrac{1}{6} = \dfrac{x}{2}$

$6\left(\dfrac{2x}{3}\right) - 6\left(\dfrac{1}{6}\right) = 6\left(\dfrac{x}{2}\right)$

$4x - 1 = 3x$

$-1 = -x$

$1 = x$

The unknown number is 1.

88. Let x be the number of days it takes them to paint the house working together.

	Days to Complete Total Job	Part of Job Completed in 1 Day
Mr. Crocker	3	$\dfrac{1}{3}$
Son	4	$\dfrac{1}{4}$
Together	x	$\dfrac{1}{x}$

$\dfrac{1}{3} + \dfrac{1}{4} = \dfrac{1}{x}$

$12x\left(\dfrac{1}{3}\right) + 12x\left(\dfrac{1}{4}\right) = 12x\left(\dfrac{1}{x}\right)$

$4x + 3x = 12$

$7x = 12$

$x = \dfrac{12}{7}$ or $1\dfrac{5}{7}$

Working together, Mr. Crocker and his son can paint the house in $1\dfrac{5}{7}$ days.

89. $\dfrac{5}{3} = \dfrac{10}{x}$

$5x = 30$

$x = 6$

The missing length is 6.

90. $\dfrac{6}{18} = \dfrac{4}{x}$

$6x = 72$

$x = 12$

The missing length is 12.

91. $\dfrac{\frac{1}{4}}{\frac{1}{3}+\frac{1}{2}} = \dfrac{12\left(\frac{1}{4}\right)}{12\left(\frac{1}{3}+\frac{1}{2}\right)} = \dfrac{12\left(\frac{1}{4}\right)}{12\left(\frac{1}{3}\right)+12\left(\frac{1}{2}\right)} = \dfrac{3}{4+6} = \dfrac{3}{10}$

92.

$$\frac{4+\frac{2}{x}}{6+\frac{3}{x}} = \frac{x\left(4+\frac{2}{x}\right)}{x\left(6+\frac{3}{x}\right)}$$

$$= \frac{x(4)+x\left(\frac{2}{x}\right)}{x(6)+x\left(\frac{3}{x}\right)}$$

$$= \frac{4x+2}{6x+3}$$

$$= \frac{2(2x+1)}{3(2x+1)}$$

$$= \frac{2}{3}$$

93.

$$\frac{y^{-2}}{1-y^{-2}} = \frac{\frac{1}{y^2}}{1-\frac{1}{y^2}} = \frac{y^2\left(\frac{1}{y^2}\right)}{y^2\left(1-\frac{1}{y^2}\right)} = \frac{1}{y^2-1}$$

94.

$$\frac{4+x^{-1}}{3+x^{-1}} = \frac{4+\frac{1}{x}}{3+\frac{1}{x}} = \frac{x\left(4+\frac{1}{x}\right)}{x\left(3+\frac{1}{x}\right)} = \frac{4x+1}{3x+1}$$

Chapter 7 Test

1. The rational expression is undefined when

$$x^2+4x+3=0$$
$$(x+3)(x+1)=0$$
$$x+3=0 \quad \text{or} \quad x+1=0$$
$$x=-3 \qquad\qquad x=-1$$

The domain is
$\{x \,|\, x \text{ is a real number}, x=-1, x \neq -3\}$.

2. a.

$$C = \frac{100x+3000}{x}$$

$$= \frac{100(200)+3000}{200}$$

$$= \frac{20,000+3000}{200}$$

$$= \frac{23,000}{200}$$

$$= 115$$

The average cost per desk is \$115.

b.

$$C = \frac{100x+3000}{x}$$

$$= \frac{100(1000)+3000}{1000}$$

$$= \frac{100,000+3000}{1000}$$

$$= \frac{103,000}{1000}$$

$$= 103$$

The average cost per desk is \$103.

3. $\dfrac{3x-6}{5x-10} = \dfrac{3(x-2)}{5(x-2)} = \dfrac{3}{5}$

4. $\dfrac{x+6}{x^2+12x+36} = \dfrac{x+6}{(x+6)^2} = \dfrac{1}{x+6}$

5. $\dfrac{x+3}{x^3+27} = \dfrac{x+3}{(x+3)(x^2-3x+9)} = \dfrac{1}{x^2-3x+9}$

6.
$$\frac{2m^3-2m^2-12m}{m^2-5m+6} = \frac{2m(m^2-m-6)}{(m-3)(m-2)}$$
$$= \frac{2m(m-3)(m+2)}{(m-3)(m-2)}$$
$$= \frac{2m(m+2)}{m-2}$$

7. $\dfrac{ay+3a+2y+6}{ay+3a+5y+15} = \dfrac{(y+3)(a+2)}{(y+3)(a+5)} = \dfrac{a+2}{a+5}$

8. $\dfrac{y-x}{x^2-y^2} = \dfrac{-(x-y)}{(x-y)(x+y)} = -\dfrac{1}{x+y}$

9. $\dfrac{3}{x-1} \cdot (5x-5) = \dfrac{3}{x-1} \cdot 5(x-1) = \dfrac{3 \cdot 5(x-1)}{x-1} = 15$

10. $\dfrac{y^2-5y+6}{2y+4} \cdot \dfrac{y+2}{2y-6} = \dfrac{(y-3)(y-2)\cdot(y+2)}{2(y+2)\cdot 2(y-3)}$
$$= \dfrac{y-2}{4}$$

11.
$$\frac{15x}{2x+5} - \frac{6-4x}{2x+5} = \frac{15x-(6-4x)}{2x+5}$$
$$= \frac{15x-6+4x}{2x+5}$$
$$= \frac{19x-6}{2x+5}$$

12. $\dfrac{5a}{a^2-a-6} - \dfrac{2}{a-3} = \dfrac{5a}{(a-3)(a+2)} - \dfrac{2(a+2)}{(a-3)(a+2)}$

$\qquad\qquad\qquad\quad = \dfrac{5a-2(a+2)}{(a-3)(a+2)}$

$\qquad\qquad\qquad\quad = \dfrac{5a-2a-4}{(a-3)(a+2)}$

$\qquad\qquad\qquad\quad = \dfrac{3a-4}{(a-3)(a+2)}$

13. $\dfrac{6}{x^2-1} + \dfrac{3}{x+1} = \dfrac{6}{(x+1)(x-1)} + \dfrac{3(x-1)}{(x+1)(x-1)}$

$\qquad\qquad\qquad\quad = \dfrac{6+3x-3}{(x+1)(x-1)}$

$\qquad\qquad\qquad\quad = \dfrac{3x+3}{(x+1)(x-1)}$

$\qquad\qquad\qquad\quad = \dfrac{3(x+1)}{(x+1)(x-1)}$

$\qquad\qquad\qquad\quad = \dfrac{3}{x-1}$

14. $\dfrac{x^2-9}{x^2-3x} \div \dfrac{xy+5x+3y+15}{2x+10} = \dfrac{x^2-9}{x^2-3x} \cdot \dfrac{2x+10}{xy+5x+3y+15}$

$\qquad\qquad\qquad\qquad\qquad = \dfrac{(x-3)(x+3)\cdot 2(x+5)}{x(x-3)\cdot(x+3)(y+5)}$

$\qquad\qquad\qquad\qquad\qquad = \dfrac{2(x+5)}{x(y+5)}$

15. $\dfrac{x+2}{x^2+11x+18} + \dfrac{5}{x^2-3x-10} = \dfrac{x+2}{(x+9)(x+2)} + \dfrac{5}{(x-5)(x+2)}$

$\qquad\qquad\qquad\qquad\qquad\quad = \dfrac{(x+2)(x-5)}{(x+9)(x+2)(x-5)} + \dfrac{5(x+9)}{(x-5)(x+2)(x+9)}$

$\qquad\qquad\qquad\qquad\qquad\quad = \dfrac{(x+2)(x-5)+5(x+9)}{(x+9)(x+2)(x-5)}$

$\qquad\qquad\qquad\qquad\qquad\quad = \dfrac{x^2-3x-10+5x+45}{(x+9)(x+2)(x-5)}$

$\qquad\qquad\qquad\qquad\qquad\quad = \dfrac{x^2+2x+35}{(x+9)(x+2)(x-5)}$

16.
$$\frac{4}{y} - \frac{5}{3} = -\frac{1}{5}$$
$$15y\left(\frac{4}{y} - \frac{5}{3}\right) = 15y\left(-\frac{1}{5}\right)$$
$$15y\left(\frac{4}{y}\right) - 15y\left(\frac{5}{3}\right) = 15y\left(-\frac{1}{5}\right)$$
$$60 - 25y = -3y$$
$$60 = 22y$$
$$\frac{60}{22} = y$$
$$y = \frac{30}{11}$$

17.
$$\frac{5}{y+1} = \frac{4}{y+2}$$
$$5(y+2) = 4(y+1)$$
$$5y + 10 = 4y + 4$$
$$y = -6$$

18.
$$\frac{a}{a-3} = \frac{3}{a-3} - \frac{3}{2}$$
$$2(a-3)\left(\frac{a}{a-3}\right) = 2(a-3)\left(\frac{3}{a-3} - \frac{3}{2}\right)$$
$$2a = 2(a-3)\left(\frac{3}{a-3}\right) - 2(a-3)\left(\frac{3}{2}\right)$$
$$2a = 6 - 3(a-3)$$
$$2a = 6 - 3a + 9$$
$$2a = 15 - 3a$$
$$5a = 15$$
$$a = 3$$

In the original equation, 3 makes a denominator 0. This equation has no solution.

19.
$$x - \frac{14}{x-1} = 4 - \frac{2x}{x-1}$$
$$(x-1)\left(x - \frac{14}{x-1}\right) = (x-1)\left(4 - \frac{2x}{x-1}\right)$$
$$x(x-1) - 14 = 4(x-1) - 2x$$
$$x^2 - x - 14 = 4x - 4 - 2x$$
$$x^2 - x - 14 = 2x - 4$$
$$x^2 - 3x - 10 = 0$$
$$(x-5)(x+2) = 0$$
$$x - 5 = 0 \quad \text{or} \quad x + 2 = 0$$
$$x = 5 \qquad\qquad x = -2$$

20.

$$\frac{10}{x^2 - 25} = \frac{3}{x+5} + \frac{1}{x-5}$$

$$\frac{10}{(x+5)(x-5)} = \frac{3}{x+5} + \frac{1}{x-5}$$

$$(x+5)(x-5)\left(\frac{10}{(x+5)(x-5)}\right) = (x+5)(x-5)\left(\frac{3}{x+5}\right) + (x+5)(x-5)\left(\frac{1}{x-5}\right)$$

$$10 = 3(x-5) + 1(x+5)$$
$$10 = 3x - 15 + x + 5$$
$$10 = 4x - 10$$
$$20 = 4x$$
$$5 = x$$

In the original equation 5 makes a denominator 0. This equation has no solution.

21. $\dfrac{\frac{5x^2}{yz^2}}{\frac{10x}{z^3}} = \dfrac{5x^2}{yz^2} \cdot \dfrac{z^3}{10x} = -\dfrac{5 \cdot x \cdot x \cdot z \cdot z^2}{y \cdot z^2 \cdot 2 \cdot 5 \cdot x} = \dfrac{xz}{2y}$

22. $\dfrac{5 - \frac{1}{y^2}}{\frac{1}{y} + \frac{2}{y^2}} = \dfrac{y^2\left(5 - \frac{1}{y^2}\right)}{y^2\left(\frac{1}{y} + \frac{2}{y^2}\right)}$

$$= \dfrac{y^2(5) - y^2\left(\frac{1}{y^2}\right)}{y^2\left(\frac{1}{y}\right) + y^2\left(\frac{2}{y^2}\right)}$$

$$= \dfrac{5y^2 - 1}{y + 2}$$

23. $\dfrac{\frac{b}{a} - \frac{a}{b}}{\frac{1}{b} + \frac{1}{a}} = \dfrac{\left(\frac{b}{a} - \frac{a}{b}\right)ab}{\left(\frac{1}{b} + \frac{1}{a}\right)ab}$

$$= \dfrac{b^2 - a^2}{a + b}$$

$$= \dfrac{(b-a)(b+a)}{a+b}$$

$$= b - a$$

24. Let x = the number of defective bulbs.

$$\frac{85}{3} = \frac{510}{x}$$
$$85x = 1530$$
$$x = 18$$

Expect to find 18 defective bulbs.

25.
$$x + 5 \cdot \frac{1}{x} = 6$$
$$x + \frac{5}{x} = 6$$
$$x\left(x + \frac{5}{x}\right) = x(6)$$
$$x(x) + x\left(\frac{5}{x}\right) = x(6)$$
$$x^2 + 5 = 6x$$
$$x^2 - 6x + 5 = 0$$
$$(x - 5)(x - 1) = 0$$
$$x - 5 = 0 \quad \text{or} \quad x - 1 = 0$$
$$x = 5 \qquad\qquad x = 1$$
The unknown number is 5 or 1.

26. Let r be the speed of the boat in still water.

	Distance =	Rate ·	Time
Upstream	14	$r - 2$	$\frac{14}{r-2}$
Downstream	16	$r + 2$	$\frac{16}{r+2}$

$$\frac{14}{r-2} = \frac{16}{r+2}$$
$$14(r + 2) = 16(r - 2)$$
$$14r + 28 = 16r - 32$$
$$60 = 2r$$
$$r = 30$$

The speed of the boat in still water is 30 miles per hour.

27. Let x be the number of hours it takes to fill the tank using both pipes.

	Hours to Complete Total Job	Part of Job Completed in 1 Hour
1st Pipe	12	$\frac{1}{12}$
2nd Pipe	15	$\frac{1}{15}$
Together	x	$\frac{1}{x}$

$$\frac{1}{12} + \frac{1}{15} = \frac{1}{x}$$
$$60x\left(\frac{1}{12}\right) + 60x\left(\frac{1}{15}\right) = 60x\left(\frac{1}{x}\right)$$
$$5x + 4x = 60$$
$$9x = 60$$
$$x = \frac{60}{9} = \frac{20}{3} = 6\frac{2}{3}$$

Together, the pipes can fill the tank in $6\frac{2}{3}$ hours.

28.
$$\frac{8}{x} = \frac{10}{15}$$
$$8(15) = 10x$$
$$120 = 10x$$
$$12 = x$$
The missing length is 12.

Chapter 7 Cumulative Review

1. a. $\frac{15}{x} = 4$

b. $12 - 3 = x$

c. $4x + 17 \neq 21$

d. $3x < 48$

2. a. $12 - x = -45$

b. $12x = -45$

c. $x - 10 = 2x$

3. Let $x =$ the amount invested at 9% for one year.

	Principal ·	Rate =	Interest
9%	x	0.09	$0.09x$
7%	$20,000 - x$	0.07	$0.07(20,000 - x)$
Total	20,000		1550

$$0.09x + 0.07(20,000 - x) = 1550$$
$$0.09x + 1400 - 0.07x = 1550$$
$$0.02x + 1400 = 1550$$
$$0.02x = 150$$
$$x = 7500$$

$20,000 - x = 20,000 - 7500 = 12,500$
He invested $7500 at 9% and $12,500 at 7%.

4. Let x be the number of bankruptcies in 1994 then $2x - 80,000$ is the number in 2002.

$$x + 2x - 80,000 = 2,290,000$$
$$3x - 80,000 = 2,290,000$$
$$3x = 2,370,000$$
$$x = 790,000$$

$2x - 80,000 = 2(790,000) - 80,000 = 1,500,000$

There were 790,000 bankruptcies in 1994 and 1,500,000 in 2002.

5. $x - 3y = 6$

x	y
0	–2
6	0

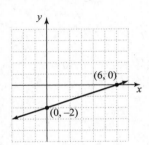

6. $7x + 2y = 9$

$$2y = -7x + 9$$
$$y = -\frac{7}{2}x + \frac{9}{2}$$
$$y = mx + b$$
$$m = -\frac{7}{2}$$

7. a. $4^2 \cdot 4^5 = 4^{2+5} = 4^7$

b. $x^4 \cdot x^6 = x^{4+6} = x^{10}$

c. $y^3 \cdot y = y^{3+1} = y^4$

d. $y^3 \cdot y^2 \cdot y^7 = y^{3+2+7} = y^{12}$

e. $(-5)^7 \cdot (-5)^8 = (-5)^{7+8} = (-5)^{15}$

f. $a^2 \cdot b^2 = a^2 b^2$

8. a. $\dfrac{x^9}{x^7} = x^{9-7} = x^2$

b. $\dfrac{x^{19} y^5}{xy} = x^{19-1} \cdot y^{5-1} = x^{18} y^4$

c. $(x^5 y^2)^3 = x^{5 \cdot 3} y^{2 \cdot 3} = x^{15} y^6$

d. $(-3a^2 b)(5a^3 b) = -15a^{2+3} b^{1+1} = -15a^5 b^2$

9. $[(8z + 11) + (9z - 2)] - (5z - 7)$
$= 8z + 11 + 9z - 2 - 5z + 7$
$= 12z + 16$

10. $(x + 1) - (9x^2 - 6x + 2) = x + 1 - 9x^2 + 6x - 2$
$\qquad\qquad\qquad\qquad\quad = -9x^2 + 7x - 1$

11. $(3a + b)^3$
$= (3a + b)(3a + b)^2$
$= (3a + b)[(3a)^2 + 2(3a)(b) + (b)^2]$
$= (3a + b)(9a^2 + 6ab + b^2)$
$= 27a^3 + 18a^2 b + 3ab^2 + 9a^2 b + 6ab^2 + b^3$
$= 27a^3 + 27a^2 b + 9ab^2 + b^3$

12. $(2x + 1)(5x^2 - x + 2)$
$= 2x(5x^2 - x + 2) + 1(5x^2 - x + 2)$
$= 10x^3 - 2x^2 + 4x + 5x^2 - x + 2$
$= 10x^3 + 3x^2 + 3x + 2$

13. a. $(t + 2)^2 = (t)^2 + 2(t)(2) + (2)^2 = t^2 + 4t + 4$

b. $(p - q)^2 = (p)^2 - 2(p)(q) + (q)^2$
$\qquad\qquad = p^2 - 2pq + q^2$

c. $(2x + 5)^2 = (2x)^2 + 2(2x)(5) + (5)^2$
$\qquad\qquad = 4x^2 + 20x + 25$

d. $(x^2 - 7y)^2 = (x^2)^2 - 2(x^2)(7y) + (7y)^2$
$\qquad\qquad\quad = x^4 - 14x^2 y + 49y^2$

14. a. $(x + 9)^2 = (x)^2 + 2(x)(9) + (9)^2$
$\qquad\qquad = x^2 + 18x + 81$

b. $(2x + 1)(2x - 1) = (2x)^2 - (1)^2 = 4x^2 - 1$

c. $8x(x^2 + 1)(x^2 - 1) = 8x[(x^2)^2 - (1)^2]$
$\qquad\qquad\qquad\qquad = 8x[x^4 - 1]$
$\qquad\qquad\qquad\qquad = 8x^5 - 8x$

15. a. $\dfrac{1}{x^{-3}} = x^3$

b. $\dfrac{1}{3^{-4}} = 3^4 = 81$

c. $\dfrac{p^{-4}}{q^{-9}} = \dfrac{q^9}{p^4}$

d. $\dfrac{5^{-3}}{2^{-5}} = \dfrac{2^5}{5^3} = \dfrac{32}{125}$

16. a. $5^{-3} = \dfrac{1}{5^3} = \dfrac{1}{125}$

b. $\dfrac{9}{x^{-7}} = 9x^7$

c. $\dfrac{11^{-1}}{7^{-2}} = \dfrac{7^2}{11^1} = \dfrac{49}{11}$

17.
$$
\begin{array}{r}
4x^2 - 4x + 6 \\
2x+3\overline{)8x^3 + 4x^2 + 0x + 7} \\
\underline{8x^3 + 12x^2} \\
-8x^2 + 0x \\
\underline{-8x^2 - 12x} \\
12x + 7 \\
\underline{12x + 18} \\
-11
\end{array}
$$

$\dfrac{4x^2 + 7 + 8x^3}{2x+3} = 4x^2 - 4x + 6 - \dfrac{11}{2x+3}$

18.
$$
\begin{array}{r}
4x^2 + 16x + 55 \\
x-4\overline{)4x^3 + 0x^2 - 9x + 2} \\
\underline{4x^3 - 16x^2} \\
16x^2 - 9x \\
\underline{16x^2 - 64x} \\
55x + 2 \\
\underline{55x - 220} \\
222
\end{array}
$$

$\dfrac{4x^3 - 9x + 2}{x-4} = 4x^2 + 16x + 55 + \dfrac{222}{x-4}$

19. a. $28 = 2 \cdot 2 \cdot 7$
$40 = 2 \cdot 2 \cdot 2 \cdot 5$
$\text{GCF} = 2^2 = 4$

b. $55 = 5 \cdot 11$
$21 = 3 \cdot 7$
$\text{GCF} = 1$

c. $15 = 3 \cdot 5$
$18 = 2 \cdot 3 \cdot 3$
$66 = 2 \cdot 3 \cdot 11$
$\text{GCF} = 3$

20. $9x^2 = 3 \cdot 3 \cdot x^2$
$6x^3 = 2 \cdot 3 \cdot x^3$
$21x^5 = 3 \cdot 7 \cdot x^5$
$\text{GCF} = 3x^2$

21. $-9a^5 + 18a^2 - 3a = -3a(3a^4 - 6a + 1)$

22. $7x^6 - 7x^5 + 7x^4 = 7x^4(x^2 - x + 1)$

23. $3m^2 - 24m - 60 = 3(m^2 - 8m - 20)$
$= 3(m^2 - 10m + 2m - 20)$
$= 3[m(m-10) + 2(m-10)]$
$= 3(m-10)(m+2)$

24. $-2a^2 + 10a + 12 = -2(a^2 - 5a - 6)$
$= -2(a+1)(a-6)$

25. $3x^2 + 11x + 6 = 3x^2 + 2x + 9x + 6$
$= x(3x+2) + 3(3x+2)$
$= (3x+2)(x+3)$

26. $10m^2 - 7m + 1 = 10m^2 - 2m - 5m + 1$
$= 2m(5m-1) - 1(5m-1)$
$= (2m-1)(5m-1)$

27. $x^2 + 12x + 36 = x^2 + 2 \cdot x \cdot 6 + 6^2 = (x+6)^2$

28. $4x^2 + 12x + 9 = (2x)^2 + 2(2x)(3) + (3)^2$
$= (2x+3)^2$

29. $x^2 + 4$ is a prime polynomial.

30. $x^2 - 4 = (x)^2 - (2)^2 = (x+2)(x-2)$

31. $x^3 + 8 = x^3 + 2^3$
$$= (x+2)(x^2 - x \cdot 2 + 2^2)$$
$$= (x+2)(x^2 - 2x + 4)$$

32. $27y^3 - 1 = (3y)^3 - (1)^3$
$$= (3y-1)[(3y)^2 + 3y(1) + (1)^2]$$
$$= (3y-1)(9y^2 + 3y + 1)$$

33. $2x^3 + 3x^2 - 2x - 3 = x^2(2x+3) - 1(2x+3)$
$$= (2x+3)(x^2 - 1)$$
$$= (2x+3)(x^2 - 1^2)$$
$$= (2x+3)(x+1)(x-1)$$

34. $3x^3 + 5x^2 - 12x - 20 = x^2(3x+5) - 4(3x+5)$
$$= (3x+5)(x^2 - 4)$$
$$= (3x+5)(x^2 - 2^2)$$
$$= (3x+5)(x+2)(x-2)$$

35. $12m^2 - 3n^2 = 3(4m^2 - n^2)$
$$= 3[(2m)^2 - (n)^2]$$
$$= 3(2m+n)(2m-n)$$

36. $x^5 - x = x(x^4 - 1)$
$$= x[(x^2)^2 - 1^2]$$
$$= x(x^2 + 1)(x^2 - 1)$$
$$= x(x^2 + 1)(x+1)(x-1)$$

37.
$$x(2x-7) = 4$$
$$2x^2 - 7x = 4$$
$$2x^2 - 7x - 4 = 0$$
$$2x^2 - 8x + x - 4 = 0$$
$$2x(x-4) + 1(x-4) = 0$$
$$(x-4)(2x+1) = 0$$
$$2x+1 = 0 \quad \text{or} \quad x-4 = 0$$
$$2x = -1 \qquad\qquad x = 4$$
$$x = -\frac{1}{2}$$

38.
$$3x^2 + 5x = 2$$
$$3x^2 + 5x - 2 = 0$$
$$3x^2 + 6x - x - 2 = 0$$
$$3x(x+2) - 1(x+2) = 0$$
$$(x+2)(3x-1) = 0$$

$$3x - 1 = 0 \quad \text{or} \quad x+2 = 0$$
$$3x = 1 \qquad\qquad x = -2$$
$$x = \frac{1}{3}$$

39. $y = x^2 - 5x + 4$
$$0 = x^2 - 5x + 4$$
$$0 = (x-4)(x-1)$$
$$x - 1 = 0 \quad \text{or} \quad x - 4 = 0$$
$$x = 1 \qquad\qquad x = 4$$
The x-intercepts are $(1, 0)$ and $(4, 0)$.

40. $y = x^2 - x - 6$
$$0 = x^2 - x - 6$$
$$0 = (x-3)(x+2)$$
$$x + 2 = 0 \quad \text{or} \quad x - 3 = 0$$
$$x = -2 \qquad\qquad x = 3$$
The x-intercepts are $(-2, 0)$ and $(3, 0)$.

41. Let x = the base and $2x - 2$ = the height.
$$A = \frac{1}{2}bh$$
$$30 = \frac{1}{2}x(2x-2)$$
$$30 = \frac{1}{2}(2x)(x-1)$$
$$30 = x(x-1)$$
$$30 = x^2 - x$$
$$0 = x^2 - x - 30$$
$$0 = (x+5)(x-6)$$
$$x - 6 = 0 \quad \text{or} \quad x + 5 = 0$$
$$x = 6 \qquad\qquad x = -5$$
Length cannot be negative, so $x = 6$.
$2x - 2 = 2(6) - 2 = 10$
The base is 6 meters and the height is 10 meters.

42. Let x = the base and $3x + 5$ = the height.
$$A = bh$$
$$182 = x(3x+5)$$
$$182 = 3x^2 + 5x$$
$$0 = 3x^2 + 5x - 182$$
$$0 = 3x^2 + 26x - 21x - 182$$
$$0 = x(3x+26) - 7(3x+26)$$
$$0 = (x-7)(3x+26)$$
$$x - 7 = 0 \quad \text{or} \quad 3x + 26 = 0$$
$$x = 7 \qquad\qquad x = -\frac{26}{3}$$
Length cannot be negative so $x = 7$.
$3x + 5 = 3(7) + 5 = 26$
The base is 7 ft and the height is 26 ft.

43. $\dfrac{18-2x^2}{x^2-2x-3} = \dfrac{2(9-x^2)}{(x+1)(x-3)}$

$= \dfrac{2(3+x)(3-x)}{(x+1)(x-3)}$

$= \dfrac{-2(3+x)(x-3)}{(x+1)(x-3)}$

$= -\dfrac{2(3+x)}{x+1}$

44. $\dfrac{2x^2-50}{4x^4-20x^3} = \dfrac{2(x^2-25)}{4x^3(x-5)}$

$= \dfrac{2(x+5)(x-5)}{4x^3(x-5)}$

$= \dfrac{x+5}{2x^3}$

45. $\dfrac{6x+2}{x^2-1} \div \dfrac{3x^2+x}{x-1} = \dfrac{6x+2}{x^2-1} \cdot \dfrac{x-1}{3x^2+x}$

$= \dfrac{2(3x+1)}{(x+1)(x-1)} \cdot \dfrac{x-1}{x(3x+1)}$

$= \dfrac{2}{x(x+1)}$

46. $\dfrac{6x^2-18x}{3x^2-2x} \cdot \dfrac{15x-10}{x^2-10} = \dfrac{6x(x-3) \cdot 5(3x-2)}{x(3x-2) \cdot (x+3)(x-3)}$

$= \dfrac{30}{x+3}$

47. $\dfrac{(2x)^{-1}+1}{2x^{-1}-1} = \dfrac{\frac{1}{2x}+1}{\frac{2}{x}-1}$

$= \dfrac{2x\left(\frac{1}{2x}+1\right)}{2x\left(\frac{2}{x}-1\right)}$

$= \dfrac{1+2x}{4-2x}$

$= \dfrac{1+2x}{2(2-x)}$

48. $\dfrac{\frac{m}{3}+\frac{n}{6}}{\frac{m+n}{12}} = \dfrac{12}{12} \cdot \dfrac{\frac{m}{3}+\frac{n}{6}}{\frac{m+n}{12}}$

$= \dfrac{12\left(\frac{m}{3}\right)+12\left(\frac{n}{6}\right)}{12\left(\frac{m+n}{12}\right)}$

$= \dfrac{4m+2n}{m+n}$ or $\dfrac{2(2m+n)}{m+n}$

Chapter 8

Section 8.1 Practice

1. $f(x) = 4x$, $g(x) = 4x - 3$

x	$f(x)$	$g(x)$
0	0	−3
−1	−4	−7
1	4	1

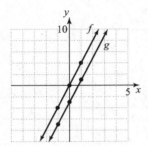

2. $f(x) = -2x$, $g(x) = -2x + 5$

x	$f(x)$	$g(x)$
0	0	5
−1	2	7
1	−2	3
2	−4	1

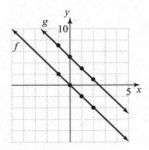

3. Use the slope-intercept form with $m = -4$ and $b = -3$.
$$y = mx + b$$
$$y = -4x + (-3)$$
$$y = -4x - 3$$
$$f(x) = -4x - 3$$

4. First find the slope.
$$m = \frac{0-2}{2-(-1)} = \frac{-2}{3} = -\frac{2}{3}$$
Use the slope and one of the points in the point-slope form. We use (2, 0).
$$y - y_1 = m(x - x_1)$$
$$y - 0 = -\frac{2}{3}(x - 2)$$
$$y = -\frac{2}{3}x + \frac{4}{3}$$
$$f(x) = -\frac{2}{3}x + \frac{4}{3}$$

5. A horizontal line has an equation of the form $y = b$. Since the line contains the point (6, −2), the equation is $y = -2$ or $f(x) = -2$.

6. Solve the given equation for y.
$$3x + 4y = 1$$
$$4y = -3x + 1$$
$$y = -\frac{3}{4}x + \frac{1}{4}$$

The slope of this line is $-\frac{3}{4}$, so the slope of any line parallel to it is also $-\frac{3}{4}$. Use this slope and the point (8, −3) in the point-slope form.
$$y - y_1 = m(x - x_1)$$
$$y - (-3) = -\frac{3}{4}(x - 8)$$
$$4(y + 3) = -3(x - 8)$$
$$4y + 12 = -3x + 24$$
$$3x + 4y = 12$$

7. Solve the given equation for y.
$$3x + 4y = 1$$
$$4y = -3x + 1$$
$$y = -\frac{3}{4}x + \frac{1}{4}$$

The slope of this line is $-\frac{3}{4}$, so the slope of any line perpendicular to it is the negative reciprocal of $-\frac{3}{4}$, or $\frac{4}{3}$. Use this slope and the point (8, −3) in the point-slope form.

$$y - y_1 = m(x - x_1)$$
$$y - (-3) = \frac{4}{3}(x - 8)$$
$$3(y + 3) = 4(x - 8)$$
$$3y + 9 = 4x - 32$$
$$3y = 4x - 41$$
$$y = \frac{4}{3}x - \frac{41}{3}$$
$$f(x) = \frac{4}{3}x - \frac{41}{3}$$

Graphing Calculator Explorations

1. $x = 3.5y$

$$y = \frac{x}{3.5}$$

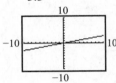

2. $-2.7y = x$

$$y = \frac{x}{-2.7} = -\frac{x}{2.7}$$

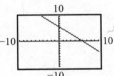

3. $5.78x + 2.31y = 10.98$

$$2.31y = -5.78x + 10.98$$
$$y = -\frac{5.78}{2.31}x + \frac{10.98}{2.31}$$

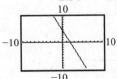

4. $-7.22x + 3.89y = 12.57$

$$3.89y = 7.22x + 12.57$$
$$y = \frac{7.22}{3.89}x + \frac{12.57}{3.89}$$

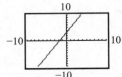

5. $y - |x| = 3.78$

$$y = |x| + 3.78$$

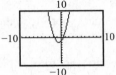

6. $3y - 5x^2 = 6x - 4$

$$3y = 5x^2 + 6x - 4$$
$$y = \frac{5}{3}x^2 + 2x - \frac{4}{3}$$

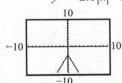

7. $y - 5.6x^2 = 7.7x + 1.5$

$$y = 5.6x^2 + 7.7x + 1.5$$

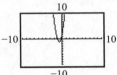

8. $y + 2.6|x| = -3.2$

$$y = -2.6|x| - 3.2$$

Vocabulary, Readiness & Video Check 8.1

1. A <u>linear</u> function can be written in the form $f(x) = mx + b$.

2. In the form $f(x) = mx + b$, the y-intercept is <u>(0, b)</u> and the slope is <u>m</u>.

3. $m = -4$, $b = 12$ so y-intercept is $(0, 12)$.

4. $m = \frac{2}{3}$, $b = -\frac{7}{2}$ so y-intercept is $\left(0, -\frac{7}{2}\right)$.

5. $m = 5$, $b = 0$ so y-intercept is $(0, 0)$.

6. $m = -1$, $b = 0$ so y-intercept is $(0, 0)$.

7. The lines both have slope 12 and they have different y-intercepts, $(0, 6)$ and $(0, -2)$, so they are parallel.

8. The lines both have slope -5 and they have different y-intercepts, $(0, 8)$ and $(0, -8)$, so they are parallel.

9. The lines have slopes -9 and $\frac{3}{2}$. The slopes are not equal and their product is not -1, so the lines are neither parallel nor perpendicular.

10. The lines have slopes 2 and $\frac{1}{2}$. The slopes are not equal and their product is not -1, so the lines are neither parallel nor perpendicular.

11. $f(x) = mx + b$, or slope-intercept form

12. The y-intercept is of the form $(0, b)$, so the b value is the y-coordinate of the y-intercept.

13. if one of the two points given is the y-intercept

14. $f(x) = \frac{1}{3}x - \frac{17}{3}$

Exercise Set 8.1

1. $f(x) = -2x$

x	0	-1	1
y	0	2	-2

Plot the points to obtain the graph.

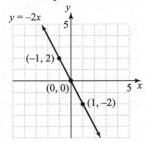

3. $f(x) = -2x + 3$

x	0	1	-1
y	3	1	5

Plot the points to obtain the graph.

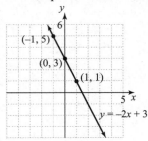

5. $f(x) = \frac{1}{2}x$

x	0	2	-2
y	0	1	-1

Plot the points to obtain the graph.

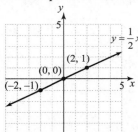

7. $f(x) = \frac{1}{2}x - 4$

x	0	2	4
y	-4	-3	-2

Plot the points to obtain the graph.

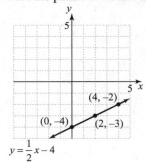

9. The graph of $f(x) = 5x - 3$ is the graph of $f(x) = 5x$ shifted down 3 units. The correct graph is C.

11. The graph of $f(x) = 5x + 1$ is the graph of $f(x) = 5x$ shifted up 1 unit. The correct graph is D.

13. $m = -1, b = 1$
$y = mx + b$
$y = -1x + 1$
$y = -x + 1$
$f(x) = -x + 1$

15. $m = 2, \ b = \dfrac{3}{4}$
$y = mx + b$
$y = 2x + \dfrac{3}{4}$
$f(x) = 2x + \dfrac{3}{4}$

17. $m = \dfrac{2}{7}, \ b = 0$
$y = mx + b$
$y = \dfrac{2}{7}x + 0$
$y = \dfrac{2}{7}x$
$f(x) = \dfrac{2}{7}x$

19. $y - y_1 = m(x - x_1)$
$y - 2 = 3(x - 1)$
$y - 2 = 3x - 3$
$y = 3x - 1$
$f(x) = 3x - 1$

21. $y - y_1 = m(x - x_1)$
$y - (-3) = -2(x - 1)$
$y + 3 = -2x + 2$
$y = -2x - 1$
$f(x) = -2x - 1$

23. $y - y_1 = m(x - x_1)$
$y - 2 = \dfrac{1}{2}[x - (-6)]$
$y - 2 = \dfrac{1}{2}(x + 6)$
$y - 2 = \dfrac{1}{2}x + 3$
$y = \dfrac{1}{2}x + 5$
$f(x) = \dfrac{1}{2}x + 5$

25. $y - y_1 = m(x - x_1)$
$y - 0 = -\dfrac{9}{10}[x - (-3)]$
$y = -\dfrac{9}{10}(x + 3)$
$y = -\dfrac{9}{10}x - \dfrac{27}{10}$
$f(x) = -\dfrac{9}{10}x - \dfrac{27}{10}$

27. $m = \dfrac{6 - 0}{4 - 2} = \dfrac{6}{2} = 3$
$y - 0 = 3(x - 2)$
$y = 3x - 6$
$f(x) = 3x - 6$

29. $m = \dfrac{13 - 5}{-6 - (-2)} = \dfrac{8}{-4} = -2$
$y - 5 = -2[x - (-2)]$
$y - 5 = -2(x + 2)$
$y - 5 = -2x - 4$
$y = -2x + 1$
$f(x) = -2x + 1$

31. $m = \dfrac{-3 - (-4)}{-4 - (-2)} = \dfrac{1}{-2} = -\dfrac{1}{2}$
$y - (-4) = -\dfrac{1}{2}[x - (-2)]$
$y + 4 = -\dfrac{1}{2}(x + 2)$
$2y + 8 = -(x + 2)$
$2y + 8 = -x - 2$
$2y = -x - 10$
$y = -\dfrac{1}{2}x - 5$
$f(x) = -\dfrac{1}{2}x - 5$

33. $m = \dfrac{-9-(-8)}{-6-(-3)} = \dfrac{-1}{-3} = \dfrac{1}{3}$

$y-(-8) = \dfrac{1}{3}[x-(-3)]$

$y+8 = \dfrac{1}{3}(x+3)$

$3y+24 = x+3$

$3y = x-21$

$y = \dfrac{1}{3}x-7$

$f(x) = \dfrac{1}{3}x-7$

35. $m = \dfrac{\frac{7}{10}-\frac{2}{5}}{-\frac{1}{5}-\frac{3}{5}} = \dfrac{\frac{7}{10}-\frac{4}{10}}{-\frac{4}{5}} = \dfrac{3}{10}\left(-\dfrac{5}{4}\right) = -\dfrac{3}{8}$

$y-\dfrac{4}{10} = -\dfrac{3}{8}\left(x-\dfrac{3}{5}\right)$

$y-\dfrac{4}{10} = -\dfrac{3}{8}x+\dfrac{9}{40}$

$y = -\dfrac{3}{8}x+\dfrac{5}{8}$

$f(x) = -\dfrac{3}{8}x+\dfrac{5}{8}$

37. $y = mx+b$

$-4 = 0(-2)+b$

$-4 = b$

$y = -4$

$f(x) = -4$

39. Every horizontal line is in the form $y = c$. Since the line passes through the point $(0, 5)$, its equation is $y = 5$ or $f(x) = 5$.

41. $y = 4x-2$ so $m = 4$

$y-8 = 4(x-3)$

$y-8 = 4x-12$

$y = 4x-4$

$f(x) = 4x-4$

43. $3y = x-6$ or $y = \dfrac{1}{3}x-2$ so

$m = \dfrac{1}{3}$ and $m_\perp = -3$

$y-(-5) = -3(x-2)$

$y+5 = -3x+6$

$y = -3x+1$

$f(x) = -3x+1$

45. $3x+2y = 5$

$2y = -3x+5$

$y = -\dfrac{3}{2}x+\dfrac{5}{2}$ so $m = -\dfrac{3}{2}$

$y-(-3) = -\dfrac{3}{2}[x-(-2)]$

$2(y+3) = -3(x+2)$

$2y+6 = -3(x+2)$

$2y+6 = -3x-6$

$y = -\dfrac{3}{2}x-6$

$f(x) = -\dfrac{3}{2}x-6$

47. $y-3 = 2[x-(-2)]$

$y-3 = 2(x+2)$

$y-3 = 2x+4$

$2x-y = -7$

49. $m = \dfrac{2-6}{5-1} = \dfrac{-4}{4} = -1$

$y-6 = -1(x-1)$

$y-6 = -x+1$

$y = -x+7$

$f(x) = -x+7$

51. $y = -\dfrac{1}{2}x+11$

$f(x) = -\dfrac{1}{2}x+11$

53. $m = \dfrac{-6-(-4)}{0-(-7)} = \dfrac{-2}{7} = -\dfrac{2}{7}$

$y = -\dfrac{2}{7}x-6$

$7y = -2x-42$

$2x+7y = -42$

55. $y-0 = -\dfrac{4}{3}[x-(-5)]$

$3y = -4(x+5)$

$3y = -4x-20$

$4x+3y = -20$

57. Every horizontal line is in the form $y = c$. Since the line passes through the point $(-2, -10)$, its equation is $y = -10$ or $f(x) = -10$.

59. $2x + 4y = 8$

$\quad 4y = -2x + 8$

$\quad\quad y = -\dfrac{1}{2}x + 2 \ \text{ so } \ m = -\dfrac{1}{2}$

$\quad y - (-2) = -\dfrac{1}{2}(x - 6)$

$\quad 2(y + 2) = -(x - 6)$

$\quad 2y + 4 = -x + 6$

$\quad x + 2y = 2$

61. Lines with slopes of 0 are horizontal. Every horizontal line is in the form $y = c$. Since the line passes through $(-9, 12)$, its equation is $y = 12$ or $f(x) = 12$.

63. $8x - y = 9$

$\quad y = 8x - 9 \ \text{ so } \ m = 8$

$\quad y - 1 = 8(x - 6)$

$\quad y - 1 = 8x - 48$

$\quad 8x - y = 47$

65. A line perpendicular to $y = 9$ will have the form $x = c$. Since the line passes through the point $(5, -6)$, its equation is $x = 5$.

67. $m = \dfrac{-5 - (-8)}{-6 - 2} = \dfrac{3}{-8} = -\dfrac{3}{8}$

$\quad y - (-8) = -\dfrac{3}{8}(x - 2)$

$\quad 8(y + 8) = -3(x - 2)$

$\quad 8y + 64 = -3x + 6$

$\quad y = -\dfrac{3}{8}x - \dfrac{29}{4}$

$\quad f(x) = -\dfrac{3}{8}x - \dfrac{29}{4}$

69. $f(x) = 2.7x + 4.1$

$\quad f(9) = 2.7(9) + 4.1 = 24.3 + 4.1 = 28.4$

In 2009, about 28.4% of students took at least one online course.

71. $g(x) = 0.07x^2 + 1.9x + 5.9$

$\quad g(16) = 0.07(16)^2 + 1.9(16) + 5.9$

$\quad\quad\quad = 17.92 + 30.4 + 5.9$

$\quad\quad\quad = 54.22$

In 2016, we predict that 54.22% of students will take at least one online course.

73. answers may vary

75. $(0, 3), (1, 1)$

$\quad m = \dfrac{1 - 3}{1 - 0} = \dfrac{-2}{1} = -2$

$\quad b = 3$

$\quad\quad y = -2x + 3$

$\quad f(x) = -2x + 3$

77. $(-2, 1), (4, 5)$

$\quad m = \dfrac{5 - 1}{4 - (-2)} = \dfrac{4}{6} = \dfrac{2}{3}$

$\quad y - 1 = \dfrac{2}{3}(x + 2)$

$\quad y - 1 = \dfrac{2}{3}x + \dfrac{4}{3}$

$\quad\quad y = \dfrac{2}{3}x + \dfrac{7}{3}$

$\quad f(x) = \dfrac{2}{3}x + \dfrac{7}{3}$

79. a. Use the points $(1, 32)$ and $(3, 96)$.

$\quad m = \dfrac{96 - 32}{3 - 1} = \dfrac{64}{2} = 32$

$\quad y - y_1 = m(x - x_1)$

$\quad y - 32 = 32(x - 1)$

$\quad y - 32 = 32x - 32$

$\quad\quad\quad y = 32x$

    b. Let $x = 4$. $y = 32x = 32(4) = 128$

The rock is traveling 128 feet per second at 4 seconds.

81. a. Use the points $(6, 2000)$ and $(8, 1500)$.

$\quad m = \dfrac{1500 - 2000}{8 - 6} = \dfrac{-500}{2} = -250$

$\quad y - y_1 = m(x - x_1)$

$\quad y - 2000 = -250(x - 6)$

$\quad y - 2000 = -250x + 1500$

$\quad\quad\quad y = -250x + 3500$

    b. Let $x = 7.50$.

$\quad y = -250x + 3500$

$\quad y = -250(7.50) + 3500 = 1625$

The daily sales of Frisbees at \$7.50 each is predicted to be 1625 Frisbees.

83. a. Use the points (0, 2619) and (10, 3200).

$$m = \frac{3200 - 2619}{10 - 0} = \frac{581}{10} = 58.1$$

$$y - y_1 = m(x - x_1)$$
$$y - 2619 = 58.1(x - 0)$$
$$y = 58.1x + 2619$$

b. Let $x = 2012 - 2008 = 4$.
$y = 58.1(4) + 2619 = 2851.4$
The estimated number of people who will be employed as registered nurses in 2012 is 2851.4 thousand.

85. a. Use the points (0, 297,000) and (5, 272,900).

$$m = \frac{272,900 - 297,000}{5 - 0} = \frac{-24,100}{5} = -4820$$

$$y - y_1 = m(x - x_1)$$
$$y - 297,000 = -4820(x - 0)$$
$$y = -4820x + 297,000$$

b. Let $x = 2013 - 2005 = 8$.
$y = -4820(8) + 297,000 = 258,440$
The average price of a new home in 2013 is predicted to be \$258,440.

87. $m = \dfrac{1 - (-1)}{-5 - 3} = \dfrac{2}{-8} = -\dfrac{1}{4}$ so $m_\perp = 4$

$$M((3, -1), (5, 1)) = \left(\frac{3 - 5}{2}, \frac{-1 + 1}{2} \right) = (1, 0)$$

$$y - 0 = 4[x - (-1)]$$
$$y = 4(x + 1)$$
$$y = 4x + 4$$
$$-4x + y = 4$$

89. $m = \dfrac{-4 - 6}{-22 - (-2)} = \dfrac{-10}{-20} = \dfrac{1}{2}$ so $m_\perp = -2$

$$M((-2, 6), (-22, -4)) = \left(\frac{-2 - 22}{2}, \frac{6 - 4}{2} \right)$$
$$= (-12, 1)$$

$$y - 1 = -2[x - (-12)]$$
$$y - 1 = -2(x + 12)$$
$$y - 1 = -2x - 24$$
$$2x + y = -23$$

91. $m = \dfrac{7 - 3}{-4 - 2} = \dfrac{4}{-6} = -\dfrac{2}{3}$ so $m_\perp = \dfrac{3}{2}$

$$M((2, 3), (-4, 7)) = \left(\frac{2 - 4}{2}, \frac{3 + 7}{2} \right)$$
$$= (-1, 5)$$

$$y - 5 = \frac{3}{2}[x - (-1)]$$
$$2(y - 5) = 3(x + 1)$$
$$2y - 10 = 3x + 3$$
$$3x - 2y = -13$$

93. answers may vary

Section 8.2 Practice

1. a. To find $f(1)$, find the y-value when $x = 1$. We see from the graph that when $x = 1$, y or $f(x) = -3$. Thus, $f(1) = -3$.

b. $f(0) = -2$ from the ordered pair (0, −2).

c. $g(-2) = 3$ from the ordered pair (−2, 3).

d. $g(0) = 1$ from the ordered pair (0, 1).

e. To find x-values such that $f(x) = 1$, we are looking for any ordered pairs on the graph of f whose $f(x)$ or y-value is 1. They are (−1, 1) and (3, 1). Thus, $f(-1) = 1$ and $f(3) = 1$. The x-values are −1 and 3.

f. Find ordered pairs on the graph of g whose $g(x)$ or y-value is −2. There is one such ordered pair, (−3, −2). Thus, $g(-3) = -2$. The only x-value is −3.

2. Find the semester fall 2003 and move upward until you reach the graph representing total enrollment. From the point on the graph, move horizontally to the left until the vertical axis is reached. In fall 2003, approximately 17 million or 17,000,000 students were enrolled.

3. 2010 is 10 years after 2000, so find $f(10)$.
$$f(x) = 0.55x + 0.3$$
$$f(10) = 0.55(10) + 0.3 = 5.8$$
We predict that the online enrollment was 5.8 million or 5,800,000 students in 2010.

4. a. $\sqrt{121} = 11$ since 11 is positive and $11^2 = 121$.

b. $\sqrt{\dfrac{1}{16}} = \dfrac{1}{4}$ since $\left(\dfrac{1}{4}\right)^2 = \dfrac{1}{16}$.

c. $-\sqrt{64} = -8$

d. $\sqrt{-64}$ is not a real number.

e. $\sqrt{100} = 10$ since $10^2 = 100$.

5. $f(x) = 2x^2$

This equation is not linear because of the x^2 term. Its graph is not a line.

If $x = -3$, then $f(-3) = 2(-3)^2$, or 18.

If $x = -2$, then $f(-2) = 2(-2)^2$, or 8.

If $x = -1$, then $f(-1) = 2(-1)^2$, or 2.

If $x = 0$, then $f(0) = 2(0)^2$, or 0.

If $x = 1$, then $f(1) = 2(1)^2$, or 2.

If $x = 2$, then $f(2) = 2(2)^2$, or 8.

If $x = 3$, then $f(3) = 2(3)^2$, or 18.

x	y or $f(x)$
-3	18
-2	8
-1	2
0	0
1	2
2	8
3	18

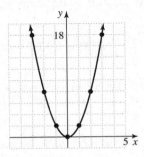

6. $f(x) = -|x|$

This equation is not linear because it cannot be written in the form $Ax + By = C$. Its graph is not a line.

If $x = -3$, then $f(-3) = -|-3|$, or -3.

If $x = -2$, then $f(-2) = -|-2|$, or -2.

If $x = -1$, then $f(-1) = -|-1|$, or -1.

If $x = 0$, then $f(0) = -|0|$, or 0.

If $x = 1$, then $f(1) = -|1|$, or -1.

If $x = 2$, then $f(2) = -|2|$, or -2.

If $x = 3$, then $f(3) = -|3|$, or -3.

x	y or $f(x)$
-3	-3
-2	-2
-1	-1
0	0
1	-1
2	-2
3	-3

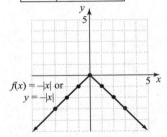

$f(x) = -|x|$ or $y = -|x|$

7. $f(x) = \sqrt{x} + 1$

This equation is not linear because it cannot be written in the form $Ax + By = C$. Its graph is not a line.

If $x = 0$, then $f(0) = \sqrt{0} + 1$, or 1.

If $x = 1$, then $f(1) = \sqrt{1} + 1$, or 2.

If $x = 4$, then $f(4) = \sqrt{4} + 1$, or 3.

If $x = 9$, then $f(9) = \sqrt{9} + 1$, or 4.

x	y or $f(x)$
0	1
1	2
4	3
9	4

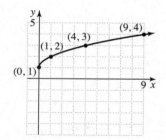

Graphing Calculator Explorations

1.

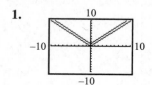

2.

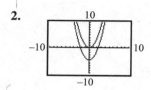

3.

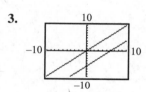

4.

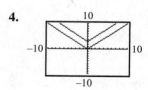

5.

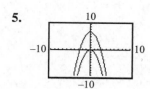

6.

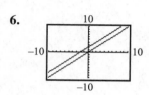

Vocabulary, Readiness & Video Check 8.2

1. The graph of $y = |x|$ looks <u>V-shaped</u>.

2. The graph of $y = x^2$ is a <u>parabola</u>.

3. If $f(-2) = 1.7$, the corresponding ordered pair is <u>$(-2, 1.7)$</u>.

4. If $f(x) = x^2$, then $f(-3) = \underline{9}$.

5. Using function notation, the replacement value for x and the resulting $f(x)$ or y-value corresponds to an ordered pair (x, y) solution to the function.

6. There is no real number whose square is -100.

7. When graphing a nonlinear equation, first recognize it as a nonlinear equation and know that the graph is <u>not</u> a line. If you don't know the <u>shape</u> of the graph, plot enough points until you see a pattern.

Exercise Set 8.2

1. To find $f(1)$, find the y-value when $x = 1$. We see from the graph that when $x = 1$, y or $f(x)$ is 0. Thus, $f(1) = 0$.

3. To find $f(-1)$, find the y-value when $x = -1$. We see from the graph that when $x = -1$, y or $f(x)$ is -4. Thus, $f(-1) = -4$.

5. To find x-values such that $f(x) = 4$, find any ordered pairs on the graph with $f(x)$- or y-value of 4. The only such point is $(3, 4)$. Thus, $f(3) = 4$ and the x-value is 3.

7. If $f(1) = -10$, then $y = -10$ when $x = 1$. The ordered pair is $(1, -10)$.

9. If $g(4) = 56$, then $y = 56$ when $x = 4$. The ordered pair is $(4, 56)$.

11. The ordered pair $(-1, -2)$ is on the graph of f. Thus, $f(-1) = -2$.

13. The ordered pair $(2, 0)$ is on the graph of g. Thus, $g(2) = 0$.

15. There are two ordered pairs on the graph of f with a y-value of -5, $(-4, -5)$ and $(0, -5)$. The x-values are -4 and 0.

17. To the right of the y-axis, there is one ordered pair on the graph of g with a y-value of 4, $(3, 4)$. The x-value is 3.

19. $\sqrt{49} = 7$, since $7^2 = 49$.

21. $-\sqrt{\dfrac{4}{9}} = -\dfrac{2}{3}$, since $\left(\dfrac{2}{3}\right)^2 = \dfrac{4}{9}$.

23. $\sqrt{64} = 8$, since $8^2 = 64$.

25. $\sqrt{81} = 9$, since $9^2 = 81$.

27. $\sqrt{-100}$ is not a real number.

29. $f(x) = x^2 + 3$

x	y or $f(x)$
-2	7
-1	4
0	3
1	4
2	7

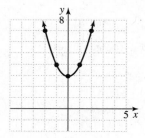

31. $h(x) = |x| - 2$

x	y or $h(x)$
-2	0
-1	-1
0	-2
1	-1
2	0

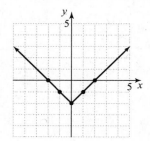

33. $g(x) = 2x^2$

x	y or $g(x)$
-2	8
-1	2
0	0
1	2
2	8

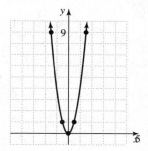

35. $f(x) = 5x - 1$

x	y or $f(x)$
-1	-6
0	-1
1	4

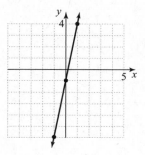

37. $f(x) = \sqrt{x+1}$

x	y or $f(x)$
-1	0
0	1
3	2
8	3

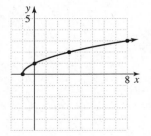

39. $g(x) = -2|x|$

x	y or $g(x)$
-2	-4
-1	-2
0	0
1	-2
2	-4

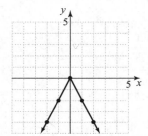

41. $h(x) = \sqrt{x} + 2$

x	y or $h(x)$
0	2
1	3
4	4
9	5

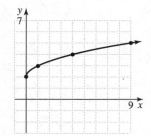

43. a. Find the semester fall 2009 and move upward until you reach the graph representing online enrollment. From the point on the graph, move horizontally to the left until the other axis is reached. The online enrollment in fall 2009 was between 5 and 6 million.

    b. Find $f(9)$.
$f(x) = 0.55x + 0.3$
$f(9) = 0.55(9) + 0.3 = 5.25$
The online enrollment in fall 2009 was approximately 5.25 million or 5,250,000.

45. Since 2012 is 12 years after 2000, find $f(12)$.
$f(x) = 0.42x + 10.5$
$\quad f(12) = 0.42(12) + 10.5$
$\quad\quad\quad = 15.54$
We predict that diamond production was $15.54 billion in 2012.

47. $A(r) = \pi r^2$
$A(5) = \pi(5)^2 = 25\pi$ square centimeters

49. $V(x) = x^3$
$V(14) = (14)^3 = 2744$ cubic inches

51. $H(f) = 2.59f + 47.24$
$H(46) = 2.59(46) + 47.24$
$\quad\quad\quad = 166.38$ centimeters

53. $D(x) = \dfrac{136}{25} x$
$D(30) = \dfrac{136}{25}(30) = 163.2$ milligrams

55. Infinite number
The reason is that a graph is a function as long as it passes the vertical line test. So it does not matter if the equation of the graph takes on the value 0 many times.

57. $3(x-2) + 5x = 6x - 16$
$\quad 3x - 6 + 5x = 6x - 16$
$\quad\quad\quad 8x - 6 = 6x - 16$
$\quad\quad\quad\quad\quad 2x = -10$
$\quad\quad\quad\quad\quad\ x = -5$
The solution is -5.

59. $3x + \dfrac{2}{5} = \dfrac{1}{10}$

$30x + 4 = 1$

$30x = -3$

$x = -\dfrac{1}{10}$

The solution is $-\dfrac{1}{10}$.

61. Look for the graph where the only nonzero y-values are 40 and 60. The answer is b.

63. Look for the graph where all the y-values are between 10 and 30. The answer is c.

65. The first segment in the graph with y-coordinate greater than 5.0 begins in 1997. Thus, 1997 is the first year that the minimum hourly wage rose above $5.00.

67. answers may vary

69. $y = x^2 - 4x + 7$

x	y
0	7
1	4
2	3
3	4
4	7

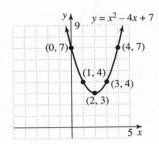

71. $f(x) = [x]$

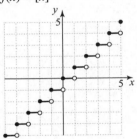

Integrated Review

1. $f(x) = 3x - 5$

$y = 3x - 5$

$y = mx + b$

The slope is $m = 3$; the y-intercept is $(0, b) = (0, -5)$.

2. $f(x) = \dfrac{5}{2}x - \dfrac{7}{2}$

$y = \dfrac{5}{2}x - \dfrac{7}{2}$

$y = mx + b$

The slope is $m = \dfrac{5}{2}$; the y-intercept is

$(0, b) = \left(0, -\dfrac{7}{2}\right)$.

3. $f(x) = 8x - 6$: slope 8; y-intercept $(0, -6)$

$g(x) = 8x + 6$: slope 8; y-intercept $(0, 6)$

The lines have the same slope and different y-intercepts, so they are parallel.

4. $f(x) = \dfrac{2}{3}x + 1$: slope $\dfrac{2}{3}$; y-intercept $(0, 1)$

$2y + 3x = 1$

$2y = -3x + 1$

$y = -\dfrac{3}{2}x + \dfrac{1}{2}$

slope $-\dfrac{3}{2}$; y-intercept $\left(0, \dfrac{1}{2}\right)$

The slopes of the lines are not equal and their product is -1, so the lines are perpendicular.

5. $(1, 6), (5, 2)$

$m = \dfrac{2 - 6}{5 - 1} = \dfrac{-4}{4} = -1$

$y - y_1 = m(x - x_1)$

$y - 6 = -1(x - 1)$

$y - 6 = -x + 1$

$y = -x + 7$

$f(x) = -x + 7$

6. $(2, -8), (-6, -5)$

$$m = \frac{-5 - (-8)}{-6 - 2} = \frac{3}{-8} = -\frac{3}{8}$$

$$y - y_1 = m(x - x_1)$$

$$y - (-8) = -\frac{3}{8}(x - 2)$$

$$y + 8 = -\frac{3}{8}x + \frac{3}{4}$$

$$y = -\frac{3}{8}x - \frac{29}{4}$$

$$f(x) = -\frac{3}{8}x - \frac{29}{4}$$

7. $3x - y = 5$

$3x - 5 = y$

The line $3x - y = 5$ has slope 3. A parallel line will also have slope 3.

$$y - y_1 = m(x - x_1); \; (-1, -5)$$

$$y - (-5) = 3[x - (-1)]$$

$$y + 5 = 3(x + 1)$$

$$y + 5 = 3x + 3$$

$$y = 3x - 2$$

$$f(x) = 3x - 2$$

8. $4x - 5y = 10$

$$-5y = -4x + 10$$

$$y = \frac{4}{5}x - 2$$

The line $4x - 5y = 10$ has slope $\frac{4}{5}$.

A perpendicular line has slope $-\frac{5}{4}$.

$$y = mx + b; \; (0, 4)$$

$$y = -\frac{5}{4}x + 4$$

$$f(x) = -\frac{5}{4}x + 4$$

9. $4x + y = \frac{2}{3}$

$$y = -4x + \frac{2}{3}$$

The line $4x + y = \frac{2}{3}$ has slope -4. A

perpendicular line has slope $\frac{1}{4}$.

$$y - y_1 = m(x - x_1); \; (2, -3)$$

$$y - (-3) = \frac{1}{4}(x - 2)$$

$$y + 3 = \frac{1}{4}x - \frac{1}{2}$$

$$y = \frac{1}{4}x - \frac{7}{2}$$

$$f(x) = \frac{1}{4}x - \frac{7}{2}$$

10. $5x + 2y = 2$

$$2y = -5x + 2$$

$$y = -\frac{5}{2}x + 1$$

The line $5x + 2y = 2$ has slope $-\frac{5}{2}$. A parallel

line will also have slope $-\frac{5}{2}$.

$$y - y_1 = m(x - x_1); \; (-1, 0)$$

$$y - 0 = -\frac{5}{2}[x - (-1)]$$

$$y = -\frac{5}{2}(x + 1)$$

$$y = -\frac{5}{2}x - \frac{5}{2}$$

$$f(x) = -\frac{5}{2}x - \frac{5}{2}$$

11. $f(x) = 4x - 2$

Linear

x	0	$\frac{1}{2}$	1
y or $f(x)$	-2	0	2

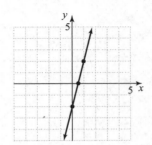

Copyright © 2013 Pearson Education, Inc.

12. $f(x) = 6x - 5$
Linear

x	0	$\frac{1}{2}$	1
y or $f(x)$	-5	-2	1

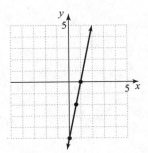

13. $g(x) = |x| + 3$
Not linear

x	-2	-1	0	1	2
y or $g(x)$	5	4	3	4	5

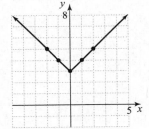

14. $h(x) = |x| + 2$
Not linear

x	-2	-1	0	1	2
y or $h(x)$	4	3	2	3	4

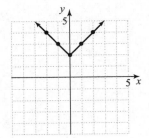

15. $f(x) = 2x^2$
Not linear

x	-2	-1	0	1	2
y or $f(x)$	8	2	0	2	8

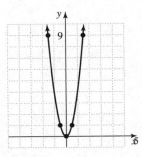

16. $F(x) = 3x^2$
Not linear

x	-2	-1	0	1	2
y or $F(x)$	12	3	0	3	12

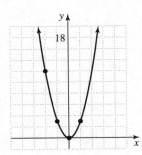

17. $h(x) = x^2 - 3$
Not linear

x	-2	-1	0	1	2
y or $h(x)$	1	-2	-3	-2	1

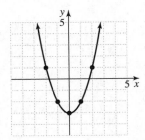

18. $G(x) = x^2 + 3$
Not linear

x	-2	-1	0	1	2
y or $G(x)$	7	4	3	4	7

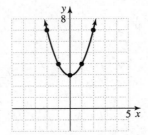

19. $F(x) = -2x$
Linear

x	-1	0	1
y or $F(x)$	2	0	-2

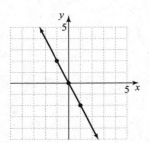

20. $H(x) = -3x$
Linear

x	-1	0	1
y or $H(x)$	3	0	-3

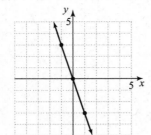

21. $G(x) = |x + 2|$
Not linear

x	-4	-3	-2	-1	0
y or $G(x)$	2	1	0	1	2

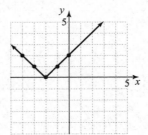

22. $g(x) = |x - 1|$
Not linear

x	-1	0	1	2	3
y or $g(x)$	2	1	0	1	2

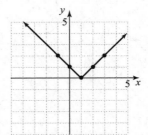

23. $f(x) = \dfrac{1}{3}x - 1$
Linear

x	-3	0	3
y or $f(x)$	-2	-1	0

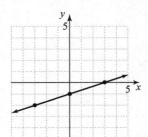

24. $f(x) = \dfrac{1}{2}x - 3$

Linear

x	−2	0	2
y or f(x)	−4	−3	−2

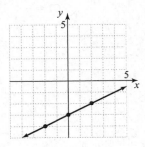

25. $g(x) = -\dfrac{3}{2}x + 1$

Linear

x	−2	0	2
y or g(x)	4	1	−2

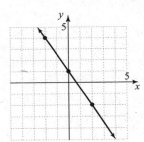

26. $G(x) = -\dfrac{2}{3}x + 1$

Linear

x	−3	0	3
y or G(x)	3	1	−1

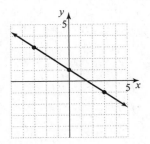

Section 8.3 Practice

1. $f(x) = \begin{cases} -4x - 2 & \text{if } x \le 0 \\ x + 1 & \text{if } x > 0 \end{cases}$

Since $4 > 0$, $f(4) = 4 + 1 = 5$.

Since $-2 \le 0$, $f(-2) = -4(-2) - 2 = 8 - 2 = 6$.

Since $0 \le 0$, $f(0) = -4(0) - 2 = 0 - 2 = -2$.

The ordered pairs are $(4, 5)$, $(-2, 6)$, and $(0, -2)$.

2. $f(x) = \begin{cases} -4x - 2 & \text{if } x \le 0 \\ x + 1 & \text{if } x > 0 \end{cases}$

For $x \le 0$:

x	f(x)
−2	6
−1	2
0	−2

For $x > 0$:

x	f(x)
1	2
2	3
3	4

Graph a closed circle at $(0, -2)$. Graph an open circle at $(0, 1)$, which is found by substituting 0 for x in $f(x) = x + 1$.

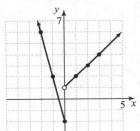

3. $f(x) = x^2$ and $g(x) = x^2 - 3$

The graph of $g(x) = x^2 - 3$ is the graph of $f(x) = x^2$ moved downward 3 units.

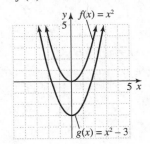

4. $f(x) = \sqrt{x}$ and $g(x) = \sqrt{x} + 1$

The graph of $g(x) = \sqrt{x} + 1$ is the graph of

$f(x) = \sqrt{x}$ moved upward 1 unit.

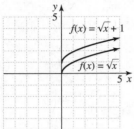

5. $f(x) = |x|$ and $g(x) = |x - 3|$

x	$f(x)$	$g(x)$
-2	2	5
-1	1	4
0	0	3
1	1	2
2	2	1
3	3	0
4	4	1
5	5	2

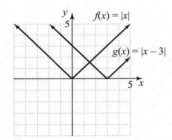

6. $f(x) = |x|$ and $g(x) = |x - 2| + 3$

The graph of $g(x)$ is the same as the graph of $f(x)$
shifted 2 units to the right and 3 units up.

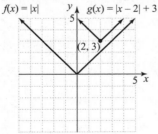

7. $h(x) = -(x + 2)^2 - 1$

The graph of $h(x) = -(x + 2)^2 - 1$ is the same as

the graph of $f(x) = x^2$ reflected about the x-axis,
then moved 2 units to the left and 1 unit
downward.

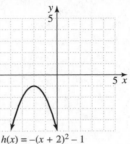

$$h(x) = -(x + 2)^2 - 1$$

Vocabulary, Readiness & Video Check 8.3

1. The graph that corresponds to $y = \sqrt{x}$ is C.

2. The graph that corresponds to $y = x^2$ is B.

3. The graph that corresponds to $y = x$ is D.

4. The graph that corresponds to $y = |x|$ is A.

5. Although $f(x) = x + 3$ isn't defined for $x = -1$, we
need to clearly indicate the point where this
piece of the graph ends. Therefore, we find this
point and graph it as an open circle.

6. Once you know the shapes, use the shifting rules
to tell you where to move the vertex or starting
point for each graph, then you can easily draw in
the appropriate basic shape.

7. The graph of $f(x) = -\sqrt{x + 6}$ has the same

shape as the graph of $f(x) = \sqrt{x + 6}$ but it is
reflected about the <u>x-axis</u>.

Exercise Set 8.3

1. $f(x) = \begin{cases} 2x & \text{if } x < 0 \\ x + 1 & \text{if } x \geq 0 \end{cases}$

For $x < 0$:　　　　For $x \geq 0$:

x	$f(x)$
-3	-6
-2	-4
-1	-2

x	$f(x)$
0	1
1	2
2	3

Graph a closed circle at (0, 1). Graph an open circle at (0, 0), which is found by substituting 0 for x in $f(x) = 2x$.

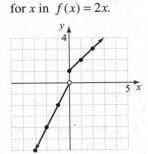

3. $f(x) = \begin{cases} 4x+5 & \text{if } x \le 0 \\ \frac{1}{4}x+2 & \text{if } x > 0 \end{cases}$

For $x \le 0$:   For $x > 0$:

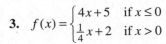

x	$f(x)$
-2	-3
-1	1
0	5

x	$f(x)$
1	$2\frac{1}{4}$
2	$2\frac{1}{2}$
4	3

Graph a closed circle at (0, 5). Graph an open circle at (0, 2), which is found by substituting 0 for x in $f(x) = \frac{1}{4}x + 2$.

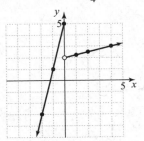

5. $g(x) = \begin{cases} -x & \text{if } x \le 1 \\ 2x+1 & \text{if } x > 1 \end{cases}$

For $x \le 1$:   For $x > 1$:

x	$g(x)$
-1	1
0	0
1	-1

x	$g(x)$
2	5
3	7
4	9

Graph a closed circle at (1, −1). Graph an open circle at (1, 3), which is found by substituting 1 for x in $g(x) = 2x + 1$.

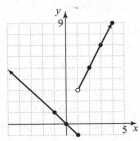

7. $f(x) = \begin{cases} 5 & \text{if } x < -2 \\ 3 & \text{if } x \ge -2 \end{cases}$

For $x < -2$:   For $x \ge -2$:

x	$f(x)$
-5	5
-4	5
-3	5

x	$f(x)$
-2	3
-1	3
0	3

Graph a closed circle at (−2, 3). Graph an open circle at (−2, 5), which is found by substituting −2 for x in $f(x) = 5$.

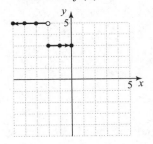

9. $f(x) = \begin{cases} -2x & \text{if } x \le 0 \\ 2x+1 & \text{if } x > 0 \end{cases}$

For $x \le 0$:   For $x > 0$:

x	$f(x)$
-1	2
0	0

x	$f(x)$
1	3
2	5

Graph a closed circle at (0, 0). Graph an open circle at (0, 1), which is found by substituting 0 for x in $f(x) = 2x + 1$.

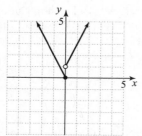

The function is defined for all real numbers, so the domain is $(-\infty, \infty)$. The function takes on all y-values greater than or equal to 0, so the range is $[0, \infty)$.

11. $h(x) = \begin{cases} 5x-5 & \text{if } x < 2 \\ -x+3 & \text{if } x \geq 2 \end{cases}$

For $x < 2$:   For $x \geq 2$:

x	$h(x)$
0	-5
1	0

x	$h(x)$
2	1
3	0

Graph a closed circle at $(2, 1)$. Graph an open circle at $(2, 5)$, which is found by substituting 2 for x in $h(x) = 5x-5$.

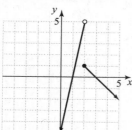

The function is defined for all real numbers, so the domain is $(-\infty, \infty)$. The function takes on all y-values less than 5, so the range is $(-\infty, 5)$.

13. $f(x) = \begin{cases} x+3 & \text{if } x < -1 \\ -2x+4 & \text{if } x \geq -1 \end{cases}$

For $x < -1$:   For $x \geq -1$:

x	$f(x)$
-4	-1
-3	0
-2	1

x	$f(x)$
-1	6
0	4
1	2

Graph a closed circle at $(-1, 6)$. Graph an open circle at $(-1, 2)$, which is found by substituting

-1 for x in $f(x) = x+3$.

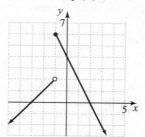

The function is defined for all real numbers, so the domain is $(-\infty, \infty)$. The function takes on all y-values less than or equal to 6, so the range is $(-\infty, 6]$.

15. $g(x) = \begin{cases} -2 & \text{if } x \leq 0 \\ -4 & \text{if } x \geq 1 \end{cases}$

For $x \leq 0$:   For $x \geq 1$:

x	$g(x)$
-2	-2
-1	-2
0	-2

x	$g(x)$
1	-4
2	-4
3	-4

Graph closed circles at $(0, -2)$ and $(1, -4)$.

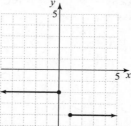

The function is defined for $x \leq 0$ or $x \geq 1$, so the domain is $(-\infty, 0] \cup [1, \infty)$. The function takes on two y-values, -2 and -4, so the range is $\{-2, -4\}$.

17. $f(x) = |x| + 3$

The graph of $f(x) = |x| + 3$ is the same as the graph of $y = |x|$ shifted up 3 units.

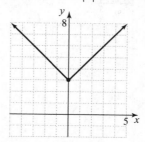

19. $f(x) = \sqrt{x} - 2$

The graph of $f(x) = \sqrt{x} - 2$ is the same as the graph of $y = \sqrt{x}$ shifted down 2 units.

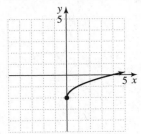

21. $f(x) = |x - 4|$

The graph of $f(x) = |x - 4|$ is the same as the graph of $y = |x|$ shifted right 4 units.

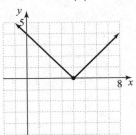

23. $f(x) = \sqrt{x + 2}$

The graph of $f(x) = \sqrt{x + 2}$ is the same as the graph of $y = \sqrt{x}$ shifted left 2 units.

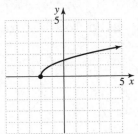

25. $y = (x - 4)^2$

The graph of $y = (x - 4)^2$ is the same as the graph of $y = x^2$ shifted right 4 units.

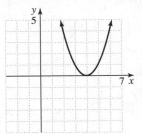

27. $f(x) = x^2 + 4$

The graph of $f(x) = x^2 + 4$ is the same as the graph of $y = x^2$ shifted up 4 units.

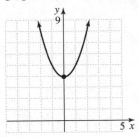

29. $f(x) = \sqrt{x - 2} + 3$

The graph of $f(x) = \sqrt{x - 2} + 3$ is the same as the graph of $y = \sqrt{x}$ shifted right 2 units and up 3 units.

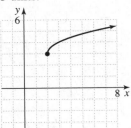

31. $f(x) = |x - 1| + 5$

The graph of $f(x) = |x - 1| + 5$ is the same as the graph of $y = |x|$ shifted right 1 unit and up 5 units.

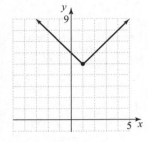

33. $f(x) = \sqrt{x+1} + 1$

The graph of $f(x) = \sqrt{x+1} + 1$ is the same as the graph of $y = \sqrt{x}$ shifted left 1 unit and up 1 unit.

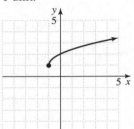

35. $f(x) = |x+3| - 1$

The graph of $f(x) = |x+3| - 1$ is the same as the graph of $y = |x|$ shifted left 3 units and down 1 unit.

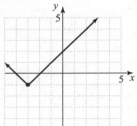

37. $g(x) = (x-1)^2 - 1$

The graph of $g(x) = (x-1)^2 - 1$ is the same as the graph of $y = x^2$ shifted right 1 unit and down 1 unit.

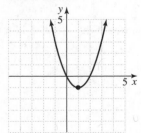

39. $f(x) = (x+3)^2 - 2$

The graph of $f(x) = (x+3)^2 - 2$ is the same as the graph of $y = x^2$ shifted left 3 units and down 2 units.

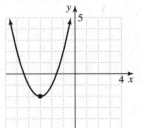

41. $f(x) = -(x-1)^2$

The graph of $f(x) = -(x-1)^2$ is the same as the graph of $y = x^2$ reflected about the x-axis and then shifted right 1 unit.

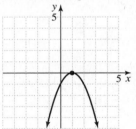

43. $h(x) = -\sqrt{x} + 3$

The graph of $h(x) = -\sqrt{x} + 3$ is the same as the graph of $y = \sqrt{x}$ reflected about the x-axis and then shifted up 3 units.

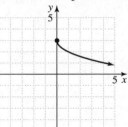

45. $h(x) = -|x+2| + 3$

The graph of $h(x) = -|x+2| + 3$ is the same as the graph of $y = |x|$ reflected about the x-axis and then shifted left 2 units and up 3 units.

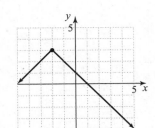

47. $f(x) = (x-3) + 2$

Since the function can be simplified to $f(x) = x - 1$, we see that its graph is a line with slope $m = 1$ and y-intercept $(0, -1)$.

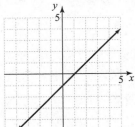

49. The graph of $y = -1$ is a horizontal line with y-intercept $(0, -1)$. The correct graph is A.

51. The graph of $x = 3$ is a vertical line with x-intercept $(3, 0)$. The correct graph is D.

53. answers may vary

55. $f(x) = \begin{cases} -\frac{1}{2}x & \text{if } x \le 0 \\ x + 1 & \text{if } 0 < x \le 2 \\ 2x - 1 & \text{if } x > 2 \end{cases}$

Some points for $x \le 0$: $(-4, 2)$, $(-2, 1)$, $(0, 0)$
Closed dot at $(0, 0)$
Some points for $0 < x \le 2$: $(1, 2)$, $(2, 3)$
Open dot at $(0, 1)$, closed dot at $(2, 3)$
Some points for $x > 2$: $(3, 5)$, $(4, 7)$
There would be an open dot at $(2, 3)$ except that it gets filled by the middle piece of the graph.

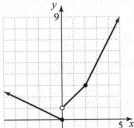

57. $f(x) = \sqrt{x - 2} + 3$

The function is defined when $x - 2 \ge 0$, or $x \ge 2$, so the domain is $[2, \infty)$. The function takes on all y-values greater than or equal to 3, so the range is $[3, \infty)$.

59. $h(x) = -|x + 2| + 3$

The function is defined for all real numbers, so the domain is $(-\infty, \infty)$. The function takes on all y-values less than or equal to 3, so the range is $(-\infty, 3]$.

61. $f(x) = 5\sqrt{x - 20} + 1$

The function is defined when $x - 20 \ge 0$, or $x \ge 20$, so the domain is $[20, \infty)$.

63. $h(x) = 5|x - 20| + 1$

The function is defined for all real numbers, so the domain is $(-\infty, \infty)$.

65. $g(x) = 9 - \sqrt{x + 103}$

The function is defined when $x + 103 \ge 0$, or $x \ge -103$, so the domain is $[-103, \infty)$.

67. $f(x) = \begin{cases} |x| & \text{if } x \le 0 \\ x^2 & \text{if } x > 0 \end{cases}$

For $x \le 0$:

x	$f(x)$
-2	2
-1	1
0	0

For $x > 0$:

x	$f(x)$
1	1
2	4
3	9

Graph a closed circle at $(0, 0)$. The graph of $f(x) = x^2$ for $x > 0$ also approaches the point $(0, 0)$.

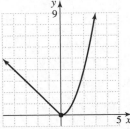

The function is defined for all real numbers, so the domain is $(-\infty, \infty)$. The function takes on all y-values greater than or equal to 0, so the range is $[0, \infty)$.

69. $g(x) = \begin{cases} |x-2| & \text{if } x < 0 \\ -x^2 & \text{if } x \geq 0 \end{cases}$

For $x < 0$:   For $x \geq 0$:

x	$g(x)$
-3	5
-2	4
-1	3

x	$g(x)$
0	0
1	-1
2	-4

Graph an open circle at (0, 2). Graph a closed circle at (0, 0).

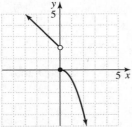

The function is defined for all real numbers, so the domain is $(-\infty, \infty)$. The function takes on all y-values such $y > 2$ or $y \leq 0$, so the range is $(-\infty, 0] \cup (2, \infty)$.

Section 8.4 Practice

1. $y = kx$

$20 = k(15)$

$\dfrac{4}{3} = k$

$k = \dfrac{4}{3}; \; y = \dfrac{4}{3}x$

2. $d = kw$

$9 = k(36)$

$\dfrac{1}{4} = k$

$d = \dfrac{1}{4}w$

$d = \dfrac{1}{4}(75)$

$d = \dfrac{75}{4}$ inches or $18\dfrac{3}{4}$ inches

3. $b = \dfrac{k}{a}$

$5 = \dfrac{k}{9}$

$k = 45; \; b = \dfrac{45}{a}$

4. $P = \dfrac{k}{V}$

$350 = \dfrac{k}{2.8}$

$980 = k$

$P = \dfrac{980}{V}$

$P = \dfrac{980}{1.5}$

$P = 653\dfrac{1}{3}$

The pressure is $653\dfrac{1}{3}$ kilopascals.

5. $A = kap$

6. $y = \dfrac{k}{x^3}$

$\dfrac{1}{2} = \dfrac{k}{2^3}$

$\dfrac{1}{2} = \dfrac{k}{8}$

$4 = k$

$k = 4; \; y = \dfrac{4}{x^3}$

7. $y = \dfrac{kz}{x^3}$

$15 = \dfrac{k \cdot 5}{3^3}$

$81 = k$

$k = 81; \; y = \dfrac{81z}{x^3}$

Vocabulary, Readiness & Video Check 8.4

1. $y = 5x$ represents direct variation.

2. $y = \dfrac{700}{x}$ represents inverse variation.

3. $y = 5xz$ represents joint variation.

4. $y = \frac{1}{2}abc$ represents joint variation.

5. $y = \frac{9.1}{x}$ represents inverse variation.

6. $y = 2.3x$ represents direct variation.

7. $y = \frac{2}{3}x$ represents direct variation.

8. $y = 3.1st$ represents joint variation.

9. linear; slope

10. When x and y vary inversely, the product of x and y is the constant of variation, k.

11. $y = ka^2b^5$

12. A combined variation problem involves combinations of direct, inverse, and/or joint variation.

Exercise Set 8.4

1. $y = kx$
$4 = k(20)$
$k = \frac{1}{5}$
$y = \frac{1}{5}x$

3. $y = kx$
$6 = k(4)$
$k = \frac{3}{2}$
$y = \frac{3}{2}x$

5. $y = kx$
$7 = k\left(\frac{1}{2}\right)$
$k = 14$
$y = 14x$

7. $y = kx$
$0.2 = k(0.8)$
$k = 0.25$
$y = 0.25x$

9. $W = kr^3$
$1.2 = k \cdot 2^3$
$k = \frac{1.2}{8} = 0.15$
$W = 0.15r^3$
$\quad = 0.15(3)^3$
$\quad = 0.15(27)$
$\quad = 4.05$
The ball weighs 4.05 pounds.

11. $P = kN$
$270,000 = k(460,000)$
$k = \frac{27}{46}$
$P = \frac{27}{46}N$
$P = \frac{27}{46}(319,000) \approx 187,239$
We expect St. Louis to produce 187,239 tons of pollution.

13. $y = \frac{k}{x}$
$6 = \frac{k}{5}$
$k = 30$
$y = \frac{30}{x}$

15. $y = \frac{k}{x}$
$100 = \frac{k}{7}$
$k = 700$
$y = \frac{700}{x}$

17. $y = \frac{k}{x}$
$\frac{1}{8} = \frac{k}{16}$
$k = 2$
$y = \frac{2}{x}$

19.
$$y = \frac{k}{x}$$
$$0.2 = \frac{k}{0.7}$$
$$k = 0.14$$
$$y = \frac{0.14}{x}$$

21.
$$R = \frac{k}{T}$$
$$45 = \frac{k}{6}$$
$$k = 270$$
$$R = \frac{270}{5} = 54$$
The car's speed is 54 mph.

23.
$$I = \frac{k}{R}$$
$$40 = \frac{k}{270}$$
$$k = 10,800$$
$$I = \frac{10,800}{R} = \frac{10,800}{150} = 72$$
The current is 72 amps.

25.
$$I_1 = \frac{k}{d^2}$$
Replace d by $2d$.
$$I_2 = \frac{k}{(2d)^2} = \frac{k}{4d^2} = \frac{1}{4}I_1$$
Thus, the intensity is divided by 4.

27. $x = kyz$

29. $r = kst^3$

31.
$$y = kx^3$$
$$9 = k(3)^3$$
$$9 = 27k$$
$$k = \frac{1}{3}$$
$$y = \frac{1}{3}x^3$$

33.
$$y = k\sqrt{x}$$
$$0.4 = k\sqrt{4}$$
$$0.4 = 2k$$
$$\frac{0.4}{2} = k$$
$$0.2 = k$$
$$y = 0.2\sqrt{x}$$

35.
$$y = \frac{k}{x^2}$$
$$0.052 = \frac{k}{5^2}$$
$$k = 1.3$$
$$y = \frac{1.3}{x^2}$$

37.
$$y = kxz^3$$
$$120 = k(5)(2^3)$$
$$120 = k(5)(8)$$
$$120 = 40k$$
$$3 = k$$
$$y = 3xz^3$$

39. $\text{Weight} = \frac{kwh^2}{l}$
$$12 = \frac{k\left(\frac{1}{2}\right)\left(\frac{1}{3}\right)^2}{10}$$
$$120 = k \cdot \frac{1}{2} \cdot \frac{1}{9}$$
$$k = 2160$$
$$\text{Weight} = \frac{2160wh^2}{l} = \frac{2160\left(\frac{2}{3}\right)\left(\frac{1}{2}\right)^2}{16} = 22.5$$
The beam can support 22.5 tons.

41.
$$V = kr^2h$$
$$32\pi = k(4)^2(6)$$
$$32\pi = k(16)(6)$$
$$32\pi = 96k$$
$$\frac{32\pi}{96} = k$$
$$\frac{\pi}{3} = k$$
$$V = \frac{\pi}{3}r^2h$$
$$V = \frac{\pi}{3}(3)^2(5)$$
$$V = 15\pi$$
The volume is 15π cubic inches.

43. $I = \dfrac{k}{x^2}$

$80 = \dfrac{k}{2^2}$

$k = 320$

$I = \dfrac{320}{x^2}$

$5 = \dfrac{320}{x^2}$

$5x^2 = 320$

$x^2 = 64$

$x = 8$

The source is 8 feet from the light source.

45. y varies directly as x is written as $y = kx$.

47. a varies inversely as b is written as $a = \dfrac{k}{b}$.

49. y varies jointly as x and z is written as $y = kxz$.

51. y varies inversely as x^3 is written as $y = \dfrac{k}{x^3}$.

53. y varies directly as x and inversely as p^2 is written as $y = \dfrac{kx}{p^2}$.

55. $r = 4$ in.

$C = 2\pi r = 2\pi(4) = 8\pi$ in.

$A = \pi r^2 = \pi(4)^2 = 16\pi$ sq in.

57. $r = 9$ cm

$C = 2\pi r = 2\pi(9) = 18\pi$ cm

$A = \pi r^2 = \pi(9)^2 = 81\pi$ sq cm

59. $|-1.2| = 1.2$

61. $-|7| = -7$

63. $-\left|-\dfrac{1}{2}\right| = -\dfrac{1}{2}$

65. $\left(\dfrac{2}{3}\right)^3 = \left(\dfrac{2}{3}\right)\left(\dfrac{2}{3}\right)\left(\dfrac{2}{3}\right) = \dfrac{8}{27}$

67. $y = \dfrac{2}{3}x$ is an example of direct variation; a.

69. $y = 9ab$ is an example of joint variation; c.

71. $H_1 = ks^3$

$H_2 = k(2s)^3 = 8(ks^3) = 8H_1$

It is multiplied by 8.

73. $y_1 = kx$

$y_2 = k(2x) = 2(kx) = 2y_1$

It is multiplied by 2.

75.

x	$\frac{1}{4}$	$\frac{1}{2}$	1	2	4
$y = \frac{3}{x}$	12	6	3	$\frac{3}{2}$	$\frac{3}{4}$

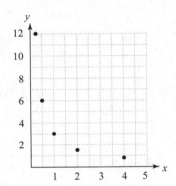

77.

x	$\frac{1}{4}$	$\frac{1}{2}$	1	2	4
$y = \dfrac{1}{2x}$	2	1	$\frac{1}{2}$	$\frac{1}{4}$	$\frac{1}{8}$

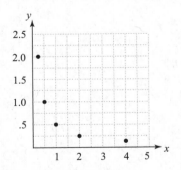

Chapter 8 Vocabulary Check

1. <u>Parallel</u> lines have the same slope and different y-intercepts.

2. <u>Slope-intercept</u> form of a linear equation in two variables is $y = mx + b$.

3. A <u>function</u> is a relation in which each first component in the ordered pairs corresponds to exactly one second component.

4. In the equation $y = 4x - 2$, the coefficient of x is the <u>slope</u> of its corresponding graph.

5. Two lines are <u>perpendicular</u> if the product of their slopes is -1.

6. A <u>linear function</u> is a function that can be written in the form $f(x) = mx + b$.

7. In the equation $y = kx$, y varies <u>directly</u> as x.

8. In the equation $y = \dfrac{k}{x}$, y varies <u>inversely</u> as x.

9. In the equation $y = kxz$, y varies <u>jointly</u> as x and z.

Chapter 8 Review

1. $f(x) = x$ or $y = x$
$m = 1,\ b = 0$

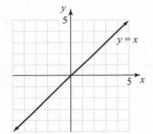

2. $f(x) = -\dfrac{1}{3}x$ or $y = -\dfrac{1}{3}x$

$m = -\dfrac{1}{3},\ b = 0$

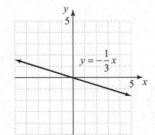

3. $g(x) = 4x - 1$ or $y = 4x - 1$
$m = 4,\ b = -1$

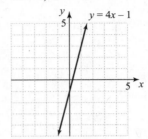

4. $F(x) = -\dfrac{2}{3}x + 2$ or $y = -\dfrac{2}{3}x + 2$

$m = -\dfrac{2}{3},\ b = 2$

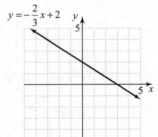

5. $f(x) = 3x + 1$
The y-intercept should be $(0, 1)$. The correct graph is C.

6. $f(x) = 3x - 2$
The y-intercept should be $(0, -2)$. The correct graph is A.

7. $f(x) = 3x + 2$
The y-intercept should be $(0, 2)$. The correct graph is B.

8. $f(x) = 3x - 5$
The y-intercept should be $(0, -5)$. The correct graph is D.

9. $f(x) = \dfrac{2}{5}x - \dfrac{4}{3}$

slope $m = \dfrac{2}{5}$; y-intercept $(0, b) = \left(0, -\dfrac{4}{3}\right)$

10. $f(x) = -\dfrac{2}{7}x + \dfrac{3}{2}$

slope $m = -\dfrac{2}{7}$; y-intercept $(0, b) = \left(0, \dfrac{3}{2}\right)$

11.
$$y - y_1 = m(x - x_1)$$
$$y - (-2) = 2(x - 5)$$
$$y + 2 = 2x - 10$$
$$2x - y = 12$$

12.
$$y - y_1 = m(x - x_1)$$
$$y - 5 = 3[x - (-3)]$$
$$y - 5 = 3(x + 3)$$
$$y - 5 = 3x + 9$$
$$3x - y = -14$$

13.
$$m = \frac{-8 - 3}{-4 - (-5)} = \frac{-11}{1} = -11$$
$$y - y_1 = m(x - x_1)$$
$$y - 3 = -11[x - (-5)]$$
$$y - 3 = -11(x + 5)$$
$$y - 3 = -11x - 55$$
$$11x + y = -52$$

14.
$$m = \frac{-2 - (-1)}{-4 - (-6)} = \frac{-1}{2} = -\frac{1}{2}$$
$$y - y_1 = m(x - x_1)$$
$$y - (-1) = -\frac{1}{2}[x - (-6)]$$
$$2(y + 1) = -(x + 6)$$
$$2y + 2 = -x - 6$$
$$x + 2y = -8$$

15. $y = 8$ has slope = 0
A line parallel to $y = 8$ has slope = 0.
$$y = -5$$

16. $x = 4$ has undefined slope.
A line perpendicular to $x = 4$ has slope = 0 and is therefore horizontal.
$$y = 3$$

17. Every horizontal line is in the form $y = c$. Since the line passes through the point $(3, -1)$, its equation is $y = -1$ or $f(x) = -1$.

18.
$$y = mx + b$$
$$y = -\frac{2}{3}x + 4$$
$$f(x) = -\frac{2}{3}x + 4$$

19.
$$y = mx + b$$
$$y = -x - 2$$
$$f(x) = -x - 2$$

20.
$$6x + 3y = 5$$
$$3y = -6x + 5$$
$$y = -2x + \frac{5}{3} \text{ so } m = -2$$
$$y - y_1 = m(x - x_1)$$
$$y - (-6) = -2(x - 2)$$
$$y + 6 = -2x + 4$$
$$y = -2x - 2$$
$$f(x) = -2x - 2$$

21.
$$3x + 2y = 8$$
$$2y = -3x + 8$$
$$y = -\frac{3}{2}x + 4 \text{ so } m = -\frac{3}{2}$$
$$y - y_1 = m(x - x_1)$$
$$y - (-2) = -\frac{3}{2}[x - (-4)]$$
$$2(y + 2) = -3(x + 4)$$
$$2y + 4 = -3x - 12$$
$$2y = -3x - 16$$
$$y = -\frac{3}{2}x - 8$$
$$f(x) = -\frac{3}{2}x - 8$$

22.
$$4x + 3y = 5$$
$$3y = -4x + 5$$
$$y = -\frac{4}{3}x + \frac{5}{3}$$
$$\text{so } m = -\frac{4}{3} \text{ and } m_\perp = \frac{3}{4}$$
$$y - y_1 = m(x - x_1)$$
$$y - (-1) = \frac{3}{4}[x - (-6)]$$
$$4(y + 1) = 3(x + 6)$$
$$4y + 4 = 3x + 18$$
$$4y = 3x + 14$$
$$y = \frac{3}{4}x + \frac{7}{2}$$
$$f(x) = \frac{3}{4}x + \frac{7}{2}$$

23. $2x - 3y = 6$

$$-3y = -2x + 6$$

$$y = \frac{2}{3}x - 2$$

so $m = \frac{2}{3}$ and $m_\perp = -\frac{3}{2}$

$$y - y_1 = m(x - x_1)$$

$$y - 5 = -\frac{3}{2}[x - (-4)]$$

$$2(y - 5) = -3(x + 4)$$

$$2y - 10 = -3x - 12$$

$$2y = -3x - 2$$

$$y = -\frac{3}{2}x - 1$$

$$f(x) = -\frac{3}{2}x - 1$$

24. Use the points (2, 17,500) and (4, 14,300).

a. $m = \dfrac{14,300 - 17,500}{4 - 2} = \dfrac{-3200}{2} = -1600$

$$y - y_1 = m(x - x_1)$$

$$y - 17,500 = -1600(x - 2)$$

$$y - 17,500 = -1600x + 3200$$

$$y = -1600x + 20,700$$

b. Let $x = 6$, since 2012 is 6 years after 2006.

$$y = -1600x + 20,700$$

$$y = -1600(6) + 20,700$$

$$= -9600 + 20,700$$

$$= 11,100$$

In 2012 the value of the automobile was $11,100.

25. a. Use the points (7, 210,000) and (12, 270,000).

$$m = \frac{270,000 - 210,000}{12 - 7} = \frac{60,000}{5} = 12,000$$

$$y - y_1 = m(x - x_1)$$

$$y - 210,000 = 12,000(x - 7)$$

$$y - 210,000 = 12,000x - 84,000$$

$$y = 12,000x + 126,000$$

b. Let $x = 18$ since 2018 is 18 years after 2000.

$$y = 12,000x + 126,000$$

$$y = 12,000(18) + 126,000$$

$$= 216,000 + 126,000$$

$$= 342,000$$

In 2018, the value of the building is estimated to be $342,000.

26. $-x + 3y = 2$

$$3y = x + 2$$

$$y = \frac{1}{3}x + \frac{2}{3}$$

$$m = \frac{1}{3}, \, b = \frac{2}{3}$$

$$6x - 18y = 3$$

$$-18y = -6x + 3$$

$$y = \frac{1}{3}x - \frac{1}{6}$$

$$m = \frac{1}{3}, \, b = -\frac{1}{6}$$

The slopes are the same and the *y*-intercepts are different, so the lines are parallel.

27. When $x = -1$, *y* or *f*(*x*) is 0, so $f(-1) = 0$.

28. When $x = 1$, *y* or *f*(*x*) is -2, so $f(1) = -2$.

29. The *x*-values that correspond to a *y*-value of 1 are -2 and 4, so $f(x) = 1$ for $x = -2$ and $x = 4$.

30. The *x*-values that correspond to a *y*-value of -1 are 0 and 2, so $f(x) = -1$ for $x = 0$ and $x = 2$.

31. $f(x) = 3x$; Linear

x	-1	0	1
y	-3	0	3

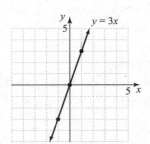

32. $f(x) = 5x$; Linear

x	-1	0	1
y	-5	0	5

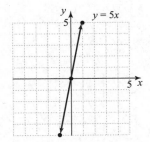

33. $g(x) = |x| + 4$; Not linear

x	–3	–2	–1	0	1	2	3
y	7	6	5	4	5	6	7

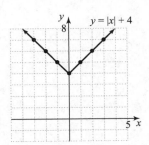

34. $h(x) = x^2 + 4$; Not linear

x	–3	–2	–1	0	1	2	3
y	13	8	5	4	5	8	13

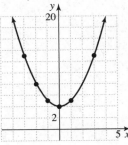

35. $F(x) = -\dfrac{1}{2}x + 2$; Linear

Find three ordered pair solutions, or find *x*- and *y*-intercepts, or find *m* and *b*.

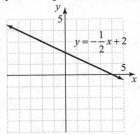

36. $G(x) = -x + 5$; Linear
Find three ordered pair solutions, or find *x*- and *y*-intercepts, or find *m* and *b*.

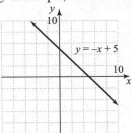

37. $y = -1.36x$; Linear
Find three ordered pair solutions, or find *x*- and *y*-intercepts, or find *m* and *b*.

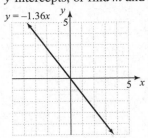

38. $y = 2.1x + 5.9$; Linear

Find three ordered pair solutions, or find *x*- and *y*-intercepts, or find *m* and *b*.

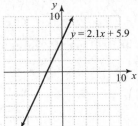

39. $H(x) = (x-2)^2$; Not linear

x	0	1	2	3	4
y	4	1	0	1	4

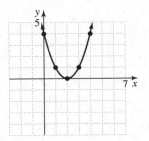

40. $f(x) = -|x - 3|$; Not linear

x	1	2	3	4	5
y	−2	−1	0	−1	−2

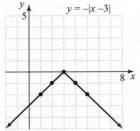

41. $g(x) = \begin{cases} -\dfrac{1}{5}x & \text{if } x \le -1 \\ -4x+2 & \text{if } x > -1 \end{cases}$

For $x \le -1$:   For $x > -1$:

x	$g(x)$
−5	1
−3	$\dfrac{3}{5}$
−1	$\dfrac{1}{5}$

x	$g(x)$
0	2
1	−2
2	−6

Graph a closed circle at $\left(-1, \dfrac{1}{5}\right)$. Graph an open circle at $(-1, 6)$, which is found by substituting −1 for x in $g(x) = -4x + 2$.

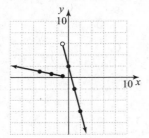

42. $f(x) = \begin{cases} -3x & \text{if } x < 0 \\ x-3 & \text{if } x \ge 0 \end{cases}$

For $x < 0$:   For $x \ge 0$:

x	$f(x)$
−3	9
−2	6
−1	3

x	$f(x)$
0	−3
1	−2
2	−1

Graph a closed circle at $(0, -3)$. Graph an open circle at $(0, 0)$, which is found by substituting 0 for x in $f(x) = -3x$.

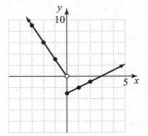

43. $f(x) = \sqrt{x-4}$

The graph of $f(x) = \sqrt{x-4}$ is the same as the graph of $y = \sqrt{x}$ shifted right 4 units.

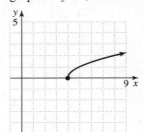

44. $y = \sqrt{x} - 4$

The graph of $f(x) = \sqrt{x} - 4$ is the same as the graph of $y = \sqrt{x}$ shifted down 4 units.

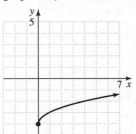

45. $h(x) = -(x+3)^2 - 1$

The graph of $h(x) = -(x+3)^2 - 1$ is the same as the graph of $y = x^2$ reflected about the x-axis and then shifted left 3 units and down 1 unit.

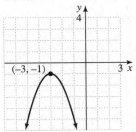

46. $g(x) = |x-2| - 2$

The graph of $g(x) = |x-2| - 2$ is the same as the graph of $y = |x|$ shifted right 2 units and down 2 units.

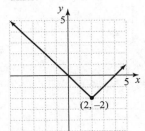

47. $A = kB$
$6 = k(14)$
$k = \dfrac{6}{14} = \dfrac{3}{7}$
$A = \dfrac{3}{7}B$
$A = \dfrac{3}{7}(21) = 9$

48. $C = \dfrac{k}{D}$
$12 = \dfrac{k}{8}$
$96 = k$
$C = \dfrac{96}{D}$
$C = \dfrac{96}{24} = 4$

49. $P = \dfrac{k}{V}$
$1250 = \dfrac{k}{2}$
$k = 2500$
$P = \dfrac{2500}{V}$
$800 = \dfrac{2500}{V}$
$800V = 2500$
$V = 3.125$

When the pressure is 800 kilopascals, the volume is 3.125 cubic meters.

50. $A = kr^2$
$36\pi = k(3)^2$
$36\pi = 9k$
$4\pi = k$
$A = 4\pi r^2$
$A = 4\pi(4)^2 = 64\pi$

When the radius is 4 inches, the surface area is 64π square inches.

51. Slope 0; through $\left(-4, \dfrac{9}{2}\right)$

A line with slope 0 is horizontal, and a horizontal line has an equation of the form $y = b$, where b is the y-coordinate of any point on the line. The equation is $y = \dfrac{9}{2}$ or $f(x) = \dfrac{9}{2}$.

52. Slope $\dfrac{3}{4}$; through $(-8, -4)$
$y - y_1 = m(x - x_1)$
$y - (-4) = \dfrac{3}{4}(x - (-8))$
$y + 4 = \dfrac{3}{4}(x + 8)$
$4(y + 4) = 3(x + 8)$
$4y + 16 = 3x + 24$
$4y = 3x + 8$
$y = \dfrac{3}{4}x + 2$
$f(x) = \dfrac{3}{4}x + 2$

53. Through $(-3, 8)$ and $(-2, 3)$
Find the slope.

$$m = \frac{3-8}{-2-(-3)} = \frac{-5}{1} = -5$$

Use the slope and one of the points in the point-slope form. We use $(-2, 3)$.

$$y - y_1 = m(x - x_1)$$
$$y - 3 = -5(x - (-2))$$
$$y - 3 = -5(x + 2)$$
$$y - 3 = -5x - 10$$
$$y = -5x - 7$$
$$f(x) = -5x - 7$$

54. Through $(-6, 1)$; parallel to $y = -\frac{3}{2}x + 11$

The slope of a line parallel to $y = -\frac{3}{2}x + 11$ will

have the same slope, $-\frac{3}{2}$.

$$y - y_1 = m(x - x_1)$$
$$y - 1 = -\frac{3}{2}(x - (-6))$$
$$y - 1 = -\frac{3}{2}(x + 6)$$
$$2(y - 1) = -3(x + 6)$$
$$2y - 2 = -3x - 18$$
$$2y = -3x - 16$$
$$y = -\frac{3}{2}x - 8$$
$$f(x) = -\frac{3}{2}x - 8$$

55. Through $(-5, 7)$; perpendicular to $5x - 4y = 10$
Find the slope of $5x - 4y = 10$.

$$5x - 4y = 10$$
$$-4y = -5x + 10$$
$$y = \frac{5}{4}x - \frac{5}{2}$$

The slope is $\frac{5}{4}$. The slope of any line

perpendicular to this line is the negative

reciprocal of $\frac{5}{4}$, or $-\frac{4}{5}$.

$$y - y_1 = m(x - x_1)$$
$$y - 7 = -\frac{4}{5}(x - (-5))$$
$$y - 7 = -\frac{4}{5}(x + 5)$$
$$5(y - 7) = -4(x + 5)$$
$$5y - 35 = -4x - 20$$
$$5y = -4x + 15$$
$$y = -\frac{4}{5}x + 3$$
$$f(x) = -\frac{4}{5}x + 3$$

56. $g(x) = \begin{cases} 4x - 3 & \text{if } x \le 1 \\ 2x & \text{if } x > 1 \end{cases}$

For $x \le 1$:   For $x > 1$:

x	$g(x)$
-1	-7
0	-3
1	1

x	$g(x)$
2	4
3	6
4	8

Graph a closed circle at $(1, 1)$. Graph an open
circle at $(1, 2)$, which is found by substituting 1
for x in $g(x) = 2x$.

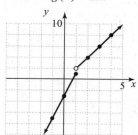

57. $f(x) = \begin{cases} x - 2 & \text{if } x \le 0 \\ -\frac{x}{3} & \text{if } x \ge 3 \end{cases}$

For $x \le 0$:   For $x \ge 3$:

x	$f(x)$
-2	-4
-1	-3
0	-2

x	$f(x)$
3	-1
4	$-\frac{4}{3}$
6	-2

Graph closed circles at $(0, -2)$ and $(3, -1)$.

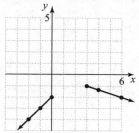

58. $f(x) = |x+1| - 3$

The graph of $f(x) = |x+1| - 3$ is the same as the graph of $y = |x|$ shifted left 1 unit and down 3 units.

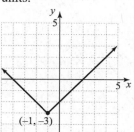

59. $f(x) = \sqrt{x-2}$

The graph of $f(x) = \sqrt{x-2}$ is the same as the graph of $y = \sqrt{x}$ shifted right 2 units.

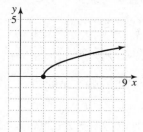

60. $y = \dfrac{k}{x}$

$14 = \dfrac{k}{6}$

$84 = k$

$y = \dfrac{84}{x}$

$y = \dfrac{84}{21} = 4$

Chapter 8 Test

1. When $x = 1$, y or $f(x)$ is 3, so $f(1) = 3$.

2. When $x = -3$, y or $f(x)$ is -5, so $f(-3) = -5$.

3. The x-values that correspond to a y-value of 0 are -2 and 2, so $f(x) = 0$ for $x = -2$ and $x = 2$.

4. The x-value that corresponds to a y-value of 4 is 0, so $f(x) = 4$ for $x = 0$.

5. $2x - 3y = -6$

    $-3y = -2x - 6$

    $y = \dfrac{2}{3}x + 2$

    $m = \dfrac{2}{3}, \ b = 2$

    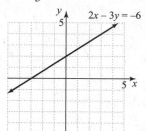

6. $f(x) = \dfrac{2}{3}x$ or $y = \dfrac{2}{3}x$

    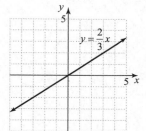

7. Horizontal; through $(2, -8)$

    A horizontal line has an equation of the form $y = b$, where b is the y-coordinate of any point on the line. The equation is $y = -8$.

8. $y - y_1 = m(x - x_1)$

    $y - (-1) = -3(x - 4)$

    $y + 1 = -3x + 12$

    $3x + y = 11$

9. $y - y_1 = m(x - x_1)$

    $y - (-2) = 5(x - 0)$

    $y + 2 = 5x$

    $5x - y = 2$

10. $m = \dfrac{-3 - (-2)}{6 - 4} = \dfrac{-1}{2} = -\dfrac{1}{2}$

$y - y_1 = m(x - x_1)$

$y - (-2) = -\dfrac{1}{2}(x - 4)$

$2(y + 2) = -(x - 4)$

$2y + 4 = -x + 4$

$2y = -x$

$y = -\dfrac{1}{2}x$

$f(x) = -\dfrac{1}{2}x$

11. $3x - y = 4$

$y = 3x - 4$

$m = 3$ so $m_\perp = -\dfrac{1}{3}$

$y - y_1 = m(x - x_1)$

$y - 2 = -\dfrac{1}{3}[x - (-1)]$

$3(y - 2) = -(x + 1)$

$3y - 6 = -x - 1$

$3y = -x + 5$

$y = -\dfrac{1}{3}x + \dfrac{5}{3}$

$f(x) = -\dfrac{1}{3}x + \dfrac{5}{3}$

12. $2y + x = 3$

$2y = -x + 3$

$y = -\dfrac{1}{2}x + 3$ so $m = -\dfrac{1}{2}$

$y - y_1 = m(x - x_1)$

$y - (-2) = -\dfrac{1}{2}(x - 3)$

$2(y + 2) = -(x - 3)$

$2y + 4 = -x + 3$

$2y = -x - 1$

$y = -\dfrac{1}{2}x - \dfrac{1}{2}$

$f(x) = -\dfrac{1}{2}x - \dfrac{1}{2}$

13. $2x - 5y = 8$

$-5y = -2x + 8$

$y = \dfrac{2}{5}x - \dfrac{8}{5}$ so $m_1 = \dfrac{2}{5}$

$m_2 = \dfrac{-1 - 4}{-1 - 1} = \dfrac{-5}{-2} = \dfrac{5}{2}$

Therefore, lines L_1 and L_2 are neither parallel nor perpendicular since their slopes are not equal and the product of their slopes is not -1.

14. Domain: $(-\infty, \infty)$
Range: $\{5\}$
Function since it passes the vertical line test.

15. Domain: $\{-2\}$
Range: $(-\infty, \infty)$
Not a function since it fails the vertical line test.

16. Domain: $(-\infty, \infty)$
Range: $[0, \infty)$
Function since it passes the vertical line test.

17. Domain: $(-\infty, \infty)$
Range: $(-\infty, \infty)$
Function since it passes the vertical line test.

18. $f(x) = 0.096x + 72.81$

   a. Let $x = 90$.
$f(90) = 0.096(90) + 72.81 = 81.45 \approx 81$
81 games would be won in the 2009 season.

   b. Let $x = 67$.
$f(67) = 0.096(67) + 72.81 = 79.242 \approx 79$
79 games would be won in the 2009 season.

   c. Let $f(x) = 95$ and solve for x.
$95 = 0.096x + 72.81$
$22.19 = 0.096x$
$231 \approx x$
A payroll of \$231 million would be necessary.

   d. The slope is 0.096. Every million dollars spent on payroll increases winnings by 0.096 game.

19. $f(x) = \begin{cases} -\dfrac{1}{2}x & \text{if } x \leq 0 \\ 2x - 3 & \text{if } x > 0 \end{cases}$

For $x \leq 0$:

x	$f(x)$
-4	2
-2	1
0	0

For $x > 0$:

x	$f(x)$
1	-1
2	1
3	3

Graph a closed circle at (0, 0). Graph an open circle at (0, −3), which is found by substituting 0 for x in $f(x) = 2x - 3$.

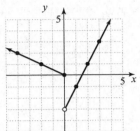

Domain: $(-\infty, \infty)$;
range: $(-3, \infty)$

20. $f(x) = (x-4)^2$

The graph of $f(x) = (x-4)^2$ is the same as the graph of $y = x^2$ shifted right 4 units.

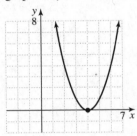

21. $g(x) = -|x+2| - 1$

The graph of $g(x) = -|x+2| - 1$ is the same as the graph of $y = |x|$ reflected about the x-axis and then shifted left 2 units and down 1 unit.

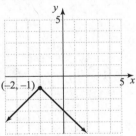

Domain: $(-\infty, \infty)$;
range: $(-\infty, -1]$

22. $h(x) = \sqrt{x} - 1$

The graph of $h(x) = \sqrt{x} - 1$ is the same as the graph of $y = \sqrt{x}$ shifted down 1 unit.

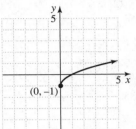

23. $W = \dfrac{k}{V}$

$20 = \dfrac{k}{12}$

$240 = k$

$W = \dfrac{240}{V}$

$W = \dfrac{240}{15} = 16$

24. $Q = kRS^2$

$24 = k(3)(4)^2$

$24 = 48k$

$\dfrac{1}{2} = k$

$Q = \dfrac{1}{2}RS^2$

$Q = \dfrac{1}{2}(2)(3)^2 = \dfrac{1}{2}(2)(9) = 9$

25. $s = k\sqrt{d}$

$160 = k\sqrt{400}$

$160 = 20k$

$8 = k$

$s = 8\sqrt{d}$

$128 = 8\sqrt{d}$

$16 = \sqrt{d}$

$16^2 = \left(\sqrt{d}\right)^2$

$256 = d$

The cliff is 256 feet tall.

Chapter 8 Cumulative Review

1. $3[4 + 2(10 - 1)] = 3[4 + 2(9)]$
$$= 3[4 + 18]$$
$$= 3[22]$$
$$= 66$$

2. $5[3 + 6(8 - 5)] = 5[3 + 6(3)]$
$$= 5[3 + 18]$$
$$= 5[21]$$
$$= 105$$

3. Let $x = 2$ and $y = -5$.

   a. $\dfrac{x - y}{12 + x} = \dfrac{2 - (-5)}{12 + 2} = \dfrac{2 + 5}{14} = \dfrac{7}{14} = \dfrac{1}{2}$

   b. $x^2 - 3y = 2^2 - 3(-5) = 4 + 15 = 19$

4. Let $x = 2$ and $y = -5$.

   a. $\dfrac{x + y}{3y} = \dfrac{2 + (-5)}{3(-5)} = \dfrac{-3}{-15} = \dfrac{1}{5}$

   b. $y^2 - x = (-5)^2 - 2 = 25 - 2 = 23$

5. $-3x = 33$
$$\frac{-3x}{-3} = \frac{33}{-3}$$
$$x = -11$$

6. $\dfrac{2}{3}y = 7$
$$\frac{3}{2}\left(\frac{2}{3}y\right) = \frac{3}{2}(7)$$
$$y = \frac{21}{2}$$

7. $8(2 - t) = -5t$
$$16 - 8t = -5t$$
$$16 - 8t + 8t = -5t + 8t$$
$$16 = 3t$$
$$\frac{16}{3} = \frac{3t}{3}$$
$$\frac{16}{3} = t$$

8. $5x - 9 = 5x - 29$
$$5x - 5x - 9 = 5x - 5x - 29$$
$$-9 = -29$$
This is a false statement, so the equation has no solution.

9. $y = mx + b$
$$y - b = mx$$
$$\frac{y - b}{m} = x \quad \text{or} \quad x = \frac{y - b}{m}$$

10. $y = 7x - 2$
$$y + 2 = 7x$$
$$\frac{y + 2}{7} = x \quad \text{or} \quad x = \frac{y + 2}{7}$$

11. $-4x + 7 \geq -9$
$$-4x + 7 - 7 \geq -9 - 7$$
$$-4x \geq -16$$
$$\frac{-4x}{-4} \leq \frac{-16}{-4}$$
$$x \leq 4$$
$(-\infty, 4]$

12. $-5x - 6 < 3x + 1$
$$-5x - 3x - 6 < 3x - 3x + 1$$
$$-8x - 6 < 1$$
$$-8x - 6 + 6 < 1 + 6$$
$$-8x < 7$$
$$\frac{-8x}{-8} > \frac{7}{-8}$$
$$x > -\frac{7}{8}$$
$\left(-\dfrac{7}{8}, \infty\right)$

13. $y = mx + b$
$y = -1 = 0x - 1$
The slope is $m = 0$.

14. $(0, 7), (-1, 0)$
$$m = \frac{0 - 7}{-1 - 0} = \frac{-7}{-1} = 7$$
The slope of a parallel line is 7.

15. $g(x) = x^2 - 3$

    a. $g(2) = 2^2 - 3 = 4 - 3 = 1$
       The ordered pair is (2, 1).

    b. $g(-2) = (-2)^2 - 3 = 4 - 3 = 1$
       The ordered pair is (−2, 1).

    c. $g(0) = 0^2 - 3 = 0 - 3 = -3$
       The ordered pair is (0, −3).

16. $f(x) = 3 - x^2$

    a. $f(2) = 3 - 2^2 = 3 - 4 = -1$
       The ordered pair is (2, −1).

    b. $f(-2) = 3 - (-2)^2 = 3 - 4 = -1$
       The ordered pair is (−2, −1).

    c. $f(0) = 3 - 0^2 = 3 - 0 = 3$
       The ordered pair is (0, 3).

17. $\begin{cases} 2x + y = 10 \\ x = y + 2 \end{cases}$

Substitute $y + 2$ for x in the first equation.
$2(y+2) + y = 10$
$2y + 4 + y = 10$
$4 + 3y = 10$
$3y = 6$
$y = 2$
Let $y = 2$ in the second equation.
$x = 2 + 2 = 4$
The solution is (4, 2).

18. $\begin{cases} 3y = x + 10 \\ 2x + 5y = 24 \end{cases}$

Solve the first equation for x.
$x = 3y - 10$
Substitute $3y - 10$ for x in the second equation.
$2(3y - 10) + 5y = 24$
$6y - 20 + 5y = 24$
$11y - 20 = 24$
$11y = 44$
$y = 4$
Let $y = 4$ in $x = 3y - 10$.
$x = 3(4) - 10 = 12 - 10 = 2$
The solution is (2, 4).

19. $\begin{cases} -x - \dfrac{y}{2} = \dfrac{5}{2} \\ \dfrac{x}{6} - \dfrac{y}{2} = 0 \end{cases}$

Multiply the first equation by 2 and the second equation by 6 to clear fractions.
$\begin{cases} -2x - y = 5 & (1) \\ x - 3y = 0 & (2) \end{cases}$
Multiply equation (2) by 2, then add.
$\begin{array}{r} -2x - y = 5 \\ 2x - 6y = 0 \\ \hline -7y = 5 \end{array}$
$y = -\dfrac{5}{7}$
Multiply equation (1) by −3, then add.
$\begin{array}{r} 6x + 3y = -15 \\ x - 3y = 0 \\ \hline 7x \quad\quad = -15 \end{array}$
$x = -\dfrac{15}{7}$
The solution is $\left(-\dfrac{15}{7}, -\dfrac{5}{7} \right)$.

20. $\begin{cases} \dfrac{x}{2} + y = \dfrac{5}{6} \\ 2x - y = \dfrac{5}{6} \end{cases}$

Multiply both equations by 6, then add.
$\begin{array}{r} 3x + 6y = 5 \\ 12x - 6y = 5 \\ \hline 15x \quad\quad = 10 \end{array}$
$x = \dfrac{2}{3}$
Let $x = \dfrac{2}{3}$ in the first equation.
$\dfrac{\frac{2}{3}}{2} + y = \dfrac{5}{6}$
$\dfrac{1}{3} + y = \dfrac{5}{6}$
$y = \dfrac{5}{6} - \dfrac{2}{6}$
$y = \dfrac{3}{6} = \dfrac{1}{2}$
The solution is $\left(\dfrac{2}{3}, \dfrac{1}{2} \right)$.

21.
$$x+3\overline{)x^2+7x+12}$$
$$\underline{x^2+3x}$$
$$4x+12$$
$$\underline{4x+12}$$
$$0$$

quotient $x+4$

$$\frac{x^2+7x+12}{x+3}=x+4$$

22.
$$\frac{5x^2y-6xy+2}{6xy}=\frac{5x^2y}{6xy}-\frac{6xy}{6xy}+\frac{2}{6xy}$$
$$=\frac{5x}{6}-1+\frac{1}{3xy}$$

23. a. $6t+18=6\cdot t+6\cdot 3=6(t+3)$

b. $y^5-y^7=y^5\cdot 1-y^5\cdot y^2=y^5(1-y^2)$

24. a. $5y-20=5\cdot y-5\cdot 4=5(y-4)$

b. $z^{10}-z^3=z^3\cdot z^7-z^3\cdot 1=z^3(z^7-1)$

25. $x^2+4x-12=(x-2)(x+6)$

26. $x^2-10x+21=(x-7)(x-3)$

27. $10x^2-13xy-3y^2=10x^2-15xy+2xy-3y^2$
$$=5x(2x-3y)+y(2x-3y)$$
$$=(2x-3y)(5x+y)$$

28. $12a^2+5ab-2b^2=12a^2-3ab+8ab-2b^2$
$$=3a(4a-b)+2b(4a-b)$$
$$=(4a-b)(3a+2b)$$

29. $x^3+8=x^3+2^3$
$$=(x+2)(x^2-x\cdot 2+2^2)$$
$$=(x+2)(x^2-2x+4)$$

30. $y^3-27=y^3-3^3$
$$=(y-3)(y^2+y\cdot 3+3^2)$$
$$=(y-3)(y^2+3y+9)$$

31. $x^2-9x-22=0$
$$(x+2)(x-11)=0$$
$$x+2=0\quad\text{or}\quad x-11=0$$
$$x=-2\qquad\qquad x=11$$
The solutions are -2 and 11.

32. $y^2-5y=-6$
$$y^2-5y+6=0$$
$$(y-2)(y-3)=0$$
$$y-2=0\quad\text{or}\quad y-3=0$$
$$y=2\qquad\qquad y=3$$
The solutions are 2 and 3.

33. a. $\dfrac{2x^2}{10x^3-2x^2}=\dfrac{2x^2}{2x^2(5x-1)}=\dfrac{1}{5x-1}$

b. $\dfrac{9x^2+13x+4}{8x^2+x-7}=\dfrac{(9x+4)(x+1)}{(8x-7)(x+1)}=\dfrac{9x+4}{8x-7}$

34. a. $\dfrac{33x^4y^2}{3xy}=\dfrac{33}{3}x^{4-1}y^{2-1}=11x^3y^1=11x^3y$

b. $\dfrac{9y}{90y^2+9y}=\dfrac{9y}{9y(10y+1)}=\dfrac{1}{10y+1}$

35. $\dfrac{3x+3}{5x-5x^2}\cdot\dfrac{2x^2+x-3}{4x^2-9}$
$$=\frac{3(x+1)\cdot(2x+3)(x-1)}{-5x(x-1)\cdot(2x+3)(2x-3)}$$
$$=-\frac{3(x+1)}{5x(2x-3)}$$

36. $\dfrac{2x}{x-6}-\dfrac{x+6}{x-6}=\dfrac{2x-(x+6)}{x-6}$
$$=\frac{2x-x-6}{x-6}$$
$$=\frac{x-6}{x-6}$$
$$=1$$

37.

$$\frac{3x^2+2x}{x-1}-\frac{10x-5}{x-1}=\frac{3x^2+2x-(10x-5)}{x-1}$$

$$=\frac{3x^2+2x-10x+5}{x-1}$$

$$=\frac{3x^2-8x+5}{x-1}$$

$$=\frac{(3x-5)(x-1)}{x-1}$$

$$=3x-5$$

38. $\frac{9}{y^2}-4y=\frac{9}{y^2}-\frac{4y}{1}\cdot\frac{y^2}{y^2}=\frac{9}{y^2}-\frac{4y^3}{y^2}=\frac{9-4y^3}{y^2}$

39.

$$3-\frac{6}{x}=x+8$$

$$x\left(3-\frac{6}{x}\right)=x(x+8)$$

$$3x-6=x^2+8x$$

$$0=x^2+5x+6$$

$$0=(x+3)(x+2)$$

$$x+3=0 \quad\text{or}\quad x+2=0$$

$$x=-3 \qquad\qquad x=-2$$

The solutions are -3 and -2.

40.

$$\frac{x}{2}+\frac{x}{5}=\frac{x-7}{20}$$

$$20\left(\frac{x}{2}+\frac{x}{5}\right)=20\left(\frac{x-7}{20}\right)$$

$$10x+4x=x-7$$

$$14x=x-7$$

$$13x=-7$$

$$x=-\frac{7}{13}$$

The solution is $-\frac{7}{13}$.

41. $(4, 0), (-4, -5)$

$$m=\frac{-5-0}{-4-4}=\frac{-5}{-8}=\frac{5}{8}$$

$$y-y_1=m(x-x_1)$$

$$y-0=\frac{5}{8}(x-4)$$

$$y=\frac{5}{8}x-\frac{5}{2}$$

$$f(x)=\frac{5}{8}x-\frac{5}{2}$$

42. $(-1, 3), (-2, 7)$

$$m=\frac{7-3}{-2-(-1)}=\frac{4}{-2+1}=\frac{4}{-1}=-4$$

$$y-y_1=m(x-x_1)$$

$$y-3=-4[x-(-1)]$$

$$y-3=-4(x+1)$$

$$y-3=-4x-4$$

$$y=-4x-1$$

$$f(x)=-4x-1$$

Chapter 9

Section 9.1 Practice

1. $A = \{1, 3, 5, 7, 9\}$ and $B = \{1, 2, 3, 4\}$
The numbers 1 and 3 are in sets A and B.
The intersection is $\{1, 3\}$. $A \cap B = \{1, 3\}$.

2. $\quad x + 3 < 8 \quad$ and $\quad 2x - 1 < 3$
$\qquad x < 5 \quad$ and $\qquad 2x < 4$
$\qquad x < 5 \quad$ and $\qquad x < 2$
$\{x | x < 5\}, (-\infty, 5)$

$\{x | x < 2\}, (-\infty, 2)$

$\{x | x < 5 \text{ and } x < 2\} = \{x | x < 2\}$

The solution set is $(-\infty, 2)$.

3. $\quad 4x \le 0 \quad$ and $\quad 3x + 2 > 8$
$\qquad x \le 0 \quad$ and $\qquad 3x > 6$
$\qquad x \le 0 \quad$ and $\qquad x > 2$
$\{x | x \le 0\}, (-\infty, 0]$

$\{x | x > 2\}, (2, \infty)$

$\{x | 4x \le 0 \text{ and } 3x + 2 > 8\} = \{\quad\}$ or $\varnothing$

4. $\qquad 3 < 5 - x < 9$
$\qquad 3 - 5 < 5 - x - 5 < 9 - 5$
$\qquad -2 < -x < 4$
$\qquad \dfrac{-2}{-1} > \dfrac{-x}{-1} > \dfrac{4}{-1}$
$\qquad 2 > x > -4$
or $-4 < x < 2$
The solution set is $(-4, 2)$.

5. $\qquad -4 \le \dfrac{x}{2} - 1 \le 3$
$\quad 2(-4) \le 2\left(\dfrac{x}{2} - 1\right) \le 2(3)$
$\qquad -8 \le x - 2 \le 6$
$\quad -8 + 2 \le x - 2 + 2 \le 6 + 2$
$\qquad -6 \le x \le 8$
The solution set is $[-6, 8]$.

6. $A = \{1, 3, 5, 7, 9\}$ and $B = \{2, 3, 4, 5, 6\}$.
The numbers that are in either set or both sets are
$\{1, 2, 3, 4, 5, 6, 7, 9\}$. This set is the union,
$A \cup B$.

7. $\quad 8x + 5 \le 8 \quad$ or $\quad x - 1 \ge 2$
$\qquad 8x \le 3 \quad$ or $\qquad x \ge 3$
$\qquad x \le \dfrac{3}{8} \quad$ or $\qquad x \ge 3$
$\left\{x \,\middle|\, x \le \dfrac{3}{8}\right\}, \left(-\infty, \dfrac{3}{8}\right]$

$\{x | x \ge 3\}, [3, \infty)$

$\left\{x \,\middle|\, x \le \dfrac{3}{8} \text{ or } x \ge 3\right\} = \left(-\infty, \dfrac{3}{8}\right] \cup [3, \infty)$

The solution set is $\left(-\infty, \dfrac{3}{8}\right] \cup [3, \infty)$.

8. $\quad -3x - 2 > -8 \quad$ or $\quad 5x > 0$
$\qquad -3x > -6 \quad$ or $\qquad x > 0$
$\qquad x < 2 \quad$ or $\qquad x > 0$
$\{x | x < 2\}, (-\infty, 2)$

$\{x | x > 0\}, (0, \infty)$

$\{x | x < 2 \text{ or } x > 0\}, (-\infty, \infty)$

The solution set is $(-\infty, \infty)$.

Vocabulary, Readiness & Video Check 9.1

1. Two inequalities joined by the words "and" or "or" are called <u>compound</u> inequalities.

2. The word <u>and</u> means intersection.

3. The word <u>or</u> means union.

4. The symbol $\cap$ means intersection.

5. The symbol $\cup$ represents union.

6. The symbol $\varnothing$ is the empty set.

7. For an element to be in the intersection of sets A and B, the element must be in set A <u>and</u> in set B.

8. Graph the two intervals, each on its own number line, so you can see their intersection. Graph this intersection on the third number line—this intersection is the solution set.

9. For an element to be in the union of sets A and B, the element must be in set A <u>or</u> set B.

10. Graph the two intervals, each on its own number line, so you can see their union. Graph this union on the third number line—this union is the solution set.

Exercise Set 9.1

1. $C \cup D = \{2, 3, 4, 5, 6, 7\}$

3. $A \cap D = \{4, 6\}$

5. $A \cup B = \{..., -2, -1, 0, 1, ...\}$

7. $B \cap D = \{5, 7\}$

9. $B \cup C = \{x \mid x \text{ is an odd integer or } x = 2 \text{ or } x = 4\}$

11. $A \cap C = \{2, 4\}$

13. $x < 1$ and $x > -3$
$-3 < x < 1$
$(-3, 1)$

15. $x \le -3$ and $x \ge -2$
$\varnothing$

17. $x < -1$ and $x < 1$
$x < -1$
$(-\infty, -1)$

19. $x + 1 \ge 7$ and $3x - 1 \ge 5$
    $x \ge 6$ and $3x \ge 6$
                        $x \ge 2$

$x \ge 6$
$[6, \infty)$

21. $4x + 2 \le -10$ and $2x \le 0$
   $4x \le -12$ and $x \le 0$
     $x \le -3$

$x \le -3$
$(-\infty, -3]$

23. $-2x < -8$ and $x - 5 < 5$
    $x > 4$ and $x < 10$
$(4, 10)$

25. $5 < x - 6 < 11$
$11 < x < 17$
$(11, 17)$

27. $-2 \le 3x - 5 \le 7$
   $3 \le 3x \le 12$
   $1 \le x \le 4$
$[1, 4]$

29. $1 \le \dfrac{2}{3}x + 3 \le 4$

   $-2 \le \dfrac{2}{3}x \le 1$

   $-3 \le x \le \dfrac{3}{2}$

$\left[-3, \dfrac{3}{2} \right]$

31. $-5 \le \dfrac{-3x + 1}{4} \le 2$

$4(-5) \le 4\left(\dfrac{-3x + 1}{4} \right) \le 4(2)$

   $-20 \le -3x + 1 \le 8$
   $-21 \le -3x \le 7$

     $7 \ge x \ge -\dfrac{7}{3}$

   $-\dfrac{7}{3} \le x \le 7$

$\left[-\dfrac{7}{3}, 7 \right]$

33. $x < 4$ or $x < 5$
$(-\infty, 5)$

35. $x \le -4$ or $x \ge 1$
$(-\infty, -4] \cup [1, \infty)$

37. $x > 0$ or $x < 3$

$(-\infty, \infty)$

39. $-2x \le -4$ or $5x - 20 \ge 5$

    $x \ge 2$ or $5x \ge 25$

                        $x \ge 5$

$x \ge 2$

$[2, \infty)$

41. $x + 4 < 0$ or $6x > -12$

    $x < -4$ or $x > -2$

$(-\infty, -4) \cup (-2, \infty)$

43. $3(x - 1) < 12$ or $x + 7 > 10$

    $x - 1 < 4$ or $x > 3$

       $x < 5$

$(-\infty, \infty)$

45. $x < \dfrac{2}{3}$ and $x > -\dfrac{1}{2}$

$-\dfrac{1}{2} < x < \dfrac{2}{3}$

$\left(-\dfrac{1}{2}, \dfrac{2}{3}\right)$

47. $x < \dfrac{2}{3}$ or $x > -\dfrac{1}{2}$

$(-\infty, \infty)$

49. $0 \le 2x - 3 \le 9$

   $3 \le 2x \le 12$

$\dfrac{3}{2} \le x \le 6$

$\left[\dfrac{3}{2}, 6\right]$

51. $\dfrac{1}{2} < x - \dfrac{3}{4} < 2$

$4\left(\dfrac{1}{2}\right) < 4\left(x - \dfrac{3}{4}\right) < 4(2)$

    $2 < 4x - 3 < 8$

    $5 < 4x < 11$

    $\dfrac{5}{4} < x < \dfrac{11}{4}$

$\left(\dfrac{5}{4}, \dfrac{11}{4}\right)$

53. $x + 3 \ge 3$ and $x + 3 \le 2$

     $x \ge 0$ and $x \le -1$

No solution exists.

$\varnothing$

55. $3x \ge 5$ or $-\dfrac{5}{8}x - 6 > 1$

    $x \ge \dfrac{5}{3}$ or $-\dfrac{5}{8}x > 7$

                         $x < -\dfrac{56}{5}$

$\left(-\infty, -\dfrac{56}{5}\right) \cup \left[\dfrac{5}{3}, \infty\right)$

57. $0 < \dfrac{5 - 2x}{3} < 5$

    $0 < 5 - 2x < 15$

   $\dfrac{-5}{-2} > \dfrac{-2x}{-2} > \dfrac{10}{-2}$

    $\dfrac{5}{2} > x > -5$

   $-5 < x < \dfrac{5}{2}$

$\left(-5, \dfrac{5}{2}\right)$

59. $-6 < 3(x - 2) \le 8$

   $-6 < 3x - 6 \le 8$

    $0 < 3x \le 14$

    $0 < x \le \dfrac{14}{3}$

$\left(0, \dfrac{14}{3}\right]$

61. $-x + 5 > 6$ and $1 + 2x \le -5$

    $-x > 1$ and $2x \le -6$

      $x < -1$ and $x \le -3$

$x \le -3$

$(-\infty, -3]$

63. $3x + 2 \le 5$ or $7x > 29$

    $3x \le 3$ or $x > \dfrac{29}{7}$

     $x \le 1$ or $x > \dfrac{29}{7}$

$(-\infty, 1] \cup \left(\dfrac{29}{7}, \infty\right)$

65. $5 - x > 7$ and $2x + 3 \geq 13$
     $-x > 2$ and $2x \geq 10$
     $x < -2$ and $x \geq 5$
   No solution exists.
   $\varnothing$

67. $-\dfrac{1}{2} \leq \dfrac{4x-1}{6} < \dfrac{5}{6}$

   $6\left(-\dfrac{1}{2}\right) \leq 6\left(\dfrac{4x-1}{6}\right) < 6\left(\dfrac{5}{6}\right)$

     $-3 \leq 4x - 1 < 5$
     $-2 \leq 4x < 6$
     $-\dfrac{1}{2} \leq x < \dfrac{3}{2}$

   $\left[-\dfrac{1}{2}, \dfrac{3}{2}\right)$

69. $\dfrac{1}{15} < \dfrac{8-3x}{15} < \dfrac{4}{5}$

   $15\left(\dfrac{1}{15}\right) < 15\left(\dfrac{8-3x}{15}\right) < 15\left(\dfrac{4}{5}\right)$

     $1 < 8 - 3x < 12$
     $-7 < -3x < 4$
     $-\dfrac{4}{3} < x < \dfrac{7}{3}$

   $\left(-\dfrac{4}{3}, \dfrac{7}{3}\right)$

71. $0.3 < 0.2x - 0.9 < 1.5$
     $1.2 < 0.2x < 2.4$
       $6 < x < 12$
     $(6, 12)$

73. $|-7| - |19| = 7 - 19 = -12$

75. $-(-6) - |-10| = 6 - 10 = -4$

77. $|x| = 7$
     $x = -7, 7$

79. $|x| = 0$
     $x = 0$

81. Both lines are above the level representing 1500 for the years 2004 and 2005.

83. answers may vary

85. $2x - 3 < 3x + 1 < 4x - 5$
     $2x - 3 < 3x + 1$ and $3x + 1 < 4x - 5$
       $-x < 4$ and $-x < -6$
       $x > -4$ and $x > 6$
   $x > 6$
   $(6, \infty)$

87. $-3(x - 2) \leq 3 - 2x \leq 10 - 3x$
     $-3x + 6 \leq 3 - 2x$ and $3 - 2x \leq 10 - 3x$
       $-x \leq -3$ and $x \leq 7$
       $x \geq 3$
   $3 \leq x \leq 7$
   $[3, 7]$

89. $5x - 8 < 2(2 + x) < -2(1 + 2x)$
     $5x - 8 < 4 + 2x$ and $4 + 2x < -2 - 4x$
       $3x < 12$ and $6x < -6$
       $x < 4$ and $x < -1$
   $x < -1$
   $(-\infty, -1)$

91. $-29 \leq C \leq 35$
     $-29 \leq \dfrac{5}{9}(F - 32) \leq 35$
     $-52.5 \leq F - 32 \leq 63$
     $-20.2 \leq F \leq 95$
     $-20.2° \leq F \leq 95°$

93. $70 \leq \dfrac{68 + 65 + 75 + 78 + 2x}{6} \leq 79$
     $420 \leq 286 + 2x \leq 474$
     $134 \leq 2x \leq 188$
       $67 \leq x \leq 94$
   If Christian scores between 67 and 94 inclusive on his final exam, he will receive a C in the course.

Section 9.2 Practice

1. $|q| = 3$
   $q = 3$ or $q = -3$
   The solution set is $\{-3, 3\}$.

2. $|2x - 3| = 5$
     $2x - 3 = 5$ or $2x - 3 = -5$
       $2x = 8$ or $2x = -2$
       $x = 4$ or $x = -1$
   The solution set is $\{-1, 4\}$.

409

3. $\left| \dfrac{x}{5} + 1 \right| = 15$

$\dfrac{x}{5} + 1 = 15$ or $\dfrac{x}{5} + 1 = -15$

$\dfrac{x}{5} = 14$ or $\dfrac{x}{5} = -16$

$x = 70$ or $x = -80$

The solutions are -80 and 70.

4. $|3x| + 8 = 14$

$|3x| = 6$

$3x = 6$ or $3x = -6$

$x = 2$ or $x = -2$

The solutions are -2 and 2.

5. $|z| = 0$

The solution is 0.

6. $3|z| + 9 = 7$

$3|z| = -2$

$|z| = -\dfrac{2}{3}$

The absolute value of a number is never negative, so there is no solution. The solution set is { } or $\varnothing$.

7. $\left| \dfrac{5x + 3}{4} \right| = -8$

The absolute value of a number is never negative, so there is no solution. The solution set is { } or $\varnothing$.

8. $|2x + 4| = |3x - 1|$

$2x + 4 = 3x - 1$ or $2x + 4 = -(3x - 1)$

$-x + 4 = -1$ $\qquad$ $2x + 4 = -3x + 1$

$-x = -5$ $\qquad$ $5x + 4 = 1$

$x = 5$ $\qquad$ $5x = -3$

$\qquad\qquad\qquad\qquad x = -\dfrac{3}{5}$

The solutions are $-\dfrac{3}{5}$ and 5.

9. $|x - 2| = |8 - x|$

$x - 2 = 8 - x$ or $x - 2 = -(8 - x)$

$2x - 2 = 8$ $\qquad$ $x - 2 = -8 + x$

$2x = 10$ $\qquad$ $-2 = -8$ False

$x = 5$

The solution is 5.

Vocabulary, Readiness & Video Check 9.2

1. $|x - 2| = 5$
C. $x - 2 = 5$ or $x - 2 = -5$

2. $|x - 2| = 0$
A. $x - 2 = 0$

3. $|x - 2| = |x + 3|$
B. $x - 2 = x + 3$ or $x - 2 = -(x + 3)$

4. $|x + 3| = 5$
E. $x + 3 = 5$ or $x + 3 = -5$

5. $|x + 3| = -5$
D. $\varnothing$

6. If a is negative, $|X| = a$ has no solution. (Also, if a is 0, we solve $X = 0$.)

Exercise Set 9.2

1. $|x| = 7$
$x = 7$ or $x = -7$

3. $|3x| = 12.6$
$3x = 12.6$ or $3x = -12.6$
$x = 4.2$ or $x = -4.2$

5. $|2x - 5| = 9$
$2x - 5 = 9$ or $2x - 5 = -9$
$2x = 14$ or $2x = -4$
$x = 7$ or $x = -2$

7. $\left| \dfrac{x}{2} - 3 \right| = 1$

$\dfrac{x}{2} - 3 = 1$ $\qquad$ or $\qquad$ $\dfrac{x}{2} - 3 = -1$

$2\left(\dfrac{x}{2} - 3 \right) = 2(1)$ or $2\left(\dfrac{x}{2} - 3 \right) = 2(-1)$

$x - 6 = 2$ $\qquad$ or $\qquad$ $x - 6 = -2$

$x = 8$ $\qquad$ or $\qquad$ $x = 4$

9. $|z| + 4 = 9$
$|z| = 5$
$z = -5$ or $z = -5$

11. $|3x| + 5 = 14$
$|3x| = 9$
$3x = 9$ or $3x = -9$
$x = 3$ or $x = -3$

13. $|2x| = 0$
$2x = 0$
$x = 0$

15. $|4n+1| + 10 = 4$
$|4n+1| = -6$ which is impossible.
The solution set is $\varnothing$.

17. $|5x-1| = 0$
$5x - 1 = 0$
$5x = 1$
$x = \dfrac{1}{5}$

19. $|5x-7| = |3x+11|$
$5x - 7 = 3x + 11$ or $5x - 7 = -(3x+11)$
$2x = 18$ or $5x - 7 = -3x - 11$
$x = 9$ or $8x = -4$
$x = -\dfrac{1}{2}$

21. $|z+8| = |z-3|$
$z + 8 = z - 3$ or $z + 8 = -(z-3)$
$8 = -3$ or $z + 8 = -z + 3$
$2z = -5$
$z = -\dfrac{5}{2}$

The only solution is $-\dfrac{5}{2}$.

23. $|x| = 4$
$x = 4$ or $x = -4$

25. $|y| = 0;\ y = 0$

27. $|z| = -2$ is impossible. The solution set is $\varnothing$.

29. $|7 - 3x| = 7$
$7 - 3x = 7$ or $7 - 3x = -7$
$-3x = 0$ or $-3x = -14$
$x = 0$ or $x = \dfrac{14}{3}$

31. $|6x| - 1 = 11$
$|6x| = 12$
$6x = 12$ or $6x = -12$
$x = 2$ or $x = -2$

33. $|4p| = -8$ is impossible. The solution set is $\varnothing$.

35. $|x-3| + 3 = 7$
$|x-3| = 4$
$x - 3 = 4$ or $x - 3 = -4$
$x = 7$ or $x = -1$

37. $\left|\dfrac{z}{4} + 5\right| = -7$ is impossible. The solution set is $\varnothing$.

39. $|9v - 3| = -8$ is impossible. The solution set is $\varnothing$.

41. $|8n+1| = 0$
$8n + 1 = 0$
$8n = -1$
$n = -\dfrac{1}{8}$

43. $|1 - 6c| - 7 = -3$
$|1 - 6c| = 4$
$1 - 6c = 4$ or $1 - 6c = -4$
$6c = 3$ or $6c = -5$
$c = \dfrac{1}{2}$ or $c = -\dfrac{5}{6}$

45. $|5x + 1| = 11$
$5x + 1 = 11$ or $5x + 1 = -11$
$5x = 10$ or $5x = -12$
$x = 2$ or $x = -\dfrac{12}{5}$

47. $|4x - 2| = |-10|$
$|4x - 2| = 10$
$4x - 2 = 10$ or $4x - 2 = -10$
$4x = 12$ or $4x = -8$
$x = 3$ or $x = -2$

49. $|5x + 1| = |4x - 7|$
$5x + 1 = 4x - 7$ or $5x + 1 = -(4x-7)$
$x = -8$ or $5x + 1 = -4x + 7$
$9x = 6$
$x = \dfrac{2}{3}$

51. $|6 + 2x| = -|-7|$
$|6 + 2x| = -7$ which is impossible. The solution set is $\varnothing$.

53. $|2x - 6| = |10 - 2x|$
$$2x - 6 = 10 - 2x \quad \text{or} \quad 2x - 6 = -(10 - 2x)$$
$$4x = 16 \quad\quad \text{or} \quad 2x - 6 = -10 + 2x$$
$$x = 4 \quad\quad\quad \text{or} \quad\quad\quad -6 = -10$$
$-6 = -10$ is impossible. The only solution is 4.

55. $\left|\dfrac{2x - 5}{3}\right| = 7$

$$\dfrac{2x - 5}{3} = 7 \quad \text{or} \quad \dfrac{2x - 5}{3} = -7$$
$$2x - 5 = 21 \quad \text{or} \quad 2x - 5 = -21$$
$$2x = 26 \quad \text{or} \quad\quad\quad 2x = -16$$
$$x = 13 \quad \text{or} \quad\quad\quad\quad x = -8$$

57. $2 + |5n| = 17$
$$|5n| = 15$$
$$5n = 15 \quad \text{or} \quad 5n = -15$$
$$n = 3 \quad \text{or} \quad\quad n = -3$$

59. $\left|\dfrac{2x - 1}{3}\right| = |-5|$

$$\left|\dfrac{2x - 1}{3}\right| = 5$$

$$\dfrac{2x - 1}{3} = 5 \quad \text{or} \quad \dfrac{2x - 1}{3} = -5$$
$$2x - 1 = 15 \quad \text{or} \quad 2x - 1 = -15$$
$$2x = 16 \quad \text{or} \quad\quad\quad 2x = -14$$
$$x = 8 \quad \text{or} \quad\quad\quad\quad x = -7$$

61. $|2y - 3| = |9 - 4y|$
$$2y - 3 = 9 - 4y \quad \text{or} \quad 2y - 3 = -(9 - 4y)$$
$$6y = 12 \quad\quad \text{or} \quad 2y - 3 = -9 + 4y$$
$$y = 2 \quad\quad\quad \text{or} \quad\quad -2y = -6$$
$$\phantom{y = 2 \quad\quad\quad \text{or} \quad\quad} y = 3$$

63. $\left|\dfrac{3n + 2}{8}\right| = |-1|$

$$\left|\dfrac{3n + 2}{8}\right| = 1$$

$$\dfrac{3n + 2}{8} = 1 \quad \text{or} \quad \dfrac{3n + 2}{8} = -1$$
$$3n + 2 = 8 \quad \text{or} \quad 3n + 2 = -8$$
$$3n = 6 \quad \text{or} \quad\quad\quad 3n = -10$$
$$n = 2 \quad \text{or} \quad\quad\quad\quad n = -\dfrac{10}{3}$$

65. $|x + 4| = |7 - x|$
$$x + 4 = 7 - x \quad \text{or} \quad x + 4 = -(7 - x)$$
$$2x = 3 \quad\quad \text{or} \quad x + 4 = -7 + x$$
$$x = \dfrac{3}{2} \quad\quad \text{or} \quad\quad\quad 4 = -7$$

$4 = -7$ is impossible. The only solution is $\dfrac{3}{2}$.

67. $\left|\dfrac{8c - 7}{3}\right| = -|-5|$

$$\left|\dfrac{8c - 7}{3}\right| = -5 \text{ which is impossible.}$$

The solution set is $\varnothing$.

69. 31% of cheese production came from cheddar cheese.

71. 9% of 360° is $0.09(360°) = 32.4°$

73. answers may vary

75. answers may vary

77. Since absolute value is never negative, the solution set is $\varnothing$.

79. All numbers whose distance from 0 is 5 units is written as $|x| = 5$.

81. answers may vary

83. $|x - 1| = 5$

85. answers may vary

87. $|x| = 6$

89. $|x - 2| = |3x - 4|$

Section 9.3 Practice

1. $|x| < 5$
The solution set of this inequality contains all numbers whose distance from 0 is less than 5. The solution set is $(-5, 5)$.

2. $|b + 1| < 3$
$$-3 < b + 1 < 3$$
$$-3 - 1 < b + 1 - 1 < 3 - 1$$
$$-4 < b < 2$$
$$(-4, 2)$$

3. $|3x - 2| + 5 \le 9$

$|3x - 2| \le 9 - 5$

$|3x - 2| \le 4$

$-4 \le 3x - 2 \le 4$

$-4 + 2 \le 3x - 2 + 2 \le 4 + 2$

$-2 \le 3x \le 6$

$-\dfrac{2}{3} \le x \le 2$

$\left[-\dfrac{2}{3}, 2 \right]$

4. $\left| 3x + \dfrac{5}{8} \right| < -4$

The absolute value of a number is always nonnegative and can never be less than −4. The solution set is { } or $\varnothing$.

5. $\left| \dfrac{3(x-2)}{5} \right| \le 0$

$\dfrac{3(x-2)}{5} = 0$

$5 \left[\dfrac{3(x-2)}{5} \right] = 5(0)$

$3(x - 2) = 0$

$3x - 6 = 0$

$3x = 6$

$x = 2$

The solution set is {2}.

6. $|y + 4| \ge 6$

$y + 4 \le -6 \qquad$ or $\qquad y + 4 \ge 6$

$y + 4 - 4 \le -6 - 4 \quad$ or $\quad y + 4 - 4 \ge 6 - 4$

$y \le -10 \qquad$ or $\qquad y \ge 2$

$(-\infty, -10] \cup [2, \infty)$

7. $|4x + 3| + 5 > 3$

$|4x + 3| + 5 - 5 > 3 - 5$

$|4x + 3| > -2$

The absolute value of any number is always nonnegative and thus is always greater than −2.

$(-\infty, \infty)$

8. $\left| \dfrac{x}{2} - 3 \right| - 5 > -2$

$\left| \dfrac{x}{2} - 3 \right| - 5 + 5 > -2 + 5$

$\left| \dfrac{x}{2} - 3 \right| > 3$

$\dfrac{x}{2} - 3 < -3 \qquad$ or $\qquad \dfrac{x}{2} - 3 > 3$

$2 \left(\dfrac{x}{2} - 3 \right) < 2(-3) \quad$ or $\quad 2 \left(\dfrac{x}{2} - 3 \right) > 2(3)$

$x - 6 < -6 \qquad$ or $\qquad x - 6 > 6$

$x < 0 \qquad$ or $\qquad x > 12$

$(-\infty, 0) \cup (12, \infty)$

Vocabulary, Readiness & Video Check 9.3

1. D

2. E

3. C

4. B

5. A

6. The left side of the inequality is an absolute value, which must be nonnegative—it must be 0 or positive. therefore, there is no value of x that can make the value of this absolute value be less than the negative value on the right side of the inequality.

7. The solution set involves "or" and "or" means "union."

Exercise Set 9.3

1. $|x| \le 4$

$-4 \le x \le 4$

$[-4, 4]$

3. $|x - 3| < 2$

$-2 < x - 3 < 2$

$1 < x < 5$

$(1, 5)$

5. $|x + 3| < 2$

$-2 < x + 3 < 2$

$-5 < x < -1$

$(-5, -1)$

7. $|2x + 7| \leq 3$

$-13 \leq 2x + 7 \leq 13$

$-20 \leq 2x \leq 6$

$-10 \leq x \leq 3$

$[-10, 3]$

9. $|x| + 7 \leq 12$

$|x| \leq 5$

$-5 \leq x \leq 5$

$[-5, 5]$

11. $|3x - 1| < -5$

No real solutions; $\varnothing$

13. $|x - 6| - 7 \leq -1$

$|x - 6| \leq 6$

$-6 \leq x - 6 \leq 6$

$0 \leq x \leq 12$

$[0, 12]$

15. $|x| > 3$

$x < -3 \quad \text{or} \quad x > 3$

$(-\infty, -3) \cup (3, \infty)$

17. $|x + 10| \geq 14$

$x + 10 \leq -14 \quad \text{or} \quad x + 10 \geq 14$

$x \leq -24 \quad \text{or} \quad x \geq 4$

$(-\infty, -24] \cup [4, \infty)$

19. $|x| + 2 > 6$

$|x| > 4$

$x < -4 \quad \text{or} \quad x > 4$

$(-\infty, -4) \cup (4, \infty)$

21. $|5x| > -4$

All real numbers

$(-\infty, \infty)$

23. $|6x - 8| + 3 > 7$

$|6x - 8| > 4$

$6x - 8 < -4 \quad \text{or} \quad 6x - 8 > 4$

$6x < 4 \quad \text{or} \quad 6x > 12$

$x < \dfrac{2}{3} \quad \text{or} \quad x > 2$

$\left(-\infty, \dfrac{2}{3}\right) \cup (2, \infty)$

25. $|x| \leq 0$

$|x| = 0$

$x = 0$

27. $|8x + 3| > 0$ only excludes $|8x + 3| = 0$

$8x + 3 = 0$

$8x = -3$

$x = -\dfrac{3}{8}$

All real numbers except $-\dfrac{3}{8}$.

$\left(-\infty, -\dfrac{3}{8}\right) \cup \left(-\dfrac{3}{8}, \infty\right)$

29. $|x| \leq 2$

$-2 \leq x \leq 2$

$[-2, 2]$

31. $|y| > 1$

$y < -1 \quad \text{or} \quad y > 1$

$(-\infty, -1) \cup (1, \infty)$

33. $|x - 3| < 8$

$-8 < x - 3 < 8$

$-5 < x < 11$

$(-5, 11)$

35. $|0.6x - 3| > 0.6$

$0.6x - 3 < -0.6 \quad \text{or} \quad 0.6x - 3 > 0.6$

$0.6x < 2.4 \quad \text{or} \quad 0.6x > 3.6$

$x < 4 \quad \text{or} \quad x > 6$

$(-\infty, 4) \cup (6, \infty)$

37. $5 + |x| \le 2$

$|x| \le -3$

No real solution

$\varnothing$

39. $|x| > -4$

All real numbers

$(-\infty, \infty)$

41. $|2x - 7| \le 11$

$-11 \le 2x - 7 \le 11$

$-4 \le 2x \le 18$

$-2 \le x \le 9$

$[-2, 9]$

43. $|x + 5| + 2 \ge 8$

$|x + 5| \ge 6$

$x + 5 \le -6 \quad \text{or} \quad x + 5 \ge 6$

$x \le -11 \quad \text{or} \quad x \ge 1$

$(-\infty, -11] \cup [1, \infty)$

45. $|x| > 0$ only excludes $|x| = 0$, or $x = 0$.

All real numbers except $x = 0$

$(-\infty, 0) \cup (0, \infty)$

47. $9 + |x| > 7$

$|x| > -2$

All real numbers

$(-\infty, \infty)$

49. $6 + |4x - 1| \le 9$

$|4x - 1| \le 3$

$-3 \le 4x - 1 \le 3$

$-2 \le 4x \le 4$

$-\dfrac{1}{2} \le x \le 1$

$\left[-\dfrac{1}{2}, 1\right]$

51. $\left|\dfrac{2}{3}x + 1\right| > 1$

$\dfrac{2}{3}x + 1 < -1 \quad \text{or} \quad \dfrac{2}{3}x + 1 > 1$

$\dfrac{2}{3}x < -2 \quad \text{or} \quad \dfrac{2}{3}x > 0$

$x < -3 \quad \text{or} \quad x > 0$

$(-\infty, -3) \cup (0, \infty)$

53. $|5x + 3| < -6$

No real solution

$\varnothing$

55. $\left|\dfrac{8x - 3}{4}\right| \le 0$

$\dfrac{8x - 3}{4} = 0$

$8x - 3 = 0$

$8x = 3$

$x = \dfrac{3}{8}$

$\left\{\dfrac{3}{8}\right\}$

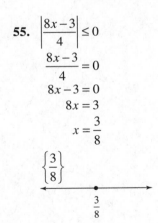

    415

57. $|1 + 3x| + 4 < 5$

$\qquad |1 + 3x| < 1$

$\quad -1 < 1 + 3x < 1$

$\quad -2 < 3x < 0$

$\quad -\dfrac{2}{3} < x < 0$

$\quad \left(-\dfrac{2}{3}, 0\right)$

59. $\left|\dfrac{x + 6}{3}\right| > 2$

$\quad \dfrac{x + 6}{3} < -2 \quad$ or $\quad \dfrac{x + 6}{3} > 2$

$\quad x + 6 < -6 \quad$ or $\quad x + 6 > 6$

$\quad x < -12 \quad$ or $\qquad x > 0$

$\quad (-\infty, -12) \cup (0, \infty)$

61. $-15 + |2x - 7| \le -6$

$\qquad |2x - 7| \le 9$

$\quad -9 \le 2x - 7 \le 9$

$\quad -2 \le 2x \le 16$

$\quad -1 \le x \le 8$

$\quad [-1, 8]$

63. $\left|2x + \dfrac{3}{4}\right| - 7 \le -2$

$\qquad \left|2x + \dfrac{3}{4}\right| \le 5$

$\quad -5 \le 2x + \dfrac{3}{4} \le 5$

$\quad -20 \le 8x + 3 \le 20$

$\quad -23 \le 8x \le 17$

$\quad -\dfrac{23}{8} \le x \le \dfrac{17}{8}$

$\quad \left[-\dfrac{23}{8}, \dfrac{17}{8}\right]$

65. $|2x - 3| < 7$

$\quad -7 < 2x - 3 < 7$

$\quad -4 < 2x < 10$

$\quad -2 < x < 5$

$\quad (-2, 5)$

67. $|2x - 3| = 7$

$\quad 2x - 3 = 7 \quad$ or $\quad 2x - 3 = -7$

$\qquad 2x = 10 \quad$ or $\qquad 2x = -4$

$\qquad x = 5 \quad$ or $\qquad x = -2$

69. $|x - 5| \ge 12$

$\quad x - 5 \le -12 \quad$ or $\quad x - 5 \ge 12$

$\qquad x \le -7 \quad$ or $\qquad x \ge 17$

$\quad (-\infty, -7] \cup [17, \infty)$

71. $|9 + 4x| = 0$

$\quad 9 + 4x = 0$

$\qquad 4x = -9$

$\qquad x = -\dfrac{9}{4}$

73. $|2x + 1| + 4 < 7$

$\qquad |2x + 1| < 3$

$\quad -3 < 2x + 1 < 3$

$\quad -4 < 2x < 2$

$\quad -2 < x < 1$

$\quad (-2, 1)$

75. $|3x - 5| + 4 = 5$

$\qquad |3x - 5| = 1$

$\quad 3x - 5 = 1 \quad$ or $\quad 3x - 5 = -1$

$\qquad 3x = 6 \quad$ or $\qquad 3x = 4$

$\qquad x = 2 \quad$ or $\qquad x = \dfrac{4}{3}$

77. $|x + 11| = -1$ is impossible. The solution set is $\varnothing$.

79. $\left|\dfrac{2x - 1}{3}\right| = 6$

$\quad \dfrac{2x - 1}{3} = 6 \quad$ or $\quad \dfrac{2x - 1}{3} = -6$

$\quad 2x - 1 = 18 \quad$ or $\quad 2x - 1 = -18$

$\qquad 2x = 19 \quad$ or $\qquad 2x = -17$

$\qquad x = \dfrac{19}{2} \quad$ or $\qquad x = -\dfrac{17}{2}$

81. $\left|\dfrac{3x-5}{6}\right| > 5$

$\dfrac{3x-5}{6} < -5 \quad$ or $\quad \dfrac{3x-5}{6} > 5$

$3x-5 < -30 \quad$ or $\quad 3x-5 > 30$

$3x < -25 \quad$ or $\quad 3x > 35$

$x < -\dfrac{25}{3} \quad$ or $\quad x > \dfrac{35}{3}$

$\left(-\infty, -\dfrac{25}{3}\right) \cup \left(\dfrac{35}{3}, \infty\right)$

83. $(2x-21) + \left(\dfrac{5}{2}x+2\right) + (3x+24) = 290$

$2x + \dfrac{5}{2}x + 3x - 21 + 2 + 24 = 290$

$\dfrac{15}{2}x + 5 = 290$

$\dfrac{15}{2}x = 285$

$x = 38$

$2x - 21 = 2(38) - 21 = 55$

$\dfrac{5}{2}x + 2 = \dfrac{5}{2}(38) + 2 = 97$

$3x + 24 = 3(38) + 24 = 138$

Year	Increase in Wi-Fi-Enabled Cell Phones	Estimated Number
2010	$2x - 21$	55 million
2012	$\dfrac{5}{2}x + 2$	97 million
2014	$3x + 24$	138 million
Total	290 million	

85. $3x - 4y = 12$

$3(2) - 4y = 12$

$6 - 4y = 12$

$-4y = 6$

$y = -\dfrac{3}{2} = -1.5$

87. $3x - 4y = 12$

$3x - 4(-3) = 12$

$3x + 12 = 12$

$3x = 0$

$x = 0$

89. $|x| < 7$

91. $|x| \le 5$

93. answers may vary

95. $|3.5 - x| < 0.05$

$-0.05 < 3.5 - x < 0.05$

$-3.55 < -x < -3.45$

$3.55 > x > 3.45$

$3.45 < x < 3.55$

Integrated Review

1. $x < 7$ and $x > -5$ is $-5 < x < 7$. The solution set is $(-5, 7)$.

2. $x < 7$ or $x > -5$
The solution set is $(-\infty, \infty)$.

3. $|4x - 3| = 1$

$4x - 3 = 1 \quad$ or $\quad 4x - 3 = -1$

$4x = 4 \quad$ or $\quad 4x = 2$

$x = 1 \quad$ or $\quad x = \dfrac{1}{2}$

The solutions are 1 and $\dfrac{1}{2}$.

4. $|2x + 1| < 5$

$-5 < 2x + 1 < 5$

$-6 < 2x < 4$

$-3 < x < 2$

The solution set is $(-3, 2)$.

5. $|6x| - 9 \ge -3$

$|6x| \ge 6$

$6x \le -6 \quad$ or $\quad 6x \ge 6$

$x \le -1 \quad$ or $\quad x \ge 1$

The solution set is $(-\infty, -1] \cup [1, \infty)$.

6. $|x - 7| = |2x + 11|$

$x - 7 = 2x + 11 \quad$ or $\quad x - 7 = -(2x + 11)$

$-7 = x + 11 \quad$ or $\quad x - 7 = -2x - 11$

$-18 = x \quad$ or $\quad 3x - 7 = -11$

$3x = -4$

$-18 = x \quad$ or $\quad x = -\dfrac{4}{3}$

The solutions are -18 and $-\dfrac{4}{3}$.

7. $-5 \le \dfrac{3x-8}{2} \le 2$

$-10 \le 3x - 8 \le 4$

$-2 \le 3x \le 12$

$-\dfrac{2}{3} \le x \le 4$

The solution set is $\left[-\dfrac{2}{3}, 4\right]$.

8. $|9x - 1| = -3$

The absolute value of a number cannot be negative. There is no solution, or $\varnothing$.

9. $3x + 2 \le 5$ or $-3x \ge 0$

     $3x \le 3$ or $\dfrac{-3x}{-3} \le \dfrac{0}{-3}$

       $x \le 1$ or $x \le 0$

The solution set is $(-\infty, 1]$.

10. $3x + 2 \le 5$ and $-3x \ge 0$

     $3x \le 3$ and $\dfrac{-3x}{-3} \le \dfrac{0}{-3}$

       $x \le 1$ and $x \le 0$

The solution set is $(-\infty, 0]$.

11. $|3 - x| - 5 \le -2$

     $|3 - x| \le 3$

   $-3 \le 3 - x \le 3$

   $-6 \le -x \le 0$

   $\dfrac{-6}{-1} \ge \dfrac{-x}{-1} \ge \dfrac{0}{-1}$

     $6 \ge x \ge 0$

     $0 \le x \le 6$

The solution set is $[0, 6]$.

12. $\left|\dfrac{4x+1}{5}\right| = |-1|$

  $\dfrac{4x+1}{5} = 1$ or $\dfrac{4x+1}{5} = -1$

  $4x + 1 = 5$ or $4x + 1 = -5$

    $4x = 4$ or $4x = -6$

     $x = 1$ or $x = \dfrac{-6}{4} = -\dfrac{3}{2}$

The solutions are 1 and $-\dfrac{3}{2}$.

13. $|2x + 1| = 5$

$2x + 1 = 5$ or $2x + 1 = -5$

This is statement B.

14. $|2x + 1| < 5$

$-5 < 2x + 1 < 5$

This is statement E.

15. $|2x + 1| > 5$

$2x + 1 < -5$ or $2x + 1 > 5$

This is statement A.

16. $x < 3$ or $x < 5$ is $x < 5$. This is statement C.

17. $x < 3$ and $x < 5$ is $x < 3$. This is statement D.

Section 9.4 Practice

1.

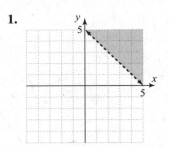

2.

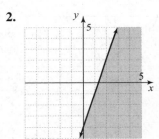

3.

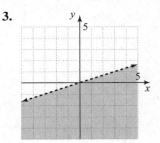

4.

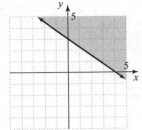

5.

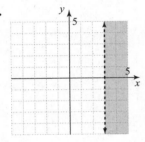

6. $\begin{cases} 4x \le y \\ x + 3y \ge 9 \end{cases}$

Graph $4x \le y$ with a solid line.
Test (1, 0)

$4(1) \overset{?}{\le} 0$

False
Shade above.
Graph $x + 3y \ge 9$ with a solid line.
Test (0, 0)

$0 + 3(0) \overset{?}{\ge} 9$

False
Shade above.
The solution of the system is the darker shaded region and includes parts of both boundary lines.

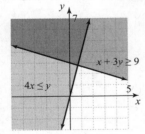

7. $\begin{cases} x - y > 4 \\ x + 3y < -4 \end{cases}$

Graph both inequalities using dashed lines. The solution of the system is the darker shaded region which does not include any of the boundary lines.

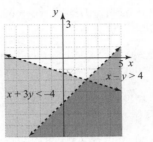

8. $\begin{cases} y \le 6 \\ -2x + 5y > 10 \end{cases}$

Graph both inequalities. The solution of the system is the darker shaded region.

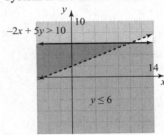

Vocabulary, Readiness & Video Check 9.4

1. The statement $5x - 6y < 7$ is an example of a
   <u>linear inequality in two variables</u>.

2. A boundary line divides a plane into two regions
   called <u>half-planes</u>.

3. The graph of $5x - 6y < 7$ does not include its
   corresponding boundary line. The statement is
   false.

4. When graphing a linear inequality to determine
   which side of the boundary line to shade, choose
   a point *not* on the boundary line. The statement
   is true.

5. The boundary line for the inequality $5x - 6y < 7$
   is the graph of $5x - 6y = 7$. The statement is true.

6. The graph of $y < 3$ is

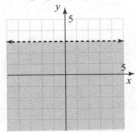

7. Yes, since the inequality is ≥, the graph includes the boundary line.

8. No, since the inequality is >, the graph does not include the boundary line.

9. Yes, since the inequality is ≥, the graph includes the boundary line.

10. No, since the inequality is >, the graph does not include the boundary line.

11. We find the boundary line equation by replacing the inequality symbol with =. The points on this line are solutions (line is solid) if the inequality symbol is ≥ or ≤; they are not solutions (line is dashed) if the inequality is > or <.

12. No; we can choose any test point except a point on the second inequality's own boundary line.

Exercise Set 9.4

1. $x - y > 3$

   $(2, -1),\ 2 - (-1) \overset{?}{>} 3$

   $2 + 1 \overset{?}{>} 3$

   $3 \overset{?}{>} 3,\ \text{False}$

   $(2, -1)$ is not a solution.

   $(5, 1),\ 5 - 1 \overset{?}{>} 3$

   $4 > 3,\ \text{True}$

   $(5, 1)$ is a solution.

3. $3x - 5y \le -4$

   $(-1, -1),\ 3(-1) - 5(-1) \overset{?}{\le} -4$

   $-3 + 5 \overset{?}{\le} -4$

   $2 \le -4,\ \text{False}$

   $(-1, -1)$ is not a solution.

   $(4, 0),\ 3(4) - 5(0) \overset{?}{\le} -4$

   $12 - 0 \overset{?}{\le} -4$

   $12 \le -4,\ \text{False}$

   $(4, 0)$ is not a solution.

5. $x < -y$

   $(0, 2),\ 0 \overset{?}{<} -2,\ \text{False}$

   $(0, 2)$ is not a solution.

   $(-5, 1),\ -5 \overset{?}{<} -1,\ \text{True}$

   $(-5, 1)$ is a solution.

7. $x + y \le 1$
   Test $(0, 0)$

   $0 + 0 \overset{?}{\le} 1,\ \text{True}$
   Shade below.

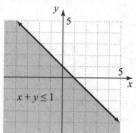

9. $2x + y > -4$
   Test $(0, 0)$

   $2(0) + 0 \overset{?}{>} -4$
   True
   Shade above.

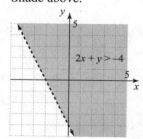

11. $x + 6y \le -6$
    Test $(0, 0)$

    $0 + 6(0) \overset{?}{\le} -6$
    False
    Shade below.

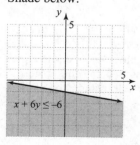

13. $2x + 5y > -10$
Test $(0, 0)$
$$2(0) + 5(0) \overset{?}{>} -10$$
True
Shade above.

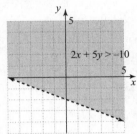

15. $x + 2y \leq 3$
Test $(0, 0)$
$$0 + 2(0) \overset{?}{\leq} 3$$
True
Shade below.

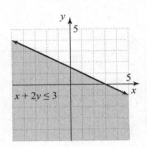

17. $2x + 7y > 5$
Test $(0, 0)$
$$2(0) + 7(0) \overset{?}{>} 5$$
False
Shade above.

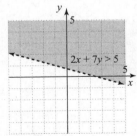

19. $x - 2y \geq 3$
Test $(0, 0)$
$$(0) - 2(0) \overset{?}{\geq} 3$$
False
Shade below.

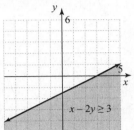

21. $5x + y < 3$
Test $(0, 0)$
$$5(0) + 0 \overset{?}{<} 3$$
True
Shade below.

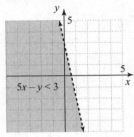

23. $4x + y < 8$
Test $(0, 0)$
$$4(09) + 0 \overset{?}{<} 8$$
True
Shade below.

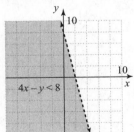

421

25. $y \geq 2x$
Test $(1, 0)$
$$0 \overset{?}{\geq} 2(1)$$
False
Shade above.

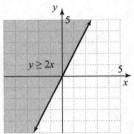

27. $x \geq 0$
Shade right.

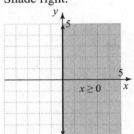

29. $y \leq -3$
Shade below.

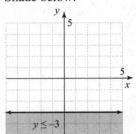

31. $2x - 7y > 0$
Test $(1, 0)$
$$2(1) - 7(0) \overset{?}{>} 0$$
True
Shade below.

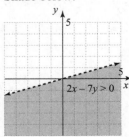

33. $3x - 7y \geq 0$
Test $(1, 0)$
$$3(1) - 7(0) \overset{?}{\geq} 0$$
True
Shade below.

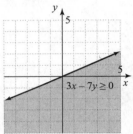

35. $x > y$
Test $(0, 1)$
$$0 \overset{?}{>} 1$$
False
Shade below.

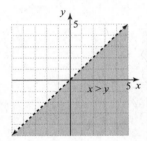

37. $x - y \leq 6$
Test $(0, 0)$
$$0 - 0 \overset{?}{\leq} 6$$
True
Shade above.

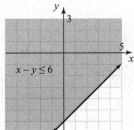

39. $-\dfrac{1}{4}y + \dfrac{1}{3}x > 1$

Test $(0, 0)$

$-\dfrac{1}{4}(0) + \dfrac{1}{3}(0) \overset{?}{>} 1$

False

Shade below.

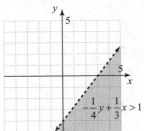

41. $-x < 0.4y$

Test $(1, 0)$

$-(1) \overset{?}{<} 0$

True

Shade above.

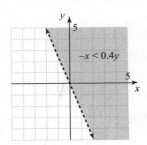

43. e

45. c

47. f

49. $\begin{cases} y \geq x+1 \\ y \geq 3-x \end{cases}$

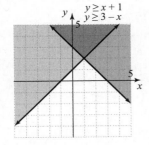

51. $\begin{cases} y < 3x-4 \\ y \leq x+2 \end{cases}$

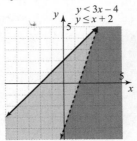

53. $\begin{cases} y \leq -2x-2 \\ y \geq x+4 \end{cases}$

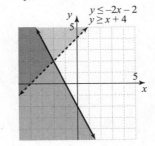

55. $\begin{cases} y \geq -x+2 \\ y \leq 2x+5 \end{cases}$

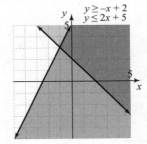

57. $\begin{cases} x \geq 3y \\ x+3y \leq 6 \end{cases}$

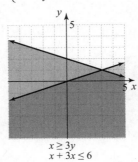

59. $\begin{cases} y + 2x \geq 0 \\ 5x - 3y \leq 12 \end{cases}$

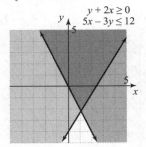

61. $\begin{cases} 3x - 4y \geq -6 \\ 2x + y \leq 7 \end{cases}$

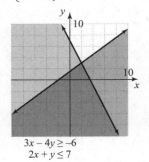

63. $\begin{cases} x \leq 2 \\ y \geq -3 \end{cases}$

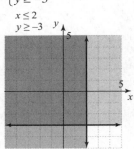

65. $\begin{cases} y \geq 1 \\ x < -3 \end{cases}$

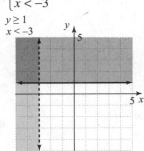

67. $\begin{cases} 2x + 3y < -8 \\ x \geq -4 \end{cases}$

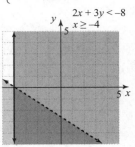

69. $\begin{cases} 2x - 5y \leq 9 \\ y \leq -3 \end{cases}$

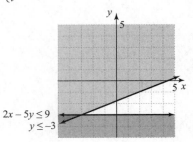

71. $\begin{cases} y \geq \dfrac{1}{2}x + 2 \\ y \leq \dfrac{1}{2}x - 3 \end{cases}$

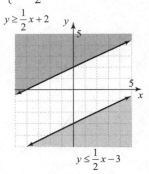

73. Let $x = -5$.

$x^2 = (-5)(-5) = 25$

75. Let $x = -1$.

$2x^3 = 2(-1)(-1)(-1) = -2$

77. $3x + 4y < 8;\ (1,1)$

$3(1) + 4(1) < 8$

$3 + 4 < 8$

$7 < 8$ True

$(1, 1)$ is included in the graph.

79. $y \geq -\frac{1}{2}x$; (1, 1)

$y \geq -\frac{1}{2}(1)$

$y \geq -\frac{1}{2}$ True

(1, 1) is included in the graph.

81. The inequality is $x + y \geq 13$.

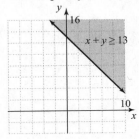

83. answers may vary

85.

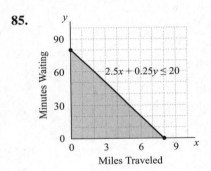

87. answers may vary

89. a. $30x + 0.15y \leq 500$

b.

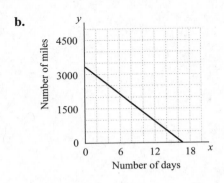

c. answers may vary

91. C

93. D

95. $\begin{cases} 2x - y \leq 6 \\ x \geq 3 \\ y > 2 \end{cases}$

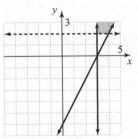

97. answers may vary

Chapter 9 Vocabulary Check

1. The statement "$x < 5$ or $x > 7$" is called a <u>compound inequality</u>.

2. The <u>intersection</u> of two sets is the set of all elements common to both sets.

3. The <u>union</u> of two sets is the set of all elements that belong to either of the sets.

4. A number's distance from 0 is called its <u>absolute value</u>.

5. When a variable in an equation is replaced by a number and the resulting equation is true, then that number is called a <u>solution</u> of the equation.

6. Two or more linear inequalities are called a <u>system of linear inequalities</u>.

Chapter 9 Review

1. $-3 < 4(2x - 1) < 12$

$-3 < 8x - 4 < 12$

$1 < 8x < 16$

$\frac{1}{8} < x < 2$

$\left(\frac{1}{8}, 2 \right)$

2. $-2 \leq 8 + 5x < -1$

$-10 \leq 5x \leq -9$

$-2 \leq x \leq -\frac{9}{5}$

$\left[-2, \frac{9}{5} \right)$

3.
$$\frac{1}{6} < \frac{4x-3}{3} \le \frac{4}{5}$$
$$30\left(\frac{1}{6}\right) < 30\left(\frac{4x-3}{3}\right) \le 30\left(\frac{4}{5}\right)$$
$$5 < 10(4x-3) \le 24$$
$$5 < 40x-30 \le 24$$
$$35 < 40x < 54$$
$$\frac{7}{8} < x \le \frac{27}{20}$$
$$\left(\frac{7}{8}, \frac{27}{20}\right]$$

4.
$$-6 < x-(3-4x) < -3$$
$$-6 < x-3+4x < -3$$
$$-6 < 5x-3 < -3$$
$$-3 < 5x < 0$$
$$-\frac{3}{5} < x < 0$$
$$\left(-\frac{3}{5}, 0\right)$$

5.
$$3x-5 > 6 \quad \text{or} \quad -x < -5$$
$$3x > 11 \quad \text{or} \quad x > 5$$
$$x > \frac{11}{3} \quad \text{or} \quad x > 5$$
$$x > \frac{11}{3}$$
$$\left(\frac{11}{3}, \infty\right)$$

6. $x \le 2 \quad \text{and} \quad x > -5$
$$-5 < x \le 2$$
$$(-5, 2]$$

7. $|8 - x| = 3$
$$8-x=3 \quad \text{or} \quad 8-x=-3$$
$$-x=-5 \quad \text{or} \quad -x=-11$$
$$x=5 \quad \text{or} \quad x=11$$

8. $|x - 7| = 9$
$$x-7=9 \quad \text{or} \quad x-7=-9$$
$$x=16 \quad \text{or} \quad x=-2$$

9. $|-3x + 4| = 7$
$$-3x+4=7 \quad \text{or} \quad -3x+4=-7$$
$$-3x=3 \quad \text{or} \quad -3x=-11$$
$$x=-1 \quad \text{or} \quad x=\frac{11}{3}$$

10. $|2x + 9| = 9$
$$2x+9=9 \quad \text{or} \quad 2x+9=-9$$
$$2x=0 \quad \text{or} \quad 2x=-18$$
$$x=0 \quad \text{or} \quad x=-9$$

11. $5 + |6x + 1| = 5$
$$|6x+1|=0$$
$$6x+1=0$$
$$6x=-1$$
$$x=-\frac{1}{6}$$

12. $|3x - 2| + 6 = 10$
$$|3x-2|=4$$
$$3x-2=4 \quad \text{or} \quad 3x-2=-4$$
$$3x=6 \quad \text{or} \quad 3x=-2$$
$$x=2 \quad \text{or} \quad x=-\frac{2}{3}$$

13. $|5 - 6x| + 8 = 3$
$$|5-6x|=-5$$
The solution set is $\varnothing$.

14. $-5 = |4x - 3|$
The solution set is $\varnothing$.

15. $\left|\frac{3x-7}{4}\right| = 2$
$$\frac{3x-7}{4}=2 \quad \text{or} \quad \frac{3x-7}{4}=-2$$
$$3x-7=8 \quad \text{or} \quad 3x-7=-8$$
$$3x=15 \quad \text{or} \quad 3x=-1$$
$$x=5 \quad \text{or} \quad x=-\frac{1}{3}$$

16. $-4 = \left|\frac{x-3}{2}\right| - 5$
$$1 = \left|\frac{x-3}{2}\right|$$
$$\frac{x-3}{2}=1 \quad \text{or} \quad \frac{x-3}{2}=-1$$
$$x-3=2 \qquad x-3=-2$$
$$x=5 \qquad x=1$$

17. $|6x + 1| = |15 + 4x|$

$6x + 1 = 15 + 4x$ or $6x + 1 = -(15 + 4x)$

$2x = 14$ or $6x + 1 = -15 - 4x$

$x = 7$ or $10x = -16$

$$x = -\frac{8}{5}$$

18. $|x - 3| = |x + 5|$

$x - 3 = x + 5$ or $x - 3 = -(x + 5)$

$-3 = 5$ False   or $x - 3 = -x - 5$

$2x - 3 = -5$

$2x = -2$

$x = -1$

19. $|5x - 1| < 9$

$-9 < 5x - 1 < 9$

$-8 < 5x < 10$

$$-\frac{8}{5} < x < 2$$

$$\left(-\frac{8}{5}, 2\right)$$

20. $|6 + 4x| \ge 10$

$6 + 4x \le -10$ or $6 + 4x \ge 10$

$4x \le -16$ or $4x \ge 4$

$x \le -4$ or $x \ge 1$

$(-\infty, -4] \cup [1, \infty)$

21. $|3x| - 8 > 1$

$|3x| > 9$

$3x < -9$ or $3x > 9$

$x < -3$ or $x > 3$

$(-\infty, -3) \cup (3, \infty)$

22. $9 + |5x| < 24$

$|5x| < 15$

$-15 < 5x < 15$

$-3 < x < 3$

$(-3, 3)$

23. $|6x - 5| \le -1$

The solution set is $\varnothing$.

24. $|6x - 5| \le 5$

$-5 \le 6x - 5 \le 5$

$0 \le 6x \le 10$

$$\frac{0}{6} \le x \le \frac{10}{6}$$

$$0 \le x \le \frac{5}{3}$$

$$\left[0, \frac{5}{3}\right]$$

25. $\left|3x + \dfrac{2}{5}\right| \ge 4$

$3x + \dfrac{2}{5} \le -4$ or $3x + \dfrac{2}{5} \ge 4$

$5\left(3x + \dfrac{2}{5}\right) \le 5(-4)$ or $5\left(3x + \dfrac{2}{5}\right) \ge 5(4)$

$15x + 2 \le -20$ or $15x + 2 \ge 20$

$15x \le -22$ or $15x \ge 18$

$x \le -\dfrac{22}{15}$ or $x \ge \dfrac{6}{5}$

$$\left(-\infty, -\frac{22}{15}\right] \cup \left[\frac{6}{5}, \infty\right)$$

26. $|5x - 3| > 2$

$5x - 3 < -2$ or $5x - 3 > 2$

$5x < 1$ or $5x > 5$

$x < \dfrac{1}{5}$ or $x > 1$

$$\left(-\infty, \frac{1}{5}\right) \cup (1, \infty)$$

27. $\left|\dfrac{x}{3}+6\right|-8>-5$

$\left|\dfrac{x}{3}+6\right|>3$

$\dfrac{x}{3}+6<-3 \quad$ or $\quad \dfrac{x}{3}+6>3$

$\dfrac{x}{3}<-9 \quad$ or $\quad \dfrac{x}{3}>-3$

$x<-27 \quad$ or $\quad x>-9$

$(-\infty,-27)\cup(-9,\infty)$

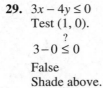

28. $\left|\dfrac{4(x-1)}{7}\right|+10<2$

$\left|\dfrac{4(x-1)}{7}\right|<-8$

The solution set is $\varnothing$.

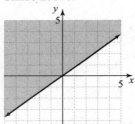

29. $3x-4y\le 0$
Test $(1, 0)$.

$3-0\overset{?}{\le}0$

False
Shade above.

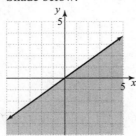

30. $3x-4y\ge 0$
Test $(1, 0)$.

$3-0\overset{?}{\ge}0$

True
Shade below.

31. $x+6y<6$
Test $(0, 0)$

$0+6(0)\overset{?}{<}6$

True
Shade below.

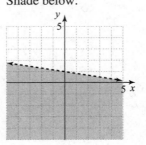

32. $y\le -4$
Shade below.

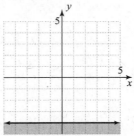

33. $y\ge -7$
Shade above.

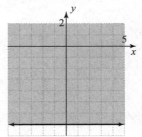

34. $x\ge -y$
Test $(1, 0)$

$1\overset{?}{\ge}0$

True
Shade above.

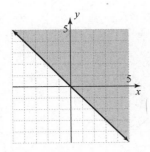

35. $\begin{cases} y \ge 2x - 3 \\ y \le -2x + 1 \end{cases}$

$y \ge 2x - 3$
$y \le -2x + 1$

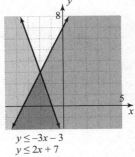

36. $\begin{cases} y \le -3x - 3 \\ y \le 2x + 7 \end{cases}$

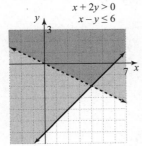

$y \le -3x - 3$
$y \le 2x + 7$

37. $\begin{cases} x + 2y > 0 \\ x - y \le 6 \end{cases}$

$x + 2y > 0$
$x - y \le 6$

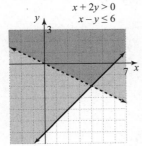

38. $\begin{cases} 4x - y \le 0 \\ 3x - 2y \ge -5 \end{cases}$

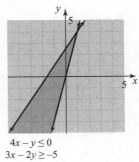

$4x - y \le 0$
$3x - 2y \ge -5$

39. $\begin{cases} 3x - 2y \le 4 \\ 2x + y \ge 5 \end{cases}$

$3x - 2y \le 4$
$2x + y \ge 5$

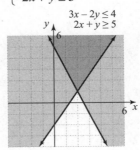

40. $\begin{cases} -2x + 3y > -7 \\ x \ge -2 \end{cases}$

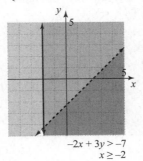

$-2x + 3y > -7$
$x \ge -2$

41. $0 \le \dfrac{2(3x + 4)}{5} \le 3$

$5(0) \le 5\left[\dfrac{2(3x + 4)}{5}\right] \le 5(3)$

$0 \le 2(3x + 4) \le 15$

$0 \le 6x + 8 \le 15$

$-8 \le 6x \le 7$

$-\dfrac{4}{3} \le x \le \dfrac{7}{6}$

$\left[-\dfrac{4}{3}, \dfrac{7}{6}\right]$

42. $x \le 2$ or $x > -5$
$(-\infty, \infty)$

43. $-2x \le 6$ and $-2x + 3 < -7$
$\quad x \ge -3$ and $-2x < -10$
$\quad x \ge -3$ and $x > 5$
$x > 5$
$(5, \infty)$

44. $|7x| - 26 = -5$
$\qquad |7x| = 21$

$7x = 21$ or $7x = -21$
$x = 3$ or $x = -3$

45. $\left|\dfrac{9 - 2x}{5}\right| = -3$
The solution set is $\varnothing$.

46. $|x - 3| = |7 + 2x|$
$x - 3 = 7 + 2x$ or $x - 3 = -(7 + 2x)$
$-10 = x$ or $x - 3 = -7 - 2x$
$\qquad\qquad\qquad\qquad 3x = -4$
$\qquad\qquad\qquad\qquad\quad x = -\dfrac{4}{3}$

47. $|6x - 5| \ge -1$
Since $|6x - 5|$ is nonnegative for all numbers x, the solution set is $(-\infty, \infty)$.

48. $\left|\dfrac{4x - 3}{5}\right| < 1$

$-1 < \dfrac{4x - 3}{5} < 1$
$-5 < 4x - 3 < 5$
$-2 < 4x < 8$
$-\dfrac{1}{2} < x < 2$
$\left(-\dfrac{1}{2}, 2\right)$

49. $-x \le y$
Test $(1, 0)$
$\qquad ?$
$-1 \le 0$
True
Shade above.

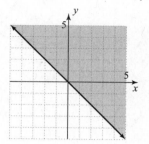

50. $x + y > -2$
Test $(0, 0)$
$\qquad ?$
$0 + 0 > -2$
True
Shade above.

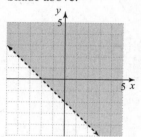

51. $\begin{cases} -3x + 2y > -1 \\ y < -2 \end{cases}$

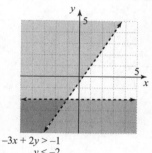

$-3x + 2y > -1$
$y < -2$

52. $\begin{cases} x - 2y \ge 7 \\ x + y \le -5 \end{cases}$

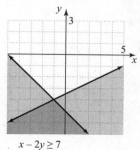

$x - 2y \ge 7$
$x + y \le -5$

Chapter 9 Test

1. $|6x - 5| - 3 = -2$

$\quad |6x - 5| = 1$

$\quad 6x - 5 = 1 \quad$ or $\quad 6x - 5 = -1$

$\qquad 6x = 6 \quad$ or $\qquad 6x = 4$

$\qquad x = 1 \quad$ or $\qquad x = \dfrac{2}{3}$

2. $|8 - 2t| = -6$
No solution, $\varnothing$

3. $|x - 5| = |x + 2|$

$\quad x - 5 = x + 2 \qquad$ or $\quad x - 5 = -(x + 2)$

$\quad -5 = 2 \quad$ False $\quad$ or $\quad x - 5 = -x - 2$

$\qquad\qquad\qquad\qquad\qquad\qquad 2x = 3$

$\qquad\qquad\qquad\qquad\qquad\qquad x = \dfrac{3}{2}$

Since $-5 = 2$ is not possible, the only solution is $\dfrac{3}{2}$.

4. $-3 < 2(x - 3) \le 4$

$\quad -3 < 2x - 6 \le 4$

$\quad\; 3 < 2x \le 10$

$\quad \dfrac{3}{2} < x \le 5$

$\quad \left(\dfrac{3}{2}, 5 \right]$

5. $|3x + 1| > 5$

$\quad 3x + 1 < -5 \quad$ or $\quad 3x + 1 > 5$

$\qquad 3x < -6 \quad$ or $\qquad 3x > 4$

$\qquad x < -2 \quad$ or $\qquad x > \dfrac{4}{3}$

$\quad (-\infty, -2) \cup \left(\dfrac{4}{3}, \infty \right)$

6. $|x - 5| - 4 < -2$

$\quad\; |x - 5| < 2$

$\quad -2 < x - 5 < 2$

$\qquad 3 < x < 7$

$\quad (3, 7)$

7. $x \le -2 \;$ and $\; x \le -5$
$(-\infty, -5]$

8. $x \le -2 \;$ or $\; x \le -5$
$(-\infty, -2]$

9. $-x > 1 \qquad$ and $\quad 3x + 3 \ge x - 3$

$\quad \dfrac{-x}{-1} < \dfrac{1}{-1} \quad$ and $\qquad 2x \ge -6$

$\qquad x < -1 \quad$ and $\qquad x \ge -3$

$\quad -3 \le x < -1$
$[-3, -1)$

10. $6x + 1 > 5x + 4 \quad$ or $\quad 1 - x > -4$

$\qquad\quad x > 3 \qquad$ or $\qquad 5 > x$

$\quad (-\infty, \infty)$

11. $\left| \dfrac{5x - 7}{2} \right| = 4$

$\quad \dfrac{5x - 7}{2} = 4 \quad$ or $\quad \dfrac{5x - 7}{2} = -4$

$\quad 5x - 7 = 8 \quad$ or $\quad 5x - 7 = -8$

$\qquad 5x = 15 \quad$ or $\qquad 5x = -1$

$\qquad x = 3 \quad$ or $\qquad x = -\dfrac{1}{5}$

12. $\left| 17x - \dfrac{1}{5} \right| > -2$

The solution set is $(-\infty, \infty)$ since an absolute value is never negative.

13. $\quad -1 \le \dfrac{2x - 5}{3} < 2$

$\quad 3(-1) \le 3\left(\dfrac{2x - 5}{3} \right) < 3(2)$

$\qquad -3 \le 2x - 5 < 6$

$\quad -3 + 5 \le 2x - 5 + 5 < 6 + 5$

$\qquad\quad 2 \le 2x < 11$

$\qquad\quad \dfrac{2}{2} \le \dfrac{2x}{2} < \dfrac{11}{2}$

$\qquad\quad 1 \le x < \dfrac{11}{2}$

$\quad \left[1, \dfrac{11}{2} \right)$

14. $y > -4x$
Test $(1, 0)$

$\qquad \overset{?}{0 > -4(1)}$

True
Shade above.

431

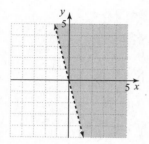

15. $2x - 3y > -6$
Test $(0, 0)$

$$2(0) - 3(0) \overset{?}{>} -6$$
True
Shade below.

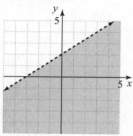

16. $\begin{cases} y + 2x \le 4 \\ y \ge 2 \end{cases}$

$y + 2x \le 4$
$y \ge 2$

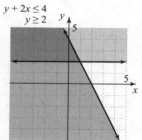

17. $\begin{cases} 2y - x \ge 1 \\ x + y \ge -4 \end{cases}$

$2y - x \ge 1$
$x + y \ge -4$

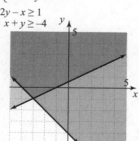

Chapter 9 Cumulative Review

1. Let $x = 2$ and $y = -5$.

    a. $\dfrac{x - y}{12 + x} = \dfrac{2 - (-5)}{12 + 2} = \dfrac{2 + 5}{14} = \dfrac{7}{14} = \dfrac{1}{2}$

    b. $x^2 - 3y = 2^2 - 3(-5) = 4 + 15 = 19$

2. Let $x = -4$ and $y = 7$.

    a. $\dfrac{x - y}{7 - x} = \dfrac{-4 - 7}{7 - (-4)} = \dfrac{-11}{7 + 4} = \dfrac{-11}{11} = -1$

    b. $x^2 + 2y = (-4)^2 + 2 \cdot 7 = 16 + 14 = 30$

3. **a.** $\dfrac{(-12)(-3) + 3}{-7 - (-2)} = \dfrac{36 + 3}{-7 + 2} = \dfrac{39}{-5} = -\dfrac{39}{5}$

    b. $\dfrac{2(-3)^2 - 20}{-5 + 4} = \dfrac{2 \cdot 9 - 20}{-1} = \dfrac{18 - 20}{-1} = \dfrac{-2}{-1} = 2$

4. **a.** $\dfrac{4(-3) - (-6)}{-8 + 4} = \dfrac{-12 + 6}{-8 + 4} = \dfrac{-6}{-4} = \dfrac{3}{2}$

    b. $\dfrac{3 + (-3)(-2)^3}{-1 - (-4)} = \dfrac{3 + (-3)(-8)}{-1 + 4}$

$$= \dfrac{3 + 24}{3}$$
$$= \dfrac{27}{3}$$
$$= 9$$

5. **a.** $2x + 3x + 5 + 2 = (2 + 3)x + (5 + 2) = 5x + 7$

    b. $-5a - 3 + a + 2 = -5a + a - 3 + 2 = -4a - 1$

    c. $4y - 3y^2$ cannot be simplified.

    d. $2.3x + 5x - 6 = 7.3x - 6$

    e. $-\dfrac{1}{2}b + b = \left(-\dfrac{1}{2} + 1\right)b = \dfrac{1}{2}b$

6. **a.** $4x - 3 + 7 - 5x = 4x - 5x - 3 + 7 = -x + 4$

    b. $-6y + 3y - 8 + 8y = -6y + 3y + 8y - 8$
$$= 5y - 8$$

    c. $2 + 8.1a + a - 6 = 8.1a + a + 2 - 6 = 9.1a - 4$

   d. $2x^2 - 2x$ cannot be simplified.

7. $2x + 3x - 5 + 7 = 10x + 3 - 6x - 4$
$$5x + 2 = 4x - 1$$
$$x + 2 = -1$$
$$x = -3$$

8. $6y - 11 + 4 + 2y = 8 + 15y - 8y$
$$8y - 7 = 7y + 8$$
$$y - 7 = 8$$
$$y = 15$$

9. $y = 3x$
$x = -1$: $y = 3(-1) = -3$
$y = 0$: $0 = 3x$
$$\frac{0}{3} = \frac{3x}{3}$$
$$0 = x$$
$y = -9$: $-9 = 3x$
$$\frac{-9}{3} = \frac{3x}{3}$$
$$-3 = x$$

x	y
-1	-3
0	0
-3	-9

10. $2x + y = 6$
$x = 0$: $2(0) + y = 6$
$$y = 6$$
$y = -2$: $2x + (-2) = 6$
$$2x - 2 = 6$$
$$2x = 8$$
$$x = 4$$
$x = 3$: $2(3) + y = 6$
$$6 + y = 6$$
$$y = 0$$

x	y
0	6
4	-2
3	0

11.   **a.** x-intercept: $(-3, 0)$; y-intercept: $(0, 2)$

     b. x-intercepts: $(-4, 0)$ and $(-1, 0)$;
         y-intercept: $(0, 1)$

     c. x-intercept: $(0, 0)$; y-intercept: $(0, 0)$

     d. x-intercept: $(2, 0)$; no y-intercept

     e. x-intercepts: $(-1, 0)$ and $(3, 0)$;
         y-intercepts: $(0, -1)$ and $(0, 2)$

12.   **a.** x-intercept: $(4, 0)$; y-intercept: $(0, 1)$

     b. x-intercepts: $(-2, 0)$, $(0, 0)$, and $(3, 0)$;
         y-intercept: $(0, 0)$

     c.   no x-intercept; y-intercept: $(0, -3)$

     d. x-intercepts: $(-3, 0)$ and $(3, 0)$;
         y-intercepts: $(0, -3)$ and $(0, 3)$

13. $y = -\dfrac{1}{5}x + 1$: $m = -\dfrac{1}{5}$, $b = 1$
$$2x + 10y = 30$$
$$10y = -2x + 30$$
$$y = -\frac{1}{5}x + 3: m = -\frac{1}{5}, b = 3$$
The slopes are the same, but the y-intercepts are different, so the lines are parallel.

14. $y = 3x + 7$: $m = 3$, $b = 7$
$$x + 3y = -15$$
$$3y = -x - 15$$
$$y = -\frac{1}{3}x - 5: m = -\frac{1}{3}, b = -5$$
The product of the slopes is -1, so the lines are perpendicular.

15. y-intercept $(0, -3)$: $b = -3$; $m = \dfrac{1}{4}$
$$y = mx + b$$
$$y = \frac{1}{4}x + (-3) \text{ or } y = \frac{1}{4}x - 3$$

16. y-intercept $(0, 4)$: $b = 4$; $m = -2$
$$y = mx + b$$
$$y = -2x + 4$$

17.   The line $y = 5$ is vertical. A parallel line will also be vertical. The vertical line passing through $(-2, -3)$ has equation $y = -3$.

18. $y = 2x + 4$: $m = 2$

A perpendicular line has slope $m = -\dfrac{1}{2}$.

$(x_1, y_1) = (1, 5)$

$y - y_1 = m(x - x_1)$

$y - 5 = -\dfrac{1}{2}(x - 1)$

$y - 5 = -\dfrac{1}{2}x + \dfrac{1}{2}$

$y = -\dfrac{1}{2}x + \dfrac{11}{2}$

19. a, b, and c are functions since they represent non-vertical lines.

20. a, c, and d are functions since they represent non-vertical lines.

21. $\begin{cases} 2x - 3y = 6 \\ x = 2y \end{cases}$

   a. Let $x = 12$ and $y = 6$.

      $2x - 3y = 6$ $x = 2y$

      $2(12) - 3(6) \overset{?}{=} 6$ $12 \overset{?}{=} 2(6)$

        $24 - 18 \overset{?}{=} 6$ $12 = 12$ True

            $6 = 6$ True

   (12, 6) is a solution.

   b. Let $x = 0$ and $y = -2$.

      $2x - 3y = 6$ $x = 2y$

      $2(0) - 3(-2) \overset{?}{=} 6$ $0 \overset{?}{=} 2(-2)$

        $0 + 6 \overset{?}{=} 6$ $0 = -4$ False

           $6 = 6$ True

   (0, −2) is not a solution.

22. $\begin{cases} 2x + y = 4 \\ x + y = 2 \end{cases}$

   a. Let $x = 1$ and $y = 1$.

      $2x + y = 4$

      $2(1) + 1 \overset{?}{=} 4$

        $2 + 1 \overset{?}{=} 4$

          $3 = 4$ False

   (1, 1) is not a solution.

   b. Let $x = 2$ and $y = 0$.

      $2x + y = 4$ $x + y = 2$

      $2(2) + 0 \overset{?}{=} 4$ $2 + 0 \overset{?}{=} 2$

        $4 + 0 \overset{?}{=} 4$ $2 = 2$ True

          $4 = 4$ True

   (2, 0) is a solution.

23. $(11x^3 - 12x^2 + x - 3) + (x^3 - 10x + 5)$

$= 11x^3 + x^3 - 12x^2 + x - 10x - 3 + 5$

$= 12x^3 - 12x^2 - 9x + 2$

24. $4a^2 + 3a - 2a^2 + 7a - 5$

$= 4a^2 - 2a^2 + 3a + 7a - 5$

$= 2a^2 + 10a - 5$

25. $x^2 + 7yx + 6y^2 = (x + 6y)(x + y)$

26. $3x^2 + 15x + 18 = 3(x^2 + 5x + 6) = 3(x + 2)(x + 3)$

27. $\dfrac{3x^3 y^7}{40} \div \dfrac{4x^3}{y^2} = \dfrac{3x^3 y^7}{40} \cdot \dfrac{y^2}{4x^3} = \dfrac{3y^9}{160}$

28. $\dfrac{12x^2 y^3}{5} \div \dfrac{3y^2}{x} = \dfrac{12x^2 y^3}{5} \cdot \dfrac{x}{3y^2} = \dfrac{4x^3 y}{5}$

29. $\dfrac{2y}{2y - 7} - \dfrac{7}{2y - 7} = \dfrac{2y - 7}{2y - 7} = 1$

30. $\dfrac{-4x^2}{x + 1} - \dfrac{4x}{x + 1} = \dfrac{-4x^2 - 4x}{x + 1} = \dfrac{-4x(x + 1)}{x + 1} = -4x$

31. $\dfrac{2x}{x^2 + 2x + 1} + \dfrac{x}{x^2 - 1} = \dfrac{2x}{(x + 1)^2} + \dfrac{x}{(x + 1)(x - 1)}$

$= \dfrac{2x(x - 1) + x(x + 1)}{(x + 1)^2 (x - 1)}$

$= \dfrac{2x^2 - 2x + x^2 + x}{(x + 1)^2 (x - 1)}$

$= \dfrac{3x^2 - x}{(x + 1)^2 (x - 1)}$

$= \dfrac{x(3x - 1)}{(x + 1)^2 (x - 1)}$

32. $\dfrac{3x}{x^2+5x+6}+\dfrac{1}{x^2+2x-3}$

$=\dfrac{3x}{(x+2)(x+3)}+\dfrac{1}{(x+3)(x-1)}$

$=\dfrac{3x(x-1)+1(x+2)}{(x+2)(x+3)(x-1)}$

$=\dfrac{3x^2-3x+x+2}{(x+2)(x+3)(x-1)}$

$=\dfrac{3x^2-2x+2}{(x+2)(x+3)(x-1)}$

33. $\dfrac{x}{2}+\dfrac{8}{3}=\dfrac{1}{6}$

$6\left(\dfrac{x}{2}+\dfrac{8}{3}\right)=6\left(\dfrac{1}{6}\right)$

$3x+16=1$

$3x=-15$

$x=-5$

34. $\dfrac{1}{21}+\dfrac{x}{7}=\dfrac{5}{3}$

$21\left(\dfrac{1}{21}+\dfrac{x}{7}\right)=21\left(\dfrac{5}{3}\right)$

$1+3x=35$

$3x=34$

$x=\dfrac{34}{3}$

35. $\begin{cases} 2x+y=7 \\ 2y=-4x \end{cases}$

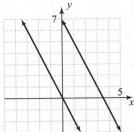

The system has no solution. The solution set is $\{\ \}$ or $\varnothing$.

36. $\begin{cases} y=x+2 \\ 2x+y=5 \end{cases}$

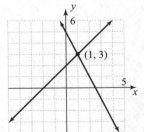

The solution is (1, 3).

37. $\begin{cases} 7x-3y=-14 \\ -3x+y=6 \end{cases}$

Solve the second equation for y.

$y=3x+6$

Substitute $3x+6$ for y in the first equation.

$7x-3(3x+6)=-14$

$7x-9x-18=-14$

$-2x-18=-14$

$-2x=4$

$x=-2$

Let $x=-2$ in $y=3x+6$.

$y=3(-2)+6=-6+6=0$

The solution is (−2, 0).

38. $\begin{cases} 5x+y=3 \\ y=-5x \end{cases}$

Substitute $-5x$ for y in the first equation.

$5x+(-5x)=3$

$0=3$

This is a false statement, so the system has no solution. The solution set is $\{\ \}$ or $\varnothing$.

39. $\begin{cases} 3x-2y=2 \\ -9x+6y=-6 \end{cases}$

Multiply the first equation by 3, then add.

$9x-6y=6$

$\underline{-9x+6y=-6}$

$0=0$

This is a true statement, so the system has an infinite number of solutions, $\{(x,y)|3x-2y=2\}$ or $\{(x,y)|-9x+6y=-6\}$.

40. $\begin{cases} -2x+y=7 \\ 6x-3y=-21 \end{cases}$

Multiply the first equation by 3, then add.

$-6x+3y=21$

$\underline{6x-3y=-21}$

$0=0$

This is a true statement, so the system has an infinite number of solutions, $\{(x, y) | -2x + y = 7\}$ or $\{(x, y) | 6x - 3y = -21\}$.

41. $\begin{cases} -3x + 4y < 12 \\ \quad\quad x \geq 2 \end{cases}$

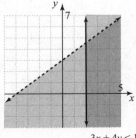

$$-3x + 4y < 12$$
$$x \geq 2$$

42. $\begin{cases} 2x - y \leq 6 \\ \quad\quad y \geq 2 \end{cases}$

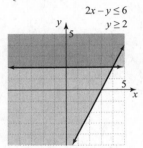

$$2x - y \leq 6$$
$$y \geq 2$$

43. a. $x^7 \cdot x^4 = x^{7+4} = x^{11}$

b. $\left(\dfrac{t}{2}\right)^4 = \dfrac{t^4}{2^4} = \dfrac{t^4}{16}$

c. $(9y^5)^2 = 9^2(y^5)^2 = 81y^{10}$

44. a. $\left(\dfrac{-6x}{y^3}\right)^3 = \dfrac{(-6)^3 x^3}{(y^3)^3} = \dfrac{-216x^3}{y^9} = -\dfrac{216x^3}{y^9}$

b. $\dfrac{(2b^2)^5}{a^2 b^7} = \dfrac{2^5(b^2)^5}{a^2 b^7}$

$$= \dfrac{32b^{10}}{a^2 b^7}$$

$$= \dfrac{32b^{10-7}}{a^2}$$

$$= \dfrac{32b^3}{a^2}$$

c. $\dfrac{(3y)^2}{y^2} = \dfrac{3^2 y^2}{y^2} = 9y^{2-2} = 9y^0 = 9$

d. $\dfrac{(x^2 y^4)^2}{xy^3} = \dfrac{(x^2)^2(y^4)^2}{xy^3}$

$$= \dfrac{x^4 y^8}{xy^3}$$

$$= x^{4-1} y^{8-3}$$

$$= x^3 y^5$$

45. $(5x - 1)(2x^2 + 15x + 18) = 0$
$(5x - 1)(2x + 3)(x + 6) = 0$
$5x - 1 = 0$ or $2x + 3 = 0$ or $x + 6 = 0$
   $5x = 1$ $2x = -3$ $x = -6$
   $x = \dfrac{1}{5}$ $x = -\dfrac{3}{2}$

The solutions are -6, $-\dfrac{3}{2}$, and $\dfrac{1}{5}$.

46. $(x + 1)(2x^2 - 3x - 5) = 0$
$(x + 1)(2x - 5)(x + 1) = 0$
$x + 1 = 0$ or $2x - 5 = 0$
   $x = -1$ $2x = 5$
                    $x = \dfrac{5}{2}$

The solutions are -1 and $\dfrac{5}{2}$.

47. $\dfrac{45}{x} = \dfrac{5}{7}$
$45 \cdot 7 = 5x$
$315 = 5x$
$\dfrac{315}{5} = \dfrac{5x}{5}$
$63 = x$

48. $\dfrac{2x + 7}{3} = \dfrac{x - 6}{2}$
$2(2x + 7) = 3(x - 6)$
$4x + 14 = 3x - 18$
$x + 14 = -18$
$x = -32$

Chapter 10

1. a. $\sqrt{49} = 7$ because $7^2 = 49$ and 7 is not negative.

b. $\sqrt{\dfrac{0}{1}} = \sqrt{0} = 0$ because $0^2 = 0$ and 0 is not negative.

c. $\sqrt{\dfrac{16}{81}} = \dfrac{4}{9}$ because $\left(\dfrac{4}{9}\right)^2 = \dfrac{16}{81}$ and $\dfrac{4}{9}$ is not negative.

d. $\sqrt{0.64} = 0.8$ because $(0.8)^2 = 0.64$.

e. $\sqrt{z^8} = z^4$ because $(z^4)^2 = z^8$.

f. $\sqrt{16b^4} = 4b^2$ because $(4b^2)^2 = 16b^4$.

g. $-\sqrt{36} = -6$. The negative in front of the radical indicates the negative square root of 36.

h. $\sqrt{-36}$ is not a real number.

2. $\sqrt{45} \approx 6.708$

Since $36 < 45 < 49$, then $\sqrt{36} < \sqrt{45} < \sqrt{49}$, or $6 < \sqrt{45} < 7$. The approximation is between 6 and 7 and thus is reasonable.

3. a. $\sqrt[3]{-1} = -1$ because $(-1)^3 = -1$.

b. $\sqrt[3]{27} = 3$ because $3^3 = 27$.

c. $\sqrt[3]{\dfrac{27}{64}} = \dfrac{3}{4}$ because $\left(\dfrac{3}{4}\right)^3 = \dfrac{27}{64}$.

d. $\sqrt[3]{x^{12}} = x^4$ because $(x^4)^3 = x^{12}$.

e. $\sqrt[3]{-8x^3} = -2x$ because $(-2x)^3 = -8x^3$.

4. a. $\sqrt[4]{10,000} = 10$ because $10^4 = 10,000$ and 10 is positive.

b. $\sqrt[5]{-1} = -1$ because $(-1)^5 = -1$.

c. $-\sqrt{81} = -9$ because -9 is the opposite of $\sqrt{81}$.

d. $\sqrt[4]{-625}$ is not a real number. There is no real number that, when raised to the fourth power, is -625.

e. $\sqrt[3]{27x^9} = 3x^3$ because $(3x^3)^3 = 27x^9$.

5. a. $\sqrt{(-4)^2} = |-4| = 4$

b. $\sqrt{x^{14}} = \left|x^7\right|$

c. $\sqrt[4]{(x+7)^4} = |x+7|$

d. $\sqrt[3]{(-7)^3} = -7$

e. $\sqrt[5]{(3x-5)^5} = 3x-5$

f. $\sqrt{49x^2} = 7|x|$

g. $\sqrt{x^2 + 16x + 64} = \sqrt{(x+8)^2} = |x+8|$

6. $f(x) = \sqrt{x+5}$, $g(x) = \sqrt[3]{x-3}$

a. $f(11) = \sqrt{11+5} = \sqrt{16} = 4$

b. $f(-1) = \sqrt{-1+5} = \sqrt{4} = 2$

c. $g(11) = \sqrt[3]{11-3} = \sqrt[3]{8} = 2$

d. $g(-6) = \sqrt[3]{-6-3} = \sqrt[3]{-9}$

7. $h(x) = \sqrt{x+2}$
Find the domain.
$x+2 \geq 0$
$x \geq -2$
The domain of $h(x)$ is $\{x \mid x \geq -2\}$.

x	$h(x) = \sqrt{x+2}$
-2	0
-1	1
1	$\sqrt{1+2} = \sqrt{3} \approx 1.7$
2	2
7	3

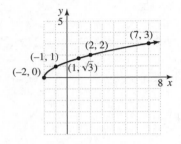

8. $f(x) = \sqrt[3]{x} - 4$

The domain is the set of all real numbers.

x	$f(x) = \sqrt[3]{x} - 4$
0	-4
1	-3
-1	-5
6	$\sqrt[3]{6} - 4 \approx 1.8 - 4 = -2.2$
-6	$\sqrt[3]{-6} - 4 \approx -1.8 - 4 = -5.8$
8	-2
-8	-6

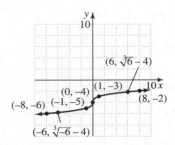

Vocabulary, Readiness & Video Check 10.1

1. In the expression $\sqrt[n]{a}$, the n is called the index, the $\sqrt{}$ is called the radical sign, and a is called the radicand.

2. If $\sqrt{a}$ is the positive square root of a, $a \neq 0$, then $-\sqrt{a}$ is the negative square root of a.

3. The square root of a negative number is not a real number.

4. Numbers such as 1, 4, 9, and 25 are called perfect squares, whereas numbers such as 1, 8, 27, and 125 are called perfect cubes.

5. The domain of the function $f(x) = \sqrt{x}$ is $[0, \infty)$.

6. The domain of the function $f(x) = \sqrt[3]{x}$ is $(-\infty, \infty)$.

7. If $f(16) = 4$, the corresponding ordered pair is (16, 4).

8. If $g(-8) = -2$, the corresponding ordered pair is (−8, −2).

9. Divide the index into each exponent in the radicand.

10. Find the nearest perfect squares both less than and greater than the radicand. The square root of the radicand falls between the square roots of these two perfect squares.

11. The square root of a negative number is not a real number, but the cube root of a negative number is a real number.

12. The even root of a negative number is not a real number.

13. For odd roots, there's only one root/answer whether the radicand is positive or negative, so absolute value bars aren't needed.

14. Since the variable x is the radicand of a square root and a square root cannot be negative, then we know that x cannot be negative.

Exercise Set 10.1

1. $\sqrt{100} = 10$ because $10^2 = 100$.

3. $\sqrt{\dfrac{1}{4}} = \dfrac{1}{2}$ because $\left(\dfrac{1}{2}\right)^2 = \dfrac{1}{4}$.

5. $\sqrt{0.0001} = 0.01$ because $(0.01)^2 = 0.0001$.

7. $-\sqrt{36} = -1 \cdot \sqrt{36} = -1 \cdot 6 = -6$

9. $\sqrt{x^{10}} = x^5$ because $(x^5)^2 = x^{10}$.

11. $\sqrt{16y^6} = 4y^3$ because $(4y^3)^2 = 16y^6$.

13. $\sqrt{7} \approx 2.646$
    Since $4 < 7 < 9$, then $\sqrt{4} < \sqrt{7} < \sqrt{9}$, or
    $2 < \sqrt{7} < 3$. The approximation is between 2 and
    3 and thus is reasonable.

15. $\sqrt{38} \approx 6.164$
    Since $36 < 38 < 49$, then $\sqrt{36} < \sqrt{38} < \sqrt{49}$, or
    $6 < \sqrt{38} < 7$. The approximation is between 6
    and 7 and thus is reasonable.

17. $\sqrt{200} \approx 14.142$
    Since $196 < 200 < 225$, then
    $\sqrt{196} < \sqrt{200} < \sqrt{225}$, or $14 < \sqrt{200} < 15$. The
    approximation is between 14 and 15 and thus is
    reasonable.

19. $\sqrt[3]{64} = 4$ because $4^3 = 64$.

21. $\sqrt[3]{\dfrac{1}{8}} = \dfrac{1}{2}$ because $\left(\dfrac{1}{2}\right)^3 = \dfrac{1}{8}$.

23. $\sqrt[3]{-1} = -1$ because $(-1)^3 = -1$.

25. $\sqrt[3]{x^{12}} = x^4$ because $(x^4)^3 = x^{12}$.

27. $\sqrt[3]{-27x^9} = -3x^3$ because $(-3x^3)^3 = -27x^9$.

29. $-\sqrt[4]{16} = -2$ because $2^4 = 16$.

31. $\sqrt[4]{-16}$ is not a real number. There is no real
    number that, when raised to the fourth power, is
    -16.

33. $\sqrt[5]{-32} = -2$ because $(-2)^5 = -32$.

35. $\sqrt[5]{x^{20}} = x^4$ because $(x^4)^5 = x^{20}$.

37. $\sqrt[6]{64x^{12}} = 2x^2$ because $(2x^2)^6 = 64x^{12}$.

39. $\sqrt{81x^4} = 9x^2$ because $(9x^2)^2 = 81x^4$.

41. $\sqrt[4]{256x^8} = 4x^2$ because $(4x^2)^4 = 256x^8$.

43. $\sqrt{(-8)^2} = |-8| = 8$

45. $\sqrt[3]{(-8)^3} = -8$

47. $\sqrt{4x^2} = |2x| = 2|x|$

49. $\sqrt[3]{x^3} = x$

51. $\sqrt{(x-5)^2} = |x-5|$

53. $\sqrt{x^2 + 4x + 4} = \sqrt{(x+2)^2} = |x+2|$

55. $-\sqrt{121} = -11$

57. $\sqrt[3]{8x^3} = 2x$

59. $\sqrt{y^{12}} = y^6$

61. $\sqrt{25a^2 b^{20}} = 5ab^{10}$

63. $\sqrt[3]{-27x^{12} y^9} = -3x^4 y^3$

65. $\sqrt[4]{a^{16} b^4} = a^4 b$

67. $\sqrt[5]{-32x^{10} y^5} = -2x^2 y$

69. $\sqrt{\dfrac{25}{49}} = \dfrac{5}{7}$

71. $\sqrt{\dfrac{x^{20}}{4y^2}} = \dfrac{x^{10}}{2y}$

73. $-\sqrt[3]{\dfrac{z^{21}}{27x^3}} = -\dfrac{z^7}{3x}$

75. $\sqrt[4]{\dfrac{x^4}{16}} = \dfrac{x}{2}$

77. $f(x) = \sqrt{2x+3}$
$f(0) = \sqrt{2(0)+3} = \sqrt{3}$

79. $g(x) = \sqrt[3]{x-8}$
$g(7) = \sqrt[3]{7-8} = \sqrt[3]{-1} = -1$

81. $g(x) = \sqrt[3]{x-8}$
$g(-19) = \sqrt[3]{-19-8} = \sqrt[3]{-27} = -3$

83. $f(x) = \sqrt{2x+3}$
$f(2) = \sqrt{2(2)+3} = \sqrt{7}$

85. $f(x) = \sqrt{x}+2$
$x \geq 0$
Domain: $[0, \infty)$

x	$f(x) = \sqrt{x}+2$
0	$\sqrt{0}+2 = 2$
1	$\sqrt{1}+2 = 3$
3	$\sqrt{3}+2 \approx 3.7$
4	$\sqrt{4}+2 = 4$

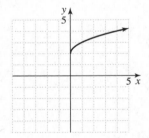

87. $f(x) = \sqrt{x-3}$
$x-3 \geq 0$
$x \geq 3$
Domain: $[3, \infty)$

x	$f(x) = \sqrt{x-3}$
3	$\sqrt{3-3} = \sqrt{0} = 0$
4	$\sqrt{4-3} = \sqrt{1} = 1$
7	$\sqrt{7-3} = \sqrt{4} = 2$
12	$\sqrt{12-3} = \sqrt{9} = 3$

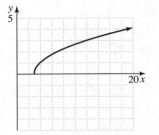

89. $f(x) = \sqrt[3]{x}+1$
Domain: $(-\infty, \infty)$

x	$f(x) = \sqrt[3]{x}+1$
-4	$\sqrt[3]{-4}+1 \approx -0.6$
-1	$\sqrt[3]{-1}+1 = 0$
0	$\sqrt[3]{0}+1 = 1$
1	$\sqrt[3]{1}+1 = 2$
4	$\sqrt[3]{4}+1 \approx 2.6$

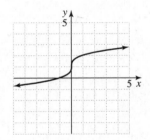

91. $g(x) = \sqrt[3]{x-1}$
Domain: $(-\infty, \infty)$

x	$g(x) = \sqrt[3]{x-1}$
1	$\sqrt[3]{1-1} = \sqrt[3]{0} = 0$
2	$\sqrt[3]{2-1} = \sqrt[3]{1} = 1$
0	$\sqrt[3]{0-1} = \sqrt[3]{-1} = -1$
9	$\sqrt[3]{9-1} = \sqrt[3]{8} = 2$
-7	$\sqrt[3]{-7-1} = \sqrt[3]{-8} = -2$

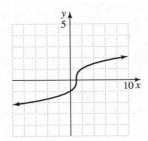

93. $(-2x^3y^2)^5 = (-2)^5 x^{3\cdot5} y^{2\cdot5} = -32x^{15}y^{10}$

95. $(-3x^2y^3z^5)(20x^5y^7) = -3(20)x^{2+5}y^{3+7}z^5$
$\qquad\qquad\qquad\qquad\qquad = -60x^7y^{10}z^5$

97. $\dfrac{7x^{-1}y}{14(x^5y^2)^{-2}} = \dfrac{7x^{-1}y}{14x^{-10}y^{-4}} = \dfrac{x^9y^5}{2}$

99. $\sqrt{-17}$ is not a real number.

101. $\sqrt[10]{-17}$ is not a real number.

103. The radical that is not a real number is $\sqrt{-10}$, choice **d**.

105. The radical that simplifies to -3 is $\sqrt[3]{-27}$, choice **d**.

107. $144 < 160 < 169$ so $\sqrt{144} < \sqrt{160} < \sqrt{169}$, or $12 < \sqrt{160} < 13$. Thus $\sqrt{160}$ is between 12 and 13. Therefore, the answer is **b**.

109. $\sqrt{30} \approx 5$, $\sqrt{10} \approx 3$, and $\sqrt{90} \approx 10$ so
$P = \sqrt{30} + \sqrt{10} + \sqrt{90} \approx 5+3+10 = 18$.
Therefore, the answer is **b**.

111. answers may vary

113. $B = \sqrt{\dfrac{hw}{3131}} = \sqrt{\dfrac{66\cdot135}{3131}}$
$\qquad = \sqrt{\dfrac{8910}{3131}}$
$\qquad \approx 1.69$ sq meters

115. $v = \sqrt{\dfrac{2Gm}{r}} = \sqrt{\dfrac{2(6.67\times10^{-11})(5.97\times10^{24})}{6.37\times10^6}}$

$\qquad = \sqrt{\dfrac{2(3.98199\times10^{14})}{6.37\times10^6}}$

$\qquad = \sqrt{\dfrac{7.96398\times10^{14}}{6.37\times10^6}}$

$\qquad \approx \sqrt{125,023,233.9}$

$\qquad \approx 11,181$

The escape velocity is 11,181 meters per second.

117. answers may vary

119. $f(x) = \sqrt{x} + 2$

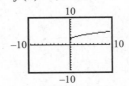

Domain: $[0, \infty)$

121. $f(x) = \sqrt[3]{x} + 1$

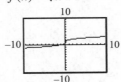

Domain: $(-\infty, \infty)$

Section 10.2 Practice

1. a. $36^{1/2} = \sqrt{36} = 6$

    b. $1000^{1/3} = \sqrt[3]{1000} = 10$

    c. $x^{1/3} = \sqrt[3]{x}$

    d. $1^{1/4} = \sqrt[4]{1} = 1$

    e. $-64^{1/2} = -\sqrt{64} = -8$

    f. $(125x^9)^{1/3} = \sqrt[3]{125x^9} = 5x^3$

    g. $3x^{1/4} = 3\sqrt[4]{x}$

2. a. $16^{3/2} = \left(\sqrt{16}\right)^3 = 4^3 = 64$

b. $-1^{3/5} = -\left(\sqrt[5]{1}\right)^3 = -(1)^3 = -1$

c. $-(81)^{3/4} = -\left(\sqrt[4]{81}\right)^3 = -(3)^3 = -27$

d. $\left(\dfrac{1}{25}\right)^{3/2} = \left(\sqrt{\dfrac{1}{25}}\right)^3 = \left(\dfrac{1}{5}\right)^3 = \dfrac{1}{125}$

e. $(3x+2)^{5/9} = \sqrt[9]{(3x+2)^5}$

3. a. $9^{-3/2} = \dfrac{1}{9^{3/2}} = \dfrac{1}{\left(\sqrt{9}\right)^3} = \dfrac{1}{3^3} = \dfrac{1}{27}$

b. $(-64)^{-2/3} = \dfrac{1}{(-64)^{2/3}} = \dfrac{1}{\left(\sqrt[3]{-64}\right)^2} = \dfrac{1}{(-4)^2} = \dfrac{1}{16}$

4. a. $y^{2/3} \cdot y^{8/3} = y^{(2/3+8/3)} = y^{10/3}$

b. $x^{3/5} \cdot x^{1/4} = x^{3/5+1/4} = x^{12/20+5/20} = x^{17/20}$

c. $\dfrac{9^{2/7}}{9^{9/7}} = 9^{2/7-9/7} = 9^{-7/7} = 9^{-1} = \dfrac{1}{9}$

d. $b^{4/9} \cdot b^{-2/9} = b^{4/9+(-2/9)} = b^{2/9}$

e. $\dfrac{\left(3x^{1/4}y^{-2/3}\right)^4}{x^4 y} = \dfrac{3^4 (x^{1/4})^4 (y^{-2/3})^4}{x^4 y}$

$= \dfrac{81xy^{-8/3}}{x^4 y}$

$= 81x^{1-4}y^{-8/3-3/3}$

$= 81x^{-3}y^{-11/3}$

$= \dfrac{81}{x^3 y^{11/3}}$

5. a. $x^{3/5}(x^{1/3} - x^2) = x^{3/5}x^{1/3} - x^{3/5}x^2$

$= x^{(3/5+1/3)} - x^{(3/5+2)}$

$= x^{(9/15+5/15)} - x^{(3/5+10/5)}$

$= x^{14/15} - x^{13/5}$

b. $(x^{1/2} + 6)(x^{1/2} - 2)$

$= x^{2/2} - 2x^{1/2} + 6x^{1/2} - 12$

$= x + 4x^{1/2} - 12$

6. $2x^{-1/5} - 7x^{4/5} = (x^{-1/5})(2) - (x^{-1/5})(7x^{5/5})$

$= x^{-1/5}(2 - 7x)$

7. a. $\sqrt[9]{x^3} = x^{3/9} = x^{1/3} = \sqrt[3]{x}$

b. $\sqrt[4]{36} = 36^{1/4} = (6^2)^{1/4} = 6^{2/4} = 6^{1/2} = \sqrt{6}$

c. $\sqrt[8]{a^4 b^2} = (a^4 b^2)^{1/8}$

$= a^{4/8}b^{2/8}$

$= a^{2/4}b^{1/4}$

$= (a^2 b)^{1/4}$

$= \sqrt[4]{a^2 b}$

8. a. $\sqrt[3]{x} \cdot \sqrt[4]{x} = x^{1/3} \cdot x^{1/4}$

$= x^{1/3+1/4}$

$= x^{4/12+3/12}$

$= x^{7/12}$

$= \sqrt[12]{x^7}$

b. $\dfrac{\sqrt[3]{y}}{\sqrt[5]{y}} = \dfrac{y^{1/3}}{y^{1/5}}$

$= y^{1/3-1/5}$

$= y^{5/15-3/15}$

$= y^{2/15}$

$= \sqrt[15]{y^2}$

c. $\sqrt[3]{5} \cdot \sqrt{3} = 5^{1/3} \cdot 3^{1/2}$

$= 5^{2/6} \cdot 3^{3/6}$

$= (5^2 \cdot 3^3)^{1/6}$

$= \sqrt[6]{5^2 \cdot 3^3}$

$= \sqrt[6]{675}$

Vocabulary, Readiness & Video Check 10.2

1. It is true that $9^{-1/2}$ is a positive number.

2. It is false that $9^{-1/2}$ is a whole number.

3. It is true that $\dfrac{1}{a^{-m/n}} = a^{m/n}$ (where $a^{m/n}$ is a nonzero real number).

4. To simplify $x^{2/3} \cdot x^{1/5}$, <u>add</u> the exponents; a.

5. To simplify $\left(x^{2/3}\right)^{1/5}$, <u>multiply</u> the exponents; c.

6. To simplify $\dfrac{x^{2/3}}{x^{1/5}}$, <u>subtract</u> the exponents; b.

7. $-\sqrt[5]{3x}$

8. The numerator is the power; the denominator is the index.

9. A negative fractional exponent will move a base from the numerator to the <u>denominator</u> with the fractional exponent becoming <u>positive</u>.

10. If applying the product rule of exponents, you <u>add</u> the exponents. If applying the quotient rule of exponents, you <u>subtract</u> the exponents. If applying the power rule of exponents, you <u>multiply</u> the exponents.

11. Write the radical using an equivalent fractional exponent form, simplify the fraction, then write as a radical again.

Exercise Set 10.2

1. $49^{1/2} = \sqrt{49} = 7$

3. $27^{1/3} = \sqrt[3]{27} = 3$

5. $\left(\dfrac{1}{16}\right)^{1/4} = \sqrt[4]{\dfrac{1}{16}} = \dfrac{1}{2}$

7. $169^{1/2} = \sqrt{169} = 13$

9. $2m^{1/3} = 2\sqrt[3]{m}$

11. $\left(9x^4\right)^{1/2} = \sqrt{9x^4} = 3x^2$

13. $(-27)^{1/3} = \sqrt[3]{-27} = -3$

15. $-16^{1/4} = -\sqrt[4]{16} = -2$

17. $16^{3/4} = \left(\sqrt[4]{16}\right)^3 = 2^3 = 8$

19. $(-64)^{2/3} = \left(\sqrt[3]{-64}\right)^2 = (-4)^2 = 16$

21. $(-16)^{3/4} = \left(\sqrt[4]{-16}\right)^3$ is not a real number.

23. $(2x)^{3/5} = \sqrt[5]{(2x)^3}$ or $\left(\sqrt[5]{2x}\right)^3$

25. $(7x+2)^{2/3} = \sqrt[3]{(7x+2)^2}$ or $\left(\sqrt[3]{7x+2}\right)^2$

27. $\left(\dfrac{16}{9}\right)^{3/2} = \left(\sqrt{\dfrac{16}{9}}\right)^3 = \left(\dfrac{4}{3}\right)^3 = \dfrac{64}{27}$

29. $8^{-4/3} = \dfrac{1}{8^{4/3}} = \dfrac{1}{\left(\sqrt[3]{8}\right)^4} = \dfrac{1}{2^4} = \dfrac{1}{16}$

31. $(-64)^{-2/3} = \dfrac{1}{(-64)^{2/3}} = \dfrac{1}{\left(\sqrt[3]{-64}\right)^2} = \dfrac{1}{(-4)^2} = \dfrac{1}{16}$

33. $(-4)^{-3/2} = \dfrac{1}{(-4)^{3/2}} = \dfrac{1}{\left(\sqrt{-4}\right)^3}$ is not a real number.

35. $x^{-1/4} = \dfrac{1}{x^{1/4}}$

37. $\dfrac{1}{a^{-2/3}} = a^{2/3}$

39. $\dfrac{5}{7x^{-3/4}} = \dfrac{5x^{3/4}}{7}$

41. $a^{2/3}a^{5/3} = a^{2/3+5/3} = a^{7/3}$

43. $x^{-2/5} \cdot x^{7/5} = x^{-\frac{2}{5}+\frac{7}{5}} = x^{5/5} = x$

45. $3^{1/4} \cdot 3^{3/8} = 3^{\frac{1}{4}+\frac{3}{8}} = 3^{\frac{2}{8}+\frac{3}{8}} = 3^{5/8}$

47. $\dfrac{y^{1/3}}{y^{1/6}} = y^{\frac{1}{3}-\frac{1}{6}} = y^{\frac{2}{6}-\frac{1}{6}} = y^{1/6}$

49. $(4u^2)^{3/2} = 4^{3/2} u^{2(3/2)}$

$\qquad = \left(\sqrt{4}\right)^3 u^3$

$\qquad = 2^3 u^3$

$\qquad = 8u^3$

51. $\dfrac{b^{1/2} b^{3/4}}{-b^{1/4}} = -b^{\frac{1}{2}+\frac{3}{4}-\frac{1}{4}} = -b^{\frac{2}{4}+\frac{3}{4}-\frac{1}{4}} = -b^1 = -b$

53. $\dfrac{(x^3)^{1/2}}{x^{7/2}} = \dfrac{x^{3/2}}{x^{7/2}}$

$\qquad = x^{3/2-7/2}$

$\qquad = x^{-2}$

$\qquad = \dfrac{1}{x^2}$

55. $\dfrac{(3x^{1/4})^3}{x^{1/12}} = \dfrac{3^3 x^{3/4}}{x^{1/12}}$

$\qquad = 27 x^{\frac{3}{4}-\frac{1}{12}}$

$\qquad = 27 x^{\frac{9}{12}-\frac{1}{12}}$

$\qquad = 27 x^{8/12}$

$\qquad = 27 x^{2/3}$

57. $\dfrac{(y^3 z)^{1/6}}{y^{-1/2} z^{1/3}} = \dfrac{y^{3/6} z^{1/6}}{y^{-1/2} z^{1/3}}$

$\qquad = y^{3/6-(-1/2)} z^{1/6-1/3}$

$\qquad = y^{1/2+1/2} z^{1/6-2/6}$

$\qquad = y^1 z^{-1/6}$

$\qquad = \dfrac{y}{z^{1/6}}$

59. $\dfrac{(x^3 y^2)^{1/4}}{(x^{-5} y^{-1})^{-1/2}} = \dfrac{x^{3/4} y^{2/4}}{x^{5/2} y^{1/2}}$

$\qquad = x^{\frac{3}{4}-\frac{5}{2}} y^{\frac{2}{4}-\frac{1}{2}}$

$\qquad = x^{\frac{3}{4}-\frac{10}{4}} y^{\frac{1}{2}-\frac{1}{2}}$

$\qquad = x^{-7/4}$

$\qquad = \dfrac{1}{x^{7/4}}$

61. $y^{1/2}(y^{1/2} - y^{2/3}) = y^{1/2} y^{1/2} - y^{1/2} y^{2/3}$

$\qquad = y^{1/2+1/2} - y^{1/2+2/3}$

$\qquad = y^1 - y^{7/6}$

$\qquad = y - y^{7/6}$

63. $x^{2/3}(x-2) = x \cdot x^{2/3} - 2x^{2/3}$

$\qquad = x^{1+2/3} - 2x^{2/3}$

$\qquad = x^{5/3} - 2x^{2/3}$

65. $(2x^{1/3} + 3)(2x^{1/3} - 3) = (2x^{1/3})^2 - 3^2$

$\qquad = 2^2 (x^{1/3})^2 - 9$

$\qquad = 4x^{2/3} - 9$

67. $x^{8/3} + x^{10/3} = x^{8/3}(1) + x^{8/3}(x^{2/3})$

$\qquad = x^{8/3}(1 + x^{2/3})$

69. $x^{2/5} - 3x^{1/5} = x^{1/5}(x^{1/5}) - x^{1/5}(3)$

$\qquad = x^{1/5}(x^{1/5} - 3)$

71. $5x^{-1/3} + x^{2/3} = x^{-1/3}(5) + x^{-1/3}(x^{3/3})$

$\qquad = x^{-1/3}(5 + x)$

73. $\sqrt[6]{x^3} = x^{3/6} = x^{1/2} = \sqrt{x}$

75. $\sqrt[6]{4} = 4^{1/6} = (2^2)^{1/6} = 2^{1/3} = \sqrt[3]{2}$

77. $\sqrt[4]{16x^2} = (16x^2)^{1/4}$

$\qquad = 16^{1/4} x^{2/4} = 2x^{1/2} = 2\sqrt{x}$

79. $\sqrt[8]{x^4 y^4} = (x^4 y^4)^{1/8}$

$\qquad = x^{4/8} y^{4/8}$

$\qquad = x^{1/2} y^{1/2}$

$\qquad = (xy)^{1/2}$

$\qquad = \sqrt{xy}$

81. $\sqrt[12]{a^8 b^4} = a^{8/12} b^{4/12}$

$\qquad = a^{2/3} b^{1/3}$

$\qquad = (a^2 b)^{1/3}$

$\qquad = \sqrt[3]{a^2 b}$

83. $\sqrt[4]{(x+3)^2} = (x+3)^{2/4} = (x+3)^{1/2} = \sqrt{x+3}$

85. $\sqrt[3]{y} \cdot \sqrt[5]{y^2} = y^{1/3} \cdot y^{2/5}$

$= y^{\frac{1}{3}+\frac{2}{5}}$

$= y^{\frac{5}{15}+\frac{6}{15}}$

$= y^{11/15}$

$= \sqrt[15]{y^{11}}$

87. $\dfrac{\sqrt[3]{b^2}}{\sqrt[4]{b}} = \dfrac{b^{2/3}}{b^{1/4}} = b^{\frac{2}{3}-\frac{1}{4}} = b^{\frac{8}{12}-\frac{3}{12}} = b^{5/12} = \sqrt[12]{b^5}$

89. $\sqrt[3]{x} \cdot \sqrt[4]{x} \cdot \sqrt[8]{x^3} = x^{1/3} \cdot x^{1/4} \cdot x^{3/8}$

$= x^{8/24} \cdot x^{6/24} \cdot x^{9/24}$

$= x^{23/24}$

$= \sqrt[24]{x^{23}}$

91. $\dfrac{\sqrt[3]{a^2}}{\sqrt[6]{a}} = \dfrac{a^{2/3}}{a^{1/6}}$

$= a^{\frac{2}{3}-\frac{1}{6}} = a^{\frac{4}{6}-\frac{1}{6}} = a^{3/6} = a^{1/2} = \sqrt{a}$

93. $\sqrt{3} \cdot \sqrt[3]{4} = 3^{1/2} \cdot 4^{1/3}$

$= 3^{3/6} \cdot 4^{2/6}$

$= (3^3 \cdot 4^2)^{1/6}$

$= (27 \cdot 16)^{1/6}$

$= (432)^{1/6}$

$= \sqrt[6]{432}$

95. $\sqrt[5]{7} \cdot \sqrt[3]{y} = 7^{1/5} \cdot y^{1/3}$

$= 7^{3/15} \cdot y^{5/15}$

$= (7^3 \cdot y^5)^{1/15}$

$= (343y^5)^{1/15}$

$= \sqrt[15]{343y^5}$

97. $\sqrt{5r} \cdot \sqrt[3]{s} = (5r)^{1/2} \cdot s^{1/3}$

$= (5r)^{3/6} \cdot s^{2/6}$

$= [(5r)^3 \cdot s^2]^{1/6}$

$= (125r^3 s^2)^{1/6}$

$= \sqrt[6]{125r^3 s^2}$

99. $75 = 25 \cdot 3$ where 25 is a perfect square.

101. $48 = 4 \cdot 12$ or $16 \cdot 3$ where both 4 and 16 are perfect squares.

103. $16 = 8 \cdot 2$ where 8 is a perfect cube.

105. $54 = 27 \cdot 2$ where 27 is a perfect cube.

107. $4^{1/2} = 2$, A

109. $(-4)^{1/2}$ is not a real number, C

111. $-8^{1/3} = -2$, B

113. $B(w) = 70w^{3/4}$

$B(60) = 70(60)^{3/4}$

≈ 1509

The BMR is 1509 calories.

115. 2010 is 15 years after 1995.

$f(x) = 25x^{23/25}$

$f(15) = 25(15)^{23/25}$

≈ 302.0

The model predicts 302.0 million subscriptions in 2010.

117. answers may vary

119. $\square \cdot a^{2/3} = a^{3/3}$

$\square = \dfrac{a^{3/3}}{a^{2/3}}$

$\square = a^{3/3-2/3}$

$\square = a^{1/3}$

121. $\dfrac{\square}{x^{-2/5}} = x^{3/5}$

$x^{-2/5}\left(\dfrac{\square}{x^{-2/5}}\right) = x^{3/5} \cdot x^{-2/5}$

$\square = x^{3/5-2/5}$

$\square = x^{1/5}$

123. $8^{1/4} \approx 1.6818$

125. $18^{3/5} \approx 5.6645$

127. $\dfrac{\sqrt{t}}{\sqrt{u}} = \dfrac{t^{1/2}}{u^{1/2}}$

Section 10.3 Practice

1. a. $\sqrt{5} \cdot \sqrt{7} = \sqrt{5 \cdot 7} = \sqrt{35}$

b. $\sqrt{13} \cdot \sqrt{z} = \sqrt{13z}$

c. $\sqrt[4]{125} \cdot \sqrt[4]{5} = \sqrt[4]{125 \cdot 5} = \sqrt[4]{625} = 5$

d. $\sqrt[3]{5y} \cdot \sqrt[3]{3x^2} = \sqrt[3]{5y \cdot 3x^2} = \sqrt[3]{15x^2 y}$

e. $\sqrt{\dfrac{5}{m}} \cdot \sqrt{\dfrac{t}{2}} = \sqrt{\dfrac{5}{m} \cdot \dfrac{t}{2}} = \sqrt{\dfrac{5t}{2m}}$

2. a. $\sqrt{\dfrac{36}{49}} = \dfrac{\sqrt{36}}{\sqrt{49}} = \dfrac{6}{7}$

b. $\sqrt{\dfrac{z}{16}} = \dfrac{\sqrt{z}}{\sqrt{16}} = \dfrac{\sqrt{z}}{4}$

c. $\sqrt[3]{\dfrac{125}{8}} = \dfrac{\sqrt[3]{125}}{\sqrt[3]{8}} = \dfrac{5}{2}$

d. $\sqrt[4]{\dfrac{5}{81x^8}} = \dfrac{\sqrt[4]{5}}{\sqrt[4]{81x^8}} = \dfrac{\sqrt[4]{5}}{3x^2}$

3. a. $\sqrt{98} = \sqrt{49 \cdot 2} = \sqrt{49} \cdot \sqrt{2} = 7\sqrt{2}$

b. $\sqrt[3]{54} = \sqrt[3]{27 \cdot 2} = \sqrt[3]{27} \cdot \sqrt[3]{2} = 3\sqrt[3]{2}$

c. The largest perfect square factor of 35 is 1, so $\sqrt{35}$ cannot be simplified further.

d. $\sqrt[4]{243} = \sqrt[4]{81 \cdot 3} = \sqrt[4]{81} \cdot \sqrt[4]{3} = 3\sqrt[4]{3}$

4. a. $\sqrt{36z^7} = \sqrt{36z^6 \cdot z} = \sqrt{36z^6} \cdot \sqrt{z} = 6z^3\sqrt{z}$

b. $\sqrt[3]{32p^4q^7} = \sqrt[3]{8 \cdot 4 \cdot p^3 \cdot p \cdot q^6 \cdot q}$

$= \sqrt[3]{8p^3q^6 \cdot 4pq}$

$= \sqrt[3]{8p^3q^6} \cdot \sqrt[3]{4pq}$

$= 2pq^2\sqrt[3]{4pq}$

c. $\sqrt[4]{16x^{15}} = \sqrt[4]{16 \cdot x^{12} \cdot x^3}$

$= \sqrt[4]{16x^{12}} \cdot \sqrt[4]{x^3}$

$= 2x^3\sqrt[4]{x^3}$

5. a. $\dfrac{\sqrt{80}}{\sqrt{5}} = \sqrt{\dfrac{80}{5}} = \sqrt{16} = 4$

b. $\dfrac{\sqrt{98z}}{3\sqrt{2}} = \dfrac{1}{3} \cdot \sqrt{\dfrac{98z}{2}}$

$= \dfrac{1}{3} \cdot \sqrt{49z}$

$= \dfrac{1}{3} \cdot \sqrt{49} \cdot \sqrt{z}$

$= \dfrac{1}{3} \cdot 7 \cdot \sqrt{z}$

$= \dfrac{7}{3}\sqrt{z}$

c. $\dfrac{5\sqrt[3]{40x^5y^7}}{\sqrt[3]{5y}} = 5 \cdot \sqrt[3]{\dfrac{40x^5y^7}{5y}}$

$= 5 \cdot \sqrt[3]{8x^5y^6}$

$= 5 \cdot \sqrt[3]{8x^3y^6 \cdot x^2}$

$= 5 \cdot \sqrt[3]{8x^3y^6} \cdot \sqrt[3]{x^2}$

$= 5 \cdot 2xy^2 \cdot \sqrt[3]{x^2}$

$= 10xy^2\sqrt[3]{x^2}$

d. $\dfrac{3\sqrt[5]{64x^9y^8}}{\sqrt[5]{x^{-1}y^2}} = 3 \cdot \sqrt[5]{\dfrac{64x^9y^8}{x^{-1}y^2}}$

$= 3 \cdot \sqrt[5]{64x^{10}y^6}$

$= 3 \cdot \sqrt[5]{32 \cdot x^{10} \cdot y^5 \cdot 2 \cdot y}$

$= 3 \cdot \sqrt[5]{32x^{10}y^5} \cdot \sqrt[5]{2y}$

$= 3 \cdot 2x^2y \cdot \sqrt[5]{2y}$

$= 6x^2y\sqrt[5]{2y}$

6. Let $(x_1,\ y_1) = (-3, 7)$ and $(x_2,\ y_2) = (-2, 3)$.

$d = \sqrt{(x_2 - x_1)^2 + (y_2 - y_1)^2}$

$= \sqrt{[-2 - (-3)]^2 + (3 - 7)^2}$

$= \sqrt{(1)^2 + (-4)^2}$

$= \sqrt{1 + 16}$

$= \sqrt{17} \approx 4.123$

The distance between the two points is exactly $\sqrt{17}$ units, or approximately 4.123 units.

7. Let $(x_1, y_1) = (5, -2)$ and $(x_2, y_2) = (8, -6)$.

$$\text{midpoint} = \left(\frac{x_1 + x_2}{2}, \frac{y_1 + y_2}{2} \right)$$

$$= \left(\frac{5+8}{2}, \frac{-2+(-6)}{2} \right)$$

$$= \left(\frac{13}{2}, \frac{-8}{2} \right)$$

$$= \left(\frac{13}{2}, -4 \right)$$

The midpoint of the segment is $\left(\frac{13}{2}, -4 \right)$.

Vocabulary, Readiness & Video Check 10.3

1. The <u>midpoint</u> of a line segment is a <u>point</u> exactly halfway between the two endpoints of the line segment.

2. The <u>distance</u> between two points is a distance, measured in units.

3. The <u>distance</u> formula is

$$d = \sqrt{(x_2 - x_1)^2 + (y_2 - y_1)^2}.$$

4. The <u>midpoint</u> formula is $\left(\frac{x_1 + x_2}{2}, \frac{y_1 + y_2}{2} \right)$.

5. the indexes must be the same

6. If you see that simplifying can be done by separating a fraction radical into separate numerator and denominator radicands or by combining separate numerator and denominator radicands under one radical.

7. The power must be 1. Any even power is a perfect square and will leave no factor in the radicand; any higher odd power can have an even power factored from it, leaving one factor remaining in the radicand.

8. Be careful of signs since you're dealing with subtraction.

9. The x-value of the midpoint is the <u>average</u> of the x-values of the endpoints and the y-value of the midpoint is the <u>average</u> of the y-values of the endpoints.

Exercise Set 10.3

1. $\sqrt{7} \cdot \sqrt{2} = \sqrt{7 \cdot 2} = \sqrt{14}$

3. $\sqrt[4]{8} \cdot \sqrt[4]{2} = \sqrt[4]{8 \cdot 2} = \sqrt[4]{16} = 2$

5. $\sqrt[3]{4} \cdot \sqrt[3]{9} = \sqrt[3]{4 \cdot 9} = \sqrt[3]{36}$

7. $\sqrt{2} \cdot \sqrt{3x} = \sqrt{2 \cdot 3x} = \sqrt{6x}$

9. $\sqrt{\frac{7}{x}} \cdot \sqrt{\frac{2}{y}} = \sqrt{\frac{7}{x} \cdot \frac{2}{y}} = \sqrt{\frac{14}{xy}}$

11. $\sqrt[4]{4x^3} \cdot \sqrt[4]{5} = \sqrt[4]{4x^3 \cdot 5} = \sqrt[4]{20x^3}$

13. $\sqrt{\frac{6}{49}} = \frac{\sqrt{6}}{\sqrt{49}} = \frac{\sqrt{6}}{7}$

15. $\sqrt{\frac{2}{49}} = \frac{\sqrt{2}}{\sqrt{49}} = \frac{\sqrt{2}}{7}$

17. $\sqrt[4]{\frac{x^3}{16}} = \frac{\sqrt[4]{x^3}}{\sqrt[4]{16}} = \frac{\sqrt[4]{x^3}}{2}$

19. $\sqrt[3]{\frac{4}{27}} = \frac{\sqrt[3]{4}}{\sqrt[3]{27}} = \frac{\sqrt[3]{4}}{3}$

21. $\sqrt[4]{\frac{8}{x^8}} = \frac{\sqrt[4]{8}}{\sqrt[4]{x^8}} = \frac{\sqrt[4]{8}}{x^2}$

23. $\sqrt[3]{\frac{2x}{81y^{12}}} = \frac{\sqrt[3]{2x}}{\sqrt[3]{81y^{12}}}$

$$= \frac{\sqrt[3]{2x}}{\sqrt[3]{27y^{12}} \cdot \sqrt[3]{3}}$$

$$= \frac{\sqrt[3]{2x}}{3y^4 \sqrt[3]{3}}$$

25. $\sqrt{\frac{x^2 y}{100}} = \frac{\sqrt{x^2 y}}{\sqrt{100}} = \frac{\sqrt{x^2} \sqrt{y}}{10} = \frac{x\sqrt{y}}{10}$

27. $\sqrt{\frac{5x^2}{4y^2}} = \frac{\sqrt{5x^2}}{\sqrt{4y^2}} = \frac{\sqrt{5}\sqrt{x^2}}{2y} = \frac{x\sqrt{5}}{2y}$

29. $-\sqrt[3]{\frac{z^7}{27x^3}} = -\frac{\sqrt[3]{z^7}}{\sqrt[3]{27x^3}} = -\frac{\sqrt[3]{z^6 \cdot z}}{3x} = -\frac{z^2 \sqrt[3]{z}}{3x}$

31. $\sqrt{32} = \sqrt{16 \cdot 2} = \sqrt{16} \cdot \sqrt{2} = 4\sqrt{2}$

33. $\sqrt[3]{192} = \sqrt[3]{64 \cdot 3} = \sqrt[3]{64} \cdot \sqrt[3]{3} = 4\sqrt[3]{3}$

35. $5\sqrt{75} = 5\sqrt{25 \cdot 3} = 5\sqrt{25} \cdot \sqrt{3} = 5(5)\sqrt{3} = 25\sqrt{3}$

37. $\sqrt{24} = \sqrt{4 \cdot 6} = \sqrt{4} \cdot \sqrt{6} = 2\sqrt{6}$

39. $\sqrt{100x^5} = \sqrt{100x^4 \cdot x} = \sqrt{100x^4} \cdot \sqrt{x} = 10x^2\sqrt{x}$

41. $\sqrt[3]{16y^7} = \sqrt[3]{8y^6 \cdot 2y} = \sqrt[3]{8y^6} \cdot \sqrt[3]{2y} = 2y^2\sqrt[3]{2y}$

43. $\sqrt[4]{a^8b^7} = \sqrt[4]{a^8b^4 \cdot b^3} = \sqrt[4]{a^8b^4} \cdot \sqrt[4]{b^3} = a^2b\sqrt[4]{b^3}$

45. $\sqrt{y^5} = \sqrt{y^4 \cdot y} = \sqrt{y^4} \cdot \sqrt{y} = y^2\sqrt{y}$

47. $\sqrt{25a^2b^3} = \sqrt{25a^2b^2 \cdot b}$
$= \sqrt{25a^2b^2} \cdot \sqrt{b}$
$= 5ab\sqrt{b}$

49. $\sqrt[5]{-32x^{10}y} = \sqrt[5]{-32x^{10} \cdot y}$
$= \sqrt[5]{-32x^{10}} \cdot \sqrt[5]{y}$
$= -2x^2\sqrt[5]{y}$

51. $\sqrt[3]{50x^{14}} = \sqrt[3]{x^{12} \cdot 50x^2}$
$= \sqrt[3]{x^{12}} \cdot \sqrt[3]{50x^2}$
$= x^4\sqrt[3]{50x^2}$

53. $-\sqrt{32a^8b^7} = -\sqrt{16a^8b^6 \cdot 2b}$
$= -\sqrt{16a^8b^6} \cdot \sqrt{2b}$
$= -4a^4b^3\sqrt{2b}$

55. $\sqrt{9x^7y^9} = \sqrt{9x^6y^8 \cdot xy}$
$= \sqrt{9x^6y^8} \cdot \sqrt{xy}$
$= 3x^3y^4\sqrt{xy}$

57. $\sqrt[3]{125r^9s^{12}} = 5r^3s^4$

59. $\sqrt[4]{32x^{12}y^5} = \sqrt[4]{16x^{12}y^4} \cdot \sqrt[4]{2y} = 2x^3y\sqrt[4]{2y}$

61. $\dfrac{\sqrt{14}}{\sqrt{7}} = \sqrt{\dfrac{14}{7}} = \sqrt{2}$

63. $\dfrac{\sqrt[3]{24}}{\sqrt[3]{3}} = \sqrt[3]{\dfrac{24}{3}} = \sqrt[3]{8} = 2$

65. $\dfrac{5\sqrt[4]{48}}{\sqrt[4]{3}} = 5\sqrt[4]{\dfrac{48}{3}} = 5\sqrt[4]{16} = 5(2) = 10$

67. $\dfrac{\sqrt{x^5y^3}}{\sqrt{xy}} = \sqrt{\dfrac{x^5y^3}{xy}} = \sqrt{x^4y^2} = x^2y$

69. $\dfrac{8\sqrt[3]{54m^7}}{\sqrt[3]{2m}} = 8\sqrt[3]{\dfrac{54m^7}{2m}}$
$= 8\sqrt[3]{27m^6}$
$= 8(3m^2)$
$= 24m^2$

71. $\dfrac{3\sqrt{100x^2}}{2\sqrt{2x^{-1}}} = \dfrac{3}{2}\sqrt{\dfrac{100x^2}{2x^{-1}}}$
$= \dfrac{3}{2}\sqrt{50x^3}$
$= \dfrac{3}{2}\sqrt{25x^2 \cdot 2x}$
$= \dfrac{3}{2}(5x)\sqrt{2x}$
$= \dfrac{15x}{2}\sqrt{2x}$

73. $\dfrac{\sqrt[4]{96a^{10}b^3}}{\sqrt[4]{3a^2b^3}} = \sqrt[4]{\dfrac{96a^{10}b^3}{3a^2b^3}}$
$= \sqrt[4]{32a^8}$
$= \sqrt[4]{16a^8 \cdot 2}$
$= 2a^2\sqrt[4]{2}$

75. $\dfrac{\sqrt[5]{64x^{10}y^3}}{\sqrt[5]{2x^3y^{-7}}} = \sqrt[5]{\dfrac{64x^{10}y^3}{2x^3y^{-7}}}$
$= \sqrt[5]{32x^7y^{10}}$
$= \sqrt[5]{32x^5y^{10}} \cdot \sqrt[5]{x^2}$
$= 2xy^2\sqrt[5]{x^2}$

77. (5, 1), (8, 5)

$d = \sqrt{(8-5)^2 + (5-1)^2}$

$= \sqrt{3^2 + 4^2}$

$= \sqrt{9 + 16}$

$= \sqrt{25}$

$= 5$ units

79. (−3, 2), (1, −3)

$d = \sqrt{[1-(-3)]^2 + (-3-2)^2}$

$= \sqrt{4^2 + (-5)^2}$

$= \sqrt{16 + 25}$

$= \sqrt{41} \approx 6.403$ units

81. (−9, 4), (−8, 1)

$d = \sqrt{[-8-(-9)]^2 + (1-4)^2}$

$= \sqrt{1^2 + (-3)^2}$

$= \sqrt{1 + 9}$

$= \sqrt{10} \approx 3.162$ units

83. $\left(0, -\sqrt{2}\right), \left(\sqrt{3}, 0\right)$

$d = \sqrt{\left(\sqrt{3}-0\right)^2 + \left[0-\left(-\sqrt{2}\right)\right]^2}$

$= \sqrt{\left(\sqrt{3}\right)^2 + \left(\sqrt{2}\right)^2}$

$= \sqrt{3 + 2}$

$= \sqrt{5} \approx 2.236$ units

85. (1.7, −3.6), (−8.6, 5.7)

$d = \sqrt{(-8.6-1.7)^2 + [5.7-(-3.6)]^2}$

$= \sqrt{(-10.3)^2 + (9.3)^2}$

$= \sqrt{192.58} \approx 13.877$ units

87. (6, −8), (2, 4)

$\left(\dfrac{6+2}{2}, \dfrac{-8+4}{2}\right) = \left(\dfrac{8}{2}, \dfrac{-4}{2}\right) = (4, -2)$

The midpoint of the segment is (4, −2).

89. (−2, −1), (−8, 6)

$\left(\dfrac{-2+(-8)}{2}, \dfrac{-1+6}{2}\right) = \left(\dfrac{-10}{2}, \dfrac{5}{2}\right) = \left(-5, \dfrac{5}{2}\right)$

The midpoint of the segment is $\left(-5, \dfrac{5}{2}\right)$.

91. (7, 3), (−1, −3)

$\left(\dfrac{7+(-1)}{2}, \dfrac{3+(-3)}{2}\right) = \left(\dfrac{6}{2}, \dfrac{0}{2}\right) = (3, 0)$

The midpoint of the segment is (3, 0).

93. $\left(\dfrac{1}{2}, \dfrac{3}{8}\right), \left(-\dfrac{3}{2}, \dfrac{5}{8}\right)$

$\left(\dfrac{\frac{1}{2}+\left(-\frac{3}{2}\right)}{2}, \dfrac{\frac{3}{8}+\frac{5}{8}}{2}\right) = \left(\dfrac{-1}{2}, \dfrac{1}{2}\right)$

The midpoint of the segment is $\left(-\dfrac{1}{2}, \dfrac{1}{2}\right)$.

95. $\left(\sqrt{2}, 3\sqrt{5}\right), \left(\sqrt{2}, -2\sqrt{5}\right)$

$\left(\dfrac{\sqrt{2}+\sqrt{2}}{2}, \dfrac{3\sqrt{5}+\left(-2\sqrt{5}\right)}{2}\right) = \left(\dfrac{2\sqrt{2}}{2}, \dfrac{\sqrt{5}}{2}\right)$

$= \left(\sqrt{2}, \dfrac{\sqrt{5}}{2}\right)$

The midpoint of the segment is $\left(\sqrt{2}, \dfrac{\sqrt{5}}{2}\right)$.

97. (4.6, −3.5), (7.8, −9.8)

$\left(\dfrac{4.6+7.8}{2}, \dfrac{-3.5+(-9.8)}{2}\right) = \left(\dfrac{12.4}{2}, \dfrac{-13.3}{2}\right)$

$= (6.2, -6.65)$

The midpoint of the segment is (6.2, −6.65).

99. $6x + 8x = (6+8)x = 14x$

101. $(2x+3)(x-5) = 2x^2 - 10x + 3x - 15$

$= 2x^2 - 7x - 15$

103. $9y^2 - 8y^2 = (9-8)y^2 = 1y^2 = y^2$

105. $-3(x+5) = -3x - 3(5) = -3x - 15$

107. $(x-4)^2 = x^2 - 2(x)(4) + 4^2$

$= x^2 - 8x + 16$

109. The statement $\sqrt[n]{a} \cdot \sqrt[n]{b} = \sqrt[n]{ab}$ is <u>true</u>.

111. The statement $\sqrt[3]{7} \cdot \sqrt{11} = \sqrt{77}$ is <u>false</u>.

113. The statement $\dfrac{\sqrt[n]{a}}{\sqrt[n]{b}} = \sqrt[n]{\dfrac{a}{b}}$ is <u>true</u>.

115. $\dfrac{\sqrt[3]{64}}{\sqrt{64}} = \dfrac{4}{8} = \dfrac{1}{2}$

117. $\sqrt[5]{x^{35}} = x^7$

119. $\sqrt[4]{a^{12}b^4c^{20}} = a^3bc^5$

121. $\sqrt[3]{z^{32}} = \sqrt[3]{z^{30} \cdot z^2} = \sqrt[3]{z^{30}} \cdot \sqrt[3]{z^2} = z^{10}\sqrt[3]{z^2}$

123. $\sqrt[7]{q^{17}r^{40}s^7} = \sqrt[7]{q^{14} \cdot q^3 \cdot r^{35} \cdot r^5 \cdot s^7}$
$\qquad = \sqrt[7]{q^{14}r^{35}s^7 \cdot q^3r^5}$
$\qquad = q^2r^5s\sqrt[7]{q^3r^5}$

125. $r = \sqrt{\dfrac{A}{4\pi}} = \sqrt{\dfrac{32.17}{4\pi}} \approx \sqrt{2.56} = 1.6$
The radius of a standard zorb is 1.6 meters.

127. $A = \pi r\sqrt{r^2 + h^2}$

   a. $A = \pi(4)\sqrt{4^2 + 3^2}$
$\qquad = 4\pi\sqrt{16 + 9}$
$\qquad = 4\pi\sqrt{25}$
$\qquad = 4\pi(5)$
$\qquad = 20\pi$
The area is 20π square centimeters.

   b. $A = \pi(6.8)\sqrt{(6.8)^2 + (7.2)^2}$
$\qquad = 6.8\pi\sqrt{46.24 + 51.84}$
$\qquad \doteq 6.8\pi\sqrt{98.08}$
$\qquad \approx 211.57$
The area is approximately 211.57 square feet.

Section 10.4 Practice

 1. a. $3\sqrt{17} + 5\sqrt{17} = (3+5)\sqrt{17} = 8\sqrt{17}$

   b. $7\sqrt[3]{5z} - 12\sqrt[3]{5z} = (7-12)\sqrt[3]{5z} = -5\sqrt[3]{5z}$

   c. $3\sqrt{2} + 5\sqrt[3]{2}$
This expression cannot be simplified since $3\sqrt{2}$ and $5\sqrt[3]{2}$ do not contain like radicals.

2. a. $\sqrt{24} + 3\sqrt{54} = \sqrt{4 \cdot 6} + 3\sqrt{9 \cdot 6}$
$\qquad = \sqrt{4} \cdot \sqrt{6} + 3 \cdot \sqrt{9} \cdot \sqrt{6}$
$\qquad = 2 \cdot \sqrt{6} + 3 \cdot 3 \cdot \sqrt{6}$
$\qquad = 2\sqrt{6} + 9\sqrt{6}$
$\qquad = 11\sqrt{6}$

   b. $\sqrt[3]{24} - 4\sqrt[3]{81} + \sqrt[3]{3}$
$\qquad = \sqrt[3]{8} \cdot \sqrt[3]{3} - 4 \cdot \sqrt[3]{27} \cdot \sqrt[3]{3} + \sqrt[3]{3}$
$\qquad = 2 \cdot \sqrt[3]{3} - 4 \cdot 3 \cdot \sqrt[3]{3} + \sqrt[3]{3}$
$\qquad = 2\sqrt[3]{3} - 12\sqrt[3]{3} + \sqrt[3]{3}$
$\qquad = -9\sqrt[3]{3}$

   c. $\sqrt{75x} - 3\sqrt{27x} + \sqrt{12x}$
$\qquad = \sqrt{25} \cdot \sqrt{3x} - 3 \cdot \sqrt{9} \cdot \sqrt{3x} + \sqrt{4} \cdot \sqrt{3x}$
$\qquad = 5 \cdot \sqrt{3x} - 3 \cdot 3 \cdot \sqrt{3x} + 2 \cdot \sqrt{3x}$
$\qquad = 5\sqrt{3x} - 9\sqrt{3x} + 2\sqrt{3x}$
$\qquad = -2\sqrt{3x}$

   d. $\sqrt{40} + \sqrt[3]{40} = \sqrt{4} \cdot \sqrt{10} + \sqrt[3]{8} \cdot \sqrt[3]{5}$
$\qquad\qquad\qquad = 2\sqrt{10} + 2\sqrt[3]{5}$

   e. $\sqrt[3]{81x^4} + \sqrt[3]{3x^4} = \sqrt[3]{27x^3} \cdot \sqrt[3]{3x} + \sqrt[3]{x^3} \cdot \sqrt[3]{3x}$
$\qquad\qquad\qquad = 3x\sqrt[3]{3x} + x\sqrt[3]{3x}$
$\qquad\qquad\qquad = 4x\sqrt[3]{3x}$

3. a. $\dfrac{\sqrt{28}}{3} - \dfrac{\sqrt{7}}{4} = \dfrac{2\sqrt{7}}{3} - \dfrac{\sqrt{7}}{4}$
$\qquad\qquad = \dfrac{2\sqrt{7} \cdot 4}{3 \cdot 4} - \dfrac{\sqrt{7} \cdot 3}{4 \cdot 3}$
$\qquad\qquad = \dfrac{8\sqrt{7}}{12} - \dfrac{3\sqrt{7}}{12}$
$\qquad\qquad = \dfrac{5\sqrt{7}}{12}$

   b. $\sqrt[3]{\dfrac{6y}{64}} + 3\sqrt[3]{6y} = \dfrac{\sqrt[3]{6y}}{\sqrt[3]{64}} + 3\sqrt[3]{6y}$
$\qquad\qquad\qquad = \dfrac{\sqrt[3]{6y}}{4} + 3\sqrt[3]{6y}$
$\qquad\qquad\qquad = \dfrac{\sqrt[3]{6y}}{4} + \dfrac{3\sqrt[3]{6y} \cdot 4}{4}$
$\qquad\qquad\qquad = \dfrac{\sqrt[3]{6y}}{4} + \dfrac{12\sqrt[3]{6y}}{4}$
$\qquad\qquad\qquad = \dfrac{13\sqrt[3]{6y}}{4}$

4. a.
$$\sqrt{5}(2+\sqrt{15}) = \sqrt{5}(2) + \sqrt{5}(\sqrt{15})$$
$$= 2\sqrt{5} + \sqrt{5 \cdot 15}$$
$$= 2\sqrt{5} + \sqrt{5 \cdot 5 \cdot 3}$$
$$= 2\sqrt{5} + 5\sqrt{3}$$

b.
$$(\sqrt{2} - \sqrt{5})(\sqrt{6} + 2)$$
$$= \sqrt{2} \cdot \sqrt{6} + \sqrt{2} \cdot 2 - \sqrt{5} \cdot \sqrt{6} - \sqrt{5} \cdot 2$$
$$= \sqrt{2 \cdot 2 \cdot 3} + 2\sqrt{2} - \sqrt{30} - 2\sqrt{5}$$
$$= 2\sqrt{3} + 2\sqrt{2} - \sqrt{30} - 2\sqrt{5}$$

c.
$$(3\sqrt{z} - 4)(2\sqrt{z} + 3)$$
$$= 3\sqrt{z}(2\sqrt{z}) + 3\sqrt{z}(3) - 4(2\sqrt{z}) - 4(3)$$
$$= 6 \cdot z + 9\sqrt{z} - 8\sqrt{z} - 12$$
$$= 6z + \sqrt{z} - 12$$

d.
$$(\sqrt{6} - 3)^2 = (\sqrt{6} - 3)(\sqrt{6} - 3)$$
$$= \sqrt{6}(\sqrt{6}) - \sqrt{6}(3) - 3(\sqrt{6}) - 3(-3)$$
$$= 6 - 3\sqrt{6} - 3\sqrt{6} + 9$$
$$= 6 - 6\sqrt{6} + 9$$
$$= 15 - 6\sqrt{6}$$

e.
$$(\sqrt{5x} + 3)(\sqrt{5x} - 3)$$
$$= \sqrt{5x} \cdot \sqrt{5x} - 3\sqrt{5x} + 3\sqrt{5x} - 3 \cdot 3$$
$$= 5x - 9$$

f.
$$(\sqrt{x+2} + 3)^2 = (\sqrt{x+2})^2 + 2 \cdot \sqrt{x+2} \cdot 3 + 3^2$$
$$= x + 2 + 6\sqrt{x+2} + 9$$
$$= x + 11 + 6\sqrt{x+2}$$

Vocabulary, Readiness & Video Check 10.4

1. The terms $\sqrt{7}$ and $\sqrt[3]{7}$ are <u>unlike</u> terms.

2. The terms $\sqrt[3]{x^2 y}$ and $\sqrt[3]{yx^2}$ are <u>like</u> terms.

3. The terms $\sqrt[3]{abc}$ and $\sqrt[3]{cba}$ are <u>like</u> terms.

4. The terms $2x\sqrt{5}$ and $2x\sqrt{10}$ are <u>unlike</u> terms.

5. $2\sqrt{3} + 4\sqrt{3} = \underline{6\sqrt{3}}$

6. $5\sqrt{7} + 3\sqrt{7} = \underline{8\sqrt{7}}$

7. $8\sqrt{x} - \sqrt{x} = \underline{7\sqrt{x}}$

8. $3\sqrt{y} - \sqrt{y} = \underline{2\sqrt{y}}$

9. $7\sqrt[3]{x} + \sqrt[3]{x} = \underline{8\sqrt[3]{x}}$

10. $8\sqrt[3]{z} + \sqrt[3]{z} = \underline{9\sqrt[3]{z}}$

11. Sometimes you can't see that there are like radicals until you simplify, so you may incorrectly think you cannot add or subtract if you don't simplify first.

12. The square root of a positive number times the square root of the same positive number is that positive number.

Exercise Set 10.4

1.
$$\sqrt{8} - \sqrt{32} = \sqrt{4 \cdot 2} - \sqrt{16 \cdot 2}$$
$$= \sqrt{4} \cdot \sqrt{2} - \sqrt{16} \cdot \sqrt{2}$$
$$= 2\sqrt{2} - 4\sqrt{2}$$
$$= -2\sqrt{2}$$

3.
$$2\sqrt{2x^3} + 4x\sqrt{8x} = 2\sqrt{x^2 \cdot 2x} + 4x\sqrt{4 \cdot 2x}$$
$$= 2\sqrt{x^2} \cdot \sqrt{2x} + 4x\sqrt{4} \cdot \sqrt{2x}$$
$$= 2x\sqrt{2x} + 4x(2)\sqrt{2x}$$
$$= 2x\sqrt{2x} + 8x\sqrt{2x}$$
$$= 10x\sqrt{2x}$$

5.
$$2\sqrt{50} - 3\sqrt{125} + \sqrt{98}$$
$$= 2\sqrt{25 \cdot 2} - 3\sqrt{25 \cdot 5} + \sqrt{49 \cdot 2}$$
$$= 2\sqrt{25} \cdot \sqrt{2} - 3\sqrt{25} \cdot \sqrt{5} + \sqrt{49} \cdot \sqrt{2}$$
$$= 2(5)\sqrt{2} - 3(5)\sqrt{5} + 7\sqrt{2}$$
$$= 10\sqrt{2} - 15\sqrt{5} + 7\sqrt{2}$$
$$= 17\sqrt{2} - 15\sqrt{5}$$

7.
$$\sqrt[3]{16x} - \sqrt[3]{54x} = \sqrt[3]{8 \cdot 2x} - \sqrt[3]{27 \cdot 2x}$$
$$= \sqrt[3]{8} \cdot \sqrt[3]{2x} - \sqrt[3]{27} \cdot \sqrt[3]{2x}$$
$$= 2\sqrt[3]{2x} - 3\sqrt[3]{2x}$$
$$= -\sqrt[3]{2x}$$

9.
$$\sqrt{9b^3} - \sqrt{25b^3} + \sqrt{49b^3}$$
$$= \sqrt{9b^2 \cdot b} - \sqrt{25b^2 \cdot b} + \sqrt{49b^2 \cdot b}$$
$$= \sqrt{9b^2} \cdot \sqrt{b} - \sqrt{25b^2} \cdot \sqrt{b} + \sqrt{49b^2} \cdot \sqrt{b}$$
$$= 3b\sqrt{b} - 5b\sqrt{b} + 7b\sqrt{b}$$
$$= 5b\sqrt{b}$$

11. $\dfrac{5\sqrt{2}}{3}+\dfrac{2\sqrt{2}}{5}=\dfrac{5\left(5\sqrt{2}\right)+3\left(2\sqrt{2}\right)}{3(5)}$

$=\dfrac{25\sqrt{2}+6\sqrt{2}}{15}$

$=\dfrac{31\sqrt{2}}{15}$

13. $\sqrt[3]{\dfrac{11}{8}}-\dfrac{\sqrt[3]{11}}{6}=\dfrac{\sqrt[3]{11}}{\sqrt[3]{8}}-\dfrac{\sqrt[3]{11}}{6}$

$=\dfrac{\sqrt[3]{11}}{2}-\dfrac{\sqrt[3]{11}}{6}$

$=\dfrac{3\sqrt[3]{11}-\sqrt[3]{11}}{6}$

$=\dfrac{2\sqrt[3]{11}}{6}$

$=\dfrac{\sqrt[3]{11}}{3}$

15. $\dfrac{\sqrt{20x}}{9}+\sqrt{\dfrac{5x}{9}}=\dfrac{\sqrt{4\cdot5x}}{9}+\dfrac{\sqrt{5x}}{\sqrt{9}}$

$=\dfrac{2\sqrt{5x}}{9}+\dfrac{\sqrt{5x}}{3}$

$=\dfrac{2\sqrt{5x}+3\sqrt{5x}}{9}$

$=\dfrac{5\sqrt{5x}}{9}$

17. $7\sqrt{9}-7+\sqrt{3}=7(3)-7+\sqrt{3}$

$=21-7+\sqrt{3}$

$=14+\sqrt{3}$

19. $2+3\sqrt{y^2}-6\sqrt{y^2}+5=2+3y-6y+5$

$=7-3y$

21. $3\sqrt{108}-2\sqrt{18}-3\sqrt{48}$

$=3\sqrt{36\cdot3}-2\sqrt{9\cdot2}-3\sqrt{16\cdot3}$

$=3\sqrt{36}\cdot\sqrt{3}-2\sqrt{9}\cdot\sqrt{2}-3\sqrt{16}\cdot\sqrt{3}$

$=3(6)\sqrt{3}-2(3)\sqrt{2}-3(4)\sqrt{3}$

$=18\sqrt{3}-6\sqrt{2}-12\sqrt{3}$

$=6\sqrt{3}-6\sqrt{2}$

23. $-5\sqrt[3]{625}+\sqrt[3]{40}=-5\sqrt[3]{125\cdot5}+\sqrt[3]{8\cdot5}$

$=-5(5)\sqrt[3]{5}+2\sqrt[3]{5}$

$=-25\sqrt[3]{5}+2\sqrt[3]{5}$

$=-23\sqrt[3]{5}$

25. $a^3\sqrt{9ab^3}-\sqrt{25a^7b^3}+\sqrt{16a^7b^3}$

$=a^3\sqrt{9b^2\cdot ab}-\sqrt{25a^6b^2\cdot ab}+\sqrt{16a^6b^2\cdot ab}$

$=a^3\cdot3b\sqrt{ab}-5a^3b\sqrt{ab}+4a^3b\sqrt{ab}$

$=3a^3b\sqrt{ab}-5a^3b\sqrt{ab}+4a^3b\sqrt{ab}$

$=2a^3b\sqrt{ab}$

27. $5y\sqrt{8y}+2\sqrt{50y^3}=5y\sqrt{4\cdot2y}+2\sqrt{25y^2\cdot2y}$

$=5y(2)\sqrt{2y}+2(5y)\sqrt{2y}$

$=10y\sqrt{2y}+10y\sqrt{2y}$

$=20y\sqrt{2y}$

29. $\sqrt[3]{54xy^3}-5\sqrt[3]{2xy^3}+y\sqrt[3]{128x}$

$=\sqrt[3]{27y^3\cdot2x}-5\sqrt[3]{y^3\cdot2x}+y\sqrt[3]{64\cdot2x}$

$=3y\sqrt[3]{2x}-5y\sqrt[3]{2x}+4y\sqrt[3]{2x}$

$=2y\sqrt[3]{2x}$

31. $6\sqrt[3]{11}+8\sqrt{11}-12\sqrt{11}=6\sqrt[3]{11}-4\sqrt{11}$

33. $-2\sqrt[4]{x^7}+3\sqrt[4]{16x^7}-x\sqrt[4]{x^3}$

$=-2\sqrt[4]{x^4\cdot x^3}+3\sqrt[4]{16x^4\cdot x^3}-x\sqrt[4]{x^3}$

$=-2x\sqrt[4]{x^3}+3(2x)\sqrt[4]{x^3}-x\sqrt[4]{x^3}$

$=-2x\sqrt[4]{x^3}+6x\sqrt[4]{x^3}-x\sqrt[4]{x^3}$

$=3x\sqrt[4]{x^3}$

35. $\dfrac{4\sqrt{3}}{3}-\dfrac{\sqrt{12}}{3}=\dfrac{4\sqrt{3}}{3}-\dfrac{\sqrt{4\cdot3}}{3}$

$=\dfrac{4\sqrt{3}-2\sqrt{3}}{3}$

$=\dfrac{2\sqrt{3}}{3}$

37. $\dfrac{\sqrt[3]{8x^4}}{7}+\dfrac{3x\sqrt[3]{x}}{7}=\dfrac{\sqrt[3]{8x^3\cdot x}}{7}+\dfrac{3x\sqrt[3]{x}}{7}$

$=\dfrac{2x\sqrt[3]{x}+3x\sqrt[3]{x}}{7}$

$=\dfrac{5x\sqrt[3]{x}}{7}$

39. $\sqrt{\dfrac{28}{x^2}} + \sqrt{\dfrac{7}{4x^2}} = \dfrac{\sqrt{28}}{\sqrt{x^2}} + \dfrac{\sqrt{7}}{\sqrt{4x^2}}$

$\qquad = \dfrac{\sqrt{4\cdot7}}{x} + \dfrac{\sqrt{7}}{2x}$

$\qquad = \dfrac{2\sqrt{7}}{x} + \dfrac{\sqrt{7}}{2x}$

$\qquad = \dfrac{2\left(2\sqrt{7}\right) + \sqrt{7}}{2x}$

$\qquad = \dfrac{4\sqrt{7} + \sqrt{7}}{2x}$

$\qquad = \dfrac{5\sqrt{7}}{2x}$

41. $\sqrt[3]{\dfrac{16}{27}} - \dfrac{\sqrt[3]{54}}{6} = \dfrac{\sqrt[3]{8\cdot2}}{\sqrt[3]{27}} - \dfrac{\sqrt[3]{27\cdot2}}{6}$

$\qquad = \dfrac{2\sqrt[3]{2}}{3} - \dfrac{3\sqrt[3]{2}}{6}$

$\qquad = \dfrac{2\left(2\sqrt[3]{2}\right) - 3\sqrt[3]{2}}{6}$

$\qquad = \dfrac{4\sqrt[3]{2} - 3\sqrt[3]{2}}{6}$

$\qquad = \dfrac{\sqrt[3]{2}}{6}$

43. $-\dfrac{\sqrt[3]{2x^4}}{9} + \sqrt[3]{\dfrac{250x^4}{27}} = -\dfrac{\sqrt[3]{x^3\cdot2x}}{9} + \dfrac{\sqrt[3]{125x^3\cdot2x}}{\sqrt[3]{27}}$

$\qquad = \dfrac{-x\sqrt[3]{2x}}{9} + \dfrac{5x\sqrt[3]{2x}}{3}$

$\qquad = \dfrac{-x\sqrt[3]{2x} + 3\left(5x\sqrt[3]{2x}\right)}{9}$

$\qquad = \dfrac{-x\sqrt[3]{2x} + 15x\sqrt[3]{2x}}{9}$

$\qquad = \dfrac{14x\sqrt[3]{2x}}{9}$

45. $P = 2\sqrt{12} + \sqrt{12} + 2\sqrt{27} + 3\sqrt{3}$

$\qquad = 2\sqrt{4\cdot3} + \sqrt{4\cdot3} + 2\sqrt{9\cdot3} + 3\sqrt{3}$

$\qquad = 2(2)\sqrt{3} + 2\sqrt{3} + 2(3)\sqrt{3} + 3\sqrt{3}$

$\qquad = 4\sqrt{3} + 2\sqrt{3} + 6\sqrt{3} + 3\sqrt{3}$

$\qquad = 15\sqrt{3}$ inches

47. $\sqrt{7}\left(\sqrt{5} + \sqrt{3}\right) = \sqrt{7}\sqrt{5} + \sqrt{7}\sqrt{3}$

$\qquad = \sqrt{35} + \sqrt{21}$

49. $\left(\sqrt{5} - \sqrt{2}\right)^2 = \left(\sqrt{5}\right)^2 - 2\sqrt{5}\sqrt{2} + \left(\sqrt{2}\right)^2$

$\qquad = 5 - 2\sqrt{10} + 2$

$\qquad = 7 - 2\sqrt{10}$

51. $\sqrt{3x}\left(\sqrt{3} - \sqrt{x}\right) = \sqrt{3x}\sqrt{3} - \sqrt{3x}\sqrt{x}$

$\qquad = \sqrt{9x} - \sqrt{3x^2}$

$\qquad = 3\sqrt{x} - x\sqrt{3}$

53. $\left(2\sqrt{x} - 5\right)\left(3\sqrt{x} + 1\right)$

$\qquad = 2\sqrt{x}\left(3\sqrt{x}\right) + 2\sqrt{x}\cdot1 - 5\left(3\sqrt{x}\right) - 5(1)$

$\qquad = 6x + 2\sqrt{x} - 15\sqrt{x} - 5$

$\qquad = 6x - 13\sqrt{x} - 5$

55. $\left(\sqrt[3]{a} - 4\right)\left(\sqrt[3]{a} + 5\right)$

$\qquad = \sqrt[3]{a}\left(\sqrt[3]{a}\right) + \sqrt[3]{a}\cdot5 - 4\sqrt[3]{a} - 4(5)$

$\qquad = \sqrt[3]{a^2} + 5\sqrt[3]{a} - 4\sqrt[3]{a} - 20$

$\qquad = \sqrt[3]{a^2} + \sqrt[3]{a} - 20$

57. $6\left(\sqrt{2} - 2\right) = 6\sqrt{2} - 6(2) = 6\sqrt{2} - 12$

59. $\sqrt{2}\left(\sqrt{2} + x\sqrt{6}\right) = \sqrt{2}\sqrt{2} + \sqrt{2}\left(x\sqrt{6}\right)$

$\qquad = 2 + x\sqrt{12}$

$\qquad = 2 + x\sqrt{4\cdot3}$

$\qquad = 2 + 2x\sqrt{3}$

61. $\left(2\sqrt{7} + 3\sqrt{5}\right)\left(\sqrt{7} - 2\sqrt{5}\right)$

$\qquad = 2\sqrt{7}\sqrt{7} + 2\sqrt{7}\left(-2\sqrt{5}\right) + 3\sqrt{5}\sqrt{7} + 3\sqrt{5}\left(-2\sqrt{5}\right)$

$\qquad = 2(7) - 4\sqrt{35} + 3\sqrt{35} - 6(5)$

$\qquad = 14 - \sqrt{35} - 30$

$\qquad = -16 - \sqrt{35}$

63. $\left(\sqrt{x} - y\right)\left(\sqrt{x} + y\right) = \left(\sqrt{x}\right)^2 - y^2 = x - y^2$

65. $\left(\sqrt{3} + x\right)^2 = \left(\sqrt{3}\right)^2 + 2\sqrt{3}\cdot x + x^2$

$\qquad = 3 + 2x\sqrt{3} + x^2$

67. $\left(\sqrt{5x}-2\sqrt{3x}\right)\left(\sqrt{5x}-3\sqrt{3x}\right)$

$=\left(\sqrt{5x}\right)^2 -\sqrt{5x}\left(3\sqrt{3x}\right)-2\sqrt{3x}\left(\sqrt{5x}\right)$
$\qquad\qquad\qquad -2\sqrt{3x}\left(-3\sqrt{3x}\right)$

$=5x-3x\sqrt{15}-2x\sqrt{15}+6\cdot 3x$

$=23x-5x\sqrt{15}$

69. $\left(\sqrt[3]{4}+2\right)\left(\sqrt[3]{2}-1\right)$

$=\sqrt[3]{4}\left(\sqrt[3]{2}\right)+\sqrt[3]{4}\cdot(-1)+2\sqrt[3]{2}+2(-1)$

$=\sqrt[3]{8}-\sqrt[3]{4}+2\sqrt[3]{2}-2$

$=2-\sqrt[3]{4}+2\sqrt[3]{2}-2$

$=2\sqrt[3]{2}-\sqrt[3]{4}$

71. $\left(\sqrt[3]{x}+1\right)\left(\sqrt[3]{x^2}-\sqrt[3]{x}+1\right)$

$=\sqrt[3]{x}\left(\sqrt[3]{x^2}\right)-\sqrt[3]{x}\left(\sqrt[3]{x}\right)+\sqrt[3]{x}(1)$

$\qquad +1\left(\sqrt[3]{x^2}\right)-1\left(\sqrt[3]{x}\right)+1(1)$

$=\sqrt[3]{x^3}-\sqrt[3]{x^2}+\sqrt[3]{x}+\sqrt[3]{x^2}-\sqrt[3]{x}+1$

$=x+1$

73. $\left(\sqrt{x-1}+5\right)^2 =\left(\sqrt{x-1}\right)^2 +2\sqrt{x-1}\cdot 5+5^2$

$\qquad\qquad =(x-1)+10\sqrt{x-1}+25$

$\qquad\qquad =x+10\sqrt{x-1}+24$

75. $\left(\sqrt{2x+5}-1\right)^2 =\left(\sqrt{2x+5}\right)^2 -2\sqrt{2x+5}\cdot 1+1^2$

$\qquad\qquad =(2x+5)-2\sqrt{2x+5}+1$

$\qquad\qquad =2x-2\sqrt{2x+5}+6$

77. $\dfrac{2x-14}{2}=\dfrac{2(x-7)}{2}=x-7$

79. $\dfrac{7x-7y}{x^2-y^2}=\dfrac{7(x-y)}{(x+y)(x-y)}=\dfrac{7}{x+y}$

81. $\dfrac{6a^2b-9ab}{3ab}=\dfrac{3ab(2a-3)}{3ab}=2a-3$

83. $\dfrac{-4+2\sqrt{3}}{6}=\dfrac{2\left(-2+\sqrt{3}\right)}{6}=\dfrac{-2+\sqrt{3}}{3}$

85. $P=2l+2w$

$=2\left(3\sqrt{20}\right)+2\left(\sqrt{125}\right)$

$=6\sqrt{4\cdot 5}+2\sqrt{25\cdot 5}$

$=6(2)\sqrt{5}+2(5)\sqrt{5}$

$=12\sqrt{5}+10\sqrt{5}$

$=22\sqrt{5}$ feet

$A=lw$

$=\left(3\sqrt{20}\right)\left(\sqrt{125}\right)$

$=3\sqrt{4\cdot 5}\sqrt{25\cdot 5}$

$=3(2)\sqrt{5}\cdot 5\sqrt{5}$

$=30\cdot 5$

$=150$ square feet

87. a. $\sqrt{3}+\sqrt{3}=2\sqrt{3}$

   b. $\sqrt{3}\cdot\sqrt{3}=\sqrt{9}=3$

   c. answers may vary

89. $\left(\sqrt{2}+\sqrt{3}-1\right)^2$

$=\left[\left(\sqrt{2}+\sqrt{3}\right)-1\right]^2$

$=\left(\sqrt{2}+\sqrt{3}\right)^2 -2\left(\sqrt{2}+\sqrt{3}\right)+1^2$

$=\left(\sqrt{2}\right)^2 +2\sqrt{2}\sqrt{3}+\left(\sqrt{3}\right)^2 -2\sqrt{2}-2\sqrt{3}+1$

$=2+2\sqrt{6}+3-2\sqrt{2}-2\sqrt{3}+1$

$=6+2\sqrt{6}-2\sqrt{2}-2\sqrt{3}$

91. answer may vary

Section 10.5 Practice

1. a. $\dfrac{5}{\sqrt{3}}=\dfrac{5\cdot\sqrt{3}}{\sqrt{3}\cdot\sqrt{3}}=\dfrac{5\sqrt{3}}{3}$

   b. $\dfrac{3\sqrt{25}}{\sqrt{4x}}=\dfrac{3(5)}{2\sqrt{x}}=\dfrac{15}{2\sqrt{x}}=\dfrac{15\cdot\sqrt{x}}{2\sqrt{x}\cdot\sqrt{x}}=\dfrac{15\sqrt{x}}{2x}$

   c. $\sqrt[3]{\dfrac{2}{9}}=\dfrac{\sqrt[3]{2}}{\sqrt[3]{9}}=\dfrac{\sqrt[3]{2}\cdot\sqrt[3]{3}}{\sqrt[3]{3^2}\cdot\sqrt[3]{3}}=\dfrac{\sqrt[3]{6}}{3}$

2. $\sqrt{\dfrac{3z}{5y}}=\dfrac{\sqrt{3z}}{\sqrt{5y}}=\dfrac{\sqrt{3z}\cdot\sqrt{5y}}{\sqrt{5y}\cdot\sqrt{5y}}=\dfrac{\sqrt{15yz}}{5y}$

3. $\dfrac{\sqrt[3]{z^2}}{\sqrt[3]{27x^4}} = \dfrac{\sqrt[3]{z^2}}{\sqrt[3]{27x^3} \cdot \sqrt[3]{x}}$

$\qquad\qquad = \dfrac{\sqrt[3]{z^2}}{3x\sqrt[3]{x}}$

$\qquad\qquad = \dfrac{\sqrt[3]{z^2} \cdot \sqrt[3]{x^2}}{3x\sqrt[3]{x} \cdot \sqrt[3]{x^2}}$

$\qquad\qquad = \dfrac{\sqrt[3]{z^2 x^2}}{3x\sqrt[3]{x^3}}$

$\qquad\qquad = \dfrac{\sqrt[3]{x^2 z^2}}{3x^2}$

4. a. $\dfrac{5}{3\sqrt{5}+2} = \dfrac{5\left(3\sqrt{5}-2\right)}{\left(3\sqrt{5}+2\right)\left(3\sqrt{5}-2\right)}$

$\qquad\qquad = \dfrac{5\left(3\sqrt{5}-2\right)}{\left(3\sqrt{5}\right)^2 - 2^2}$

$\qquad\qquad = \dfrac{5\left(3\sqrt{5}-2\right)}{45-4}$

$\qquad\qquad = \dfrac{5\left(3\sqrt{5}-2\right)}{41}$

b. $\dfrac{\sqrt{2}+5}{\sqrt{3}-\sqrt{5}} = \dfrac{\left(\sqrt{2}+5\right)\left(\sqrt{3}+\sqrt{5}\right)}{\left(\sqrt{3}-\sqrt{5}\right)\left(\sqrt{3}+\sqrt{5}\right)}$

$\qquad\qquad = \dfrac{\sqrt{2}\sqrt{3}+\sqrt{2}\sqrt{5}+5\sqrt{3}+5\sqrt{5}}{\left(\sqrt{3}\right)^2 - \left(\sqrt{5}\right)^2}$

$\qquad\qquad = \dfrac{\sqrt{6}+\sqrt{10}+5\sqrt{3}+5\sqrt{5}}{3-5}$

$\qquad\qquad = \dfrac{\sqrt{6}+\sqrt{10}+5\sqrt{3}+5\sqrt{5}}{-2}$

c. $\dfrac{3\sqrt{x}}{2\sqrt{x}+\sqrt{y}} = \dfrac{3\sqrt{x}\left(2\sqrt{x}-\sqrt{y}\right)}{\left(2\sqrt{x}+\sqrt{y}\right)\left(2\sqrt{x}-\sqrt{y}\right)}$

$\qquad\qquad = \dfrac{6\sqrt{x^2}-3\sqrt{xy}}{\left(2\sqrt{x}\right)^2 - \left(\sqrt{y}\right)^2}$

$\qquad\qquad = \dfrac{6x-3\sqrt{xy}}{4x-y}$

5. $\dfrac{\sqrt{32}}{\sqrt{80}} = \dfrac{\sqrt{16\cdot2}}{\sqrt{16\cdot5}} = \dfrac{4\sqrt{2}}{4\sqrt{5}} = \dfrac{\sqrt{2}}{\sqrt{5}} = \dfrac{\sqrt{2}\cdot\sqrt{2}}{\sqrt{5}\cdot\sqrt{2}} = \dfrac{2}{\sqrt{10}}$

6. $\dfrac{\sqrt[3]{5b}}{\sqrt[3]{2a}} = \dfrac{\sqrt[3]{5b}\cdot\sqrt[3]{25b^2}}{\sqrt[3]{2a}\cdot\sqrt[3]{25b^2}} = \dfrac{\sqrt[3]{125b^3}}{\sqrt[3]{50ab^2}} = \dfrac{5b}{\sqrt[3]{50ab^2}}$

7. $\dfrac{\sqrt{x}-3}{4} = \dfrac{\left(\sqrt{x}-3\right)\left(\sqrt{x}+3\right)}{4\left(\sqrt{x}+3\right)}$

$\qquad\qquad = \dfrac{\left(\sqrt{x}\right)^2 - (3)^2}{4\left(\sqrt{x}+3\right)}$

$\qquad\qquad = \dfrac{x-9}{4\left(\sqrt{x}+3\right)}$

Vocabulary, Readiness & Video Check 10.5

1. The <u>conjugate</u> of $a + b$ is $a - b$.

2. The process of writing an equivalent expression, but without a radical in the denominator, is called <u>rationalizing the denominator</u>.

3. The process of writing an equivalent expression, but without a radical in the numerator, is called <u>rationalizing the numerator</u>.

4. To rationalize the denominator of $\dfrac{5}{\sqrt{3}}$, we

multiply by $\dfrac{\sqrt{3}}{\sqrt{3}}$.

5. To write an equivalent expression without a radical in the denominator.

6. Using the FOIL order to multiply, the Outer product and the Inner product are opposites and they will subtract out.

7. No, except for the fact you're working with numerators, the process is the same.

Exercise Set 10.5

1. $\dfrac{\sqrt{2}}{\sqrt{7}} = \dfrac{\sqrt{2}\cdot\sqrt{7}}{\sqrt{7}\cdot\sqrt{7}} = \dfrac{\sqrt{14}}{\sqrt{49}} = \dfrac{\sqrt{14}}{7}$

3. $\sqrt{\dfrac{1}{5}} = \dfrac{\sqrt{1}}{\sqrt{5}} = \dfrac{1\cdot\sqrt{5}}{\sqrt{5}\cdot\sqrt{5}} = \dfrac{\sqrt{5}}{5}$

5. $\sqrt{\dfrac{4}{x}} = \dfrac{\sqrt{4}}{\sqrt{x}} = \dfrac{2\cdot\sqrt{x}}{\sqrt{x}\cdot\sqrt{x}} = \dfrac{2\sqrt{x}}{\sqrt{x^2}} = \dfrac{2\sqrt{x}}{x}$

7. $\dfrac{4}{\sqrt[3]{3}} = \dfrac{4 \cdot \sqrt[3]{9}}{\sqrt[3]{3} \cdot \sqrt[3]{9}} = \dfrac{4\sqrt[3]{9}}{\sqrt[3]{27}} = \dfrac{4\sqrt[3]{9}}{3}$

9. $\dfrac{3}{\sqrt{8x}} = \dfrac{3 \cdot \sqrt{2x}}{\sqrt{8x} \cdot \sqrt{2x}} = \dfrac{3\sqrt{2x}}{\sqrt{16x^2}} = \dfrac{3\sqrt{2x}}{4x}$

11. $\dfrac{3}{\sqrt[3]{4x^2}} = \dfrac{3 \cdot \sqrt[3]{2x}}{\sqrt[3]{4x^2} \cdot \sqrt[3]{2x}} = \dfrac{3\sqrt[3]{2x}}{\sqrt[3]{8x^3}} = \dfrac{3\sqrt[3]{2x}}{2x}$

13. $\dfrac{9}{\sqrt{3a}} = \dfrac{9 \cdot \sqrt{3a}}{\sqrt{3a} \cdot \sqrt{3a}} = \dfrac{9\sqrt{3a}}{3a} = \dfrac{3\sqrt{3a}}{a}$

15. $\dfrac{3}{\sqrt[3]{2}} = \dfrac{3 \cdot \sqrt[3]{4}}{\sqrt[3]{2} \cdot \sqrt[3]{4}} = \dfrac{3\sqrt[3]{4}}{\sqrt[3]{8}} = \dfrac{3\sqrt[3]{4}}{2}$

17. $\dfrac{2\sqrt{3}}{\sqrt{7}} = \dfrac{2\sqrt{3} \cdot \sqrt{7}}{\sqrt{7} \cdot \sqrt{7}} = \dfrac{2\sqrt{21}}{\sqrt{49}} = \dfrac{2\sqrt{21}}{7}$

19. $\sqrt{\dfrac{2x}{5y}} = \dfrac{\sqrt{2x}}{\sqrt{5y}} = \dfrac{\sqrt{2x} \cdot \sqrt{5y}}{\sqrt{5y} \cdot \sqrt{5y}} = \dfrac{\sqrt{10xy}}{5y}$

21. $\sqrt[3]{\dfrac{3}{5}} = \dfrac{\sqrt[3]{3}}{\sqrt[3]{5}} \cdot \dfrac{\sqrt[3]{25}}{\sqrt[3]{25}} = \dfrac{\sqrt[3]{75}}{5}$

23. $\sqrt{\dfrac{3x}{50}} = \dfrac{\sqrt{3x}}{\sqrt{50}}$

$= \dfrac{\sqrt{3x}}{5\sqrt{2}}$

$= \dfrac{\sqrt{3x} \cdot \sqrt{2}}{5\sqrt{2} \cdot \sqrt{2}}$

$= \dfrac{\sqrt{6x}}{5 \cdot 2}$

$= \dfrac{\sqrt{6x}}{10}$

25. $\dfrac{1}{\sqrt{12z}} = \dfrac{1}{\sqrt{4 \cdot 3z}} = \dfrac{1}{2\sqrt{3z}} \cdot \dfrac{\sqrt{3z}}{\sqrt{3z}} = \dfrac{\sqrt{3z}}{6z}$

27. $\dfrac{\sqrt[3]{2y^2}}{\sqrt[3]{9x^2}} = \dfrac{\sqrt[3]{2y^2} \cdot \sqrt[3]{3x}}{\sqrt[3]{9x^2} \cdot \sqrt[3]{3x}} = \dfrac{\sqrt[3]{6xy^2}}{3x}$

29. $\sqrt[4]{\dfrac{81}{8}} = \dfrac{\sqrt[4]{81}}{\sqrt[4]{8}} = \dfrac{3 \cdot \sqrt[4]{2}}{\sqrt[4]{8} \cdot \sqrt[4]{2}} = \dfrac{3\sqrt[4]{2}}{\sqrt[4]{16}} = \dfrac{3\sqrt[4]{2}}{2}$

31. $\sqrt[4]{\dfrac{16}{9x^7}} = \dfrac{\sqrt[4]{16}}{\sqrt[4]{9x^7}} = \dfrac{2 \cdot \sqrt[4]{9x}}{\sqrt[4]{9x^7} \cdot \sqrt[4]{9x}} = \dfrac{2\sqrt[4]{9x}}{\sqrt[4]{81x^8}} = \dfrac{2\sqrt[4]{9x}}{3x^2}$

33. $\dfrac{5a}{\sqrt[5]{8a^9b^{11}}} = \dfrac{5a \cdot \sqrt[5]{4ab^4}}{\sqrt[5]{8a^9b^{11}} \cdot \sqrt[5]{4ab^4}}$

$= \dfrac{5a\sqrt[5]{4ab^4}}{\sqrt[5]{32a^{10}b^{15}}}$

$= \dfrac{5a\sqrt[5]{4ab^4}}{2a^2b^3}$

$= \dfrac{5\sqrt[5]{4ab^4}}{2ab^3}$

35. The conjugate of $\sqrt{2} + x$ is $\sqrt{2} - x$.

37. The conjugate of $5 - \sqrt{a}$ is $5 + \sqrt{a}$.

39. The conjugate of $-7\sqrt{5} + 8\sqrt{x}$ is $-7\sqrt{5} - 8\sqrt{x}$.

41. $\dfrac{6}{2 - \sqrt{7}} = \dfrac{6\left(2 + \sqrt{7}\right)}{\left(2 - \sqrt{7}\right)\left(2 + \sqrt{7}\right)}$

$= \dfrac{6\left(2 + \sqrt{7}\right)}{2^2 - \left(\sqrt{7}\right)^2}$

$= \dfrac{6\left(2 + \sqrt{7}\right)}{4 - 7}$

$= \dfrac{6\left(2 + \sqrt{7}\right)}{-3}$

$= -2\left(2 + \sqrt{7}\right)$

43. $\dfrac{-7}{\sqrt{x} - 3} = \dfrac{-7\left(\sqrt{x} + 3\right)}{\left(\sqrt{x} - 3\right)\left(\sqrt{x} + 3\right)}$

$= \dfrac{-7\left(\sqrt{x} + 3\right)}{\left(\sqrt{x}\right)^2 - (3)^2}$

$= \dfrac{-7\left(\sqrt{x} + 3\right)}{x - 9}$ or $\dfrac{7\left(\sqrt{x} + 3\right)}{9 - x}$

45. $\dfrac{\sqrt{2}-\sqrt{3}}{\sqrt{2}+\sqrt{3}} = \dfrac{\left(\sqrt{2}-\sqrt{3}\right)\left(\sqrt{2}-\sqrt{3}\right)}{\left(\sqrt{2}+\sqrt{3}\right)\left(\sqrt{2}-\sqrt{3}\right)}$

$\quad = \dfrac{\left(\sqrt{2}\right)^2 - 2\sqrt{2}\sqrt{3} + \left(\sqrt{3}\right)^2}{\left(\sqrt{2}\right)^2 - \left(\sqrt{3}\right)^2}$

$\quad = \dfrac{2 - 2\sqrt{6} + 3}{2 - 3}$

$\quad = \dfrac{5 - 2\sqrt{6}}{-1}$

$\quad = -5 + 2\sqrt{6}$

47. $\dfrac{\sqrt{a}+1}{2\sqrt{a}-\sqrt{b}}$

$\quad = \dfrac{\left(\sqrt{a}+1\right)\left(2\sqrt{a}+\sqrt{b}\right)}{\left(2\sqrt{a}-\sqrt{b}\right)\left(2\sqrt{a}+\sqrt{b}\right)}$

$\quad = \dfrac{\sqrt{a}\cdot 2\sqrt{a} + \sqrt{a}\sqrt{b} + 1\cdot 2\sqrt{a} + 1\cdot\sqrt{b}}{\left(2\sqrt{a}\right)^2 - \left(\sqrt{b}\right)^2}$

$\quad = \dfrac{2a + \sqrt{ab} + 2\sqrt{a} + \sqrt{b}}{4a - b}$

49. $\dfrac{8}{1+\sqrt{10}} = \dfrac{8\left(1-\sqrt{10}\right)}{\left(1+\sqrt{10}\right)\left(1-\sqrt{10}\right)}$

$\quad = \dfrac{8\left(1-\sqrt{10}\right)}{1^2 - \left(\sqrt{10}\right)^2}$

$\quad = \dfrac{8\left(1-\sqrt{10}\right)}{1 - 10}$

$\quad = -\dfrac{8\left(1-\sqrt{10}\right)}{9}$

51. $\dfrac{\sqrt{x}}{\sqrt{x}+\sqrt{y}} = \dfrac{\sqrt{x}\left(\sqrt{x}-\sqrt{y}\right)}{\left(\sqrt{x}+\sqrt{y}\right)\left(\sqrt{x}-\sqrt{y}\right)}$

$\quad = \dfrac{\sqrt{x}\left(\sqrt{x}-\sqrt{y}\right)}{\left(\sqrt{x}\right)^2 - \left(\sqrt{y}\right)^2}$

$\quad = \dfrac{\sqrt{x}\left(\sqrt{x}-\sqrt{y}\right)}{x - y}$

$\quad = \dfrac{\sqrt{x}\sqrt{x} - \sqrt{x}\sqrt{y}}{x - y}$

$\quad = \dfrac{x - \sqrt{xy}}{x - y}$

53. $\dfrac{2\sqrt{3}+\sqrt{6}}{4\sqrt{3}-\sqrt{6}} = \dfrac{\left(2\sqrt{3}+\sqrt{6}\right)\left(4\sqrt{3}+\sqrt{6}\right)}{\left(4\sqrt{3}-\sqrt{6}\right)\left(4\sqrt{3}+\sqrt{6}\right)}$

$\quad = \dfrac{8\cdot 3 + 2\sqrt{18} + 4\sqrt{18} + 6}{\left(4\sqrt{3}\right)^2 - \left(\sqrt{6}\right)^2}$

$\quad = \dfrac{30 + 6\sqrt{18}}{16\cdot 3 - 6}$

$\quad = \dfrac{30 + 6(3)\sqrt{2}}{42}$

$\quad = \dfrac{30 + 18\sqrt{2}}{42}$

$\quad = \dfrac{6\left(5 + 3\sqrt{2}\right)}{42}$

$\quad = \dfrac{5 + 3\sqrt{2}}{7}$

55. $\sqrt{\dfrac{5}{3}} = \dfrac{\sqrt{5}}{\sqrt{3}} = \dfrac{\sqrt{5}\cdot\sqrt{5}}{\sqrt{3}\cdot\sqrt{5}} = \dfrac{\sqrt{25}}{\sqrt{15}} = \dfrac{5}{\sqrt{15}}$

57. $\sqrt{\dfrac{18}{5}} = \dfrac{\sqrt{18}}{\sqrt{5}}$

$\quad = \dfrac{\sqrt{9}\cdot\sqrt{2}}{\sqrt{5}}$

$\quad = \dfrac{3\sqrt{2}}{\sqrt{5}}$

$\quad = \dfrac{3\sqrt{2}\cdot\sqrt{2}}{\sqrt{5}\cdot\sqrt{2}}$

$\quad = \dfrac{3\cdot 2}{\sqrt{10}}$

$\quad = \dfrac{6}{\sqrt{10}}$

59. $\dfrac{\sqrt{4x}}{7} = \dfrac{2\sqrt{x}}{7} = \dfrac{2\sqrt{x} \cdot \sqrt{x}}{7 \cdot \sqrt{x}} = \dfrac{2\sqrt{x^2}}{7\sqrt{x}} = \dfrac{2x}{7\sqrt{x}}$

61. $\dfrac{\sqrt[3]{5y^2}}{\sqrt[3]{4x}} = \dfrac{\sqrt[3]{5y^2} \cdot \sqrt[3]{5^2\,y}}{\sqrt[3]{4x} \cdot \sqrt[3]{5^2\,y}} = \dfrac{\sqrt[3]{5^3\,y^3}}{\sqrt[3]{100xy}} = \dfrac{5y}{\sqrt[3]{100xy}}$

63. $\sqrt{\dfrac{2}{5}} = \dfrac{\sqrt{2}}{\sqrt{5}} = \dfrac{\sqrt{2} \cdot \sqrt{2}}{\sqrt{5} \cdot \sqrt{2}} = \dfrac{\sqrt{4}}{\sqrt{10}} = \dfrac{2}{\sqrt{10}}$

65. $\dfrac{\sqrt{2x}}{11} = \dfrac{\sqrt{2x} \cdot \sqrt{2x}}{11 \cdot \sqrt{2x}} = \dfrac{\sqrt{4x^2}}{11\sqrt{2x}} = \dfrac{2x}{11\sqrt{2x}}$

67. $\sqrt[3]{\dfrac{7}{8}} = \dfrac{\sqrt[3]{7}}{\sqrt[3]{8}}$

$\qquad = \dfrac{\sqrt[3]{7}}{2}$

$\qquad = \dfrac{\sqrt[3]{7} \cdot \sqrt[3]{7^2}}{2 \cdot \sqrt[3]{7^2}}$

$\qquad = \dfrac{\sqrt[3]{7^3}}{2\sqrt[3]{49}}$

$\qquad = \dfrac{7}{2\sqrt[3]{49}}$

69. $\dfrac{\sqrt[3]{3x^5}}{10} = \dfrac{\sqrt[3]{x^3 \cdot 3x^2}}{10}$

$\qquad = \dfrac{x\sqrt[3]{3x^2}}{10}$

$\qquad = \dfrac{x\sqrt[3]{3x^2} \cdot \sqrt[3]{3^2\,x}}{10 \cdot \sqrt[3]{3^2\,x}}$

$\qquad = \dfrac{x\sqrt[3]{3^3\,x^3}}{10\sqrt[3]{9x}}$

$\qquad = \dfrac{x \cdot 3x}{10\sqrt[3]{9x}}$

$\qquad = \dfrac{3x^2}{10\sqrt[3]{9x}}$

71. $\sqrt{\dfrac{18x^4y^6}{3z}} = \dfrac{\sqrt{18x^4y^6}}{\sqrt{3z}}$

$\qquad = \dfrac{\sqrt{9x^4y^6 \cdot 2}}{\sqrt{3z}}$

$\qquad = \dfrac{3x^2y^3\sqrt{2}}{\sqrt{3z}}$

$\qquad = \dfrac{3x^2y^3\sqrt{2} \cdot \sqrt{2}}{\sqrt{3z} \cdot \sqrt{2}}$

$\qquad = \dfrac{3x^2y^3 \cdot 2}{\sqrt{6z}}$

$\qquad = \dfrac{6x^2y^3}{\sqrt{6z}}$

73. $\dfrac{2-\sqrt{11}}{6} = \dfrac{\left(2-\sqrt{11}\right)\left(2+\sqrt{11}\right)}{6\left(2+\sqrt{11}\right)}$

$\qquad = \dfrac{4-11}{12+6\sqrt{11}}$

$\qquad = \dfrac{-7}{12+6\sqrt{11}}$

75. $\dfrac{2-\sqrt{7}}{-5} = \dfrac{\left(2-\sqrt{7}\right)\left(2+\sqrt{7}\right)}{-5\left(2+\sqrt{7}\right)}$

$\qquad = \dfrac{4-7}{-5\left(2+\sqrt{7}\right)}$

$\qquad = \dfrac{-3}{-5\left(2+\sqrt{7}\right)}$

$\qquad = \dfrac{3}{5\left(2+\sqrt{7}\right)}$

$\qquad = \dfrac{3}{10+5\sqrt{7}}$

77. $\dfrac{\sqrt{x}+3}{\sqrt{x}} = \dfrac{\left(\sqrt{x}+3\right)\left(\sqrt{x}-3\right)}{\sqrt{x}\left(\sqrt{x}-3\right)}$

$\qquad = \dfrac{\sqrt{x^2}-9}{\sqrt{x^2}-3\sqrt{x}}$

$\qquad = \dfrac{x-9}{x-3\sqrt{x}}$

79. $\dfrac{\sqrt{2}-1}{\sqrt{2}+1} = \dfrac{\left(\sqrt{2}-1\right)\left(\sqrt{2}+1\right)}{\left(\sqrt{2}+1\right)\left(\sqrt{2}+1\right)}$

$= \dfrac{\sqrt{4}-1}{\sqrt{4}+2\sqrt{2}+1}$

$= \dfrac{2-1}{2+2\sqrt{2}+1}$

$= \dfrac{1}{3+2\sqrt{2}}$

81. $\dfrac{\sqrt{x}+1}{\sqrt{x}-1} = \dfrac{\left(\sqrt{x}+1\right)\left(\sqrt{x}-1\right)}{\left(\sqrt{x}-1\right)\left(\sqrt{x}-1\right)}$

$= \dfrac{\sqrt{x^2}-1}{\sqrt{x^2}-2\sqrt{x}+1}$

$= \dfrac{x-1}{x-2\sqrt{x}+1}$

83. $2x-7 = 3(x-4)$
$2x-7 = 3x-12$
$-x-7 = -12$
$-x = -5$
$x = 5$
The solution is 5.

85. $(x-6)(2x+1) = 0$
$x-6 = 0 \quad \text{or} \quad 2x+1 = 0$
$x = 6 \quad \text{or} \qquad 2x = -1$
$$x = -\dfrac{1}{2}$$
The solutions are $-\dfrac{1}{2}, 6$.

87. $\quad x^2 - 8x = -12$
$x^2 - 8x + 12 = 0$
$(x-6)(x-2) = 0$
$x-6 = 0 \quad \text{or} \quad x-2 = 0$
$x = 6 \quad \text{or} \qquad x = 2$
The solutions are 2, 6.

89. $r = \sqrt{\dfrac{A}{4\pi}}$

$= \dfrac{\sqrt{A}}{\sqrt{4\pi}}$

$= \dfrac{\sqrt{A}}{2\sqrt{\pi}}$

$= \dfrac{\sqrt{A} \cdot \sqrt{\pi}}{2\sqrt{\pi} \cdot \sqrt{\pi}}$

$= \dfrac{\sqrt{A\pi}}{2\pi}$

91. a. $\dfrac{\sqrt{5y^3} \cdot \sqrt{12x^3}}{\sqrt{12x^3} \cdot \sqrt{12x^3}} = \dfrac{\sqrt{60x^3y^3}}{\sqrt{(12x^3)^2}}$

$= \dfrac{2xy\sqrt{15xy}}{12x^3}$

$= \dfrac{y\sqrt{15xy}}{6x^2}$

 b. $\dfrac{\sqrt{5y^3} \cdot \sqrt{3x}}{\sqrt{12x^3} \cdot \sqrt{3x}} = \dfrac{\sqrt{15xy^3}}{\sqrt{36x^4}} = \dfrac{y\sqrt{15xy}}{6x^2}$

 c. answers may vary

93. $\dfrac{9}{\sqrt[3]{5}} = \dfrac{9}{\sqrt[3]{5}} \cdot \dfrac{\sqrt[3]{25}}{\sqrt[3]{25}} = \dfrac{9\sqrt[3]{25}}{\sqrt[3]{125}} = \dfrac{9\sqrt[3]{25}}{5}$

The smallest number is $\sqrt[3]{25}$.

95. answers may vary

97. answers may vary

Integrated Review

 1. $\sqrt{81} = 9$ because $9^2 = 81$.

 2. $\sqrt[3]{-8} = -2$ because $(-2)^3 = -8$.

 3. $\sqrt[4]{\dfrac{1}{16}} = \dfrac{1}{2}$ because $\left(\dfrac{1}{2}\right)^4 = \dfrac{1}{16}$.

 4. $\sqrt{x^6} = x^3$ because $(x^3)^2 = x^6$.

 5. $\sqrt[3]{y^9} = y^3$ because $(y^3)^3 = y^9$.

 6. $\sqrt{4y^{10}} = 2y^5$ because $(2y^5)^2 = 4y^{10}$.

7. $\sqrt[5]{-32y^5} = -2y$ because $(-2y)^5 = -32y^5$.

8. $\sqrt[4]{81b^{12}} = 3b^3$ because $(3b^3)^4 = 81b^{12}$.

9. $36^{1/2} = \sqrt{36} = 6$

10. $(3y)^{1/4} = \sqrt[4]{3y}$

11. $64^{-2/3} = \dfrac{1}{\left(\sqrt[3]{64}\right)^2} = \dfrac{1}{4^2} = \dfrac{1}{16}$

12. $(x+1)^{3/5} = \sqrt[5]{(x+1)^3}$

13. $y^{-1/6} \cdot y^{7/6} = y^{-\frac{1}{6}+\frac{7}{6}} = y^{6/6} = y$

14. $\dfrac{(2x^{1/3})^4}{x^{5/6}} = 16x^{4/3}x^{-5/6}$

$\qquad = 16x^{\frac{8}{6}-\frac{5}{6}}$

$\qquad = 16x^{3/6}$

$\qquad = 16x^{1/2}$

15. $\dfrac{x^{1/4}x^{3/4}}{x^{-1/4}} = x^{\frac{1}{4}+\frac{3}{4}+\frac{1}{4}} = x^{5/4}$

16. $4^{1/3} \cdot 4^{2/5} = 4^{\frac{1}{3}+\frac{2}{5}} = 4^{\frac{5}{15}+\frac{6}{15}} = 4^{11/15}$

17. $\sqrt[3]{8x^6} = (8x^6)^{1/3} = (2^3 x^6)^{1/3} = 2^{3/3}x^{6/3} = 2x^2$

18. $\sqrt[12]{a^9 b^6} = (a^9 b^6)^{1/12}$

$\qquad = a^{9/12}b^{6/12}$

$\qquad = a^{3/4}b^{1/2}$

$\qquad = a^{3/4}b^{2/4}$

$\qquad = (a^3 b^2)^{1/4}$

$\qquad = \sqrt[4]{a^3 b^2}$

19. $\sqrt[4]{x} \cdot \sqrt{x} = x^{1/4} \cdot x^{1/2} = x^{\frac{1}{4}+\frac{2}{4}} = x^{3/4} = \sqrt[4]{x^3}$

20. $\sqrt{5} \cdot \sqrt[3]{2} = 5^{1/2} \cdot 2^{1/3}$

$\qquad = 5^{3/6} \cdot 2^{2/6}$

$\qquad = (5^3 \cdot 2^2)^{1/6}$

$\qquad = \sqrt[6]{5^3 \cdot 2^2}$

$\qquad = \sqrt[6]{500}$

21. $\sqrt{40} = \sqrt{4}\sqrt{10} = 2\sqrt{10}$

22. $\sqrt[4]{16x^7 y^{10}} = \sqrt[4]{16x^4 y^8}\sqrt[4]{x^3 y^2} = 2xy^2\sqrt[4]{x^3 y^2}$

23. $\sqrt[3]{54x^4} = \sqrt[3]{27x^3}\sqrt[3]{2x} = 3x\sqrt[3]{2x}$

24. $\sqrt[5]{-64b^{10}} = \sqrt[5]{-32b^{10}}\sqrt[5]{2} = -2b^2\sqrt[5]{2}$

25. $\sqrt{5} \cdot \sqrt{x} = \sqrt{5x}$

26. $\sqrt[3]{8x} \cdot \sqrt[3]{8x^2} = \sqrt[3]{64x^3} = 4x$

27. $\dfrac{\sqrt{98y^6}}{\sqrt{2y}} = \sqrt{\dfrac{98y^6}{2y}}$

$\qquad = \sqrt{49y^5}$

$\qquad = \sqrt{49y^4} \cdot \sqrt{y}$

$\qquad = 7y^2\sqrt{y}$

28. $\dfrac{\sqrt[4]{48a^9 b^3}}{\sqrt[4]{ab^3}} = \sqrt[4]{\dfrac{48a^9 b^3}{ab^3}}$

$\qquad = \sqrt[4]{48a^8}$

$\qquad = \sqrt[4]{16a^8} \cdot \sqrt[4]{3}$

$\qquad = 2a^2\sqrt[4]{3}$

29. $\sqrt{20} - \sqrt{75} + 5\sqrt{7} = \sqrt{4}\sqrt{5} - \sqrt{25}\sqrt{3} + 5\sqrt{7}$

$\qquad = 2\sqrt{5} - 5\sqrt{3} + 5\sqrt{7}$

30. $\sqrt[3]{54y^4} - y\sqrt[3]{16y} = \sqrt[3]{27y^3}\sqrt[3]{2y} - y\sqrt[3]{8}\sqrt[3]{2y}$

$\qquad = 3y\sqrt[3]{2y} - 2y\sqrt[3]{2y}$

$\qquad = y\sqrt[3]{2y}$

31. $\sqrt{3}\left(\sqrt{5} - \sqrt{2}\right) = \sqrt{3}\sqrt{5} - \sqrt{3}\sqrt{2} = \sqrt{15} - \sqrt{6}$

32. $\left(\sqrt{7} + \sqrt{3}\right)^2 = \left(\sqrt{7}\right)^2 + 2\sqrt{7}\sqrt{3} + \left(\sqrt{3}\right)^2$

$\qquad = 7 + 2\sqrt{21} + 3$

$\qquad = 10 + 2\sqrt{21}$

33. $\left(2x - \sqrt{5}\right)\left(2x + \sqrt{5}\right) = (2x)^2 - \left(\sqrt{5}\right)^2$

$\qquad = 4x^2 - 5$

34. $\left(\sqrt{x+1}-1\right)^2 = \left(\sqrt{x+1}\right)^2 - 2\left(\sqrt{x+1}\right) + 1^2$

$\qquad = x+1-2\sqrt{x+1}+1$

$\qquad = x+2-2\sqrt{x+1}$

35. $\sqrt{\dfrac{7}{3}} = \dfrac{\sqrt{7}}{\sqrt{3}} = \dfrac{\sqrt{7}}{\sqrt{3}} \cdot \dfrac{\sqrt{3}}{\sqrt{3}} = \dfrac{\sqrt{21}}{3}$

36. $\dfrac{5}{\sqrt[3]{2x^2}} = \dfrac{5}{\sqrt[3]{2x^2}} \cdot \dfrac{\sqrt[3]{4x}}{\sqrt[3]{4x}} = \dfrac{5\sqrt[3]{4x}}{\sqrt[3]{8x^3}} = \dfrac{5\sqrt[3]{4x}}{2x}$

37. $\dfrac{\sqrt{3}-\sqrt{7}}{2\sqrt{3}+\sqrt{7}}$

$\qquad = \dfrac{\sqrt{3}-\sqrt{7}}{2\sqrt{3}+\sqrt{7}} \cdot \dfrac{\left(2\sqrt{3}-\sqrt{7}\right)}{\left(2\sqrt{3}-\sqrt{7}\right)}$

$\qquad = \dfrac{\sqrt{3}\left(2\sqrt{3}\right)-\sqrt{3}\sqrt{7}-\sqrt{7}\left(2\sqrt{3}\right)+\sqrt{7}\sqrt{7}}{\left(2\sqrt{3}\right)^2 - \left(\sqrt{7}\right)^2}$

$\qquad = \dfrac{6-\sqrt{21}-2\sqrt{21}+7}{12-7}$

$\qquad = \dfrac{13-3\sqrt{21}}{5}$

38. $\sqrt{\dfrac{7}{3}} = \dfrac{\sqrt{7}}{\sqrt{3}} = \dfrac{\sqrt{7}}{\sqrt{3}} \cdot \dfrac{\sqrt{7}}{\sqrt{7}} = \dfrac{7}{\sqrt{21}}$

39. $\sqrt[3]{\dfrac{9y}{11}} = \dfrac{\sqrt[3]{9y}}{\sqrt[3]{11}} = \dfrac{\sqrt[3]{9y}}{\sqrt[3]{11}} \cdot \dfrac{\sqrt[3]{3y^2}}{\sqrt[3]{3y^2}} = \dfrac{\sqrt[3]{27y^3}}{\sqrt[3]{31y^2}} = \dfrac{3y}{\sqrt[3]{33y^2}}$

40. $\dfrac{\sqrt{x}-2}{\sqrt{x}} = \dfrac{\sqrt{x}-2}{\sqrt{x}} \cdot \dfrac{\sqrt{x}+2}{\sqrt{x}+2}$

$\qquad = \dfrac{\left(\sqrt{x}\right)^2 - 2^2}{\sqrt{x}\sqrt{x}+2\sqrt{x}}$

$\qquad = \dfrac{x-4}{x+2\sqrt{x}}$

Section 10.6 Practice

1. $\sqrt{3x-5} = 7$

$\left(\sqrt{3x-5}\right)^2 = 7^2$

$\qquad 3x-5 = 49$

$\qquad\quad 3x = 54$

$\qquad\quad\ x = 18$

Check:

$\sqrt{3x-5} = 7$

$\sqrt{3(18)-5} \stackrel{?}{=} 7$

$\sqrt{54-5} \stackrel{?}{=} 7$

$\sqrt{49} \stackrel{?}{=} 7$

$\qquad 7 = 7$

The solution is 18.

2. $\sqrt{16x-3} - 4x = 0$

$\sqrt{16x-3} = 4x$

$\left(\sqrt{16x-3}\right)^2 = (4x)^2$

$\qquad 16x-3 = 16x^2$

$\qquad\quad 0 = 16x^2 - 16x + 3$

$\qquad\quad 0 = (4x-1)(4x-3)$

$4x-1 = 0 \quad\text{or}\quad 4x-3 = 0$

$\quad x = \dfrac{1}{4} \quad\text{or}\qquad x = \dfrac{3}{4}$

Check $\dfrac{1}{4}$:

$\sqrt{16\cdot\dfrac{1}{4}-3} - 4\left(\dfrac{1}{4}\right) \stackrel{?}{=} 0$

$\sqrt{4-3} - 1 \stackrel{?}{=} 0$

$\sqrt{1} - 1 \stackrel{?}{=} 0$

$1 - 1 \stackrel{?}{=} 0$

$0 = 0$

Check $\dfrac{3}{4}$:

$\sqrt{16\cdot\dfrac{3}{4}-3} - 4\left(\dfrac{3}{4}\right) \stackrel{?}{=} 0$

$\sqrt{12-3} - 4\left(\dfrac{3}{4}\right) \stackrel{?}{=} 0$

$\sqrt{9} - 3 \stackrel{?}{=} 0$

$3 - 3 \stackrel{?}{=} 0$

$0 = 0$

The solutions are $\dfrac{1}{4}$ and $\dfrac{3}{4}$.

3. $\sqrt[3]{x-2} + 1 = 3$

$\sqrt[3]{x-2} = 2$

$\left(\sqrt[3]{x-2}\right)^3 = 2^3$

$\qquad x-2 = 8$

$\qquad\quad x = 10$

Check:
$$\sqrt[3]{x-2}+1=3$$
$$\sqrt[3]{10-2}+1\overset{?}{=}3$$
$$\sqrt[3]{8}+1\overset{?}{=}3$$
$$2+1=3$$
The solution is 10.

4. $$\sqrt{16+x}=x-4$$
$$\left(\sqrt{16+x}\right)^2=(x-4)^2$$
$$16+x=x^2-8x+16$$
$$x^2-9x=0$$
$$x(x-9)=0$$
$$x=0 \text{ or } x-9=0$$
$$x=9$$

Check 0:
$$\sqrt{16+x}=x-4$$
$$\sqrt{16+0}\overset{?}{=}0-4$$
$$\sqrt{16}\overset{?}{=}-4$$
$$4\neq-4$$
Check 9:
$$\sqrt{16+x}=x-4$$
$$\sqrt{16+9}\overset{?}{=}9-4$$
$$\sqrt{25}\overset{?}{=}5$$
$$5=5$$
0 does not check, so the only solution is 9.

5. $$\sqrt{8x+1}+\sqrt{3x}=2$$
$$\sqrt{8x+1}=2-\sqrt{3x}$$
$$\left(\sqrt{8x+1}\right)^2=\left(2-\sqrt{3x}\right)^2$$
$$8x+1=4-4\sqrt{3x}+3x$$
$$4\sqrt{3x}=3-5x$$
$$\left(4\sqrt{3x}\right)^2=(3-5x)^2$$
$$16(3x)=9-30x+25x^2$$
$$25x^2-78x+9=0$$
$$(25x-3)(x-3)=0$$
$$25x-3=0 \text{ or } x-3=0$$
$$x=\frac{3}{25} \text{ or } x=3$$

Check $\frac{3}{25}$:
$$\sqrt{8x+1}+\sqrt{3x}=2$$
$$\sqrt{8\left(\frac{3}{25}\right)+1}+\sqrt{3\left(\frac{3}{25}\right)}\overset{?}{=}2$$
$$\sqrt{\frac{24}{25}+\frac{25}{25}}+\sqrt{\frac{9}{25}}\overset{?}{=}2$$
$$\sqrt{\frac{49}{25}}+\sqrt{\frac{9}{25}}\overset{?}{=}2$$
$$\frac{7}{5}+\frac{3}{5}\overset{?}{=}2$$
$$\frac{10}{5}=2$$
Check 3:
$$\sqrt{8x+1}+\sqrt{3x}=2$$
$$\sqrt{8(3)+1}+\sqrt{3(3)}\overset{?}{=}2$$
$$\sqrt{25}+\sqrt{9}\overset{?}{=}2$$
$$5+3\neq2$$

3 does not check, so the only solution is $\frac{3}{25}$.

6. $$a^2+b^2=c^2$$
$$a^2+6^2=12^2$$
$$a^2+36=144$$
$$a^2=108$$
$$a=\pm\sqrt{108}=\pm\sqrt{36\cdot3}=\pm6\sqrt{3}$$
Since a is a length, we will use the positive value only. The unknown leg is $6\sqrt{3}$ meters long.

7.   Consider the base of the tank, and the plastic divider in the diagonal. Use the Pythagorean theorem to find l.

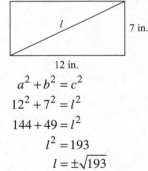

12 in.

$$a^2+b^2=c^2$$
$$12^2+7^2=l^2$$
$$144+49=l^2$$
$$l^2=193$$
$$l=\pm\sqrt{193}$$
We will use the positive value because l represents length. The divider must be $\sqrt{193}\approx13.89$ inches long.

Graphing Calculator Explorations

1.

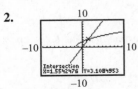

The solution is 3.19.

2.

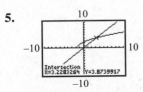

The solution is 1.55.

3.

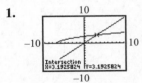

There is no solution. The solution set is $\varnothing$.

4.

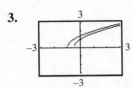

The solution is 0.34.

5.

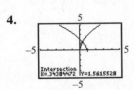

The solution is 3.23.

6.

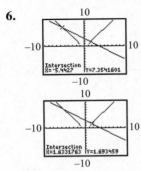

The solutions are -5.44 and 7.35.

Vocabulary, Readiness & Video Check 10.6

1. A proposed solution that is not a solution of the original equation is called an <u>extraneous solution</u>.

2. The Pythagorean Theorem states that $a^2 + b^2 = c^2$ where a and b are the lengths of the <u>legs</u> of a <u>right</u> triangle and c is the length of the <u>hypotenuse</u>.

3. The square of $x - 5$, or $(x-5)^2 = \underline{x^2 - 10x + 25}$.

4. The square of $4 - \sqrt{7x}$, or
    $$\left(4 - \sqrt{7x}\right)^2 = \underline{16 - 8\sqrt{7x} + 7x}.$$

5. Applying the power rule can result in an equation with more solutions than the original equation, so you need to check all proposed solutions in the original equation.

6. The Pythagorean theorem works for a right triangle only and the side opposite the right angle is the hypotenuse, which is c in the formula $a^2 + b^2 = c^2$.

7. Our answer is either a positive square root of a value or a negative square root of a value. We're looking for a length, which must be positive, so our answer must be the positive square root.

Exercise Set 10.6

1. $\sqrt{2x} = 4$
    $\left(\sqrt{2x}\right)^2 = 4^2$
    $2x = 16$
    $x = 8$
    The solution is 8.

3. $\sqrt{x-3} = 2$
    $\left(\sqrt{x-3}\right)^2 = 2^2$
    $x - 3 = 4$
    $x = 7$
    The solution is 7.

5. $\sqrt{2x} = -4$
    No solution since a principle square root does not yield a negative number.

7. $\sqrt{4x-3} - 5 = 0$
$\sqrt{4x-3} = 5$
$\left(\sqrt{4x-3}\right)^2 = 5^2$
$4x - 3 = 25$
$4x = 28$
$x = 7$
The solution is 7.

9. $\sqrt{2x-3} - 2 = 1$
$\sqrt{2x-3} = 3$
$\left(\sqrt{2x-3}\right)^2 = 3^2$
$2x - 3 = 9$
$2x = 12$
$x = 6$
The solution is 6.

11. $\sqrt[3]{6x} = -3$
$\left(\sqrt[3]{6x}\right)^3 = (-3)^3$
$6x = -27$
$x = -\dfrac{27}{6} = -\dfrac{9}{2}$
The solution is $-\dfrac{9}{2}$.

13. $\sqrt[3]{x-2} - 3 = 0$
$\sqrt[3]{x-2} = 3$
$\left(\sqrt[3]{x-2}\right)^3 = 3^3$
$x - 2 = 27$
$x = 29$
The solution is 29.

15. $\sqrt{13-x} = x - 1$
$\left(\sqrt{13-x}\right)^2 = (x-1)^2$
$13 - x = x^2 - 2x + 1$
$0 = x^2 - x - 12$
$0 = (x-4)(x+3)$
$x - 4 = 0$ or $x + 3 = 0$
$x = 4$ or $x = -3$
We discard –3 as extraneous. The solution is 4.

17. $x - \sqrt{4-3x} = -8$
$x + 8 = \sqrt{4-3x}$
$(x+8)^2 = \left(\sqrt{4-3x}\right)^2$
$x^2 + 16x + 64 = 4 - 3x$
$x^2 + 19x + 60 = 0$
$(x+4)(x+15) = 0$
$x + 4 = 0$ or $x + 15 = 0$
$x = -4$ or $x = -15$
We discard –15 as extraneous. The solution is –4.

19. $\sqrt{y+5} = 2 - \sqrt{y-4}$
$\left(\sqrt{y+5}\right)^2 = \left(2 - \sqrt{y-4}\right)^2$
$y + 5 = 4 - 4\sqrt{y-4} + (y-4)$
$y + 5 = y - 4\sqrt{y-4}$
$5 = -4\sqrt{y-4}$
$5^2 = \left(-4\sqrt{y-4}\right)^2$
$25 = 16(y-4)$
$25 = 16y - 64$
$89 = 16y$
$\dfrac{89}{16} = y$
We discard $\dfrac{89}{16}$ as extraneous. There is no solution.

21. $\sqrt{x-3} + \sqrt{x+2} = 5$
$\sqrt{x-3} = 5 - \sqrt{x+2}$
$\left(\sqrt{x-3}\right)^2 = \left(5 - \sqrt{x+2}\right)^2$
$x - 3 = 25 - 10\sqrt{x+2} + (x+2)$
$x - 3 = 27 - 10\sqrt{x+2} + x$
$-30 = -10\sqrt{x+2}$
$3 = \sqrt{x+2}$
$3^2 = \left(\sqrt{x+2}\right)^2$
$9 = x + 2$
$7 = x$
The solution is 7.

23.
$$\sqrt{3x-2} = 5$$
$$\left(\sqrt{3x-2}\right)^2 = 5^2$$
$$3x - 2 = 25$$
$$3x = 27$$
$$x = 9$$
The solution is 9.

25.
$$-\sqrt{2x} + 4 = -6$$
$$10 = \sqrt{2x}$$
$$10^2 = \left(\sqrt{2x}\right)^2$$
$$100 = 2x$$
$$50 = x$$
The solution is 50.

27.
$$\sqrt{3x+1} + 2 = 0$$
$$\sqrt{3x+1} = -2$$
No solution since a principle square root does not yield a negative number.

29.
$$\sqrt[4]{4x+1} - 2 = 0$$
$$\sqrt[4]{4x+1} = 2$$
$$\left(\sqrt[4]{4x+1}\right)^4 = 2^4$$
$$4x + 1 = 16$$
$$4x = 15$$
$$x = \frac{15}{4}$$
The solution is $\frac{15}{4}$.

31.
$$\sqrt{4x-3} = 7$$
$$\left(\sqrt{4x-3}\right)^2 = 7^2$$
$$4x - 3 = 49$$
$$4x = 52$$
$$x = 13$$
The solution is 13.

33.
$$\sqrt[3]{6x-3} - 3 = 0$$
$$\sqrt[3]{6x-3} = 3$$
$$\left(\sqrt[3]{6x-3}\right)^3 = 3^3$$
$$6x - 3 = 27$$
$$6x = 30$$
$$x = 5$$
The solution is 5.

35.
$$\sqrt[3]{2x-3} - 2 = -5$$
$$\sqrt[3]{2x-3} = -3$$
$$\left(\sqrt[3]{2x-3}\right)^3 = (-3)^3$$
$$2x - 3 = -27$$
$$2x = -24$$
$$x = -12$$
The solution is -12.

37.
$$\sqrt{x+4} = \sqrt{2x-5}$$
$$\left(\sqrt{x+4}\right)^2 = \left(\sqrt{2x-5}\right)^2$$
$$x + 4 = 2x - 5$$
$$-x = -9$$
$$x = 9$$
The solution is 9.

39.
$$x - \sqrt{1-x} = -5$$
$$x + 5 = \sqrt{1-x}$$
$$(x+5)^2 = \left(\sqrt{1-x}\right)^2$$
$$x^2 + 10x + 25 = 1 - x$$
$$x^2 + 11x + 24 = 0$$
$$(x+8)(x+3) = 0$$
$$x + 8 = 0 \quad \text{or} \quad x + 3 = 0$$
$$x = -8 \quad \text{or} \qquad x = -3$$
We discard -8 as extraneous. The solution is -3.

41.
$$\sqrt[3]{-6x-1} = \sqrt[3]{-2x-5}$$
$$\left(\sqrt[3]{-6x-1}\right)^3 = \left(\sqrt[3]{-2x-5}\right)^3$$
$$-6x - 1 = -2x - 5$$
$$-4x = -4$$
$$x = 1$$
The solution is 1.

43. $\sqrt{5x-1} - \sqrt{x} + 2 = 3$

$\sqrt{5x-1} = \sqrt{x} + 1$

$\left(\sqrt{5x-1}\right)^2 = \left(\sqrt{x}+1\right)^2$

$5x - 1 = x + 2\sqrt{x} + 1$

$4x - 2 = 2\sqrt{x}$

$2x - 1 = \sqrt{x}$

$(2x-1)^2 = \left(\sqrt{x}\right)^2$

$4x^2 - 4x + 1 = x$

$4x^2 - 5x + 1 = 0$

$(4x-1)(x-1) = 0$

$4x - 1 = 0$ or $x - 1 = 0$

$4x = 1$ or $\quad x = 1$

$x = \dfrac{1}{4}$

We discard $\dfrac{1}{4}$ as extraneous. The solution is 1.

45. $\sqrt{2x-1} = \sqrt{1-2x}$

$\left(\sqrt{2x-1}\right)^2 = \left(\sqrt{1-2x}\right)^2$

$2x - 1 = 1 - 2x$

$4x = 2$

$x = \dfrac{2}{4} = \dfrac{1}{2}$

The solution is $\dfrac{1}{2}$.

47. $\sqrt{3x+4} - 1 = \sqrt{2x+1}$

$\sqrt{3x+4} = \sqrt{2x+1} + 1$

$\left(\sqrt{3x+4}\right)^2 = \left(\sqrt{2x+1}+1\right)^2$

$3x + 4 = (2x+1) + 2\sqrt{2x+1} + 1$

$3x + 4 = 2x + 2 + 2\sqrt{2x+1}$

$x + 2 = 2\sqrt{2x+1}$

$(x+2)^2 = \left(2\sqrt{2x+1}\right)^2$

$x^2 + 4x + 4 = 4(2x+1)$

$x^2 + 4x + 4 = 8x + 4$

$x^2 - 4x = 0$

$x(x-4) = 0$

$x = 0$ or $x - 4 = 0$

$\quad x = 4$

The solutions are 0 and 4.

49. $\sqrt{y+3} - \sqrt{y-3} = 1$

$\sqrt{y+3} = 1 + \sqrt{y-3}$

$\left(\sqrt{y+3}\right)^2 = \left(1+\sqrt{y-3}\right)^2$

$y + 3 = 1 + 2\sqrt{y-3} + (y-3)$

$y + 3 = -2 + 2\sqrt{y-3} + y$

$5 = 2\sqrt{y-3}$

$(5)^2 = \left(2\sqrt{y-3}\right)^2$

$25 = 4(y-3)$

$25 = 4y - 12$

$37 = 4y$

$\dfrac{37}{4} = y$

The solution is $\dfrac{37}{4}$.

51. Let c = length of the hypotenuse.

$6^2 + 3^2 = c^2$

$36 + 9 = c^2$

$45 = c^2$

$\sqrt{45} = \sqrt{c^2}$

$\sqrt{9 \cdot 5} = c$

$3\sqrt{5} = c$ so $c = 3\sqrt{5}$ feet

53. Let b = length of the unknown leg.

$3^2 + b^2 = 7^2$

$9 + b^2 = 49$

$b^2 = 40$

$\sqrt{b^2} = \sqrt{40}$

$b = \sqrt{4 \cdot 10}$

$b = 2\sqrt{10}$ meters

55. Let b = length of the unknown leg.

$9^2 + b^2 = \left(11\sqrt{5}\right)^2$

$81 + b^2 = 121 \cdot 5$

$81 + b^2 = 605$

$b^2 = 524$

$\sqrt{b^2} = \sqrt{524}$

$b = \sqrt{4 \cdot 131}$

$b = 2\sqrt{131} \approx 22.9$ meters

57. Let c = length of the hypotenuse.
$$7^2 + (7.2)^2 = c^2$$
$$49 + 51.84 = c^2$$
$$100.84 = c^2$$
$$\sqrt{100.84} = \sqrt{c^2}$$
$$c = \sqrt{100.84} \approx 10.0 \text{ mm}$$

59. Let x = amount of cable needed.
$$15^2 + 8^2 = x^2$$
$$225 + 64 = x^2$$
$$289 = x^2$$
$$\sqrt{289} = \sqrt{x^2}$$
$$17 = x$$
Thus, 17 feet of cable is needed.

61. Let c = length of the ladder.
$$12^2 + 5^2 = c^2$$
$$144 + 25 = c^2$$
$$169 = c^2$$
$$\sqrt{169} = \sqrt{c^2}$$
$$13 = c$$
A 13-foot ladder is needed.

63.
$$r = \sqrt{\frac{A}{4\pi}}$$
$$1080 = \sqrt{\frac{A}{4\pi}}$$
$$(1080)^2 = \left(\sqrt{\frac{A}{4\pi}}\right)^2$$
$$1{,}166{,}400 = \frac{A}{4\pi}$$
$$14{,}657{,}415 \approx A$$
The surface area is approximately 14,657,415 square miles.

65.
$$v = \sqrt{2gh}$$
$$80 = \sqrt{2(32)h}$$
$$(80)^2 = \left(\sqrt{64h}\right)^2$$
$$6400 = 64h$$
$$100 = h$$
The object fell 100 feet.

67.
$$S = 2\sqrt{I} - 9$$
$$11 = 2\sqrt{I} - 9$$
$$20 = 2\sqrt{I}$$
$$10 = \sqrt{I}$$
$$10^2 = \left(\sqrt{I}\right)^2$$
$$100 = I$$
The estimated IQ is 100.

69.
$$P = 2\pi\sqrt{\frac{l}{32}}$$
$$= 2\pi\sqrt{\frac{2}{32}}$$
$$= 2\pi\sqrt{\frac{1}{16}}$$
$$= 2\pi\left(\frac{1}{4}\right)$$
$$= \frac{\pi}{2} \text{ sec} \approx 1.57 \text{ sec}$$

71.
$$P = 2\pi\sqrt{\frac{l}{32}}$$
$$4 = 2\pi\sqrt{\frac{l}{32}}$$
$$\frac{4}{2\pi} = \sqrt{\frac{l}{32}}$$
$$\left(\frac{2}{\pi}\right)^2 = \left(\sqrt{\frac{l}{32}}\right)^2$$
$$\frac{4}{\pi^2} = \frac{l}{32}$$
$$l = 32\left(\frac{4}{\pi^2}\right) \approx 12.97 \text{ feet}$$

73. Answers may vary

75. $s = \dfrac{1}{2}(6 + 10 + 14) = \dfrac{1}{2}(30) = 15$
$$A = \sqrt{s(s-a)(s-b)(s-c)}$$
$$= \sqrt{15(15-6)(15-10)(15-14)}$$
$$= \sqrt{15(9)(5)(1)}$$
$$= \sqrt{675}$$
$$= \sqrt{225 \cdot 3}$$
$$= 15\sqrt{3} \text{ sq mi} \approx 25.98 \text{ sq mi.}$$

77. Answers may vary

79.
$$D(h) = 111.7\sqrt{h}$$
$$80 = 111.7\sqrt{h}$$
$$\frac{80}{111.7} = \sqrt{h}$$
$$\left(\frac{80}{111.7}\right)^2 = \left(\sqrt{h}\right)^2$$
$$0.5129483389 \approx h$$
$$h \approx 0.51 \text{ km}$$

81. Function; no vertical line intersects the graph more than one time.

83. Function; no vertical line intersects the graph more than one time.

85. Not a function; the y-axis is an example of a vertical line that intersects the graph more than one time.

87. $\dfrac{\frac{x}{6}}{\frac{2x}{3} + \frac{1}{2}} = \dfrac{\left(\frac{x}{6}\right)6}{\left(\frac{2x}{3} + \frac{1}{2}\right)6} = \dfrac{x}{4x+3}$

89. $\dfrac{\frac{z}{5} + \frac{1}{10}}{\frac{z}{20} - \frac{z}{5}} = \dfrac{\left(\frac{z}{5} + \frac{1}{10}\right)20}{\left(\frac{z}{20} - \frac{z}{5}\right)20}$
$$= \dfrac{4z+2}{z-4z}$$
$$= \dfrac{4z+2}{-3z}$$
$$= -\dfrac{4z+2}{3z}$$

91.
$$\sqrt{5x-1} + 4 = 7$$
$$\sqrt{5x-1} = 3$$
$$\left(\sqrt{5x-1}\right)^2 = 3^2$$
$$5x-1 = 9$$
$$5x = 10$$
$$x = 2$$

93.
$$\sqrt{\sqrt{x+3} + \sqrt{x}} = \sqrt{3}$$
$$\left(\sqrt{\sqrt{x+3} + \sqrt{x}}\right)^2 = \left(\sqrt{3}\right)^2$$
$$\sqrt{x+3} + \sqrt{x} = 3$$
$$\sqrt{x+3} = 3 - \sqrt{x}$$
$$\left(\sqrt{x+3}\right)^2 = \left(3 - \sqrt{x}\right)^2$$
$$x+3 = 9 - 6\sqrt{x} + x$$
$$-6 = -6\sqrt{x}$$
$$(-6)^2 = \left(-6\sqrt{x}\right)^2$$
$$36 = 36x$$
$$1 = x$$

95. a. answers may vary

   b. answers may vary

97. $3\sqrt{x^2 - 8x} = x^2 - 8x$

Let $t = x^2 - 8x$. Then
$$3\sqrt{t} = t$$
$$\left(3\sqrt{t}\right)^2 = t^2$$
$$9t = t^2$$
$$0 = t^2 - 9t$$
$$0 = t(t-9)$$
$$t = 0 \quad \text{or} \quad t = 9$$
Replace t with $x^2 - 8x$.

$x^2 - 8x = 0$ or $x^2 - 8x = 9$
$x(x-8) = 0$ $x^2 - 8x - 9 = 0$
$x = 0$ or $x = 8$ $(x-9)(x+1) = 0$
                    $x = 9$ or $x = -1$

The solutions are -1, 0, 8, and 9.

99. $7 - (x^2 - 3x) = \sqrt{(x^2 - 3x) + 5}$

Let $t = x^2 - 3x$. Then
$$7 - t = \sqrt{t+5}$$
$$(7-t)^2 = \left(\sqrt{t+5}\right)^2$$
$$49 - 14t + t^2 = t + 5$$
$$t^2 - 15t + 44 = 0$$
$$(t-11)(t-4) = 0$$
$$t = 11 \text{ or } t = 4$$
Replace t with $x^2 - 3x$.

$x^2 - 3x = 11$ or $x^2 - 3x = 4$

$x^2 - 3x - 11 = 0$ $x^2 - 3x - 4 = 0$

Can't factor $(x - 4)(x + 1) = 0$

                       $x = 4$ or $x = -1$

The solutions are -1 and 4.

Section 10.7 Practice

1. a. $\sqrt{-4} = \sqrt{-1 \cdot 4} = \sqrt{-1} \cdot \sqrt{4} = i \cdot 2$, or $2i$

   b. $\sqrt{-7} = \sqrt{-1(7)} = \sqrt{-1} \cdot \sqrt{7} = i\sqrt{7}$

   c. $-\sqrt{-18} = -\sqrt{-1 \cdot 18}$

               $= -\sqrt{-1} \cdot \sqrt{9 \cdot 2}$

               $= -i \cdot 3\sqrt{2}$

               $= -3i\sqrt{2}$

2. a. $\sqrt{-5} \cdot \sqrt{-6} = i\sqrt{5}\left(i\sqrt{6}\right)$

                  $= i^2\sqrt{30}$

                  $= -1\sqrt{30}$

                  $= -\sqrt{30}$

   b. $\sqrt{-9} \cdot \sqrt{-1} = 3i \cdot i = 3i^2 = 3(-1) = -3$

   c. $\sqrt{125} \cdot \sqrt{-5} = 5\sqrt{5}\left(i\sqrt{5}\right)$

                 $= 5i\left(\sqrt{5}\sqrt{5}\right)$

                 $= 5i(5)$

                 $= 25i$

   d. $\dfrac{\sqrt{-27}}{\sqrt{3}} = \dfrac{i\sqrt{27}}{\sqrt{3}} = i\sqrt{9} = 3i$

3. a. $(3 - 5i) + (-4 + i) = (3 - 4) + (-5 + 1)i$

                           $= -1 - 4i$

   b. $4i - (3 - i) = 4i - 3 + i$

                 $= -3 + (4 + 1)i$

                 $= -3 + 5i$

   c. $(-5 - 2i) - (-8) = -5 - 2i + 8$

                    $= (-5 + 8) - 2i$

                    $= 3 - 2i$

4. a. $-4i \cdot 5i = -20i^2 = -20(-1) = 20 = 20 + 0i$

   b. $5i(2 + i) = 5i \cdot 2 + 5i \cdot i$

                  $= 10i + 5i^2$

                  $= 10i + 5(-1)$

                  $= 10i - 5$

                  $= -5 + 10i$

   c. $(2 + 3i)(6 - i) = 2(6) - 2(i) + 3i(6) - 3i(i)$

                      $= 12 - 2i + 18i - 3i^2$

                      $= 12 + 16i - 3(-1)$

                      $= 12 + 16i + 3$

                      $= 15 + 16i$

   d. $(3 - i)^2 = (3 - i)(3 - i)$

               $= 3(3) - 3(i) - 3(i) + i^2$

               $= 9 - 6i + (-1)$

               $= 8 - 6i$

   e. $(9 + 2i)(9 - 2i) = 9(9) - 9(2i) + 2i(9) - 2i(2i)$

                       $= 81 - 18i + 18i - 4i^2$

                       $= 81 - 4(-1)$

                       $= 81 + 4$

                       $= 85$

                       $= 85 + 0i$

5. a. $\dfrac{4 - i}{3 + i} = \dfrac{(4 - i)(3 - i)}{(3 + i)(3 - i)}$

           $= \dfrac{4(3) - 4(i) - 3(i) + i^2}{3^2 - i^2}$

           $= \dfrac{12 - 7i - 1}{9 + 1}$

           $= \dfrac{11 - 7i}{10}$

           $= \dfrac{11}{10} - \dfrac{7i}{10}$ or $\dfrac{11}{10} - \dfrac{7}{10}i$

   b. $\dfrac{5}{2i} = \dfrac{5(-2i)}{2i(-2i)}$

        $= \dfrac{-10i}{-4i^2}$

        $= \dfrac{-10i}{-4(-1)}$

        $= \dfrac{-10i}{4}$

        $= \dfrac{-5i}{2}$

        $= 0 - \dfrac{5i}{2}$ or $0 - \dfrac{5}{2}i$

6. a. $i^9 = i^4 \cdot i^4 \cdot i = 1 \cdot 1 \cdot i = i$

b. $i^{16} = (i^4)^4 = 1^4 = 1$

c. $i^{34} = i^{32} \cdot i^2 = (i^4)^8 \cdot i^2 = 1^8(-1) = -1$

d. $i^{-24} = \dfrac{1}{i^{24}} = \dfrac{1}{(i^4)^6} = \dfrac{1}{(1)^6} = \dfrac{1}{1} = 1$

Vocabulary, Readiness & Video Check 10.7

1. A <u>complex</u> number is one that can be written in the form $a + bi$ where a and b are real numbers.

2. In the complex number system, i denotes the <u>imaginary unit</u>.

3. $i^2 = \underline{-1}$

4. $i = \sqrt{-1}$

5. A complex number, $a + bi$, is a <u>real</u> number if $b = 0$.

6. A complex number, $a + bi$, is a <u>pure imaginary</u> number if $a = 0$ and $b \neq 0$.

7. The product rule for radicals; you need to first simplify each separate radical and have nonnegative radicands before applying the product rule.

8. combining like terms; i is *not* a variable, but a constant, $\sqrt{-1}$

9. The fact that $i^2 = -1$.

10. using conjugates to rationalize denominators with two terms

11. $i, i^2 = -1, i^3 = -i, i^4 = 1$

Exercise Set 10.7

1. $\sqrt{-81} = \sqrt{-1 \cdot 81} = \sqrt{-1}\sqrt{81} = 9i$

3. $\sqrt{-7} = \sqrt{-1 \cdot 7} = \sqrt{-1}\sqrt{7} = i\sqrt{7}$

5. $-\sqrt{16} = -4$

7. $\sqrt{-64} = \sqrt{-1 \cdot 64} = \sqrt{-1}\sqrt{64} = 8i$

9. $\sqrt{-24} = \sqrt{-1 \cdot 24} = \sqrt{-1}\sqrt{4 \cdot 6} = i \cdot 2\sqrt{6} = 2i\sqrt{6}$

11. $-\sqrt{-36} = -\sqrt{-1 \cdot 36} = -\sqrt{-1}\sqrt{36} = -i \cdot 6 = -6i$

13. $8\sqrt{-63} = 8\sqrt{-1 \cdot 63}$
$= 8\sqrt{-1}\sqrt{9 \cdot 7}$
$= 8i \cdot 3\sqrt{7}$
$= 24i\sqrt{7}$

15. $-\sqrt{54} = -\sqrt{9 \cdot 6} = -3\sqrt{6} = -3\sqrt{6} + 0i$

17. $\sqrt{-2} \cdot \sqrt{-7} = i\sqrt{2} \cdot i\sqrt{7}$
$= i^2\sqrt{14}$
$= (-1)\sqrt{14}$
$= -\sqrt{14}$

19. $\sqrt{-5} \cdot \sqrt{-10} = i\sqrt{5} \cdot i\sqrt{10}$
$= i^2\sqrt{50}$
$= (-1)\sqrt{25 \cdot 2}$
$= -5\sqrt{2}$

21. $\sqrt{16} \cdot \sqrt{-1} = 4i$

23. $\dfrac{\sqrt{-9}}{\sqrt{3}} = \dfrac{i\sqrt{9}}{\sqrt{3}} = i\sqrt{\dfrac{9}{3}} = i\sqrt{3}$

25. $\dfrac{\sqrt{-80}}{\sqrt{-10}} = \dfrac{i\sqrt{80}}{i\sqrt{10}} = \sqrt{\dfrac{80}{10}} = \sqrt{8} = \sqrt{4 \cdot 2} = 2\sqrt{2}$

27. $(4 - 7i) + (2 + 3i) = (4 + 2) + (-7 + 3)i$
$= 6 + (-4)i$
$= 6 - 4i$

29. $(6 + 5i) - (8 - i) = 6 + 5i - 8 + i$
$= (6 - 8) + (5 + 1)i$
$= -2 + 6i$

31. $6 - (8 + 4i) = 6 - 8 - 4i$
$= (6 - 8) - 4i$
$= -2 - 4i$

33. $-10i \cdot -4i = 40i^2 = 40(-1) = -40 = -40 + 0i$

35. $6i(2 - 3i) = 12i - 18i^2$
$= 12i - 18(-1)$
$= 18 + 12i$

37. $\left(\sqrt{3}+2i\right)\left(\sqrt{3}-2i\right)$
$= \sqrt{3}\cdot\sqrt{3}-\sqrt{3}\cdot 2i+\sqrt{3}\cdot 2i-4i^2$
$= 3-4(-1)+0i$
$= 3+4+0i$
$= 7+0i$

39. $\left(4-2i\right)^2 = (4-2i)(4-2i)$
$= 16-4\cdot 2i-4\cdot 2i+4i^2$
$= 16-8i-8i+4(-1)$
$= 16-16i-4$
$= 12-16i$

41. $\dfrac{4}{i} = \dfrac{4(-i)}{i(-i)} = \dfrac{-4i}{-i^2} = \dfrac{-4i}{-(-1)} = -4i = 0-4i$

43. $\dfrac{7}{4+3i} = \dfrac{7(4-3i)}{(4+3i)(4-3i)}$
$= \dfrac{28-21i}{4^2-9i^2}$
$= \dfrac{28-21i}{16+9}$
$= \dfrac{28-21i}{25}$
$= \dfrac{28}{25}-\dfrac{21}{25}i$

45. $\dfrac{3+5i}{1+i} = \dfrac{(3+5i)(1-i)}{(1+i)(1-i)}$
$= \dfrac{3-3i+5i-5i^2}{1^2-i^2}$
$= \dfrac{3+2i+5}{1+1}$
$= \dfrac{8+2i}{2}$
$= \dfrac{8}{2}+\dfrac{2}{2}i$
$= 4+i$

47. $\dfrac{5-i}{3-2i} = \dfrac{(5-i)(3+2i)}{(3-2i)(3+2i)}$
$= \dfrac{15+10i-3i-2i^2}{3^2-4i^2}$
$= \dfrac{15+7i+2}{9+4}$
$= \dfrac{17+7i}{13}$
$= \dfrac{17}{13}+\dfrac{7}{13}i$

49. $(7i)(-9i) = -63i^2 = -63(-1) = 63 = 63+0i$

51. $(6-3i)-(4-2i) = 6-3i-4+2i = 2-i$

53. $-3i(-1+9i) = 3i-27i^2$
$= 3i-27(-1)$
$= 27+3i$

55. $\dfrac{4-5i}{2i} = \dfrac{4-5i}{2i}\cdot\dfrac{-2i}{-2i}$
$= \dfrac{-8i+10i^2}{-4i^2}$
$= \dfrac{-10-8i}{4}$
$= \dfrac{-10}{4}-\dfrac{8}{4}i$
$= -\dfrac{5}{2}-2i$

57. $(4+i)(5+2i) = 20+8i+5i+2i^2$
$= 20+13i+2(-1)$
$= 20+13i-2$
$= 18+13i$

59. $(6-2i)(3+i) = 18+6i-6i-2i^2$
$= 18+2+0i$
$= 20+0i$

61. $(8-3i)+(2+3i) = 8-3i+2+3i = 10+0i$

63. $(1-i)(1+i) = 1+i-i-i^2 = 1+1+0i = 2+0i$

65. $\dfrac{16+15i}{-3i} = \dfrac{(16+15i)(3i)}{-3i(3i)}$
$= \dfrac{48i+45i^2}{-9i^2}$
$= \dfrac{-45+48i}{9}$
$= \dfrac{-45}{9}+\dfrac{48}{9}i$
$= -5+\dfrac{16}{3}i$

67. $(9+8i)^2 = 9^2+2(9)(8i)+(8i)^2$
$= 81+144i+64i^2$
$= 81+144i-64$
$= 17+144i$

69.
$$\frac{2}{3+i} = \frac{2(3-i)}{(3+i)(3-i)}$$
$$= \frac{6-2i}{3^2-i^2}$$
$$= \frac{6-2i}{9+1}$$
$$= \frac{6-2i}{10}$$
$$= \frac{6}{10} - \frac{2}{10}i$$
$$= \frac{3}{5} - \frac{1}{5}i$$

71. $(5-6i) - 4i = 5 - 6i - 4i = 5 - 10i$

73.
$$\frac{2-3i}{2+i} = \frac{(2-3i)(2-i)}{(2+i)(2-i)}$$
$$= \frac{4-2i-6i+3i^2}{2^2-i^2}$$
$$= \frac{4-8i-3}{4+1}$$
$$= \frac{1-8i}{5}$$
$$= \frac{1}{5} - \frac{8}{5}i$$

75. $(2+4i) + (6-5i) = 2 + 4i + 6 - 5i = 8 - i$

77.
$$\left(\sqrt{6}+i\right)\left(\sqrt{6}-i\right) = \left(\sqrt{6}\right)^2 - i^2$$
$$= 6 - (-1)$$
$$= 6 + 1$$
$$= 7$$
$$= 7 + 0i$$

79.
$$4(2-i)^2 = 4(2^2 - 2 \cdot 2i + i^2)$$
$$= 4(4 - 4i - 1)$$
$$= 4(3 - 4i)$$
$$= 12 - 16i$$

81. $i^8 = (i^4)^2 = 1^2 = 1$

83. $i^{21} = i^{20} \cdot i = (i^4)^5 \cdot i = 1^5 \cdot i = i$

85. $i^{11} = i^8 \cdot i^3 = (i^4)^2 \cdot i^3 = 1^2 \cdot (-i) = -i$

87. $i^{-6} = \dfrac{1}{i^6} = \dfrac{1}{i^4 \cdot i^2} = \dfrac{1}{1 \cdot (-1)} = -1$

89. $(2i)^6 = 2^6 i^6 = 64i^4 \cdot i^2 = 64(1)(-1) = -64$

91. $(-3i)^5 = (-3)^5 i^5 = -243i^4 \cdot i = -243(1)i = -243i$

93.
$$x + 50° + 90° = 180°$$
$$x + 140° = 180°$$
$$x = 40°$$

95.

$$\begin{array}{r|rrrr} 1\!\!\!| & 1 & -6 & 3 & -4 \\ & & 1 & -5 & -2 \\ \hline & 1 & -5 & -2 & -6 \end{array}$$

Answer: $x^2 - 5x - 2 - \dfrac{6}{x-1}$

97. 5 people

99. 5 + 9 = 14 people

101.
$$\frac{5 \text{ people}}{30 \text{ people}} = \frac{1}{6} \approx 0.1666$$
About 16.7% of the people reported an average checking balance of \$201 to \$300.

103. $i^3 - i^4 = -i - 1 = -1 - i$

105.
$$i^6 + i^8 = i^4 \cdot i^2 + (i^4)^2$$
$$= 1(-1) + 1^2$$
$$= -1 + 1$$
$$= 0$$
$$= 0 + 0i$$

107. $2 + \sqrt{-9} = 2 + i\sqrt{9} = 2 + 3i$

109.
$$\frac{6 + \sqrt{-18}}{3} = \frac{6 + i\sqrt{9 \cdot 2}}{3}$$
$$= \frac{6 + 3i\sqrt{2}}{3}$$
$$= \frac{6}{3} + \frac{3\sqrt{2}}{3}i$$
$$= 2 + i\sqrt{2}$$

111.
$$\frac{5-\sqrt{-75}}{10} = \frac{5-i\sqrt{25\cdot 3}}{10}$$
$$= \frac{5-5i\sqrt{3}}{10}$$
$$= \frac{5}{10} - \frac{5\sqrt{3}}{10}i$$
$$= \frac{1}{2} - \frac{\sqrt{3}}{2}i$$

113. answers may vary

115.
$$\left(8-\sqrt{-3}\right) - \left(2+\sqrt{-12}\right)$$
$$= 8 - i\sqrt{3} - 2 - 2i\sqrt{3}$$
$$= 6 - 3i\sqrt{3}$$

117.
$$x^2 + 4 = 0$$
$$(2i)^2 + 4 = 0$$
$$4i^2 + 4 = 0$$
$$4(-1) + 4 = 0$$
$$-4 + 4 = 0, \text{ which is true.}$$
Yes, $2i$ is a solution.

Chapter 10 Vocabulary Check

1. The <u>conjugate</u> of $\sqrt{3}+2$ is $\sqrt{3}-2$.

2. The <u>principal square root</u> of a nonnegative number a is written as $\sqrt{a}$.

3. The process of writing a radical expression as an equivalent expression but without a radical in the denominator is called <u>rationalizing</u> the denominator.

4. The <u>imaginary unit</u>, written i, is the number whose square is -1.

5. The <u>cube root</u> of a number is written as $\sqrt[3]{a}$.

6. In the notation $\sqrt[n]{a}$, n is called the <u>index</u> and a is called the <u>radicand</u>.

7. Radicals with the same index and the same radicand are called <u>like radicals</u>.

8. A <u>complex number</u> is a number that can be written in the form $a + bi$, where a and b are real numbers.

9. The <u>distance</u> formula is
$$d = \sqrt{(x_2 - x_1)^2 + (y_2 - y_1)^2}.$$

10. The <u>midpoint</u> formula is $\left(\dfrac{x_1 + x_2}{2}, \dfrac{y_1 + y_2}{2}\right)$.

Chapter 10 Review

1. $\sqrt{81} = 9$ because $9^2 = 81$.

2. $\sqrt[4]{81} = 3$ because $3^4 = 81$.

3. $\sqrt[3]{-8} = -2$ because $(-2)^3 = -8$.

4. $\sqrt[4]{-16}$ is not a real number.

5. $-\sqrt{\dfrac{1}{49}} = -\dfrac{1}{7}$ because $\left(\dfrac{1}{7}\right)^2 = \dfrac{1}{49}$.

6. $\sqrt{x^{64}} = x^{32}$ because $(x^{32})^2 = x^{32\cdot 2} = x^{64}$.

7. $-\sqrt{36} = -6$ because $6^2 = 36$.

8. $\sqrt[3]{64} = 4$ because $4^3 = 64$.

9.
$$\sqrt[3]{-a^6 b^9} = \sqrt[3]{-1}\sqrt[3]{a^6}\sqrt[3]{b^9}$$
$$= -1a^2 b^3$$
$$= -a^2 b^3$$

10. $\sqrt{16a^4 b^{12}} = \sqrt{16}\sqrt{a^4}\sqrt{b^{12}} = 4a^2 b^6$

11. $\sqrt[5]{32a^5 b^{10}} = \sqrt[5]{32}\sqrt[5]{a^5}\sqrt[5]{b^{10}} = 2ab^2$

12. $\sqrt[5]{-32x^{15}y^{20}} = \sqrt[5]{-32}\sqrt[5]{x^{15}}\sqrt[5]{y^{20}} = -2x^3 y^4$

13. $\sqrt{\dfrac{x^{12}}{36y^2}} = \dfrac{\sqrt{x^{12}}}{\sqrt{36y^2}} = \dfrac{x^6}{6y}$

14. $\sqrt[3]{\dfrac{27y^3}{z^{12}}} = \dfrac{\sqrt[3]{27y^3}}{\sqrt[3]{z^{12}}} = \dfrac{3y}{z^4}$

15. $\sqrt{(-x)^2} = |-x|$

16. $\sqrt[4]{(x^2 - 4)^4} = |x^2 - 4|$

17. $\sqrt[3]{(-27)^3} = -27$

18. $\sqrt[5]{(-5)^5} = -5$

19. $-\sqrt[5]{x^5} = -x$

20. $-\sqrt[3]{x^3} = -x$

21. $\sqrt[4]{16(2y+z)^4} = 2|2y+z|$

22. $\sqrt{25(x-y)^2} = 5|x-y|$

23. $\sqrt[5]{y^5} = y$

24. $\sqrt[6]{x^6} = |x|$

25. a. $f(x) = \sqrt{x} + 3$
$f(0) = \sqrt{0} + 3 = 0 + 3 = 3$
$f(9) = \sqrt{9} + 3 = 3 + 3 = 6$

   b. $f(x) = \sqrt{x} + 3$
   $x \geq 0$
   Domain: $[0, \infty)$

   c.

x	0	1	4	9
$f(x)$	3	4	5	6

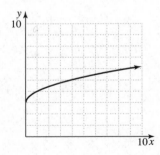

26. a. $g(x) = \sqrt[3]{x-3}$
$g(11) = \sqrt[3]{11-3} = \sqrt[3]{8} = 2$
$g(20) = \sqrt[3]{20-3} = \sqrt[3]{17}$

   b. $g(x) = \sqrt[3]{x-3}$
   Domain: $(-\infty, \infty)$

   c.

x	−5	2	3	4	11
$g(x)$	−2	−1	0	1	2

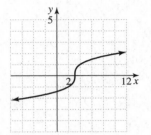

27. $\left(\dfrac{1}{81}\right)^{1/4} = \dfrac{\sqrt[4]{1}}{\sqrt[4]{81}} = \dfrac{1}{3}$

28. $\left(-\dfrac{1}{27}\right)^{1/3} = \dfrac{\sqrt[3]{-1}}{\sqrt[3]{27}} = -\dfrac{1}{3}$

29. $(-27)^{-1/3} = \dfrac{1}{\sqrt[3]{-27}} = -\dfrac{1}{3}$

30. $(-64)^{-1/3} = \dfrac{1}{\sqrt[3]{-64}} = -\dfrac{1}{4}$

31. $-9^{3/2} = -\left(\sqrt{9}\right)^3 = -(3)^3 = -27$

32. $64^{-1/3} = \dfrac{1}{\sqrt[3]{64}} = \dfrac{1}{4}$

33. $(-25)^{5/2} = \left(\sqrt{-25}\right)^5$ is not a real number.

34. $\left(\dfrac{25}{49}\right)^{-3/2} = \dfrac{1}{\left(\frac{25}{49}\right)^{3/2}}$
$= \left(\dfrac{49}{25}\right)^{3/2}$
$= \left(\dfrac{\sqrt{49}}{\sqrt{25}}\right)^3$
$= \left(\dfrac{7}{5}\right)^3$
$= \dfrac{343}{125}$

35. $\left(\dfrac{8}{27}\right)^{-2/3} = \dfrac{1}{\left(\dfrac{8}{27}\right)^{2/3}}$

$\qquad = \left(\dfrac{27}{8}\right)^{2/3}$

$\qquad = \left(\dfrac{\sqrt[3]{27}}{\sqrt[3]{8}}\right)^2$

$\qquad = \left(\dfrac{3}{2}\right)^2$

$\qquad = \dfrac{9}{4}$

36. $\left(-\dfrac{1}{36}\right)^{-1/4}$ is not a real number.

37. $\sqrt[3]{x^2} = x^{2/3}$

38. $\sqrt[5]{5x^2 y^3} = 5^{1/5} x^{2/5} y^{3/5}$

39. $y^{4/5} = \sqrt[5]{y^4}$

40. $5(xy^2 z^5)^{1/3} = 5\sqrt[3]{xy^2 z^5}$

41. $(x+2)^{-1/3} = \dfrac{1}{\sqrt[3]{x+2}}$

42. $(x+2y)^{-1/2} = \dfrac{1}{\sqrt{x+2y}}$

43. $a^{1/3} a^{4/3} a^{1/2} = a^{2/6} a^{8/6} a^{3/6} = a^{13/6}$

44. $\dfrac{b^{1/3}}{b^{4/3}} = b^{\frac{1}{3}-\frac{4}{3}} = b^{-3/3} = b^{-1} = \dfrac{1}{b}$

45. $(a^{1/2} a^{-2})^3 = a^{3/2} a^{-6} = a^{\frac{3}{2}-\frac{12}{2}} = a^{-9/2} = \dfrac{1}{a^{9/2}}$

46. $(x^{-3} y^6)^{1/3} = x^{-3/3} y^{6/3} = x^{-1} y^2 = \dfrac{y^2}{x}$

47. $\left(\dfrac{b^{3/4}}{a^{-1/2}}\right)^8 = \dfrac{b^{24/4}}{a^{-8/2}} = \dfrac{b^6}{a^{-4}} = a^4 b^6$

48. $\dfrac{x^{1/4} x^{-1/2}}{x^{2/3}} = x^{\frac{1}{4}-\frac{1}{2}-\frac{2}{3}} = x^{\frac{3}{12}-\frac{6}{12}-\frac{8}{12}} = x^{-\frac{11}{12}} = \dfrac{1}{x^{11/12}}$

49. $\left(\dfrac{49c^{5/3}}{a^{-1/4} b^{5/6}}\right)^{-1} = \left(\dfrac{a^{-1/4} b^{5/6}}{49c^{5/3}}\right) = \dfrac{b^{5/6}}{49 a^{1/4} c^{5/3}}$

50. $a^{-1/4}(a^{5/4} - a^{9/4}) = a^{-\frac{1}{4}+\frac{5}{4}} - a^{-\frac{1}{4}+\frac{9}{4}}$

$\qquad = a^{\frac{4}{4}} - a^{\frac{8}{4}}$

$\qquad = a - a^2$

51. $\sqrt{20} \approx 4.472$

52. $\sqrt[3]{-39} \approx -3.391$

53. $\sqrt[4]{726} \approx 5.191$

54. $56^{1/3} \approx 3.826$

55. $-78^{3/4} \approx -26.246$

56. $105^{-2/3} \approx 0.045$

57. $\sqrt[3]{2} \cdot \sqrt{7} = 2^{1/3} \cdot 7^{1/2}$

$\qquad = 2^{2/6} \cdot 7^{3/6}$

$\qquad = \sqrt[6]{2^2 \cdot 7^3}$

$\qquad = \sqrt[6]{1372}$

58. $\sqrt[3]{3} \cdot \sqrt[4]{x} = 3^{1/3} x^{1/4}$

$\qquad = 3^{4/12} \cdot x^{3/12}$

$\qquad = \sqrt[12]{3^4 x^3}$

$\qquad = \sqrt[12]{81x^3}$

59. $\sqrt{3} \cdot \sqrt{8} = \sqrt{24} = \sqrt{4}\sqrt{6} = 2\sqrt{6}$

60. $\sqrt[3]{7y} \cdot \sqrt[3]{x^2 z} = \sqrt[3]{7x^2 yz}$

61. $\dfrac{\sqrt{44x^3}}{\sqrt{11x}} = \sqrt{\dfrac{44x^3}{11x}} = \sqrt{4x^2} = 2x$

62. $\dfrac{\sqrt[4]{a^6 b^{13}}}{\sqrt[4]{a^2 b}} = \sqrt[4]{\dfrac{a^6 b^{13}}{a^2 b}} = \sqrt[4]{a^4 b^{12}} = ab^3$

63. $\sqrt{60} = \sqrt{4}\sqrt{15} = 2\sqrt{15}$

64. $-\sqrt{75} = -\sqrt{25}\sqrt{3} = -5\sqrt{3}$

65. $\sqrt[3]{162} = \sqrt[3]{27}\sqrt[3]{6} = 3\sqrt[3]{6}$

66. $\sqrt[3]{-32} = \sqrt[3]{-8}\sqrt[3]{4} = -2\sqrt[3]{4}$

67. $\sqrt{36x^7} = \sqrt{36x^6}\sqrt{x} = 6x^3\sqrt{x}$

68. $\sqrt[3]{24a^5b^7} = \sqrt[3]{8a^3b^6}\sqrt[3]{3a^2b^1} = 2ab^2\sqrt[3]{3a^2b}$

69. $\sqrt{\dfrac{p^{17}}{121}} = \dfrac{\sqrt{p^{16}}\sqrt{p}}{\sqrt{121}} = \dfrac{p^8\sqrt{p}}{11}$

70. $\sqrt[3]{\dfrac{y^5}{27x^6}} = \dfrac{\sqrt[3]{y^3}\sqrt[3]{y^2}}{\sqrt[3]{27x^6}} = \dfrac{y\sqrt[3]{y^2}}{3x^2}$

71. $\sqrt[4]{\dfrac{xy^6}{81}} = \dfrac{\sqrt[4]{y^4}\sqrt[4]{xy^2}}{\sqrt[4]{81}} = \dfrac{y\sqrt[4]{xy^2}}{3}$

72. $\sqrt{\dfrac{2x^3}{49y^4}} = \dfrac{\sqrt{x^2}\sqrt{2x}}{\sqrt{49y^4}} = \dfrac{x\sqrt{2x}}{7y^2}$

73. $r = \sqrt{\dfrac{A}{\pi}}$

$= \sqrt{\dfrac{25}{\pi}}$

$= \dfrac{\sqrt{25}}{\sqrt{\pi}}$

$= \dfrac{5}{\sqrt{\pi}}$ meters or $\dfrac{5\sqrt{\pi}}{\pi}$ meters

74. $r = \sqrt{\dfrac{A}{\pi}} = \sqrt{\dfrac{104}{\pi}} = 5.75$ inches

75. $(x_1,\ y_1) = (-6, 3),\ (x_2,\ y_2) = (8, 4)$

$d = \sqrt{(x_2 - x_1)^2 + (y_2 - y_1)^2}$

$= \sqrt{(8+6)^2 + (4-3)^2}$

$= \sqrt{196 + 1}$

$= \sqrt{197}$ units ≈ 14.036 units

76. $(x_1,\ y_1) = (-4, -6),\ (x_2,\ y_2) = (-1, 5)$

$d = \sqrt{(x_2 - x_1)^2 + (y_2 - y_1)^2}$

$= \sqrt{(-1+4)^2 + (5+6)^2}$

$= \sqrt{9 + 121}$

$= \sqrt{130}$ units ≈ 11.402 units

77. $(x_1,\ y_1) = (-1, 5),\ (x_2,\ y_2) = (2, -3)$

$d = \sqrt{(x_2 - x_1)^2 + (y_2 - y_1)^2}$

$= \sqrt{(2+1)^2 + (-3-5)^2}$

$= \sqrt{9 + 64}$

$= \sqrt{73}$ units ≈ 8.544 units

78. $(x_1,\ y_1) = \left(-\sqrt{2}, 0\right),\ (x_2,\ y_2) = \left(0, -4\sqrt{6}\right)$

$d = \sqrt{(x_2 - x_1)^2 + (y_2 - y_1)^2}$

$= \sqrt{\left(0 + \sqrt{2}\right)^2 + \left(-4\sqrt{6} - 0\right)^2}$

$= \sqrt{2 + 96}$

$= \sqrt{98}$

$= 7\sqrt{2}$ units ≈ 9.899 units

79. $(x_1,\ y_1) = \left(-\sqrt{5}, -\sqrt{11}\right),$

$(x_2,\ y_2) = \left(-\sqrt{5}, -3\sqrt{11}\right)$

$d = \sqrt{(x_2 - x_1)^2 + (y_2 - y_1)^2}$

$= \sqrt{\left(-\sqrt{5} + \sqrt{5}\right)^2 + \left(-3\sqrt{11} + \sqrt{11}\right)^2}$

$= \sqrt{0 + 44}$

$= \sqrt{44}$

$= 2\sqrt{11}$ units ≈ 6.633 units

80. $(x_1,\ y_1) = (7.4, -8.6),\ (x_2,\ y_2) = (-1.2, 5.6)$

$d = \sqrt{(-1.2 - 7.4)^2 + (5.6 + 8.6)^2}$

$= \sqrt{(-8.6)^2 + (14.2)^2}$

$= \sqrt{73.96 + 201.64}$

$= \sqrt{275.6}$ units ≈ 16.601 units

81. $(x_1, y_1) = (2, 6), (x_2, y_2) = (-12, 4)$

$\text{midpoint} = \left(\dfrac{x_1 + x_2}{2}, \dfrac{y_1 + y_2}{2} \right)$

$= \left(\dfrac{2 - 12}{2}, \dfrac{6 + 4}{2} \right)$

$= \left(\dfrac{-10}{2}, \dfrac{10}{2} \right)$

$= (-5, 5)$

82. $(x_1, y_1) = (-6, -5), (x_2, y_2) = (-9, 7)$

$\text{midpoint} = \left(\dfrac{x_1 + x_2}{2}, \dfrac{y_1 + y_2}{2} \right)$

$= \left(\dfrac{-6 - 9}{2}, \dfrac{-5 + 7}{2} \right)$

$= \left(-\dfrac{15}{2}, 1 \right)$

83. $(x_1, y_1) = (4, -6), (x_2, y_2) = (-15, 2)$

$\text{midpoint} = \left(\dfrac{x_1 + x_2}{2}, \dfrac{y_1 + y_2}{2} \right)$

$= \left(\dfrac{4 - 15}{2}, \dfrac{-6 + 2}{2} \right)$

$= \left(-\dfrac{11}{2}, -2 \right)$

84. $(x_1, y_1) = \left(0, -\dfrac{3}{8} \right), (x_2, y_2) = \left(\dfrac{1}{10}, 0 \right)$

$\text{midpoint} = \left(\dfrac{x_1 + x_2}{2}, \dfrac{y_1 + y_2}{2} \right)$

$= \left(\dfrac{0 + \frac{1}{10}}{2}, \dfrac{-\frac{3}{8} + 0}{2} \right)$

$= \left(\dfrac{1}{20}, -\dfrac{3}{16} \right)$

85. $(x_1, y_1) = \left(\dfrac{3}{4}, -\dfrac{1}{7} \right), (x_2, y_2) = \left(-\dfrac{1}{4}, -\dfrac{3}{7} \right)$

$\text{midpoint} = \left(\dfrac{\frac{3}{4} - \frac{1}{4}}{2}, \dfrac{-\frac{1}{7} - \frac{3}{7}}{2} \right)$

$= \left(\dfrac{\frac{1}{2}}{2}, \dfrac{-\frac{11}{7}}{2} \right)$

$= \left(\dfrac{1}{4}, -\dfrac{2}{7} \right)$

86. $(x_1, y_1) = \left(\sqrt{3}, -2\sqrt{6} \right), (x_2, y_2) = \left(\sqrt{3}, -4\sqrt{6} \right)$

$\text{midpoint} = \left(\dfrac{x_1 + x_2}{2}, \dfrac{y_1 + y_2}{2} \right)$

$= \left(\dfrac{\sqrt{3} + \sqrt{3}}{2}, \dfrac{-2\sqrt{6} - 4\sqrt{6}}{2} \right)$

$= \left(\dfrac{2\sqrt{3}}{2}, \dfrac{-6\sqrt{6}}{2} \right)$

$= \left(\sqrt{3}, -3\sqrt{6} \right)$

87. $\sqrt{20} + \sqrt{45} - 7\sqrt{5} = 2\sqrt{5} + 3\sqrt{5} - 7\sqrt{5} = -2\sqrt{5}$

88. $x\sqrt{75x} - \sqrt{27x^3} = 5x\sqrt{3x} - 3x\sqrt{3x} = 2x\sqrt{3x}$

89. $\sqrt[3]{128} + \sqrt[3]{250} = 4\sqrt[3]{2} + 5\sqrt[3]{2} = 9\sqrt[3]{2}$

90. $3\sqrt[4]{32a^5} - a\sqrt[4]{162a} = 6a\sqrt[4]{2a} - 3a\sqrt[4]{2a} = 3a\sqrt[4]{2a}$

91. $\dfrac{5}{\sqrt{4}} + \dfrac{\sqrt{3}}{3} = \dfrac{5}{2} + \dfrac{\sqrt{3}}{3} = \dfrac{15}{6} + \dfrac{2\sqrt{3}}{6} = \dfrac{15 + 2\sqrt{3}}{6}$

92. $\sqrt{\dfrac{8}{x^2}} - \sqrt{\dfrac{50}{16x^2}} = \dfrac{2\sqrt{2}}{x} - \dfrac{5\sqrt{2}}{4x}$

$= \dfrac{8\sqrt{2} - 5\sqrt{2}}{4x}$

$= \dfrac{3\sqrt{2}}{4x}$

93. $2\sqrt{50} - 3\sqrt{125} + \sqrt{98} = 10\sqrt{2} - 15\sqrt{5} + 7\sqrt{2}$

$= 17\sqrt{2} - 15\sqrt{5}$

94. $2a\sqrt[4]{32b^5} - 3b\sqrt[4]{162a^4b} + \sqrt[4]{2a^4b^5}$

$= 4ab\sqrt[4]{2b} - 9ab\sqrt[4]{2b} + ab\sqrt[4]{2b}$

$= -4ab\sqrt[4]{2b}$

95. $\sqrt{3}\left(\sqrt{27} - \sqrt{3} \right) = \sqrt{3}\sqrt{27} - \sqrt{3}\sqrt{3}$

$= \sqrt{81} - \sqrt{9}$

$= 9 - 3$

$= 6$

96. $\left(\sqrt{x} - 3 \right)^2 = \left(\sqrt{x} \right)^2 - (2)(3)\sqrt{x} + 9 = x - 6\sqrt{x} + 9$

477

97. $\left(\sqrt{5}-5\right)\left(2\sqrt{5}+2\right)$
$=\sqrt{5}\left(2\sqrt{5}\right)+2\sqrt{5}-10\sqrt{5}-10$
$=10-8\sqrt{5}-10$
$=-8\sqrt{5}$

98. $\left(2\sqrt{x}-3\sqrt{y}\right)\left(2\sqrt{x}+3\sqrt{y}\right)=\left(2\sqrt{x}\right)^2-\left(3\sqrt{y}\right)^2$
$=4x-9y$

99. $\left(\sqrt{a}+3\right)\left(\sqrt{a}-3\right)=\left(\sqrt{a}\right)^2-(3)^2=a-9$

100. $\left(\sqrt[3]{a}+2\right)^2=\left(\sqrt[3]{a}\right)^2+2(2)\left(\sqrt[3]{a}\right)+2^2$
$=\sqrt[3]{a^2}+4\sqrt[3]{a}+4$

101. $\left(\sqrt[3]{5x}+9\right)\left(\sqrt[3]{5x}-9\right)=\left(\sqrt[3]{5x}\right)^2-9^2$
$=\sqrt[3]{25x^2}-81$

102. $\left(\sqrt[3]{a}+4\right)\left(\sqrt[3]{a^2}-4\sqrt[3]{a}+16\right)$
$=\sqrt[3]{a}\sqrt[3]{a^2}-\sqrt[3]{a}\left(4\sqrt[3]{a}\right)+16\sqrt[3]{a}+4\sqrt[3]{a^2}-16\sqrt[3]{a}+64$
$=a-4\sqrt[3]{a^2}+4\sqrt[3]{a^2}+64$
$=a+64$

103. $\dfrac{3}{\sqrt{7}}=\dfrac{3}{\sqrt{7}}\cdot\dfrac{\sqrt{7}}{\sqrt{7}}=\dfrac{3\sqrt{7}}{7}$

104. $\sqrt{\dfrac{x}{12}}=\dfrac{\sqrt{x}}{\sqrt{12}}=\dfrac{\sqrt{x}}{2\sqrt{3}}\cdot\dfrac{\sqrt{3}}{\sqrt{3}}=\dfrac{\sqrt{3x}}{6}$

105. $\dfrac{5}{\sqrt[3]{4}}=\dfrac{5}{\sqrt[3]{4}}\cdot\dfrac{\sqrt[3]{2}}{\sqrt[3]{2}}=\dfrac{5\sqrt[3]{2}}{2}$

106. $\sqrt{\dfrac{24x^5}{3y}}=\dfrac{\sqrt{24x^5}}{\sqrt{3y}}$
$=\dfrac{2x^2\sqrt{6x}}{\sqrt{3y}}\cdot\dfrac{\sqrt{3y}}{\sqrt{3y}}$
$=\dfrac{2x^2\sqrt{9xy}}{3y}$
$=\dfrac{6x^2\sqrt{2xy}}{3y}$
$=\dfrac{2x^2\sqrt{2xy}}{y}$

107. $\sqrt[3]{\dfrac{15x^6y^7}{z^2}}=\dfrac{\sqrt[3]{15x^6y^7}}{\sqrt[3]{z^2}}$
$=\dfrac{x^2y^2\sqrt[3]{15y}}{\sqrt[3]{z^2}}\cdot\dfrac{\sqrt[3]{z}}{\sqrt[3]{z}}$
$=\dfrac{x^2y^2\sqrt[3]{15yz}}{z}$

108. $\sqrt[4]{\dfrac{81}{8x^{10}}}=\dfrac{\sqrt[4]{81}}{\sqrt[4]{8x^{10}}}=\dfrac{3}{x^2\sqrt[4]{8x^2}}\cdot\dfrac{\sqrt[4]{2x^2}}{\sqrt[4]{2x^2}}=\dfrac{3\sqrt[4]{2x^2}}{2x^3}$

109. $\dfrac{3}{\sqrt{y}-2}=\dfrac{3}{\sqrt{y}-2}\cdot\dfrac{\sqrt{y}+2}{\sqrt{y}+2}=\dfrac{3\sqrt{y}+6}{y-4}$

110. $\dfrac{\sqrt{2}-\sqrt{3}}{\sqrt{2}+\sqrt{3}}=\dfrac{\sqrt{2}-\sqrt{3}}{\sqrt{2}+\sqrt{3}}\cdot\dfrac{\sqrt{2}-\sqrt{3}}{\sqrt{2}-\sqrt{3}}$
$=\dfrac{2-2\sqrt{6}+3}{2-3}$
$=-5+2\sqrt{6}$

111. $\dfrac{\sqrt{11}}{3}=\dfrac{\sqrt{11}}{3}\cdot\dfrac{\sqrt{11}}{\sqrt{11}}=\dfrac{11}{3\sqrt{11}}$

112. $\sqrt{\dfrac{18}{y}}=\dfrac{\sqrt{18}}{\sqrt{y}}=\dfrac{3\sqrt{2}}{\sqrt{y}}\cdot\dfrac{\sqrt{2}}{\sqrt{2}}=\dfrac{6}{\sqrt{2y}}$

113. $\dfrac{\sqrt[3]{9}}{7}=\dfrac{\sqrt[3]{9}}{7}\cdot\dfrac{\sqrt[3]{3}}{\sqrt[3]{3}}=\dfrac{3}{7\sqrt[3]{3}}$

114. $\sqrt{\dfrac{24x^5}{3y^2}} = \dfrac{\sqrt{24x^5}}{\sqrt{3y^2}}$

$\qquad = \dfrac{2x^2\sqrt{6x}}{y\sqrt{3}} \cdot \dfrac{\sqrt{6x}}{\sqrt{6x}}$

$\qquad = \dfrac{12x^3}{3y\sqrt{2x}}$

$\qquad = \dfrac{4x^3}{y\sqrt{2x}}$

115. $\sqrt[3]{\dfrac{xy^2}{10z}} = \dfrac{\sqrt[3]{xy^2}}{\sqrt[3]{10z}} = \dfrac{\sqrt[3]{xy^2}}{\sqrt[3]{10z}} \cdot \dfrac{\sqrt[3]{x^2 y}}{\sqrt[3]{x^2 y}} = \dfrac{xy}{\sqrt[3]{10x^2 yz}}$

116. $\dfrac{\sqrt{x}+5}{-3} = \dfrac{\sqrt{x}+5}{-3} \cdot \dfrac{\sqrt{x}-5}{\sqrt{x}-5} = \dfrac{x-25}{-3\sqrt{x}+15}$

117. $\sqrt{y-7} = 5$

$\qquad y - 7 = 25$

$\qquad y = 32$

The solution is 32.

118. $\sqrt{2x} + 10 = 4$

$\qquad \sqrt{2x} = -6$

No solution since a principal square root does not yield a negative number. The solution set is $\varnothing$.

119. $\sqrt[3]{2x-6} = 4$

$\qquad \left(\sqrt[3]{2x-6}\right)^3 = 4^3$

$\qquad 2x - 6 = 64$

$\qquad 2x = 70$

$\qquad x = 35$

The solution is 35.

120. $\sqrt{x+6} = \sqrt{x+2}$

$\qquad \left(\sqrt{x+6}\right)^2 = \left(\sqrt{x+2}\right)^2$

$\qquad x + 6 = x + 2$

$\qquad 6 = 2$ False

The solution set is $\varnothing$.

121. $2x - 5\sqrt{x} = 3$

$\qquad 2x - 3 = 5\sqrt{x}$

$\qquad 4x^2 - 12x + 9 = 25x$

$\qquad 4x^2 - 37x + 9 = 0$

$\qquad (4x-1)(x-9) = 0$

$\qquad 4x - 1 = 0 \quad \text{or} \quad x - 9 = 0$

$\qquad 4x = 1 \qquad\qquad\qquad x = 9$

$\qquad x = \dfrac{1}{4}$

We discard the $\dfrac{1}{4}$ as extraneous, leaving $x = 9$ as the only solution. The solution is 9.

122. $\sqrt{x+9} = 2 + \sqrt{x-7}$

$\qquad x + 9 = 4 + (2)2\sqrt{x-7} + x - 7$

$\qquad 12 = 4\sqrt{x-7}$

$\qquad 3 = \sqrt{x-7}$

$\qquad 9 = x - 7$

$\qquad 16 = x$

The solution is 16.

123. $a^2 + b^2 = c^2$

$\qquad 3^2 + 3^2 = c^2$

$\qquad 9 + 9 = c^2$

$\qquad 18 = c^2$

$\qquad \sqrt{18} = c$

$\qquad 3\sqrt{2} = c$

The unknown length is $3\sqrt{2}$ centimeters.

124. $a^2 + b^2 = c^2$

$\qquad 7^2 + \left(8\sqrt{3}\right)^2 = c^2$

$\qquad 49 + 192 = c^2$

$\qquad 241 = c^2$

$\qquad \sqrt{241} = c$

The unknown length is $\sqrt{241}$ feet.

125. $a^2 + b^2 = c^2$

$\qquad a^2 + 40^2 = 65^2$

$\qquad a^2 + 1600 = 4225$

$\qquad a^2 = 2625$

$\qquad a = 51.2$

The width of the pond is 51.2 feet.

126.
$$a^2 + b^2 = c^2$$
$$3^2 + 3^2 = c^2$$
$$9 + 9 = c^2$$
$$18 = c^2$$
$$\sqrt{18} = c$$
$3\sqrt{2}$ or $4.24 = c$
The length is 4.24 feet.

127. $\sqrt{-8} = \sqrt{-1} \cdot \sqrt{8} = 2i\sqrt{2} = 0 + 2i\sqrt{2}$

128. $-\sqrt{-6} = -\sqrt{-1} \cdot \sqrt{6} = -i\sqrt{6} = 0 - i\sqrt{6}$

129.
$$\sqrt{-4} + \sqrt{-16} = \sqrt{-1}\sqrt{4} + \sqrt{-1}\sqrt{16}$$
$$= 2i + 4i$$
$$= 6i$$
$$= 0 + 6i$$

130.
$$\sqrt{-2} \cdot \sqrt{-5} = \sqrt{-1} \cdot \sqrt{2} \cdot \sqrt{-1} \cdot \sqrt{5}$$
$$= i\sqrt{2} \cdot i\sqrt{5}$$
$$= i^2\sqrt{10}$$
$$= -\sqrt{10}$$
$$= -\sqrt{10} + 0i$$

131. $(12 - 6i) + (3 + 2i) = (12 + 3) + (-6 + 2)i = 15 - 4i$

132. $(-8 - 7i) - (5 - 4i) = (-8 - 5) + [-7 - (-4)]i$
$$= -13 - 3i$$

133. $(2i)^6 = 2^6 \cdot i^6 = 64 \cdot (-1) = -64$

134. $(3i)^4 = 3^4 \cdot i^4 = 81 \cdot 1 = 81$

135. $-3i(6 - 4i) = -18i + 12(-1) = -12 - 18i$

136. $(3 + 2i)(1 + i) = 3 + 3i + 2i - 2 = 1 + 5i$

137. $(2 - 3i)^2 = 2^2 - 2(2)(3i) + (3i)^2$
$$= 4 - 12i - 9$$
$$= -5 - 12i$$

138. $\left(\sqrt{6} - 9i\right)\left(\sqrt{6} + 9i\right) = \left(\sqrt{6}\right)^2 - (9i)^2 = 6 + 81 = 87$

139. $\dfrac{2 + 3i}{2i} = \dfrac{2 + 3i}{2i} \cdot \dfrac{-2i}{-2i} = \dfrac{-4i + 6}{4} = \dfrac{3}{2} - i$

140. $\dfrac{1 + i}{-3i} = \dfrac{1 + i}{-3i} \cdot \dfrac{3i}{3i} = \dfrac{3i - 3}{9} = \dfrac{-1 + i}{3} = -\dfrac{1}{3} + \dfrac{1}{3}i$

141. $\sqrt[3]{x^3} = x$

142. $\sqrt{(x + 2)^2} = |x + 2|$

143. $-\sqrt{100} = -10$

144. $\sqrt[3]{-x^{12}y^3} = -x^4y$

145. $\sqrt[4]{\dfrac{y^{20}}{16x^{12}}} = \dfrac{\sqrt[4]{y^{20}}}{\sqrt[4]{16x^{12}}} = \dfrac{y^5}{2x^3}$

146. $9^{1/2} = \sqrt{9} = 3$

147. $64^{-1/2} = \dfrac{1}{64^{1/2}} = \dfrac{1}{\sqrt{64}} = \dfrac{1}{8}$

148. $\left(\dfrac{27}{64}\right)^{-2/3} = \left(\dfrac{64}{27}\right)^{2/3} = \left(\sqrt[3]{\dfrac{64}{27}}\right)^2 = \left(\dfrac{4}{3}\right)^2 = \dfrac{16}{9}$

149.
$$\dfrac{(x^{2/3}x^{-3})^3}{x^{-1/2}} = \dfrac{x^{6/3}x^{-9}}{x^{-1/2}}$$
$$= x^{2 - 9 + \frac{1}{2}}$$
$$= x^{-13/2}$$
$$= \dfrac{1}{x^{13/2}}$$

150. $\sqrt{200x^9} = 10x^4\sqrt{2x}$

151. $\sqrt{\dfrac{3n^3}{121m^{10}}} = \dfrac{\sqrt{3n^3}}{\sqrt{121m^{10}}} = \dfrac{n\sqrt{3n}}{11m^5}$

152.
$$3\sqrt{20} - 7x\sqrt[3]{40} + 3\sqrt[3]{5x^3}$$
$$= 3\sqrt{4}\sqrt{5} - 7x\sqrt[3]{8}\sqrt[3]{54} + 3\sqrt[3]{x^3}\sqrt[3]{5}$$
$$= 6\sqrt{5} - 14x\sqrt[3]{5} + 3x\sqrt[3]{5}$$
$$= 6\sqrt{5} - 11x\sqrt[3]{5}$$

153. $\left(2\sqrt{x} - 5\right)^2 = \left(2\sqrt{x}\right)^2 - 2(5)\left(2\sqrt{x}\right) + 5^2$
$$= 4x - 20\sqrt{x} + 25$$

154. $(x_1, y_1) = (-3, 5)$, $(x_2, y_2) = (-8, 9)$

$$d = \sqrt{(x_2 - x_1)^2 + (y_2 - y_1)^2}$$
$$= \sqrt{(-8 + 3)^2 + (9 - 5)^2}$$
$$= \sqrt{(-5)^2 + (4)^2}$$
$$= \sqrt{25 + 16}$$
$$= \sqrt{41}$$

The distance is $\sqrt{41}$ units.

155. $(x_1, y_1) = (-3, 8)$, $(x_2, y_2) = (11, 24)$

$$\text{midpoint} = \left(\frac{x_1 + x_2}{2}, \frac{y_1 + y_2}{2} \right)$$
$$= \left(\frac{-3 + 11}{2}, \frac{8 + 24}{2} \right)$$
$$= \left(\frac{8}{2}, \frac{32}{2} \right)$$
$$= (4, 16)$$

156. $\dfrac{7}{\sqrt{13}} = \dfrac{7}{\sqrt{13}} \cdot \dfrac{\sqrt{13}}{\sqrt{13}} = \dfrac{7\sqrt{13}}{13}$

157. $\dfrac{2}{\sqrt{x} + 3} = \dfrac{2}{\sqrt{x} + 3} \cdot \dfrac{\sqrt{x} - 3}{\sqrt{x} - 3} = \dfrac{2\sqrt{x} - 6}{x - 9}$

158. $\sqrt{x} + 2 = x$

$$\sqrt{x} = x - 2$$
$$\left(\sqrt{x} \right)^2 = (x - 2)^2$$
$$x = x^2 - 4x + 4$$
$$0 = x^2 - 5x + 4$$
$$0 = (x - 4)(x - 1)$$
$$x - 4 = 0 \quad \text{or} \quad x - 1 = 0$$
$$x = 4 \qquad\qquad x = 1$$

Discard the extraneous solution $x = 1$. The solution is 4.

159. $\sqrt{2x - 1} + 2 = x$

$$\sqrt{2x - 1} = x - 2$$
$$\left(\sqrt{2x - 1} \right)^2 = (x - 2)^2$$
$$2x - 1 = x^2 - 4x + 4$$
$$0 = x^2 - 6x + 5$$
$$0 = (x - 5)(x - 1)$$
$$x - 5 = 0 \quad \text{or} \quad x - 1 = 0$$
$$x = 5 \quad \text{or} \qquad x = 1$$

Discard the extraneous solution $x = 1$. The solution is 5.

Chapter 10 Test

1. $\sqrt{216} = \sqrt{36 \cdot 6} = 6\sqrt{6}$

2. $-\sqrt[4]{x^{64}} = -x^{16}$

3. $\left(\dfrac{1}{125} \right)^{1/3} = \dfrac{1}{125^{1/3}} = \dfrac{1}{\sqrt[3]{125}} = \dfrac{1}{5}$

4. $\left(\dfrac{1}{125} \right)^{-1/3} = \dfrac{1}{\left(\frac{1}{125} \right)^{1/3}} = \dfrac{1}{\frac{1}{5}} = 5$

5. $\left(\dfrac{8x^3}{27} \right)^{2/3} = \dfrac{(8x^3)^{2/3}}{27^{2/3}}$

$$= \dfrac{\left(\sqrt[3]{8x^3} \right)^2}{\left(\sqrt[3]{27} \right)^2}$$
$$= \dfrac{(2x)^2}{3^2}$$
$$= \dfrac{4x^2}{9}$$

6. $\sqrt[3]{-a^{18}b^9} = \sqrt[3]{-1 a^{18} b^9} = (-1)a^6 b^3 = -a^6 b^3$

7. $\left(\dfrac{64c^{4/3}}{a^{-2/3}b^{5/6}} \right)^{1/2} = \left(\dfrac{64a^{2/3}c^{4/3}}{b^{5/6}} \right)^{1/2}$

$$= \dfrac{64^{1/2}(a^{2/3})^{1/2}(c^{4/3})^{1/2}}{(b^{5/6})^{1/2}}$$
$$= \dfrac{\sqrt{64}a^{1/3}c^{2/3}}{b^{5/12}}$$
$$= \dfrac{8a^{1/3}c^{2/3}}{b^{5/12}}$$

8. $a^{-2/3}(a^{5/4} - a^3) = a^{-2/3}a^{5/4} - a^{-2/3}a^3$

$$= a^{-\frac{2}{3} + \frac{5}{4}} - a^{-\frac{2}{3} + 3}$$
$$= a^{-\frac{8}{12} + \frac{15}{12}} - a^{-\frac{2}{3} + \frac{9}{3}}$$
$$= a^{7/12} - a^{7/3}$$

9. $\sqrt[4]{(4xy)^4} = |4xy| = 4|xy|$

10. $\sqrt[3]{(-27)^3} = -27$

11. $\sqrt{\dfrac{9}{y}} = \dfrac{\sqrt{9}}{\sqrt{y}} = \dfrac{3}{\sqrt{y}} = \dfrac{3 \cdot \sqrt{y}}{\sqrt{y} \cdot \sqrt{y}} = \dfrac{3\sqrt{y}}{y}$

12. $\dfrac{4-\sqrt{x}}{4+2\sqrt{x}} = \dfrac{4-\sqrt{x}}{2(2+\sqrt{x})}$

$= \dfrac{(4-\sqrt{x})(2-\sqrt{x})}{2(2+\sqrt{x})(2-\sqrt{x})}$

$= \dfrac{8-4\sqrt{x}-2\sqrt{x}+x}{2\left[2^2-(\sqrt{x})^2\right]}$

$= \dfrac{8-6\sqrt{x}+x}{2(4-x)}$ or $\dfrac{8-6\sqrt{x}+x}{8-2x}$

13. $\dfrac{\sqrt[3]{ab}}{\sqrt[3]{ab^2}} = \sqrt[3]{\dfrac{ab}{ab^2}}$

$= \sqrt[3]{\dfrac{1}{b}}$

$= \dfrac{1}{\sqrt[3]{b}}$

$= \dfrac{1 \cdot \sqrt[3]{b^2}}{\sqrt[3]{b} \cdot \sqrt[3]{b^2}}$

$= \dfrac{\sqrt[3]{b^2}}{b}$

14. $\dfrac{\sqrt{6}+x}{8} = \dfrac{(\sqrt{6}+x)(\sqrt{6}-x)}{8(\sqrt{6}-x)}$

$= \dfrac{(\sqrt{6})^2 - x^2}{8(\sqrt{6}-x)}$

$= \dfrac{6-x^2}{8(\sqrt{6}-x)}$

15. $\sqrt{125x^3} - 3\sqrt{20x^3} = \sqrt{25x^2 \cdot 5x} - 3\sqrt{4x^2 \cdot 5x}$

$= 5x\sqrt{5x} - 3 \cdot 2x\sqrt{5x}$

$= 5x\sqrt{5x} - 6x\sqrt{5x}$

$= -x\sqrt{5x}$

16. $\sqrt{3}\left(\sqrt{16}-\sqrt{2}\right) = \sqrt{3}\left(4-\sqrt{2}\right)$

$= 4\sqrt{3} - \sqrt{3}\sqrt{2}$

$= 4\sqrt{3} - \sqrt{6}$

17. $\left(\sqrt{x}+1\right)^2 = \left(\sqrt{x}\right)^2 + 2\sqrt{x} + 1^2$

$= x + 2\sqrt{x} + 1$

18. $\left(\sqrt{2}-4\right)\left(\sqrt{3}+1\right) = \sqrt{2}\sqrt{3} + 1 \cdot \sqrt{2} - 4\sqrt{3} - 4$

$= \sqrt{6} + \sqrt{2} - 4\sqrt{3} - 4$

19. $\left(\sqrt{5}+5\right)\left(\sqrt{5}-5\right) = \left(\sqrt{5}\right)^2 - 5^2$

$= 5 - 25$

$= -20$

20. $\sqrt{561} \approx 23.685$

21. $386^{-2/3} \approx 0.019$

22. $x = \sqrt{x-2} + 2$

$x - 2 = \sqrt{x-2}$

$(x-2)^2 = \left(\sqrt{x-2}\right)^2$

$x^2 - 4x + 4 = x - 2$

$x^2 - 5x + 6 = 0$

$(x-2)(x-3) = 0$

$x = 2$ or $x = 3$

The solutions are 2 and 3.

23. $\sqrt{x^2-7} + 3 = 0$

$\sqrt{x^2-7} = -3$

No solution exists since the principle square root of a number is not negative.

24. $\sqrt[3]{x+5} = \sqrt[3]{2x-1}$

$\left(\sqrt[3]{x+5}\right)^3 = \left(\sqrt[3]{2x-1}\right)^3$

$x + 5 = 2x - 1$

$-x = -6$

$x = 6$

The solution is 6.

25. $\sqrt{-2} = i\sqrt{2} = 0 + i\sqrt{2}$

26. $-\sqrt{-8} = -i\sqrt{4 \cdot 2} = -2i\sqrt{2} = 0 - 2i\sqrt{2}$

27. $(12-6i) - (12-3i) = 12 - 6i - 12 + 3i = 0 - 3i$

28. $(6-2i)(6+2i) = 6^2 - (2i)^2$
$$= 36 - 4i^2$$
$$= 36 + 4$$
$$= 40$$
$$= 40 + 0i$$

29. $(4+3i)^2 = 4^2 + 2\cdot4\cdot3i + (3i)^2$
$$= 16 + 24i + 9i^2$$
$$= 16 + 24i - 9$$
$$= 7 + 24i$$

30. $\dfrac{1+4i}{1-i} = \dfrac{(1+4i)(1+i)}{(1-i)(1+i)}$
$$= \dfrac{1+i+4i+4i^2}{1^2 - i^2}$$
$$= \dfrac{1+5i-4}{1-(-1)}$$
$$= \dfrac{-3+5i}{2}$$
$$= -\dfrac{3}{2} + \dfrac{5}{2}i$$

31. $x^2 + x^2 = 5^2$
$$2x^2 = 25$$
$$x^2 = \dfrac{25}{2}$$
$$\sqrt{x^2} = \sqrt{\dfrac{25}{2}}$$
$$x = \dfrac{5}{\sqrt{2}} = \dfrac{5\cdot\sqrt{2}}{\sqrt{2}\cdot\sqrt{2}} = \dfrac{5\sqrt{2}}{2}$$

32. $g(x) = \sqrt{x+2}$
$$x+2 \geq 0$$
$$x \geq -2$$
Domain: $[-2, \infty)$

x	-2	-1	2	7
$g(x)$	0	1	2	3

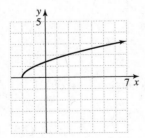

33. $(x_1, y_1) = (-6, 3), (x_2, y_2) = (-8, -7)$
$$d = \sqrt{(-8-(-6))^2 + (-7-3)^2}$$
$$= \sqrt{(-2)^2 + (-10)^2}$$
$$= \sqrt{4+100}$$
$$= \sqrt{104}$$
$$= \sqrt{4\cdot26}$$
$$= 2\sqrt{26}$$
The distance is $2\sqrt{26}$ units.

34. $(x_1, y_1) = \left(-2\sqrt{5}, \sqrt{10}\right),$
$(x_2, y_2) = \left(-\sqrt{5}, 4\sqrt{10}\right)$
$$d = \sqrt{(x_2-x_1)^2 + (y_2-y_1)^2}$$
$$= \sqrt{\left(-\sqrt{5}+2\sqrt{5}\right)^2 + \left(4\sqrt{10}-\sqrt{10}\right)^2}$$
$$= \sqrt{\left(\sqrt{5}\right)^2 + \left(3\sqrt{10}\right)^2}$$
$$= \sqrt{5+90}$$
$$= \sqrt{95}$$
The distance is $\sqrt{95}$ units.

35. $(x_1, y_1) = (-2, -5), (x_2, y_2) = (-6, 12)$
$$\text{midpoint} = \left(\dfrac{x_1+x_2}{2}, \dfrac{y_1+y_2}{2}\right)$$
$$= \left(\dfrac{-2-6}{2}, \dfrac{-5+12}{2}\right)$$
$$= \left(-\dfrac{8}{2}, \dfrac{7}{2}\right)$$
$$= \left(-4, \dfrac{7}{2}\right)$$

36. $(x_1, y_1) = \left(-\dfrac{2}{3}, -\dfrac{1}{5}\right), (x_2, y_2) = \left(-\dfrac{1}{3}, \dfrac{4}{5}\right)$
$$\text{midpoint} = \left(\dfrac{x_1+x_2}{2}, \dfrac{y_1+y_2}{2}\right)$$
$$= \left(\dfrac{-\frac{2}{3}-\frac{1}{3}}{2}, \dfrac{-\frac{1}{5}+\frac{4}{5}}{2}\right)$$
$$= \left(\dfrac{-\frac{3}{3}}{2}, \dfrac{\frac{3}{5}}{2}\right)$$
$$= \left(-\dfrac{1}{2}, \dfrac{3}{10}\right)$$

37. $V(r) = \sqrt{2.5r}$

$V(300) = \sqrt{2.5(300)} = \sqrt{750} \approx 27$ mph

38. $V(r) = \sqrt{2.5r}$

$30 = \sqrt{2.5r}$

$30^2 = \left(\sqrt{2.5r}\right)^2$

$900 = 2.5r$

$r = \dfrac{900}{2.5} = 360$ feet

Chapter 10 Cumulative Review

1. a. $\dfrac{(-12)(-3)+3}{-7-(-2)} = \dfrac{36+3}{-7+2} = \dfrac{39}{-5} = -\dfrac{39}{5}$

b. $\dfrac{2(-3)^2 - 20}{-5+4} = \dfrac{2\cdot 9 - 20}{-1} = \dfrac{18-20}{-1} = \dfrac{-2}{-1} = 2$

2. a. $2(x-3) + (5x+3) = 2x - 6 + 5x + 3$
$= 7x - 3$

b. $4(3x+2) - 3(5x-1) = 12x + 8 - 15x + 3$
$= -3x + 11$

c. $7x + 2(x-7) - 3x = 7x + 2x - 14 - 3x$
$= 6x - 14$

3. $\dfrac{x}{2} - 1 = \dfrac{2}{3}x - 3$

$6\left(\dfrac{x}{2} - 1\right) = 6\left(\dfrac{2}{3}x - 3\right)$

$3x - 6 = 4x - 18$

$-6 = x - 18$

$12 = x$

4. $\dfrac{a-1}{2} + a = 2 - \dfrac{2a+7}{8}$

$8\left(\dfrac{a-1}{2} + a\right) = 8\left(2 - \dfrac{2a+7}{8}\right)$

$4(a-1) + 8a = 16 - (2a+7)$

$4a - 4 + 8a = 16 - 2a - 7$

$12a - 4 = 9 - 2a$

$14a = 13$

$a = \dfrac{13}{14}$

5. Let x = the length of the shorter piece. Then the longer piece has length $3x$.

$x + 3x = 48$

$4x = 48$

$\dfrac{4x}{4} = \dfrac{48}{4}$

$x = 12$

$3x = 3(12) = 36$

The pieces are 12 inches and 36 inches long.

6. Let r = their average speed. Use $r \cdot t = d$.
The round-trip distance is $2(121.5) = 243$ miles.

$r \cdot 4.5 = 243$

$r = \dfrac{243}{4.5} = 54$

Their average speed was 54 mph.

7. $\begin{cases} 3x - y = 4 \\ x + 2y = 8 \end{cases}$

$3x - y = 4$

$-y = -3x + 4$

$y = 3x - 4$: $m = 3$

$x + 2y = 8$

$2y = -x + 8$

$y = -\dfrac{1}{2}x + 4$: $m = -\dfrac{1}{2}$

Since the slopes are different, the lines intersect in one point. The system has one solution.

8. $|3x - 2| + 5 = 5$

$|3x - 2| = 0$

$3x - 2 = 0$

$3x = 2$

$x = \dfrac{2}{3}$

9. $\begin{cases} x + 2y = 7 \\ 2x + 2y = 13 \end{cases}$

Solve the first equation for x.

$x = -2y + 7$

Substitute $-2y + 7$ for x in the second equation.

$2(-2y+7) + 2y = 13$

$-4y + 14 + 2y = 13$

$-2y + 14 = 13$

$-2y = -1$

$y = \dfrac{1}{2}$

Let $y = \dfrac{1}{2}$ in the equation $x = -2y + 7$.

$$x = -2y + 7 = -2\left(\frac{1}{2}\right) + 7 = -1 + 7 = 6$$

The solution is $\left(6, \frac{1}{2}\right)$.

10. $\left|\frac{x}{2} - 1\right| \le 0$

$$\frac{x}{2} - 1 = 0$$

$$\frac{x}{2} = 1$$

$$x = 2$$

11. $\begin{cases} 2x - y = 7 \\ 8x - 4y = 1 \end{cases}$

Multiply the first equation by -4, then add.

$$-8x + 4y = -28$$
$$\underline{8x - 4y = 1}$$
$$0 = -27$$

This is a false statement, so the system has no solution. The solution set is { } or $\varnothing$.

12. $y = |x - 2|$

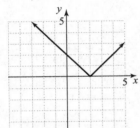

13. Let x be the amount of 30% alcohol solution and y the amount of 80% solution.

$$\begin{cases} x + y = 70 \\ 0.30x + 0.80y = 0.50(70) \end{cases}$$

$$\begin{cases} x + y = 70 \\ 3x + 8y = 350 \end{cases}$$

Multiply the first equation by -3, then add.

$$-3x - 3y = -210$$
$$\underline{3x + 8y = 350}$$
$$5y = 140$$
$$y = 28$$

Let $y = 28$ in the first equation.

$$x + y = 70$$
$$x + 28 = 70$$
$$x = 42$$

She should mix 42 liters of 30% solution with 28 liters of 80% solution.

14. a. Domain: $(-\infty, 0]$, Range: $(-\infty, \infty)$
not a function

b. Domain: $(-\infty, \infty)$, Range: $(-\infty, \infty)$
function

c. Domain: $(-\infty, -2] \cup [2, \infty)$
Range: $(-\infty, \infty)$
not a function

15. $P(x) = 3x^2 - 2x - 5$

a. $P(1) = 3(1)^2 - 2(1) - 5$
$= 3(1) - 2(1) - 5$
$= 3 - 2 - 5$
$= -4$

b. $P(-2) = 3(-2)^2 - 2(-2) - 5$
$= 3(4) - (-4) - 5$
$= 12 + 4 - 5$
$= 11$

16. $f(x) = -2$
This is a horizontal line passing through $(0, -2)$.

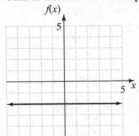

17. $\dfrac{6m^2 + 2m}{2m} = \dfrac{2m(3m + 1)}{2m} = 3m + 1$

18. $y = -3$ is a horizontal line. The slope is 0.

19. $\begin{array}{r|rrrr} 3 & 2 & -1 & -13 & 1 \\ & & 6 & 15 & 6 \\ \hline & 2 & 5 & 2 & 7 \end{array}$

Answer: $2x^2 + 5x + 2 + \dfrac{7}{x - 3}$

20. $\begin{cases} \dfrac{x}{6} - \dfrac{y}{2} = 1 \\ \dfrac{x}{3} - \dfrac{y}{4} = 2 \end{cases}$ or $\begin{cases} x - 3y = 6 \quad (1) \\ 4x - 3y = 24 \quad (2) \end{cases}$

Solve equation (1) for x.

485

$x - 3y = 6$

$\qquad x = 3y + 6$

Replace x with $3y + 6$ in equation (2).

$4(3y + 6) - 3y = 24$

$12y + 24 - 3y = 24$

$\qquad 9y + 24 = 24$

$\qquad\quad 9y = 0$

$\qquad\quad y = 0$

Substitute 0 for y in $x = 3y + 6$.

$x = 3(0) + 6 = 0 + 6 = 6$

The solution is $(6, 0)$.

21. $x^2 + 7yx + 6y^2 = (x + 6y)(x + y)$

22. Let $x =$ number of tee-shirts and
$y =$ number of shorts.
$$\begin{cases} x + y = 9 & (1) \\ 3.50x + 4.25y = 33.75 & (2) \end{cases}$$
Solve equation (1) for y.

$x + y = 9$

$\qquad y = 9 - x$

Substitute $9 - x$ for y in equation (2).

$\quad 3.50x + 4.25(9 - x) = 33.75$

$\quad 3.50x + 38.25 - 4.25x = 33.75$

$\qquad -0.75x + 38.25 = 33.75$

$\qquad\qquad -0.75x = -4.5$

$\qquad\qquad\quad x = \dfrac{-4.5}{-0.75} = 6$

Replace x with 6 in $y = 9 - x$.

$y = 9 - 6 = 3$

Nana bought 6 shirts and 3 shorts.

23. a. $\dfrac{x^3 + 8}{2 + x} = \dfrac{x^3 + 2^3}{x + 2}$

$\qquad\quad = \dfrac{(x + 2)(x^2 - 2x + 4)}{x + 2}$

$\qquad\quad = x^2 - 2x + 4$

b. $\dfrac{2y^2 + 2}{y^3 - 5y^2 + y - 5} = \dfrac{2(y^2 + 1)}{y^2(y - 5) + 1(y - 5)}$

$\qquad\qquad\qquad\quad = \dfrac{2(y^2 + 1)}{(y - 5)(y^2 + 1)}$

$\qquad\qquad\qquad\quad = \dfrac{2}{y - 5}$

24. $\dfrac{0.0000035 \times 4000}{0.28} = \dfrac{(3.5 \times 10^{-6}) \times (4 \times 10^3)}{2.8 \times 10^{-1}}$

$\qquad\qquad\qquad\quad = \dfrac{3.5 \times 4}{2.8} \times 10^{-6 + 3 - (-1)}$

$\qquad\qquad\qquad\quad = 5 \times 10^{-2}$

25. $\dfrac{3x^3 y^7}{40} \div \dfrac{4x^3}{y^2} = \dfrac{3x^3 y^7}{40} \cdot \dfrac{y^2}{4x^3} = \dfrac{3y^9}{160}$

26. $[(5x^2 - 3x + 6) + (4x^2 + 5x - 3)] - (2x - 5)$

$\quad = 5x^2 - 3x + 6 + 4x^2 + 5x - 3 - 2x + 5$

$\quad = 9x^2 + 8$

27. $\dfrac{2y}{2y - 7} - \dfrac{7}{2y - 7} = \dfrac{2y - 7}{2y - 7} = 1$

28. a. $(y - 2)(3y + 4) = 3y^2 + 4y - 6y - 8$

$\qquad\qquad\qquad\quad = 3y^2 - 2y - 8$

b. $(3y - 1)(2y^2 + 3y - 1)$

$\quad = 6y^3 + 9y^2 - 3y - 2y^2 - 3y + 1$

$\quad = 6y^3 + 7y^2 - 6y + 1$

29. $\dfrac{2x}{x^2 + 2x + 1} + \dfrac{x}{x^2 - 1} = \dfrac{2x}{(x + 1)^2} + \dfrac{x}{(x + 1)(x - 1)}$

$\qquad\qquad\qquad\qquad\quad = \dfrac{2x(x - 1) + x(x + 1)}{(x + 1)^2(x - 1)}$

$\qquad\qquad\qquad\qquad\quad = \dfrac{2x^2 - 2x + x^2 + x}{(x + 1)^2(x - 1)}$

$\qquad\qquad\qquad\qquad\quad = \dfrac{3x^2 - x}{(x + 1)^2(x - 1)}$

$\qquad\qquad\qquad\qquad\quad = \dfrac{x(3x - 1)}{(x + 1)^2(x - 1)}$

30. $x^3 - x^2 + 4x - 4 = (x^3 - x^2) + (4x - 4)$

$\qquad\qquad\qquad\qquad = x^2(x - 1) + 4(x - 1)$

$\qquad\qquad\qquad\qquad = (x - 1)(x^2 + 4)$

31. a. $\dfrac{\frac{5x}{x+2}}{\frac{10}{x-2}} = \dfrac{5x}{x+2} \cdot \dfrac{x-2}{10} = \dfrac{x(x-2)}{2(x+2)}$

b. $\dfrac{\frac{x}{y^2}+\frac{1}{y}}{\frac{y}{x^2}+\frac{1}{x}} = \dfrac{\left(\frac{x}{y^2}+\frac{1}{y}\right)x^2y^2}{\left(\frac{y}{x^2}+\frac{1}{x}\right)x^2y^2}$

$= \dfrac{x^3+x^2y}{y^3+xy^2}$

$= \dfrac{x^2(x+y)}{y^2(y+x)}$

$= \dfrac{x^2}{y^2}$

32. a. $\dfrac{a^3-8}{2-a} = \dfrac{a^3-2^3}{2-a}$

$= \dfrac{(a-2)(a^2+2a+4)}{-1(a-2)}$

$= -1(a^2+2a+4)$

$= -a^2-2a-4$

b. $\dfrac{3a^2-3}{a^3+5a^2-a-5} = \dfrac{3(a^2-1)}{a^2(a+5)-1(a+5)}$

$= \dfrac{3(a^2-1)}{(a+5)(a^2-1)}$

$= \dfrac{3}{a+5}$

33. $\dfrac{x}{2}+\dfrac{8}{3} = \dfrac{1}{6}$

$6\left(\dfrac{x}{2}+\dfrac{8}{3}\right) = 6\left(\dfrac{1}{6}\right)$

$3x+16 = 1$

$3x = -15$

$x = -5$

34. a. $\dfrac{3}{xy^2}-\dfrac{2}{3x^2y} = \dfrac{3 \cdot 3x}{xy^2 \cdot 3x}-\dfrac{2 \cdot y}{3x^2y \cdot y}$

$= \dfrac{9x-2y}{3x^2y^2}$

b. $\dfrac{5x}{x+3}-\dfrac{2x}{x-3} = \dfrac{5x(x-3)-2x(x+3)}{(x+3)(x-3)}$

$= \dfrac{5x^2-15x-2x^2-6x}{(x+3)(x-3)}$

$= \dfrac{3x^2-21x}{(x+3)(x-3)}$

$\text{or } \dfrac{3x(x-7)}{(x+3)(x-3)}$

c. $\dfrac{x}{x-2}-\dfrac{5}{2-x} = \dfrac{x}{x-2}+\dfrac{5}{x-2} = \dfrac{x+5}{x-2}$

35. $\dfrac{x}{10} = \dfrac{3}{2}$

$2x = 10 \cdot 3$

$2x = 30$

$x = 15$

The missing length is 15 yards.

36. a. $\dfrac{\frac{y-2}{16}}{\frac{2y+3}{12}} = \dfrac{y-2}{16} \cdot \dfrac{12}{2y+3} = \dfrac{3(y-2)}{4(2y+3)}$

b. $\dfrac{\frac{x}{16}-\frac{1}{x}}{1-\frac{4}{x}} = \dfrac{\left(\frac{x}{16}-\frac{1}{x}\right)16x}{\left(1-\frac{4}{x}\right)16x}$

$= \dfrac{x^2-16}{16x-64}$

$= \dfrac{(x+4)(x-4)}{16(x-4)}$

$= \dfrac{x+4}{16}$

37. a. $\sqrt[3]{1} = 1$, since $1^3 = 1$.

b. $\sqrt[3]{-64} = \sqrt[3]{(-4)^3} = -4$

c. $\sqrt[3]{\dfrac{8}{125}} = \sqrt[3]{\left(\dfrac{2}{5}\right)^3} = \dfrac{2}{5}$

d. $\sqrt[3]{x^6} = \sqrt[3]{(x^2)^3} = x^2$

e. $\sqrt[3]{-27x^9} = \sqrt[3]{(-3x^3)^3} = -3x^3$

38.

$$x-2\overline{\smash{\big)}\,x^3-2x^2+3x-6}$$

$$\,x^2+3$$

$$\underline{x^3-2x^2}$$

$$3x-6$$

$$\underline{3x-6}$$

$$0$$

Answer: x^2+3

39. a. $16^{-3/4}=\dfrac{1}{16^{3/4}}=\dfrac{1}{\left(\sqrt[4]{16}\right)^3}=\dfrac{1}{(2)^3}=\dfrac{1}{8}$

b. $(-27)^{-2/3}=\dfrac{1}{(-27)^{2/3}}$

$$=\dfrac{1}{\left(\sqrt[3]{-27}\right)^2}$$

$$=\dfrac{1}{(-3)^2}$$

$$=\dfrac{1}{9}$$

40.

$$\begin{array}{r|rrrr} 3 & 4 & -12 & -1 & 12 \\ & & 12 & 0 & -3 \\ \hline & 4 & 0 & -1 & 9 \end{array}$$

Answer: $4y^2-1+\dfrac{9}{y-3}$

41. $\dfrac{\sqrt{x}+2}{5}=\dfrac{\left(\sqrt{x}+2\right)\left(\sqrt{x}-2\right)}{5\left(\sqrt{x}-2\right)}$

$$=\dfrac{\left(\sqrt{x}\right)^2-2^2}{5\left(\sqrt{x}-2\right)}$$

$$=\dfrac{x-4}{5\left(\sqrt{x}-2\right)}$$

42.

$$\frac{28}{9-a^2} = \frac{2a}{a-3} + \frac{6}{a+3}$$

$$\frac{28}{-(a^2-9)} = \frac{2a}{a-3} + \frac{6}{a+3}$$

$$\frac{-28}{(a+3)(a-3)} = \frac{2a}{a-3} + \frac{6}{a+3}$$

$$(a+3)(a-3) \cdot \frac{-28}{(a+3)(a-3)} = (a+3)(a-3) \cdot \left(\frac{2a}{a-3} + \frac{6}{a+3}\right)$$

$$-28 = 2a(a+3) + 6(a-3)$$

$$-28 = 2a^2 + 6a + 6a - 18$$

$$0 = 2a^2 + 12a + 10$$

$$0 = 2(a^2 + 6a + 5)$$

$$0 = 2(a+5)(a+1)$$

$a = -5$ or $a = -1$

The solutions are -5 and -1.

43. $u = \dfrac{k}{w}$

$3 = \dfrac{k}{5}$

$k = 3(5) = 15$

$u = \dfrac{15}{w}$

44. $y = kx$

$0.51 = k(3)$

$k = \dfrac{0.51}{3} = 0.17$

$y = 0.17x$

Chapter 11

Section 11.1 Practice

1. $x^2 = 32$

$x = \pm\sqrt{32}$

$x = \pm 4\sqrt{2}$

Check:

Let $x = 4\sqrt{2}$.   Let $x = -4\sqrt{2}$.

$x^2 = 32$ $x^2 = 32$

$\left(4\sqrt{2}\right)^2 \overset{?}{=} 32$ $\left(-4\sqrt{2}\right)^2 \overset{?}{=} 32$

$16 \cdot 2 \overset{?}{=} 32$ $16 \cdot 2 \overset{?}{=} 32$

$32 = 32$ True $32 = 32$ True

The solutions are $4\sqrt{2}$ and $-4\sqrt{2}$, or the

solution set is $\left\{-4\sqrt{2}, 4\sqrt{2}\right\}$.

2. First we get the squared variable alone on one side of the equation.

$5x^2 - 50 = 0$

$5x^2 = 50$

$x^2 = 10$

$x = \pm\sqrt{10}$

The solutions are $\sqrt{10}$ and $-\sqrt{10}$, or the

solution set is $\left\{-\sqrt{10}, \sqrt{10}\right\}$.

3. $(x+3)^2 = 20$

$x + 3 = \pm\sqrt{20}$

$x + 3 = \pm 2\sqrt{5}$

$x = -3 \pm 2\sqrt{5}$

Check:

$(x+3)^2 = 20$

$\left(-3 + 2\sqrt{5} + 3\right)^2 \overset{?}{=} 20$

$\left(2\sqrt{5}\right)^2 \overset{?}{=} 20$

$4 \cdot 5 \overset{?}{=} 20$

$20 = 20$ True

$(x+3)^2 = 20$

$\left(-3 - 2\sqrt{5} + 3\right)^2 \overset{?}{=} 20$

$\left(-2\sqrt{5}\right)^2 \overset{?}{=} 20$

$4 \cdot 5 \overset{?}{=} 20$

$20 = 20$ True

The solutions are $-3 + 2\sqrt{5}$ and $-3 - 2\sqrt{5}$.

4. $(5x-2)^2 = -9$

$5x - 2 = \pm\sqrt{-9}$

$5x - 2 = \pm 3i$

$5x = 2 \pm 3i$

$x = \dfrac{2 \pm 3i}{5} = \dfrac{2}{5} \pm \dfrac{3}{5}i$

The solutions are $\dfrac{2+3i}{5}$ and $\dfrac{2-3i}{5}$ or $\dfrac{2}{5} + \dfrac{3}{5}i$

and $\dfrac{2}{5} - \dfrac{3}{5}i$.

5. $b^2 + 4b = 3$

Add the square of half the coefficient of b to both sides.

$b^2 + 4b + \left(\dfrac{4}{2}\right)^2 = 3 + \left(\dfrac{4}{2}\right)^2$

$b^2 + 4b + 4 = 7$

$(b+2)^2 = 7$

$b + 2 = \pm\sqrt{7}$

$b = -2 \pm \sqrt{7}$

The solutions are $-2 + \sqrt{7}$ and $-2 - \sqrt{7}$.

6. $p^2 - 3p + 1 = 0$

Subtract 1 from both sides.

$p^2 - 3p = -1$

Add the square of half the coefficient of p to both sides.

Copyright © 2013 Pearson Education, Inc.

$$p^2 - 3p + \left(\frac{-3}{2}\right)^2 = -1 + \left(\frac{-3}{2}\right)^2$$

$$p^2 - 3p + \frac{9}{4} = -1 + \frac{9}{4} = \frac{5}{4}$$

$$\left(p - \frac{3}{2}\right)^2 = \frac{5}{4}$$

$$p - \frac{3}{2} = \pm\frac{\sqrt{5}}{2}$$

$$p = \frac{3 \pm \sqrt{5}}{2}$$

The solutions are $\dfrac{3+\sqrt{5}}{2}$ and $\dfrac{3-\sqrt{5}}{2}$.

7. $3x^2 - 12x + 1 = 0$
Divide both sides by 3.

$$3x^2 - 12x + 1 = 0$$

$$x^2 - 4x + \frac{1}{3} = 0$$

$$x^2 - 4x = -\frac{1}{3}$$

Find the square of half of -4.

$$\left(\frac{-4}{2}\right)^2 = (-2)^2 = 4$$

Add 4 to both sides of the equation.

$$x^2 - 4x + 4 = -\frac{1}{3} + 4$$

$$(x-2)^2 = -\frac{1}{3} + \frac{12}{3} = \frac{11}{3}$$

$$x - 2 = \pm\sqrt{\frac{11}{3}} = \pm\frac{\sqrt{33}}{3}$$

$$x = \frac{6}{3} \pm \frac{\sqrt{33}}{3} = \frac{6 \pm \sqrt{33}}{3}$$

The solutions are $\dfrac{6+\sqrt{33}}{3}$ and $\dfrac{6-\sqrt{33}}{3}$.

8. $2x^2 - 5x + 7 = 0$

$$2x^2 - 5x = -7$$

$$x^2 - \frac{5}{2}x = -\frac{7}{2}$$

Since $\dfrac{1}{2}\left(-\dfrac{5}{2}\right) = -\dfrac{5}{4}$ and $\left(-\dfrac{5}{4}\right)^2 = \dfrac{25}{16}$, we add

$\dfrac{25}{16}$ to both sides of the equation.

$$x^2 - \frac{5}{2}x + \frac{25}{16} = -\frac{7}{2} + \frac{25}{16}$$

$$\left(x - \frac{5}{4}\right)^2 = -\frac{56}{16} + \frac{25}{16} = -\frac{31}{16}$$

$$x - \frac{5}{4} = \pm\sqrt{-\frac{31}{16}}$$

$$x = \frac{5}{4} \pm \frac{i\sqrt{31}}{4} = \frac{5 \pm i\sqrt{31}}{4}$$

The solutions are $\dfrac{5 + i\sqrt{31}}{4}$ and $\dfrac{5 - i\sqrt{31}}{4}$ or

$\dfrac{5}{4} + \dfrac{\sqrt{31}}{4}i$ and $\dfrac{5}{4} - \dfrac{\sqrt{31}}{4}i$.

9. $A = P(1+r)^t$; $A = 5618$, $P = 5000$, $t = 2$

$$A = P(1+r)^t$$

$$5618 = 5000(1+r)^2$$

$$1.1236 = (1+r)^2$$

$$\pm\sqrt{1.1236} = 1 + r$$

$$-1 \pm 1.06 = r$$

$0.06 = r$ or $-2.06 = r$

The rate cannot be negative, so we reject -2.06.

Check: $A = 5000(1 + 0.06)^2$

$$= 5000(1.06)^2$$

$$= 5000 \cdot 1.1236$$

$$= 5618$$

The interest rate is 6% compounded annually.

Graphing Calculator Explorations

1. $-1.27, 6.27$

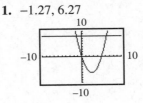

2. $-3.45, 1.45$

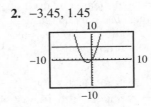

3. $-1.10, 0.90$

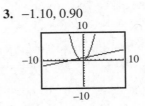

4. -1.54, 1.94

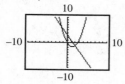

5. No real solutions, or $\varnothing$

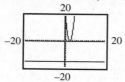

6. answers may vary

Vocabulary, Readiness & Video Check 11.1

1. By the square root property, if b is a real number, and $a^2 = b$, then $a = \underline{\pm\sqrt{b}}$.

2. A <u>quadratic</u> equation can be written in the form $ax^2 + bx + c = 0$, $a \neq 0$.

3. The process of writing a quadratic equation so that one side is a perfect square trinomial is called <u>completing the square</u>.

4. A perfect square trinomial is one that can be factored as a <u>binomial</u> squared.

5. To solve $x^2 + 6x = 10$ by completing the square, add $\underline{9}$ to both sides.

6. To solve $x^2 + bx = c$ by completing the square, add $\underline{\left(\dfrac{b}{2}\right)^2}$ to both sides.

7. We need a quantity shown squared by itself on one side of the equation. The only quantity squared is x, so divide both sides by 2 before applying the square root property.

8. The coefficient of y^2 is 3. To use the completing the square method, the coefficient of the squared variable must be 1, so we first divide through by 3.

9. We're looking for an interest rate so a negative value does not make sense.

Exercise Set 11.1

1. $x^2 = 16$
   $x = \pm\sqrt{16}$
   $x = \pm 4$
   The solutions are -4 and 4.

3. $x^2 - 7 = 0$
   $x^2 = 7$
   $x = \pm\sqrt{7}$
   The solutions are $\sqrt{7}$ and $-\sqrt{7}$.

5. $x^2 = 18$
   $x = \pm\sqrt{18}$
   $x = \pm\sqrt{9 \cdot 2}$
   $x = \pm 3\sqrt{2}$
   The solutions are $3\sqrt{2}$ and $-3\sqrt{2}$.

7. $3z^2 - 30 = 0$
   $3z^2 = 30$
   $z^2 = 10$
   $z = \pm\sqrt{10}$
   The solutions are $\sqrt{10}$ and $-\sqrt{10}$.

9. $(x+5)^2 = 9$
   $x + 5 = \pm\sqrt{9}$
   $x + 5 = \pm 3$
   $x = -5 \pm 3$
   $x = -8$ or $x = -2$
   The solutions are -8 and -2.

11. $(z-6)^2 = 18$
    $z - 6 = \pm\sqrt{18}$
    $z - 6 = \pm 3\sqrt{2}$
    $z = 6 \pm 3\sqrt{2}$
    The solutions are $6 + 3\sqrt{2}$ and $6 - 3\sqrt{2}$.

13. $(2x-3)^2 = 8$
    $2x - 3 = \pm\sqrt{8}$
    $2x - 3 = \pm 2\sqrt{2}$
    $2x = 3 \pm 2\sqrt{2}$
    $x = \dfrac{3 \pm 2\sqrt{2}}{2}$
    The solutions are $\dfrac{3 + 2\sqrt{2}}{2}$ and $\dfrac{3 - 2\sqrt{2}}{2}$.

15. $x^2 + 9 = 0$

$x^2 = -9$

$x = \pm\sqrt{-9}$

$x = \pm 3i$

The solutions are $3i$ and $-3i$.

17. $x^2 - 6 = 0$

$x^2 = 6$

$x = \pm\sqrt{6}$

The solutions are $\sqrt{6}$ and $-\sqrt{6}$.

19. $2z^2 + 16 = 0$

$2z^2 = -16$

$z^2 = -8$

$z = \pm\sqrt{-8}$

$z = \pm i\sqrt{8}$

$z = \pm 2i\sqrt{2}$

The solutions are $2i\sqrt{2}$ and $-2i\sqrt{2}$.

21. $(3x - 1)^2 = -16$

$3x - 1 = \pm\sqrt{-16}$

$3x - 1 = \pm 4i$

$3x = 1 \pm 4i$

$x = \dfrac{1 \pm 4i}{3}$

The solutions are $\dfrac{1 + 4i}{3}$ and $\dfrac{1 - 4i}{3}$ or $\dfrac{1}{3} + \dfrac{4}{3}i$

and $\dfrac{1}{3} - \dfrac{4}{3}i$.

23. $(z + 7)^2 = 5$

$z + 7 = \pm\sqrt{5}$

$z = -7 \pm \sqrt{5}$

The solutions are $-7 + \sqrt{5}$ and $-7 - \sqrt{5}$.

25. $(x + 3)^2 + 8 = 0$

$(x + 3)^2 = -8$

$x + 3 = \pm\sqrt{-8}$

$x + 3 = \pm i\sqrt{8}$

$x + 3 = \pm 2i\sqrt{2}$

$x = -3 \pm 2i\sqrt{2}$

The solutions are $-3 + 2i\sqrt{2}$ and $-3 - 2i\sqrt{2}$.

27. $x^2 + 16x + \left(\dfrac{16}{2}\right)^2 = x^2 + 16x + 64$

$= (x + 8)^2$

29. $z^2 - 12z + \left(\dfrac{-12}{2}\right)^2 = z^2 - 12z + 36$

$= (z - 6)^2$

31. $p^2 + 9p + \left(\dfrac{9}{2}\right)^2 = p^2 + 9p + \dfrac{81}{4}$

$= \left(p + \dfrac{9}{2}\right)^2$

33. $x^2 + x + \left(\dfrac{1}{2}\right)^2 = x^2 + 16x + \dfrac{1}{4}$

$= \left(x + \dfrac{1}{2}\right)^2$

35. $x^2 + 8x = -15$

$x^2 + 8x + \left(\dfrac{8}{2}\right)^2 = -15 + 16$

$x^2 + 8x + 16 = 1$

$(x + 4)^2 = 1$

$x + 4 = \pm\sqrt{1}$

$x = -4 \pm 1$

$x = -5 \ \text{or} \ x = -3$

The solutions are -5 and -3.

37. $x^2 + 6x + 2 = 0$

$x^2 + 6x = -2$

$x^2 + 6x + \left(\dfrac{6}{2}\right)^2 = -2 + 9$

$x^2 + 6x + 9 = 7$

$(x + 3)^2 = 7$

$x + 3 = \pm\sqrt{7}$

$x = -3 \pm \sqrt{7}$

The solutions are $-3 + \sqrt{7}$ and $-3 - \sqrt{7}$.

39.
$$x^2 + x - 1 = 0$$
$$x^2 + x = 1$$
$$x^2 + x + \left(\frac{1}{2}\right)^2 = 1 + \frac{1}{4}$$
$$x^2 + x + \frac{1}{4} = \frac{5}{4}$$
$$\left(x + \frac{1}{2}\right)^2 = \frac{5}{4}$$
$$x + \frac{1}{2} = \pm\sqrt{\frac{5}{4}}$$
$$x = -\frac{1}{2} \pm \frac{\sqrt{5}}{2} = \frac{-1 \pm \sqrt{5}}{2}$$

The solutions are $\dfrac{-1+\sqrt{5}}{2}$ and $\dfrac{-1-\sqrt{5}}{2}$.

41.
$$x^2 + 2x - 5 = 0$$
$$x^2 + 2x = 5$$
$$x^2 + 2x + \left(\frac{2}{2}\right)^2 = 5 + 1$$
$$x^2 + 2x + 1 = 6$$
$$(x+1)^2 = 6$$
$$x + 1 = \pm\sqrt{6}$$
$$x = -1 \pm \sqrt{6}$$

The solutions are $-1+\sqrt{6}$ and $-1-\sqrt{6}$.

43.
$$y^2 + y - 7 = 0$$
$$y^2 + y = 7$$
$$y^2 + y + \left(\frac{1}{2}\right)^2 = 7 + \frac{1}{4}$$
$$y^2 + y + \frac{1}{4} = \frac{29}{4}$$
$$\left(y + \frac{1}{2}\right)^2 = \frac{29}{4}$$
$$y + \frac{1}{2} = \pm\sqrt{\frac{29}{4}}$$
$$y = -\frac{1}{2} \pm \frac{\sqrt{29}}{2} = \frac{-1 \pm \sqrt{29}}{2}$$

The solutions are $\dfrac{-1+\sqrt{29}}{2}$ and $\dfrac{-1-\sqrt{29}}{2}$.

45.
$$3p^2 - 12p + 2 = 0$$
$$p^2 - 4p + \frac{2}{3} = 0$$
$$p^2 - 4p = -\frac{2}{3}$$
$$p^2 - 4p + \left(\frac{-4}{2}\right)^2 = -\frac{2}{3} + 4$$
$$(p-2)^2 = \frac{10}{3}$$
$$p - 2 = \pm\sqrt{\frac{10}{3}}$$
$$p - 2 = \pm\frac{\sqrt{10} \cdot \sqrt{3}}{\sqrt{3} \cdot \sqrt{3}}$$
$$p - 2 = \pm\frac{\sqrt{30}}{3}$$
$$p = 2 \pm \frac{\sqrt{30}}{3} = \frac{6 \pm \sqrt{30}}{3}$$

The solutions are $\dfrac{6+\sqrt{30}}{3}$ and $\dfrac{6-\sqrt{30}}{3}$.

47.
$$4y^2 - 2 = 12y$$
$$4y^2 - 12y - 2 = 0$$
$$y^2 - 3y - \frac{1}{2} = 0$$
$$y^2 - 3y = \frac{1}{2}$$
$$y^2 - 3y + \left(\frac{-3}{2}\right)^2 = \frac{1}{2} + \frac{9}{4}$$
$$y^2 - 3y + \frac{9}{4} = \frac{11}{4}$$
$$\left(y - \frac{3}{2}\right)^2 = \frac{11}{4}$$
$$y - \frac{3}{2} = \pm\sqrt{\frac{11}{4}}$$
$$y = \frac{3}{2} \pm \frac{\sqrt{11}}{2} = \frac{3 \pm \sqrt{11}}{2}$$

The solutions are $\dfrac{3+\sqrt{11}}{2}$ and $\dfrac{3-\sqrt{11}}{2}$.

49.

$$2x^2 + 7x = 4$$

$$x^2 + \frac{7}{2}x = 2$$

$$x^2 + \frac{7}{2}x + \left(\frac{\frac{7}{2}}{2}\right)^2 = 2 + \frac{49}{16}$$

$$x^2 + \frac{7}{2}x + \frac{49}{16} = \frac{81}{16}$$

$$\left(x + \frac{7}{4}\right)^2 = \frac{81}{16}$$

$$x + \frac{7}{4} = \pm\sqrt{\frac{81}{16}}$$

$$x = -\frac{7}{4} \pm \frac{9}{4} = \frac{-7 \pm 9}{4}$$

$$x = -4 \text{ or } \frac{1}{2}$$

The solutions are -4 and $\frac{1}{2}$.

51.

$$x^2 + 8x + 1 = 0$$

$$x^2 + 8x = -1$$

$$x^2 + 8x + \left(\frac{8}{2}\right)^2 = -1 + 16$$

$$x^2 + 8x + 16 = 15$$

$$(x + 4)^2 = 15$$

$$x + 4 = \pm\sqrt{15}$$

$$x = -4 \pm \sqrt{15}$$

The solutions are $-4 + \sqrt{15}$ and $-4 - \sqrt{15}$.

53.

$$3y^2 + 6y - 4 = 0$$

$$y^2 + 2y - \frac{4}{3} = 0$$

$$y^2 + 2y = \frac{4}{3}$$

$$y^2 + 2y + \left(\frac{2}{2}\right)^2 = \frac{4}{3} + 1$$

$$y^2 + 2y + 1 = \frac{7}{3}$$

$$(y + 1)^2 = \frac{7}{3}$$

$$y + 1 = \pm\sqrt{\frac{7}{3}}$$

$$y + 1 = \pm\frac{\sqrt{7} \cdot \sqrt{3}}{\sqrt{3} \cdot \sqrt{3}}$$

$$y + 1 = \pm\frac{\sqrt{21}}{3}$$

$$y = -1 \pm \frac{\sqrt{21}}{3} = \frac{-3 \pm \sqrt{21}}{3}$$

The solutions are $\dfrac{-3 + \sqrt{21}}{3}$ and $\dfrac{-3 - \sqrt{21}}{3}$.

55.

$$2x^2 - 3x - 5 = 0$$

$$x^2 - \frac{3}{2}x - \frac{5}{2} = 0$$

$$x^2 - \frac{3}{2}x = \frac{5}{2}$$

$$x^2 - \frac{3}{2}x + \left(\frac{\frac{3}{2}}{2}\right)^2 = \frac{5}{2} + \frac{9}{16}$$

$$x^2 - \frac{3}{2}x + \frac{9}{16} = \frac{49}{16}$$

$$\left(x - \frac{3}{4}\right)^2 = \frac{49}{16}$$

$$x - \frac{3}{4} = \pm\sqrt{\frac{49}{16}}$$

$$x = \frac{3}{4} \pm \frac{7}{4} = \frac{3 \pm 7}{4}$$

$$x = -1 \text{ or } \frac{5}{2}$$

The solutions are -1 and $\frac{5}{2}$.

57.
$$y^2 + 2y + 2 = 0$$
$$y^2 + 2y = -2$$
$$y^2 + 2y + \left(\frac{2}{2}\right)^2 = -2 + 1$$
$$y^2 + 2y + 1 = -1$$
$$(y+1)^2 = -1$$
$$y + 1 = \pm\sqrt{-1}$$
$$y = -1 \pm i$$

The solutions are $-1 + i$ and $-1 - i$.

59.
$$y^2 + 6y - 8 = 0$$
$$y^2 + 6y = 8$$
$$y^2 + 6y + \left(\frac{6}{2}\right)^2 = 8 + 9$$
$$y^2 + 6y + 9 = 17$$
$$(y+3)^2 = 17$$
$$y + 3 = \pm\sqrt{17}$$
$$y = -3 \pm \sqrt{17}$$

The solutions are $-3 + \sqrt{17}$ and $-3 - \sqrt{17}$.

61.
$$2a^2 + 8a = -12$$
$$a^2 + 4a = -6$$
$$a^2 + 4a + \left(\frac{4}{2}\right)^2 = -6 + 4$$
$$a^2 + 4a + 4 = -2$$
$$(a+2)^2 = -2$$
$$a + 2 = \pm\sqrt{-2}$$
$$a + 2 = \pm i\sqrt{2}$$
$$a = -2 \pm i\sqrt{2}$$

The solutions are $-2 + i\sqrt{2}$ and $-2 - i\sqrt{2}$.

63.
$$5x^2 + 15x - 1 = 0$$
$$x^2 + 3x - \frac{1}{5} = 0$$
$$x^2 + 3x = \frac{1}{5}$$
$$x^2 + 3x + \left(\frac{3}{2}\right)^2 = \frac{1}{5} + \frac{9}{4}$$
$$x^2 + 3x + \frac{9}{4} = \frac{49}{20}$$
$$\left(x + \frac{3}{2}\right)^2 = \frac{49}{20}$$
$$x + \frac{3}{2} = \pm\sqrt{\frac{49}{20}}$$
$$x + \frac{3}{2} = \pm\frac{7}{\sqrt{20}}$$
$$x + \frac{3}{2} = \pm\frac{7}{2\sqrt{5}}$$
$$x + \frac{3}{2} = \pm\frac{7 \cdot \sqrt{5}}{2\sqrt{5} \cdot \sqrt{5}}$$
$$x + \frac{3}{2} = \pm\frac{7\sqrt{5}}{10}$$
$$x = -\frac{3}{2} \pm \frac{7\sqrt{5}}{10} = \frac{-15 \pm 7\sqrt{5}}{10}$$

The solutions are $\dfrac{-15 + 7\sqrt{5}}{10}$ and $\dfrac{-15 - 7\sqrt{5}}{10}$.

65.
$$2x^2 - x + 6 = 0$$
$$x^2 - \frac{1}{2}x + 3 = 0$$
$$x^2 - \frac{1}{2}x = -3$$
$$x^2 - \frac{1}{2}x + \left(\frac{-\frac{1}{2}}{2}\right)^2 = -3 + \frac{1}{16}$$
$$x^2 - \frac{1}{2}x + \frac{1}{16} = -\frac{47}{16}$$
$$\left(x - \frac{1}{4}\right)^2 = -\frac{47}{16}$$
$$x - \frac{1}{4} = \pm\sqrt{-\frac{47}{16}}$$
$$x - \frac{1}{4} = \pm i\frac{\sqrt{47}}{4}$$
$$x = \frac{1}{4} \pm i\frac{\sqrt{47}}{4} = \frac{1 \pm i\sqrt{47}}{4}$$

The solutions are $\dfrac{1 + i\sqrt{47}}{4}$ and $\dfrac{1 - i\sqrt{47}}{4}$ or $\dfrac{1}{4} + \dfrac{\sqrt{47}}{4}i$ and $\dfrac{1}{4} - \dfrac{\sqrt{47}}{4}i$.

67.
$$x^2 + 10x + 28 = 0$$
$$x^2 + 10x = -28$$
$$x^2 + 10x + \left(\frac{10}{2}\right)^2 = -28 + 25$$
$$(x + 5)^2 = -3$$
$$x + 5 = \pm\sqrt{-3}$$
$$x = -5 \pm i\sqrt{3}$$

The solutions are $-5 + i\sqrt{3}$ and $-5 - i\sqrt{3}$.

69.
$$z^2 + 3z - 4 = 0$$
$$z^2 + 3z = 4$$
$$z^2 + 3z + \left(\frac{3}{2}\right)^2 = 4 + \frac{9}{4}$$
$$z^2 + 3z + \frac{9}{4} = \frac{25}{4}$$
$$\left(z + \frac{3}{2}\right)^2 = \frac{25}{4}$$
$$z + \frac{3}{2} = \pm\sqrt{\frac{25}{4}}$$
$$z = -\frac{3}{2} \pm \frac{5}{2} = \frac{-3 \pm 5}{2}$$
$$z = -4 \text{ or } 1$$

The solutions are -4 and 1.

71.
$$2x^2 - 4x = -3$$
$$x^2 - 2x = -\frac{3}{2}$$
$$x^2 - 2x + \left(\frac{-2}{2}\right)^2 = -\frac{3}{2} + 1$$
$$x^2 - 2x + 1 = -\frac{1}{2}$$
$$(x - 1)^2 = -\frac{1}{2}$$
$$x - 1 = \pm\sqrt{-\frac{1}{2}}$$
$$x - 1 = \pm i\frac{1}{\sqrt{2}}$$
$$x - 1 = \pm i\frac{1 \cdot \sqrt{2}}{\sqrt{2} \cdot \sqrt{2}}$$
$$x - 1 = \pm i\frac{\sqrt{2}}{2}$$
$$x = 1 \pm i\frac{\sqrt{2}}{2} = \frac{2 \pm i\sqrt{2}}{2}$$

The solutions are $\dfrac{2 + i\sqrt{2}}{2}$ and $\dfrac{2 - i\sqrt{2}}{2}$ or $1 + \dfrac{\sqrt{2}}{2}i$ and $1 - \dfrac{\sqrt{2}}{2}i$.

73.
$$3x^2 + 3x = 5$$
$$x^2 + x = \frac{5}{3}$$
$$x^2 + x + \left(\frac{1}{2}\right)^2 = \frac{5}{3} + \frac{1}{4}$$
$$x^2 + x + \frac{1}{4} = \frac{23}{12}$$
$$\left(x + \frac{1}{2}\right)^2 = \frac{23}{12}$$
$$x + \frac{1}{2} = \pm\sqrt{\frac{23}{12}}$$
$$x + \frac{1}{2} = \pm\frac{\sqrt{23}}{2\sqrt{3}}$$
$$x + \frac{1}{2} = \pm\frac{\sqrt{23} \cdot \sqrt{3}}{2\sqrt{3} \cdot \sqrt{3}}$$
$$x + \frac{1}{2} = \pm\frac{\sqrt{69}}{6}$$
$$x = -\frac{1}{2} \pm \frac{\sqrt{69}}{6} = \frac{-3 \pm \sqrt{69}}{6}$$
The solutions are $\dfrac{-3 + \sqrt{69}}{6}$ and $\dfrac{-3 - \sqrt{69}}{6}$.

75.
$$A = P(1 + r)^t$$
$$4320 = 3000(1 + r)^2$$
$$\frac{4320}{3000} = (1 + r)^2$$
$$1.44 = (1 + r)^2$$
$$\pm\sqrt{1.44} = 1 + r$$
$$\pm 1.2 = 1 + r$$
$$-1 \pm 1.2 = r$$
$$-2.2 = r \quad \text{or} \quad 0.2 = r$$
Rate cannot be negative, so the rate is
$r = 0.2 = 20\%$.

77.
$$A = P(1 + r)^t$$
$$16{,}224 = 15{,}000(1 + r)^2$$
$$\frac{16{,}224}{15{,}000} = (1 + r)^2$$
$$\pm\sqrt{1.0816} = 1 + r$$
$$\pm 1.04 = 1 + r$$
$$-1 \pm 1.04 = r$$
$$0.04 = r \quad \text{or} \quad -2.04 = r$$
Rate cannot be negative, so the rate is $r = 0.04$,
or 4%.

79.
$$s(t) = 16t^2$$
$$1483 = 16t^2$$
$$t^2 = \frac{1483}{16}$$
$$t = \pm\sqrt{\frac{1483}{16}}$$
$$t \approx 9.63 \text{ or } -9.63 \text{ (disregard)}$$
It would take 9.63 seconds.

81.
$$s(t) = 16t^2$$
$$1100 = 16t^2$$
$$\frac{1100}{16} = t^2$$
$$\frac{275}{4} = t^2$$
$$\pm\sqrt{\frac{275}{4}} = t^2$$
$$\pm\frac{5\sqrt{11}}{2} = t$$
The time cannot be negative, so $-\dfrac{5\sqrt{11}}{2}$ is
rejected.
It takes the object $\dfrac{5\sqrt{11}}{2} \approx 8.29$ seconds to fall
from the top to the base of the dam.

83. Let x be the length of one side.
$$x^2 = 225$$
$$x = \pm\sqrt{225}$$
$$x = \pm 15$$
The length cannot be negative, so -15 is
rejected. The dimensions are 15 feet by 15 feet.

85.
$$a^2 + b^2 = c^2$$
$$x^2 + x^2 = 20^2$$
$$2x^2 = 400$$
$$x^2 = 200$$
$$x = \pm\sqrt{200}$$
$$x = \pm 10\sqrt{2}$$
The length cannot be negative, so $-10\sqrt{2}$ is
rejected. The length of each leg is
$10\sqrt{2}$ centimeters.

87. $\dfrac{1}{2} - \sqrt{\dfrac{9}{4}} = \dfrac{1}{2} - \dfrac{3}{2} = -\dfrac{2}{2} = -1$

89. $\dfrac{6+4\sqrt{5}}{2} = \dfrac{6}{2} + \dfrac{4\sqrt{5}}{2} = 3 + 2\sqrt{5}$

91. $\dfrac{3-9\sqrt{2}}{6} = \dfrac{3}{6} - \dfrac{9\sqrt{2}}{6} = \dfrac{1}{2} - \dfrac{3\sqrt{2}}{2} = \dfrac{1-3\sqrt{2}}{2}$

93. $\sqrt{b^2 - 4ac} = \sqrt{(4)^2 - 4(2)(-1)}$
$\qquad\qquad\quad = \sqrt{16+8}$
$\qquad\qquad\quad = \sqrt{24}$
$\qquad\qquad\quad = 2\sqrt{6}$

95. $\sqrt{b^2 - 4ac} = \sqrt{(-1)^2 - 4(3)(-2)}$
$\qquad\qquad\quad = \sqrt{1+24}$
$\qquad\qquad\quad = \sqrt{25}$
$\qquad\qquad\quad = 5$

97. The solutions of $(x+1)^2 = -1$ are complex, but not real numbers.

99. The solutions of $3z^2 = 10$ are real numbers.

101. The solutions of $(2y-5)^2 + 7 = 3$ are complex, but not real numbers.

103. $x^2 + \underline{\quad} + 16$
$\qquad \left(\dfrac{b}{2}\right)^2 = 16$
$\qquad\quad \dfrac{b}{2} = \pm\sqrt{16}$
$\qquad\quad \dfrac{b}{2} = \pm 4$
$\qquad\quad\ b = \pm 8$
Answer: $\pm 8x$

105. $z^2 + \underline{\quad} + \dfrac{25}{4}$
$\qquad \left(\dfrac{b}{2}\right)^2 = \dfrac{25}{4}$
$\qquad\quad \dfrac{b}{2} = \pm\sqrt{\dfrac{25}{4}}$
$\qquad\quad \dfrac{b}{2} = \pm\dfrac{5}{2}$
$\qquad\quad\ b = \pm 5$
Answer: $\pm 5z$

107. answers may vary

109. compound interest is preferable; answers may vary

111. $p = -x^2 + 47$
$\quad 11 = -x^2 + 47$
$\quad\ x^2 = 36$
$\qquad x = \pm\sqrt{36}$
$\qquad x = \pm 6$
Demand cannot be negative. Therefore, the demand is 6 thousand scissors.

Section 11.2 Practice

1. $3x^2 - 5x - 2 = 0$
$a = 3, b = -5, c = -2$
$x = \dfrac{-b \pm \sqrt{b^2 - 4ac}}{2a}$
$ = \dfrac{-(-5) \pm \sqrt{(-5)^2 - 4(3)(-2)}}{2(3)}$
$ = \dfrac{5 \pm \sqrt{25 + 24}}{6}$
$ = \dfrac{5 \pm \sqrt{49}}{6}$
$ = \dfrac{5 \pm 7}{6}$
$x = \dfrac{5+7}{6} = \dfrac{12}{6} = 2$ or $x = \dfrac{5-7}{6} = \dfrac{-2}{6} = -\dfrac{1}{3}$

The solutions are $-\dfrac{1}{3}$ and 2, or the solution set is $\left\{-\dfrac{1}{3}, 2\right\}$.

2. $3x^2 - 8x = 2$

Write in standard form.

$3x^2 - 8x - 2 = 0$

$a = 3,\ b = -8,\ c = -2$

$$x = \frac{-b \pm \sqrt{b^2 - 4ac}}{2a}$$

$$= \frac{-(-8) \pm \sqrt{(-8)^2 - 4(3)(-2)}}{2(3)}$$

$$= \frac{8 \pm \sqrt{64 + 24}}{6}$$

$$= \frac{8 \pm \sqrt{88}}{6}$$

$$= \frac{8 \pm 2\sqrt{22}}{6}$$

$$= \frac{4 \pm \sqrt{22}}{3}$$

The solutions are $\dfrac{4 + \sqrt{22}}{3}$ and $\dfrac{4 - \sqrt{22}}{3}$, or the

solution set is $\left\{ \dfrac{4 + \sqrt{22}}{3},\ \dfrac{4 - \sqrt{22}}{3} \right\}$.

3. $\dfrac{1}{8}x^2 - \dfrac{1}{4}x - 2 = 0$

Multiply both sides of the equation by 8.

$$8\left(\frac{1}{8}x^2 - \frac{1}{4}x - 2 \right) = 8 \cdot 0$$

$$x^2 - 2x - 16 = 0$$

Substitute $a = 1$, $b = -2$, and $c = -16$ into the quadratic formula and simplify.

$$x = \frac{-(-2) \pm \sqrt{(-2)^2 - 4(1)(-16)}}{2(1)}$$

$$= \frac{2 \pm \sqrt{4 + 64}}{2}$$

$$= \frac{2 \pm \sqrt{68}}{2}$$

$$= \frac{2 \pm 2\sqrt{17}}{2}$$

$$= 1 \pm \sqrt{17}$$

The solutions are $1 + \sqrt{17}$ or $1 - \sqrt{17}$.

4. $x = -2x^2 - 2$

The equation in standard form is

$2x^2 + x + 2 = 0$. Thus, let $a = 2$, $b = 1$, and $c = 2$ in the quadratic formula.

$$x = \frac{-1 \pm \sqrt{1^2 - 4(2)(2)}}{2(2)}$$

$$= \frac{-1 \pm \sqrt{1 - 16}}{4}$$

$$= \frac{-1 \pm \sqrt{-15}}{4}$$

$$= \frac{-1 \pm i\sqrt{15}}{4}$$

The solutions are $\dfrac{-1 + i\sqrt{15}}{4}$ and $\dfrac{-1 - i\sqrt{15}}{4}$ or

$-\dfrac{1}{4} + \dfrac{\sqrt{15}}{4}i$ and $-\dfrac{1}{4} - \dfrac{\sqrt{15}}{4}i$.

5. a. $x^2 - 6x + 9 = 0$

In $x^2 - 6x + 9$, $a = 1$, $b = -6$, and $c = 9$. Thus,

$b^2 - 4ac = (-6)^2 - 4(1)(9) = 36 - 36 = 0$

Since $b^2 - 4ac = 0$, this equation has one real solution.

b. $x^2 - 3x - 1 = 0$

In this equation, $a = 1$, $b = -3$, and $c = -1$.

$b^2 - 4ac = (-3)^2 - 4(1)(-1) = 9 + 4 = 13 > 0$

Since $b^2 - 4ac$ is positive, this equation has two real solutions.

c. $7x^2 + 11 = 0$

In this equation, $a = 7$, $b = 0$, and $c = 11$.

$b^2 - 4ac = 0^2 - 4(7)(11) = -308 < 0$

Since $b^2 - 4ac$ is negative, this equation has two complex but not real solutions.

6. By the Pythagorean theorem, we have

$$x^2 + (x + 3)^2 = 15^2$$

$$x^2 + x^2 + 6x + 9 = 225$$

$$2x^2 + 6x - 216 = 0$$

$$x^2 + 3x - 108 = 0$$

Here, $a = 1$, $b = 3$, and $c = -108$. By the quadratic formula,

$$x = \frac{-3 \pm \sqrt{3^2 - 4(1)(-108)}}{2(1)}$$

$$= \frac{-3 \pm \sqrt{9 + 432}}{2}$$

$$= \frac{-3 \pm \sqrt{441}}{2}$$

$$= \frac{-3 \pm 21}{2}$$

$$x = \frac{-3 + 21}{2} = \frac{18}{2} = 9 \text{ or}$$

$$x = \frac{-3 - 21}{2} = \frac{-24}{2} = -12$$

The length can't be negative, so reject -12. The distance along the sidewalk is
$x + (x + 3) = 2x + 3 = 2(9) + 3 = 18 + 3 = 21$ feet
A person can save $21 - 15 = 6$ feet by cutting across the lawn.

7. $h = -16t^2 + 20t + 45$
    At the ground, $h = 0$.
    $0 = -16t^2 + 20t + 45$
    Here, $a = -16$, $b = 20$, and $c = 45$. By the quadratic formula,

$$t = \frac{-20 \pm \sqrt{20^2 - 4(-16)(45)}}{2(-16)}$$

$$= \frac{-20 \pm \sqrt{400 + 2880}}{-32}$$

$$= \frac{-20 \pm \sqrt{3280}}{-32}$$

$$= \frac{20 \pm \sqrt{16 \cdot 205}}{32}$$

$$= \frac{20 \pm 4\sqrt{205}}{32}$$

$$= \frac{5 \pm \sqrt{205}}{8}$$

$$t = \frac{5 + \sqrt{205}}{8} \approx 2.4 \text{ or } t = \frac{5 - \sqrt{205}}{8} \approx -1.2$$

Since the time won't be negative, we reject -1.2. The rocket will strike the ground 2.4 seconds after launch.

Vocabulary, Readiness & Video Check 11.2

1. The quadratic formula is $x = \dfrac{-b \pm \sqrt{b^2 - 4ac}}{2a}$.

2. For $2x^2 + x + 1 = 0$, if $a = 2$, then $b = \underline{1}$ and $c = \underline{1}$.

3. For $5x^2 - 5x - 7 = 0$, if $a = 5$, then $b = \underline{-5}$ and $c = \underline{-7}$.

4. For $7x^2 - 4 = 0$, if $a = 7$, then $b = \underline{0}$ and $c = \underline{-4}$.

5. For $x^2 + 9 = 0$, if $c = 9$, then $a = \underline{1}$ and $b = \underline{0}$.

6. The correct simplified form of $\dfrac{5 \pm 10\sqrt{2}}{5}$ is $\underline{1 \pm 2\sqrt{2}}$. The answer is **c**.

7. **a.** Yes, in order to make sure we have correct values for a, b, and c.

    b. No; clearing fractions makes the work less tedious, but it's not a necessary step.

8. The discriminant is the <u>radicand</u> in the quadratic formula and can be used to find the number and type of solutions of a quadratic equation without <u>solving</u> the equation. To use the discriminant, the quadratic equation needs to be written in <u>standard</u> form.

9. With applications, we need to make sure we answer the question(s) asked. Here we're asked how much distance is saved, so once the dimensions of the triangle are known, further calculations are needed to answer this question and solve the problem.

Exercise Set 11.2

1. $m^2 + 5m - 6 = 0$
    $a = 1, b = 5, c = -6$

$$m = \frac{-5 \pm \sqrt{(5)^2 - 4(1)(-6)}}{2(1)}$$

$$= \frac{-5 \pm \sqrt{25 + 24}}{2}$$

$$= \frac{-5 \pm \sqrt{49}}{2}$$

$$= \frac{-5 \pm 7}{2}$$

$$= -6 \text{ or } 1$$

The solutions are -6 and 1.

3.
$$2y = 5y^2 - 3$$
$$5y^2 - 2y - 3 = 0$$
$$a = 5, \, b = -2, \, c = -3$$
$$y = \frac{2 \pm \sqrt{(-2)^2 - 4(5)(-3)}}{2(5)}$$
$$= \frac{2 \pm \sqrt{4 + 60}}{10}$$
$$= \frac{2 \pm \sqrt{64}}{10}$$
$$= \frac{2 \pm 8}{10}$$
$$= -\frac{3}{5} \text{ or } 1$$

The solutions are $-\dfrac{3}{5}$ and 1.

5. $x^2 - 6x + 9 = 0$
$$a = 1, \, b = -6, \, c = 9$$
$$x = \frac{6 \pm \sqrt{(-6)^2 - 4(1)(9)}}{2(1)}$$
$$= \frac{6 \pm \sqrt{36 - 36}}{2}$$
$$= \frac{6 \pm \sqrt{0}}{2}$$
$$= \frac{6}{2}$$
$$= 3$$
The solution is 3.

7. $x^2 + 7x + 4 = 0$
$$a = 1, \, b = 7, \, c = 4$$
$$x = \frac{-7 \pm \sqrt{(7)^2 - 4(1)(4)}}{2(1)}$$
$$= \frac{-7 \pm \sqrt{49 - 16}}{2}$$
$$= \frac{-7 \pm \sqrt{33}}{2}$$

The solutions are $\dfrac{-7 + \sqrt{33}}{2}$ and $\dfrac{-7 - \sqrt{33}}{2}$.

9.
$$8m^2 - 2m = 7$$
$$8m^2 - 2m - 7 = 0$$
$$a = 8, \, b = -2, \, c = -7$$
$$m = \frac{2 \pm \sqrt{(-2)^2 - 4(8)(-7)}}{2(8)}$$
$$= \frac{2 \pm \sqrt{4 + 224}}{16}$$
$$= \frac{2 \pm \sqrt{228}}{16}$$
$$= \frac{2 \pm \sqrt{4 \cdot 57}}{16}$$
$$= \frac{2 \pm 2\sqrt{57}}{16}$$
$$= \frac{1 \pm \sqrt{57}}{8}$$

The solutions are $\dfrac{1 + \sqrt{57}}{8}$ and $\dfrac{1 - \sqrt{57}}{8}$.

11.
$$3m^2 - 7m = 3$$
$$3m^2 - 7m - 3 = 0$$
$$a = 3, \, b = -7, \, c = -3$$
$$m = \frac{7 \pm \sqrt{(-7)^2 - 4(3)(-3)}}{2(3)}$$
$$= \frac{7 \pm \sqrt{49 + 36}}{6}$$
$$= \frac{7 \pm \sqrt{85}}{6}$$

The solutions are $\dfrac{7 + \sqrt{85}}{6}$ and $\dfrac{7 - \sqrt{85}}{6}$.

13. $\dfrac{1}{2}x^2 - x - 1 = 0$
$$x^2 - 2x - 2 = 0$$
$$a = 1, \, b = -2, \, c = -2$$
$$x = \frac{2 \pm \sqrt{(-2)^2 - 4(1)(-2)}}{2(1)}$$
$$= \frac{2 \pm \sqrt{4 + 8}}{2}$$
$$= \frac{2 \pm \sqrt{12}}{2}$$
$$= \frac{2 \pm 2\sqrt{3}}{2}$$
$$= 1 \pm \sqrt{3}$$
The solutions are $1 + \sqrt{3}$ and $1 - \sqrt{3}$.

15. $\dfrac{2}{5}y^2 + \dfrac{1}{5}y = \dfrac{3}{5}$

$2y^2 + y - 3 = 0$

$a = 2, b = 1, c = -3$

$y = \dfrac{-1 \pm \sqrt{(1)^2 - 4(2)(-3)}}{2(2)}$

$= \dfrac{-1 \pm \sqrt{1 + 24}}{4}$

$= \dfrac{-1 \pm \sqrt{25}}{4}$

$= \dfrac{-1 \pm 5}{4}$

$= -\dfrac{3}{2}$ or 1

The solutions are $-\dfrac{3}{2}$ and 1.

17. $\dfrac{1}{3}y^2 = y + \dfrac{1}{6}$

$\dfrac{1}{3}y^2 - y - \dfrac{1}{6} = 0$

$2y^2 - 6y - 1 = 0$

$a = 2, b = -6, c = -1$

$y = \dfrac{6 \pm \sqrt{(-6)^2 - 4(2)(-1)}}{2(2)}$

$= \dfrac{6 \pm \sqrt{36 + 8}}{4}$

$= \dfrac{6 \pm \sqrt{44}}{4}$

$= \dfrac{6 \pm 2\sqrt{11}}{4}$

$= \dfrac{3 \pm \sqrt{11}}{2}$

The solutions are $\dfrac{3 + \sqrt{11}}{2}$ and $\dfrac{3 - \sqrt{11}}{2}$.

19. $x^2 + 5x = -2$

$x^2 + 5x + 2 = 0$

$a = 1, b = 5, c = 2$

$x = \dfrac{-5 \pm \sqrt{(5)^2 - 4(1)(2)}}{2(1)}$

$= \dfrac{-5 \pm \sqrt{25 - 8}}{2}$

$= \dfrac{-5 \pm \sqrt{17}}{2}$

The solutions are $\dfrac{-5 + \sqrt{17}}{2}$ and $\dfrac{-5 - \sqrt{17}}{2}$.

21. $(m + 2)(2m - 6) = 5(m - 1) - 12$

$2m^2 - 6m + 4m - 12 = 5m - 5 - 12$

$2m^2 - 7m + 5 = 0$

$a = 2, b = -7, c = 5$

$m = \dfrac{7 \pm \sqrt{(-7)^2 - 4(2)(5)}}{2(2)}$

$= \dfrac{7 \pm \sqrt{49 - 40}}{4}$

$= \dfrac{7 \pm \sqrt{9}}{4}$

$= \dfrac{7 \pm 3}{4}$

$= 1$ or $\dfrac{5}{2}$

The solutions are 1 and $\dfrac{5}{2}$.

23. $x^2 + 6x + 13 = 0$

$a = 1, b = 6, c = 13$

$x = \dfrac{-6 \pm \sqrt{(6)^2 - 4(1)(13)}}{2(1)}$

$= \dfrac{-6 \pm \sqrt{36 - 52}}{2}$

$= \dfrac{-6 \pm \sqrt{-16}}{2}$

$= \dfrac{-6 \pm 4i}{2}$

$= -3 \pm 2i$

The solutions are $-3 + 2i$ and $-3 - 2i$.

25.
$$(x+5)(x-1) = 2$$
$$x^2 + 4x - 5 = 2$$
$$x^2 + 4x - 7 = 0$$
$$a = 1, b = 4, c = -7$$
$$x = \frac{-4 \pm \sqrt{(4)^2 - 4(1)(-7)}}{2(1)}$$
$$= \frac{-4 \pm \sqrt{16 + 28}}{2}$$
$$= \frac{-4 \pm \sqrt{44}}{2}$$
$$= \frac{-4 \pm 2\sqrt{11}}{2}$$
$$= -2 \pm \sqrt{11}$$
The solutions are $-2 + \sqrt{11}$ and $-2 - \sqrt{11}$.

27.
$$6 = -4x^2 + 3x$$
$$4x^2 - 3x + 6 = 0$$
$$a = 4, b = -3, c = 6$$
$$x = \frac{3 \pm \sqrt{(-3)^2 - 4(4)(6)}}{2(4)}$$
$$= \frac{3 \pm \sqrt{9 - 96}}{8}$$
$$= \frac{3 \pm \sqrt{-87}}{8}$$
$$= \frac{3 \pm i\sqrt{87}}{8}$$

The solutions are $\dfrac{3 + i\sqrt{87}}{8}$ and $\dfrac{3 - i\sqrt{87}}{8}$ or

$\dfrac{3}{8} + \dfrac{\sqrt{87}}{8}i$ and $\dfrac{3}{8} - \dfrac{\sqrt{87}}{8}i$.

29.
$$\frac{x^2}{3} - x = \frac{5}{3}$$
$$x^2 - 3x = 5$$
$$x^2 - 3x - 5 = 0$$
$$a = 1, b = -3, c = -5$$
$$x = \frac{3 \pm \sqrt{(-3)^2 - 4(1)(-5)}}{2(1)}$$
$$= \frac{3 \pm \sqrt{9 + 20}}{2}$$
$$= \frac{3 \pm \sqrt{29}}{2}$$

The solutions are $\dfrac{3 + \sqrt{29}}{2}$ and $\dfrac{3 - \sqrt{29}}{2}$.

31.
$$10y^2 + 10y + 3 = 0$$
$$a = 10, b = 10, c = 3$$
$$y = \frac{-10 \pm \sqrt{(10)^2 - 4(10)(3)}}{2(10)}$$
$$= \frac{-10 \pm \sqrt{100 - 120}}{20}$$
$$= \frac{-10 \pm \sqrt{-20}}{20}$$
$$= \frac{-10 \pm i\sqrt{4 \cdot 5}}{20}$$
$$= \frac{-10 \pm 2i\sqrt{5}}{20}$$
$$= \frac{-5 \pm i\sqrt{5}}{10}$$

The solutions are $\dfrac{-5 + i\sqrt{5}}{10}$ and $\dfrac{-5 - i\sqrt{5}}{10}$ or

$-\dfrac{1}{2} + \dfrac{\sqrt{5}}{10}i$ and $-\dfrac{1}{2} - \dfrac{\sqrt{5}}{10}i$.

33.
$$x(6x + 2) = 3$$
$$x(6x + 2) - 3 = 0$$
$$6x^2 + 2x - 3 = 0$$
$$a = 6, b = 2, c = -3$$
$$x = \frac{-2 \pm \sqrt{(2)^2 - 4(6)(-3)}}{2(6)}$$
$$= \frac{-2 \pm \sqrt{4 + 72}}{12}$$
$$= \frac{-2 \pm \sqrt{76}}{12}$$
$$= \frac{-2 \pm \sqrt{4 \cdot 19}}{12}$$
$$= \frac{-2 \pm 2\sqrt{19}}{12}$$
$$= \frac{-1 \pm \sqrt{19}}{6}$$

The solutions are $\dfrac{-1 + \sqrt{19}}{6}$ and $\dfrac{-1 - \sqrt{19}}{6}$.

35. $\dfrac{2}{5}y^2 + \dfrac{1}{5}y + \dfrac{3}{5} = 0$

$2y^2 + y + 3 = 0$

$a = 2, b = 1, c = 3$

$y = \dfrac{-1 \pm \sqrt{(1)^2 - 4(2)(3)}}{2(2)}$

$= \dfrac{-1 \pm \sqrt{1 - 24}}{4}$

$= \dfrac{-1 \pm \sqrt{-23}}{4}$

$= \dfrac{-1 \pm i\sqrt{23}}{4}$

The solutions are $\dfrac{-1 + i\sqrt{23}}{4}$ and $\dfrac{-1 - i\sqrt{23}}{4}$ or

$-\dfrac{1}{4} + \dfrac{\sqrt{23}}{4}i$ and $-\dfrac{1}{4} - \dfrac{\sqrt{23}}{4}i$.

37. $\dfrac{1}{2}y^2 = y - \dfrac{1}{2}$

$y^2 = 2y - 1$

$y^2 - 2y + 1 = 0$

$a = 1, b = -2, c = 1$

$y = \dfrac{2 \pm \sqrt{(-2)^2 - 4(1)(1)}}{2(1)}$

$= \dfrac{2 \pm \sqrt{4 - 4}}{2}$

$= \dfrac{2 \pm \sqrt{0}}{2}$

$= \dfrac{2}{2}$

$= 1$

The solution is 1.

39. $(n - 2)^2 = 2n$

$n^2 - 4n + 4 = 2n$

$n^2 - 6n + 4 = 0$

$a = 1, b = -6, c = 4$

$n = \dfrac{6 \pm \sqrt{(-6)^2 - 4(1)(4)}}{2(1)}$

$= \dfrac{6 \pm \sqrt{36 - 16}}{2}$

$= \dfrac{6 \pm \sqrt{20}}{2}$

$= \dfrac{6 \pm 2\sqrt{5}}{2}$

$= 3 \pm \sqrt{5}$

The solutions are $3 + \sqrt{5}$ and $3 - \sqrt{5}$.

41. $x^2 - 5 = 0$

$a = 1, b = 0, c = -5$

$b^2 - 4ac = 0^2 - 4(1)(-5) = 20 > 0$

Therefore, there are two real solutions.

43. $4x^2 + 12x = -9$

$4x^2 - 12x + 9 = 0$

$a = 4, b = -12, c = 9$

$b^2 - 4ac = (-12)^2 - 4(4)(9)$

$= 144 - 144$

$= 0$

Therefore, there is one real solution.

45. $3x = -2x^2 + 7$

$2x^2 + 3x - 7 = 0$

$a = 2, b = 3, c = -7$

$b^2 - 4ac = 3^2 - 4(2)(-7)$

$= 9 + 56$

$= 65 > 0$

Therefore, there are two real solutions.

47. $6 = 4x - 5x^2$

$5x^2 - 4x + 6 = 0$

$a = 5, b = -4, c = 6$

$b^2 - 4ac = (-4)^2 - 4(5)(6)$

$= 16 - 120$

$= -104 < 0$

Therefore, there are two complex but not real solutions.

49. $9x - 2x^2 + 5 = 0$

$-2x^2 + 9x + 5 = 0$

$a = -2, b = 9, c = 5$

$b^2 - 4ac = 9^2 - 4(-2)(5)$

$\qquad = 81 + 40$

$\qquad = 121 > 0$

Therefore, there are two real solutions.

51. $(x+8)^2 + x^2 = 36^2$

$(x^2 + 16x + 64) + x^2 = 1296$

$2x^2 + 16x - 1232 = 0$

$a = 2, b = 16, c = -1232$

$x = \dfrac{-16 \pm \sqrt{(16)^2 - 4(2)(-1232)}}{2(2)}$

$ = \dfrac{-16 \pm \sqrt{10{,}112}}{4}$

$x \approx 21$ or $x \approx -29$ (disregard)

$x + (x+8) = 21 + 21 + 8 = 50$

$50 - 36 = 14$

They save about 14 feet of walking distance.

53. Let x = length of leg. Then $x + 2$ = length of hypotenuse.

$x^2 + x^2 = (x+2)^2$

$2x^2 = x^2 + 4x + 4$

$x^2 - 4x - 4 = 0$

$a = 1, b = -4, c = -4$

$x = \dfrac{4 \pm \sqrt{(-4)^2 - 4(1)(-4)}}{2(1)}$

$ = \dfrac{4 \pm \sqrt{32}}{2}$

$ = \dfrac{4 \pm 4\sqrt{2}}{2}$

$ = 2 \pm 2\sqrt{2}$ (disregard the negative)

$ = 2 + 2\sqrt{2}$

The sides measure $\left(2 + 2\sqrt{2}\right)$ cm, $\left(2 + 2\sqrt{2}\right)$ cm, and $\left(4 + 2\sqrt{2}\right)$ cm.

55. Let x = width; then $x + 10$ = length.

Area = length · width

$400 = (x+10)x$

$0 = x^2 + 10x - 400$

$a = 1, b = 10, c = -400$

$x = \dfrac{-10 \pm \sqrt{(10)^2 - 4(1)(-400)}}{2(1)}$

$ = \dfrac{-10 \pm \sqrt{1700}}{2}$

$ = \dfrac{-10 \pm 10\sqrt{17}}{2}$

$ = -5 \pm 5\sqrt{17}$

Disregard the negative length. The width is $\left(-5 + 5\sqrt{17}\right)$ ft and the length is $\left(5 + 5\sqrt{17}\right)$ ft.

57. a. Let x = length.

$x^2 + x^2 = 100^2$

$2x^2 - 10{,}000 = 0$

$a = 2, b = 0, c = -10{,}000$

$x = \dfrac{0 \pm \sqrt{(0)^2 - 4(2)(-10{,}000)}}{2(2)}$

$ = \dfrac{\pm\sqrt{80{,}000}}{4}$

$ = \dfrac{\pm 200\sqrt{2}}{4}$

$ = \pm 50\sqrt{2}$

Disregard the negative length. The side measures $50\sqrt{2}$ meters.

b. Area $= s^2$

$ = \left(50\sqrt{2}\right)^2$

$ = 2500(2)$

$ = 5000$

The area is 5000 square meters.

59. Let w = width; then $w + 1.1$ = height.

Area = length · width

$1439.9 = (w + 1.1)w$

$0 = w^2 + 1.1w - 1439.9$

$a = 1, b = 1.1, c = -1439.9$

$w = \dfrac{-1.1 \pm \sqrt{(1.1)^2 - 4(1)(-1439.9)}}{2(1)}$

$ = \dfrac{-1.1 \pm \sqrt{5760.81}}{2}$

$ = 37.4$ or -38.5 (disregard)

Its width is 37.4 ft and its height is 38.5 ft.

61. Let $h =$ height. Then $2h + 4 =$ base.

$\text{Area} = \dfrac{1}{2} \text{base} \cdot \text{height}$

$42 = \dfrac{1}{2}(2h + 4)h$

$42 = h^2 + 2h$

$0 = h^2 + 2h - 42$

$a = 1, b = 2, c = -42$

$h = \dfrac{-2 \pm \sqrt{(2)^2 - 4(1)(-42)}}{2(1)}$

$= \dfrac{-2 \pm \sqrt{172}}{2}$

$= \dfrac{-2 \pm 2\sqrt{43}}{2}$

$= -1 \pm \sqrt{43}$ (disregard the negative)

$\text{base} = 2\left(-1 + \sqrt{43}\right) + 4 = 2 + 2\sqrt{43}$

Height: $\left(-1 + \sqrt{43}\right)$ cm

Base: $\left(2 + 2\sqrt{43}\right)$ cm

63. $h = -16t^2 + 20t + 1100$

$0 = -16t^2 + 20t + 1100$

$a = -16, b = 20, c = 1100$

$t = \dfrac{-20 \pm \sqrt{(20)^2 - 4(-16)(1100)}}{2(-16)}$

$= \dfrac{-20 \pm \sqrt{70,800}}{-32}$

≈ 8.9 or -7.7 (disregard)

It will take about 8.9 seconds.

65. $h = -16t^2 - 20t + 180$

$0 = -16t^2 - 20t + 180$

$a = -16, b = -20, c = 180$

$t = \dfrac{20 \pm \sqrt{(-20)^2 - 4(-16)(180)}}{2(-16)}$

$= \dfrac{20 \pm \sqrt{11,920}}{-32}$

≈ 2.8 or -4.0 (disregard)

It will take about 2.8 seconds.

67. $\sqrt{5x - 2} = 3$

$\left(\sqrt{5x - 2}\right)^2 = 3^2$

$5x - 2 = 9$

$5x = 11$

$x = \dfrac{11}{5}$

69. $\dfrac{1}{x} + \dfrac{2}{5} = \dfrac{7}{x}$

$5x\left(\dfrac{1}{x} + \dfrac{2}{5}\right) = 5x\left(\dfrac{7}{x}\right)$

$5 + 2x = 35$

$2x = 30$

$x = 15$

71. $x^4 + x^2 - 20 = (x^2 + 5)(x^2 - 4)$

$\qquad\qquad\qquad = (x^2 + 5)(x + 2)(x - 2)$

73. $z^4 - 13z^2 + 36 = (z^2 - 9)(z^2 - 4)$

$\qquad\qquad\qquad\quad = (z + 3)(z - 3)(z + 2)(z - 2)$

75. $x^2 = -10$

$x^2 + 10 = 0$

$a = 1, b = 0, c = 10$

The correct substitution is **b.**

77. $m^2 + 5m - 6 = 0$

$(m + 6)(m - 1) = 0$

$m + 6 = 0 \quad$ or $\quad m - 1 = 0$

$\qquad m = -6$ or $\qquad m = 1$

The results are the same. answers may vary.

79. $2x^2 - 6x + 3 = 0$

$a = 2, b = -6, c = 3$

$x = \dfrac{6 \pm \sqrt{(-6)^2 - 4(2)(3)}}{2(2)}$

$= \dfrac{6 \pm \sqrt{12}}{4}$

≈ 0.6 or 2.4

81. From Sunday to Monday

83. Wednesday

85. $f(x) = 3x^2 - 18x + 56$

$f(4) = 3(4)^2 - 18(4) + 56 = 32$

This answers appears to agree with the graph.

87. a. 2010 is 10 years after 2000, so $x = 10$.

$$f(x) = 22x^2 + 274x + 15,628$$

$$f(10) = 22(10)^2 + 274(10) + 15,628$$
$$= 20,568$$

The number of college students in the United States in 2010 was 20,568 thousand students.

b. $24,500 = 22x^2 + 274x + 15,628$

$$0 = 22x^2 + 274x - 8872$$

$$0 = 2(11x^2 + 137x - 4436)$$

$$a = 11, \ b = 137, \ c = -4436$$

$$x = \frac{-b \pm \sqrt{b^2 - 4ac}}{2a}$$

$$x = \frac{-137 \pm \sqrt{137^2 - 4(11)(-4436)}}{2(11)}$$

$$x \approx \frac{-137 \pm 462.55}{22}$$

$x \approx 15$ (reject a negative time because we are not concerned with the past)

The model predicts 24,500 thousand students 15 years after 2000, or in 2015.

89. a. 2004 is 4 years after 2000, so $x = 4$.

$$y = -10x^2 + 193x + 8464$$

$$y = -10(4)^2 + 193(4) + 8464 = 9076$$

The average total daily supply in 2004 was 9076 thousand barrels per day.

b. $9325 = -10x^2 + 193x + 8464$

$$10x^2 - 193x + 861 = 0$$

$$a = 10, \ b = -193, \ c = 861$$

$$x = \frac{-b \pm \sqrt{b^2 - 4ac}}{2a}$$

$$x = \frac{193 \pm \sqrt{(-193)^2 - 4(10)(861)}}{2(10)}$$

$$= \frac{193 \pm \sqrt{2809}}{20}$$

$$= \frac{193 \pm 53}{20}$$

$x = 12.3$ or $x = 7$

The first time the supply reaches this level is 7 years after 2000, or in 2007. The supply was 9325 thousand barrels per day in 2007.

c. 12.3 years after 2000, or in 2012, the supply will be 9325 thousand barrels per day.

91. $\dfrac{-b + \sqrt{b^2 - 4ac}}{2a} + \dfrac{-b - \sqrt{b^2 - 4ac}}{2a}$

$$= \frac{-b + \sqrt{b^2 - 4ac} - b - \sqrt{b^2 - 4ac}}{2a}$$

$$= \frac{-2b}{2a}$$

$$= -\frac{b}{a}$$

93. $3x^2 - \sqrt{12}x + 1 = 0$

$$a = 3, \ b = -\sqrt{12}, \ c = 1$$

$$x = \frac{\sqrt{12} \pm \sqrt{\left(-\sqrt{12}\right)^2 - 4(3)(1)}}{2(3)}$$

$$= \frac{\sqrt{12} \pm \sqrt{12 - 12}}{6}$$

$$= \frac{\sqrt{4 \cdot 3} \pm \sqrt{0}}{6}$$

$$= \frac{2\sqrt{3}}{6}$$

$$= \frac{\sqrt{3}}{3}$$

The solution is $\dfrac{\sqrt{3}}{3}$.

95. $x^2 + \sqrt{2}x + 1 = 0$

$$a = 1, \ b = \sqrt{2}, \ c = 1$$

$$x = \frac{-\sqrt{2} \pm \sqrt{\left(\sqrt{2}\right)^2 - 4(1)(1)}}{2(1)}$$

$$= \frac{-\sqrt{2} \pm \sqrt{2 - 4}}{2}$$

$$= \frac{-\sqrt{2} \pm \sqrt{-2}}{2}$$

$$= \frac{-\sqrt{2} \pm i\sqrt{2}}{2}$$

The solutions are $\dfrac{-\sqrt{2} + i\sqrt{2}}{2}$ and $\dfrac{-\sqrt{2} - i\sqrt{2}}{2}$ or

$-\dfrac{\sqrt{2}}{2} + \dfrac{\sqrt{2}}{2}i$ and $-\dfrac{\sqrt{2}}{2} - \dfrac{\sqrt{2}}{2}i$.

97. $2x^2 - \sqrt{3}x - 1 = 0$

$a = 2, b = -\sqrt{3}, c = -1$

$$x = \frac{\sqrt{3} \pm \sqrt{\left(-\sqrt{3}\right)^2 - 4(2)(-1)}}{2(2)}$$

$$= \frac{\sqrt{3} \pm \sqrt{3 + 8}}{4}$$

$$= \frac{\sqrt{3} \pm \sqrt{11}}{4}$$

The solutions are $\dfrac{\sqrt{3} + \sqrt{11}}{4}$ and $\dfrac{\sqrt{3} - \sqrt{11}}{4}$.

99. Exercise 63:

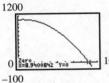

Exercise 65:

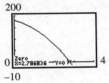

101. $y = 9x - 2x^2 + 5$

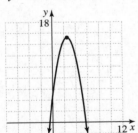

There are two *x*-intercepts. There are two real solutions.

Section 11.3 Practice

1. $x - \sqrt{x+1} - 5 = 0$

Get the radical alone on one side of the equation. Then square both sides.

$$x - \sqrt{x+1} - 5 = 0$$
$$x - 5 = \sqrt{x+1}$$
$$(x-5)^2 = x+1$$
$$x^2 - 10x + 25 = x + 1$$
$$x^2 - 11x + 24 = 0$$
$$(x-8)(x-3) = 0$$

$x - 8 = 0$ or $x - 3 = 0$

$x = 8$ or $x = 3$

Check:

Let $x = 3$.

$x - \sqrt{x+1} - 5 = 0$

$3 - \sqrt{3+1} - 5 \stackrel{?}{=} 0$

$-2 - \sqrt{4} \stackrel{?}{=} 0$

$-2 - 2 \stackrel{?}{=} 0$

$-4 = 0$ False

Let $x = 8$.

$x - \sqrt{x+1} - 5 = 0$

$8 - \sqrt{8+1} - 5 \stackrel{?}{=} 0$

$3 - \sqrt{9} \stackrel{?}{=} 0$

$3 - 3 \stackrel{?}{=} 0$

$0 = 0$ True

The solution is 8 or the solution set is $\{8\}$.

2. $\dfrac{5x}{x+1} - \dfrac{x+4}{x} = \dfrac{3}{x(x+1)}$

x cannot be either -1 or 0, because these values cause denominators to equal zero. Multiply both sides of the equation by $x(x + 1)$.

$x(x+1)\left(\dfrac{5x}{x+1}\right) - x(x+1)\left(\dfrac{x+4}{x}\right) = x(x+1)\left[\dfrac{3}{x(x+1)}\right]$

$5x^2 - (x+1)(x+4) = 3$

$5x^2 - x^2 - 5x - 4 = 3$

$4x^2 - 5x - 7 = 0$

Use the quadratic formula with $a = 4$, $b = -5$, and $c = -7$.

$x = \dfrac{-(-5) \pm \sqrt{(-5)^2 - 4(4)(-7)}}{2(4)} = \dfrac{5 \pm \sqrt{25 + 112}}{8} = \dfrac{5 \pm \sqrt{137}}{8}$

Neither proposed solution will make denominators 0. The solutions are $\dfrac{5+\sqrt{137}}{8}$ and $\dfrac{5-\sqrt{137}}{8}$ or the solution

set is $\left\{\dfrac{5+\sqrt{137}}{8}, \dfrac{5-\sqrt{137}}{8}\right\}$.

3. $p^4 - 7p^2 - 144 = 0$

$(p^2 + 9)(p^2 - 16) = 0$

$(p^2 + 9)(p + 4)(p - 4) = 0$

$p^2 + 9 = 0$ or $p + 4 = 0$ or $p - 4 = 0$

$p^2 = -9$ $p = -4$ $p = 4$

$p = \pm\sqrt{-9}$

$p = \pm 3i$

The solutions are 4, -4, $3i$, and $-3i$.

4. $(x+2)^2 - 2(x+2) - 3 = 0$

Let $y = x + 2$.

$y^2 - 2y - 3 = 0$

$(y-3)(y+1) = 0$

$y - 3 = 0$ or $y + 1 = 0$

     $y = 3$ $y = -1$

Substitute $x + 2$ for y.

$x + 2 = 3$ or $x + 2 = -1$

    $x = 1$ $x = -3$

Both 1 and -3 check. The solutions are 1 and -3.

5. $x^{2/3} - 5x^{1/3} + 4 = 0$

Let $m = x^{1/3}$.

$m^2 - 5m + 4 = 0$

$(m-4)(m-1) = 0$

$m - 4 = 0$ or $m - 1 = 0$

     $m = 4$ $m = 1$

Since $m = x^{1/3}$, we have

$x^{1/3} = 4$ or $x^{1/3} = 1$

   $x = 4^3 = 64$ $x = 1^3 = 1$

Both 64 and 1 check. The solutions are 64 and 1.

6. Let $x =$ the time in hours it takes Steve to groom all the dogs. Then,

$x - 1 =$ the time it takes Katy to groom all the dogs.

The part of the job completed in one hour by Steve is $\dfrac{1}{x}$, and the part completed by Katy in one hour is $\dfrac{1}{x-1}$. In

one hour, $\dfrac{1}{4}$ of the job is completed. We have,

$$\frac{1}{x} + \frac{1}{x-1} = \frac{1}{4}$$

$$4x(x-1)\left(\frac{1}{x}\right) + 4x(x-1)\left(\frac{1}{x-1}\right) = 4x(x-1)\left(\frac{1}{4}\right)$$

$$4(x-1) + 4x = x(x-1)$$

$$4x - 4 + 4x = x^2 - x$$

$$0 = x^2 - 9x + 4$$

Use the quadratic formula with $a = 1$, $b = -9$, and $c = 4$.

$$x = \frac{-(-9) \pm \sqrt{(-9)^2 - 4(1)(4)}}{2(1)}$$

$$x = \frac{9 \pm \sqrt{81 - 16}}{2} = \frac{9 \pm \sqrt{65}}{2}$$

$x \approx 8.53$ or $x \approx 0.47$

Since $x - 1 = 0.47 - 1 = -0.53 < 0$, representing negative time worked, we reject 0.47. It takes Steve

$\dfrac{9 + \sqrt{65}}{2} \approx 8.5$ hours and Katy $\dfrac{9 + \sqrt{65}}{2} - 1 = \dfrac{7 + \sqrt{65}}{2} \approx 7.5$ hours to groom all the dogs when working alone.

7. Let x = the speed driven to Shanghai. Then $x + 50$ = the speed driven to Ningbo.

	distance =	rate	·	time
To Shanghai	36	x		$\frac{36}{x}$
To Ningbo	36	$x + 50$		$\frac{36}{x+50}$

The total travel time was 1.3 hours, so

$$\frac{36}{x} + \frac{36}{x+50} = 1.3$$

$$x(x+50)\left(\frac{36}{x}\right) + x(x+50)\left(\frac{36}{x+50}\right) = 1.3x(x+50)$$

$$36(x+50) + 36x = 1.3x^2 + 65x$$

$$36x + 1800 + 36x = 1.3x^2 + 65x$$

$$0 = 1.3x^2 - 7x - 1800$$

Use the quadratic formula with $a = 1.3$, $b = -7$, and $c = -1800$.

$$x = \frac{-(-7) \pm \sqrt{(-7)^2 - 4(1.3)(-1800)}}{2(1.3)} = \frac{7 \pm \sqrt{9409}}{2.6}$$

$$x = \frac{7 + \sqrt{9409}}{2.6} = 40 \quad \text{or} \quad x = \frac{7 - \sqrt{9409}}{2.6} \approx -34.6$$

The speed is not negative, so reject −34.6. The speed to Shanghai was 40 km/hr and to Ningbo it was 40 + 50 = 90 km/hr.

Vocabulary, Readiness & Video Check 11.3

1. The values we get for the substituted variable are *not* our final answers. Remember to always substitute back to the original variable and solve for it if necessary.

2. The rational equation simplifies to a quadratic equation once you multiply through by the LCD to rid the equation of fractions.

Exercise Set 11.3

1.
$$2x = \sqrt{10 + 3x}$$
$$4x^2 = 10 + 3x$$
$$4x^2 - 3x - 10 = 0$$
$$(4x + 5)(x - 2) = 0$$
$$4x + 5 = 0 \quad \text{or} \quad x - 2 = 0$$
$$x = -\frac{5}{4} \quad \text{or} \quad x = 2$$

Discard $-\frac{5}{4}$. The solution is 2.

3.
$$x - 2\sqrt{x} = 8$$
$$x - 8 = 2\sqrt{x}$$
$$(x-8)^2 = \left(2\sqrt{x}\right)^2$$
$$x^2 - 16x + 64 = 4x$$
$$x^2 - 20x + 64 = 0$$
$$(x-16)(x-4) = 0$$
$$x - 16 = 0 \quad \text{or} \quad x - 4 = 0$$
$$x = 16 \quad \text{or} \quad x = 4 \text{ (discard)}$$
The solution is 16.

5.
$$\sqrt{9x} = x + 2$$
$$\left(\sqrt{9x}\right)^2 = (x+2)^2$$
$$9x = x^2 + 4x + 4$$
$$0 = x^2 - 5x + 4$$
$$0 = (x-4)(x-1)$$
$$x - 4 = 0 \quad \text{or} \quad x - 1 = 0$$
$$x = 4 \quad \text{or} \quad x = 1$$
The solutions are 1 and 4.

7. $\dfrac{2}{x} + \dfrac{3}{x-1} = 1$

Multiply each term by $x(x-1)$.
$$2(x-1) + 3x = x(x-1)$$
$$2x - 2 + 3x = x^2 - x$$
$$0 = x^2 - 6x + 2$$
$$x = \frac{6 \pm \sqrt{(-6)^2 - 4(1)(2)}}{2(1)}$$
$$= \frac{6 \pm \sqrt{28}}{2}$$
$$= \frac{6 \pm 2\sqrt{7}}{2} = 3 \pm \sqrt{7}$$

The solutions are $3 + \sqrt{7}$ and $3 - \sqrt{7}$.

9. $\dfrac{3}{x} + \dfrac{4}{x+2} = 2$

Multiply each term by $x(x+2)$.
$$3(x+2) + 4x = 2x(x+2)$$
$$3x + 6 + 4x = 2x^2 + 4x$$
$$0 = 2x^2 - 3x - 6$$

$$x = \frac{3 \pm \sqrt{(-3)^2 - 4(2)(-6)}}{2(2)}$$
$$= \frac{3 \pm \sqrt{57}}{4}$$
The solutions are $\dfrac{3+\sqrt{57}}{4}$ and $\dfrac{3-\sqrt{57}}{4}$.

11. $\dfrac{7}{x^2 - 5x + 6} = \dfrac{2x}{x-3} - \dfrac{x}{x-2}$

$$\frac{7}{(x-3)(x-2)} = \frac{2x}{x-3} - \frac{x}{x-2}$$
Multiply each term by $(x-3)(x-2)$.
$$7 = 2x(x-2) - x(x-3)$$
$$7 = 2x^2 - 4x - x^2 + 3x$$
$$0 = x^2 - x - 7$$
$$x = \frac{1 \pm \sqrt{(-1)^2 - 4(1)(-7)}}{2(1)}$$
$$= \frac{1 \pm \sqrt{29}}{2}$$

The solutions are $\dfrac{1+\sqrt{29}}{2}$ and $\dfrac{2-\sqrt{29}}{2}$.

13.
$$p^4 - 16 = 0$$
$$(p^2 - 4)(p^2 + 4) = 0$$
$$(p+2)(p-2)(p^2 + 4) = 0$$
$$p + 2 = 0 \quad \text{or} \quad p - 2 = 0 \text{ or } p^2 + 4 = 0$$
$$p = -2 \text{ or} \quad p = 2 \text{ or} \quad p^2 = -4$$
$$p = \pm\sqrt{-4}$$
$$p = \pm 2i$$
The solutions are -2, 2, $-2i$, and $2i$.

15.
$$4x^4 + 11x^2 = 3$$
$$4x^4 + 11x^2 - 3 = 0$$
$$(4x^2 - 1)(x^2 + 3) = 0$$
$$(2x+1)(2x-1)(x^2 + 3) = 0$$
$$2x + 1 = 0 \quad \text{or} \quad 2x - 1 = 0 \text{ or } x^2 + 3 = 0$$
$$x = -\frac{1}{2} \text{ or} \quad x = \frac{1}{2} \text{ or} \quad x^2 = -3$$
$$x = \pm\sqrt{-3}$$
$$x = \pm i\sqrt{3}$$

The solutions are $-\dfrac{1}{2}, \dfrac{1}{2}, -i\sqrt{3}$, and $i\sqrt{3}$.

17.
$$z^4 - 13z^2 + 36 = 0$$
$$(z^2 - 9)(z^2 - 4) = 0$$
$$(z+3)(z-3)(z+2)(z-2) = 0$$
$$z = -3, z = 3, z = -2, z = 2$$
The solutions are –3, 3, –2, and 2.

19. $x^{2/3} - 3x^{1/3} - 10 = 0$
Let $y = x^{1/3}$. Then $y^2 = x^{2/3}$ and
$$y^2 - 3y - 10 = 0$$
$$(y-5)(y+2) = 0$$
$$y - 5 = 0 \quad \text{or} \quad y + 2 = 0$$
$$y = 5 \quad \text{or} \quad y = -2$$
$$x^{1/3} = 5 \quad \text{or} \quad x^{1/3} = -2$$
$$x = 125 \text{ or} \quad x = -8$$
The solutions are –8 and 125.

21. $(5n+1)^2 + 2(5n+1) - 3 = 0$
Let $y = 5n + 1$. Then $y^2 = (5n+1)^2$ and
$$y^2 + 2y - 3 = 0$$
$$(y+3)(y-1) = 0$$
$$y + 3 = 0 \quad \text{or} \quad y - 1 = 0$$
$$y = -3 \text{ or} \quad y = 1$$
$$5n + 1 = -3 \text{ or } 5n + 1 = 1$$
$$5n = -4 \text{ or} \quad 5n = 0$$
$$n = -\frac{4}{5} \quad \text{or} \quad n = 0$$

The solutions are $-\frac{4}{5}$ and 0.

23. $2x^{2/3} - 5x^{1/3} = 3$
Let $y = x^{1/3}$. Then $y^2 = x^{2/3}$ and
$$2y^2 - 5y = 3$$
$$2y^2 - 5y - 3 = 0$$
$$(2y+1)(y-3) = 0$$
$$2y + 1 = 0 \quad \text{or} \quad y - 3 = 0$$
$$y = -\frac{1}{2} \quad \text{or} \quad y = 3$$
$$x^{1/3} = -\frac{1}{2} \quad \text{or} \quad x^{1/3} = 3$$
$$x = -\frac{1}{8} \quad \text{or} \quad x = 27$$

The solutions are $-\frac{1}{8}$ and 27.

25.
$$1 + \frac{2}{3t-2} = \frac{8}{(3t-2)^2}$$
$$(3t-2)^2 + 2(3t-2) = 8$$
$$(3t-2)^2 + 2(3t-2) - 8 = 0$$
Let $y = 3t - 2$. Then $y^2 = (3t-2)^2$ and
$$y^2 + 2y - 8 = 0$$
$$(y+4)(y-2) = 0$$
$$y + 4 = 0 \quad \text{or} \quad y - 2 = 0$$
$$y = -4 \text{ or} \quad y = 2$$
$$3t - 2 = -4 \text{ or } 3t - 2 = 2$$
$$3t = -2 \text{ or} \quad 3t = 4$$
$$t = -\frac{2}{3} \quad \text{or} \quad t = \frac{4}{3}$$

The solutions are $-\frac{2}{3}$ and $\frac{4}{3}$.

27. $20x^{2/3} - 6x^{1/3} - 2 = 0$
Let $y = x^{1/3}$. Then $y^2 = x^{2/3}$ and
$$20y^2 - 6y - 2 = 0$$
$$2(10y^2 - 3y - 1) = 0$$
$$2(5y+1)(2y-1) = 0$$
$$5y + 1 = 0 \quad \text{or} \quad 2y - 1 = 0$$
$$y = -\frac{1}{5} \quad \text{or} \quad y = \frac{1}{2}$$
$$x^{1/3} = -\frac{1}{5} \quad \text{or} \quad x^{1/3} = \frac{1}{2}$$
$$x = -\frac{1}{125} \quad \text{or} \quad x = \frac{1}{8}$$

The solutions are $\frac{1}{8}$ and $-\frac{1}{125}$.

29.
$$a^4 - 5a^2 + 6 = 0$$
$$(a^2 - 3)(a^2 - 2) = 0$$
$$a^2 - 3 = 0 \quad \text{or } a^2 - 2 = 0$$
$$a^2 = 3 \quad \text{or} \quad a^2 = 2$$
$$a = \pm\sqrt{3} \text{ or} \quad a = \pm\sqrt{2}$$
The solutions are $-\sqrt{3}, \sqrt{3}, -\sqrt{2}$, and $\sqrt{2}$.

31. $\frac{2x}{x-2} + \frac{x}{x+3} = -\frac{5}{x+3}$
Multiply each term by $(x+3)(x-2)$.
$$2x(x+3) + x(x-2) = -5(x-2)$$
$$2x^2 + 6x + x^2 - 2x = -5x + 10$$
$$3x^2 + 9x - 10 = 0$$

$$x = \frac{-9 \pm \sqrt{(9)^2 - 4(3)(-10)}}{2(3)}$$

$$= \frac{-9 \pm \sqrt{201}}{6}$$

The solutions are $\dfrac{-9 + \sqrt{201}}{6}$ and $\dfrac{-9 - \sqrt{201}}{6}$.

33.
$$(p+2)^2 = 9(p+2) - 20$$
$$(p+2)^2 - 9(p+2) + 20 = 0$$

Let $x = p + 2$. Then $x^2 = (p+2)^2$ and

$$x^2 - 9x + 20 = 0$$
$$(x-5)(x-4) = 0$$
$$x = 5 \quad \text{or} \quad x = 4$$
$$p + 2 = 5 \quad \text{or} \quad p + 2 = 4$$
$$p = 3 \quad \text{or} \quad p = 2$$

The solutions are 2 and 3.

35.
$$2x = \sqrt{11x + 3}$$
$$(2x)^2 = \left(\sqrt{11x+3}\right)^2$$
$$4x^2 = 11x + 3$$
$$4x^2 - 11x - 3 = 0$$
$$(4x+1)(x-3) = 0$$
$$x = -\frac{1}{4} \text{ (discard)} \quad \text{or} \quad x = 3$$

The solution is 3.

37. $x^{2/3} - 8x^{1/3} + 15 = 0$

Let $y = x^{1/3}$. Then $y^2 = x^{2/3}$ and

$$y^2 - 8y + 15 = 0$$
$$(y-5)(y-3) = 0$$
$$y = 5 \quad \text{or} \quad y = 3$$
$$x^{1/3} = 5 \quad \text{or} \quad x^{1/3} = 3$$
$$x = 125 \quad \text{or} \quad x = 27$$

The solutions are 27 and 125.

39.
$$y^3 + 9y - y^2 - 9 = 0$$
$$y(y^2 + 9) - 1(y^2 + 9) = 0$$
$$(y^2 + 9)(y - 1) = 0$$
$$y^2 + 9 = 0 \quad \text{or } y - 1 = 0$$
$$y^2 = -9 \quad \text{or} \quad y = 1$$
$$y = \pm\sqrt{-9}$$
$$y = \pm 3i$$

The solutions are 1, $-3i$, and $3i$.

41. $2x^{2/3} + 3x^{1/3} - 2 = 0$

Let $y = x^{1/3}$. Then $y^2 = x^{2/3}$ and

$$2y^2 + 3y - 2 = 0$$
$$(2y - 1)(y + 2) = 0$$
$$y = \frac{1}{2} \quad \text{or} \quad y = -2$$
$$x^{1/3} = \frac{1}{2} \quad \text{or } x^{1/3} = -2$$
$$x = \frac{1}{8} \quad \text{or} \quad x = -8$$

The solutions are -8 and $\dfrac{1}{8}$.

43. $x^{-2} - x^{-1} - 6 = 0$

Let $y = x^{-1}$. Then $y^2 = x^{-2}$ and

$$y^2 - y - 6 = 0$$
$$(y - 3)(y + 2) = 0$$
$$y = 3 \quad \text{or} \quad y = -2$$
$$x^{-1} = 3 \quad \text{or} \quad x^{-1} = -2$$
$$\frac{1}{x} = 3 \quad \text{or} \quad \frac{1}{x} = -2$$
$$x = \frac{1}{3} \quad \text{or} \quad x = -\frac{1}{2}$$

The solutions are $-\dfrac{1}{2}$ and $\dfrac{1}{3}$.

45.
$$x - \sqrt{x} = 2$$
$$x - 2 = \sqrt{x}$$
$$(x - 2)^2 = x$$
$$x^2 - 4x + 4 = x$$
$$x^2 - 5x + 4 = 0$$
$$(x - 4)(x - 1) = 0$$
$$x = 4 \text{ or } x = 1 \text{ (discard)}$$

The solution is 4.

47.
$$\frac{x}{x-1} + \frac{1}{x+1} = \frac{2}{x^2 - 1}$$
$$\frac{x}{x-1} + \frac{1}{x+1} = \frac{2}{(x+1)(x-1)}$$
$$x(x+1) + (x-1) = 2$$
$$x^2 + x + x - 1 = 2$$
$$x^2 + 2x - 3 = 0$$
$$(x+3)(x-1) = 0$$
$$x = -3 \text{ or } x = 1 \text{ (discard)}$$

The solution is -3.

49.

$$p^4 - p^2 - 20 = 0$$
$$(p^2 - 5)(p^2 + 4) = 0$$
$$p^2 - 5 = 0 \quad \text{or} \quad p^2 + 4 = 0$$
$$p^2 = 5 \quad \text{or} \quad p^2 = -4$$
$$p = \pm\sqrt{5} \quad \text{or} \quad p = \pm 2i$$

The solutions are $-\sqrt{5}, \sqrt{5}, -2i,$ and $2i$.

51. $(x+3)(x^2 - 3x + 9) = 0$

$$x + 3 = 0 \quad \text{or} \quad x^2 - 3x + 9 = 0$$
$$x = -3 \quad \text{or}$$
$$x = \frac{3 \pm \sqrt{(-3)^2 - 4(1)(9)}}{2(1)}$$
$$= \frac{3 \pm \sqrt{-27}}{2}$$
$$= \frac{3 \pm 3i\sqrt{3}}{2}$$

The solutions are $-3, \dfrac{3 + 3i\sqrt{3}}{2},$ and $\dfrac{3 - 3i\sqrt{3}}{2}$ or $-3, \dfrac{3}{2} + \dfrac{3\sqrt{3}}{2}i,$ and $\dfrac{3}{2} - \dfrac{3\sqrt{3}}{2}i.$

53.

$$1 = \frac{4}{x-7} + \frac{5}{(x-7)^2}$$

$$(x-7)^2 - 4(x-7) - 5 = 0$$

Let $y = x - 7$. Then $y^2 = (x-7)^2$ and

$$y^2 - 4y - 5 = 0$$
$$(y-5)(y+1) = 0$$
$$y = 5 \quad \text{or} \quad y = -1$$
$$x - 7 = 5 \quad \text{or} \quad x - 7 = -1$$
$$x = 12 \quad \text{or} \quad x = 6$$

The solutions are 6 and 12.

55.

$$27y^4 + 15y^2 = 2$$
$$27y^4 + 15y^2 - 2 = 0$$
$$(9y^2 - 1)(3y^2 + 2) = 0$$
$$(3y+1)(3y-1)(3y^2 + 2) = 0$$
$$y = -\frac{1}{3} \quad \text{or} \quad y = \frac{1}{3} \quad \text{or} \quad y^2 = -\frac{2}{3}$$
$$y = \pm\sqrt{-\frac{2}{3}}$$
$$y = \pm\frac{i\sqrt{6}}{3}$$

The solutions are $-\dfrac{1}{3}, \dfrac{1}{3}, -\dfrac{i\sqrt{6}}{3},$ and $\dfrac{i\sqrt{6}}{3}.$

57. $x - \sqrt{19 - 2x} - 2 = 0$

$$x - 2 = \sqrt{19 - 2x}$$

$$(x - 2)^2 = \left(\sqrt{19 - 2x}\right)^2$$

$$x^2 - 4x + 4 = 19 - 2x$$

$$x^2 - 2x - 15 = 0$$

$$(x - 5)(x + 3) = 0$$

$x - 5 = 0$ or $x + 3 = 0$

$x = 5$ or $x = -3$

Reject $x = -3$ as an extraneous solution. The solution is 5.

59. Let x be the rate to Tucson. Then $x + 11$ is the rate returning home.

	distance	=	rate	·	time
To Tucson	330		x		$\frac{320}{x}$
Return trip	330		$x + 11$		$\frac{330}{x-11}$

The LCD is $x(x + 11)$.

$$\frac{330}{x} = \frac{330}{x + 11} + 1$$

$$\frac{330}{x} - \frac{330}{x + 11} = 1$$

$$x(x+11)\left(\frac{330}{x}\right) - x(x+11)\left(\frac{330}{x+11}\right) = x(x+11)(1)$$

$$330(x+11) - 330x = x(x+11)$$

$$330x + 3630 - 330x = x^2 + 11x$$

$$0 = x^2 + 11x - 3630$$

$$0 = (x + 66)(x - 55)$$

$x + 66 = 0$ or $x - 55 = 0$

$x = -66$ $x = 55$

Reject -66 since rate cannot be negative. The speed to Tucson is 55 mph and the speed returning is $x + 11 = 55 + 11 = 66$ mph.

61. Let $x = $ speed on the first part. Then $x - 1 = $ speed on the second part.

$$d = rt \implies t = \frac{d}{r}$$

$$t_{\text{on first part}} + t_{\text{on second part}} = 1\frac{3}{5}$$

$$\frac{3}{x} + \frac{4}{x - 1} = \frac{8}{5}$$

$$3 \cdot 5(x - 1) + 4 \cdot 5x = 8x(x - 1)$$

$$15x - 15 + 20x = 8x^2 - 8x$$

$$0 = 8x^2 - 43x + 15$$

$$0 = (8x - 3)(x - 5)$$

$8x - 3 = 0$ or $x - 5 = 0$

$x = \dfrac{3}{8}$ or $x = 5$

$x - 1 = 4$

Discard $\dfrac{3}{8}$. Her speeds were 5 mph then 4 mph.

63. Let x = time for hose alone. Then
$x - 1$ = time for the inlet pipe alone.

$$\dfrac{1}{x} + \dfrac{1}{x-1} = \dfrac{1}{8}$$
$$8(x-1) + 8x = x(x-1)$$
$$8x - 8 + 8x = x^2 - x$$
$$0 = x^2 - 17x + 8$$
$$x = \dfrac{17 \pm \sqrt{(-17)^2 - 4(1)(8)}}{2(1)}$$
$$= \dfrac{17 \pm \sqrt{257}}{2}$$

$x \approx 0.5$ (discard)   or $x \approx 16.5$

$x - 1 \approx 15.5$

Hose: 16.5 hrs; Inlet pipe: 15.5 hrs

65. Let x = time for son alone. Then
$x - 1$ = time for dad alone.

$$\dfrac{1}{x} + \dfrac{1}{x-1} = \dfrac{1}{4}$$
$$4(x-1) + 4x = x(x-1)$$
$$4x - 4 + 4x = x^2 - x$$
$$0 = x^2 - 9x + 4$$
$$x = \dfrac{9 \pm \sqrt{(-9)^2 - 4(1)(4)}}{2(1)}$$
$$= \dfrac{9 \pm \sqrt{65}}{2}$$

≈ 0.5 (discard)   or   8.5

It takes his son about 8.5 hours.

67. Let x = the number.
$$x(x - 4) = 96$$
$$x^2 - 4x - 96 = 0$$
$$(x - 12)(x + 8) = 0$$
$$x = 12 \text{ or } x = -8$$
The number is 12 or –8.

69. a. length $= x - 3 - 3 = x - 6 = (x - 6)$ in.

    b. $V = lwh$
$$300 = (x - 6)(x - 6) \cdot 3$$

c. $300 = 3(x - 6)^2$
$$100 = x^2 - 12x + 36$$
$$0 = x^2 - 12x - 64$$
$$0 = (x - 16)(x + 4)$$
$$x = 16 \text{ or } x = -4 \text{ (discard)}$$
The sheet is 16 in. by 16 in.
Check: $V = 3(x - 6)(x - 6)$
$$= 3(16 - 6)(16 - 6)$$
$$= 3(10)(10)$$
$$= 300 \text{ cubic in.}$$

71. Let x = length of the side of the square.

    Area $= x^2$
$$920 = x^2$$
$$\sqrt{920} = x$$

Adding another radial line to a different corner would yield a right triangle with legs r and hypotenuse x.

$$r^2 + r^2 = x^2$$
$$2r^2 = \left(\sqrt{920}\right)^2$$
$$2r^2 = 920$$
$$r^2 = 460$$
$$r = \pm\sqrt{460} = \pm 21.4476$$

Disregard the negative. The smallest radius would be 22 feet.

73. $\dfrac{5x}{3} + 2 \le 7$

$$\dfrac{5x}{3} \le 5$$
$$5x \le 15$$
$$x \le 3$$
$(-\infty, 3]$

75. $\dfrac{y-1}{15} > -\dfrac{2}{5}$

$$15\left(\dfrac{y-1}{15}\right) > 15\left(-\dfrac{2}{5}\right)$$
$$y - 1 > -6$$
$$y > -5$$
$(-5, \infty)$

77. Domain: $\{x \mid x \text{ is a real number}\}$ or $(-\infty, \infty)$
Range: $\{y \mid y \text{ is a real number}\}$ or $(-\infty, \infty)$
It is a function.

79. Domain: $\{x \mid x \text{ is a real number}\}$ or $(-\infty, \infty)$

Range: $\{y \mid y \geq -1\}$ or $[-1, \infty)$

It is a function.

81. $\quad 5y^3 - 45y - 5y^2 - 45 = 0$

$\quad 5y(y^2 + 9) - 5(y^2 + 9) = 0$

$\qquad (5y - 5)(y^2 + 9) = 0$

$\qquad 5(y - 1)(y^2 + 9) = 0$

$y - 1 = 0 \quad \text{or} \quad y^2 + 9 = 0$

$\quad y = 1 \qquad\qquad y^2 = -9$

$\qquad\qquad\qquad\qquad y = \pm\sqrt{-9} = \pm 3i$

The solutions are 1, $3i$, and $-3i$.

83. Let $u = x^{-1}$, then $x^{-2} = u^2$.

$\quad 3x^{-2} - 3x^{-1} - 18 = 0$

$\qquad 3u^2 - 3u - 18 = 0$

$\qquad 3(u^2 - u - 6) = 0$

$\qquad 3(u - 3)(u + 2) = 0$

$u - 3 = 0 \quad \text{or} \quad u + 2 = 0$

$\quad u = 3 \qquad\qquad u = -2$

$\quad x^{-1} = 3 \qquad\quad x^{-1} = -2$

$\quad x = \dfrac{1}{3} \qquad\qquad x = -\dfrac{1}{2}$

The solutions are $\dfrac{1}{3}$ and $-\dfrac{1}{2}$.

85. $\qquad\qquad\qquad 2x^3 = -54$

$\qquad\qquad\qquad\quad x^3 = -27$

$\qquad\qquad\qquad x^3 + 27 = 0$

$\qquad (x + 3)(x^2 - 3x + 9) = 0$

$\quad x + 3 = 0 \quad \text{or} \quad x^2 - 3x + 9 = 0$

$\qquad x = -3$

$\quad x = \dfrac{-(-3) \pm \sqrt{(-3)^2 - 4(1)(9)}}{2(1)}$

$\quad x = \dfrac{3 \pm \sqrt{9 - 36}}{2} = \dfrac{3 \pm \sqrt{-27}}{2}$

$\quad x = \dfrac{3 \pm 3i\sqrt{3}}{2}$

The solutions are -3, $\dfrac{3 + 3i\sqrt{3}}{2}$, and $\dfrac{3 - 3i\sqrt{3}}{2}$ or -3, $\dfrac{3}{2} + \dfrac{3\sqrt{3}}{2}i$, and $\dfrac{3}{2} - \dfrac{3\sqrt{3}}{2}i$.

87. answers may vary

89. a. Let x be Voeckler's average speed. Using $d = r \cdot t$ or $t = \dfrac{d}{r}$, Voeckler's time for that stage was

$\dfrac{119,000}{x}$ seconds. Contador's average speed is then $(x - 0.0034)$ meters per second and the time for his stage

was $\dfrac{119,000}{x - 0.0034}$ seconds, which is also $\dfrac{119,000}{x} + 3$.

The LCD is $x(x - 0.0034)$.

$$\frac{119,000}{x} + 3 = \frac{119,000}{x - 0.0034}$$

$$x(x - 0.0034)\left(\frac{119,000}{x}\right) + x(x - 0.0034)(3) = x(x - 0.0034)\left(\frac{119,000}{x - 0.0034}\right)$$

$$119,000(x - 0.0034) + 3x(x - 0.0034) = x(119,000)$$

$$119,000x - 404.6 + 3x^2 - 0.0102x = 119,000x$$

$$3x^2 - 0.0102x - 404.6 = 0$$

$a = 3,\ b = -0.0102,\ c = -404.6$

$$x = \frac{-b \pm \sqrt{b^2 - 4ac}}{2a}$$

$$x = \frac{-(-0.0102) \pm \sqrt{(-0.0102)^2 - 4(3)(-404.6)}}{2(3)}$$

$x \approx -11.612$ or $x \approx 11.615$

Since the speed cannot be negative, reject -11.612.
Voeckler's average speed was approximately 11.615 meters per second.

b. Contador's average speed was $11.615 - 0.0034 \approx 11.612$ meters per second.

c. $\dfrac{11.615 \text{ m}}{1 \text{ sec}} \cdot \dfrac{3600 \text{ sec}}{\text{hr}} \cdot \dfrac{1 \text{ km}}{1000 \text{ m}} \cdot \dfrac{0.62 \text{ mi}}{1 \text{ km}} \approx 25.925$ mi/hr

Voeckler's average speed was about 25.925 miles per hour.

Integrated Review

1. $x^2 - 10 = 0$

$\qquad x^2 = 10$

$\qquad x = \pm\sqrt{10}$

The solutions are $\sqrt{10}$ and $-\sqrt{10}$.

2. $x^2 - 14 = 0$

$\qquad x^2 = 14$

$\qquad x = \pm\sqrt{14}$

The solutions are $\sqrt{14}$ and $-\sqrt{14}$.

3. $(x - 1)^2 = 8$

$\qquad x - 1 = \pm\sqrt{8}$

$\qquad x - 1 = \pm 2\sqrt{2}$

$\qquad\quad x = 1 \pm 2\sqrt{2}$

The solutions are $1 + 2\sqrt{2}$ and $1 - 2\sqrt{2}$.

4. $(x+5)^2 = 12$

$\quad x+5 = \pm\sqrt{12}$

$\quad x+5 = \pm 2\sqrt{3}$

$\quad x = -5 \pm 2\sqrt{3}$

The solutions are $-5+2\sqrt{3}$ and $-5-2\sqrt{3}$.

5. $\quad x^2 + 2x - 12 = 0$

$x^2 + 2x + \left(\dfrac{2}{2}\right)^2 = 12 + 1$

$\quad x^2 + 2x + 1 = 13$

$\quad (x+1)^2 = 13$

$\quad x+1 = \pm\sqrt{13}$

$\quad x = -1 \pm \sqrt{13}$

The solutions are $-1+\sqrt{13}$ and $-1-\sqrt{13}$.

6. $\quad x^2 - 12x + 11 = 0$

$x^2 - 12x + \left(\dfrac{-12}{2}\right)^2 = -11 + 36$

$\quad x^2 - 12x + 36 = 25$

$\quad (x-6)^2 = \pm\sqrt{25}$

$\quad x - 6 = \pm 5$

$\quad x = 6 \pm 5$

$\quad x = 1 \text{ or } x = 11$

The solutions are 1 and 11.

7. $\quad 3x^2 + 3x = 5$

$\quad x^2 + x = \dfrac{5}{3}$

$x^2 + x + \left(\dfrac{1}{2}\right)^2 = \dfrac{5}{3} + \dfrac{1}{4}$

$\quad x^2 + x + \dfrac{1}{4} = \dfrac{23}{12}$

$\quad \left(x+\dfrac{1}{2}\right)^2 = \dfrac{23}{12}$

$\quad x+\dfrac{1}{2} = \pm\sqrt{\dfrac{23}{12}}$

$\quad x+\dfrac{1}{2} = \pm\dfrac{\sqrt{23}}{2\sqrt{3}}$

$\quad x+\dfrac{1}{2} = \pm\dfrac{\sqrt{23}\cdot\sqrt{3}}{2\sqrt{3}\cdot\sqrt{3}}$

$\quad x+\dfrac{1}{2} = \pm\dfrac{\sqrt{69}}{6}$

$\quad x = -\dfrac{1}{2} \pm \dfrac{\sqrt{69}}{6} = \dfrac{-3\pm\sqrt{69}}{6}$

The solutions are $\dfrac{-3+\sqrt{69}}{6}$ and $\dfrac{-3-\sqrt{69}}{6}$.

8. $\quad 16y^2 + 16y = 1$

$\quad y^2 + y = \dfrac{1}{16}$

$y^2 + y + \left(\dfrac{1}{2}\right)^2 = \dfrac{1}{16} + \dfrac{1}{4}$

$\quad y^2 + y + \dfrac{1}{4} = \dfrac{5}{16}$

$\quad \left(y+\dfrac{1}{2}\right)^2 = \dfrac{5}{16}$

$\quad y+\dfrac{1}{2} = \pm\sqrt{\dfrac{5}{16}}$

$\quad y+\dfrac{1}{2} = \pm\dfrac{\sqrt{5}}{4}$

$\quad y = -\dfrac{1}{2} \pm \dfrac{\sqrt{5}}{4} = \dfrac{-2\pm\sqrt{5}}{4}$

The solutions are $\dfrac{-2+\sqrt{5}}{4}$ and $\dfrac{-2-\sqrt{5}}{4}$.

9. $2x^2 - 4x + 1 = 0$

$a = 2, b = -4, c = 1$

$$x = \frac{4 \pm \sqrt{(-4)^2 - 4(2)(1)}}{2(2)}$$

$$= \frac{4 \pm \sqrt{8}}{4}$$

$$= \frac{4 \pm 2\sqrt{2}}{4} = \frac{2 \pm \sqrt{2}}{2}$$

The solutions are $\dfrac{2 + \sqrt{2}}{2}$ and $\dfrac{2 - \sqrt{2}}{2}$.

10. $\dfrac{1}{2}x^2 + 3x + 2 = 0$

$x^2 + 6x + 4 = 0$

$a = 1, b = 6, c = 4$

$$x = \frac{-6 \pm \sqrt{(6)^2 - 4(1)(4)}}{2(1)}$$

$$= \frac{-6 \pm \sqrt{20}}{2}$$

$$= \frac{-6 \pm 2\sqrt{5}}{2} = -3 \pm \sqrt{5}$$

The solutions are $-3 + \sqrt{5}$ and $-3 - \sqrt{5}$.

11. $x^2 + 4x = -7$

$x^2 + 4x + 7 = 0$

$a = 1, b = 4, c = 7$

$$x = \frac{-4 \pm \sqrt{(4)^2 - 4(1)(7)}}{2(1)}$$

$$= \frac{-4 \pm \sqrt{-12}}{2}$$

$$= \frac{-4 \pm i\sqrt{4 \cdot 3}}{2}$$

$$= \frac{-4 \pm 2i\sqrt{3}}{2} = -2 \pm i\sqrt{3}$$

The solutions are $-2 + i\sqrt{3}$ and $-2 - i\sqrt{3}$.

12. $x^2 + x = -3$

$x^2 + x + 3 = 0$

$a = 1, b = 1, c = 3$

$$x = \frac{-1 \pm \sqrt{(1)^2 - 4(1)(3)}}{2(1)}$$

$$= \frac{-1 \pm \sqrt{-11}}{2}$$

$$= \frac{-1 \pm i\sqrt{11}}{2}$$

The solutions are $\dfrac{-1 + i\sqrt{11}}{2}$ and $\dfrac{-1 - i\sqrt{11}}{2}$ or

$-\dfrac{1}{2} + \dfrac{\sqrt{11}}{2}i$ and $-\dfrac{1}{2} - \dfrac{\sqrt{11}}{2}i$.

13. $x^2 + 3x + 6 = 0$

$a = 1, b = 3, c = 6$

$$x = \frac{-3 \pm \sqrt{(3)^2 - 4(1)(6)}}{2(1)}$$

$$= \frac{-3 \pm \sqrt{-15}}{2}$$

$$= \frac{-3 \pm i\sqrt{15}}{2}$$

The solutions are $\dfrac{-3 + i\sqrt{15}}{2}$ and $\dfrac{-3 - i\sqrt{15}}{2}$ or

$-\dfrac{3}{2} + \dfrac{\sqrt{15}}{2}i$ and $-\dfrac{3}{2} - \dfrac{\sqrt{15}}{2}i$.

14. $2x^2 + 18 = 0$

$2x^2 = -18$

$x^2 = -9$

$x = \pm\sqrt{-9}$

$x = \pm 3i$

The solutions are $3i$ and $-3i$.

15. $x^2 + 17x = 0$

$x(x + 17) = 0$

$x = 0$ or $x + 17 = 0$

$\qquad\qquad\qquad x = -17$

$x = 0, -17$

The solutions are 0 and -17.

16. $4x^2 - 2x - 3 = 0$

$a = 4, b = -2, c = -3$

$$x = \frac{2 \pm \sqrt{(-2)^2 - 4(4)(-3)}}{2(4)}$$

$$= \frac{2 \pm \sqrt{52}}{8}$$

$$= \frac{2 \pm 2\sqrt{13}}{8}$$

$$= \frac{1 \pm \sqrt{13}}{4}$$

The solutions are $\dfrac{1 + \sqrt{13}}{4}$ and $\dfrac{1 - \sqrt{13}}{4}$.

17. $(x - 2)^2 = 27$

$x - 2 = \pm\sqrt{27}$

$x - 2 = \pm 3\sqrt{3}$

$x = 2 \pm 3\sqrt{3}$

The solutions are $2 + 3\sqrt{3}$ and $2 - 3\sqrt{3}$.

18. $\dfrac{1}{2}x^2 - 2x + \dfrac{1}{2} = 0$

$x^2 - 4x + 1 = 0$

$x^2 - 4x + \left(\dfrac{-4}{2}\right)^2 = -1 + 4$

$x^2 - 4x + 4 = 3$

$(x - 2)^2 = 3$

$x - 2 = \pm\sqrt{3}$

$x = 2 \pm \sqrt{3}$

The solutions are $2 + \sqrt{3}$ and $2 - \sqrt{3}$.

19. $3x^2 + 2x = 8$

$3x^2 + 2x - 8 = 0$

$(3x - 4)(x + 2) = 0$

$3x - 4 = 0$ or $x + 2 = 0$

$x = \dfrac{4}{3}$ or $x = -2$

The solutions are $\dfrac{4}{3}$ and -2.

20. $2x^2 = -5x - 1$

$2x^2 + 5x + 1 = 0$

$a = 2, b = 5, c = 1$

$$x = \frac{-5 \pm \sqrt{(5)^2 - 4(2)(1)}}{2(2)}$$

$$= \frac{-5 \pm \sqrt{17}}{4}$$

The solutions are $\dfrac{-5 + \sqrt{17}}{4}$ and $\dfrac{-5 - \sqrt{17}}{4}$.

21. $x(x - 2) = 5$

$x^2 - 2x = 5$

$x^2 - 2x + \left(\dfrac{-2}{2}\right)^2 = 5 + 1$

$x^2 - 2x + 1 = 6$

$(x - 1)^2 = 6$

$x - 1 = \pm\sqrt{6}$

$x = 1 \pm \sqrt{6}$

The solutions are $1 + \sqrt{6}$ and $1 - \sqrt{6}$.

22. $x^2 - 31 = 0$

$x^2 = 31$

$x = \pm\sqrt{31}$

The solutions are $\sqrt{31}$ and $-\sqrt{31}$.

23. $5x^2 - 55 = 0$

$5x^2 = 55$

$x^2 = 11$

$x = \pm\sqrt{11}$

The solutions are $\sqrt{11}$ and $-\sqrt{11}$.

24. $5x^2 + 55 = 0$

$5x^2 = -55$

$x^2 = -11$

$x = \pm\sqrt{-11}$

$x = \pm i\sqrt{11}$

The solutions are $i\sqrt{11}$ and $-i\sqrt{11}$.

25.

$$x(x+5) = 66$$
$$x^2 + 5x = 66$$
$$x^2 + 5x - 66 = 0$$
$$(x+11)(x-6) = 0$$
$$x+11 = 0 \quad \text{or} \quad x-6 = 0$$
$$x = -11 \quad \text{or} \quad x = 6$$

The solutions are −11 and 6.

26. $5x^2 + 6x - 2 = 0$

$a = 5, b = 6, c = -2$

$$x = \frac{-6 \pm \sqrt{(6)^2 - 4(5)(-2)}}{2(5)}$$

$$= \frac{-6 \pm \sqrt{76}}{10}$$

$$= \frac{-6 \pm \sqrt{4 \cdot 19}}{10}$$

$$= \frac{-6 \pm 2\sqrt{19}}{10}$$

$$= \frac{-3 \pm \sqrt{19}}{5}$$

The solutions are $\dfrac{-3 + \sqrt{19}}{5}$ and $\dfrac{-3 - \sqrt{19}}{5}$.

27.

$$2x^2 + 3x = 1$$
$$2x^2 + 3x - 1 = 0$$
$$a = 2, b = 3, c = -1$$

$$x = \frac{-3 \pm \sqrt{(3)^2 - 4(2)(-1)}}{2(2)}$$

$$= \frac{-3 \pm \sqrt{17}}{4}$$

The solutions are $\dfrac{-3 + \sqrt{17}}{4}$ and $\dfrac{-3 - \sqrt{17}}{4}$.

28. $x - \sqrt{13 - 3x} - 3 = 0$

$$x - 3 = \sqrt{13 - 3x}$$
$$(x-3)^2 = \left(\sqrt{13 - 3x}\right)^2$$
$$x^2 - 6x + 9 = 13 - 3x$$
$$x^2 - 3x - 4 = 0$$
$$(x-4)(x+1) = 0$$
$$x - 4 = 0 \quad \text{or} \quad x + 1 = 0$$
$$x = 4 \qquad \qquad x = -1$$

The value −1 does not check, so the solution is 4.

29. The LCD is $x(x-2)$.

$$\frac{5x}{x-2} - \frac{x+1}{x} = \frac{3}{x(x-2)}$$

$$x(x-2)\left(\frac{5x}{x-2}\right) - x(x-2)\left(\frac{x+1}{x}\right) = x(x-2)\left[\frac{3}{x(x-2)}\right]$$

$$x(5x) - (x-2)(x+1) = 3$$

$$5x^2 - (x^2 - x - 2) = 3$$

$$5x^2 - x^2 + x + 2 = 3$$

$$4x^2 + x - 1 = 0$$

$a = 4,\ b = 1,\ c = -1$

$$x = \frac{-b \pm \sqrt{b^2 - 4ac}}{2a}$$

$$x = \frac{-1 \pm \sqrt{1^2 - 4(4)(-1)}}{2(4)}$$

$$= \frac{-1 \pm \sqrt{1 + 16}}{8}$$

$$= \frac{-1 \pm \sqrt{17}}{8}$$

The solutions are $\dfrac{-1-\sqrt{17}}{8}$ and $\dfrac{-1+\sqrt{17}}{8}$.

30. $a^2 + b^2 = c^2$

$$x^2 + x^2 = 20^2$$

$$2x^2 = 400$$

$$x^2 = 200$$

$$x = \pm\sqrt{200}$$

$$= \pm 10\sqrt{2} \approx 14.1421$$

Disregard the negative. A side of the room is $10\sqrt{2}$ feet ≈ 14.1 feet.

31. Let x = time for Jack alone. Then
$x - 2$ = time for Lucy alone.

$$\frac{1}{x} + \frac{1}{x-2} = \frac{1}{4}$$

$$4(x-2) + 4x = x(x-2)$$

$$4x - 8 + 4x = x^2 - 2x$$

$$0 = x^2 - 10x + 8$$

$$x = \frac{10 \pm \sqrt{(-10)^2 - 4(1)(8)}}{2(1)}$$

$$= \frac{10 \pm \sqrt{68}}{2}$$

$$\approx 9.1 \text{ or } 0.9 \text{ (disregard)}$$

$x - 2 = 9.1 - 2 = 7.1$

It would take Jack 9.1 hours and Lucy 7.1 hours.

32. Let x = initial speed on treadmill. Then $x + 1$ = speed increased.

$$t_{initial} + t_{increased} = \frac{4}{3}$$

$$\frac{5}{x} + \frac{2}{x+1} = \frac{4}{3}$$

$$5 \cdot 3(x+1) + 2 \cdot 3x = 4x(x+1)$$

$$15x + 15 + 6x = 4x^2 + 4x$$

$$0 = 4x^2 - 17x - 15$$

$$0 = (4x+3)(x-5)$$

$$x = -\frac{4}{3} \text{ (disregard) or } x = 5$$

$$x + 1 = 5 + 1 = 6$$

Initial speed: 5 mph
Increased speed: 6 mph

Section 11.4 Practice

1. $(x - 4)(x + 3) > 0$

Solve the related equation, $(x - 4)(x + 3) = 0$.

$$(x - 4)(x + 3) = 0$$
$$x - 4 = 0 \quad \text{or} \quad x + 3 = 0$$
$$x = 4 \qquad\qquad x = -3$$

Test points in the three regions separated by $x = 4$ and $x = -3$.

Region	Test Point	$(x - 4)(x + 3) > 0$ Result
A: $(-\infty, -3)$	-4	$(-8)(-1) > 0$ True
B: $(-3, 4)$	0	$(-4)(3) > 0$ False
C: $(4, \infty)$	5	$(1)(8) > 0$ True

The points in regions A and C satisfy the inequality. The numbers 4 and -3 are not included in the solution since the inequality symbol is $>$. The solution set is $(-\infty, -3) \cup (4, \infty)$.

2. $x^2 - 8x \leq 0$

Solve the related equation, $x^2 - 8x = 0$.

$$x^2 - 8x = 0$$
$$x(x - 8) = 0$$
$$x = 0 \quad \text{or} \quad x - 8 = 0$$
$$x = 8$$

The numbers 0 and 8 separate the number line into three regions, A, B, and C. Test a point in each region.

Region	Test Point	$x^2 - 8x \leq 0$ Result
A: $(-\infty, 0]$	-1	$1 + 8 \leq 0$ False
B: $[0, 8]$	1	$1 - 8 \leq 0$ True
C: $[8, \infty)$	9	$81 - 72 \leq 0$ False

Values in region B satisfy the inequality. The numbers 0 and 8 are included in the solution since the inequality symbol is $\leq$. The solution set is $[0, 8]$.

3. $(x + 3)(x - 2)(x + 1) \leq 0$

Solve $(x + 3)(x - 2)(x + 1) = 0$ by inspection.

$$x = -3 \quad \text{or} \quad x = 2 \quad \text{or} \quad x = -1$$

These separate the number line into four regions. Test points in each region.

Region	Test Point	$(x + 3)(x - 2)(x + 1) \leq 0$ Result
A: $(-\infty, -3]$	-4	$(-1)(-6)(-3) \leq 0$ True
B: $[-3, -1]$	-2	$(1)(-4)(-1) \leq 0$ False
C: $[-1, 2]$	0	$(3)(-2)(1) \leq 0$ True
D: $[2, \infty)$	3	$(6)(1)(4) \leq 0$ False

The solution set is $(-\infty, -3] \cup [-1, 2]$. We include the numbers -3, -1, and 2 because the inequality symbol is $\leq$.

4. $\dfrac{x-5}{x+4} \le 0$

$x+4=0$

$\quad x=-4$

$x=-4$ makes the denominator zero. Solve the

related equation $\dfrac{x-5}{x+4}=0$.

$\dfrac{x-5}{x+4}=0$

$x-5=0$

$\quad x=5$

Test points in the three regions separated by $x=-4$ and $x=5$.

Region	Test Point	$\dfrac{x-5}{x+4} \le 0$ Result
A: $(-\infty, -4)$	-5	$\dfrac{-10}{-1} \le 0$ False
B: $(-4, 5]$	0	$\dfrac{-5}{4} \le 0$ True
C: $[5, \infty)$	6	$\dfrac{1}{10} \le 0$ False

The solution set is $(-4, 5]$. The interval includes 5 because 5 satisfies the original inequality. This interval does not include -4, because -4 would make the denominator zero.

5. $\dfrac{7}{x+3} < 5$

$x+3=0$

$\quad x=-3$

$x=-3$ makes the denominator zero.

Solve $\dfrac{7}{x+3}=5$.

$(x+3)\left(\dfrac{7}{x+3}\right)=5(x+3)$

$\quad\quad\quad 7 = 5x+15$

$\quad\quad -8 = 5x$

$\quad\quad -\dfrac{8}{5}=x$

We use these two solutions to divide the number line into three regions and choose test points.

Region	Test Point	$\dfrac{7}{x+3} < 5$ Result
A: $(-\infty, -3)$	-4	$\dfrac{7}{-1} < 5$ True
B: $\left(-3, -\dfrac{8}{5}\right)$	-2	$\dfrac{7}{1} < 5$ False
C: $\left(-\dfrac{8}{5}, \infty\right)$	0	$\dfrac{7}{3} < 5$ True

The solution set is $(-\infty, -3) \cup \left(-\dfrac{8}{5}, \infty\right)$.

Vocabulary, Readiness & Video Check 11.4

1. $[-7, 3)$

2. $(-1, 5]$

3. $(-\infty, 0]$

4. $(-\infty, -8]$

5. $(-\infty, -12) \cup [-10, \infty)$

6. $(-\infty, -3] \cup (4, \infty)$

7. We use the solutions to the related equation to divide the number line into regions that either entirely are or entirely are not solution regions; the solutions to the related equation are solutions to the inequality only if the inequality symbol is $\le$ or $\ge$.

8. The solution set cannot include values that make the denominator zero.

Exercise Set 11.4

1. $(x + 1)(x + 5) > 0$
$\quad x + 1 = 0 \quad$ or $\quad x + 5 = 0$
$\qquad x = -1 \quad$ or $\qquad x = -5$

Region	Test Point	$(x + 1)(x + 5) > 0$ Result
A: $(-\infty, -5)$	-6	$(-5)(-1) > 0$ True
B: $(-5, -1)$	-2	$(-1)(3) > 0$ False
C: $(-1, \infty)$	0	$(1)(5) > 0$ True

Solution: $(-\infty, -5) \cup (-1, \infty)$

3. $(x - 3)(x + 4) \le 0$
$\quad x - 3 = 0 \quad$ or $\quad x + 4 = 0$
$\qquad x = 3 \quad$ or $\qquad x = -4$

Region	Test Point	$(x - 3)(x + 4) \le 0$ Result
A: $(-\infty, -4]$	-5	$(-8)(-1) \le 0$ False
B: $[-4, 3]$	0	$(-3)(4) \le 0$ True
C: $[3, \infty)$	4	$(1)(8) \le 0$ False

Solution: $[-4, 3]$

5. $x^2 - 7x + 10 \le 0$
$\quad (x - 5)(x - 2) \le 0$
$\quad x - 5 = 0 \quad$ or $\quad x - 2 = 0$
$\qquad x = 5 \quad$ or $\qquad x = 2$

Region	Test Point	$(x - 5)(x - 2) \le 0$ Result
A: $(-\infty, 2]$	0	$(-5)(-2) \le 0$ False
B: $[2, 5]$	3	$(-2)(1) \le 0$ True
C: $[5, \infty)$	6	$(1)(4) \le 0$ False

Solution: $[2, 5]$

7. $\qquad 3x^2 + 16 < -5$
$\quad 3x^2 + 16x + 5 < 0$
$\quad (3x + 1)(x + 5) < 0$
$\quad 3x + 1 = 0 \quad$ or $\quad x + 5 = 0$
$\qquad x = -\dfrac{1}{3} \quad$ or $\qquad x = -5$

Region	Test Point	$(3x + 1)(x + 5) < 0$ Result
A: $(-\infty, -5)$	-6	$(-17)(-1) < 0$ False
B: $\left(-5, -\dfrac{1}{3}\right)$	-1	$(-2)(4) < 0$ True
C: $\left(-\dfrac{1}{3}, \infty\right)$	0	$(1)(5) < 0$ False

Solution: $\left(-5, -\dfrac{1}{3}\right)$

9. $(x - 6)(x - 4)(x - 2) > 0$
$\quad x - 6 = 0 \quad$ or $\quad x - 4 = 0 \quad$ or $\quad x - 2 = 0$
$\qquad x = 6 \quad$ or $\qquad x = 4 \quad$ or $\qquad x = 2$

Region	Test Point	$(x - 6)(x - 4)(x - 2) > 0$ Result
A: $(-\infty, 2)$	0	$(-6)(-4)(-2) > 0$ False
B: $(2, 4)$	3	$(-3)(-1)(1) > 0$ True
C: $(4, 6)$	5	$(-1)(1)(3) > 0$ False
D: $(6, \infty)$	7	$(1)(3)(5) > 0$ True

Solution: $(2, 4) \cup (6, \infty)$

11. $x(x-1)(x+4) \leq 0$

$x = 0$ or $x - 1 = 0$ or $x + 4 = 0$

             $x = 1$ or $x = -4$

Region	Test Point	$x(x-1)(x+4) \leq 0$ Result
A: $(-\infty, -4]$	-5	$-5(-6)(-1) \leq 0$ True
B: $[-4, 0]$	-1	$-1(-2)(3) \leq 0$ False
C: $[0, 1]$	$\dfrac{1}{2}$	$\dfrac{1}{2}\left(-\dfrac{1}{2}\right)\left(\dfrac{9}{2}\right) \leq 0$ True
D: $[1, \infty)$	2	$2(1)(6) \leq 0$ False

Solution: $(-\infty, -4] \cup [0, 1]$

13. $(x^2 - 9)(x^2 - 4) > 0$

$(x+3)(x-3)(x+2)(x-2) > 0$

$x + 3 = 0$ or $x - 3 = 0$ or $x + 2 = 0$ or $x - 2 = 0$

  $x = -3$ or $x = 3$ or $x = -2$ or $x = 2$

Region	Test Point	$(x+3)(x-3)(x+2)(x-2) > 0$ Result
A: $(-\infty, -3)$	-4	$(-1)(-7)(-2)(-6) > 0$ True
B: $(-3, -2)$	$-\dfrac{5}{2}$	$\left(\dfrac{1}{2}\right)\left(-\dfrac{11}{2}\right)\left(-\dfrac{1}{2}\right)\left(-\dfrac{9}{2}\right) > 0$ False
C: $(-2, 2)$	0	$(3)(-3)(2)(-2) > 0$ True
D: $(2, 3)$	$\dfrac{5}{2}$	$\left(\dfrac{11}{2}\right)\left(-\dfrac{1}{2}\right)\left(\dfrac{9}{2}\right)\left(\dfrac{1}{2}\right) > 0$ False
E: $(3, \infty)$	4	$(7)(1)(6)(2) > 0$ True

Solution: $(-\infty, -3) \cup (-2, 2) \cup (3, \infty)$

15. $\dfrac{x+7}{x-2} < 0$

$x + 7 = 0 \quad$ or $\quad x - 2 = 0$

$\quad x = -7 \quad$ or $\qquad x = 2$

Region	Test Point	$\dfrac{x+7}{x-2} < 0$ False
$A: (-\infty, -7)$	-8	$\dfrac{-1}{-10} < 0$ False
$B: (-7, 2)$	0	$\dfrac{7}{-2} < 0$ True
$C: (2, \infty)$	3	$\dfrac{10}{1} < 0$ False

Solution: $(-7, 2)$

17. $\dfrac{5}{x+1} > 0$

$x + 1 = 0$

$\quad x = -1$

Region	Test Point	$\dfrac{5}{x+1} > 0$ Result
$A: (-\infty, -1)$	-2	$\dfrac{5}{-1} > 0$ False
$B: (-1, \infty)$	0	$\dfrac{5}{1} > 0$ True

Solution: $(-1, \infty)$

19. $\dfrac{x+1}{x-4} \geq 0$

$x + 1 = 0 \quad$ or $\quad x - 4 = 0$

$\quad x = -1 \quad$ or $\qquad x = 4$

Region	Test Point	$\dfrac{x+1}{x-4} \geq 0$ Result
$A: (-\infty, -1]$	-2	$\dfrac{-1}{-6} \geq 0$ True
$B: [-1, 4)$	0	$\dfrac{1}{-4} \geq 0$ False
$C: (4, \infty)$	5	$\dfrac{6}{1} \geq 0$ True

Solution: $(-\infty, -1] \cup (4, \infty)$

21. $\dfrac{3}{x-2} < 4$

The denominator is equal to 0 when $x - 2 = 0$, or $x = 2$.

$\dfrac{3}{x-2} = 4$

$\quad\quad 3 = 4x - 8$

$\quad 11 = 4x$

$\dfrac{11}{4} = x$

Region	Test Point	$\dfrac{3}{x-2} < 4$ Result
$A: (-\infty, 2)$	0	$\dfrac{3}{-2} < 4$ True
$B: \left(2, \dfrac{11}{4}\right)$	$\dfrac{5}{2}$	$\dfrac{3}{\frac{1}{2}} = 6 < 4$ False
$C: \left(\dfrac{11}{4}, \infty\right)$	4	$\dfrac{3}{2} < 4$ True

Solution: $(-\infty, 2) \cup \left(\dfrac{11}{4}, \infty\right)$

23. $\dfrac{x^2+6}{5x} \ge 1$

The denominator is equal to 0 when $5x = 0$, or $x = 0$.

$$\frac{x^2+6}{5x} = 1$$
$$x^2+6 = 5x$$
$$x^2-5x+6 = 0$$
$$(x-2)(x-3) = 0$$
$$x-2 = 0 \quad \text{or} \quad x-3 = 0$$
$$x = 2 \quad \text{or} \quad x = 3$$

Region	Test Point	$\dfrac{x^2+6}{5x} \ge 1$ Result
A: $(-\infty, 0)$	-1	$\dfrac{7}{-5} \ge 1$ False
B: $(0, 2]$	1	$\dfrac{7}{5} \ge 1$ True
C: $[2, 3]$	$\dfrac{5}{2}$	$\dfrac{\frac{49}{4}}{\frac{25}{2}} = \dfrac{49}{50} \ge 1$ False
D: $[3, \infty)$	4	$\dfrac{22}{20} \ge 1$ True

Solution: $(0, 2] \cup [3, \infty)$

25. $\dfrac{x+2}{x-3} < 1$

$x - 3 = 0$

$x = 3$ makes the denominator 0.

$$\frac{x+2}{x-3} = 1$$
$$x+2 = x-3$$
$$2 = -3 \quad \text{False}$$

The related equation has no solution.

Region	Test Point Value	$\dfrac{x+2}{x-3} < 1$ Result
$(-\infty, 3)$	0	$\dfrac{2}{-3} < 1$ True
$(3, \infty)$	4	$\dfrac{6}{1} < 1$ False

Solution: $(-\infty, 3)$

27. $(2x-3)(4x+5) \le 0$

$$2x-3 = 0 \quad \text{or} \quad 4x+5 = 0$$
$$x = \frac{3}{2} \quad \text{or} \quad x = -\frac{5}{4}$$

Region	Test Point	$(2x-3)(4x+5) \le 0$ Result
A: $\left(-\infty, -\dfrac{5}{4}\right]$	-2	$(-7)(-3) \le 0$ False
B: $\left[-\dfrac{5}{4}, \dfrac{3}{2}\right]$	0	$(-3)(5) \le 0$ True
C: $\left[\dfrac{3}{2}, \infty\right)$	2	$(1)(13) \le 0$ False

Solution: $\left[-\dfrac{5}{4}, \dfrac{3}{2}\right]$

29. $x^2 > x$

$$x^2 - x > 0$$
$$x(x-1) > 0$$
$$x = 0 \quad \text{or} \quad x-1 = 0$$
$$x = 1$$

Region	Test Point	$x(x-1) > 0$ Result
A: $(-\infty, 0)$	-1	$-1(-2) > 0$ True
B: $(0, 1)$	$\dfrac{1}{2}$	$\dfrac{1}{2}\left(-\dfrac{1}{2}\right) > 0$ False
C: $(1, \infty)$	2	$2(1) > 0$ True

Solution: $(-\infty, 0) \cup (1, \infty)$

31. $(2x - 8)(x + 4)(x - 6) \le 0$

$2x - 8 = 0$ or $x + 4 = 0$ or $x - 6 = 0$

     $x = 4$ or $x = -4$ or $x = 6$

Region	Test Point	$(2x - 8)(x + 4)(x - 6) \le 0$ Result
A: $(-\infty, -4]$	-5	$(-18)(-1)(-11) \le 0$ True
B: $[-4, 4]$	0	$(-8)(4)(-6) \le 0$ False
C: $[4, 6]$	5	$(2)(9)(-1) \le 0$ True
D: $[6, \infty)$	7	$(6)(11)(1) \le 0$ False

Solution: $(-\infty, -4] \cup [4, 6]$

33. $6x^2 - 5x \ge 6$

       $6x^2 - 5x - 6 \ge 0$

   $(3x + 2)(2x - 3) \ge 0$

   $3x + 2 = 0$ or $2x - 3 = 0$

       $x = -\dfrac{2}{3}$ or $x = \dfrac{3}{2}$

Region	Test Point	$(3x + 2)(2x - 3) \ge 0$ Result
A: $\left(-\infty, -\dfrac{2}{3}\right]$	-1	$(-1)(-5) \ge 0$ True
B: $\left[-\dfrac{2}{3}, \dfrac{3}{2}\right]$	0	$(2)(-3) \ge 0$ False
C: $\left[\dfrac{3}{2}, \infty\right)$	2	$(8)(1) \ge 0$ True

Solution: $\left(-\infty, -\dfrac{2}{3}\right] \cup \left[\dfrac{3}{2}, \infty\right)$

35.
$$4x^3 + 16x^2 - 9x - 36 > 0$$
$$4x^2(x+4) - 9(x+4) > 0$$
$$(x+4)(4x^2 - 9) > 0$$
$$(x+4)(2x+3)(2x-3) > 0$$

$$x + 4 = 0 \quad \text{or} \quad 2x + 3 = 0 \quad \text{or} \quad 2x - 3 = 0$$
$$x = -4 \quad \text{or} \quad x = -\frac{3}{2} \quad \text{or} \quad x = \frac{3}{2}$$

Region	Test Point	$(x+4)(2x+3)(2x-3) > 0$
$A: (-\infty, -4)$	-5	$(-1)(-7)(-13) > 0$ False
$B: \left(-4, -\frac{3}{2}\right)$	-3	$(1)(-3)(-9) > 0$ True
$C: \left(-\frac{3}{2}, \frac{3}{2}\right)$	0	$(4)(3)(-3) > 0$ False
$D: \left(\frac{3}{2}, \infty\right)$	4	$(8)(11)(5) > 0$ True

Solution: $\left(-4, -\dfrac{3}{2}\right) \cup \left(\dfrac{3}{2}, \infty\right)$

37.
$$x^4 - 26x^2 + 25 \geq 0$$
$$(x^2 - 25)(x^2 - 1) \geq 0$$
$$(x+5)(x-5)(x+1)(x-1) \geq 0$$
$$x = -5 \quad \text{or} \quad x = 5 \quad \text{or} \quad x = -1 \quad \text{or} \quad x = 1$$

Region	Test Point	$(x+5)(x-5)(x+1)(x-1) \geq 0$ Result
$A: (-\infty, -5]$	-6	$(-1)(-11)(-5)(-7) \geq 0$ True
$B: [-5, -1]$	-2	$(3)(-7)(-1)(-3) \geq 0$ False
$C: [-1, 1]$	0	$(5)(-5)(1)(-1) \geq 0$ True
$D: [1, 5]$	2	$(7)(-3)(3)(1) \geq 0$ False
$E: [5, \infty)$	6	$(11)(1)(7)(5) \geq 0$ True

Solution: $(-\infty, -5] \cup [-1, 1] \cup [5, \infty)$

39. $(2x - 7)(3x + 5) > 0$

$$2x - 7 = 0 \quad \text{or} \quad 3x + 5 = 0$$

$$x = \frac{7}{2} \quad \text{or} \quad x = -\frac{5}{3}$$

Region	Test Point	$(2x - 7)(3x + 5) > 0$
$A: \left(-\infty, -\frac{5}{3}\right)$	-2	$(-11)(-1) > 0$ True
$B: \left(-\frac{5}{3}, \frac{7}{2}\right)$	0	$(-7)(5) > 0$ False
$C: \left(\frac{7}{2}, \infty\right)$	4	$(1)(17) > 0$ True

Solution: $\left(-\infty, -\frac{5}{3}\right) \cup \left(\frac{7}{2}, \infty\right)$

41. $\dfrac{x}{x - 10} < 0$

$$x = 0 \quad \text{or} \quad x - 10 = 0$$

$$x = 10$$

Region	Test Point	$\dfrac{x}{x - 10} < 0$ Result
$A: (-\infty, 0)$	-1	$\dfrac{-1}{-11} < 0$ False
$B: (0, 10)$	5	$\dfrac{5}{-5} < 0$ True
$C: (10, \infty)$	11	$\dfrac{11}{1} < 0$ False

Solution: $(0, 10)$

43. $\dfrac{x - 5}{x + 4} \geq 0$

$$x - 5 = 0 \quad \text{or} \quad x + 4 = 0$$

$$x = 5 \quad \text{or} \quad x = -4$$

Region	Test Point	$\dfrac{x-5}{x+4} \geq 0$ Result
A: $(-\infty, -4)$	-5	$\dfrac{-10}{-1} \geq 0$ True
B: $(-4, 5]$	0	$\dfrac{-5}{4} \geq 0$ False
C: $[5, \infty)$	6	$\dfrac{1}{10} \geq 0$ True

Solution: $(-\infty, -4) \cup [5, \infty)$

45. $\dfrac{x(x+6)}{(x-7)(x+1)} \geq 0$

$x = 0$ or $x + 6 = 0$ or $x - 7 = 0$ or $x + 1 = 0$
$x = -6$ or $x = 7$ or $x = -1$

Region	Test Point	$\dfrac{x(x+6)}{(x-7)(x+1)} \geq 0$ Result
A: $(-\infty, -6]$	-7	$\dfrac{-7(-1)}{(-14)(-6)} \geq 0$ True
B: $[-6, -1)$	-3	$\dfrac{-3(3)}{(-10)(-2)} \geq 0$ False
C: $(-1, 0]$	$-\dfrac{1}{2}$	$\dfrac{-\frac{1}{2}\left(\frac{11}{2}\right)}{\left(-\frac{15}{2}\right)\left(\frac{1}{2}\right)} \geq 0$ True
D: $[0, 7)$	2	$\dfrac{2(8)}{(-5)(3)} \geq 0$ False
E: $(7, \infty)$	8	$\dfrac{8(14)}{(1)(9)} \geq 0$ True

Solution: $(-\infty, -6] \cup (-1, 0] \cup (7, \infty)$

47. $\dfrac{-1}{x-1} > -1$

The denominator is equal to 0 when $x - 1 = 0$, or $x = 1$.

$$\dfrac{-1}{x-1} = -1$$
$$-1 = -1(x-1)$$
$$-1 = -x+1$$
$$x = 2$$

Region	Test Point	$\dfrac{-1}{x-1} > -1$ Result
A: $(-\infty, 1)$	0	$\dfrac{-1}{-1} > -1$ True
B: $(1, 2)$	$\dfrac{3}{2}$	$\dfrac{-1}{\frac{1}{2}} = -2 > -1$ False
C: $(2, \infty)$	3	$\dfrac{-1}{2} > -1$ True

Solution: $(-\infty, 1) \cup (2, \infty)$

49. $\dfrac{x}{x+4} \le 2$

The denominator is equal to 0 when $x + 4 = 0$, or $x = -4$.

$$\dfrac{x}{x+4} = 2$$
$$x = 2x + 8$$
$$-x = 8$$
$$x = -8$$

Region	Test Point	$\dfrac{x}{x+4} \le 2$ Result
A: $(-\infty, -8]$	-9	$\dfrac{-9}{-5} \le 2$ True
B: $[-8, -4)$	-6	$\dfrac{-6}{-2} \le 2$ False
C: $(-4, \infty)$	0	$\dfrac{0}{4} \le 2$ True

Solution: $(-\infty, -8] \cup (-4, \infty)$

51. $\dfrac{z}{z-5} \geq 2z$

The denominator is equal to 0 when $z - 5 = 0$, or $z = 5$.

$$\dfrac{z}{z-5} = 2z$$
$$z = 2z(z-5)$$
$$z = 2z^2 - 10z$$
$$0 = 2z^2 - 11z$$
$$0 = z(2z - 11)$$
$$z = 0 \quad \text{or} \quad 2z - 11 = 0$$
$$z = \dfrac{11}{2}$$

Region	Test Point	$\dfrac{z}{z-5} \geq 2z$ Result
A: $(-\infty, 0]$	-1	$\dfrac{-1}{-6} \geq -2$ True
B: $[0, 5)$	1	$\dfrac{1}{-4} \geq 2$ False
C: $\left(5, \dfrac{11}{2}\right]$	$\dfrac{21}{4}$	$\dfrac{\left(\frac{21}{4}\right)}{\left(\frac{1}{4}\right)} \geq \dfrac{21}{2}$ $21 \geq \dfrac{21}{2}$ True
D: $\left[\dfrac{11}{2}, \infty\right)$	6	$\dfrac{6}{1} \geq 12$ False

Solution: $(-\infty, 0] \cup \left(5, \dfrac{11}{2}\right]$

53. $\dfrac{(x+1)^2}{5x} > 0$

The denominator is equal to 0 when $5x = 0$, or $x = 0$.

$$\dfrac{(x+1)^2}{5x} = 0$$
$$(x+1)^2 = 0$$
$$x + 1 = 0$$
$$x = -1$$

Region	Test Point	$\dfrac{(x+1)^2}{5x} > 0$ Result
A: $(-\infty, -1)$	-2	$\dfrac{1}{-10} > 0$ False
B: $(-1, 0)$	$-\dfrac{1}{2}$	$\dfrac{\left(\frac{1}{4}\right)}{\left(-\frac{5}{2}\right)} > 0$ False
C: $(0, \infty)$	1	$\dfrac{4}{5} > 0$ True

Solution: $(0, \infty)$

55. $g(x) = |x| + 2$

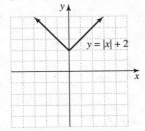

57. $F(x) = |x| - 1$

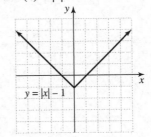

59. $F(x) = x^2 - 3$

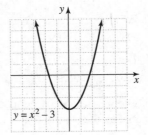

61. $H(x) = x^2 + 1$

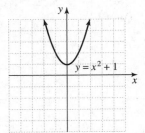

$y = x^2 + 1$

63. answers may vary

65. Let x = the number. Then

$\dfrac{1}{x}$ = the reciprocal of the number.

$$x - \frac{1}{x} < 0$$

$$\frac{x^2 - 1}{x} < 0$$

$$\frac{(x+1)(x-1)}{x} < 0$$

$x + 1 = 0$ or $x - 1 = 0$ or $x = 0$

$x = -1$ or $x = 1$

Region	Test Point	$\dfrac{(x+1)(x-1)}{x} < 0$ Result
A: $(-\infty, -1)$	-2	$\dfrac{(-1)(-3)}{-2} < 0$ True
B: $(-1, 0)$	$-\dfrac{1}{2}$	$\dfrac{\left(\frac{1}{2}\right)\left(-\frac{3}{2}\right)}{\left(-\frac{1}{2}\right)} < 0$ False
C: $(0, 1)$	$\dfrac{1}{2}$	$\dfrac{\left(\frac{3}{2}\right)\left(-\frac{1}{2}\right)}{\left(\frac{1}{2}\right)} < 0$ True
D: $(1, \infty)$	2	$\dfrac{(3)(1)}{2} < 0$ False

$(-\infty, -1) \cup (0, 1)$ or any number less than -1 or between 0 and 1 satisfies the conditions.

67. $P(x) = -2x^2 + 26x - 44$

$$-2x^2 + 26x - 44 > 0$$
$$-2(x^2 + 13x - 22) > 0$$
$$-2(x - 11)(x - 2) > 0$$
$$x - 11 = 0 \quad \text{or} \quad x - 2 = 0$$
$$x = 11 \quad \text{or} \quad x = 2$$

Region	Test Point	$-2(x - 11)(x - 2) > 0$ Result
A: $(0, 2)$	1	$-2(-10)(-3) > 0$ False
B: $(2, 11)$	3	$-2(-8)(1) > 0$ True
C: $(11, \infty)$	12	$-2(1)(10) > 0$ False

The company makes a profit when x is between 2 and 11.

69.
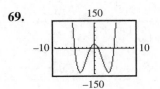

71.
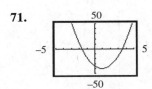

Section 11.5 Practice

1. $f(x) = x^2$ and $g(x) = x^2 - 4$

Construct a table of values for $f(x)$ and $g(x)$.

x	$f(x) = x^2$	$g(x) = x^2 - 4$
-2	4	0
-1	1	-3
0	0	-4
1	1	-3
2	4	0

538

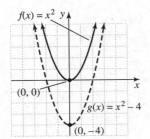

2. a. $f(x) = x^2 - 5$

The graph of $f(x)$ is obtained by shifting the graph of $y = x^2$ downward 5 units.

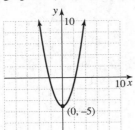

b. $g(x) = x^2 + 3$

The graph of $g(x)$ is obtained by shifting the graph of $y = x^2$ upward 3 units.

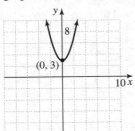

3. $f(x) = x^2$ and $g(x) = (x+6)^2$

Plot points. Notice that the graph of $g(x)$ is the graph of $f(x)$ shifted 6 units to the left.

x	$f(x) = x^2$	x	$g(x) = (x+6)^2$
−2	4	−8	4
−1	1	−7	1
0	0	−6	0
1	1	−5	1
2	4	−4	4

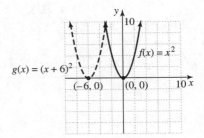

4. a. $G(x) = (x+4)^2$

The graph of $G(x)$ is obtained by shifting the graph of $y = x^2$ to the left 4 units.

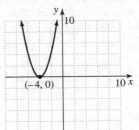

b. $H(x) = (x-7)^2$

The graph of $H(x)$ is obtained by shifting the graph of $y = x^2$ to the right 7 units.

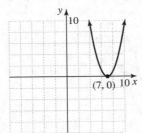

5. $f(x) = (x+2)^2 + 2$

The graph of $f(x)$ is the graph of $y = x^2$ shifted 2 units to the left and 2 units upward. The vertex is then (−2, 2), and the axis of symmetry is $x = -2$.

x	$f(x) = (x+2)^2 + 2$
−4	6
−3	3
−1	3
0	6

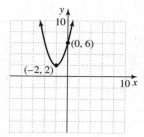

6. $f(x) = x^2$, $g(x) = 4x^2$, and $h(x) = \frac{1}{4}x^2$

Comparing tables of values, we see that for each x-value, the corresponding value of $g(x)$ is four times that of $f(x)$. Similarly, the value of $h(x)$ is one quarter the value of $f(x)$.

x	$f(x) = x^2$	$g(x) = 4x^2$	$h(x) = \frac{1}{4}x^2$
-2	4	16	1
-1	1	4	$\frac{1}{4}$
0	0	0	0
1	1	4	$\frac{1}{4}$
2	4	16	1

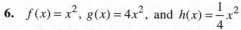
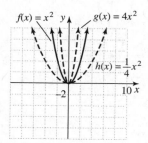

7. $f(x) = -\frac{1}{2}x^2$

Because $a = -\frac{1}{2}$, a negative value, this parabola opens downward. Since $\left| -\frac{1}{2} \right| = \frac{1}{2} < 1$, the parabola is wider than the graph of $y = x^2$. The vertex is $(0, 0)$, and the axis of symmetry is the y-axis.

x	$f(x) = -\frac{1}{2}x^2$
-2	-2
-1	$-\frac{1}{2}$
0	0
1	$-\frac{1}{2}$
2	-2

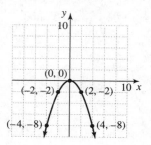

8. $h(x) = \frac{1}{3}(x-4)^2 - 3$

This graph is the same as $y = x^2$ shifted 4 units to the right and 3 units downward, and it is wider because a is $\frac{1}{3}$. The vertex is $(4, -3)$, and the axis of symmetry is $x = 4$.

x	$h(x) = \frac{1}{3}(x-4)^2 - 3$
2	$-\frac{5}{3}$
3	$-\frac{8}{3}$
4	-3
5	$-\frac{8}{3}$
6	$-\frac{5}{3}$

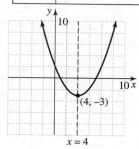

Graphing Calculator Explorations

1.
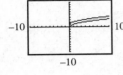

2.
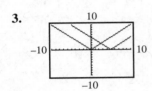

3.

4.
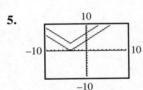

5.
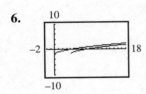

6.
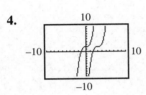

Vocabulary, Readiness & Video Check 11.5

1. A <u>quadratic</u> function is one that can be written in the form $f(x) = ax^2 + bx + c$, $a \neq 0$.

2. The graph of a quadratic function is a <u>parabola</u> opening <u>upward</u> or <u>downward</u>.

3. If $a > 0$, the graph of the quadratic function opens <u>upward</u>.

4. If $a < 0$, the graph of the quadratic function opens <u>downward</u>.

5. The vertex of a parabola is the <u>lowest</u> point if $a > 0$.

6. The vertex of a parabola is the <u>highest</u> point if $a < 0$.

7. $f(x) = x^2$; vertex: (0, 0)

8. $f(x) = -5x^2$; vertex: (0, 0)

9. $g(x) = (x - 2)^2$; vertex: (2, 0)

10. $g(x) = (x + 5)^2$; vertex: (−5, 0)

11. $f(x) = 2x^2 + 3$; vertex: (0, 3)

12. $h(x) = x^2 - 1$; vertex: (0, −1)

13. $g(x) = (x + 1)^2 + 5$; vertex: (−1, 5)

14. $h(x) = (x - 10)^2 - 7$; vertex: (10, −7)

15. Graphs of the form $f(x) = x^2 + k$ shift up or down the y-axis k units from $y = x^2$; the y-intercept.

16. Graphs of the form $f(x) = (x - h)^2$ shift right or left on the x-axis h units from $y = x^2$; the x-intercept.

17. The vertex, (h, k) and the axis of symmetry, $x = h$; the basic shape of $y = x^2$ does not change.

18. whether the graph is wider or narrower than $y = x^2$

19. the coordinates of the vertex, whether the graph opens upward or downward, whether the graph is narrower or wider than $y = x^2$, and the graph's axis of symmetry

Exercise Set 11.5

1. $f(x) = x^2 - 1$

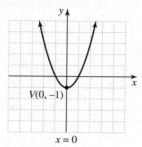

3. $h(x) = x^2 + 5$

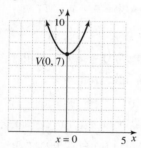

5. $g(x) = x^2 + 7$

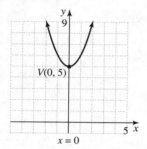

7. $f(x) = (x-5)^2$

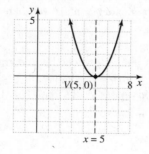

9. $h(x) = (x+2)^2$

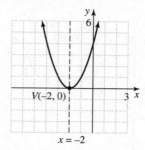

11. $G(x) = (x+3)^2$

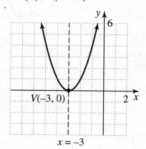

13. $f(x) = (x-2)^2 + 5$

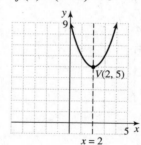

15. $h(x) = (x+1)^2 + 4$

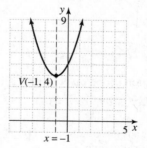

17. $g(x) = (x+2)^2 - 5$

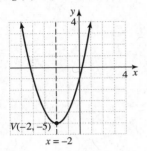

19. $H(x) = 2x^2$

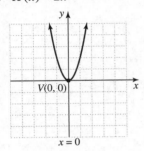

21. $h(x) = \frac{1}{3}x^2$

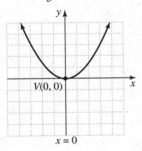

23. $g(x) = -x^2$

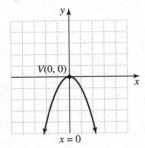

25. $f(x) = 2(x-1)^2 + 3$

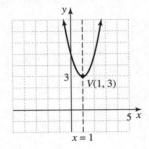

27. $h(x) = -3(x+3)^2 + 1$

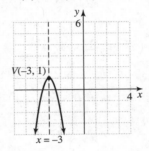

29. $H(x) = \frac{1}{2}(x-6)^2 - 3$

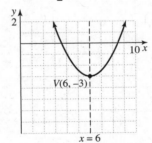

31. $f(x) = -(x-2)^2$

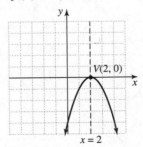

33. $F(x) = -x^2 + 4$

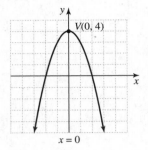

35. $F(x) = 2x^2 - 5$

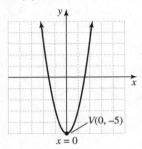

37. $h(x) = (x-6)^2 + 4$

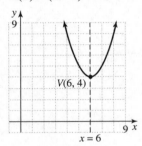

39. $F(x) = \left(x + \dfrac{1}{2}\right)^2 - 2$

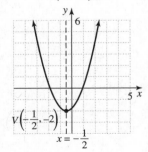

41. $F(x) = \dfrac{3}{2}(x+7)^2 + 1$

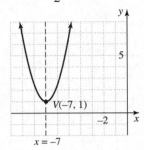

43. $f(x) = \dfrac{1}{4}x^2 - 9$

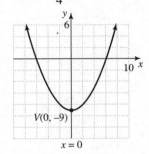

45. $G(x) = 5\left(x + \dfrac{1}{2}\right)^2$

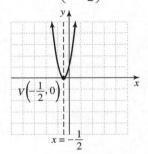

47. $h(x) = -(x-1)^2 - 1$

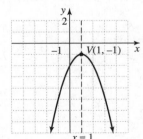

49. $g(x) = \sqrt{3}(x+5)^2 + \dfrac{3}{4}$

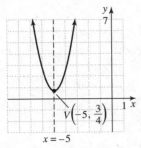

51. $h(x) = 10(x+4)^2 - 6$

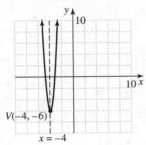

53. $f(x) = -2(x-4)^2 + 5$

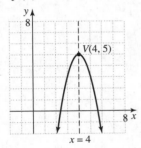

55. $x^2 + 8x$

$$\left[\frac{1}{2}(8)\right]^2 = (4)^2 = 16$$

$$x^2 + 8x + 16$$

57. $z^2 - 16z$

$$\left[\frac{1}{2}(-16)\right]^2 = (-8)^2 = 64$$

$$z^2 - 16z + 64$$

59. $y^2 + y$

$$\left[\frac{1}{2}(1)\right]^2 = \left(\frac{1}{2}\right)^2 = \frac{1}{4}$$

$$y^2 + y + \frac{1}{4}$$

61.
$$x^2 + 4x = 12$$
$$x^2 + 4x + \left(\frac{4}{2}\right)^2 = 12 + 4$$
$$x^2 + 4x + 4 = 16$$
$$(x+2)^2 = 16$$
$$x + 2 = \pm\sqrt{16}$$
$$x + 2 = \pm 4$$
$$x = -2 \pm 4$$
$$x = -6 \text{ or } 2$$

63.
$$z^2 + 10z - 1 = 0$$
$$z^2 + 10z = 1$$
$$z^2 + 10z + \left(\frac{10}{2}\right)^2 = 1 + 25$$
$$z^2 + 10z + 25 = 26$$
$$(z+5)^2 = 26$$
$$z + 5 = \pm\sqrt{26}$$
$$z = -5 \pm \sqrt{26}$$

65.
$$z^2 - 8z = 2$$
$$z^2 - 8z + \left(\frac{-8}{2}\right)^2 = 2 + 16$$
$$z^2 - 8z + 16 = 18$$
$$(z-4)^2 = 18$$
$$z - 4 = \pm\sqrt{18}$$
$$z - 4 = \pm 3\sqrt{2}$$
$$z = 4 \pm 3\sqrt{2}$$

67. $f(x) = -213(x-0.1)^2 + 3.6$
$a = -213 < 0$, so $f(x)$ opens downward.
The vertex is (0.1, 3.6). The correct answer is **c**.

69. $f(x) = 5(x-2)^2 + 3$

71. $f(x) = 5[x-(-3)]^2 + 6$
$$= 5(x+3)^2 + 6$$

73. $y = f(x) + 1$

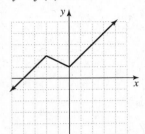

75. $y = f(x-3)$

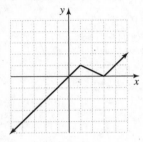

77. $y = f(x+2) + 2$

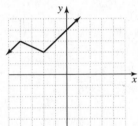

79. $f(x) = 2158x^2 - 10{,}339x + 6731$

   a. $x = 2010 - 2000 = 10$

$$f(10) = 2158(10)^2 - 10{,}339(10) + 6731$$
$$= 119{,}141$$

There were approximately 119,141 million text messages sent.

   b. $x = 2014 - 2000 = 14$

$$f(14) = 2158(14)^2 - 10{,}339(14) + 6731$$
$$= 284{,}953$$

The model predicts approximately 284,953 million text messages sent in 2014.

Section 11.6 Practice

1. $g(x) = x^2 - 2x - 3$

Write in the form $y = (x-h)^2 + k$ by completing the square.

$$y = x^2 - 2x - 3$$
$$y + 3 = x^2 - 2x$$
$$y + 3 + \left(\frac{-2}{2}\right)^2 = x^2 - 2x + \left(\frac{-2}{2}\right)^2$$
$$y + 4 = x^2 - 2x + 1$$
$$y = (x-1)^2 - 4$$

The vertex is at (1, −4).
Let $g(x) = 0$.

$$0 = x^2 - 2x - 3$$
$$0 = (x-3)(x+1)$$
$$x - 3 = 0 \quad \text{or} \quad x + 1 = 0$$
$$x = 3 \qquad\qquad x = -1$$

The *x*-intercepts are (3, 0) and (−1, 0).
Let $x = 0$.

$$g(0) = 0^2 - 2(0) - 3 = -3$$

The *y*-intercept is (0, −3).

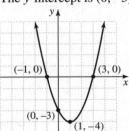

2. $g(x) = 4x^2 + 4x + 3$

Replace $g(x)$ with y and complete the square to write the equation in the form $y = a(x-h)^2 + k$.

$$y = 4x^2 + 4x + 3$$
$$y - 3 = 4x^2 + 4x = 4(x^2 + x)$$
$$y - 3 + 4\left(\frac{1}{2}\right)^2 = 4\left[x^2 + x + \left(\frac{1}{2}\right)^2\right]$$
$$y - 3 + 1 = 4\left(x^2 + x + \frac{1}{4}\right)$$
$$y = 4\left(x + \frac{1}{2}\right)^2 + 2$$

$a = 4$, $h = -\frac{1}{2}$, and $k = 2$.

The parabola opens upward with vertex

$\left(-\dfrac{1}{2}, 2\right)$, and has an axis of symmetry $x = -\dfrac{1}{2}$.

Let $x = 0$.

$g(0) = 4(0)^2 + 4(0) + 3 = 3$

The y-intercept is $(0, 3)$. There are no x-intercepts.

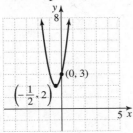

3. $g(x) = -x^2 + 5x + 6$

Write $g(x)$ in the form $a(x - h)^2 + k$ by completing the square. Replace $g(x)$ with y.

$$y = -x^2 + 5x + 6$$

$$y - 6 = -x^2 + 5x$$

$$y - 6 = -1(x^2 - 5x)$$

$$y - 6 - \left(\frac{-5}{2}\right)^2 = -1\left[x^2 - 5x + \left(\frac{-5}{2}\right)^2\right]$$

$$y - 6 - \frac{25}{4} = -1\left(x^2 - 5x + \frac{25}{4}\right)$$

$$y - \frac{49}{4} = -\left(x - \frac{5}{2}\right)^2$$

$$y = -\left(x - \frac{5}{2}\right)^2 + \frac{49}{4}$$

Since $a = -1$, the parabola opens downward with vertex $\left(\dfrac{5}{2}, \dfrac{49}{4}\right)$ and axis of symmetry $x = \dfrac{5}{2}$.

Let $x = 0$.

$y = -0^2 + 5(0) + 6 = 6$

The y-intercept is $(0, 6)$. Let $y = 0$.

$0 = -x^2 + 5x + 6$

$0 = x^2 - 5x - 6$

$0 = (x - 6)(x + 1)$

$x - 6 = 0 \quad$ or $\quad x + 1 = 0$

$\quad x = 6 \qquad\qquad x = -1$

The x-intercepts are $(6, 0)$ and $(-1, 0)$.

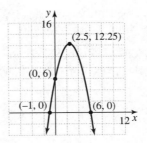

4. $g(x) = x^2 - 2x - 3$

$a = 1$, $b = -2$, and $c = -3$

$$\frac{-b}{2a} = \frac{-(-2)}{2(1)} = \frac{2}{2} = 1$$

The x-value of the vertex is 1.

$g(1) = 1^2 - 2(1) - 3 = 1 - 2 - 3 = -4$

The vertex is $(1, -4)$.

5. $h(t) = -16t^2 + 24t$

Find the vertex of $h(t)$ to find its maximum value.

$a = -16$, $b = 24$, and $c = 0$

$$\frac{-b}{2a} = \frac{-24}{2(-16)} = \frac{3}{4}$$

The t-value of the vertex is $\dfrac{3}{4}$.

$$h\left(\frac{3}{4}\right) = -16\left(\frac{3}{4}\right)^2 + 24\left(\frac{3}{4}\right)$$

$$= -16\left(\frac{9}{16}\right) + 18$$

$$= -9 + 18$$

$$= 9$$

The vertex is $\left(\dfrac{3}{4}, 9\right)$. Thus, the ball reaches its maximum height of 9 feet in $\dfrac{3}{4}$ second.

Vocabulary, Readiness & Video Check 11.6

1.   If a quadratic function is in the form $f(x) = a(x - h)^2 + k$, the vertex of its graph is <u>(h, k)</u>.

2.   The graph of $f(x) = ax^2 + bx + c$, $a \ne 0$, is a parabola whose vertex has x-value $\dfrac{-b}{2a}$.

3. We can immediately identify the vertex (h, k), whether the parabola opens upward or downward, and know its axis of symmetry; completing the square.

4. This information tells us whether or not the graph has x-intercepts. For example, if the vertex is in quadrant III or IV and the parabola opens downward, then there aren't any x-intercepts and there's no need to go through the steps to locate any.

5. the vertex

Exercise Set 11.6

	Parabola Opens	*Vertex Location*	*Number of x-intercept(s)*	*Number of y-intercept(s)*
1.	up	Q I	0	1
3.	down	Q II	2	1
5.	up	x-axis	1	1
7.	down	Q III	0	
9.	up	Q IV	2	

11. $f(x) = x^2 + 8x + 7$

    $-\dfrac{b}{2a} = \dfrac{-8}{2(1)} = -4$ and

    $\begin{aligned} f(-4) &= (-4)^2 + 8(-4) + 7 \\ &= 16 - 32 + 7 \\ &= -9 \end{aligned}$

    Thus, the vertex is $(-4, -9)$.

13. $f(x) = -x^2 + 10x + 5$

    $-\dfrac{b}{2a} = \dfrac{-10}{2(-1)} = 5$ and

    $\begin{aligned} f(5) &= -(5)^2 + 10(5) + 5 \\ &= -25 + 50 + 5 \\ &= 30 \end{aligned}$

    Thus, the vertex is $(5, 30)$.

15. $f(x) = 5x^2 - 10x + 3$

    $-\dfrac{b}{2a} = \dfrac{-(-10)}{2(5)} = 1$ and

    $\begin{aligned} f(1) &= 5(1)^2 - 10(1) + 3 \\ &= 5 - 10 + 3 \\ &= -2 \end{aligned}$

    Thus, the vertex is $(1, -2)$.

17. $f(x) = -x^2 + x + 1$

$-\dfrac{b}{2a} = \dfrac{-1}{2(-1)} = \dfrac{1}{2}$ and

$f\left(\dfrac{1}{2}\right) = -\left(\dfrac{1}{2}\right)^2 + \left(\dfrac{1}{2}\right) + 1$

$= -\dfrac{1}{4} + \dfrac{1}{2} + 1$

$= \dfrac{5}{4}$

Thus, the vertex is $\left(\dfrac{1}{2}, \dfrac{5}{4}\right)$.

19. $f(x) = x^2 - 4x + 3$

$-\dfrac{b}{2a} = \dfrac{-(-4)}{2(1)} = 2$ and

$f(2) = (2)^2 - 4(2) + 3 = -1$

The vertex is (2, –1), so the graph is D.

21. $f(x) = x^2 - 2x - 3$

$-\dfrac{b}{2a} = \dfrac{-(-2)}{2(1)} = 1$ and

$f(1) = (1)^2 - 2(1) - 3 = -4$

The vertex is (1, –4), so the graph is B.

23. $f(x) = x^2 + 4x - 5$

$-\dfrac{b}{2a} = \dfrac{-4}{2(1)} = -2$ and

$f(-2) = (-2)^2 + 4(-2) - 5 = -9$

Thus, the vertex is (–2, –9).

The graph opens upward ($a = 1 > 0$).

$x^2 + 4x - 5 = 0$

$(x + 5)(x - 1) = 0$

$x + 5 = 0$ or $x - 1 = 0$

$x = -5$ or $x = 1$

x-intercepts: (–5, 0) and (1, 0).

$f(0) = -5$, so the y-intercept is (0, –5).

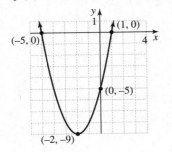

25. $f(x) = -x^2 + 2x - 1$

$-\dfrac{b}{2a} = \dfrac{-2}{2(-1)} = 1$ and

$f(1) = -(1)^2 + 2(1) - 1 = 0$

Thus, the vertex is (1, 0).

The graph opens downward ($a = -1 < 0$).

$-x^2 + 2x - 1 = 0$

$x^2 - 2x + 1 = 0$

$(x - 1)^2 = 0$

$x - 1 = 0$

$x = 1$

x-intercept: (1, 0).

$f(0) = -1$, so the y-intercept is (0, –1).

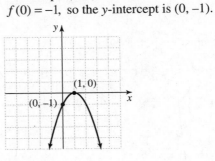

27. $f(x) = x^2 - 4$

$-\dfrac{b}{2a} = \dfrac{-0}{2(1)} = 0$ and

$f(0) = (0)^2 - 4 = -4$

Thus, the vertex is (0, –4).

The graph opens upward ($a = 1 > 0$).

$x^2 - 4 = 0$

$(x + 2)(x - 2) = 0$

$x + 2 = 0$ or $x - 2 = 0$

$x = -2$ or $x = 2$

x-intercepts: (–2, 0) and (2, 0).

$f(0) = -4$, so the y-intercept is (0, –4).

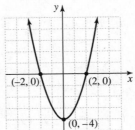

29. $f(x) = 4x^2 + 4x - 3$

$-\dfrac{b}{2a} = -\dfrac{4}{2(4)} = -\dfrac{1}{2}$ and

$f\left(-\dfrac{1}{2}\right) = 4\left(-\dfrac{1}{2}\right)^2 + 4\left(-\dfrac{1}{2}\right) - 3 = -4$

Thus, the vertex is $\left(-\dfrac{1}{2}, -4\right)$.

The graph opens upward ($a = 4 > 0$).

$4x^2 + 4x - 3 = 0$

$(2x + 3)(2x - 1) = 0$

$2x + 3 = 0$ or $2x - 1 = 0$

$x = -\dfrac{3}{2}$ or $x = \dfrac{1}{2}$

x-intercepts: $\left(-\dfrac{3}{2}, 0\right)$ and $\left(\dfrac{1}{2}, 0\right)$.

$f(0) = -3$, so the y-intercept is $(0, -3)$.

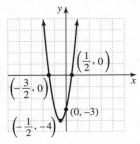

31. $f(x) = \dfrac{1}{2}x^2 + 4x + \dfrac{15}{2}$

$-\dfrac{b}{2a} = -\dfrac{4}{2\left(\frac{1}{2}\right)} = -\dfrac{4}{1} = -4$ and

$f(-4) = \dfrac{1}{2}(-4)^2 + 4(-4) + \dfrac{15}{2} = -\dfrac{1}{2}$.

Thus, the vertex is $\left(-4, -\dfrac{1}{2}\right)$.

The graph opens upward $\left(a = \dfrac{1}{2} > 0\right)$.

$\dfrac{1}{2}x^2 + 4x + \dfrac{15}{2} = 0$

$2\left(\dfrac{1}{2}x^2 + 4x + \dfrac{15}{2}\right) = 2(0)$

$x^2 + 8x + 15 = 0$

$(x + 5)(x + 3) = 0$

$x + 5 = 0$ or $x + 3 = 0$

$x = -5$ or $x = -3$

x-intercepts: $(-5, 0)$ and $(-3, 0)$.

$f(0) = \dfrac{15}{2}$, so the y-intercept is $\left(0, \dfrac{15}{2}\right)$.

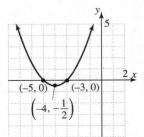

33. $f(x) = x^2 - 6x + 5$

$y = x^2 - 6x + 5$

$y - 5 = x^2 - 6x$

$y - 5 + 9 = x^2 - 6x + 9$

$y + 4 = (x - 3)^2$

$y = (x - 3)^2 - 4$

$f(x) = (x - 3)^2 - 4$

Thus, the vertex is $(3, -4)$.

The graph opens upward ($a = 1 > 0$).

$x^2 - 6x + 5 = 0$

$(x - 5)(x - 1) = 0$

$x = 5$ or $x = 1$

x-intercepts: $(5, 0)$ and $(1, 0)$.

$f(0) = 5$, so the y-intercept is $(0, 5)$.

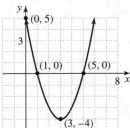

35. $f(x) = x^2 - 4x + 5$

$y = x^2 - 4x + 5$

$y - 5 = x^2 - 4x$

$y - 5 + 4 = x^2 - 4x + 4$

$y - 1 = (x - 2)^2$

$y = (x - 2)^2 + 1$

$f(x) = (x - 2)^2 + 1$

Thus, the vertex is $(2, 1)$.

The graph opens upward ($a = 1 > 0$).

$$x^2 - 4x + 5 = 0$$

$$x = \frac{4 \pm \sqrt{(-4)^2 - 4(1)(5)}}{2(1)} = \frac{4 \pm \sqrt{-4}}{2}$$

which give non-real solutions.

Hence, there are no x-intercepts.

$f(0) = 5$, so the y-intercept is $(0, 5)$.

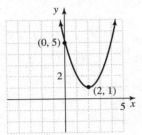

37.
$$f(x) = 2x^2 + 4x + 5$$
$$y = 2x^2 + 4x + 5$$
$$y - 5 = 2(x^2 + 2x)$$
$$y - 5 + 2(1) = 2(x^2 + 2x + 1)$$
$$y - 3 = 2(x+1)^2$$
$$y = 2(x+1)^2 + 3$$
$$f(x) = 2(x+1)^2 + 3$$

Thus, the vertex is $(-1, 3)$.

The graph opens upward ($a = 2 > 0$).

$$2x^2 + 4x + 5 = 0$$

$$x = \frac{-4 \pm \sqrt{(4)^2 - 4(2)(5)}}{2(2)} = \frac{-4 \pm \sqrt{-24}}{4}$$

which give non-real solutions.

Hence, there are no x-intercepts.

$f(0) = 5$, so the y-intercept is $(0, 5)$.

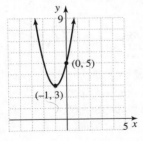

39.
$$f(x) = -2x^2 + 12x$$
$$y = -2(x^2 - 6x)$$
$$y + [-2(9)] = -2(x^2 - 6x + 9)$$
$$y - 18 = -2(x-3)^2$$
$$y = -2(x-3)^2 + 18$$
$$f(x) = -2(x-3)^2 + 18$$

Thus, the vertex is $(3, 18)$.

The graph opens downward ($a = -2 < 0$).

$$-2x^2 + 12x = 0$$
$$-2x(x-6) = 0$$
$$x = 0 \text{ or } x - 6 = 0$$
$$x = 6$$

x-intercepts: $(0, 0)$ and $(6, 0)$

$f(0) = 0$, so the y-intercept is $(0, 0)$.

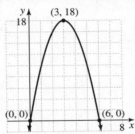

41. $f(x) = x^2 + 1$

$$x = -\frac{b}{2a} = -\frac{0}{2(1)} = 0$$

$$f(0) = (0)^2 + 1 = 1$$

Thus, the vertex is $(0, 1)$.

The graph opens upward ($a = 1 > 0$).

$$x^2 + 1 = 0$$
$$x^2 = -1$$

which give non-real solutions.

Hence, there are no x-intercepts.

$f(0) = 1$, so the y-intercept is $(0, 1)$.

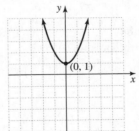

43.

$$f(x) = x^2 - 2x - 15$$
$$y = x^2 - 2x - 15$$
$$y + 15 = x^2 - 2x$$
$$y + 15 + 1 = x^2 - 2x + 1$$
$$y + 16 = (x-1)^2$$
$$y = (x-1)^2 - 16$$
$$f(x) = (x-1)^2 - 16$$

Thus, the vertex is (1, −16).
The graph opens upward ($a = 1 > 0$).

$$x^2 - 2x - 15 = 0$$
$$(x-5)(x+3) = 0$$
$$x = 5 \text{ or } x = -3$$

x-intercepts: (−3, 0) and (5, 0).
$f(0) = -15$ so the y-intercept is (0, −15).

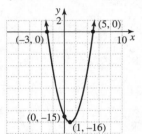

45. $f(x) = -5x^2 + 5x$

$$x = -\frac{b}{2a} = \frac{-5}{2(-5)} = \frac{1}{2} \text{ and}$$

$$f\left(\frac{1}{2}\right) = -5\left(\frac{1}{2}\right)^2 + 5\left(\frac{1}{2}\right) = -\frac{5}{4} + \frac{5}{2} = \frac{5}{4}$$

Thus, the vertex is $\left(\frac{1}{2}, \frac{5}{4}\right)$.

The graph opens downward ($a = -5 < 0$).

$$-5x^2 + 5x = 0$$
$$-5x(x-1) = 0$$
$$x = 0 \text{ or } x - 1 = 0$$
$$x = 1$$

x-intercepts: (0, 0) and (1, 0)
$f(0) = 0$, so the y-intercept is (0, 0).

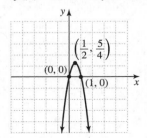

47. $f(x) = -x^2 + 2x - 12$

$$x = -\frac{b}{2a} = \frac{-2}{2(-1)} = 1 \text{ and}$$

$$f(1) = -(1)^2 + 2(1) - 12 = -11$$

Thus, the vertex is (1, −11).
The graph opens downward ($a = -1 < 0$).

$$-x^2 + 2x - 12 = 0$$
$$x^2 - 2x + 12 = 0$$

$$x = \frac{2 \pm \sqrt{(-2)^2 - 4(1)(12)}}{2(1)} = \frac{2 \pm \sqrt{-44}}{2}$$

which yields non-real solutions.
Hence, there are no x-intercepts.
$f(0) = -12$ so the y-intercept is (0, −12).

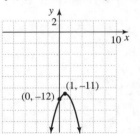

49. $f(x) = 3x^2 - 12x + 15$

$$x = -\frac{b}{2a} = \frac{-(-12)}{2(3)} = \frac{12}{6} = 2 \text{ and}$$

$$f(2) = 3(2)^2 - 12(2) + 15$$
$$= 12 - 24 + 15$$
$$= 3$$

Thus, the vertex is (2, 3).
The graph opens upward ($a = 3 > 0$).

$$3x^2 - 12x + 15 = 0$$
$$x^2 - 4x + 5 = 0$$

$$x = \frac{4 \pm \sqrt{(-4)^2 - 4(1)(5)}}{2(1)} = \frac{4 \pm \sqrt{-4}}{2}$$

which yields non-real solutions.
Hence, there are no x-intercepts.
$f(0) = 15$, so the y-intercept is (0, 15).

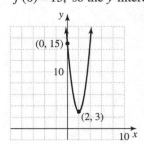

51. $f(x) = x^2 + x - 6$

$x = -\dfrac{b}{2a} = \dfrac{-1}{2(1)} = -\dfrac{1}{2}$ and

$$f\left(-\dfrac{1}{2}\right) = \left(-\dfrac{1}{2}\right)^2 + \left(-\dfrac{1}{2}\right) - 6$$
$$= \dfrac{1}{4} - \dfrac{1}{2} - 6$$
$$= -\dfrac{25}{4}$$

Thus, the vertex is $\left(-\dfrac{1}{2}, -\dfrac{25}{4}\right)$.

The graph opens upward ($a = 1 > 0$).

$x^2 + x - 6 = 0$

$(x+3)(x-2) = 0$

$x = -3$ or $x = 2$

x-intercepts: $(-3, 0)$ and $(2, 0)$.

$f(0) = -6$ so the y-intercept is $(0, -6)$.

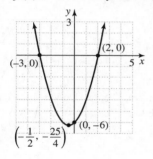

53. $f(x) = -2x^2 - 3x + 35$

$x = -\dfrac{b}{2a} = \dfrac{-(-3)}{2(-2)} = -\dfrac{3}{4}$ and

$$f\left(-\dfrac{3}{4}\right) = -2\left(-\dfrac{3}{4}\right)^2 - 3\left(-\dfrac{3}{4}\right) + 35$$
$$= -\dfrac{9}{8} + \dfrac{9}{4} + 35$$
$$= \dfrac{289}{8}$$

Thus, the vertex is $\left(-\dfrac{3}{4}, \dfrac{289}{8}\right)$.

The graph opens downward ($a = -2 < 0$).

$-2x^2 - 3x + 35 = 0$

$2x^2 + 3x - 35 = 0$

$(2x-7)(x+5) = 0$

$2x - 7 = 0$ or $x + 5 = 0$

$x = \dfrac{7}{2}$ or $x = -5$

x-intercepts: $(-5, 0)$ and $\left(\dfrac{7}{2}, 0\right)$.

$f(0) = 35$ so the y-intercept is $(0, 35)$.

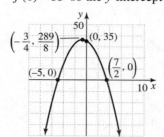

55. $h(t) = -16t^2 + 96t$

$t = -\dfrac{b}{2a} = \dfrac{-96}{2(-16)} = \dfrac{96}{32} = 3$ and

$h(3) = -16(3)^2 + 96(3)$
$= -144 + 288$
$= 144$

The maximum height is 144 feet.

57. $C(x) = 2x^2 - 800x + 92,000$

  a. $x = -\dfrac{b}{2a} = \dfrac{-(-800)}{2(2)} = 200$

       200 bicycles are needed to minimize the cost.

  b. $C(200) = 2(200)^2 - 800(200) + 92,000$
              $= 12,000$

       The minimum cost is \$12,000.

59. Let $x = $ one number. Then
$60 - x = $ the other number.
$f(x) = x(60 - x)$
     $= 60x - x^2$
     $= -x^2 + 60x$

The maximum will occur at the vertex.

$x = -\dfrac{b}{2a} = \dfrac{-60}{2(-1)} = 30$

$60 - x = 60 - 30 = 30$

The numbers are 30 and 30.

61. Let $x = $ one number. Then
$10 + x = $ the other number.
$f(x) = x(10 + x)$
     $= 10x + x^2$
     $= x^2 + 10x$

The minimum will occur at the vertex.

$$x = -\frac{b}{2a} = \frac{-10}{2(1)} = -5$$

$$10 + x = 10 + (-5) = 5$$

The numbers are –5 and 5.

63. Let x = width. Then $40 - x$ = the length.

Area = length · width

$$A(x) = (40 - x)x$$
$$= 40x - x^2$$
$$= -x^2 + 40x$$

The maximum will occur at the vertex.

$$x = -\frac{b}{2a} = \frac{-40}{2(-1)} = 20$$

$$40 - x = 40 - 20 = 20$$

The maximum area will occur when the length and width are 20 units each.

65. $f(x) = x^2 + 2$

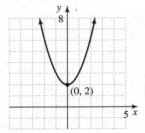

67. $g(x) = x + 2$

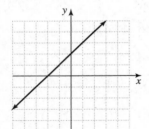

69. $f(x) = (x + 5)^2 + 2$

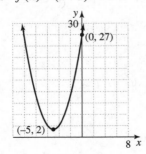

71. $f(x) = 3(x - 4)^2 + 1$

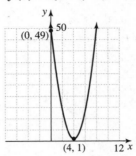

73. $f(x) = -(x - 4)^2 + \frac{3}{2}$

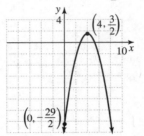

75. $f(x) = 2x^2 - 5$

Since $a = 2 > 0$, the graph opens upward; thus, $f(x)$ has a minimum value.

77. $F(x) = 3 - \frac{1}{2}x^2$

Since $a = -\frac{1}{2} < 0$, the graph opens downward; thus, $F(x)$ has a maximum value.

79. $f(x) = x^2 + 10x + 15$

$$x = -\frac{b}{2a} = \frac{-10}{2(1)} = -5 \text{ and}$$

$$f(-5) = (-5)^2 + 10(-5) + 15 = -10$$

Thus, the vertex is $(-5, -10)$.

The graph opens upward ($a = 1 > 0$).

$f(0) = 15$ so the y-intercept is $(0, 15)$.

$$x^2 + 10x + 15 = 0$$

$$x = \frac{-10 \pm \sqrt{(10)^2 - 4(1)(15)}}{2(1)}$$

$$= \frac{-10 \pm \sqrt{40}}{2}$$

$$\approx -8.2 \text{ or } -1.8$$

The x-intercepts are approximately $(-8.2, 0)$ and $(-1.8, 0)$.

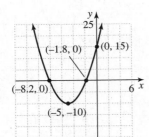

81. $f(x) = 3x^2 - 6x + 7$

$$x = -\frac{b}{2a} = \frac{-(-6)}{2(3)} = 1 \text{ and}$$

$$f(1) = 3(1)^2 - 6(1) + 7 = 4$$

Thus, the vertex is $(1, 4)$.

The graph opens upward $(a = 3 > 0)$.

$f(0) = 7$ so the y-intercept is $(0, 7)$.

$$3x^2 - 6x + 7 = 0$$

$$x = \frac{6 \pm \sqrt{(-6)^2 - 4(3)(7)}}{2(3)} = \frac{6 \pm \sqrt{-48}}{6}$$

which yields non-real solutions.

Hence, there are no x-intercepts.

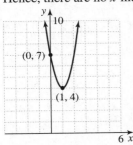

83. $f(x) = 2.3x^2 - 6.1x + 3.2$

$$x = \frac{-(-6.1)}{2(2.3)} \approx 1.33$$

$f(1.33) \approx -0.84$

minimum ≈ -0.84

Alternative solution:

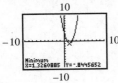

85. $f(x) = -1.9x^2 + 5.6x - 2.7$

$$x = \frac{-5.6}{2(-1.9)} \approx 1.47$$

$f(1.47) \approx 1.43$

maximum ≈ 1.43

Alternate solution:

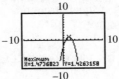

87. a. The function will have a maximum value; answers may vary.

b. $c(x) = -0.4x^2 + 21x + 35$

$a = -0.4$, $b = 21$, $c = 35$

$$-\frac{b}{2a} = -\frac{21}{2(-0.4)} = 26.25$$

The maximum will be reached 26.25 years after 2009 or in 2009 + 26 = 2035.

c. $c(26.25) = -0.4(26.25)^2 + 21(26.25) + 35$

≈ 310.6

The maximum number of Wi-Fi-enabled cell phones is predicted to be about 311 million.

89.

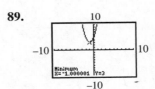

91.

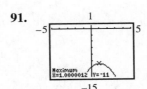

Chapter 11 Vocabulary Check

1. The <u>discriminant</u> helps us find the number and type of solutions of a quadratic equation.

2. If $a^2 = b$, then $a = \underline{\pm\sqrt{b}}$.

3. The graph of $f(x) = ax^2 + bx + c$, where a is not 0, is a parabola whose vertex has x-value $\underline{\dfrac{-b}{2a}}$.

4. A <u>quadratic inequality</u> is an inequality that can be written so that one side is a quadratic expression and the other side is 0.

5. The process of writing a quadratic equation so that one side is a perfect square trinomial is called <u>completing the square</u>.

6. The graph of $f(x) = x^2 + k$ has vertex <u>(0, k)</u>.

7. The graph of $f(x) = (x - h)^2$ has vertex <u>(h, 0)</u>.

8. The graph of $f(x) = (x - h)^2 + k$ has vertex <u>(h, k)</u>.

9. The formula $x = \dfrac{-b \pm \sqrt{b^2 - 4ac}}{2a}$ is called the <u>quadratic formula</u>.

10. A <u>quadratic</u> equation is one that can be written in the form $ax^2 + bx + c = 0$ where a, b, and c are real numbers and a is not 0.

Chapter 11 Review

1. $x^2 - 15x + 14 = 0$
$(x - 14)(x - 1) = 0$
$x - 14 = 0$ or $x - 1 = 0$
$\quad x = 14$ or $\quad x = 1$
The solutions are 1 and 14.

2. $\qquad 7a^2 = 29a + 30$
$7a^2 - 29a - 30 = 0$
$(7a + 6)(a - 5) = 0$
$7a + 6 = 0$ or $a - 5 = 0$
$\quad 7a = -6$ or $\quad a = 5$
$\quad a = -\dfrac{6}{7}$

The solutions are $-\dfrac{6}{7}$ and 5.

3. $4m^2 = 196$
$\quad m^2 = 49$
$\quad m = \pm\sqrt{49}$
$\quad m = \pm 7$
The solutions are -7 and 7.

4. $(5x - 2)^2 = 2$
$5x - 2 = \pm\sqrt{2}$
$\quad 5x = 2 \pm \sqrt{2}$
$\quad\quad x = \dfrac{2 \pm \sqrt{2}}{5}$

The solutions are $\dfrac{2 + \sqrt{2}}{5}$ and $\dfrac{2 - \sqrt{2}}{5}$.

5. $\qquad z^2 + 3z + 1 = 0$
$\qquad z^2 + 3z = -1$
$z^2 + 3z + \left(\dfrac{3}{2}\right)^2 = -1 + \dfrac{9}{4}$
$\qquad \left(z + \dfrac{3}{2}\right)^2 = \dfrac{5}{4}$
$\qquad\quad z + \dfrac{3}{2} = \pm\sqrt{\dfrac{5}{4}}$
$\qquad\quad z + \dfrac{3}{2} = \pm\dfrac{\sqrt{5}}{2}$
$\qquad\qquad z = -\dfrac{3}{2} \pm \dfrac{\sqrt{5}}{2} = \dfrac{-3 \pm \sqrt{5}}{2}$

The solutions are $\dfrac{-3 + \sqrt{5}}{2}$ and $\dfrac{-3 - \sqrt{5}}{2}$.

6. $\qquad (2x + 1)^2 = x$
$\quad 4x^2 + 4x + 1 = x$
$\qquad 4x^2 + 3x = -1$
$\qquad x^2 + \dfrac{3}{4}x = -\dfrac{1}{4}$
$x^2 + \dfrac{3}{4}x + \left(\dfrac{\frac{3}{4}}{2}\right)^2 = -\dfrac{1}{4} + \dfrac{9}{64}$
$\qquad \left(x + \dfrac{3}{8}\right)^2 = -\dfrac{7}{64}$
$\qquad\quad x + \dfrac{3}{8} = \pm\sqrt{-\dfrac{7}{64}}$
$\qquad\quad x + \dfrac{3}{8} = \pm\dfrac{i\sqrt{7}}{8}$
$\qquad\qquad x = -\dfrac{3}{8} \pm \dfrac{i\sqrt{7}}{8} = \dfrac{-3 \pm i\sqrt{7}}{8}$

The solutions are $\dfrac{-3 + i\sqrt{7}}{8}$ and $\dfrac{-3 - i\sqrt{7}}{8}$.

7.
$$A = P(1+r)^2$$
$$2717 = 2500(1+r)^2$$
$$\frac{2717}{2500} = (1+r)^2$$
$$(1+r)^2 = 1.0868$$
$$1+r = \pm\sqrt{1.0868}$$
$$1+r = \pm 1.0425$$
$$r = -1 \pm 1.0425$$
$$= 0.0425 \text{ or } -2.0425 \text{ (disregard)}$$
The interest rate is 4.25%.

8. Let x = distance traveled.
$$a^2 + b^2 = c^2$$
$$x^2 + x^2 = (150)^2$$
$$2x^2 = 22{,}500$$
$$x^2 = 11{,}250$$
$$x = \pm 75\sqrt{2} \approx \pm 106.1$$
Disregard the negative. The ships each traveled $75\sqrt{2} \approx 106.1$ miles.

9. Two complex but not real solutions exist.

10. Two real solutions exist.

11. Two real solutions exist.

12. One real solution exists.

13. $x^2 - 16x + 64 = 0$
$$a = 1, b = -16, c = 64$$
$$x = \frac{16 \pm \sqrt{(-16)^2 - 4(1)(64)}}{2(1)}$$
$$= \frac{16 \pm \sqrt{256 - 256}}{2}$$
$$= \frac{16 \pm \sqrt{0}}{2}$$
$$= 8$$
The solution is 8.

14. $x^2 + 5x = 0$
$$a = 1, b = 5, c = 0$$
$$x = \frac{-5 \pm \sqrt{(5)^2 - 4(1)(0)}}{2(1)}$$
$$= \frac{-5 \pm \sqrt{25}}{2}$$
$$= \frac{-5 \pm 5}{2}$$
$$= 0 \text{ or } -5$$
The solutions are −5 and 0.

15. $2x^2 + 3x = 5$
$$2x^2 + 3x - 5 = 0$$
$$a = 2, b = 3, c = -5$$
$$x = \frac{-3 \pm \sqrt{(3)^2 - 4(2)(-5)}}{2(2)}$$
$$= \frac{-3 \pm \sqrt{49}}{4}$$
$$= \frac{-3 \pm 7}{4}$$
$$= 1 \text{ or } -\frac{5}{2}$$

The solutions are $-\dfrac{5}{2}$ and 1.

16. $9x^2 + 4 = 2x$
$$9x^2 - 2x + 4 = 0$$
$$x = \frac{2 \pm \sqrt{(-2)^2 - 4(9)(4)}}{2(9)}$$
$$= \frac{2 \pm \sqrt{-140}}{18}$$
$$= \frac{2 \pm i\sqrt{4 \cdot 35}}{18}$$
$$= \frac{2 \pm 2i\sqrt{35}}{18}$$
$$= \frac{1 \pm i\sqrt{35}}{9}$$

The solutions are $\dfrac{1+i\sqrt{35}}{9}$ and $\dfrac{1-i\sqrt{35}}{9}$.

17.
$$6x^2 + 7 = 5x$$
$$6x^2 - 5x + 7 = 0$$
$$a = 6, b = -5, c = 7$$
$$x = \frac{5 \pm \sqrt{(-5)^2 - 4(6)(7)}}{2(6)}$$
$$= \frac{5 \pm \sqrt{25 - 168}}{12}$$
$$= \frac{5 \pm \sqrt{-143}}{12}$$
$$= \frac{5 \pm i\sqrt{143}}{12}$$

The solutions are $\dfrac{5 + i\sqrt{143}}{12}$ and $\dfrac{5 - i\sqrt{143}}{12}$.

18.
$$(2x - 3)^2 = x$$
$$4x^2 - 12x + 9 - x = 0$$
$$4x^2 - 13x + 9 = 0$$
$$a = 4, b = -13, c = 9$$
$$x = \frac{13 \pm \sqrt{(-13)^2 - 4(4)(9)}}{2(4)}$$
$$= \frac{13 \pm \sqrt{169 - 144}}{8}$$
$$= \frac{13 \pm \sqrt{25}}{8}$$
$$= \frac{13 \pm 5}{8}$$
$$= \frac{9}{4} \text{ or } 1$$

The solutions are 1 and $\dfrac{9}{4}$.

19. $d(t) = -16t^2 + 30t + 6$

a. $d(1) = -16(1)^2 + 30(1) + 6$
$$= -16 + 30 + 6$$
$$= 20 \text{ feet}$$

b.
$$-16t^2 + 30t + 6 = 0$$
$$8t^2 - 15t - 3 = 0$$
$$a = 8, b = -15, c = -3$$
$$t = \frac{15 \pm \sqrt{(-15)^2 - 4(8)(-3)}}{2(8)}$$
$$= \frac{15 \pm \sqrt{225 + 96}}{16}$$
$$= \frac{15 \pm \sqrt{321}}{16}$$

Disregarding the negative, we have
$$t = \frac{15 + \sqrt{321}}{16} \text{ seconds}$$
$$\approx 2.1 \text{ seconds.}$$

20. Let x = length of the legs. Then
$x + 6$ = length of the hypotenuse.
$$x^2 + x^2 = (x + 6)^2$$
$$2x^2 = x^2 + 12x + 36$$
$$x^2 - 12x - 36 = 0$$
$$a = 1, b = -12, c = -36$$
$$x = \frac{12 \pm \sqrt{(-12)^2 - 4(1)(-36)}}{2(1)}$$
$$= \frac{12 \pm \sqrt{144 + 144}}{2}$$
$$= \frac{12 \pm \sqrt{144 \cdot 2}}{2}$$
$$= \frac{12 \pm 12\sqrt{2}}{2}$$
$$= 6 \pm 6\sqrt{2}$$

Disregard the negative. The length of each leg is $\left(6 + 6\sqrt{2}\right)$ cm.

21.
$$x^3 = 27$$
$$x^3 - 27 = 0$$
$$(x-3)(x^2 + 3x + 9) = 0$$
$$x - 3 = 0 \text{ or } x^2 + 3x + 9 = 0$$
$$x = 3 \qquad a = 1, b = 3, c = 9$$
$$x = \frac{-3 \pm \sqrt{(3)^2 - 4(1)(9)}}{2(1)}$$
$$= \frac{-3 \pm \sqrt{9 - 36}}{2}$$
$$= \frac{-3 \pm \sqrt{-27}}{2}$$
$$= \frac{-3 \pm 3i\sqrt{3}}{2}$$

The solutions are 3, $\dfrac{-3 + 3i\sqrt{3}}{2}$, and

$\dfrac{-3 - 3i\sqrt{3}}{2}$.

22.
$$y^3 = -64$$
$$y^3 + 64 = 0$$
$$(y+4)(y^2 - 4y + 16) = 0$$
$$y + 4 = 0 \text{ or } y^2 - 4y + 16 = 0$$
$$y = -4 \qquad a = 1, b = -4, c = 16$$
$$y = \frac{4 \pm \sqrt{(-4)^2 - 4(1)(16)}}{2(1)}$$
$$= \frac{4 \pm \sqrt{16 - 64}}{2}$$
$$= \frac{4 \pm \sqrt{-48}}{2}$$
$$= \frac{4 \pm 4i\sqrt{3}}{2}$$
$$= 2 \pm 2i\sqrt{3}$$

The solutions are -4, $2 + 2i\sqrt{3}$, and
$2 - 2i\sqrt{3}$.

23.
$$\frac{5}{x} + \frac{6}{x-2} = 3$$
$$x(x-2)\left(\frac{5}{x} + \frac{6}{x-2}\right) = 3x(x-2)$$
$$5(x-2) + 6x = 3x^2 - 6x$$
$$5x - 10 + 6x = 3x^2 - 6x$$
$$0 = 3x^2 - 17x + 10$$
$$0 = (3x - 2)(x - 5)$$
$$3x - 2 = 0 \text{ or } x - 5 = 0$$
$$x = \frac{2}{3} \text{ or } \qquad x = 5$$

The solutions are $\dfrac{2}{3}$ and 5.

24.
$$x^4 - 21x^2 - 100 = 0$$
$$(x^2 - 25)(x^2 + 4) = 0$$
$$(x+5)(x-5)(x^2 + 4) = 0$$
$$x + 5 = 0 \text{ or } x - 5 = 0 \text{ or } x^2 + 4 = 0$$
$$x = -5 \text{ or } \qquad x = 5 \text{ or } \qquad x^2 = -4$$
$$x = \pm 2i$$

The solutions are -5, 5 $-2i$, and $2i$.

25. $x^{2/3} - 6x^{1/3} + 5 = 0$

Let $y = x^{1/3}$. Then $y^2 = x^{2/3}$ and
$$y^2 - 6y + 5 = 0$$
$$(y-5)(y-1) = 0$$
$$y - 5 = 0 \text{ or } y - 1 = 0$$
$$y = 5 \text{ or } \qquad y = 1$$
$$x^{1/3} = 5 \text{ or } x^{1/3} = 1$$
$$x = 125 \text{ or } \qquad x = 1$$

The solutions are 1 and 125.

26. $5(x+3)^2 - 19(x+3) = 4$

$5(x+3)^2 - 19(x+3) - 4 = 0$

Let $y = x + 3$. Then $y^2 = (x+3)^2$ and

$5y^2 - 19y - 4 = 0$

$(5y+1)(y-4) = 0$

$5y+1 = 0$ or $y - 4 = 0$

$y = -\dfrac{1}{5}$ or $y = 4$

$x + 3 = -\dfrac{1}{5}$ or $x + 3 = 4$

$x = -\dfrac{16}{5}$ or $x = 1$

The solutions are $-\dfrac{16}{5}$ and 1.

27.

$a^6 - a^2 = a^4 - 1$

$a^6 - a^4 - a^2 + 1 = 0$

$a^4(a^2 - 1) - 1(a^2 - 1) = 0$

$(a^2 - 1)(a^4 - 1) = 0$

$(a+1)(a-1)(a^2 + 1)(a^2 - 1) = 0$

$(a+1)(a-1)(a^2 + 1)(a+1)(a-1) = 0$

$(a+1)^2(a-1)^2(a^2 + 1) = 0$

$(a+1)^2 = 0$ or $(a-1)^2 = 0$ or $a^2 + 1 = 0$

$a + 1 = 0$ or $a - 1 = 0$ or $a^2 = -1$

$a = -1$ or $a = 1$ or $a = \pm i$

The solutions are -1, 1, $-i$, and i.

28. $y^{-2} + y^{-1} = 20$

$\dfrac{1}{y^2} + \dfrac{1}{y} = 20$

$1 + y = 20y^2$

$0 = 20y^2 - y - 1$

$0 = (5y+1)(4y-1)$

$5y+1 = 0$ or $4y - 1 = 0$

$y = -\dfrac{1}{5}$ or $y = \dfrac{1}{4}$

The solutions are $-\dfrac{1}{5}$ and $\dfrac{1}{4}$.

29. Let x = time for Jerome alone. Then
$x - 1$ = time for Tim alone.

$\dfrac{1}{x} + \dfrac{1}{x-1} = \dfrac{1}{5}$

$5(x-1) + 5x = x(x-1)$

$5x - 5 + 5x = x^2 - x$

$0 = x^2 - 11x + 5$

$a = 1, b = -11, c = 5$

$x = \dfrac{11 \pm \sqrt{(-11)^2 - 4(1)(5)}}{2(1)}$

$= \dfrac{11 \pm \sqrt{101}}{2}$

≈ 0.475 (disregard) or 10.525

Jerome: 10.5 hours

Tim: 9.5 hours

30. Let x = the number; then

$\dfrac{1}{x}$ = the reciprocal of the number.

$x - \dfrac{1}{x} = -\dfrac{24}{5}$

$5x\left(x - \dfrac{1}{x}\right) = 5x\left(-\dfrac{24}{5}\right)$

$5x^2 - 5 = -24x$

$5x^2 + 24x - 5 = 0$

$(5x-1)(x+5) = 0$

$5x - 1 = 0$ or $x + 5 = 0$

$x = \dfrac{1}{5}$ or $x = -5$

Disregard the positive value as extraneous. The
number is -5.

31. $2x^2 - 50 \le 0$

$2(x^2 - 25) \le 0$

$2(x+5)(x-5) \le 0$

$x + 5 = 0$ or $x - 5 = 0$

$x = -5$ or $x = 5$

Region	Test Point	$2(x + 5)(x - 5) \le 0$ Result
A: $(-\infty, -5]$	-6	$2(-1)(-11) \le 0$ False
B: $[-5, 5]$	0	$2(5)(-5) \le 0$ True
C: $[5, \infty)$	6	$2(11)(1) \le 0$ False

Solution: $[-5, 5]$

32.

$$\frac{1}{4}x^2 < \frac{1}{16}$$

$$x^2 < \frac{1}{4}$$

$$x^2 - \frac{1}{4} < 0$$

$$\left(x + \frac{1}{2}\right)\left(x - \frac{1}{2}\right) < 0$$

$$x + \frac{1}{2} = 0 \quad \text{or} \quad x - \frac{1}{2} = 0$$

$$x = -\frac{1}{2} \quad \text{or} \quad x = \frac{1}{2}$$

Region	Test Point	$\left(x + \frac{1}{2}\right)\left(x - \frac{1}{2}\right) < 0$ Result
A: $\left(-\infty, -\frac{1}{2}\right)$	-1	$\left(-\frac{1}{2}\right)\left(-\frac{3}{2}\right) < 0$ False
B: $\left(-\frac{1}{2}, \frac{1}{2}\right)$	0	$\left(\frac{1}{2}\right)\left(-\frac{1}{2}\right) < 0$ True
C: $\left(\frac{1}{2}, \infty\right)$	1	$\left(\frac{3}{2}\right)\left(\frac{1}{2}\right) < 0$ False

Solution: $\left(-\frac{1}{2}, \frac{1}{2}\right)$

33.
$$(x^2 - 4)(x^2 - 25) \leq 0$$
$$(x+2)(x-2)(x+5)(x-5) \leq 0$$
$$x+2 = 0 \quad \text{or} \quad x-2 = 0 \quad \text{or} \quad x+5 = 0 \quad \text{or} \quad x-5 = 0$$
$$x = -2 \quad \text{or} \quad x = 2 \quad \text{or} \quad x = -5 \quad \text{or} \quad x = 5$$

Region	Test Point	$(x^2 - 4)(x^2 - 25) \leq 0$ Result
A: $(-\infty, -5)$	-6	$352 \leq 0$ False
B: $(-5, -2)$	-3	$-80 \leq 0$ True
C: $(-2, 2)$	0	$100 \leq 0$ False
D: $(2, 5)$	3	$-80 \leq 0$ True
E: $(5, \infty)$	6	$352 \leq 0$ False

Solution: $[-5, -2] \cup [2, 5]$

34.
$$(x^2 - 16)(x^2 - 1) > 0$$
$$(x+4)(x-4)(x+1)(x-1) > 0$$
$$x+4 = 0 \quad \text{or} \quad x-4 = 0 \quad \text{or} \quad x+1 = 0 \quad \text{or} \quad x-1 = 0$$
$$x = -4 \quad \text{or} \quad x = 4 \quad \text{or} \quad x = -1 \quad \text{or} \quad x = 1$$

Region	Test Point	$(x+4)(x-4)(x+1)(x-1) > 0$ Result
A: $(-\infty, -4)$	-5	$(-1)(-9)(-4)(-6) > 0$ True
B: $(-4, -1)$	-2	$(2)(-6)(-1)(-3) > 0$ False
C: $(-1, 1)$	0	$(4)(-4)(1)(-1) > 0$ True
D: $(1, 4)$	2	$(6)(-2)(3)(1) > 0$ False
E: $(4, \infty)$	5	$(9)(1)(6)(4) > 0$ True

Solution: $(-\infty, -4) \cup (-1, 1) \cup (4, \infty)$

35. $\dfrac{x-5}{x-6} < 0$

$x - 5 = 0$ or $x - 6 = 0$

$x = 5$ or $x = 6$

Region	Test Point	$\dfrac{x-5}{x-6} < 0$ Result
A: $(-\infty, 5)$	0	$\dfrac{-5}{-6} < 0$ False
B: $(5, 6)$	$\dfrac{11}{2}$	$\dfrac{\frac{1}{2}}{-\frac{1}{2}} < 0$ True
C: $(6, \infty)$	7	$\dfrac{2}{1} < 0$ False

Solution: $(5, 6)$

36. $\dfrac{(4x+3)(x-5)}{x(x+6)} > 0$

$4x + 3 = 0,\ x - 5 = 0,\ x = 0,\ \text{or } x + 6 = 0$

$x = -\dfrac{3}{4},\ x = 5,\ x = 0,\ \text{or } x = -6$

Region	Test Point	$\dfrac{(4x+3)(x-5)}{x(x+6)} > 0$ Result
A: $(-\infty, -6)$	-7	$\dfrac{(-25)(-12)}{-7(-1)} > 0$ True
B: $\left(-6, -\dfrac{3}{4}\right)$	-3	$\dfrac{(-9)(-8)}{-3(3)} > 0$ False
C: $\left(-\dfrac{3}{4}, 0\right)$	$-\dfrac{1}{2}$	$\dfrac{(1)\left(-\frac{11}{2}\right)}{-\frac{1}{2}\left(\frac{11}{2}\right)} > 0$ True
D: $(0, 5)$	1	$\dfrac{(7)(-4)}{1(7)} > 0$ False
E: $(5, \infty)$	6	$\dfrac{(27)(1)}{6(12)} > 0$ True

Solution: $(-\infty, -6) \cup \left(-\dfrac{3}{4}, 0\right) \cup (5, \infty)$

37. $(x+5)(x-6)(x+2) \le 0$

$x + 5 = 0$ or $x - 6 = 0$ or $x + 2 = 0$

$x = -5$ or $x = 6$ or $x = -2$

Region	Test Point	$(x+5)(x-6)(x+2) \le 0$ Result
A: $(-\infty, -5]$	-6	$(-1)(-12)(-4) \le 0$ True
B: $[-5, -2]$	-3	$(2)(-9)(-1) \le 0$ False
C: $[-2, 6]$	0	$(5)(-6)(2) \le 0$ True
D: $[6, \infty)$	7	$(12)(1)(9) \le 0$ False

Solution: $(-\infty, -5] \cup [-2, 6]$

38. $x^3 + 3x^2 - 25x - 75 > 0$

$x^2(x+3) - 25(x+3) > 0$

$(x+3)(x^2 - 25) > 0$

$(x+3)(x+5)(x-5) > 0$

$x+3 = 0$ or $x+5 = 0$ or $x-5 = 0$

$x = -3$ or $x = -5$ or $x = 5$

Region	Test Point	$(x+3)(x+5)(x-5) > 0$ Result
A: $(-\infty, -5)$	-6	$(-3)(-1)(-11) > 0$ False
B: $(-5, -3)$	-4	$(-1)(1)(-9) > 0$ True
C: $(-3, 5)$	0	$(3)(5)(-5) > 0$ False
D: $(5, \infty)$	6	$(9)(11)(1) > 0$ True

Solution: $(-5, -3) \cup (5, \infty)$

39. $\dfrac{x^2 + 4}{3x} \le 1$

The denominator equals 0 when $3x = 0$, or $x = 0$.

$$\frac{x^2 + 4}{3x} = 1$$

$$x^2 + 4 = 3x$$

$$x^2 - 3x + 4 = 0$$

$$x = \frac{3 \pm \sqrt{(-3)^2 - 4(1)(4)}}{2(1)} = \frac{3 \pm \sqrt{-7}}{2}$$

which yields non-real solutions.

Region	Test Point	$\dfrac{x^2 + 4}{3x} \le 1$ Result
A: $(-\infty, 0)$	-1	$\dfrac{5}{-3} \le 1$ True
B: $(0, \infty)$	1	$\dfrac{5}{3} \le 1$ False

Solution: $(\infty, 0)$

40. $\dfrac{3}{x-2} > 2$

The denominator is equal to 0 when $x - 2 = 0$, or $x = 2$.

$$\frac{3}{x-2} = 2$$

$$3 = 2(x-2)$$

$$3 = 2x - 4$$

$$7 = 2x$$

$$\frac{7}{2} = x$$

Region	Test Point	$\dfrac{3}{x-2} > 2$ Result
A: $(-\infty, 2)$	0	$\dfrac{3}{-2} > 2$ False
B: $\left(2, \dfrac{7}{2}\right)$	3	$\dfrac{3}{1} > 2$ True
C: $\left(\dfrac{7}{2}, \infty\right)$	5	$\dfrac{3}{3} > 2$ False

Solution: $\left(2, \dfrac{7}{2}\right)$

41. $f(x) = x^2 - 4$

Vertex: $(0, -4)$

Axis of symmetry: $x = 0$

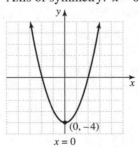

42. $g(x) = x^2 + 7$

Vertex: $(0, 7)$

Axis of symmetry: $x = 0$

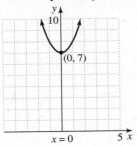

43. $H(x) = 2x^2$

Vertex: $(0, 0)$

Axis of symmetry: $x = 0$

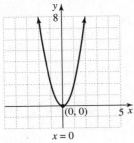

44. $h(x) = -\dfrac{1}{3}x^2$

Vertex: $(0, 0)$

Axis of symmetry: $x = 0$

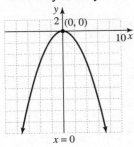

45. $F(x) = (x-1)^2$

Vertex: $(1, 0)$

Axis of symmetry: $x = 1$

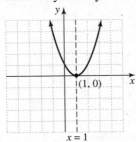

46. $G(x) = (x+5)^2$

Vertex: $(-5, 0)$

Axis of symmetry: $x = -5$

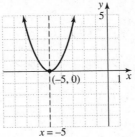

47. $f(x) = (x-4)^2 - 2$

Vertex: $(4, -2)$

Axis of symmetry: $x = 4$

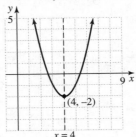

48. $f(x) = -3(x-1)^2 + 1$

Vertex: $(1, 1)$

Axis of symmetry: $x = 1$

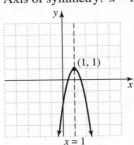

49. $f(x) = x^2 + 10x + 25$

$x = -\dfrac{b}{2a} = \dfrac{-10}{2(1)} = -5$

$f(-5) = (-5)^2 + 10(-5) + 25 = 0$

Vertex: $(-5, 0)$

$x^2 + 10x + 25 = 0$

$(x+5)^2 = 0$

$x + 5 = 0$

$x = -5$

x-intercept: $(-5, 0)$

$f(0) = 25$ so the y-intercept is $(0, 25)$.

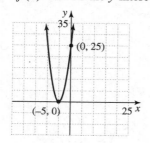

50. $f(x) = -x^2 + 6x - 9$

$x = -\dfrac{b}{2a} = \dfrac{-6}{2(-1)} = 3$

$f(3) = -(3)^2 + 6(3) - 9 = 0$

Vertex: $(3, 0)$

x-intercept: $(3, 0)$

$f(0) = -9$

y-intercept: $(0, -9)$

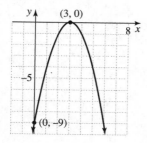

51. $f(x) = 4x^2 - 1$

$x = -\dfrac{b}{2a} = \dfrac{-0}{2(4)} = 0$

$f(0) = 4(0)^2 - 1 = -1$

Vertex: $(0, -1)$

$4x^2 - 1 = 0$

$(2x+1)(2x-1) = 0$

$x = -\dfrac{1}{2}$ or $x = \dfrac{1}{2}$

x-intercepts: $\left(-\dfrac{1}{2}, 0\right), \left(\dfrac{1}{2}, 0\right)$

$f(0) = -1$

y-intercept: $(0, -1)$

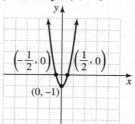

52. $f(x) = -5x^2 + 5$

$x = -\dfrac{b}{2a} = \dfrac{-0}{2(-5)} = 0$

$f(0) = -5(0)^2 + 5 = 5$

Vertex: $(0, 5)$

$-5x^2 + 5 = 0$

$-5x^2 = -5$

$x^2 = 1$

$x = \pm 1$

x-intercepts: $(-1, 0), (1, 0)$

$f(0) = 5$

y-intercept: $(0, 5)$

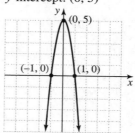

53. $f(x) = -3x^2 - 5x + 4$

$x = -\dfrac{b}{2a} = \dfrac{-(-5)}{2(-3)} = -\dfrac{5}{6}$

$f\left(-\dfrac{5}{6}\right) = -3\left(-\dfrac{5}{6}\right)^2 - 5\left(-\dfrac{5}{6}\right) + 4 = \dfrac{73}{12}$

Vertex: $\left(-\dfrac{5}{6}, \dfrac{73}{12}\right)$

The graph opens downward ($a = -3 < 0$).

$f(0) = 4 \Rightarrow y$-intercept: $(0, 4)$

$-3x^2 - 5x + 4 = 0$

$$x = \frac{5 \pm \sqrt{(-5)^2 - 4(-3)(4)}}{2(-3)}$$

$$= \frac{5 \pm \sqrt{73}}{-6}$$

$$\approx -2.2573 \text{ or } 0.5907$$

x-intercepts: $(-2.3, 0), (0.6, 0)$

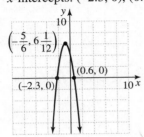

54. $h(t) = -16t^2 + 120t + 300$

a.
$$350 = -16t^2 + 120t + 300$$
$$16t^2 - 120t + 50 = 0$$
$$8t^2 - 60t + 25 = 0$$
$$a = 8, b = -60, c = 25$$
$$t = \frac{60 \pm \sqrt{(-60)^2 - 4(8)(25)}}{2(8)}$$
$$= \frac{60 \pm \sqrt{2800}}{16}$$
$$\approx 0.4 \text{ second and } 7.1 \text{ seconds}$$

b. The object will be at 350 feet on the way up and on the way down.

55. Let $x =$ one number; then
$420 - x =$ the other number.
Let $f(x)$ represent their product.
$$f(x) = x(420 - x)$$
$$= 420x - x^2$$
$$= -x^2 + 420x$$
$$x = -\frac{b}{2a} = \frac{-420}{2(-1)} = 210;$$
$$420 - x = 420 - 210 = 210$$
Therefore, the numbers are both 210.

56. $y = a(x - h)^2 + k$

vertex $(-3, 7)$ with $a = -\frac{7}{9}$ gives

$$y = -\frac{7}{9}(x + 3)^2 + 7 \ .$$

57. $x^2 - x - 30 = 0$
$(x + 5)(x - 6) = 0$
$x + 5 = 0 \quad$ or $\quad x - 6 = 0$
$\quad x = -5 \quad$ or $\quad\quad x = 6$
The solutions are -5 and 6.

58.
$$10x^2 = 3x + 4$$
$$10x^2 - 3x - 4 = 0$$
$$(5x - 4)(2x + 1) = 0$$
$$5x - 4 = 0 \text{ or } 2x + 1 = 0$$
$$5x = 4 \text{ or } \quad 2x = -1$$
$$x = \frac{4}{5} \text{ or } \quad x = -\frac{1}{2}$$

The solutions are $-\frac{1}{2}$ and $\frac{4}{5}$.

59. $9y^2 = 36$
$$y^2 = 4$$
$$y = \pm\sqrt{4}$$
$$y = \pm 2$$
The solutions are -2 and 2.

60. $(9n + 1)^2 = 9$
$$9n + 1 = \pm\sqrt{9}$$
$$9n + 1 = \pm 3$$
$$9n = -1 \pm 3$$
$$n = \frac{-1 \pm 3}{9} = \frac{2}{9}, -\frac{4}{9}$$

The solutions are $-\frac{4}{9}$ and $\frac{2}{9}$.

61.
$$x^2 + x + 7 = 0$$
$$x^2 + x = -7$$
$$x^2 + x + \left(\frac{1}{2}\right)^2 = -7 + \frac{1}{4}$$
$$\left(x + \frac{1}{2}\right)^2 = -\frac{27}{4}$$
$$x + \frac{1}{2} = \pm\sqrt{-\frac{27}{4}}$$
$$x + \frac{1}{2} = \pm\frac{i\sqrt{9 \cdot 3}}{2}$$
$$x + \frac{1}{2} = \pm\frac{3i\sqrt{3}}{2}$$
$$x = -\frac{1}{2} \pm \frac{3i\sqrt{3}}{2} = \frac{-1 \pm 3i\sqrt{3}}{2}$$

The solutions are $\dfrac{-1 + 3i\sqrt{3}}{2}$ and $\dfrac{-1 - 3i\sqrt{3}}{2}$.

62.
$$(3x - 4)^2 = 10x$$
$$9x^2 - 24x + 16 = 10x$$
$$9x^2 - 34x = -16$$
$$x^2 - \frac{34}{9}x = -\frac{16}{9}$$
$$x^2 - \frac{34}{9}x + \left(\frac{-\frac{34}{9}}{2}\right)^2 = -\frac{16}{9} + \frac{289}{81}$$
$$\left(x - \frac{17}{9}\right)^2 = \frac{145}{81}$$
$$x - \frac{17}{9} = \pm\sqrt{\frac{145}{81}}$$
$$x - \frac{17}{9} = \pm\frac{\sqrt{145}}{9}$$
$$x = \frac{17 \pm \sqrt{145}}{9}$$

The solutions are $\dfrac{17 + \sqrt{145}}{9}$ and $\dfrac{17 - \sqrt{145}}{9}$.

63.
$$x^2 + 11 = 0$$
$$a = 1, b = 0, c = 11$$
$$x = \frac{0 \pm \sqrt{(0)^2 - 4(1)(11)}}{2(1)}$$
$$= \frac{\pm\sqrt{-44}}{2}$$
$$= \frac{\pm 2i\sqrt{11}}{2}$$
$$= \pm i\sqrt{11}$$

The solutions are $-i\sqrt{11}$ and $i\sqrt{11}$.

64.
$$x^2 + 7 = 0$$
$$x^2 = -7$$
$$\sqrt{x^2} = \pm\sqrt{-7}$$
$$x = \pm i\sqrt{7}$$

The solutions are $-i\sqrt{7}$ and $i\sqrt{7}$.

65.
$$(5a - 2)^2 - a = 0$$
$$25a^2 - 20a + 4 - a = 0$$
$$25a^2 - 21a + 4 = 0$$
$$a = \frac{21 \pm \sqrt{(-21)^2 - 4(25)(4)}}{2(25)}$$
$$= \frac{21 \pm \sqrt{441 - 400}}{50}$$
$$= \frac{21 \pm \sqrt{41}}{50}$$

The solutions are $\dfrac{21 + \sqrt{41}}{50}$ and $\dfrac{21 - \sqrt{41}}{50}$.

66.
$$\frac{7}{8} = \frac{8}{x^2}$$
$$7x^2 = 64$$
$$x^2 = \frac{64}{7}$$
$$x = \pm\sqrt{\frac{64}{7}}$$
$$x = \pm\frac{8}{\sqrt{7}} = \pm\frac{8 \cdot \sqrt{7}}{\sqrt{7} \cdot \sqrt{7}} = \pm\frac{8\sqrt{7}}{7}$$

The solutions are $-\dfrac{8\sqrt{7}}{7}$ and $\dfrac{8\sqrt{7}}{7}$.

67. $x^{2/3} - 6x^{1/3} = -8$

$x^{2/3} - 6x^{1/3} + 8 = 0$

Let $y = x^{1/3}$. Then $y^2 = x^{2/3}$ and

$y^2 - 6y + 8 = 0$

$(y - 4)(y - 2) = 0$

$y - 4 = 0$ or $y - 2 = 0$

    $y = 4$ or $y = 2$

   $x^{1/3} = 4$ or $x^{1/3} = 2$

    $x = 64$ or $x = 8$

The solutions are 8 and 64.

68. $(2x - 3)(4x + 5) \geq 0$

$2x - 3 = 0$ or $4x + 5 = 0$

    $x = \dfrac{3}{2}$ or $x = -\dfrac{5}{4}$

Region	Test Point	$(2x - 3)(4x + 5) \geq 0$ Result
$A: \left(-\infty, -\dfrac{5}{4}\right]$	-2	$(-7)(-3) \geq 0$ True
$B: \left[-\dfrac{5}{4}, \dfrac{3}{2}\right]$	0	$(-3)(5) \geq 0$ False
$C: \left[\dfrac{3}{2}, \infty\right)$	3	$(3)(17) \geq 0$ True

Solution: $\left(-\infty, -\dfrac{5}{4}\right] \cup \left[\dfrac{3}{2}, \infty\right)$

69. $\dfrac{x(x+5)}{4x - 3} \geq 0$

$x = 0$ or $x + 5 = 0$ or $4x - 3 = 0$

            $x = -5$ or $x = \dfrac{3}{4}$

Region	Test Point	$\dfrac{x(x+5)}{4x-3} \geq 0$ Result
$A: (-\infty, -5]$	-6	$\dfrac{-6(-1)}{-27} \geq 0$ False
$B: [-5, 0]$	-1	$\dfrac{-1(4)}{-7} \geq 0$ True
$C: \left[0, \dfrac{3}{4}\right)$	$\dfrac{1}{2}$	$\dfrac{\frac{1}{2}\left(\frac{11}{2}\right)}{-1} \geq 0$ False
$D: \left(\dfrac{3}{4}, \infty\right)$	1	$\dfrac{1(6)}{1} \geq 0$ True

Solution: $[-5, 0] \cup \left(\dfrac{3}{4}, \infty\right)$

70. $\dfrac{3}{x - 2} > 2$

The denominator is equal to 0 when $x - 2 = 0$, or $x = 2$.

$\dfrac{3}{x - 2} = 2$

    $3 = 2(x - 2)$

    $3 = 2x - 4$

    $7 = 2x$

   $\dfrac{7}{2} = x$

Region	Test Point	$\dfrac{3}{x-2} > 2$ Result
A: $(-\infty, 2)$	0	$\dfrac{3}{-2} > 2$ False
B: $\left(2, \dfrac{7}{2}\right)$	3	$\dfrac{3}{1} > 2$ True
C: $\left(\dfrac{7}{2}, \infty\right)$	5	$\dfrac{3}{3} > 2$ False

Solution: $\left(2, \dfrac{7}{2}\right)$

71. $y = -32x^2 + 1733x + 76{,}362$

    a. $x = 2015 - 2000 = 15$

$$y = -32(15)^2 + 1733(15) + 76{,}362$$
$$= 95{,}175$$

The passenger traffic will be approximately 95,157 thousand.

    b. Let $y = 99{,}000$.

$$99{,}000 = -32x^2 + 1733x + 76{,}362$$
$$0 = -32x^2 + 1733x - 22{,}638$$

$a = -32$, $b = 1733$, $c = -22{,}638$

$$x = \frac{-b \pm \sqrt{b^2 - 4ac}}{2a}$$

$$x = \frac{-1733 \pm \sqrt{1733^2 - 4(-32)(-22{,}638)}}{2(-32)}$$

$$= \frac{-1733 \pm \sqrt{105{,}625}}{-64}$$

$$= \frac{-1733 \pm 325}{-64}$$

$$= 22 \text{ or approximately } 32$$

Choosing the first root, $x = 22$, we see that there will first be 99,000 thousand passengers in 2022.

Chapter 11 Test

1. $5x^2 - 2x = 7$

$$5x^2 - 2x - 7 = 0$$
$$(5x - 7)(x + 1) = 0$$
$$5x - 7 = 0 \quad \text{or} \quad x + 1 = 0$$
$$x = \frac{7}{5} \quad \text{or} \quad x = -1$$

The solutions are -1 and $\dfrac{7}{5}$.

2. $(x+1)^2 = 10$

$$x + 1 = \pm\sqrt{10}$$
$$x = -1 \pm \sqrt{10}$$

The solutions are $-1 + \sqrt{10}$ and $-1 - \sqrt{10}$.

3. $m^2 - m + 8 = 0$

$a = 1$, $b = -1$, $c = 8$

$$m = \frac{1 \pm \sqrt{(-1)^2 - 4(1)(8)}}{2(1)}$$

$$= \frac{1 \pm \sqrt{1 - 32}}{2}$$

$$= \frac{1 \pm \sqrt{-31}}{2}$$

$$= \frac{1 \pm i\sqrt{31}}{2}$$

The solutions are $\dfrac{1 + i\sqrt{31}}{2}$ and $\dfrac{1 - i\sqrt{31}}{2}$.

4. $u^2 - 6u + 2 = 0$

$a = 1$, $b = -6$, $c = 2$

$$u = \frac{-(-6) \pm \sqrt{(-6)^2 - 4(1)(2)}}{2(1)}$$

$$= \frac{6 \pm \sqrt{36 - 8}}{2}$$

$$= \frac{6 \pm \sqrt{28}}{2}$$

$$= \frac{6 \pm 2\sqrt{7}}{2}$$

$$= 3 \pm \sqrt{7}$$

The solutions are $3 + \sqrt{7}$ and $3 - \sqrt{7}$.

5. $7x^2 + 8x + 1 = 0$

$(7x+1)(x+1) = 0$

$7x + 1 = 0$ or $x + 1 = 0$

$\quad 7x = -1 \qquad\qquad\quad x = -1$

$\quad\quad x = -\dfrac{1}{7}$

The solutions are $-\dfrac{1}{7}$ and -1.

6. $y^2 - 3y = 5$

$y^2 - 3y - 5 = 0$

$a = 1, b = -3, c = -5$

$y = \dfrac{3 \pm \sqrt{(-3)^2 - 4(1)(-5)}}{2(1)}$

$ = \dfrac{3 \pm \sqrt{9 + 20}}{2}$

$ = \dfrac{3 \pm \sqrt{29}}{2}$

The solutions are $\dfrac{3 + \sqrt{29}}{2}$ and $\dfrac{3 - \sqrt{29}}{2}$.

7. $\dfrac{4}{x+2} + \dfrac{2x}{x-2} = \dfrac{6}{x^2 - 4}$

$\dfrac{4}{x+2} + \dfrac{2x}{x-2} = \dfrac{6}{(x+2)(x-2)}$

$4(x-2) + 2x(x+2) = 6$

$4x - 8 + 2x^2 + 4x = 6$

$2x^2 + 8x - 14 = 0$

$x^2 + 4x - 7 = 0$

$a = 1, b = 4, c = -7$

$x = \dfrac{-4 \pm \sqrt{(4)^2 - 4(1)(-7)}}{2(1)}$

$ = \dfrac{-4 \pm \sqrt{16 + 28}}{2}$

$ = \dfrac{-4 \pm \sqrt{44}}{2}$

$ = \dfrac{-4 \pm 2\sqrt{11}}{2}$

$ = -2 \pm \sqrt{11}$

The solutions are $-2 + \sqrt{11}$ and $-2 - \sqrt{11}$.

8. $x^5 + 3x^4 = x + 3$

$x^5 + 3x^4 - x - 3 = 0$

$x^4(x+3) - 1(x+3) = 0$

$(x+3)(x^4 - 1) = 0$

$(x+3)(x^2 + 1)(x^2 - 1) = 0$

$x + 3 = 0$ or $x^2 + 1 = 0$ or $x^2 - 1 = 0$

$\quad x = -3$ or $\quad x^2 = -1$ or $\quad x^2 = 1$

$\qquad\qquad\qquad x = \pm i$ or $\qquad x = \pm 1$

The solutions are $-3, -1, 1, -i,$ and i.

9. $x^6 + 1 = x^4 + x^2$

$x^6 - x^4 - x^2 + 1 = 0$

$x^4(x^2 - 1) - (x^2 - 1) = 0$

$(x^4 - 1)(x^2 - 1) = 0$

$(x^2 + 1)(x^2 - 1)(x+1)(x-1) = 0$

$(x^2 + 1)(x+1)^2(x-1)^2 = 0$

$x^2 + 1 = 0$ or $x + 1 = 0$ or $x - 1 = 0$

$\quad x^2 = -1 \qquad\qquad\quad x = -1 \qquad\qquad\quad x = 1$

$\quad x = \pm i$

The solutions are $-i, i, -1,$ and 1.

10. $(x+1)^2 - 15(x+1) + 56 = 0$

Let $y = x + 1$. Then $y^2 = (x+1)^2$ and

$y^2 - 15y + 56 = 0$

$(y-8)(y-7) = 0$

$y = 8$ or $y = 7$

$x + 1 = 8$ or $x + 1 = 7$

$\quad x = 7$ or $x = 6$

The solutions are 6 and 7.

11. $x^2 - 6x = -2$

$x^2 - 6x + \left(\dfrac{-6}{2}\right)^2 = -2 + 9$

$x^2 - 6x + 9 = 7$

$(x-3)^2 = 7$

$x - 3 = \pm\sqrt{7}$

$x = 3 \pm \sqrt{7}$

The solutions are $3 + \sqrt{7}$ and $3 - \sqrt{7}$.

12.
$$2a^2 + 5 = 4a$$
$$2a^2 - 4a = -5$$
$$a^2 - 2a = -\frac{5}{2}$$
$$a^2 - 2a + \left(\frac{-2}{2}\right)^2 = -\frac{5}{2} + 1$$
$$a^2 - 2a + 1 = -\frac{3}{2}$$
$$(a-1)^2 = -\frac{3}{2}$$
$$a - 1 = \pm\sqrt{-\frac{3}{2}} = \pm\frac{i\sqrt{3}}{\sqrt{2}}$$
$$a - 1 = \pm\frac{i\sqrt{6}}{2}$$
$$a = 1 \pm \frac{i\sqrt{6}}{2} \quad \text{or} \quad \frac{2 \pm i\sqrt{6}}{2}$$

The solutions are $\dfrac{2 + i\sqrt{6}}{2}$ and $\dfrac{2 - i\sqrt{6}}{2}$.

13.
$$2x^2 - 7x > 15$$
$$2x^2 - 7x - 15 > 0$$
$$(2x+3)(x-5) > 0$$
$$2x + 3 = 0 \quad \text{or} \quad x - 5 = 0$$
$$x = -\frac{3}{2} \quad \text{or} \quad x = 5$$

Region	Test Point	$(2x + 3)(x - 5) > 0$ Result
A: $\left(-\infty, -\dfrac{3}{2}\right)$	-2	$(-1)(-7) > 0$ True
B: $\left(-\dfrac{3}{2}, 5\right)$	0	$(3)(-5) > 0$ False
C: $(5, \infty)$	6	$(15)(1) > 0$ True

Solution: $\left(-\infty, -\dfrac{3}{2}\right) \cup (5, \infty)$

14.
$$(x^2 - 16)(x^2 - 25) \geq 0$$
$$(x+4)(x-4)(x+5)(x-5) \geq 0$$
$$x+4 = 0 \quad \text{or} \quad x-4 = 0 \quad \text{or} \quad x+5 = 0 \quad \text{or} \quad x-5 = 0$$
$$x = -4 \quad \text{or} \quad x = 4 \quad \text{or} \quad x = -5 \quad \text{or} \quad x = 5$$

Region	Test Point	$(x+4)(x-4)(x+5)(x-5) \geq 0$ Result
A: $(-\infty, -5]$	-6	$(-2)(-10)(-1)(-11) \geq 0$ True
B: $[-5, -4]$	$-\dfrac{9}{2}$	$\left(-\dfrac{1}{2}\right)\left(-\dfrac{17}{2}\right)\left(\dfrac{1}{2}\right)\left(-\dfrac{19}{2}\right) \geq 0$ False
C: $[-4, 4]$	0	$(4)(-4)(5)(-5) \geq 0$ True
D: $[4, 5]$	$\dfrac{9}{2}$	$\left(\dfrac{17}{2}\right)\left(\dfrac{1}{2}\right)\left(\dfrac{19}{2}\right)\left(-\dfrac{1}{2}\right) \geq 0$ False
E: $[5, \infty)$	6	$(10)(2)(11)(1) \geq 0$ True

Solution: $(-\infty, -5] \cup [-4, 4] \cup [5, \infty)$

15. $\dfrac{5}{x+3} < 1$

The denominator is equal to 0 when $x + 3 = 0$, or $x = -3$.

$$\dfrac{5}{x+3} = 1$$
$$5 = x + 3 \quad \text{so} \quad x = 2$$

Region	Test Point	$\dfrac{5}{x+3} < 1$ Result
A: $(-\infty, -3)$	-4	$\dfrac{5}{-1} < 1$ True
B: $(-3, 2)$	0	$\dfrac{5}{3} < 1$ False
C: $(2, \infty)$	3	$\dfrac{5}{6} < 1$ True

Solution: $(-\infty, -3) \cup (2, \infty)$

Chapter 11: Quadratic Equations and Functions

SSM: Beginning and Intermediate Algebra

16.
$$\frac{7x-14}{x^2-9}\le 0$$
$$\frac{7(x-2)}{(x+3)(x-3)}\le 0$$
$x-2=0$ or $x+3=0$ or $x-3=0$
$x=2$ or $x=-3$ or $x=3$

Region	Test Point	$\dfrac{7(x-2)}{(x+3)(x-3)}\le 0$ Result
A: $(-\infty, -3)$	-4	$\dfrac{7(-6)}{(-1)(-7)}\le 0$ True
B: $(-3, 2]$	0	$\dfrac{7(-2)}{(3)(-3)}\le 0$ False
C: $[2, 3)$	$\dfrac{5}{2}$	$\dfrac{7\left(\frac12\right)}{\left(\frac{11}{2}\right)\left(-\frac12\right)}\le 0$ True
D: $(3, \infty)$	4	$\dfrac{7(2)}{(7)(1)}\le 0$ False

Solution: $(-\infty, -3)\cup[2, 3)$

17. $f(x)=3x^2$
Vertex: $(0, 0)$
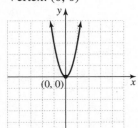

18. $G(x)=-2(x-1)^2+5$
Vertex: $(1, 5)$
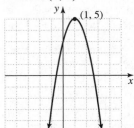

19. $h(x)=x^2-4x+4$
$x=-\dfrac{b}{2a}=\dfrac{-(-4)}{2(1)}=2$
$h(2)=(2)^2-4(2)+4=0$
Vertex: $(2, 0)$
$h(0)=4\Rightarrow$ y-intercept: $(0, 4)$
x-intercept: $(2, 0)$
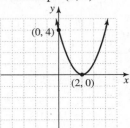

20. $F(x)=2x^2-8x+9$
$x=-\dfrac{b}{2a}=\dfrac{-(-8)}{2(2)}=2$
$F(2)=2(2)^2-8(2)+9=1$
Vertex: $(2, 1)$
$F(0)=9\Rightarrow$ y-intercept: $(0, 9)$
$2x^2-8x+9=0$
$a=2, b=-8, c=9$
$$x=\frac{8\pm\sqrt{(-8)^2-4(2)(9)}}{2(2)}$$
$$=\frac{8\pm\sqrt{-8}}{4}$$
which yields non-real solutons.
Therefore, there are no x-intercepts.

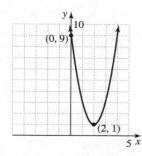

21. Let t = time for Sandy alone. Then
$t - 2$ = time for Dave alone.

$$\frac{1}{t} + \frac{1}{t-2} = \frac{1}{4}$$

$$4(t-2) + 4t = t(t-2)$$

$$4t - 8 + 4t = t^2 - 2t$$

$$0 = t^2 - 10t + 8$$

$$a = 1, b = -10, c = 8$$

$$t = \frac{10 \pm \sqrt{(-10)^2 - 4(1)(8)}}{2(1)}$$

$$= \frac{10 \pm \sqrt{68}}{2}$$

$$= \frac{10 \pm 2\sqrt{17}}{2}$$

$$= 5 \pm \sqrt{17}$$

$$\approx 9.12 \text{ or } 0.88 \text{ (discard)}$$

It takes her about 9.12 hours.

22. $s(t) = -16t^2 + 32t + 256$

   a. $t = -\dfrac{b}{2a} = \dfrac{-32}{2(-16)} = 1$

       $s(1) = -16(1)^2 + 32(1) + 256 = 272$

       Vertex: (1, 272)

       The maximum height is 272 feet.

   b. $-16t^2 + 32t + 256 = 0$

$$t^2 - 2t - 16 = 0$$

$$a = 1, b = -2, c = -16$$

$$t = \frac{2 \pm \sqrt{(-2)^2 - 4(1)(-16)}}{2(1)}$$

$$= \frac{2 \pm \sqrt{68}}{2}$$

$$= \frac{2 \pm 2\sqrt{17}}{2}$$

$$= 1 \pm \sqrt{17}$$

$$\approx -3.12 \text{ and } 5.12$$

Disregard the negative. The stone will hit
the water in about 5.12 seconds.

23.
$$a^2 + b^2 = c^2$$

$$x^2 + (x+8)^2 = (20)^2$$

$$x^2 + (x^2 + 16x + 64) = 400$$

$$2x^2 + 16x - 336 = 0$$

$$x^2 + 8x - 168 = 0$$

$$a = 1, b = 8, c = -168$$

$$x = \frac{-8 \pm \sqrt{(8)^2 - 4(1)(-168)}}{2(1)}$$

$$= \frac{-8 \pm \sqrt{736}}{2}$$

$$\approx -17.565 \text{ or } 9.565$$

Disregard the negative.

$$x \approx 9.6$$

$$x + 8 \approx 9.6 + 8 = 17.6$$

$$17.6 + 9.6 = 27.2$$

$$27.2 - 20 = 7.2$$

They would save about 7.2 feet.

Chapter 11 Cumulative Review

1. Let $x = 2$ and $y = -5$.

   a. $\dfrac{x-y}{12+x} = \dfrac{2-(-5)}{12+2} = \dfrac{2+5}{14} = \dfrac{7}{14} = \dfrac{1}{2}$

   b. $x^2 - 3y = 2^2 - 3(-5) = 4 + 15 = 19$

2. $|3x - 2| = -5$ which is impossible. Thus, there is
no solution, or $\varnothing$.

3. **a.** $2x + 3x + 5 + 2 = (2+3)x + (5+2) = 5x + 7$

    b. $-5a - 3 + a + 2 = -5a + a - 3 + 2 = -4a - 1$

c. $4y - 3y^2$ cannot be simplified.

d. $2.3x + 5x - 6 = 7.3x - 6$

e. $-\dfrac{1}{2}b + b = \left(-\dfrac{1}{2} + 1\right)b = \dfrac{1}{2}b$

4. $\begin{cases} -6x + y = 5 \ \ (1) \\ 4x - 2y = 6 \ \ (2) \end{cases}$

Multiply E1 by 2 and add to E2.
$-12x + 2y = 10$
$\underline{4x - 2y = 6}$
$-8x = 16$
$ x = -2$

Replace x with -2 in E1.
$-6(-2) + y = 5$
$12 + y = 5$
$ y = -7$

The solution is $(-2, -7)$.

5. $\begin{cases} 2x + y = 7 \\ 2y = -4x \end{cases}$

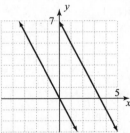

The system has no solution. The solution set is $\{ \ \}$ or $\varnothing$.

6. a. $(a^{-2}bc^3)^{-3} = (a^{-2})^{-3}b^{-3}(c^3)^{-3}$
$\phantom{(a^{-2}bc^3)^{-3}} = a^6 b^{-3} c^{-9}$
$\phantom{(a^{-2}bc^3)^{-3}} = \dfrac{a^6}{b^3 c^9}$

b. $\left(\dfrac{a^{-4}b^2}{c^3}\right)^{-2} = \dfrac{(a^{-4})^{-2}(b^2)^{-2}}{(c^3)^{-2}}$
$\phantom{\left(\dfrac{a^{-4}b^2}{c^3}\right)^{-2}} = \dfrac{a^8 b^{-4}}{c^{-6}}$
$\phantom{\left(\dfrac{a^{-4}b^2}{c^3}\right)^{-2}} = \dfrac{a^8 c^6}{b^4}$

c. $\left(\dfrac{3a^8 b^2}{12a^5 b^5}\right)^{-2} = \left(\dfrac{a^3}{4b^3}\right)^{-2}$
$\phantom{\left(\dfrac{3a^8 b^2}{12a^5 b^5}\right)^{-2}} = \dfrac{(a^3)^{-2}}{4^{-2}(b^3)^{-2}}$
$\phantom{\left(\dfrac{3a^8 b^2}{12a^5 b^5}\right)^{-2}} = \dfrac{4^2 a^{-6}}{b^{-6}}$
$\phantom{\left(\dfrac{3a^8 b^2}{12a^5 b^5}\right)^{-2}} = \dfrac{16b^6}{a^6}$

7. $\begin{cases} 7x - 3y = -14 \\ -3x + y = 6 \end{cases}$

Solve the second equation for y.
$y = 3x + 6$
Substitute $3x + 6$ for y in the first equation.
$7x - 3(3x + 6) = -14$
$7x - 9x - 18 = -14$
$-2x - 18 = -14$
$-2x = 4$
$x = -2$
Let $x = -2$ in $y = 3x + 6$.
$y = 3x + 6 = 3(-2) + 6 = -6 + 6 = 0$
The solution is $(-2, 0)$.

8. a. $(4a - 3)(7a - 2) = 28a^2 - 8a - 21a + 6$
$ = 28a^2 - 29a + 6$

b. $(2a + b)(3a - 5b)$
$= 6a^2 - 10ab + 3ab - 5b^2$
$= 6a^2 - 7ab - 5b^2$

9. a. $\dfrac{x^5}{x^2} = x^{5-2} = x^3$

b. $\dfrac{4^7}{4^3} = 4^{7-3} = 4^4 = 256$

c. $\dfrac{(-3)^5}{(-3)^2} = (-3)^{5-2} = (-3)^3 = -27$

d. $\dfrac{s^2}{t^3}$ cannot be simplified.

e. $\dfrac{2x^5 y^2}{xy} = 2x^{5-1}y^{2-1} = 2x^4 y^1 = 2x^4 y$

10. a. $9x^3 + 27x^2 - 15x = 3x(3x^2 + 9x - 5)$

b. $2x(3y-2) - 5(3y-2)$
$= (3y-2)(2x-5)$

c. $2xy + 6x - y - 3 = 2x(y+3) - 1(y+3)$
$= (y+3)(2x-1)$

11. $P(x) = 2x^3 - 4x^2 + 5$

a. $P(2) = 2(2)^3 - 4(2)^2 + 5$
$= 2(8) - 4(4) + 5$
$= 16 - 16 + 5$
$= 5$

b.
$$\begin{array}{r|rrrr} 2 & 2 & -4 & 0 & 5 \\ & & 4 & 0 & 0 \\ \hline & 2 & 0 & 0 & 5 \end{array}$$
Thus, $P(2) = 5$.

12. $x^2 - 2x - 48 = (x+6)(x-8)$

13. $(5x-1)(2x^2 + 15x + 18) = 0$
$(5x-1)(2x+3)(x+6) = 0$
$5x-1 = 0$ or $2x+3 = 0$ or $x+6 = 0$
$5x = 1$ $2x = -3$ $x = -6$
$x = \dfrac{1}{5}$ $x = -\dfrac{3}{2}$

The solutions are -6, $-\dfrac{3}{2}$, and $\dfrac{1}{5}$.

14. $2ax^2 - 12axy + 18ay^2 = 2a(x^2 - 6xy + 9y^2)$
$= 2a(x-3y)(x-3y)$
$= 2a(x-3y)^2$

15. $\dfrac{2x^2}{10x^3 - 2x^2} = \dfrac{2x^2}{2x^2(5x-1)} = \dfrac{1}{5x-1}$

16. $2(a^2+2) - 8 = -2a(a-2) - 5$
$2a^2 + 4 - 8 = -2a^2 + 4a - 5$
$4a^2 - 4a + 1 = 0$
$(2a-1)^2 = 0$
$2a - 1 = 0$
$2a = 1$
$a = \dfrac{1}{2}$
The solution is $\dfrac{1}{2}$.

17. $\dfrac{x^{-1} + 2xy^{-1}}{x^{-2} - x^{-2}y^{-1}} = \dfrac{\dfrac{1}{x} + \dfrac{2x}{y}}{\dfrac{1}{x^2} - \dfrac{1}{x^2 y}}$
$= \dfrac{\left(\dfrac{1}{x} + \dfrac{2x}{y}\right)x^2 y}{\left(\dfrac{1}{x^2} - \dfrac{1}{x^2 y}\right)x^2 y}$
$= \dfrac{xy + 2x^3}{y-1}$

18. $f(x) = x^2 + x - 12$
$x = -\dfrac{b}{2a} = -\dfrac{1}{2(1)} = -\dfrac{1}{2}$
$f\left(-\dfrac{1}{2}\right) = \left(-\dfrac{1}{2}\right)^2 + \left(-\dfrac{1}{2}\right) - 12$
$= \dfrac{1}{4} - \dfrac{1}{2} - 12$
$= -\dfrac{49}{4}$
Vertex: $\left(-\dfrac{1}{2}, -\dfrac{49}{4}\right)$
$x^2 + x - 12 = 0$
$(x+4)(x-3) = 0$
$x+4 = 0$ or $x-3 = 0$
$x = -4$ $x = 3$
x-intercepts: $(-4, 0)$, $(3, 0)$
$f(0) = 0^2 + 0 - 12 = -12$
y-intercept: $(0, -12)$

19. $4m^2 - 4m + 1 = (2m)^2 - 2 \cdot 2m \cdot 1 + 1^2 = (2m-1)^2$

20. $\dfrac{x^2 - 4x + 4}{2-x} = \dfrac{(x-2)^2}{-(x-2)} = \dfrac{x-2}{-1} = 2-x$

21. Let $x =$ the number.

$$x^2 + 3x = 70$$

$$x^2 + 3x - 70 = 0$$

$$(x+10)(x-7) = 0$$

$$x + 10 = 0 \quad \text{or} \quad x - 7 = 0$$

$$x = -10 \qquad \qquad x = 7$$

The number is -10 or 7.

22. $\dfrac{a+1}{a^2 - 6a + 8} - \dfrac{3}{16 - a^2}$

$$= \frac{a+1}{(a-4)(a-2)} - \frac{3}{(4+a)(4-a)}$$

$$= \frac{a+1}{(a-4)(a-2)} + \frac{3}{(4+a)(a-4)}$$

$$= \frac{(a+1)(a+4) + 3(a-2)}{(a-4)(a-2)(a+4)}$$

$$= \frac{(a^2 + 4a + a + 4) + 3a - 6}{(a-4)(a-2)(a+4)}$$

$$= \frac{a^2 + 8a - 2}{(a-4)(a-2)(a+4)}$$

23. a. $\sqrt{25x^3} = \sqrt{25x^2 \cdot x} = 5x\sqrt{x}$

b. $\sqrt[3]{54x^6 y^8} = \sqrt[3]{27x^6 y^6 \cdot 2y^2}$

$$= 3x^2 y^2 \sqrt[3]{2y^2}$$

c. $\sqrt[4]{81z^{11}} = \sqrt[4]{81z^8 \cdot z^3} = 3z^2 \sqrt[4]{z^3}$

24. $\dfrac{(2a)^{-1} + b^{-1}}{a^{-1} + (2b)^{-1}} = \dfrac{\dfrac{1}{2a} + \dfrac{1}{b}}{\dfrac{1}{a} + \dfrac{1}{2b}}$

$$= \frac{\left(\dfrac{1}{2a} + \dfrac{1}{b}\right)2ab}{\left(\dfrac{1}{a} + \dfrac{1}{2b}\right)2ab}$$

$$= \frac{b + 2a}{2b + a}$$

$$= \frac{2a + b}{a + 2b}$$

25. a. $\dfrac{2}{\sqrt{5}} = \dfrac{2 \cdot \sqrt{5}}{\sqrt{5} \cdot \sqrt{5}} = \dfrac{2\sqrt{5}}{5}$

b. $\dfrac{2\sqrt{16}}{\sqrt{9x}} = \dfrac{2 \cdot 4}{3\sqrt{x}} = \dfrac{8 \cdot \sqrt{x}}{3\sqrt{x} \cdot \sqrt{x}} = \dfrac{8\sqrt{x}}{3x}$

c. $\sqrt[3]{\dfrac{1}{2}} = \dfrac{\sqrt[3]{1}}{\sqrt[3]{2}} = \dfrac{1}{\sqrt[3]{2}} = \dfrac{1 \cdot \sqrt[3]{2^2}}{\sqrt[3]{2} \cdot \sqrt[3]{2^2}} = \dfrac{\sqrt[3]{4}}{2}$

26.

$$
\begin{array}{r}
x^2 - 6x + 8 \\
x+3 \overline{)\, x^3 - 3x^2 - 10x + 24} \\
\underline{x^3 + 3x^2} \\
-6x^2 - 10x \\
\underline{-6x^2 - 18x} \\
8x + 24 \\
\underline{8x + 24} \\
0
\end{array}
$$

Answer: $x^2 - 6x + 8$

27. $\sqrt{2x+5} + \sqrt{2x} = 3$

$$\sqrt{2x+5} = 3 - \sqrt{2x}$$

$$\left(\sqrt{2x+5}\right)^2 = \left(3 - \sqrt{2x}\right)^2$$

$$2x + 5 = 9 - 6\sqrt{2x} + 2x$$

$$-4 = -6\sqrt{2x}$$

$$(-4)^2 = \left(-6\sqrt{2x}\right)^2$$

$$16 = 36(2x)$$

$$16 = 72x$$

$$x = \frac{16}{72} = \frac{2}{9}$$

The solution is $\dfrac{2}{9}$.

28. $P(x) = 4x^3 - 2x^2 + 3$

a. $P(-2) = 4(-2)^3 - 2(-2)^2 + 3$

$$= 4(-8) - 2(4) + 3$$

$$= -32 - 8 + 3$$

$$= -37$$

b.

$$
\begin{array}{r|rrrr}
-2 & 4 & -2 & 0 & 3 \\
   &   & -8 & 20 & -40 \\
\hline
   & 4 & -10 & 20 & -37
\end{array}
$$

Thus, $P(-2) = -37$.

29. $\dfrac{x}{2} + \dfrac{8}{3} = \dfrac{1}{6}$

$6\left(\dfrac{x}{2} + \dfrac{8}{3}\right) = 6\left(\dfrac{1}{6}\right)$

$3x + 16 = 1$

$3x = -15$

$x = -5$

30. $\dfrac{x+3}{x^2 + 5x + 6} = \dfrac{3}{2x+4} - \dfrac{1}{x+3}$

$\dfrac{x+3}{(x+3)(x+2)} = \dfrac{3}{2(x+2)} - \dfrac{1}{x+3}$

$2(x+3) = 3(x+3) - 2(x+2)$

$2x + 6 = 3x + 9 - 2x - 4$

$2x + 6 = x + 5$

$x = -1$

31. Let $x =$ the number.

$\dfrac{x}{6} - \dfrac{5}{3} = \dfrac{x}{2}$

$6\left(\dfrac{x}{6} - \dfrac{5}{3}\right) = 6\left(\dfrac{x}{2}\right)$

$x - 10 = 3x$

$-10 = 2x$

$-5 = x$

The number is -5.

32. Let $t =$ time to roof the house together.

$\dfrac{1}{24} + \dfrac{1}{40} = \dfrac{1}{t}$

$120t\left(\dfrac{1}{24} + \dfrac{1}{40}\right) = 120t\left(\dfrac{1}{t}\right)$

$5t + 3t = 120$

$8t = 120$

$t = \dfrac{120}{8} = 15$

It would take them 15 hours to roof the house working together.

33. $y = kx$

$5 = k(30)$

$k = \dfrac{5}{30} = \dfrac{1}{6}$ and $y = \dfrac{1}{6}x$

34. $y = \dfrac{k}{x}$

$8 = \dfrac{k}{24}$

$k = 8(24) = 192$ and $y = \dfrac{192}{x}$

35. a. $\sqrt{(-3)^2} = |-3| = 3$

    b. $\sqrt{x^2} = |x|$

    c. $\sqrt[4]{(x-2)^4} = |x-2|$

    d. $\sqrt[3]{(-5)^3} = -5$

    e. $\sqrt[5]{(2x-7)^5} = 2x-7$

    f. $\sqrt{25x^2} = \sqrt{25} \cdot \sqrt{x^2} = 5|x|$

    g. $\sqrt{x^2 + 2x + 1} = \sqrt{(x+1)^2} = |x+1|$

36. a. $\sqrt{(-2)^2} = |-2| = 2$

    b. $\sqrt{y^2} = |y|$

    c. $\sqrt[4]{(a-3)^4} = |a-3|$

    d. $\sqrt[3]{(-6)^3} = -6$

    e. $\sqrt[5]{(3x-1)^5} = 3x-1$

37. a. $\sqrt[8]{x^4} = x^{4/8} = x^{1/2} = \sqrt{x}$

    b. $\sqrt[6]{25} = (25)^{1/6}$
$= (5^2)^{1/6} = 5^{2/6} = 5^{1/3} = \sqrt[3]{5}$

    c. $\sqrt[4]{r^2 s^6} = (r^2 s^6)^{1/4}$
$= r^{2/4} s^{6/4}$
$= r^{1/2} s^{3/2}$
$= (rs^3)^{1/2} = \sqrt{rs^3}$

38. a. $\sqrt[4]{5^2} = 5^{2/4} = 5^{1/2} = \sqrt{5}$

    b. $\sqrt[12]{x^3} = x^{3/12} = x^{1/4} = \sqrt[4]{x}$

c. $\sqrt[6]{x^2 y^4} = (x^2 y^4)^{1/6}$

$\qquad = x^{2/6} y^{4/6}$

$\qquad = x^{1/3} y^{2/3}$

$\qquad = (xy^2)^{1/3} = \sqrt[3]{xy^2}$

39. a. $\dfrac{2+i}{1-i} = \dfrac{(2+i)\cdot(1+i)}{(1-i)\cdot(1+i)}$

$\qquad = \dfrac{2+2i+1i+i^2}{1^2 - i^2}$

$\qquad = \dfrac{2+3i-1}{1+1}$

$\qquad = \dfrac{1+3i}{2}$ or $\dfrac{1}{2} + \dfrac{3}{2}i$

b. $\dfrac{7}{3i} = \dfrac{7\cdot(-3i)}{3i\cdot(-3i)} = \dfrac{-21i}{-9i^2} = \dfrac{-21i}{9} = -\dfrac{7}{3}i = 0 - \dfrac{7}{3}i$

40. a. $3i(5-2i) = 15i - 6i^2$

$\qquad = 15i + 6$

$\qquad = 6 + 15i$

b. $(6-5i)^2 = 6^2 - 2(6)(5i) + (5i)^2$

$\qquad = 36 - 60i + 25i^2$

$\qquad = 36 - 60i - 25$

$\qquad = 11 - 60i$

c. $\left(\sqrt{3}+2i\right)\left(\sqrt{3}-2i\right) = \left(\sqrt{3}\right)^2 - (2i)^2$

$\qquad = 3 - 4i^2$

$\qquad = 3 + 4$

$\qquad = 7$

41. $(x+1)^2 = 12$

$\qquad x+1 = \pm\sqrt{12}$

$\qquad x+1 = \pm 2\sqrt{3}$

$\qquad x = -1 \pm 2\sqrt{3}$

The solutions are $-1+2\sqrt{3}$ and $-1-2\sqrt{3}$.

42. $(y-1)^2 = 24$

$\qquad y-1 = \pm\sqrt{24}$

$\qquad y-1 = \pm 2\sqrt{6}$

$\qquad y = 1 \pm 2\sqrt{6}$

The solutions are $1+2\sqrt{6}$ and $1-2\sqrt{6}$.

43. $x - \sqrt{x} - 6 = 0$

Let $y = \sqrt{x}$. Then $y^2 = x$ and $\quad \begin{aligned} y^2 - y - 6 &= 0 \\ (y-3)(y+2) &= 0 \end{aligned}$

$\qquad y - 3 = 0$ or $y + 2 = 0$

$\qquad\quad y = 3$ or $y = -2$

$\qquad \sqrt{x} = 3$ or $\sqrt{x} = -2$ (can't happen)

$\qquad\quad x = 9$

The solution is 9.

44. $m^2 = 4m + 8$

$\qquad m^2 - 4m - 8 = 0$

$\qquad a = 1, b = -4, c = -8$

$\qquad x = \dfrac{4 \pm \sqrt{(-4)^2 - 4(1)(-8)}}{2(1)}$

$\qquad = \dfrac{4 \pm \sqrt{16 + 32}}{2}$

$\qquad = \dfrac{4 \pm \sqrt{48}}{2}$

$\qquad = \dfrac{4 \pm 4\sqrt{3}}{2}$

$\qquad = 2 \pm 2\sqrt{3}$

The solutions are $2+2\sqrt{3}$ and $2-2\sqrt{3}$.

Chapter 12

Section 12.1 Practice

1. $f(x) = x + 2$; $g(x) = 3x + 5$

   a. $(f + g)(x) = f(x) + g(x)$
$$= (x + 2) + (3x + 5)$$
$$= 4x + 7$$

   b. $(f - g)(x) = f(x) - g(x)$
$$= (x + 2) - (3x + 5)$$
$$= x + 2 - 3x - 5$$
$$= -2x - 3$$

   c. $(f \cdot g)(x) = f(x) \cdot g(x)$
$$= (x + 2)(3x + 5)$$
$$= 3x^2 + 6x + 5x + 10$$
$$= 3x^2 + 11x + 10$$

   d. $\left(\dfrac{f}{g}\right)(x) = \dfrac{f(x)}{g(x)} = \dfrac{x+2}{3x+5}$, where $x \neq -\dfrac{5}{3}$.

2. $f(x) = x^2 + 1$; $g(x) = 3x - 5$

   a. $(f \circ g)(4) = f(g(4)) = f(7) = 50$
$$(g \circ f)(4) = g(f(4)) = g(17) = 46$$

   b. $(f \circ g)(x) = f(g(x))$
$$= f(3x - 5)$$
$$= (3x - 5)^2 + 1$$
$$= 9x^2 - 30x + 26$$
$$(g \circ f)(x) = g(f(x))$$
$$= g(x^2 + 1)$$
$$= 3(x^2 + 1) - 5$$
$$= 3x^2 - 2$$

3. $f(x) = x^2 + 5$; $g(x) = x + 3$

   a. $(f \circ g)(x) = f(g(x))$
$$= f(x + 3)$$
$$= (x + 3)^2 + 5$$
$$= x^2 + 6x + 14$$

   b. $(g \circ f)(x) = g(f(x))$
$$= g(x^2 + 5)$$
$$= (x^2 + 5) + 3$$
$$= x^2 + 8$$

4. $f(x) = 3x$; $g(x) = x - 4$; $h(x) = |x|$

   a. $F(x) = |x - 4|$
$$F(x) = (h \circ g)(x)$$
$$= h(g(x))$$
$$= h(x - 4)$$
$$= |x - 4|$$

   b. $G(x) = 3x - 4$
$$G(x) = (g \circ f)(x)$$
$$= g(f(x))$$
$$= g(3x)$$
$$= 3x - 4$$

Vocabulary, Readiness & Video Check 12.1

1. $(f \circ g)(x) = f(g(x))$; C

2. $(f \cdot g)(x) = f(x) \cdot g(x)$; E

3. $(f - g)(x) = f(x) - g(x)$; F

4. $(g \circ f)(x) = g(f(x))$; A

5. $\left(\dfrac{f}{g}\right)(x) = \dfrac{f(x)}{g(x)}$, $g(x) \neq 0$; D

6. $(f + g)(x) = f(x) + g(x)$; B

7. You can find $(f + g)(x)$ and then find $(f + g)(2)$ or you can find $f(2)$ and $g(2)$ and then add those results.

8. Yes, sometimes they can be equal.

Exercise Set 12.1

1. a. $(f + g)(x) = (x - 7) + (2x + 1) = 3x - 6$

   b. $(f - g)(x) = (x - 7) - (2x + 1)$
$$= x - 7 - 2x - 1$$
$$= -x - 8$$

   c. $(f \cdot g)(x) = (x - 7)(2x + 1) = 2x^2 - 13x - 7$

d. $\left(\dfrac{f}{g}\right)(x) = \dfrac{x-7}{2x+1}$, where $x \neq -\dfrac{1}{2}$.

3. a. $(f+g)(x) = (x^2+1)+5x = x^2+5x+1$

   b. $(f-g)(x) = (x^2+1)-5x = x^2-5x+1$

   c. $(f \cdot g)(x) = (x^2+1)(5x) = 5x^3+5x$

   d. $\left(\dfrac{f}{g}\right)(x) = \dfrac{x^2+1}{5x}$, where $x \neq 0$

5. a. $(f+g)(x) = \sqrt[3]{x}+x+5$

   b. $(f-g)(x) = \sqrt[3]{x}-(x+5) = \sqrt{x}-x-5$

   c. $(f \cdot g)(x) = \sqrt[3]{x}(x+5)$
      $= x\sqrt[3]{x}+5\sqrt[3]{x}$

   d. $\left(\dfrac{f}{g}\right)(x) = \dfrac{\sqrt[3]{x}}{x+5}$; where $x \neq -5$.

7. a. $(f+g)(x) = -3x+5x^2$ or $5x^2-3x$

   b. $(f-g)(x) = -3x-5x^2$ or $-5x^2-3x$

   c. $(f \cdot g)(x) = (-3x)(5x^2) = -15x^3$

   d. $\left(\dfrac{f}{g}\right)(x) = \dfrac{-3x}{5x^2}$
      $= -\dfrac{3}{5x}$, where $x \neq 0$.

9. $(f \circ g)(2) = f(g(2))$
   $= f(-4)$
   $= (-4)^2 - 6(-4) + 2$
   $= 16 + 24 + 2$
   $= 42$

11. $(g \circ f)(-1) = g(f(-1))$
    $= g(9)$
    $= -2(9)$
    $= -18$

13. $(g \circ h)(0) = g(h(0))$
    $= g(0)$
    $= -2(0)$
    $= 0$

15. $(f \circ g)(x) = f(g(x))$
    $= f(5x)$
    $= (5x)^2 + 1$
    $= 25x^2 + 1$
    $(g \circ f)(x) = g(f(x))$
    $= g(x^2+1)$
    $= 5(x^2+1)$
    $= 5x^2 + 5$

17. $(f \circ g)(x) = f(g(x))$
    $= f(x+7)$
    $= 2(x+7) - 3$
    $= 2x + 14 - 3$
    $= 2x + 11$
    $(g \circ f)(x) = g(f(x))$
    $= g(2x-3)$
    $= (2x-3) + 7$
    $= 2x + 4$

19. $(f \circ g)(x) = f(g(x))$
    $= f(-2x)$
    $= (-2x)^3 + (-2x) - 2$
    $= -8x^3 - 2x - 2$
    $(g \circ f)(x) = g(f(x))$
    $= g(x^3+x-2)$
    $= -2(x^3+x-2)$
    $= -2x^3 - 2x + 4$

21. $(f \circ g)(x) = f(g(x))$
    $= f(10x-3)$
    $= |10x-3|$
    $(g \circ f)(x) = g(f(x)) = g(|x|) = 10|x| - 3$

23. $(f \circ g)(x) = f(g(x)) = f(-5x+2) = \sqrt{-5x+2}$
    $(g \circ f)(x) = g(f(x)) = g(\sqrt{x}) = -5\sqrt{x} + 2$

25. $H(x) = (g \circ h)(x)$
    $= g(h(x))$
    $= g(x^2+2)$
    $= \sqrt{x^2+2}$

27. $F(x) = (h \circ f)(x)$
         $= h(f(x))$
         $= h(3x)$
         $= (3x)^2 + 2$
         $= 9x^2 + 2$

29. $G(x) = (f \circ g)(x)$
         $= f(g(x))$
         $= f\left(\sqrt{x}\right)$
         $= 3\sqrt{x}$

31. answers may vary; for example, $g(x) = x + 2$ and $f(x) = x^2$

33. answers may vary; for example, $g(x) = x + 5$ and $f(x) = \sqrt{x} + 2$

35. answers may vary; for example, $g(x) = 2x - 3$ and $f(x) = \dfrac{1}{x}$

37. $x = y + 2$
     $y = x - 2$

39. $x = 3y$
     $y = \dfrac{x}{3}$

41. $x = -2y - 7$
     $2y = -x - 7$
     $y = -\dfrac{x + 7}{2}$

43. $(f + g)(2) = f(2) + g(2) = 7 + (-1) = 6$

45. $(f \circ g)(2) = f(g(2)) = f(-1) = 4$

47. $(f \cdot g)(7) = f(7) \cdot g(7) = 1 \cdot 4 = 4$

49. $\left(\dfrac{f}{g}\right)(-1) = \dfrac{f(-1)}{g(-1)} = \dfrac{4}{-4} = -1$

51. answers may vary.

53. Profit is equal to the revenue minus the cost; $P(x) = R(x) - C(x)$

Section 12.2 Practice

1. a. $f = \{(4, -3), (3, -4), (2, 7), (5, 0)\}$
       f is one-to-one since each y-value corresponds to only one x-value.

     b. $g = \{(8, 4), (-2, 0), (6, 4), (2, 6)\}$
       g is not one-to-one because the y-value 4 in (8, 4) and (6, 4) corresponds to two different x-values.

c. $h = \{(2, 4), (1, 3), (4, 6), (-2, 4)\}$

h is not one-to-one because the y-value 4 in (2, 4) and (−2, 4) corresponds to two different x-values.

d.

Year	1950	1963	1968	1975	1997	2008
Federal Minimum Wage	$0.75	$1.25	$1.60	$2.10	$5.15	$6.55

This function is one-to-one because each wage corresponds to only one year.

e. The function represented by the graph is not one-to-one because the y-value 2 in (2, 2) and (3, 2) corresponds to two different x-values.

f. The function represented by the diagram is not one-to-one because the score 509 corresponds to two different states.

2. Graphs **a**, **b**, and **c** all pass the vertical line test, so only these graphs are functions. But, of these, only **b** and **c** pass the horizontal line test, so only **b** and **c** are graphs of one-to-one functions.

3. $f = \{(3, 4), (-2, 0), (2, 8), (6, 6)\}$

Switching the coordinates of each ordered pair gives $f^{-1} = \{(4, 3), (0, -2), (8, 2), (6, 6)\}$

4. $f(x) = 6 - x$

Replace $f(x)$ with y.

$y = 6 - x$

Interchange x and y.

$x = 6 - y$

Solve for y.

$x = 6 - y$

$y = 6 - x$

Replace y with $f^{-1}(x)$.

$f^{-1}(x) = 6 - x$

5. $f(x) = 5x + 2$

Replace $f(x)$ with y.

$y = 5x + 2$

Interchange x and y.

$x = 5y + 2$

Solve for y.

$x = 5y + 2$

$x - 2 = 5y$

$\dfrac{x - 2}{5} = y$

Replace y with $f^{-1}(x)$.

$f^{-1}(x) = \dfrac{x - 2}{5}$

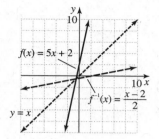

$f(x) = 5x + 2$

$f^{-1}(x) = \dfrac{x-2}{2}$

$y = x$

6. a.

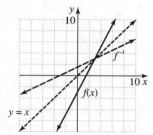

f^{-1}

$y = x$

$f(x)$

b.

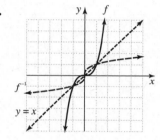

f

f^{-1}

$y = x$

7. $f(x) = 4x - 1; \ f^{-1}(x) = \dfrac{x+1}{4}$

$$(f \circ f^{-1})(x) = f(f^{-1}(x))$$

$$= f\left(\dfrac{x+1}{4}\right)$$

$$= 4\left(\dfrac{x+1}{4}\right) - 1$$

$$= x + 1 - 1$$

$$= x$$

$$(f^{-1} \circ f)(x) = f^{-1}(f(x))$$

$$= f^{-1}(4x - 1)$$

$$= \dfrac{(4x-1)+1}{4}$$

$$= \dfrac{4x-1+1}{4}$$

$$= \dfrac{4x}{4}$$

$$= x$$

Since $f \circ f^{-1} = x$ and $f^{-1} \circ f = x$, if

$f(x) = 4x - 1, \ f^{-1}(x) = \dfrac{x+1}{4}$.

Vocabulary, Readiness & Video Check 12.2

1. If $f(2) = 11$, the corresponding ordered pair is (2, 11).

2. If (7, 3) is an ordered pair solution of $f(x)$, and $f(x)$ has an inverse, then an ordered pair solution of $f^{-1}(x)$ is (3, 7).

3. The symbol f^{-1} means the inverse of f.

4. True or false: The function notation $f^{-1}(x)$ means $\dfrac{1}{f(x)}$. false

5. To tell whether a graph is the graph of a function, use the vertical line test.

6. To tell whether the graph of a function is also a one-to-one function, use the horizontal line test.

7. The graphs of f and f^{-1} are symmetric about the line $y = x$.

8. Two functions are inverse of each other if $(f \circ f^{-1})(x) = x$ and $(f^{-1} \circ f)(x) = x$.

9. Every function must have each x-value correspond to only one y-value. A one-to-one function must also have each y-value correspond to only one x-value.

10. No, a graph must pass the vertical line test to even be a function—a graph must pass *both* the vertical and horizontal line tests to be a one-to-one function.

11. Yes; by the definition of an inverse function.

12. The definition of inverse function tells us that f^{-1} consists of the ordered pairs (y, x) when (x, y) belongs to f. So it makes sense that switching x and y in the equations would result in switching the x and y values in the ordered pairs.

13. Once you know some points of the original equation or graph, you can switch the x's and y's of these points to find points that satisfy the inverse and then graph it. You can also check that the two graphs (the original and the inverse) are symmetric about the line $y = x$.

14. You must show that $f(f^{-1}(x))$ and $f^{-1}(f(x))$ both equal x in order to prove they are inverses of each other.

Exercise Set 12.2

1. $f = \{(-1,-1),(1,1),(0,2),(2,0)\}$ is a one-to-one function.
$f^{-1} = \{(-1,-1),(1,1),(2,0),(0,2)\}$

3. $h = \{(10,10)\}$ is a one-to-one function.
$h^{-1} = \{(10,10)\}$

5. $f = \{(11,12),(4,3),(3,4),(6,6)\}$ is a one-to-one function.
$f^{-1} = \{(12,11),(3,4),(4,3),(6,6)\}$

7. This function is not one-to-one because the months November and December have the same output, 10.0.

9. This function is one-to-one.

Rank in Population (input)	1	47	16	25	36	7
State (output)	California	Alaska	Indiana	Louisiana	New Mexico	Ohio

11. $f(x) = x^3 + 2$

    a. $f(1) = 1^3 + 2 = 3$

    b. $f^{-1}(3) = 1$

13. $f(x) = x^3 + 2$

    a. $f(-1) = (-1)^3 + 2 = 1$

    b. $f^{-1}(1) = -1$

15. The graph represents a one-to-one function because it passes the horizontal line test.

17. The graph does not represent a one-to-one function because it does not pass the horizontal line test.

19. The graph represents a one-to-one function because it passes the horizontal line test.

21. The graph does not represent a one-to-one function because it does not pass the horizontal line test.

23. $f(x) = x + 4$
$y = x + 4$
$x = y + 4$
$y = x - 4$
$f^{-1}(x) = x - 4$

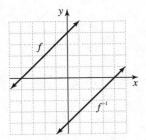

25. $f(x) = 2x - 3$

$y = 2x - 3$

$x = 2y - 3$

$2y = x + 3$

$y = \dfrac{x+3}{2}$

$f^{-1}(x) = \dfrac{x+3}{2}$

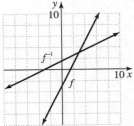

27. $f(x) = \dfrac{1}{2}x - 1$

$y = \dfrac{1}{2}x - 1$

$x = \dfrac{1}{2}y - 1$

$\dfrac{1}{2}y = x + 1$

$y = 2x + 2$

$f^{-1}(x) = 2x + 2$

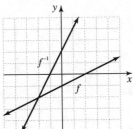

29. $f(x) = x^3$

$y = x^3$

$x = y^3$

$y = \sqrt[3]{x}$

$f^{-1}(x) = \sqrt[3]{x}$

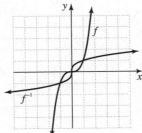

31. $f(x) = 5x + 2$

$y = 5x + 2$

$x = 5y + 2$

$5y = x - 2$

$y = \dfrac{x-2}{5}$

$f^{-1}(x) = \dfrac{x-2}{5}$

33. $f(x) = \dfrac{x-2}{5}$

$y = \dfrac{x-2}{5}$

$x = \dfrac{y-2}{5}$

$5x = y - 2$

$y = 5x + 2$

$f^{-1}(x) = 5x + 2$

35. $f(x) = \sqrt[3]{x}$

$y = \sqrt[3]{x}$

$x = \sqrt[3]{y}$

$x^3 = y$

$f^{-1}(x) = x^3$

37. $f(x) = \dfrac{5}{3x+1}$

$y = \dfrac{5}{3x+1}$

$x = \dfrac{5}{3y+1}$

$3y + 1 = \dfrac{5}{x}$

$3y = \dfrac{5}{x} - 1$

$3y = \dfrac{5-x}{x}$

$y = \dfrac{5-x}{3x}$

$f^{-1}(x) = \dfrac{5-x}{3x}$

39. $f(x) = (x+2)^3$

$y = (x+2)^3$

$x = (y+2)^3$

$\sqrt[3]{x} = y + 2$

$\sqrt[3]{x} - 2 = y$

$f^{-1}(x) = \sqrt[3]{x} - 2$

41.

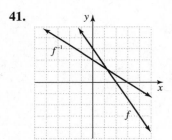

43.

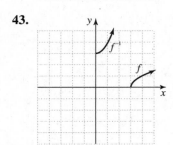

45.

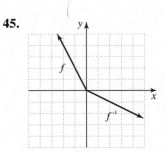

47. $(f \circ f^{-1})(x) = f(f^{-1}(x))$

$\qquad = f\left(\dfrac{x-1}{2}\right)$

$\qquad = 2\left(\dfrac{x-1}{2}\right) + 1$

$\qquad = x - 1 + 1$

$\qquad = x$

$(f^{-1} \circ f)(x) = f^{-1}(f(x))$

$\qquad = f^{-1}(2x+1)$

$\qquad = \dfrac{(2x+1)-1}{2}$

$\qquad = \dfrac{2x}{2}$

$\qquad = x$

49. $(f \circ f^{-1})(x) = f(f^{-1}(x))$

$\qquad = f\left(\sqrt[3]{x-6}\right)$

$\qquad = \left(\sqrt[3]{x-6}\right)^3 + 6$

$\qquad = x - 6 + 6$

$\qquad = x$

$(f^{-1} \circ f)(x) = f^{-1}(f(x))$

$\qquad = f^{-1}(x^3 + 6)$

$\qquad = \sqrt[3]{(x^3 + 6) - 6}$

$\qquad = \sqrt[3]{x^3}$

$\qquad = x$

51. $25^{1/2} = \sqrt{25} = 5$

53. $16^{3/4} = \left(\sqrt[4]{16}\right)^3 = 2^3 = 8$

55. $9^{-3/2} = \dfrac{1}{9^{3/2}} = \dfrac{1}{\left(\sqrt{9}\right)^3} = \dfrac{1}{3^3} = \dfrac{1}{27}$

57. $f(x) = 3^x$

$f(2) = 3^2 = 9$

　　Copyright © 2013 Pearson Education, Inc.

59. $f(x) = 3^x$

$f\left(\dfrac{1}{2}\right) = 3^{1/2} \approx 1.73$

61. $f(2) = 9$

    a. $(2, 9)$

    b. $(9, 2)$

63. a. $\left(-2, \dfrac{1}{4}\right), \left(-1, \dfrac{1}{2}\right), (0,1), (1,2), (2,5)$

    b. $\left(\dfrac{1}{4}, -2\right), \left(\dfrac{1}{2}, -1\right), (1,0), (2,1), (5,2)$

    c, d.

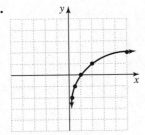

65. answers may vary.

67.
$$f(x) = 3x + 1$$
$$y = 3x + 1$$
$$x = 3y + 1$$
$$x - 1 = 3y$$
$$y = \dfrac{x-1}{3}$$
$$f^{-1}(x) = \dfrac{x-1}{3}$$

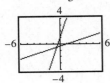

69.

$$f(x) = \sqrt[3]{x+1}$$
$$y = \sqrt[3]{x+1}$$
$$x = \sqrt[3]{y+1}$$
$$x^3 = y+1$$
$$y = x^3 - 1$$
$$f^{-1}(x) = x^3 - 1$$

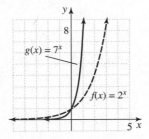

Section 12.3 Practice

1.

$f(x) = 2^x$	x	0	1	2	3	−1	−2
	$f(x)$	1	2	4	8	$\frac{1}{2}$	$\frac{1}{4}$

$g(x) = 7^x$	x	0	1	2	3	−1	−2
	$g(x)$	1	7	49	343	$\frac{1}{7}$	$\frac{1}{49}$

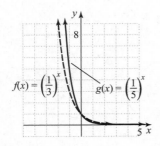

2.

$f(x) = \left(\dfrac{1}{3}\right)^x$	x	0	1	.2	3	−1	−2
	$f(x)$	1	$\frac{1}{3}$	$\frac{1}{9}$	$\frac{1}{27}$	3	9

$g(x) = \left(\dfrac{1}{5}\right)^x$	x	0	1	2	3	−1	−2
	$g(x)$	1	$\frac{1}{5}$	$\frac{1}{25}$	$\frac{1}{125}$	5	25

3. $f(x) = 2^{x-3}$

$f(x) = 2^{x-3}$	x	6	5	4	3	2	1	0
	$f(x)$	8	4	2	1	$\frac{1}{2}$	$\frac{1}{4}$	$\frac{1}{8}$

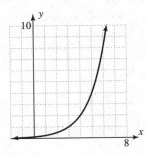

4. a. $3^x = 9$

Write 9 as a power of 3, $9 = 3^2$.

$3^x = 3^2$, thus, $x = 2$.

b. $8^x = 16$

Write 8 and 16 as powers of 2.

$8 = 2^3$ and $16 = 2^4$.

$$8^x = 16$$
$$(2^3)^x = 2^4$$
$$2^{3x} = 2^4$$
$$3x = 4$$
$$x = \frac{4}{3}$$

c. $125^x = 25^{x-2}$

Write 125 and 25 as powers of 5.

$125 = 5^3$ and $25 = 5^2$.

$$125^x = 25^{x-2}$$
$$(5^3)^x = (5^2)^{x-2}$$
$$5^{3x} = 5^{2x-4}$$
$$3x = 2x - 4$$
$$x = -4$$

5. $P = \$3000$, $r = 7\% = 0.07$, $n = 2$, and $t = 4$.

$$A = P\left(1 + \frac{r}{n}\right)^{nt}$$

$$A = 3000\left(1 + \frac{0.07}{2}\right)^{2(4)}$$

$$= 3000(1.035)^8$$

$$\approx 3950.43$$

Thus, the amount owed is approximately $3950.43.

6. a. $p(n) = 100(2.7)^{-0.05n}$, $n = 2$

$$p(2) = 100(2.7)^{-0.05(2)}$$

$$= 100(2.7)^{-0.1}$$

$$\approx 90.54$$

Thus, approximately 90.54% of the light passes through.

b. $p(n) = 100(2.7)^{-0.05n}$, $n = 10$

$$p(10) = 100(2.7)^{-0.05(10)}$$

$$= 100(2.7)^{-0.5}$$

$$\approx 60.86$$

Thus, approximately 60.86% of the light passes through.

Graphing Calculator Explorations

1.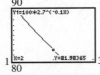

The expected percent after 2 days is 81.98%.

2.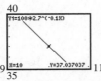

The expected percent after 10 days is 37.04%.

3.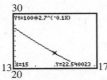

The expected percent after 15 days is 22.54%.

4.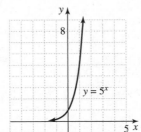

The expected percent after 25 days is 8.35%.

Vocabulary, Readiness & Video Check 12.3

1. A function such as $f(x) = 2^x$ is an <u>exponential</u> function; **C**.

2. If $7^x = 7^y$, then <u>$x = y$</u>; **B**.

3. Yes, the graph passes the vertical line test.

4. Yes, the function passes both the vertical and horizontal line tests.

5. The function has no x-intercept.

6. The function has a y-intercept of <u>(0, 1)</u>.

7. The domain of this function, in interval notation, is <u>$(-\infty, \infty)$</u>.

8. The range of this function, in interval notation, is <u>$(0, \infty)$</u>.

9. In a polynomial function, the base is the variable and the exponent is the constant; in an exponential function, the base is the constant and the exponent is the variable.

10. Rewrite the equation so the bases are the same.

11. $y = 30(0.996)^{101} \approx 20.0$ lb

Exercise Set 12.3

1. $y = 5^x$

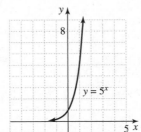

3. $y = 2^x + 1$

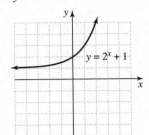

5. $y = \left(\dfrac{1}{4}\right)^x$

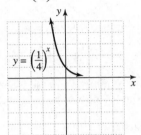

7. $y = \left(\dfrac{1}{2}\right)^x - 2$

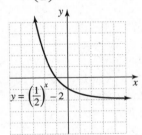

9. $y = -2^x$

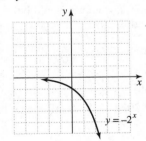

11. $y = -\left(\dfrac{1}{4}\right)^x$

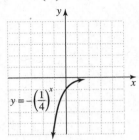

13. $f(x) = 2^{x+1}$

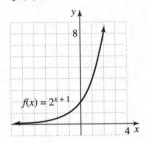

15. $f(x) = 4^{x-2}$

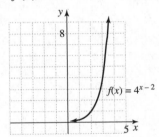

17. C

19. B

21. $3^x = 27$
$3^x = 3^3$
$x = 3$
The solution is 3.

23. $16^x = 8$
$(2^4)^x = 2^3$
$2^{4x} = 2^3$
$4x = 3$
$x = \dfrac{3}{4}$

The solution is $\dfrac{3}{4}$.

25.

$$32^{2x-3} = 2$$
$$(2^5)^{2x-3} = 2^1$$
$$2^{10x-15} = 2^1$$
$$10x - 15 = 1$$
$$10x = 16$$
$$x = \frac{8}{5}$$

The solution is $\frac{8}{5}$.

27.

$$\frac{1}{4} = 2^{3x}$$
$$2^{-2} = 2^{3x}$$
$$3x = -2$$
$$x = -\frac{2}{3}$$

The solution is $-\frac{2}{3}$.

29.

$$5^x = 625$$
$$5^x = 5^4$$
$$x = 4$$

The solution is 4.

31.

$$4^x = 8$$
$$(2^2)^x = 2^3$$
$$2^{2x} = 2^3$$
$$2x = 3$$
$$x = \frac{3}{2}$$

The solution is $\frac{3}{2}$.

33.

$$27^{x+1} = 9$$
$$(3^3)^{x+1} = 3^2$$
$$3^{3x+3} = 3^2$$
$$3x + 3 = 2$$
$$3x = -1$$
$$x = -\frac{1}{3}$$

The solution is $-\frac{1}{3}$.

35.

$$81^{x-1} = 27^{2x}$$
$$(3^4)^{x-1} = (3^3)^{2x}$$
$$3^{4x-4} = 3^{6x}$$
$$4x - 4 = 6x$$
$$-4 = 2x$$
$$x = -2$$

The solution is -2.

37.

$$y = 30(0.996)^x$$
$$y = 30(0.996)^{50} \approx 24.6$$

There will be about 24.6 pounds left after 50 days.

39. a. 2007 is 4 years after 2003, so $x = 4$.

$$y = 8.6(1.03)^x$$
$$y = 8.6(1.03)^4 \approx 9.7$$

The total cheese production in the United States was about 9.7 billion pounds in 2007.

b. 2015 is 12 years after 2003, so $x = 12$.

$$y = 8.6(1.03)^{12} \approx 12.3$$

The total cheese production is predicted to be 12.3 billion pounds in 2015.

41. $y = 140,242(1.083)^x$

a. $x = 2004 - 2000 = 4$

$$y = 140,242(1.083)^4 \approx 192,927$$

192,927 American students studied abroad in 2004.

b. $x = 2015 - 2000 = 15$

$$y = 140,242(1.083)^{15} \approx 463,772$$

463,722 American students would be studying abroad in 2015.

43. 2012 is 10 years after 2002, so $x = 10$.

$$y = 136.76(1.107)^x$$
$$y = 136.76(1.107)^{10} \approx 378.0$$

The prediction for 2012 is 378.0 million cellular phone users.

45. $y = 200,000(1.08)^x$, $x = 13$

$$y = 200,000(1.08)^{13} \approx 544,000$$

There will be approximately 544,000 mosquitoes on May 25.

47. $A = P\left(1 + \dfrac{r}{n}\right)^{nt}$

$t = 3$, $P = 6000$, $r = 0.08$, and $n = 12$

$A = 6000\left(1 + \dfrac{0.08}{12}\right)^{12(3)}$

$\quad = 6000\left(1 + \dfrac{0.08}{12}\right)^{36}$

$\quad \approx 7621.42$

Erica would owe \$7621.42 after 3 years.

49. $A = P\left(1 + \dfrac{r}{n}\right)^{nt}$

$P = 2000$

$r = 0.06$, $n = 2$, and $t = 12$

$A = 2000\left(1 + \dfrac{0.06}{2}\right)^{2(12)}$

$\quad = 2000(1.03)^{24}$

$\quad \approx 4065.59$

Janina has approximately \$4065.59 in her savings account.

51. $5x - 2 = 18$

$\quad 5x = 20$

$\quad\;\; x = 4$

The solution is 4.

53. $3x - 4 = 3(x + 1)$

$\quad 3x - 4 = 3x + 3$

$\quad\;\; -4 = 3$

This is a false statement. The solution set is $\varnothing$.

55. $\quad\quad x^2 + 6 = 5x$

$\quad x^2 - 5x + 6 = 0$

$\quad (x - 2)(x - 3) = 0$

$x = 2 \;$ or $\; x = 3$

The solutions are 2 and 3.

57. $2^x = 8$

$\quad 2^3 = 8$

$\quad\;\; x = 3$

59. $5^x = \dfrac{1}{5}$

$\quad 5^{-1} = \dfrac{1}{5}$

$\quad\;\; x = -1$

61. Since there are no variables in the exponent in $f(x) = 1.5x^2$, it is not an exponential function.

63. Since there are no variables in the exponent in $h(x) = \left(\dfrac{1}{2}x\right)^2$, it is not an exponential function.

65. $f(x) = 2^{-x}$

$f(1) = 2^{-1} = \dfrac{1}{2}$

This is graph C.

67. $f(x) = 4^{-x}$

$f(1) = 4^{-1} = \dfrac{1}{4}$

This is graph D.

69. answers may vary

71. $y = \left|3^x\right|$

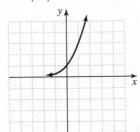

73. $y = 3^{|x|}$

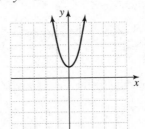

75.

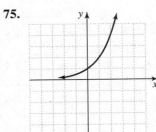

The graphs are the same, since $\left(\dfrac{1}{2}\right)^{-x} = 2^x$.

77. The result is the same, 24.55 pounds.

79. $y = 30(0.996)^x$

$y = 30(0.996)^{100} \approx 20.09$

After 100 days, the estimate is 20.09 pounds left.

Section 12.4 Practice

1. $C = 25{,}000, r = 0.12, x = 2015 - 2000 = 15$

$y = C(1+r)^x$

$y = 25{,}000(1 + 0.12)^{15}$

$\quad = 25{,}000(1.12)^{15}$

$\quad \approx 136{,}839$

In 2015, the predicted population is 136,839.

2. $C = 800, r = 0.30, x = 9$

$y = C(1-r)^x$

$y = 800(1 - 0.30)^9 = 800(0.70)^9 \approx 32$

After 9 rounds, there are 32 players remaining.

3. $C = 500, r = 0.50, x = \dfrac{51}{15} = 3.4$

$y = C(1-r)^x$

$y = 500(1 - 0.50)^{3.4} = 500(0.50)^{3.4} \approx 47.4$

In 51 years, 47.4 grams of DDT remain.

Vocabulary, Readiness & Video Check 12.4

1. For Example 1, the growth rate is given as 5% per year. Since this is "per year," the number of time intervals is the "number of years," or 8.

2. The number of employees is decreasing and not increasing. It is exponential decay because the decrease is the same percent per year.

3. time intervals = years/half-life; the decay rate is 50% or $\dfrac{1}{2}$ because half-life is the amount of time it takes half of a substance to decay.

Exercise Set 12.4

	Original Amount	Growth Rate Per Year	Number of Years, x	Final Amount after x Years of Growth
1.	305	5%	8	$y = 305(1 + 0.05)^8 \approx 451$
3.	2000	11%	41	$y = 2000(1 + 0.11)^{41} \approx 144{,}302$
5.	17	29%	28	$y = 17(1 + 0.29)^{28} \approx 21{,}231$

	Original Amount	Decay Rate per Year	Number of Years, x	Final Amount after x Years of Decay
7.	305	5%	8	$y = 305(1 - 0.05)^8 \approx 202$
9.	10,000	12%	15	$y = 10,000(1 - 0.12)^{15} \approx 1470$
11.	207,000	32%	25	$y = 207,000(1 - 0.32)^{25} \approx 13$

13. $C = 500,000$, $r = 0.03$, $x = 12$

$y = C(1 + r)^x = 500,000(1 + 0.03)^{12} \approx 712,880$

In 12 years, the population is predicted to be 712,880.

15. $C = 640$, $r = 0.05$, $x = 10$

$y = C(1 - r)^x = 640(1 - 0.05)^{10} \approx 383$

In 10 years, the number of employees is predicted to be 383.

17. $C = 260$, $r = 0.025$, $x = 10$

$y = C(1 + r)^x = 260(1 + 0.025)^{10} \approx 333$

In 10 years, the predicted number of bison is 333.

19. $C = 5$, $r = 0.15$, $x = 10$

$y = C(1 - r)^x = 5(1 - 0.15)^{10} \approx 1$

After 10 seconds, there will be 1 gram of the isotope.

	Original Amount	Half-Life (in years)	Number of Years	Time intervals, x	Final Amount after x Time Intervals	Is amount reasonable?
21. a.	40	7	14	$\frac{14}{7} = 2$	$40(1 - 0.5)^2 = 10$	yes
b.	40	7	11	$\frac{11}{7} \approx 1.6$	$40(1 - 0.5)^{1.6} \approx 13.2$	yes
23.	21	152	500	$\frac{500}{152} \approx 3.3$	$21(1 - 0.5)^{3.3} \approx 2.1$	yes

25. $C = 30$, $r = 0.5$, $x = \dfrac{250}{96}$

$y = 30(1 - 0.5)^{250/96}$

$y \approx 4.9$

In 250 years, the predicted amount is 4.9 grams.

27. $2^x = 8$

$2^x = 2^3$

$x = 3$

29. $5^x = \dfrac{1}{5}$

$5^x = 5^{-1}$

$x = -1$

31. no; answers may vary.

Section 12.5 Practice

1. a. $\log_3 81 = 4$ means $3^4 = 81$.

  b. $\log_5 \dfrac{1}{5} = -1$ means $5^{-1} = \dfrac{1}{5}$.

  c. $\log_7 \sqrt{7} = \dfrac{1}{2}$ means $7^{1/2} = \sqrt{7}$.

  d. $\log_{13} y = 4$ means $13^4 = y$.

2. a. $4^3 = 64$ means $\log_4 64 = 3$.

  b. $6^{1/3} = \sqrt[3]{6}$ means $\log_6 \sqrt[3]{6} = \dfrac{1}{3}$.

  c. $5^{-3} = \dfrac{1}{125}$ means $\log_5 \dfrac{1}{125} = -3$.

  d. $\pi^7 = z$ means $\log_\pi z = 7$.

3. a. $\log_3 9 = 2$ because $3^2 = 9$.

  b. $\log_2 \dfrac{1}{8} = -3$ because $2^{-3} = \dfrac{1}{8}$.

  c. $\log_{49} 7 = \dfrac{1}{2}$ because $49^{1/2} = 7$.

4. a. $\log_5 \dfrac{1}{25} = x$

  $\log_5 \dfrac{1}{25} = x$ means $5^x = \dfrac{1}{25}$. Solve

  $5^x = \dfrac{1}{25}$.

  $5^x = \dfrac{1}{25}$

  $5^x = 5^{-2}$

  Since the bases are the same, by the uniqueness of b^x, we have that $x = -2$. The solution is -2 or the solution set is $\{-2\}$.

  b. $\log_x 8 = 3$

  $x^3 = 8$

  $x^3 = 2^3$

  $x = 2$

  c. $\log_6 x = 2$

  $6^2 = x$

  $36 = x$

  d. $\log_{13} 1 = x$

  $13^x = 1$

  $13^x = 13^0$

  $x = 0$

  e. $\log_h 1 = x$

  $h^x = 1$

  $h^x = h^0$

  $x = 0$

5. a. From Property 2, $\log_5 5^4 = 4$.

  b. From Property 2, $\log_9 9^{-2} = -2$.

  c. From Property 3, $6^{\log_6 5} = 5$.

  d. From Property 3, $7^{\log_7 4} = 4$.

6. $y = \log_9 x$ means that $9^y = x$. Find some ordered pair solutions that satisfy $9^y = x$.

$x = 9^y$	y
1	0
9	1
$\dfrac{1}{9}$	-1
$\dfrac{1}{81}$	-2

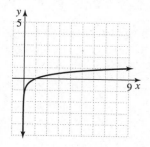

7. $y = \log_{1/4} x$ means that $\left(\dfrac{1}{4}\right)^y = x$. Find some

ordered-pair solutions that satisfy $\left(\dfrac{1}{4}\right)^y = x$.

$x = \left(\dfrac{1}{4}\right)^y$	y
1	0
$\dfrac{1}{4}$	1
4	−1
16	−2

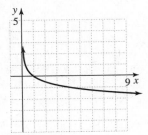

Vocabulary, Readiness & Video Check 12.5

1. A function such as $y = \log_2 x$ is a <u>logarithmic</u> function; **B**.

2. If $y = \log_2 x$, then <u>$2^y = x$</u>; **C**.

3. Yes, the function passes both the horizontal- and vertical-line tests.

4. The function has an *x*-intercept of <u>(1, 0)</u>.

5. The function has no *y*-intercept.

6. The domain of this function, in interval notation, is <u>(0, ∞)</u>.

7. The range of this function, in interval notation, is <u>(−∞, ∞)</u>.

8. Logarithms are exponents.

9. First write the equation as an equivalent exponential equation. Then solve.

10. The exponential is solved for *x*, and *y* is the exponent in the equation. Since the exponential equation is solved for *x*, it is easier to choose a *y*-value and simplify the expression containing *y*, which is then the *x*-value.

Exercise Set 12.5

1. $\log_6 36 = 2$
   $6^2 = 36$

3. $\log_3 \dfrac{1}{27} = -3$
   $3^{-3} = \dfrac{1}{27}$

5. $\log_{10} 1000 = 3$
   $10^3 = 1000$

7. $\log_9 x = 4$
   $9^4 = x$

9. $\log_\pi \dfrac{1}{\pi^2} = -2$
   $\pi^{-2} = \dfrac{1}{\pi^2}$

11. $\log_7 \sqrt{7} = \dfrac{1}{2}$
    $7^{1/2} = \sqrt{7}$

13. $\log_{0.7} 0.343 = 3$
    $0.7^3 = 0.343$

15. $\log_3 \dfrac{1}{81} = -4$
    $3^{-4} = \dfrac{1}{81}$

17. $2^4 = 16$
    $\log_2 16 = 4$

19. $10^2 = 100$
    $\log_{10} 100 = 2$

21. $\pi^3 = x$
    $\log_\pi x = 3$

23. $10^{-1} = \dfrac{1}{10}$

$\log_{10} \dfrac{1}{10} = -1$

25. $4^{-2} = \dfrac{1}{16}$

$\log_4 \dfrac{1}{16} = -2$

27. $5^{1/2} = \sqrt{5}$

$\log_5 \sqrt{5} = \dfrac{1}{2}$

29. $\log_2 8 = 3$ since $2^3 = 8$.

31. $\log_3 \dfrac{1}{9} = -2$ since $3^{-2} = \dfrac{1}{9}$.

33. $\log_{25} 5 = \dfrac{1}{2}$ since $25^{1/2} = 5$.

35. $\log_{1/2} 2 = -1$ since $\left(\dfrac{1}{2}\right)^{-1} = 2$.

37. $\log_6 1 = 0$ since $6^0 = 1$.

39. $\log_{10} 100 = \log_{10} 10^2 = 2$

41. $\log_3 81 = \log_3 3^4 = 4$

43. $\log_4 \dfrac{1}{64} = \log_4 4^{-3} = -3$

45. $\log_3 9 = x$

$3^x = 9$

$3^x = 3^2$

$x = 2$

47. $\log_3 x = 4$

$x = 3^4 = 81$

49. $\log_x 49 = 2$

$x^2 = 49$

$x = \pm 7$

We discard the negative base.

$x = 7$

51. $\log_2 \dfrac{1}{8} = x$

$2^x = \dfrac{1}{8}$

$2^x = 2^{-3}$

$x = -3$

53. $\log_3 \dfrac{1}{27} = x$

$\dfrac{1}{27} = 3^x$

$3^{-3} = 3^x$

$-3 = x$

55. $\log_8 x = \dfrac{1}{3}$

$x = 8^{1/3} = 2$

57. $\log_4 16 = x$

$4^x = 16$

$4^x = 4^2$

$x = 2$

59. $\log_{3/4} x = 3$

$\left(\dfrac{3}{4}\right)^3 = x$

$\dfrac{27}{64} = x$

61. $\log_x 100 = 2$

$x^2 = 100$

$x = \pm 10$

We discard the negative base.

$x = 10$

63. $\log_2 2^4 = x$

$2^x = 2^4$

$x = 4$

65. $3^{\log_3 5} = x$

$5 = x$

67. $\log_x \dfrac{1}{7} = \dfrac{1}{2}$

$x^{1/2} = \dfrac{1}{7}$

$x = \dfrac{1}{49}$

69. $\log_5 5^3 = 3$

71. $2^{\log_2 3} = 3$

73. $\log_9 9 = 1$

75. $\log_8 (8)^{-1} = -1$

77. $y = \log_3 x$

$y = 0:$

$\log_3 x = 0$

$\quad x = 3^0 = 1$

(1, 0) is the only x-intercept. No y-intercept exists.

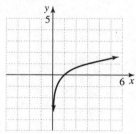

79. $f(x) = \log_{1/4} x$

$y = 0:\ 0 = \log_{1/4} x$

$\quad x = \left(\dfrac{1}{4}\right)^0 = 1$

(1, 0) is the x-intercept. No y-intercept exists.

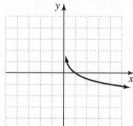

81. $f(x) = \log_5 x$

$y = 0:\ 0 = \log_5 x$

$\quad x = 5^0 = 1$

(1, 0) is the x-intercept. No y-intercept exists.

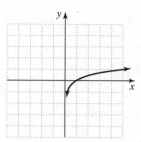

83. $f(x) = \log_{1/16} x$

$y = 0:$

$0 = \log_{1/6} x$

$\quad x = \left(\dfrac{1}{6}\right)^0 = 1$

(1, 0) is the x-intercept. No y-intercept exists.

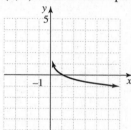

85. $\dfrac{x+3}{3+x} = \dfrac{x+3}{x+3} = 1$

87. $\dfrac{x^2 - 8x + 16}{2x - 8} = \dfrac{(x-4)^2}{2(x-4)} = \dfrac{x-4}{2}$

89. $\dfrac{2}{x} + \dfrac{3}{x^2} = \dfrac{2x}{x^2} + \dfrac{3}{x^2} = \dfrac{2x+3}{x^2}$

91. $\dfrac{3x}{x+3} + \dfrac{9}{x+3} = \dfrac{3x+9}{x+3} = \dfrac{3(x+3)}{x+3} = 3$

93. $f(x) = \log_5 x;\ f^{-1}(x) = g(x) = 5^x$

  a. (2, 25) implies $g(2) = 25$.

  b. Since $f^{-1}(x) = g(x)$, (25, 2) is a solution of $f(x)$.

  c. (25, 2) implies $f(25) = 2$.

95. answers may vary

97. $\log_7(5x-2)=1$
$$5x-2=7^1$$
$$5x=9$$
$$x=\frac{9}{5}$$

99. $\log_3\left(\log_5 125\right)=\log_3(3)=1$

101. $y=4^x$; $y=\log_4 x$

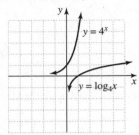

103. $y=\left(\dfrac{1}{3}\right)^x$; $y=\log_{1/3} x$

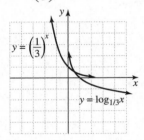

105. answers may vary

107. $\log_{10}(1-k)=\dfrac{-0.3}{H}$, $H=8$

$$\log_{10}(1-k)=\frac{-0.3}{8}=-0.0375$$
$$1-k=10^{-0.0375}$$
$$1-10^{-0.0375}=k$$
$$k\approx 0.0827$$
The rate of decay is 0.0827.

Section 12.6 Practice

1. a. $\log_8 5+\log_8 3=\log_8(5\cdot 3)=\log_8 15$

   b. $\log_2\dfrac{1}{3}+\log_2 18=\log_2\left(\dfrac{1}{3}\cdot 18\right)=\log_2 6$

   c. $\log_5(x-1)+\log_5(x+1)=\log_5[(x+1)(x+1)]$
$$=\log_5(x^2-1)$$

2. a. $\log_5 18-\log_5 6=\log_5\dfrac{18}{6}=\log_5 3$

   b. $\log_6 x-\log_6 3=\log_6\dfrac{x}{3}$

   c. $\log_4(x^2+1)-\log_4(x^2+3)=\log_4\dfrac{x^2+1}{x^2+3}$

3. a. $\log_7 x^8=8\log_7 x$

   b. $\log_5\sqrt[4]{7}=\log_5 7^{1/4}=\dfrac{1}{4}\log_5 7$

4. a. $2\log_5 4+5\log_5 2=\log_5 4^2+\log_5 2^5$
$$=\log_5 16+\log_5 32$$
$$=\log_5(16\cdot 32)$$
$$=\log_5 512$$

   b. $2\log_8 x-\log_8(x+3)=\log_8 x^2-\log_8(x+3)$
$$=\log_8\frac{x^2}{x+3}$$

   c. $\log_7 12+\log_7 5-\log_7 4$
$$=\log_7(12\cdot 5)-\log_7 4$$
$$=\log_7 60-\log_7 4$$
$$=\log_7\frac{60}{4}$$
$$=\log_7 15$$

5. a. $\log_5\dfrac{4\cdot 3}{7}=\log_5(4\cdot 3)-\log_5 7$
$$=\log_5 4+\log_5 3-\log_5 7$$

   b. $\log_4\dfrac{a^2}{b^5}=\log_4 a^2-\log_4 b^5$
$$=2\log_4 a-5\log_4 b$$

6. $\log_b 5=0.83$ and $\log_b 3=0.56$

   a. $\log_b 15=\log_b(3\cdot 5)$
$$=\log_b 3+\log_b 5$$
$$=0.56+0.83$$
$$=1.39$$

b. $\log_b 25 = \log_b 5^2 = 2\log_b 5 = 2(0.83) = 1.66$

c. $\log_b \sqrt{3} = \log_b 3^{1/2}$

$\qquad\qquad = \dfrac{1}{2}\log_b 3$

$\qquad\qquad = \dfrac{1}{2}(0.56)$

$\qquad\qquad = 0.28$

Vocabulary, Readiness & Video Check 12.6

1. $\log_b 12 + \log_b 3 = \log_b (12 \cdot 3) = \log_b \underline{36}$;   **a.**

2. $\log_b 12 - \log_b 3 = \log_b \dfrac{12}{3} = \log_b \underline{4}$;   **c.**

3. $7\log_b 2 = \underline{\log_b 2^7}$;   **b.**

4. $\log_b 1 = \underline{0}$;   **c.**

5. $b^{\log_b x} = \underline{x}$;   **a.**

6. $\log_5 5^2 = \underline{2}$;   **b.**

7.   No, the product property says the logarithm of a product can be written as a sum of logarithms—the expression in Example 2 is a logarithm of a sum.

8.   The bases must be the same.

9.   Since $\dfrac{1}{x} = x^{-1}$, this gives us $\log_2 x^{-1}$. Using the power property, we get $-1\log_2 x$ or $-\log_2 x$.

10.   From writing logarithms as equivalent exponents and then using the rules for exponents.

Exercise Set 12.6

1. $\log_5 2 + \log_5 7 = \log_5 (2 \cdot 7) = \log_5 14$

3. $\log_4 9 + \log_4 x = \log_4 9x$

5. $\log_6 x + \log_6 (x+1) = \log_6 [x(x+1)]$

$\qquad\qquad\qquad\qquad = \log_6 (x^2 + x)$

7. $\log_{10} 5 + \log_{10} 2 + \log_{10}(x^2 + 2)$

$\qquad = \log_{10}\left[5 \cdot 2\left(x^2 + 2\right)\right]$

$\qquad = \log_{10}\left(10x^2 + 20\right)$

9. $\log_5 12 - \log_5 4 = \log_5 \dfrac{12}{4} = \log_5 3$

11. $\log_3 8 - \log_3 2 = \log_3 \dfrac{8}{2} = \log_3 4$

13. $\log_2 x - \log_2 y = \log_2 \dfrac{x}{y}$

15. $\log_2(x^2 + 6) - \log(x^2 + 1) = \log_2 \dfrac{x^2 + 6}{x^2 + 1}$

17. $\log_3 x^2 = 2\log_3 x$

19. $\log_4 5^{-1} = -\log_4 5$

21. $\log_5 \sqrt{y} = \log_5 y^{1/2} = \dfrac{1}{2}\log_5 y$

23. $\log_2 5 + \log_2 x^3 = \log_2 5x^3$

25. $3\log_4 2 + \log_4 6 = \log_4 2^3 + \log_4 6$

$\qquad\qquad\qquad\quad = \log_4 8 + \log_4 6$

$\qquad\qquad\qquad\quad = \log_4 (8 \cdot 6)$

$\qquad\qquad\qquad\quad = \log_4 48$

27. $3\log_5 x + 6\log_5 z = \log_5 x^3 + \log_5 z^6$

$\qquad\qquad\qquad\quad = \log_5 x^3 z^6$

29. $\log_4 2 + \log_4 10 - \log_4 5 = \log_4 (2 \cdot 10) - \log_4 5$

$\qquad\qquad\qquad\qquad\qquad = \log_4 \dfrac{20}{5}$

$\qquad\qquad\qquad\qquad\qquad = \log_4 4$

$\qquad\qquad\qquad\qquad\qquad = 1$

31. $\log_7 6 + \log_7 3 - \log_7 4 = \log_7 (6 \cdot 3) - \log_7 4$

$\qquad\qquad\qquad\qquad\qquad = \log_7 \dfrac{18}{4}$

$\qquad\qquad\qquad\qquad\qquad = \log_7 \dfrac{9}{2}$

33. $\log_{10} x - \log_{10}(x+1) + \log_{10}(x^2 - 2)$

$= \log_{10}\dfrac{x}{x+1} + \log_{10}(x^2 - 2)$

$= \log_{10}\dfrac{x(x^2 - 2)}{x+1}$

$= \log_{10}\dfrac{x^3 - 2x}{x+1}$

35. $3\log_2 x + \dfrac{1}{2}\log_2 x - 2\log_2(x+1)$

$= \log_2 x^3 + \log_2 x^{1/2} - \log_2(x+1)^2$

$= \log_2(x^3 \cdot x^{1/2}) - \log_2(x+1)^2$

$= \log_2 x^{7/2} - \log_2(x+1)^2$

$= \log_2\dfrac{x^{7/2}}{(x+1)^2}$

37. $2\log_8 x - \dfrac{2}{3}\log_8 x + 4\log_8 x = \left(2 - \dfrac{2}{3} + 4\right)\log_8 x$

$= \dfrac{16}{3}\log_8 x$

$= \log_8 x^{16/3}$

39. $\log_3\dfrac{4y}{5} = \log_3 4y - \log_3 5$

$= \log_3 4 + \log_3 y - \log_3 5$

41. $\log_4\dfrac{5}{9z} = \log_4 5 - \log_4 9z$

$= \log_4 5 - (\log_4 9 + \log_4 z)$

$= \log_4 5 - \log_4 9 - \log_4 z$

43. $\log_2\dfrac{x^3}{y} = \log_2 x^3 - \log_2 y$

$= 3\log_2 x - \log_2 y$

45. $\log_b \sqrt{7x} = \log_b(7x)^{1/2}$

$= \dfrac{1}{2}\log_b(7x)$

$= \dfrac{1}{2}\big[\log_b 7 + \log_b x\big]$

$= \dfrac{1}{2}\log_b 7 + \dfrac{1}{2}\log_b x$

47. $\log_6 x^4 y^5 = \log_6 x^4 + \log_6 y^5$

$= 4\log_6 x + 5\log_6 y$

49. $\log_5 x^3(x+1) = \log_5 x^3 + \log_5(x+1)$

$= 3\log_5 x + \log_5(x+1)$

51. $\log_6\dfrac{x^2}{x+3} = \log_6 x^2 - \log_6(x+3)$

$= 2\log_6 x - \log_6(x+3)$

53. $\log_b 15 = \log_b(5 \cdot 3)$

$= \log_b 5 + \log_b 3$

$= 0.7 + 0.5$

$= 1.2$

55. $\log_b\dfrac{5}{3} = \log_b 5 - \log_b 3 = 0.7 - 0.5 = 0.2$

57. $\log_b \sqrt{5} = \log_b 5^{1/2} = \dfrac{1}{2}\log_b 5 = \dfrac{1}{2}(0.7) = 0.35$

59. $\log_b 8 = \log_b 2^3 = 3\log_b 2 = 3(0.43) = 1.29$

61. $\log_b\dfrac{3}{9} = \log_b 3 - \log_b 9$

$= \log_b 3 - \log_b 3^2$

$= \log_b 3 - 2\log_b 3$

$= -\log_b 3$

$= -0.68$

63. $\log_b \sqrt{\dfrac{2}{3}} = \log_b\left(\dfrac{2}{3}\right)^{1/2}$

$= \dfrac{1}{2}\log_b\dfrac{2}{3}$

$= \dfrac{1}{2}\big(\log_b 2 - \log_b 3\big)$

$= \dfrac{1}{2}(0.43 - 0.68)$

$= \dfrac{1}{2}(-0.25)$

$= -0.125$

65. $y = 10^x$ and $y = \log_{10} x$

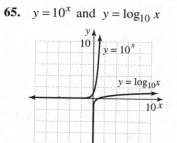

67. $\log_{10} 100 = \log_{10} 10^2 = 2$

69. $\log_7 7^2 = 2$

71. $\log_3 \dfrac{14}{11} = \log_3 14 - \log_3 11$; **b**

73. $\log_2 x^3 = 3\log_2 x$ is true.

75. $\dfrac{\log_7 10}{\log_7 5} = \log_7 2$ is false.

77. $\dfrac{\log_7 x}{\log_7 y} = (\log_7 x) - (\log_7 y)$ is false.

79. $\log_b 8$ equals $\log_b 8 + \log_b 1$ because $\log_b 1 = 0$.

Integrated Review

1. $(f+g)(x) = x - 6 + x^2 + 1 = x^2 + x - 5$

2. $(f-g)(x) = x - 6 - (x^2 + 1) = -x^2 + x - 7$

3. $(f \cdot g)(x) = (x-6)(x^2+1) = x^3 - 6x^2 + x - 6$

4. $\left(\dfrac{f}{g}\right)(x) = \dfrac{x-6}{x^2+1}$

5. $(f \circ g)(x) = f(g(x)) = f(3x-1) = \sqrt{3x-1}$

6. $(g \circ f)(x) = g(f(x)) = g\left(\sqrt{x}\right) = 3\sqrt{x} - 1$

7. one-to-one; inverse:
$\{(6,-2),(8,4),(-6,2),(3,3)\}$

8. not one-to-one

9. not one-to-one

10. one-to-one

11. not one-to-one

12. $f(x) = 3x$
         $y = 3x$

         $x = 3y$
         $y = \dfrac{x}{3}$
       $f^{-1}(x) = \dfrac{x}{3}$

13. $f(x) = x + 4$
         $y = x + 4$

         $x = y + 4$
         $y = x - 4$
       $f^{-1}(x) = x - 4$

14. $f(x) = 5x - 1$
         $y = 5x - 1$

         $x = 5y - 1$
         $5y = x + 1$
         $y = \dfrac{x+1}{5}$
       $f^{-1}(x) = \dfrac{x+1}{5}$

15. $f(x) = 3x + 2$
         $y = 3x + 2$

         $x = 3y + 2$
         $3y = x - 2$
         $y = \dfrac{x-2}{3}$
       $f^{-1}(x) = \dfrac{x-2}{3}$

16. $y = \left(\dfrac{1}{2}\right)^x$

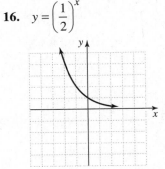

17. $y = 2^x + 1$

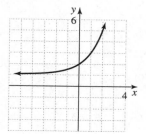

18. $y = \log_3 x$

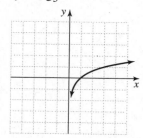

19. $y = \log_{1/3} x$

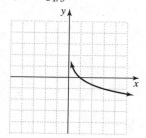

20. $2^x = 8$
$2^x = 2^3$
$x = 3$
The solution is 3.

21. $9 = 3^{x-5}$
$3^2 = 3^{x-5}$
$2 = x - 5$
$7 = x$
The solution is 7.

22. $4^{x-1} = 8^{x+2}$
$(2^2)^{x-1} = (2^3)^{x+2}$
$2^{2x-2} = 2^{3x+6}$
$2x - 2 = 3x + 6$
$-8 = x$
The solution is -8.

23. $25^x = 125^{x-1}$
$(5^2)^x = (5^3)^{x-1}$
$5^{2x} = 5^{3x-3}$
$2x = 3x - 3$
$3 = x$
The solution is 3.

24. $\log_4 16 = x$
$4^x = 16$
$4^x = 4^2$
$x = 2$
The solution is 2.

25. $\log_{49} 7 = x$
$49^x = 7$
$(7^2)^x = 7$
$7^{2x} = 7$
$2x = 1$
$x = \dfrac{1}{2}$

The solution is $\dfrac{1}{2}$.

26. $\log_2 x = 5$
$2^5 = x$
$32 = x$
The solution is 32.

27. $\log_x 64 = 3$
$x^3 = 64$
$x^3 = 4^3$
$x = 4$
The solution is 4.

28. $\log_x \dfrac{1}{125} = -3$
$x^{-3} = \dfrac{1}{125}$
$x^{-3} = 5^{-3}$
$x = 5$
The solution is 5.

29. $\log_3 x = -2$
$3^{-2} = x$
$x = \dfrac{1}{3^2} = \dfrac{1}{9}$

The solution is $\dfrac{1}{9}$.

30. $5\log_2 x = \log_2 x^5$

31. $x\log_2 5 = \log_2 5^x$

32. $3\log_5 x - 5\log_5 y = \log_5 x^3 - \log_5 y^5 = \log_5 \dfrac{x^3}{y^5}$

33. $9\log_5 x + 3\log_5 y = \log_5 x^9 + \log_5 y^3$
$= \log_5 x^9 y^3$

34. $\log_2 x + \log_2 (x-3) - \log_2 (x^2 + 4)$
$= \log_2 [x(x-3)] - \log_2 (x^2 + 4)$
$= \log_2 (x^2 - 3x) - \log_2 (x^2 + 4)$
$= \log_2 \dfrac{x^2 - 3x}{x^2 + 4}$

35. $\log_3 y - \log_3 (y+2) + \log_3 (y^3 + 11)$
$= \log_3 \dfrac{y}{y+2} + \log_3 (y^3 + 11)$
$= \log_3 \dfrac{y(y^3 + 11)}{y+2}$
$= \log_3 \dfrac{y^4 + 11y}{y+2}$

36. $\log_7 \dfrac{9x^2}{y} = \log_7 9x^2 - \log_7 y$
$= \log_7 9 + \log_7 x^2 - \log_7 y$
$= \log_7 9 + 2\log_7 x - \log_7 y$

37. $\log_6 \dfrac{5y}{z^2} = \log_6 5y - \log_6 z^2$
$= \log_6 5 + \log_6 y - 2\log_6 z$

38. $C = 100{,}000,\ r = 6\% = 0.06,\ x = 17 - 1 = 16$
$y = C(1+r)^x$
$= 100{,}000(1+0.06)^{16}$
$\approx 254{,}000$
There will be approximately 254,000 mosquitoes on April 17.

Section 12.7 Practice

1. To four decimal places, $\log 15 \approx 1.1761$.

2. a. $\log \dfrac{1}{100} = \log 10^{-2} = 2$

b. $\log 100{,}000 = \log 10^5 = 5$

c. $\log \sqrt[5]{10} = \log 10^{1/5} = \dfrac{1}{5}$

d. $\log 0.001 = \log 10^{-3} = -3$

3. $\log x = 3.4$
$x = 10^{3.4}$
$x \approx 2511.8864$

4. $a = 450$ micrometers
$T = 4.2$ seconds
$B = 3.6$
$R = \log\left(\dfrac{a}{T}\right) + B$
$= \log\left(\dfrac{450}{4.2}\right) + 3.6$
$\approx 2.0 + 3.6$
$= 5.6$
The earthquake had a magnitude of 5.6 on the Richter scale.

5. To four decimal places, $\ln 13 \approx 2.5649$.

6. a. $\ln e^4 = 4$

b. $\ln \sqrt[3]{e} = \ln e^{1/3} = \dfrac{1}{3}$

7. $\ln 5x = 8$
$e^8 = 5x$
$\dfrac{e^8}{5} = x$
$x = \dfrac{1}{5}e^8 \approx 596.1916$

8. $P = \$2400$
$r = 6\% = 0.06$
$t = 4$ years
$A = Pe^{rt} = 2400e^{0.06(4)} = 2400e^{0.24} \approx 3051.00$
The total amount of money owed is $3051.00.

9. $\log_8 5 = \dfrac{\log 5}{\log 8} \approx \dfrac{0.6989700043}{0.903089987} \approx 0.773976$
To four decimal places, $\log_8 5 \approx 0.7740$.

Vocabulary, Readiness & Video Check 12.7

1. The base of $\log 7$ is $\underline{10}$; **c.**

2. The base of $\ln 7$ is $\underline{e}$; **a.**

3. $\log_{10} 10^7 = \underline{7}$; **b.**

4. $\log_7 1 = \underline{0}$; **d.**

5. $\log_e e^5 = \underline{5}$; **b.**

6. $\ln e^5 = \underline{5}$; **b.**

7. $\log_2 7 = \dfrac{\log 7}{\log 2}$ or $\dfrac{\ln 7}{\ln 2}$; **a** and **b.**

8. 10

9. The understood base of a common logarithm is 10. If you're finding the common logarithm of a known power of 10, then the common logarithm is the known power of 10.

10. e

11. $\log_b b^x = x$

12. $\dfrac{\ln 4}{\ln 6}$ or $\dfrac{\log 4}{\log 6}$; also $\dfrac{\ln 4}{\ln 6} = \dfrac{\log 4}{\log 6}$

Exercise Set 12.7

1. $\log 8 \approx 0.9031$

3. $\log 2.31 \approx 0.3636$

5. $\ln 2 \approx 0.6931$

7. $\ln 0.0716 \approx -2.6367$

9. $\log 12.6 \approx 1.1004$

11. $\ln 5 \approx 1.6094$

13. $\log 41.5 \approx 1.6180$

15. $\log 100 = \log 10^2 = 2$

17. $\log \dfrac{1}{1000} = \log 10^{-3} = -3$

19. $\ln e^2 = 2$

21. $\ln \sqrt[4]{e} = \ln e^{1/4} = \dfrac{1}{4}$

23. $\log 10^3 = 3$

25. $\ln e^{-7} = -7$

27. $\log 0.0001 = \log 10^{-4} = -4$

29. $\ln \sqrt{e} = \ln e^{1/2} = \dfrac{1}{2}$

31. $\ln 2x = 7$
   $2x = e^7$
   $x = \dfrac{1}{2}e^7 \approx 548.3166$

33. $\log x = 1.3$
   $x = 10^{1.3} \approx 19.9526$

35. $\log 2x = 1.1$
   $2x = 10^{1.1}$
   $x = \dfrac{10^{1.1}}{2} \approx 6.2946$

37. $\ln x = 1.4$
   $x = e^{1.4} \approx 4.0552$

39. $\ln(3x - 4) = 2.3$
   $3x - 4 = e^{2.3}$
   $3x = 4 + e^{2.3}$
   $x = \dfrac{4 + e^{2.3}}{3} \approx 4.6581$

41. $\log x = 2.3$
   $x = 10^{2.3} \approx 199.5262$

43. $\ln x = -2.3$
   $x = e^{-2.3} \approx 0.1003$

45. $\log(2x + 1) = -0.5$
   $2x + 1 = 10^{-0.5}$
   $2x = 10^{-0.5} - 1$
   $x = \dfrac{10^{-0.5} - 1}{2} \approx -0.3419$

47. $\ln 4x = 0.18$

$4x = e^{0.18}$

$x = \dfrac{e^{0.18}}{4} \approx 0.2993$

49. $\log_2 3 = \dfrac{\log 3}{\log 2} \approx 1.5850$

51. $\log_{1/2} 5 = \dfrac{\ln 5}{\ln\left(\frac{1}{2}\right)} \approx -2.3219$

53. $\log_4 9 = \dfrac{\ln 9}{\ln 4} \approx 1.5850$

55. $\log_3\left(\dfrac{1}{6}\right) = \dfrac{\log\left(\frac{1}{6}\right)}{\log 3} \approx -1.6309$

57. $\log_8 6 = \dfrac{\log 6}{\log 8} \approx 0.8617$

59. $R = \log\left(\dfrac{a}{T}\right) + B,\ a = 200,\ T = 1.6$

$B = 2.1$

$R = \log\left(\dfrac{200}{1.6}\right) + 2.1 \approx 4.2$

The earthquake measures 4.2 on the Richter scale.

61. $R = \log\left(\dfrac{a}{T}\right) + B,\ a = 400,\ T = 2.6$

$B = 3.1$

$R = \log\left(\dfrac{400}{2.6}\right) + 3.1 \approx 5.3$

The earthquake measures 5.3 on the Richter scale.

63. $A = Pe^{rt},\ t = 12,\ P = 1400, r = 0.08$

$A = 1400e^{(0.08)12} = 1400e^{0.96} \approx 3656.38$

Dana has \$3656.38 after 12 years.

65. $A = Pe^{rt},\ t = 4,\ P = 2000, r = 0.06$

$A = 2000e^{(0.06)4} = 2000e^{0.24} \approx 2542.50$

Barbara owes \$2542.50 at the end of 4 years.

67. $6x - 3(2 - 5x) = 6$

$6x - 6 + 15x = 6$

$21x = 12$

$x = \dfrac{12}{21} = \dfrac{4}{7}$

The solution is $\dfrac{4}{7}$.

69. $2x + 3y = 6x$

$3y = 4x$

$x = \dfrac{3y}{4}$

71. $x^2 + 7x = -6$

$x^2 + 7x + 6 = 0$

$(x + 6)(x + 1) = 0$

$x + 6 = 0 \quad$ or $\quad x + 1 = 0$

$x = -6 \quad$ or $\qquad x = -1$

The solutions are -6 and -1.

73. $\begin{cases} x + 2y = -4 \\ 3x - y = 9 \end{cases}$

Multiply the second equation by 2, then add.

$x + 2y = -4$

$\dfrac{6x - 2y = 18}{7x \qquad = 14}$

$x = 2$

Replace x with 2 in the first equation.

$x + 2y = -4$

$2 + 2y = -4$

$2y = -6$

$y = -3$

The solution is $(2, -3)$.

75. answers may vary

77. ln 50 is larger. answers may vary

79. $f(x) = e^x$

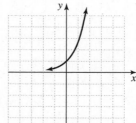

81. $f(x) = e^{-3x}$

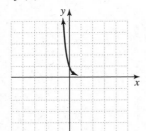

83. $f(x) = e^x + 2$

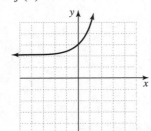

85. $f(x) = e^{x-1}$

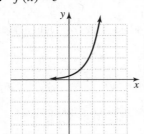

87. $f(x) = 3e^x$

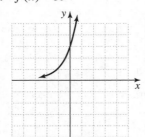

89. $f(x) = \ln x$

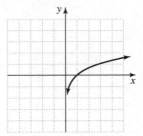

91. $f(x) = -2\log x$

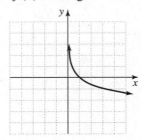

93. $f(x) = \log(x+2)$

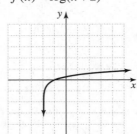

95. $f(x) = \ln x - 3$

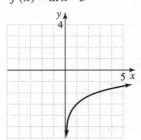

97. $f(x) = e^x$

$\quad\quad f(x) = e^x + 2$

$\quad\quad f(x) = e^x - 3$

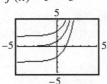

answers may vary

Section 12.8 Practice

1. $5^x = 9$

$$\log 5^x = \log 9$$
$$x \log 5 = \log 9$$
$$x = \frac{\log 9}{\log 5} \approx 1.3652$$

The solution is $\dfrac{\log 9}{\log 5}$, or approximately 1.3652.

2. $\log_2(x-1) = 5$

$$2^5 = x-1$$
$$32 = x-1$$
$$33 = x$$

Check: $\log_2(x-1) = 5$
$$\log_2(33-1) \overset{?}{=} 5$$
$$\log_2 32 \overset{?}{=} 5$$
$$2^5 = 32 \quad \text{True}$$

The solution is 33.

3. $\log_5 x + \log_5(x+4) = 1$
$$\log_5 x(x+4) = 1$$
$$\log_5(x^2+4x) = 1$$

$$5^1 = x^2 + 4x$$
$$0 = x^2 + 4x - 5$$
$$0 = (x+5)(x-1)$$
$$x + 5 = 0 \quad \text{or} \quad x - 1 = 0$$
$$x = -5 \qquad\qquad x = 1$$

Since $\log_5(-5)$ is undefined, -5 is rejected. The solution is 1.

4. $\log(x+3) - \log x = 1$
$$\log \frac{x+3}{x} = 1$$
$$10^1 = \frac{x+3}{x}$$
$$10x = x+3$$
$$9x = 3$$
$$x = \frac{1}{3}$$

The solution is $\dfrac{1}{3}$.

5. $y_0 = 60; \ t = 3$

$$y = y_0 e^{0.916t}$$
$$y = 60 e^{0.916(3)} = 60 e^{2.748} \approx 937$$

The population will be approximately 937 rabbits.

6. $P = \$3000; \ r = 7\% = 0.07; \ n = 12;$
$A = 2P = \$6000$

$$A = P\left(1 + \frac{r}{n}\right)^{nt}$$
$$6000 = 3000\left(1 + \frac{0.07}{12}\right)^{12t}$$
$$2 = \left(1 + \frac{0.07}{12}\right)^{12t}$$
$$\log 2 = \log\left(1 + \frac{0.07}{12}\right)^{12t}$$
$$\log 2 = 12t \log\left(1 + \frac{0.07}{12}\right)$$
$$\frac{\log 2}{12 \log\left(1 + \frac{0.07}{12}\right)} = t$$
$$9.9 \approx t$$

It takes nearly 10 years to double.

Graphing Calculator Explorations

1. $Y_1 = 5000\left(1 + \dfrac{0.05}{4}\right)^{4x}$, $Y_2 = 6000$

It takes 3.67 years, or 3 years and 8 months.

2. $Y_1 = 1000\left(1 + \dfrac{0.045}{365}\right)^{365x}$, $Y_2 = 2000$

It takes 15.40 years or 15 years and 5 months.

3. $Y_1 = 10,000\left(1 + \dfrac{0.06}{12}\right)^{12x}$, $Y_2 = 40,000$

It takes 23.16 years or 23 years and 2 months.

4. $Y_1 = 500\left(1+\dfrac{0.04}{2}\right)^{2x}$, $Y_2 = 800$

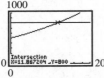

It takes 11.87 years or 11 years and 10 months.

Vocabulary, Readiness & Video Check 12.8

1. $\ln(4x-2) = \ln 3$ is the same as $\log_e(4x-2) = \log_e 3$. Therefore, from the logarithm property of equality, we know that $4x - 2 = 3$.

2. Substituting 8 in the original equation gives us the logarithm of a negative number, which does not exist—we can only take the logarithm of a positive number.

3. $2000 = 1000\left(1+\dfrac{0.07}{12}\right)^{12 \cdot t}$

$t \approx 9.9$

As long as the interest rate and compounding are the same, it takes any amount of money the same time to double.

Exercise Set 12.8

1. $3^x = 6$

$\log 3^x = \log 6$

$x \log 3 = \log 6$

$x = \dfrac{\log 6}{\log 3} \approx 1.6309$

3. $3^{2x} = 3.8$

$\log 3^{2x} = \log 3.8$

$2x \log 3 = \log 3.8$

$x = \dfrac{\log 3.8}{2 \log 3} \approx 0.6076$

5. $2^{x-3} = 5$

$\log 2^{x-3} = \log 5$

$(x-3) \log 2 = \log 5$

$x \log 2 - 3 \log 2 = \log 5$

$x \log 2 = 3 \log 2 + \log 5$

$x = \dfrac{3 \log 2 + \log 5}{\log 2}$

or

$x = 3 + \dfrac{\log 5}{\log 2}$

$x \approx 5.3219$

7. $9^x = 5$

$\log 9^x = \log 5$

$x \log 9 = \log 5$

$x = \dfrac{\log 5}{\log 9} \approx 0.7325$

9. $4^{x+7} = 3$

$\log 4^{x+7} = \log 3$

$(x+7) \log 4 = \log 3$

$x \log 4 + 7 \log 4 = \log 3$

$x \log 4 = \log 3 - 7 \log 4$

$x = \dfrac{\log 3 - 7 \log 4}{\log 4}$

or

$x = \dfrac{\log 3}{\log 4} - 7$

$x \approx -6.2075$

11. $\log_2(x+5) = 4$

$x + 5 = 2^4$

$x + 5 = 16$

$x = 11$

13. $\log_4 2 + \log_4 x = 0$

$\log_4(2x) = 0$

$2x = 4^0$

$2x = 1$

$x = \dfrac{1}{2}$

15. $\log_2 6 - \log_2 x = 3$

$$\log_2\left(\frac{6}{x}\right) = 3$$

$$\frac{6}{x} = 2^3$$

$$\frac{6}{x} = 8$$

$$8x = 6$$

$$x = \frac{3}{4}$$

17. $\log_6(x^2 - x) = 1$

$$6^1 = x^2 - x$$

$$0 = x^2 - x - 6$$

$$0 = (x - 3)(x + 2)$$

$x = 3$ or $x = -2$

19. $\log_4 x + \log_4 (x + 6) = 2$

$$\log_4 x(x + 6) = 2$$

$$x(x + 6) = 4^2$$

$$x^2 + 6x = 16$$

$$x^2 + 6x - 16 = 0$$

$$(x + 8)(x - 2) = 0$$

$x = -8$ or $x = 2$

We discard -8 as extraneous, the solution is 2.

21. $\log_5(x + 3) - \log_5 x = 2$

$$\log_5\left(\frac{x + 3}{x}\right) = 2$$

$$\frac{x + 3}{x} = 5^2$$

$$\frac{x + 3}{x} = 25$$

$$x + 3 = 25x$$

$$3 = 24x$$

$$x = \frac{1}{8}$$

23. $7^{3x-4} = 11$

$$\log 7^{3x-4} = \log 11$$

$$(3x - 4)\log 7 = \log 11$$

$$3x \log 7 - 4 \log 7 = \log 11$$

$$3x \log 7 = 4 \log 7 + \log 11$$

$$x = \frac{4 \log 7 + \log 11}{3 \log 7}$$

or

$$x = \frac{1}{3}\left(4 + \frac{\log 11}{\log 7}\right)$$

$x \approx 1.7441$

25. $\log_4(x^2 - 3x) = 1$

$$x^2 - 3x = 4$$

$$x^2 - 3x - 4 = 0$$

$$(x - 4)(x + 1) = 0$$

$x = 4$ or $x = -1$

27. $e^{6x} = 5$

$$\ln e^{6x} = \ln 5$$

$$6x = \ln 5$$

$$x = \frac{\ln 5}{6} \approx 0.2682$$

29. $\log_3 x^2 = 4$

$$x^2 = 3^4$$

$$x^2 = 81$$

$$x = \pm 9$$

31. $\ln 5 + \ln x = 0$

$$\ln(5x) = 0$$

$$e^0 = 5x$$

$$1 = 5x$$

$$\frac{1}{5} = x$$

33. $3 \log x - \log x^2 = 2$

$$3 \log x - 2 \log x = 2$$

$$\log x = 2$$

$$x = 10^2$$

$$x = 100$$

35. $\log_4 x - \log_4(2x-3) = 3$

$$\log_4\left(\frac{x}{2x-3}\right) = 3$$

$$\frac{x}{2x-3} = 4^3$$

$$x = 64(2x-3)$$

$$x = 128x - 192$$

$$192 = 127x$$

$$x = \frac{192}{127}$$

37. $\log_2 x + \log_2(3x+1) = 1$

$$\log_2 x(3x+1) = 1$$

$$x(3x+1) = 2$$

$$3x^2 + x - 2 = 0$$

$$(3x-2)(x+1) = 0$$

$$3x - 2 = 0 \quad \text{or} \quad x + 1 = 0$$

$$x = \frac{2}{3} \quad \text{or} \qquad x = -1$$

We discard -1 as extraneous, the solution is $\frac{2}{3}$.

39. $\log_2 x + \log_2(x+5) = 1$

$$\log_2 x(x+5) = 1$$

$$x(x+5) = 2$$

$$x^2 + 5x - 2 = 0$$

$$a = 1, b = 5, c = -2$$

$$x = \frac{-5 \pm \sqrt{5^2 - 4(1)(-2)}}{2(1)}$$

$$x = \frac{-5 \pm \sqrt{33}}{2}$$

Discard $\frac{-5 - \sqrt{33}}{2}$, the solution is $\frac{-5 + \sqrt{33}}{2}$.

41. Let $y_0 = 83$ and $t = 5$.

$$y = y_0 e^{0.043t}$$

$$y = 83e^{0.043(5)} = 83e^{0.215} \approx 103$$

The population is estimated to be 103 wolves in 5 years.

43. Let $y_0 = 11,488$ and $t = 15$.

$$y = y_0 e^{-0.0277t}$$

$$y = 11,488e^{-0.0277(15)} = 11,488e^{-0.4155} \approx 7582$$

The population of the Cook Islands is predicted to be 7582 in 2025.

45. $A = P\left(1 + \frac{r}{n}\right)^{nt}$, $P = 600$,

$$A = 2(600) = 1200, r = 0.07, n = 12$$

$$1200 = 600\left(1 + \frac{0.07}{12}\right)^{12t}$$

$$2 = \left(1 + \frac{0.07}{12}\right)^{12t}$$

$$\log 2 = \log\left(1 + \frac{0.07}{12}\right)^{12t}$$

$$\log 2 = 12t \log\left(1 + \frac{0.07}{12}\right)$$

$$\frac{\log 2}{12 \log\left(1 + \frac{0.07}{12}\right)} = t$$

$$9.9 \approx t$$

It takes approximately 9.9 years for the $600 to double.

47. $A = P\left(1 + \frac{r}{n}\right)^{nt}$, $P = 1200$,

$$A = P + I = 1200 + 200 = 1400$$

$$r = 0.09, n = 4$$

$$1400 = 1200\left(1 + \frac{0.09}{4}\right)^{4t}$$

$$\frac{7}{6} = (1.0225)^{4t}$$

$$\log\frac{7}{6} = \log 1.0225^{4t}$$

$$\log\frac{7}{6} = 4t \log 1.0225$$

$$t = \frac{\log\frac{7}{6}}{4 \log 1.0225}$$

$$t \approx 1.7$$

It would take the investment approximately 1.7 years to earn $200.

49. $A = P\left(1 + \dfrac{r}{n}\right)^{nt}$, $P = 1000$

$A = 2(1000) = 2000$, $r = 0.08$, $n = 2$

$2000 = 1000\left(1 + \dfrac{0.08}{2}\right)^{2t}$

$2 = (1.04)^{2t}$

$\log 2 = \log 1.04^{2t}$

$\log 2 = 2t \log 1.04$

$t = \dfrac{\log 2}{2 \log 1.04}$

$t \approx 8.8$

It takes 8.8 years to double.

51. $w = 0.00185 h^{2.67}$, and $h = 35$

$w = 0.00185(35)^{2.67} \approx 24.5$

The expected weight of a boy 35 inches tall is 24.5 pounds.

53. $w = 0.00185 h^{2.67}$, and $w = 85$

$85 = 0.00185 h^{2.67}$

$\dfrac{85}{0.00185} = h^{2.67}$

$h = \left(\dfrac{85}{0.00185}\right)^{1/2.67} \approx 55.7$

The expected height of the boy is 55.7 inches.

55. $P = 14.7 e^{-0.21x}$, $x = 1$

$P = 14.7 e^{-0.21(1)}$

$\quad = 14.7 e^{-0.21}$

$\quad \approx 11.9$

The average atmospheric pressure in Denver is approximately 11.9 pounds per square inch.

57. $P = 14.7 e^{-0.21x}$, $P = 7.5$

$7.5 = 14.7 e^{-0.21x}$

$\dfrac{7.5}{14.7} = e^{-0.21x}$

$-0.21x = \ln\left(\dfrac{7.5}{14.7}\right)$

$x = -\dfrac{1}{0.21} \ln\left(\dfrac{7.5}{14.7}\right) \approx 3.2$

The elevation of the jet is approximately 3.2 miles.

59. $t = \dfrac{1}{c} \ln\left(\dfrac{A}{A - N}\right)$

$t = \dfrac{1}{0.09} \ln\left(\dfrac{75}{75 - 50}\right)$

$t = \dfrac{1}{0.09} \ln(3)$

$t \approx 12.21$

It will take 12 weeks.

61. $t = \dfrac{1}{c} \ln\left(\dfrac{A}{A - N}\right)$

$t = \dfrac{1}{0.07} \ln\left(\dfrac{210}{210 - 150}\right)$

$t = \dfrac{1}{0.07} \ln(3.5)$

$t \approx 17.9$

It will take 18 weeks.

63. $\dfrac{x^2 - y + 2z}{3x} = \dfrac{(-2)^2 - 0 + 2(3)}{3(-2)}$

$\qquad = \dfrac{4 + 6}{-6}$

$\qquad = \dfrac{10}{-6}$

$\qquad = -\dfrac{5}{3}$

65. $\dfrac{3z - 4x + y}{x + 2z} = \dfrac{3(3) - 4(-2) + 0}{-2 + 2(3)} = \dfrac{9 + 8}{-2 + 6} = \dfrac{17}{4}$

67. $f(x) = 5x + 2$

$y = 5x + 2$

$x = 5y + 2$

$\dfrac{x - 2}{5} = y$

$f^{-1}(x) = \dfrac{x - 2}{5}$

69. $y = y_0 e^{-0.003t}$

$y = 9,500,000, \; y_0 = 9,939,000$

$9,500,000 = 9,939,000 e^{-0.003t}$

$\dfrac{9,500,000}{9,939,000} = e^{-0.003t}$

$\ln \dfrac{9500}{9939} = -0.003t$

$t = \dfrac{\ln \frac{9500}{9939}}{-0.003} \approx 15.1$

The population of Michigan will be 9,500,000 after approximately 15 years.

71. answers may vary

73. $Y_1 = e^{0.3x}, \; Y_2 = 8$

$x \approx 6.93$

75. $Y_1 = 2\log(-5.6x + 1.3) + x + 1, \; Y_2 = 0$

$x \approx -3.68, \, 0.19$

77. $Y_1 = 7^{3x-4} - 11, \; Y_2 = 0$

$x \approx 1.74$

79. $Y_1 = \ln 5 + \ln x, \; Y_2 = 0$

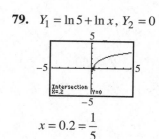

$x = 0.2 = \dfrac{1}{5}$

Chapter 12 Vocabulary Check

1. For a one-to-one function, we can find its <u>inverse</u> function by switching the coordinates of the ordered pairs of the function.

2. The <u>composition</u> of functions f and g is $(f \circ g)(x) = f(g(x))$.

3. A function of the form $f(x) = b^x$ is called an <u>exponential</u> function if $b > 0$, b is not 1, and x is a real number.

4. The graphs of f and f^{-1} are <u>symmetric</u> about the line $y = x$.

5. <u>Natural</u> logarithms are logarithms to base e.

6. <u>Common</u> logarithms are logarithms to base 10.

7. To see whether a graph is the graph of a one-to-one function, apply the <u>vertical</u> line test to see whether it is a function, and then apply the <u>horizontal</u> line test to see whether it is a one-to-one function.

8. A <u>logarithmic</u> function is a function that can be defined by $f(x) = \log_b x$ where x is a positive real number, b is a constant positive real number, and b is not 1.

9. <u>Half-life</u> is the amount of time it takes for half of the amount of a substance to decay.

10. A quantity that grows or decays by the same percent at regular time periods is said to have <u>exponential</u> growth or decay.

Chapter 12 Review

1. $(f + g)(x) = f(x) + g(x)$
$\qquad = (x - 5) + (2x + 1)$
$\qquad = x - 5 + 2x + 1$
$\qquad = 3x - 4$

2. $(f-g)(x) = f(x) - g(x)$
$= (x-5) - (2x+1)$
$= x - 5 - 2x - 1$
$= -x - 6$

3. $(f \cdot g)(x) = f(x) \cdot g(x)$
$= (x-5)(2x+1)$
$= 2x^2 + x - 10x - 5$
$= 2x^2 - 9x - 5$

4. $\left(\dfrac{g}{f}\right)(x) = \dfrac{g(x)}{f(x)} = \dfrac{2x+1}{x-5}, x \neq 5$

5. $(f \circ g)(x) = f(g(x))$
$= f(x+1)$
$= (x+1)^2 - 2$
$= x^2 + 2x - 1$

6. $(g \circ f)(x) = g(f(x))$
$= g(x^2 - 2)$
$= x^2 - 2 + 1$
$= x^2 - 1$

7. $(h \circ g)(2) = h(g(2)) = h(3) = 3^3 - 3^2 = 18$

8. $(f \circ f)(x) = f(f(x))$
$= f(x^2 - 2)$
$= (x^2 - 2)^2 - 2$
$= x^4 - 4x^2 + 4 - 2$
$= x^4 - 4x^2 + 2$

9. $(f \circ g)(-1) = f(g(-1)) = f(0) = 0^2 - 2 = -2$

10. $(h \circ h)(2) = h(h(2)) = h(4) = 4^3 - 4^2 = 48$

11. The function is one-to-one.
$h^{-1} = \{(14, -9), (8, 6), (12, -11), (15, 15)\}$

12. The function is not one-to-one.

13. The function is one-to-one.

Rank in Housing Starts for 2009 (Input)	4	3	1	2
U.S. Region (Output)	Northeast	Midwest	South	West

14. The function is not one-to-one.

15. $f(x) = \sqrt{x+2}$

　　a. $f(7) = \sqrt{7+2} = \sqrt{9} = 3$

　　b. $f^{-1}(3) = 7$

16. $f(x) = \sqrt{x+2}$

　　a. $f(-1) = \sqrt{-1+2} = \sqrt{1} = 1$

　　b. $f^{-1}(1) = -1$

17. The graph does not represent a one-to-one function.

18. The graph does not represent a one-to-one function.

19. The graph does not represent a one-to-one function.

20. The graph represents a one-to-one function.

21. $f(x) = x - 9$
$y = x - 9$
$x = y - 9$
$y = x + 9$
$f^{-1}(x) = x + 9$

22. $f(x) = x + 8$
$y = x + 8$
$x = y + 8$
$y = x - 8$
$f^{-1}(x) = x - 8$

23.
$$f(x) = 6x + 11$$
$$y = 6x + 11$$
$$x = 6y + 11$$
$$6y = x - 11$$
$$y = \frac{x - 11}{6}$$
$$f^{-1}(x) = \frac{x - 11}{6}$$

24.
$$f(x) = 12x$$
$$y = 12x$$
$$x = 12y$$
$$y = \frac{x}{12}$$
$$f^{-1}(x) = \frac{x}{12}$$

25.
$$f(x) = x^3 - 5$$
$$y = x^3 - 5$$
$$x = y^3 - 5$$
$$y^3 = x + 5$$
$$y = \sqrt[3]{x + 5}$$
$$f^{-1}(x) = \sqrt[3]{x + 5}$$

26.
$$f(x) = \sqrt[3]{x + 2}$$
$$y = \sqrt[3]{x + 2}$$
$$x = \sqrt[3]{y + 2}$$
$$x^3 = y + 2$$
$$y = x^3 - 2$$
$$f^{-1}(x) = x^3 - 2$$

27.
$$g(x) = \frac{12x - 7}{6}$$
$$y = \frac{12x - 7}{6}$$
$$x = \frac{12y - 7}{6}$$
$$6x = 12y - 7$$
$$12y = 6x + 7$$
$$y = \frac{6x + 7}{12}$$
$$g^{-1}(x) = \frac{6x + 7}{12}$$

28.
$$r(x) = \frac{13x - 5}{2}$$
$$y = \frac{13x - 5}{2}$$
$$x = \frac{13y - 5}{2}$$
$$2x = 13y - 5$$
$$2x + 5 = 13y$$
$$y = \frac{2x + 5}{13}$$
$$r^{-1}(x) = \frac{2x + 5}{13}$$

29.
$$f(x) = -2x + 3$$
$$y = -2x + 3$$
$$x = -2y + 3$$
$$x - 3 = -2y$$
$$-\frac{x - 3}{2} = y$$
$$f^{-1}(x) = -\frac{x - 3}{2}$$

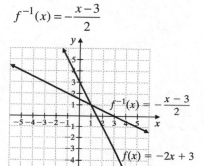

30.
$$f(x) = 5x - 5$$
$$y = 5x - 5$$
$$x = 5y - 5$$
$$x + 5 = 5y$$
$$\frac{x + 5}{5} = y$$
$$f^{-1}(x) = \frac{x + 5}{5}$$

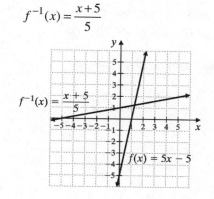

31. $4^x = 64$
$4^x = 4^3$
$x = 3$

32. $3^x = \dfrac{1}{9}$
$3^x = 3^{-2}$
$x = -2$

33. $2^{3x} = \dfrac{1}{16}$
$2^{3x} = 2^{-4}$
$3x = -4$
$x = -\dfrac{4}{3}$

34. $5^{2x} = 125$
$5^{2x} = 5^3$
$2x = 3$
$x = \dfrac{3}{2}$

35. $9^{x+1} = 243$
$(3^2)^{x+1} = 3^5$
$3^{2x+2} = 3^5$
$2x + 2 = 5$
$2x = 3$
$x = \dfrac{3}{2}$

36. $8^{3x-2} = 4$
$(2^3)^{3x-2} = 2^2$
$2^{9x-6} = 2^2$
$9x - 6 = 2$
$9x = 8$
$x = \dfrac{8}{9}$

37. $y = 3^x$

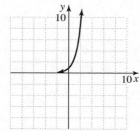

38. $y = \left(\dfrac{1}{3}\right)^x$

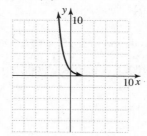

39. $y = 2^{x-4}$

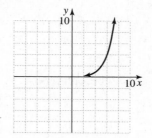

40. $y = 2^x + 4$

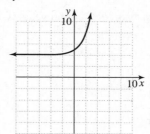

41. $A = P\left(1 + \dfrac{r}{n}\right)^{nt}$

$A = 1600\left(1 + \dfrac{0.09}{2}\right)^{(2)(7)}$
$A \approx 2963.11$
The amount accrued is $2963.11.

42. $A = P\left(1 + \dfrac{r}{n}\right)^{nt}$

$A = 800\left(1 + \dfrac{0.07}{4}\right)^{(4)(5)}$
$A \approx 1131.82$
The certificate is worth $1131.82 at the end of 5 years.

43. $C = 79{,}087,\ r = 0.044,\ x = 2020 - 2000 = 20$

$y = C(1+r)^x$

$y = 79{,}087(1+0.044)^{20}$

$y \approx 187{,}118$

The predicted population in 2020 is 187,118.

44. $C = 287{,}370,\ r = 0.042,\ x = 2018 - 2000 = 18$

$y = C(1+r)^x$

$y = 287{,}370(1+0.042)^{18}$

$y \approx 602{,}643$

The predicted population in 2018 is 602,643.

45. $C = 1024,\ r = 0.5,\ x = 7$

$y = C(1-r)^x$

$y = 1024(1-0.5)^7$

$y = 8$

After 7 rounds there will be 8 players.

46. $C = 1280,\ r = 0.11,\ x = 6$

$y = C(1-r)^x$

$y = 1280(1-0.11)^6$

$y \approx 636$

The predicted bear population in 6 years is 636.

47. $\qquad 49 = 7^2$

$\log_7 49 = 2$

48. $\qquad 2^{-4} = \dfrac{1}{16}$

$\log_2 \dfrac{1}{16} = -4$

49. $\log_{1/2} 16 = -4$

$\left(\dfrac{1}{2}\right)^{-4} = 16$

50. $\log_{0.4} 0.064 = 3$

$0.4^3 = 0.064$

51. $\log_4 x = -3$

$x = 4^{-3} = \dfrac{1}{64}$

52. $\log_3 x = 2$

$x = 3^2 = 9$

53. $\log_3 1 = x$

$3^x = 1$

$3^x = 3^0$

$x = 0$

54. $\log_4 64 = x$

$4^x = 64$

$4^x = 4^3$

$x = 3$

55. $\log_4 4^5 = x$

$x = 5$

56. $\log_7 7^{-2} = x$

$x = -2$

57. $5^{\log_5 4} = x$

$x = 4$

58. $2^{\log_2 9} = x$

$9 = x$

59. $\log_2(3x-1) = 4$

$3x - 1 = 2^4$

$3x - 1 = 16$

$3x = 17$

$x = \dfrac{17}{3}$

60. $\log_3(2x+5) = 2$

$2x + 5 = 3^2$

$2x + 5 = 9$

$2x = 4$

$x = 2$

61. $\log_4(x^2 - 3x) = 1$

$x^2 - 3x = 4$

$x^2 - 3x - 4 = 0$

$(x+1)(x-4) = 0$

$x = -1$ or $x = 4$

62. $\log_8(x^2 + 7x) = 1$

$x^2 + 7x = 8$

$x^2 + 7x - 8 = 0$

$(x+8)(x-1) = 0$

$x = -8$ or $x = 1$

63. $y = 2^x$ and $y = \log_2 x$

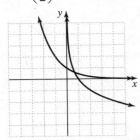

64. $y = \left(\dfrac{1}{2}\right)^x$ and $y = \log_{1/2} x$

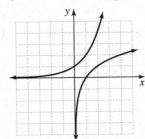

65. $\log_3 8 + \log_3 4 = \log_3(8 \cdot 4) = \log_3 32$

66. $\log_2 6 + \log_2 3 = \log_2(6 \cdot 3) = \log_2 18$

67. $\log_7 15 - \log_7 20 = \log_7 \dfrac{15}{20} = \log_7 \dfrac{3}{4}$

68. $\log 18 - \log 12 = \log \dfrac{18}{12} = \log \dfrac{3}{2}$

69. $\log_{11} 8 + \log_{11} 3 - \log_{11} 6 = \log_{11} \dfrac{(8)(3)}{6}$
$= \log_{11} 4$

70. $\log_5 14 + \log_5 3 - \log_5 21$
$= \log_5 (14 \cdot 3) - \log_5 21$
$= \log_5 \dfrac{42}{21}$
$= \log_5 2$

71. $2\log_5 x - 2\log_5(x+1) + \log_5 x$
$= \log_5 x^2 - \log_5(x+1)^2 + \log_5 x$
$= \log_5 \dfrac{(x^2)(x)}{(x+1)^2}$
$= \log_5 \dfrac{x^3}{(x+1)^2}$

72. $4\log_3 x - \log_3 x + \log_3(x+2)$
$= 3\log_3 x + \log_3(x+2)$
$= \log_3 x^3 + \log_3(x+2)$
$= \log_3 \left[x^3(x+2) \right]$
$= \log_3(x^4 + 2x^3)$

73. $\log_3 \dfrac{x^3}{x+2} = \log_3 x^3 - \log_3(x+2)$
$= 3\log_3 x - \log_3(x+2)$

74. $\log_4 \dfrac{x+5}{x^2} = \log_4(x+5) - \log_4 x^2$
$= \log_4(x+5) - 2\log_4 x$

75. $\log_2 \dfrac{3x^2 y}{z} = \log_2(3x^2 y) - \log_2 z$
$= \log_2 3 + \log_2 x^2 + \log_2 y - \log_2 z$
$= \log_2 3 + 2\log_2 x + \log_2 y - \log_2 z$

76. $\log_7 \dfrac{yz^3}{x} = \log_7(yz^3) - \log_7 x$
$= \log_7 y + \log_7 z^3 - \log_7 x$
$= \log_7 y + 3\log_7 z - \log_7 x$

77. $\log_b 50 = \log_b (5)(5)(2)$
$= \log_b(5) + \log_b(5) + \log_b(2)$
$= 0.83 + 0.83 + 0.36$
$= 2.02$

78. $\log_b \dfrac{4}{5} = \log_b 4 - \log_b 5$
$= \log_b 2^2 - \log_b 5$
$= 2\log_b 2 - \log_b 5$
$= 2(0.36) - 0.83$
$= 0.72 - 0.83$
$= -0.11$

79. $\log 3.6 \approx 0.5563$

80. $\log 0.15 \approx -0.8239$

81. $\ln 1.25 \approx 0.2231$

82. $\ln 4.63 \approx 1.5326$

83. $\log 1000 = \log 10^3 = 3$

84. $\log\dfrac{1}{10} = \log 10^{-1} = -1$

85. $\ln\dfrac{1}{e} = \ln e^{-1} = -1$

86. $\ln e^4 = 4$

87. $\ln(2x) = 2$

$2x = e^2$

$x = \dfrac{e^2}{2}$

88. $\ln(3x) = 1.6$

$3x = e^{1.6}$

$x = \dfrac{e^{1.6}}{3}$

89. $\ln(2x-3) = -1$

$2x - 3 = e^{-1}$

$x = \dfrac{e^{-1}+3}{2}$

90. $\ln(3x+1) = 2$

$3x + 1 = e^2$

$3x = e^2 - 1$

$x = \dfrac{e^2 - 1}{3}$

91. $\ln\dfrac{I}{I_0} = -kx$

$\ln\dfrac{0.03 I_0}{I_0} = -2.1x$

$\ln 0.03 = -2.1x$

$\dfrac{\ln 0.03}{-2.1} = x$

$x \approx 1.67$

The depth is 1.67 millimeters.

92. $\ln\dfrac{I}{I_0} = -kx$

$\ln\dfrac{0.02 I_0}{I_0} = -3.2x$

$\ln 0.02 = -3.2x$

$\dfrac{\ln 0.02}{-3.2} = x$

$x \approx 1.22$

2% of the original radioactivity will penetrate at a depth of approximately 1.22 millimeters.

93. $\log_5 1.6 = \dfrac{\log 1.6}{\log 5} \approx 0.2920$

94. $\log_3 4 = \dfrac{\log 4}{\log 3} \approx 1.2619$

95. $A = Pe^{rt}$

$A = 1450 e^{(0.03)(5)}$

$A \approx 1684.66$

The accrued amount is \$1684.66.

96. $A = Pe^{rt}$

$A = 940 e^{0.04(3)} = 940 e^{0.12} \approx 1059.85$

The investment grows to \$1059.85.

97. $3^{2x} = 7$

$\log 3^{2x} = \log 7$

$2x \log 3 = \log 7$

$x = \dfrac{\log 7}{2\log 3} \approx 0.8856$

98. $6^{3x} = 5$

$\log 6^{3x} = \log 5$

$3x \log 6 = \log 5$

$x = \dfrac{\log 5}{3\log 6} \approx 0.2994$

99.
$$3^{2x+1} = 6$$
$$\log 3^{2x+1} = \log 6$$
$$(2x+1)\log 3 = \log 6$$
$$2x \log 3 + \log 3 = \log 6$$
$$2x \log 3 = \log 6 - \log 3$$
$$x = \frac{\log 6 - \log 3}{2 \log 3}$$
or
$$x = \frac{1}{2}\left(\frac{\log 6}{\log 3} - 1\right)$$
$$x \approx 0.3155$$

100.
$$4^{3x+2} = 9$$
$$\log 4^{3x+2} = \log 9$$
$$(3x+2)\log 4 = \log 9$$
$$3x \log 4 + 2 \log 4 = \log 9$$
$$3x \log 4 = \log 9 - 2 \log 4$$
$$x = \frac{\log 9 - 2 \log 4}{3 \log 4}$$
or
$$x = \frac{1}{3}\left(\frac{\log 9}{\log 4} - 2\right)$$
$$x \approx -0.1383$$

101.
$$5^{3x-5} = 4$$
$$\log 5^{3x-5} = \log 4$$
$$(3x-5)\log 5 = \log 4$$
$$3x \log 5 - 5 \log 5 = \log 4$$
$$3x \log 5 = \log 4 + 5 \log 5$$
$$x = \frac{\log 4 + 5 \log 5}{3 \log 5}$$
or
$$x = \frac{1}{3}\left(\frac{\log 4}{\log 5} + 5\right)$$
$$x \approx 1.9538$$

102.
$$8^{4x-2} = 3$$
$$\log 8^{4x-2} = \log 3$$
$$(4x-2)\log 8 = \log 3$$
$$4x \log 8 - 2 \log 8 = \log 3$$
$$4x \log 8 = \log 3 + 2 \log 8$$
$$x = \frac{\log 3 + 2 \log 8}{4 \log 8}$$
or
$$x = \frac{1}{4}\left(\frac{\log 3}{\log 8} + 2\right)$$
$$x \approx 0.6321$$

103.
$$5^{x-1} = \frac{1}{2}$$
$$\log 5^{x-1} = \log \frac{1}{2}$$
$$(x-1)\log 5 = \log \frac{1}{2}$$
$$x \log 5 - \log 5 = \log \frac{1}{2}$$
$$x \log 5 = \log \frac{1}{2} + \log 5$$
$$x = \frac{\log \frac{1}{2} + \log 5}{\log 5}$$
or
$$x = -\frac{\log 2}{\log 5} + 1$$
$$x \approx 0.5693$$

104.
$$4^{x+5} = \frac{2}{3}$$
$$\log 4^{x+5} = \log \frac{2}{3}$$
$$(x+5)\log 4 = \log \frac{2}{3}$$
$$x \log 4 + 5 \log 4 = \log \frac{2}{3}$$
$$x \log 4 = \log \frac{2}{3} - 5 \log 4$$
$$x = \frac{\log \frac{2}{3} - 5 \log 4}{\log 4}$$
or
$$x = \frac{\log \frac{2}{3}}{\log 4} - 5$$
$$x \approx -5.2925$$

105. $\log_5 2 + \log_5 x = 2$
$$\log_5 2x = 2$$
$$2x = 5^2$$
$$2x = 25$$
$$x = \frac{25}{2}$$

106. $\log_3 x + \log_3 10 = 2$
$$\log_3(10x) = 2$$
$$10x = 3^2$$
$$10x = 9$$
$$x = \frac{9}{10}$$

107. $\log(5x) - \log(x+1) = 4$
$$\log\frac{5x}{x+1} = 4$$
$$\frac{5x}{x+1} = 10^4$$
$$\frac{5x}{x+1} = 10,000$$
$$5x = 10,000x + 10,000$$
$$x = -1.0005$$

no solution, or $\varnothing$

108. $-\log_6(4x+7) + \log_6 x = 1$
$$\log_6\frac{x}{4x+7} = 1$$
$$\frac{x}{4x+7} = 6$$
$$x = 6(4x+7)$$
$$x = 24x + 42$$
$$x = -\frac{42}{23}$$

$-\frac{42}{23}$ is rejected since $\log_6\left(-\frac{42}{23}\right)$ is undefined.
There is no solution, or $\varnothing$.

109. $\log_2 x + \log_2 2x - 3 = 1$
$$\log_2(x \cdot 2x) = 4$$
$$2x^2 = 2^4$$
$$2x^2 = 16$$
$$x^2 = 8$$
$$x = \pm 2\sqrt{2}$$

$-2\sqrt{2}$ is rejected since $\log_2\left(-2\sqrt{2}\right)$ is undefined. The solution is $2\sqrt{2}$.

110. $\log_3(x^2 - 8x) = 2$
$$3^2 = x^2 - 8x$$
$$0 = x^2 - 8x - 9$$
$$0 = (x-9)(x+1)$$
$x = 9$ or $x = -1$

111. Let $y_0 = 27$, $y = 347$, $r = 11.4\% = 0.114$.
$$y = y_0 e^{kt}$$
$$347 = 27e^{(0.114)t}$$
$$\frac{347}{27} = e^{0.114t}$$
$$\ln\left(\frac{347}{27}\right) = 0.114t$$
$$\frac{1}{0.114}\ln\left(\frac{347}{27}\right) = t$$
$$22.4 \approx t$$
It took approximately 22.4 years.

112.
$$y = y_0 e^{kt}$$
$$70,000,000 = 65,822,000e^{0.004t}$$
$$\frac{70,000,000}{65,822,000} = e^{0.004t}$$
$$\ln\frac{70,000,000}{65,822,000} = 0.004t$$
$$t = \frac{1}{0.004}\ln\frac{70,000,000}{65,822,000} \approx 15.4$$
It will take approximately 15.4 years.

113.
$$y = y_0 e^{kt}$$
$$2(22,600,000) = 22,600,000e^{0.007t}$$
$$2 = e^{0.007t}$$
$$\ln 2 = 0.007t$$
$$t = \frac{\ln 2}{0.007} \approx 99.0$$
It will take approximately 99.0 years.

114.
$$y = y_0 t^{kt}$$
$$2(7,746,400) = 7,746,400e^{0.016t}$$
$$2 = e^{0.016t}$$
$$\ln 2 = 0.016t$$
$$t = \frac{\ln 2}{0.016} \approx 43.3$$
It will take approximately 43.3 years.

115.
$$A = P\left(1 + \frac{r}{n}\right)^{nt}$$
$$10{,}000 = 5000\left(1 + \frac{0.08}{4}\right)^{4t}$$
$$2 = (1.02)^{4t}$$
$$\log 2 = \log 1.02^{4t}$$
$$\log 2 = 4t \log 1.02$$
$$t = \frac{\log 2}{4 \log 1.02} \approx 8.8$$

It will take 8.8 years.

116.
$$A = P\left(1 + \frac{r}{n}\right)^{nt}$$
$$10{,}000 = 6000\left(1 + \frac{0.06}{12}\right)^{12t}$$
$$\frac{5}{3} = (1.005)^{12t}$$
$$\log \frac{5}{3} = \log 1.005^{12t}$$
$$\log \frac{5}{3} = 12t \log 1.005$$
$$t = \frac{1}{12}\left(\frac{\log\left(\frac{5}{3}\right)}{\log(1.005)}\right) \approx 8.5$$

It was invested for approximately 8.5 years.

117. $Y_1 = e^x,\ Y_2 = 2$

$x \approx 0.69$

118. $Y_1 = 10^{0.3x},\ Y_2 = 7$

$x \approx 2.82$

119. $3^x = \dfrac{1}{81}$
$$3^x = 3^{-4}$$
$$x = -4$$

120. $7^{4x} = 49$
$$7^{4x} = 7^2$$
$$4x = 2$$
$$x = \frac{1}{2}$$

121. $8^{3x-2} = 32$
$$(2^3)^{(3x-2)} = 2^5$$
$$2^{9x-6} = 2^5$$
$$9x - 6 = 5$$
$$9x = 11$$
$$x = \frac{11}{9}$$

122. $9^{x-2} = 27$
$$(3^2)^{(x-2)} = 3^3$$
$$3^{2x-4} = 3^3$$
$$2x - 4 = 3$$
$$2x = 7$$
$$x = \frac{7}{2}$$

123. $\log_4 4 = x$
$$4^x = 4^1$$
$$x = 1$$

124. $\log_3 x = 4$
$$3^4 = x$$
$$81 = x$$

125. $\log_5(x^2 - 4x) = 1$
$$5^1 = x^2 - 4x$$
$$0 = x^2 - 4x - 5$$
$$0 = (x - 5)(x + 1)$$
$$x - 5 = 0 \quad \text{or} \quad x + 1 = 0$$
$$x = 5 \qquad\qquad x = -1$$

Both check, so the solutions are 5 and -1.

126. $\log_4(3x - 1) = 2$
$$4^2 = 3x - 1$$
$$16 + 1 = 3x$$
$$\frac{17}{3} = x$$

127. $\ln x = -3.2$
$$e^{\ln x} = e^{-3.2}$$
$$x = e^{-3.2}$$

625

128. $\log_5 x + \log_5 10 = 2$

$\qquad \log_5 (10x) = 2$

$\qquad\qquad 5^2 = 10x$

$\qquad\qquad \dfrac{25}{10} = x$

$\qquad\qquad \dfrac{5}{2} = x$

129. $\ln x - \ln 2 = 1$

$\qquad \ln \dfrac{x}{2} = 1$

$\qquad e^{\ln \frac{x}{2}} = e^1$

$\qquad \dfrac{x}{2} = e$

$\qquad x = 2e$

130. $\log_6 x - \log_6 (4x+7) = 1$

$\qquad \log_6 \dfrac{x}{4x+7} = 1$

$\qquad\qquad 6^1 = \dfrac{x}{4x+7}$

$\qquad 24x + 42 = x$

$\qquad\qquad 23x = -42$

$\qquad\qquad x = -\dfrac{42}{23}$

$-\dfrac{42}{23}$ is rejected since $\log_6 \left(-\dfrac{42}{23}\right)$ is undefined.

There is no solution, or $\varnothing$.

Chapter 12 Test

1. $f(x) = x$ and $g(x) = 2x - 3$

$(f \cdot g)(x) = f(x) \cdot g(x) = x(2x-3) = 2x^2 - 3x$

2. $f(x) = x$ and $g(x) = 2x - 3$

$(f - g)(x) = f(x) - g(x)$

$\qquad\qquad = x - (2x-3)$

$\qquad\qquad = -x + 3$

$\qquad\qquad = 3 - x$

3. $(f \circ h)(0) = f(h(0)) = f(5) = 5$

4. $(g \circ f)(x) = g(f(x)) = g(x) = x - 7$

5. $(g \circ h)(x) = g(h(x))$

$\qquad\qquad = g(x^2 - 6x + 5)$

$\qquad\qquad = x^2 - 6x + 5 - 7$

$\qquad\qquad = x^2 - 6x - 2$

6. $f(x) = 7x - 14$, $f^{-1}(x) = \dfrac{x+14}{7}$

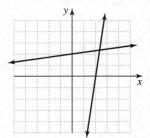

7. The graph represents a one-to-one function.

8. The graph does not represent a one-to-one function.

9. $y = 6 - 2x$ is one-to-one.

$\qquad x = 6 - 2y$

$\qquad 2y = -x + 6$

$\qquad y = \dfrac{-x+6}{2}$

$\qquad f^{-1}(x) = \dfrac{-x+6}{2}$

10. $f = \{(0,0), (2,3), (-1,5)\}$ is one-to-one.

$f^{-1} = \{(0,0), (3,2), (5,-1)\}$

11. The function is not one-to-one.

12. $\log_3 6 + \log_3 4 = \log_3 (6 \cdot 4) = \log_3 24$

13. $\log_5 x + 3\log_5 x - \log_5 (x+1)$

$= 4\log_5 x - \log_5 (x+1)$

$= \log_5 x^4 - \log_5 (x+1)$

$= \log_5 \dfrac{x^4}{x+1}$

14. $\log_6 \dfrac{2x}{y^3} = \log_6 2x - \log_6 y^3$

$\qquad\qquad = \log_6 2 + \log_6 x - 3\log_6 y$

15. $\log_b \left(\dfrac{3}{25}\right) = \log_b 3 - \log_b 25$

$\qquad\qquad = \log_b 3 - \log_b 5^2$

$\qquad\qquad = \log_b 3 - 2\log_b 5$

$\qquad\qquad = 0.79 - 2(1.16)$

$\qquad\qquad = -1.53$

16. $\log_7 8 = \dfrac{\ln 8}{\ln 7} \approx 1.0686$

17. $8^{x-1} = \dfrac{1}{64}$

$8^{x-1} = 8^{-2}$

$x - 1 = -2$

$x = -1$

18. $3^{2x+5} = 4$

$\log 3^{2x+5} = \log 4$

$(2x+5)\log 3 = \log 4$

$2x = \dfrac{\log 4}{\log 3} - 5$

$x = \dfrac{1}{2}\left(\dfrac{\log 4}{\log 3} - 5\right)$

$x \approx -1.8691$

19. $\log_3 x = -2$

$x = 3^{-2}$

$x = \dfrac{1}{9}$

20. $\ln \sqrt{e} = x$

$\ln e^{1/2} = x$

$\dfrac{1}{2} = x$

21. $\log_8(3x-2) = 2$

$3x - 2 = 8^2$

$3x - 2 = 64$

$3x = 66$

$x = \dfrac{66}{3} = 22$

22. $\log_5 x + \log_5 3 = 2$

$\log_5(3x) = 2$

$3x = 5^2$

$3x = 25$

$x = \dfrac{25}{3}$

23. $\log_4(x+1) - \log_4(x-2) = 3$

$\log_4 \dfrac{x+1}{x-2} = 3$

$\dfrac{x+1}{x-2} = 4^3$

$\dfrac{x+1}{x-2} = 64$

$x + 1 = 64x - 128$

$129 = 63x$

$\dfrac{129}{63} = x$

$\dfrac{43}{21} = x$

24. $\ln(3x+7) = 1.31$

$3x + 7 = e^{1.31}$

$3x = e^{1.31} - 7$

$x = \dfrac{e^{1.31} - 7}{3} \approx -1.0979$

25. $y = \left(\dfrac{1}{2}\right)^x + 1$

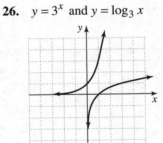

26. $y = 3^x$ and $y = \log_3 x$

27. $A = \left(1 + \dfrac{r}{n}\right)^{nt}, P = 4000, t = 3, r = 0.09,$

    and $n = 12$

    $A = 4000\left(1 + \dfrac{0.09}{12}\right)^{12(3)}$

    $= 4000(1.0075)^{36}$

    ≈ 5234.58

    \$5234.58 will be in the account.

28. $A = \left(1 + \dfrac{r}{n}\right)^{nt}, P = 2000, A = 3000$

    $r = 0.07, n = 2$

    $3000 = 2000\left(1 + \dfrac{0.07}{2}\right)^{2t}$

    $1.5 = (1.035)^{2t}$

    $\log 1.5 = \log 1.035^{2t}$

    $\log 1.5 = 2t \log 1.035$

    $t = \dfrac{\log 1.5}{2 \log 1.035} \approx 5.9$

    It would take 6 years.

29. Let $P = 3000, t = 10$.

    Semiannually: Let $n = 2, r = 0.065$.

    $A = P\left(1 + \dfrac{r}{n}\right)^{nt}$

    $A = 3000\left(1 + \dfrac{0.065}{2}\right)^{2(10)}$

    $A \approx 5688$

    Monthly: Let $n = 12, r = 0.06$.

    $A = 3000\left(1 + \dfrac{0.06}{12}\right)^{12(10)}$

    $A \approx 5458$

    The better investment is the 6.5% investment by
    \$5688 − \$5458 = \$230.

30. $C = 150{,}000, r = 0.02, x = 20$

    $y = C(1 - r)^x$

    $y = 150{,}000(1 - 0.02)^{20}$

    $y \approx 100{,}141$

    The predicted population in 20 years is 100,141.

31. $C = 57{,}000, r = 0.026, x = 5$

    $y = C(1 + r)^x$

    $y = 57{,}000(1 + 0.026)^5$

    $y \approx 64{,}805$

    In 5 years, the predicted population is
    64,805 animals.

32. $C = 400, r = 0.062, y = 1000$

    $y = C(1 + r)^x$

    $1000 = 400(1 + 0.062)^x$

    $\dfrac{1000}{400} = 1.062^x$

    $\ln\left(\dfrac{1000}{400}\right) = x \ln 1.062$

    $x = \dfrac{\ln\left(\frac{1000}{400}\right)}{\ln 1.062}$

    $x \approx 15$

    It will take about 15 years to reach their goal.

33. Let $t = 0.5$.

    $R(t) = 2.7^{-(1/3)t}$

    $R(0.5) = 2.7^{-(1/3)(0.5)} \approx 0.85$

    The probability is about 85%.

34. Let $t = 2$.

    $R(t) = 2.7^{-(1/3)(2)} \approx 0.52$

    The probability is about 52%.

Chapter 12 Cumulative Review

1. a. $\dfrac{4}{5} \div \dfrac{5}{16} = \dfrac{4}{5} \cdot \dfrac{16}{5} = \dfrac{4 \cdot 16}{5 \cdot 5} = \dfrac{64}{25}$

   b. $\dfrac{7}{10} \div 14 = \dfrac{7}{10} \div \dfrac{14}{1} = \dfrac{7}{10} \cdot \dfrac{1}{14} = \dfrac{7 \cdot 1}{10 \cdot 2 \cdot 7} = \dfrac{1}{20}$

   c. $\dfrac{3}{8} \div \dfrac{3}{10} = \dfrac{3}{8} \cdot \dfrac{10}{3} = \dfrac{3 \cdot 2 \cdot 5}{2 \cdot 4 \cdot 3} = \dfrac{5}{4}$

2. $\dfrac{1}{3}(x - 2) = \dfrac{1}{4}(x + 1)$

   $4(x - 2) = 3(x + 1)$

   $4x - 8 = 3x + 3$

   $x = 11$

3. $f(x) = x^2$

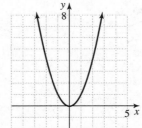

4. $y = f(x) = -3x + 4$, $m = -3$

Perpendicular line: $m = \dfrac{1}{3}$, through $(-2, 6)$

$$y - y_1 = m(x - x_1)$$
$$y - 6 = \frac{1}{3}[x - (-2)]$$
$$y - 6 = \frac{1}{3}x + \frac{2}{3}$$
$$y = \frac{1}{3}x + \frac{20}{3}$$
$$f(x) = \frac{1}{3}x + \frac{20}{3}$$

5. Equation 2 is twice the opposite of equation 1 and equation 3 is one-half of equation 1. Therefore, the system is dependent. The solution is $\{(x, y, z) | x - 5y - 2z = 6\}$.

6. The angles labeled $y°$ and $(x - 40)°$ are alternate interior angles, so $y = x - 40$. The angles labeled $x°$ and $y°$ are supplementary, so $x + y = 180$.

$$\begin{cases} y = x - 40 \\ x + y = 180 \end{cases}$$

Replace y with $x - 40$ in the second equation.
$$x + (x - 40) = 180$$
$$2x = 220$$
$$x = 110$$
$$y = x - 40 = 110 - 40 = 70$$

7. a. $-3 + [(-2 - 5) - 2] = -3 + [-7 - 2]$
$$= -3 + [-9]$$
$$= -12$$

    b. $2^3 - |10| + [-6 - (-5)] = 2^3 - |10| + [-6 + 5]$
$$= 2^3 - |10| + (-1)$$
$$= 8 - 10 - 1$$
$$= -2 - 1$$
$$= -3$$

8. a. $(4a^3)^2 = 4^2(a^3)^2 = 16a^6$

    b. $\left(-\dfrac{2}{3}\right)^3 = \dfrac{(-2)^3}{3^3} = \dfrac{-8}{27} = -\dfrac{8}{27}$

    c. $\left(\dfrac{4a^5}{b^3}\right)^3 = \dfrac{4^3(a^5)^3}{(b^3)^3} = \dfrac{64a^{15}}{b^9}$

    d. $\left(\dfrac{3^{-2}}{x}\right)^{-3} = \dfrac{(3^{-2})^{-3}}{x^{-3}} = \dfrac{3^6}{x^{-3}} = 729x^3$

    e. $(a^{-2}b^3c^{-4})^{-2} = (a^{-2})^{-2}(b^3)^{-2}(c^{-4})^{-2}$
$$= a^4 b^{-6} c^8$$
$$= \frac{a^4 c^8}{b^6}$$

9. a. $C(100) = \dfrac{2.6(100) + 10,000}{100}$
$$= 102.60$$
The cost is $102.60 per disc for 100 discs.

    b. $C(1000) = \dfrac{2.6(1000) + 10,000}{1000}$
$$= 12.60$$
The cost is $12.60 per disc for 1000 discs.

10. a. $(3x - 1)^2 = (3x)^2 - 2(3x)(1) + 1^2$
$$= 9x^2 - 6x + 1$$

    b. $\left(\dfrac{1}{2}x + 3\right)\left(\dfrac{1}{2}x - 3\right) = \left(\dfrac{1}{2}x\right)^2 - 3^2$
$$= \frac{1}{4}x^2 - 9$$

    c. $(2x - 5)(6x + 7) = 12x^2 + 14x - 30x - 35$
$$= 12x^2 - 16x - 35$$

11. $12a - 8a = 10 + 2a - 13 - 7$
$$4a = 2a - 10$$
$$2a = -10$$
$$a = -5$$

12. $\dfrac{5}{x - 2} + \dfrac{3}{x^2 + 4x + 4} - \dfrac{6}{x + 2}$

$$= \frac{5}{x - 2} + \frac{3}{(x + 2)^2} - \frac{6}{x + 2}$$

$$= \frac{5(x + 2)^2 + 3(x - 2) - 6(x - 2)(x + 2)}{(x - 2)(x + 2)(x + 2)}$$

$$= \frac{-x^2 + 23x + 38}{(x - 2)(x + 2)^2}$$

13. $\dfrac{8x^2y^2 - 16xy + 2x}{4xy} = \dfrac{8x^2y^2}{4xy} - \dfrac{16xy}{4xy} + \dfrac{2x}{4xy}$

$\qquad\qquad\qquad\qquad = 2xy - 4 + \dfrac{1}{2y}$

14. a. $\dfrac{\frac{a}{5}}{\frac{a-1}{10}} = \dfrac{a}{5} \cdot \dfrac{10}{a-1} = \dfrac{2a}{a-1}$

b. $\dfrac{\frac{3}{2+a} + \frac{6}{2-a}}{\frac{5}{a+2} - \frac{1}{a-2}} = \dfrac{\frac{3}{a+2} - \frac{6}{a-2}}{\frac{5}{a+2} - \frac{1}{a-2}}$

Multiply the numerator and the denominator by $(a+2)(a-2)$.

$\dfrac{3(a-2) - 6(a+2)}{5(a-2) - 1(a+2)} = \dfrac{3a - 6 - 6a - 12}{5a - 10 - a - 2}$

$\qquad\qquad\qquad\qquad = \dfrac{-3a - 18}{4a - 12}$

c. $\dfrac{x^{-1} + y^{-1}}{xy} = \dfrac{\frac{1}{x} + \frac{1}{y}}{xy} = \dfrac{\left(\frac{1}{x} + \frac{1}{y}\right)xy}{(xy)(xy)} = \dfrac{y+x}{x^2y^2}$

15. $3m^2 - 24m - 60 = 3(m^2 - 8m - 20)$

$\qquad\qquad\qquad\quad = 3(m+2)(m-10)$

16. $5x^2 - 85x + 350 = 5(x^2 - 17x + 70)$

$\qquad\qquad\qquad\quad = 5(x-10)(x-7)$

17. $\dfrac{3x^2 + 2x}{x-1} - \dfrac{10x - 5}{x-1} = \dfrac{3x^2 + 2x - (10x - 5)}{x-1}$

$\qquad\qquad\qquad\qquad = \dfrac{3x^2 + 2x - 10x + 5}{x-1}$

$\qquad\qquad\qquad\qquad = \dfrac{3x^2 - 8x + 5}{x-1}$

$\qquad\qquad\qquad\qquad = \dfrac{(3x-5)(x-1)}{x-1}$

$\qquad\qquad\qquad\qquad = 3x - 5$

18. $\begin{array}{r|rrr} 2 & 8 & -12 & -7 \\ & & 16 & 8 \\ \hline & 8 & 4 & 1 \end{array}$

Solution: $8x + 4 + \dfrac{1}{x-2}$

19. a. $\sqrt[4]{81} = \sqrt[4]{3^4} = 3$

b. $\sqrt[5]{-243} = \sqrt[5]{(-3)^5} = -3$

c. $-\sqrt{25} = -\sqrt{5^2} = -5$

d. $\sqrt[4]{-81}$ is not a real number.

e. $\sqrt[3]{64x^3} = \sqrt[3]{4^3 x^3} = 4x$

20. $\dfrac{1}{a+5} = \dfrac{1}{3a+6} - \dfrac{a+2}{a^2 + 7a + 10}$

$\dfrac{1}{a+5} = \dfrac{1}{3(a+2)} - \dfrac{a+2}{(a+2)(a+5)}$

$3(a+2) = a + 5 - 3(a+2)$

$3a + 6 = a + 5 - 3a - 6$

$5a = -7$

$a = -\dfrac{7}{5}$

21. a. $\sqrt{x} \cdot \sqrt[4]{x} = x^{1/2} \cdot x^{1/4} = x^{3/4} = \sqrt[4]{x^3}$

b. $\dfrac{\sqrt{x}}{\sqrt[3]{x}} = \dfrac{x^{1/2}}{x^{1/3}} = x^{\frac{1}{2} - \frac{1}{3}} = x^{1/6} = \sqrt[6]{x}$

c. $\sqrt[3]{3} \cdot \sqrt{2} = 3^{1/3} \cdot 2^{1/2}$

$\qquad\qquad\quad = 3^{2/6} \cdot 2^{3/6}$

$\qquad\qquad\quad = 9^{1/6} \cdot 8^{1/6}$

$\qquad\qquad\quad = 72^{1/6}$

$\qquad\qquad\quad = \sqrt[6]{72}$

22. $y = kx$

$\dfrac{1}{2} = 12k$

$k = \dfrac{1}{24},\; y = \dfrac{1}{24}x$

23. a. $\sqrt{3}\left(5 + \sqrt{30}\right) = 5\sqrt{3} + \sqrt{90} = 5\sqrt{3} + 3\sqrt{10}$

b. $\left(\sqrt{5} - \sqrt{6}\right)\left(\sqrt{7} + 1\right) = \sqrt{35} + \sqrt{5} - \sqrt{42} - \sqrt{6}$

c. $\left(7\sqrt{x} + 5\right)\left(3\sqrt{x} - \sqrt{5}\right)$

$\qquad = 21x - 7\sqrt{5x} + 15\sqrt{x} - 5\sqrt{5}$

d. $\left(4\sqrt{3} - 1\right)^2$

$\qquad = \left(4\sqrt{3}\right)^2 - 2\left(4\sqrt{3}\right)(1) + 1^2$

$\qquad = 16 \cdot 3 - 8\sqrt{3} + 1$

$\qquad = 49 - 8\sqrt{3}$

e. $\left(\sqrt{2x}-5\right)\left(\sqrt{2x}+5\right)=\left(\sqrt{2x}\right)^2-5^2$
$$=2x-25$$

f. $\left(\sqrt{x-3}+5\right)^2=\left(\sqrt{x-3}\right)^2+2\sqrt{x-3}(5)+5^2$
$$=x-3+10\sqrt{x-3}+25$$
$$=x+22+10\sqrt{x-3}$$

24. a. $\sqrt{9}=\sqrt{3^2}=3$

b. $\sqrt[3]{-27}=\sqrt[3]{(-3)^3}=-3$

c. $\sqrt{\dfrac{9}{64}}=\sqrt{\left(\dfrac{3}{8}\right)^2}=\dfrac{3}{8}$

d. $\sqrt[4]{x^{12}}=x^3$

e. $\sqrt[3]{-125y^6}=-5y^2$

25. $\dfrac{\sqrt[4]{x}}{\sqrt[4]{81y^5}}=\dfrac{\sqrt[4]{x}}{\sqrt[4]{81y^5}}\cdot\dfrac{\sqrt[4]{y^3}}{\sqrt[4]{y^3}}=\dfrac{\sqrt[4]{xy^3}}{3y^2}$

26. a. $a^{1/4}(a^{3/4}-a^8)=a^{4/4}-a^{33/4}=a-a^{39/4}$

b. $(x^{1/2}-3)(x^{1/2}+5)$
$$=x^{2/2}+5x^{1/2}-3x^{1/2}-15$$
$$=x+2x^{1/2}-15$$

27. $\sqrt{4-x}=x-2$
$$\left(\sqrt{4-x}\right)^2=(x-2)^2$$
$$4-x=x^2-4x+4$$
$$0=x^2-3x$$
$$0=x(x-3)$$
$$x=0 \quad \text{or} \quad x-3=0$$
$$x=3$$
$x=0$ does not check, so the only solution is $x=3$.

28. a. $\dfrac{\sqrt{54}}{\sqrt{6}}=\sqrt{\dfrac{54}{6}}=\sqrt{9}=3$

b. $\dfrac{\sqrt{108a^2}}{3\sqrt{3}}=\dfrac{1}{3}\sqrt{\dfrac{108a^2}{3}}$
$$=\dfrac{1}{3}\sqrt{36a^2}$$
$$=\dfrac{1}{3}(6a)$$
$$=2a$$

c. $\dfrac{3\sqrt[3]{81a^5b^{10}}}{\sqrt[3]{3b^4}}=3\sqrt[3]{\dfrac{81a^5b^{10}}{3b^4}}$
$$=3\sqrt[3]{27a^5b^6}$$
$$=9ab^2\sqrt[3]{a^2}$$

29. $3x^2-9x+8=0$
$$x^2-3x+\dfrac{8}{3}=0$$
$$x^2-3x=-\dfrac{8}{3}$$
$$x^2-3x+\left(\dfrac{-3}{2}\right)^2=-\dfrac{8}{3}+\left(\dfrac{-3}{2}\right)^2$$
$$x^2-3x+\dfrac{9}{4}=-\dfrac{8}{3}+\dfrac{9}{4}$$
$$\left(x-\dfrac{3}{2}\right)^2=-\dfrac{5}{12}$$
$$x-\dfrac{3}{2}=\pm\sqrt{-\dfrac{5}{12}}$$
$$x-\dfrac{3}{2}=\pm\dfrac{i\sqrt{5}}{2\sqrt{3}}$$
$$x-\dfrac{3}{2}=\pm\dfrac{i\sqrt{15}}{6}$$
$$x=\dfrac{3}{2}\pm\dfrac{i\sqrt{15}}{6}$$
$$=\dfrac{9}{6}\pm\dfrac{i\sqrt{15}}{6}$$
$$=\dfrac{9\pm i\sqrt{15}}{6}$$

The solutions are $\dfrac{9+i\sqrt{15}}{6}$ and $\dfrac{9-i\sqrt{15}}{6}$ or

$\dfrac{3}{2}+\dfrac{\sqrt{15}}{6}i$ and $\dfrac{3}{2}-\dfrac{\sqrt{15}}{6}i$.

30. a.

$$\frac{\sqrt{20}}{3} + \frac{\sqrt{5}}{4} = \frac{2\sqrt{5}}{3} + \frac{\sqrt{5}}{4}$$
$$= \frac{8\sqrt{5} + 3\sqrt{5}}{12}$$
$$= \frac{11\sqrt{5}}{12}$$

b.

$$\sqrt[3]{\frac{24x}{27}} - \frac{\sqrt[3]{3x}}{2} = \frac{2\sqrt[3]{3x}}{3} - \frac{\sqrt[3]{3x}}{2}$$
$$= \frac{4\sqrt[3]{3x} - 3\sqrt[3]{3x}}{6}$$
$$= \frac{\sqrt[3]{3x}}{6}$$

31.

$$\frac{3x}{x-2} - \frac{x+1}{x} = \frac{6}{x(x-2)}$$
$$3x(x) - (x+1)(x-2) = 6$$
$$3x^2 - x^2 + x + 2 = 6$$
$$2x^2 + x - 4 = 0$$
$$a = 2, b = 1, c = -4$$
$$x = \frac{-1 \pm \sqrt{1^2 - 4(2)(-4)}}{2(2)} = \frac{-1 \pm \sqrt{33}}{4}$$

32.

$$\sqrt[3]{\frac{27}{m^4 n^8}} = \frac{\sqrt[3]{27}}{\sqrt[3]{m^4 n^8}}$$
$$= \frac{3}{mn^2 \sqrt[3]{mn^2}}$$
$$= \frac{3 \cdot \sqrt[3]{m^2 n}}{mn^2 \sqrt[3]{mn^2} \cdot \sqrt[3]{m^2 n}}$$
$$= \frac{3\sqrt[3]{m^2 n}}{m^2 n^3}$$

33. $x^2 - 4x \le 0$
$$x(x-4) = 0$$
$$x = 0, x = 4$$

Region	Test Point	$x(x-4) \le 0$	Result
$x < 0$	$x = -1$	$(-1)(-5) \le 0$	False
$0 < x < 4$	$x = 2$	$2(-2) \le 0$	True
$x > 4$	$x = 5$	$5(1) \le 0$	False

Solution: [0, 4]

34.

$$c^2 = a^2 + b^2$$
$$8^2 = 4^2 + b^2$$
$$64 = 16 + b^2$$
$$48 = b^2$$
$$\pm 4\sqrt{3} = b$$

$b > 0$ so the length is $4\sqrt{3}$ inches.

35. $F(x) = (x-3)^2 + 1$

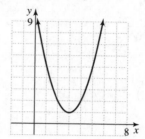

36. a. $i^8 = (i^2)^4 = (-1)^4 = 1$

b. $i^{21} = i(i^{20}) = i$

c. $i^{42} = i^2(i^{40}) = i^2 = -1$

d. $i^{-13} = \frac{1}{i^{13}} = \frac{1}{i(i^{12})} = \frac{1}{i} = \frac{i}{i^2} = -i$

37.

$$\frac{45}{x} = \frac{5}{7}$$
$$45 \cdot 7 = 5x$$
$$315 = 5x$$
$$63 = x$$

38. $4x^2 + 8x - 1 = 0$

$x^2 + 2x - \dfrac{1}{4} = 0$

$x^2 + 2x = \dfrac{1}{4}$

$x^2 + 2x + \left(\dfrac{2}{2}\right)^2 = \dfrac{1}{4} + \left(\dfrac{2}{2}\right)^2$

$x^2 + 2x + 1 = \dfrac{1}{4} + 1$

$(x+1)^2 = \dfrac{5}{4}$

$x + 1 = \pm\sqrt{\dfrac{5}{4}}$

$x + 1 = \pm\dfrac{\sqrt{5}}{2}$

$x = -1 \pm \dfrac{\sqrt{5}}{2}$

$\quad = \dfrac{-2 \pm \sqrt{5}}{2}$

The solutions are $\dfrac{-2 + \sqrt{5}}{2}$ and $\dfrac{-2 - \sqrt{5}}{2}$.

39. $f(x) = x + 3$

$y = x + 3$

$x = y + 3$

$y = x - 3$

$f^{-1}(x) = x - 3$

40. $\left(x - \dfrac{1}{2}\right)^2 = \dfrac{x}{2}$

$x^2 - x + \dfrac{1}{4} = \dfrac{1}{2}x$

$x^2 - \dfrac{3}{2}x + \dfrac{1}{4} = 0$

$4x^2 - 6x + 1 = 0$

$a = 4, \, b = -6, \, c = 1$

$x = \dfrac{-(-6) \pm \sqrt{(-6)^2 - 4(4)(1)}}{2(4)}$

$\quad = \dfrac{6 \pm \sqrt{20}}{8}$

$\quad = \dfrac{6 \pm 2\sqrt{5}}{8}$

$\quad = \dfrac{3 \pm \sqrt{5}}{4}$

The solutions are $\dfrac{3 + \sqrt{5}}{4}$ and $\dfrac{3 - \sqrt{5}}{4}$.

41. a. $\log_4 16 = \log_4 4^2 = 2$

b. $\log_{10} \dfrac{1}{10} = \log_{10} 10^{-1} = -1$

c. $\log_9 3 = \log_9 9^{1/2} = \dfrac{1}{2}$

42. $f(x) = -(x+1)^2 + 1$

Vertex: $(-1, 1)$

Axis of symmetry: $x = -1$

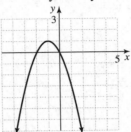

Chapter 13

1. $x = \dfrac{1}{2}y^2$; $a = \dfrac{1}{2}$, $h = 0$, $k = 0$; vertex: (0, 0)

x	y
2	−2
$\dfrac{1}{2}$	−1
0	0
$\dfrac{1}{2}$	1
2	2

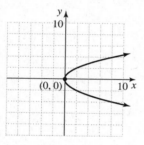

2. $x = -2(y+4)^2 - 1$; $a = -2$, $h = -1$, $k = -4$; vertex: (−1, −4)

x	y
−9	−6
3	−5
−1	−4
−3	−3
−9	−2

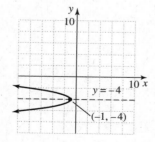

3.
$$y = -x^2 + 4x + 6$$
$$y - 6 = -x^2 + 4x$$
$$y - 6 = -(x^2 - 4x)$$
$$y - 6 - (+4) = -(x^2 - 4x + 4)$$
$$y - 10 = -(x - 2)^2$$
$$y = -(x - 2)^2 + 10$$

$a = -1$, $h = 2$, $k = 10$
vertex: (2, 10)

x	y
−1	1
0	6
1	9
2	10
3	9
4	6
5	1

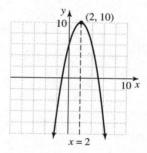

4. $x = 3y^2 + 6y + 4$
Find the vertex.
$$y = \frac{-b}{2a} = \frac{-6}{2(3)} = -1$$

$x = 3(-1)^2 + 6(-1) + 4 = 3 - 6 + 4 = 1$
vertex: (1, −1)
The axis of symmetry is the line $y = -1$.
Since $a > 0$, the parabola opens to the right.
$x = 3(0)^2 + 6(0) + 4 = 4$
The x-intercept is (4, 0).

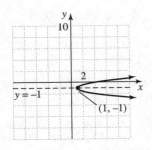

$y = -1$

$(1, -1)$

5.
$$x^2 + y^2 = 25$$
$$(x-0)^2 + (y-0)^2 = 5^2$$
center: (0, 0); radius = 5

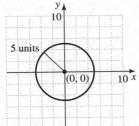

5 units

$(0, 0)$

6. $(x-3)^2 + (y+2)^2 = 4$

$h = 3, k = -2, r = \sqrt{4} = 2$

center: (3, –2)

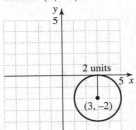

2 units

$(3, -2)$

7.
$$x^2 + y^2 + 6x - 2y = 6$$
$$(x^2 + 6x) + (y^2 - 2y) = 6$$
$$(x^2 + 6x + 9) + (y^2 - 2y + 1) = 6 + 9 + 1$$
$$(x+3)^2 + (y-1)^2 = 16$$

Center: (–3, 1); radius = $\sqrt{16} = 4$

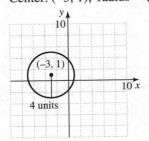

$(-3, 1)$

4 units

8. Center: (–2, –5); radius = 9

$$(x-h)^2 + (y-k)^2 = r^2$$

$h = -2, k = -5$, and $r = 9$.

The equation is $(x+2)^2 + (y+5)^2 = 81$.

Graphing Calculator Explorations

1. $x^2 + y^2 = 55$
$$y^2 = 55 - x^2$$
$$y = \pm\sqrt{55 - x^2}$$

2. $x^2 + y^2 = 20$
$$y^2 = 20 - x^2$$
$$y = \pm\sqrt{20 - x^2}$$

3. $5x^2 + 5y^2 = 50$
$$5y^2 = 50 - 5x^2$$
$$y^2 = 10 - x^2$$
$$y = \pm\sqrt{10 - x^2}$$

4. $6x^2 + 6y^2 = 105$
$$6y^2 = 105 - 6x^2$$
$$y^2 = 17.5 - x^2$$
$$y = \pm\sqrt{17.5 - x^2}$$

5. $2x^2 + 2y^2 - 34 = 0$

$$2y^2 = 34 - 2x^2$$
$$y^2 = 17 - x^2$$
$$y = \pm\sqrt{17 - x^2}$$

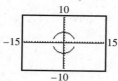

6. $4x^2 + 4y^2 - 48 = 0$

$$4y^2 = 48 - 4x^2$$
$$y^2 = 12 - x^2$$
$$y = \pm\sqrt{12 - x^2}$$

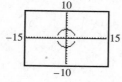

7. $7x^2 + 7y^2 - 89 = 0$

$$7y^2 = 89 - 7x^2$$
$$y^2 = \frac{89 - 7x^2}{7}$$
$$y = \pm\sqrt{\frac{89 - 7x^2}{7}}$$

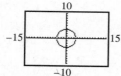

8. $3x^2 + 3y^2 - 35 = 0$

$$3y^2 = 35 - 3x^2$$
$$y^2 = \frac{35 - 3x^2}{3}$$
$$y = \pm\sqrt{\frac{35 - 3x^2}{3}}$$

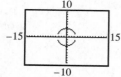

Vocabulary, Readiness & Video Check 13.1

1. The circle, parabola, ellipse, and hyperbola are called the <u>conic sections</u>.

2. For a parabola that opens upward the lowest point is the <u>vertex</u>.

3. A <u>circle</u> is the set of all points in a plane that are the same distance from a fixed point. The fixed point is called the <u>center</u>.

4. The midpoint of a diameter of a circle is the <u>center</u>.

5. The distance from the center of a circle to any point of the circle is called the <u>radius</u>.

6. Twice a circle's radius is its <u>diameter</u>.

7. No, their graphs don't pass the vertical line test.

8. $x^2 + y^2 = r^2$

9. The formula for the standard form of a circle identifies the center and radius, so you just need to substitute these values into this formula and simplify.

10. Since the standard form of a circle involves a squared binomial for both x and y, we need to complete the square on both x and y.

Exercise Set 13.1

1. $y = x^2 - 7x + 5$; $a = 1$, upward

3. $x = -y^2 - y + 2$; $a = -1$, to the left

5. $y = -x^2 + 2x + 1$; $a = -1$, downward

7. $x = 3y^2$

$$x = 3(y - 0)^2 + 0$$

$V(0, 0)$; opens to the right

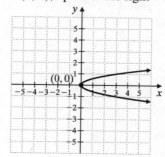

9. $x = -2y^2$

$x = -2(y-0)^2 + 0$

$V(0, 0)$; opens to the left

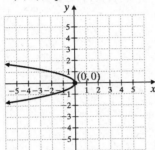

11. $y = -4x^2 = -4(x-0)^2 + 0$

$V(0, 0)$; opens downward

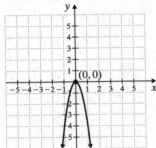

13. $x = (y-2)^2 + 3$

$V(3, 2)$; opens to the right

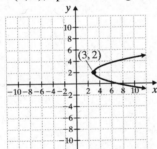

15. $y = -3(x-1)^2 + 5$

$V(1, 5)$; opens downward

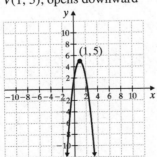

17. $x = y^2 + 6y + 8$

$x = y^2 + 6y + 9 + 8 - 9$

$x = (y+3)^2 - 1$

$V(-1, -3)$; opens to the right

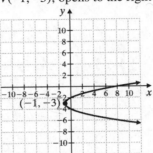

19. $y = x^2 + 10x + 20$

$y = x^2 + 10x + 25 + 20 - 25$

$y = (x+5)^2 - 5$

$V(-5, -5)$; opens upward

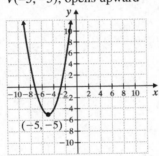

21. $x = -2y^2 + 4y + 6$

$x = -2(y^2 - 2y) + 6$

$x = -2(y^2 - 2y + 1) + 6 + 2$

$x = -2(y-1)^2 + 8$

$V(8, 1)$; opens to the left

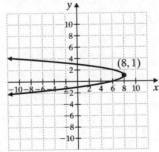

23.
$$x^2 + y^2 = 9$$
$$(x-0)^2 + (y-0)^2 = 3^2$$
$C(0, 0)$ and $r = 3$

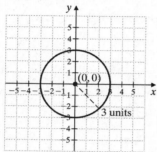

25.
$$x^2 + (y-2)^2 = 1$$
$$(x-0)^2 + (y-2)^2 = 1^2$$
$C(0, 2)$ and $r = 1$

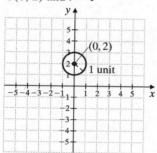

27.
$$(x-5)^2 + (y+2)^2 = 1$$
$$(x-5)^2 + (y+2)^2 = 1^2$$
$C(5, -2)$ and $r = 1$

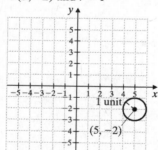

29.
$$x^2 + y^2 + 6y = 0$$
$$x^2 + y^2 + 6y + 9 = 0 + 9$$
$$(x-0)^2 + (y+3)^2 = 3^2$$
$C(0, -3)$ and $r = 3$

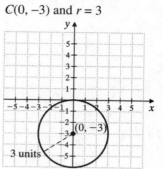

31.
$$x^2 + y^2 + 2x - 4y = 4$$
$$x^2 + 2x + 1 + y^2 - 4y + 4 = 4 + 1 + 4$$
$$(x+1)^2 + (y-2)^2 = 9$$
$$(x+1)^2 + (y-2)^2 = 3^2$$
$C(-1, 2)$ and $r = 3$

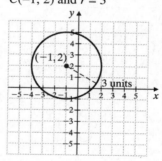

33.
$$(x+2)^2 + (y-3)^2 = 7$$
$$(x+2)^2 + (y-3)^2 = \left(\sqrt{7}\right)^2$$
$C(-2, 3)$ and $r = \sqrt{7}$

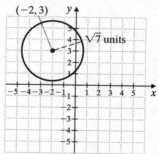

35.
$$x^2 + y^2 - 4x - 8y - 2 = 0$$
$$x^2 - 4x + 4 + y^2 - 8y + 16 = 2 + 4 + 16$$
$$(x-2)^2 + (y-4)^2 = 22$$
$$(x-2)^2 + (y-4)^2 = \left(\sqrt{22}\right)^2$$
$C(2, 4)$ and $r = \sqrt{22}$

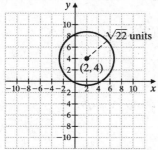

37.
$$3x^2 + 3y^2 = 75$$
$$x^2 + y^2 = 25$$
$$(x-0)^2 + (y-0)^2 = 5^2$$
$C(0, 0)$ and $r = 5$

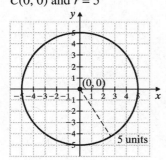

39.
$$6(x-4)^2 + 6(y-1)^2 = 24$$
$$(x-4)^2 + (y-1)^2 = 4$$
$$(x-4)^2 + (y-1)^2 = 2^2$$
$C(4, 1)$ and $r = 2$

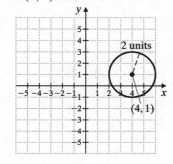

41.
$$4(x+1)^2 + 4(y-3)^2 = 12$$
$$(x+1)^2 + (y-3)^2 = 3$$
$$(x+1)^2 + (y-3)^2 = \left(\sqrt{3}\right)^2$$
$C(-1, 3)$ and $r = \sqrt{3}$

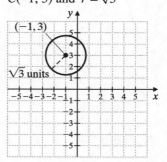

43. Center $(h, k) = (2, 3)$ and radius $r = 6$.
$$(x-h)^2 + (y-k)^2 = r^2$$
$$(x-2)^2 + (y-3)^2 = 6^2$$
$$(x-2)^2 + (y-3)^2 = 36$$

45. Center $(h, k) = (0, 0)$ and radius $r = \sqrt{3}$.
$$(x-h)^2 + (y-k)^2 = r^2$$
$$(x-0)^2 + (y-0)^2 = \left(\sqrt{3}\right)^2$$
$$x^2 + y^2 = 3$$

47. Center $(h, k) = (-5, 4)$ and radius $r = 3\sqrt{5}$.
$$(x-h)^2 + (y-k)^2 = r^2$$
$$[x-(-5)]^2 + (y-4)^2 = \left(3\sqrt{5}\right)^2$$
$$(x+5)^2 + (y-4)^2 = 45$$

49.
$$x = y^2 - 3$$
$$x = (y-0)^2 - 3$$
Vertex: $(-3, 0)$

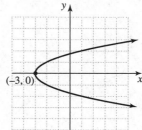

51. $y = (x-2)^2 - 2$

Vertex: (2, –2)

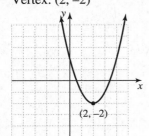

53. $x^2 + y^2 = 1$

Center: (0, 0), radius $r = \sqrt{1} = 1$

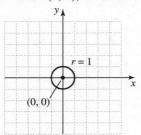

55. $x = (y+3)^2 - 1$

Vertex: (–1, –3)

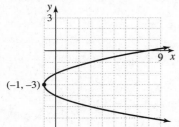

57. $(x-2)^2 + (y-2)^2 = 16$

Center: (2, 2), radius $r = \sqrt{16} = 4$

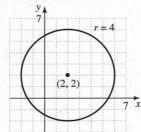

59. $x = -(y-1)^2$

Vertex: (0, 1)

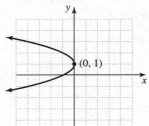

61. $(x-4)^2 + y^2 = 7$

Center: (4, 0), radius $r = \sqrt{7}$

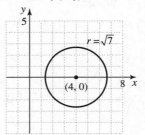

63. $y = 5(x+5)^2 + 3$

Vertex: (–5, 3)

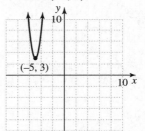

65. $\dfrac{x^2}{8} + \dfrac{y^2}{8} = 2$

$8\left(\dfrac{x^2}{8} + \dfrac{y^2}{8}\right) = 8(2)$

$x^2 + y^2 = 16$

Center: (0, 0), radius $r = \sqrt{16} = 4$

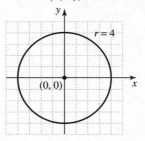

67.
$$y = x^2 + 7x + 6$$
$$y - 6 = x^2 + 7x$$
$$y - 6 + \frac{49}{4} = x^2 + 7x + \frac{49}{4}$$
$$y + \frac{25}{4} = \left(x + \frac{7}{2}\right)^2$$
$$y = \left(x + \frac{7}{2}\right)^2 - \frac{25}{4}$$

Vertex: $\left(-\frac{7}{2}, -\frac{25}{4}\right)$

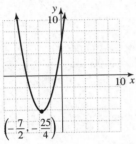

69.
$$x^2 + y^2 + 2x + 12y - 12 = 0$$
$$(x^2 + 2x) + (y^2 + 12y) = 12$$
$$(x^2 + 2x + 1) + (y^2 + 12y + 36) = 12 + 1 + 36$$
$$(x + 1)^2 + (y + 6)^2 = 49$$

Center: (–1, –6), radius $r = \sqrt{49} = 7$

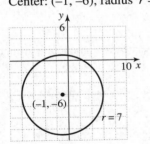

71.
$$x = y^2 + 8y - 4$$
$$x + 4 = y^2 + 8y$$
$$x + 4 + 16 = y^2 + 8y + 16$$
$$x + 20 = (y + 4)^2$$
$$x = (y + 4)^2 - 20$$

Vertex: (–20, –4)

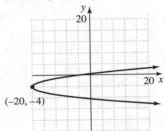

73.
$$x^2 - 10y + y^2 + 4 = 0$$
$$x^2 + (y^2 - 10y) = -4$$
$$x^2 + (y^2 - 10y + 25) = -4 + 25$$
$$x^2 + (y - 5)^2 = 21$$

Center: (0, 5), radius $r = \sqrt{21}$

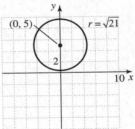

75.
$$x = -3y^2 + 30y$$
$$x = -3(y^2 - 10y)$$
$$x + [-3(25)] = -3(y^2 - 10y + 25)$$
$$x - 75 = -3(y - 5)^2$$
$$x = -3(y - 5)^2 + 75$$

Vertex: (75, 5)

77. $5x^2 + 5y^2 = 25$

$x^2 + y^2 = 5$

Center: (0, 0), radius $r = \sqrt{5}$

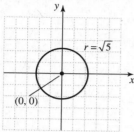

79.
$$y = 5x^2 - 20x + 16$$
$$y - 16 = 5(x^2 - 4x)$$
$$y - 16 + 5(4) = 5(x^2 - 4x + 4)$$
$$y + 4 = (x - 2)^2$$
$$y = (x - 2)^2 - 4$$

Vertex: (2, –4)

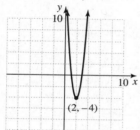

81. $y = 2x + 5$

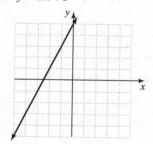

83. $y = 3$

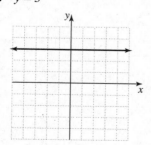

85. $\dfrac{1}{\sqrt{3}} = \dfrac{1 \cdot \sqrt{3}}{\sqrt{3} \cdot \sqrt{3}} = \dfrac{\sqrt{3}}{3}$

87. $\dfrac{4\sqrt{7}}{\sqrt{6}} = \dfrac{4\sqrt{7} \cdot \sqrt{6}}{\sqrt{6} \cdot \sqrt{6}} = \dfrac{4\sqrt{42}}{6} = \dfrac{2\sqrt{42}}{3}$

89. The vertex is (1, –5).

91. a. radius $= \dfrac{1}{2}$(diameter)

$= \dfrac{1}{2}$(33 meters)

$= 16.5$ meters

b. circumference $= \pi$(diameter)

$= \pi$(33 meters)

≈ 103.67 meters

c. $\dfrac{103.67}{30} \approx 3.5$ meters apart

d. center: (0, 16.5)

e. $(x - 0)^2 + (y - 16.5)^2 = 16.5^2$

$x^2 + (y - 16.5)^2 = 16.5^2$

93. a. The radius was one-half of the diameter, or 125 feet.

b. $264 - 250 = 14$
The wheel was 14 feet above the ground.

c. $125 + 14 = 139$
The center of the wheel was 139 feet from the ground.

d. From the drawing, the center was at (0, 139).

e. $(x - 0)^2 + (y - 139)^2 = 125^2$

$x^2 + (y - 139)^2 = 125^2$

95. answers may vary

97. Using A as $(0, 0)$, point B is at $(3, 1)$ and point C is at $(19, 13)$.

$$d(B, C) = \sqrt{(x_2 - x_1)^2 + (y_2 - y_1)^2}$$
$$= \sqrt{(19-3)^2 + (13-1)^2}$$
$$= \sqrt{16^2 + 12^2}$$
$$= \sqrt{256 + 144}$$
$$= \sqrt{400}$$
$$= 20$$

The distance across the lake is 20 meters.

99. $5x^2 + 5y^2 = 25$
$$5y^2 = 25 - 5x^2$$
$$y^2 = 5 - x^2$$
$$y = \pm\sqrt{5 - x^2}$$

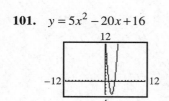

101. $y = 5x^2 - 20x + 16$

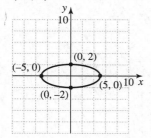

Section 13.2 Practice

1. $\dfrac{x^2}{25} + \dfrac{y^2}{4} = 1$

The equation is an ellipse with $a = 5$ and $b = 2$. The center is $(0, 0)$. The x-intercepts are $(5, 0)$ and $(-5, 0)$. The y-intercepts are $(2, 0)$ and $(-2, 0)$.

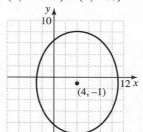

2. $9x^2 + 4y^2 = 36$
$$\frac{9x^2}{36} + \frac{4y^2}{36} = \frac{36}{36}$$
$$\frac{x^2}{4} + \frac{y^2}{9} = 1$$

This is an equation of an ellipse with $a = 2$ and $b = 3$. The ellipse has center $(0, 0)$, x-intercepts $(2, 0)$ and $(-2, 0)$, and y-intercepts $(3, 0)$ and $(-3, 0)$.

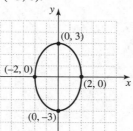

3. $\dfrac{(x-4)^2}{49} + \dfrac{(y+1)^2}{81} = 1$

This ellipse has center $(4, -1)$.
$a = 7$ and $b = 9$.
Find four points on the ellipse.
$(4 + 7, -1) = (11, -1)$
$(4 - 7, -1) = (-3, -1)$
$(4, -1 + 9) = (4, 8)$
$(4, -1 - 9) = (4, -10)$

4. $\dfrac{x^2}{9} - \dfrac{y^2}{16} = 1$

This is a hyperbola with $a = 3$ and $b = 4$. It has center $(0, 0)$ and x-intercepts $(3, 0)$ and $(-3, 0)$. The asymptotes pass through $(3, 4)$, $(3, -4)$, $(-3, 4)$, and $(-3, -4)$.

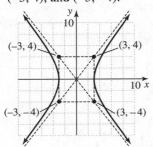

5. $9y^2 - 25x^2 = 225$

$$\dfrac{9y^2}{225} - \dfrac{25x^2}{225} = \dfrac{225}{225}$$

$$\dfrac{y^2}{25} - \dfrac{x^2}{9} = 1$$

This is a hyperbola with $a = 3$ and $b = 5$. The center is at $(0, 0)$ with y-intercepts $(0, 5)$ and $(0, -5)$. The asymptotes pass through $(3, 5)$, $(3, -5)$, $(-3, 5)$, and $(-3, -5)$.

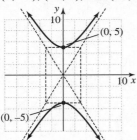

Graphing Calculator Explorations

1. $10x^2 + y^2 = 32$

$$y^2 = 32 - 10x^2$$

$$y = \pm\sqrt{32 - 10x^2}$$

2. $x^2 + 6y^2 = 35$

$$6y^2 = 35 - x^2$$

$$y^2 = \dfrac{35 - x^2}{6}$$

$$y = \pm\sqrt{\dfrac{35 - x^2}{6}}$$

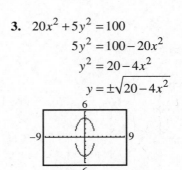

3. $20x^2 + 5y^2 = 100$

$$5y^2 = 100 - 20x^2$$

$$y^2 = 20 - 4x^2$$

$$y = \pm\sqrt{20 - 4x^2}$$

4. $4y^2 + 12x^2 = 48$

$$4y^2 = 48 - 12x^2$$

$$y^2 = 12 - 3x^2$$

$$y = \pm\sqrt{12 - 3x^2}$$

5. $7.3x^2 + 15.5y^2 = 95.2$

$$15.5y^2 = 95.2 - 7.3x^2$$

$$y^2 = \dfrac{95.2 - 7.3x^2}{15.5}$$

$$y = \pm\sqrt{\dfrac{95.2 - 7.3x^2}{15.5}}$$

6. $18.8x^2 + 36.1y^2 = 205.8$

$$36.1y^2 = 205.8 - 18.8x^2$$

$$y^2 = \frac{205.8 - 18.8x^2}{36.1}$$

$$y = \pm\sqrt{\frac{205.8 - 18.8x^2}{36.1}}$$

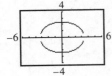

Vocabulary, Readiness & Video Check 13.2

1. A <u>hyperbola</u> is the set of points in a plane such that the absolute value of the differences of their distances from two fixed points is constant.

2. An <u>ellipse</u> is the set of points in a plane such that the sum of their distances from two fixed points is constant.

3. The two fixed points are each called a <u>focus</u>.

4. The point midway between the foci is called the <u>center</u>.

5. The graph of $\frac{x^2}{a^2} - \frac{y^2}{b^2} = 1$ is a <u>hyperbola</u> with center <u>(0, 0)</u> and x-intercepts of <u>(a, 0) and (−a, 0)</u>.

6. The graph of $\frac{x^2}{a^2} + \frac{y^2}{b^2} = 1$ is an <u>ellipse</u> with center <u>(0, 0)</u> and x-intercepts of <u>(a, 0) and (−a, 0)</u>.

7. *a* and *b* give us the location of 4 intercepts—

$(a, 0)$, $(−a, 0)$, $(0, b)$ and $(0, −b)$ for $\frac{x^2}{a^2} + \frac{y^2}{b^2} = 1$

with center (0, 0). For Example 2, the values of *a* and *b* give us 4 points of the graph, just not intercepts. Here we move a distance of *a* units horizontally to the left and right of the center and *b* units above and below.

8. We use these points to draw asymptotes (also not part of the graph) which help us draw the correct shape of the hyperbola. The graph of a hyperbola gets closer and closer to the asymptotes without crossing them.

Exercise Set 13.2

1. $\frac{x^2}{16} + \frac{y^2}{4} = 1$ is an ellipse.

3. $x^2 - 5y^2 = 3$ is a hyperbola.

5. $-\frac{y^2}{25} + \frac{x^2}{36} = 1$ or

$\frac{x^2}{36} - \frac{y^2}{25} = 1$ is a hyperbola.

7. $\frac{x^2}{4} + \frac{y^2}{25} = 1$

$\frac{x^2}{2^2} + \frac{y^2}{5^2} = 1$

Center: (0, 0)
x-intercepts: (−2, 0), (2, 0)
y-intercepts: (0, −5), (0, 5)

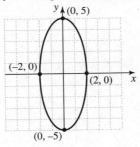

9. $\frac{x^2}{9} + y^2 = 1$

$\frac{x^2}{3^2} + \frac{y^2}{1^2} = 1$

Center: (0, 0)
x-intercepts: (−3, 0), (3, 0)
y-intercepts: (0, −1), (0, 1)

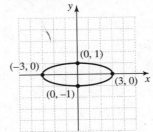

11. $9x^2 + y^2 = 36$

$$\frac{x^2}{4} + \frac{y^2}{36} = 1$$

$$\frac{x^2}{2^2} + \frac{y^2}{6^2} = 1$$

Center: (0, 0)
x-intercepts: (−2, 0), (2, 0)
y-intercepts: (0, −6), (0, 6)

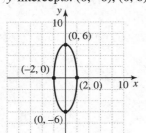

13. $4x^2 + 25y^2 = 100$

$$\frac{x^2}{25} + \frac{y^2}{4} = 1$$

$$\frac{x^2}{5^2} + \frac{y^2}{2^2} = 1$$

Center: (0, 0)
x-intercepts: (−5, 0), (5, 0)
y-intercepts: (0, −2), (0, 2)

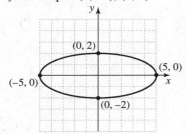

15. $\dfrac{(x+1)^2}{36} + \dfrac{(y-2)^2}{49} = 1$

$$\frac{(x+1)^2}{6^2} + \frac{(y-2)^2}{7^2} = 1$$

Center: (−1, 2)
Other points:

$(-1-6, 2) = (-7, 2)$
$(-1+6, 2) = (5, 2)$
$(-1, 2-7) = (-1, -5)$
$(-1, 2+7) = (-1, 9)$

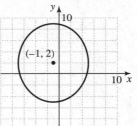

17. $\dfrac{(x-1)^2}{4} + \dfrac{(y-1)^2}{25} = 1$

$$\frac{(x-1)^2}{2^2} + \frac{(y-1)^2}{5^2} = 1$$

Center: (1, 1)
Other points:
$(1-2, 1) = (-1, 1)$
$(1+2, 1) = (3, 1)$
$(1, 1-5) = (1, -4)$
$(1, 1+5) = (1, 6)$

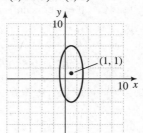

19. $\dfrac{x^2}{4} - \dfrac{y^2}{9} = 1$

$$\frac{x^2}{2^2} - \frac{y^2}{3^2} = 1$$

$a = 2, b = 3$

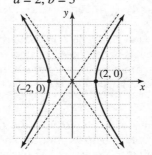

21. $\dfrac{y^2}{25} - \dfrac{x^2}{16} = 1$

$\dfrac{y^2}{5^2} - \dfrac{x^2}{4^2} = 1$

$a = 4, \ b = 5$

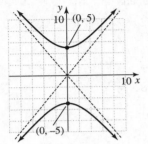

23. $x^2 - 4y^2 = 16$

$\dfrac{x^2}{16} - \dfrac{y^2}{4} = 1$

$\dfrac{x^2}{4^2} - \dfrac{y^2}{2^2} = 1$

$a = 4, \ b = 2$

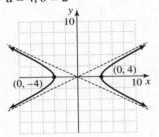

25. $16y^2 - x^2 = 16$

$\dfrac{y^2}{1} - \dfrac{x^2}{16} = 1$

$\dfrac{y^2}{1^2} - \dfrac{x^2}{4^2} = 1$

$a = 4, \ b = 1$

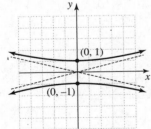

27. $\dfrac{y^2}{36} = 1 - x^2$

$x^2 + \dfrac{y^2}{36} = 1$

$\dfrac{x^2}{1^2} + \dfrac{y^2}{6^2} = 1$

$C(0, 0)$

x-intercepts: $(-1, 0), (1, 0)$

y-intercepts: $(0, -6), (0, 6)$

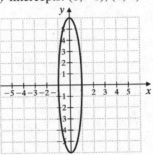

29. $4(x-1)^2 + 9(y+2)^2 = 36$

$\dfrac{(x-1)^2}{9} + \dfrac{(y+2)^2}{4} = 1$

$\dfrac{(x-1)^2}{3^2} + \dfrac{(y+2)^2}{2^2} = 1$

$C(1, -2)$

other points:

$(1 - 3, -2)$, or $(-2, -2)$

$(1 + 3, -2)$, or $(4, -2)$

$(1, -2 - 2)$, or $(1, -4)$

$(1, -2 + 2)$, or $(1, 0)$

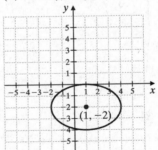

31. $8x^2 + 2y^2 = 32$

$$\frac{8x^2}{32} + \frac{2y^2}{32} = 1$$

$$\frac{x^2}{4} + \frac{y^2}{16} = 1$$

$$\frac{x^2}{2^2} + \frac{y^2}{4^2} = 1$$

$C(0, 0)$

x-intercepts: $(-2, 0)$, $(2, 0)$

y-intercepts: $(0, -4)$, $(0, 4)$

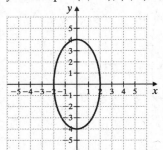

33. $25x^2 - y^2 = 25$

$$x^2 - \frac{y^2}{25} = 1$$

$$\frac{x^2}{1^2} - \frac{y^2}{5^2} = 1$$

$a = 1, b = 5$

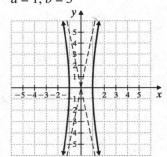

35. $(x-7)^2 + (y-2)^2 = 4$

Circle; center $(7, 2)$, radius $r = \sqrt{4} = 2$

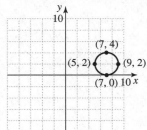

37. $y = x^2 + 12x + 36$

Parabola; $x = \dfrac{-b}{2a} = \dfrac{-12}{2(1)} = -6$

$y = (-6)^2 + 12(-6) + 36 = 0$

Vertex: $(-6, 0)$, opens upward

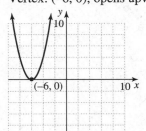

39. $\dfrac{y^2}{9} - \dfrac{x^2}{9} = 1$

$$\frac{y^2}{3^2} - \frac{x^2}{3^2} = 1$$

Hyperbola; center: $(0, 0)$

$a = 3, b = 3$

y-intercepts $(0, -3)$, $(0, 3)$

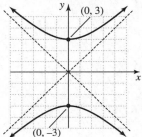

41. $\dfrac{x^2}{16} + \dfrac{y^2}{4} = 1$

$$\frac{x^2}{4^2} + \frac{y^2}{2^2} = 1$$

Ellipse; center: $(0, 0)$, $a = 4, b = 2$

x-intercepts $(-4, 0)$, $(4, 0)$

y-intercepts $(0, -2)$, $(0, 2)$

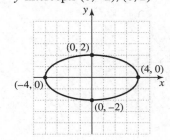

43. $x = y^2 + 4y - 1$

Parabola: $y = \dfrac{-b}{2a} = \dfrac{-4}{2(1)} = -2$

$x = (-2)^2 + 4(-2) - 1 = -5$

Vertex: $(-5, -2)$, opens to the right.

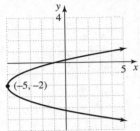

45. $9x^2 - 4y^2 = 36$

$\dfrac{x^2}{4} - \dfrac{y^2}{9} = 1$

$\dfrac{x^2}{2^2} - \dfrac{y^2}{3^2} = 1$

Hyperbola: center = $(0, 0)$
$a = 2, b = 3$
x-intercepts: $(2, 0), (-2, 0)$

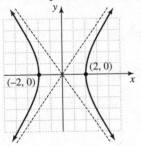

47. $\dfrac{(x-1)^2}{49} + \dfrac{(y+2)^2}{25} = 1$

$\dfrac{(x-1)^2}{7^2} + \dfrac{(y+2)^2}{5^2} = 1$

Ellipse; center: $(1, -2)$
$a = 7, b = 5$

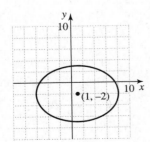

49. $\left(x + \dfrac{1}{2}\right)^2 + \left(y - \dfrac{1}{2}\right)^2 = 1$

Circle; center: $\left(-\dfrac{1}{2}, \dfrac{1}{2}\right)$, radius $r = 1$

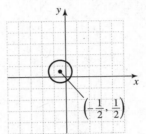

51. $(2x^3)(-4x^2) = -8x^5$

53. $-5x^2 + x^2 = -4x^2$

55. $\dfrac{x^2}{16} + \dfrac{y^2}{25} = 1$

$\sqrt{16} = 4$, so the distance between the
x-intercepts is $4 + 4 = 8$ units.
$\sqrt{25} = 5$, so the distance between the
y-intercepts is $5 + 5 = 10$ units.
The distance between the y-intercepts is longer
by $10 - 8 = 2$ units.

57. $4x^2 + y^2 = 16$

$\dfrac{x^2}{4} + \dfrac{y^2}{16} = 1$

$\sqrt{4} = 2$, so the distance between the x-intercepts
is $2 + 2 = 4$ units.
$\sqrt{16} = 4$, so the distance between the
y-intercepts is $4 + 4 = 8$ units.
The distance between the y-intercepts is longer
by $8 - 4 = 4$ units.

59. answers may vary

61. Circles: B, F
Ellipses: C, E, H
Hyperbolas: A, D, G

63. A: $c^2 = 36 + 13 = 49$; $c = \sqrt{49} = 7$

B: $c^2 = 4 - 4 = 0$; $c = \sqrt{0} = 0$

C: $c^2 = |25 - 16| = 9$; $c = \sqrt{9} = 3$

D: $c^2 = 39 + 25 = 64$; $c = \sqrt{64} = 8$

E: $c^2 = |81 - 17| = 64; \; c = \sqrt{64} = 8$

F: $c^2 = |36 - 36| = 0; \; c = \sqrt{0} = 0$

G: $c^2 = 65 + 16 = 81; \; c = \sqrt{81} = 9$

H: $c^2 = |144 - 140| = 4; \; c = \sqrt{4} = 2$

65. A: $e = \dfrac{7}{6}$

B: $e = \dfrac{0}{2} = 0$

C: $e = \dfrac{3}{5}$

D: $e = \dfrac{8}{5}$

E: $e = \dfrac{8}{9}$

F: $e = \dfrac{0}{6} = 0$

G: $e = \dfrac{9}{4}$

H: $e = \dfrac{2}{12} = \dfrac{1}{6}$

67. They are equal to 0.

69. answers may vary

71. $a = 130{,}000{,}000 \Rightarrow a^2 = (130{,}000{,}000)^2$
$$= 1.69 \times 10^{16}$$
$b = 125{,}000{,}000 \Rightarrow b^2 = (125{,}000{,}000)^2$
$$= 1.5625 \times 10^{16}$$

Thus, the equation is
$$\dfrac{x^2}{1.69 \times 10^{16}} + \dfrac{y^2}{1.5625 \times 10^{16}} = 1.$$

73. $9x^2 + 4y^2 = 36$
$$4y^2 = 36 - 9x^2$$
$$y^2 = \dfrac{36 - 9x^2}{4}$$
$$y = \pm\sqrt{\dfrac{36 - 9x^2}{4}} = \pm\dfrac{\sqrt{36 - 9x^2}}{2}$$

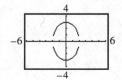

75. $\dfrac{(x-1)^2}{4} - \dfrac{(y+1)^2}{25} = 1$

Center: $(1, -1)$

$a = 2, \; b = 5$

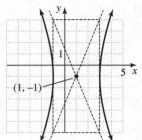

77. $\dfrac{y^2}{16} - \dfrac{(x+3)^2}{9} = 1$

Center: $(-3, 0)$

$a = 3, \; b = 4$

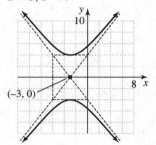

79. $\dfrac{(x+5)^2}{16} - \dfrac{(y+2)^2}{25} = 1$

Center: $(-5, -2)$

$a = 4, \; b = 5$

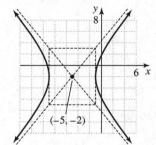

Integrated Review

1. $(x-7)^2 + (y-2)^2 = 4$

Circle; center: $(7, 2)$,

radius: $r = \sqrt{4} = 2$

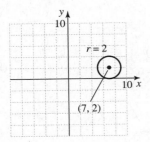

2. $y = x^2 + 4$

Parabola; vertex: $(0, 4)$

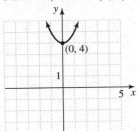

3. $y = x^2 + 12x + 36$

Parabola; $x = \dfrac{-b}{2a} = \dfrac{-12}{2(1)} = -6$

$y = (-6)^2 + 12(-6) + 36 = 0$

Vertex: $(-6, 0)$

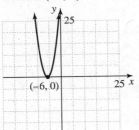

4. $\dfrac{x^2}{4} + \dfrac{y^2}{9} = 1$

Ellipse; center: $(0, 0)$

$a = 2, b = 3$

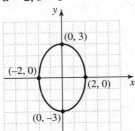

5. $\dfrac{y^2}{9} - \dfrac{x^2}{9} = 1$

Hyperbola; center: $(0, 0)$

$a = 3, b = 3$

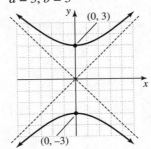

6. $\dfrac{x^2}{16} - \dfrac{y^2}{4} = 1$

Hyperbola; center: $(0, 0)$

$a = 4, b = 2$

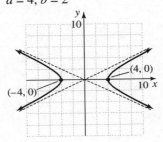

7. $\dfrac{x^2}{16} + \dfrac{y^2}{4} = 1$

Ellipse; center: $(0, 0)$

$a = 4, b = 2$

8. $x^2 + y^2 = 16$

Circle; center: (0, 0)

radius: $r = \sqrt{16} = 4$

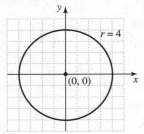

9. $x = y^2 + 4y - 1$

Parabola; $y = \dfrac{-b}{2a} = \dfrac{-4}{2(1)} = -2$

$x = (-2)^2 + 4(-2) - 1 = -5$

Vertex: (−5, −2)

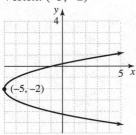

10. $x = -y^2 + 6y$

Parabola; $y = \dfrac{-b}{2a} = \dfrac{-6}{2(-1)} = 3$

$x = -(3)^2 + 6(3) = 9$

Vertex: (9, 3)

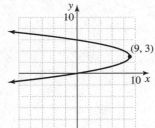

11. $9x^2 - 4y^2 = 36$

$\dfrac{x^2}{4} - \dfrac{y^2}{9} = 1$

Hyperbola; center: (0, 0)

$a = 2, b = 3$

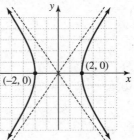

12. $9x^2 + 4y^2 = 36$

$\dfrac{x^2}{4} + \dfrac{y^2}{9} = 1$

Ellipse; center: (0, 0)

$a = 2, b = 3$

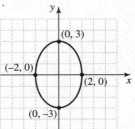

13. $\dfrac{(x-1)^2}{49} + \dfrac{(y+2)^2}{25} = 1$

Ellipse; center: (1, −2),

$a = 7, b = 5$

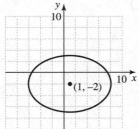

14.
$$y^2 = x^2 + 16$$
$$y^2 - x^2 = 16$$
$$\frac{y^2}{16} - \frac{x^2}{16} = 1$$
Hyperbola; center: $(0, 0)$
$a = 4, b = 4$

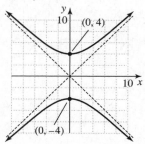

15. $\left(x + \frac{1}{2}\right)^2 + \left(y - \frac{1}{2}\right)^2 = 1$

Circle; center: $\left(-\frac{1}{2}, \frac{1}{2}\right)$, radius: $r = \sqrt{1} = 1$

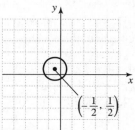

Section 13.3 Practice

1. $\begin{cases} x^2 - 4y = 4 \\ x + y = -1 \end{cases}$

Solve $x + y = -1$ for y.
$y = -x - 1$
Replace y with $-x - 1$ in the first equation and solve for x.
$$x^2 - 4(-x - 1) = 4$$
$$x^2 + 4x + 4 = 4$$
$$x^2 + 4x = 0$$
$$x(x + 4) = 0$$
$x = 0$ or $x = -4$
Let $x = 0$,   Let $x = -4$,
$y = -0 - 1 = -1$ $y = -(-4) - 1 = 3$
The solutions are $(0, -1)$ and $(-4, 3)$.

2. $\begin{cases} y = -\sqrt{x} \\ x^2 + y^2 = 20 \end{cases}$

Substitute $-\sqrt{x}$ for y in the second equation.
$$x^2 + \left(-\sqrt{x}\right)^2 = 20$$
$$x^2 + x = 20$$
$$x^2 + x - 20 = 0$$
$$(x + 5)(x - 4) = 0$$
$x = -5$ or $x = 4$
Let $x = -5$.
$y = -\sqrt{-5}$ Not a real number
Let $x = 4$.
$y = -\sqrt{4} = -2$
The solution is $(4, -2)$.

3. $\begin{cases} x^2 + y^2 = 9 \\ x - y = 5 \end{cases}$

Solve the second equation for x.
$x = y + 5$
Let $x = y + 5$ in the first equation.
$$(y + 5)^2 + y^2 = 9$$
$$y^2 + 10y + 25 + y^2 = 9$$
$$2y^2 + 10y + 16 = 0$$
$$y^2 + 5y + 8 = 0$$
By the quadratic formula,
$$y = \frac{-5 \pm \sqrt{5^2 - 4(1)(8)}}{2(1)} = \frac{-5 \pm \sqrt{-7}}{2}$$
$\sqrt{-7}$ is not a real number. There is no real solution, or $\varnothing$.

4. $\begin{cases} x^2 + 4y^2 = 16 \\ x^2 - y^2 = 1 \end{cases}$

Add the opposite of the second equation to the first.
$$x^2 + 4y^2 = 16$$
$$\underline{-x^2 + y^2 = -1}$$
$$0 + 5y^2 = 15$$
$$y^2 = 3$$
$$y = \pm\sqrt{3}$$

Let $y=\sqrt{3}.$ Let $y=-\sqrt{3}.$

$x^2-\left(\sqrt{3}\right)^2=1$ $x^2-\left(-\sqrt{3}\right)^2=1$

$x^2-3=1$ $x^2-3=1$

$x^2=4$ $x^2=4$

$x=\pm2$ $x=\pm2$

The solutions are $\left(2,\sqrt{3}\right),\left(2,-\sqrt{3}\right),\left(-2,\sqrt{3}\right),$ and $\left(-2,-\sqrt{3}\right).$

Vocabulary, Readiness & Video Check 13.3

1. Solving for y would either introduce tedious fractions (2nd equation) or a square root (1st equation) into the calculations.

2. When you multiply the left side of the equation by this number, do not forget to also multiply the right side.

Exercise Set 13.3

1. $\begin{cases} x^2+y^2=25 & (1) \\ 4x+3y=0 & (2) \end{cases}$

Solve E2 for y.

$3y=-4x$

$y=-\dfrac{4x}{3}$

Substitute into E1.

$x^2+\left(-\dfrac{4x}{3}\right)^2=25$

$x^2+\dfrac{16x^2}{9}=25$

$9\left(x^2+\dfrac{16x^2}{9}\right)=9(25)$

$9x^2+16x^2=225$

$25x^2=225$

$x^2=9$

$x=\pm\sqrt{9}=\pm3$

$x=3:y=-\dfrac{4(3)}{3}=-4$

$x=-3:y=-\dfrac{4(-3)}{3}=4$

The solutions are $(3,-4)$ and $(-3,4)$.

3. $\begin{cases} x^2+4y^2=10 & (1) \\ y=x & (2) \end{cases}$

Substitute x for y in E1.

$x^2+4x^2=10$

$5x^2=10$

$x^2=2$

$x=\pm\sqrt{2}$

Substitute these values into E2.

$x=\sqrt{2}:y=x=\sqrt{2}$

$x=-\sqrt{2}:y=x=-\sqrt{2}$

The solutions are $\left(\sqrt{2},\sqrt{2}\right)$ and $\left(-\sqrt{2},-\sqrt{2}\right).$

5. $\begin{cases} y^2=4-x & (1) \\ x-2y=4 & (2) \end{cases}$

Solve E2 for x.

$x=2y+4$

Substitute into E1.

$y^2=4-(2y+4)$

$y^2=-2y$

$y^2+2y=0$

$y(y+2)=0$

$y=0$ or $y+2=0$

$y=-2$

Substitute these values into the equation $x=2y+4$.

$y=0:x=2(0)+4=4$

$y=-2:x=2(-2)+4=0$

The solutions are $(4,0)$ and $(0,-2)$.

7. $\begin{cases} x^2+y^2=9 & (1) \\ 16x^2-4y^2=64 & (2) \end{cases}$

Multiply E1 by 4 and add to E2.

$4x^2+4y^2=36$

$\underline{16x^2-4y^2=64}$

$20x^2\qquad=100$

$x^2=5$

$x=\pm\sqrt{5}$

Substitute 5 for x^2 into E1.

$5+y^2=9$

$y^2=4$

$y=\pm2$

The solutions are $\left(-\sqrt{5},-2\right),\left(-\sqrt{5},2\right),\left(\sqrt{5},-2\right),$ and $\left(\sqrt{5},2\right).$

9. $\begin{cases} x^2 + 2y^2 = 2 & (1) \\ x - y = 2 & (2) \end{cases}$

Solve E2 for x: $x = y + 2$

Substitute into E1.

$(y+2)^2 + 2y^2 = 2$

$y^2 + 4y + 4 + 2y^2 = 2$

$3y^2 + 4y + 2 = 0$

$y = \dfrac{-4 \pm \sqrt{(4)^2 - 4(3)(2)}}{2(3)} = \dfrac{-4 \pm \sqrt{-8}}{6}$

There are no real solutions. The solution is $\varnothing$.

11. $\begin{cases} y = x^2 - 3 & (1) \\ 4x - y = 6 & (2) \end{cases}$

Substitute $x^2 - 3$ for y in E2.

$4x - (x^2 - 3) = 6$

$4x - x^2 + 3 = 6$

$0 = x^2 - 4x + 3$

$0 = (x-3)(x-1)$

$x - 3 = 0$ or $x - 1 = 0$

$x = 3$ or $ x = 1$

Substitute these values into E1.

$x = 3 : y = (3)^2 - 3 = 6$

$x = 1 : y = (1)^2 - 3 = -2$

The solutions are $(3, 6)$ and $(1, -2)$.

13. $\begin{cases} y = x^2 & (1) \\ 3x + y = 10 & (2) \end{cases}$

Substitute x^2 for y in E2.

$3x + x^2 = 10$

$x^2 + 3x - 10 = 0$

$(x+5)(x-2) = 0$

$x + 5 = 0$ or $x - 2 = 0$

$x = -5$ or $ x = 2$

Substitute these values into E1.

$x = -5 : y = (-5)^2 = 25$

$x = 2 : y = (2)^2 = 4$

The solutions are $(-5, 25)$ and $(2, 4)$.

15. $\begin{cases} y = 2x^2 + 1 & (1) \\ x + y = -1 & (2) \end{cases}$

Substitute $2x^2 + 1$ for y in E2.

$x + 2x^2 + 1 = -1$

$2x^2 + x + 2 = 0$

$x = \dfrac{-1 \pm \sqrt{(1)^2 - 4(2)(2)}}{2(2)} = \dfrac{-1 \pm \sqrt{-15}}{4}$

There are no real solutions. The solution is $\varnothing$.

17. $\begin{cases} y = x^2 - 4 & (1) \\ y = x^2 - 4x & (2) \end{cases}$

Substitute $x^2 - 4$ for y in E2.

$x^2 - 4 = x^2 - 4x$

$-4 = -4x$

$1 = x$

Substitute this value into E1.

$y = (1)^2 - 4 = -3$

The solution is $(1, -3)$.

19. $\begin{cases} 2x^2 + 3y^2 = 14 & (1) \\ -x^2 + y^2 = 3 & (2) \end{cases}$

Multiply E2 by 2 and add to E1.

$\begin{aligned} 2x^2 + 3y^2 &= 14 \\ \underline{-2x^2 + 2y^2} &= \underline{6} \\ 5y^2 &= 20 \\ y^2 &= 4 \\ y &= \pm 2 \end{aligned}$

Substitute 4 for y^2 into E2.

$-x^2 + 4 = 3$

$-x^2 = -1$

$x^2 = 1$

$x = \pm 1$

The solutions are $(-1, -2)$, $(-1, 2)$, $(1, -2)$, and $(1, 2)$.

21. $\begin{cases} x^2 + y^2 = 1 & (1) \\ x^2 + (y+3)^2 = 4 & (2) \end{cases}$

Multiply E1 by -1 and add to E2.

$$-x^2 - y^2 = -1$$

$$\underline{x^2 + (y+3)^2 = 4}$$

$$(y+3)^3 - y^2 = 3$$

$$y^2 + 6y + 9 - y^2 = 3$$

$$6y = -6$$

$$y = -1$$

Replace y with -1 in E1.

$$x^2 + (-1)^2 = 1$$

$$x^2 = 0$$

$$x = 0$$

The solution is $(0, -1)$.

23. $\begin{cases} y = x^2 + 2 & (1) \\ y = -x^2 + 4 & (2) \end{cases}$

Add E1 and E2.

$$y = x^2 + 2$$

$$\underline{y = -x^2 + 4}$$

$$2y = 6$$

$$y = 3$$

Substitute this value into E1.

$$3 = x^2 + 2$$

$$1 = x^2$$

$$\pm 1 = x$$

The solutions are $(-1, 3)$ and $(1, 3)$.

25. $\begin{cases} 3x^2 + y^2 = 9 & (1) \\ 3x^2 - y^2 = 9 & (2) \end{cases}$

Add E1 and E2.

$$3x^2 + y^2 = 9$$

$$\underline{3x^2 - y^2 = 9}$$

$$6x^2 \qquad = 18$$

$$x^2 = 3$$

$$x = \pm\sqrt{3}$$

Substitute 3 for x^2 in E1.

$$3(3) + y^2 = 9$$

$$y^2 = 0$$

$$y = 0$$

The solutions are $\left(-\sqrt{3}, 0\right)$, $\left(\sqrt{3}, 0\right)$.

27. $\begin{cases} x^2 + 3y^2 = 6 & (1) \\ x^2 - 3y^2 = 10 & (2) \end{cases}$

Solve E2 for x^2: $x^2 = 3y^2 + 10$.

Substitute into E1.

$$(3y^2 + 10) + 3y^2 = 6$$

$$6y^2 = -4$$

$$y^2 = -\frac{2}{3}$$

There are no real solutions. The solution is $\varnothing$.

29. $\begin{cases} x^2 + y^2 = 36 & (1) \\ y = \dfrac{1}{6}x^2 - 6 & (2) \end{cases}$

Solve E1 for x^2: $x^2 = 36 - y^2$.

Substitute into E2.

$$y = \frac{1}{6}(36 - y^2) - 6$$

$$y = 6 - \frac{1}{6}y^2 - 6$$

$$6y = -y^2$$

$$y^2 + 6y = 0$$

$$y(y + 6) = 0$$

$$y = 0 \text{ or } y = -6$$

Substitute these values into the equation $x^2 = 36 - y^2$.

$$y = 0: x^2 = 36 - (0)^2$$

$$x^2 = 36$$

$$x = \pm 6$$

$$y = -6: x^2 = 36 - (6)^2$$

$$x^2 = 0$$

$$x = 0$$

The solutions are $(-6, 0)$, $(6, 0)$ and $(0, -6)$.

31. $\begin{cases} y = \sqrt{x} \\ x^2 + y^2 = 12 \end{cases}$

Substitute.

$$x^2 + \left(\sqrt{x}\right)^2 = 12$$

$$x^2 + x = 12$$

$$x^2 + x - 12 = 0$$

$$(x + 4)(x - 3) = 0$$

$$x + 4 = 0 \quad \text{or} \quad x - 3 = 0$$

$$x = -4 \qquad\qquad x = 3$$

$$x = -4: \; y = \sqrt{-4}$$

$x = 3$: $y = \sqrt{3}$

Since $\sqrt{-4}$ is not a real number, the only

solution is $\left(3, \sqrt{3}\right)$.

33. $x > -3$

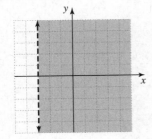

35. $y < 2x - 1$

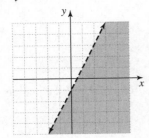

37. $P = x + (2x - 5) + (5x - 20) = (8x - 25)$ inches

39. $P = 2(x^2 + 3x + 1) + 2(x^2)$

$= 2x^2 + 6x + 2 + 2x^2$

$= (4x^2 + 6x + 2)$ meters

41. answers may vary

43. There are 0, 1, 2, 3, or 4 possible real solutions. answers may vary

45. Let x and y represent the numbers.

$\begin{cases} x^2 + y^2 = 130 \\ x^2 - y^2 = 32 \end{cases}$

Add the equations.

$x^2 + y^2 = 130$

$\underline{x^2 - y^2 = 32}$

$2x^2 \qquad = 162$

$x^2 = 81$

$x = \pm 9$

Replace x^2 with 81 in the first equation.

$81 + y^2 = 130$

$y^2 = 49$

$y = \pm 7$

The numbers are –9 and –7, –9 and 7, 9 and –7, and 9 and 7.

47. Let x and y be the length and width.

$\begin{cases} xy = 285 \\ 2x + 2y = 68 \end{cases}$

Solve the first equation for y: $y = \dfrac{285}{x}$.

Substitute into the second equation.

$2x + 2\left(\dfrac{285}{x}\right) = 68$

$x + \dfrac{285}{x} = 34$

$x^2 + 285 = 34x$

$x^2 - 34x + 285 = 0$

$(x - 19)(x - 15) = 0$

$x = 19$ or $x = 15$

Using $x = 19$, $y = \dfrac{285}{x} = \dfrac{285}{19} = 15$.

Using $x = 15$, $y = \dfrac{285}{x} = \dfrac{285}{15} = 19$.

The dimensions are 19 cm by 15 cm.

49. $\begin{cases} p = -0.01x^2 - 0.2x + 9 \\ p = 0.01x^2 - 0.1x + 3 \end{cases}$

Substitute.

$-0.01x^2 - 0.2x + 9 = 0.01x^2 - 0.1x + 3$

$0 = 0.02x^2 + 0.1x - 6$

$0 = x^2 + 5x - 300$

$0 = (x + 20)(x - 15)$

$x + 20 = 0$ or $x - 15 = 0$

$x = -20$ or $x = 15$

Disregard the negative.

$p = -0.01(15)^2 - 0.2(15) + 9$

$p = 3.75$

The equilibrium quantity is 15,000 compact discs, and the corresponding price is \$3.75.

51. $\begin{cases} x^2 + 4y^2 = 10 \\ y = x \end{cases}$

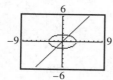

53. $\begin{cases} y = x^2 + 2 \\ y = -x^2 + 4 \end{cases}$

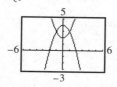

Section 13.4 Practice

1. $\dfrac{x^2}{36} + \dfrac{y^2}{16} \geq 1$

First graph the ellipse $\dfrac{x^2}{36} + \dfrac{y^2}{16} = 1$ as a solid

curve. Choose (0, 0) as a test point.

$$\dfrac{x^2}{36} + \dfrac{y^2}{16} \geq 1$$

$$\dfrac{0^2}{36} + \dfrac{0^2}{16} \geq 1$$

$$0 \geq 1 \quad \text{False}$$

The solution set is the region that does not contain (0, 0).

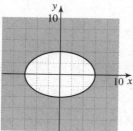

2. $16y^2 > 9x^2 + 144$

The related equation is $16y^2 = 9x^2 + 144$.

$$16y^2 - 9x^2 = 144$$

$$\dfrac{y^2}{9} - \dfrac{x^2}{16} = 1$$

Graph the hyperbola as a dashed curve.

Choose (0, 0), (0, 4), and (0, −4) as test points.

(0, 0): $16(0)^2 > 9(0)^2 + 144$

$0 > 144 \quad$ False

(0, 4): $16(4)^2 > 9(0)^2 + 144$

$256 > 144 \quad$ True

(0, −4): $16(-4)^2 > 9(0)^2 + 144$

$256 > 144 \quad$ True

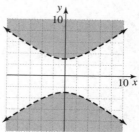

3. $\begin{cases} y \geq x^2 \\ y \leq -3x + 2 \end{cases}$

Solve the related system $\begin{cases} y = x^2 \\ y = -3x + 2 \end{cases}$.

Substitute $-3x + 2$ for y in the first equation.

$$x^2 = -3x + 2$$

$$x^2 + 3x - 2 = 0$$

$$x = \dfrac{-3 \pm \sqrt{3^2 - 4(1)(-2)}}{2(1)}$$

$$= \dfrac{-3 \pm \sqrt{17}}{2}$$

$$\approx 0.56 \text{ or } -3.56$$

$y = -3x + 2 \approx -3(0.56) + 2 = 0.32$

$y \approx -3(-3.56) + 2 = 12.68$

The points of intersection are approximately (0.56, 0.32) and (−3.56, 12.68).

Graph $y = x^2$ and $y = -3x + 2$ as solid curves.

The region of the solution set is above the parabola but below the line.

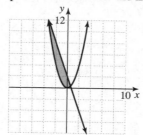

4. $\begin{cases} x^2 + y^2 < 16 \\ \dfrac{x^2}{4} - \dfrac{y^2}{9} < 1 \\ y < x + 3 \end{cases}$

Graph $x^2 + y^2 = 16$, $\dfrac{x^2}{4} - \dfrac{y^2}{9} = 1$, and $y = x + 3$.

The test point $(0, 0)$ gives true statements for all three inequalities; thus, the innermost region is the solution set.

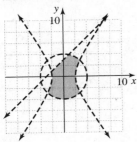

Vocabulary, Readiness & Video Check 13.4

1. For both, we graph the related equation to find the boundary and sketch it as a solid boundary for ≤ or ≥ and a dashed boundary for < or >; also we choose a test point not on the boundary and shade that region if the test point is a solution of the original inequality or shade the other region if not.

2. A circle within a circle (either circle solid or dashed) where the inner circle is shaded inside and the outer circle is shaded outside; also, two non-intersecting circles (either circle solid or dashed), both shaded inside just to name a few examples.

Exercise Set 13.4

1. $y < x^2$

First graph the parabola as a dashed curve.

Test Point	$y < x^2$; Result
$(0, 1)$	$1 < 0^2$; False

Shade the region which does not contain $(0, 1)$.

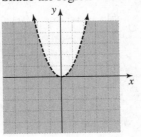

3. $x^2 + y^2 \geq 16$

First graph the circle as a solid curve.

Test Point	$x^2 + y^2 \geq 16$; Result
$(0, 0)$	$0^2 + 0^2 \geq 16$; False

Shade the region which does not contain $(0, 0)$.

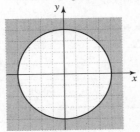

5. $\dfrac{x^2}{4} - y^2 < 1$

First graph the hyperbola as a dashed curve.

Test Point	$\dfrac{x^2}{4} - y^2 < 1$; Result
$(-4, 0)$	$\dfrac{(-4)^2}{4} - 0^2 < 1$; False
$(0, 0)$	$\dfrac{(0)^2}{4} - 0^2 < 1$; True
$(4, 0)$	$\dfrac{(4)^2}{4} - 0^2 < 1$; False

Shade the region containing $(0, 0)$.

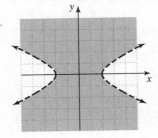

7. $y > (x-1)^2 - 3$

First graph the parabola as a dashed curve.

Test Point	$y > (x-1)^2 - 3$; Result
$(0, 0)$	$0 > (0-1)^2 - 3$; True

Shade the region containing $(0, 0)$.

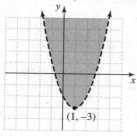

9. $x^2 + y^2 \leq 9$

First graph the circle as a solid curve.

Test Point	$x^2 + y^2 \leq 9$; Result
$(0, 0)$	$0^2 + 0^2 \leq 9$; True

Shade the region containing $(0, 0)$.

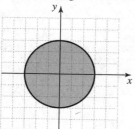

11. $y > -x^2 + 5$

First graph the parabola as a dashed curve.

Test Point	$y > -x^2 + 5$; Result
$(0, 0)$	$0 > -(0)^2 + 5$; False

Shade the region which does not contain $(0, 0)$.

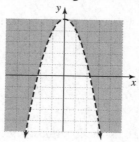

13. $\dfrac{x^2}{4} + \dfrac{y^2}{9} \leq 1$

First graph the ellipse as a solid curve.

Test Point	$\dfrac{x^2}{4} + \dfrac{y^2}{9} \leq 1$; Result
$(0, 0)$	$\dfrac{(0)^2}{4} + \dfrac{(0)^2}{9} \leq 1$; True

Shade the region containing $(0, 0)$.

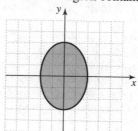

15. $\dfrac{y^2}{4} - x^2 \leq 1$

First graph the hyperbola as solid curves.

Test Point	$\dfrac{y^2}{4} - x^2 \leq 1$; Result
$(0, -4)$	$\dfrac{(-4)^2}{4} - 0^2 \leq 1$; False
$(0, 0)$	$\dfrac{(0)^2}{4} - 0^2 \leq 1$; True
$(0, 4)$	$\dfrac{(4)^2}{4} - 0^2 \leq 1$; False

Shade the region containing (0, 0).

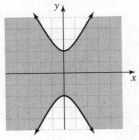

17. $y < (x-2)^2 + 1$

First graph the parabola as a dashed curve.

Test Point	$y < (x-2)^2 + 1$; Result
(0, 0)	$0 < (0-2)^2 + 1$; True

Shade the region containing (0, 0).

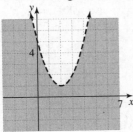

19. $y \le x^2 + x - 2$

First graph the parabola as a solid curve.

Test Point	$y \le x^2 + x - 2$; Result
(0, 0)	$0 \le (0)^2 + (0) - 2$; False

Shade the region which does not contain (0, 0).

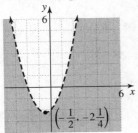

21. $\begin{cases} 4x + 3y \ge 12 \\ x^2 + y^2 < 16 \end{cases}$

First graph $4x + 3y = 12$ as a solid line.

Test Point	$4x + 3y \ge 12$; Result
(0, 0)	$4(0) + 3(0) \ge 12$; False

Shade the region which does not contain (0, 0).
Next, graph the circle $x^2 + y^2 = 16$ as a dashed curve.

Test Point	$x^2 + y^2 < 16$; Result
(0, 0)	$0^2 + 0^2 < 16$; True

Shade the region containing (0, 0). The solution to the system is the intersection.

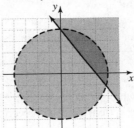

23. $\begin{cases} x^2 + y^2 \le 9 \\ x^2 + y^2 \ge 1 \end{cases}$

First graph the circle with radius 3 as a solid curve.

Test Point	$x^2 + y^2 \le 9$; Result
(0, 0)	$0^2 + 0^2 \le 9$; True

Shade the region containing (0, 0). Next, graph the circle with 1 as a dashed curve.

Test Point	$x^2 + y^2 \ge 1$; Result
(0, 0)	$0^2 + 0^2 \ge 1$; False

Shade the region which does not contain (0, 0). The solution to the system is the intersection.

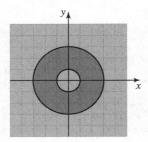

25. $\begin{cases} y > x^2 \\ y \geq 2x+1 \end{cases}$

First graph the parabola as a dashed curve.

Test Point	$y > x^2$; Result
(0, 1)	$1 > 0^2$; True

Shade the region containing (0, 1). Next, graph $y = 2x+1$ as a solid line.

Test Point	$y \geq 2x + 1$; Result
(0, 0)	$0 \geq 2(0) + 1$; False

Shade the region which does not contain (0, 0). The solution to the system is the intersection.

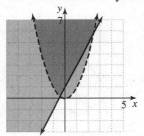

27. $\begin{cases} x^2 + y^2 > 9 \\ y > x^2 \end{cases}$

First graph the circle as a dashed curve.

Test Point	$x^2 + y^2 > 9$; Result
(0, 0)	$0^2 + 0^2 > 9$; False

Shade the region which does not contain (0, 0). Next, graph the parabola as a dashed curve.

Test Point	$y > x^2$; Result
(0, 1)	$1 > 0^2$; True

Shade the region containing (0, 1). The solution to the system is the intersection.

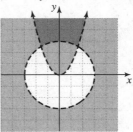

29. $\begin{cases} \dfrac{x^2}{4} + \dfrac{y^2}{9} \geq 1 \\ x^2 + y^2 \geq 4 \end{cases}$

First graph the ellipse as a solid curve.

Test Point	$\dfrac{x^2}{4} + \dfrac{y^2}{9} \geq 1$; Result
(0, 0)	$\dfrac{0^2}{4} + \dfrac{0^2}{9} \geq 1$; False

Shade the region which does not contain (0, 0). Next, graph the circle as a solid curve.

Test Point	$x^2 + y^2 \geq 4$; Result
(0, 0)	$0^2 + 0^2 \geq 4$; False

Shade the region which does not contain (0, 0). The solution to the system is the intersection.

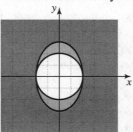

31. $\begin{cases} x^2 - y^2 \geq 1 \\ y \geq 0 \end{cases}$

First graph the hyperbola as solid curves.

Test Point	$x^2 - y^2 \geq 1$; Result
$(-2, 0)$	$(-2)^2 - 0^2 \geq 1$; True
$(0, 0)$	$0^2 - 0^2 \geq 1$; False
$(2, 0)$	$2^2 - 0^2 \geq 1$; True

Shade the region which does not contain $(0, 0)$. Next, graph $y = 0$ as a solid line.

Test Point	$y > 0$; Result
$(0, 1)$	$1 \geq 0$; True

Shade the region containing $(0, 1)$. The solution to the system is the intersection.

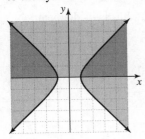

33. $\begin{cases} x + y \geq 1 \\ 2x + 3y < 1 \\ x > -3 \end{cases}$

First graph $x + y = 1$ as a solid line.

Test Point	$x + y \geq 1$; Result
$(0, 0)$	$0 + 0 \geq 1$; False

Shade the region which does not contain $(0, 0)$. Next, graph $2x + 3y = 1$ as a dashed line.

Test Point	$2x + 3y < 1$; Result
$(0, 0)$	$2(0) 1 + 3(0) < 1$; True

Shade the region containing $(0, 0)$. Now graph the line $x = -3$ as a dashed line.

Test Point	$x > -3$; Result
$(0, 0)$	$0 > -3$; True

Shade the region containing $(0, 0)$. The solution to the system is the intersection.

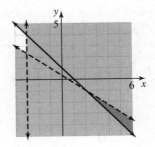

35. $\begin{cases} x^2 - y^2 < 1 \\ \dfrac{x^2}{16} + y^2 \leq 1 \\ x \geq -2 \end{cases}$

First graph the hyperbola as dashed curves.

Test Point	$x^2 - y^2 < 1$; Result
$(-2, 0)$	$(-2)^2 - 0^2 < 1$; False
$(0, 0)$	$0^2 - 0^2 < 1$; True
$(2, 0)$	$2^2 - 0^2 < 1$; False

Shade the region containing $(0, 0)$. Next, graph the ellipse as a solid curve.

Test Point	$\dfrac{x^2}{16} + y^2 \leq 1$; Result
$(0, 0)$	$\dfrac{0^2}{16} + 0^2 \leq 1$; True

Shade the region containing $(0, 0)$. Now graph the line $x = -2$ as a solid line.

Test Point	$x \geq -2$; Result
$(0, 0)$	$0 \geq -2$; True

Shade the region containing $(0, 0)$. The solution to the system is the intersection.

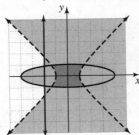

37. This is not a function because a vertical line can cross the graph in more than one place.

39. This is a function because a vertical line can cross the graph in no more than one place.

41. $f(x) = 3x^2 - 2$
$f(-1) = 3(-1)^2 - 2 = 3 - 2 = 1$

43. $f(x) = 3x^2 - 2$
$f(a) = 3(a)^2 - 2 = 3a^2 - 2$

45. answers may vary

47. $\begin{cases} y \le x^2 \\ y \ge x + 2 \\ x \ge 0 \\ y \ge 0 \end{cases}$

First graph $y = x^2$ as a solid curve.

Test Point	$y \le x^2$; Result
(0, 1)	$1 \le 0^2$; False

Shade the region which does not contain (0, 1). Next, graph $y = x + 2$ as a solid line.

Test Point	$y \ge x + 2$; Result
(0, 0)	$0 \ge 0 + 2$; False

Shade the region which does not contain (0, 0). Next graph the line $x = 0$ as a solid line, and shade to the right. Now graph the line $y = 0$ as a solid line, and shade above. The solution to the system is the intersection.

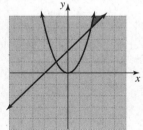

Chapter 13 Vocabulary Check

1. A <u>circle</u> is the set of all points in a plane that are the same distance from a fixed point, called the <u>center</u>.

2. A <u>nonlinear system of equations</u> is a system of equations at least one of which is not linear.

3. An <u>ellipse</u> is the set of points on a plane such that the sum of the distances of those points from two fixed points is a constant.

4. In a circle, the distance from the center to a point of the circle is called its <u>radius</u>.

5. A <u>hyperbola</u> is the set of points in a plane such that the absolute value of the difference of the distance from two fixed points is constant.

6. The circle, parabola, ellipse, and hyperbola are called the <u>conic sections</u>.

7. For a parabola that opens upward, the lowest point is the <u>vertex</u>.

8. Twice a circle's radius is its <u>diameter</u>.

Chapter 13 Review

1. center (–4, 4), radius 3
$[x - (-4)]^2 + (y - 4)^2 = 3^2$
$(x + 4)^2 + (y - 4)^2 = 9$

2. center (5, 0), radius 5
$(x - 5)^2 + (y - 0)^2 = 5^2$
$(x - 5)^2 + y^2 = 25$

3. center (–7, –9), radius $\sqrt{11}$
$[x - (-7)]^2 + [y - (-9)]^2 = \left(\sqrt{11}\right)^2$
$(x + 7)^2 + (y + 9)^2 = 11$

4. center (0, 0), radius $\dfrac{7}{2}$
$(x - 0)^2 + (y - 0)^2 = \left(\dfrac{7}{2}\right)^2$
$x^2 + y^2 = \dfrac{49}{4}$

5. $x^2 + y^2 = 7$

Circle; center (0, 0), radius $r = \sqrt{7}$

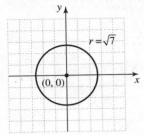

6. $x = 2(y-5)^2 + 4$

Parabola; vertex: (4, 5)

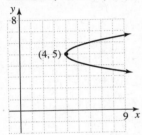

7. $x = -(y+2)^2 + 3$

Parabola; vertex: (3, –2)

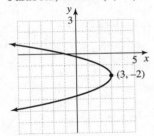

8. $(x-1)^2 + (y-2)^2 = 4$

Circle; center (1, 2), radius $r = \sqrt{4} = 2$

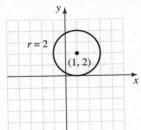

9. $y = -x^2 + 4x + 10$

Parabola; $x = \dfrac{-b}{2a} = \dfrac{-4}{2(-1)} = 2$

$y = -(2)^2 + 4(2) + 10 = 14$

Vertex: (2, 14)

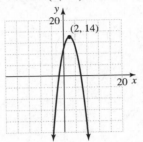

10. $x = -y^2 - 4y + 6$

Parabola; $y = \dfrac{-b}{2a} = \dfrac{-(-4)}{2(-1)} = -2$

$x = -(-2)^2 - 4(-2) + 6 = 10$

Vertex: (10, –2)

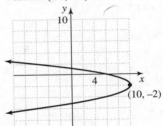

11. $x = \dfrac{1}{2}y^2 + 2y + 1$

Parabola; $y = \dfrac{-b}{2a} = \dfrac{-2}{2\left(\frac{1}{2}\right)} = -2$

$x = \dfrac{1}{2}(-2)^2 + 2(-2) + 1 = -1$

Vertex: (–1, –2)

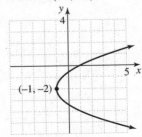

12. $y = -3x^2 + \dfrac{1}{2}x + 4$

Parabola; $x = \dfrac{-b}{2a} = \dfrac{-\frac{1}{2}}{2(-3)} = \dfrac{1}{12}$

$y = -3\left(\dfrac{1}{12}\right)^2 + \dfrac{1}{2}\left(\dfrac{1}{12}\right) + 4 = \dfrac{193}{48}$

Vertex: $\left(\dfrac{1}{12}, \dfrac{193}{48}\right)$

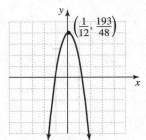

13.

$x^2 + y^2 + 2x + y = \dfrac{3}{4}$

$(x^2 + 2x) + (y^2 + y) = \dfrac{3}{4}$

$(x^2 + 2x + 1) + \left(y^2 + y + \dfrac{1}{4}\right) = \dfrac{3}{4} + 1 + \dfrac{1}{4}$

$(x+1)^2 + \left(y + \dfrac{1}{2}\right)^2 = 2$

Circle; center $\left(-1, -\dfrac{1}{2}\right)$, radius $r = \sqrt{2}$

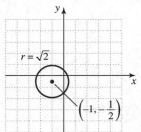

14.

$x^2 + y^2 - 3y = \dfrac{7}{4}$

$x^2 + \left(y^2 - 3y + \dfrac{9}{4}\right) = \dfrac{7}{4} + \dfrac{9}{4}$

$x^2 + \left(y - \dfrac{3}{2}\right)^2 = 4$

Circle; center $\left(0, \dfrac{3}{2}\right)$, radius $r = \sqrt{4} = 2$

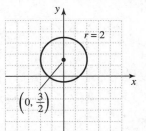

15.

$4x^2 + 4y^2 + 16x + 8y = 1$

$(x^2 + 4x) + (y^2 + 2y) = \dfrac{1}{4}$

$(x^2 + 4x + 4) + (y^2 + 2y + 1) = \dfrac{1}{4} + 4 + 1$

$(x+2)^2 + (y+1)^2 = \dfrac{21}{4}$

Circle; center $(-2, -1)$, radius $r = \sqrt{\dfrac{21}{4}} = \dfrac{\sqrt{21}}{2}$

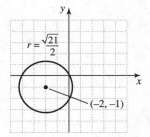

16.

$3x^2 + 3y^2 + 18x - 12y = -12$

$x^2 + y^2 + 6x - 4y = -4$

$x^2 + 6x + 9 + y^2 - 4y + 4 = -4 + 9 + 4$

$(x+3)^2 + (y-2)^2 = 9$

$(x+3)^2 + (y-2)^2 = 3^2$

$C(-3, 2), r = 3$

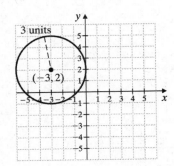

17. $x^2 - \dfrac{y^2}{4} = 1$

$\dfrac{x^2}{1^2} - \dfrac{y^2}{2^2} = 1$

$a = 1,\ b = 2$

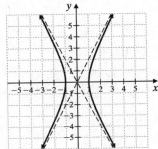

18. $x^2 + \dfrac{y^2}{4} = 1$

$\dfrac{x^2}{1^2} + \dfrac{y^2}{2^2} = 1$

$C(0, 0)$

x-intercepts: $(-1, 0),\ (1, 0)$

y-intercepts: $(0, -2),\ (0, 2)$

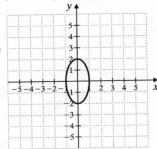

19. $4y^2 + 9x^2 = 36$

$\dfrac{y^2}{9} + \dfrac{x^2}{4} = 1$

$\dfrac{x^2}{2^2} + \dfrac{y^2}{3^2} = 1$

$C(0, 0)$

x-intercepts: $(-2, 0),\ (2, 0)$

y-intercepts: $(0, -3),\ (0, 3)$

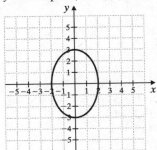

20. $-5x^2 + 25y^2 = 125$

$-\dfrac{x^2}{25} + \dfrac{y^2}{5} = 1$

$\dfrac{y^2}{\left(\sqrt{5}\right)^2} - \dfrac{x^2}{5^2} = 1$

$a = 5,\ b = \sqrt{5}$

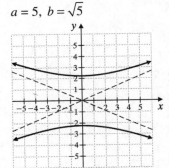

21. $x^2 - y^2 = 1$

$\dfrac{x^2}{1^2} - \dfrac{y^2}{1^2} = 1$

$a = 1,\ b = 1$

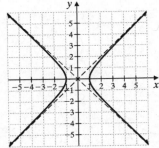

22. $\dfrac{(x+3)^2}{9}+\dfrac{(y-4)^2}{25}=1$

$\dfrac{(x+3)^2}{3^2}+\dfrac{(y-4)^2}{5^2}=1$

$C(-3, 4)$
other points:
$(-3 + 3, 4)$, or $(0, 4)$
$(-3 - 3, 4)$, or $(-6, 4)$
$(-3, 4 - 5)$, or $(-3, -1)$
$(-3, 4 + 5)$, or $(-3, 9)$

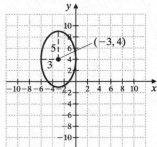

23. $y = x^2 + 9$

$V(0, 9)$
The parabola opens upward.

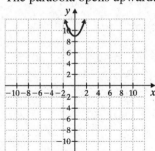

24. $36y^2 - 49x^2 = 1764$

$\dfrac{y^2}{49}-\dfrac{x^2}{36}=1$

$\dfrac{y^2}{7^2}-\dfrac{x^2}{6^2}=1$

$a = 6, b = 7$

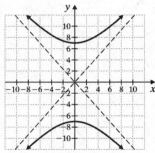

25. $x = 4y^2 - 16$

$V(-16, 0)$
The parabola opens to the right.

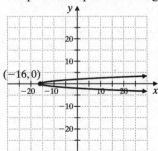

26. $y = x^2 + 4x + 6$

$y = (x^2 + 4x + 4) + 6 - 4$

$y = (x+2)^2 + 2$

$V(-2, 2)$
The parabola opens upward.

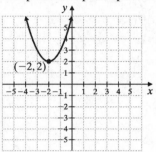

27. $y^2 + 2(x-1)^2 - 8 = 0$

$y^2 + 2(x-1)^2 = 8$

$\dfrac{y^2}{8}+\dfrac{(x-1)^2}{4}=1$

$\dfrac{y^2}{\left(\sqrt{8}\right)^2}+\dfrac{(x-1)^2}{2^2}=1$

$C(1, 0)$

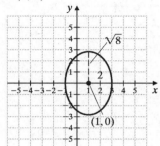

28. $x - 4y = y^2$

$x = y^2 + 4y$

$x = y^2 + 4y + 4 - 4$

$x = (y+2)^2 - 4$

$V(-4, -2)$

The parabola opens to the right.

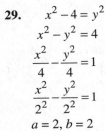

$(-4, -2)$

29. $x^2 - 4 = y^2$

$x^2 - y^2 = 4$

$\dfrac{x^2}{4} - \dfrac{y^2}{4} = 1$

$\dfrac{x^2}{2^2} - \dfrac{y^2}{2^2} = 1$

$a = 2, b = 2$

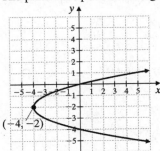

30. $x^2 = 4 - y^2$

$x^2 + y^2 = 4$

$x^2 + y^2 = 2^2$

$C(0, 0), r = 2$

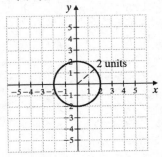

2 units

31. $36y^2 = 576 + 16x^2$

$36y^2 - 16x^2 = 576$

$\dfrac{y^2}{16} - \dfrac{x^2}{36} = 1$

$\dfrac{y^2}{4^2} - \dfrac{x^2}{6^2} = 1$

$a = 6, b = 4$

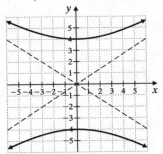

32. $3(x-7)^2 + 3(y+4)^2 = 1$

$(x-7)^2 + (y+4)^2 = \dfrac{1}{3}$

$(x-7)^2 + (y+4)^2 = \left(\sqrt{\dfrac{1}{3}}\right)^2$

$(x-7)^2 + (y+4)^2 = \left(\dfrac{\sqrt{3}}{3}\right)^2$

$C(7, -4)$

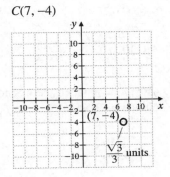

$(7, -4)$

$\dfrac{\sqrt{3}}{3}$ units

33. $\begin{cases} y = 2x - 4 \\ y^2 = 4x \end{cases}$

Substitute.

$$y^2 = 4x$$
$$(2x - 4)^2 = 4x$$
$$4x^2 - 16x + 16 = 4x$$
$$4x^2 - 16x + 16 = 4x$$
$$4x^2 - 20x + 16 = 0$$
$$4(x^2 - 5x + 4) = 0$$
$$4(x - 4)(x - 1) = 0$$
$$x - 4 = 0 \quad \text{or} \quad x - 1 = 0$$
$$x = 4 \qquad\qquad x = 1$$

$x = 4$: $y = 2(4) - 4 = 4$
$x = 1$: $y = 2(1) - 4 = -2$
The solutions are $(1, -2)$ and $(4, 4)$.

34. $\begin{cases} x^2 + y^2 = 4 \\ x - y = 4 \end{cases}$

Solve equation 2 for x.
$x = 4 + y$
Substitute.

$$x^2 + y^2 = 4$$
$$(4 + y)^2 + y^2 = 4$$
$$16 + 8y + y^2 + y^2 = 4$$
$$2y^2 + 8y + 12 = 0$$
$$a = 2, \, b = 8, \, c = 12$$
$$y = \frac{-8 \pm \sqrt{8^2 - 4(2)(12)}}{2(2)} = \frac{-8 \pm \sqrt{-32}}{4}$$

Since $\sqrt{-32}$ is not a real number, there is no solution. The solution set is $\varnothing$.

35. $\begin{cases} y = x + 2 \\ y = x^2 \end{cases}$

Substitute.

$$x + 2 = x^2$$
$$0 = x^2 - x - 2$$
$$0 = (x - 2)(x + 1)$$
$$x - 2 = 0 \quad \text{or} \quad x + 1 = 0$$
$$x = 2 \qquad\qquad x = -1$$

$x = 2$: $y = 2^2 = 4$

$x = -1$: $y = (-1)^2 = 1$

The solutions are $(-1, 1)$ and $(2, 4)$.

36. $\begin{cases} 4x - y^2 = 0 \\ 2x^2 + y^2 = 16 \end{cases}$

Add the equations.

$$2x^2 + 4x = 16$$
$$2x^2 + 4x - 16 = 0$$
$$2(x^2 + 2x - 8) = 0$$
$$2(x + 4)(x - 2) = 0$$
$$x + 4 = 0 \quad \text{or} \quad x - 2 = 0$$
$$x = -4 \qquad\qquad x = 2$$

$x = -4$: $4(-4) - y^2 = 0$
$$-16 = y^2$$
$$\pm\sqrt{-16} = y$$

$x = 2$: $4(2) - y^2 = 0$
$$8 = y^2$$
$$\pm\sqrt{8} = y$$
$$\pm 2\sqrt{2} = y$$

Since $\sqrt{-16}$ is not a real number, the solutions are $\left(2, 2\sqrt{2}\right)$ and $\left(2, -2\sqrt{2}\right)$.

37. $\begin{cases} x^2 + 4y^2 = 16 \\ x^2 + y^2 = 4 \end{cases}$

Multiply equation 2 by -1. Add the results.

$$x^2 + 4y^2 = 16$$
$$\underline{-x^2 - y^2 = -4}$$
$$3y^2 = 12$$
$$y^2 = 4$$
$$y = \pm\sqrt{4}$$
$$y = \pm 2$$

$y = -2$: $x^2 + (-2)^2 = 4$
$$x^2 = 0$$
$$x = 0$$

$y = 2$: $x^2 + (2)^2 = 4$
$$x^2 = 0$$
$$x = 0$$

The solutions are $(0, 2)$ and $(0, -2)$.

38. $\begin{cases} x^2 + 2y = 9 \\ 5x - 2y = 5 \end{cases}$

Add the equations.

$$x^2 + 5x = 14$$
$$x^2 + 5x - 14 = 0$$
$$(x + 7)(x - 2) = 0$$

$$x + 7 = 0 \quad \text{or} \quad x - 2 = 0$$
$$x = -7 \qquad\qquad x = 2$$
$$x = -7: \ 5(-7) - 2y = 5$$
$$-2y = 40$$
$$y = -20$$
$$x = 2: \ 5(2) - 2y = 5$$
$$-2y = -5$$
$$y = \frac{5}{2}$$

The solutions are $\left(2, \dfrac{5}{2}\right)$ and $(-7, -20)$.

39. $\begin{cases} y = 3x^2 + 5x - 4 \\ y = 3x^2 - x + 2 \end{cases}$

Substitute.
$$3x^2 + 5x - 4 = 3x^2 - x + 2$$
$$5x - 4 = -x + 2$$
$$6x = 6$$
$$x = 1$$
$$x = 1: \ y = 3(1)^2 + 5(1) - 4 = 4$$

The solution is $(1, 4)$.

40. $\begin{cases} x^2 - 3y^2 = 1 \\ 4x^2 + 5y^2 = 21 \end{cases}$

Solve equation 1 for x^2.
$$x^2 = 1 + 3y^2$$

Substitute.
$$4(1 + 3y^2) + 5y^2 = 21$$
$$4 + 12y^2 + 5y^2 = 21$$
$$17y^2 = 17$$
$$y^2 = 1$$
$$y = \pm 1$$
$$y = -1: \ x^2 = 1 + 3(-1)^2$$
$$x^2 = 4$$
$$x = \pm 2$$
$$y = 1: \ x^2 = 1 + 3(1)^2$$
$$x^2 = 4$$
$$x = \pm 2$$

The solutions are $(-2, -1)$, $(-2, 1)$, $(2, -1)$, and $(2, 1)$.

41. Let x be the width and y be the length.

$\begin{cases} xy = 150 \\ 2x + 2y = 50 \end{cases}$

Solve equation 2 for x.
$$2x + 2y = 50$$
$$x + y = 25$$
$$x = 25 - y$$

Substitute.
$$xy = 150$$
$$(25 - y)y = 150$$
$$25y - y^2 = 150$$
$$0 = y^2 - 25y + 150$$
$$0 = (y - 15)(y - 10)$$
$$y - 15 = 0 \quad \text{or} \quad y - 10 = 0$$
$$y = 15 \qquad\qquad y = 10$$
$$y = 15: \ x = 25 - 15 = 10$$
$$y = 10: \ x = 25 - 10 = 15$$

The dimensions are length = 15 feet and width = 10 feet.

42. An ellipse and a hyperbola can intersect at a maximum of 4 points. Therefore, there are a maximum of 4 real number solutions.

43. $y \le -x^2 + 3$

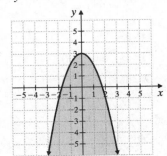

44. $x < y^2 - 1$

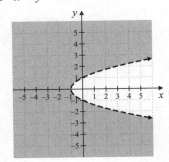

45. $x^2 + y^2 < 9$

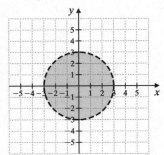

46. $\dfrac{x^2}{4} + \dfrac{y^2}{9} \geq 1$

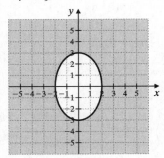

47. $\begin{cases} 3x + 4y \leq 12 \\ x - 2y > 6 \end{cases}$

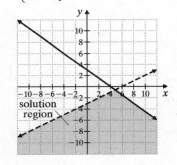

48. $\begin{cases} x^2 + y^2 \leq 16 \\ x^2 + y^2 \geq 4 \end{cases}$

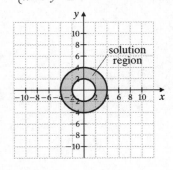

49. $\begin{cases} x^2 + y^2 < 4 \\ x^2 - y^2 \leq 1 \end{cases}$

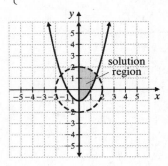

50. $\begin{cases} x^2 + y^2 < 4 \\ y \geq x^2 - 1 \\ x \geq 0 \end{cases}$

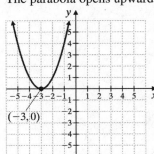

51. $C(-7, 8)$, $r = 5$

$[x - (-7)]^2 + (y - 8)^2 = 5^2$

$(x + 7)^2 + (y - 8)^2 = 25$

52. $y = x^2 + 6x + 9$

$y = (x + 3)^2$

$V(-3, 0)$

The parabola opens upward.

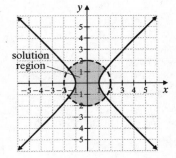

53. $x = y^2 + 6y + 9$

$x = (y + 3)^2$

$V(0, -3)$

The parabola opens to the right.

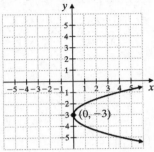

54. $\dfrac{y^2}{4} - \dfrac{x^2}{16} = 1$

$\dfrac{y^2}{2^2} - \dfrac{x^2}{4^2} = 1$

$a = 4, b = 2$

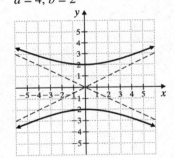

55. $\dfrac{y^2}{4} + \dfrac{x^2}{16} = 1$

$\dfrac{y^2}{2^2} + \dfrac{x^2}{4^2} = 1$

$C(0, 0)$

x-intercepts: $(-4, 0), (4, 0)$

y-intercepts: $(0, -2), (0, 2)$

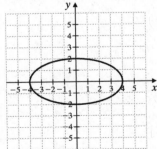

56. $\dfrac{(x-2)^2}{4} + (y - 1)^2 = 1$

$\dfrac{(x-2)^2}{2^2} + \dfrac{(y-1)^2}{1^2} = 1$

$C(2, 1)$

other points:

$(2 - 2, 1)$, or $(0, 1)$

$(2 + 2, 1)$, or $(4, 1)$

$(2, 1 - 1)$, or $(2, 0)$

$(2, 1 + 1)$, or $(2, 2)$

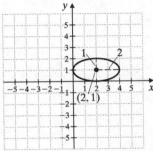

57. $y^2 = x^2 + 6$

$y^2 - x^2 = 6$

$\dfrac{y^2}{6} - \dfrac{x^2}{6} = 1$

$\dfrac{y^2}{\left(\sqrt{6}\right)^2} - \dfrac{x^2}{\left(\sqrt{6}\right)^2} = 1$

$a = \sqrt{6}, b = \sqrt{6}$

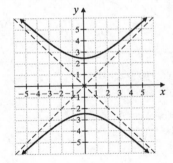

58. $y^2 + (x-2)^2 = 10$

$(x-2)^2 + y^2 = \left(\sqrt{10}\right)^2$

$C(2, 0),\ r = \sqrt{10}$

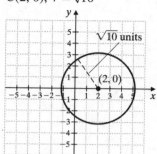

59. $3x^2 + 6x + 3y^2 = 9$

$x^2 + 2x + y^2 = 3$

$(x^2 + 2x + 1) + y^2 = 3 + 1$

$(x+1)^2 + y^2 = 4$

$(x+1)^2 + (y-0)^2 = 2^2$

$C(-1, 0),\ r = 2$

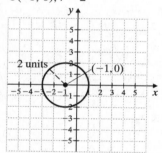

60. $x^2 + y^2 - 8y = 0$

$x^2 + (y^2 - 8y + 16) = 0 + 16$

$x^2 + (y-4)^2 = 16$

$(x-0)^2 + (y-4)^2 = 4^2$

$C(0, 4),\ r = 4$

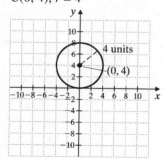

61. $6(x-2)^2 + 9(y+5)^2 = 36$

$\dfrac{(x-2)^2}{6} + \dfrac{(y+5)^2}{4} = 1$

$\dfrac{(x-2)^2}{\left(\sqrt{6}\right)^2} + \dfrac{(y+5)^2}{2^2} = 1$

$C(2, -5)$
other points:

$\left(2 - \sqrt{6}, -5\right)$

$\left(2 + \sqrt{6}, -5\right)$

$(2, -5 - 2),$ or $(2, -7)$
$(2, -5 + 2),$ or $(2, -3)$

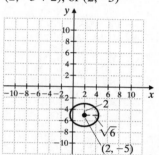

62. $\dfrac{x^2}{16} - \dfrac{y^2}{25} = 1$

$\dfrac{x^2}{4^2} - \dfrac{y^2}{5^2} = 1$

$a = 4,\ b = 5$

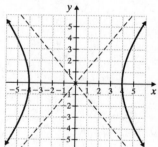

63. $\begin{cases} y = x^2 - 5x + 1 \\ y = -x + 6 \end{cases}$

Substitute.

$x^2 - 5x + 1 = -x + 6$

$x^2 - 4x - 5 = 0$

$(x-5)(x+1) = 0$

$x - 5 = 0 \quad$ or $\quad x + 1 = 0$

$\qquad x = 5 \qquad\qquad\quad x = -1$

$x = 5$: $y = -5 + 6 = 1$
$x = -1$: $y = -(-1) + 6 = 7$
The solutions are (5, 1) and (−1, 7).

64. $\begin{cases} x^2 + y^2 = 10 \\ 9x^2 + y^2 = 18 \end{cases}$

Multiply equation 1 by −1. Add the results.

$-x^2 - y^2 = -10$

$\dfrac{9x^2 + y^2 = 18}{8x^2 \qquad = 8}$

$x^2 = 1$

$x = \pm\sqrt{1}$

$x = \pm 1$

$x = -1$: $(-1)^2 + y^2 = 10$

$y^2 = 9$

$y = \pm 3$

$x = 1$: $(1)^2 + y^2 = 10$

$y^2 = 9$

$y = \pm 3$

The solutions are (−1, 3), (−1, −3), (1, 3), and (1, −3).

65. $x^2 - y^2 < 1$

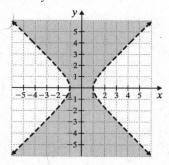

66. $\begin{cases} y > x^2 \\ x + y \geq 3 \end{cases}$

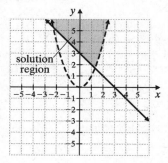

Chapter 13 Test

1. $x^2 + y^2 = 36$

Circle; center: (0, 0), radius $r = \sqrt{36} = 6$

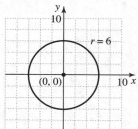

2. $x^2 - y^2 = 36$

$\dfrac{x^2}{36} - \dfrac{y^2}{36} = 1$

Hyperbola; center: (0, 0), $a = 6$, $b = 6$

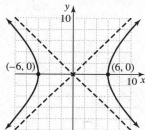

3. $16x^2 + 9y^2 = 144$

$\dfrac{x^2}{9} + \dfrac{y^2}{16} = 1$

Ellipse; center: (0, 0), $a = 3$, $b = 4$

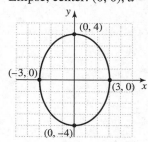

4. $y = x^2 - 8x + 16$

$y = (x-4)^2$

Parabola; vertex: (4, 0)

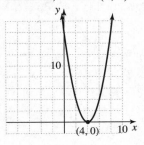

5. $x^2 + y^2 + 6x = 16$

$(x^2 + 6x) + y^2 = 16$

$(x^2 + 6x + 9) + y^2 = 16 + 9$

$(x+3)^2 + y^2 = 25$

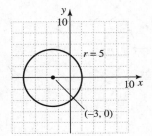

6. $x = y^2 + 8y - 3$

$x + 16 = (y^2 + 8y + 16) - 3$

$x = (y+4)^2 - 19$

Parabola; vertex: (−4, −19)

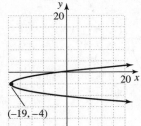

7. $\dfrac{(x-4)^2}{16} + \dfrac{(y-3)^2}{9} = 1$

Ellipse: center: (4, 3), $a = 4$, $b = 3$

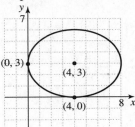

8. $y^2 - x^2 = 1$

Hyperbola: center: (0, 0), $a = 1$, $b = 1$

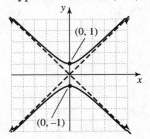

9. $\begin{cases} x^2 + y^2 = 169 \\ 5x + 12y = 0 \end{cases}$

$12y = -5x$

$y = -\dfrac{5x}{12}$

Substitute.

$x^2 + \left(-\dfrac{5x}{12}\right)^2 = 169$

$x^2 + \dfrac{25x^2}{144} = 169$

$\dfrac{169x^2}{144} = 169$

$\dfrac{x^2}{144} = 1$

$x^2 = 144$ so $x = \pm 12$.

Substitute back.

$x = 12$: $y = -\dfrac{5}{12}(12) = -5$

$x = -12$: $y = -\dfrac{5}{12}(-12) = 5$

The solutions are (12, −5) and (−12, 5).

10. $\begin{cases} x^2 + y^2 = 26 \\ x^2 - 2y^2 = 23 \end{cases}$

Multiply the second equation by -1. Add the results.

$x^2 + y^2 = 26$

$\underline{-x^2 + 2y^2 = -23}$

$3y^2 = 3$

$y^2 = 1$

$y = \pm 1$

$y = -1: \quad x^2 + (-1)^2 = 26$

$\qquad\qquad\quad x^2 = 25$

$\qquad\qquad\qquad x = \pm 5$

$y = 1: \quad x^2 + 1^2 = 26$

$\qquad\qquad x^2 = 25$

$\qquad\qquad\quad x = \pm 5$

The solutions are $(-5, -1)$, $(-5, 1)$, $(5, -1)$, and $(5, 1)$.

11. $\begin{cases} y = x^2 - 5x + 6 \\ y = 2x \end{cases}$

Substitute.

$x^2 - 5x + 6 = 2x$

$x^2 - 7x + 6 = 0$

$(x - 6)(x - 1) = 0$

$x - 6 = 0 \quad \text{or} \quad x - 1 = 0$

$\quad x = 6 \qquad\qquad\quad x = 1$

$x = 6: \; y = 2(6) = 12$

$x = 1: \; y = 2(1) = 2$

The solutions are $(6, 12)$ and $(1, 2)$.

12. $\begin{cases} x^2 + 4y^2 = 5 \\ y = x \end{cases}$

Substitute.

$x^2 + 4x^2 = 5$

$\quad 5x^2 = 5$

$\quad\; x^2 = 1$

$\qquad x = \pm 1$

$x = -1: y = -1$

$x = 1: y = 1$

The solutions are $(-1, -1)$ and $(1, 1)$.

13. $\begin{cases} 2x + 5y \geq 10 \\ y \geq x^2 + 1 \end{cases}$

First graph $\begin{cases} 2x + 5y = 10 \\ \qquad y = x^2 + 1 \end{cases}$ or

$\begin{cases} y = -\dfrac{2}{5}x + 2 \\ y = 1 \cdot (x - 0)^2 + 1 \end{cases}$

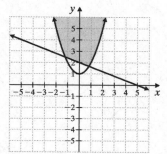

14. $\begin{cases} \dfrac{x^2}{4} + y^2 \leq 1 \\ \quad x + y > 1 \end{cases}$

First graph $\begin{cases} \dfrac{x^2}{2^2} + \dfrac{y^2}{1^2} = 1 \\ \quad x + y = 1 \end{cases}$

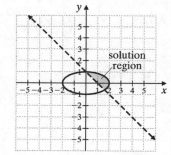

15. $\begin{cases} x^2 + y^2 > 1 \\ \dfrac{x^2}{4} - y^2 \geq 1 \end{cases}$

First graph $\begin{cases} x^2 + y^2 = 1 \\ \dfrac{x^2}{2^2} - \dfrac{y^2}{1^2} = 1 \end{cases}$

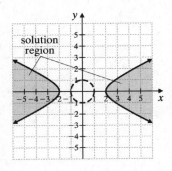

16. $\begin{cases} x^2 + y^2 \ge 4 \\ x^2 + y^2 < 16 \\ y \ge 0 \end{cases}$

First graph $\begin{cases} x^2 + y^2 = 2^2 \\ x^2 + y^2 = 4^2 \\ y = 0 \end{cases}$

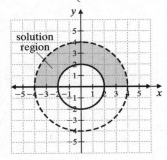

17. Graph B; vertex in second quadrant, opens to the right.

18. $100x^2 + 225y^2 = 22{,}500$

$$\frac{x^2}{225} + \frac{y^2}{100} = 1$$

$$\frac{x^2}{15^2} + \frac{y^2}{10^2} = 1$$

Height = 10 feet
Width = 2(15) = 30 feet

Chapter 13 Cumulative Review

1. $2x \ge 0$ and $4x - 1 \le -9$
$\quad x \ge 0$ and $4x \le -8$
$\quad x \ge 0$ and $x \le -2$

There is no solution, or $\varnothing$.

2. $3x + 4 > 1$ and $2x - 5 \le 9$
$\quad\quad 3x > -3$ and $\quad 2x \le 14$
$\quad\quad\; x > -1$ and $\quad\quad x \le 7$
$-1 < x \le 7$
$(-1, 7]$

3. $5x - 3 \le 10$ or $x + 1 \ge 5$
$\quad\; 5x \le 13$ or $\quad x \ge 4$
$\quad\;\; x \le \dfrac{13}{5}$ or $\quad x \ge 4$

The solution set is $\left(-\infty, \dfrac{13}{5}\right] \cup [4, \infty)$.

4. $(3, 2), (1, -4)$
$$m = \frac{-4 - 2}{1 - 3} = \frac{-6}{-2} = 3$$

5. $|5w + 3| = 7$
$5w + 3 = 7$ or $5w + 3 = -7$
$\quad 5w = 4$ or $\quad\; 5w = -10$
$\quad\;\; w = \dfrac{4}{5}$ or $\quad\quad w = -2$

The solutions are -2 and $\dfrac{4}{5}$.

6. Let x = speed of one plane. Then
$x + 25$ = speed of the other plane.
$d_{\text{plane 1}} + d_{\text{plane 2}} = 650$ miles
$\quad 2x + 2(x + 25) = 650$
$\quad\quad 2x + 2x + 50 = 650$
$\quad\quad\quad\quad\quad 4x = 600$
$\quad\quad\quad\quad\quad\; x = 150$
$x + 25 = 150 + 25 = 175$
The planes are traveling at 150 mph and 175 mph.

7. $\left|\dfrac{x}{2} - 1\right| = 11$

$\dfrac{x}{2} - 1 = 11$ or $\dfrac{x}{2} - 1 = -11$

$\quad\;\; \dfrac{x}{2} = 12$ or $\quad\;\; \dfrac{x}{2} = -10$

$\quad\quad x = 24$ or $\quad\quad x = -20$

The solutions are -20 and 24.

8. a. $\dfrac{4^8}{4^3} = 4^{8-3} = 4^5$

b. $\dfrac{y^{11}}{y^5} = y^{11-5} = y^6$

c. $\dfrac{32x^7}{4x^6} = \dfrac{32}{4}x^{7-6} = 8x$

d. $\dfrac{18a^{12}b^6}{12a^8b^6} = \dfrac{18}{12}a^{12-8}b^{6-6} = \dfrac{3}{2}a^4b^0 = \dfrac{3a^4}{2}$

9. $|3x + 2| = |5x - 8|$

$\begin{array}{llll} 3x+2 = 5x-8 & \text{or} & 3x+2 = -(5x-8) \\ 2 = 2x-8 & \text{or} & 3x+2 = -5x+8 \\ 10 = 2x & \text{or} & 8x+2 = 8 \\ 5 = x & \text{or} & 8x = 6 \\ & & \quad\ x = \dfrac{3}{4} \end{array}$

The solutions are $\dfrac{3}{4}$ and 5.

10. a. $3y^2 + 14y + 15 = (3y+5)(y+3)$

b. $20a^5 + 54a^4 + 10a^3$
$= 2a^3(10a^2 + 27a + 5)$
$= 2a^3(2a+5)(5a+1)$

c. $(y-3)^2 - 2(y-3) - 8$

Let $u = y - 3$. Then $u^2 = (y-3)^2$ and

$\begin{aligned} u^2 - 2u - 8 &= (u-4)(u+2) \\ &= [(y-3)-4][(y-3)+2] \\ &= (y-7)(y-1) \end{aligned}$

11. $|m - 6| < 2$
$-2 < m-6 < 2$
$4 < m < 8$
The solution set is (4, 8).

12. $\dfrac{2}{3a-15} - \dfrac{a}{25-a^2}$

$= \dfrac{2}{3(a-5)} + \dfrac{a}{a^2-25}$

$= \dfrac{2}{3(a-5)} + \dfrac{a}{(a+5)(a-5)}$

$= \dfrac{2(a+5)+3a}{3(a-5)(a+5)}$

$= \dfrac{2a+10+3a}{3(a-5)(a+5)}$

$= \dfrac{5a+10}{3(a-5)(a+5)}$

13. $\dfrac{x^{-1}+2xy^{-1}}{x^{-2}-x^{-2}y^{-1}} = \dfrac{\dfrac{1}{x}+\dfrac{2x}{y}}{\dfrac{1}{x^2}-\dfrac{1}{x^2y}}$

$= \dfrac{x^2y\left(\dfrac{1}{x}+\dfrac{2x}{y}\right)}{x^2y\left(\dfrac{1}{x^2}-\dfrac{1}{x^2y}\right)}$

$= \dfrac{xy+2x^3}{y-1}$

14. a. $(a^{-1}-b^{-1})^{-1} = \left(\dfrac{1}{a}-\dfrac{1}{b}\right)^{-1}$

$= \left(\dfrac{b-a}{ab}\right)^{-1}$

$= \dfrac{ab}{b-a}$

b. $\dfrac{2-\dfrac{1}{x}}{4x-\dfrac{1}{x}} = \dfrac{\left(2-\dfrac{1}{x}\right)x}{\left(4x-\dfrac{1}{x}\right)x}$

$= \dfrac{2x-1}{4x^2-1}$

$= \dfrac{2x-1}{(2x+1)(2x-1)}$

$= \dfrac{1}{2x+1}$

15. $|2x+9| + 5 > 3$
$|2x+9| > -2$

The absolute value is never negative, so all real numbers are solutions. The solution set is $(-\infty, \infty)$.

16.
$$\frac{2}{x+3} = \frac{1}{x^2-9} - \frac{1}{x-3}$$
$$\frac{2}{x+3} = \frac{1}{(x+3)(x-3)} - \frac{1}{x-3}$$
$$2(x-3) = 1 - 1(x+3)$$
$$2x - 6 = 1 - x - 3$$
$$2x - 6 = -x - 2$$
$$3x = 4$$
$$x = \frac{4}{3}$$

17.

$$\begin{array}{r|rrrrrrr} 4 & 4 & -25 & 35 & 0 & 17 & 0 & 0 \\ & & 16 & -36 & -4 & -16 & 4 & 16 \\ \hline & 4 & -9 & -1 & -4 & 1 & 4 & 16 \end{array}$$

Thus, $P(4) = 16$.

18.
$$y = \frac{k}{x}$$
$$3 = \frac{k}{\frac{2}{3}}$$
$$k = 3\left(\frac{2}{3}\right) = 2$$

Thus, the equation is $y = \frac{2}{x}$.

19. a. $\sqrt[3]{1} = 1$

b. $\sqrt[3]{-64} = -4$

c. $\sqrt[3]{\frac{8}{125}} = \frac{\sqrt[3]{8}}{\sqrt[3]{125}} = \frac{2}{5}$

d. $\sqrt[3]{x^6} = x^2$

e. $\sqrt[3]{-27x^9} = -3x^3$

20. a. $\sqrt{5}\left(2 + \sqrt{15}\right) = 2\sqrt{5} + \sqrt{5} \cdot \sqrt{15}$
$$= 2\sqrt{5} + \sqrt{75}$$
$$= 2\sqrt{5} + 5\sqrt{3}$$

b. $\left(\sqrt{3} - \sqrt{5}\right)\left(\sqrt{7} - 1\right)$
$$= \sqrt{3} \cdot \sqrt{7} - \sqrt{3} \cdot 1 - \sqrt{5} \cdot \sqrt{7} + \sqrt{5} \cdot 1$$
$$= \sqrt{21} - \sqrt{3} - \sqrt{35} + \sqrt{5}$$

c. $\left(2\sqrt{5} - 1\right)^2 = \left(2\sqrt{5}\right)^2 - 2 \cdot 2\sqrt{5} \cdot 1 + 1^2$
$$= 4(5) - 4\sqrt{5} + 1$$
$$= 21 - 4\sqrt{5}$$

d. $\left(3\sqrt{2} + 5\right)\left(3\sqrt{2} - 5\right) = \left(3\sqrt{2}\right)^2 - 5^2$
$$= 9(2) - 25$$
$$= 18 - 25$$
$$= -7$$

21. a. $z^{2/3}(z^{1/3} - z^5) = z^{2/3+1/3} - z^{2/3+5}$
$$= z^{3/3} - z^{2/3+15/3}$$
$$= z - z^{17/3}$$

b. $(x^{1/3} - 5)(x^{1/3} + 2)$
$$= x^{1/3} \cdot x^{1/3} + 2x^{1/3} - 5x^{1/3} - 5(2)$$
$$= x^{2/3} - 3x^{1/3} - 10$$

22.
$$\frac{-2}{\sqrt{3}+3} = \frac{-2\left(\sqrt{3}-3\right)}{\left(\sqrt{3}+3\right)\left(\sqrt{3}-3\right)}$$
$$= \frac{-2\left(\sqrt{3}-3\right)}{\left(\sqrt{3}\right)^2 - 3^2}$$
$$= \frac{-2\left(\sqrt{3}-3\right)}{3-9}$$
$$= \frac{-2\left(\sqrt{3}-3\right)}{-6}$$
$$= \frac{\sqrt{3}-3}{3}$$

23. a. $\frac{\sqrt{20}}{\sqrt{5}} = \sqrt{\frac{20}{5}} = \sqrt{4} = 2$

b. $\frac{\sqrt{50x}}{2\sqrt{2}} = \frac{1}{2}\sqrt{\frac{50x}{2}} = \frac{1}{2}\sqrt{25x} = \frac{5\sqrt{x}}{2}$

c. $\frac{7\sqrt[3]{48x^4y^8}}{\sqrt[3]{6y^2}} = 7\sqrt[3]{\frac{48x^4y^8}{6y^2}}$
$$= 7\sqrt[3]{8x^4y^6}$$
$$= 7\sqrt[3]{8x^3y^6 \cdot x}$$
$$= 7 \cdot 2xy^2\sqrt[3]{x}$$
$$= 14xy^2\sqrt[3]{x}$$

d.
$$\frac{2\sqrt[4]{32a^8b^6}}{\sqrt[4]{a^{-1}b^2}} = 2\sqrt[4]{\frac{32a^8b^6}{a^{-1}b^2}}$$
$$= 2\sqrt[4]{32a^9b^4}$$
$$= 2\sqrt[4]{16a^8b^4 \cdot 2a}$$
$$= 2 \cdot 2a^2b\sqrt[4]{2a}$$
$$= 4a^2b\sqrt[4]{2a}$$

24.
$$\sqrt{2x-3} = x-3$$
$$\left(\sqrt{2x-3}\right)^2 = (x-3)^2$$
$$2x-3 = x^2 - 6x + 9$$
$$0 = x^2 - 8x + 12$$
$$0 = (x-6)(x-2)$$
$$x-6 = 0 \text{ or } x-2 = 0$$
$$x = 6 \text{ or } \qquad x = 2$$
Discard 2 as an extraneous solution. The solution is 6.

25. a.
$$\frac{\sqrt{45}}{4} - \frac{\sqrt{5}}{3} = \frac{3\sqrt{5}}{4} - \frac{\sqrt{5}}{3}$$
$$= \frac{9\sqrt{5} - 4\sqrt{5}}{12}$$
$$= \frac{5\sqrt{5}}{12}$$

b.
$$\sqrt[3]{\frac{7x}{8}} + 2\sqrt[3]{7x} = \frac{\sqrt[3]{7x}}{2} + 2\sqrt[3]{7x}$$
$$= \frac{\sqrt[3]{7x}}{2} + \frac{4\sqrt[3]{7x}}{2}$$
$$= \frac{5\sqrt[3]{7x}}{2}$$

26.
$$9x^2 - 6x = -4$$
$$9x^2 - 6x + 4 = 0$$
$$a = 9, b = -6, c = 4$$
$$b^2 - 4ac = (-6)^2 - 4(9)(4)$$
$$= 36 - 144$$
$$= -108$$
Two complex but not real solutions

27.　$\sqrt{\dfrac{7x}{3y}} = \dfrac{\sqrt{7x}}{\sqrt{3y}} = \dfrac{\sqrt{7x} \cdot \sqrt{3y}}{\sqrt{3y} \cdot \sqrt{3y}} = \dfrac{\sqrt{21xy}}{3y}$

28.
$$\frac{4}{x-2} - \frac{x}{x+2} = \frac{16}{x^2-4}$$
$$\frac{4}{x-2} - \frac{x}{x+2} = \frac{16}{(x+2)(x-2)}$$
$$4(x+2) - x(x-2) = 16$$
$$4x + 8 - x^2 + 2x = 16$$
$$0 = x^2 - 6x + 8$$
$$0 = (x-4)(x-2)$$
$$x-4 = 0 \text{ or } x-2 = 0$$
$$x = 4 \text{ or } \qquad x = 2$$
Discard the solution 2 as extraneous. The solution is 4.

29.
$$\sqrt{2x-3} = 9$$
$$\left(\sqrt{2x-3}\right)^2 = 9^2$$
$$2x-3 = 81$$
$$2x = 84$$
$$x = 42$$
The solution is 42.

30.
$$x^3 + 2x^2 - 4x \geq 8$$
$$x^3 + 2x^2 - 4x - 8 \geq 0$$
$$x^2(x+2) - 4(x+2) \geq 0$$
$$(x+2)(x^2-4) \geq 0$$
$$(x+2)(x+2)(x-2) \geq 0$$
$$(x+2)^2(x-2) \geq 0$$
$$(x+2)^2 = 0 \text{ or } x-2 = 0$$
$$x+2 = 0 \text{ or } \qquad x = 2$$
$$x = -2$$

Region	Test Point	$(x+2)^2(x-2) \geq 0$ Result
A: $(-\infty, -2)$	-3	$(-1)^2(-5) \geq 0$ False
B: $(-2, 2)$	0	$(2)^2(-2) \geq 0$ False
C: $(2, \infty)$	3	$(5)^2(1) \geq 0$ True

Solution: $[2, \infty)$

31. a. $i^7 = i^4 \cdot i^3 = 1 \cdot (-i) = -i$

b. $i^{20} = (i^4)^5 = 1^5 = 1$

c. $i^{46} = i^{44} \cdot i^2 = (i^4)^{11} \cdot (-1) = 1^{11}(-1) = -1$

d. $i^{-12} = \dfrac{1}{i^{12}} = \dfrac{1}{(i^4)^3} = \dfrac{1}{1^3} = 1$

32. $f(x) = (x+2)^2 - 1$

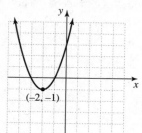

$(-2, -1)$

33.
$$p^2 + 2p = 4$$
$$p^2 + 2p + \left(\frac{2}{2}\right)^2 = 4 + 1$$
$$p^2 + 2p + 1 = 5$$
$$(p+1)^2 = 5$$
$$p + 1 = \pm\sqrt{5}$$
$$p = -1 \pm \sqrt{5}$$
The solutions are $-1 + \sqrt{5}$ and $-1 - \sqrt{5}$.

34. $f(x) = -x^2 - 6x + 4$

The maximum will occur at the vertex.
$$x = \frac{-b}{2a} = \frac{-(-6)}{2(-1)} = -3$$
$$f(-3) = -(-3)^2 - 6(-3) + 4 = 13$$
The maximum value is 13.

35.
$$\frac{1}{4}m^2 - m + \frac{1}{2} = 0$$
$$4\left(\frac{1}{4}m^2 - m + \frac{1}{2}\right) = 4(0)$$
$$m^2 - 4m + 2 = 0$$
$$a = 1, b = -4, c = 2$$
$$m = \frac{-(-4) \pm \sqrt{(-4)^2 - 4(1)(2)}}{2(1)}$$
$$= \frac{4 \pm \sqrt{16-8}}{2}$$
$$= \frac{4 \pm \sqrt{8}}{2}$$
$$= \frac{4 \pm 2\sqrt{2}}{2}$$
$$= 2 \pm \sqrt{2}$$
The solutions are $2 + \sqrt{2}$ and $2 - \sqrt{2}$.

36.
$$f(x) = \frac{x+1}{2}$$
$$y = \frac{x+1}{2}$$
$$x = \frac{y+1}{2}$$
$$2x = y + 1$$
$$2x - 1 = y$$
$$f^{-1}(x) = 2x - 1$$

37.
$$p^4 - 3p^2 - 4 = 0$$
$$(p^2 - 4)(p^2 + 1) = 0$$
$$(p+2)(p-2)(p^2 + 1) = 0$$
$$p + 2 = 0 \quad \text{or} \quad p - 2 = 0 \quad \text{or} \quad p^2 + 1 = 0$$
$$p = -2 \quad \text{or} \quad p = 2 \quad \text{or} \quad p^2 = -1$$
$$p = \pm i$$
The solutions are -2, 2, $-i$, and i.

38. a. $\dfrac{\sqrt{32}}{\sqrt{4}} = \sqrt{\dfrac{32}{4}} = \sqrt{8} = \sqrt{4 \cdot 2} = 2\sqrt{2}$

b. $\dfrac{\sqrt[3]{240y^2}}{5\sqrt[3]{3y^{-4}}} = \dfrac{1}{5}\sqrt[3]{\dfrac{240y^2}{3y^{-4}}}$

$= \dfrac{1}{5}\sqrt[3]{80y^6}$

$= \dfrac{1}{5}\sqrt[3]{8y^6 \cdot 10}$

$= \dfrac{2y^3\sqrt[3]{10}}{5}$

c. $\dfrac{\sqrt[5]{64x^9y^2}}{\sqrt[5]{2x^2y^{-8}}} = \sqrt[5]{\dfrac{64x^9y^2}{2x^2y^{-8}}}$

$= \sqrt[5]{32x^7y^{10}}$

$= \sqrt[5]{32x^5y^{10} \cdot x^2}$

$= 2xy^2\sqrt[5]{x^2}$

39. $\dfrac{x+2}{x-3} \le 0$

$x+2=0 \quad \text{or} \quad x-3=0$

$x=-2 \quad \text{or} \qquad x=3$

Region	Test Point	$\dfrac{x+2}{x-3} \le 0$ Result
A: $(-\infty, -2)$	-3	$\dfrac{-1}{-6} \le 0$; False
B: $(-2, 3)$	0	$\dfrac{2}{-3} \le 0$; True
C: $(3, \infty)$	4	$\dfrac{6}{1} \le 0$; False

Solution: $[-2, 3)$

40. $4x^2 + 9y^2 = 36$

$\dfrac{x^2}{9} + \dfrac{y^2}{4} = 1$

Ellipse: center $(0, 0)$, $a = 3$, $b = 2$

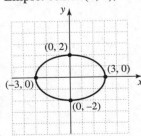

41. $g(x) = \dfrac{1}{2}(x+2)^2 + 5$

Vertex: $(-2, 5)$, axis: $x = -2$

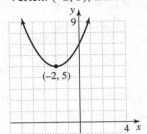

42. a. $64^x = 4$

$(4^2)^x = 4$

$4^{2x} = 4$

$2x = 1$

$x = \dfrac{1}{2}$

b. $125^{x-3} = 25$

$(5^3)^{x-3} = 5^2$

$5^{3x-9} = 5^2$

$3x - 9 = 2$

$3x = 11$

$x = \dfrac{11}{3}$

c. $\dfrac{1}{81} = 3^{2x}$

$3^{-4} = 3^{2x}$

$-4 = 2x$

$-\dfrac{4}{2} = x$

$-2 = x$

43. $f(x) = x^2 - 4x - 12$

$x = \dfrac{-b}{2a} = \dfrac{-(-4)}{2(1)} = 2$

$f(2) = (2)^2 - 4(2) - 12 = -16$

Vertex: $(2, -16)$

44. $\begin{cases} x + 2y < 8 \\ \quad\ y \geq x^2 \end{cases}$

First, graph $x + 2y = 8$ as a dashed line.

Test Point	$x + 2y < 8$; Result
(0, 0)	$0 + 2(0) < 8$; True

Shade the region containing (0, 0). Next, graph the parabola $y = x^2$ as a solid curve.

Test Point	$y \geq x^2$; Result
(0, 1)	$1 \geq 0^2$; True

Shade the region containing (0, 1). The solution to the system is the intersection.

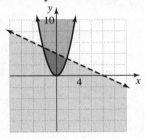

45. (2, –5), (1, –4)

$$d = \sqrt{[-4 - (-5)]^2 + (1 - 2)^2}$$
$$= \sqrt{1^2 + (-1)^2}$$
$$= \sqrt{2} \approx 1.414$$

46. $\begin{cases} x^2 + y^2 = 36 \quad (1) \\ \quad\quad y = x + 6 \quad (2) \end{cases}$

Substitute $x + 6$ for y in E1.

$$x^2 + (x + 6)^2 = 36$$
$$x^2 + (x^2 + 12x + 36) = 36$$
$$2x^2 + 12x = 0$$
$$2x(x + 6) = 0$$

$2x = 0$ or $x + 6 = 0$

$\ x = 0$ or $x = -6$

Use these values in E2 to find y.

$x = 0: y = 0 + 6 = 6$

$x = -6: y = -6 + 6 = 0$

The solutions are (0, 6) and (–6, 0).

Chapter 14

Section 14.1 Practice

1. $a_n = 5 + n^2$

   $a_1 = 5 + 1^2 = 5 + 1 = 6$

   $a_2 = 5 + 2^2 = 5 + 4 = 9$

   $a_3 = 5 + 3^2 = 5 + 9 = 14$

   $a_4 = 5 + 4^2 = 5 + 16 = 21$

   $a_5 = 5 + 5^2 = 5 + 25 = 30$

   Thus, the first five terms of the sequence are 6, 9, 14, 21, and 30.

2. $a_n = \dfrac{(-1)^n}{5n}$

   a. $a_1 = \dfrac{(-1)^1}{5(1)} = -\dfrac{1}{5}$

   b. $a_4 = \dfrac{(-1)^4}{5(4)} = \dfrac{1}{20}$

   c. $a_{30} = \dfrac{(-1)^{30}}{5(30)} = \dfrac{1}{150}$

   d. $a_{19} = \dfrac{(-1)^{19}}{5(19)} = -\dfrac{1}{95}$

3. **a.** 1, 3, 5, 7, …

   These numbers are the first four odd natural numbers, so a general term might be $a_n = 2n - 1$.

   b. 3, 9, 27, 81, …

   These numbers are all powers of 3 ($3 = 3^1$, $9 = 3^2$, $27 = 3^3$, and $81 = 3^4$), so a general term might be $a_n = 3^n$.

   c. $\dfrac{1}{2}, \dfrac{2}{3}, \dfrac{3}{4}, \dfrac{4}{5}, \ldots$

   The numerators are the first four natural numbers and each denominator is one greater than the numerator, so a general term might be $a_n = \dfrac{n}{n+1}$.

   d. $-\dfrac{1}{2}, -\dfrac{1}{3}, -\dfrac{1}{4}, -\dfrac{1}{5}, \ldots$

   The denominators are consecutive natural numbers beginning with 2 and each term is negative, so a general term might be $a_n = -\dfrac{1}{n+1}$.

4. $v_n = 3950(0.8)^n$

   $v_3 = 3950(0.8)^3$

   $\quad = 3950(0.512)$

   $\quad = 2022.4$

   The value of the copier after three years is $2022.40.

Vocabulary, Readiness & Video Check 14.1

1. The nth term of the sequence a_n is called the <u>general</u> term.

2. A <u>finite</u> sequence is a function whose domain is $\{1, 2, 3, 4, \ldots, n\}$ where n is some natural number.

3. An <u>infinite</u> sequence is a function whose domain is $\{1, 2, 3, 4, \ldots\}$.

4. $a_n = 7^n$

   $a_1 = 7^1 = 7$

5. $a_n = \dfrac{(-1)^n}{n}$

   $a_1 = \dfrac{(-1)^1}{1} = -1$

6. $a_n = (-1)^n \cdot n^4$

   $a_1 = (-1)^1 \cdot 1^4 = -1$

7. A sequence is a <u>function</u> whose <u>domain</u> is the set of natural numbers. We use $\underline{a_n}$ to mean the general term of a sequence.

8. If the negative is inside the parentheses, such as $(-2)^n$, it is also raised to the power. Since a_2 and a_4 would then have even powers, they would then be positive terms—but all original terms are negative.

9. $a_9 = 0.10(2)^{9-1} = \$25.60$

Exercise Set 14.1

1. $a_n = n + 4$
$a_1 = 1 + 4 = 5$
$a_2 = 2 + 4 = 6$
$a_3 = 3 + 4 = 7$
$a_4 = 4 + 4 = 8$
$a_5 = 5 + 4 = 9$
Thus, the first five terms of the sequence
$a_n = n + 4$ are 5, 6, 7, 8, 9.

3. $a_n = (-1)^n$

$a_1 = (-1)^1 = -1$
$a_2 = (-1)^2 = 1$
$a_3 = (-1)^3 = -1$
$a_4 = (-1)^4 = 1$
$a_5 = (-1)^5 = -1$
Thus, the first five terms of the sequence
$a_n = (-1)^n$ are $-1, 1, -1, 1, -1$.

5. $a_n = \dfrac{1}{n+3}$

$a_1 = \dfrac{1}{1+3} = \dfrac{1}{4}$
$a_2 = \dfrac{1}{2+3} = \dfrac{1}{5}$
$a_3 = \dfrac{1}{3+3} = \dfrac{1}{6}$
$a_4 = \dfrac{1}{4+3} = \dfrac{1}{7}$
$a_5 = \dfrac{1}{5+3} = \dfrac{1}{8}$
Thus, the first five terms of the sequence
$a_n = \dfrac{1}{n+3}$ are $\dfrac{1}{4}, \dfrac{1}{5}, \dfrac{1}{6}, \dfrac{1}{7}, \dfrac{1}{8}$.

7. $a_n = 2n$
$a_1 = 2(1) = 2$
$a_2 = 2(2) = 4$
$a_3 = 2(3) = 6$
$a_4 = 2(4) = 8$
$a_5 = 2(5) = 10$
Thus, the first five terms of the sequence
$a_n = 2n$ are 2, 4, 6, 8, 10.

9. $a_n = -n^2$
$a_1 = -1^2 = -1$
$a_2 = -2^2 = -4$
$a_3 = -3^2 = -9$
$a_4 = -4^2 = -16$
$a_5 = -5^2 = -25$
Thus, the first five terms of the sequence
$a_n = n^2$ are $-1, -4, -8, -16, -25$.

11. $a_n = 2^n$
$a_1 = 2^1 = 2$
$a_2 = 2^2 = 4$
$a_3 = 2^3 = 8$
$a_4 = 2^4 = 16$
$a_5 = 2^5 = 32$
Thus, the first five terms of the sequence
$a_n = 2^n$ are 2, 4, 8, 16, 32.

13. $a_n = 2n + 5$
$a_1 = 2(1) + 5 = 2 + 5 = 7$
$a_2 = 2(2) + 5 = 4 + 5 = 9$
$a_3 = 2(3) + 5 = 6 + 5 = 11$
$a_4 = 2(4) + 5 = 8 + 5 = 13$
$a_5 = 2(5) + 5 = 10 + 5 = 15$
Thus, the first five terms of the sequence
$a_n = 2n + 5$ are 7, 9, 11, 13, 15.

15. $a_n = (-1)^n n^2$
$a_1 = (-1)^1(1)^2 = -1(1) = -1$
$a_2 = (-1)^2(2)^2 = 1(4) = 4$
$a_3 = (-1)^3(3)^2 = -1(9) = -9$
$a_4 = (-1)^4(4)^2 = 1(16) = 16$
$a_5 = (-1)^5(5)^2 = -1(25) = -25$
Thus, the first five terms of the sequence
$a_n = (-1)^n n^2$ are $-1, 4, -9, 16, -25$.

17. $a_n = 3n^2$
$a_5 = 3(5)^2 = 3(25) = 75$

19. $a_n = 6n - 2$
$a_{20} = 6(20) - 2 = 120 - 2 = 118$

21. $a_n = \dfrac{n+3}{n}$

$a_{15} = \dfrac{15+3}{15} = \dfrac{18}{15} = \dfrac{6}{5}$

23. $a_n = (-3)^n$

$a_6 = (-3)^6 = 729$

25. $a_n = \dfrac{n-2}{n+1}$

$a_6 = \dfrac{6-2}{6+1} = \dfrac{4}{7}$

27. $a_n = \dfrac{(-1)^n}{n}$

$a_8 = \dfrac{(-1)^8}{8} = \dfrac{1}{8}$

29. $a_n = -n^2 + 5$

$a_{10} = -10^2 + 5 = -100 + 5 = -95$

31. $a_n = \dfrac{(-1)^n}{n+6}$

$a_{19} = \dfrac{(-1)^{19}}{19+6} = -\dfrac{1}{25}$

33. 3, 7, 11, 15, or 4(1) – 1, 4(2) – 1, 4(3) – 1, 4(4) – 1. In general, $a_n = 4n - 1$.

35. –2, –4, –8, –16, or –2, -2^2, -2^3, -2^4.

In general, $a_n = -2^n$.

37. $\dfrac{1}{3}, \dfrac{1}{9}, \dfrac{1}{27}, \dfrac{1}{81}$, or $\dfrac{1}{3}, \dfrac{1}{3^2}, \dfrac{1}{3^3}, \dfrac{1}{3^4}$

In general, $a_n = \dfrac{1}{3^n}$.

39. $a_n = 32n - 16$

$a_2 = 32(2) - 16 = 64 - 16 = 48$ ft
$a_3 = 32(3) - 16 = 96 - 16 = 80$ ft
$a_4 = 32(4) - 16 = 128 - 16 = 112$ ft

41. 0.10, 0.20, 0.40, or 0.10, 0.10(2), $0.10(2)^2$

In general, $a_n = 0.10(2)^{n-1}$

$a_{14} = 0.10(2)^{13} = \$819.20$

43. $a_n = 75(2)^{n-1}$

$a_6 = 75(2)^5 = 75(32) = 2400$ cases
$a_1 = 75(2)^0 = 75(1) = 75$ cases

45. $a_n = \dfrac{1}{2}a_{n-1}$ for $n > 1, a_1 = 800$

In 2000, $n = 1$ and $a_1 = 800$.

In 2001, $n = 2$ and $a_2 = \dfrac{1}{2}(800) = 400$.

In 2002, $n = 3$ and $a_3 = \dfrac{1}{2}(400) = 200$.

In 2003, $n = 4$ and $a_4 = \dfrac{1}{2}(200) = 100$.

In 2004, $n = 5$ and $a_5 = \dfrac{1}{2}(100) = 50$.

The population estimate for 2004 is 50 sparrows. Continuing the sequence:

In 2005, $n = 6$ and $a_6 = \dfrac{1}{2}(50) = 25$.

In 2006, $n = 7$ and $a_7 = \dfrac{1}{2}(25) \approx 12$.

In 2007, $n = 8$ and $a_8 = \dfrac{1}{2}(12) = 6$.

In 2008, $n = 9$ and $a_9 = \dfrac{1}{2}(6) = 3$.

In 2009, $n = 10$ and $a_{10} = \dfrac{1}{2}(3) \approx 1$.

In 2010, $n = 11$ and $a_{11} = \dfrac{1}{2}(1) \approx 0$.

The population is estimated to become extinct in 2010.

47. $f(x) = (x-1)^2 + 3$

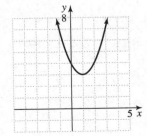

49. $f(x) = 2(x+4)^2 + 2$

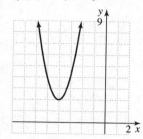

51. $(-4, -1)$ and $(-7, -3)$

$$d = \sqrt{[-7-(-4)]^2 + [-3-(-1)]^2}$$
$$= \sqrt{(-7+4)^2 + (-3+1)^2}$$
$$= \sqrt{(-3)^2 + (-2)^2}$$
$$= \sqrt{9+4}$$
$$= \sqrt{13} \text{ units}$$

53. $(2, -7)$ and $(-3, -3)$

$$d = \sqrt{(-3-2)^2 + [-3-(-7)]^2}$$
$$= \sqrt{(-5)^2 + (-3+7)^2}$$
$$= \sqrt{(-5)^2 + (4)^2}$$
$$= \sqrt{25+16}$$
$$= \sqrt{41} \text{ units}$$

55. $a_n = \dfrac{1}{\sqrt{n}}$

$$a_1 = \frac{1}{\sqrt{1}} = \frac{1}{1} = 1$$
$$a_2 = \frac{1}{\sqrt{2}} \approx 0.7071$$
$$a_3 = \frac{1}{\sqrt{3}} \approx 0.5774$$
$$a_4 = \frac{1}{\sqrt{4}} = \frac{1}{2} = 0.5$$
$$a_5 = \frac{1}{\sqrt{5}} \approx 0.4472$$

Thus, the first five terms of the sequence

$a_n = \dfrac{1}{\sqrt{n}}$ are 1, 0.7071, 0.5774, 0.5, 0.4472.

57. $a_n = \left(1 + \dfrac{1}{n}\right)^n$

$$a_1 = \left(1 + \frac{1}{1}\right)^1 = (2)^1 = 2$$
$$a_2 = \left(1 + \frac{1}{2}\right)^2 = \left(\frac{3}{2}\right)^2 = 2.25$$
$$a_3 = \left(1 + \frac{1}{3}\right)^3 = \left(\frac{4}{3}\right)^3 \approx 2.3704$$
$$a_4 = \left(1 + \frac{1}{4}\right)^4 = \left(\frac{5}{4}\right)^4 \approx 2.4414$$
$$a_5 = \left(1 + \frac{1}{5}\right)^5 = \left(\frac{6}{5}\right)^5 \approx 2.4883$$

Thus, the first five terms of the sequence

$a_n = \left(1 + \dfrac{1}{n}\right)^n$ are 2, 2.25, 2.3704, 2.4414, 2.4883.

Section 14.2 Practice

1. $a_1 = 4$
$a_2 = 4 + 5 = 9$
$a_3 = 9 + 5 = 14$
$a_4 = 14 + 5 = 19$
$a_5 = 19 + 5 = 24$
The first five terms are 4, 9, 14, 19, 24.

2. a. $a_n = a_1 + (n-1)d$
Here, $a_1 = 2$ and $d = -3$.
$a_n = 2 + (n-1)(-3) = 2 - 3n + 3 = 5 - 3n$

b. $a_n = 5 - 3n$
$a_{12} = 5 - 3 \cdot 12 = 5 - 36 = -31$

3. Since the sequence is arithmetic, the ninth term is $a_9 = a_1 + (9-1)d = a_1 + 8d$.
a_1 is the first term of the sequence, so $a_1 = 3$. d is the constant difference, so
$d = a_2 - a_1 = 9 - 3 = 6$. Thus,
$a_9 = a_1 + 8d = 3 + 8 \cdot 6 = 51$.

4. We need to find a_1 and d. The given facts, $a_3 = 23$ and $a_8 = 63$, lead to a system of linear equations.

$$\begin{cases} a_3 = a_1 + (3-1)d \\ a_8 = a_1 + (8-1)d \end{cases} \text{ or } \begin{cases} 23 = a_1 + 2d \\ 63 = a_1 + 7d \end{cases}$$

We solve the system by elimination. Multiply both sides of the second equation by -1.

$$\begin{cases} 23 = a_1 + 2d \\ -1(63) = -1(a_1 + 7d) \end{cases} \text{ or } \begin{cases} 23 = a_1 + 2d \\ \underline{-63 = -a_1 - 7d} \\ -40 = -5d \\ 8 = d \end{cases}$$

To find a_1, let $d = 8$ in $23 = a_1 + 2d$.

$23 = a_1 + 2(8)$
$23 = a_1 + 16$
$7 = a_1$

Thus, $a_1 = 7$ and $d = 8$, so

$a_n = 7 + (n-1)(8) = 7 + 8n - 8 = -1 + 8n$ and
$a_6 = -1 + 8 \cdot 6 = 47$.

5. The first term, a_1, is 57,000, and d is 2200.

$a_n = 57,000 + (n-1)(2200)$
$\quad = 54,800 + 2200n$

$a_3 = 54,800 + 2200 \cdot 3 = 61,400$

The salary for the third year is $61,400.

6. $a_1 = 8$

$a_2 = 8(-3) = -24$
$a_3 = -24(-3) = 72$
$a_4 = 72(-3) = -216$

The first four terms are 8, −24, 72, and −216.

7. $a_n = a_1 r^{n-1}$

Here, $a_1 = 64$ and $r = \dfrac{1}{4}$.

Evaluate a_n for $n = 7$.

$a_7 = 64 \left(\dfrac{1}{4} \right)^{7-1}$

$\quad = 64 \left(\dfrac{1}{4} \right)^6$

$\quad = 64 \left(\dfrac{1}{4096} \right)$

$\quad = \dfrac{1}{64}$

8. Since the sequence is geometric and $a_1 = -3$, the seventh term must be $a_1 r^{7-1}$, or $-3r^6$. r is the common ratio of terms, so r must be $\dfrac{6}{-3}$, or −2.

$a_7 = -3r^6$
$a_7 = -3(-2)^6 = -192$

9. Notice that $\dfrac{27}{4} \div \dfrac{9}{2} = \dfrac{3}{2}$, so $r = \dfrac{3}{2}$.

$a_2 = a_1 \left(\dfrac{3}{2} \right)^{2-1}$

$\dfrac{9}{2} = a_1 \left(\dfrac{3}{2} \right)^1$, or $a_1 = 3$

The first term is 3, and the common ratio is $\dfrac{3}{2}$.

10. Since the culture is reduced by one-half each day, the population sizes are modeled by a geometric sequence. Here, $a_1 = 4800$ and $r = \dfrac{1}{2}$.

$a_n = a_1 r^{n-1} = 4800 \left(\dfrac{1}{2} \right)^{n-1}$

$a_7 = 4800 \left(\dfrac{1}{2} \right)^{7-1} = 75$

The bacterial culture should measure 75 units at the beginning of day 7.

Vocabulary, Readiness & Video Check 14.2

1. A geometric sequence is one in which each term (after the first) is obtained by multiplying the preceding term by a constant r. The constant r is called the common ratio.

2. An arithmetic sequence is one in which each term (after the first) differs from the preceding term by a constant amount d. The constant d is called the common difference.

3. The general term of an arithmetic sequence is $a_n = a_1 + (n-1)d$ where a_1 is the first term and d is the common difference.

4. The general term of a geometric sequence is $a_n = a_1 r^{n-1}$ where a_1 is the first term and r is the common ratio.

5. If there is a common difference between each term and its preceding term in a sequence, it's an arithmetic sequence.

6. An arithmetic sequence has a common difference between terms—you add the same number to go from term to term. A geometric sequence has a common ratio between terms—you multiply by the same number to go from term to term.

Exercise Set 14.2

1. $a_n = a_1 + (n-1)d$
$a_1 = 4; d = 2$
$a_1 = 4$
$a_2 = 4 + (2-1)2 = 6$
$a_3 = 4 + (3-1)2 = 8$
$a_4 = 4 + (4-1)2 = 10$
$a_5 = 4 + (5-1)2 = 12$
The first five terms are 4, 6, 8, 10, 12.

3. $a_n = a_1 + (n-1)d$
$a_1 = 6, d = -2$
$a_1 = 6$
$a_2 = 6 + (2-1)(-2) = 4$
$a_3 = 6 + (3-1)(-2) = 2$
$a_4 = 6 + (4-1)(-2) = 0$
$a_5 = 6 + (5-1)(-2) = -2$
The first five terms are 6, 4, 2, 0, -2.

5. $a_n = a_1 r^{n-1}$
$a_1 = 1, r = 3$
$a_1 = 1(3)^{1-1} = 1$
$a_2 = 1(3)^{2-1} = 3$
$a_3 = 1(3)^{3-1} = 9$
$a_4 = 1(3)^{4-1} = 27$
$a_5 = 1(3)^{5-1} = 81$
The first five terms are 1, 3, 9, 27, 81.

7. $a_n = a_1 r^{n-1}$
$a_1 = 48, r = \dfrac{1}{2}$
$a_1 = 48\left(\dfrac{1}{2}\right)^{1-1} = 48$
$a_2 = 48\left(\dfrac{1}{2}\right)^{2-1} = 24$
$a_3 = 48\left(\dfrac{1}{2}\right)^{3-1} = 12$
$a_4 = 48\left(\dfrac{1}{2}\right)^{4-1} = 6$
$a_5 = 48\left(\dfrac{1}{2}\right)^{5-1} = 3$
The first five terms are 48, 24, 12, 6, 3.

9. $a_n = a_1 + (n-1)d$
$a_1 = 12, d = 3$
$a_n = 12 + (n-1)3$
$a_8 = 12 + 7(3) = 12 + 21 = 33$

11. $a_n = a_1 r^{n-1}$
$a_1 = 7, d = -5$
$a_n = a_1 r^{n-1}$
$a_4 = 7(-5)^3 = 7(-125) = -875$

13. $a_n = a_1 + (n-1)d$
$a_1 = -4, d = -4$
$a_n = -4 + (n-1)(-4)$
$a_{15} = -4 + 14(-4) = -4 - 56 = -60$

15. 0, 12, 24
$a_1 = 0$ and $d = 12$
$a_n = 0 + (n-1)12$
$a_9 = 8(12) = 96$

17. 20, 18, 16
$a_1 = 20$ and $d = -2$
$a_n = 20 + (n-1)(-2)$
$a_{25} = 20 + 24(-2) = 20 - 48 = -28$

19. 2, -10, 50
$a_1 = 2$ and $r = -5$
$a_n = 2(-5)^{n-1}$
$a_5 = 2(-5)^4 = 2(625) = 1250$

21. $a_4 = 19, a_{15} = 52$
$\begin{cases} a_4 = a_1 + (4-1)d \\ a_{15} = a_1 + (15-1)d \end{cases}$ or
$\begin{cases} 19 = a_1 + 3d \\ 52 = a_1 + 14d \end{cases}$
$\begin{cases} -19 = -a_1 - 3d \\ 52 = a_1 + 14d \end{cases}$
Adding yields $33 = 11d$ or $d = 3$. Then
$a_1 = 19 - 3(3) = 10$.
$a_n = 10 + (n-1)3$
$= 10 + 3n - 3$
$= 7 + 3n$
and $a_8 = 7 + 3(8)$
$= 7 + 24$
$= 31$

23. $a_2 = -1$, $a_4 = 5$

$\begin{cases} a_2 = a_1 + (2-1)d \\ a_4 = a_1 + (4-1)d \end{cases}$ or

$\begin{cases} -1 = a_1 + d \\ 5 = a_1 + 3d \end{cases}$

$\begin{cases} 1 = -a_1 - d \\ 5 = a_1 + 3d \end{cases}$

Adding yields $6 = 2d$ or $d = 3$. Then
$a_1 = -1 - 3 = -4$.

$\begin{aligned} a_n &= -4 + (n-1)3 \\ &= -4 + 3n - 3 \\ &= -7 + 3n \end{aligned}$

and $\begin{aligned} a_9 &= -7 + 3(9) \\ &= -7 + 27 \\ &= 20 \end{aligned}$

25. $a_2 = -\dfrac{4}{3}$ and $a_3 = \dfrac{8}{3}$

Notice that $\dfrac{8}{3} \div \dfrac{-4}{3} = \dfrac{8}{3} \cdot -\dfrac{3}{4} = -2$, so $r = -2$.

Then

$a_2 = a_1(-2)^{2-1}$

$-\dfrac{4}{3} = a_1(-2)$

$\dfrac{2}{3} = a_1.$

The first term is $\dfrac{2}{3}$ and the common ratio is -2.

27. answers may vary

29. $2, 4, 6$ is an arithmetic sequence.
$a_1 = 2$ and $d = 2$

31. $5, 10, 20$ is a geometric sequence.
$a_1 = 5$ and $r = 2$

33. $\dfrac{1}{2}, \dfrac{1}{10}, \dfrac{1}{50}$ is a geometric sequence.

$a_1 = \dfrac{1}{2}$ and $r = \dfrac{1}{5}$

35. $x, 5x, 25x$ is a geometric sequence.
$a_1 = x$ and $r = 5$

37. $p, p + 4, p + 8$ is an arithmetic sequence.
$a_1 = p$ and $d = 4$

39. $a_1 = 14$ and $d = \dfrac{1}{4}$

$a_n = 14 + (n-1)\dfrac{1}{4}$

$a_{21} = 14 + 20\left(\dfrac{1}{4}\right) = 14 + 5 = 19$

41. $a_1 = 3$ and $r = -\dfrac{2}{3}$

$a_n = 3\left(-\dfrac{2}{3}\right)^{n-1}$

$a_4 = 3\left(-\dfrac{2}{3}\right)^3 = 3\left(-\dfrac{8}{27}\right) = -\dfrac{8}{9}$

43. $\dfrac{3}{2}, 2, \dfrac{5}{2}, ...$

$a_1 = \dfrac{3}{2}$ and $d = \dfrac{1}{2}$

$a_n = \dfrac{3}{2} + (n-1)\dfrac{1}{2}$

$a_{15} = \dfrac{3}{2} + 14\left(\dfrac{1}{2}\right) = \dfrac{17}{2}$

45. $24, 8, \dfrac{8}{3}, ...$

$a_1 = 24$ and $r = \dfrac{1}{3}$

$a_n = 24\left(\dfrac{1}{3}\right)^{n-1}$

$a_6 = 24\left(\dfrac{1}{3}\right)^5 = 24\left(\dfrac{1}{243}\right) = \dfrac{8}{81}$

47. $a_3 = 2$, $a_{17} = -40$

$\begin{cases} a_3 = a_1 + (3-1)d \\ a_{17} = a_1 + (17-1)d \end{cases}$ or

$\begin{cases} 2 = a_1 + 2d \\ -40 = a_1 + 16d \end{cases}$

$\begin{cases} -2 = -a_1 - 2d \\ -40 = a_1 + 16d \end{cases}$

Adding yields $-42 = 14d$ or $d = -3$. Then
$a_1 = 2 - 2(-3) = 8$.

$a_n = 8 + (n-1)(-3) = 8 - 3n + 3 = 11 - 3n$

and

$a_{10} = 11 - 3(10) = 11 - 30 = -19$

49. $54, 58, 62$

$a_1 = 54$ and $d = 4$

$a_n = 54 + (n-1)4$

$a_{20} = 54 + 19(4) = 54 + 76 = 130$

The general term of the sequence is $a_n = 4n + 50$. There are 130 seats in the twentieth row.

51. $a_1 = 6$ and $r = 3$

$a_n = 6(3)^{n-1} = 2 \cdot 3 \cdot (3)^{n-1} = 2(3)^n$

The general term of the sequence is $a_n = 6(3)^{n-1}$ or $a_n = 2(3)^n$.

53. $a_1 = 486$ and $r = \dfrac{1}{3}$

Initial Height $= a_1 = 486\left(\dfrac{1}{3}\right)^{1-1} = 486$

Rebound 1 $= a_2 = 486\left(\dfrac{1}{3}\right)^{2-1} = 162$

Rebound 2 $= a_3 = 486\left(\dfrac{1}{3}\right)^{3-1} = 54$

Rebound 3 $= a_4 = 486\left(\dfrac{1}{3}\right)^{4-1} = 18$

Rebound 4 $= a_5 = 486\left(\dfrac{1}{3}\right)^{5-1} = 6$

The first five terms of the sequence are 486, 162, 54, 18, 6.

The general term is $a_n = 486\left(\dfrac{1}{3}\right)^{n-1}$ or

$a_n = \dfrac{486}{3^{n-1}}$. Since $a_6 = 2$ and $a_7 = \dfrac{2}{3}$, a_7 is the

first term less than 1. Since a_7 corresponds to the 6th bounce, it takes 6 bounces for the ball to rebound less than 1 foot.

55. $a_1 = 4000$ and $d = 125$

$a_n = 4000 + (n-1)125$ or

$a_n = 3875 + 125n$

$a_{12} = 4000 + 11(125) = 5375$

His salary for his last month of training is $5375.

57. $a_1 = 400$ and $r = \dfrac{1}{2}$

12 hours $= 4(3$ hours$)$, so we seek the fourth term after a_1, namely a_5.

$a_n = a_1 r^{n-1}$

$a_5 = 400\left(\dfrac{1}{2}\right)^4 = \dfrac{400}{16} = 25$

25 grams of the radioactive material remain after 12 hours.

59. $\dfrac{1}{3(1)} + \dfrac{1}{3(2)} + \dfrac{1}{3(3)} = \dfrac{1}{3} + \dfrac{1}{6} + \dfrac{1}{9}$

$= \dfrac{6}{18} + \dfrac{3}{18} + \dfrac{2}{18}$

$= \dfrac{11}{18}$

61. $3^0 + 3^1 + 3^2 + 3^3 = 1 + 3 + 9 + 27 = 40$

63. $\dfrac{8-1}{8+1} + \dfrac{8-2}{8+2} + \dfrac{8-3}{8+3} = \dfrac{7}{9} + \dfrac{6}{10} + \dfrac{5}{11}$

$= \dfrac{770}{990} + \dfrac{594}{990} + \dfrac{450}{990}$

$= \dfrac{1814}{990}$

$= \dfrac{907}{495}$

65. $a_1 = \$11,782.40$

$r = 0.5$

$a_2 = (11,782.40)(0.5) = \$5891.20$

$a_3 = (5891.20)(0.5) = \$2945.60$

$a_4 = (2945.60)(0.5) = \$1472.80$

The first four terms of the sequence are $11,782.40, $5891.20, $2945.60, $1472.80.

67. $a_1 = 19.652$ and $d = -0.034$

$a_2 = 19.652 - 0.034 = 19.618$

$a_3 = 19.618 - 0.034 = 19.584$

$a_4 = 19.584 - 0.034 = 19.550$

69. answers may vary

Section 14.3 Practice

1. a. $\displaystyle\sum_{i=0}^{4} \dfrac{i-3}{4} = \dfrac{0-3}{4} + \dfrac{1-3}{4} + \dfrac{2-3}{4} + \dfrac{3-3}{4} + \dfrac{4-3}{4}$

$= \left(-\dfrac{3}{4}\right) + \left(-\dfrac{2}{4}\right) + \left(-\dfrac{1}{4}\right) + 0 + \dfrac{1}{4}$

$= -\dfrac{5}{4}$ or $-1\dfrac{1}{4}$

b. $\displaystyle\sum_{i=2}^{5} 3^i = 3^2 + 3^3 + 3^4 + 3^5$

$$= 9 + 27 + 81 + 243$$
$$= 360$$

2. a. Since the difference of each term and the preceding term is 5, the terms correspond to the first six terms of the arithmetic sequence $a_n = 5 + (n-1)5 = 5n$. Thus, in summation notation,

$$5 + 10 + 15 + 20 + 25 + 30 = \sum_{i=1}^{6} 5i.$$

b. Since each term is the product of the preceding term and $\dfrac{1}{5}$, these terms correspond to the first four terms of the geometric sequence $a_n = \dfrac{1}{5}\left(\dfrac{1}{5}\right)^{n-1} = \left(\dfrac{1}{5}\right)^n$.

In summation notation,

$$\frac{1}{5} + \frac{1}{25} + \frac{1}{125} + \frac{1}{625} = \sum_{i=1}^{4} \left(\frac{1}{5}\right)^i.$$

3. $S_4 = \displaystyle\sum_{i=1}^{4} \dfrac{2+3i}{i^2}$

$$= \frac{2+3\cdot 1}{1^2} + \frac{2+3\cdot 2}{2^2} + \frac{2+3\cdot 3}{3^2} + \frac{2+3\cdot 4}{4^2}$$

$$= \frac{5}{1} + \frac{8}{4} + \frac{11}{9} + \frac{14}{16}$$

$$= 5 + 2 + \frac{11}{9} + \frac{7}{8}$$

$$= \frac{655}{72} \text{ or } 9\frac{7}{72}$$

4. $S_5 = \displaystyle\sum_{i=1}^{5} i(2i-1)$

$$= 1(2\cdot 1 - 1) + 2(2\cdot 2 - 1) + 3(2\cdot 3 - 1)$$
$$+ 4(2\cdot 4 - 1) + 5(2\cdot 5 - 1)$$

$$= 1 + 6 + 15 + 28 + 45$$

$$= 95$$

There are 95 plants after 5 years.

Vocabulary, Readiness & Video Check 14.3

1. A series is an <u>infinite</u> series if it is the sum of all the terms of an infinite sequence.

2. A series is a <u>finite</u> series if it is the sum of a finite number of terms.

3. A shorthand notation for denoting a series when the general term of the sequence is known is called <u>summation</u> notation.

4. In the notation $\displaystyle\sum_{i=1}^{7} (5i-2)$, the Σ is the Greek uppercase letter <u>sigma</u> and the i is called the <u>index of summation</u>.

5. The sum of the first n terms of a sequence is a finite series known as a <u>partial sum</u>.

6. For the notation in Exercise 4 above, the beginning value of i is $\underline{1}$ and the ending value of i is $\underline{7}$.

7. sigma/sum, index of summation, beginning value of i, ending value of i, and general term of the sequence

8. the sum of the first 7 terms of the sequence

Exercise Set 14.3

1. $\displaystyle\sum_{i=1}^{4} (i-3) = (1-3) + (2-3) + (3-3) + (4-3)$

$$= -2 + (-1) + 0 + 1$$
$$= -2$$

3. $\displaystyle\sum_{i=4}^{7} (2i+4) = [2(4)+4] + [2(5)+4] + [2(6)+4]$

$$+ [2(7)+4]$$

$$= 12 + 14 + 16 + 18$$
$$= 60$$

5. $\displaystyle\sum_{i=2}^{4} (i^2 - 3) = (2^2 - 3) + (3^2 - 3) + (4^2 - 3)$

$$= 1 + 6 + 13$$
$$= 20$$

7. $\displaystyle\sum_{i=1}^{3} \left(\dfrac{1}{i+5}\right) = \dfrac{1}{1+5} + \dfrac{1}{2+5} + \dfrac{1}{3+5}$

$$= \frac{1}{6} + \frac{1}{7} + \frac{1}{8}$$

$$= \frac{28}{168} + \frac{24}{168} + \frac{21}{168}$$

$$= \frac{73}{168}$$

9. $\displaystyle\sum_{i=1}^{3}\frac{1}{6i}=\frac{1}{6(1)}+\frac{1}{6(2)}+\frac{1}{6(3)}$

$\qquad\quad =\dfrac{1}{6}+\dfrac{1}{12}+\dfrac{1}{18}$

$\qquad\quad =\dfrac{6+3+2}{36}$

$\qquad\quad =\dfrac{11}{36}$

11. $\displaystyle\sum_{i=2}^{6}3i=3(2)+3(3)+3(4)+3(5)+3(6)$

$\qquad\quad =6+9+12+15+18$

$\qquad\quad =60$

13. $\displaystyle\sum_{i=3}^{5}i(i+2)=3(3+2)+4(4+2)+5(5+2)$

$\qquad\qquad\quad =15+24+35$

$\qquad\qquad\quad =74$

15. $\displaystyle\sum_{i=1}^{5}2^{i}=2^{1}+2^{2}+2^{3}+2^{4}+2^{5}$

$\qquad\quad =2+4+8+16+32$

$\qquad\quad =62$

17. $\displaystyle\sum_{i=1}^{4}\frac{4i}{i+3}=\frac{4(1)}{1+3}+\frac{4(2)}{2+3}+\frac{4(3)}{3+3}+\frac{4(4)}{4+3}$

$\qquad\qquad =1+\dfrac{8}{5}+2+\dfrac{16}{7}$

$\qquad\qquad =\dfrac{105}{35}+\dfrac{56}{35}+\dfrac{80}{35}$

$\qquad\qquad =\dfrac{241}{35}$

19. $1+3+5+7+9$

$\quad a_1=1,\ d=2$

$\quad a_n=1+(n-1)2=2n-1$

$\quad \displaystyle\sum_{i=1}^{5}(2i-1)$

21. $4+12+36+108=4+4(3)+4(3)^{2}+4(3)^{3}$

$\qquad\qquad\qquad\quad =\displaystyle\sum_{i=1}^{4}4(3)^{i-1}$

23. $12+9+6+3+0+(-3)$

$\quad a_1=12,\ d=-3$

$\quad a_n=12+(n-1)(-3)=-3n+15$

$\quad \displaystyle\sum_{i=1}^{6}(-3i+15)$

25. $12+4+\dfrac{4}{3}+\dfrac{4}{9}=\dfrac{4}{3^{-1}}+\dfrac{4}{3^{0}}+\dfrac{4}{3}+\dfrac{4}{3^{2}}$

$\qquad\qquad\qquad\quad =\displaystyle\sum_{i=1}^{4}\frac{4}{3^{i-2}}$

27. $1+4+9+16+25+36+49$

$\quad =1^{2}+2^{2}+3^{2}+4^{2}+5^{2}+6^{2}+7^{2}$

$\quad =\displaystyle\sum_{i=1}^{7}i^{2}$

29. $a_n=(n+2)(n-5)$

$\quad S_2=\displaystyle\sum_{i=1}^{2}(i+2)(i-5)$

$\qquad =(1+2)(1-5)+(2+2)(2-5)$

$\qquad =3(-4)+4(-3)$

$\qquad =-12-12$

$\qquad =-24$

31. $a_n=(-1)^{n}$

$\quad S_6=\displaystyle\sum_{i=1}^{6}(-1)^{i}$

$\qquad =(-1)^{1}+(-1)^{2}+(-1)^{3}+(-1)^{4}+(-1)^{5}$

$\qquad\qquad +(-1)^{6}$

$\qquad =-1+1+(-1)+1+(-1)+1$

$\qquad =0$

33. $a_n=(n+3)(n+1)$

$\quad S_4=\displaystyle\sum_{i=1}^{4}(i+3)(i+1)$

$\qquad =(1+3)(1+1)+(2+3)(2+1)+(3+3)(3+1)$

$\qquad\qquad +(4+3)(4+1)$

$\qquad =4(2)+5(3)+6(4)+7(5)$

$\qquad =8+15+24+35$

$\qquad =82$

35. $a_n = -2n$

$$S_4 = \sum_{i=1}^{4}(-2i)$$
$$= -2(1) + (-2)(2) + (-2)(3) + (-2)(4)$$
$$= -2 - 4 - 6 - 8$$
$$= -20$$

37. $a_n = -\dfrac{n}{3}$

$$S_3 = \sum_{i=1}^{3} -\frac{i}{3} = -\frac{1}{3} - \frac{2}{3} - \frac{3}{3} = -2$$

39. $1, 2, 3, \ldots, 10$

$a_n = n$

$$S_{10} = \sum_{i=1}^{10} i = 1 + 2 + 3 + \ldots + 10 = 55$$

A total of 55 trees were planted.

41. $a_n = 6 \cdot 2^{n-1}$

$$S_5 = \sum_{i=1}^{5} 6 \cdot 2^{i-1}$$
$$= 6(2)^0 + 6(2)^1 + 6(2)^2 + 6(2)^3 + 6(2)^4$$
$$= 6 + 12 + 24 + 48 + 96$$
$$= 186$$

Adding in the original 6 units, there will be 192 fungus units at the end of the fifth day.

43. $a_4 = (4+1)(4+3) = 5(7) = 35$

In the fourth year, 35 species were born.

$$S_4 = \sum_{i=1}^{4}(i+1)(i+3)$$
$$= 2(4) + 3(5) + 4(6) + 5(7)$$
$$= 8 + 15 + 24 + 12$$
$$= 82$$

There were 82 species born in the first four years.

45. $a_n = (n+1)(n+2)$

$a_4 = (4+1)(4+2) = 5(6) = 30$

30 opossums were killed in the fourth month.

$$S_4 = \sum_{i=1}^{4}(i+1)(i+2)$$
$$= 2(3) + (3)(4) + (4)(5) + (5)(6)$$
$$= 6 + 12 + 20 + 30$$
$$= 68$$

68 opossums were killed in the four months.

47. $a_n = 100(0.5)^n$

$a_4 = 100(0.5)^4 = 6.25$

The decay in the fourth year is 6.25 pounds.

$$S_4 = \sum_{i=1}^{4} 100(0.5)^i$$
$$= 100(0.5)^1 + 100(0.5)^2 + 100(0.5)^3$$
$$\quad + 100(0.5)^4$$
$$= 100(0.5) + 100(0.25) + 100(0.125)$$
$$\quad + 100(0.0625)$$
$$= 50 + 25 + 12.5 + 6.25$$
$$= 93.75$$

The decay over the four years is 93.75 pounds.

49. $a_1 = 40$ and $r = \dfrac{4}{5}$

$$a_5 = 40\left(\frac{4}{5}\right)^4 = 16.384$$

The length of the fifth swing is approximately 16.4 inches.

$$S_5 = \sum_{i=1}^{5} 40\left(\frac{4}{5}\right)^{i-1}$$
$$= 40\left(\frac{4}{5}\right)^0 + 40\left(\frac{4}{5}\right)^1 + 40\left(\frac{4}{5}\right)^2 + 40\left(\frac{4}{5}\right)^3$$
$$\quad + 40\left(\frac{4}{5}\right)^4$$
$$= 40 + 32 + 25.6 + 20.48 + 16.384$$
$$= 134.464$$

The pendulum swings about 134.5 inches in five swings.

51. $\dfrac{5}{1-\frac{1}{2}} = \dfrac{5}{\frac{1}{2}} = 5 \cdot \dfrac{2}{1} = 10$

53. $\dfrac{\frac{1}{3}}{1-\frac{1}{10}} = \dfrac{\frac{1}{3}}{\frac{9}{10}} = \dfrac{1}{3} \cdot \dfrac{10}{9} = \dfrac{10}{27}$

55. $\dfrac{3(1-2^4)}{1-2} = \dfrac{3(1-16)}{-1} = \dfrac{3(-15)}{-1} = \dfrac{-45}{-1} = 45$

57. $\dfrac{10}{2}(3+15) = \dfrac{10}{2}(18) = \dfrac{180}{2} = 90$

59. a. $\displaystyle\sum_{i=1}^{7}(i+i^2)$

$= (1+1^2) + (2+2^2) + (3+3^2) + (4+4^2)$
$\quad + (5+5^2) + (6+6^2) + (7+7^2)$
$= 2+6+12+20+30+42+56$

b. $\displaystyle\sum_{i=1}^{7}i + \sum_{i=1}^{7}i^2$

$= (1+2+3+4+5+6+7)$
$\quad + (1+4+9+16+25+36+49)$

c. answers may vary

d. True; answers may vary.

Integrated Review

1. $a_n = n-3$
$a_1 = 1-3 = -2$
$a_2 = 2-3 = -1$
$a_3 = 3-3 = 0$
$a_4 = 4-3 = 1$
$a_5 = 5-3 = 2$
Therefore, the first five terms are −2, −1, 0, 1, 2.

2. $a_n = \dfrac{7}{1+n}$
$a_1 = \dfrac{7}{1+1} = \dfrac{7}{2}$
$a_2 = \dfrac{7}{1+2} = \dfrac{7}{3}$
$a_3 = \dfrac{7}{1+3} = \dfrac{7}{4}$
$a_4 = \dfrac{7}{1+4} = \dfrac{7}{5}$
$a_5 = \dfrac{7}{1+5} = \dfrac{7}{6}$
The first five terms are $\dfrac{7}{2}, \dfrac{7}{3}, \dfrac{7}{4}, \dfrac{7}{5}$, and $\dfrac{7}{6}$.

3. $a_n = 3^{n-1}$
$a_1 = 3^{1-1} = 3^0 = 1$
$a_2 = 3^{2-1} = 3^1 = 3$
$a_3 = 3^{3-1} = 3^2 = 9$
$a_4 = 3^{4-1} = 3^3 = 27$
$a_5 = 3^{5-1} = 3^4 = 81$
The first five terms are 1, 3, 9, 27, and 81.

4. $a_n = n^2 - 5$
$a_1 = 1^2 - 5 = 1 - 5 = -4$
$a_2 = 2^2 - 5 = 4 - 5 = -1$
$a_3 = 3^2 - 5 = 9 - 5 = 4$
$a_4 = 4^2 - 5 = 16 - 5 = 11$
$a_5 = 5^2 - 5 = 25 - 5 = 20$
The first five terms are −4, −1, 4, 11, and 20.

5. $(-2)^n;\ a_6$
$a_6 = (-2)^6 = 64$

6. $-n^2 + 2;\ a_4$
$a_4 = -(4)^2 + 2 = -16 + 2 = -14$

7. $\dfrac{(-1)^n}{n};\ a_{40}$
$a_{40} = \dfrac{(-1)^{40}}{40} = \dfrac{1}{40}$

8. $\dfrac{(-1)^n}{2n};\ a_{41}$
$a_{41} = \dfrac{(-1)^{41}}{2(41)} = \dfrac{-1}{82} = -\dfrac{1}{82}$

9. $a_1 = 7;\ d = -3$
$a_1 = 7$
$a_2 = 7 - 3 = 4$
$a_3 = 4 - 3 = 1$
$a_4 = 1 - 3 = -2$
$a_5 = -2 - 3 = -5$
The first five terms are 7, 4, 1, −2, −5.

10. $a_1 = -3;\ r = 5$

$a_1 = -3$

$a_2 = -3(5) = -15$

$a_3 = -15(5) = -75$

$a_4 = -75(5) = -375$

$a_5 = -375(5) = -1875$

The first five terms are $-3, -15, -75, -375,$ $-1875.$

11. $a_1 = 45;\ r = \dfrac{1}{3}$

$a_1 = 45$

$a_2 = 45\left(\dfrac{1}{3}\right) = 15$

$a_3 = 15\left(\dfrac{1}{3}\right) = 5$

$a_4 = 5\left(\dfrac{1}{3}\right) = \dfrac{5}{3}$

$a_5 = \dfrac{5}{3}\left(\dfrac{1}{3}\right) = \dfrac{5}{9}$

The first five terms are $45, 15, 5, \dfrac{5}{3}, \dfrac{5}{9}.$

12. $a_1 = -12;\ d = 10$

$a_1 = -12$

$a_2 = -12 + 10 = -2$

$a_3 = -2 + 10 = 8$

$a_4 = 8 + 10 = 18$

$a_5 = 18 + 10 = 28$

The first five terms are $-12, -2, 8, 18, 28.$

13. $a_1 = 20;\ d = 9$

$a_n = a_1 + (n-1)d$

$\begin{aligned} a_{10} &= 20 + (10-1)9 \\ &= 20 + 81 \\ &= 101 \end{aligned}$

14. $a_1 = 64;\ r = \dfrac{3}{4}$

$a_n = a_1 r^{n-1}$

$\begin{aligned} a_6 &= 64\left(\dfrac{3}{4}\right)^{6-1} \\ &= 64\left(\dfrac{3}{4}\right)^{5} \\ &= 64\left(\dfrac{243}{1024}\right) \\ &= \dfrac{243}{16} \end{aligned}$

15. $a_1 = 6;\ r = \dfrac{-12}{6} = -2$

$a_n = a_1 r^{n-1}$

$a_7 = 6(-2)^{7-1} = 6(-2)^6 = 6(64) = 384$

16. $a_1 = -100;\ d = -85 - (-100) = 15$

$a_n = a_1 + (n-1)d$

$\begin{aligned} a_{20} &= -100 + (20-1)(15) \\ &= -100 + (19)(15) \\ &= -100 + 285 \\ &= 185 \end{aligned}$

17. $a_4 = -5,\ a_{10} = -35$

$a_n = a_1 + (n-1)d$

$\begin{cases} a_4 = a_1 + (4-1)d \\ a_{10} = a_1 + (10-1)d \end{cases}$

$\begin{cases} -5 = a_1 + 3d \\ -35 = a_1 + 9d \end{cases}$

Multiply eq. 2 by -1, then add the equations.

$\begin{cases} -5 = a_1 + 3d \\ (-1)(-35) = -1(a_1 + 9d) \end{cases}$

$\begin{cases} -5 = a_1 + 3d \\ 35 = -a_1 - 9d \end{cases}$

$30 = -6d$

$-5 = d$

To find a_1, let $d = -5$ in

$-5 = a_1 + 3d$

$-5 = a_1 + 3(-5)$

$10 = a_1$

Thus, $a_1 = 10$ and $d = -5$, so

$a_n = 10 + (n-1)(-5) = -5n + 15$

$a_5 = -5(5) + 15 = -10$

18. $a_4 = 1$; $a_7 = \dfrac{1}{8}$

$a_n = a_1 r^{n-1}$

$a_4 = a_1 r^{4-1}$ so $1 = a_1 r^3$

$a_7 = a_1 r^{7-1}$ so $\dfrac{1}{8} = a_1 r^6$

Since $a_1 r^6 = (a_1 r^3) r^3$, $\dfrac{1}{8} = 1 \cdot r^3$ and $r = \dfrac{1}{2}$.

$a_5 = a_4 \cdot r$ so $a_5 = 1 \cdot \dfrac{1}{2} = \dfrac{1}{2}$

19. $\displaystyle\sum_{i=1}^{4} 5i = 5(1) + 5(2) + 5(3) + 5(4)$

$= 5 + 10 + 15 + 20$

$= 50$

20. $\displaystyle\sum_{i=1}^{7} (3i + 2)$

$= (3(1) + 2) + (3(2) + 2) + (3(3) + 2)$
$\quad + (3(4) + 2) + (3(5) + 2) + (3(6) + 2)$
$\quad + (3(7) + 2)$

$= 5 + 8 + 11 + 14 + 17 + 20 + 23$

$= 98$

21. $\displaystyle\sum_{i=3}^{7} 2^{i-4}$

$= 2^{3-4} + 2^{4-4} + 2^{5-4} + 2^{6-4} + 2^{7-4}$

$= 2^{-1} + 2^0 + 2^1 + 2^2 + 2^3$

$= \dfrac{1}{2} + 1 + 2 + 4 + 8$

$= 15\dfrac{1}{2}$

$= \dfrac{31}{2}$

22. $\displaystyle\sum_{i=2}^{5} \dfrac{i}{i+1} = \dfrac{2}{2+1} + \dfrac{3}{3+1} + \dfrac{4}{4+1} + \dfrac{5}{5+1}$

$= \dfrac{2}{3} + \dfrac{3}{4} + \dfrac{4}{5} + \dfrac{5}{6}$

$= \dfrac{61}{20}$

23. $S_3 = \displaystyle\sum_{i=1}^{3} i(i-4)$

$= 1(1-4) + 2(2-4) + 3(3-4)$

$= -3 - 4 - 3$

$= -10$

24. $S_{10} = \displaystyle\sum_{i=1}^{10} (-1)^i (i+1)$

$= (-1)^1 (1+1) + (-1)^2 (2+1)$
$\quad + (-1)^3 (3+1) + (-1)^4 (4+1)$
$\quad + (-1)^5 (5+1) + (-1)^6 (6+1)$
$\quad + (-1)^7 (7+1) + (-1)^8 (8+1)$
$\quad + (-1)^9 (9+1) + (-1)^{10} (10+1)$

$= -2 + 3 - 4 + 5 - 6 + 7 - 8 + 9 - 10 + 11$

$= 5$

Section 14.4 Practice

1. 2, 9, 16, 23, 30, ...
Use the formula for S_n of an arithmetic sequence, replacing n with 5, a_1 with 2, and a_n with 30.

$S_n = \dfrac{n}{2}(a_1 + a_n)$

$S_5 = \dfrac{5}{2}(2 + 30) = \dfrac{5}{2}(32) = 80$

2. Because 1, 2, 3, …, 50 is an arithmetic sequence, use the formula for S_n with $n = 50$, $a_1 = 1$, and $a_n = 50$.

$S_n = \dfrac{n}{2}(a_1 + a_n)$

$S_5 = \dfrac{50}{2}(1 + 50) = 25(51) = 1275$

3. The list 6, 7, …, 15 is the first 10 terms of an arithmetic sequence. Use the formula for S_n with $n = 10$, $a_1 = 6$, and $a_n = 15$.

$S_{10} = \dfrac{10}{2}(6 + 15) = 5(21) = 105$

There are a total of 105 blocks of ice.

4. 32, 8, 2, $\dfrac{1}{2}$, $\dfrac{1}{8}$

Use the formula for the partial sum S_n of the terms of a geometric sequence. Here, $n = 5$, the first term $a_1 = 32$, and the common ratio $r = \dfrac{1}{4}$.

$$S_n = \frac{a_1(1-r^n)}{1-r}$$

$$S_5 = \frac{32\left[1-\left(\frac{1}{4}\right)^5\right]}{1-\frac{1}{4}}$$

$$= \frac{32\left(1-\frac{1}{1024}\right)}{\frac{3}{4}}$$

$$= \frac{32-\frac{1}{32}}{\frac{3}{4}}$$

$$= \frac{\frac{1023}{32}}{\frac{3}{4}}$$

$$= \frac{1023}{32} \cdot \frac{4}{3}$$

$$= \frac{341}{8} \text{ or } 42\frac{5}{8}$$

5. The donations are modeled by the first seven terms of a geometric sequence. Evaluate S_n when $n = 7$, $a_1 = 250,000$, and $r = 0.8$.

$$S_7 = \frac{250,000[1-(0.8)^7]}{1-0.8} = 987,856$$

The total amount donated during the seven years is \$987,856.

6. $7, \frac{7}{4}, \frac{7}{16}, \frac{7}{64}, \ldots$

For this geometric sequence $r = \frac{1}{4}$. Since $|r| < 1$, use the formula for S_∞ of a geometric sequence with $a_1 = 7$ and $r = \frac{1}{4}$.

$$S_\infty = \frac{a_1}{1-r} = \frac{7}{1-\frac{1}{4}} = \frac{7}{\frac{3}{4}} = \frac{28}{3} = 9\frac{1}{3}$$

7. We must find the sum of the terms of an infinite geometric sequence whose first term, a_1, is 36 and whose common ratio, r, is 0.96. Since $|r| < 1$, we may use the formula for S_∞.

$$S_\infty = \frac{a_1}{1-r} = \frac{36}{1-0.96} = \frac{36}{0.04} = 900$$

The ball travels a total distance of 900 inches before it comes to a rest.

Vocabulary, Readiness & Video Check 14.4

1. Each term after the first is 5 more than the preceding term; the sequence is <u>arithmetic</u>.

2. Each term after the first is 2 times the preceding term; the sequence is <u>geometric</u>.

3. Each term after the first is −3 times the preceding term; the sequence is <u>geometric</u>.

4. Each term after the first is 2 more than the preceding term; the sequence is <u>arithmetic</u>.

5. Each term after the first is 7 more than the preceding term; the sequence is <u>arithmetic</u>.

6. Each term after the first is −1 times the preceding term; the sequence is <u>geometric</u>.

7. Use the general term formula from Section 11.2 for the general term of an arithmetic sequence: $a_n = a_1 + (n-1)d$.

8. It would be a sequence in which every number is the same since you would multiply each term by 1 to get the next term; to find a partial sum of n terms, just multiply the first term by n since that term, which just repeats, will be added n times.

9. The common ratio r is 3 for this sequence so that $|r| \geq 1$, or $|3| \geq 1$; S_∞ doesn't exist if $|r| \geq 1$.

Exercise Set 14.4

1. $1, 3, 5, 7, \ldots$

   $d = 2$; $a_6 = 1 + (6-1)(2) = 11$

   $$S_6 = \frac{6}{2}(1+11) = 3(12) = 36$$

3. $4, 12, 36, \ldots$

   $a_1 = 4$, $r = 3$, $n = 5$

   $$S_5 = \frac{4(1-3^5)}{1-3} = 484$$

5. $3, 6, 9, \ldots$

   $d = 3$; $a_6 = 3 + (6-1)(3) = 18$

   $$S_6 = \frac{6}{2}(3+18) = 3(21) = 63$$

7. $2, \dfrac{2}{5}, \dfrac{2}{25}, \ldots$

$a_1 = 2, \ r = \dfrac{1}{5}, \ n = 4$

$S_4 = \dfrac{2\left[1-\left(\frac{1}{5}\right)^4\right]}{1-\frac{1}{5}} = \dfrac{\frac{1248}{625}}{\frac{4}{5}} = \dfrac{312}{125}$

9. $1, 2, 3, \ldots, 10$

The first term is 1 and the tenth term is 10.

$S_{10} = \dfrac{10}{2}(1+10) = 5(11) = 55$

11. $1, 2, 3, 7$

The first term is 1 and the fourth term is 7.

$S_4 = \dfrac{4}{2}(1+7) = 2(8) = 16$

13. $12, 6, 3, \ldots$

$a_1 = 12, \ r = \dfrac{1}{2}$

$S_\infty = \dfrac{12}{1-\frac{1}{2}} = \dfrac{12}{\frac{1}{2}} = 12 \cdot \dfrac{2}{1} = 24$

15. $\dfrac{1}{10}, \dfrac{1}{100}, \dfrac{1}{1000}, \ldots$

$a_1 = \dfrac{1}{10}, \ r = \dfrac{1}{10}$

$S_\infty = \dfrac{\frac{1}{10}}{1-\frac{1}{10}} = \dfrac{\frac{1}{10}}{\frac{9}{10}} = \dfrac{1}{10} \cdot \dfrac{10}{9} = \dfrac{1}{9}$

17. $-10, -5, -\dfrac{5}{2}, \ldots$

$a_1 = -10, \ r = \dfrac{1}{2}$

$S_\infty = \dfrac{-10}{1-\frac{1}{2}} = \dfrac{-10}{\frac{1}{2}} = -10 \cdot \dfrac{2}{1} = -20$

19. $2, -\dfrac{1}{4}, \dfrac{1}{32}, \ldots$

$a_1 = 2, \ r = -\dfrac{1}{8}$

$S_\infty = \dfrac{2}{1-\left(-\frac{1}{8}\right)} = \dfrac{2}{\frac{9}{8}} = 2 \cdot \dfrac{8}{9} = \dfrac{16}{9}$

21. $\dfrac{2}{3}, -\dfrac{1}{3}, \dfrac{1}{6}, \ldots$

$a_1 = \dfrac{2}{3}, \ r = -\dfrac{1}{2}$

$S_\infty = \dfrac{\frac{2}{3}}{1-\left(-\frac{1}{2}\right)} = \dfrac{\frac{2}{3}}{\frac{3}{2}} = \dfrac{2}{3} \cdot \dfrac{2}{3} = \dfrac{4}{9}$

23. $-4, 1, 6, \ldots, 41$

The first term is -4 and the tenth term is 41.

$S_{10} = \dfrac{10}{2}(-4+41) = 5(37) = 185$

25. $3, \dfrac{3}{2}, \dfrac{3}{4}, \ldots$

$a_1 = 3, \ r = \dfrac{1}{2}, \ n = 7$

$S_7 = \dfrac{3\left[1-\left(\frac{1}{2}\right)^7\right]}{1-\frac{1}{2}} = \dfrac{381}{64}$

27. $-12, 6, -3, \ldots$

$a_1 = -12, \ r = -\dfrac{1}{2}, \ n = 5$

$S_5 = \dfrac{-12\left[1-\left(-\frac{1}{2}\right)^5\right]}{1-\left(-\frac{1}{2}\right)} = -\dfrac{33}{4} = -8.25$

29. $\dfrac{1}{2}, \dfrac{1}{4}, 0, \ldots, -\dfrac{17}{4}$

The first term is $\dfrac{1}{2}$ and the twentieth term is $-\dfrac{17}{4}$.

$S_{20} = \dfrac{20}{2}\left(\dfrac{1}{2} - \dfrac{17}{4}\right) = 10\left(\dfrac{-15}{4}\right) = -\dfrac{75}{2}$

31. $a_1 = 8, \ r = -\dfrac{2}{3}, \ n = 3$

$S_3 = \dfrac{8\left[1-\left(-\frac{2}{3}\right)^3\right]}{1-\left(-\frac{2}{3}\right)} = \dfrac{56}{9}$

33. The first five terms are 4000, 3950, 3900, 3850, 3800.

$a_1 = 4000$, $d = -50$, $n = 12$

$a_{12} = 4000 + 11(-50) = 3450$

3450 cars will be sold in month 12.

$S_{12} = \dfrac{12}{2}(4000 + 3450) = 44,700$

44,700 cars will be sold in the first year.

35. Firm *A*:
The first term is 22,000 and the tenth term is 31,000.

$S_{10} = \dfrac{10}{2}(22,000 + 31,000)$

$\quad = \$265,000$

Firm *B*:
The first term is 20,000 and the tenth term is 30,800.

$S_{10} = \dfrac{10}{2}(20,000 + 30,800)$

$\quad = \$254,000$

Thus, Firm *A* is making the more profitable offer.

37. $a_1 = 30,000$, $r = 1.10$, $n = 4$

$a_4 = 30,000(1.10)^{4-1} = 39,930$
She made $39,930 during her fourth year of business.

$S_4 = \dfrac{30,000(1-1.10^4)}{1-1.10} = 139,230$

She made $139,230 during the first four years of business.

39. $a_1 = 30$, $r = 0.9$, $n = 5$

$a_5 = 30(0.9)^{5-1} = 19.683$

Approximately 20 minutes to assemble the first computer.

$S_5 = \dfrac{30(1-0.9^5)}{1-0.9} = 122.853$

Approximately 123 minutes to assemble the first 5 computers.

41. $a_1 = 20$, $r = \dfrac{4}{5}$

$S_\infty = \dfrac{20}{1-\frac{4}{5}} = 100$

We double the number (to account for the flight up as well as down) and subtract 20 (since the first bounce was preceded by only a downward flight). Thus, the ball travels
$2(100) - 20 = 180$ feet.

43. Player *A*:
The first term is 1 and the ninth term is 9.

$S_9 = \dfrac{9}{2}(1+9) = 45$ points

Player *B*:
The first term is 10 and the sixth term is 15.

$S_6 = \dfrac{6}{2}(10+15) = 75$ points

45. The first term is 200 and the twentieth is
$200 - 19(5) = 105$.

$S_{20} = \dfrac{20}{2}(200 + 105) = 3050$

Thus, $3050 rent is paid for 20 days during the holiday rush.

47. $a_1 = 0.01$, $r = 2$, $n = 30$

$S_3 = \dfrac{0.01\left[1-2^{30}\right]}{1-2} = 10,737,418.23$

He would pay $10,737,418.23 in room and board for the 30 days.

49. $6 \cdot 5 \cdot 4 \cdot 3 \cdot 2 \cdot 1 = 720$

51. $\dfrac{3 \cdot 2 \cdot 1}{2 \cdot 1} = \dfrac{3 \cdot \cancel{2} \cdot \cancel{1}}{\cancel{2} \cdot \cancel{1}} = 3$

53. $(x+5)^2 = x^2 + 2 \cdot x \cdot 5 + 5^2 = x^2 + 10x + 25$

55. $(2x-1)^3 = (2x-1)^2(2x-1)$

$\quad = (4x^2 - 4x + 1)(2x - 1)$

$\quad = 8x^3 - 4x^2 + 2x - 8x^2 + 4x - 1$

$\quad = 8x^3 - 12x^2 + 6x - 1$

57. $0.\overline{888} = 0.8 + 0.08 + 0.008 + \cdots$

$\quad = \dfrac{8}{10} + \dfrac{8}{100} + \dfrac{8}{1000} + \cdots$

This is a geometric series with $a_1 = \dfrac{8}{10}$, $r = \dfrac{1}{10}$.

$S_\infty = \dfrac{\frac{8}{10}}{1-\frac{1}{10}} = \dfrac{\frac{8}{10}}{\frac{9}{10}} = \dfrac{8}{10} \cdot \dfrac{10}{9} = \dfrac{8}{9}$

59. answers may vary

Section 14.5 Practice

1. $(p+r)^7$

    The $n = 7$ row of Pascal's triangle is
    1   7   21   35   35   21   7   1
    Using the $n = 7$ row of Pascal's triangle as the coefficients, $(p + r)^7$ can be expanded as
    $$p^7 + 7p^6 r + 21p^5 r^2 + 35p^4 r^3 + 35p^3 r^4 + 21p^2 r^5 + 7pr^6 + r^7$$

2. **a.** $\dfrac{6!}{7!} = \dfrac{6 \cdot 5 \cdot 4 \cdot 3 \cdot 2 \cdot 1}{7 \cdot 6 \cdot 5 \cdot 4 \cdot 3 \cdot 2 \cdot 1} = \dfrac{1}{7}$

    b. $\begin{aligned} \dfrac{8!}{4!2!} &= \dfrac{8 \cdot 7 \cdot 6 \cdot 5 \cdot 4!}{4! \cdot 2 \cdot 1} \\ &= \dfrac{8 \cdot 7 \cdot 6 \cdot 5}{2 \cdot 1} \\ &= 4 \cdot 7 \cdot 6 \cdot 5 \\ &= 840 \end{aligned}$

    c. $\dfrac{5!}{4!1!} = \dfrac{5 \cdot 4 \cdot 3 \cdot 2 \cdot 1}{4 \cdot 3 \cdot 2 \cdot 1 \cdot 1} = 5$

    d. $\dfrac{9!}{9!0!} = \dfrac{9!}{9! \cdot 1} = 1$

3. $(a+b)^9$

    Let $n = 9$ in the binomial formula.
    $$(a+b)^9 = a^9 + \frac{9}{1!}a^8 b + \frac{9 \cdot 8}{2!}a^7 b^2 + \frac{9 \cdot 8 \cdot 7}{3!}a^6 b^3 + \frac{9 \cdot 8 \cdot 7 \cdot 6}{4!}a^5 b^4 + \frac{9 \cdot 8 \cdot 7 \cdot 6 \cdot 5}{5!}a^4 b^5 + \frac{9 \cdot 8 \cdot 7 \cdot 6 \cdot 5 \cdot 4}{6!}a^3 b^6$$
    $$+ \frac{9 \cdot 8 \cdot 7 \cdot 6 \cdot 5 \cdot 4 \cdot 3}{7!}a^2 b^7 + \frac{9 \cdot 8 \cdot 7 \cdot 6 \cdot 5 \cdot 4 \cdot 3 \cdot 2}{8!}ab^8 + b^9$$
    $$= a^9 + 9a^8 b + 36a^7 b^2 + 84a^6 b^3 + 126a^5 b^4 + 126a^4 b^5 + 84a^3 b^6 + 36a^2 b^7 + 9ab^8 + b^9$$

4. $(a+5b)^3$

    Replace b with $5b$ in the binomial formula.
    $$\begin{aligned} (a+5b)^3 &= a^3 + \frac{3}{1!}a^2(5b) + \frac{3 \cdot 2}{2!}a(5b)^2 + (5b)^3 \\ &= a^3 + 3a^2(5b) + 3a(25b^2) + 125b^3 \\ &= a^3 + 15a^2 b + 75ab^2 + 125b^3 \end{aligned}$$

5. $(3x-2y)^3$

    Let $a = 3x$ and $b = -2y$ in the binomial formula.
    $$\begin{aligned} (3x-2y)^3 &= (3x)^3 + \frac{3}{1!}(3x)^2(-2y) + \frac{3 \cdot 2}{2!}(3x)(-2y)^2 + (-2y)^3 \\ &= 27x^3 + 3(9x^2)(-2y) + 3(3x)(4y^2) - 8y^3 \\ &= 27x^3 - 54x^2 y + 36xy^2 - 8y^3 \end{aligned}$$

6. $(x-4y)^{11}$

     Use the formula with $n = 11$, $a = x$, $b = -4y$, and $r + 1 = 7$. Notice that, since $r + 1 = 7$, $r = 6$.

$$\frac{n!}{r!(n-r)!}a^{n-r}b^r = \frac{11!}{6!5!}x^5(-4y)^6$$
$$= 462x^5(4096y^6)$$
$$= 1,892,352x^5y^6$$

Vocabulary, Readiness & Video Check 14.5

1. $0! = \underline{1}$

2. $1! = \underline{1}$

3. $4! = 4 \cdot 3 \cdot 2 \cdot 1 = \underline{24}$

4. $2! = 2 \cdot 1 = \underline{2}$

5. $3!0! = 3 \cdot 2 \cdot 1 \cdot 1 = \underline{6}$

6. $0!2! = 1 \cdot 2 \cdot 1 = \underline{2}$

7. Pascal's triangle gives you the coefficients of the terms of the expanded binomial; also, the power tells you how many terms the expansion has (1 more than the power on the binomial).

8. $4 \cdot 3 \cdot 2 \cdot 1 = 24$; $0! = 1$

9. The theorem is in terms of $(a+b)^n$, so if your binomial is in the form $(a-b)^n$, then remember to think of it as $(a+(-b))^n$, so your second term is $-b$.

10. We are using the formula for the $(r + 1)$st term in a binomial—so if $r + 1 = 4$ then $r = 3$.

Exercise Set 14.5

1. $(m+n)^3 = m^3 + 3m^2n + 3mn^2 + n^3$

3. $(c+d)^5 = c^5 + 5c^4d + 10c^3d^2 + 10c^2d^3 + 5cd^4 + d^5$

5. $(y-x)^5 = \left[y + (-x)\right]^5$
$$= y^5 - 5y^4x + 10y^3x^2 - 10y^2x^3 + 5yx^4 - x^5$$

7. answers may vary

9. $\dfrac{8!}{7!} = \dfrac{8 \cdot 7!}{7!} = 8$

11. $\dfrac{7!}{5!} = \dfrac{7 \cdot 6 \cdot 5!}{5!} = 7 \cdot 6 = 42$

13. $\dfrac{10!}{7!2!} = \dfrac{10 \cdot 9 \cdot 8 \cdot 7!}{7!2!} = \dfrac{10 \cdot 9 \cdot 8}{2 \cdot 1} = 360$

15. $\dfrac{8!}{6!0!} = \dfrac{8 \cdot 7 \cdot 6!}{6!1} = 8 \cdot 7 = 56$

17. Let $n = 7$ in the binomial theorem.

$$(a+b)^7 = a^7 + \frac{7}{1!}a^6 b + \frac{7 \cdot 6}{2!}a^5 b^2 + \frac{7 \cdot 6 \cdot 5}{3!}a^4 b^3 + \frac{7 \cdot 6 \cdot 5 \cdot 4}{4!}a^3 b^4 + \frac{7 \cdot 6 \cdot 5 \cdot 4 \cdot 3}{5!}a^2 b^5 + \frac{7 \cdot 6 \cdot 5 \cdot 4 \cdot 3 \cdot 2}{6!}ab^6 + b^7$$
$$= a^7 + 7a^6 b + 21a^5 b^2 + 35a^4 b^3 + 35a^3 b^4 + 21a^2 b^5 + 7ab^6 + b^7$$

19. Let $b = 2b$ and $n = 5$ in the binomial theorem.

$$(a+2b)^5 = a^5 + \frac{5}{1!}a^4(2b) + \frac{5 \cdot 4}{2!}a^3(2b)^2 + \frac{5 \cdot 4 \cdot 3}{3!}a^2(2b)^3 + \frac{5 \cdot 4 \cdot 3 \cdot 2}{4!}a(2b)^4 + (2b)^5$$
$$= a^5 + 10a^4 b + 40a^3 b^2 + 80a^2 b^3 + 80ab^4 + 32b^5$$

21. Let $a = q$, $b = r$, and $n = 9$ in the binomial theorem.

$$(q+r)^2 = q^9 + \frac{9}{1!}q^8 r + \frac{9 \cdot 8}{2!}q^7 r^2 + \frac{9 \cdot 8 \cdot 7}{3!}q^6 r^3 + \frac{9 \cdot 8 \cdot 7 \cdot 6}{4!}q^5 r^4 + \frac{9 \cdot 8 \cdot 7 \cdot 6 \cdot 5}{5!}q^4 r^5 + \frac{9 \cdot 8 \cdot 7 \cdot 6 \cdot 5 \cdot 4}{6!}q^3 r^6$$
$$+ \frac{9 \cdot 8 \cdot 7 \cdot 6 \cdot 5 \cdot 4 \cdot 3}{7!}q^2 r^7 + \frac{9 \cdot 8 \cdot 7 \cdot 6 \cdot 5 \cdot 4 \cdot 3 \cdot 2}{8!}qr^8 + r^9$$
$$= q^9 + 9q^8 r + 36q^7 r^2 + 84q^6 r^3 + 126q^5 r^4 + 126q^4 r^5 + 84q^3 r^6 + 36q^2 r^7 + 9qr^8 + r^9$$

23. Let $a = 4a$ and $n = 5$ in the binomial theorem.

$$(4a+b)^5 = (4a)^5 + \frac{5}{1!}(4a)^4 b + \frac{5 \cdot 4}{2!}(4a)^3 b^2 + \frac{5 \cdot 4 \cdot 3}{3!}(4a)^2 b^3 + \frac{5 \cdot 4 \cdot 3 \cdot 2}{4!}(4a)b^4 + b^5$$
$$= 1024a^5 + 1280a^4 b + 640a^3 b^2 + 160a^2 b^3 + 20ab^4 + b^5$$

25. Let $a = 5a$, $b = -2b$, and $n = 4$ in the binomial theorem.

$$(5a-2b)^4 = (5a)^4 + \frac{4}{1!}(5a)^3(-2b) + \frac{4 \cdot 3}{2!}(5a)^2(-2b)^2 + \frac{4 \cdot 3 \cdot 2}{3!}(5a)(-2b)^3 + (-2b)^4$$
$$= 625a^4 - 1000a^3 b + 600a^2 b^2 - 160ab^3 + 16b^4$$

27. Let $a = 2a$, $b = 3b$, and $n = 3$ in the binomial theorem.

$$(2a+3b)^3 = (2a)^3 + \frac{3}{1!}(2a)^2(3b) + \frac{3 \cdot 2}{2!}(2a)(3b)^2 + (3b)^3$$
$$= 8a^3 + 36a^2 b + 54ab^2 + 27b^3$$

29. Let $a = x$, $b = 2$, and $n = 5$ in the binomial theorem.

$$(x+2)^5 = x^5 + \frac{5}{1!}x^4(2) + \frac{5 \cdot 4}{2!}x^3(2)^2 + \frac{5 \cdot 4 \cdot 3}{3!}x^2(2)^3 + \frac{5 \cdot 4 \cdot 3 \cdot 2}{4!}x(2)^4 + (2)^5$$
$$= x^5 + 10x^4 + 40x^3 + 80x^2 + 80x + 32$$

31. 5th term of $(c-d)^5$ corresponds to $r = 4$:

$$\frac{5!}{4!(5-4)!}c^{5-4}(-d)^4 = 5cd^4$$

33. 8th term of $(2c+d)^7$ corresponds to $r = 7$:

$$\frac{7!}{7!(7-7)!}(2c)^{7-7}(d)^7 = d^7$$

35. 4th term of $(2r-s)^5$ corresponds to $r = 3$:

$$\frac{5!}{3!(5-3)!}(2r)^{5-3}(-s)^3 = -40r^2s^3$$

37. 3rd term of $(x+y)^4$ corresponds to $r = 2$: $\dfrac{4!}{2!(4-2)!}(x)^{4-2}(y)^2 = 6x^2y^2$

39. 2nd term of $(a+3b)^{10}$ corresponds to $r = 1$: $\dfrac{10!}{1!(10-1)!}(a)^{10-1}(3b)^1 = 30a^9b$

41. $f(x) = |x|$

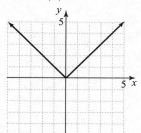

Not one-to-one

43. $H(x) = 2x + 3$

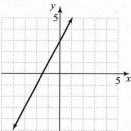

One-to-one

45. $f(x) = x^2 + 3$

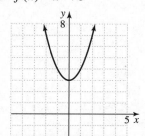

Not one-to-one

47. $(\sqrt{x} + \sqrt{3})^5$

Use the binomial theorem with $n = 5$, $a = \sqrt{x}$, , and $b = \sqrt{3}$.

$$\left(\sqrt{x}+\sqrt{3}\right)^5 = \left(\sqrt{x}\right)^5 + 5\left(\sqrt{x}\right)^4\left(\sqrt{3}\right) + 10\left(\sqrt{x}\right)^3\left(\sqrt{3}\right)^2 + 10\left(\sqrt{x}\right)^2\left(\sqrt{3}\right)^3 + 5\left(\sqrt{x}\right)\left(\sqrt{3}\right)^4 + \left(\sqrt{3}\right)^5$$

$$= x^2\sqrt{x} + 5\sqrt{3}x^2 + 30x\sqrt{x} + 30\sqrt{3}x + 45\sqrt{x} + 9\sqrt{3}$$

49. $\binom{9}{5} = \dfrac{9!}{5!(9-5)!}$

$= \dfrac{9!}{5!4!}$

$= \dfrac{9 \cdot 8 \cdot 7 \cdot 6 \cdot 5 \cdot 4 \cdot 3 \cdot 2 \cdot 1}{(5 \cdot 4 \cdot 3 \cdot 2 \cdot 1) \cdot (4 \cdot 3 \cdot 2 \cdot 1)}$

$= 126$

51. $\binom{8}{2} = \dfrac{8!}{2!(8-2)!}$

$= \dfrac{8!}{2!6!}$

$= \dfrac{8 \cdot 7 \cdot 6 \cdot 5 \cdot 4 \cdot 3 \cdot 2 \cdot 1}{(2 \cdot 1) \cdot (6 \cdot 5 \cdot 4 \cdot 3 \cdot 2 \cdot 1)}$

$= 28$

53. answers may vary.

Chapter 14 Vocabulary Check

1. A <u>finite sequence</u> is a function whose domain is the set of natural numbers $\{1, 2, 3, ..., n\}$, where n is some natural number.

2. The <u>factorial of n</u>, written $n!$, is the product of the first n consecutive natural numbers.

3. An <u>infinite sequence</u> is a function whose domain is the set of natural numbers.

4. A <u>geometric sequence</u> is a sequence in which each term (after the first) is obtained by multiplying the preceding term by a constant amount r. The constant r is called the <u>common ratio</u> of the sequence.

5. A sum of the terms of a sequence is called a <u>series</u>.

6. The nth term of the sequence a_n is called the <u>general term</u>.

7. An <u>arithmetic sequence</u> is a sequence in which each term (after the first) differs from the preceding term by a constant amount d. The constant d is called the <u>common difference</u> of the sequence.

8. A triangular array of the coefficients of the terms of the expansions of $(a+b)^n$ is called <u>Pascal's triangle</u>.

Chapter 14 Review

1. $a_n = -3n^2$

$a_1 = -3(1)^2 = -3$
$a_2 = -3(2)^2 = -12$
$a_3 = -3(3)^2 = -27$
$a_4 = -3(4)^2 = -48$
$a_5 = -3(5)^2 = -75$

2. $a_n = n^2 + 2n$

$a_1 = 1^2 + 2(1) = 3$
$a_2 = 2^2 + 2(2) = 8$
$a_3 = 3^2 + 2(3) = 15$
$a_4 = 4^2 + 2(4) = 24$
$a_5 = 5^2 + 2(5) = 35$

3. $a_n = \dfrac{(-1)^n}{100}$

$a_{100} = \dfrac{(-1)^{100}}{100} = \dfrac{1}{100}$

4. $a_n = \dfrac{2n}{(-1)^n}$

$a_{50} = \dfrac{2(50)}{(-1)^{50}} = 100$

5. $\dfrac{1}{6 \cdot 1}, \dfrac{1}{6 \cdot 2}, \dfrac{1}{6 \cdot 3}, \cdots$

In general, $a_n = \dfrac{1}{6n}$.

6. $-1, 4, -9, 16, ...$

$a_n = (-1)^n n^2$

7. $a_n = 32n - 16$
$a_5 = 32(5) - 16 = 144$ feet
$a_6 = 32(6) - 16 = 176$ feet
$a_7 = 32(7) - 16 = 208$ feet

8. Since the measure is 80 at the end of the first day is 80, the measures at the ends of the first 5 days are:
80
$2 \cdot 80 = 160$
$2 \cdot 160 = 320$
$2 \cdot 320 = 640$
$2 \cdot 640 = 1280$

The general terms for the measure at the end of the nth day is $80 \cdot 2^{n-1}$.

$$80 \cdot 2^{n-1} \geq 10,000$$
$$2^{n-1} \geq 125$$

Since $2^6 = 64$ and $2^7 = 128$, $n - 1 = 7$, so $n = 8$. It will take 8 days before the yeast culture measures at least 10,000.

9. 2006: $a_1 = 660,000$

    2007: $a_2 = 660,000(2) = 1,320,000$

    2008: $a_3 = 1,320,000(2) = 2,640,000$

    2009: $a_4 = 2,640,000(2) = 5,280,000$

    2010: $a_5 = 5,280,000(2) = 10,560,000$

    There will be 10,560,000 acres of infested trees in 2010.

10. $a_n = 50 + (n-1)8$

    $a_1 = 50$

    $a_2 = 50 + 8 = 58$

    $a_3 = 50 + 2(8) = 66$

    $a_4 = 50 + 3(8) = 74$

    $a_5 = 50 + 4(8) = 82$

    $a_6 = 50 + 5(8) = 90$

    $a_7 = 50 + 6(8) = 98$

    $a_8 = 50 + 7(8) = 106$

    $a_9 = 50 + 8(8) = 114$

    $a_{10} = 50 + 9(8) = 122$

    There are 122 seats in the tenth row.

11. $a_1 = -2$, $r = \dfrac{2}{3}$

    $a_1 = -2$

    $a_2 = -2\left(\dfrac{2}{3}\right) = -\dfrac{4}{3}$

    $a_3 = \left(-\dfrac{4}{3}\right)\left(\dfrac{2}{3}\right) = -\dfrac{8}{9}$

    $a_4 = \left(-\dfrac{8}{9}\right)\left(\dfrac{2}{3}\right) = -\dfrac{16}{27}$

    $a_5 = \left(-\dfrac{16}{27}\right)\left(\dfrac{2}{3}\right) = -\dfrac{32}{81}$

12. $a_n = 12 + (n-1)(-1.5)$

    $a_1 = 12$

    $a_2 = 12 + (1)(-1.5) = 10.5$

    $a_3 = 12 + 2(-1.5) = 9$

    $a_4 = 12 + 3(-1.5) = 7.5$

    $a_5 = 12 + 4(-1.5) = 6$

13. $a_n = -5 + (n-1)^4$

    $a_{30} = 5 + (30-1)4 = 111$

14. $a_n = 2 + (n-1)\dfrac{3}{4}$

    $a_{11} = 2 + 10\left(\dfrac{3}{4}\right) = \dfrac{19}{2}$

15. 12, 7, 2,...

    $a_1 = 12$, $d = -5$, $n = 20$

    $a_{20} = 12 + (20-1)(-5) = -83$

16. $a_n = a_1 r^{n-1}$, $a_1 = 4$, $r = \dfrac{3}{2}$

    $a_6 = 4\left(\dfrac{3}{2}\right)^{6-1} = \dfrac{243}{8}$

17. $a_4 = 18$, $a_{20} = 98$

    $\begin{cases} a_4 = a_1 + (4-1)d \\ a_{20} = a_1 + (20-1)d \end{cases}$

    $\begin{cases} 18 = a_1 + 3d \\ 98 = a_1 + 19d \end{cases}$

    $\begin{cases} -18 = -a_1 - 3d \\ 98 = a_1 + 19d \end{cases}$

    Adding yields $80 = 16d$ or $d = 5$. Then $a_1 = 18 - 3(5) = 3$.

18. $a_3 = -48$, $a_4 = 192$

    $r = \dfrac{a_4}{a_3} = \dfrac{192}{-48} = -4$

    $a_3 = a_1 r^{3-1}$

    $-48 = a_1(-4)^2$

    $-48 = 16a_1$

    $-3 = a_1$

    $r = -4$, $a_1 = -3$

19. $\dfrac{3}{10}, \dfrac{3}{10^2}, \dfrac{3}{10^3},\ldots$

    In general, $a_n = \dfrac{3}{10^n}$

20. 50, 58, 66, ...

    $a_n = 50 + (n-1)8$ or $a_n = 42 + 8n$

21. $\dfrac{8}{3}$, 4, 6, ...

Geometric; $a_1 = \dfrac{8}{3}$,

$r = \dfrac{4}{\frac{8}{3}} = 4 \cdot \dfrac{3}{8} = \dfrac{12}{8} = \dfrac{3}{2}$

22. -10.5, -6.1, -1.7
Arithmetic; $a_1 = -10.5$,
$d = -6.1 - (-10.5) = 4.4$

23. $7x$, $-14x$, $28x$
Geometric; $a_1 = 7x$, $r = -2$

24. neither

25. $a_1 = 8$, $r = 0.75$
$a_1 = 8$
$a_2 = 8(0.75) = 6$
$a_3 = 8(0.75)^2 = 4.5$
$a_4 = 8(0.75)^3 \approx 3.4$
$a_5 = 8(0.75)^4 \approx 2.5$
$a_6 = 8(0.75)^5 \approx 1.9$
Yes, a ball that rebounds to a height of 2.5 feet after the fifth bounce is good, since $2.5 \geq 1.9$.

26. $a_1 = 25$, $d = -4$
$a_n = a_1 + (n-1)d$
$a_n = 25 + (n-1)(-4) = 29 - 4n$
$a_7 = 25 + 6(-4) = 1$
Continuing the progression as far as possible leaves 1 can in the top row.

27. $a_1 = 1$, $r = 2$
$a_n = 2^{n-1}$
$a_{10} = 2^9 = 512$
$a_{30} = 2^{29} = 536{,}870{,}912$
You save $512 on the tenth day and $536,870,912 on the thirtieth day.

28. $a_n = a_1 r^{n-1}$, $a_1 = 30$, $r = 0.7$
$a_5 = 30(0.7)^4 = 7.203$
The length is 7.203 inches on the fifth swing.

29. $a_1 = 900$, $d = 150$
$a_n = 900 + (n-1)150 = 150_n + 750$
$a_6 = 900 + (6-1)150 = 1650$
Her salary is $1650 per month at the end of training.

30. $\dfrac{1}{512}$, $\dfrac{1}{256}$, $\dfrac{1}{128}$, ...

first fold: $a_1 = \dfrac{1}{256}$, $r = 2$

$a_{15} = \dfrac{1}{256}(2)^{15-1} = 64$

After 15 folds, the thickness is 64 inches.

31. $\displaystyle\sum_{i=1}^{5}(2i-1) = \left[2(1)-1\right] + \left[2(2)-1\right] + \left[2(3)-1\right]$
$\qquad\qquad + \left[2(4)-1\right] + \left[2(5)-1\right]$
$\qquad = 1 + 3 + 5 + 7 + 9$
$\qquad = 25$

32. $\displaystyle\sum_{i=1}^{5} i(i+2) = 1(1+2) + 2(2+2) + 3(3+2)$
$\qquad\qquad + 4(4+2) + 5(5+2)$
$\qquad = 3 + 8 + 15 + 24 + 35$
$\qquad = 85$

33. $\displaystyle\sum_{i=2}^{4}\dfrac{(-1)^i}{2i} = \dfrac{(-1)^2}{2(2)} + \dfrac{(-1)^3}{2(3)} + \dfrac{(-1)^4}{2(4)}$
$\qquad = \dfrac{1}{4} - \dfrac{1}{6} + \dfrac{1}{8}$
$\qquad = \dfrac{5}{24}$

34. $\displaystyle\sum_{i=3}^{5} 5(-1)^{i-1} = 5(-1)^{3-1} + 5(-1)^{4-1} + 5(-1)^{5-1}$
$\qquad = 5(1) + 5(-1) + 5(1)$
$\qquad = 5 - 5 + 5$
$\qquad = 5$

35. $1 + 3 + 9 + 27 + 81 + 243$
$= 3^0 + 3^1 + 3^2 + 3^3 + 3^4 + 3^5$
$= \displaystyle\sum_{i=1}^{6} 3^{i-1}$

36. $6+2+(-2)+(-6)+(-10)+(-14)+(-18)$

$a_1 = 6, \; d = -4$

$a_n = 6+(n-1)(-4)$

$\displaystyle\sum_{i=1}^{7}[6+(i-1)(-4)]$

37. $\dfrac{1}{4}+\dfrac{1}{16}+\dfrac{1}{64}+\dfrac{1}{256}=\dfrac{1}{4^1}+\dfrac{1}{4^2}+\dfrac{1}{4^3}+\dfrac{1}{4^4}$

$\qquad\qquad = \displaystyle\sum_{i=1}^{4}\dfrac{1}{4^i}$

38. $1+\left(-\dfrac{3}{2}\right)+\dfrac{9}{4}=\left(-\dfrac{3}{2}\right)^0+\left(-\dfrac{3}{2}\right)^1+\left(-\dfrac{3}{2}\right)^2$

$\qquad\qquad = \displaystyle\sum_{i=1}^{3}\left(-\dfrac{3}{2}\right)^{i-1}$

39. $a_1 = 20, \; r = 2$

$a_n = 20(2)^n$ represents the number of yeast, where n represents the number of 8-hour periods. Since $48 = 6(8)$ here, $n = 6$.

$a_6 = 20(2)^6 = 1280$

There are 1280 yeast after 48 hours.

40. $a_n = n^2 + 2n - 1$

$a_4 = (4)^2 + 2(4) - 1 = 23$

$S_4 = \displaystyle\sum_{i=1}^{4}(i^2+2i-1)$

$\quad = (1+2-1)+(4+4-1)+(9+6-1)$
$\qquad +(16+8-1)$

$\quad = 46$

23 cranes are born in the fourth year and 46 cranes are born in the first four years.

41. For Job A: $a_1 = 39,500, \; d = 2200;$

$a_5 = 39,500+(5-1)2200 = \$48,330$

For Job B: $a_1 = 41,000, \; d = 1400$

$a_5 = 41,000+(5-1)1400 = \$46,600$

For the fifth year, Job A has a higher salary.

42. $a_n = 200(0.5)^n$

$a_3 = 200(0.5)^3 = 25$

$S_3 = \displaystyle\sum_{i=1}^{3}200(0.5)^i$

$\quad = 200(0.5)+200(0.5)^2+200(0.5)^3$

$\quad = 175$

25 kilograms decay in the third year and 175 kilograms decay in the first three years.

43. $a_n = (n-3)(n+2)$

$S_4 = \displaystyle\sum_{i=1}^{4}(i-3)(i+2)$

$\quad = (1-3)(1+2)+(2-3)(2+2)$
$\qquad +(3-3)(3+2)+(4-3)(4+2)$

$\quad = -6-4+0+6$

$\quad = -4$

44. $a_n = n^2$

$S_6 = \displaystyle\sum_{i=1}^{6}i^2$

$\quad = (1)^2+(2)^2+(3)^2+(4)^2+(5)^2+(6)^2$

$\quad = 91$

45. $a_n = -8+(n-1)3 = 3n-11$

$S_5 = \displaystyle\sum_{i=1}^{5}(3i-11)$

$\quad = [3(1)-11]+[3(2)-11]+[3(3)-11]$
$\qquad +[3(4)-11]+[3(5)-11]$

$\quad = -8-5-2+1+4$

$\quad = -10$

46. $a_n = 5(4)^{n-1}$

$S_3 = \displaystyle\sum_{i=1}^{3}5(4)^{i-1} = 5(4)^0+5(4)^1+5(4)^2 = 105$

47. $15, 19, 23, \ldots$

$a_1 = 15, \; d = 4, \; a_6 = 15+(6-1)4 = 35$

$S_6 = \dfrac{6}{2}[15+35] = 150$

48. $5, -10, 20, \ldots$

$a_1 = 5, r = -2$

$S_n = \dfrac{a_1(1 - r^n)}{1 - r}$

$S_9 = \dfrac{5(1 - (-2)^9)}{1 - (-2)} = 855$

49. $a_1 = 1, d = 2, n = 30, a_{30} = 1 + (30 - 1)2 = 59$

$S_{30} = \dfrac{30}{2}[1 + 59] = 900$

50. $7, 14, 21, 28, \ldots$

$a_n = 7 + (n - 1)7$

$a_{20} = 7 + (20 - 1)7 = 140$

$S_{20} = \dfrac{20}{2}(7 + 140) = 1470$

51. $8, 5, 2, \ldots$

$a_1 = 8, d = -3, n = 20$

$a_{20} = 8 + (20 - 1)(-3) = -49$

$S_{20} = \dfrac{20}{2}[8 + (-49)]$

$\quad\;\; = -410$

52. $\dfrac{3}{4}, \dfrac{9}{4}, \dfrac{27}{4}, \ldots$

$a_1 = \dfrac{3}{4}, \; r = 3$

$S_8 = \dfrac{\dfrac{3}{4}(1 - 3^8)}{1 - 3} = 2460$

53. $a_1 = 6, r = 5$

$S_4 = \dfrac{6(1 - 5^4)}{1 - 5} = 936$

54. $a_1 = -3, d = -6$

$a_n = -3 + (n - 1)(-6)$

$a_{100} = -3 + (100 - 1)(-6) = -597$

$S_{100} = \dfrac{100}{2}(-3 + (-597)) = -30,000$

55. $5, \dfrac{5}{2}, \dfrac{5}{4}, \ldots$

$a_1 = 5, \; r = \dfrac{1}{2}$

$S_\infty = \dfrac{5}{1 - \frac{1}{2}} = 10$

56. $18, -2, \dfrac{2}{9}, \ldots$

$a_1 = 18, \; r = -\dfrac{1}{9}$

$S_\infty = \dfrac{18}{1 + \frac{1}{9}} = \dfrac{81}{5}$

57. $-20, -4, -\dfrac{4}{5}, \ldots$

$a_1 = -20, \; r = \dfrac{1}{5}$

$S_\infty = \dfrac{-20}{1 - \frac{1}{5}} = -25$

58. $0.2, 0.02, 0.002, \ldots$

$a_1 = 0.2 = \dfrac{1}{5}, \; r = \dfrac{1}{10}$

$S_\infty = \dfrac{\frac{1}{5}}{1 - \frac{1}{10}} = \dfrac{2}{9}$

59. $a_1 = 20,000, r = 1.15, n = 4$

$a_4 = 20,000(1.15)^{4-1} = 30,418$

$S_4 = \dfrac{20,000(1 - 1.15^4)}{1 - 1.15} = 99,868$

He earned \$30,418 during the fourth year and \$99,868 over the four years.

60. $a_n = 40(0.8)^{n-1}$

$a_4 = 40(0.8)^{4-1} = 20.48$

$S_4 = \dfrac{40(1 - 0.8^4)}{1 - 0.8} = 118.08$

He takes 20 minutes to assemble the fourth television and 118 minutes to assemble the first four televisions.

61. $a_1 = 100, d = -7, n = 7$

$a_7 = 100 + (7 - 1)(-7) = 58$

$S_7 = \dfrac{7}{2}(100 + 58) = 553$

The rent for the seventh day is \$58 and the rent for 7 days is \$553.

62. $a_1 = 15$, $r = 0.8$

$S_\infty = \dfrac{15}{1-0.8} = 75$ feet downward

$a_1 = 15(0.8) = 12$, $r = 0.8$

$S_\infty = \dfrac{12}{1-0.8} = 60$ feet upward

The total distance is 135 feet.

63. $1800, 600, 200, \ldots$

$a_1 = 1800$, $r = \dfrac{1}{3}$, $n = 6$

$S_6 = 1800 \dfrac{\left(1 - \left(\frac{1}{3}\right)^6\right)}{1 - \frac{1}{3}} \approx 2696$

Approximately 2696 mosquitoes were killed during the first six days after the spraying.

64. $1800, 600, 200, \ldots$

For which n is $a_n < 1$?

$a_n = 1800 \left(\dfrac{1}{3}\right)^{n-1} < 1$

$\left(\dfrac{1}{3}\right)^{n-1} < \dfrac{1}{1800}$

$(n-1)\log \dfrac{1}{3} < \log \dfrac{1}{1800}$

$(n-1)\log 3^{-1} < \log 1800^{-1}$

$(n-1)(-\log 3) < -\log 1800$

$n - 1 > \dfrac{-\log 1800}{-\log 3}$

$n > 1 + \dfrac{\log 1800}{\log 3}$

$n > 7.8$

No longer effective on the 8th day

$S_8 = \dfrac{1800 \left(1 - \left(\frac{1}{3}\right)^8\right)}{1 - \frac{1}{3}} \approx 2700$

About 2700 mosquitoes were killed.

65. $0.55\overline{5} = 0.5 + 0.05 + 0.005 + \cdots$

$a_1 = 0.5$, $r = 0.1$

$S_\infty = \dfrac{0.5}{1 - 0.1} = \dfrac{0.5}{0.9} = \dfrac{5}{9}$

66. 27, 30, 33, …

$$a_n = 27 + (n-1)(3)$$
$$a_{20} = 27 + (20-1)(3) = 84$$
$$S_{20} = \frac{20}{2}(27+84) = 1110$$

There are 1110 seats in the theater.

67. $(x+z)^5 = x^5 + 5x^4 z + 10x^3 z^2 + 10x^2 z^3 + 5xz^4 + z^5$

68. $(y-r)^6 = y^6 + 6y^5(-r) + 15y^4(-r)^2 + 20y^3(-r)^3 + 15y^2(-r)^4 + 6y(-r)^5 + (-r)^6$
$$= y^6 - 6y^5 r + 15y^4 r^2 - 20y^3 r^3 + 15y^2 r^4 - 6yr^5 + r^6$$

69. $(2x+y)^4 = (2x)^4 + 4(2x)^3 y + 6(2x)^2 y^2 + 4(2x)y^3 + y^4$
$$= 16x^4 + 32x^3 y + 24x^2 y^2 + 8xy^3 + y^4$$

70. $(3y-z)^4 = (3y)^4 + 4(3y)^3(-z) + 6(3y)^2(-z)^2 + 4(3y)(-z)^3 + (-z)^4$
$$= 81y^4 - 108y^3 z + 54y^2 z^2 - 12yz^3 + z^4$$

71. $(b+c)^8 = b^8 + \dfrac{8}{1!}b^7 c + \dfrac{8 \cdot 7}{2!}b^6 c^2 + \dfrac{8 \cdot 7 \cdot 6}{3!}b^5 c^3 + \dfrac{8 \cdot 7 \cdot 6 \cdot 5}{4!}b^4 c^4 + \dfrac{8 \cdot 7 \cdot 6 \cdot 5 \cdot 4}{5!}b^3 c^5$
$$+ \frac{8 \cdot 7 \cdot 6 \cdot 5 \cdot 4 \cdot 3}{6!}b^2 c^6 + \frac{8 \cdot 7 \cdot 6 \cdot 5 \cdot 4 \cdot 3 \cdot 2}{7!}bc^7 + c^8$$
$$= b^8 + 8b^7 c + 28b^6 c^2 + 56b^5 c^3 + 70b^4 c^4 + 56b^3 c^5 + 28b^2 c^6 + 8bc^7 + c^8$$

72. $(x-w)^7 = x^7 + \dfrac{7}{1!}x^6(-w) + \dfrac{7 \cdot 6}{2!}x^5(-w)^2 + \dfrac{7 \cdot 6 \cdot 5}{3!}x^4(-w)^3 + \dfrac{7 \cdot 6 \cdot 5 \cdot 4}{4!}x^3(-w)^4 + \dfrac{7 \cdot 6 \cdot 5 \cdot 4 \cdot 3}{5!}x^2(-w)^5$
$$+ \frac{7 \cdot 6 \cdot 5 \cdot 4 \cdot 3 \cdot 2}{6!}x(-w)^6 + (-w)^7$$
$$= x^7 - 7x^6 w + 21x^5 w^2 - 35x^4 w^3 + 35x^3 w^4 - 21x^2 w^5 + 7xw^6 - w^7$$

73. $(4m-n)^4 = (4m)^4 + \dfrac{4}{1!}(4m)^3(-n) + \dfrac{4 \cdot 3}{2!}(4m)^2(-n)^2 + \dfrac{4 \cdot 3 \cdot 2}{3!}(4m)(-n)^3 + (-n)^4$
$$= 256m^4 - 256m^3 n + 96m^2 n^2 - 16mn^3 + n^4$$

74. $(p-2r)^5 = p^5 + \dfrac{5}{1!}p^4(-2r) + \dfrac{5 \cdot 4}{2!}p^3(-2r)^2 + \dfrac{5 \cdot 4 \cdot 3}{3!}p^2(-2r)^3 + \dfrac{5 \cdot 4 \cdot 3 \cdot 2}{4!}p(-2r)^4 + (-2r)^5$
$$= p^5 - 10p^4 r + 40p^3 r^2 - 80p^2 r^3 + 80pr^4 - 32r^5$$

75. The 4th term corresponds to $r = 3$.
$$\frac{7!}{3!(7-3)!}a^{7-3}b^3 = 35a^4 b^3$$

76. The 11th term corresponds to $r = 10$.
$$\frac{10!}{10!0!}y^{10-10}(2z)^{10} = 1024z^{10}$$

Chapter 14 Test

1. $a_n = \dfrac{(-1)^n}{n+4}$

   $a_1 = \dfrac{(-1)^1}{1+4} = -\dfrac{1}{5}$

   $a_2 = \dfrac{(-1)^2}{2+4} = \dfrac{1}{6}$

   $a_3 = \dfrac{(-1)^3}{3+4} = -\dfrac{1}{7}$

   $a_4 = \dfrac{(-1)^4}{4+4} = \dfrac{1}{8}$

   $a_5 = \dfrac{(-1)^5}{5+4} = -\dfrac{1}{9}$

2. $a_n = 10 + 3(n-1)$

   $a_{80} = 10 + 3(80-1) = 247$

3. $\dfrac{2}{5}, \dfrac{2}{25}, \dfrac{2}{125}, \dots$

   In general, $a_n = \dfrac{2}{5}\left(\dfrac{1}{5}\right)^{n-1}$ or $a_n = \dfrac{2}{5^n}$.

4. $(-1)^1 9 \cdot 1,\ (-1)^2 9 \cdot 2, \dots, a_n = (-1)^n 9n$

5. $a_n = 5(2)^{n-1}, S_5 = \dfrac{5(1-2^5)}{1-2} = 155$

6. $a_n = 18 + (n-1)(-2)$

   $a_1 = 18,\ a_{30} = 18 + (30-1)(-2) = -40$

   $S_{30} = \dfrac{30}{2}[18-40] = -330$

7. $a_1 = 24,\ r = \dfrac{1}{6}$

   $S_\infty = \dfrac{24}{1-\frac{1}{6}} = \dfrac{144}{5}$

8. $\dfrac{3}{2}, -\dfrac{3}{4}, \dfrac{3}{8}, \dots$

   $a_1 = \dfrac{3}{2},\ r = -\dfrac{1}{2}$

   $S_\infty = \dfrac{\frac{3}{2}}{1-\left(-\frac{1}{2}\right)} = 1$

9. $\displaystyle\sum_{i=1}^{4} i(i-2) = 1(1-2) + 2(2-2) + 3(3-2) + 4(4-2)$
$= -1 + 0 + 3 + 8 - 20 + 40 - 80$
$= 10$

10. $\displaystyle\sum_{i=2}^{4} 5(2)^i (-1)^{i-1} = 5(2)^2 (-1)^{2-1} + 5(2)^3 (-1)^{3-1} + 5(2)^4 (-1)^{4-1} = -20 + 40 - 80 = -60$

11. $(a-b)^6 = a^6 - 6a^5 b + 15a^4 b^2 - 20a^3 b^3 + 15a^2 b^4 - 6ab^5 + b^6$

12. $(2x+y)^5 = (2x)^5 + \dfrac{5}{1!}(2x)^4 y + \dfrac{5 \cdot 4}{2!}(2x)^3 y^2 + \dfrac{5 \cdot 4 \cdot 3}{3!}(2x)^2 y^3 + \dfrac{5 \cdot 4 \cdot 3 \cdot 2}{4!}(2x)y^4 + y^5$
$= 32x^5 + 80x^4 y + 80x^3 y^2 + 40x^2 y^3 + 10xy^4 + y^5$

13. $a_n = 250 + 75(n-1)$
$a_{10} = 250 + 75(10-1) = 925$
There were 925 people in the town at the beginning of the tenth year.
$a_1 = 250 + 75(1-1) = 250$
There were 250 people in the town at the beginning of the first year.

14. $1, 3, 5, \ldots$
$a_1 = 1,\ d = 2,\ n = 8$
$a_8 = 1 + (8-1)2 = 15$
$1 + 3 + 5 + 7 + 9 + 11 + 13 + 15$
$S_8 = \dfrac{8}{2}[1+15] = 64$
There were 64 shrubs planted in the 8 rows.

15. $a_1 = 80,\ r = \dfrac{3}{4},\ n = 4$
$a_4 = 80\left(\dfrac{3}{4}\right)^{4-1} = 33.75$
The arc length is 33.75 cm on the 4th swing.
$S_4 = \dfrac{80\left(1 - \left(\frac{3}{4}\right)^4\right)}{1 - \frac{3}{4}} = 218.75$
The total of the arc lengths is 218.75 cm for the first 4 swings.

16. $a_1 = 80,\ r = \dfrac{3}{4}$
$S_\infty = \dfrac{80}{1 - \frac{3}{4}} = 320$
The total of the arc lengths is 320 cm before the pendulum comes to rest.

17. 16, 48, 80,...

$a_{10} = 16 + (10-1)32 = 304$

He falls 304 feet during the 10th second.

$S_{10} = \dfrac{10}{2}[16 + 304] = 1600$

He falls 1600 feet during the first 10 seconds.

18. $0.42\overline{42} = 0.42 + 0.0042 + 0.000042$

$a_1 = 0.42 = \dfrac{42}{100}, \; r = 0.01 = \dfrac{1}{100}$

$S_\infty = \dfrac{\frac{42}{100}}{1 - \frac{1}{100}} = \dfrac{42}{100} \cdot \dfrac{100}{99} = \dfrac{14}{33}$

Thus, $0.42\overline{42} = \dfrac{14}{33}$.

Chapter 14 Cumulative Review

1. a. $(-2)^3 = (-2)(-2)(-2) = -8$

    b. $-2^3 = -(2)(2)(2) = -8$

    c. $(-3)^2 = (-3)(-3) = 9$

    d. $-3^2 = -(3)(3) = -9$

2. a. $3a - (4a + 3) = 3a - 4a - 3 = -a - 3$

    b. $(5x - 3) + (2x + 6) = 7x + 3$

    c. $4(2x - 5) - 3(5x + 1) = 8x - 20 - 15x - 3$
$= -7x - 23$

3. $(2x - 3) - (4x - 2) = 2x - 3 - 4x + 2 = -2x - 1$

4. Let $x =$ the price before taxes, then

$x + 0.06x = 344.50$

$\quad\quad 1.06x = 344.50$

$\quad\quad\quad\quad x = 325$

The price before taxes was $325.

5. $y = mx + b; \; m = \dfrac{1}{4}, \; b = -3$

$y = \dfrac{1}{4}x - 3$

6. If the line is to be parallel, then the slope has to be the same as the slope of the given line.

Therefore, $m = \dfrac{3}{2}$.

$(y - (-2)) = \dfrac{3}{2}(x - 3)$

$y + 2 = \dfrac{3}{2}(x - 3)$

$y = \dfrac{3}{2}x - \dfrac{13}{2}$

$f(x) = \dfrac{3}{2}x - \dfrac{13}{2}$

7. $(2, 5), (-3, 4)$

$m = \dfrac{y_2 - y_1}{x_2 - x_1} = \dfrac{4 - 5}{-3 - 2} = \dfrac{-1}{-5} = \dfrac{1}{5}$

$y - y_1 = m(x - x_1)$

$y - 5 = \dfrac{1}{5}(x - 2)$

$5y - 25 = x - 2$

$-23 = x - 5y$ or $x - 5y = -23$

8.
$$y^3 + 5y^2 - y = 5$$
$$y^3 + 5y^2 - y - 5 = 0$$
$$(y^3 + 5y^2) + (-y - 5) = 0$$
$$y^2(y + 5) - 1(y + 5) = 0$$
$$(y^2 - 1)(y + 5) = 0$$
$$(y + 1)(y - 1)(y + 5) = 0$$
$$y = -5, -1, 1$$

9.

$$
\begin{array}{r|rrrrr}
-2 & 1 & -2 & -11 & 5 & 34 \\
   &   & -2 & 8 & 6 & -22 \\
\hline
   & 1 & -4 & -3 & 11 & 12 \\
\end{array}
$$

Answer: $x^3 - 4x^2 - 3x + 11 + \dfrac{12}{x + 2}$

10. $\dfrac{5}{3a - 6} - \dfrac{a}{a - 2} + \dfrac{3 + 2a}{5a - 10}$

$= \dfrac{5}{3(a - 2)} - \dfrac{a}{a - 2} + \dfrac{3 + 2a}{5(a - 2)}$

$= \dfrac{5 \cdot 5 - 3 \cdot 5a + 3(3 + 2a)}{3 \cdot 5(a - 2)}$

$= \dfrac{25 - 15a + 9 + 6a}{15(a - 2)}$

$= \dfrac{34 - 9a}{15(a - 2)}$

11. a. $\sqrt{50} = \sqrt{2}\sqrt{25} = 5\sqrt{2}$

 b. $\sqrt[3]{24} = \sqrt[3]{8}\sqrt[3]{3} = 2\sqrt[3]{3}$

 c. $\sqrt{26} = \sqrt{26}$

 d. $\sqrt[4]{32} = \sqrt[4]{16}\sqrt[4]{2} = 2\sqrt[4]{2}$

12. $\sqrt{3x+6} - \sqrt{7x-6} = 0$

$$\sqrt{3x+6} = \sqrt{7x-6}$$
$$\left(\sqrt{3x+6}\right)^2 = \left(\sqrt{7x-6}\right)^2$$
$$3x+6 = 7x-6$$
$$-4x = -12$$
$$x = 3$$

13. $2420 = 2000(1+r)^2$

$$\frac{2420}{2000} = (1+r)^2$$
$$\frac{121}{100} = (1+r)^2$$
$$\pm\sqrt{\frac{121}{100}} = 1+r$$
$$\pm\frac{11}{10} = 1+r$$
$$-1 \pm \frac{11}{10} = r$$

Discard the negative value.

$$r = -1 + \frac{11}{10} = \frac{1}{10} = 0.10$$

The interest rate is 10%.

14. a. $\sqrt[3]{\dfrac{4}{3x}} = \dfrac{\sqrt[3]{4}}{\sqrt[3]{3x}} = \left(\dfrac{\sqrt[3]{9x^2}}{\sqrt[3]{9x^2}}\right) = \dfrac{\sqrt[3]{36x^2}}{3x}$

 b. $\dfrac{\sqrt{2}+1}{\sqrt{2}-1} = \dfrac{\sqrt{2}+1}{\sqrt{2}-1} \cdot \left(\dfrac{\sqrt{2}+1}{\sqrt{2}+1}\right)$

$$= \dfrac{2+2\sqrt{2}+1}{2-1}$$
$$= 3+2\sqrt{2}$$

15. $(x-3)^2 - 3(x-3) - 4 = 0$

$$x^2 - 6x + 9 - 3x + 9 - 4 = 0$$
$$x^2 - 9x + 14 = 0$$
$$(x-2)(x-7) = 0$$
$$x = 2, 7$$

16. $\dfrac{10}{(2x+4)^2} - \dfrac{1}{2x+4} = 3$

$$10 - (2x+4) = 3(2x+4)^2$$
$$10 - 2x - 4 = 3(4x^2 + 16x + 16)$$
$$-2x + 6 = 12x^2 + 48x + 48$$
$$12x^2 + 50x + 42 = 0$$
$$6x^2 + 25x + 21 = 0$$
$$(6x+7)(x+3) = 0$$
$$x = -\frac{7}{6}, -3$$

17. $\dfrac{5}{x+1} < -2$

$$x + 1 = 0$$
$$x = -1$$

Solve $\dfrac{5}{x+1} = -2$.

$$(x+1)\frac{5}{x+1} = (x+1)(-2)$$
$$5 = -2x - 2$$
$$7 = -2x$$
$$-\frac{7}{2} = x$$

Region	Test Point	$\dfrac{5}{x+1} < -2$; Result
$\left(-\infty, -\dfrac{7}{2}\right)$	$x = -6$	$\dfrac{5}{-5} < -2$; False
$\left(-\dfrac{7}{2}, -1\right)$	$x = -2$	$\dfrac{5}{-1} < -2$; True
$(-1, \infty)$	$x = 4$	$\dfrac{5}{5} < -2$; False

The solution set is $\left(-\dfrac{7}{2}, -1\right)$.

18. $f(x) = (x+2)^2 - 6$

Axis of symmetry: $x = -2$

vertex: $(-2, -6)$

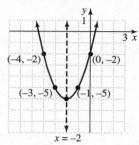

19. $f(t) = -16t^2 + 20t$

The maximum height occurs at the vertex.

$t = \dfrac{-20}{2(-16)} = \dfrac{5}{8}$

$f\left(\dfrac{5}{8}\right) = -16\left(\dfrac{5}{8}\right)^2 + 20\left(\dfrac{5}{8}\right) = \dfrac{25}{4}$

The maximum height of $\dfrac{25}{4}$ feet occurs at

$\dfrac{5}{8}$ second.

20. $f(x) = x^2 + 3x - 18$

$a = 1, b = 3, c = -18$

$x = \dfrac{-3}{2(1)} = -\dfrac{3}{2}$

$f\left(-\dfrac{3}{2}\right) = \left(-\dfrac{3}{2}\right)^2 + 3\left(\dfrac{3}{2}\right) - 18 = -\dfrac{81}{4}$

The vertex is $\left(-\dfrac{3}{2}, -\dfrac{81}{4}\right)$.

21. a. $(f \circ g)(2) = f(g(2)) = f(5) = 5^2 = 25$

$(g \circ f)(2) = g(f(2)) = g(4) = 4 + 3 = 7$

   b. $(f \circ g)(x) = f(x+3)$

$= (x+3)^2$

$= x^2 + 6x + 9$

$(g \circ f)(x) = g(x^2) = x^2 + 3$

22. $f(x) = -2x + 3$

$y = -2x + 3$

$x = -2y + 3$

$x - 3 = -2y$

$\dfrac{x-3}{-2} = y$

$f^{-1}(x) = -\dfrac{x-3}{2}$ or $f^{-1}(x) = \dfrac{3-x}{2}$

23. $f^{-1} = \{(1,0),\ (7,-2),\ (-6,3),\ (4,4)\}$

24. a. $(f \circ g)(2) = f(g(2)) = f(3) = 3^2 - 2 = 7$

$(g \circ f)(2) = g(f(2)) = g(2) = 2 + 1 = 3$

   b. $(f \circ g)(x) = f(x+1)$

$= (x+1)^2 - 2$

$= x^2 + 2x - 1$

$(g \circ f)(x) = g(x^2 - 2) = x^2 - 2 + 1 = x^2 - 1$

25. a. $2^x = 16$

$2^x = 2^4$

$x = 4$

   b. $9^x = 27$

$(3^2)^x = 3^3$

$2x = 3$

$x = \dfrac{3}{2}$

   c. $4^{x+3} = 8^x$

$(2^2)^{x+3} = (2^3)^x$

$2^{2x+6} = 2^{3x}$

$2x + 6 = 3x$

$x = 6$

26. a. $\log_2 32 = x$

$2^x = 32$

$2^x = 2^5$

$x = 5$

   b. $\log_4 \dfrac{1}{64} = x$

$4^x = \dfrac{1}{64}$

$4^x = 4^{-3}$

$x = -3$

c. $\log_{\frac{1}{2}} x = 5$

$\left(\frac{1}{2}\right)^5 = x$

$x = \frac{1}{32}$

27. a. $\log_3 3^2 = 2$

b. $\log_7 7^{-1} = -1$

c. $5^{\log_5 3} = 3$

d. $2^{\log_2 6} = 6$

28. a. $4^x = 64$

$\left(2^2\right)^x = 2^6$

$2x = 6$

$x = 3$

b. $8^x = 32$

$\left(2^3\right)^x = 2^5$

$3x = 5$

$x = \frac{5}{3}$

c. $9^{x+4} = 243^x$

$(3^2)^{x+4} = (3^5)^x$

$3^{2x+8} = 3^{5x}$

$2x + 8 = 5x$

$8 = 3x$

$x = \frac{8}{3}$

29. a. $\log_{11} 10 + \log_{11} 3 = \log_{11}(10 \cdot 3) = \log_{11} 30$

b. $\log_3 \frac{1}{2} + \log_3 12 = \log_3 \left(\frac{1}{2} \cdot 12\right) = \log_3 6$

c. $\log_2(x+2) + \log_2 x = \log_2[(x+2)x]$

$= \log_2(x^2 + 2x)$

30. a. $\log 100,000 = \log_{10} 10^5 = 5$

b. $\log 10^{-3} = \log_{10} 10^{-3} = -3$

c. $\ln \sqrt[5]{e} = \ln e^{1/5} = \frac{1}{5}$

d. $\ln e^4 = 4$

31. $A = Pe^{rt}$

$A = 1600e^{0.09(5)} \approx 2509.30$

$2509.30 is owed after 5 years.

32. a. $\log_6 5 + \log_6 4 = \log_6(5 \cdot 4) = \log_6 20$

b. $\log_8 12 - \log_8 4 = \log_8 \frac{12}{4} = \log_8 3$

c. $2\log_2 x + 3\log_2 x - 2\log_2(x-1)$

$= 5\log_2 x - \log_2(x-1)^2$

$= \log_2 x^5 - \log_2(x-1)^2$

$= \log_2 \frac{x^5}{(x-1)^2}$

33. $3^x = 7$

$\log 3^x = \log 7$

$x\log 3 = \log 7$

$x = \frac{\log 7}{\log 3} \approx 1.7712$

34. $10,000 = 5000\left(1 + \frac{0.02}{4}\right)^{4t}$

$2 = (1.005)^{4t}$

$\ln 2 = \ln 1.005^{4t}$

$\ln 2 = 4t \ln(1.005)$

$t = \frac{\ln 2}{4\ln 1.005} \approx 34.7$

It takes 34.7 years.

35. $\log_4(x-2) = 2$

$4^2 = x - 2$

$x - 2 = 16$

$x = 18$

36. $\log_4 10 - \log_4 x = 2$

$$\log_4 \frac{10}{x} = 2$$

$$4^2 = \frac{10}{x}$$

$$16 = \frac{10}{x}$$

$$16x = 10$$

$$x = \frac{5}{8}$$

37. $\dfrac{x^2}{16} - \dfrac{y^2}{25} = 1$

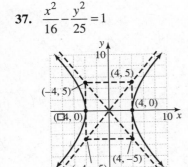

38. $(8, 5), (-2, 4)$

$$d = \sqrt{(-2-8)^2 + (4-5)^2} = \sqrt{101} \text{ units}$$

39. $\begin{cases} y = \sqrt{x} \\ x^2 + y^2 = 6 \end{cases}$

Replace y with $\sqrt{x}$ in the first equation.

$$(x)^2 + \left(\sqrt{x}\right)^2 = 6$$

$$x^2 + x - 6 = 0$$

$$(x+3)(x-2) = 0$$

$$x = -3 \text{(discard) or } x = 2$$

$$x = 2: \ y = \sqrt{x} = \sqrt{2}$$

$$\left(2, \sqrt{2}\right)$$

40. $\begin{cases} x^2 + y^2 = 36 \\ x - y = 6 \Rightarrow x = y + 6 \end{cases}$

Replace x with $y + 6$ in the first equation.

$$(y+6)^2 + y^2 = 36$$

$$2y^2 + 12y = 0$$

$$2y(y+6) = 0$$

$$y = 0 \qquad \text{or} \qquad y = -6$$

$$x = 0 + 6 = 6 \qquad x = -6 + 6 = 0$$

$$(0, -6); (6, 0)$$

41. $\dfrac{x^2}{9} + \dfrac{y^2}{16} \le 1$

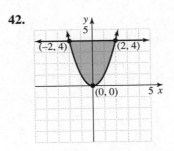

$$\frac{x^2}{9} + \frac{y^2}{16} = 1$$

42.

43. $a_n = n^2 - 1$

$$a_1 = 1^2 - 1 = 0$$

$$a_2 = 2^2 - 1 = 3$$

$$a_3 = 3^2 - 1 = 8$$

$$a_4 = 4^2 - 1 = 15$$

$$a_5 = 5^2 - 1 = 24$$

44. $a_n = \dfrac{n}{n+4}$

$$a_8 = \frac{8}{8+4} = \frac{8}{12} = \frac{2}{3}$$

45. $a_1 = 2, \ d = 9 - 2 = 7$

$$a_{11} = 2 + (11-1)(7) = 72$$

46. $a_1 = 2, \ r = \dfrac{10}{2} = 5$

$$a_6 = 2(5)^{6-1} = 2(5)^5 = 6250$$

47. a. $\displaystyle\sum_{i=0}^{6} \frac{i-2}{2} = \frac{0-2}{2} + \frac{1-2}{2} + \frac{2-2}{2} + \frac{3-2}{2} + \frac{4-2}{2} + \frac{5-2}{2} + \frac{6-2}{2}$

$$= -1 - \frac{1}{2} + 0 + \frac{1}{2} + 1 + \frac{3}{2} + 2$$

$$= \frac{7}{2}$$

b. $\displaystyle\sum_{i=3}^{5} 2^{i} = 2^3 + 2^4 + 2^5 = 8 + 16 + 32 = 56$

48. a. $\displaystyle\sum_{i=0}^{4} i(i+1) = 0(0+1) + 1(1+1) + 2(2+1) + 3(3+1) + 4(4+1)$

$$= 0 + 2 + 6 + 12 + 20$$

$$= 40$$

b. $\displaystyle\sum_{i=0}^{3} 2^{i} = 2^0 + 2^1 + 2^2 + 2^3 = 1 + 2 + 4 + 8 = 15$

49. $a_1 = 1,\ a_{30} = 30$

$$S_n = \frac{n}{2}(a_1 + a_n) = \frac{30}{2}(1 + 30) = 465$$

50. $(x-y)^6$ where $a = x$, $b = -y$, $n = 6$, and $r = 2$.

$$\frac{6!}{2!(6-2)!} x^{6-2} y^2 = 15x^4 y^2$$

The third term in the expansion of $(x-y)^6$ is $15x^4 y^2$.

Appendix A

Appendix A Exercise Set

1. $\begin{array}{r} 9.076 \\ + 8.004 \\ \hline 17.080 \end{array}$

3. $\begin{array}{r} 27.004 \\ - 14.200 \\ \hline 12.804 \end{array}$

5. $\begin{array}{r} 107.92 \\ + 3.04 \\ \hline 110.96 \end{array}$

7. $\begin{array}{r} 10.0 \\ - 7.6 \\ \hline 2.4 \end{array}$

9. $\begin{array}{r} 126.32 \\ - 97.89 \\ \hline 28.43 \end{array}$

11. $\begin{array}{r} 3.25 \\ \times 70 \\ \hline 227.50 \end{array}$

13. $\begin{array}{r} 2.7 \\ 3\overline{)8.1} \\ \underline{6} \\ 2\,1 \\ \underline{2\,1} \\ 0 \end{array}$

15. $\begin{array}{r} 55.4050 \\ - 6.1711 \\ \hline 49.2339 \end{array}$

17. $0.75\overline{)60}$ becomes $\begin{array}{r} 80 \\ 75\overline{)6000} \\ \underline{600} \\ 00 \\ \underline{00} \end{array}$

19. $7.612 \div 100 = 0.07612$

21. $2.7\overline{)12.312}$ becomes $\begin{array}{r} 4.56 \\ 27\overline{)123.12} \\ \underline{108} \\ 15\,1 \\ \underline{13\,5} \\ 1\,62 \\ \underline{1\,62} \\ 0 \end{array}$

23. $\begin{array}{r} 569.20 \\ 71.25 \\ + 8.01 \\ \hline 648.46 \end{array}$

25. $\begin{array}{r} 768.00 \\ - 0.17 \\ \hline 767.83 \end{array}$

27. $\begin{array}{r} 12.000 \\ + 0.062 \\ \hline 12.062 \end{array}$

29. $\begin{array}{r} 76.00 \\ - 14.52 \\ \hline 61.48 \end{array}$

31. $0.43\overline{)3.311}$ becomes $\begin{array}{r} 7.7 \\ 43\overline{)331.1} \\ \underline{301} \\ 30\,1 \\ \underline{30\,1} \\ 0 \end{array}$

33. $\begin{array}{r} 762.12 \\ 89.70 \\ + 11.55 \\ \hline 863.37 \end{array}$

35. $\begin{array}{r} 23.400 \\ - 0.821 \\ \hline 22.579 \end{array}$

37. $\begin{array}{r} 476.12 \\ - 112.97 \\ \hline 363.15 \end{array}$

39. $\begin{array}{r} 0.007 \\ + 7.000 \\ \hline 7.007 \end{array}$

Appendix B

Appendix B.1 Practice

1.
$$3(x-5) = 6x-3$$
$$3x-15 = 6x-3$$
$$3x-15-6x = 6x-3-6x$$
$$-3x-15 = -3$$
$$-3x-15+15 = -3+15$$
$$-3x = 12$$
$$\frac{-3x}{-3} = \frac{12}{-3}$$
$$x = -4$$

2.
$$\frac{y}{2} - \frac{y}{5} = \frac{1}{4}$$
$$20\left(\frac{y}{2} - \frac{y}{5}\right) = 20\left(\frac{1}{4}\right)$$
$$20\left(\frac{y}{2}\right) - 20\left(\frac{y}{5}\right) = 5$$
$$10y - 4y = 5$$
$$6y = 5$$
$$\frac{6y}{6} = \frac{5}{6}$$
$$y = \frac{5}{6}$$

3.
$$8(x^2+3)+4 = -8x(x+3)+19$$
$$8x^2+24+4 = -8x^2-24x+19$$
$$16x^2+24x+9 = 0$$
$$(4x+3)(4x+3) = 0$$
$$4x+3 = 0 \quad \text{or} \quad 4x+3 = 0$$
$$4x = -3 \quad \text{or} \quad 4x = -3$$
$$x = -\frac{3}{4} \quad \text{or} \quad x = -\frac{3}{4}$$

The solution is $-\dfrac{3}{4}$.

4.
$$x - \frac{x-2}{12} = \frac{x+3}{4} + \frac{1}{4}$$
$$12\left(x - \frac{x-2}{12}\right) = 12\left(\frac{x+3}{4} + \frac{1}{4}\right)$$
$$12 \cdot x - 12\left(\frac{x-2}{12}\right) = 12\left(\frac{x+3}{4}\right) + 12 \cdot \frac{1}{4}$$
$$12x - (x-2) = 3(x+3)+3$$
$$12x - x + 2 = 3x+9+3$$
$$11x+2 = 3x+12$$
$$11x+2-3x = 3x+12-3x$$
$$8x+2 = 12$$
$$8x+2-2 = 12-2$$
$$8x = 10$$
$$\frac{8x}{8} = \frac{10}{8}$$
$$x = \frac{5}{4}$$

5.
$$4x^2 = \frac{15}{2}x+1$$
$$2(4x^2) = 2\left(\frac{15}{2}x+1\right)$$
$$8x^2 = 15x+2$$
$$8x^2-15x-2 = 0$$
$$(8x+1)(x-2) = 0$$
$$8x+1 = 0 \quad \text{or} \quad x-2 = 0$$
$$8x = -1$$
$$x = -\frac{1}{8} \quad \text{or} \quad x = 2$$

The solutions are $-\dfrac{1}{8}$ and 2.

Appendix B.1 Exercise Set

1.
$$x^2+11x+24 = 0$$
$$(x+3)(x+8) = 0$$
$$x+3 = 0 \quad \text{or} \quad x+8 = 0$$
$$x = -3 \qquad x = -8$$
The solutions are -3 and -8.

3.
$$3x-4-5x = x+4+x$$
$$-2x-4 = 2x+4$$
$$-4 = 4x+4$$
$$-8 = 4x$$
$$-2 = x$$
The solution is -2.

5. $12x^2 + 5x - 2 = 0$

$(4x - 1)(3x + 2) = 0$

$4x - 1 = 0$ or $3x + 2 = 0$

$4x = 1$ $3x = -2$

$x = \dfrac{1}{4}$ $x = -\dfrac{2}{3}$

The solutions are $-\dfrac{2}{3}$ and $\dfrac{1}{4}$.

7. $z^2 + 9 = 10z$

$z^2 - 10z + 9 = 0$

$(z - 1)(z - 9) = 0$

$z - 1 = 0$ or $z - 9 = 0$

$z = 1$ $z = 9$

The solutions are 1 and 9.

9. $5(y + 4) = 4(y + 5)$

$5y + 20 = 4y + 20$

$y + 20 = 20$

$y = 0$

The solution is 0.

11. $0.6x - 10 = 1.4x - 14$

$-10 = 0.8x - 14$

$4 = 0.8x$

$40 = 8x$

$5 = x$

The solution is 5.

13. $x(5x + 2) = 3$

$5x^2 + 2x = 3$

$5x^2 + 2x - 3 = 0$

$(5x - 3)(x + 1) = 0$

$5x - 3 = 0$ or $x + 1 = 0$

$5x = 3$ $x = -1$

$x = \dfrac{3}{5}$

The solutions are -1 and $\dfrac{3}{5}$.

15. $6x - 2(x - 3) = 4(x + 1) + 4$

$6x - 2x + 6 = 4x + 4 + 4$

$4x + 6 = 4x + 8$

$6 = 8$

This is a false statement, so the equation has no solution.

17. $\dfrac{3}{8} + \dfrac{b}{3} = \dfrac{5}{12}$

$24\left(\dfrac{3}{8} + \dfrac{b}{3}\right) = 24\left(\dfrac{5}{12}\right)$

$9 + 8b = 10$

$8b = 1$

$b = \dfrac{1}{8}$

The solution is $\dfrac{1}{8}$.

19. $x^2 - 6x = x(8 + x)$

$x^2 - 6x = 8x + x^2$

$-6x = 8x$

$0 = 14x$

$0 = x$

The solution is 0.

21. $\dfrac{z^2}{6} - \dfrac{z}{2} - 3 = 0$

$z^2 - 3z - 18 = 0$

$(z + 3)(z - 6) = 0$

$z + 3 = 0$ or $z - 6 = 0$

$z = -3$ $z = 6$

The solutions are -3 and 6.

23. $z + 3(2 + 4z) = 6(z + 1) + 5z$

$z + 6 + 12z = 6z + 6 + 5z$

$6 + 13z = 11z + 6$

$6 + 2z = 6$

$2z = 0$

$z = 0$

The solution is 0.

25. $\dfrac{x^2}{2} + \dfrac{x}{20} = \dfrac{1}{10}$

$10x^2 + x = 2$

$10x^2 + x - 2 = 0$

$(2x + 1)(5x - 2) = 0$

$2x + 1 = 0$ or $5x - 2 = 0$

$2x = -1$ $5x = 2$

$x = -\dfrac{1}{2}$ $x = \dfrac{2}{5}$

The solutions are $-\dfrac{1}{2}$ and $\dfrac{2}{5}$.

27.
$$\frac{4t^2}{5} = \frac{t}{5} + \frac{3}{10}$$
$$8t^2 = 2t + 3$$
$$8t^2 - 2t - 3 = 0$$
$$(2t+1)(4t-3) = 0$$
$$2t+1 = 0 \quad \text{or} \quad 4t-3 = 0$$
$$2t = -1 \qquad\qquad 4t = 3$$
$$t = -\frac{1}{2} \qquad\qquad t = \frac{3}{4}$$

The solutions are $-\frac{1}{2}$ and $\frac{3}{4}$.

29.
$$\frac{3t+1}{8} = \frac{5+2t}{7} + 2$$
$$56\left(\frac{3t+1}{8}\right) = 56\left(\frac{5+2t}{7} + 2\right)$$
$$7(3t+1) = 8(5+2t) + 112$$
$$21t + 7 = 40 + 16t + 112$$
$$21t + 7 = 16t + 152$$
$$5t + 7 = 152$$
$$5t = 145$$
$$t = 29$$
The solution is 29.

31.
$$\frac{m-4}{3} - \frac{3m-1}{5} = 1$$
$$15\left(\frac{m-4}{3} - \frac{3m-1}{5}\right) = 15 \cdot 1$$
$$5(m-4) - 3(3m-1) = 15$$
$$5m - 20 - 9m + 3 = 15$$
$$-4m - 17 = 15$$
$$-4m = 32$$
$$m = -8$$
The solution is -8.

33.
$$3x^2 = -x$$
$$3x^2 + x = 0$$
$$x(3x+1) = 0$$
$$x = 0 \quad \text{or} \quad 3x+1 = 0$$
$$3x = -1$$
$$x = -\frac{1}{3}$$

The solutions are $-\frac{1}{3}$ and 0.

35.
$$x(x-3) = x^2 + 5x + 7$$
$$x^2 - 3x = x^2 + 5x + 7$$
$$-3x = 5x + 7$$
$$-8x = 7$$
$$x = -\frac{7}{8}$$
The solution is $-\frac{7}{8}$.

37.
$$3(t-8) + 2t = 7 + t$$
$$3t - 24 + 2t = 7 + t$$
$$5t - 24 = 7 + t$$
$$4t - 24 = 7$$
$$4t = 31$$
$$t = \frac{31}{4}$$
The solution is $\frac{31}{4}$.

39.
$$-3(x-4) + x = 5(3-x)$$
$$-3x + 12 + x = 15 - 5x$$
$$-2x + 12 = 15 - 5x$$
$$3x + 12 = 15$$
$$3x = 3$$
$$x = 1$$
The solution is 1.

41.
$$(x-1)(x+4) = 24$$
$$x^2 + 3x - 4 = 24$$
$$x^2 + 3x - 28 = 0$$
$$(x+7)(x-4) = 0$$
$$x + 7 = 0 \quad \text{or} \quad x - 4 = 0$$
$$x = -7 \qquad\qquad x = 4$$
The solutions are -7 and 4.

43.
$$\frac{x^2}{4} - \frac{5}{2}x + 6 = 0$$
$$x^2 - 10x + 24 = 0$$
$$(x-6)(x-4) = 0$$
$$x - 6 = 0 \quad \text{or} \quad x - 4 = 0$$
$$x = 6 \qquad\qquad x = 4$$
The solutions are 4 and 6.

45.

$$y^2 + \frac{1}{4} = -y$$

$$4y^2 + 1 = -4y$$

$$4y^2 + 4y + 1 = 0$$

$$(2y+1)^2 = 0$$

$$2y + 1 = 0$$

$$2y = -1$$

$$y = -\frac{1}{2}$$

The solution is $-\frac{1}{2}$.

47. a. Incorrect; answers may vary

b. Correct; answers may vary

c. Correct; answers may vary

d. Incorrect; answers may vary

49.

$$3.2x + 4 = 5.4x - 7$$

$$3.2x + 4 - 4 = 5.4x - 7 - 4$$

$$3.2x = 5.4x - 11$$

$$K = -11$$

51.

$$\frac{x}{6} + 4 = \frac{x}{3}$$

$$6\left(\frac{x}{6} + 4\right) = 6\left(\frac{x}{3}\right)$$

$$x + 24 = 2x$$

$$K = 24$$

53.

$$2.569x = -12.48534$$

$$\frac{2.569x}{2.569} = \frac{-12.48534}{2.569}$$

$$x = -4.86$$

Check: $2.569x = -12.48534$

$2.569(-4.86) \stackrel{?}{=} -12.48534$

$-12.48534 = -12.48534$

The solution is -4.86.

55.
$$2.86z - 8.1258 = -3.75$$
$$2.86z - 8.1258 + 8.1258 = -3.75 + 8.1258$$
$$2.86z = 4.3758$$
$$\frac{2.86z}{2.86} = \frac{4.3758}{2.86}$$
$$z = 1.53$$

Check:
$$2.86z - 8.1258 = -3.75$$
$$2.86 \cdot 1.53 - 8.1258 \stackrel{?}{=} -3.75$$
$$4.3758 - 8.1258 \stackrel{?}{=} -3.75$$
$$-3.75 = -3.75$$

The solution is 1.53.

57. The quotient of 8 and a number is $\dfrac{8}{x}$.

59. The product of 8 and a number is $8x$.

61. 2 more than three times a number is $3x + 2$.

Appendix B.2 Practice

1. a. In words:

first integer	plus	second odd integer	plus	third odd integer
↓	↓	↓	↓	↓

Translate: $\quad x \qquad + \qquad (x+2) \qquad + \qquad (x+4)$

Then $x + (x+2) + (x+4) = x + x + 2 + x + 4 = 3x + 6$

b. In words:

side	+	side	+	side	+	side
↓	↓	↓	↓	↓	↓	↓

Translate: $\quad x \qquad + \qquad 2x \qquad + \qquad (x+2) \qquad + \qquad (2x-3)$

Then $x + 2x + (x+2) + (2x-3) = x + 2x + x + 2 + 2x - 3 = 6x - 1$

2. If x = number of arrivals and departures at Frankfurt airport, then $x + 12.9$ = number at London, and $x + 5.2$ = number at Paris.

In words:

number at Frankfurt	+	number at London	+	number at Paris
↓	↓	↓	↓	↓

Translate: $\quad x \qquad + \qquad (x+12.9) \qquad + \qquad (x+5.2)$

Then $x + (x + 12.9) + (x + 5.2) = x + x + 12.9 + x + 5.2 = 3x + 18.1$

3. Let x = the first number, then $3x - 8$ = the second number, and $5x$ = the third number.
The sum of the three numbers is 118.
$$x + (3x - 8) + 5x = 118$$
$$x + 3x + 5x - 8 = 118$$
$$9x - 8 = 118$$
$$9x = 126$$
$$x = 14$$
The numbers are 14, $3x - 8 = 3(14) - 8 = 34$, and $5x = 5(14) = 70$.

4. Let x = the original price. Then $0.4x$ = the discount. The original price, minus the discount, is equal to $270.
$$x - 0.4x = 270$$
$$0.6x = 270$$
$$x = \frac{270}{0.6} = 450$$
The original price was $450.

Vocabulary & Readiness Check

1. 130% of a number $\underline{>}$ the number.

2. 70% of a number $\underline{<}$ the number.

3. 100% of a number $\underline{=}$ the number.

4. 200% of a number $\underline{>}$ the number.

		First Integer	All Described Integers
5.	Four consecutive integers	31	31, 32, 33, 34
6.	Three consecutive odd integers	31	31, 33, 35
7.	Three consecutive even integers	18	18, 20, 22
8.	Four consecutive even integers	92	92, 94, 96, 98
9.	Three consecutive integers	y	$y, y + 1, y + 2$
10.	Three consecutive even integers	z (z is even)	$z, z + 2, z + 4$
11.	Four consecutive integers	p	$p, p + 1, p + 2, p + 3$
12.	Three consecutive odd integers	s (s is odd)	$s, s + 2, s + 4$

Appendix B.2 Exercise Set

1. The perimeter is the sum of the lengths of the four sides.
$$y + y + y + y = 4y$$

3. Let z = first integer, then $z + 1$ = second integer, and $z + 2$ = third integer.
$$z + (z + 1) + (z + 2) = z + z + z + 1 + 2 = 3z + 3$$

5. Find the sum of x nickels worth 5¢ each, and $(x + 3)$ dimes worth 10¢ each, and $2x$ quarters worth 25¢ each.
$$5x + 10(x + 3) + 25(2x) = 5x + 10x + 30 + 50x$$
$$= 65x + 30$$
The total amount is $(65x + 30)$ cents.

7. $4x + 3(2x + 1) = 4x + 6x + 3 = 10x + 3$

9. The length of the side denoted by ? is $10 - 2 = 8$. Similarly, the length of the unmarked side is $(x - 3) - (x - 10) = x - 3 - x + 10 = 7$. Thus the perimeter of the floor plan is given by $(x - 10) + 2 + 7 + 8 + (x - 3) + 10 = 2x + 14$.

11. Let x = the number.
$$4(x - 2) = 2 + 4x + 2x$$
$$4x - 8 = 2 + 6x$$
$$-2x = 10$$
$$x = -5$$
The number is -5.

13. Let x = the first number, then $5x$ = the second number, and $x + 100$ = the third number.
$$x + 5x + (x + 100) = 415$$
$$7x + 100 = 415$$
$$7x = 315$$
$$x = 45$$
$5x = 225$
$x + 100 = 145$
The numbers are 45, 225, and 145.

15. 29% of $2271 = 0.29 \cdot 2271 = 658.59$;
$2271 - 658.59 = 1612.41$.
Approximately 1612.41 million acres are not federally owned.

17. 91.4% of $8476 = 0.914 \cdot 8476 \approx 7747$
Approximately 7747 minor earthquakes occurred in the United States in 2010.

19. 15% of $1500 = 0.15 \cdot 1500 = 225$
$1500 - 225 = 1275$
1275 are willing to do business with any size retailer.

21. Let x be the percent of email users that spend less than 15 minutes on email at work per day.
$$x + 50 + 8 + 10 + 9 = 100$$
$$x + 77 = 100$$
$$x = 23$$
23% of email users spend less than 15 minutes on email each day.

23. Let x be the number of employees who use email for more than 3 hours per day. Then x is 9% of 4633.
$x = 0.09(4633) = 416.97$
417 employees would be expected to use email for more than 3 hours per day.

25. $x + 4x + (x + 6) = 180$
$$6x + 6 = 180$$
$$6x = 174$$
$$x = 29$$
$4x = 4(29) = 116$
$x + 6 = 29 + 6 = 35$
The angles measure 29°, 35°, and 116°.

27. $(4x) + (5x + 1) + (5x + 3) = 102$
$$14x + 4 = 102$$
$$14x = 98$$
$$x = 7$$
$4x = 4(7) = 28$
$5x + 1 = 5(7) + 1 = 36$
$5x + 3 = 5(7) + 3 = 38$
The sides measure 28 meters, 36 meters, and 38 meters.

29. $x + (2.5x - 9) + x + 1.5x = 99$
$$6x - 9 = 99$$
$$6x = 108$$
$$x = 18$$
$1.5x = 1.5(18) = 27$
$2.5x - 9 = 2.5(18) - 9 = 36$
The sides measure 18 inches, 18 inches, 27 inches, and 36 inches.

31. Let x = first integer; then $x + 1$ = next integer and $x + 2$ = third integer.
$$x + (x + 1) + (x + 2) = 228$$
$$3x + 3 = 228$$
$$3x = 225$$
$$x = 75$$
$x + 1 = 75 + 1 = 76$
$x + 2 = 75 + 2 = 77$
The integers are 75, 76, and 77.

Copyright © 2013 Pearson Education, Inc.

33. Let x = first even integer, then
$x + 2$ = second even integer, and
$x + 4$ = third even integer.
$$2x + (x + 4) = 268,222$$
$$3x + 4 = 268,222$$
$$3x = 268,218$$
$$x = 89,406$$
$x + 2 = 89,408$
$x + 4 = 89,410$
Fallon's ZIP code is 89406, Fernley's ZIP code
is 89408, and Gardnerville Ranchos's ZIP code
is 89410.

35. $(2x - 21) + \left(\dfrac{5}{2}x + 2 \right) + (3x + 24) = 290$
$$2x + \dfrac{5}{2}x + 3x - 21 + 2 + 24 = 290$$
$$\dfrac{15}{2}x + 5 = 290$$
$$\dfrac{15}{2}x = 285$$
$$x = 38$$
$2x - 21 = 2(38) - 21 = 55$
$\dfrac{5}{2}x + 2 = \dfrac{5}{2}(38) + 2 = 97$
$3x + 24 = 3(38) + 24 = 138$

Year	Increase in Wi-Fi-Enabled Cell Phones	Predicted Number
2010	$2x - 21$	55 million
2012	$\dfrac{5}{2}x + 2$	97 million
2014	$3x + 24$	138 million
Total	290 million	

37. Let x be the growth in the number of biomedical engineer jobs (in thousands). Then $3x - 7$ is the growth in the number of physician assistant jobs, and $\dfrac{1}{2}x + 9$ is the growth in the number of skin care specialist jobs.

$$x + (3x - 7) + \left(\dfrac{1}{2}x + 9 \right) = 56$$
$$x + 3x + \dfrac{1}{2}x - 7 + 9 = 56$$
$$\dfrac{9}{2}x + 2 = 56$$
$$\dfrac{9}{2}x = 54$$
$$x = 12$$
$3x - 7 = 3(12) - 7 = 29$
$\dfrac{1}{2}x + 9 = \dfrac{1}{2}(12) + 9 = 15$
The growth predictions are as follows:
biomedical engineer: 12 thousand
skin care specialist: 15 thousand
physician assistant: 29 thousand

39. Let x be the number of seats in a B737-200 aircraft. Then the number of seats in a B767-300ER is $x + 88$, and the number of seats in a F-100 is $x - 32$.
$$x + (x + 88) + (x - 32) = 413$$
$$3x + 56 = 413$$
$$3x = 357$$
$$x = 119$$
$x + 88 = 119 + 88 = 207$
$x - 32 = 119 - 32 = 87$
The B737-200 has 119 seats, the B767-300ER has 207 seats, and the F-100 has 87 seats.

41. Let x be the price of the fax machine before tax.
$$x + 0.08x = 464.40$$
$$1.08x = 464.40$$
$$x = 430$$
The fax machine cost $430 before tax.

43. The new salary is 1.023 times the current salary.
$436,000(1.023) = 446,028$
The new salary is $446,028.

45. Let x be the expected population of Swaziland in 2050. Then x is 1,200,000 plus the increase of 50% of 1,200,000.
$$x = 1,200,000 + 0.50(1,200,000)$$
$$= 1,200,000 + 600,000$$
$$= 1,800,000$$
The population of Swaziland is expected to be 1,800,000 in 2050.

47. Let x = measure of the angle; then
$180 - x$ = measure of its supplement.
$$x = 3(180 - x) + 20$$
$$x = 540 - 3x + 20$$
$$4x = 560$$
$$x = 140$$
$$180 - x = 180 - 140 = 40$$
The angles measure 140° and 40°.

49. Let x = measure of second angle; then
$2x$ = measure of first angle and
$3x - 12$ = measure of third angle.
$$x + 2x + (3x - 12) = 180$$
$$6x - 12 = 180$$
$$6x = 192$$
$$x = 32$$
$$2x = 2(32) = 64$$
$$3x - 12 = 3(32) - 12 = 84$$
The angles measure 64°, 32°, and 84°.

51. Let x = the length of a side of the square. Then
$x + 6$ = the length of a side of the triangle.
$$4x = 3(x + 6)$$
$$4x = 3x + 18$$
$$x = 18$$
The sides of the square are 18 cm and the sides of the triangle are 24 cm.

53. Let x = first even integer, then
$x + 2$ = second even integer, and
$x + 4$ = third even integer.
$$x + (x + 4) = 156$$
$$2x + 4 = 156$$
$$2x = 152$$
$$x = 76$$
$$x + 2 = 78$$
$$x + 4 = 80$$
The integers are 76, 78, and 80.

55. Let x be the number of grandstand seats at Darlington Motor Raceway. Then $2x + 37,000$ is the number of grandstand seats at Daytona International Speedway.
$$x + (2x + 37,000) = 220,000$$
$$3x + 37,000 = 220,000$$
$$3x = 183,000$$
$$x = 61,000$$
$$2x + 37,000 = 2(61,000) + 37,000 = 159,000$$
Darlington has 61,000 grandstand seats and Daytona has 159,000 grandstand seats.

57. Let x be the population of the New York metropolitan region. Then Mexico City's population is $x + 0.03$, and Tokyo's population is $2x - 2.19$.
$$x + (x + 0.03) + (2x - 2.19) = 75.56$$
$$4x - 2.16 = 75.56$$
$$4x = 77.72$$
$$x = 19.43$$
$$x + 0.03 = 19.46$$
$$2x - 2.19 = 36.67$$
The population of New York is 19.43 million, the population of Mexico City is 19.46 million, and the population of Tokyo is 36.67 million.

59.
$$x + 5x + (6x - 3) = 483$$
$$12x - 3 = 483$$
$$12x = 486$$
$$x = 40.5$$
$$5x = 5(40.5) = 202.5$$
$$6x - 3 = 6(40.5) - 3 = 240$$
The sides measure 40.5 feet, 202.5 feet, and 240 feet.

61. Let x be the number of bulb hours for a halogen bulb. Then $25x$ is the number of bulb hours for a fluorescent bulb, and $x - 2500$ is the number of bulb hours for an incandescent bulb.
$$x + 25x + (x - 2500) = 105,500$$
$$27x - 2500 = 105,500$$
$$27x = 108,000$$
$$x = 4000$$
$$25x = 25(4000) = 100,000$$
$$x - 2500 = 4000 - 2500 = 1500$$
A halogen bulb lasts 4000 bulb hours. A fluorescent bulb lasts 100,000 bulb hours. An incandescent bulb lasts 1500 bulb hours.

63. Let x = number of wins for the Chicago Cubs. Then $x + 1$ = number of wins for the Houston Astros, and $x + 2$ = number of wins for the Milwaukee Brewers.
$$x + (x + 1) + (x + 2) = 228$$
$$3x + 3 = 228$$
$$3x = 225$$
$$x = 75$$
$$x + 1 = 76$$
$$x + 2 = 77$$
The number of wins are as follows:
Chicago Cubs: 75 wins
Houston Astros: 76 wins
Milwaukee Brewers: 77 wins

65. Let x = height of Galter Pavilion; then
$x + 67$ = height of Guy's Tower and
$x + 47$ = height of Queen Mary

$$x + (x + 67) + (x + 47) = 1320$$
$$3x + 114 = 1320$$
$$3x = 1206$$
$$x = 402$$

$x + 67 = 402 + 67 = 469$
$x + 47 = 402 + 47 = 449$
Galter Pavilion: 402 ft
Guy's Tower: 469 ft
Queen Mary: 449 ft

Appendix B.3 Practice

1. The six points are graphed as shown.

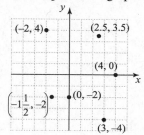

a. $(3, -4)$ lies in quadrant IV.

b. $(0, -2)$ is on the y-axis.

c. $(-2, 4)$ lies in quadrant II.

d. $(4, 0)$ is on the x-axis.

e. $\left(-1\frac{1}{2},\ -2\right)$ is in quadrant III.

f. $(2.5, 3.5)$ is in quadrant I.

2. $y = -3x - 2$

This is a linear equation. (In standard form, it is $3x + y = -2$.) Since the equation is solved for y, we choose three x-values.
Let $x = 0$.
$y = -3x - 2$
$y = -3 \cdot 0 - 2$
$y = -2$
Let $x = -1$.
$y = -3x - 2$
$y = -3(-1) - 2$
$y = 1$
Let $x = -2$.

$y = -3x - 2$
$y = -3(-2) - 2$
$y = 4$

The three ordered pairs $(0, -2)$, $(-1, 1)$, and $(-2, 4)$ are listed in the table.

x	y
0	-2
-1	1
-2	4

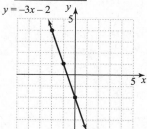

3. $y = -\dfrac{1}{2}x$

To avoid fractions, we choose x-values that are multiples of 2. To find the y-intercept, we let $x = 0$.

If $x = 0$, then $y = -\dfrac{1}{2}(0)$, or 0.

If $x = 2$, then $y = -\dfrac{1}{2}(2)$, or -1.

If $x = -2$, then $y = -\dfrac{1}{2}(-2)$, or 1.

x	y
0	0
2	-1
-2	1

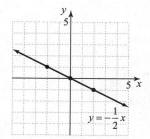

Appendix B.3 Exercise Set

1. Point A is $(5, 2)$.

3. Point C is $(3, 0)$.

5. Point E is $(-5, -2)$.

7. Point G is $(-1, 0)$.

9. $(2, 3)$; QI

11. $(-2, 7)$; QII

13. $(-1, -4)$; QIII

15. $(0, -100)$; y-axis

17. $(-10, -30)$; QIII

19. $(-87, 0)$; x-axis

21. $(x, -y)$ lies in quadrant IV.

23. $(x, 0)$ lies on the x-axis.

25. $(-x, -y)$ lies in quadrant III.

27. $y = -x - 2$
Let $x = 0$.
$y = -0 - 2 = -2$
Let $x = -1$.
$y = -(-1) - 2 = 1 - 2 = -1$
Let $x = 1$.
$y = -1 - 2 = -3$

x	y
0	-2
-1	-1
1	-3

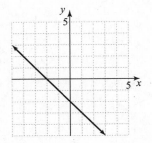

29. $3x - 4y = 8$
Let $x = 0$.
$3(0) - 4y = 8$
$-4y = 8$
$y = -2$
Let $y = 0$.
$3x - 4(0) = 8$
$3x = 8$
$x = \dfrac{8}{3}$
Let $x = 4$.
$3(4) - 4y = 8$
$12 - 4y = 8$
$-4y = -4$
$y = 1$

x	y
0	-2
$\dfrac{8}{3}$	0
4	1

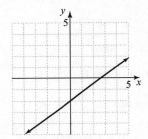

31. $y = \dfrac{1}{3}x$
Let $x = 0$.
$y = \dfrac{1}{3}(0) = 0$
Let $x = 3$.
$y = \dfrac{1}{3}(3) = 1$
Let $x = -3$.
$y = \dfrac{1}{3}(-3) = -1$

x	y
0	0
3	1
−3	−1

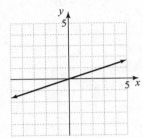

33. $y + 4 = 0$

$\qquad y = -4$

This is a horizontal line.

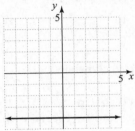

35. The point $(4, 1)$ is on the graph of f.
Thus, $f(4) = 1$.

37. The point $(0, -4)$ is on the graph of g.
Thus, $g(0) = -4$.

39. The points on the graph of f with y-value of 0 are
$(1, 0)$ and $(3, 0)$. Thus, $f(x) = 0$ when $x = 1$ and
$x = 3$.

41. $g(-1) = -2$

Appendix B.4 Practice

1. a. $12x^2 y - 3xy = 3xy(4x) + 3xy(-1)$
$\qquad\qquad\qquad = 3xy(4x - 1)$

b. $49x^2 - 4 = (7x)^2 - 2^2 = (7x + 2)(7x - 2)$

c. $5x^2 + 2x - 3 = (5x - 3)(x + 1)$

d. $3x^2 + 6 + x^3 + 2x = 3(x^2 + 2) + x(x^2 + 2)$
$\qquad\qquad\qquad\qquad = (x^2 + 2)(3 + x)$

e. $4x^2 + 20x + 25 = (2x)^2 + 2 \cdot 2x \cdot 5 + 5^2$
$\qquad\qquad\qquad\qquad = (2x + 5)^2$

f. $b^2 + 100$ cannot be factored.

2. a. $64x^3 + y^3 = (4x)^3 + y^3$
$\qquad\qquad\qquad = (4x + y)[(4x)^2 - 4x \cdot y + y^2]$
$\qquad\qquad\qquad = (4x + y)(16x^2 - 4xy + y^2)$

b. $7x^2 y^2 - 63y^4 = 7y^2(x^2 - 9y^2)$
$\qquad\qquad\qquad\quad = 7y^2[x^2 - (3y)^2]$
$\qquad\qquad\qquad\quad = 7y^2(x - 3y)(x + 3y)$

c. $3x^2 + 12x + 12 - 3b^2$
$\quad = 3(x^2 + 4x + 4 - b^2)$
$\quad = 3[(x + 2)^2 - b^2]$
$\quad = 3(x + 2 + b)(x + 2 - b)$

d. $x^5 y^4 + 27x^2 y$
$\quad = x^2 y(x^3 y^3 + 27)$
$\quad = x^2 y[(xy)^3 + 3^3]$
$\quad = x^2 y(xy + 3)(x^2 y^2 - 3xy + 9)$

e. $(x + 7)^2 - 81y^2 = (x + 7)^2 - (9y)^2$
$\qquad\qquad\qquad\quad = (x + 7 + 9y)(x + 7 - 9y)$

Appendix B.4 Exercise Set

1. $(-y^2 + 6y - 1) + (3y^2 - 4y - 10)$
$\quad = -y^2 + 6y - 1 + 3y^2 - 4y - 10$
$\quad = 2y^2 + 2y - 11$

3. $(x^2 - 6x + 2) - (x - 5) = x^2 - 6x + 2 - x + 5$
$\qquad\qquad\qquad\qquad\qquad = x^2 - 7x + 7$

5. $(5x - 3)^2 = (5x)^2 - 2(5x)(3) + 3^2$
$\qquad\qquad\quad = 25x^2 - 30x + 9$

7.

$$\begin{array}{r} 2x^3 - 4x^2 + 5x - 3 \\ x+2\overline{\smash{\big)}\ 2x^4 + 0x^3 - 3x^2 + 5x - 2} \\ \underline{2x^4 + 4x^3} \\ -4x^3 - 3x^2 \\ \underline{-4x^3 - 8x^2} \\ 5x^2 + 5x \\ \underline{5x^2 + 10x} \\ -5x - 2 \\ \underline{-5x - 10} \\ 8 \end{array}$$

$$(2x^4 - 3x^2 + 5x - 2) \div (x+2)$$
$$= 2x^3 - 4x^2 + 5x - 5 + \frac{8}{x+2}$$

9. $x^2 - 8x + 16 - y^2 = (x-4)^2 - y^2$
$$= (x-4+y)(x-4-y)$$

11. $x^4 - x = x(x^3 - 1) = x(x-1)(x^2 + x + 1)$

13. $14x^2 y - 2xy = 2xy(7x - 1)$

15. $4x^2 - 16 = 4(x^2 - 4) = 4(x+2)(x-2)$

17. $3x^2 - 8x - 11 = (3x - 11)(x + 1)$

19. $4x^2 + 8x - 12 = 4(x^2 + 2x - 3)$
$$= 4(x+3)(x-1)$$

21. $4x^2 + 36x + 81 = (2x)^2 + 2 \cdot 2x \cdot 9 + 9^2$
$$= (2x + 9)^2$$

23. $8x^3 + 125y^3 = (2x)^3 + (5y)^3$
$$= (2x + 5y)(4x^2 - 10xy + 25y^2)$$

25. $64x^2 y^3 - 8x^2 = 8x^2(8y^3 - 1)$
$$= 8x^2[(2y)^3 - 1^3]$$
$$= 8x^2(2y - 1)(4y^2 + 2y + 1)$$

27. $(x+5)^3 + y^3$
$$= [(x+5) + y][(x+5)^2 - (x+5)y + y^2]$$
$$= (x + y + 5)(x^2 + 10x + 25 - xy - 5y + y^2)$$
$$= (x + y + 5)(x^2 + 10x - xy - 5y + y^2 + 25)$$

29. Let $y = 5a - 3$. Then
$$(5a - 3)^2 - 6(5a - 3) + 9 = y^2 - 6y + 9$$
$$= (y - 3)(y - 3)$$
$$= (y - 3)^2$$
$$= [(5a - 3) - 3]^2$$
$$= (5a - 6)^2$$

31. $7x^2 - 63x = 7x(x - 9)$

33. $ab - 6a + 7b - 42 = a(b - 6) + 7(b - 6)$
$$= (a + 7)(b - 6)$$

35. $x^4 - 1 = (x^2)^2 - 1^2$
$$= (x^2 + 1)(x^2 - 1)$$
$$= (x^2 + 1)(x + 1)(x - 1)$$

37. $10x^2 - 7x - 33 = (5x - 11)(2x + 3)$

39. $5a^3 b^3 - 50a^3 b = 5a^3 b(b^2 - 10)$

41. $16x^2 + 25$ is a prime polynomial.

43. $10x^3 - 210x^2 + 1100x = 10x(x^2 - 21x + 110)$
$$= 10x(x - 11)(x - 10)$$

45. $64a^3 b^4 - 27a^3 b$
$$= a^3 b(64b^3 - 27)$$
$$= a^3 b[(4b)^3 - 3^3]$$
$$= a^3 b(4b - 3)(16b^2 + 12b + 9)$$

47. $2x^3 - 54 = 2(x^3 - 27)$
$$= 2(x^3 - 3^3)$$
$$= 2(x - 3)(x^2 + 3x + 9)$$

49. $3y^5 - 5y^4 + 6y - 10 = y^4(3y - 5) + 2(3y - 5)$
$$= (y^4 + 2)(3y - 5)$$

51. $100z^3 + 100 = 100(z^3 + 1)$
$$= 100(z + 1)(z^2 - z + 1)$$

53. $4b^2 - 36b + 81 = (2b)^2 - 2 \cdot 2b \cdot 9 + 9^2$
$$= (2b - 9)^2$$

55. Let $x = y - 6$. Then

$$(y-6)^2 + 3(y-6) + 2 = x^2 + 3x + 2$$
$$= (x+2)(x+1)$$
$$= [(y-6)+2][(y-6)+1]$$
$$= (y-4)(y-5)$$

57. Area $= 3^2 - 4x^2 = 3^2 - (2x)^2 = (3+2x)(3-2x)$

Appendix B.5 Practice

1. a.

$$\frac{2+5n}{3n} \cdot \frac{6n+3}{5n^2 - 3n - 2}$$
$$= \frac{2+5n}{3n} \cdot \frac{3(2n+1)}{(5n+2)(n-1)}$$
$$= \frac{2n+1}{n(n-1)}$$

b.

$$\frac{x^3 - 8}{-6x + 12} \cdot \frac{6x^2}{x^2 + 2x + 4}$$
$$= \frac{(x-2)(x^2+2x+4)}{-6(x-2)} \cdot \frac{6x^2}{x^2+2x+4}$$
$$= \frac{(x-2)(x^2+2x+4)\cdot 6 \cdot x^2}{-1\cdot 6(x-2)(x^2+2x+4)}$$
$$= \frac{x^2}{-1}$$
$$= -x^2$$

2. a.

$$\frac{6y^3}{3y^2 - 27} \div \frac{42}{3-y} = \frac{6y^3}{3y^2 - 27} \cdot \frac{3-y}{42}$$
$$= \frac{6y^3(3-y)}{3(y+3)(y-3)\cdot 42}$$
$$= \frac{6y^3 \cdot (-1)(y-3)}{3(y+3)(y-3)\cdot 6 \cdot 7}$$
$$= -\frac{y^3}{21(y+3)}$$

b.

$$\frac{10x^2 + 23x - 5}{5x^2 - 51x + 10} \div \frac{2x^2 + 9x + 10}{7x^2 - 68x - 20}$$
$$= \frac{10x^2 + 23x - 5}{5x^2 - 51x + 10} \cdot \frac{7x^2 - 68x - 20}{2x^2 + 9x + 10}$$
$$= \frac{(5x-1)(2x+5)}{(5x-1)(x-10)} \cdot \frac{(7x+2)(x-10)}{(2x+5)(x+2)}$$
$$= \frac{7x+2}{x+2}$$

3. a. The LCD is $5p^4 q$.

$$\frac{4}{p^3 q} + \frac{3}{5p^4 q} = \frac{4 \cdot 5p}{p^3 q \cdot 5p} + \frac{3}{5p^4 q}$$
$$= \frac{20p}{5p^4 q} + \frac{3}{5p^4 q}$$
$$= \frac{20p+3}{5p^4 q}$$

b. The LCD is the product of the two denominators: $(y+3)(y-3)$.

$$\frac{4}{y+3} + \frac{5y}{y-3}$$
$$= \frac{4\cdot (y-3)}{(y+3)\cdot (y-3)} + \frac{5y\cdot (y+3)}{(y-3)\cdot (y+3)}$$
$$= \frac{4y-12}{(y+3)(y-3)} + \frac{5y^2 + 15y}{(y+3)(y-3)}$$
$$= \frac{4y-12+5y^2+15y}{(y+3)(y-3)}$$
$$= \frac{5y^2 + 19y - 12}{(y+3)(y-3)}$$

c. The LCD is either $z - 5$ or $5 - z$.

$$\frac{3z-18}{z-5} - \frac{3}{5-z} = \frac{3z-18}{z-5} - \frac{3}{-1(z-5)}$$
$$= \frac{3z-18}{z-5} - \frac{-1\cdot 3}{z-5}$$
$$= \frac{3z-18-(-3)}{z-5}$$
$$= \frac{3z-18+3}{z-5}$$
$$= \frac{3z-15}{z-5}$$
$$= \frac{3(z-5)}{z-5}$$
$$= 3$$

4. $x^2 - 4 = (x+2)(x-2)$

The LCD is $(x+2)(x-2)$.

$$\frac{2}{x-2} - \frac{5+2x}{x^2-4} = \frac{x}{x+2}$$

$$(x+2)(x-2) \cdot \frac{2}{x-2} - (x+2)(x-2) \cdot \frac{5+2x}{(x+2)(x-2)} = (x+2)(x-2) \cdot \frac{x}{x+2}$$

$$2(x+2) - (5+2x) = x(x-2)$$

$$2x+4-5-2x = x^2-2x$$

$$x^2 - 2x + 1 = 0$$

$$(x-1)(x-1) = 0$$

$$x-1 = 0$$

$$x = 1$$

Since 1 does not make any denominator 0, the solution is 1.

Appendix B.5 Exercise Set

1. $\dfrac{x}{2} = \dfrac{1}{8} + \dfrac{x}{4}$

The LCD is 8.

$$8 \cdot \frac{x}{2} = 8 \cdot \frac{1}{8} + 8 \cdot \frac{x}{4}$$

$$4x = 1 + 2x$$

$$2x = 1$$

$$x = \frac{1}{2}$$

The solution is $\dfrac{1}{2}$.

3. $\dfrac{1}{8} + \dfrac{x}{4} = \dfrac{1}{8} + \dfrac{x}{4} \cdot \dfrac{2}{2} = \dfrac{1}{8} + \dfrac{2x}{8} = \dfrac{1+2x}{8}$

5. $\dfrac{4}{x+2} - \dfrac{2}{x-1} = \dfrac{4}{x+2} \cdot \dfrac{x-1}{x-1} - \dfrac{2}{x-1} \cdot \dfrac{x+2}{x+2}$

$$= \frac{4(x-1)}{(x+2)(x-1)} - \frac{2(x+2)}{(x+2)(x+1)}$$

$$= \frac{4x-4-2x-4}{(x+2)(x-1)}$$

$$= \frac{2x-8}{(x+2)(x-1)}$$

$$= \frac{2(x-4)}{(x+2)(x-1)}$$

7. $\dfrac{4}{x+2} = \dfrac{2}{x-1}$

The LCD is $(x + 2)(x - 1)$.

$$(x+2)(x-1) \cdot \dfrac{4}{x+2} = (x+2)(x-1) \cdot \dfrac{2}{x-1}$$
$$4(x-1) = 2(x+2)$$
$$4x-4 = 2x+4$$
$$4x = 2x+8$$
$$2x = 8$$
$$x = 4$$

The solution is 4.

9. $x^2 - 4 = (x+2)(x-2)$

The LCD is $(x + 2)(x - 2)$.

$$\dfrac{2}{x^2-4} = \dfrac{1}{x+2} - \dfrac{3}{x-2}$$
$$(x+2)(x-2) \cdot \dfrac{2}{x^2-4} = (x+2)(x-2) \cdot \dfrac{1}{x+2} - (x+2)(x-2) \cdot \dfrac{3}{x-2}$$
$$2 = (x-2) - 3(x+2)$$
$$2 = x-2-3x-6$$
$$2 = -2x-8$$
$$2x = -10$$
$$x = -5$$

The solution is −5.

11. $\dfrac{5}{x^2-3x} + \dfrac{4}{2x-6} = \dfrac{5}{x(x-3)} + \dfrac{4}{2(x-3)}$

$$= \dfrac{5}{x(x-3)} \cdot \dfrac{2}{2} + \dfrac{4}{2(x-3)} \cdot \dfrac{x}{x}$$
$$= \dfrac{10}{2x(x-3)} + \dfrac{4x}{2x(x-3)}$$
$$= \dfrac{4x+10}{2x(x-3)}$$
$$= \dfrac{2(2x+5)}{2x(x-3)}$$
$$= \dfrac{2x+5}{x(x-3)}$$

13. $x^2 - 1 = (x-1)(x+1)$

    The LCD is $(x - 1)(x + 1)$.

    $$\frac{x-1}{x+1} + \frac{x+7}{x-1} = \frac{4}{x^2-1}$$

    $$(x-1)(x+1) \cdot \frac{x-1}{x+1} + (x-1)(x+1) \cdot \frac{x+7}{x-1} = (x-1)(x+1) \cdot \frac{4}{(x-1)(x+1)}$$

    $$(x-1)(x-1) + (x+1)(x+7) = 4$$

    $$x^2 - 2x + 1 + x^2 + 8x + 7 = 4$$

    $$2x^2 + 6x + 8 = 4$$

    $$2x^2 + 6x + 4 = 0$$

    $$2(x^2 + 3x + 2) = 0$$

    $$2(x+1)(x+2) = 0$$

    $x + 1 = 0 \quad$ or $\quad x + 2 = 0$

    $\quad x = -1 \quad$ or $\quad\quad x = -2$

    The number -1 makes the denominator $x + 1$ equal to 0, so it is not a solution. The solution is -2.

15. $\dfrac{a^2-9}{a-6} \cdot \dfrac{a^2-5a-6}{a^2-a-6} = \dfrac{(a+3)(a-3)}{a-6} \cdot \dfrac{(a-6)(a+1)}{(a-3)(a+2)}$

    $\qquad\qquad\qquad\qquad\quad = \dfrac{(a+3)(a+1)}{a+2}$

17. $\dfrac{2x+3}{3x-2} = \dfrac{4x+1}{6x+1}$

    The LCD is $(3x - 2)(6x + 1)$.

    $$(3x-2)(6x+1) \cdot \frac{2x+3}{3x-2} = (3x-2)(6x+1) \cdot \frac{4x+1}{6x+1}$$

    $$(6x+1)(2x+3) = (3x-2)(4x+1)$$

    $$12x^2 + 18x + 2x + 3 = 12x^2 + 3x - 8x - 2$$

    $$12x^2 + 20x + 3 = 12x^2 - 5x - 2$$

    $$20x + 3 = -5x - 2$$

    $$25x + 3 = -2$$

    $$25x = -5$$

    $$x = -\frac{5}{25}$$

    $$x = -\frac{1}{5}$$

    The solution is $-\dfrac{1}{5}$.

19. $\dfrac{a}{9a^2-1}+\dfrac{2}{6a-2}$

$=\dfrac{a}{(3a-1)(3a+1)}+\dfrac{2}{2(3a-1)}$

$=\dfrac{a}{(3a-1)(3a+1)}\cdot\dfrac{2}{2}+\dfrac{2}{2(3a-1)}\cdot\dfrac{(3a+1)}{(3a+1)}$

$=\dfrac{2a}{2(3a-1)(3a+1)}+\dfrac{6a+2}{2(3a-1)(3a+1)}$

$=\dfrac{8a+2}{2(3a-1)(3a+1)}$

$=\dfrac{2(4a+1)}{2(3a-1)(3a+1)}$

$=\dfrac{4a+1}{(3a-1)(3a+1)}$

21. The LCD is x^2.

$$-\dfrac{3}{x^2}-\dfrac{1}{x}+2=0$$

$$x^2\cdot-\dfrac{3}{x^2}-x^2\cdot\dfrac{1}{x}+x^2\cdot2=0$$

$$-3-x+2x^2=0$$

$$2x^2-x-3=0$$

$$(2x-3)(x+1)=0$$

$2x-3=0$ or $x+1=0$

$2x=3$ or $x=-1$

$x=\dfrac{3}{2}$ or $x=-1$

The solutions are -1 and $\dfrac{3}{2}$.

23. $\dfrac{x-8}{x^2-x-2}+\dfrac{2}{x-2}=\dfrac{x-8}{(x-2)(x+1)}+\dfrac{2}{x-2}$

$=\dfrac{x-8}{(x-2)(x+1)}+\dfrac{2}{x-2}\cdot\dfrac{x+1}{x+1}$

$=\dfrac{x-8}{(x-2)(x+1)}+\dfrac{2x+2}{(x-2)(x+1)}$

$=\dfrac{x-8+2x+2}{(x-2)(x+1)}$

$=\dfrac{3x-6}{(x-2)(x+1)}$

$=\dfrac{3(x-2)}{(x-2)(x+1)}$

$=\dfrac{3}{x+1}$

25. The LCD is a.

$$\dfrac{3}{a}-5=\dfrac{7}{a}-1$$

$$a\cdot\dfrac{3}{a}-a\cdot5=a\cdot\dfrac{7}{a}-a\cdot1$$

$$3-5a=7-a$$

$$3=7+4a$$

$$-4=4a$$

$$-1=a$$

The solution is -1.

27. a. $\dfrac{x}{5}-\dfrac{x}{4}+\dfrac{1}{10}$ is an expression.

b. The first step to simplify this expression is to write each rational expression term so that the denominator is the LCD, 20.

c. $\dfrac{x}{5}-\dfrac{x}{4}+\dfrac{1}{10}=\dfrac{x}{5}\cdot\dfrac{4}{4}-\dfrac{x}{4}\cdot\dfrac{5}{5}+\dfrac{1}{10}\cdot\dfrac{2}{2}$

$=\dfrac{4x}{20}-\dfrac{5x}{20}+\dfrac{2}{20}$

$=\dfrac{4x-5x+2}{20}$

$=\dfrac{-x+2}{20}$

29. $\dfrac{\triangle+\square}{\triangle}=\dfrac{\triangle}{\triangle}+\dfrac{\square}{\triangle}=1+\dfrac{\square}{\triangle}$

b is the correct answer.

31. $\dfrac{\triangle}{\square}\cdot\dfrac{\bigcirc}{\square}=\dfrac{\triangle\bigcirc}{\square\square}$

d is the correct answer.

33. $\dfrac{\frac{\triangle+\square}{\bigcirc}}{\frac{\triangle}{\bigcirc}}=\dfrac{\triangle+\square}{\bigcirc}\div\dfrac{\triangle}{\bigcirc}=\dfrac{\triangle+\square}{\bigcirc}\cdot\dfrac{\bigcirc}{\triangle}=\dfrac{\triangle+\square}{\triangle}$

d is the correct answer.

Appendix C

Viewing Window and Interpreting Window Settings Exercise Set

1. Yes, since every coordinate is between −10 and 10.

3. No, since −11 is less than −10.

5. Answers may vary. Any values such that Xmin < −90, Ymin < −80, Xmax > 55, and Ymax > 80.

7. Answers may vary. Any values such that Xmin < −11, Ymin < −5, Xmax > 7, and Ymax > 2.

9. Answers may vary. Any values such that Xmin < 50, Ymin < −50, Xmax > 200, and Ymax > 200.

11. Xmin = −12   Ymin = −12
Xmax = 12   Ymax = 12
Xscl = 3   Yscl = 3

13. Xmin = −9   Ymin = −12
Xmax = 9   Ymax = 12
Xscl = 1   Yscl = 2

15. Xmin = −10   Ymin = −25
Xmax = 10   Ymax = 25
Xscl = 2   Yscl = 5

17. Xmin = −10   Ymin = −30
Xmax = 10   Ymax = 30
Xscl = 1   Yscl = 3

19. Xmin = −20   Ymin = −30
Xmax = 30   Ymax = 50
Xscl = 5   Yscl = 10

Graphing Equations and Square Viewing Window Exercise Set

1. Setting A:

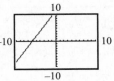

Setting B:

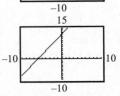

Setting B shows all intercepts.

3. Setting A:

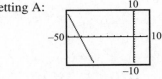

Setting B:

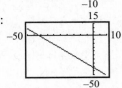

Setting B shows all intercepts.

5. Setting A:

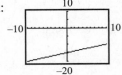

Setting B:

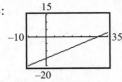

Setting B shows all intercepts.

7. $3x = 5y$

$y = \dfrac{3}{5}x$

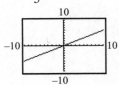

9. $9x - 5y = 30$
$$-5y = -9x + 30$$
$$y = \frac{9}{5}x - 6$$

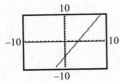

11. $y = -7$

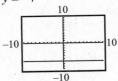

13. $x + 10y = -5$
$$10y = -x - 5$$
$$y = -\frac{1}{10}x - \frac{1}{2}$$

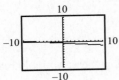

15. $y = \sqrt{x}$

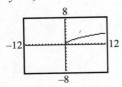

17. $y = x^2 + 2x + 1$

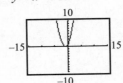

19. $y = |x|$

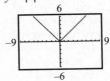

21. $x + 2y = 30$
$$2y = -x + 30$$
$$y = -\frac{1}{2}x + 15$$

Standard window:

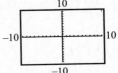

Adjusted window:

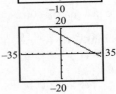

Appendix D

Appendix D Practice

1. $\begin{cases} x + 4y = -2 \\ 3x - y = 7 \end{cases}$

The corresponding matrix is $\begin{bmatrix} 1 & 4 & | & -2 \\ 3 & -1 & | & 7 \end{bmatrix}$. The element in the first row, first column is already 1. Multiply row 1 by -3 and add to row 2 to get a 0 below the 1.

$\begin{bmatrix} 1 & 4 & | & -2 \\ -3(1)+3 & -3(4)+(-1) & | & -3(-2)+7 \end{bmatrix}$

$\begin{bmatrix} 1 & 4 & | & -2 \\ 0 & -13 & | & 13 \end{bmatrix}$

We change -13 to a 1 by dividing row 2 by -13.

$\begin{bmatrix} 1 & 4 & | & -2 \\ 0 & \frac{-13}{-13} & | & \frac{13}{-13} \end{bmatrix}$

$\begin{bmatrix} 1 & 4 & | & -2 \\ 0 & 1 & | & -1 \end{bmatrix}$

The last matrix corresponds to $\begin{cases} x + 4y = -2 \\ y = -1 \end{cases}$

To find x, we let $y = -1$ in the first equation.

$x + 4y = -2$
$x + 4(-1) = -2$
$x - 4 = -2$
$x = 2$

The solution is $(2, -1)$.

2. $\begin{cases} x - 3y = 3 \\ -2x + 6y = 4 \end{cases}$

The corresponding matrix is $\begin{bmatrix} 1 & -3 & | & 3 \\ -2 & 6 & | & 4 \end{bmatrix}$. The element in the first row, first column is already 1. Multiply row 1 by 2 and add to row 2 to get a 0 below the 1.

$\begin{bmatrix} 1 & -3 & | & 3 \\ 2(1)+(-2) & 2(-3)+6 & | & 2(3)+4 \end{bmatrix}$

$\begin{bmatrix} 1 & -3 & | & 3 \\ 0 & 0 & | & 10 \end{bmatrix}$

The corresponding system is $\begin{cases} x - 3y = 3 \\ 0 = 10 \end{cases}$

The equation $0 = 10$ is false. Hence, the system is inconsistent and has no solution. The solution set is $\varnothing$.

3. $\begin{cases} x + 3y - z = 0 \\ 2x + y + 3z = 5 \\ -x - 2y + 4z = 7 \end{cases}$

The corresponding matrix is $\begin{bmatrix} 1 & 3 & -1 & | & 0 \\ 2 & 1 & 3 & | & 5 \\ -1 & -2 & 4 & | & 7 \end{bmatrix}$.

The element in the first row, first column is already 1. Multiply row 1 by -2 and add to row 2 to get a 0 below the 1 in row 2. Add row 1 to row 3 to get a 0 below the 1 in row 3.

$\begin{bmatrix} 1 & 3 & -1 & | & 0 \\ -2(1)+2 & -2(3)+1 & -2(-1)+3 & | & -2(0)+5 \\ 1+(-1) & 3+(-2) & -1+4 & | & 0+7 \end{bmatrix}$

$\begin{bmatrix} 1 & 3 & -1 & | & 0 \\ 0 & -5 & 5 & | & 5 \\ 0 & 1 & 3 & | & 7 \end{bmatrix}$

Now we want a 1 where the -5 is now. Interchange rows 2 and 3.

$\begin{bmatrix} 1 & 3 & -1 & | & 0 \\ 0 & 1 & 3 & | & 7 \\ 0 & -5 & 5 & | & 5 \end{bmatrix}$

Now we want a 0 below the 1. Multiply row 2 by 5 and add to row 3.

$\begin{bmatrix} 1 & 3 & -1 & | & 0 \\ 0 & 1 & 3 & | & 7 \\ 5(0)+0 & 5(1)+(-5) & 5(3)+5 & | & 5(7)+5 \end{bmatrix}$

$\begin{bmatrix} 1 & 3 & -1 & | & 0 \\ 0 & 1 & 3 & | & 7 \\ 0 & 0 & 20 & | & 40 \end{bmatrix}$

Finally, divide row 3 by 20.

$\begin{bmatrix} 1 & 3 & -1 & | & 0 \\ 0 & 1 & 3 & | & 7 \\ 0 & 0 & \frac{20}{20} & | & \frac{40}{20} \end{bmatrix}$

$\begin{bmatrix} 1 & 3 & -1 & | & 0 \\ 0 & 1 & 3 & | & 7 \\ 0 & 0 & 1 & | & 2 \end{bmatrix}$

This matrix corresponds to the system

$\begin{cases} x + 3y - z = 0 \\ y + 3z = 7 \\ z = 2 \end{cases}$

The z-coordinate is 2. Replace z with 2 in the second equation and solve for y.

$$y + 3z = 7$$
$$y + 3(2) = 7$$
$$y + 6 = 7$$
$$y = 1$$

To find x, we let $z = 2$ and $y = 1$ in the first equation.

$$x + 3y - z = 0$$
$$x + 3(1) - 2 = 0$$
$$x + 1 = 0$$
$$x = -1$$

The solution is $(-1, 1, 2)$.

Appendix D Exercise Set

1. $\begin{cases} x + y = 1 \\ x - 2y = 4 \end{cases}$

$$\begin{bmatrix} 1 & 1 & | & 1 \\ 1 & -2 & | & 4 \end{bmatrix}$$

Multiply R1 by -1 and add to R2.

$$\begin{bmatrix} 1 & 1 & | & 1 \\ 0 & -3 & | & 3 \end{bmatrix}$$

Divide R2 by -3.

$$\begin{bmatrix} 1 & 1 & | & 1 \\ 0 & 1 & | & -1 \end{bmatrix}$$

This corresponds to $\begin{cases} x + y = 1 \\ y = -1 \end{cases}$.

$$x + (-1) = 1$$
$$x - 1 = 1$$
$$x = 2$$

The solution is $(2, -1)$.

3. $\begin{cases} x + 3y = 2 \\ x + 2y = 0 \end{cases}$

$$\begin{bmatrix} 1 & 3 & | & 2 \\ 1 & 2 & | & 0 \end{bmatrix}$$

Multiply R1 by -1 and add to R2.

$$\begin{bmatrix} 1 & 3 & | & 2 \\ 0 & -1 & | & -2 \end{bmatrix}$$

Multiply R2 by -1.

$$\begin{bmatrix} 1 & 3 & | & 2 \\ 0 & 1 & | & 2 \end{bmatrix}$$

This corresponds to $\begin{cases} x + 3y = 2 \\ y = 2 \end{cases}$.

$$x + 3(2) = 2$$
$$x + 6 = 2$$
$$x = -4$$

The solution is $(-4, 2)$.

5. $\begin{cases} x - 2y = 4 \\ 2x - 4y = 4 \end{cases}$

$$\begin{bmatrix} 1 & -2 & | & 4 \\ 2 & -4 & | & 4 \end{bmatrix}$$

Multiply R1 by -2 and add to R2.

$$\begin{bmatrix} 1 & -2 & | & 4 \\ 0 & 0 & | & -4 \end{bmatrix}$$

This corresponds to $\begin{cases} x - 2y = 4 \\ 0 = -4 \end{cases}$.

This is an inconsistent system. The solution is $\varnothing$.

7. $\begin{cases} 3x - 3y = 9 \\ 2x - 2y = 6 \end{cases}$

$$\begin{bmatrix} 3 & -3 & | & 9 \\ 2 & -2 & | & 6 \end{bmatrix}$$

Divide R1 by 3.

$$\begin{bmatrix} 1 & -1 & | & 3 \\ 2 & -2 & | & 6 \end{bmatrix}$$

Multiply R1 by -2 and add to R2.

$$\begin{bmatrix} 1 & -1 & | & 3 \\ 0 & 0 & | & 0 \end{bmatrix}$$

This corresponds to $\begin{cases} x - y = 3 \\ 0 = 0 \end{cases}$.

This is a dependent system. The solution is $\{(x, y) | x - y = 3\}$.

9. $\begin{cases} x + y = 3 \\ 2y = 10 \\ 3x + 2y - 4z = 12 \end{cases}$

$$\begin{bmatrix} 1 & 1 & 0 & | & 3 \\ 0 & 2 & 0 & | & 10 \\ 3 & 2 & -4 & | & 12 \end{bmatrix}$$

Multiply R1 by -3 and add to R3.

$$\begin{bmatrix} 1 & 1 & 0 & | & 3 \\ 0 & 2 & 0 & | & 10 \\ 0 & -1 & -4 & | & 3 \end{bmatrix}$$

Divide R2 by 2.

$$\begin{bmatrix} 1 & 1 & 0 & | & 3 \\ 0 & 1 & 0 & | & 5 \\ 0 & -1 & -4 & | & 3 \end{bmatrix}$$

Add R2 to R3.

$$\begin{bmatrix} 1 & 1 & 0 & | & 3 \\ 0 & 1 & 0 & | & 5 \\ 0 & 0 & -4 & | & 8 \end{bmatrix}$$

Divide R3 by -4.

$$\begin{bmatrix} 1 & 1 & 0 & | & 3 \\ 0 & 1 & 0 & | & 5 \\ 0 & 0 & 1 & | & -2 \end{bmatrix}$$

This corresponds to $\begin{cases} x + y = 3 \\ \quad\;\; y = 5 \\ \quad\quad\;\; z = -2 \end{cases}$.

$x + 5 = 3$
$\quad x = -2$
The solution is $(-2, 5, -2)$.

11. $\begin{cases} \quad\;\; 2y - \;\; z = -7 \\ x + 4y + \;\; z = -4 \\ 5x - y + 2z = 13 \end{cases}$

$$\begin{bmatrix} 0 & 2 & -1 & | & -7 \\ 1 & 4 & 1 & | & -4 \\ 5 & -1 & 2 & | & 13 \end{bmatrix}$$

Interchange R1 and R2.

$$\begin{bmatrix} 1 & 4 & 1 & | & -4 \\ 0 & 2 & -1 & | & -7 \\ 5 & -1 & 2 & | & 13 \end{bmatrix}$$

Multiply R1 by -5 and add to R3.

$$\begin{bmatrix} 1 & 4 & 1 & | & -4 \\ 0 & 2 & -1 & | & -7 \\ 0 & -21 & -3 & | & 33 \end{bmatrix}$$

Divide R2 by 2.

$$\begin{bmatrix} 1 & 4 & 1 & | & -4 \\ 0 & 1 & -\frac{1}{2} & | & -\frac{7}{2} \\ 0 & -21 & -3 & | & 33 \end{bmatrix}$$

Multiply R2 by 21 and add to R3.

$$\begin{bmatrix} 1 & 4 & 1 & | & -4 \\ 0 & 1 & -\frac{1}{2} & | & -\frac{7}{2} \\ 0 & 0 & -\frac{27}{2} & | & -\frac{81}{2} \end{bmatrix}$$

Multiply R2 by $-\dfrac{2}{27}$.

$$\begin{bmatrix} 1 & 4 & 1 & | & -4 \\ 0 & 1 & -\frac{1}{2} & | & -\frac{7}{2} \\ 0 & 0 & 1 & | & 3 \end{bmatrix}$$

This corresponds to $\begin{cases} x + 4y + z = -4 \\ \quad\;\; y - \dfrac{1}{2}z = -\dfrac{7}{2} \\ \quad\quad\quad\; z = 3 \end{cases}$.

$y - \dfrac{1}{2}(3) = -\dfrac{7}{2}$
$\quad\; y - \dfrac{3}{2} = -\dfrac{7}{2}$
$\quad\quad\; y = -2$
$x + 4(-2) + 3 = -4$
$\quad x - 8 + 3 = -4$
$\quad\quad\quad x = 1$
The solution is $(1, -2, 3)$.

13. $\begin{cases} x - 4 = 0 \\ x + y = 1 \end{cases}$ or $\begin{cases} x \quad\quad = 4 \\ x + y = 1 \end{cases}$

$$\begin{bmatrix} 1 & 0 & | & 4 \\ 1 & 1 & | & 1 \end{bmatrix}$$

Multiply R1 by -1 and add to R2.

$$\begin{bmatrix} 1 & 0 & | & 4 \\ 0 & 1 & | & -3 \end{bmatrix}$$

This corresponds to $\begin{cases} x = 4 \\ y = -3 \end{cases}$

The solution is $(4, -3)$.

15. $\begin{cases} x + y + z = 2 \\ 2x \quad\;\; - z = 5 \\ \quad\; 3y + z = 2 \end{cases}$

$$\begin{bmatrix} 1 & 1 & 1 & | & 2 \\ 2 & 0 & -1 & | & 5 \\ 0 & 3 & 1 & | & 2 \end{bmatrix}$$

Multiply R1 by -2 and add to R2.

$$\begin{bmatrix} 1 & 1 & 1 & | & 2 \\ 0 & -2 & -3 & | & 1 \\ 0 & 3 & 1 & | & 2 \end{bmatrix}$$

Divide R2 by -2.

$$\begin{bmatrix} 1 & 1 & 1 & | & 2 \\ 0 & 1 & \frac{3}{2} & | & -\frac{1}{2} \\ 0 & 3 & 1 & | & 2 \end{bmatrix}$$

Multiply R2 by -3 and add to R3.

$$\begin{bmatrix} 1 & 1 & 1 & | & 2 \\ 0 & 1 & \frac{3}{2} & | & -\frac{1}{2} \\ 0 & 0 & -\frac{7}{2} & | & \frac{7}{2} \end{bmatrix}$$

Multiply R3 by $-\dfrac{2}{7}$.

$$\begin{bmatrix} 1 & 1 & 1 & | & 2 \\ 0 & 1 & \frac{3}{2} & | & -\frac{1}{2} \\ 0 & 0 & 1 & | & -1 \end{bmatrix}$$

This corresponds to $\begin{cases} x+y+z=2 \\ y+\dfrac{3}{2}z=-\dfrac{1}{2} \\ z=-1 \end{cases}$.

$$y+\frac{3}{2}(-1)=-\frac{1}{2}$$
$$y-\frac{3}{2}=-\frac{1}{2}$$
$$y=1$$
$$x+1+(-1)=2$$
$$x=2$$

The solution is $(2, 1, -1)$.

17. $\begin{cases} 5x-2y=27 \\ -3x+5y=18 \end{cases}$

$$\begin{bmatrix} 5 & -2 & | & 27 \\ -3 & 5 & | & 18 \end{bmatrix}$$

Divide R1 by 5.

$$\begin{bmatrix} 1 & -\frac{2}{5} & | & \frac{27}{5} \\ -3 & 5 & | & 18 \end{bmatrix}$$

Multiply R1 by 3 and add to R2.

$$\begin{bmatrix} 1 & -\frac{2}{5} & | & \frac{27}{5} \\ 0 & \frac{19}{5} & | & \frac{171}{5} \end{bmatrix}$$

Multiply R2 by $\dfrac{5}{19}$.

$$\begin{bmatrix} 1 & -\frac{2}{5} & | & \frac{27}{5} \\ 0 & 1 & | & 9 \end{bmatrix}$$

This corresponds to $\begin{cases} x-\dfrac{2}{5}y=\dfrac{27}{5} \\ y=9 \end{cases}$.

$$x-\frac{2}{5}(9)=\frac{27}{5}$$
$$x-\frac{18}{5}=\frac{27}{5}$$
$$x=9$$

The solution is $(9, 9)$.

19. $\begin{cases} 4x-7y=7 \\ 12x-21y=24 \end{cases}$

$$\begin{bmatrix} 4 & -7 & | & 7 \\ 12 & -21 & | & 24 \end{bmatrix}$$

Divide R1 by 4.

$$\begin{bmatrix} 1 & -\frac{7}{4} & | & \frac{7}{4} \\ 12 & -21 & | & 24 \end{bmatrix}$$

Multiply R1 by -12 and add to R2.

$$\begin{bmatrix} 1 & -\frac{7}{4} & | & \frac{7}{4} \\ 0 & 0 & | & 3 \end{bmatrix}$$

This corresponds to $\begin{cases} x-\dfrac{7}{4}y=\dfrac{7}{4} \\ 0=3 \end{cases}$.

This is an inconsistent system. The solution set is $\varnothing$.

21. $\begin{cases} 4x-y+2z=5 \\ 2y+z=4 \\ 4x+y+3z=10 \end{cases}$

$$\begin{bmatrix} 4 & -1 & 2 & | & 5 \\ 0 & 2 & 1 & | & 4 \\ 4 & 1 & 3 & | & 10 \end{bmatrix}$$

Divide R1 by 4.

$$\begin{bmatrix} 1 & -\frac{1}{4} & \frac{1}{2} & | & \frac{5}{4} \\ 0 & 2 & 1 & | & 4 \\ 4 & 1 & 3 & | & 10 \end{bmatrix}$$

Multiply R1 by -4 and add to R3.

$$\begin{bmatrix} 1 & -\frac{1}{4} & \frac{1}{2} & | & \frac{5}{4} \\ 0 & 2 & 1 & | & 4 \\ 0 & 2 & 1 & | & 5 \end{bmatrix}$$

Divide R2 by 2.

$$\begin{bmatrix} 1 & -\frac{1}{4} & \frac{1}{2} & | & \frac{5}{4} \\ 0 & 1 & \frac{1}{2} & | & 2 \\ 0 & 2 & 1 & | & 5 \end{bmatrix}$$

Multiply R2 by -2 and add to R3.

$$\begin{bmatrix} 1 & -\frac{1}{4} & \frac{1}{2} & | & \frac{5}{4} \\ 0 & 1 & \frac{1}{2} & | & 2 \\ 0 & 0 & 0 & | & 1 \end{bmatrix}$$

This corresponds to $\begin{cases} x-\dfrac{1}{4}y+\dfrac{1}{2}z=\dfrac{5}{4} \\ y+\dfrac{1}{2}z=2 \\ 0=1 \end{cases}$.

This is an inconsistent system. The solution set is $\varnothing$.

23. $\begin{cases} 4x + y + z = 3 \\ -x + y - 2z = -11 \\ x + 2y + 2z = -1 \end{cases}$

$$\begin{bmatrix} 4 & 1 & 1 & | & 3 \\ -1 & 1 & -2 & | & -11 \\ 1 & 2 & 2 & | & -1 \end{bmatrix}$$

Interchange R1 and R3.

$$\begin{bmatrix} 1 & 2 & 2 & | & -1 \\ -1 & 1 & -2 & | & -11 \\ 4 & 1 & 1 & | & 3 \end{bmatrix}$$

Add R1 to R2. Multiply R1 by –4 and add to R3.

$$\begin{bmatrix} 1 & 2 & 2 & | & -1 \\ 0 & 3 & 0 & | & -12 \\ 0 & -7 & -7 & | & 7 \end{bmatrix}$$

Divide R2 by 3.

$$\begin{bmatrix} 1 & 2 & 2 & | & -1 \\ 0 & 1 & 0 & | & -4 \\ 0 & -7 & -7 & | & 7 \end{bmatrix}$$

Multiply R2 by 7 and add to R3.

$$\begin{bmatrix} 1 & 2 & 2 & | & -1 \\ 0 & 1 & 0 & | & -4 \\ 0 & 0 & -7 & | & -21 \end{bmatrix}$$

Divide R3 by –7.

$$\begin{bmatrix} 1 & 2 & 2 & | & -1 \\ 0 & 1 & 0 & | & -4 \\ 0 & 0 & 1 & | & 3 \end{bmatrix}$$

This corresponds to $\begin{cases} x + 2y + 2z = -1 \\ \quad y \quad\quad = -4. \\ \quad\quad\quad z = 3 \end{cases}$

$x + 2(-4) + 2(3) = -1$
$\quad\quad x - 8 + 6 = -1$
$\quad\quad\quad\quad\quad x = 1$

The solution is (1, –4, 3).

25. The matrix should have four columns, so (a) is not the correct matrix. The matrix should have a 0 in the first column, second row since the coefficient of x in the second equation is 0, so (b) is not the correct matrix. The correct matrix is (c).

Appendix E

1. $\begin{vmatrix} 3 & 5 \\ -1 & 7 \end{vmatrix} = ad - bc = 3(7) - 5(-1) = 21 + 5 = 26$

3. $\begin{vmatrix} 9 & -2 \\ 4 & -3 \end{vmatrix} = ad - bc$

$= 9(-3) - (-2)(4)$

$= -27 + 8$

$= -19$

5. $\begin{vmatrix} -2 & 9 \\ 4 & -18 \end{vmatrix} = ad - bc$

$= -2(-18) - 9(4)$

$= 36 - 36$

$= 0$

7. $\begin{cases} 2y - 4 = 0 \\ x + 2y = 5 \end{cases}$ or $\begin{cases} 0x + 2y = 4 \\ 1x + 2y = 5 \end{cases}$

$D = \begin{vmatrix} 0 & 2 \\ 1 & 2 \end{vmatrix} = 0 - 2 = -2$

$D_x = \begin{vmatrix} 4 & 2 \\ 5 & 2 \end{vmatrix} = 8 - 10 = -2$

$D_y = \begin{vmatrix} 0 & 4 \\ 1 & 5 \end{vmatrix} = 0 - 4 = -4$

$x = \dfrac{D_x}{D} = \dfrac{-2}{-2} = 1$

$y = \dfrac{D_y}{D} = \dfrac{-4}{-2} = 2$

The solution is (1, 2).

9. $\begin{cases} 3x + y = 1 \\ 2y = 2 - 6x \end{cases}$ or $\begin{cases} 3x + 1y = 1 \\ 6x + 2y = 2 \end{cases}$

$D = \begin{vmatrix} 3 & 1 \\ 6 & 2 \end{vmatrix} = 6 - 6 = 0$

Since $D = 0$, Cramer's Rule cannot be used.
Notice that equation (2) is equation (1)
multiplied by 2.
The solution is $\{(x, y) | 3x + y = 1\}$.

11. $\begin{cases} 5x - 2y = 27 \\ -3x + 5y = 18 \end{cases}$

$D = \begin{vmatrix} 5 & -2 \\ -3 & 5 \end{vmatrix} = 25 - 6 = 19$

$D_x = \begin{vmatrix} 27 & -2 \\ 18 & 5 \end{vmatrix} = 135 + 36 = 171$

$D_y = \begin{vmatrix} 5 & 27 \\ -3 & 18 \end{vmatrix} = 90 + 81 = 171$

$x = \dfrac{D_x}{D} = \dfrac{171}{19} = 9$ and $y = \dfrac{D_y}{D} = \dfrac{171}{19} = 9$

The solution is (9, 9).

13. Expand by first row.

$\begin{vmatrix} 2 & 1 & 0 \\ 0 & 5 & -3 \\ 4 & 0 & 2 \end{vmatrix} = 2\begin{vmatrix} 5 & -3 \\ 0 & 2 \end{vmatrix} - 1\begin{vmatrix} 0 & -3 \\ 4 & 2 \end{vmatrix} + 0\begin{vmatrix} 0 & 5 \\ 4 & 0 \end{vmatrix}$

$= 2(10 - 0) - 1(0 + 12) + 0$

$= 20 - 12$

$= 8$

15. Expand by third column.

$\begin{vmatrix} 4 & -6 & 0 \\ -2 & 3 & 0 \\ 4 & -6 & 1 \end{vmatrix} = 0\begin{vmatrix} -2 & 3 \\ 4 & -6 \end{vmatrix} - 0\begin{vmatrix} 4 & -6 \\ 4 & -6 \end{vmatrix} + 1\begin{vmatrix} 4 & -6 \\ -2 & 3 \end{vmatrix}$

$= 0 - 0 + 1(12 - 12)$

$= 0$

17. Expand by first row.

$\begin{vmatrix} 3 & 6 & -3 \\ -1 & -2 & 3 \\ 4 & -1 & 6 \end{vmatrix} = 3\begin{vmatrix} -2 & 3 \\ -1 & 6 \end{vmatrix} - 6\begin{vmatrix} -1 & 3 \\ 4 & 6 \end{vmatrix} - 3\begin{vmatrix} -1 & -2 \\ 4 & -1 \end{vmatrix}$

$= 3(-12 + 3) - 6(-6 - 12) - 3(1 + 8)$

$= -27 + 108 - 27$

$= 54$

19. $\begin{cases} 3x \quad\quad + z = -1 \\ -x - 3y + z = 7 \\ \quad\quad 3y + z = 5 \end{cases}$

$D = \begin{vmatrix} 3 & 0 & 1 \\ -1 & -3 & 1 \\ 0 & 3 & 1 \end{vmatrix} = 3\begin{vmatrix} -3 & 1 \\ 3 & 1 \end{vmatrix} - 0\begin{vmatrix} -1 & 1 \\ 0 & 1 \end{vmatrix} + 1\begin{vmatrix} -1 & -3 \\ 0 & 3 \end{vmatrix}$

$= 3(-3 - 3) - 0 + 1(-3 - 0)$

$= -18 - 3$

$= -21$

$$D_x = \begin{vmatrix} -1 & 0 & 1 \\ 7 & -3 & 1 \\ 5 & 3 & 1 \end{vmatrix}$$

$$= -1\begin{vmatrix} -3 & 1 \\ 3 & 1 \end{vmatrix} - 0\begin{vmatrix} 7 & 1 \\ 5 & 1 \end{vmatrix} + 1\begin{vmatrix} 7 & -3 \\ 5 & 3 \end{vmatrix}$$

$$= -1(-3-3) - 0 + 1|21+15|$$

$$= 6 + 36$$

$$= 42$$

$$D_y = \begin{vmatrix} 3 & -1 & 1 \\ -1 & 7 & 1 \\ 0 & 5 & 1 \end{vmatrix} = 3\begin{vmatrix} 7 & 1 \\ 5 & 1 \end{vmatrix} + 1\begin{vmatrix} -1 & 1 \\ 5 & 1 \end{vmatrix} + 0\begin{vmatrix} -1 & 1 \\ 7 & 1 \end{vmatrix}$$

$$= 3(7-5) + 1(-1-5) + 0$$

$$= 6 - 6$$

$$= 0$$

$$D_z = \begin{vmatrix} 3 & 0 & -1 \\ -1 & -3 & 7 \\ 0 & 3 & 5 \end{vmatrix}$$

$$= 3\begin{vmatrix} -3 & 7 \\ 3 & 5 \end{vmatrix} - 0\begin{vmatrix} -1 & 7 \\ 0 & 5 \end{vmatrix} - 1\begin{vmatrix} -1 & -3 \\ 0 & 3 \end{vmatrix}$$

$$= 3(-15-21) - 0 - 1(-3-0)$$

$$= -108 + 3$$

$$= -105$$

$$x = \frac{D_x}{D} = \frac{42}{-21} = -2, \quad y = \frac{D_y}{D} = \frac{0}{-21} = 0,$$

$$z = \frac{D_z}{D} = \frac{-105}{-21} = 5$$

The solution is (−2, 0, 5).

21. $\begin{cases} x + y + z = 8 \\ 2x - y - z = 10 \\ x - 2y + 3z = 22 \end{cases}$

$$D = \begin{vmatrix} 1 & 1 & 1 \\ 2 & -1 & -1 \\ 1 & -2 & 3 \end{vmatrix}$$

$$= 1\begin{vmatrix} -1 & -1 \\ -2 & 3 \end{vmatrix} - 1\begin{vmatrix} 2 & -1 \\ 1 & 3 \end{vmatrix} + 1\begin{vmatrix} 2 & -1 \\ 1 & -2 \end{vmatrix}$$

$$= 1(-3-2) - 1(6+1) + 1(-4+1)$$

$$= -5 - 7 - 3$$

$$= -15$$

$$D_x = \begin{vmatrix} 8 & 1 & 1 \\ 10 & -1 & -1 \\ 22 & -2 & 3 \end{vmatrix}$$

$$= 8\begin{vmatrix} -1 & -1 \\ -2 & 3 \end{vmatrix} - 10\begin{vmatrix} 1 & 1 \\ -2 & 3 \end{vmatrix} + 22\begin{vmatrix} 1 & 1 \\ -1 & -1 \end{vmatrix}$$

$$= 8(-3-2) - 10(3+2) + 22(-1+1)$$

$$= -40 - 50 + 0$$

$$= -90$$

$$D_y = \begin{vmatrix} 1 & 8 & 1 \\ 2 & 10 & -1 \\ 1 & 22 & 3 \end{vmatrix}$$

$$= 1\begin{vmatrix} 10 & -1 \\ 22 & 3 \end{vmatrix} - 8\begin{vmatrix} 2 & -1 \\ 1 & 3 \end{vmatrix} + 1\begin{vmatrix} 2 & 10 \\ 1 & 22 \end{vmatrix}$$

$$= 1(30+22) - 8(6+1) + 1(44-10)$$

$$= 52 - 56 + 34$$

$$= 30$$

$$D_z = \begin{vmatrix} 1 & 1 & 8 \\ 2 & -1 & 10 \\ 1 & -2 & 22 \end{vmatrix}$$

$$= 1\begin{vmatrix} -1 & 10 \\ -2 & 22 \end{vmatrix} - 1\begin{vmatrix} 2 & 10 \\ 1 & 22 \end{vmatrix} + 8\begin{vmatrix} 2 & -1 \\ 1 & -2 \end{vmatrix}$$

$$= 1(-22+20) - 1(44-10) + 8(-4+1)$$

$$= -2 - 34 - 24$$

$$= -60$$

$$x = \frac{D_x}{D} = \frac{-90}{-15} = 6, \quad y = \frac{D_y}{D} = \frac{30}{-15} = -2,$$

$$z = \frac{D_z}{D} = \frac{-60}{-15} = 4$$

The solution is (6, −2, 4).

23. $\begin{vmatrix} 10 & -1 \\ -4 & 2 \end{vmatrix} = 10(2) - (-1)(-4) = 20 - 4 = 16$

25. Expand by first row.

$$\begin{vmatrix} 1 & 0 & 4 \\ 1 & -1 & 2 \\ 3 & 2 & 1 \end{vmatrix} = 1\begin{vmatrix} -1 & 2 \\ 2 & 1 \end{vmatrix} - 0\begin{vmatrix} 1 & 2 \\ 3 & 1 \end{vmatrix} + 4\begin{vmatrix} 1 & -1 \\ 3 & 2 \end{vmatrix}$$

$$= 1(-1-4) - 0 + 4(2+3)$$

$$= -5 + 20$$

$$= 15$$

27. $\begin{vmatrix} \frac{3}{4} & \frac{5}{2} \\ -\frac{1}{6} & \frac{7}{3} \end{vmatrix} = \left(\frac{3}{4}\right)\left(\frac{7}{3}\right) - \left(\frac{5}{2}\right)\left(-\frac{1}{6}\right)$

$$= \frac{7}{4} + \frac{5}{12}$$

$$= \frac{21}{12} + \frac{5}{12}$$

$$= \frac{26}{12}$$

$$= \frac{13}{6}$$

29. Expand by first row.

$$\begin{vmatrix} 4 & -2 & 2 \\ 6 & -1 & 3 \\ 2 & 1 & 1 \end{vmatrix} = 4\begin{vmatrix} -1 & 3 \\ 1 & 1 \end{vmatrix} - (-2)\begin{vmatrix} 6 & 3 \\ 2 & 1 \end{vmatrix} + 2\begin{vmatrix} 6 & -1 \\ 2 & 1 \end{vmatrix}$$
$$= 4(-1-3) + 2(6-6) + 2(6+2)$$
$$= -16 + 0 + 16$$
$$= 0$$

31. Expand by first row.

$$\begin{vmatrix} -2 & 5 & 4 \\ 5 & -1 & 3 \\ 4 & 1 & 2 \end{vmatrix} = -2\begin{vmatrix} -1 & 3 \\ 1 & 2 \end{vmatrix} - 5\begin{vmatrix} 5 & 3 \\ 4 & 2 \end{vmatrix} + 4\begin{vmatrix} 5 & -1 \\ 4 & 1 \end{vmatrix}$$
$$= -2(-2-3) - 5(10-12) + 4(5+4)$$
$$= 10 + 10 + 36$$
$$= 56$$

33. $\begin{cases} 2x - 5y = 4 \\ x + 2y = -7 \end{cases}$

$$D = \begin{vmatrix} 2 & -5 \\ 1 & 2 \end{vmatrix} = 4 + 5 = 9$$

$$D_x = \begin{vmatrix} 4 & -5 \\ -7 & 2 \end{vmatrix} = 8 - 35 = -27$$

$$D_y = \begin{vmatrix} 2 & 4 \\ 1 & -7 \end{vmatrix} = -14 - 4 = -18$$

$$x = \frac{D_x}{D} = \frac{-27}{9} = -3 \text{ and } y = \frac{D_y}{D} = \frac{-18}{9} = -2$$

The solution is $(-3, -2)$.

35. $\begin{cases} 4x + 2y = 5 \\ 2x + y = -1 \end{cases}$

$$D = \begin{vmatrix} 4 & 2 \\ 2 & 1 \end{vmatrix} = 4 - 4 = 0$$

Since $D = 0$, Cramer's rule cannot be used.
Multiply the second equation by -2, then add.

$$\begin{array}{r} 4x + 2y = 5 \\ \underline{-4x - 2y = 2} \\ 0 = 7 \end{array}$$

This is a false statement, so the system has no solution, or $\varnothing$.

37. $\begin{cases} 2x + 2y + z = 1 \\ -x + y + 2z = 3 \\ x + 2y + 4z = 0 \end{cases}$

$$D = \begin{vmatrix} 2 & 2 & 1 \\ -1 & 1 & 2 \\ 1 & 2 & 4 \end{vmatrix} = 2\begin{vmatrix} 1 & 2 \\ 2 & 4 \end{vmatrix} - 2\begin{vmatrix} -1 & 2 \\ 1 & 4 \end{vmatrix} + 1\begin{vmatrix} -1 & 1 \\ 1 & 2 \end{vmatrix}$$
$$= 2(4-4) - 2(-4-2) + 1(-2-1)$$
$$= 0 + 12 - 3$$
$$= 9$$

$$D_x = \begin{vmatrix} 1 & 2 & 1 \\ 3 & 1 & 2 \\ 0 & 2 & 4 \end{vmatrix} = 1\begin{vmatrix} 1 & 2 \\ 2 & 4 \end{vmatrix} - 3\begin{vmatrix} 2 & 1 \\ 2 & 4 \end{vmatrix} + 0$$
$$= 1(4-4) - 3(8-2)$$
$$= -18$$

$$D_y = \begin{vmatrix} 2 & 1 & 1 \\ -1 & 3 & 2 \\ 1 & 0 & 4 \end{vmatrix} = 2\begin{vmatrix} 3 & 2 \\ 0 & 4 \end{vmatrix} - 1\begin{vmatrix} -1 & 2 \\ 1 & 4 \end{vmatrix} + 1\begin{vmatrix} -1 & 3 \\ 1 & 0 \end{vmatrix}$$
$$= 2(12-0) - 1(-4-2) + 1(0-3)$$
$$= 24 + 6 - 3$$
$$= 27$$

$$D_z = \begin{vmatrix} 2 & 2 & 1 \\ -1 & 1 & 3 \\ 1 & 2 & 0 \end{vmatrix} = 2\begin{vmatrix} 1 & 3 \\ 2 & 0 \end{vmatrix} - 2\begin{vmatrix} -1 & 3 \\ 1 & 0 \end{vmatrix} + 1\begin{vmatrix} -1 & 1 \\ 1 & 2 \end{vmatrix}$$
$$= 2(0-6) - 2(0-3) + 1(-2-1)$$
$$= -12 + 6 - 3$$
$$= -9$$

$$x = \frac{D_x}{D} = \frac{-18}{9} = -2, \quad y = \frac{D_y}{D} = \frac{27}{9} = 3,$$

$$z = \frac{D_z}{D} = \frac{-9}{9} = -1$$

The solution is $(-2, 3, -1)$.

39. $\begin{cases} \dfrac{2}{3}x - \dfrac{3}{4}y = -1 \\ -\dfrac{1}{6}x + \dfrac{3}{4}y = \dfrac{5}{2} \end{cases}$

$$D = \begin{vmatrix} \frac{2}{3} & -\frac{3}{4} \\ -\frac{1}{6} & \frac{3}{4} \end{vmatrix} = \frac{1}{2} - \frac{1}{8} = \frac{3}{8}$$

$$D_x = \begin{vmatrix} -1 & -\frac{3}{4} \\ \frac{5}{2} & \frac{3}{4} \end{vmatrix} = -\frac{3}{4} + \frac{15}{8} = \frac{9}{8}$$

$$D_y = \begin{vmatrix} \frac{2}{3} & -1 \\ -\frac{1}{6} & \frac{5}{2} \end{vmatrix} = \frac{10}{6} - \frac{1}{6} = \frac{9}{6}$$

$x = \dfrac{D_x}{D} = \dfrac{\frac{9}{8}}{\frac{3}{8}} = 3$ and $y = \dfrac{D_y}{D} = \dfrac{\frac{9}{6}}{\frac{3}{8}} = 4$

The solution is (3, 4).

41. $\begin{cases} 0.7x - 0.2y = -1.6 \\ 0.2x \quad - y = -1.4 \end{cases}$

$D = \begin{vmatrix} 0.7 & -0.2 \\ 0.2 & -1 \end{vmatrix} = -0.7 + 0.04 = -0.66$

$D_x = \begin{vmatrix} -1.6 & -0.2 \\ -1.4 & -1 \end{vmatrix} = 1.6 - 0.28 = 1.32$

$D_y = \begin{vmatrix} 0.7 & -1.6 \\ 0.2 & -1.4 \end{vmatrix} = -0.98 + 0.32 = -0.66$

$x = \dfrac{D_x}{D} = \dfrac{1.32}{-0.66} = -2$

$y = \dfrac{D_y}{D} = \dfrac{-0.66}{-0.66} = 1$

The solution is (−2, 1).

43. $\begin{cases} -2x + 4y - 2z = 6 \\ x - 2y + z = -3 \\ 3x - 6y + 3z = -9 \end{cases}$

$D = \begin{vmatrix} -2 & 4 & -2 \\ 1 & -2 & 1 \\ 3 & -6 & 3 \end{vmatrix}$

$= -2\begin{vmatrix} -2 & 1 \\ -6 & 3 \end{vmatrix} - 4\begin{vmatrix} 1 & 1 \\ 3 & 3 \end{vmatrix} - 2\begin{vmatrix} 1 & -2 \\ 3 & -6 \end{vmatrix}$

$= -2(-6 + 6) - 4(3 - 3) - 2(-6 + 6)$

$= 0 - 0 - 0$

$= 0$

Since $D = 0$, Cramer's rule cannot be used. Note that the first equation is −2 times the second, and the third equation is 3 times the second. Thus, the system is dependent and the solution set is $\{(x, y, z) \mid x - 2y + z = -3\}$.

45. $\begin{cases} x - 2y + z = -5 \\ 3y + 2z = 4 \\ 3x - y = -2 \end{cases}$

$D = \begin{vmatrix} 1 & -2 & 1 \\ 0 & 3 & 2 \\ 3 & -1 & 0 \end{vmatrix} = 1\begin{vmatrix} 3 & 2 \\ -1 & 0 \end{vmatrix} + 2\begin{vmatrix} 0 & 2 \\ 3 & 0 \end{vmatrix} + 1\begin{vmatrix} 0 & 3 \\ 3 & -1 \end{vmatrix}$

$= 1(0 + 2) + 2(0 - 6) + 1(0 - 9)$

$= 2 - 12 - 9$

$= -19$

$D_x = \begin{vmatrix} -5 & -2 & 1 \\ 4 & 3 & 2 \\ -2 & -1 & 0 \end{vmatrix}$

$= -5\begin{vmatrix} 3 & 2 \\ -1 & 0 \end{vmatrix} + 2\begin{vmatrix} 4 & 2 \\ -2 & 0 \end{vmatrix} + 1\begin{vmatrix} 4 & 3 \\ -2 & -1 \end{vmatrix}$

$= -5(0 + 2) + 2(0 + 4) + 1(-4 + 6)$

$= -10 + 8 + 2$

$= 0$

$D_y = \begin{vmatrix} 1 & -5 & 1 \\ 0 & 4 & 2 \\ 3 & -2 & 0 \end{vmatrix} = 1\begin{vmatrix} 4 & 2 \\ -2 & 0 \end{vmatrix} + 5\begin{vmatrix} 0 & 2 \\ 3 & 0 \end{vmatrix} + 1\begin{vmatrix} 0 & 4 \\ 3 & -2 \end{vmatrix}$

$= 1(0 + 4) + 5(0 - 6) + 1(0 - 12)$

$= 4 - 30 - 12$

$= -38$

$D_z = \begin{vmatrix} 1 & -2 & -5 \\ 0 & 3 & 4 \\ 3 & -1 & -2 \end{vmatrix}$

$= 1\begin{vmatrix} 3 & 4 \\ -1 & -2 \end{vmatrix} + 2\begin{vmatrix} 0 & 4 \\ 3 & -2 \end{vmatrix} - 5\begin{vmatrix} 0 & 3 \\ 3 & -1 \end{vmatrix}$

$= 1(-6 + 4) + 2(0 - 12) - 5(0 - 9)$

$= -2 - 24 + 45$

$= 19$

$x = \dfrac{D_x}{D} = \dfrac{0}{-19} = 0$, $y = \dfrac{D_y}{D} = \dfrac{-38}{-19} = 2$,

$z = \dfrac{D_z}{D} = \dfrac{19}{-19} = -1$

The solution is (0, 2, −1).

47. $\begin{vmatrix} 1 & x \\ 2 & 7 \end{vmatrix} = -3$

$(1)(7) - x \cdot 2 = -3$

$7 - 2x = -3$

$-2x = -10$

$x = 5$

49. 0; answers may vary

51. $\begin{matrix} + & - & + & - \\ - & + & - & + \\ + & - & + & - \\ - & + & - & + \end{matrix}$

53. Expand by first row.

$$\begin{vmatrix} 5 & 0 & 0 & 0 \\ 0 & 4 & 2 & -1 \\ 1 & 3 & -2 & 0 \\ 0 & -3 & 1 & 2 \end{vmatrix} = 5\begin{vmatrix} 4 & 2 & -1 \\ 3 & -2 & 0 \\ -3 & 1 & 2 \end{vmatrix} - 0\begin{vmatrix} 0 & 2 & -1 \\ 1 & -2 & 0 \\ 0 & 1 & 2 \end{vmatrix} + 0\begin{vmatrix} 0 & 4 & -1 \\ 1 & 3 & 0 \\ 0 & -3 & 2 \end{vmatrix} - 0\begin{vmatrix} 0 & 4 & 2 \\ 1 & 3 & -2 \\ 0 & -3 & 1 \end{vmatrix}$$

$$= 5\left[4\begin{vmatrix} -2 & 0 \\ 1 & 2 \end{vmatrix} - 2\begin{vmatrix} 3 & 0 \\ -3 & 2 \end{vmatrix} + (-1)\begin{vmatrix} 3 & -2 \\ -3 & 1 \end{vmatrix} \right] - 0 + 0 - 0$$

$$= 5[4(-4-0) - 2(6+0) - 1(3-6)]$$

$$= 5(-16-12+3)$$

$$= 5(-25)$$

$$= -125$$

55. Expand by first column.

$$\begin{vmatrix} 4 & 0 & 2 & 5 \\ 0 & 3 & -1 & 1 \\ 0 & 0 & 2 & 0 \\ 0 & 0 & 0 & 1 \end{vmatrix} = 4\begin{vmatrix} 3 & -1 & 1 \\ 0 & 2 & 0 \\ 0 & 0 & 1 \end{vmatrix} - 0\begin{vmatrix} 0 & 2 & 5 \\ 0 & 2 & 0 \\ 0 & 0 & 1 \end{vmatrix} + 0\begin{vmatrix} 0 & 2 & 5 \\ 3 & -1 & 1 \\ 0 & 0 & 1 \end{vmatrix} - 0\begin{vmatrix} 0 & 2 & 5 \\ 3 & -1 & 1 \\ 0 & 2 & 0 \end{vmatrix}$$

$$= 4\left[3\begin{vmatrix} 2 & 0 \\ 0 & 1 \end{vmatrix} - 0\begin{vmatrix} -1 & 1 \\ 0 & 1 \end{vmatrix} + 0\begin{vmatrix} -1 & 1 \\ 2 & 0 \end{vmatrix} \right] - 0 + 0 - 0$$

$$= 4[3(2-0) - 0 + 0]$$

$$= 4(6)$$

$$= 24$$

Appendix F

Appendix F Exercise Set

1. 21, 28, 16, 42, 38

$$\overline{x} = \frac{21+28+16+42+38}{5} = \frac{145}{5} = 29$$

16, 21, 28, 38, 42

median = 28

no mode

3. 7.6, 8.2, 8.2, 9.6, 5.7, 9.1

$$\overline{x} = \frac{7.6+8.2+8.2+9.6+5.7+9.1}{6} = \frac{48.4}{6} = 8.1$$

5.7, 7.6, 8.2, 8.2, 9.1, 9.6

$$\text{median} = \frac{8.2+8.2}{2} = 8.2$$

mode = 8.2

5. 0.2, 0.3, 0.5, 0.6, 0.6, 0.9, 0.2, 0.7, 1.1

$$\overline{x} = \frac{0.2+0.3+0.5+0.6+0.6+0.9+0.2+0.7+1.1}{9}$$

$$= \frac{5.1}{9}$$

$$= 0.6$$

median = 0.6

mode = 0.2 and 0.6

7. 231, 543, 601, 293, 588, 109, 334, 268

$$\overline{x} = \frac{231+543+601+293+588+109+334+268}{8}$$

$$= \frac{2967}{8}$$

$$= 370.9$$

109, 231, 268, 293, 334, 543, 588, 601

$$\text{median} = \frac{293+334}{2} = 313.5$$

no mode

9. 1454, 1250, 1136, 1127, 1107

$$\overline{x} = \frac{1454+1250+1136+1127+1107}{5}$$

$$= \frac{6074}{5}$$

$$= 1214.8 \text{ feet}$$

11. 1454, 1250, 1136, 1127, 1107, 1046, 1023, 1002

$$\text{median} = \frac{1127+1107}{2} = 1117 \text{ feet}$$

13. $$\overline{x} = \frac{7.8+6.9+7.5+4.7+6.9+7.0}{6}$$

$$= \frac{40.8}{6}$$

$$= 6.8 \text{ seconds}$$

15. 4.7, 6.9, 6.9, 7.0, 7.5, 7.8

mode = 6.9

17. 74, 77, 85, 86, 91, 95

$$\text{median} = \frac{85+86}{2} = 85.5$$

19. Sum = 78 + 80 + 66 + 68 + 71 + 64 + 82 + 71
$$+ \ 70 + 65 + 70 + 75 + 77 + 86 + 72$$
$$= 1095$$

$$\overline{x} = \frac{1095}{15} = 73$$

21. 64, 65, 66, 68, 70, 70, 71, 71, 72, 75, 77, 78, 80, 82, 86

mode = 70 and 71

23. 64, 65, 66, 68, 70, 70, 71, 71, 72, 75, 77, 78, 80, 82, 86
$$\uparrow$$
$$\text{mean} = 73$$

9 rates were lower than the mean.

25. __, __, 16, 18, __;

Since the mode is 21, at least two of the missing numbers must be 21. The mean is 20. Let the one unknown number be x.

$$\overline{x} = \frac{21+21+16+18+x}{5} = 20$$

$$\frac{76+x}{5} = 20$$

$$76 + x = 100$$

$$x = 24$$

The missing numbers are 21, 21, 24.

Appendix G

1. $90° - 19° = 71°$

3. $90° - 70.8° = 19.2°$

5. $90° - 11\frac{1}{4}° = 78\frac{3}{4}°$

7. $180° - 150° = 30°$

9. $180° - 30.2° = 149.8°$

11. $180° - 79\frac{1}{2}° = 100\frac{1}{2}°$

13. $m\angle 1 = 110°$
$m\angle 2 = 180° - 110° = 70°$
$m\angle 3 = m\angle 2 = 70°$
$m\angle 4 = m\angle 2 = 70°$
$m\angle 5 = m\angle 1 = 110°$
$m\angle 6 = m\angle 4 = 70°$
$m\angle 7 = m\angle 5 = 110°$

15. $180° - 11° - 79° = 90°$

17. $180° - 25° - 65° = 90°$

19. $180° - 30° - 60° = 90°$

21. $90° - 45° = 45°$
$45°, 90°$

23. $90° - 17° = 73°$
$73°, 90°$

25. $90° - 39\frac{3}{4}° = 50\frac{1}{4}°$
$50\frac{1}{4}°, \ 90°$

27. $\dfrac{12}{4} = \dfrac{18}{x}$
$4x\left(\dfrac{12}{4}\right) = 4x\left(\dfrac{18}{x}\right)$
$12x = 72$
$x = 6$

29. $\dfrac{6}{9} = \dfrac{3}{x}$
$9x\left(\dfrac{6}{9}\right) = 9x\left(\dfrac{3}{x}\right)$
$6x = 27$
$x = 4.5$

31. $a^2 + b^2 = c^2$
$6^2 + 8^2 = c^2$
$36 + 64 = c^2$
$100 = c^2$
$10 = c$

33. $a^2 + b^2 = c^2$
$5^2 + b^2 = 13^2$
$25 + b^2 = 169$
$b^2 = 144$
$b = 12$

Practice Final Exam

1. $6[5+2(3-8)-3] = 6[5+2(-5)-3]$
$$= 6[5+(-10)-3]$$
$$= 6[-5-3]$$
$$= 6[-8]$$
$$= -48$$

2. $-3^4 = -(3^4) = -81$

3. $4^{-3} = \dfrac{1}{4^3} = \dfrac{1}{64}$

4. $\dfrac{1}{2} - \dfrac{5}{6} = \dfrac{1}{2} \cdot \dfrac{3}{3} - \dfrac{5}{6} = \dfrac{3}{6} - \dfrac{5}{6} = \dfrac{-2}{6} = -\dfrac{1}{3}$

5. $5x^3 + x^2 + 5x - 2 - (8x^3 - 4x^2 + x - 7)$
$$= 5x^3 + x^2 + 5x - 2 - 8x^3 + 4x^2 - x + 7$$
$$= 5x^3 - 8x^3 + x^2 + 4x^2 + 5x - x - 2 + 7$$
$$= -3x^3 + 5x^2 + 4x + 5$$

6. $(4x-2)^2 = (4x)^2 - 2(4x)(2) + 2^2$
$$= 16x^2 - 16x + 4$$

7. $(3x+7)(x^2+5x+2)$
$$= 3x(x^2+5x+2) + 7(x^2+5x+2)$$
$$= 3x^3 + 15x^2 + 6x + 7x^2 + 35x + 14$$
$$= 3x^3 + 22x^2 + 41x + 14$$

8. $y^2 - 8y - 48 = (y-12)(y+4)$

9. $9x^3 + 39x^2 + 12x = 3x(3x^2 + 13x + 4)$
$$= 3x(3x+1)(x+4)$$

10. $180 - 5x^2 = 5(36 - x^2)$
$$= 5(6^2 - x^2)$$
$$= 5(6+x)(6-x)$$

11. $3a^2 + 3ab - 7a - 7b = 3a(a+b) - 7(a+b)$
$$= (a+b)(3a-7)$$

12. $8y^3 - 64 = 8(y^3 - 8)$
$$= 8(y^3 - 2^3)$$
$$= 8(y-2)(y^2 + y \cdot 2 + 2^2)$$
$$= 8(y-2)(y^2 + 2y + 4)$$

13. $\left(\dfrac{x^2 y^3}{x^3 y^{-4}}\right)^2 = \left(\dfrac{y^{3-(-4)}}{x^{3-2}}\right)^2 = \left(\dfrac{y^7}{x^1}\right)^2 = \dfrac{y^{7 \cdot 2}}{x^{1 \cdot 2}} = \dfrac{y^{14}}{x^2}$

14. $-4(a+1) - 3a = -7(2a-3)$
$$-4a - 4 - 3a = -14a + 21$$
$$-7a - 4 = -14a + 21$$
$$7a - 4 = 21$$
$$7a = 25$$
$$a = \dfrac{25}{7}$$

15. $3x - 5 \geq 7x + 3$
$$-5 \geq 4x + 3$$
$$-8 \geq 4x$$
$$-2 \geq x$$
$$x \leq -2$$
$$(-\infty, -2]$$

16. $\quad x(x+6) = 7$
$$x^2 + 6x = 7$$
$$x^2 + 6x - 7 = 0$$
$$(x+7)(x-1) = 0$$
$$x + 7 = 0 \quad \text{or} \quad x - 1 = 0$$
$$x = -7 \qquad\qquad x = 1$$

17. $5x - 7y = 10$

x	y
2	0
0	$-\dfrac{10}{7}$

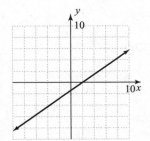

Copyright © 2013 Pearson Education, Inc.

18. $x - 3 = 0$

    $x = 3$

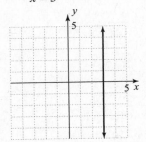

19. $m = \dfrac{y_2 - y_1}{x_2 - x_1}$

    $m = \dfrac{2 - (-5)}{-1 - 6} = \dfrac{2 + 5}{-7} = \dfrac{7}{-7} = -1$

20. $-3x + y = 5$

    $y = 3x + 5$

    $m = 3$

21. $(x_1,\, y_1) = (2,\, -5),\ (x_2,\, y_2) = (1,\, 3)$

    $m = \dfrac{y_2 - y_1}{x_2 - x_1} = \dfrac{3 - (-5)}{1 - 2} = \dfrac{3 + 5}{-1} = \dfrac{8}{-1} = -8$

    $y - y_1 = m(x - x_1)$

    $y - (-5) = -8(x - 2)$

    $y + 5 = -8x + 16$

    $8x + y = 11$

22. A line parallel to $x = 7$ is a vertical line. The vertical line through $(-5,\, -1)$ has equation $x = -5$.

23. $\begin{cases} \dfrac{1}{2}x + 2y = -\dfrac{15}{4} \\ 4x = -y \end{cases}$

Solve the second equation for y.

$y = -4x$

Substitute $-4x$ for y in the first equation.

$\dfrac{1}{2}x + 2(-4x) = -\dfrac{15}{4}$

$\dfrac{1}{2}x - 8x = -\dfrac{15}{4}$

$\dfrac{1}{2}x - \dfrac{16x}{2} = -\dfrac{15}{4}$

$-\dfrac{15}{2}x = -\dfrac{15}{4}$

$-\dfrac{2}{15}\left(-\dfrac{15}{2}x\right) = -\dfrac{2}{15}\left(-\dfrac{15}{4}\right)$

$x = \dfrac{1}{2}$

Let $x = \dfrac{1}{2}$ in $y = -4x$.

$y = -4x = -4\left(\dfrac{1}{2}\right) = -2$

The solution is $\left(\dfrac{1}{2},\, -2\right)$.

24. $\begin{cases} 4x - 6y = 7 \\ -2x + 3y = 0 \end{cases}$

Multiply the second equation by 2 and add the result to the first equation.

$\begin{array}{r} 4x - 6y = 7 \\ -4x + 6y = 0 \\ \hline 0 = 7 \end{array}$

The statement $0 = 7$ is false, so the system has no solution. The solution set is $\{\ \}$ or $\varnothing$.

25.
$$3x + 2 \overline{) 27x^3 + 0x^2 + 0x - 8 }$$

$$\begin{array}{r} 9x^2 - 6x + 4 \\ 3x+2\,\overline{)\,27x^3 + 0x^2 + 0x - 8} \\ \underline{27x^3 + 18x^2} \\ -18x^2 + 0x \\ \underline{-18x^2 - 12x} \\ 12x - 8 \\ \underline{12x + 8} \\ -16 \end{array}$$

$\dfrac{27x^3 - 8}{3x + 2} = 9x^2 - 6x + 4 - \dfrac{16}{3x + 2}$

26. $h(x) = x^3 - x$

    a. $h(-1) = (-1)^3 - (-1) = -1 + 1 = 0$

    b. $h(0) = 0^3 - 0 = 0 - 0 = 0$

c. $h(4) = 4^3 - 4 = 64 - 4 = 60$

27. x-intercepts: $(0, 0)$, $(4, 0)$
y-intercept: $(0, 0)$
Domain: $(-\infty, \infty)$
Range: $(-\infty, 4]$

28. Let x be the smaller area code. Then the other area code is $2x$.
$x + 2x = 1203$
$3x = 1203$
$x = 401$
$2x = 2(401) = 802$
The area codes are 401 (Rhode Island) and 802 (Vermont).

29. Let x be the number of hours since the trains left. One train will have traveled $50x$ miles and the other will have traveled $64x$ miles.
$50x + 64x = 285$
$114x = 285$
$x = \dfrac{285}{114} = \dfrac{5}{2}$ or $2\dfrac{1}{2}$

The trains are 285 miles apart after $2\dfrac{1}{2}$ hours.

30. Let x be the amount of 12% solution.

Amount	Percent	Total Saline
x	$12\% = 0.12$	$0.12x$
80	$22\% = 0.22$	$0.22 \cdot 80 = 17.6$
$80 + x$	$16\% = 0.16$	$0.16(80 + x)$ $= 12.8 + 0.16x$

$0.12x + 17.6 = 12.8 + 0.16x$
$4.8 = 0.04x$
$120 = x$
120 cc of 12% saline solution should be added.

31. $x^2 + 4x + 3 = 0$
$(x + 1)(x + 3) = 0$
$x + 1 = 0$ or $x + 3 = 0$
$x = -1$ $x = -3$
The domain of $g(x)$ is
$\{x \mid x$ is a real number, $x \neq -1, x \neq -3\}$.

32. $\dfrac{15x}{2x+5} - \dfrac{6-4x}{2x+5} = \dfrac{15x - (6-4x)}{2x+5}$
$= \dfrac{15x - 6 + 4x}{2x+5}$
$= \dfrac{19x - 6}{2x+5}$

33. $\dfrac{x^2 - 9}{x^2 - 3x} \div \dfrac{xy + 5x + 3y + 15}{2x + 10}$
$= \dfrac{x^2 - 9}{x^2 - 3x} \cdot \dfrac{2x + 10}{xy + 5x + 3y + 15}$
$= \dfrac{(x+3)(x-3) \cdot 2(x+5)}{x(x-3) \cdot (y+5)(x+3)}$
$= \dfrac{2(x+5)}{x(y+5)}$

34. $a^2 - a - 6 = (a-3)(a+2)$
LCD $= (a-3)(a+2)$
$\dfrac{5a}{a^2 - a - 6} - \dfrac{2}{a-3}$
$= \dfrac{5a}{(a-3)(a+2)} - \dfrac{2 \cdot (a+2)}{(a-3) \cdot (a+2)}$
$= \dfrac{5a - 2(a+2)}{(a-3)(a+2)}$
$= \dfrac{5a - 2a - 4}{(a-3)(a+2)}$
$= \dfrac{3a - 4}{(a-3)(a+2)}$

35. $\dfrac{5 - \frac{1}{y^2}}{\frac{1}{y} + \frac{2}{y^2}} = \dfrac{y^2\left(5 - \frac{1}{y^2}\right)}{y^2\left(\frac{1}{y} + \frac{2}{y^2}\right)} = \dfrac{5y^2 - 1}{y + 2}$

36. $\dfrac{4}{y} - \dfrac{5}{3} = -\dfrac{1}{5}$
$15y\left(\dfrac{4}{y}\right) - 15y\left(\dfrac{5}{3}\right) = 15y\left(-\dfrac{1}{5}\right)$
$60 - 25y = -3y$
$60 = 22y$
$\dfrac{60}{22} = y$
$\dfrac{30}{11} = y$

37.
$$\frac{5}{y+1} = \frac{4}{y+2}$$
$$5(y+2) = 4(y+1)$$
$$5y+10 = 4y+4$$
$$y+10 = 4$$
$$y = -6$$

38.
$$\frac{a}{a-3} = \frac{3}{a-3} - \frac{3}{2}$$
$$2(a-3)\left(\frac{a}{a-3}\right) = 2(a-3)\left(\frac{3}{a-3}\right) - 2(a-3)\left(\frac{3}{2}\right)$$
$$2a = 6 - 3(a-3)$$
$$2a = 6 - 3a + 9$$
$$2a = 15 - 3a$$
$$5a = 15$$
$$a = 3$$

In the original equation, $a = 3$ makes a denominator 0, so it is extraneous. The equation has no solution.

39. Let x be the number.
$$x + 5 \cdot \frac{1}{x} = 6$$
$$x\left(x + \frac{5}{x}\right) = x \cdot 6$$
$$x^2 + 5 = 6x$$
$$x^2 - 6x + 5 = 0$$
$$(x-1)(x-5) = 0$$
$$x - 1 = 0 \quad \text{or} \quad x - 5 = 0$$
$$x = 1 \qquad\qquad x = 5$$
The number is 1 or 5.

40. $\sqrt{216} = \sqrt{36 \cdot 6} = \sqrt{36} \cdot \sqrt{6} = 6\sqrt{6}$

41. $\left(\dfrac{1}{125}\right)^{-1/3} = 125^{1/3} = \sqrt[3]{125} = \sqrt[3]{5^3} = 5$

42.
$$\left(\frac{64c^{4/3}}{a^{-2/3}b^{5/6}}\right)^{1/2} = \frac{64^{1/2}c^{\frac{4}{3}\cdot\frac{1}{2}}}{a^{-\frac{2}{3}\cdot\frac{1}{2}}b^{\frac{5}{6}\cdot\frac{1}{2}}}$$
$$= \frac{8c^{2/3}}{a^{-1/3}b^{5/12}}$$
$$= \frac{8a^{1/3}c^{2/3}}{b^{5/12}}$$

43.
$$\sqrt{125x^3} - 3\sqrt{20x^3} = \sqrt{25x^2 \cdot 5x} - 3\sqrt{4x^2 \cdot 5x}$$
$$= 5x\sqrt{5x} - 3 \cdot 2x\sqrt{5x}$$
$$= 5x\sqrt{5x} - 6x\sqrt{5x}$$
$$= (5x - 6x)\sqrt{5x}$$
$$= -x\sqrt{5x}$$

44. $\left(\sqrt{5}+5\right)\left(\sqrt{5}-5\right) = \left(\sqrt{5}\right)^2 - 5^2 = 5 - 25 = -20$

45.
$$|6x-5| - 3 = -2$$
$$|6x-5| = 1$$
$$6x - 5 = -1 \quad \text{or} \quad 6x - 5 = 1$$
$$6x = 4 \quad \text{or} \quad 6x = 6$$
$$x = \frac{4}{6} \quad \text{or} \quad x = 1$$
$$x = \frac{2}{3}$$
Both solutions check.

46.
$$-3 < 2(x-3) \le 4$$
$$-3 < 2x - 6 \le 4$$
$$-3+6 < 2x - 6 + 6 \le 4 + 6$$
$$3 < 2x \le 10$$
$$\frac{3}{2} < \frac{2x}{2} \le \frac{10}{2}$$
$$\frac{3}{2} < x \le 5$$
$$\left(\frac{3}{2}, 5\right]$$

47.
$$|3x+1| > 5$$
$$3x + 1 < -5 \quad \text{or} \quad 3x + 1 > 5$$
$$3x < -6 \qquad\qquad 3x > 4$$
$$x < -2 \qquad\qquad x > \frac{4}{3}$$
$$(-\infty, -2) \cup \left(\frac{4}{3}, \infty\right)$$

48.
$$y^2 - 3y = 5$$
$$y^2 - 3y - 5 = 0$$
$$y = \frac{-(-3) \pm \sqrt{(-3)^2 - 4(1)(-5)}}{2(1)}$$
$$y = \frac{3 \pm \sqrt{9 + 20}}{2}$$
$$y = \frac{3 \pm \sqrt{29}}{2}$$

49.
$$x = \sqrt{x-2} + 2$$
$$x - 2 = \sqrt{x-2}$$
$$(x-2)^2 = \left(\sqrt{x-2}\right)^2$$
$$x^2 - 4x + 4 = x - 2$$
$$x^2 - 5x + 6 = 0$$
$$(x-2)(x-3) = 0$$
$$x - 2 = 0 \quad \text{or} \quad x - 3 = 0$$
$$x = 2 \quad \text{or} \quad x = 3$$

50.
$$2x^2 - 7x > 15$$
$$2x^2 - 7x - 15 > 0$$
$$(2x+3)(x-5) > 0$$
$$2x + 3 = 0 \quad \text{or} \quad x - 5 = 0$$
$$x = -\frac{3}{2} \quad \text{or} \quad x = 5$$

51. $y > -4x$

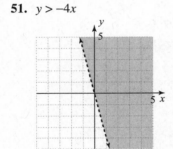

52. $g(x) = -|x + 2| - 1$

x	$g(x) = -|x+2| - 1$	$g(x)$
-5	$-|-5+2| - 1 = -4$	-4
-4	$-|-4+2| - 1 = -3$	-3
-3	$-|-3+2| - 1 = -2$	-2
-2	$-|-2+2| - 1 = -1$	-1
-1	$-|-1+2| - 1 = -2$	-2
0	$-|0+2| - 1 = -3$	-3
1	$-|1+2| - 1 = -4$	-4

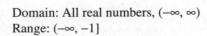

Domain: All real numbers, $(-\infty, \infty)$
Range: $(-\infty, -1]$

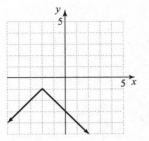

53. $h(x) = x^2 - 4x + 4$
x-intercept: Let $h(x) = 0$ and solve for x.
$$0 = x^2 - 4x + 4$$
$$0 = (x-2)^2$$
$$x - 2 = 0$$
$$x = 2$$
x-intercept: $(2, 0)$
y-intercept: Let $x = 0$.
$$h(0) = 0^2 - 4(0) + 4 = 4$$
y-intercept: $(0, 4)$
x-coordinate of vertex:
$$-\frac{b}{2a} = -\frac{-4}{2(1)} = 2$$
vertex: $(2, 0)$

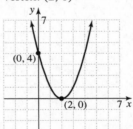

54. $f(x) = \begin{cases} -\dfrac{1}{2}x & \text{if} \quad x \le 0 \\ 2x - 3 & \text{if} \quad x > 0 \end{cases}$

If $x \le 0$

x	$-\frac{1}{2}x$	$f(x)$
0	$-\frac{1}{2}(0)$	0
-2	$-\frac{1}{2}(-2)$	1
-4	$-\frac{1}{2}(-4)$	2

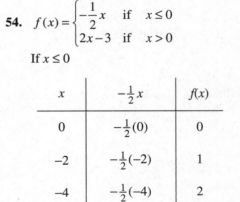

If $x > 0$

x	$2x - 3$	$f(x)$
1	$2(1) - 3$	-1
2	$2(2) - 3$	1
3	$2(3) - 3$	3

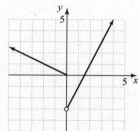

Domain: $(-\infty, \infty)$
Range: $(-3, \infty)$

55. through $(4, -2)$ and $(6, -3)$

$$\text{slope} = m = \frac{y_2 - y_1}{x_2 - x_1} = \frac{-3 - (-2)}{6 - 4} = \frac{-1}{2}$$

$$y - y_1 = m(x - x_1)$$

$$y - (-2) = -\frac{1}{2}(x - 4)$$

$$y + 2 = -\frac{1}{2}x + 2$$

$$y = -\frac{1}{2}x$$

$$f(x) = -\frac{1}{2}x$$

56. through $(-1, 2)$ and perpendicular to $3x - y = 4$
Find the slope of $3x - y = 4$ by writing the equation in slope-intercept form.

$$3x - y = 4$$

$$-y = -3x + 4$$

$$y = 3x - 4$$

The slope is 3. The slope of a line perpendicular

to this line is $-\frac{1}{3}$.

Substitute $m = -\frac{1}{3}$ and $(x_1, y_1) = (-1, 2)$ in the

equation:

$$y - y_1 = m(x - x_1)$$

$$y - 2 = -\frac{1}{3}[x - (-1)]$$

$$y - 2 = -\frac{1}{3}(x + 1)$$

$$y - 2 = -\frac{1}{3}x - \frac{1}{3}$$

$$y = -\frac{1}{3}x + \frac{5}{3}$$

$$f(x) = -\frac{1}{3}x + \frac{5}{3}$$

57. $(x_1, y_1) = (-6, 3);\ (x_2, y_2) = (-8, -7)$

$$d = \sqrt{(x_2 - x_1)^2 + (y_2 - y_1)^2}$$

$$= \sqrt{[-8 - (-6)]^2 + (-7 - 3)^2}$$

$$= \sqrt{(-2)^2 + (-10)^2}$$

$$= \sqrt{4 + 100}$$

$$= \sqrt{104}$$

$$= 2\sqrt{26} \text{ units}$$

58. $(x_1, y_1) = (-2, -5);\ (x_2, y_2) = (-6, 12)$

$$\text{midpoint} = \left(\frac{x_1 + x_2}{2}, \frac{y_1 + y_2}{2}\right)$$

$$= \left(\frac{-2 + (-6)}{2}, \frac{-5 + 12}{2}\right)$$

$$= \left(\frac{-8}{2}, \frac{7}{2}\right)$$

$$= \left(-4, \frac{7}{2}\right)$$

59. $\displaystyle \sqrt{\frac{9}{y}} = \frac{\sqrt{9}}{\sqrt{y}} = \frac{\sqrt{9}}{\sqrt{y}} \cdot \frac{\sqrt{y}}{\sqrt{y}} = \frac{\sqrt{9} \cdot \sqrt{y}}{\sqrt{y} \cdot \sqrt{y}} = \frac{3\sqrt{y}}{y}$

60.

$$\frac{4 - \sqrt{x}}{4 + 2\sqrt{x}} = \frac{4 - \sqrt{x}}{4 + 2\sqrt{x}} \cdot \frac{4 - 2\sqrt{x}}{4 - 2\sqrt{x}}$$

$$= \frac{\left(4 - \sqrt{x}\right)\left(4 - 2\sqrt{x}\right)}{\left(4 + 2\sqrt{x}\right)\left(4 - 2\sqrt{x}\right)}$$

$$= \frac{16 - 12\sqrt{x} + 2x}{16 - 4x}$$

$$= \frac{2\left(8 - 6\sqrt{x} + x\right)}{2(8 - 2x)}$$

$$= \frac{8 - 6\sqrt{x} + x}{8 - 2x}$$

61. $W = \dfrac{k}{V}$

Find k by substituting $W = 20$ and $V = 12$.

$20 = \dfrac{k}{12}$

$240 = k$

Write the inverse relation equation.

$W = \dfrac{240}{V}$

Let $V = 15$ and find W.

$W = \dfrac{240}{15}$

$W = 16$

62. Use the Pythagorean Theorem.

$c^2 = a^2 + b^2$

$20^2 = x^2 + (x+8)^2$

$400 = x^2 + x^2 + 16x + 64$

$0 = 2x^2 + 16x - 336$

$0 = 2(x^2 + 8x - 168)$

$x = \dfrac{-8 \pm \sqrt{8^2 - 4(1)(-168)}}{2(1)}$

$x = \dfrac{-8 \pm \sqrt{736}}{2}$

$x \approx -17.6$ or $x \approx 9.6$

Discard a negative distance.

$x + 8 + x = 9.6 + 8 + 9.6 = 27.2$

$27.2 - 20 = 7.2$ or about 7

A person saves about 7 feet.

63. a. Find the vertex.

$s(t) = -16t^2 + 32t + 256$

t-value: $\dfrac{-b}{2a} = \dfrac{-32}{2(-16)} = 1$

$s(t)$-value:

$s(1) = -16(1)^2 + 32(1) + 256 = 272$

The maximum height is 272 feet.

b. Let $s(t) = 0$ and solve for t.

$0 = -16t^2 + 32t + 256$

$0 = -16(t^2 - 2t - 16)$

$t = \dfrac{-(-2) \pm \sqrt{(-2)^2 - 4(1)(-16)}}{2(1)}$

$t = \dfrac{2 \pm \sqrt{68}}{2}$

$t = \dfrac{2 \pm 2\sqrt{17}}{2}$

$t = 1 \pm \sqrt{17}$

$t \approx -3.12$ or $t \approx 5.12$

Discard a negative time.

The stone will hit the water in approximately 5.12 seconds.

64. $-\sqrt{-8} = -\sqrt{4 \cdot (-1) \cdot 2}$

$= -\sqrt{4} \cdot \sqrt{-1} \cdot \sqrt{2}$

$= -2i\sqrt{2}$

$= 0 - 2i\sqrt{2}$

65. $(12 - 6i) - (12 - 3i) = 12 - 6i - 12 + 3i$

$= 12 - 12 - 6i + 3i$

$= 0 - 3i$

66. $(4 + 3i)^2 = (4 + 3i)(4 + 3i)$

$= 16 + 12i + 12i + 9i^2$

$= 16 + 24i - 9$

$= 7 + 24i$

67. $\dfrac{1 + 4i}{1 - i} = \dfrac{1 + 4i}{1 - i} \cdot \dfrac{1 + i}{1 + i}$

$= \dfrac{(1 + 4i)(1 + i)}{(1 - i)(1 + i)}$

$= \dfrac{1 + 5i + 4i^2}{1 - i^2}$

$= \dfrac{1 + 5i - 4}{1 - (-1)}$

$= \dfrac{-3 + 5i}{2}$

$= -\dfrac{3}{2} + \dfrac{5}{2}i$

68. $g(x) = x - 7$ and $h(x) = x^2 - 6x + 5$

$(g \circ h)(x) = (x^2 - 6x + 5) - 7 = x^2 - 6x - 2$

69. $f(x) = 6 - 2x$ is a one-to-one function since there is only one $f(x)$ value for each x-value.

Inverse:

$y = 6 - 2x$ $\Rightarrow$ $x = 6 - 2y$

$x + 2y = 6$

$2y = -x + 6$

$y = \dfrac{-x + 6}{2}$

$f^{-1}(x) = \dfrac{-x + 6}{2}$

70. $\log_5 x + 3\log_5 x - \log_5(x+1)$

$= \log_5 x + \log_5 x^3 - \log_5(x+1)$

$= \log_5 x \cdot x^3 - \log_5(x+1)$

$= \log_5 x^4 - \log_5(x+1)$

$= \log_5 \dfrac{x^4}{x+1}$

71. $8^{x-1} = \dfrac{1}{64}$

$(2^3)^{x-1} = \dfrac{1}{2^6}$

$2^{3(x-1)} = 2^{-6}$

$3(x-1) = -6$

$3x - 3 = -6$

$3x = -3$

$x = -1$

72. $3^{2x+5} = 4$

$\log 3^{2x+5} = \log 4$

$(2x+5)\log 3 = \log 4$

$2x + 5 = \dfrac{\log 4}{\log 3}$

$2x = \dfrac{\log 4}{\log 3} - 5$

$x = \dfrac{1}{2}\left(\dfrac{\log 4}{\log 3} - 5\right)$

$x \approx -1.8691$

73. $\log_8(3x-2) = 2$

$8^2 = 3x - 2$

$64 = 3x - 2$

$66 = 3x$

$22 = x$

74. $\log_4(x+1) - \log_4(x-2) = 3$

$\log_4 \dfrac{x+1}{x-2} = 3$

$4^3 = \dfrac{x+1}{x-2}$

$64 = \dfrac{x+1}{x-2}$

$64(x-2) = x+1$

$64x - 128 = x + 1$

$63x = 129$

$x = \dfrac{129}{63} = \dfrac{43}{21}$

75. $\ln \sqrt{e} = x$

$\ln e^{1/2} = x$

$\dfrac{1}{2}\ln e = x$

$\dfrac{1}{2} = x$

76. $y = \left(\dfrac{1}{2}\right)^x + 1$

x	$\left(\frac{1}{2}\right)^x + 1$	y
-3	$\left(\frac{1}{2}\right)^{-3} + 1 = 9$	9
-2	$\left(\frac{1}{2}\right)^{-2} + 1 = 5$	5
-1	$\left(\frac{1}{2}\right)^{-1} + 1 = 3$	3
0	$\left(\frac{1}{2}\right)^{0} + 1 = 2$	2
1	$\left(\frac{1}{2}\right)^{1} + 1 = 1\frac{1}{2}$	$1\frac{1}{2}$
2	$\left(\frac{1}{2}\right)^{2} + 1 = 1\frac{1}{4}$	$1\frac{1}{4}$
3	$\left(\frac{1}{2}\right)^{3} + 1 = 1\frac{1}{8}$	$1\frac{1}{8}$

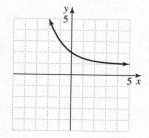

77. Use $y = C(1+r)^x$ with $C = 57{,}000$,
$r = 2.6\% = 0.026$, and $x = 5$.
$y = 57{,}000(1 + 0.026)^5 \approx 64{,}805$
There will be 64,805 prairie dogs in 5 years.

78. $x^2 - y^2 = 36$
$x^2 - y^2 = 6^2$
hyperbola, with x-intercepts $(-6, 0)$, $(6, 0)$

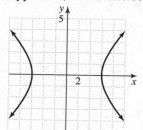

79. $16x^2 + 9y^2 = 144$
$\dfrac{16x^2}{144} + \dfrac{9y^2}{144} = \dfrac{144}{144}$
$\dfrac{x^2}{9} + \dfrac{y^2}{16} = 1$
Ellipse, x-intercepts $(-3, 0)$, $(3, 0)$
y-intercepts $(0, -4)$, $(0, 4)$

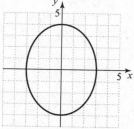

80. $x^2 + y^2 + 6x = 16$
$(x^2 + 6x + 9) + y^2 = 16 + 9$
$(x+3)^2 + y^2 = 25$
$[x - (-3)]^2 + (y - 0)^2 = 5^2$
circle with center $(-3, 0)$ and radius 5

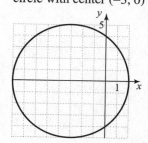

81. $\begin{cases} x^2 + y^2 = 26 \\ x^2 - 2y^2 = 23 \end{cases}$

Multiply equation (2) by -1 and add the equations.

$\begin{aligned} x^2 + y^2 &= 26 \\ \underline{-x^2 + 2y^2} &= \underline{-23} \\ 3y^2 &= 3 \\ y^2 &= 1 \\ y &= \pm 1 \end{aligned}$

Substitute $y = -1$ and $y = 1$ into equation (1).
$x^2 + (-1)^2 = 26$
$x^2 = 25$
$x = \pm 5$
$x^2 + 1^2 = 26$
$x^2 = 25$
$x = \pm 5$
The solutions are $(-5, -1)$, $(-5, 1)$, $(5, -1)$, $(5, 1)$.

82. $a_n = \dfrac{(-1)^n}{n+4}$

$a_1 = \dfrac{(-1)^1}{1+4} = -\dfrac{1}{5}$

$a_2 = \dfrac{(-1)^2}{2+4} = \dfrac{1}{6}$

$a_3 = \dfrac{(-1)^3}{3+4} = -\dfrac{1}{7}$

$a_4 = \dfrac{(-1)^4}{4+4} = \dfrac{1}{8}$

$a_5 = \dfrac{(-1)^5}{5+4} = -\dfrac{1}{9}$

The first five terms are $-\dfrac{1}{5}, \dfrac{1}{6}, -\dfrac{1}{7}, \dfrac{1}{8}, -\dfrac{1}{9}$.

83. $a_n = 5(2)^{n-1}$

$a_1 = 5(2)^{1-1} = 5(2)^0 = 5$

$r = 2$

$n = 5$

$S_n = \dfrac{a_1(1-r^n)}{1-r}$

$S_5 = \dfrac{5(1-2^5)}{1-2} = \dfrac{5(1-32)}{-1} = 155$

84. Sequence $\dfrac{3}{2}, -\dfrac{3}{4}, \dfrac{3}{8}, \dots$

$a_1 = \dfrac{3}{2}, r = -\dfrac{1}{2}$

$S_\infty = \dfrac{a_1}{1-r} = \dfrac{\frac{3}{2}}{1-\left(-\frac{1}{2}\right)} = \dfrac{\frac{3}{2}}{\frac{3}{2}} = 1$

85. $\displaystyle\sum_{i=1}^{4} i(i-2) = 1(1-2) + 2(2-2) + 3(3-2) + 4(4-2)$

$\qquad\qquad = 1(-1) + 2(0) + 3(1) + 4(2)$

$\qquad\qquad = -1 + 0 + 3 + 8$

$\qquad\qquad = 10$

86. $(2x+y)^5 = \binom{5}{0}(2x)^5 + \binom{5}{1}(2x)^4(y) + \binom{5}{2}(2x)^3(y)^2 + \binom{5}{3}(2x)^2(y)^3 + \binom{5}{4}(2x)^1(y)^4 + \binom{5}{5}y^5$

$\qquad\qquad = 2^5 x^5 + 5 \cdot 2^4 x^4 y + 10 \cdot 2^3 x^3 y^2 + 10 \cdot 2^2 x^2 y^3 + 5 \cdot 2xy^4 + y^5$

$\qquad\qquad = 32x^5 + 80x^4 y + 80x^3 y^2 + 40x^2 y^3 + 10xy^4 + y^5$